A GEOLOGIC TIME SCALE 2004

An international team of over 40 stratigraphic experts, many actively involved in the International Commission of Stratigraphy (ICS), have helped to build the most up-to-date international stratigraphic framework for the Precambrian and Phanerozoic. This successor to *A Geologic Time Scale 1989* by W. Brian Harland *et al.* (Cambridge, 1989) begins with an introduction to the theory and methodology behind the construction of the new time scale. The main part of the book is devoted to the scale itself, systematically presenting the standard subdivisions at all levels using a variety of correlation markers. Extensive use is made of stable and unstable isotope geochronology, geomathematics, and orbital tuning to produce a standard geologic scale of unprecedented detail and accuracy with a full error analysis. A wallchart summarizing the whole time scale, with paleogeographic reconstructions throughout the Phanerozoic is included in the back of the book. The time scale will be an invaluable reference source for academic and professional researchers and students.

FELIX GRADSTEIN is currently Professor of Stratigraphy and Micropaleontology at the Geological Museum of the University of Oslo where he leads the offshore relational stratigraphic database funded by a petroleum consortium. He is the current chair of the International Commission on Stratigraphy, which is working on the formal classification of the global Precambrian and Phanerozoic rock record and the international time scale.

JAMES OGG is Professor of Stratigraphy at Purdue University, West Lafayette, Indiana, and has research interests in magnetochronology, cyclostratigraphy, sedimentology, and stratigraphic databases. He is presently Secretary General of the International Commission on Stratigraphy (ICS).

ALAN SMITH is Reader in Geology at the University of Cambridge and a Fellow of St. John's College. His principal research interests are paleogeographic reconstruction and related software development.

A Geologic Time Scale 2004

Edited by
Felix M. Gradstein, James G. Ogg, and Alan G. Smith

PUBLISHED BY THE PRESS SYNDICATE OF THE UNIVERSITY OF CAMBRIDGE
The Pitt Building, Trumpington Street, Cambridge, United Kingdom

CAMBRIDGE UNIVERSITY PRESS
The Edinburgh Building, Cambridge CB2 2RU, UK
40 West 20th Street, New York, NY 10011-4211, USA
477 Williamstown Road, Port Melbourne, VIC 3207, Australia
Ruiz de Alarcón 13, 28014 Madrid, Spain
Dock House, The Waterfront, Cape Town 8001, South Africa

http://www.cambridge.org

First published 2004

Printed in the United Kingdom at the University Press, Cambridge

Typeface Ehrhardt 10/14 pt. *System* LaTeX 2ε [TB]

A catalog record for this book is available from the British Library

Library of Congress Cataloging in Publication data

A Geologic Time Scale 2004 / Felix M. Gradstein . . . [et al.].
p. cm.
Includes bibliographical references and index.
ISBN 0 521 78142 6 ISBN 0 521 78673 8 (paperback)
1. Geological time. I. Gradstein, F. M.
QE508.G3956 2004
551.7′01–dc22 2004043586

ISBN 0 521 78142 6 hardback
ISBN 0 521 78673 8 paperback

Dedication

We dedicate this third edition of the Geologic Time Scale book to W. B. (Brian) Harland[†]. He was an inspiring leader in practical stratigraphy, its philosophical roots, and its prime product: The Geologic Time Scale!

[†]Deceased.

With the acceptance of a reliable time scale, geology will have gained an invaluable key to further discovery. In every branch of science its mission will be to unify and correlate, and with its help a fresh light will be thrown on the more fascinating problems of the Earth and its Past.

Arthur Holmes, 1913, *The Age of the Earth*

Contents

Contributors

FELIX M. GRADSTEIN
Geological Museum
University of Oslo
PO Box 1172 Blindern
N-0318 Oslo
Norway
felix.gradstein@nhm.uio.no

JAMES G. OGG
Department of Earth and Atmospheric Sciences
550 Stadium Mall Drive
Purdue University
West Lafayette, IN 47907-2051
USA
jogg@purdue.edu

ALAN G. SMITH
Department of Earth Sciences
University of Cambridge
Downing Street
Cambridge CB2 3EQ
UK
ags1@esc.cam.ac.uk

FRITS P. AGTERBERG
Geological Survey of Canada
601 Booth Street
Ottawa, Ontario K1A OE8
Canada

WOUTER BLEEKER
Geological Survey of Canada
601 Booth Street
Ottawa, Ontario K1A OE8
Canada

ROGER A. COOPER
Geological Time Section
Institute of Geological and Nuclear Sciences
PO Box 30368
Lower Hutt
New Zealand

VLADIMIR DAVYDOV
Permian Research Institute
Boise State University
1910 University Drive
Boise, ID 83725-1535
USA

PHIL GIBBARD
Godwin Institute of Quaternary Research
Department of Geography
University of Cambridge
Downing Street
Cambridge CB2 3EN
UK

LINDA A. HINNOV
Department of Earth and Planetary Sciences
The Johns Hopkins University
Baltimore, MD 21218
USA

MICHAEL R. HOUSE[†]
Department of Geology
Southampton Oceanographic Centre
Southampton
UK

LUCAS LOURENS
Faculty of Earth Sciences
Department of Geology
Utrecht University
Budapestlaan 4
3508 TA Utrecht
The Netherlands

HANS-PETER LUTERBACHER
Institut und Museum für Geologie und Palaontologie
Eberhard-Karls Universität
Sigwartstrasse 10
D-72076 Tubingen
Germany

[†]Deceased.

JOHN MCARTHUR
Institute of Geological Sciences
University College London
Gower Street
London WC1E 6BT
UK

MIKE J. MELCHIN
Department of Earth Sciences
St. Francis Xavier University
PO Box 5000
Antigonish, NS B2G2W5
Canada

LAURENCE J. ROBB
Economic Geology Research Institute
Hugh Allsopp Laboratory
University of the Witwatersrand
Private Bag 3, Wits 2050
South Africa

JOHN SHERGOLD
La Freunie
Benayes
19510 Masseret
France

MIKE VILLENEUVE
Geological Survey of Canada
601 Booth Street
Ottawa, ON K1A OE8
Canada

BRUCE R. WARDLAW
US Geological Survey
926A National Center
Reston, VA 20192-0001
USA

JASON ALI
Department of Earth Sciences
University of Hong Kong
Pokfulam Road
Hong Kong
People's Republic of China

HENK BRINKHUIS
Laboratory of Paleobotany and Palynology
Faculty of Biology
Utrecht University
Budapestlaan 4
3584 CD Utrecht
The Netherlands

FREDERIK J. HILGEN
Faculty of Earth Sciences
Department of Geology
Utrecht University
Budapestlaan 4
3584 CD Utrecht
The Netherlands

JERRY HOOKER
The Natural History Museum, Paleontology
Cromwell Road
London SW7 5BD
UK

RICHARD J. HOWARTH
Institute of Geological Sciences
University College London
Gower Street
London WC1E 6BT
UK

ANDREW H. KNOLL
Department of Earth and Planetary Sciences
Harvard University
24 Oxford Street
Cambridge, MA 02138
USA

JACQUES LASKAR
Astronomie et Systemes Dynamiques
Bureau des Longitudes
77 Av. Denfert-Rochereau
F-75014 Paris
France

SIMONETTA MONECHI
Dipartimento di Scienze della Terra
Università di Firenze
4, Via La Pira
I-50121 Firenze
Italy

JAMES POWELL
Dinosystems
105 Albert Road
Richmond
Surrey TW10 6DJ
UK

KENNETH A. PLUMB
PO Box 102
Hawker, ACT 2614
Australia

ISABELLA RAFFI
Dipartimento di Scienze della Terra
Universitario "G. D'Annunzio"
66013 Chieti Scalo
Italy

URSULA RÖHL
Geowissenschaften
Universität Bremen
PO Box 33 04 40
D-28334 Bremen
Germany

PETER SADLER
Department of Earth Sciences
University of California at Riverside
Riverside, CA 92521
USA

ANNIKA SANFILIPPO
Scripps Institution of Oceanography
University of California at San Diego
La Jolla, CA 92033
USA

BIRGER SCHMITZ
Marine Geology, Earth Science Centre
Göteborg University
Box 7064
S-41381 Göteborg
Sweden

NICHOLAS J. SHACKLETON
Godwin Laboratory
Department of Earth Sciences
University of Cambridge
Cambridge CB2 3SA
UK

GRAHAM A. SHIELDS
School of Earth Sciences
James Cook University
Townsville, Old. 4811
Australia

HARALD STRAUSS
Geologisch-Paläontologisches Institut
Westfälische Wilhelms-Universität Münster
48149 Münster
Germany

JAN VEIZER
Ottawa–Carleton Geoscience Centre
University of Ottawa
Ottawa, Ontario K1N 6N5
Canada

J. VAN DAM
Faculty of Earth Sciences
Utrecht University
Budapestlaan 4
3584 CD Utrecht
The Netherlands

THIJS VAN KOLFSCHOTEN
Faculty of Archaeology
Leiden University
Reuvenplaats 4, 2300 RA Leiden
The Netherlands

DOUG WILSON
Department of Geological Sciences
University of California
Santa Barbara, CA
USA

Preface

This study presents the science community with a new geologic time scale for circa 3850 million years of Earth history. The scale encompasses many recent advances in stratigraphy, the science of the layering of strata on Earth. The new scale closely links radiometric and astronomical age dating, and provides comprehensive error analysis on the age of boundaries for a majority of the geologic divisions of time. Much advantage in time scale construction is gained by the concept of stage boundary definition, developed and actively pursued under the auspices of the International Commission on Stratigraphy (ICS), that co-sponsors this study.

It was in 1997 that Alan Smith approached two of us (F.M.G. and J.G.O.) with the request to undertake a new edition of *A Geologic Time Scale 1989* (GTS89) for Cambridge University Press. This was just after the "Phanerozoic Time Scale" with the A3 format time scale colour chart as insert, sponsored by Saga Petroleum in Norway, had appeared in *Episodes*. Although we realized this new request was a tall order, we optimistically accepted. A proposal was formulated and improved through peer review. As with GTS89, the new edition of the book would not necessarily give the very latest developments in any field, but would present a balanced overview designed to be educational and useful for advanced university students. In particular progress with the concept and defining of stage boundaries had delineated most international geologic stages.

Initially, a rather limited slate of specialists was engaged, and we optimistically projected completion of a revised GTS89 at the turn of the Millennium. Slightly after, F.M.G. and J.G.O. became executive officers of ICS for the 2000–2004 term, and the GTS project was incorporated in ICS's formal objectives.

Creating a new GTS in 2000, 2001, or even 2002, turned out to be rather optimistic. The more we involved ourselves in the myriad of challenges in stratigraphy and the Phanerozoic geologic time scale the more we realized that a major overhaul was in order. Rather than updating and revising chapters of GTS89 we set out to re-write the book from scratch and expand geologic period chapters along a "fixed," and ambitious format of text and figures. Advances in time scale methodology involving cycle stratigraphy, mathematics and statistics, stable isotope stratigraphy, and the formidable progress in high-resolution age dating all demanded close attention with data integration and specialists chapters.

The vast progress in Precambrian and Phanerozoic stratigraphy achieved during the last decade required intense involvement of many more geoscientists than initially envisioned. Although the more ambitious scope and bigger team did push back completion deadlines, we are confident it has enhanced the consensus value of the new geologic time scale, named GTS2004. Had we known beforehand that a total of 18 senior and 22 contributing authors, for a total of 40 geoscience specialists from 15 different countries, would work on the book and deal with the new time scale, we might have had second thoughts about our undertaking. The number of e-mails sent "criss-cross" over the globe as part of GTS2004 is in the tens of thousands. A fundamental difference between multidisciplinary studies and geologic time scale studies is that all chapters must align along the arrow of time. To put it simply: the Carboniferous cannot end at 291 Ma with the Permian starting at 299 Ma. Close agreement on type of data and standards admitted in actual time scale building is also vital. Hence, the actual data standardization and time scale calculation for each chapter was kept to a small team in which Mike Villeneuve, Frits Agterberg, F.M.G., and J.G.O. played key roles, with other senior authors as advisors. The new Neogene time scale was developed by Luc Lourens and his team of tuning specialists.

The fascination in creating a new geologic time scale is that it evokes images of creating a beautiful carpet, using many skilled hands. All stitches must conform to a pre-determined pattern, in this case the pattern of physical, chemical, and biological events on Earth aligned along the arrow of time. It is thus that this new scale is a tribute to the truly close cooperation achieved by this slate of outstanding co-authors. We also consider the new time scale a tribute to the scientific competence harbored and fostered by ICS.

We are deeply grateful to all co-authors who without reservation accepted the challenge to be part of this dedicated team, slowly (!) stitching and weaving this carpet of time and its events that are Earth's unique and splendid history.

It is with deep regret that we learned in mid 2002 that one of our most valuable scholars in Paleozoic stratigraphy, Professor Michael House, had passed away, very shortly after submission of his draft chapter on Devonian stratigraphy. It has been an honor to complete the task he set himself to create this erudite chapter of expansive and dramatic Earth history between 416 and 359 Ma. Vascular plants and forests established on Earth, exceptional high global sea level occurred, ice caps formed in the south polar region in late Devonian time, and present continents and shelves assembled on one hemisphere. Old Red Sandstone is one of the Devonian's great continental remnants.

Through the NUNA Conference in Canada in March 2003 on "New Frontiers in the Fourth Dimension: Generation, Calibration and Application of Geologic Timescales" the essay "Toward a natural Precambrian time scale" by Wouter Bleeker came to this book. Hence, this period of over 88% of Earth history is getting some more urgent attention. We thank Mike Villeneuve and his team for organizing this timely geochronology conference.

We are pleased to acknowledge the financial contribution of ExxonMobil, Statoil, ChevronTexaco, and BP. With these vital donations the elaborate graphics became possible. J. G. O. acknowledges partial support by the US National Science Foundation under Grant No. 0313524. The Geological Survey of Canada and the Network of Offshore Records in Geology and Stratigraphy (NORGES) project at the Geology Museum of the University of Oslo assisted with design and printing of the time scale wall chart.

Cambridge University Press patiently awaited the fruits of our labor, and we are much obliged to Matt Lloyd, Sally Thomas, and Lesley Thomas for their thorough editorial advice and assistance.

Figure 1.4 in the Introduction chapter illustrates the 1960 geologic time scale by its pioneer, Arthur Holmes, who introduced period scaling from observed maximum thickness. The appearance datum of this new opus in mid 2004 is nearly 90 years after Arthur Holmes's first humble geologic time scale in 1913 in search of the age of the Earth and its remarkable historic components.

This publication on the International Geologic Time Scale was produced under the auspices of the International Commission on Stratigraphy (ICS). Information on ICS, its organization, its mandate, and its wide-ranging geoscience program can be obtained from www.stratigraphy.org.

Felix M. Gradstein
James G. Ogg
Alan G. Smith

Acknowledgments

The authors and co-authors of the 23 chapters in this book are very pleased to acknowledge the many geoscientists that actively and generously gave of their time to assist with and advise on GTS2004.

F.M.G. and J.G.O. thank Frits Agterberg, Sam Bowring, David Bruton, Pierre Bultynck, Cinzia Cervato, Roger Cooper, Ferdinand Corfu, Sorin Filipescu, Stan Finney, Rich Lane, Luc Lourens, John McArthur, Ed de Mulder, Jurgen Remane, Paul Renne, Otto van Bremen, and Mike Villeneuve for general advice and/or help over the many years of book gestation.

John McArthur also checked every section dealing with geochemistry and reviewed and helped update all relevant figures; Luc Lourens assisted with the task to link the complex and detailed Neogene and Pleistocene chapters; Heiko Pälike offered insight in future trends with regard to orbital tuning of the steadily improving Paleogene time scale.

GTS2004 would not have been possible without Gabi Ogg's tireless dedication and truly hard and highly creative work with scientific design, figure and table drafting, and error checking. Virtually all of the 156 figures came from her "drawing table." We also thank Jane Dolven and Gisli Sigtryggsson for their contributions with drafting, revising, and printing of some figures.

F.M.G. and J.G.O. thank the International Union of Geological Sciences (IUGS) and the Commission for the Geological Map of the World (CGMW) for advice and support with the GTS project.

Linda Hinnov is very grateful to Marie-France Loutre and Timothy Herbert for taking the time to comment on specific issues discussed in her chapter on Earth's orbital parameters and cycle stratigraphy.

Mike Villeneuve thanks Richard Stern and Otto van Bremen for assistance with Chapter 6, Radiogenic isotope geochronolology.

Frits P. Agterberg, author of Chapter 8, Geomathematics, expresses thanks to Roger Cooper, Felix Gradstein, Jim Ogg, Peter Sadler, and Alan Smith for helpful discussions and information; to Mike Villeneuve for discussions regarding geochronological aspects; to Graeme Bonham-Carter and Danny Wright for help with software; and to Gabi Ogg for help with constructing the many diagrams.

Wouter Bleeker reviewed Chapter 9, The Precambrian: the Archean and Proterozoic Eons; his valuable comments improved the manuscript. In Chapter 10, Toward a "Natural" Precambrian time scale, W. Bleeker discussed some of the ideas expressed by Euan Nisbet, who together with Preston Cloud has been a vocal critic of the current Precambrian time scale. Discussions with Yuri Amelin and Richard Stern helped clarify the magnitude of the limitations due to uncertainties in the decay constant. He thanks Felix Gradstein and Mike Villeneuve for their encouragement to formulate this study, and for their critical comments on an early draft of the manuscript. Chapter 10 is Geological Survey of Canada contribution 2 003 066.

Chapter 11, The Cambrian Period, was reviewed by Soren Jensen, Graham Shields, and David Bruton; senior chapter author John Shergold thanks these contributors.

The authors of Chapter 12, The Ordovician Period, Roger Cooper and Peter Sadler, thank Mike Villeneuve, Sam Bowring, and Bill Compston for their comments on radioisotopic dating methods. However, the choice of dates used for calibration of the time scale was theirs alone. Frits Agterberg and Felix Gradstein performed the time scale calculations.

Chapter 13, The Silurian Period, benefited from helpful comments and review by David Bruton, and discussions with Alfred Lenz, Tatiana Koren, David Loydell, Godfrey Nowlan, Gary Mullins, and John Beck. Mike Villeneuve discussed radiometric dating methodology with Roger Cooper, and Frits Agterberg and Felix Gradstein performed age data standardization and time scale and uncertainty calculations. Mike Melchin, who is senior author of the chapter, also gratefully acknowledges financial support from the Natural Sciences and Engineering Research Council of Canada.

Completion of Chapter 14, The Devonian Period, undertaken by F.M.G. after Professor Michael House sadly passed away, would not have been possible without the vital help of Pierre Bultynck, Bernd Kaufmann, and David Bruton, all of whom provided valuable advice and information; their comments improved the text and figures. F.M.G. thanks Dr Jim

House for his assistance in obtaining figures from the estate of Professor House, and thanks E. A. Williams for permission to utilize his illustration of discrepancies between dating methods. Frits P. Agterberg kindly executed the statistical–mathematical routines.

The Carboniferous Period, Chapter 15, was undertaken by senior author Vladimir Davydov. He thanks many individuals for their input toward this chapter on this complex subject. In particular, Peter Jones, John Groves, Paul Brenckle, and Phil Heckel reviewed the text and made several significant suggestions that improved both style and content. The authors are grateful to E. Trapp and B. Kaufmann for providing new TIMS dates from Germany, and thank Sam Bowring and Mike Villeneuve for advice on radiometric dates, Chris Klootwijk for advice on magnetostratigraphy, Frits Agterberg for help with the numerical analysis, and Gabi Ogg for drafting the figures.

The senior author of Chapter 16, The Permian Period, Bruce Wardlaw, thanks Sam Bowring and Frits Agterberg for advice and assistance, and Gabi Ogg for drafting of the many figures.

Chapter 17, The Triassic Period, was authored by J.G.O. He is most grateful to Tim Tozer, Linda Hinnov, Nereo Preto, John McArthur, and, especially, Mike Orchard for providing valuable insights into the intricate history and current debates on Triassic subdivisions and stratigraphy and/or reviewing earlier versions of this summary. The reviewers cautiously advise that the Triassic time scale may evolve in unforeseen ways in the future as GSSPs are formalized and better global correlations are achieved. Josef Pálfy, Sam Bowring, Roland Mundil, and Paul Renne enlightened the author about the constraints and disagreements with radioisotopic ages. Gabi Ogg drafted the final versions of the diagrams.

As author of The Jurassic Period, Chapter 18, J.G.O. acknowledges the countless Jurassic experts who have contributed their expertise on the intricacies of the Jurassic world. In particular, he thanks Josef Pálfy, Nicol Morton, Angela Coe, Bill Wimbledon, and Simon Kelly for reviewing drafts of this chapter; and John McArthur for updating the geochemistry. Gabi Ogg drafted the final versions of the diagrams.

Drafts of The Cretaceous Period, Chapter 19, were reviewed by Peter Rawson, Stephane Reboullet, Jorg Mutterlose, Jurgen Thurow, and Linda Hinnov; and John McArthur updated the geochemistry. Gabi Ogg drafted the final versions of the detailed diagrams. Senior author J.G.O. thanks all these contributors for their vital assistance in elucidating the stratigraphy of the longest period (80 myr) in the Phanerozoic.

The Paleogene Period, Chapter 20, was authored by Hans-Peter Luterbacher and many contributors. Heiko Pälike reviewed and advised on the actual time scale, Jan Hardenbol advised on an early draft of the manuscript, and Frits Agterberg performed the scaling and uncertainty calculations.

The Neogene Period, Chapter 21, came to light under the senior authorship of Luc Lourens; he and his co-authors like to express their thanks to Hayfaa Abdul-Aziz, Torsten Bickert, Katharina Billups, Henk Brinkhuis, Peter DeMenocal, Wout Krijgsman, Klaudia Kuiper, Alan Mix, Heiko Pälike, Jamie Powell, Isabella Raffi, Javier Sierro, Ralf Tiedemann, Elena Turco, Jan Van Dam, Erwin Van der Laan, and Jan-Willem Zachariasse for their indispensable contributions. The authors acknowledge Bill Berggren, whose studies inspired them in writing this chapter. Davide Castradori and Linda Hinnov provided very helpful and constructive comments on an earlier version of this manuscript.

A fruitful discussion on the nettlesome Quaternary issue between F.M.G. and Phil Gibbard during the *First Conference on Future Directions in Stratigraphy* in 2002 in Urbino, Italy, led to Chapter 22, The Pleistocene and Holocene Epochs. Luc Lourens is thanked for a careful review, and Brad Pillans for advice. The authors thank Steve Boreham and Gabi Ogg for help in the compilation of the chart.

Wendy Green and Lawrence Rush helped with getting the references into EndNote. The GSSP maps were drawn with the ATLAS program of Cambridge Paleomap Services, written by Lawrence Rush.

Abbreviations and acronyms

ORGANIZATIONS

CGMW	Commission for the Geological Map of the World
DNAG	Decade of North American Geology
DSDP	Deep Sea Drilling Project
GSC	Geological Survey of Canada
ICS	International Commission of Stratigraphy
IGC	International Geological Congress
IGCP	International Geological Correlation Project
INQUA	International Quaternary Association
IUGS	International Union of Geological Sciences
ODP	Ocean Drilling Project
SNS	Subcommision (of ICS) on Neogene Stratigraphy
SQS	Subcommission (of ICS) on Quaternary Stratigraphy
SOS	Subcommission (of ICS) on Ordovician Stratigraphy
UNESCO	United Nations Education, Scientific, and Cultural Organization
USGS	United States Geological Survey

TIME SCALE PUBLICATIONS (see References for details)

NDS82	*Numerical Dating in Stratigraphy* (Odin *et al.*, 1982)
GTS82	*A Geologic Time Scale* (Harland *et al.*, 1982)
DNAG83	*Geologic Time Scale, Decade of North American Geology* (Palmer, 1983)
KG85	Kent and Gradstein (1985)
EX88	Exxon 1988 (Haq *et al.*, 1988)
GTS89	*A Geologic Time Scale 1989* (Harland *et al.*, 1990)
OB93	Obradovich (1993)
JGR94	*Journal of Geophysical Research* 1994 (Gradstein *et al.*, 1994)
SEPM95	Society for Sedimentary Geology 1995 (Gradstein *et al.*, 1995)
GO96	Gradstein and Ogg (1996)

GEOSCIENTIFIC CONCEPTS

FAD	First appearance datum
FCT	Fish Canyon Tuff sanidine monitor standard (in Ar–Ar dating)
GPTS	Geomagnetic polarity time scale
GSSP	Global Stratotype Section and Point
GSSA	Global Standard Stratigraphic Age (in Precambrian)
HR–SIMS	High-resolution secondary ion mass spectrometry (in U–Pb dating)
LAD	Last appearance datum
LA2003	Laskar 2003 numerical solution of orbital periodicities
MMhb-1	McClure Mountain hornblende monitor standard (in Ar–Ar dating)
SL13	Sri Lanka 13 monitor zircon standard (in HR–SIMS dating)
TIMS	Thermal ionization mass spectrometry (in U–Pb dating)
TCR	Taylor Creek Rhyolite Sanidine monitor standard (in Ar–Ar dating)

SYMBOLS

ka	10^3 years ago (kilo annum)
kyr	10^3 years duration
Ga	10^9 years ago (giga annum)
Ma	10^6 years ago (mega annum)
myr	10^6 years duration
SI	Système Internationale d'Unités
yr	year duration

Part I • Introduction

1 • Introduction

F. M. GRADSTEIN

The development of new dating methods and the extension of existing methods has stimulated the need for a comprehensive review of the geologic time scale. The construction of geologic time scales evolved as a result of applying new ideas, methods, and data.

1.1 A GEOLOGIC TIME SCALE 2004

The geologic time scale is the framework for deciphering the history of the Earth. Since the time scale is the tool "par excellence" of the geological trade, insight in its construction, strengths, and limitations greatly enhances its function and its utility. All earth scientists should understand how the evolving time scales are constructed and calibrated, rather than merely using the numbers in them.

This calibration to linear time of the succession of events recorded in the rock record has three components:

1. the international stratigraphic divisions and their correlation in the global rock record,
2. the means of measuring linear time or elapsed durations from the rock record, and
3. the methods of effectively joining the two scales.

For convenience in international communication, the rock record of Earth's history is subdivided into a "chronostratigraphic" scale of standardized global stratigraphic units, such as "Jurassic," "Eocene," "*Harpoceras falciferum* ammonite zone," or "polarity Chron C24r." Unlike the continuous ticking clock of the "chronometric" scale (measured in years before present), the chronostratigraphic scale is based on relative time units, in which global reference points at boundary stratotypes define the limits of the main formalized units, such as "Devonian." The chronostratigraphic scale is an agreed convention, whereas its calibration to linear time is a matter for discovery or estimation (Fig. 1.1).

A Geologic Time Scale 2004, eds. Felix M. Gradstein, James G. Ogg, and Alan G. Smith. Published by Cambridge University Press.

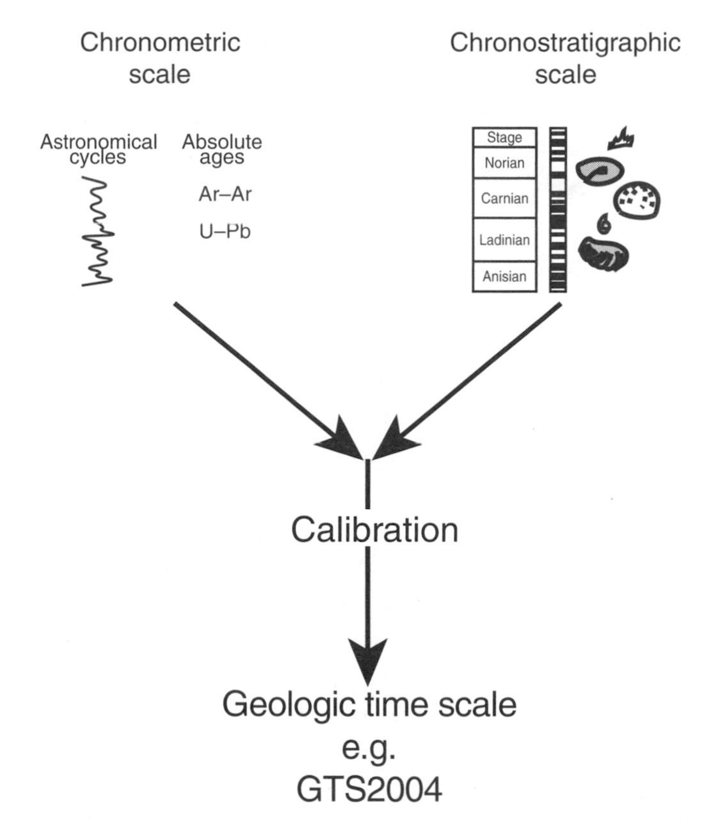

Figure 1.1 The construction of a geologic time scale is the merger of a chronometric scale (measured in years) and a chronostratigraphic scale (formalized definitions of geologic stages, biostratigraphic zonation units, magnetic polarity zones, and other subdivisions of the rock record).

By contrast, Precambrian stratigraphy is formally classified chronometrically (see Chapter 9), i.e. the base of each Precambrian eon, era, and period is assigned a numerical age (Table 1.1).

Continual improvements in data coverage, methodology, and standardization of chronostratigraphic units imply that no geologic time scale can be final. *A Geologic Time Scale 2004* (GTS2004) provides an overview of the status of the geological time scale and is the successor to GTS1989 (Harland *et al.*, 1990), which in turn was preceded by GTS1982 (Harland *et al.*, 1982).

Table 1.1 *Current framework for subdividing terrestrial stratigraphy*

Eon	Era	Definition of base	Age in (Ma)
Phanerozoic	Cenozoic Mesozoic Paleozoic	Boundaries defined in rock (chronostratigraphically) by GSSPs	To be discovered by correlation from GSSPs and dating. Base of Phanerozoic dated at 542 Ma
Proterozoic	Neoproterozoic Mesoproterozoic Paleoproterozoic	Boundaries defined in time (chronometrically) by arbitrary assignment of numerical age	Age of basal Proterozoic defined as 2500 Ma
Archean	Neoarchean Mesoarchean Paleoarchean Eoarchean	Boundaries defined in time (chronometrically) by arbitrary assignment of numerical age	Age of basal Archean not defined

Since 1989, there have been several major developments:

1. Stratigraphic standardization through the work of the International Commission on Stratigraphy (ICS) has greatly refined the international chronostratigraphic scale. In some cases, traditional European-based geological stages have been replaced with new subdivisions that allow global correlation.
2. New or enhanced methods of extracting linear time from the rock record have enabled high-precision age assignments. An abundance of high-resolution radiometric dates have been generated that has led to improved age assignments of key geologic stage boundaries. The use of global geochemical variations, Milankovitch climate cycles, and magnetic reversals have become important calibration tools.
3. Statistical techniques of extrapolating ages and associated uncertainties to stratigraphic events have evolved to meet the challenge of more accurate age dates and more precise zonal assignments. Fossil event databases with multiple stratigraphic sections through the globe can be integrated into composite standards.

The compilation of GTS2004 has involved a large number of specialists, including contributions by past and present chairs of different subcommissions of ICS, geochemists working with radiogenic and stable isotopes, stratigraphers using diverse tools from traditional fossils to astronomical cycles to database programming, and geomathematicians.

The set of chronostratigraphic units (stages, eras) and their computed ages which constitute the main framework for *A Geologic Time Scale 2004* are summarized as Fig. 1.2, with detailed descriptions and stratigraphic scales in appropriate chapters.

1.2 HOW THIS BOOK IS ARRANGED

The foundation of the geologic time scale is the standardized system of international stratigraphic units. Chapter 2 summarizes the philosophy of the construction of this international standard, gives selected examples of defining boundaries, and reviews the origin of the main divisions of eons and eras.

Biostratigraphy, or the use of fossils in the rock record for assigning relative ages, has merged with mathematical and statistical methods to enable scaled composites of global succession of events. Chapter 3 on biostratigraphy summarizes these quantitative methods, which were used to construct the primary standard for most of the Paleozoic time scale (from 542 to 251 Ma).

Periodic multi-thousand-year oscillations in the Earth's orbit and tilt relative to the Sun produce cyclic environmental changes that are recorded in sediments. Chapter 4 summarizes how these astronomical signals are extracted from the sediments and used to construct a very high-resolution time scale that can be tied to the present orbital condition (linear time) or to measure actual elapsed time. Cycle stratigraphy has calibrated the time scales for most of the Neogene Period (i.e. for the past 23 million years), and for portions of the Paleogene Period (from 65 to 23 Ma) and Mesozoic Era (from 251 to 65 Ma).

Reversals of the Earth's geomagnetic field are recorded by sediments, by volcanic rocks, and by the oceanic crust. Chapter 5 explains how the oceanic magnetic anomalies are calibrated with spreading models to produce a powerful correlation tool for sediments deposited during the past 160 million years. These calibrated C-sequence and M-sequence polarity time scales enable assignment of ages to stage boundaries and to

GEOLOGIC TIME SCALE

PHANEROZOIC

CENOZOIC

Period	Epoch		Stage	AGE (Ma)
Quaternary	Holocene			
	Pleistocene			1.81
Neogene	Pliocene	L	Gelasian	2.59
			Piacenzian	3.60
		E	Zanclean	5.33
	Miocene	L	Messinian	7.25
			Tortonian	11.61
		M	Serravallian	13.65
			Langhian	15.97
		E	Burdigalian	20.43
			Aquitanian	23.03
Paleogene	Oligocene	L	Chattian	28.4
		E	Rupelian	33.9
	Eocene	L	Priabonian	37.2
		M	Bartonian	40.4
			Lutetian	48.6
		E	Ypresian	55.8
	Paleocene	L	Thanetian	58.7
		M	Selandian	61.7
		E	Danian	65.5

MESOZOIC

Period	Epoch	Stage	AGE (Ma)
Cretaceous	Late		65.5
		Maastrichtian	70.6
		Campanian	83.5
		Santonian	85.8
		Coniacian	89.3
		Turonian	93.5
		Cenomanian	99.6
	Early	Albian	112.0
		Aptian	125.0
		Barremian	130.0
		Hauterivian	136.4
		Valanginian	140.2
		Berriasian	145.5
Jurassic	Late	Tithonian	150.8
		Kimmeridgian	155.7
		Oxfordian	161.2
	Middle	Callovian	164.7
		Bathonian	167.7
		Bajocian	171.6
		Aalenian	175.6
	Early	Toarcian	183.0
		Pliensbachian	189.6
		Sinemurian	196.5
		Hettangian	199.6
Triassic	Late	Rhaetian	203.6
		Norian	216.5
		Carnian	228.0
	Middle	Ladinian	237.0
		Anisian	245.0
	Early	Olenekian	249.7
		Induan	251.0

PALEOZOIC

Period	Epoch		Stage	AGE (Ma)
Permian	Lopingian			251.0
			Changhsingian	253.8
			Wuchiapingian	260.4
	Guadalupian		Capitanian	265.8
			Wordian	268.0
			Roadian	270.6
	Cisuralian		Kungurian	275.6
			Artinskian	284.4
			Sakmarian	294.6
			Asselian	299.0
Carboniferous	Pennsylvanian	Late	Gzhelian	303.9
			Kasimovian	306.5
		Middle	Moscovian	311.7
		Early	Bashkirian	318.1
	Mississippian	Late	Serpukhovian	326.4
		Middle	Visean	345.3
		Early	Tournaisian	359.2
Devonian	Late		Famennian	374.5
			Frasnian	385.3
	Middle		Givetian	391.8
			Eifelian	397.5
	Early		Emsian	407.0
			Pragian	411.2
			Lochkovian	416.0
Silurian	Pridoli			418.7
	Ludlow		Ludfordian	421.3
			Gorstian	422.9
	Wenlock		Homerian	426.2
			Sheinwoodian	428.2
	Llandovery		Telychian	436.0
			Aeronian	439.0
			Rhuddanian	443.7
Ordovician	Late		Hirnantian	445.6
				455.8
				460.9
	Middle		Darriwilian	468.1
				471.8
	Early			478.6
			Tremadocian	488.3
Cambrian	Furongian			
			Paibian	501
	Middle			513
	Early			542.0

PRECAMBRIAN

Eon	Era	Period	AGE (Ma)
Proterozoic	Neoproterozoic		542
		Ediacaran	~630
		Cryogenian	850
		Tonian	1000
	Mesoproterozoic	Stenian	1200
		Ectasian	1400
		Calymmian	1600
	Paleoproterozoic	Statherian	1800
		Orosirian	2050
		Rhyacian	2300
		Siderian	2500
Archean	Neoarchean		2800
	Mesoarchean		3200
	Paleoarchean		3600
	Eoarchean	Lower limit is not defined	

Figure 1.2 Summary of *A Geologic Time Scale 2004*.

biostratigraphic and other stratigraphic events through much of that interval.

Chapter 6 on radiogenic isotopes summarizes the evolving techniques used to acquire high-precision ages from the rock record. However, high precision does not always imply accuracy, and this chapter explains some of the pitfalls induced by geological distortions or laboratory standards.

Stable isotopes of strontium reveal a wealth of information about past environmental conditions and geochemical cycling. Chapter 7 explains the use of trends in the strontium isotope ratios of past seawater for global correlation and for relative scaling of stratigraphic events, and presents these trends for the past 600 million years.

Assembling the array of radiometric, biostratigraphic, cycles, magnetic, and other data into a unified geologic time scale, and extrapolating the ages and uncertainties on stratigraphic boundaries is the topic of Chapter 8, Geomathematics. This chapter also details construction methods and results for GTS2004.

The Precambrian encompasses the 4 billion years from the formation of the Earth to the evolution of multicellular life. In addition to summarizing major geologic and geochemical trends, the two chapters on the Precambrian highlight the philosophical difference in establishing chronostratigraphic subdivisions based on pure linear age versus identifying significant global events.

The Phanerozoic (the past 542 million years of Earth history) is subdivided into 11 periods. Each of the "period chapters" has three principal parts: an explanation of the formal subdivision into stages using global boundary stratotypes associated with primary and secondary correlation markers; a summary of the biostratigraphy, cycle stratigraphy, magnetic stratigraphy, and geochemical stratigraphy features that are applied to construct high-resolution relative time scales; and the methods of calibration to a linear time framework. Each period chapter includes a detailed graphic presentation of its integrated geologic time scale, and these are drawn at a uniform scale among all chapters and in the color plates section to allow visual comparison of rates.

The summary of GTS2004 (Fig. 1.2) in Chapter 23 reviews the entire geologic time scale, summarizes its construction and uncertainties, and outlines potentially rewarding directions for future time scale research.

1.3 CONVENTIONS AND STANDARDS

Ages are given in years before "Present" (BP). To avoid a constantly changing datum, "Present" was fixed as AD 1950 (as in ^{14}C determinations), the point in time at which modern isotope dating research began in laboratories around the world. For most geologists, this offset of official "Present" from "today" is not important. However, for archeologists and researchers into events during the Holocene (the past 11500 years), the current offset (50 years) between the "BP" convention from radiometric laboratories and actual total elapsed calendar years becomes significant.

For clarity, the linear age in years is abbreviated as "*a*" (for annum), and ages are generally measured in *ka* or *Ma*, for thousands, millions, or billions of years before present. The elapsed time or duration is abbreviated as "*yr*" (for year), and durations are generally in *kyr* or *myr*. Therefore, the Cenozoic began at 65.5 Ma, and spans 65.5 myr (to the present day).

The uncertainties on computed ages or durations are expressed as standard deviation (1-sigma or 68% confidence) or 2-sigma (95% confidence). The uncertainty is indicated by "±" and will have implied units of thousands or millions of years as appropriate to the magnitude of the age. Therefore, an age cited as "124.6 ± 0.3 Ma" implies a 0.3 myr uncertainty (1-sigma, unless specified as 2-sigma) on the 124.6 Ma date. We present the uncertainties (±) on summary graphics of the geologic time scale as 2-sigma (95% confidence) values.

Geologic time is measured in years, but the standard unit for time is the second *s*. Because the Earth's rotation is not uniform, this "second" is not defined as a fraction (1/86 400) of a solar day, but as the *atomic second*. The basic principle of the atomic clock is that electromagnetic waves of a particular frequency are emitted when an atomic transition occurs. In 1967, the Thirteenth General Conference on Weights and Measures defined the *atomic second* as the duration of 9 192 631 770 periods of the radiation corresponding to the transition between two hyperfine levels of the ground state of cesium-133. This value was established to agree as closely as possible with the solar-day second. The frequency of 9 192 631 770 hertz (Hz), which the definition assigns to cesium radiation was carefully chosen to make it impossible, by any existing experimental evidence, to distinguish the atomic second from the ephemeris second based on the Earth's motion. The advantage of having the atomic second as the unit of time in the International System of Units is the relative ease, in theory, for anyone to build and calibrate an atomic clock with a precision of 1 part per 10^{11} (or better). In practice, clocks are calibrated against broadcast time signals, with the frequency oscillations in hertz being the "pendulum" of the atomic time keeping device.

1 year is approximately 31.56 mega seconds (1 a = ~31.56 Ms).

The Système Internationale d'Unités (SI) conventions at 10^3 intervals that are relevant for spans of geologic time through sizes of microfossils are:

10^9	giga G
10^6	mega M
10^3	kilo k
10^0	unity 1
10^{-3}	milli m
10^{-6}	micro μ
10^{-9}	nanno n

Although dates assigned in the geologic time scale are measured in multiples of the atomic second as unit of time (year), there are two other types of seconds: *mean solar* second and *ephemeris* second.

1.3.1 Universal time

Universal time is utilized in the application of astronomy to navigation. Measurement of universal time is made directly from observing the times of transits of stars; since the Earth's rotation is not uniform, corrections are applied to obtain a more uniform time system. In essence, universal time is the mean solar time on the Greenwich meridian, reckoned in days of 24 mean solar hours beginning with zero hour at midnight, and derives from the average rate of the daily motion of the Sun relative to the meridian of Greenwich. The *mean solar second* is 1/86 400 of the mean solar day, but because of non-uniformity this unit is no longer the standard of international time.

1.3.2 Ephemeris time

Ephemeris time (ET) is uniform and obtained from observation by directly comparing positions of the Sun, Moon, and the planets with calculated ephemerides of their coordinates. Webster's dictionary defines ephemeris as any tabular statement of the assigned places of a celestial body for regular intervals. Ephemeris time is based on the *ephemeris second* defined as 1/31 556 925.9447 of the tropical year for 1900 January 0 day 12 hour ET. The ephemeris day is 86 400 ephemeris seconds, which unit in 1957 was adopted by the International Astronomical Union as the fundamental invariable unit of time.

1.4 HISTORICAL OVERVIEW OF GEOLOGIC TIME SCALES

Stitching together the many data points on the loom of time requires an elaborate combination of Earth science and mathematical/statistical methods. Hence, the time and effort involved in constructing a new geologic time scale and assembling all relevant information is considerable. Because of this, and because continuous updating in small measure with new information is not advantageous to the stability of any common standard, new geologic time scales spanning the entire Phanerozoic tend to come out sparsely (e.g. Harland *et al.*, 1982, 1990; Gradstein and Ogg, 1996; Remane, 2000).

In the absence of accepted accurate dates at each stage boundary, extrapolating the ages of geologic stages is a major challenge in time scale building and various methods have been employed by different compilations, including this GTS2004 version. A major challenge in itself is to try to understand the precision of radiometric ages, including calibrations between different radiometric methods, now that analytical errors are greatly reduced.

Figure 1.3 summarizes those and some others in terms of 12 methods applied since 1937. Radiometric age dating, stratigraphic reasoning, and biostratigraphic/geomagnetic calibrations are three corner stones of time scale building. Stratigraphic reasoning, although fuzzy, evaluates the complex web of correlations around stage boundaries or other key levels, and is paramount in the science of stratigraphy. Geomathematical methods involve mathematical/statistical routines and interpolations that can estimate margins of error on limits of stratigraphic units; such errors are of two main types, stratigraphic and analytical (see Chapter 8). Tuning of cyclic sequences to orbital time scales, either counting back from an anchor level such as the "present," or tuning individual cyclic segments with orbital periodicities (floating time scale), has the potential to be the most accurate calibration of the geologic time scale (see Chapters 4 and 21). Such an orbitally tuned time scale can also calibrate the standards and decay constants of radiometric methods.

1.4.1 Arthur Holmes and age–thickness interpolations

Arthur Holmes (1890–1965) was the first to combine radiometric ages with geologic formations in order to create a geologic time scale. His book, *The Age of the Earth* (1913, 2nd edition 1937), written when he was only 22, had a major impact on those interested in geochronology. For his pioneering scale, Holmes carefully plotted four radiometric dates, one in the Eocene and three in the Paleozoic from radiogenic helium and lead in uranium minerals, against estimates of the accumulated maximum thickness of Phanerozoic sediments. If we ignore sizable error margins, the base of Cambrian interpolates at 600 Ma, curiously close to modern estimates. The new approach was

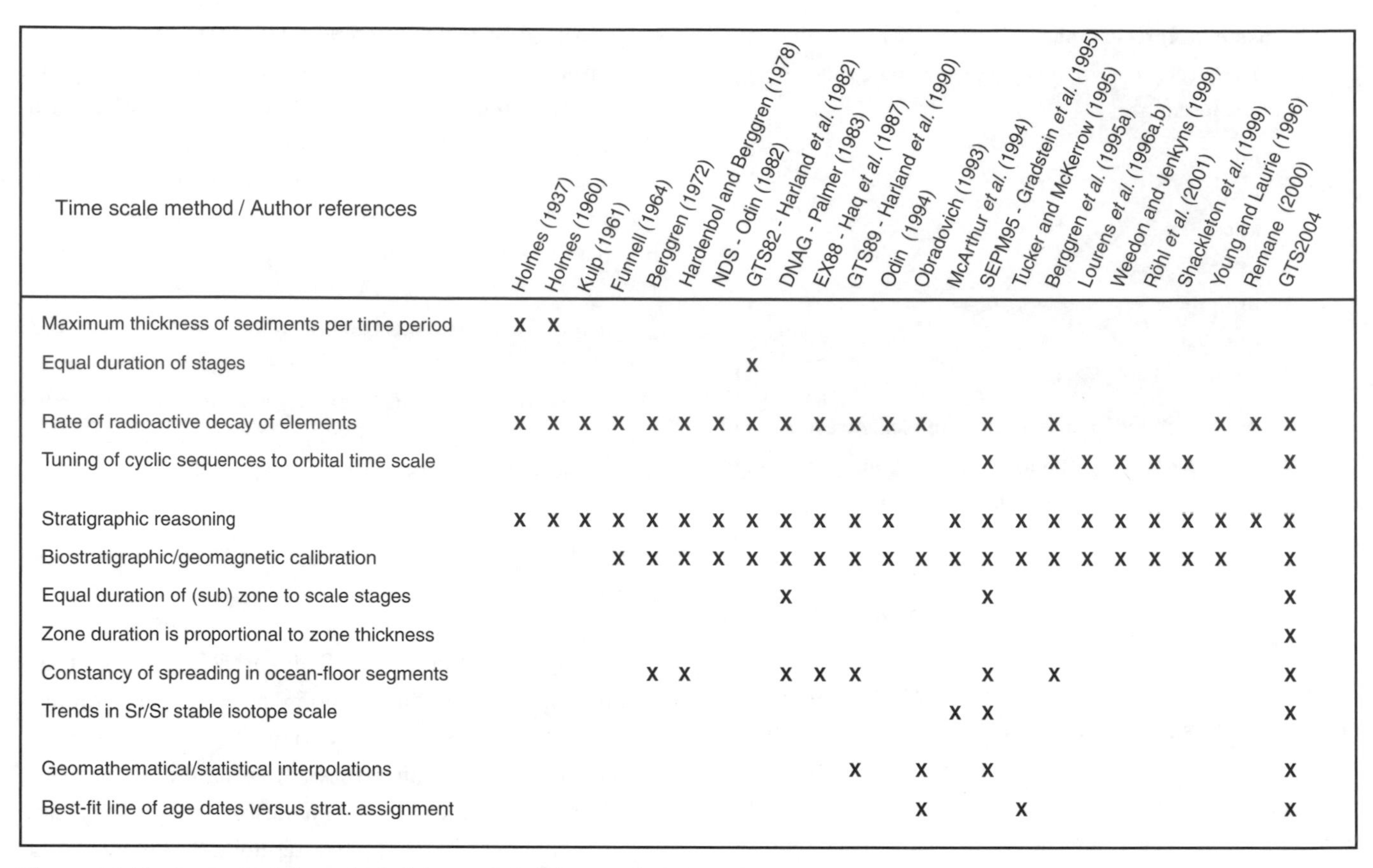

Time scale method / Author references	Holmes (1937)	Holmes (1960)	Kulp (1961)	Funnell (1964)	Berggren (1972)	Hardenbol and Berggren (1978)	NDS - Odin (1982)	GTS82 - Harland *et al.* (1982)	DNAG - Palmer (1983)	EX88 - Haq *et al.* (1987)	GTS89 - Harland *et al.* (1990)	Odin (1994)	Obradovich (1993)	McArthur *et al.* (1994)	SEPM95 - Gradstein *et al.* (1995)	Tucker and McKerrow (1995)	Berggren *et al.* (1995a)	Lourens *et al.* (1996a,b)	Weedon and Jenkyns (1999)	Röhl *et al.* (2001)	Shackleton *et al.* (1999)	Young and Laurie (1996)	Remane (2000)	GTS2004
Maximum thickness of sediments per time period	X	X																						
Equal duration of stages								X																
Rate of radioactive decay of elements	X	X	X	X	X	X	X	X	X	X	X	X	X		X		X					X	X	X
Tuning of cyclic sequences to orbital time scale															X		X	X	X	X	X			X
Stratigraphic reasoning	X	X	X	X	X	X	X	X	X	X	X	X		X	X	X	X	X	X	X	X	X	X	X
Biostratigraphic/geomagnetic calibration				X	X	X	X	X	X	X	X	X	X	X	X	X	X	X	X	X	X	X		X
Equal duration of (sub) zone to scale stages									X						X									X
Zone duration is proportional to zone thickness																								X
Constancy of spreading in ocean-floor segments					X	X			X	X	X				X		X							X
Trends in Sr/Sr stable isotope scale														X	X									X
Geomathematical/statistical interpolations											X		X		X									X
Best-fit line of age dates versus strat. assignment													X			X								X

Figure 1.3 Twelve methods in geological time scale building applied since 1937.

a major improvement over a previous "hour-glass" method that tried to estimate maximum thickness of strata per period to determine their relative duration, but had no way of estimating rates of sedimentation independently. As late as 1960, Holmes, being well aware of limitations, elegantly phrased it thus (p. 184):

> The [now obsolete] 1947 scale was tied to the five dates listed. . . . In order to estimate dates for the beginning and end of each period by interpolation, I adopted a modification of Samuel Haughton's celebrated principle of 1878 that "the proper relative measure of geological periods is the maximum thickness of strata formed during those periods", and plotted the five dates against the cumulative sums of the maximum thicknesses in what were thought to be their most probable positions. I am fully aware that this method of interpolation has obvious weaknesses, but at least it provides an objective standard, and so far as I know, no one has suggested a better one.

In 1960, Holmes compiled a revised version of the age-versus-thickness scale (Fig. 1.4). Compared with the initial 1913 scale, the projected durations of the Jurassic and Permian are more or less doubled, the Triassic and Carboniferous are extended about 50%, and the Cambrian gains 20 myr at the expense of the Ordovician.

1.4.2 Phanerozoic radiometric databases, statistical scales, and compilations

W. B. Harland and E. H. Francis as part of a Phanerozoic time scale symposium coordinated a systematic, numbered radiometric database with critical evaluations. Items 1–337 in *The Phanerozoic Time-Scale: A Symposium* (Harland *et al.*, 1964) were listed in the order as received by the editors. Supplements of items 338–366 were assembled by the Geological Society's Phanerozoic Time-scale Sub-Committee from publications omitted from the previous volume or published between 1964 and 1968, and of items 367–404 relating specifically to the Pleistocene most were provided by N. J. Shackleton. The compilation of these additional items with critical evaluations was included in *The Phanerozoic Time-Scale: A Supplement* (Harland and Francis, 1971). In 1978, R. L. Armstrong published a re-evalution and continuation of *The Phanerozoic Time-Scale* database (Armstrong, 1978). This publication did not include abstracting and critical commentary. These catalogs of items 1–404 and of Armstrong's continuation of items

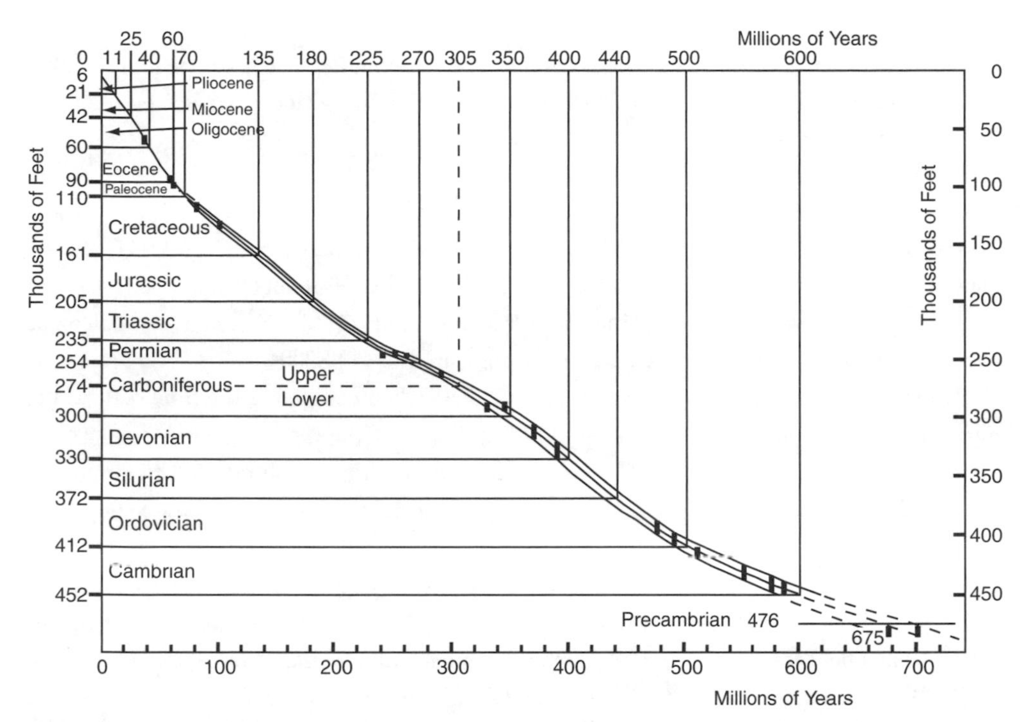

Figure 1.4 Scaling concept employed by Arthur Holmes in the first half of the previous century to construct the geologic time scale. The cumulative sum of maximum thicknesses of strata in thousands of feet per stratigraphic unit is plotted along the vertical axis and selected radiometric dates from volcanic tuffs, glauconites, and magmatic intrusives along the horizontal linear axis. This version (Holmes, 1960) incorporated an uncertainty envelope from the errors on the radiometric age constraints.

404–522 were denoted "PTS" and "A," respectively, in later publications.

In 1976, the Subcommision on Geochronology recommended an intercalibrated set of decay constants and isotopic abundances for the U–Th–Pb, Rb–Sr, and K–Ar systems with the uranium decay constants by Jaffey *et al.* (1971) as the mainstay for the standard set (Steiger and Jaeger, 1978). This new set of decay constants necessitated systematic upward or downward revisions of previous radiometric ages by 1–2%.

In *A Geological Time Scale* (Cambridge University Press, 1982), Harland *et al.* standardized the Mesozoic–Paleozoic portion of the previous PTS–A series to the new decay constants and included a few additional ages published in *Contributions to the Geological Time Scale* (Cohee *et al.*, 1978) and by McKerrow *et al.* (1985). Simultaneously, G. S. Odin supervised a major compilation and critical review of 251 radiometric dating studies as Part II of *Numerical Dating in Stratigraphy* (Odin, 1982). This "NDS" compilation also re-evaluated many of the dates included in the previous "PTS–A" series. A volume of papers on *The Chronology of the Geological Record* (Snelling, 1985) from a 1982 symposium included re-assessments of the combined PTS–NDS database with additional data for different time intervals.

After applying rigorous selection criteria to the PTS–A and NDS databases and incorporating many additional studies (mainly between 1981 and 1988) in a statistical evaluation, Harland and co-workers presented *A Geological Time Scale 1989* (Cambridge University Press, 1990).

The statistical method of time scale building employed by GTS82 and refined by GTS89 derived from the marriage of the chronogram concept with the chron concept, both of which represented an original path to a more reproducible and objective scale. Having created a high-temperature radiometric age data set, the chronogram method was applied that minimizes the misfit of stratigraphically inconsistent radiometric age dates around trial boundary ages to arrive at an estimated age of stage boundaries. From the error functions a set of age/stage plots was created (Appendix 4 in GTS89) that depict the best age estimate for Paleozoic, Mesozoic, and Cenozoic stage boundaries. Because of wide errors, particularly in Paleozoic and Mesozoic dates, GTS89 plotted the chronogram ages for stage boundaries against the same stages with relative durations scaled proportionally to their component "chrons." For convenience, chrons were equated with biostratigraphic zones. The chron concept in GTS89 implied equal duration of zones in prominent biozonal schemes, such as a conodont scheme for the Devonian. In Chapter 8, the chronogram method is discussed in more detail and compared with the maximum likelihood method of interpolations using all radiometric ages, not only chronostratigraphically inconsistent ones.

The Bureau de Récherches Géologiques et Minières and the Société Géologique de France published a stratigraphic scale and time scale compiled by Odin and Odin (1990). Of the more than 90 Phanerozoic stage boundaries, 20 lacked adequate radiometric constraints, the majority of which were in the Paleozoic.

Three compilations spanning the entire Phanerozoic were published in the late 1990s. A comprehensive review of the geologic time scale by Young and Laurie (1996) was oriented toward correlating Australian strata to international standards, and is rich in detail, graphics, and zonal charts. Gradstein and Ogg (1996) assembled a composite Phanerozoic scale from various published sources, including McKerrow *et al.* (1985), Berggren *et al.* (1995a), Gradstein *et al.* (1995), Roberts *et al.* (1995a), and Tucker and McKerrow (1995). The International Stratigraphic Chart (Remane, 2000) is an important document for stratigraphic nomenclature (including Precambrian), and included a contrast of age estimates for stratigraphic boundaries modified from Odin and Odin (1990), Odin (1994), Berggren *et al.* (1995a), and individual ICS subcommissions.

During the 1990s, a series of developments in integrated stratigraphy and isotopic methodology enabled relative and linear geochronology at unprecedented high resolution. Magnetostratigraphy provided correlation of biostratigraphic datums to marine magnetic anomalies for the Late Jurassic through Cenozoic. Argon–argon dating of sanidine crystals and new techniques of uranium–lead dating of individual zircon crystals yielded ages for sediment-hosted volcanic ashes with analytical precessions less than 1%. Comparison of volcanic-derived ages to those obtained from glauconite grains in sediments indicated that the majority of glauconite grains yielded systematically younger ages (e.g. Obradovich, 1998; Gradstein *et al.*, 1994a), thereby removing a former method of obtaining direct ages on stratigraphic levels. Pelagic sediments record features from the regular climate oscillations produced by changes in the Earth's orbit, and recognition of these "Milankovitch" cycles allowed precise tuning of the associated stratigraphy to astronomical constants.

Aspects of the GTS89 compilation began a trend in which different portions of the geologic time scale were calibrated by different methods. The Paleozoic and early Mesozoic portions continued to be dominated by refinement of integrating biostratigraphy with radiometric tie points, whereas the late Mesozoic and Cenozoic also utilized oceanic magnetic anomaly patterns and astronomical tuning.

A listing of the radiometric dates and discussion of specific methods employed in building GTS 2004 can be found within individual chapters relating to specific geological periods.

1.4.3 Paleozoic scales

The Paleozoic spans 291 myr between 542 and 251 Ma. Its estimated duration has decreased about 60 myr since the scales of Holmes (1960) and Kulp (1961). Selected key Paleozoic time scales are compared to GTS2004 in Fig. 1.5*a*,*b*; historic changes stand out best when comparing the time scale at the period level in Fig. 1.5*a*.

Differences in relative estimated durations of component period and stages are substantial (e.g. the Ludlow Stage in the Silurian, or the Emsian Stage in the Devonian). Whereas most of the Cenozoic and Mesozoic have had relatively stable stage nomenclature for some decades (Figs. 1.6 and 1.7), the prior lack of an agreed nomenclature for the Permian, Carboniferous, Ordovician, and Cambrian periods complicates comparison of time scales (Fig. 1.5*a*; see also Chapters 11, 12, 15, and 16).

The 570 through 245 Ma Paleozoic time scale in GTS89 derived from the marriage of the chronogram method with the chron concept. The chron concept in GTS89 implied equal duration of zones in prominent biozonal schemes, such as a conodont scheme for the Devonian, etc. The two-way graphs for each period in the Paleozoic were interpolated by hand, weighting tie points subjectively. Error bars on stage boundaries calculated with the chronogram method were lost in the process of drawing the best-fit line. The fact that the Paleozoic suffered both from a lack of data points and relatively large uncertainties led to poorly constrained age estimates for stages; this uncertainty is readily noticeable in the chronogram/chron figures of GTS89.

The 545 through 248 Ma Paleozoic part of the Phanerozoic time scale of Gradstein and Ogg (1996) is a composite from various sources, including the well-known scales by McKerrow *et al.* (1985), Harland *et al.* (1990), Roberts *et al.* (1995a), and Tucker and McKerrow (1995).

The *International Stratigraphic Chart* (Remane, 2000) provides two different sets of ages for part of the Paleozoic stage boundaries. The column that has ages for most stages appears to slightly update Odin and Odin (1990), and Odin (1994) and is shown here.

Modern radiometric techniques that are having significant impact for Paleozoic dates include high-precision U–Pb dates from magmatic zircon crystals in tuffs (K-bentonites) wedged in marine strata that supercede older schemes with ^{40}Ar–^{39}Ar, Rb–Sr, and K–Ar dates on minerals like glauconite, and on

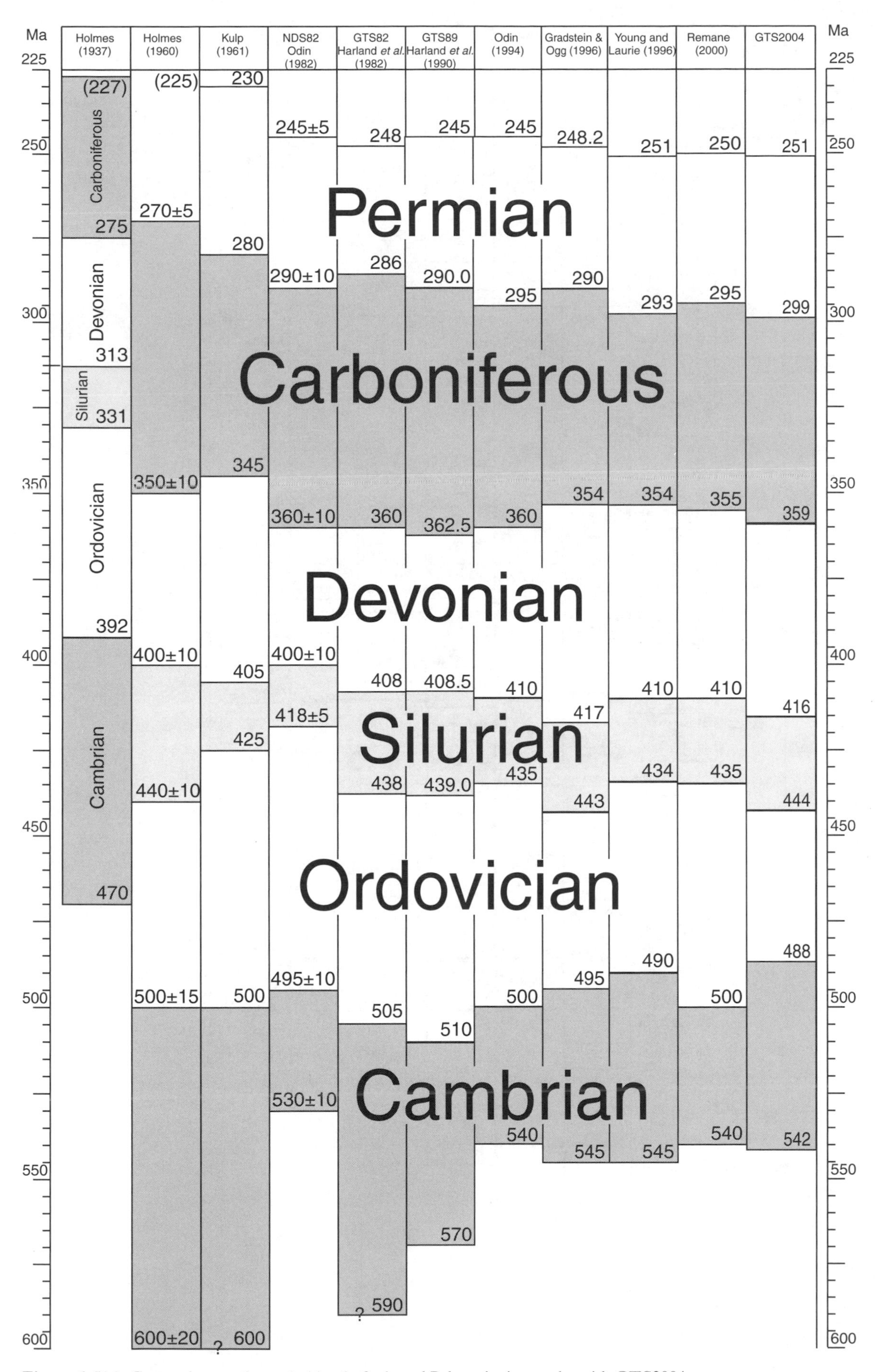

Figure 1.5(*a*) Comparison at the period level of selected Paleozoic time scales with GTS2004.

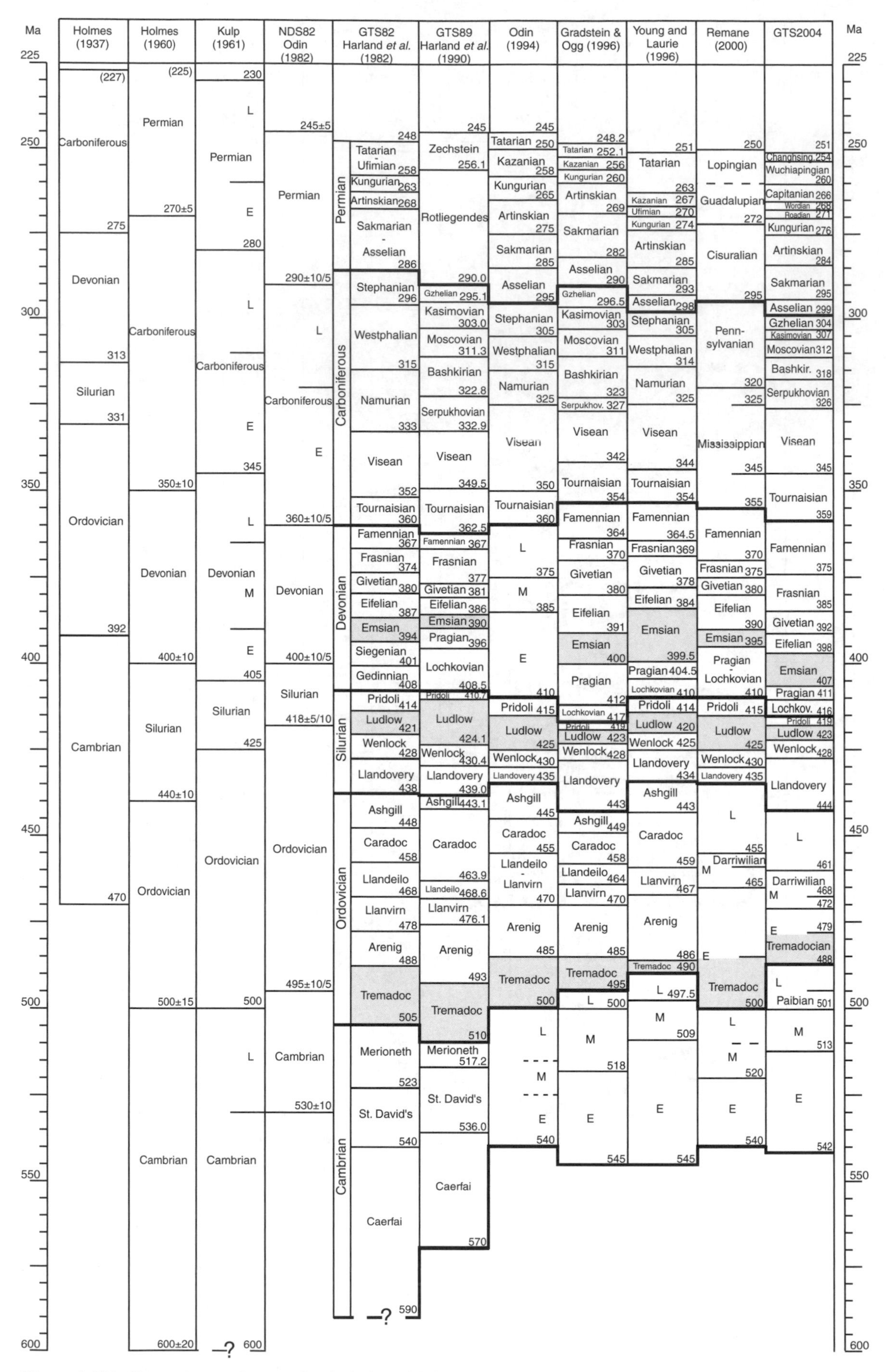

Figure 1.5(*b*) Comparison at the stage level of selected Paleozoic time scales with GTS2004. In some columns, epochs and stages are stacked together; scales 1, 2, and 3 are more detailed than shown.

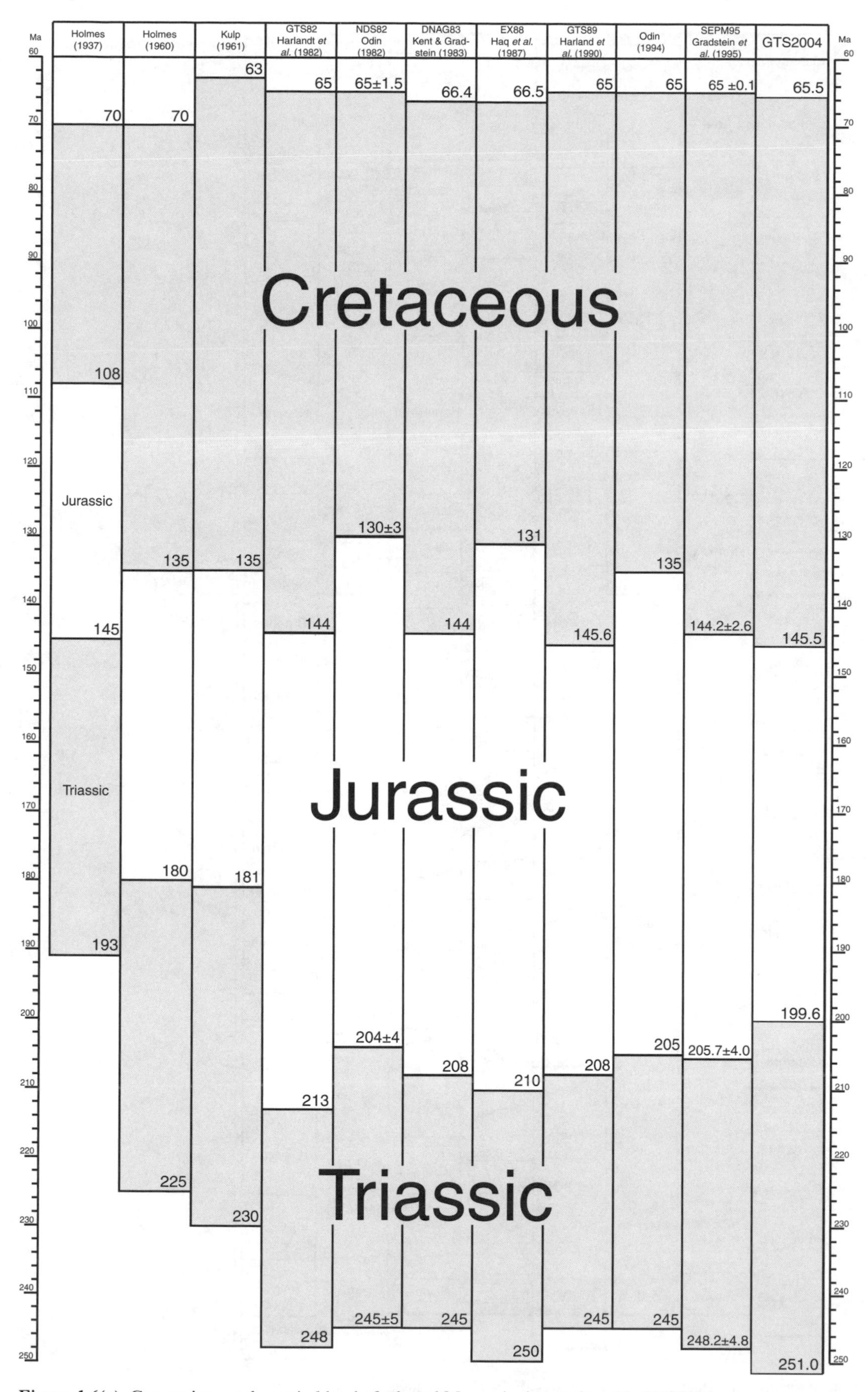

Figure 1.6(*a*) Comparison at the period level of selected Mesozoic time scales with GTS2004.

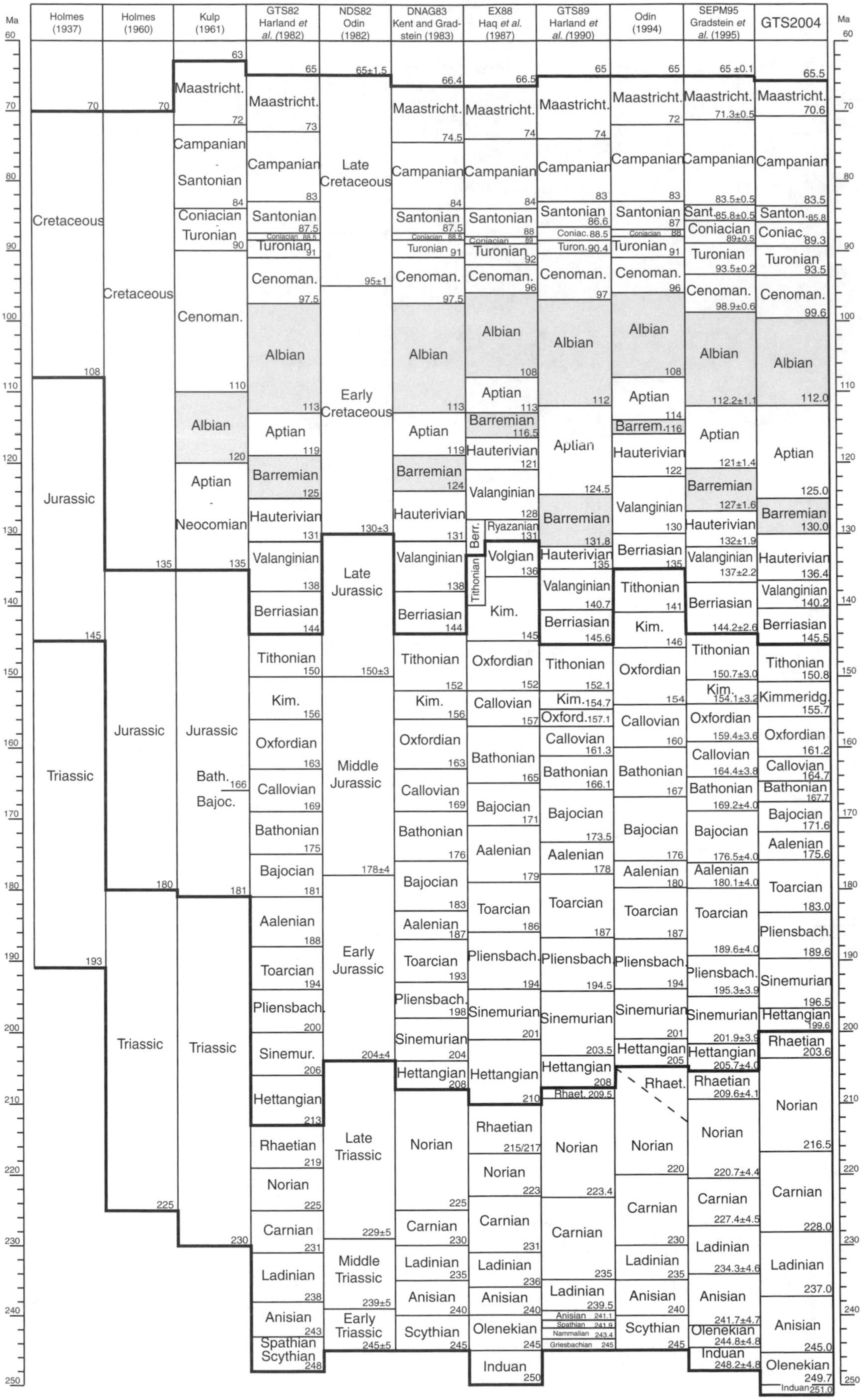

Ma 60–250

Holmes (1937): 70 · Cretaceous · 108 · Jurassic · 145 · Triassic · 193

Holmes (1960): 70 · Cretaceous · 135 · Jurassic · 180 · Triassic · 225

Kulp (1961): 63 · Maastricht. · 72 · Campanian - Santonian · 84 · Coniacian - Turonian · 90 · Cenoman. · 110 · Albian · 120 · Aptian - Neocomian · 135 · Jurassic · Bath. 166 Bajoc. · 181 · Triassic · 230

GTS82 Harland *et al.* (1982): 65 · Maastricht. · 73 · Campanian · 83 · Santonian · 87.5 · Coniacian · 88.5 · Turonian · 91 · Cenoman. · 97.5 · Albian · 113 · Aptian · 119 · Barremian · 125 · Hauterivian · 131 · Valanginian · 138 · Berriasian · 144 · Tithonian · 150 · Kim. · 156 · Oxfordian · 163 · Callovian · 169 · Bathonian · 175 · Bajocian · 181 · Aalenian · 188 · Toarcian · 194 · Pliensbach. · 200 · Sinemur. · 206 · Hettangian · 213 · Rhaetian · 219 · Norian · 225 · Carnian · 231 · Ladinian · 238 · Anisian · 243 · Spathian Scythian · 248

NDS82 Odin (1982): 65±1.5 · Late Cretaceous · 95±1 · Early Cretaceous · 130±3 · Late Jurassic · 150±3 · Middle Jurassic · 178±4 · Early Jurassic · 204±4 · Late Triassic · 229±5 · Middle Triassic · 239±5 · Early Triassic · 245±5

DNAG83 Kent and Gradstein (1983): 66.4 · Maastricht. · 74.5 · Campanian · 84 · Santonian · 87.5 · Coniacian · 88.5 · Turonian · 91 · Cenoman. · 97.5 · Albian · 113 · Aptian · 119 · Barremian · 124 · Hauterivian · 131 · Valanginian · 138 · Berriasian · 144 · Tithonian · 152 · Kim. · 156 · Oxfordian · 163 · Callovian · 169 · Bathonian · 176 · Bajocian · 183 · Aalenian · 187 · Toarcian · 193 · Pliensbach. · 198 · Sinemurian · 204 · Hettangian · 208 · Norian · 225 · Carnian · 230 · Ladinian · 235 · Anisian · 240 · Scythian · 245

EX88 Haq *et al.* (1987): 66.5 · Maastricht. · 74 · Campanian · 84 · Santonian · 88 · Coniacian · 89 · Turonian · 92 · Cenoman. · 96 · Albian · 108 · Aptian · 113 · Barremian · 116.5 · Hauterivian · 121 · Valanginian · 128 · Berr. Ryazanian · 131 · Tithonian Volgian · 136 · Kim. · 145 · Oxfordian · 152 · Callovian · 157 · Bathonian · 165 · Bajocian · 171 · Aalenian · 179 · Toarcian · 186 · Pliensbach. · 194 · Sinemurian · 201 · Hettangian · 210 · Rhaetian · 215/217 · Norian · 223 · Carnian · 231 · Ladinian · 236 · Anisian · 240 · Olenekian · 245 · Induan · 250

GTS89 Harland *et al.* (1990): 65 · Maastricht. · 74 · Campanian · 83 · Santonian · 86.6 · Coniac. · 88.5 · Turon. · 90.4 · Cenoman. · 97 · Albian · 112 · Aptian · 124.5 · Barremian · 131.8 · Hauterivian · 135 · Valanginian · 140.7 · Berriasian · 145.6 · Tithonian · 152.1 · Kim. · 154.7 · Oxford. · 157.1 · Callovian · 161.3 · Bathonian · 166.1 · Bajocian · 173.5 · Aalenian · 178 · Toarcian · 187 · Pliensbach. · 194.5 · Sinemurian · 203.5 · Hettangian · 208 · Rhaet. · 209.5 · Norian · 223.4 · Carnian · 235 · Ladinian · 239.5 · Anisian · 241.1 · Spathian · 241.9 · Nammalian · 243.4 · Griesbachian · 245

Odin (1994): 65 · Maastricht. · 72 · Campanian · 83 · Santonian · 87 · Coniacian · 88 · Turonian · 91 · Cenoman. · 96 · Albian · 108 · Aptian · 114 · Barrem. · 116 · Hauterivian · 122 · Valanginian · 130 · Berriasian · 135 · Tithonian · 141 · Kim. · 146 · Oxfordian · 154 · Callovian · 160 · Bathonian · 167 · Bajocian · 176 · Aalenian · 180 · Toarcian · 187 · Pliensbach. · 194 · Sinemurian · 201 · Hettangian · 205 · Rhaet. · Norian · 220 · Carnian · 230 · Ladinian · 235 · Anisian · 240 · Scythian · 245

SEPM95 Gradstein *et al.* (1995): 65 ±0.1 · Maastricht. · 71.3±0.5 · Campanian · 83.5±0.5 · Sant. · 85.8±0.5 · Coniacian · 89±0.5 · Turonian · 93.5±0.2 · Cenoman. · 98.9±0.6 · Albian · 112.2±1.1 · Aptian · 121±1.4 · Barremian · 127±1.6 · Hauterivian · 132±1.9 · Valanginian · 137±2.2 · Berriasian · 144.2±2.6 · Tithonian · 150.7±3.0 · Kim. · 154.1±3.2 · Oxfordian · 159.4±3.6 · Callovian · 164.4±3.8 · Bathonian · 169.2±4.0 · Bajocian · 176.5±4.0 · Aalenian · 180.1±4.0 · Toarcian · 189.6±4.0 · Pliensbach. · 195.3±3.9 · Sinemurian · 201.9±3.9 · Hettangian · 205.7±4.0 · Rhaetian · 209.6±4.1 · Norian · 220.7±4.4 · Carnian · 227.4±4.5 · Ladinian · 234.3±4.6 · Anisian · 241.7±4.7 · Olenekian · 244.8±4.8 · Induan · 248.2±4.8

GTS2004: 65.5 · Maastricht. · 70.6 · Campanian · 83.5 · Santon. · 85.8 · Coniac. · 89.3 · Turonian · 93.5 · Cenoman. · 99.6 · Albian · 112.0 · Aptian · 125.0 · Barremian · 130.0 · Hauterivian · 136.4 · Valanginian · 140.2 · Berriasian · 145.5 · Tithonian · 150.8 · Kimmeridg. · 155.7 · Oxfordian · 161.2 · Callovian · 164.7 · Bathonian · 167.7 · Bajocian · 171.6 · Aalenian · 175.6 · Toarcian · 183.0 · Pliensbach. · 189.6 · Sinemurian · 196.5 · Hettangian · 199.6 · Rhaetian · 203.6 · Norian · 216.5 · Carnian · 228.0 · Ladinian · 237.0 · Anisian · 245.0 · Olenekian · 249.7 · Induan · 251.0

Figure 1.6(*b*) Comparison at the stage level of selected Mesozoic time scales with GTS2004.

whole-rock samples. A good review in this respect for the Devonian is found in Williams *et al.* (2000), whose study points out that it is clearly desirable to combine high analytical precision with narrow biostratigraphic control to provide the most useful points for time scale calibration. These authors make a case that the Carboniferous–Devonian boundary is near 362 Ma instead of near 354 Ma or even younger, as shown in more recent scales of Fig. 1.5. The same authors point out the considerable variation in the estimated age for the Silurian–Devonian boundary from ~418 to 410 Ma, and some exceptional short estimates for stage durations, such as 1 myr for the Pridoli Stage (Tucker *et al.*, 1998) and 0.9 myr for the Pragian Stage (Compston, 2000b). The latter conflicts with the analysis of cyclicity in the limestones in the classical Devonian sections of the Barrandian (Czech Republic), which suggests the Pragian Stage is not much shorter than the underlying Lochkovian Stage (Chlupác, 2000).

Because of the relative scarcity of reliable dates with high stratigraphic precision, geomathematical/statistical techniques for direct estimation of stage boundaries are not easily applicable in the Paleozoic, and various best-fit line techniques are utilized. Tucker and McKerrow (1995, their Fig. 1) plotted selected age dates for Cambrian–Devonian from well-established stratigraphic levels against their fossil age in an iterative manner, juggling radiometric dates of selected samples against their stratigraphic age determined by fossils such that a straight fit was created relative to the adjusted stage boundaries.

An improved version of this graphical method was employed by Tucker *et al.* (1998) to arrive at a "time line" for the Devonian. First, they used graphical correlation plus biostratigraphic intuition to scale the seven Devonian stages. Then, a suite of U–Pb zircon ages using the TIMS method for six volcanic ashes closely tied to biostratigraphic zones were used to adjust and calibrate this scaling. The Devonian scale in GTS2004 uses a modified version of their biostratigraphic scaling with a calibration from additional age dates (see Chapter 14). A similar technique is applied to the Carboniferous and Permian in GTS2004.

Cooper and Sadler added a new tool to the arsenal of time scale methodology, as applied to the Early Paleozoic time scale (see Chapters 12 and 13). Using detailed graptolite sequences from over 200 sections from oceanic and slope environment basins, a robust composite fossil sequence was calculated using the constrained optimization method of compositing. The Ordovician is taken to be 44.6 myr in duration, and lasted from 488.3 to 443.7 Ma; the Silurian lasted for 27.7 myr from 443.7 to 416 Ma. Calculated uncertainties are relatively small.

1.4.4 Mesozoic scales

The Mesozoic time scale spans an interval of 186 myr, from 251 to 65.5 Ma, which is a decrease of ~60 myr since Holmes (1937) and of ~35 myr compared to the scales of Holmes (1960) and Kulp (1961). Selected key Mesozoic time scales are compared to GTS2004 in Figs. 1.6*a,b*. The geologic time scale for the Mesozoic has undergone various improvements during the last two decades. The Larson and Hilde's (1975) marine magnetic anomaly profile displayed by the Hawaiian spreading lineation was adapted for scaling of the Oxfordian through Aptian Stages in KG85 and SEPM95 to compensate for a paucity of isotope dates. Databases of radiometric ages have been statistically analyzed with various best-fit methods to estimate ages of stage boundaries (GTS89 and SEPM95). Nevertheless, there have been substantial differences in the estimated ages and durations of stages and periods among scales constructed in the last two decades. For example, GTS1989 and SEPM95 estimated the Barremian Stage to be over 6 myr long, whereas EX88 and Odin and Odin (1993) suggested a duration of 2 myr.

Age differences are particularly obvious for the Jurassic–Cretaceous transition: the Tithonian–Berriasian boundary (which lacks an international definition) is 130 Ma in NDS82, 135 Ma in Remane (2000), but ~145 Ma in GTS89 and SEPM95, both of which excluded glauconite dates.

The Jurassic scales of van Hinte (1976), NDS82, KG85, EX88, Westermann (1988), and SEPM95 relied on biochronology to interpolate the duration of stages. As a first approximation, it was assumed that the numerous ammonite zones and/or subzones of the Jurassic have approximately equal mean duration between adjacent stages. Toarcian and Bajocian Stages have double the number of ammonite subzones compared to the Aalenian, so are assumed to span twice as much time. The limited age control on the duration of the entire Jurassic indicates that the average duration of each zone is ~1 myr and each subzone is ~0.45 myr (e.g. Westermann, 1988). KG85 and SEPM95 also took into account some intra-Jurassic age control points to constrain the proportional scaling of the component stages. A smoothing spline fit was applied by F. P. Agterberg in SEPM95 that incorporates the error limits of the isotope age dates. At the individual subzone or zonal level, this equal-duration assumption is known to be incorrect. For example, McArthur *et al.* (2000) observed a dramatic variability in Pliensbachian and Toarcian ammonite zones when scaled to a linear trend in the $^{87}Sr/^{86}Sr$ ratio of the oceans (see Chapter 18). However, the average of the durations is not much off. Westermann's (1988) estimate, and application of a combined strontium trend and cycle stratigraphy to Lower Jurassic stages (Weedon *et al.*, 1999) yielded relative durations for the Hettangian, Sinemurian, and Pliensbachian that are within error limits of those of SEPM95.

The advent of $^{40}Ar/^{39}Ar$ radiometric age dates on bentonites in local ammonite zones in a large part of the US Western Interior Cretaceous was a significant improvement for Late Cretaceous chronology. With this method Obradovich (1993) calibrated a Late Cretaceous time scale. He rejected all ages derived from biotites in bentonites as too young, and considered all his previous K–Ar ages on sanidines to be obsolete. The monitor standards for $^{40}Ar/^{39}Ar$ dating have undergone revisions during the late 1990s (see detailed discussion in Chapter 6). The text of Obradovich (1993) implies that all ages were normalized to a value of 520.4 Ma for the McLure Mountain hornblende monitor MMhb-1, thereby requiring significant recalculation to the current recommendation of 523.1 ($\sim$0.5 myr older for Late Cretaceous ages). But in fact, Obradovich used the Taylor Creek (TC) rhyolite as an internal monitor standard with a value of 28.32 Ma (J. Obradovich, pers. comm. 1999), hence recalculation to the currently recommended TC monitor value of 28.34 Ma is only on the order of 0.05 myr. Correlation of the North American ammonite zonation and Obradovich's associated linear scale to Upper Cretaceous European stages and zones was partially achieved through rare interchanges of ammonite and other marine macrofauna (reviewed in Cobban, 1993) and strontium isotope curves for portions of the Campanian and Maastrichtian (e.g. McArthur *et al.*, 1993, 1994). Gradstein *et al.* (1994a, 1995) incorporated the high-precision $^{40}Ar/^{39}Ar$ data of Obradovich (1993); the authors applied a cubic-spline fit to the data set. An even more refined version of this analysis is the basis for the GTS2004 scale for Late Cretaceous (see Chapters 8 and 19). Unfortunately, except for the basal-Turonian, it is difficult to associate the ammonite zones calibrated by Obradovich (1993) with the international definitions of Late Cretaceous stage boundaries.

In 2000, Pálfy *et al.* summarized 14 U–Pb TIMS dates from the Lower and Middle Jurassic of Western Canada, calibrated to regional ammonites stratigraphy. Complex U–Pb systematics made it difficult to obtain precise ages for some of the samples, and additional uncertainties enter when calibrating the regional biostratigraphy to the European standard ammonite zonation, but this data set provides the most important constraint on the basal-Jurassic through Toarcian stages (see Chapter 18).

Cycle stratigraphy, which has become the primary method of scaling the Cenozoic time scale, has been applied to portions of the Triassic, Jurassic, and Cretaceous time scales (reviewed in Chapters 17–19). As an example, Herbert *et al.* (1995) summarized orbitally tuned cycle counts using geochemical and color data from outcrop and core studies in northern and central Italy to estimate the duration of the Cenomanian as 6.0 ± 0.5 myr, the Albian as 11.92 ± 0.2 myr, and Aptian as 10.6 ± 0.2 myr. The Cenomanian and Albian cycle-scaling results have been verified by additional studies in Italy by Fiet *et al.* (2001) and Grippo *et al.* (2004) using other proxies and methods of spectral analysis, and are within the error bars of results derived from statistical fits to the limited radiometric data (e.g. SEPM95). The main differences seem to be in the choice of the "pin" age for hanging the cycle series from the base-Turonian or base-Cenomanian, the selected marker for the yet-to-be-defined stage and substage boundaries within the Albian and Aptian, and which orbital frequency is for tuning. This cycle scaling of the Albian events, but incorporating a potential nannofossil marker for the Albian–Aptian boundary, is used in GTS2004 (see Chapter 19).

1.4.5 Cenozoic scales

The Cenozoic time scale, from 65 Ma to Recent contains stages that vary in duration from almost 8 myr for the Lutetian to less than 1 myr for the Gelasian, and with the Holocene Epoch of only 11 500 yr.

Although the Cenozoic Era is known in most detail, standardization of stage boundaries with consensus definitions and GSSPs has been slow. In the Paleocene Period, only the epoch boundaries are formally defined: base-Paleocene, base-Eocene, and base-Oligocene. All Cenozoic standard stages are originally based on European stratotypes, with the Neogene Mediterranean ones more difficult to correlate world-wide as a function of increasing provincialism and diachronism in faunal and floral events in the face of higher latitude climatic cooling. Selected key Cenozoic time scales are compared to GTS2004 in Figs. 1.7*a,b*.

Since 1964, when B. F. Funnel presented the first, relatively detailed and accurate Cenozoic time scale with radiometric age estimates, many marine time scales have been erected with a progressive enhancement of scaling methods. Berggren (1972) and NDS82 combined radiometric age dating, stratigraphic reasoning, and biostratigraphic/geomagnetic calibrations. Hardenbol and Berggren (1978), GTS82, DNAG83, and EX88 added marine magnetic reversal calibrations.

Whereas the Paleozoic and Mesozoic time scales generally lack a unifying interpolation method, the marine magnetic reversals profile provides a powerful interpolator for the

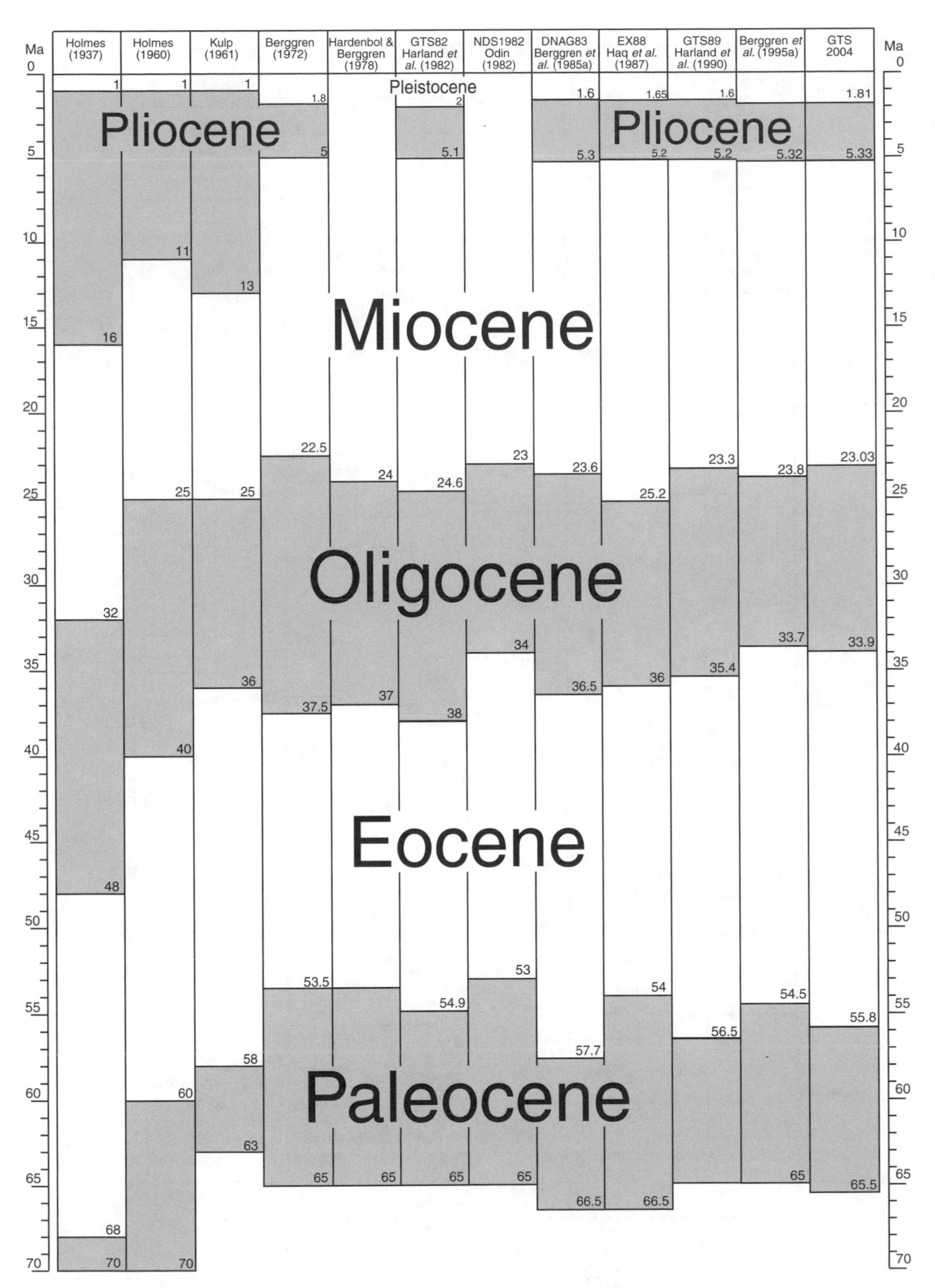

Figure 1.7(*a*) Comparison at the period level of selected Cenozoic time scales with GTS2004.

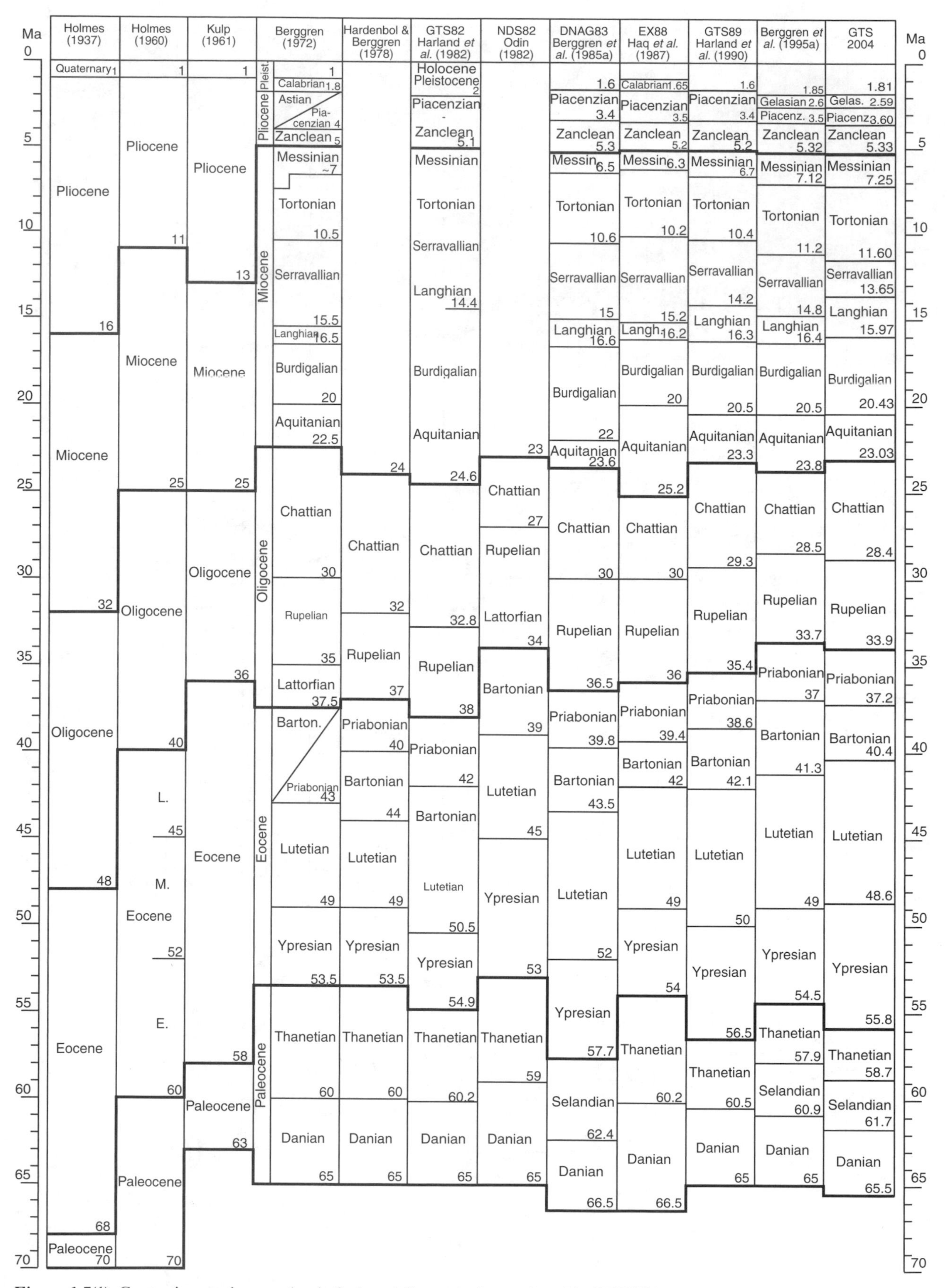

Figure 1.7(*b*) Comparison at the stage level of selected Cenozoic time scales with GTS2004.

Cenozoic time scale. The large number of geomagnetic field reversals since Late Santonian time, coupled with a wealth of seafloor magnetic profiles, and detailed knowledge of the radiometric age of selected magnetic polarity reversals in lavas and sediments provide a finely spaced scale. These are combined with a line fit or cubic spline to produce spreading-rate models for ocean basins and an associated magnetic polarity time scale (see Chapter 5). An excellent account of the method and its early applications is given by A. V. Cox in Harland *et al.* (1982).

The method itself dates back to Heirtzler *et al.* (1968), who selected a detailed profile in the Southern Atlantic from anomalies 2 through 32. The only calibrated tie point was magnetic anomaly 2A at 3.4 Ma, based on the radiometrically dated magnetic reversal scale of Cox *et al.* (1964) in Pliocene through Pleistocene lavas. Assuming that ocean-floor spreading had an invariant spreading rate of 1.9 cm/10^3 yr through the Campanian (~80 Ma), ages were assigned to the main Campanian through Pleistocene polarity chrons. This ambitious extrapolation has turned out to be within ~10% of later interpolations using a more detailed composite seafloor profile, and an improved array of age-calibrated tie points (Hardenbol and Berggren, 1978, DNAG83, EX88, and GTS89).

Cande and Kent (1992a,b, 1995) constructed a new geomagnetic reversals time scale using a composite of marine magnetic anomalies from the South Atlantic with short splices from fast-spreading Pacific and Indian Ocean segments, better estimates of anomaly width, nine age tie points, and a cubic-spline smoothing. Using an array of bio–magnetostratigraphic correlations with the Cande and Kent spreading model, Berggren *et al.* (1995a) compiled a comprehensive Cenozoic time scale.

Orbital tuning has become the dominant method for constructing detailed Neogene time scales (e.g. Shackleton *et al.*, 1990, 1999, 2000; Hilgen, 1991a; Hilgen *et al.*, 1995, 1997), and is making inroads in the Paleogene. These Milankovitch cycles of climate oscillations are recorded in nearly all oceanic and continental deposits, and have become a requirement for placement of stage-boundary stratotypes within the Neogene (see Chapter 21). Among a long list of differences we mention that the Oligocene–Miocene boundary appears 800 kyr younger, but the Tortonian–Messinian boundary is 120 kyr older than in Berggren *et al.* (1995a). In general, the Cande and Kent (1995) geomagnetic polarity time scale for the Late Neogene is slightly too young.

Cycle tuning relative to the well-dated base-Paleogene has enabled scaling of Paleocene magnetic chrons (Röhl *et al.*, 2001) and refined estimates of spreading rates for the South Atlantic profile (see Chapter 5). If the current pace of cycle stratigraphy applications continues, it is quite likely that tuning to astronomical cycles will enable detailed scaling of many more segments of the geologic time scale within the next decade.

2 • Chronostratigraphy: linking time and rock

F. M. GRADSTEIN, J. G. OGG, AND A. G. SMITH

Geologic stages and other international subdivisions of the Phanerozoic portion of the geologic scale are defined by their lower boundaries at Global Stratotype Sections and Points (GSSPs). The main criteria for a GSSP are that primary and secondary markers provide the means for global correlation. GSSP theory and criteria are outlined, the status of ratified GSSPs provided, and three examples discussed of prominent GSSPs. Subdivisions of the *International Stratigraphic Chart* are summarized and illustrated.

2.1 TIME AND ROCK

Geologic time and the observed rock record are separate but related concepts. A geologic time unit (geochronologic unit) is an abstract concept measured from the rock record by radioactive decay, Milankovitch cycles, or other means. A "rock-time" or chronostratigraphic unit consists of the total rocks formed globally during a specified interval of geologic time. The chronostratigraphic units are grouped into a hierarchy to subdivide the geologic record on Earth progressively. This chronostratigraphic scale was originally established from a combination of regional lithologic units (e.g. the Chalk of England defined the "Cretaceous," and the "Triassic" was assigned to a trio of distinctive formations in Germany) and of unique, non-recurring events provided by biological evolution.

These fragmentary chapters in the history of life and regional sediment facies gave rise to the succession of the standard geologic periods and the subdivision of periods into stages that form the chronostratigraphic time scale. In its classic usage, each geological stage was delimited at a "stratotype" to indicate the idealized extent and fossil content. The historical development of stratigraphy utilized former marginal marine to pelagic successions that are now uplifted in Europe, and these quasi-regional units still provide the basic nomenclature of most geological stages. However, the geologic record is discontinuous, and these stratotype-based chronostratigraphic units are an imperfect record of the continuum of geologic time (e.g. Paleogene stratotypes in Fig. 2.1). Therefore, a distinction between a hierarchy of material chronostratigraphic units (rock-time) and abstract geochronologic units (Earth time) units was required, and a dual nomenclature system was codified (Table 2.1). The divisions of geologic time range from an eon to the shortest formal unit of "age."*

Table 2.1 *Duality of some principle geochronologic (time) and chronostratigraphic (time–rock) units*[a]

Time units	Chronostratigraphic units
Geologic time scale intervals	
Eon	Eonothem
Era	Erathem
Period	System
Epoch	Series
Age	Stage
Non-hierarchal interval	
Chron	Zone (or chronozone)
Geomagnetic intervals	
Polarity chron	Polarity zone (or polarity chronozone)
Biostratigraphy intervals	
Biochron	Biozone (range zone, interval zone, etc.)

[a] Modified from *International Stratigraphic Guide*, 2nd edition, Salvador, 1994.

The two concepts of geochronologic and chronostratigraphic scales are now united by formally establishing markers within continuous intervals of the stratigraphic record to

* *Note:* To avoid misleading readers by using the term "age" to refer to a time span (as in the current International Stratigraphic Guide) instead of to a numerical date, we will generally use the term "stage" in this book to refer to both the time interval and the rocks deposited during that time interval. The practice of using the term "stage" for both time and for rock has the advantage of simplifying stratigraphy (as will be explained below) and liberating "age" for general use.

define the beginnings both of each successive chronostratigraphic unit and of the associated geochronologic unit. This concept of a Global Stratotype Section and Point (GSSP) to define each stage has replaced the earlier use of "stage stratotypes," and has enabled compilation of an international stratigraphic chart for the geologic time scale. In some respects, the concept of the beginning of each chronostratigraphic unit being bound by an isochronous surface defined at a GSSP has made the dual nomenclature unnecessary for the units of the geologic time scale (e.g. Walsh, 2001, 2003; Remane, 2003), as discussed later.

Prior to the evolution of metazoan life, the biological record is not generally suitable for a detailed subdivision of Precambrian time. Thus, Precambrian time is currently subdivided by the artificial assignment of numerical ages to stratigraphic boundaries. The Precambrian time scale is therefore a chronometric rather than a chronostratigraphic scale. However, it may become possible to apply the GSSP concept to some intervals of the Precambrian (see Chapter 10).

2.2 STANDARDIZATION OF THE CHRONOSTRATIGRAPHIC SCALE

2.2.1 History of geologic stratigraphic standardization*

The prodigious stratigraphic labors of the nineteenth century resulted in innumerable competing stratigraphic schemes. To impose some order, the first International Geological Congress (IGC) in Paris in 1878 set as its objective the production of a standard stratigraphic scale. Suggestions were made for standard colors (Anon., 1882, pp. 70–82), uniformity of geologic nomenclature (pp. 82–4), and the adoption of uniform subdivisions (pp. 85–7). There was also a review of several regional stratigraphic problems. In the succeeding congress at Bologna in 1881, many of the above suggestions were taken substantially further, i.e. the international geological maps were planned with standard colors for stratigraphic periods and rock types (e.g. Anon., 1882, pp. 297–411), and annexes contained national contributions toward standardization of stratigraphic classification, etc. (pp. 429–658).

In spite of this promising start, the IGCs did not have the continuing organization to carry these proposals through, except for the commissions established to produce international maps. The latter is now the Commission for the Geologic Map of the World (CGMW; see www.cgmw.org). Guides setting out the stratigraphic principles, terminology, and classificatory procedures were prepared by the International Commission on Stratigraphic Terminology, created in 1952 by the 19th IGC in Algiers, and now the International Subcommission on Stratigraphic Classification (ISSC) under the International Commission of Stratigraphy (ICS) of the International Union of Geological Sciences (IUGS). The *International Stratigraphic Guide* was published in 1976 (Hedberg, 1976), and is now in its second edition (Salvador, 1994; Murphy and Salvador, 1999).

It was not until the establishment of the International Union of Geological Sciences (IUGS) around 1960 that the goal of establishing an international chronostratigraphic scale had a means of fulfillment, through the IUGS's International Commission on Stratigraphy (ICS) and its many subcommissions. Guidelines for defining global chronostratigraphic units were established (e.g. Cowie *et al.*, 1986; enhanced by Remane *et al.*, 1996). At the occasion of the 28th IGC, the ICS published the first *Global Stratigraphic Chart* that reflects much current stratigraphic use. At the 31st IGC in Rio de Janeiro, a new edition of that chart indicated the current international standardization, and included abbreviations and colors of the stratigraphic units as adopted by the CGMW (Remane, 2000; see also Appendix 1).

2.2.2 Global boundary Stratotype Section and Point (GSSP)

How can one standardize such fragmentary and disparate material as the stratigraphic record?

Even by the first IGC in 1878, the belief that the stratigraphic systems and other divisions being described in any one place were natural chapters of Earth history was fading, and the need to adopt some conventions was widely recognized. Even so, the practice continued of treating strata divisions largely as biostratigraphic units, and even today it is an article of faith for many Earth scientists that divisions of the developing international stratigraphic scale are defined by the fossil content of the rocks. To follow this through, however, leads to difficulties: boundaries may change with new fossil discoveries; boundaries defined by particular fossils will tend to be diachronous; there will be disagreement as to which taxa shall be definitive.

As elaborated by one of the major champions of practical and rational thinking in stratigraphic standardization (Hedberg, 1976, p. 35):

> In my opinion, the first and most urgent task in connection with our present international geochronology scale is to

* This section is updated from Harland *et al.*, 1990, p. 2.

AGE (Ma)	Epoch		Stage	Boundary (Ma)
	Neogene			23.03
25	Oligocene	L	Chattian	
				28.45
30		E	Rupelian	
				33.90
35	Eocene	L	Priabonian	
				37.20
			Bartonian	
40				40.40
		M	Lutetian	
45				
				48.60
50		E	Ypresian	
55				55.80
	Paleocene	L	Thanetian	
				58.70
60		M	Selandian	
				61.70
		E	Danian	
65				65.50
	Cretaceous			

Polarity Chron: C6B, C6C, C7, C7A, C8, C9, C10, C11, C12, C13, C15, C16, C17, C18, C19, C20, C21, C22, C23, C24, C25, C26, C27, C28, C29, C30

Plankt. Foram.: M1, P22, P21 (b, a), P20, P19, P18, P17, P16, P15, P14, P13, P12, P11, P10, P9, P8, P7, P6 (b, a), P5, P4 (c, b, a), P3 (b, a), P2, P1 (c, b, a)

Calc. Nannopl.: NN1, NP25, NP24, NP23, NP22, NP21, NP20, NP19, NP18, NP17, NP16, NP15, NP14, NP13, NP12, NP11, NP10, NP9, NP8, NP7, NP6, NP5, NP4, NP3, NP2, NP1

Radio-laria: RP22, RP21, RP20, RP19, RP18, RP17, RP16, RP15, RP14, RP13, RP12, RP11, RP10, RP9, RP8, RP7, RP6 (c, b, a), unzoned

Position of Stage Stratotypes

Danian, Thanetian, Bartonian, Lattorfian, Ypresian, Lutetian, Cuisian, Wemmelian, Priabonian, Tongrian, Rupelian, Chattian, Eo (Kassel), Neo (Doberg)

achieve a better definition of its units and horizons so that each will have a standard fixed-time significance, and the same time significance for all geologists everywhere. Most of the named international chronostratigraphic (geochronology) units still lack precise globally accepted definitions and consequently their limits are controversial and variably interpreted by different workers. This is a serious and wholly unnecessary impediment to progress in global stratigraphy. What we need is simply a single permanently fixed and globally accepted standard definition for each named unit or horizon, and this is where the concept of stratotype standards (particularly boundary stratotypes and other horizon stratotypes) provides a satisfactory answer.

The standardization advocated by Hedberg and other stratigraphers has been the major task of ICS through application of the principle of boundary stratotypes; the current status of this application is actively maintained in the official website of ICS. The traditional stratigraphic scale using stage stratotypes has evolved into a standard chronostratigraphic scale in which the basal boundary of each stage is standardized at a point in a single reference section within an interval exhibiting continuous sedimentation. This precise reference point for each boundary is known as the Global Stratotype Section and Point (GSSP), and represents the point in time when that part of the rock succession began. The global chronostratigraphic scale is ultimately defined by a sequence of GSSPs.

It is now over 25 years since the first boundary stratotype or GSSP "golden spike" was defined. It fixed the lower limit of the Lochkovian Stage, the oldest stage of the Devonian, at a precise level in an outcrop with the appropriate name of Klonk in the Czech Republic (Martinsson, 1977). Paleontologically, the base of the Lochkovian Stage coincides with the first occurrence of the Devonian graptolite *Monograptus uniformis* in bed No. 20 of the Klonk Section, northeast of the village of Suchomasty (Chlupác, 1993). However, once the golden spike has been agreed, the discovery, say, of *Monograptus uniformis* below the GSSP does not require a re-definition of its position, but simply an acknowledgement that the initial level chosen was not in fact at the lowest occurrence of the particular graptolite. For this reason, multiple secondary correlation markers, including non-biostratigraphic methods, are desirable within each GSSP section.

Each GSSP must meet certain requirements and secondary desirable characteristics (Remane *et al.*, 1996; Table 2.2). The main considerations are: (1) that the boundary is recognizable outside the GSSP locality, therefore it must be tied to other events in Earth history that are documented in sediments elsewhere; and (2) the reference GSSP section is well exposed with the GSSP level within an interval of apparent continuous sedimentation.

The choice of an appropriate boundary level is of paramount importance. "Before formally defining a geochronologic boundary by a GSSP, its practical value – i.e. its correlation potential – has to be thoroughly tested. In this sense, *correlation precedes definition*" (Remane, 2003). Without correlation, stratigraphic units and their constituent boundaries are of not much use, and devoid of meaning for Earth history. Most GSSPs coincide with a single "primary marker," which is generally a biostratigraphic event, but other stratigraphic events with widespread correlation potential should coincide or bracket the GSSP level. The choice of the criteria for an international stage boundary can be a contentious issue (e.g. Fig. 2.2). Most primary markers for GSSPs have been biostratigraphic events, but some have utilized other global stratigraphic episodes (e.g. the iridium spike at the base-Cenozoic, the carbon isotope anomaly at the base-Eocene, base of magnetic polarity Chron C6Cn.2n at base-Neogene, a specific Milankovitch cycle for base-Pleistocene, etc.).

The requirement for continuous sedimentation across the GSSP level and the bracketing correlation markers is to avoid assigning a boundary to a known "gap" in the geologic record. This requirement has generally eliminated most historical stratotypes for stages, which were commonly delimited by flooding or exposure surfaces and formally represent synthems. As a result, the scope of classical stages is modified, and either the traditional nomenclature is abandoned (e.g. the revised stage nomenclature for the Ordovician and Cambrian Periods), or an historical name is given a slightly new meaning to update its practical usage.

Figure 2.1 Stratigraphic range of historical stratotypes of some Paleogene stages. The left-hand columns include the microfossil zones and polarity chrons that span the complete Paleogene according to coring of marine sediments (see Chapter 20; Martini, 1971; Martini and Muller, 1971; Roth, 1970; Roth *et al.*, 1971). The stratotypes span less than half of Paleogene time; some are simply facies equivalents rather than chronostratigraphically distinct units. Only a few of these competing stage concepts were preserved in the nomenclature of the present Paleogene geologic time scale. International stages for the Paleogene are defined at boundary stratotypes at which the basal boundary of the stage is positioned relative to primary and secondary biostratigraphic, geochemical, or magnetic polarity events for global correlation (modified from Hardenbol and Berggren, 1978).

Table 2.2 *Requirements for establishing a global stratotype section and point (GSSP)*[a]

1. **Name and stratigraphic rank of the boundary**
   Including concise statement of GSSP definition
2. **GSSP geographic and physical geology**
   Geographic location, including map coordinates
   Geologic setting (lithostratigraphy, sedimentology, paleobathymetry, post-depositional tectonics, etc.)
   Precise location and stratigraphic position of GSSP level and specific point
   Stratigraphic completeness across the GSSP level
   Adequate thickness and stratigraphic extent of section above and below
   Accessibility, including logistics, national politics, and property rights
   Provisions for conservation and protection
3. **Primary and secondary markers**
   Principal correlation event (marker) at GSSP level
   Other primary and secondary markers – biostratigraphy, magnetostratigraphy, chemical stratigraphy, sequence stratigraphy, cycle stratigraphy, other event stratigraphy, marine land correlation potential
   Potential age dating from volcanic ashes and/or orbital tuning
   Demonstration of regional and global correlation
4. **Summary of selection process**
   Relation of the GSSP to historical usage; references to historical background and adjacent (stage) units; selected publications
   Other candidates and reasons for rejection; summary of votes and received comments
   Other useful reference sections
5. **Official publication**
   Summary documentation in IUGS journal *Episodes*
   Full publication in journal *Lethaia* (ICS's official publication channel)

[a] Revised from Remane *et al.* (1996) according to current procedures and recommendations of the IUGS International Commission on Stratigraphy (ICS).

Figure 2.2 One reason that decisions on international boundaries of stages are difficult. Two experts with different paleontological specialties arguing over the suitable primary marker based on different biostratigraphic criteria. Modified from *Episodes* 8: 89, Fig. 6, 1985 (based on Birkelund *et al.*, 1983).

Difficulties in identifying global correlation criteria, problems introduced by biogeographic provincialism, and the occasional need to abandon stage concepts based on historical regional usage have slowed assignment of GSSPs in some periods, as will be elaborated in Chapters 11–22. Suitable GSSPs with full documentation are proposed by stratigraphic subcommissions or working groups under ICS; undergo approval voting through ICS and ratification by IUGS; and then are published in *Lethaia* (journal officially dedicated to ICS science), with summary documentation being published in *Episodes* (official journal of IUGS).

As of January 2004, over half of the stages have been defined by boundary stratotypes, and the criteria for most primary markers to be associated with GSSPs for other stages have been decided (Table 2.3, Figs. 2.3 and 2.4). The great majority of defined and probable GSSPs are in western Europe (Fig. 2.5). This distribution mostly reflects the historical accident that stratigraphic studies first developed in western Europe, but is also due to tectonic processes that kept western Europe in low-latitude shallow-sea environments for much of the Phanerozoic Eon and have subsequently exposed the richly fossiliferous sections that were the basis of the historical compilations of the chronostratigraphic scale.

Mesozoic - Cenozoic Stratigraphic Chart and GSSPs

Era	Period	Series/ Epoch	Stage	GSSPs	
Cenozoic	Neogene	Holocene			
		Pleistocene		GSSP	Vrica, Calabria, Italy
		Pliocene	Gelasian	GSSP	Monte San Nicola, Sicily, Italy
			Piacenzian	GSSP	Punta Picola, Sicily, Italy
			Zanclean	GSSP	Eraclea Minoa, Sicily, Italy
		Miocene	Messinian	GSSP	Oued Akrech, Rabbat, Morocco
			Tortonian	GSSP	Monte dei Corvi Beach, Ancona, Italy
			Serravallian		
			Langhian		
			Burdigalian		
			Aquitanian	GSSP	Lemme-Carrosio, N. Italy
	Paleogene	Oligocene	Chattian		
			Rupelian	GSSP	Massignano, Ancona, Italy
		Eocene	Priabonian		
			Bartonian		
			Lutetian		
			Ypresian	GSSP	Dababiya, Luxor, Egypt
		Paleocene	Thanetian		
			Selandian		
			Danian	GSSP	El Kef, Tunisia
Mesozoic	Cretaceous	Late	Maastrichtian	GSSP	Tercis-les-Bains, Landes, SW. France
			Campanian		
			Santonian		
			Coniacian		
			Turonian	GSSP	Rock Canyon, Pueblo, Colorado, USA
			Cenomanian	GSSP	Mont Risou, Rosans, Haute-Alpes, France
		Early	Albian		
			Aptian		
			Barremian		
			Hauterivian		
			Valanginian		
			Berriasian		
	Jurassic	Late	Tithonian		
			Kimmeridgian		
			Oxfordian		
		Middle	Callovian		
			Bathonian		
			Bajocian	GSSP	Cabo Mondego, W. Portugal
			Aalenian	GSSP	Fuentelsalz, Spain
		Early	Toarcian		
			Pliensbachian	GSSP	Robin Hood's Bay, Yorkshire, UK
			Sinemurian	GSSP	East Quantox Head, West Somerset, GB
			Hettangian		
	Triassic	Late	Rhaetian		
			Norian		
			Carnian		
		Middle	Ladinian		
			Anisian		
		Early	Olenekian		
			Induan	GSSP	Meishan, Zhejiang, China

Figure 2.3 Distribution of ratified GSSPs in the Mesozoic and Cenozoic Eras (status in December, 2004). The *International Stratigraphic Chart for the Precambrian and Phanerozoic* with ratified GSSPs is presented in the special color section of this book. Updated versions of the color chart are available from the website of the International Commission on Stratigraphy (www.stratigraphy.org) and can be downloaded in the colors of either the Commission for the Geologic Map of the World (CGMW) or the United States Geological Survey (USGS).

Paleozoic Stratigraphic Chart and GSSPs

Era	Period	Series/ Epoch	Stage	GSSPs	
Paleozoic	Permian	Lopingian	Changhsingian		
			Wuchiapingian	GSSP	Penglaitan, Guangxi Province, China
		Guadalupian	Capitanian	GSSP	Nipple Hill, Guadalupe Mountains, TX, USA
			Wordian	GSSP	Guadalupe Mountains, TX, USA
			Roadian	GSSP	Stratotype Canyon, Guadalupe Mountains, TX, USA
		Cisuralian	Kungurian		
			Artinskian		
			Sakmarian		
			Asselian	GSSP	Aidaralash, Ural Mountains, Kazakhstan
	Carboniferous	Pennsylvanian	Gzhelian		
			Kasimovian		
			Moscovian		
			Bashkirian	GSSP	Arrow Canyon, Nevada, USA
		Mississippian	Serpukhovian		
			Visean		
			Tournaisian	GSSP	La Serre, Montagne Noir, France
	Devonian	Late	Famennian	GSSP	Coumiac, Cessenon, Montagne Noir, France
			Frasnian	GSSP	Col du Puech, Montagne Noir, France
		Middle	Givetian	GSSP	Jebel Mech Irdane, Tafilalt, Morocco
			Eifelian	GSSP	Wetteldorf Richtschnitt, Eifel Hills, Germany
		Early	Emsian	GSSP	Zinzil'ban Gorge, Uzbekistan
			Pragian	GSSP	Velka Chuchle, SW Prague, Czech Rep.
			Lochkovian	GSSP	Klonk, SW of Prague, Czech Republic
	Silurian	Pridoli		GSSP	Pozary, Prague, Czech Republic
		Ludlow	Ludfordian	GSSP	Sunnyhill, Ludlow, UK
			Gorstian	GSSP	Pitch Coppice, Ludlow, UK
		Wenlock	Homerian	GSSP	Whitwell Coppice, Homer, UK
			Sheinwoodian	GSSP	Hughley Brook, Apedale, UK
		Llandovery	Telychian	GSSP	Cefn Cerig, Llandovery, UK
			Aeronian	GSSP	Trefawr, Llandovery, UK
			Rhuddanian	GSSP	Dob's Linn, Moffat, UK
	Ordovician	Late	Hirnantian		
				GSSP	Fagelsang, Scania, Sweden
		Middle	Darriwilian	GSSP	Huangnitang, Zhejiang Province, China
		Early		GSSP	Diasbasbrottet, Hunneberg, Sweden
			Tremadocian	GSSP	Green Point Newfoundland, Canada
	Cambrian	Furongian			
			Paibian	GSSP	Paibi, Hunan Province, China
		Middle			
		Early			
				GSSP	Fortune Head, Newfoundland, Canada

Figure 2.4 Distribution of ratified GSSPs in the Paleozoic Era (status in December 2004).

Each chapter of this book devoted to the Phanerozoic begins with a global reconstruction (Mollweide) map that displays the distribution of the GSSPs for that period. The reconstructions have been compiled from published global databases of ocean-floor spreading and paleomagnetic poles linked to tectonics (Smith, 2001).

2.2.3 Global Standard Stratigraphic Age (GSSA)

Due to the fact that most Proterozoic and Archean rocks lack adequate fossils for correlation, a different type of boundary definition was applied for subdividing these eons into eras and periods (see Chapter 9). For these two eons, the assigned boundary, called a Global Standard Stratigraphic Age (GSSA), is a chronometric boundary and is not represented by a GSSP in rocks, nor can it ever be. However, although there appears to be consensus that the division into eras is possible, the finer period subdivisions often contain no dateable rocks, which make their assignment difficult. An alternative Precambrian classification based on stages in planetary evolution with, in most cases, possible associated GSSPs is presented by W. Bleeker in Chapter 10.

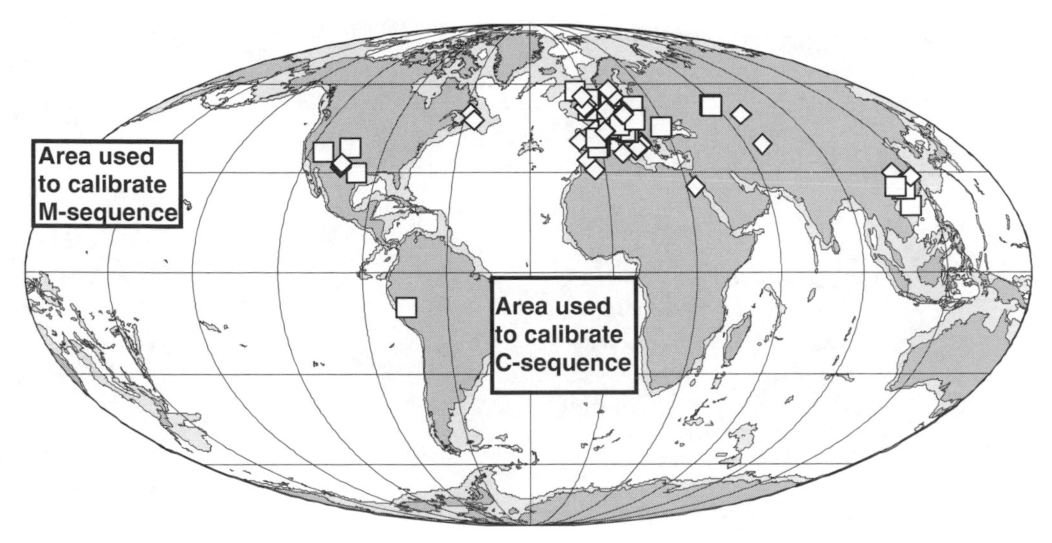

Figure 2.5 Geographic distribution of ratified (diamonds) and candidate (squares) GSSPs on a present-day (0 Ma) map (status in January 2004; see Table 2.3). Most of the GSSPs are in western Europe, where the clustering has overlapped many additional GSSPs. The approximate oceanic areas used to calibrate the C- and M-sequences of the magnetic polarity time scale are indicated by large rectangles (for details see Chapter 5).

2.2.4 Other considerations for choosing a GSSP

The basic requirements for a GSSP are that it is located in a stratigraphically continuous section; that it should be readily accessible and well exposed; and that it should ideally contain multiple markers suitable for global correlation. A GSSP is the precise definition of the base of a stratigraphic boundary in a rock sequence, but that boundary is defined only at one point on Earth. Assignment of the chronostratigraphic boundary within other stratigraphic sections requires correlation to the GSSP.

The ideal GSSP would be in a low-latitude highly fossiliferous marine section (for global biostratigraphic correlation) that contains cyclic sediments or interbedded volcanic ash or lava beds (for isotopic dating or measurement of durations), unambiguous magnetic polarity changes (for high-precision global correlation), and one or more geochemical signatures (to provide additional high-precision global correlation markers).

Surprisingly perhaps, GSSPs located in sections that have an abundant fauna may also introduce unknown correlation errors, particularly if they are in shallow-water shelf environments likely to give rise to an hiatus (Sadler, 1981; Sadler and Strauss, 1990). It is unclear what contribution such an hiatus makes to the overall global correlation uncertainties, as opposed to those of purely evolutionary origin, such as the presence or absence of a given fossil or fossils.

GSSPs are necessarily part of an outcrop exposed by uplift and erosion, and most are in relatively undistorted strata. However, some GSSPs, such as the Late Devonian GSSPs at the Montagne Noir of southeastern France and base-Maastrichtian GSSP, are tightly folded and may no longer retain magnetostratigraphic, geochemical, or other secondary markers for global correlation.

Absence of precise global markers for high-precision time scale work is a key problem that was glossed over in some GSSP decisions. If the GSSP is defined in purely biostratigraphic or lithostratigraphic terms and there are no accompanying high-precision secondary markers, such as is the case for some of the Silurian GSSPs, then the likely correlation errors are at least 0.5 myr and in some cases perhaps as high as 5 myr. Such GSSPs are unsatisfactory and will eventually need reconsideration (e.g. by working groups of the Silurian Subcommission of ICS established in 2002).

The ideal GSSP is at a horizon amenable to radiometric and/or astronomical cycle calibration or is bracketed by dateable horizons. This coincidence has been achieved for only a few GSSP placements (e.g. bases of all Pliocene and Upper Miocene stages, the base of the Turonian Stage of the Cretaceous, the base of the Triassic). If such a horizon is absent, it is essential to be able to correlate to dateable horizons elsewhere using precise global correlation markers.

The stratigraphic advances made by ocean drilling (e.g. the Deep Sea Drilling Project or the Ocean Drilling Program) was from multidisciplinary teams utilizing a wide array of shipboard and down-hole investigations. By contrast, GSSP golden spikes are most commonly placed in well-exposed continental sections where the excellence of the outcrop seems

Table 2.3 *Status of defining Global boundary Stratotype Sections and Points (GSSPs) for stage boundaries (Status in January 2004; updated tables are available from the ICS website* www.stratigraphy.org*)*

EON, Era, System, Series, Stage	Age (Ma) GTS2004	Est. ± myr	Principal correlative events	GSSP and location	Status	Publication
PHANEROZOIC						
Cenozoic Era						
Neogene System			*"Quaternary"* is traditionally considered to be the interval of oscillating climatic extremes (glacial and interglacial episodes) that was initiated at about 2.6 Ma, therefore encompasses the Holocene, Pleistocene, and uppermost Pliocene. A formal decision on its chronostratigraphic rank is pending			Ogg (2004); Pillans (2004); Pillans and Naish (2004)
Holocene Series						
base Holocene	11.5 ka	0.00	Exactly 10 000 carbon-14 years (i.e. 11.5 ka calendar years BP) at the end of the Younger Dryas cold spell		–[a]	
Pleistocene Series						
base Upper Pleistocene subseries	0.126	0.00	Base of the Eemian interglacial stage (base of marine isotope stage 5e) before final glacial episode of Pleistocene	Potentially, within sediment core under the Netherlands (Eemian type area)	–[a]	
base Middle Pleistocene subseries	0.781	0.00	Brunhes–Matuyama magnetic reversal		–[a]	
base Pleistocene Series	1.806	0.00	Just above top of magnetic polarity chronozone C2n (Olduvai) and the extinction level of calcareous nannofossil *Discoaster brouweri* (base Zone CN13). Above are lowest occurrence of calcareous nannofossil medium *Gephyrocapsa* spp. and extinction level of planktonic foraminifera *Globigerinoides extremus*	Top of sapropel layer "e", Vrica section, Calabria, Italy	1985[b]	*Episodes* **8**(2): 116–20, 1985
Pliocene Series						
base Gelasian Stage	2.588	0.00	Isotopic stage 103, base of magnetic polarity chronozone C2r (Matuyama). Above are extinction levels of calcareous nannofossil *Discoaster pentaradiatus* and *D. surculus* (base Zone CN12c)	Midpoint of sapropelic Nicola Bed ("A5"), Monte San Nicola, Gela, Sicily, Italy	1996[b]	*Episodes* **21**(2): 82–7, 1998

Table 2.3 (*cont.*)

EON, Era, System, Series, Stage	Age (Ma) GTS2004	Est. ± myr	Principal correlative events	GSSP and location	Status	Publication
base Piacenzian Stage	3.600	0.00	Base of magnetic polarity chronozone C2An (Gauss); extinction levels of planktonic foraminifera *Globorotalia margaritae* (base Zone PL3) and *Pulleniatina primalis*	Base of beige layer of carbonate cycle 77, Punta Piccola, Sicily, Italy	1997[b]	*Episodes* **21**(2): 88–93, 1998
base Zanclean Stage, base Pliocene Series	5.332	0.00	Top of magnetic polarity chronozone C3r, ~100 kyr before Thvera normal-polarity subchronozone (C3n.4n). Calcareous nannofossils–near extinction level of *Triquetrorhabdulus rugosus* (base Zone CN10b) and the lowest occurrence of *Ceratolithus acutus*	Base of Trubi Fm (base of carbonate cycle 1), Eraclea Minoa, Sicily, Italy	2000[b]	*Episodes* **23**(3): 179–87, 2000
Miocene Series						
base Messinian Stage	7.246	0.00	Astrochronology age of 7.246 Ma; middle of magnetic polarity chronozone C3Br.1r; lowest regular occurrence of the *Globorotalia conomiozea* planktonic foraminifera group	Base of red layer of carbonate cycle 15, Oued Akrech, Rabat, Morocco	2000[b]	*Episodes* **23**(3): 172–8, 2000
base Tortonian Stage	11.608	0.00	Last common occurrences of the calcareous nannofossil *Discoaster kugleri* and the planktonic foraminifera *Globigerinoides subquadratus*. Associated with the short normal-polarity subchron C5r.2n	Midpoint of sapropel 76, Monte dei Corvi beach section, Ancona, Italy	2003[b]	*Episodes* in prep.
base Serravillian Stage	13.65	0.00	Near lowest occurrence of nannofossil *Sphenolithus heteromorphus*, and within magnetic polarity chronozone C5ABr		2004[c]	
base Langhian Stage	15.97	0.0	Near first occurrence of planktonic foraminifera *Praeorbulina glomerosa* and top of magnetic polarity chronozone C5Cn.1n		2004[c]	
base Burdigalian Stage	20.43	0.0	Near lowest occurrence of planktonic foraminifera *Globigerinoides altiaperturus* or near top of magnetic polarity chronozone C6An		–[d]	

(*cont.*)

Table 2.3 (*cont.*)

EON, Era, System, Series, Stage	Age (Ma) GTS2004	Est. ± myr	Principal correlative events	GSSP and location	Status	Publication
base Aquitanian Stage, base Miocene Series, base Neogene System	23.03	0.0	Base of magnetic polarity chronozone C6Cn.2n; lowest occurrence of planktonic foraminifera *Paragloborotalia kugleri*; near extinction of calcareous nannofossil *Reticulofenestra bisecta* (base Zone NN1)	35 m from top of Lemme–Carrosio section, Carrosio village, north of Genoa, Italy	1996[b]	*Episodes* **20**(1): 23–8, 1997
Paleogene System						
Oligocene Series						
base Chattian Stage	28.4	0.1	Planktonic foraminifera, extinction of *Chiloguembelina* (base Zone P21b)	Probably in Umbria–Marche region of Italy	2004[c]	
base Rupelian Stage, base Oligocene Series	33.9	0.1	Planktonic foraminifera, extinction of *Hantkenina*	Base of marl bed at 19 m above base of Massignano quarry, Ancona, Italy	1992[b]	*Episodes* **16**(3): 379–82, 1993
Eocene Series						
base Priabonian Stage	37.2	0.1	Near lowest occurrence of calcareous nannofossil *Chiasmolithus oamaruensis* (base Zone NP18)	Probably in Umbria–Marche region of Italy		
base Bartonian Stage	40.4	0.2	Near extinction of calcareous nannofossil *Reticulofenestra reticulata*			
base Lutetian Stage	48.6	0.2	Planktonic foraminifera, lowest occurrence of *Hantkenina*	Leading candidate is Fortuna section, Murcia province, Betic Cordilleras, Spain	2004[c]	
base Ypresian Stage, base Eocene Series	55.8	0.2	Base of negative carbon isotope excursion	Dababiya section near Luxor, Egypt	2003[b]	*Episodes* in prep.
Paleocene Series						
base Thanetian Stage	58.7	0.2	Magnetic polarity chronozone, base of C26n, is a temporary assignment	Leading candidate is Zumaya section, northern Spain	–[d]	
base Selandian Stage	61.7	0.2	Boundary task group is considering a higher level–base of calcareous nannofossil zone NP5–which would be ~1 myr younger	Leading candidate is Zumaya section, northern Spain	–[d]	
base Danian Stage, base Paleogene System, base Cenozoic	65.5	0.3	Iridium geochemical anomaly. Associated with a major extinction horizon (foraminifera, calcareous nannofossils, dinosaurs, etc.)	Base of boundary clay, El Kef, Tunisia (*but deterioration may require assigning a replacement section*)	1991[b]	

Table 2.3 (*cont.*)

EON, Era, System, Series, Stage	Age (Ma) GTS2004	Est. ± myr	Principal correlative events	GSSP and location	Status	Publication
Mesozoic Era						
Cretaceous System			*Most substages of Cretaceous also have recommended GSSP criteria*			
Upper						
base Maastrichtian Stage	70.6	0.6	Mean of 12 biostratigraphic criteria of equal importance. Closely above is lowest occurrence of ammonite *Pachydiscus neubergicus*. Boreal proxy is lowest occurrence of belemnite *Belemnella lanceolata*	115.2 m level in Grande Carrière quarry, Tercis-les-Bains, Landes province, southwest France	2001[b]	*Episodes* **24**(4): 229–38, 2001; Odin (ed.) IUGS Spec. Publ. Series, V. 36, Elsevier, 910 pp.
base Campanian Stage	83.5	0.7	Crinoid, extinction of *Marsupites testudinarius*	Leading candidates are in southern England and in Texas		
base Santonian Stage	85.8	0.7	Inoceramid bivalve, lowest occurrence of *Cladoceramus undulatoplicatus*	Leading candidates are in Spain, England and Texas		
base Coniacian Stage	89.3	1.0	Inoceramid bivalve, lowest occurrence of *Cremnoceramus rotundatus* (*sensu* Tröger *non* Fiege)	Base of Bed MK47, Salzgitter–Salder Quarry, SW of Hannover, Lower Saxony, northern Germany	2004[c]	
base Turonian Stage	93.5	0.8	Ammonite, lowest occurrence of *Watinoceras devonense*	Base of Bed 86, Rock Canyon Anticline, east of Pueblo, Colorado, west–central USA	2003[b]	*Episodes* in prep.
base Cenomanian Stage	99.6	0.9	Planktonic foraminifera, lowest occurrence of *Rotalipora globotruncanoides*	36 m below top of Marnes Bleues Formation, Mont Risou, Rosans, Haute-Alpes, southeast France	2002[b]	*Episodes* **27**(1): 21–32, 2004
Lower						
base Albian Stage	112.0	1.0	Calcareous nannofossil, lowest occurrence of *Praediscosphaera columnata* (= *P. cretacea* of some earlier studies) is one potential marker		–[d]	
base Aptian Stage	125.0	1.0	Magnetic polarity chronozone, base of M0r	Leading candidate is Gorgo a Cerbara, Piobbico, Umbria–Marche, central Italy		
base Barremian Stage	130.0	1.5	Ammonite, lowest occurrence of *Spitidiscus hugii* – *Spitidiscus vandeckii* group	Leading candidate is Río Argos near Caravaca, Murcia province, Spain		

(*cont.*)

Table 2.3 (*cont.*)

EON, Era, System, Series, Stage	Age (Ma) GTS2004	Est. ± myr	Principal correlative events	GSSP and location	Status	Publication
base Hauterivian Stage	136.4	2.0	Ammonite, lowest occurrence of genus *Acanthodiscus* (especially *A. radiatus*)	Leading candidate is La Charce village, Drôme province, southeast France		
base Valanginian Stage	140.2	3.0	Calpionellid, lowest occurrence of *Calpionellites darderi* (base of Calpionellid Zone E); followed by the lowest occurrence of ammonite "*Thurmanniceras*" *pertransiens*	Leading candidate is near Montbrun-les-Bains, Drôme province, southeast France		
base Berriasian Stage, base Cretaceous System	145.5	4.0	Maybe near lowest occurrence of ammonite *Berriasella jacobi*		–[d]	
Jurassic System						
Upper						
base Tithonian Stage	150.8	4.0	Near base of *Hybonoticeras hybonotum* ammonite zone and lowest occurrence of *Gravesia* genus, and the base of magnetic polarity chronozone M22An		–[d]	
base Kimmeridgian Stage	155.7	4.0	Ammonite, near base of *Pictonia baylei* ammonite zone of Boreal realm	Leading candidates are in Scotland, southeast France, and Poland	2004[c]	
base Oxfordian Stage	161.2	4.0	Ammonite, *Brightia thuouxensis* Horizon at base of the *Cardioceras scarburgense* subzone (*Quenstedtoceras mariae* Zone)	Leading candidates are in southeast France and southern England	2004[c]	
Middle						
base Callovian Stage	164.7	4.0	Ammonite, lowest occurrence of the genus *Kepplerites* (*Kosmoceratidae*) (defines base of *Macrocephalites herveyi* Zone in sub-Boreal province of Great Britain to southwest Germany)	Leading candidate is Pfeffingen, Swabian Alb, southwest Germany	2004[c]	
base Bathonian Stage	167.7	3.5	Ammonite, lowest occurrence of *Parkinsonia* (*G.*) *convergens* (defines base of *Zigzagiceras zigzag* Zone)			
base Bajocian Stage	171.6	3.0	Ammonite, lowest occurrence of the genus *Hyperlioceras* (defines base of the *Hyperlioceras discites* Zone)	Base of Bed AB11, 77.8 m above base of Murtinheira section, Cabo Mondego, western Portugal	1996[b]	*Episodes* **20**(1): 16–22, 1997
base Aalenian Stage	175.6	2.0	Ammonite, lowest occurrence of *Leioceras* genus	Base of Bed FZ107, Fuentelsalz, central Spain	2000[b]	*Episodes* **24**(3): 166–75, 2001

Table 2.3 (*cont.*)

EON, Era, System, Series, Stage	Age (Ma) GTS2004	Est. ± myr	Principal correlative events	GSSP and location	Status	Publication
Lower						
base Toarcian Stage	183.0	1.5	Ammonite, near lowest occurrence of a diversified *Eodactylites* ammonite fauna; correlates with the northwest European *Paltus* horizon		–[d]	
base Pliensbachian Stage	189.6	1.5	Ammonite, lowest occurrences of *Bifericeras donovani* and of genera *Apoderoceras* and *Gleviceras*	Wine Haven section, Robin Hood's Bay, Yorkshire, England	2004[c]	
base Sinemurian Stage	196.5	1.0	Ammonite, lowest occurrence of arietitid genera *Vermiceras* and *Metophioceras*	0.9 m above base of Bed 145, East Quantoxhead, Watchet, West Somerset, southwest England	2000[b]	*Episodes* **25**(1): 22–6, 2002
base Hettangian Stage, base Jurassic System	199.6	0.6	Near lowest occurrence of smooth *Psiloceras planorbis* ammonite group		–[d]	
Triassic System						
Upper						
base Rhaetian Stage	203.3	1.5	Near lowest occurrence of ammonite *Cochlocera*, conodonts *Misikella* spp. and *Epigondolella mosheri*, and radiolarian *Proparvicingula moniliformis*	Key sections in Austria, British Columbia (Canada), and Turkey	–[d]	
base Norian Stage	216.5	2.0	Base of *Klamathites macrolobatus* or *Stikinoceras kerri* ammonoid zones and the *Metapolygnathus communisti* or *M. primitius* conodont zones	Leading candidates are in British Columbia (Canada), Sicily (Italy), and possibly Slovakia, Turkey (Antalya Taurus), and Oman	–[d]	
base Carnian Stage	228.0	2.0	Near first occurrence of the ammonoids *Daxatina* or *Trachyceras*, and of the conodont *Metapolygnathus polygnathiformis*	Candidate section at Prati di Stuores, Dolomites, northern Italy. Important reference sections in Spiti (India) and New Pass, Nevada	–[d]	
Middle						
base Ladinian Stage	237.0	2.0	Alternate levels are near base of *Reitzi, Secedensis*, or *Curionii* ammonite zone; near first occurrence of the conodont genus *Budurovignathus*	Leading candidates are Bagolino (Italy) and Felsoons (Hungary). Important reference sections in the Humboldt Range, Nevada	–[d]	

(*cont.*)

Table 2.3 (*cont.*)

EON, Era, System, Series, Stage	Age (Ma) GTS2004	Est. ± myr	Principal correlative events	GSSP and location	Status	Publication
base Anisian Stage	245.0	1.5	Ammonite, near lowest occurrences of genera *Japonites, Paradanubites, and Paracrochordiceras*; and of the conodont *Chiosella timorensis*	Candidate section probable at Desli Caira, Dobrogea, Romania; significant sections in Guizhou Province (China)	2004[c]	
Lower						
base Olenekian Stage	249.7	0.7	Near lowest occurrence of *Hedenstroemia* or *Meekoceras gracilitatis* ammonites, and of the conodont *Neospathodus waageni*	Candidate sections in Siberia (Russia) and probably Chaohu, Anhui Province, China. Important sections also in Spiti	–[d]	
base Induan Stage, base Triassic System, base Mesozoic	251.0	0.4	Conodont, lowest occurrence of *Hindeodus parvus*; termination of major negative carbon isotope excursion. About 1 myr after peak of Late Permian extinctions	Base of Bed 27c, Meishan, Zhejiang, China	2001[b]	*Episodes* **24**(2): 102–14, 2001
Paleozoic Era						
Permian System						
Lopingian Series						
base Changhsingian Stage	253.8	0.7	Conodont, lowest occurrence of conodont *Clarkina wangi*	Leading candidates are in China		
base Wuchiapingian Stage	260.4	0.7	Conodont, near lowest occurrence of conodont *Clarkina postbitteri*	Base of Bed 6K/115, Penglaitan section, 20 km southeast of Laibin, Guangxi Province, China	2004[b]	*Lethaia*, in prep.
Guadalupian Series						
base Capitanian Stage	265.8	0.7	Conodont, lowest occurrence of *Jinogondolella postserrata*	4.5 m above base of Pinery Limestone Member, Nipple Hill, southeast Guadalupe Mountains, Texas	2001[b]	*Episodes* in prep.
base Wordian Stage	268.0	0.7	Conodont, lowest occurrence of *Jinogondolella aserrata*	7.6 m above base of Getaway Ledge outcrop, Guadalupe Pass, southeast Guadalupe Mountains, Texas	2001[b]	*Episodes* in prep.
base Roadian Stage	270.6	0.7	Conodont, lowest occurrence of *Jinogondolella nanginkensis*	42.7 m above base of Cutoff Formation, Stratotype Canyon, southern Guadalupe Mountains, Texas	2001[b]	*Episodes* in prep.

Table 2.3 (*cont.*)

EON, Era, System, Series, Stage	Age (Ma) GTS2004	Est. ± myr	Principal correlative events	GSSP and location	Status	Publication
Cisuralian Series						
base Kungurian Stage	275.6	0.7	Conodont, near lowest occurrence of conodont *Neostreptognathus pnevi–N. exculptu*	Leading candidates are in southern Ural Mountains		
base Artinskian Stage	284.4	0.7	Conodont, lowest occurrence of conodont *Sweetognathus whitei*	Leading candidates are in southern Ural Mountains		
base Sakmarian Stage	294.6	0.8	Conodont, near lowest occurrence of conodont *Sweetognathus merrilli*	Leading candidate is at Kondurovsky, Orenburg Province, Russia		
base Asselian Stage, base Permian System	299.0	0.8	Conodont, lowest occurrence of *Streptognathodus isolatus* within the *S.* "*wabaunsensis*" conodont chronocline. 6 m higher is lowest fusilinid foraminifera *Sphaeroschwagerina vulgaris aktjubensis*	27 m above base of Bed 19, Aidaralash Creek, Aktöbe, southern Ural Mountains, northern Kazakhstan	1996[b]	*Episodes* **21**(1): 11–18, 1998
Carboniferous System						
Pennsylvanian Subsystem			*Series classification approved in 2004*			
base Gzhelian Stage	303.9	0.9	Near lowest occurrences of the fusulinids *Daixina*, *Jigulites*, and *Rugosofusulina*		–[d]	
base Kasimovian Stage, base Upper Pennsylvanian Series	306.5	1.0	Near base of *Obsoletes obsoletes* and *Protriticites pseudomontiparus* fusulinid zone, or lowest occurrence of *Parashumardites* ammonoid		–[d]	
base Moscovian Stage, base Middle Pennsylvanian Series	311.7	1.1	Near lowest occurrences of *Declinognathodus donetzianus* and/or *Idiognathoides postsulcatus* conodont species, and fusulinid species *Aljutovella aljutovica*		–[d]	
base Bashkirian Stage, base Pennsylvanian Subsystem	318.1	1.3	Conodont, lowest occurrence of *Declinognathodus nodiliferus s.l.*	82.9 m above top of Battleship Wash Fm., Arrow Canyon, southern Nevada	1996[b,e]	*Episodes* **22**(4): 272–83, 1999
Mississippian Subsystem			*Series classification approved in 2004*			
base Serpukhovian, base Upper Mississippian Series	326.4	1.6	Near lowest occurrence of conodont, *Lochriea crusiformis* or *L. Ziegleri*		–[d]	

(*cont.*)

Table 2.3 (*cont.*)

EON, Era, System, Series, Stage	Age (Ma) GTS2004	Est. ± myr	Principal correlative events	GSSP and location	Status	Publication
base Visean, base Middle Mississippian Series	345.3	2.1	Foraminifera, lineage *Eoparastaffella simplex* morphotype 1/ morphotype 2	Leading candidate is Pengchong, south China		
base Tournaisian, base Mississippian Subsystem, base Carboniferous System	359.2	2.5	Conodont, above lowest occurrence of *Siphonodella sulcata*	Base of Bed 89, La Serre, Montagne Noir, Cabrières, southern France	1990[b]	*Episodes* **14**(4): 331–6, 1991
Devonian System						
Upper						
base Famennian Stage	374.5	2.6	Just above major extinction horizon (Upper Kellwasser Event), including conodonts *Ancyrodella* and *Ozarkodina* and goniatites of *Gephuroceratidae* and *Beloceratidae*	Base of Bed 32a, upper Coumiac quarry, Cessenon, Montagne Noir, southern France	1993[b]	*Episodes* **16**(4): 433–41, 1993
base Frasnian Stage	385.3	2.6	Conodont, lowest occurrence of *Ancyrodella rotundiloba* (defines base of Lower *Polygnathus asymmetricus* conodont zone)	Base of Bed 42a', Col du Puech de la Suque section, St. Nazaire-de-Ladarez, southeast Montagne Noir, southern France	1986[b]	*Episodes* **10**(2): 97–101, 1987
Middle						
base Givetian Stage	391.8	2.7	Conodont, lowest occurrence of *Polygnathus hemiansatus*, near base of goniatite *Maenioceras Stufe*	Base of Bed 123, Jebel Mech Irdane ridge, Tafilalt, Morocco	1994[b]	*Episodes* **18**(3): 107–15, 1995
base Eifelian Stage	397.5	2.7	Conodont, lowest occurrence of *Polygnathus costatus partitus*; major faunal turnover	Base unit WP30, trench at Wetteldorf Richtschnitt, Schönecken-Wetteldorf, Eifel Hills, western Germany	1985[b]	*Episodes* **8**(2): 104–9, 1985
Lower						
base Emsian Stage	407.0	2.8	Conodont, lowest occurrence of *Polygnathus kitabicus* (= *Po. dehiscens*)	Base of Bed 9/5, Zinzil'ban Gorge, SE of Samarkand, Uzbekistan	1995[b]	*Episodes* **20**(4): 235–40, 1997
base Pragian Stage	411.2	2.8	Conodont, lowest occurrence of *Eognathodus sulcatus*	Base of Bed 12, Velká Chuchle quarry, southwest part of Prague city, Czech Republic	1989[b]	*Episodes* **12**(2): 109–13, 1989
base Lochkovian Stage, base Devonian System	416.0	2.8	Graptolite, lowest occurrence of *Monograptus uniformis*	Within Bed 20, Klonk, Barrandian area, southwest of Prague, Czech Republic	1972	Martinsson (1977)

Table 2.3 (*cont.*)

EON, Era, System, Series, Stage	Age (Ma) GTS2004	Est. ± myr	Principal correlative events	GSSP and location	Status	Publication
Silurian System						Holland and Bassett (1989)
Pridoli Series						
base Pridoli Series (*not subdivided in stages*)	418.7	2.7	Graptolite, lowest occurrence of *Monograptus parultimus*	Within Bed 96, Pozáry section near Reporje, Barrandian area, Prague, Czech Republic	1984[b]	*Episodes* **8**(2): 101–3, 1985
Ludlow Series		2.6				
base Ludfordian Stage	421.3	2.6	*Imprecise*. May be near base of *Saetograptus leintwardinensis* graptolite zone	Base of lithological unit C, Sunnyhill Quarry, Ludlow, Shropshire, southwest England	1980[b]	*Lethaia* **14**: 168, 1981; *Episodes* **5**(3): 21–3, 1982
base Gorstian Stage	422.9	2.5	*Imprecise*. Just below base of local acritarch *Leptobrachion longhopense* range zone. May be near base of *Neodiversograptus nilssoni* graptolite zone	Base of lithological unit F, Pitch Coppice quarry, Ludlow, Shropshire, southwest England	1980[b]	*Lethaia* **14**: 168, 1981; *Episodes* **5**(3): 21–3, 1982
Wenlock Series						
base Homerian Stage	426.2	2.4	Graptolite, lowest occurrence of *Cyrtograptus lundgreni* (defines base of *C. lundgreni* graptolite zone)	Graptolite biozone intersection in stream section in Whitwell Coppice, Homer, Shropshire, southwest England	1980[b]	*Lethaia* **14**: 168, 1981; *Episodes* **5**(3): 21–3, 1982
base Sheinwoodian Stage	428.2	2.3	*Imprecise*. Between the base of acritarch biozone 5 and extinction of conodont *Pterospathodus amorphognathoides*. May be near base of *Cyrtograptus centrifugus* graptolite zone	Base of lithological unit G, Hughley Brook, Apedale, Shropshire, southwest England	1980[b]	*Lethaia* **14**: 168, 1981; *Episodes* **5**(3): 21–3, 1982
Llandovery Series						
base Telychian Stage	436.0	1.9	Brachiopods, just above extinction of *Eocoelia intermedia* and below lowest succeeding species *Eocoelia curtisi*. Near base of *Monograptus turriculatus* graptolite zone	Locality 162 in transect d, Cefn Cerig Road, Llandovery area, south-central Wales	1984[b]	*Episodes* **8**(2): 101–3, 1985
base Aeronian Stage	439.0	1.8	Graptolite, lowest occurrence of *Monograptus austerus sequens* (defines base of *Monograptus triangulatus* graptolite zone)	Base of locality 72 in transect h, Trefawr forestry road, north of Cwm-coed-Aeron Farm, Llandovery area, south-central Wales	1984[b]	*Episodes* **8**(2): 101–3, 1985

(*cont.*)

Table 2.3 (*cont.*)

EON, Era, System, Series, Stage	Age (Ma) GTS2004	Est. ± myr	Principal correlative events	GSSP and location	Status	Publication
base Rhuddanian Stage, base Silurian System	443.7	1.5	Graptolites, lowest occurrences of *Parakidograptus acuminatus* and *Akidograptus ascensus*	1.6 m above base of Birkhill Shale Fm., Dob's Linn, Moffat, Scotland	1984[b]	*Episodes* **8**(2): 98–100, 1985
Ordovician System						
Upper						
base Hirnantian Stage	445.6	1.5	Potentially at base of the *Normalograptus extraordinarius–N oisuensis* graptolite biozone	Candidate section is Wangjiawan, China		
base of sixth stage (*not yet named*)	455.8	1.6	Potentially near first appearance of the graptolite *Dicellograptus caudatus*	Candidate sections ae Black Knob Ridge (Oklahoma) and Hartfell Spa (southern Scotland)		
base of fifth stage (*not yet named*)	460.9	1.6	Graptolite, lowest occurrence of *Nemagraptus gracilis*	1.4 m below phosphorite in E14a outcrop, Fågelsång, Scane, southern Sweden	2002[b]	*Episodes* **23**(2): 102–9, 2000 (*proposal; formal GSSP publication in preparation*)
Middle						
base Darriwilian Stage	468.1	1.6	Graptolite, lowest occurrence of *Undulograptus austrodentatus*	Base of Bed AEP184, 22 m below top of Ningkuo Fm., Huangnitang, Changshan, Zhejiang province, southeast China	1997[b]	*Episodes* **20**(3): 158–66, 1997
base of third stage (*not yet named*)	471.8	1.6	Conodont, potentially lowest occurrence of *Protoprioniodus aranda* or of *Baltoniodus triangularis*	Candidate sections at Niquivil (Argentina) and Huanghuachang (China)		
Lower						
base of second stage (*not yet named*)	478.6	1.7	Graptolite, lowest occurrence of *Tetragraptus approximatus*	Just above E bed, Diabasbrottet quarry, Västergötland, southern Sweden	2002[b]	*Episodes* in prep.
base of Tremadocian Stage, base Ordovician System	488.3	1.7	Conodont, lowest occurrence of *Iapetognathus fluctivagus*; just above base of *Cordylodus lindstromi* conodont Zone. Just below lowest occurrence of planktonic graptolites. Currently dated around 489 Ma	Within Bed 23 at the 101.8 m level, Green Point, western Newfoundland, Canada	2000[b]	*Episodes* **24**(1): 19–28,

Table 2.3 (*cont.*)

EON, Era, System, Series, Stage	Age (Ma) GTS2004	Est. ± myr	Principal correlative events	GSSP and location	Status	Publication
Cambrian System			Potential GSSP correlation levels include *Cordylodus proavus*, *Glyptagnostus reticulatus*, *Ptychagnostus punctuosus*, *Acidusus atavus*, and *Oryctocephalus indicus*			Overview of potential subdivisions in *Episodes* **23**(3): 188–95, 2000
Upper ("Furongian") Series						
upper stage(s) in Furongian			*Potential GSSP levels in upper Cambrian are based on trilobites and condonts*			
base Paibian Stage, base Furongian Series	501.0	2.0	Trilobite, lowest occurrence of agnostoid *Glyptagnostus reticulatus*. Coincides with base of large positive carbon isotope excursion	369.06 m above base of Huaqiao Fm., Paibi section, NW Hunan province, south China	2003[b]	*Episodes* in prep.
Middle	513.0	2.0	Potential GSSP levels in Middle Cambrian are based mainly on trilobites			
Lower			Potential GSSP levels in Lower Cambrian are based on archaeocyatha, small shelly fossils, and to a lesser extent, trilobites			
base Cambrian System, base Paleozoic, base PHANEROZOIC	542.0	1.0	Trace fossil, lowest occurrence of *Treptichnus (Phycodes) pedum*. Near base of negative carbon isotope excursion	2.4 m above base of Member 2 of Chapel Island Fm., Fortune Head, Burin Peninsula, southeast Newfoundland, Canada	1992[b]	*Episodes* **17**(1&2): 3–8, 1994
PROTEROZOIC			PreCambrian eras and systems below Ediacaran are defined by absolute ages, rather than stratigraphic points			
Neoproterozoic Era						
base Ediacaran System	630		Termination of Varanger (or Marinoan) glaciation	Base of the Nuccaleena Formation cap carbonate, immediately above the Elatina diamictite in the Enorama Creek section, Flinders Ranges, South Australia	–[f]	*Lethaia*, in prep.

(*cont.*)

Table 2.3 (*cont.*)

EON, Era, System, Series, Stage	Age (Ma) GTS2004	Est. ± myr	Principal correlative events	GSSP and location	Status	Publication
Cryogenian System	850		Base = 850 Ma			
Tonian System	1000		Base = 1000 Ma		1990[b]	*Episodes* **14**(2): 139–40, 1991
Mesoproterozoic Era						
Stenian System	1200		Base = 1200 Ma		1990[b]	*Episodes* **14**(2): 139–40, 1991
Ectasian System	1400		Base = 1400 Ma		1990[b]	*Episodes* **14**(2): 139–40, 1991
Calymmian System	1600		Base = 1600 Ma		1990[b]	*Episodes* **14**(2): 139–40, 1991
Paleoproterozoic Era						
Statherian System	1800		Base = 1800 Ma		1990[b]	*Episodes* **14**(2): 139–40, 1991
Orosirian System	2050		Base = 2050 Ma		1990[b]	*Episodes* **14**(2): 139–40, 1991
Rhyacian System	2300		Base = 2300 Ma		1990[b]	*Episodes* **14**(2): 139–40, 1991
Siderian System	2500		Base = 2500 Ma		1990[b]	*Episodes* **14**(2): 139–40, 1991
ARCHEAN						
Neoarchean Era	2800		Base = 2800 Ma			Informally in *Episodes* **15**(2): 122–3, 1992
Mesoarchean Era	3200		Base = 3200 Ma			Informally in *Episodes* **15**(2): 122–3, 1992
Paleoarchean Era	3600		Base = 3600 Ma			Informally in *Episodes* **15**(2): 122–3, 1992
Eoarchean Era						

[a] Informal working definition.
[b] Year GSSP ratified.
[c] Year in which ratification of GSSP anticipated.
[d] Guide event undecided.
[e] Subsytem rank of Mississippian and Pennsylvanian names ratified 2000.
[f] Age definition (650 Ma) ratified 1990; replaced by Australian GSSP 2004.

to preclude the need for more expensive methods of sampling such as coring. Coring and multimethod analysis of GSSP sections could provide a wealth of data for high-precision time scale work as well as providing secondary correlation markers. Previously unsuspected events, such as an hiatus or subtle rhythmic sedimentary patterns that cannot easily be detected by outcrop sampling, might also be revealed. For example, magnetic susceptibility has been fundamental for correlating Late Pleistocene sediments over a wide area of the northeast Atlantic (Robinson *et al.*, 1995) and for correlating and identifying the orbital components in latest Triassic–Jurassic successions in England (Weedon *et al.*, 1999).

2.2.5 Subdividing long stages

The emphasis on defining and dating stage boundaries tends to overlook that some Paleozoic and Mesozoic stages are rather long in duration. Ten such stages, all over 10 myr long are:

Lower Cambrian (not a stage, but a subsystem), Frasnian, Famennian, Tournaisian, Visean (almost 20 myr long), Carnian, Norian, Aptian, Albian, and Campanian. Their internal middle and upper boundaries should be subject to formal definition and GSSP standardization, just like their lower boundaries.

2.2.6 Do GSSP boundary stratotypes simplify stratigraphic classification?

Since the global chronostratigraphic scale is ultimately defined by a complete sequence of GSSPs, the limits of chronostratigraphic units (stages) are fully defined in time. Harland *et al.* (1990, p. 21) realized that the GSSP concept leads to the redundancy of a separate set of hierarchical terms for the time-rock domain:

> The terms system, series, etc. are commonly referred to as "time-rock units". It is argued here that they are now redundant and even confusing, being better replaced by reference to time divisions or rock units, as the meaning requires. It is commonly assumed that the time-rock couplets era–erathem, period–system, epoch–series, and age–stage, as well as early–lower, mid–middle, and late–upper, are precisely equivalent and should be selected only according to context in sentence. It is simpler for the chronostratigraphic scale to apply only the first term in each couplet. This is being recommended as being convenient for both thought and expression. Some others adopt the opposite simplification, i.e. Lower and Upper for time as well as rock. On that basis Lower Cambrian time (for example) may be expanded to "the time in which all Lower Cambrian rock formed (as well as intervening time not represented by rock) falling within the Early Cambrian time interval which in turn is defined by the two (initial and terminal) GSSPs in rock, each point representing an event in time". The use of time-rock terms (e.g. Lower Cambrian) predates the standardization of time terms, so it is an understandable perpetuation of an old habit that it is now nevertheless timely to replace. By referring to Early Cambrian rather than Lower Cambrian, the definition (and concept) is more direct. Early Cambrian rocks are any rocks formed in Early Cambrian time. The geologic period is defined by the initial and terminal events represented by the GSSP. The system is the rock estimated to have formed in that interval. It [the system] cannot define the period because the system boundaries are unknowable except at unconformities where the boundary rocks of uncertain age are missing. This work eliminates the use of time-rock terms such as system, without loss of meaning.

The same redundancy was elaborated by Walsh (2001). Zalasiewicz and colleagues of the Stratigraphic Commission of the Geological Society of London phrased it elegantly, by stating (J. Zalasiewicz, pers. comm., 2000):

> We consider that the practice of Chronostratigraphy today defines the time framework of Geochronology, because intervals of geological time are now being precisely defined within rock successions by GSSPs. The effect of this is that Chronostratigraphy and Geochronology should become one and the same discipline, as Harland *et al.* (1990) realized. For this one discipline we propose to keep the name Chronostratigraphy, which is the definition and application of a hierarchy of Eons, Eras, Periods, Epochs, Stages and Chrons. The formal terms Eonothem, Erathem, System, Series, Age and Chronozone thus become redundant. We include here the use of "Stage" (rather than "Age" of the standard geochronologic scale), which, as Harland *et al.*, 1990 argued, liberates "Age" for general use. The time units defined by Chronostratigraphy may be qualified by Early/Mid/Late, but not by Lower/Middle/Upper. As an example, one would not speak of "Lower January" or "Upper July". The qualifiers Lower/Middle/Upper continue to be applicable to the rock bodies of lithostratigraphy. The time units defined by Chronostratigraphy are founded within strata, but encompass all rocks on Earth. The term Geochronology reverts to its original use of referring to obtaining numerical estimates of time, through radiometric dating, the counting of Milankovitch cycles and so on.

The practice of using the term *stage* for both time and for rock has the advantage of simplifying stratigraphy, liberating *age* for general use, and avoids some ambiguity and confusion. The preferred hierarchy would be eon, era, period, epoch, and stage. One speaks of the beginning and end of a stage, or epoch or period in a time sense, and when speaking about the lower part of a stage refers to the rocks of that stage age and uses lithostratigraphic classification. However, not all co-authors of this time scale book adhere to this suggested practice, and a common new terminology has not been pursued. More philosophical discussion on the challenging issue of simplifying stratigraphy without stage stratotypes that embody time and rock as correlative units is desirable (e.g. Walsh, 2001).

2.3 CASE EXAMPLES OF GSSPs

2.3.1 Pliocene–Pleistocene boundary

The first example of a prominent GSSP highlights that different arguments lead to different preferences.

The main criterion used to place the boundary between the Pleistocene and Pliocene is changes in marine fauna, preferably in Italy since it is a classic area for marine and continental Pleistocene deposits. The base-Pleistocene GSSP ratified in 1985 is the base of sapropel layer "e" at the Vrica section in Calabria, Italy (*Episodes* 8: 116–20, 1985). It is just above top of magnetic polarity Chron C2n ("Olduvai") and the extinction level of calcareous nannofossil *Discoaster brouweri* (base of nannofossil Zone CN13). Milankovitch cyclicity of the Vrica section enables assignment of an age of 1.806 Ma.

However, after ratification of the GSSP, some stratigraphers proposed that the boundary be lowered to 2.5 Ma near the Gauss–Matuyama polarity chron boundary at the approximate onset of the rapid climate cycles that characterize the Pleistocene (or traditional "Quaternary"). Both in northwest Europe and in New Zealand this boundary is prominent in the continental rock record. The intense debate culminated in a formal vote among representatives from ICS and the International Quaternary Association (INQUA), which retained the GSSP at Vrica (Remane and Michelsen, 1999).

From a practical point of view, both boundaries are acceptable. Both can be tied to the magnetic polarity scale and linked to Milankovitch cycles to give highly precise dates for either definition. One argument for retaining the original GSSP was to promote a degree of stability in defining international chronostratigraphic units, rather than instate a new round of shifting standards.

As outlined in Chapter 21, complex historical arguments on the stratigraphic bracketing of the terms Neogene and Quaternary and its historic rock content continue to flavor "post-ratification" discussions. A young level as base for the Pleistocene near 900 ka can also thus be argued.

2.3.2 Eocene–Oligocene boundary

The second example of discussion surrounding a GSSP and its stages involves the GSSP for the base of the Oligocene that should (but appears not quite to) correspond to the boundary between the Priabonian and Rupelian Stages, accepted international stages for the Late Eocene and Early Oligocene subepochs, respectively.

One of the outcomes of the IGCP Project 174, Terminal Eocene Events, was to recommend and accept the extinction level of the low- to mid-latitude planktonic foraminiferal group of the Hantkeninidae for biostratigraphic recognition of the Eocene–Oligocene boundary. Hence, Odin and Montanari (1988) proposed that the primary marker for the Eocene–Oligocene boundary would be the extinction level of the genus *Hantkenina* (base of foraminiferal Zone P18), and that a GSSP should be placed at the base of a marl bed located 19 m above the base of Massignano quarry, Ancona, Italy. This GSSP was ratified in 1992 (*Episodes* **16**: 379–82, 1993) and is in the younger portion of polarity Chron C13r. An age of 33.7 ± 0.5 Ma is indicated for the events. It should be mentioned in this context that the Eocene–Oligocene boundary study only made passing reference to the stages adjoining the boundary, with a comment that more study would be desirable to relate to the shallow- to marginal-marine type Priabonian, Upper Eocene.

Dinoflagellate cyst correlations, in particular the boundary between the *Achomosphaera alicornu* Zone and the *Glaphyrocysta semitecta* Zone, now indicates that the GSSP for the lower limit of Oligocene at Massignano correlates to the middle part of the sequence of shallow- to marginal-marine lithostratigraphic units assigned to the Upper Eocene Priabonian Stage in northeast Italy (Brinkhuis and Visscher, 1995). A sequence stratigraphic approach suggests that the very top of the Priabonian in the type section corresponds to the base of the typical Rupelian, i.e. the base of the Oligocene in Belgium. A palynological study undertaken by Stover and Hardenbol (1993) determined that the base of beds considered Rupelian in its type area in Belgium is one dinoflagellate zone above the GSSP in Massignano; however, the Rupelian base itself is an hiatus, the result of relative sea-level fall and tectonic uplift. Thus, although formal acceptance of the Eocene–Oligocene boundary in Massignano appears to overlap slightly with Priabonian beds in a shallow- to marginal- marine facies, it appears to leave a short gap to the immediately overlying Rupelian Stage. A practical solution would be to extend the Rupelian slightly downward, at the expense of some marginal-marine Priabonian strata that have only local correlative significance. Hence, this GSSP implicitly assigns revised definitions to the historical Rupelian and the Priabonian Stages that are suitable for global correlation.

2.3.3 Permian–Triassic boundary

The third example of GSSP selection involves the Permian–Triassic boundary (Yin, 1996; Yin *et al.*, 1996). For this important boundary between the Paleozoic and the Mesozoic, a golden spike was finally hammered in early August 2001, after 18 years of study and debate by the ICS working group on this subject (1981–1999). The GSSP is at the base of Bed 27c at Meishan section "D" in the Zhejiang Province of China (Fig. 2.3; Yin *et al.*, 2001). The State Council of China approved this boundary type section as an open region to all visitors with a valid visa; free access for field study being one of the criteria of a GSSP.

For over 100 years, beds with the ammonite genus *Otoceras*, first found in the Himalayas, were widely considered to represent lowest Triassic strata. The genus is also widely known through the Arctic, but appears limited to the Permian–Triassic temperate and non-tropical realm. Generally, it was accepted that beds with *Otoceras* should be taken to define the base of the Triassic, and the base of the Griesbachian Stage. Particularly in the Arctic, *Otoceras* bearing beds may attain a thickness of over 70 m, versus only a few meters in Asia. In the Himalayas, the main phase of mass extinction of Paleozoic fauna and flora lies immediately below the level with *Otoceras*, with the latter representing a major transgressive phase.

In contrast, key Permian–Triassic boundary sections in China, Kashmir, and Tibet were found to be stratigraphically continuous, or most likely so, and have a particular good record of condodonts, a microfossil group of choice for near-global marine correlations. The species *Hindeodus parvus*, in an evolutionary lineage of several related forms, is found in China, Kashmir, Salt Range, Caucasus, Iran, Italy, Austria, Hungary, western USA, and British Columbia, Canada, and is considered to have a pan-Tethyan and circum-Pacific distribution in shallow-marine to occasionally deep-marine facies. After extensive debate and many publications, well summarized in Yin (1996), the Meishan section in Zhejiang Province of China was accepted as a GSSP, with the onset of *H. parvus* the bounding criteria for the Paleozoic–Mesozoic Era boundary, the Permian–Triassic Period boundary, and the boundary between the Changhsingian and Induan Stages (Fig. 2.6).

Important also is the sharp negative excursion of $\delta^{13}C$ just below the FO of *H. parvus* in Meishan and many sections world-wide, considered to reflect a considerable loss in microplankton productivity and biomass coincident with the end-Permian mass extinction (see Chapter 17). Thus, this stable isotope event is an excellent global auxiliary marker virtually at this important boundary.

Volcanic ashes yield radiometric ages bracketing the GSSP in Meishan. It appears that the base-Triassic, as defined by the GSSP, is significantly younger than the peak of the end-Permian mass extinctions, and re-assigns some "basal-Triassic" ammonite beds in the Arctic and Himalayas to the Permian Period.

2.4 MAJOR SUBDIVISIONS OF THE GEOLOGIC TIME SCALE

The development of the subdivisions of geologic time is a fascinating topic filled with philosophical concepts, dramatic personal disputes, and perceptive geological thinking (e.g. see the lively historical compilation by Berry, 1987). Several historical aspects still linger in informal stratigraphic divisions (e.g. the "Tertiary" was introduced by Giovanni Arduino circa 1759 (see also second color plate), a mining inspector in Tuscany, in a letter to a professor at the University of Padua about the relative age of low hills in Italy in which the sediments were commonly derived from "secondary" layered rocks; and similarly the "Dogger" was introduced from a clay-rich interval with concretions within the Jurassic of Britain). The proliferation of regional subdivisions eventually culminated in the recognition that international standardization is required, and the status of defining global intervals of geologic time is one of the main topics of this book.

Earth's history has been subdivided into three eons: the Archean, the Proterozoic, and the Phanerozoic (Fig. 2.7). The first two eons are grouped into an informal chronostratigraphic unit called the "Precambrian." The Phanerozoic Eon is subdivided into three eras: the Paleozoic, the Mesozoic, and the Cenozoic. Subdivisions of the Precambrian eons and the Phanerozoic eras are discussed in detail in Chapters 9–22, and only a brief outline is given here.

2.4.1 Archean and Proterozoic Eons (Precambrian)

"Precambrian" is an informal stratigraphic term that encompasses all geologic time and rock prior to the Cambrian Period. It was originally known as the "pre-Cambrian" and originated from the attempts of Adam Sedgwick and Roderick Murchison to subdivide the "grauwacke" strata of Wales into correlatable systems. Initially, the "Cambrian System" was applied by Sedgwick to non-fossiliferous strata below Murchison's fossil-bearing Silurian (Sedgwick and Murchison, 1835), but after suffering a contentious history, it was reinstated by Charles Lapworth (1879a) as the period with the earliest animal life forms. Thereafter, any older non-fossiliferous strata, schists, and crystalline rocks were referred to as "pre-Cambrian."

The Precambrian is currently formally divided into an older Archean Eon and a younger Proterozoic Eon, using the nomenclature suggested by the Royal Society of Canada (Alcock, 1934). The boundary between the two eons is defined chronometrically at 2500 Ma. The Archean and Proterozoic Eons are divided into four and three eras, respectively. The eras of the Proterozoic are subdivided into periods (see Chapter 9). An

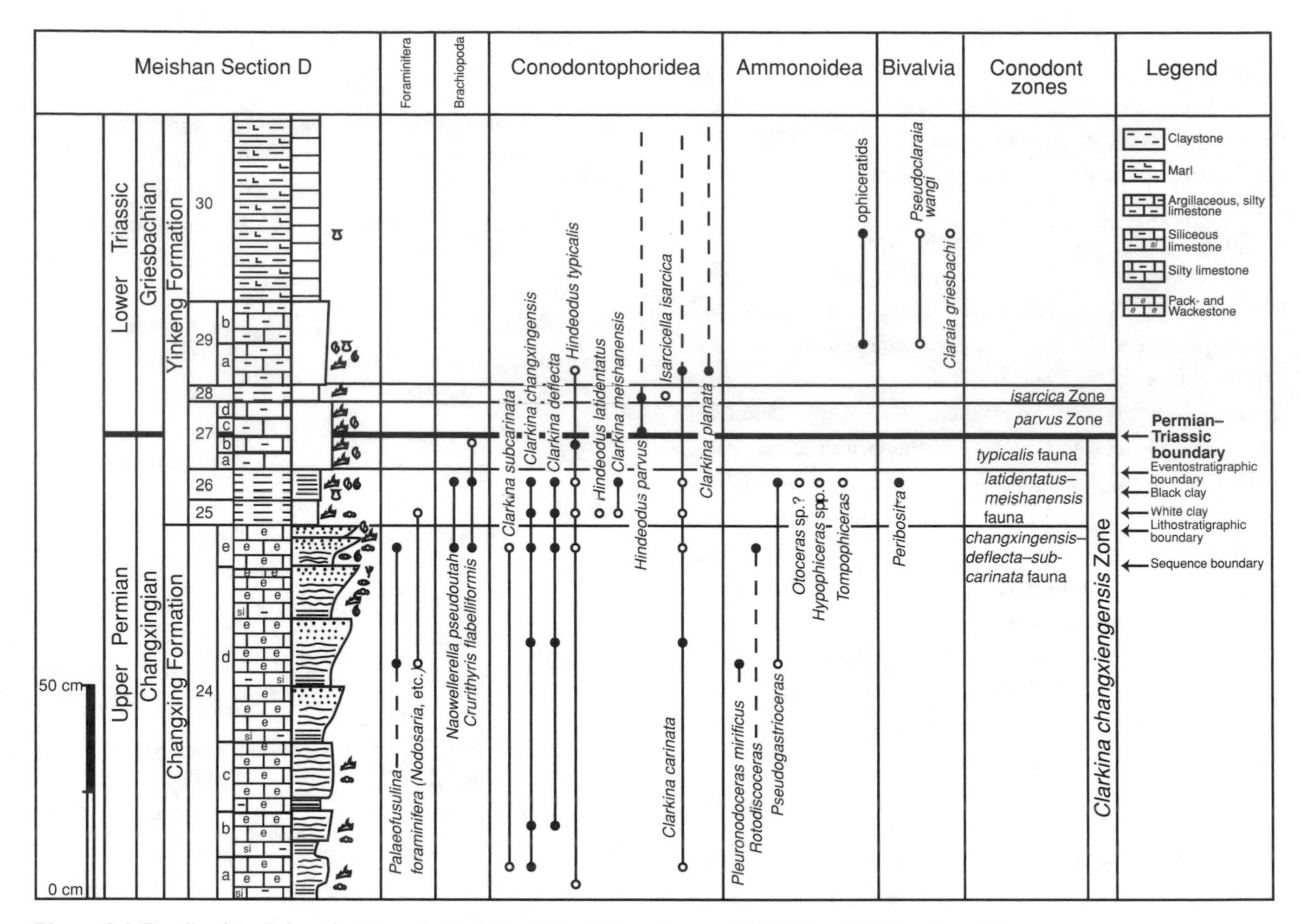

Figure 2.6 Details of the Paleozoic-Mesozoic (Permian-Triassic) boundary at the Meishan GSSP section, China.

historical review of Precambrian subdivisions is compiled in Harland *et al.* (1990, pp. 14–18). There is no defined older limit to the Archean Eon, but the term "Hadean" has been suggested for the interval from the Earth's formation to the oldest geological record (hence from ~4600 to ~3850 Ma) and is reviewed in Chapters 9 and 10.

2.4.2 Phanerozoic Eon

In an abstract, Chadwick (1930) proposed grouping geologic time into two eons: the "Phanerozoic" to encompass the Paleozoic, Mesozoic, and Cenozoic Eras; and "Cryptozoic" for the pre-Cambrian. Only the term Phanerozoic is currently in general use.

THE PALEOZOIC ERA

The "Palaeozoic" Series was proposed by Adam Sedgwick (1838) in a presentation to the Geological Society of London to encompass his Cambrian Period and Roderick Murchison's Silurian Period, and ending at the base of the Old Red Sandstone (Devonian). John Phillips (1841) extended the Paleozoic (meaning "ancient animal life") to also include the Old Red Sandstone through the Permian. The international convention is to drop the "a" from "ae" in *Palaeo*, hence it is now spelled as "Paleozoic" Era. The Paleozoic is divided into six periods, from the Cambrian at the base, through the Ordovician, Silurian, Devonian, Carboniferous, and Permian.

THE MESOZOIC ERA

Phillips (1841) introduced the term Mesozoic (meaning "middle animal life") in a brief article in *The Penny Cyclopaedia of the Society for the Diffusion of Useful Knowledge* to follow the Paleozoic, and later proposed that the Mesozoic would span the New Red Formation (the lower part was later determined to be the Permian) through the Cretaceous Chalk (Phillips, 1841). The Mesozoic Era is divided into the Triassic, Jurassic, and Cretaceous Periods.

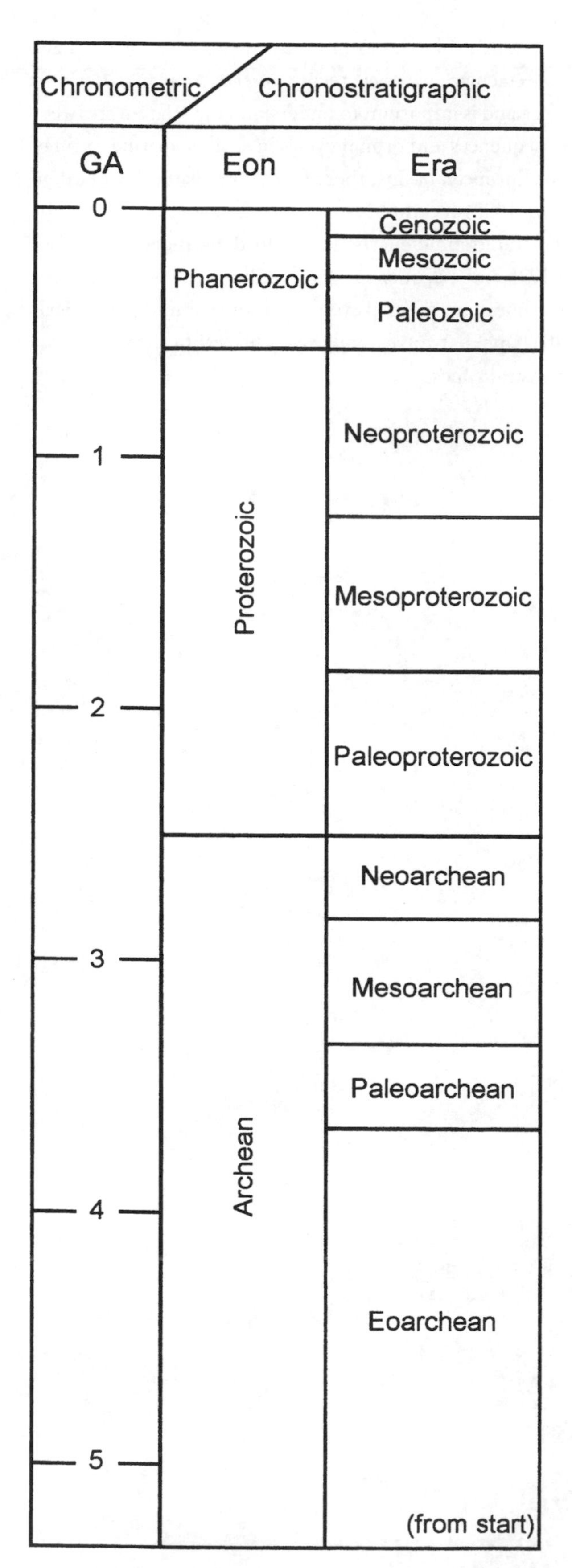

Figure 2.7 Main divisions of Earth history into eons and eras.

THE CENOZOIC (CAINOZOIC) ERA

The name "Kainozoic" (from the Greek word *kainos* meaning "recent") was coined by Phillips (1841), and later spelled by him as "Cainozoic", to apply to British strata younger than the Chalk (Cretaceous). It is now known as "Cenozoic" (see the linguistic history discussion in Harland *et al.*, 1990, p. 31).

The Cenozoic is divided into Paleogene and Neogene Periods. We refrain from using the informal terms Tertiary and Quaternary, remnants of a classification that included Primary and Secondary (e.g. by Arduino, 1760b; see also second color plate), but their origins and history are well described in Harland *et al.* (1990) and Berggren (1998). The changing concept of the term "Quaternary" is summarized in Chapter 21.

2.5 EXAMPLES OF STRATIGRAPHIC CHARTS AND TABLES

The plethora of names for time and time-rock units in local regions lends itself to the production of wall charts and stratigraphic lexicons to summarize the regional schemes and links to a standard scale. The international standard is developed by the International Commission on Stratigraphy (ICS), and is formally published in conjunction with the International Geological Congress (e.g. Remane, 2000; and an edition to appear in *Episodes* in mid 2004). The updated charts in PDF format are freely available from the ICS website at www.stratigraphy.org.

Nearly all nations, states, and/or continents have compiled regional chronostratigraphic charts or lexicons of regional stratigraphy. Some selected examples of "recent" compilations are mentioned below.

In 2002, the *Stratigraphic Table of Germany 2002* (German Stratigraphic Commission, 2002) saw the light. The wall chart is well laid out and documents the interrelation of regional German rock units through Precambrian and Phanerozoic time. Chronostratigraphic units of the standard reference scale have the suffix -ium (in English -ian, and in French -ien); this is elegant and deserves consideration in other Germanic, including Nordic, languages where orthographic principles have occasionally "muddied" stage, series, and system nomenclature. Linear time scale modifications for parts of Paleozoic and Mesozoic are documented summarily. The discrepancy between U–Pb TIMS and HR–SIMS dates in the Devonian–Carboniferous is resolved by taking the youngest possible estimate of age dates with the former method and the oldest possible estimate of age dates with the latter one. Essentially, a regression line is forced through the opposite extremes of error bar values for successive age dates to interpolate stages.

An Australian Phanerozoic Timescale (Young and Laurie, 1996) is an erudite and well-illustrated standard work for that part of the world. Detailed explanatory notes make this study a valuable compendium of bio-, magneto-, and chronostratigraphic information. Particularly, there is a wealth of data on Australian biostratigraphy, well documented in many detailed charts and a wall chart linking local and standard zonations in one scheme.

The detailed *New Zealand Geologic Time Scale* study will go to press in 2004 (R. Cooper and J. Crampton, pers. comm., 2003). Considerable progress has been made with typifying the previously biostratigraphically defined regional stages. The spectacular Neogene record of the Wanganui Basin in New Zealand is important to understand the relations between global sequences and orbital cycles in shallow marine settings.

For European basins, there is the compilation by Hardenbol *et al.* (1998) of eight Mesozoic and Cenozoic bio–magneto–chronostratigraphic charts, linked to the standard time scale anno 1995 and a sequence stratigraphic framework. The detailed calibrations of fossil events and zonal units is particularly valuable, since it involves many classical localities, classical taxa, and classical zones.

Part II • Concepts and methods

3 • Biostratigraphy: time scales from graphic and quantitative methods

F. M. GRADSTEIN, R. A. COOPER, AND P. M. SADLER

Semi-quantitative and quantitative biostratigraphy methods are assisting with scaling of stages, as exemplified in the Ordovician–Silurian and Carboniferous–Permian segments of GTS2004. This chapter focuses on some theory and practical considerations.

3.1 INTRODUCTION

The larger part of the Phanerozoic time scale in this book relies on a construction where stages are first scaled "geologically" with biostratigraphic compositing techniques, and than stretched in linear time using key radiometric dates. The advent of versatile and "clever" semi-quantitative and quantitative biostratigraphy methods is assisting with this geological scaling. The methods also add a new dimension to the construction of local or standard biochronologies, and its time scale derivatives.

In particular, three methods, each with their own PC-based programs, merit attention when it comes to scaling biostratigraphic data for standard or regional time scales:

- graphic correlation,
- constrained optimization
- ranking and scaling.

Each of these three methods aims at a particular segment of time scale building and its application, using complex and/or large microfossil data files. Constrained optimization is directly utilized in building the early Paleozoic segment of GTS2004, and graphic correlation plays a key role in building the biostratigraphic composite for the late Paleozoic. Ranking and scaling has been used in construction of local biochronologies. In this chapter more general examples will be given of the approaches; a summary of the numerical and graphic methods is presented in Table 3.1.

A Geologic Time Scale 2004, eds. Felix M. Gradstein, James G. Ogg, and Alan G. Smith. Published by Cambridge University Press.

3.2 GRAPHIC CORRELATION

Rates of sediment accumulation have been used to derive time scales. The simplest methods average fossil zone thickness in several sections and assume that thickness is directly proportional to duration (e.g. Carter *et al.*, 1980). However, because zone boundaries are defined by the stratigraphic ranges of one, or a few species, only a very small subset of the total biostratigraphic information is used in the exercise. Worse is that sedimentation rarely is constant (linear) through time, making the assumption tenuous.

Graphic correlation (Shaw, 1964; Edwards, 1984; Mann and Lane, 1995; Gradstein, 1996; see also Table 3.1) is a method that makes better use of the biostratigraphic information in sections, and is thus used for time scale construction. Graphic correlation proceeds by the pair-wise correlation of all sections to build up a composite stratigraphic section. With each successive round of correlation, biostratigraphic range-end events missing from the composite are interpolated into it via a "line of correlation" (LOC). At the same time, the stratigraphic ranges of taxa are extended to accommodate the highest range-tops and lowest range-bases recorded in any of the sections used in the analysis. This procedure is based on the assumption that, because of incomplete sampling, non-preservation, unsuitable facies, and other reasons, local sections will underestimate the true stratigraphic range of species. Isotopic dates, and other physical events can also be interpolated (Prell *et al.*, 1986). The composite section thus becomes a hypothetical section that contains all stratigraphic correlation events, and in which local taxon ranges are extended to approximate their true range in time, as recorded among all the sections.

When the composite section is based on a relatively large number of individual stratigraphic sections, it has been regarded as a good approximation of a relative time scale itself (Sweet, 1984, 1988, 1995; Kleffner, 1989; Fordham, 1992). These workers have used conodont-bearing carbonate sections to build graphic correlation time scales for the Ordovician and Silurian. It is assumed that variations in sediment accumulation rate are evened out in the composite, during the process of

Table 3.1 *Summary of graphic and numerical methods in biostratigraphy, used to assist with construction of geologic time scales*

Graphic correlation	Constrained optimization	Ranking and scaling
Programs GRAPHCOR, STRATCOR	Program CONOP	Programs RASC and CASC
Deterministic method – graphic correlation in bivariate plots. Program STRATCOR can also simulate probabilistic solutions	Mostly a deterministic method, but can also simulate probabilistic solutions. Constrained optimization with simulated annealing and penalty score.	Probabilistic method – ranking, scaling normality testing, and most likely correlation of events; error analysis
Uses event order and thickness spacing; works best with data sets having both first and last occurrences of taxa	Uses event order, event cross-over, and thickness spacing; data sets best have both first and last occurrences of taxa	Uses event order, and scores of cross-over from well to well for all event pairs in the ranked optimum sequence
Best suited for small data sets; can also operate on larger data sets	Processes medium to large data sets	Processes large data sets fast; has data input and multi-well data bookkeeper
An initial standard section is selected, after which section after section is composited in the relative standard to arrive at a final standard composite	Treats all sections and events simultaneously (operates a bit like multidimensional graphic correlation)	Treats all sections and events simultaneously
Line of correlation (LOC) fitting in section-by-section plots; technique can be partially automated	Multidimensional LOC; automated fitting; can generate several different composites depending on run options	Automated execution; generates several scaled optimum sequences per data set depending on run parameters, and tests to omit "bad" sections or "bad" events
Attempts to find maximum stratigraphic range of taxa among the sections	Attempts to find maximum or most common stratigraphic ranges of taxa	Finds average stratigraphic position of first and/or last occurrence events
Builds a composite of events by interpolation of missing events in successive section-by-section plots, via the LOC	Uses simulated annealing to find either the "best" or a good multidimensional LOC and composite sequence of events	Uses scores of event order relationships to find their most likely order, which represents the stratigraphic order found on average among the sections
Relative spacing of events is a composite of original event spacing in meters in the sections	Relative spacing of events in the composite is derived from original event spacing in meters or sample levels	Relative spacing of events in the scaled optimum sequence derives from z-transformation of cross-over frequencies
No automatic correlation of sections; composite standard can be converted in time scale	***Correlates sections automatically; zonal composite can be converted to time scale***	***Optimum sequence can be scaled to linear time; automated correlation of sections using isochrones***
No error analysis; sensitive to geological reworking and other "stratigraphic noise," and sensitive to order in which sections are composited during analysis	Numerous numerical tests and graphical analysis of stratigraphic results; finds best break points for assemblage zones	Three tests of stratigraphic normality of sections and events; calculates standard deviation of each event as a function of its stratigraphic scatter in wells
Interactive operation under DOS; graphic displays of scattergrams and best-fit lines	Batch operation under Windows®; color graphics display shows progress of run	Button operated under Windows®, fast batch runs; color graphics of output and options for interactive graphics editing

LOC fitting and extension of stratigraphic ranges. The composite units by which the composite section is scaled are assumed to be of approximately equal duration (Sweet, 1988), and therefore are time units of unspecified duration ("standard time units"). Finally, the relative scale can be calibrated with radioisotopic dates that are tied to the biostratigraphic scale. The assumptions, and some of the problems with this method, are summarized by Smith (1993).

In a variation on the method, Cooper (1992) used graphic correlation of long-ranging, deep-water, Ordovician graptolite-bearing shale sections to test for uniformity (steadiness) of depositional rate. A regional composite for Scandinavia was plotted against a composite for Newfoundland and gave a reasonable approximation to a rectilinear fit. The same two sections were then plotted against an exceptionally long-ranging section for western Canada, with the same result, and were taken to indicate that sediment accumulation rates in the three regions were approximately constant with time. The thickness scale for the Scandinavian composite (the most fossiliferous one) was then taken as a reasonable proxy for a relative time scale. This scale was then adjusted as necessary to fit the relatively sparse isotopic dates (Cooper, 1992) to give a calibration of Early Ordovician graptolite zones and stages.

For GTS2004, graphic correlation was applied for the Carboniferous and Permian time scale segment (Section 15.3.3), involving about 40 Carboniferous and 20 Permian sections, using all available fossil groups.

3.3 CONSTRAINED OPTIMIZATION

A major disadvantage of the graphic correlation method is the limited number of sections and taxa that can be used in the analysis, for practical reasons. Another disadvantage is the requirement for one of the sections to be adopted as the starting "standard section," the stratigraphic thickness measurements of which become the composite units in the composite section. Third, assumptions about relative accumulation rates may bias the sequence of events in the composite. These problems are avoided by automating correlation procedure as a constrained optimization (CONOP software, Kemple *et al.*, 1995; Sadler, 1999).

Like graphic correlation, several of the options in CONOP seek the maximum stratigraphic ranges of taxa as represented in the sections and build a composite (Sadler and Cooper, 2004). Unlike graphic correlation, the method readily enables a large number of sections and species to be used and processes all taxa and all sections simultaneously. Thus, it resembles a multidimensional graphical correlation in the sense that it considers all the local stratigraphic sections. It differs, however, in treating all sections simultaneously. A closer analogy exists between CONOP and algorithms that search for the most parsimonious cladogram.

Over 230 measured stratigraphic sections in graptolite-bearing deep-water shales from around the world and containing 1400 species were compiled in a data set. The Ordovician part alone includes 119 sections, containing 669 taxa with ranges wholly or partly in the Ordovician. The total data set was used to derive a global composite section for the Ordovician and Silurian (discussed in more detail in Chapter 12). Since graptolite specimens are rarely, if ever, found reworked, such "stratigraphic noise" is readily avoided.

The impressive graptolite composite was built in two steps. In the first step, the order of events was established by minimizing misfit between the composite and each of the individual sections in turn. The method operates heuristically, searching and discovering, in what can be a very large number of operations, which composite is best. Misfit was gauged by the net distance that range-ends had to be extended among all sections, as measured by the number of correlative biostratigraphic event levels, rather than stratigraphic thickness (as in graphic correlation). The composite is only an ordinal sequence of events. The spacing is undetermined and, unlike graphic correlation, assumptions about accumulation rate do not influence the sequence.

In the second step, the spacing of every pair of adjacent events in the composite was determined from the average of the rescaled spacing of events in the sections. The observed ranges in the individual sections were first extended to match the composite sequence. The thickness of each section was rescaled according to the number of events that it spanned in the composite sequence. The scaling of the composite is therefore derived from all of the sections, rather than from an initial "standard section" as in graphic correlation, and it is the ratio of the thicknesses between events that is used, not the absolute thickness. The influence of aberrant sections, incomplete preservation, and non-uniform depositional rates is thus minimized. Graptolite zone boundaries and stage boundaries were then located in the composite, producing a relative time scale for the Ordovician and Silurian.

Twenty-two U–Pb zircon dates that were reliably tied to the graptolite sequence and included in the compositing process were plotted against the relative time scale, and the resulting near-rectilinear fit demonstrates the reliability of the method (Fig. 3.1). These dates were then used to calibrate the relative scale, which was adjusted accordingly.

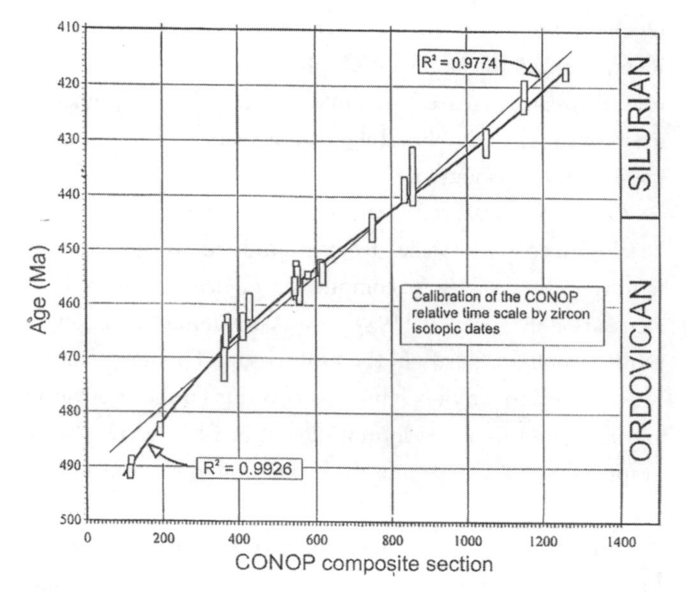

Figure 3.1 Calibration of the CONOP relative time scale in composite units by zircon isotopic dates in the Ordovician and Silurian (Sadler and Cooper, 2003, 2004). The extent to which the best fit is linear is the extent to which the CONOP relative biostratigraphic scale is linear.

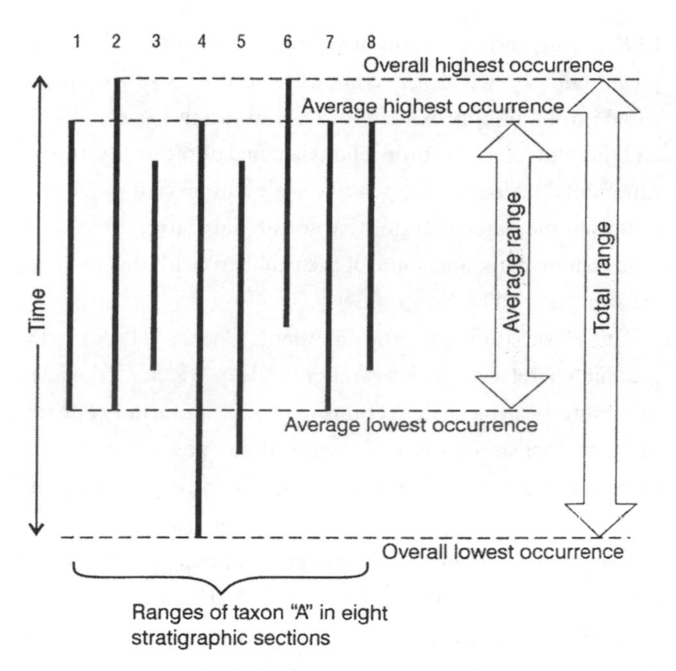

Figure 3.2 Theoretical illustration of the difference between average and maximum ranges of a species (taxon A) in eight stratigraphic sections. Probabilistic methods seek the average stratigraphic range, deterministic methods seek the total range (after Cooper *et al.*, 2001).

The result is a finely calibrated time scale. The method is applicable to any part of the time scale with suitable pelagic fossil groups, and is most suitable where isotopic dates are scarce. It provides a method for estimating the age of biostratigraphic and chronostratigraphic boundaries and events that lie stratigraphically between radiometric calibration points. Its underlying assumptions, methodology, and limitations are outlined by Sadler and Cooper (2004b). These authors demonstrate the method on the Ordovician and Silurian time scale.

In Chapters 12 and 13, the linear scaling of the CONOP graptolite composite is further refined through the use of mathematical and statistical techniques, incorporating error analysis.

3.4 RANKING AND SCALING

Both the graphic correlation and some options in the CONOP methods belong in the category of deterministic stratigraphy methods, and contrast with probabilistic methods. Deterministic methods seek the total or maximum stratigraphic range of taxa, whereas probabilistic methods estimate the most probable or average range (Fig. 3.2), to be accompanied by an estimate of stratigraphic uncertainty. Deterministic methods assume that inconsistencies in the stratigraphic range of a taxon from section to section or well to well are due to missing data. On the other hand, probabilistic methods assume that the inconsistencies are the result of random deviations from a most commonly occurring or average stratigraphic range. Or, to say it in terms of youngest occurrence events of taxa (or "tops" in exploration micropaleontology jargon): deterministic methods assume that there is a true order of events, and that inconsistencies in the relative order of tops from well to well are due to missing data. Probabilistic methods on the other hand consider such inconsistencies to be the result of random deviations from a most likely or optimum sequence of tops.

The most probable order of stratigraphic events in a sedimentary basin, with an estimate of uncertainty in event position, best predicts what order of events to expect in a new well or section. Calculation of the "true" order on the other hand would be most comparable to conventional, subjective results in range charts.

The principal method of probabilistic biostratigraphy, operating completely different from CONOP, is called RASC (Agterberg and Gradstein, 1999; Gradstein *et al.*, 1999). RASC is an acronym for *ra*nking and *sc*aling of biostratigraphic events; its sister method CASC stands for *c*orrelation *a*nd *s*tandard error *c*alculation. Data sets may vary from a few (e.g. 4) to many (25 or many more) wells or outcrop sections, and thousands of records, depending on requirement. For error analysis to have meaning, more wells are better than few.

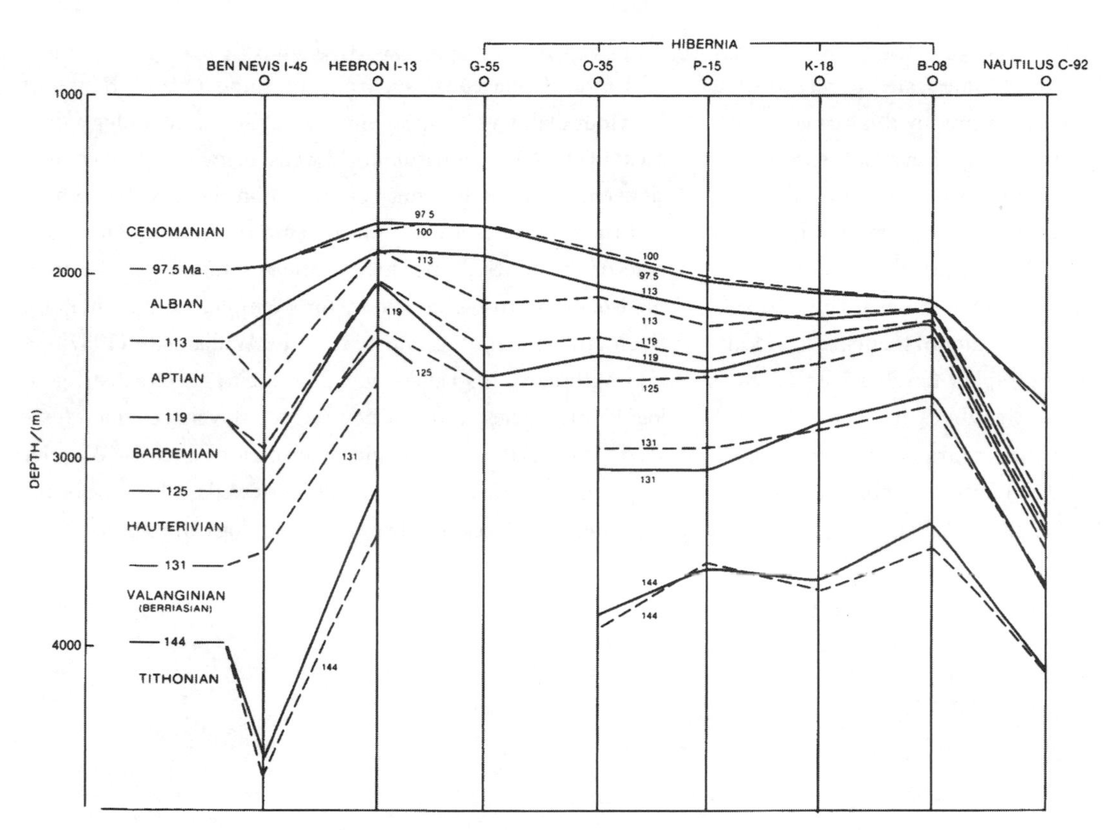

Figure 3.3 By using CASC interpolation, RASC biochronology ischrons can be correlated through the wells, as shown for the subjective (solid line) and most likely (dashed) depths of Cretaceous isochrons in northern Grand Banks wells (after Williamson, 1987). The automation of this process makes the method suitable for subsurface contouring using computer workstations.

Unlike graphic correlation, the RASC method considers the stratigraphic order of all (pairs of) fossil events in all wells simultaneously. It scores all order relationships of all event pairs in a matrix, and, using various modifications of trinomial set theory, calculates the most likely order of events. In this optimum sequence, each event position is an average of all individual positions encountered in the wells. Standard deviations of the event positions in the most likely (optimum) sequence are proportional to the amount of their stratigraphic scatter in all wells or outcrop sections.

Scaling of the optimum sequence in relative time is a function of the frequency with which events in each pair in the optimum sequence cross over their relative positions (observed records) from well to well; the more often two events cross over from well to well, the smaller their inter-fossil distance. Using a statistical model for the frequencies of cross-overs, these estimates are converted to z values of the normal distribution. Final distance estimates are expressed in dendrogram format, where tightness of clustering is a measure of nearness of events along a stratigraphic scale. The scaled version of the optimum sequence features time successive clusters, each of which bundles distinctive events. Individual bundles of events are assigned zonal status. The process of zone assignment in the scaled optimum sequence is subjective, as guided by the stratigraphic experience of the users. Large inter-fossil distances between successive dendrogram clusters agree with zonal boundaries, reflecting breaks in the fossil record due to average grouping of event extinctions. Such extinctions occur for a variety of reasons, and may reflect sequence boundaries. From a practical point of view, it suffices to say that taxa in a zone, on average, top close together in relative time.

The CASC method and program takes the RASC zonation, and calculates the most likely correlation of all events in the zonation over all wells. Interpolated event positions have error bars attached, and are compared to observed event positions in the wells examined. Since 1982, RASC and CASC have has wide stratigraphic application to a variety of microfossils, including dinoflagellate cysts, pollen/spores, diatoms, radiolarians, benthic and planktonic foraminifera, and also physical log markers inserted in zonations. A majority of applications involve well data sets from industry and from scientific ocean drilling. Published literature on the method and its uses is extensive.

Since the RASC optimum sequence has a numerical and linear scale, it may be converted to a time scale. A prerequisite is that, for an appropriate set of events in this scaled optimum sequence, absolute age estimates are available (e.g. from planktonic foraminiferal or nannofossil events in standard zonations). The more events in the scaled optimum sequence, the better the stratigraphic resolution, shrinking the gap between unevenly spaced events in estimated linear time. Next, the conversion of the RASC scaled optimum sequence to a local biochronology enables the stratigrapher familiar with CASC to trace isochrons in the same way as zones are traced. Examples of such exercises are presented in Gradstein *et al.* (1985), in Agterberg and Gradstein (1988) for the Cenozoic, offshore Labrador and Newfoundland, and in Williamson (1987) for the Jurassic–Cretaceous, offshore Newfoundland.

Figure 3.3 shows a close fit of subjective and likely traces of Lower Cretaceous isochrons in some Grand Banks of Newfoundland wells applying the CASC methodology on a most likely RASC zonation in 13 wells, using hundreds of fossil events. The dashed lines are based on the CASC method, and the solid lines are a subjective interpretation. An advantage of the CASC-type interpolation is that it can be used for isochron cross-sections at, for example, 1 myr intervals. Such cross-sections as constructed by Williamson (1987; see also Agterberg, 1990 and Fig. 9.22 therein) have realistic geological properties, and can be used to convert seismic cross-sections quickly into geologic time sections such as Wheeler diagrams, and thus to detect an hiatus in wells. This type of application enhances the role of biochronology in regional basin studies.

4 • Earth's orbital parameters and cycle stratigraphy

L. A. HINNOV

The Milankovitch theory that quasi-periodic oscillations in the Sun-Earth position have induced significant 10^4–10^6-year-scale variations in the Earth's stratigraphic record of climate is widely acknowledged. This chapter discusses the Earth's orbital parameters, the nature of orbitally forced incoming solar radiation, fossil orbital signals in Phanerozoic stratigraphy, and the use of these orbital signals in calibrating geologic time.

4.1 INTRODUCTION

Over the past century, paleoclimatological research has led to wide acceptance that quasi-periodic oscillations in the Sun–Earth position have induced significant variations in the Earth's past climate. These orbitally forced variations influenced climate-sensitive sedimentation, and thereby came to be fossilized in the Earth's cyclic stratigraphic record. The detection of orbital variations in Earth's cycle stratigraphy was progressively facilitated by advancements in celestial mechanics, which have provided more accurate models of the Earth's orbital–rotational behavior through geological time, and by improvements in data collection and analysis.

A principal outcome of the research has been the recognition that cycle stratigraphy, when shown to carry the signal specific to Earth's orbital behavior, serves as a powerful geochronometer. High-quality data collected over the past decade, in particular, have proven to have faithfully recorded all of the orbital cycles predicted by modern celestial mechanics over 0–23 Ma. Consequently, for the first time, the entire Neogene Period has been astronomically calibrated, and is reported in Chapter 21 as the Astronomically Tuned Neogene Time Scale 2004 (ATNTS2004). Cycle stratigraphy from more remote geological ages has not yet been calibrated directly to the orbital cycles, because of model limitations and greater uncertainties in determining stratigraphic age. Nonetheless, in numerous instances signal components analogous to those of the orbital variations have been detected in cycle stratigraphy, prompting the development of "floating time scales" that are calibrated to the average value of one or several model orbital frequencies. In Chapters 17–19, orbitally calibrated floating time scales are presented for intervals that extend through entire stages of the Mesozoic periods.

This chapter provides an introduction to the Earth's orbital parameters, the nature of the orbitally forced incoming solar radiation (insolation), and the discovery of orbitally forced insolation signals in cycle stratigraphy. The chapter concludes with remarks on the precision and accuracy that can be expected from orbitally calibrated cycle stratigraphy.

4.2 EARTH'S ORBITAL PARAMETERS

The Earth undergoes quasi-periodic changes in its orientation relative to the Sun as a consequence of interactions between the Earth's axial precession and variable orbit induced by motions of the other planets. These changes are described in terms of the Earth's *orbital parameters* (Fig. 4.1). Quantification of the orbital parameters has been carried out numerous times in the past with analytical approximations of the planetary motions (e.g. Milankovitch, 1941; Bretagnon, 1974; Berger, 1977a,b; Berger *et al.*, 1989; Laskar, 1990). Today, orbital models are performed largely by computerized numerical integration, and while they continue to share many of the features contained in the earlier ones, important new variables have been included, e.g. relativitistic effects, Earth's tidal braking, dynamical ellipticity, climate friction, Sun's oblateness, etc. For example, the model of Laskar *et al.* (1993) is reported as $La93_{(CMAR,FGAM)}$, where CMAR is input for the tidal effect of the Moon and FGAM is the input for the dynamical ellipticity of the Earth. The nominal model $La93_{(0,1)}$ assumes no tidal friction and present-day ellipticity; $La93_{(1,1)}$ builds in the tidal effect of Quinn *et al.* (1991). La93 provides an accurate "ephemeris" for the past 16 million years; uncertainties in the model's initial conditions and indications for chaotic motions of the planets impose an absolute accuracy limit at about 32 myr BP (Laskar, 1999).

A Geologic Time Scale 2004, eds. Felix M. Gradstein, James G. Ogg, and Alan G. Smith. Published by Cambridge University Press.

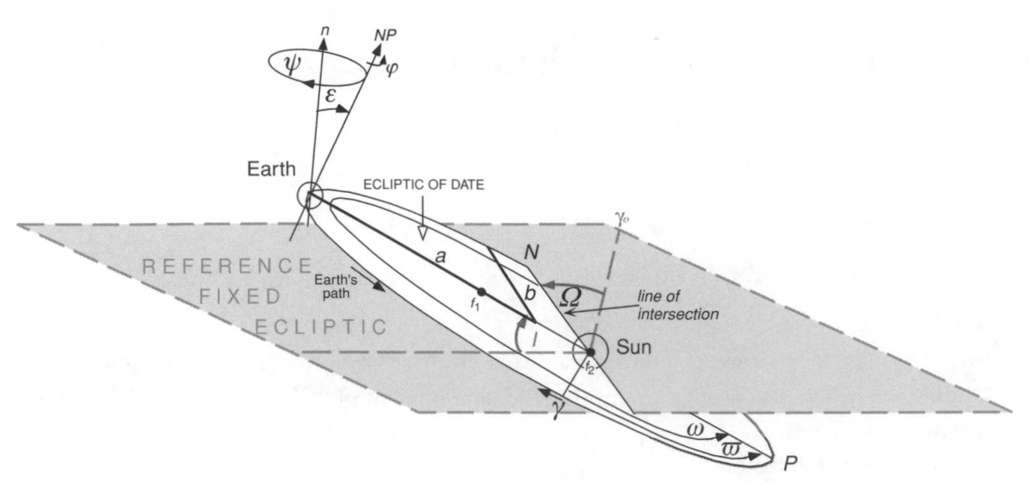

Figure 4.1 The Earth's orbital parameters seen from a view above the Earth's geographic North Pole (NP) in a configuration of northern summer solstice (NP pointed toward the Sun). The Earth's orbit is elliptical with (invariant) semi-major axis *a* and semi-minor axis *b* defining eccentricity *e*. The Sun occupies one of the two foci (f_1, f_2). Variables *e*, Π, *I* and Ω are "orbital elements," where $\Pi = \Omega + \omega$. The plane of the Earth's orbit (the "ecliptic of date") is inclined an angle *I* relative to a fixed reference ecliptic, and intersects this plane at a longitude Ω at point *N*, the ascending node, relative to fixed vernal point γ_0. (In this depiction, *I* is greatly exaggerated from its actual magnitude of only 1 to 2°.) The orbital perihelion point *P* is measured relative to γ_0 as the longitude of perihelion Π, and moves slowly anticlockwise. The Earth's figure is tilted with respect to the ecliptic of date normal *n* at obliquity angle ε. Earth's rotation φ is anticlockwise; gravitational forces along the ecliptic of date from the Moon and Sun act on the Earth's equatorial bulge and cause a clockwise precession Ψ of the rotation axis. This precession causes the vernal equinox point γ to migrate clockwise along the Earth's orbit, shifting the seasons relative to the orbit's eccentric shape; this motion constitutes the "precession of the equinoxes." The angle ω between γ and *P* is the moving longitude of perihelion and is used in the precession index $e \sin \omega$ to track Earth–Sun distance. Variations of *e*, ε, and $e \sin \omega$ are shown in Fig. 4.2.

According to $La93_{(0,1)}$, over the past 10 million years the Earth's *orbital eccentricity* varied from 0 to ~7% (Fig. 4.2*a*) with principal modes at 95, 99, 124, 131, 404, and 2360 kyr (Fig. 4.3*a*), caused by gravitational perturbations from motions of the other planets acting on Earth's orbital elements Π and *e* (Fig. 4.1). The *obliquity* variation has involved changes in the Earth's axial tilt between 22 and 24° (Fig. 4.2*b*), with a principal mode at 41 kyr, and lesser ones at 39, 54, and 29 kyr (Fig. 4.3*b*), due to planetary motions acting mainly on orbital elements *I* and Ω (Fig. 4.1). The *precession index* represents the combined effects of orbital eccentricity and the Earth's axial precession on the Sun–Earth distance (Fig. 4.2*c*), and has principal modes at 24, 22, 19, and 17 kyr (Fig. 4.3*c*). Long-period modulations of the obliquity and precession index (Figs. 4.2*b,c*) can be traced to the secular motions of individual planets (e.g. Berger and Loutre, 1990); frequency components of these modulations are summarized in Hinnov (2000).

Over geological time, dissipation of tidal energy is thought to have slowed the Earth's rotation rate (e.g. Berger *et al.*, 1989). This deceleration, accompanied by lunar recession, a declining Earth ellipticity and axial precession rate, would have resulted in a progressive lengthening of the obliquity and precession index modes toward the present (Berger *et al.*, 1992). Table 4.1 presents model predictions for the shortening of the major obliquity and precession index modes over the past 500 Ma, indicating significant changes over the Phanerozoic Eon. In addition, climate friction, i.e. episodic glacial loading of the Earth's crust, may result in "obliquity–oblateness feedback" and possibly significant deviations in the average obliquity angle (e.g. Bills, 1994; Rubincam, 1995). In contrast, orbital eccentricity is independent of geodynamical effects, and, although chaotic motions of some of the planets could destabilize some of the eccentricity modes from their current values, others, in particular the 404 kyr mode, may have remained relatively stable over much of Phanerozoic time (Laskar, 1999).

A newly updated orbital model, $La2003_{(CMAR,FGAM,CLIM)}$ exploits recent advances in numerical integration techniques and promises high-precision computations of the orbital parameters over geological times up to 40 Ma. This new model also provides a third parameter, CLIM, to include the aforementioned climate friction effect (CLIM = 0 assumes no effect). Differences between La93 and La2003 are noticeable at times previous to 20 Ma, and are discussed further in Chapter 21, where the ATNTS-2004 is calibrated to $La2003_{(1,1,0)}$ back to the Oligocene–Miocene boundary (23.03 Ma).

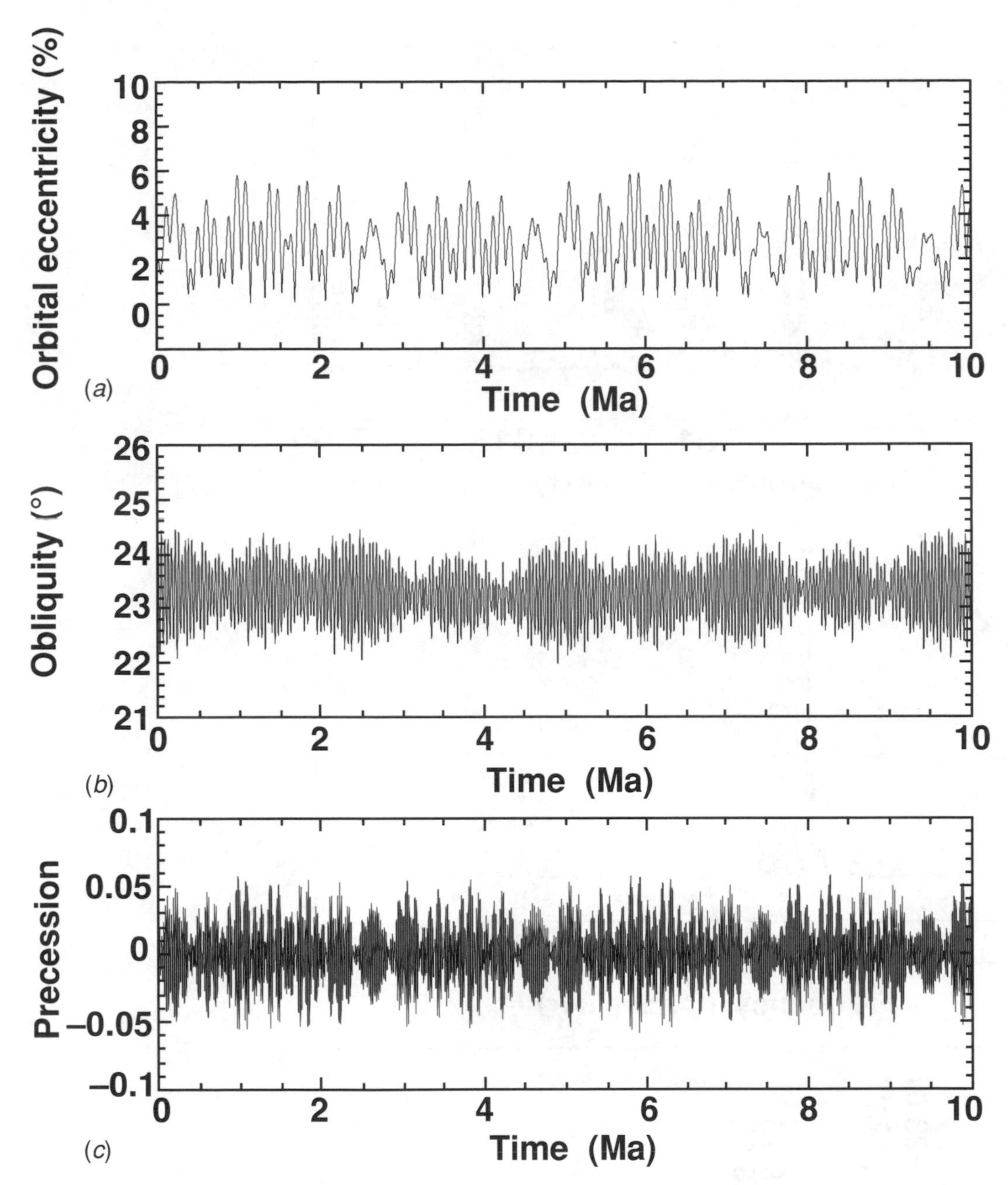

Figure 4.2 Variation of the Earth's orbital parameters over the past 10 million years according to $La93_{(0,1)}$, assuming Earth's present-day ellipticity and rotation rate. (*a*) Orbital eccentricity, in percent; (*b*) obliquity variation, in degrees of axial tilt; (*c*) precession index, in standardized units (dimensionless). All values are from *Analyseries* (Paillard *et al.*, 1996), which calculates La90($= La93_{(0,1)}$). A FORTRAN code to calculate adjustable models $La93_{(CMAR,\ FGAM)}$ can be downloaded from http://xml.gsfc.nasa.gov/archive/catalogs/616063/index_long.html.

4.3 ORBITALLY FORCED INSOLATION

The orbital parameters affect changes in the intensity and timing of the incoming solar radiation, or insolation, at all points on the Earth. These insolation changes comprise the well-known "Milankovitch cycles" (Milankovitch, 1941; re-issued in English in 1998). Geographical location, time of year, and even the time of day all determine the relative contributions of the orbital parameters to the inter-annual insolation (e.g. Berger *et al.*, 1993). For example, the insolation curves in Fig. 4.4 depict the globally available spectral power of orbitally forced daily insolation at the top of the atmosphere on June 21 (solstice) and March 21 (equinox). These curves are idealized in the sense that it is unlikely that climate responds to insolation only on one day of the year, but integrates insolation over certain times of the year and collectively over specific geographic

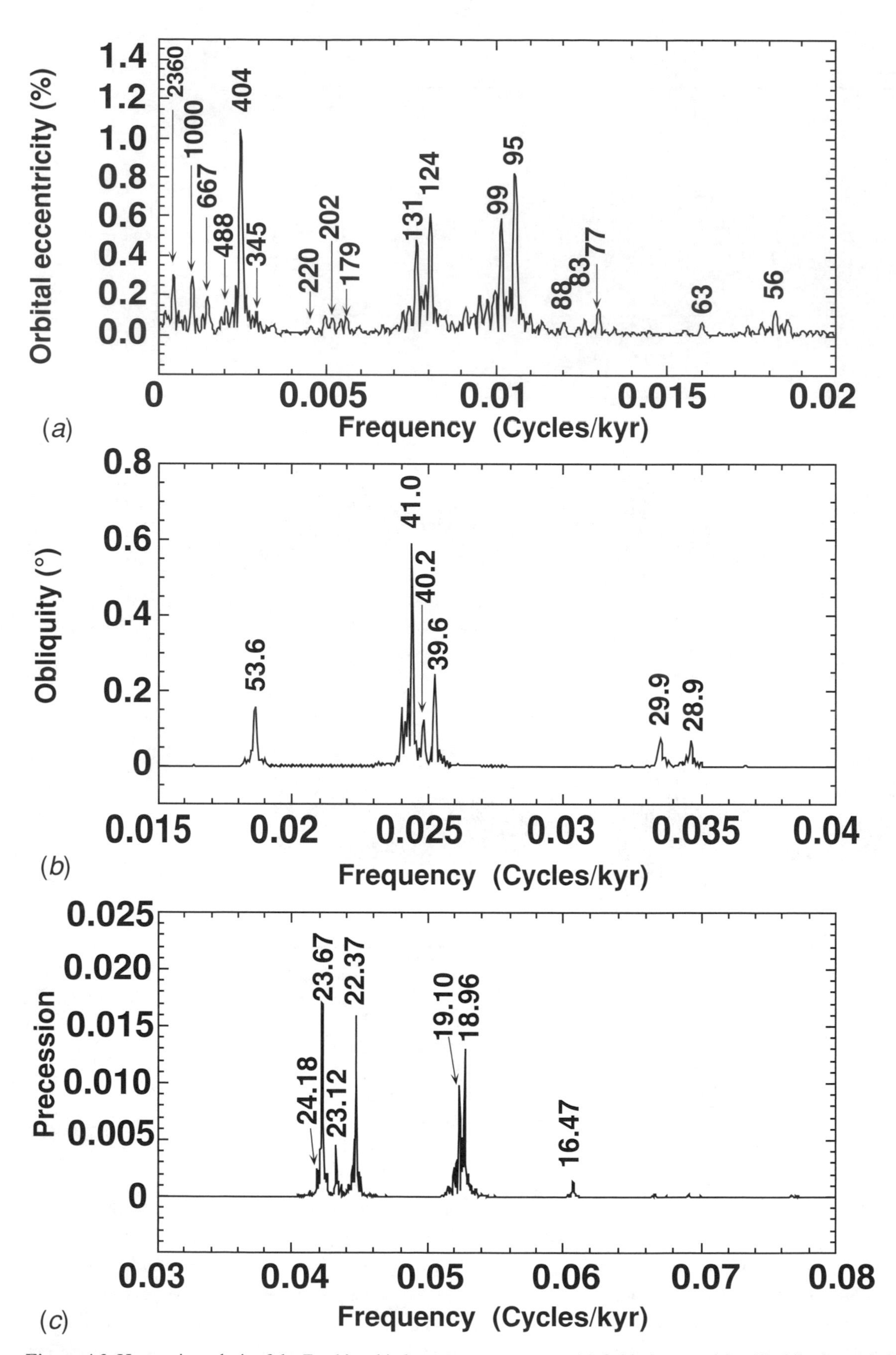

Figure 4.3 Harmonic analysis of the Earth's orbital parameters depicted in Fig. 4.2. All labels identify periodic components in thousands of years that pass the F-ratio test of Thomson (1982) for significant lines above the 95% level, using 4π multi-tapers. (*a*) Orbital eccentricity, (*b*) obliquity variation, (*c*) precession index. (Note: Due to the quasi-periodic nature of the parameters, the significance, periodicity, and amplitude of the labeled signal components will change for analyses over different time segments.)

Table 4.1 *Model of changes in the main periods of the Earth's obliquity variation and precession index*[a]

Time (Ma)	41-kyr mode	39-kyr mode	54-kyr mode	29-kyr mode
OBLIQUITY VARIATION				
0	41 057	39 663	53 805	28 929
72	39 333	38 052	50 883	28 062
270	34 820	33 812	43 577	25 687
298	34 203	33 231	42 615	25 350
380	32 426	31 550	39 891	24 360
440	31 168	30 358	38 003	23 643
500	29 916	29 169	36 158	22 916

Time (Ma)	24-kyr mode	22-kyr mode	19-kyr mode	16.5-kyr mode
PRECESSION INDEX				
0	23 708	22 394	18 966	16 470
72	23 123	21 871	18 590	16 188
270	21 485	20 401	17 517	15 368
298	21 249	20 188	17 359	15 247
380	20 549	19 555	16 889	14 884
440	20 037	19 090	16 542	14 612
500	19 512	18 613	16 182	14 332

[a] As in Hinnov (2000), based on Berger and Loutre (1994, their Table 2, Model 2). Values are in thousands of years. (Note: The precessional 16.5-kyr mode was calculated using $g = 1/(45\,865$ yr), from Laskar (1990).)

areas, possibly different areas at different times. This "climatic filtering" serves to alter the relative contributions of the orbital parameters to the total output climate response, this even prior to internal climate system responses to the insolation. Thus, it is left to the discretion of the paleoclimatologist to determine which time(s) of the year and at which location(s) a prevailing climate responded to insolation; this can require considerable insight into the infinite number of ways that one can sample insolation in space–time (Rubincam, 1994).

4.4 ORBITAL SIGNALS IN CYCLE STRATIGRAPHY

The prospect that orbital variations exerted large-scale climatic changes that could be detected in the geologic record was already being debated in the nineteenth century (e.g. Herschel, 1830; Adhémar, 1842; Lyell, 1867; Croll, 1875). Gilbert (1895) was the first to attribute the origin of limestone/shale cyclic strata of the Cretaceous Niobrara chalks (Colorado, USA) to astronomical forcing. Bradley (1929) counted varves in the lacustrine oil-shale–marl cycles of the Eocene Green River Formation (Utah, USA) estimating an average 21 630-year time scale for the cycles, and pointing to the precession of the equinoxes as a potential cause. Milankovitch (1941) was the first to attempt a quantitative correlation between astronomically calculated insolation minima and Late Quaternary Ice Age deposits of the Alps. However, later radiocarbon studies of glaciation timings in North America did not clearly corroborate Milankovitch's insolation calculations, and the orbital theory fell into disfavor (see the review in Imbrie and Imbrie, 1979; see also the update in Broecker and Denton, 1989).

At the same time, significant progress was made in understanding the origins of the prevalent rhythmic stratification of Mesozoic Alpine limestones (e.g. Schwarzacher, 1947, 1954). This research culminated in the seminal work of Fischer (1964), who found that the meter-scale beds (the Lofer cyclothems) of the Triassic Dachstein Limestone contained vertically repeating facies indicative of shallow marine environments exposed to oscillating sea levels, with an ~40 kyr timing. However, glaciations were unknown for the Triassic, raising doubts about the mechanisms by which such sea-level oscillations could have occurred; the origin of the Lofer cyclothems continues to be debated today (e.g. Schwarzacher, 1993; Enos and Samankassou, 1998).

It was not until investigation of the Late Quaternary deep-sea sedimentary record was undertaken that Milankovitch's theory of climate change was firmly validated. Emiliani (1955, 1966) explained oxygen isotope fractionation in marine calcareous microfauna as a function of ocean temperature and salinity; subsequently, Dansgaard and Tauber (1969) demonstrated that

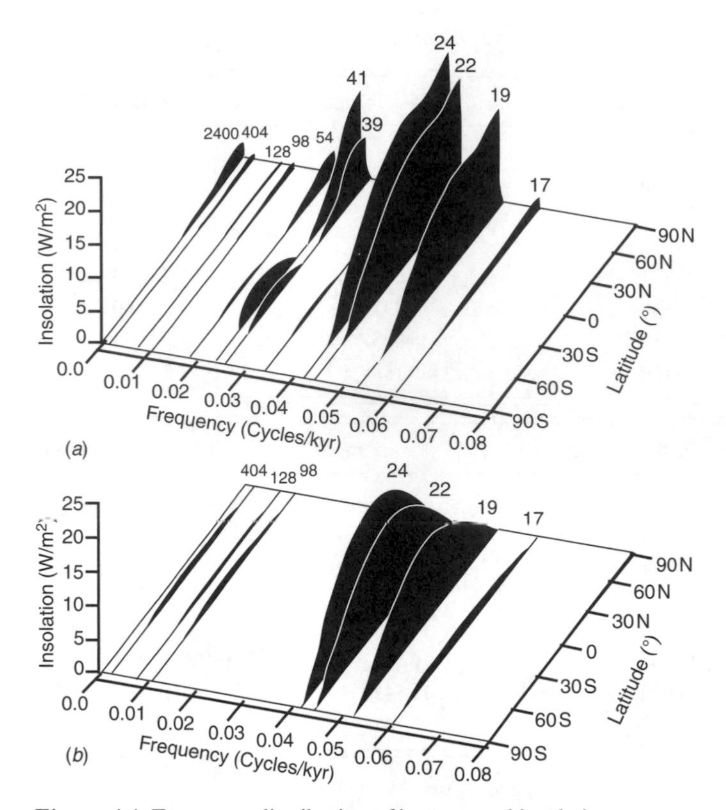

Figure 4.4 Frequency distribution of inter-annual insolation (Laskar, 1990) over 0–5 million years ago, sampled at 1-kyr intervals and displayed as amplitude spectra with respect to geographic latitude. (*a*) Daily mean insolation on June 21 (solstice). Latitudes south of ~66° S receive no insolation on this day. Maximum daily insolation occurs in the northern polar region, which experiences 24-hour exposure. The phase of the eccentricity and obliquity remains constant at all latitudes; for the precession phase shifts progressively from 0 to 180° from month to month (not depicted). (*b*) Daily mean insolation on March 21 (equinox). Insolation strength is a function of local solar altitude, highest at the Equator on this day of equal-time exposure everywhere. Contributions from the obliquity variation are absent. The phase of the eccentricity is constant at all latitudes; for the precession phase shifts progressively from 0 to 180° from pole to pole (not depicted). (Additional notes: (i) Insolation for the December 21 solstice similar to *a*, but with reversed latitudes; and the September 21 equinox is practically identical to *b*. (ii) The precession component of variation in *a* is at all locations 90° out of phase with the precession component in *b*. Additional examples are given in Berger *et al.*, 1993.)

the majority of changes in the marine oxygen isotope fractionation were linked to ocean volume. This result was followed by the landmark study of Hays *et al.* (1976) in which the oxygen isotope record was quantitatively linked to the Milankovitch cycles. Bolstered by the advent of global paleomagnetic stratigraphy, it was subsequently discovered that the same isotope signal, now encompassing the entire Brunhes chron (from 0 to 0.78 Ma), was present in all of the major oceans (Imbrie *et al.*, 1984). Finally, calibration of this proxy for global ocean volume to geological evidence for large sea-level changes (Chappell and Shackleton, 1986) established, albeit indirectly, the connection between the Quaternary Ice Ages and Milankovitch cycles. Later research into polar ice stratigraphy uncovered other isotope signals with strong orbital frequencies, providing additional, overwhelming support for the orbital forcing theory (Petit *et al.*, 1999).

These milestone studies touched off multiple initiatives to search for orbital cycles in stratigraphy back through geologic time using isotopes as well as other climate proxies, including facies stratigraphy, percent carbonate, biogenic silica, magnetic susceptibility, wireline logs, and grayscale scans (see Table 4.2). Continental Plio–Pleistocene sediments recovered from Lake Baikal revealed a strong biogenic silica signal closely mimicking that of the marine isotope record (Williams *et al.*, 1997; Prokopenko *et al.*, 2001). Additional deep-sea drilling yielded a continuous oxygen isotope signal spanning 0–6 Ma (Shackleton, 1995), and today, there is near-continuous Milankovitch coverage back through the Eocene from combinations of various marine climate proxies from both deep-sea drilling and outcrop studies (see Chapters 20 and 21). Evidence for orbital forcing has also been found in Cretaceous pelagic stratigraphy (see Chapter 19) and in Jurassic formations (see Chapter 18). The thick Upper Triassic continental lacustrine deposits of eastern North America contain a nearly perfect eccentricity signal that modulates facies successions linked to wetting–drying climate cycles at precessional time scales (see Chapter 17). It should be noted that cycle stratigraphy much older than ~20 Ma may never successfully be correlated directly to the orbital cycles, but only indirectly though comparison of average signal characteristics between data and orbital theory.

For geologic times prior to the Late Triassic, evidence for orbitally forced stratigraphy is less clear. One reason is that pre-Jurassic oceanic sediments are not composed of the abundant, continuous rain of pelagic oozes as are post-Jurassic ones. Therefore research has focused largely on the more prolific shallow-marine record, for which the primary evidence of Milankovitch forcing is systematic "interruption" rather than a continuous recording (Fischer, 1995). Definitive evidence for orbital signals in the Paleozoic is still weaker. The Permian Castile Formation, a marine evaporite sequence, shows a strong, but apparently short-lived Milankovitch signal (Anderson, 1982). The spectacular shelf carbonate cycles

Table 4.2 *Commonly measured sedimentary parameters that have been linked to orbitally forced climate change, and inferred climate conditions*

	Sedimentary parameter	Associated climate conditions
EXTRINSIC[a]	Oxygen isotopes	Temperature/salinity/precipitation/eustasy
(independent of	Carbon isotopes	Productivity/C-sequestration/redox conditions Surface
sedimentation rate)	Clay assemblages	hydrology
	Microfossil assemblages	Salinity/temperature
INTRINSIC[b]	Percent $CaCO_3$, Si, C_{org}	Productivity
(directly related to	Magnetic susceptibility	Sedimentation rate
and/or influenced by	Microfossil abundance	Productivity
sedimentation rate)	Clay/dust abundance	Surface hydrology/atmospheric circulation
	Lithofacies	Depositional environment
	Sediment color	Productivity/redox conditions
	Grain size	Erosion intensity/hydrodynamics

[a] Extrinsic parameters vary independently from sediment supply.
[b] Intrinsic parameters are directly related to sediment supply, and their signals tend to be more dramatically influenced (distorted) by changes in sedimentation rate (Herbert, 1994).

of the Pennsylvanian Paradox Basin (Utah, USA) indicate high-frequency sea-level oscillations with some orbital signal characteristics (Goldhammer *et al.*, 1994). The classic transgressive–regressive cyclothems of mid-continental USA (e.g. Heckel, 1986) and the rhythmic Mississippian hemipelagic limestones of Ireland (Schwarzacher, 1993) appear to express the dominant 404-kyr eccentricity cycle. There are at present only a few reports of orbital-scale cycles in Devonian formations (e.g. Elrick, 1995; Yang *et al.*, 1995; Bai, 1995; Crick *et al.*, 2001) and fewer still for the Silurian (Crick *et al.*, 2001; Nestor *et al.*, 2001). The thick Cambrian–Ordovician cyclic carbonate banks found world-wide, on the other hand, show some evidence of Milankovitch forcing (Bond *et al.*, 1991; Bazykin and Hinnov, 2002); however, the origins of the vast majority of these high-frequency eustatic signal proxies remain uninvestigated (e.g. Osleger, 1995).

4.5 ESTIMATING ORBITAL CHRONOLOGIES

Early on, the time predictability of the Earth's orbital parameters led to the practice of using orbitally forced cycle stratigraphy as a high-resolution geochronometer. While this application had already been considered by Gilbert (1895) and Barrell (1917), it was Milankovitch (1941) who first calibrated theoretical orbital insolation directly to the geologic record, adjusting approximately known ages of the Late Quaternary Alpine Ice Ages to the insolation minima of his calculated curves. The most significant advances in orbital chronology took place during the latter half of the twentieth century with the development of a high-resolution global marine oxygen isotope stratigraphy and magnetostratigraphy for the Pleistocene Epoch (see the review in Kent, 1999).

True orbital time calibration is possible only for cycle stratigraphy that can be connected to the "canonical" orbital variations, i.e. those quantitatively predicted by orbital theory. In GTS2004, this involves cycle stratigraphy back to the Miocene–Oligocene boundary only (0–23.03 Ma). Calibration of a sequence of cyclic strata begins with the assumption of a target orbital curve. This may take the form of an orbitally forced insolation signal, which most likely affected climate and was subsequently recorded by sedimentation (e.g. the 65° N summer insolation), or it can be as simple as the sum of the standardized orbital parameters (e.g. the ETP curve of Imbrie *et al.*, 1984). This initial assumption, however, introduces a basic source of error, because the true nature of the orbital forcing of the sediment is not known exactly. To account for this, Hilgen *et al.* (2000a), for example, calibrated ("tuned") their Miocene marl–clay deep-sea cycles to two possible target curves, the 65° N summer insolation and the precession index, correlating the mid points of the marls to the centers of the insolation maxima and precession minima. These two calibrations produced chronologies that differed from cycle to cycle by several thousand years; this was taken as a fair representation of the uncertainty of the chronology. Questions remain about which model produces the most accurate orbital cycles

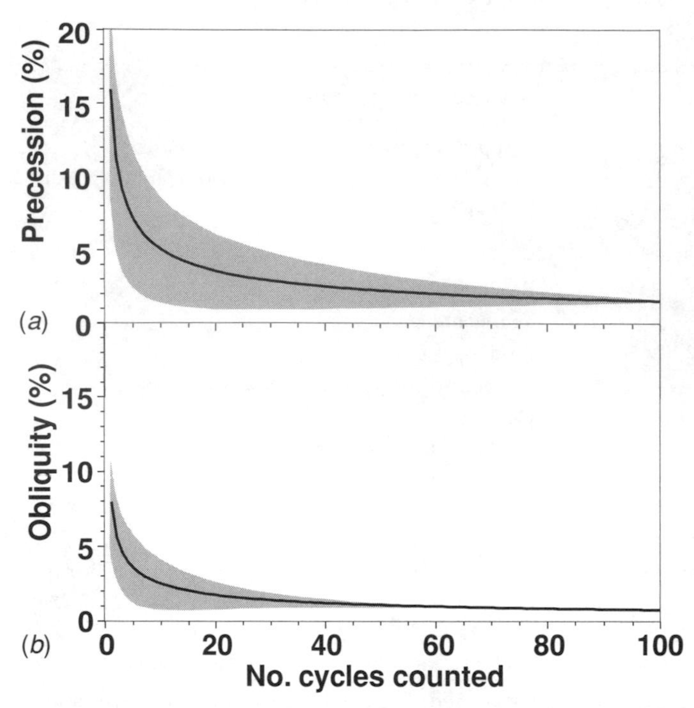

Figure 4.5 Estimated error in measuring time in using orbitally forced cycle stratigraphy when assuming that each cycle represents a duration equal to the average period of the orbital parameter that forced its formation (the "metronome" approach discussed in Herbert *et al.*, 1995). (*a*) Precession index (average period 21 kyr), (*b*) obliquity variation (average period 41 kyr). The curves indicate substantial reductions in potential error for longer time measurements over extended numbers of sequential cycles. The gray envelopes indicate the dispersion (1-sigma level) of this error to account for the high quasi-periodicity of these two orbital parameters (based upon $La93_{(0,1)}$) over short time spans.

back through time, insofar as contributions from Earth's tidal dissipation and dynamical ellipticity have been only partially explored and only in Plio–Pleistocene data (see further discussion in Chapter 21 relating to potential timing errors in ATNTS2004).

Floating orbital time scales (i.e. time scales that are disconnected from canonical orbital variations) are based upon the assumption that frequency components observed in cycle stratigraphy can be related to one or several frequencies predicted by orbital theory. This requires an additional, provisional assumption that planetary motions were stable back to the geological time represented by the data, and that current models of tidal dissipation and dynamical ellipticity which predict progressively shorter orbital periodicities back through time (e.g. Berger *et al.*, 1992) are accurate. This assumption, however, remains largely untested for times prior to the Oligocene, although some studies have shown that numerous key similarities between Earth's data and the current orbital models appear to have existed during geological times as remote as the Triassic (e.g. Olsen and Kent, 1999). Herbert *et al.* (1995) suggested that in some cases it may be possible to "lock on" to individual frequency components of the precession, for example, to obtain high-resolution orbitally calibrated time scales. Such calibrations are typically limited to counting stratigraphic cycles by visual inspection and assume an average orbital cycle period, e.g. 21 kyr for precession, for each cycle. Nevertheless, even this approach, if conducted on perfectly recorded sequences of orbitally forced cycles, can result in highly accurate floating time scales (Fig. 4.5), with uncertainties at only a few percent for only a few tens of cycles counted.

5 • The geomagnetic polarity time scale

J. G. OGG AND A. G. SMITH

The patterns of marine magnetic anomalies for the Late Cretaceous through Neogene (C-sequence) and Late Jurassic through Early Cretaceous (M-sequence) have been calibrated by magnetostratigraphic studies to biostratigraphy, cyclostratigraphy, and a few radiometrically dated levels. The geomagnetic polarity time scale for the past 160 myr has been constructed by fitting these constraints and a selected model for spreading rates. The status of the geomagnetic polarity time scale for each geological period is summarized in Chapters 11–22 as appropriate.

5.1 PRINCIPLES

5.1.1 Magnetic field reversals and magnetostratigraphy

The principal goal of magnetostratigraphy is to document and calibrate the global geomagnetic polarity sequence in stratified rocks and to apply this geomagnetic polarity time scale for high-resolution correlation of marine magnetic anomalies and of polarity zones in other sections. The basis of magnetostratigraphy is the retention by rocks of a magnetic imprint acquired in the geomagnetic field that existed when the sedimentary rock was deposited or the igneous rock underwent cooling. The imprint most useful for paleomagnetic directions and magnetostratigraphy is recorded by particles of iron oxide minerals.

Most of the material in this chapter is updated from summaries in Harland *et al.* (1990) and Ogg (1995). Excellent reviews are given in Opdyke and Channell (1996) for magnetostratigraphy and McElhinny and McFadden (2000) for general paleomagnetism.

A dynamo capable of generating a magnetic field appears to be a general property of planets and stars that possess a relatively large electrically conducting region that is rotating and convecting (Merrill *et al.*, 1996). Fluid motions in the Earth's outer core generate a global magnetic field that approximates a dipole field. This dipole has an irregular drift or secular variation about the Earth's rotational axis such that the time-averaged field of about 10 000 years roughly coincides with the Earth's rotational poles. For reasons that are still not well understood, at irregular times the currents flowing in the core reverse their direction, producing a reversal in the polarity of the dipole magnetic field.

By convention, in paleomagnetism, present-day polarity is *normal*: lines of magnetic force at the Earth's surface are directed in toward the North magnetic pole and the N-seeking pole of a compass-needle points North (its *declination*). The *inclination* of the magnetic field dips progressively steeper downward with increasing latitude in the northern hemisphere and dips upward in the southern hemisphere. When the polarity is *reversed*, the lines of force are directed in the opposite direction and a compass needle would point to the South. The sign of the inclination is then reversed in both hemispheres.

Reversals of the polarity of the main geomagnetic dipole field are geologically rapid events, typically less than 5000 years in duration, which occur at random intervals. The average frequency of geomagnetic reversals during the Cenozoic is about two or three per million years; and the most recent reversal was about 780 000 years ago. To a good approximation, geomagnetic reversals reflect a random process with durations independent of previous polarity changes. The lengths of polarity intervals vary from as little as 30 kyr to several tens of millions of years. Because polarity reversals are potentially recorded simultaneously in rocks all over the world, these magnetostratigraphic divisions, unlike lithostratigraphic and biostratigraphic divisions, are not time-transgressive. Thus, the pattern of polarity reversals provides a unique global "bar-code" for correlating polarity reversals recorded in rock strata, a usage first suggested by Khramov (1958). Therefore, magnetostratigraphy enables the correlation of rock strata among diverse depositional and faunal realms and the assignment of geologic ages to anomalies of marine magnetic intensities. Considering that the polarity reversal transition spans only about 5000 years, then paleomagnetic correlation is the most precise method available for global correlation in virtually all stratified rocks of all ages, but only directly at the recorded reversal boundary.

A Geologic Time Scale 2004, eds. Felix M. Gradstein, James G. Ogg, and Alan G. Smith. Published by Cambridge University Press.

In practice, secondary magnetizations are acquired by sediments or lavas upon compaction, lithification, diagenesis, long-term exposure to other magnetic field directions, and other processes. Therefore, various methods of demagnetization are required to separate the later secondary components from the primary magnetization directions, if such an unambiguous resolution is possible.

Marine magnetic anomalies are produced when oceanic crust cools, and are typically observed by shipboard magnetometers as an enhanced magnetic field intensity over normal-polarity anomalies and a partially cancelled magnetic field intensity over reversed-polarity anomalies. The correlation of the patterns of these marine magnetic anomalies was the initial evidence for the theory of seafloor spreading, which was later verified by drilling and dating of the oceanic basalt crust.

The geomagnetic polarity time scale is constructed from paleomagnetic analyses of various sedimentary sections having detailed biostratigraphy and by correlations to marine magnetic anomaly patterns. For Late Jurassic to Recent strata, the ocean-floor magnetic anomalies with their calibrations to biostratigraphy serve as a template against which the polarity reversals in stratigraphic sections, either on land or in deep-sea or other cores, can be identified. It is essential to have some biostratigraphic constraints on the polarity zone pattern resolved from any given section in order to propose a non-ambiguous correlation to the reference geomagnetic polarity time scale.

5.1.2 Polarity intervals, zones, and chrons

The time interval that elapses between two successive reversals in the polarity of the geomagnetic dynamo is generally referred to as a *polarity interval* (Cox, 1968); it describes a physical phenomenon, not a chronostratigraphic unit. This usage accords with the statement in the *International Stratigraphic Guide* that "interval" may refer to either time or space intervals and therefore should be used as a general term and not as a formal stratigraphic unit or division (Hedberg, 1976, p. 15).

Polarity chrons are intervals of geological time having a constant magnetic field polarity delimited by reversals (International Subcommission on Stratigraphic Classification, 1979). A *polarity zone* is the corresponding interval in a stratigraphic section deposited during the polarity chron. The chron or zone is called "normal" if the geomagnetic field orientation is similar to the present dipole polarity, and "reversed" if it is opposite in orientation (Fig. 5.1). Some publications refer to a "polarity chron" as merely a "chron." Although "chron" in the *International Stratigraphic Guide* designates a formal subdivision of a geological stage (e.g. "Planula Chron" within the Oxfordian Stage), it is usually clear from the context whether a chron refers to a magnetic polarity interval. Possible ambiguities can be avoided by retaining the prefix "polarity."

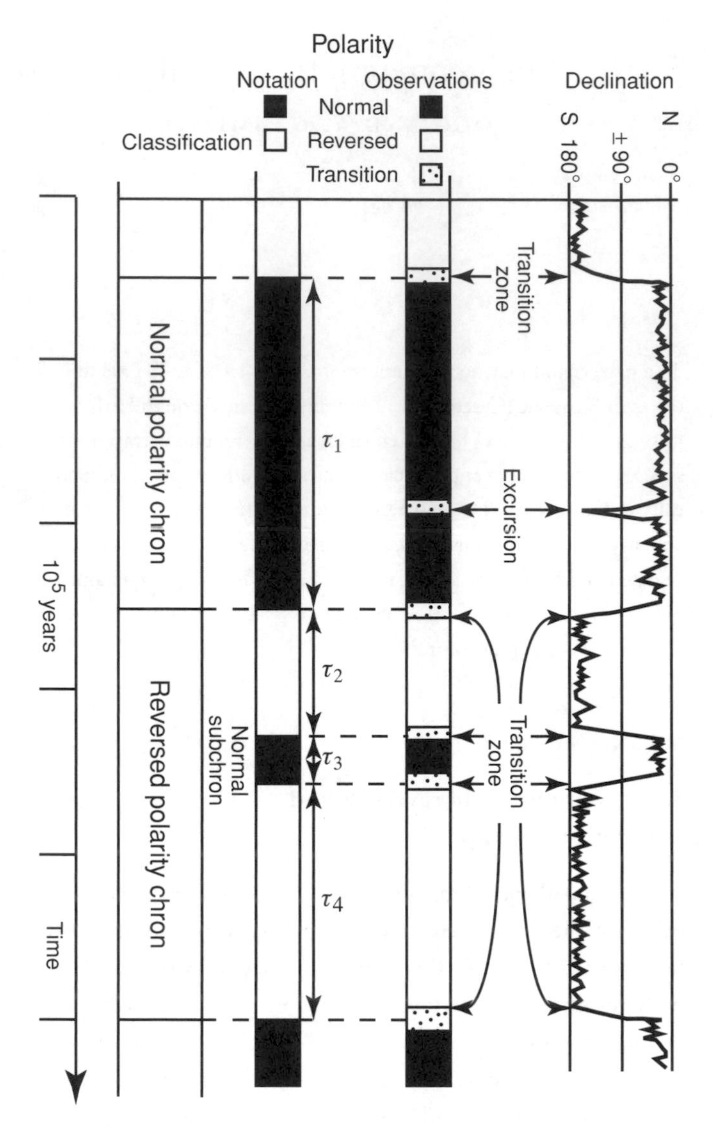

Figure 5.1 Polarity chrons, polarity subchrons, transition zones, and excursions from Harland *et al.*, 1982, their Fig. 4.1.

A hierarchical system for grouping polarity intervals together for successively longer time intervals (International Subcommission on Stratigraphic Classification, 1979) was modified by McElhinny and McFadden (2000) and further modified here (Table 5.1). Cande and Kent (1992b) suggest that anomalies with a time span of <30 kyr should be named *cryptochrons*, which makes redundant the rarely used term of "microchron." However, the distinction between "chron," "subchron," or "cryptochron" varies significantly among workers, including the current chron–subchron designations for subdivisions of the Cenozoic C-sequence geomagnetic polarity time scale (e.g. Cande and Kent, 1992a; our Table 5.2).

In the course of paleomagnetic research, each new discovery of a short polarity interval changes the local polarity

Table 5.1 *Suggested hierarchical scheme for magnetostratigraphic (rock) and polarity chron (time) units*[a]

Magnetostratigraphic polarity units (rock record)	Geochronologic unit (time equivalent)	Approximate duration (year)
Polarity megazone	Megachron	10^8–10^9
Polarity superzone	Superchron	10^7–10^8
Polarity zone	Chron	10^6–10^7
Polarity subzone	Subchron	10^5–10^6
Polarity cryptozone	Cryptochron	$<3\times10^4$

[a] Modified from McElhinny and McFadden, 2000, their Table 4.3.

structure. This may be seen in Fig. 5.1, where prior to the discovery of the short polarity interval labeled τ_3, only one reversed interval would have been recognized spanning the intervals τ_2, τ_3, and τ_4. Therefore in naming or numbering polarity intervals for stratigraphic purposes, a hierarchical set of names is needed that does not change drastically with the discovery of additional short polarity intervals. Figure 5.1 demonstrates that a scheme of simply numbering polarity intervals in the sequence of their occurrence does not provide such a system.

5.1.3 Events, excursions, magnetic anomaly wiggles, and cryptochrons

Magnetostratigraphic studies commonly reveal short-lived polarity reversals of unknown extent or duration. These are referred to by the informal term "*event*." Events have several origins; among them are excursions of the geomagnetic field, paleointensity variations, and short-lived polarity reversals.

Even when the dipole field has a steady polarity, it undergoes swings in direction with typical amplitudes of 15° and periods of 10^2–10^4 years. This *geomagnetic secular variation* is generally too small to be mistaken for the 180° changes in field direction that characterize polarity reversals. Occasionally the field appears to undergo an *excursion* characterized by a large change in direction that may approach 180°. Since excursions are thought to have durations of about 1000 years, they offer the potential of providing very precise, but rarely observable, stratigraphic markers. At least eight excursions have been identified in the past 0.78 myr, making it likely that they are a common feature of the field.

Detailed studies of magnetic anomalies from fast-spreading ridges show several tiny "wiggles" that clearly reflect changes in the paleomagnetic field (Cande and Kent, 1992a,b). Larger amplitude wiggles are probably subchrons representing short-lived field reversals, some of which are known independently from magnetostratigraphic studies. However, most of the small amplitude wiggles are not known from magnetostratigraphic work. As noted above, anomalies with a time span of <30 kyr are named *cryptochrons*. Cande and Kent (1992a) suggest that anomalies with a time span of >30 kyr are well characterized and are probably true polarity reversals. Those with a duration of <30 kyr could be either very short-lived polarity reversals, longer period variations in the intensity of the field, or geomagnetic field excursions (Cande and Kent, 1992b).

5.2 LATE CRETACEOUS–CENOZOIC GEOMAGNETIC POLARITY TIME SCALE

5.2.1 C-sequence of marine magnetic anomalies and associated chron nomenclature

The four youngest polarity chrons – Brunhes, Matuyama, Gauss, and Gilbert – span the past ~6 myr, and are named for the scientists who were important in founding the field of geomagnetism. The existence of these polarity chrons was established largely through dating lava flows on land. After the discovery of ocean-floor spreading, the marine magnetic anomaly sequence of a long traverse in the South Atlantic was taken as a marine standard for the polarity pattern spanning the latest Cretaceous through Cenozoic. The anomalies of this Cenozoic or "C-sequence" were numbered from 1 to 34 (oldest). Refinements of this C-sequence led to insertion of many additional anomalies with a complex letter–number system, and the deletion of anomaly "14."

The corresponding pair of polarity chrons (time) and polarity zones (stratigraphy) are prefaced by the letter C – for Cenozoic – before the named magnetic anomaly, with a suffix "*n*" denoting the *younger* normal polarity interval or *r* denoting the *older* reversed polarity interval, for instance Chron C15r (e.g. Tauxe *et al.*, 1984; Cande and Kent, 1992a). When a major numbered polarity chron is further subdivided, the resulting subchrons are denoted by a suffix of a corresponding numbered polarity chron. For example, "C8n.2n" is the second-oldest normal-polarity subchron comprising normal-polarity Chron C8n. For the younger part of the time scale (Pliocene–Pleistocene), the traditional names are often used to refer to the chrons and subchrons (e.g. Brunhes = C1n, Matuyama = C1r, Jaramillo subchron = C1r.1n, etc.).

Cryptochrons are designated by appending to a chron or subchron name the designation of "-1" (youngest), "-2" (next older), etc. For example, the youngest cryptochron, the Emperor cryptochron, is in C1n and designated "C1n-1."

The relative timing (position) of an event (level) within a polarity chron (zone) is "defined as the relative position in time or distance between the younger and older chronal boundaries"

Table 5.2 *C-sequence marine magnetic anomaly distances and age calibration*[a]

Geologic time scale				Distance (km) from South Atlantic spreading center ("CK92")			Age calibration (orbital tuning or spline fit)			
Period, Epoch (stage)	Age of top of stage (Ma)	Polarity chron		Young end	Old end	*Width (km)*	Top Ma	Base Ma	*Duration (myr)*	Comments
NEOGENE										
HOLOCENE	0.00									
PLEISTOCENE	11.5 ka	C1	C1n (Brunhes)	0.00	12.14	*12.14*	0	0.781	*0.781*	
Middle	0.126		C1r.1r (Matuyama)	12.14	15.37	*3.23*	0.781	0.988	*0.207*	
Early	0.78		C1r.1n (Jaramillo)	15.37	16.39	*1.02*	0.988	1.072	*0.084*	Base of Middle Pleistocene = base of Brunhes Chron
			C1r.2r				1.072	1.173	*0.101*	
			C1r.2n (Cobb Mountain)				1.173	1.185	*0.012*	Cobb Mountain cryptochron is within early part of Matuyama (C1r) Chron
			C1r.3r	16.39	27.80	*11.41*	1.185	1.778	*0.593*	
		C2	C2n (Olduvai)	27.80	31.51	*3.71*	1.778	1.945	*0.167*	
PLIOCENE	1.81									
Late (Gelasian)			C2r.1r	31.51	35.04	*3.53*	1.945	2.128	*0.183*	
			C2r.1n (Reunion)	35.04	35.57	*0.53*	2.128	2.148	*0.020*	
			C2r.2r (Matuyama)	35.57	41.75	*6.18*	2.148	2.581	*0.433*	
		C2A	C2An.1n (Gauss)	41.75	49.44	*7.69*	2.581	3.032	*0.451*	"Gauss Normal Chron" (C2An) contains two reversed intervals – Kaena (2An.1r) and Mammoth (2An.2r)
(Piacenzian)	2.59									
			C2An.1r (Kaena)	49.44	50.70	*1.26*	3.032	3.116	*0.084*	
			C2An.2n	50.70	52.31	*1.61*	3.116	3.207	*0.091*	
			C2An.2r (Mammoth)	52.31	54.10	*1.79*	3.207	3.330	*0.123*	
			C2An.3n (Gauss)	54.10	58.03	*3.93*	3.330	3.596	*0.266*	

Table 5.2 (*cont.*)

Geologic time scale		Polarity chron		Distance (km) from South Atlantic spreading center ("CK92")			Age calibration (orbital tuning or spline fit)			Comments
Period, Epoch (stage)	Age of top of stage (Ma)			Young end	Old end	*Width (km)*	Top Ma	Base Ma	*Duration (myr)*	
Early (Zanclean)	**3.60**									
			C2Ar (Gilbert)	58.03	66.44	*8.41*	3.596	4.187	*0.591*	"Gilbert Reversed Chron" spans Chrons C2Ar through C3r
		C3	C3n.1n (Cochiti)	66.44	68.23	*1.79*	4.187	4.300	*0.113*	
			C3n.1r	68.23	70.56	*2.33*	4.300	4.493	*0.193*	
			C3n.2n (Nunivak)	70.56	73.56	*3.00*	4.493	4.631	*0.138*	
			C3n.2r	73.56	76.76	*3.20*	4.631	4.799	*0.168*	
			C3n.3n (Sidufjall)	76.76	78.26	*1.50*	4.799	4.896	*0.097*	
			C3n.3r	78.26	80.40	*2.14*	4.896	4.997	*0.101*	
			C3n.4n (Thvera)	80.40	84.68	*4.28*	4.997	5.235	*0.238*	
			C3r (Gilbert)	84.68	96.87	*12.19*	5.235	6.033	*0.798*	
MIOCENE	**5.33**									
Late (Messinian)		C3A	C3An.1n	96.87	101.42	*4.55*	6.033	6.252	*0.219*	
			C3An.1r	101.42	103.92	*2.50*	6.252	6.436	*0.184*	
			C3An.2n	103.92	109.60	*5.68*	6.436	6.733	*0.297*	
			C3Ar	109.60	116.70	*7.10*	6.733	7.140	*0.407*	
		C3B	C3Bn	116.70	119.74	*3.04*	7.140	7.212	*0.072*	
			C3Br.1r	119.74	120.62	*0.88*	7.212	7.251	*0.039*	
(Tortonian)	**7.25**									
			C3Br.1n	120.62	121.30	*0.68*	7.251	7.285	*0.034*	
			C3Br.2r	121.30	124.68	*3.38*	7.285	7.454	*0.169*	
			C3Br.2n	124.68	125.35	*0.67*	7.454	7.489	*0.035*	
			C3Br.3r	125.35	126.48	*1.13*	7.489	7.528	*0.039*	
		C4	C4n.1n	126.48	129.08	*2.60*	7.528	7.642	*0.114*	
			C4n.1r	129.08	130.83	*1.75*	7.642	7.695	*0.053*	
			C4n.2n	130.83	139.37	*8.54*	7.695	8.108	*0.413*	
			C4r.1r	139.37	142.49	*3.12*	8.108	8.254	*0.146*	
			C4r.1n	142.49	143.15	*0.66*	8.254	8.300	*0.046*	
			C4r.2r	143.15	152.32	*9.17*	8.300	8.769	*0.469*	
			C4r.2r-1				*8.661*	*8.699*	*0.037*	*Cryptochron within C4r.2r*
		C4A	C4An	152.32	159.16	*6.84*	8.769	9.098	*0.329*	
			C4Ar.1r	159.16	163.49	*4.33*	9.098	9.312	*0.214*	
			C4Ar.1n	163.49	165.16	*1.67*	9.312	9.409	*0.097*	

(*cont.*)

Table 5.2 (*cont.*)

Geologic time scale				Distance (km) from South Atlantic spreading center ("CK92")			Age calibration (orbital tuning or spline fit)			
Period, Epoch (stage)	Age of top of stage (Ma)	Polarity chron		Young end	Old end	*Width (km)*	Top Ma	Base Ma	*Duration (myr)*	Comments
			C4Ar.2r	165.16	171.00	*5.84*	9.409	9.656	*0.247*	
			C4Ar.2n	171.00	172.34	*1.34*	9.656	9.717	*0.060*	
			C4Ar.3r	172.34	174.47	*2.13*	9.717	9.779	*0.063*	
		C5	C5n.1n	174.47	177.49	*3.02*	9.779	9.934	*0.155*	
			C5n.1r	177.49	178.38	*0.89*	9.934	9.987	*0.053*	
			C5n.2n	178.38	201.13	*22.75*	9.987	11.040	*1.053*	
			C5r.1r	201.13	203.44	*2.31*	11.040	11.118	*0.078*	
			C5r.1n	203.44	204.51	*1.07*	11.118	11.154	*0.036*	
			C5r.2r	204.51	213.04	*8.53*	11.154	11.554	*0.400*	
			C5r.2r-1				*11.267*	11.298	*0.031*	*Cryptochron within C5r.2r*
			C5r.2n	213.04	214.28	*1.24*	11.554	11.614	*0.060*	
Middle (Serravallian)	**11.60**									
			C5r.3r	214.28	223.52	*9.24*	11.614	12.014	*0.400*	
		C5A	C5An.1n	223.52	226.81	*3.29*	12.014	12.116	*0.102*	
			C5An.1r	226.81	229.23	*2.42*	12.116	12.207	*0.091*	
			C5An.2n	229.23	234.25	*5.02*	12.207	12.415	*0.208*	
			C5Ar.1r	234.25	240.65	*6.40*	12.415	12.730	*0.315*	
			C5Ar.1n	240.65	241.35	*0.70*	12.730	12.765	*0.035*	
			C5Ar.2r	241.35	242.90	*1.55*	12.765	12.820	*0.055*	
			C5Ar.2n	242.90	243.94	*1.04*	12.820	12.878	*0.058*	
			C5Ar.3r	243.94	247.92	*3.98*	12.878	13.015	*0.137*	
		C5AA	C5AAn	247.92	251.38	*3.46*	13.015	13.183	*0.168*	
			C5AAr	251.38	255.19	*3.81*	13.183	13.369	*0.186*	
		C5AB	C5ABn	255.19	260.03	*4.84*	13.369	13.605	*0.236*	
			C5ABr	260.03	264.53	*4.50*	13.605	13.734	*0.129*	
(Langhian)	**13.65**									
		C5AC	C5ACn	264.53	273.28	*8.75*	*13.734*	*14.095*	*0.361*	
			C5ACr	273.28	275.66	*2.38*	*14.095*	*14.194*	*0.099*	
		C5AD	C5ADn	275.66	285.80	*10.14*	*14.194*	*14.581*	*0.387*	
			C5ADr	285.80	290.17	*4.37*	*14.581*	*14.784*	*0.203*	
		C5B	C5Bn.1n	290.17	292.24	*2.07*	*14.784*	*14.877*	*0.093*	
			C5Bn.1r	292.24	295.63	*3.39*	*14.877*	*15.032*	*0.155*	
			C5Bn.2n	295.63	298.45	*2.82*	*15.032*	15.160	*0.128*	
			C5Br	298.45	318.39	*19.94*	15.160	**15.974**	*0.814*	
Early (Burdigalian)	**15.97**									Base of Langhian = base of Chron C5Br
		C5C	C5Cn.1n	318.39	324.87	*6.48*	15.974	*16.268*	*0.293*	
			C5Cn.1r	324.87	325.65	*0.78*	*16.268*	*16.303*	*0.035*	
			C5Cn.2n	325.65	329.38	*3.73*	*16.303*	*16.472*	*0.169*	
			C5Cn.2r	329.38	330.95	*1.57*	*16.472*	*16.543*	*0.071*	
			C5Cn.3n	330.95	334.88	*3.93*	*16.543*	*16.721*	*0.178*	
			C5Cr	334.88	347.64	*12.76*	*16.721*	*17.235*	*0.514*	

Table 5.2 (*cont.*)

Geologic time scale				**Distance (km) from South Atlantic spreading center ("CK92")**			**Age calibration** (orbital tuning or spline fit)			
Period, Epoch (stage)	Age of top of stage (Ma)	**Polarity chron**		Young end	Old end	*Width (km)*	Top Ma	Base Ma	*Duration (myr)*	Comments
		C5D	C5Dn	347.64	355.45	7.81	17.235	17.533	0.298	
			C5Dr	355.45	360.88	5.43	17.533	17.717	0.184	
			C5Dr.1n	360.88	361.55	0.68	17.717	17.740	0.023	
			C5Dr.2r	361.55	370.87	9.32	17.740	18.056	0.316	
		C5E	C5En	370.87	382.45	11.58	18.056	18.524	0.468	
			C5Er	382.45	388.64	6.19	18.524	18.748	0.224	
			C6n	388.64	413.88	25.24	18.748	19.722	0.974	
			C6r	413.88	422.93	9.05	19.722	20.040	0.318	
		C6A	C6An.1n	422.93	427.81	4.88	20.040	20.213	0.173	
			C6An.1r	427.81	434.18	6.37	20.213	20.439	0.226	
(Aquitanian)	**20.43**									
			C6An.2n	434.18	441.85	7.67	20.439	20.709	0.270	
			C6Ar	441.85	452.46	10.61	20.709	21.083	0.374	
		C6AA	C6AAn	452.46	454.63	2.17	21.083	21.159	0.076	
			C6AAr.1r	454.63	461.59	6.96	21.159	21.403	0.244	
			C6AAr.1n	461.59	463.92	2.33	21.403	21.483	0.080	
			C6AAr.2r	463.92	468.97	5.05	21.483	21.659	0.176	
			C6AAr.2n	468.97	469.79	0.82	21.659	21.688	0.029	
			C6AAr.3r	469.79	472.08	2.29	21.688	21.767	0.079	
		C6B	C6Bn.1n	472.08	475.99	3.91	21.767	21.936	0.169	
			C6Bn.1r	475.99	477.29	1.30	21.936	21.992	0.056	
			C6Bn.2n	477.29	483.70	6.41	21.992	22.268	0.276	
			C6Br	483.70	490.61	6.91	22.268	22.564	0.296	
		C6C	C6Cn.1n	490.61	495.05	4.44	22.564	22.754	0.190	
			C6Cn.1r	495.05	498.54	3.49	22.754	22.902	0.148	
			C6Cn.2n	498.54	501.55	3.01	22.902	**23.030**	0.128	
PALEOGENE										
OLIGOCENE	**23.03**									Base of Miocene = base of Chron C6Cn.2n
Late (Chattian)			C6Cn.2r	501.55	506.47	4.92	23.030	23.249	0.219	
			C6Cn.3n	506.47	509.41	2.94	23.249	23.375	0.125	
			C6Cr	509.41	524.64	15.23	23.375	24.044	0.670	
		C7	C7n.1n	524.64	525.92	1.28	24.044	24.102	0.057	
			C7n.1r	525.92	527.29	1.37	24.102	24.163	0.061	
			C7n.2n	527.29	536.04	8.75	24.163	24.556	0.393	
			C7r	536.04	543.97	7.93	24.556	24.915	0.359	
		C7A	C7An	543.97	547.82	3.85	24.915	25.091	0.175	
			C7Ar	547.82	552.30	4.48	25.091	25.295	0.204	
		C8	C8n.1n	552.30	555.55	3.25	25.295	25.444	0.149	
			C8n.1r	555.55	556.60	1.05	25.444	25.492	0.048	
			C8n.2n	556.60	571.04	14.44	25.492	26.154	0.662	
			C8r	571.04	583.30	12.26	26.154	26.714	0.561	

(*cont.*)

Table 5.2 (*cont.*)

Geologic time scale				Distance (km) from South Atlantic spreading center ("CK92")			Age calibration (orbital tuning or spline fit)			
Period, Epoch (stage)	Age of top of stage (Ma)	Polarity chron		Young end	Old end	*Width (km)*	Top Ma	Base Ma	*Duration (myr)*	Comments
		C9	C9n	583.30	607.96	*24.66*	26.714	27.826	*1.112*	
			C9r	607.96	616.12	*8.16*	27.826	28.186	*0.360*	
			C10n.1n	616.12	622.16	*6.04*	28.186	**28.450**	*0.264*	
Early (Rupelian)	**28.45**									Base of Chattian assigned here as base of Chron C10n.1n
			C10n.1r	622.16	623.90	*1.74*	28.450	28.525	*0.075*	
			C10n.2n	623.90	628.29	*4.39*	28.525	28.715	*0.190*	
			C10r	628.29	645.65	*17.36*	28.715	29.451	*0.736*	
			C11n.1n	645.65	652.56	*6.91*	29.451	29.740	*0.288*	
			C11n.1r	652.56	655.31	*2.75*	29.740	29.853	*0.114*	
			C11n.2n	655.31	664.15	*8.84*	29.853	30.217	*0.364*	
			C11r	664.15	674.26	*10.11*	30.217	30.627	*0.409*	
			C12n	674.26	686.50	*12.24*	30.627	31.116	*0.489*	
			C12r	686.50	742.63	*56.13*	31.116	33.266	*2.150*	
		C13	C13n	742.63	755.44	*12.81*	33.266	33.738	*0.472*	
			C13r	755.44	784.40	*28.96*	33.738	34.782	*1.044*	
EOCENE	**33.9**									"C14" does not exist
Late (Priabonian)		C15	C15n	784.40	791.78	*7.38*	34.782	35.043	*0.260*	
			C15r	791.78	802.15	*10.37*	35.043	35.404	*0.361*	
		C16	C16n.1n	802.15	806.87	*4.72*	35.404	35.567	*0.163*	
			C16n.1r	806.87	810.93	*4.06*	35.567	35.707	*0.140*	
			C16n.2n	810.93	827.67	*16.74*	35.707	36.276	*0.569*	
			C16r	827.67	834.68	*7.01*	36.276	36.512	*0.237*	
		C17	C17n.1n	834.68	856.19	*21.51*	36.512	**37.235**	*0.723*	
Middle (Bartonian)	**37.2**									Base of Priabonian assigned here as base of Chron C17n.1n
			C17n.1r	856.19	859.46	*3.27*	37.235	37.345	*0.110*	
			C17n.2n	859.46	865.54	*6.08*	37.345	37.549	*0.204*	
			C17n.2r	865.54	867.33	*1.79*	37.549	37.610	*0.060*	
			C17n.3n	867.33	872.10	*4.77*	37.610	37.771	*0.161*	
			C17r	872.10	879.83	*7.73*	37.771	38.032	*0.261*	
		C18	C18n.1n	879.83	907.31	*27.48*	38.032	38.975	*0.943*	
			C18n.1r	907.31	909.21	*1.90*	38.975	39.041	*0.066*	
			C18n.2n	909.21	921.21	*12.00*	39.041	39.464	*0.424*	
			C18r	921.21	947.96	*26.75*	39.464	**40.439**	*0.975*	

(*cont.*)

Table 5.2 (*cont.*)

Geologic time scale				Distance (km) from South Atlantic spreading center ("CK92")			Age calibration (orbital tuning or spline fit)			
Period, Epoch (stage)	Age of top of stage (Ma)	Polarity chron		Young end	Old end	*Width (km)*	Top Ma	Base Ma	*Duration (myr)*	Comments
(Lutetian)	**40.4**									Base of Bartonian assigned here as base of Chron C18r
		C19	C19n	947.96	954.12	*6.16*	40.439	40.671	*0.232*	
			C19r	954.12	977.65	*23.53*	40.671	41.590	*0.918*	
		C20	C20n	977.65	1006.06	*28.41*	41.590	42.774	*1.185*	
			C20r	1006.06	1060.24	*54.18*	42.774	45.346	*2.571*	
		C21	C21n	1060.24	1094.71	*34.47*	45.346	47.235	*1.889*	
			C21r	1094.71	1117.55	*22.84*	47.235	**48.599**	*1.364*	
Early Eocene (Ypresian)	**48.6**									Base of Lutetian assigned here as base of Chron C21r
		C22	C22n	1117.55	1130.78	*13.23*	48.599	49.427	*0.828*	
			C22r	1130.78	1150.83	*20.05*	49.427	50.730	*1.303*	
		C23	C23n.1n	1150.83	1153.90	*3.07*	50.730	50.932	*0.203*	
			C23n.1r	1153.90	1155.75	*1.85*	50.932	51.057	*0.125*	
			C23n.2n	1155.75	1168.20	*12.45*	51.057	51.901	*0.844*	
			C23r	1168.20	1178.96	*10.76*	51.901	52.648	*0.747*	
		C24	C24n.1n	1178.96	1184.03	*5.07*	52.648	53.004	*0.355*	
			C24n.1r	1184.03	1185.61	*1.58*	53.004	53.116	*0.112*	
			C24n.2n	1185.61	1186.34	*0.73*	53.116	53.167	*0.052*	
			C24n.2r	1186.34	1188.05	*1.71*	53.167	53.286	*0.119*	
			C24n.3n	1188.05	1195.35	*7.30*	53.286	53.808	*0.521*	
			C24r	1195.35	1234.51	*39.16*	53.808	56.665	*2.858*	
PALEOCENE	**55.8**									Base of Eocene is 0.98 myr above top of C25n according to cycle stratigraphy
Late (Thanetian)		C25	C25n	1234.51	1241.50	*6.99*	56.665	57.180	*0.515*	
			C25r	1241.50	1257.81	*16.31*	57.180	58.379	*1.199*	
		C26	C26n	1257.81	1262.74	*4.93*	58.379	**58.737**	*0.359*	
Selandian	**58.7**									Base of Thanetian is assigned as base of Chron C26n, as in Berggren *et al.* (1995a)
			C26r	1262.74	1303.81	*41.07*	58.737	61.650	*2.913*	

Table 5.2 (*cont.*)

Geologic time scale				Distance (km) from South Atlantic spreading center ("CK92")			Age calibration (orbital tuning or spline fit)			
Period, Epoch (stage)	Age of top of stage (Ma)	Polarity chron		Young end	Old end	*Width (km)*	Top Ma	Base Ma	*Duration (myr)*	Comments
Early (Danian)	**61.7**									Base of Selandian assigned here as Chron C27n(0.9), as in Berggren *et al.* (1995a)
		C27	C27n	1303.81	1308.70	*4.89*	61.650	61.983	*0.333*	
			C27r	1308.70	1325.71	*17.01*	61.983	63.104	*1.121*	
		C28	C28n	1325.71	1341.99	*16.28*	63.104	64.128	*1.024*	
			C28r	1341.99	1347.03	*5.04*	64.128	64.432	*0.304*	
		C29	C29n	1347.03	1358.66	*11.63*	64.432	65.118	*0.685*	
			C29r	1358.66	1371.84	*13.18*	65.118	65.861	*0.743*	
CRETACEOUS										
MAAS-TRICHTIAN	**65.5**									Mesozoic–Cenozoic boundary event is midway in Chron C29r
Late		C30	C30n	1371.84	1407.22	*35.38*	65.861	67.696	*1.835*	Ages of pre-Cenozoic portion of C-sequence are less well constrained by spreading models
			C30r	1407.22	1409.56	*2.34*	67.696	67.809	*0.113*	
		C31	C31n	1409.56	1429.14	*19.58*	67.809	68.732	*0.923*	Base of Chron C31n constrained by Ar–Ar ages to ~69.0 ± 0.5 Ma (2-sigma)
Early	*~69.3*		C31r	1429.14	1481.12	*51.98*	68.732	70.961	2.229	Base of Chron C31r constrained by Ar–Ar ages to ~ 70.45 ± 0.65 Ma (2-sigma)
CAMPANIAN Late	**70.60**	C32	C32n.1n	1481.12	1487.68	*6.56*	70.961	71.225	*0.264*	
			C32n.1r	1487.68	1493.94	*6.26*	71.225	71.474	*0.249*	
			C32n.2n	1493.94	1531.81	*37.87*	71.474	72.929	*1.456*	
			C32r.1r	1531.81	1539.94	*8.13*	72.929	73.231	*0.301*	
			C32r.1n	1539.94	1542.32	*2.38*	73.231	73.318	*0.087*	
			C32r.2r	1542.32	1549.41	*7.09*	73.318	73.577	*0.259*	

Table 5.2 (*cont.*)

Geologic time scale		Polarity chron		Distance (km) from South Atlantic spreading center ("CK92")			Age calibration (orbital tuning or spline fit)			Comments
Period, Epoch (stage)	Age of top of stage (Ma)			Young end	Old end	*Width (km)*	Top Ma	Base Ma	*Duration (myr)*	
Middle	*~76.4*	C33	C33n	1549.41	1723.76	*174.35*	73.577	79.543	*5.965*	Base of C33n constrained by Ar–Ar ages to ~79.3 ± 0.5 Ma (2-sigma)
Early	*~80.8*		C33r	1723.76	1862.32	*138.56*	79.543	~84	*4.457*	
SANTONIAN	**83.50**	C34	C34n	1862.32			~84			Base of Chron C33r is near the top of Santonian stage

[a] The Late Cretaceous through Neogene C-sequence of marine magnetic anomalies is a synthetic flow profile for the South Atlantic with relative distances from the spreading axis as compiled by Cande and Kent (1992a). Distances are tabulated to two decimal points for preserving the relative widths of subchrons and cryptochrons and for applying spreading-rate models, but the actual accuracy is much less – tables of relative uncertainties on this anomaly series are in Cande and Kent (1992a).

The age model for converting the marine magnetic anomaly pattern to absolute ages is a combination of astronomical orbital tuning and spreading-rate model derived from a spline fit to selected radiometric ages. Details and references for the assigned ages are given in Chapters 20 and 21. The associated geologic time scale framework is derived from biostratigraphic correlations to C-sequence polarity chrons or independent ages obtained by astronomical or radiometric dating of stage boundaries or zonal datums (see Chapters 20 and 21). Some Paleogene stage boundaries have not yet been defined by a GSSP or an accepted primary marker, therefore they are assigned an age according to a possible placement with respect to the C-sequence time scale (see Chapter 20). The composite biostratigraphic and geomagnetic polarity time scales for the Neogene, Paleogene, and Cretaceous are illustrated in Chapters 19–21.

(system of Hallam *et al.*, 1985, p. 126). In this proportional stratigraphic convention, the location of the Cretaceous–Paleogene boundary at Gubbio (Alvarez *et al.*, 1977) occurs at C29r(0.75), indicating that 75% of reversed-polarity zone C29r is below the boundary event. For clarity, the decimal fraction is best placed after the polarity chron name enclosed in parentheses (Cande and Kent, 1992a). It should be noted that Cande and Kent (1992a) recommended an inverted stratigraphic placement relative to present; therefore, C29r(0.3) in their notation indicated that 30% of reversed-polarity Chron C29r followed the event. Their suggested system mirrors the convention of measuring geological time and the numbering of magnetic anomalies backwards from the present.

5.2.2 Calibration and ages of the Late Cretaceous–Cenozoic geomagnetic polarity time scale

A composite C-sequence magnetic anomaly pattern for the latest Cretaceous and Cenozoic was assembled by Cande and Kent (1992a, 1995) from a composite of South Atlantic profiles with additional resolution from selected Pacific surveys. An absolute age model for this synthetic "CK92" magnetic anomaly pattern was calculated by applying a cubic-spline fit to selected radiometric age controls (Cande and Kent, 1992a,b, 1995; Fig. 5.2). Berggren *et al.* (1995a) calibrated a vast array of biostratigraphic and chronostratigraphic events to a revision ("CK95") of this geomagnetic polarity time scale to construct a detailed Cenozoic chronostratigraphic time scale.

Potassium–argon ages of lava flows whose polarity was known initially dated the youngest chrons, C1–C3. This method has been superseded by relating polarity reversals to Milankovitch cycles, thereby yielding "absolute" orbital-cycle ages with very high precision (see Chapters 20 and 21 for details). Cycle stratigraphy will eventually calibrate all the C-sequence polarity chrons to the astronomical orbital time scale and assign absolute durations to polarity chrons throughout the Phanerozoic.

We have recalibrated the "CK92" marine magnetic anomaly pattern using an array of astronomical tuning, both to the Neogene absolute-age orbital solution and to estimated durations from Paleogene Milankovitch cycles, and a suite of additional or revised radiometric age calibrations to C-sequence polarity chrons (e.g. Tables 5.2, 20.2, 20.3 and A2.1). The Late Cretaceous portion of the M-sequence is also constrained by several radiometric ages (Hicks *et al.*, 1995, 1999; Hicks and Obradovich, 1995). In the revised model of

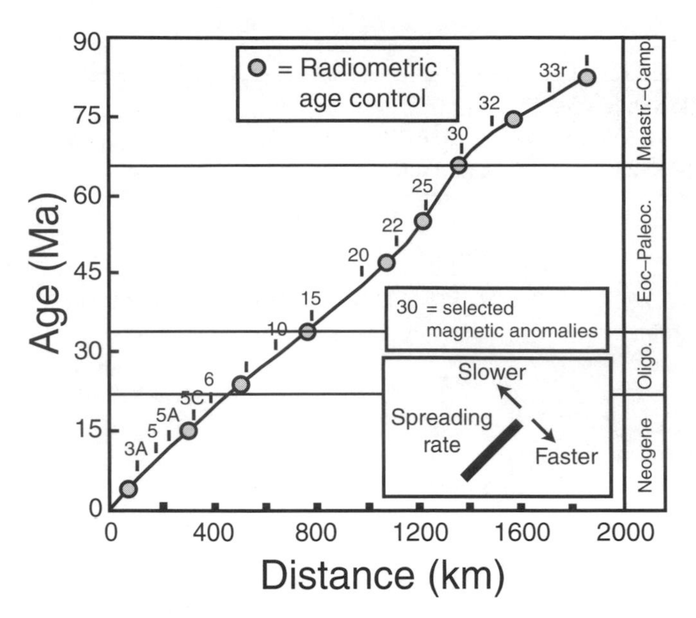

Figure 5.2 Age calibration of Cenozoic through Campanian marine magnetic anomaly pattern. Age versus relative distance from the spreading axis of a synthetic flow profile for the South Atlantic as compiled by Cande and Kent (1992a).

GTS2004, spreading rates along the synthetic South Atlantic profile decelerated smoothly through the Campanian to Danian to reach a minimum of about 13 km/myr at about 55 Ma (Röhl *et al.*, 2001), accelerated again through the Eocene to an Oligocene plateau at about 25 km/myr, and progressively slowed through the Neogene (Fig. 5.3). The apparent oscillations in spreading rates during the Oligocene may be artifacts of the spline-fitting methodology (see Chapter 8) combined with the relatively young base-Miocene age (23.0 Ma) assigned from cyclostratigraphy that requires an elevated spreading rate in Early Miocene following a slower rate in Late Oligocene.

Similar to the procedure used by Berggren *et al.* (1995a), the calibration of geological time scale boundaries and biostratigraphic datums to this geomagnetic polarity time scale generates the absolute time scale for the Paleogene and for the early part of the Neogene. The geomagnetic polarity time scale also constrains age estimates for the latest Cretaceous (Campanian and Maastrichtian stages).

5.3 MIDDLE JURASSIC–EARLY CRETACEOUS GEOMAGNETIC POLARITY TIME SCALE

5.3.1 M-sequence of marine magnetic anomalies

An extended ~35 myr interval of normal polarity, the "Cretaceous long normal-polarity Chron" or polarity Superchron C34n, extends from the Early Aptian to approximately the Santonian–Campanian boundary. Oceanic crust of late-Middle Jurassic through earliest Aptian age displays a second series of magnetic anomalies, which were named the "M" – for Mesozoic – series. The M-sequence for anomalies and corresponding chrons of M0r through M25r was derived from a block model of the Hawaiian lineations by Larson and Hilde (1975) and has undergone relatively minor refinements (e.g. Tamaki and Larson, 1988). To confuse nomenclature somewhat, the M anomalies start at M0 (*not* M1) and the younger set of "M1, M2, M3, M4, M5" designate marine anomalies of alternating polarity (Table 5.3).

After the M-sequence was numbered, three events or clusters of brief reversed-polarity excursions or subzones were reported, especially within drilling cores of deep-sea sediments, from the Aptian–Albian portion of Chron C34n. Ryan *et al.* (1978) summarized these events and suggested an "upward," hence "negative numbering" continuation to the M-sequence younger than polarity Chron M0r: (1) Chron M"-1r" in Late Aptian, with an alternate designation as the "ISEA" event (Tarduno, 1990); (2) Chron M"-2r" set of Middle Albian events; and (3) Chron M"-3r" set in Late Albian. Further details of this enigmatic "negative-numbered" set are given in Chapter 19.

Additional marine magnetic surveys in the Pacific extended the M-sequence (e.g. Cande *et al.*, 1978; Handschumacher *et al.*, 1988; Sager *et al.*, 1998). At the time of our writing, the oldest numbered M-sequence anomaly is M41 of Bathonian age (Table 5.3).

5.3.2 Constructing a composite M-sequence

Magnetostratigraphic studies in combination with constraints from drilling of oceanic crust have calibrated much of the M-sequence to ammonite and microfossil datums, and are summarized in Chapters 18 and 19. We derived a geomagnetic polarity time scale for this time interval using the same methodology as was applied to the C-sequence (Cande and Kent, 1992a, 1995; Berggren *et al.*, 1995). First, we compiled a synthetic M-sequence profile that was scaled to a single spreading center (the "Hawaiian" ridge in the Pacific). Second, we applied a spreading-rate model constrained by cycle stratigraphy and a suite of radiometric ages to this synthetic M-sequence profile to construct an absolute age scale for the M-sequence anomalies and associated chrons. Finally, we projected this geomagnetic polarity time scale onto estimated calibrations of biostratigraphic datums and zonal boundaries to M-sequence chrons to assign absolute ages to the Late Jurassic–Early Cretaceous subdivisions and events.

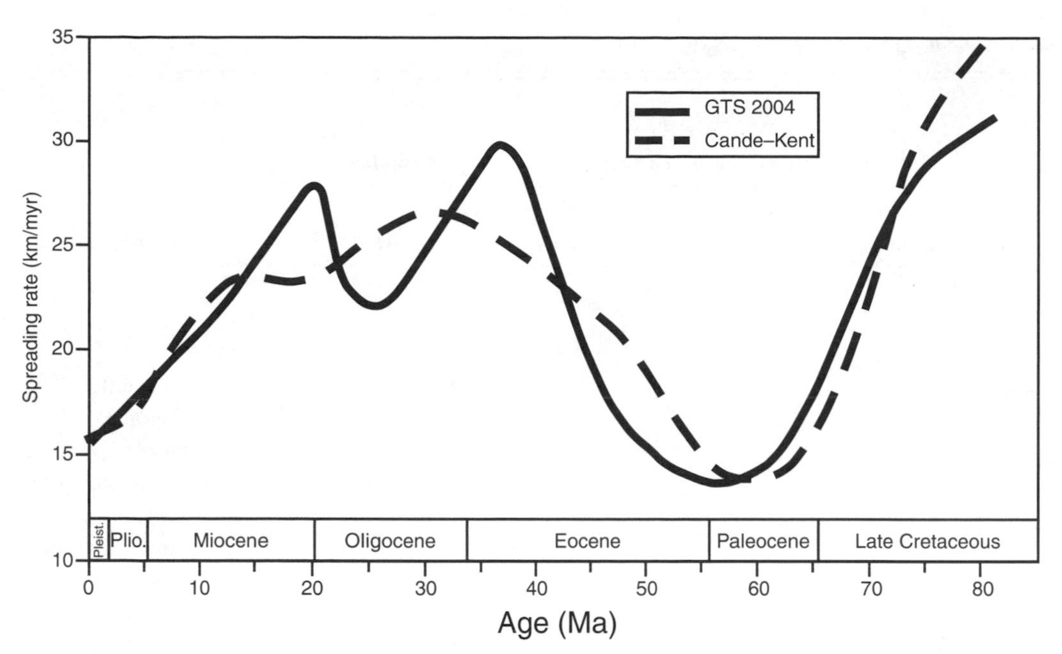

Figure 5.3 Spreading history of the South Atlantic. Variation in spreading rates (km/myr) for marine magnetic anomalies along a synthetic flow profile for the South Atlantic. The dashed line is the spreading model of Cande and Kent (1992a,b, 1995) to calibrate the Late Cretaceous through recent magnetic polarity time scale. The solid line is the GTS2004 estimates from Neogene and Paleocene cycle-stratigraphy calibration of magnetic polarity chrons and additional radiometric ages (Oligocene, Late Eocene, Maastrichtian–Campanian). A rapid slowing of spreading rates during Campanian–mid-Paleocene, a major surge during Early–Middle Eocene, and a progressive decline through the Neogene is indicated by both the original model and later independent estimates. The Neogene portion has been simplified, because the combination of cycle-tuning of individual short-duration polarity chrons and of uncertainties in the corresponding magnetic anomaly widths in the South Atlantic synthetic profile results in artifacts of spurious high-amplitude oscillations in apparent spreading rates.

A dearth of reliable high-resolution radiometric dates implies that the estimated absolute time scale and duration of ammonite zones for a disproportionately long interval of the geologic time scale, the Oxfordian–Barremian stages, or about 35 myr, depends on the calibration of the M-sequence magnetic polarity time scale. Therefore, the details of this process are elaborated below.

ANOMALIES M0R (BASE-APTIAN) TO M25N (BASE-KIMMERIDGIAN)

The M-sequence pattern is best documented at the trio of expanding spreading centers that created the current Pacific plate (e.g. Larson and Hilde, 1975). Channell *et al.* (1995b) developed a composite scaling of marine magnetic anomaly block models from approximately M0r through M29r from the individual Pacific spreading centers. This composite "CENT" scaling is a more robust estimate of the M-sequence pattern than the splicing of selected profiles on the Hawaiian lineations by Larson and Hilde (1975; Roger Larson, pers. comm. to Jim Ogg, 1997). However, the CENT composite did not resolve some of the fine-scale features in individual profiles that are also documented by magnetostratigraphy and lacks adequate resolution for anomalies older than approximately M22r. Following the CENT procedure, a series of synthetic kilometer-distances were added to replicate the fine-scale structure of the Larson–Hilde block model for several intervals and for the revised relative durations of subchrons in M11An observed from magnetostratigraphic studies (Channell *et al.*, 1995b, p. 60). Anomalies from M22r through M25n were proportioned according to the Hawaiian-lineation pattern of Larson and Hilde (1975) into the base-M22r to base-M25n distance of the CENT composite. The base of Chron M0r is the proposed GSSP primary marker of the Aptian Stage. For simplicity in computing pre-M25 magnetic anomalies, the Hawaiian-lineation kilometer scale of Channell *et al.* (1995b) was inverted to indicate distances relative to the old end (base) of anomaly M0r (upper portion of Table 5.3).

ANOMALIES M25R (BASE-KIMMERIDGIAN) TO M41 (BATHONIAN)

The continuation of the M-sequence into older oceanic crust than M25n is best documented in the magnetic anomaly

Table 5.3 *M-sequence marine magnetic anomaly distances and age calibration*[a]

Geologic time scale		Polarity chron		Distance (km) from Hawaiian spreading center			Age calibration (spreading-rate model)			
Period, Stage, Substage	Age of top of substage (Ma)			Young end	Old end	*Width (km)*	Top Ma	Base Ma	*Duration (myr)*	Comments
Early *CRETACEOUS*										
ALBIAN	*99.6*									
Middle	*106.5*	within C34n	M"-3r"				~102		*not known*	Not fully verified, and may be multiple excursions; age is projected from reported coincidence with foraminifera
Early	*~109.0*		M"-2r"				~108		*not known*	Not fully verified, and may be multiple excursions; age is projected from reported coincidence with foraminifera
APTIAN	*112.0*									
Middle	*~115*		M"-1r", or ISEA				118.5		*0.10*	Age is projected from reported coincidence with foraminifera
Early	*~121*	M0	M0r	−9.808	0.000	*9.808*	124.61	**125.00**	*0.39*	Duration of M0r is 380 kyr from cycle stratigraphy
BARREMIAN	**125.0**									Base of Aptian = Base of Chron M0r
Late			M1n	0.000	53.607	*53.607*	125.00	127.24	*2.24*	Duration of Barremian stage is ~5 myr from cycle stratigraphy
		M1	M1r (or "M1")	53.607	62.439	*8.832*	127.24	127.61	*0.37*	
		M3	M3n (or "M2,")	62.439	74.613	*12.174*	127.61	128.11	*0.51*	
Early	*128.3*		M3r (or "M3")	74.613	114.185	*39.572*	128.11	129.76	*1.65*	Base of U. Barrem = upper Chron M3r
		M5	M5n (or "M4")	114.185	136.295	*22.110*	129.76	130.80	*1.03*	

Table 5.3 (*cont.*)

Geologic time scale: *Period,* Stage, Substage	Age of top of substage (Ma)	Polarity chron		Distance (km) from Hawaiian spreading center: Young end	Old end	*Width (km)*	Age calibration (spreading-rate model): Top Ma	Base Ma	*Duration (myr)*	Comments
HAUTERIVIAN	**130.0**									Base of Barremian assigned here as Chron M5n(0.8)
Late			M5r (or "M5")	136.295	144.678	*8.383*	130.80	131.19	*0.39*	Duration of combined Hauterivian and Valanginian stages are ~11 myr from cycle stratigraphy
		M6	M6n	144.678	149.412	*4.734*	131.19	131.41	*0.22*	
			M6r	149.412	152.557	*3.145*	131.41	131.56	*0.15*	
		M7	M7n	152.557	158.782	*6.225*	131.56	131.85	*0.29*	
			M7r	158.782	166.232	*7.450*	131.85	132.20	*0.35*	
		M8	M8n	166.232	173.095	*6.863*	132.20	132.52	*0.32*	
			M8r	173.095	179.674	*6.579*	132.52	132.83	*0.31*	
		M9	M9n	179.674	186.405	*6.731*	132.83	133.14	*0.31*	
			M9r	186.405	194.062	*7.657*	133.14	133.50	*0.36*	
Early	*~133.8*	M10	M10n	194.062	202.040	*7.978*	133.50	133.87	*0.37*	
			M10r	202.040	211.178	*9.138*	133.87	134.30	*0.43*	
		M10N	M10Nn.1n	211.178	218.137	*6.959*	134.30	134.62	*0.33*	The "N" of M10N was in recognition of Fred Naugle by Larson and Hilde (1975)
			M10Nn.1r	218.137	219.117	*0.980*	134.62	134.67	*0.05*	
			M10Nn.2n	219.117	225.737	*6.620*	134.67	134.98	*0.31*	
			M10Nn.2r	225.737	226.217	*0.480*	134.98	135.00	*0.02*	
			M10Nn.3n	226.217	232.213	*5.996*	135.00	135.28	*0.28*	
			M10Nr	232.213	240.878	*8.665*	135.28	135.69	*0.40*	
		M11	M11n	240.878	257.083	*16.205*	135.69	**136.44**	*0.76*	
VALANGINIAN	**136.4**									Base of Hauterivian assigned here as base of Chron M11n
Late			M11r.1r	257.083	262.217	*5.134*	136.44	136.68	*0.24*	
			M11r.1n	262.217	262.717	*0.500*	136.68	136.71	*0.02*	
			M11r.2r	262.717	266.946	*4.229*	136.71	136.90	*0.19*	
		M11A	M11An.1n	266.946	277.897	*10.951*	136.90	137.39	*0.49*	
			M11An.1r	277.897	279.117	*1.220*	137.39	137.44	*0.05*	
			M11An.2n	279.117	280.817	*1.700*	137.44	137.51	*0.07*	
			M11Ar	280.817	282.787	*1.970*	137.51	137.60	*0.09*	

(*cont.*)

Table 5.3 (*cont.*)

Geologic time scale										
				Distance (km) from Hawaiian spreading center			**Age calibration (spreading-rate model)**			
Period, Stage, Substage	Age of top of substage (Ma)	**Polarity chron**		Young end	Old end	*Width (km)*	Top Ma	Base Ma	*Duration (myr)*	Comments
		M12	M12n	282.787	287.797	*5.010*	137.60	137.82	*0.22*	
			M12r.1r	287.797	306.148	*18.351*	137.82	138.56	*0.74*	
			M12r.1n	306.148	307.997	*1.849*	138.56	138.63	*0.07*	
			M12r.2r	307.997	311.677	*3.680*	138.63	138.78	*0.15*	
		M12A	M12An	311.677	318.047	*6.370*	138.78	139.03	*0.25*	
Early	*~139*		M12Ar	318.047	320.247	*2.200*	139.03	139.12	*0.09*	
		M13	M13n	320.247	324.815	*4.568*	139.12	139.29	*0.18*	
			M13r	324.815	331.276	*6.461*	139.29	139.53	*0.24*	
		M14	M14n	331.276	338.168	*6.892*	139.53	139.77	*0.24*	
			M14r	338.168	356.584	*18.416*	139.77	140.36	*0.59*	
BERRIASIAN	**140.0**									Base of Valanginian assigned here as Chron M14r(0.3)
Late			M15n	356.584	366.264	*9.680*	140.36	140.66	*0.29*	
		M15	M15r	366.264	379.164	*12.900*	140.66	141.05	*0.39*	
		M16	M16n	379.164	412.472	*33.308*	141.05	142.06	*1.01*	
Early	*142.3*		M16r	412.472	428.465	*15.993*	142.06	142.55	*0.49*	
		M17	M17n	428.465	437.933	*9.468*	142.55	142.84	*0.29*	
			M17r	437.933	477.493	*39.560*	142.84	144.04	*1.20*	
		M18	M18n	477.493	494.915	*17.422*	144.04	144.57	*0.53*	
			M18r	494.915	505.139	*10.224*	144.57	144.88	*0.31*	
		M19	M19n.1n	505.139	508.767	*3.628*	144.88	144.99	*0.11*	
			M19n.1r	508.767	511.217	*2.450*	144.99	145.06	*0.07*	
			M19n.2n	511.217	540.261	*29.044*	145.06	145.95	*0.88*	
Late *JURASSIC*										
TITHONIAN	**145.5**									Base of Cretaceous (base of Berriasian) assigned here as Chron M19n.2n(0.55)
Late			M19r	540.261	547.398	*7.137*	145.95	146.16	*0.22*	
		M20	M20n.1n	547.398	557.487	*10.089*	146.16	146.47	*0.31*	
			M20n.1r	557.487	559.197	*1.710*	146.47	146.52	*0.05*	
			M20n.2n	559.197	580.172	*20.975*	146.52	147.16	*0.64*	
Early	*147.2*		M20r	580.172	600.286	*20.114*	147.16	147.77	*0.61*	
		M21	M21n	600.286	625.619	*25.333*	147.77	148.54	*0.77*	
			M21r	625.619	638.098	*12.479*	148.54	148.92	*0.38*	
		M22	M22n.1n	638.098	675.017	*36.919*	148.92	150.05	*1.12*	
			M22n.1r	675.017	676.237	*1.220*	150.05	150.08	*0.04*	
			M22n.2n	676.237	677.467	*1.230*	150.08	150.12	*0.04*	

Table 5.3 (*cont.*)

Geologic time scale				Distance (km) from Hawaiian spreading center			Age calibration (spreading-rate model)			
Period, Stage, Substage	Age of top of substage (Ma)	Polarity chron		Young end	Old end	*Width (km)*	Top Ma	Base Ma	*Duration (myr)*	Comments
			M22n.2r	677.467	678.677	1.210	150.12	150.16	0.04	
			M22n.3n	678.677	680.415	1.738	150.16	150.21	0.05	
			M22r	680.415	697.476	17.061	150.21	150.73	0.52	
		M22A	M22An	697.476	701.268	3.792	150.73	**150.84**	0.12	
KIMMERID-GIAN	**150.8**									Base of Tithonian assigned here as base of Chron M22An
			M22Ar	701.268	706.577	5.309	150.84	151.01	0.16	
			M23n	706.577	717.448	10.871	151.01	151.34	0.33	
		M23	M23r.1r	717.448	725.791	8.343	151.34	151.62	0.28	
			M23r.1n	725.791	726.549	0.758	151.62	151.64	0.03	
			M23r.2r	726.549	745.004	18.455	151.64	152.26	0.62	
		M24	M24n	745.004	752.069	7.065	152.26	152.50	0.24	
			M24r.1r	752.069	765.735	13.666	152.50	152.96	0.46	
			M24r.1n	765.735	766.493	0.758	152.96	152.98	0.03	
			M24r.2r	766.493	772.561	6.067	152.98	153.18	0.20	
		M24A	M24An	772.561	776.353	3.792	153.18	153.31	0.13	
			M24Ar	776.353	784.190	7.837	153.31	153.58	0.26	
		M24B	M24Bn	784.190	794.808	10.618	153.58	153.93	0.36	
			M24Br	794.808	799.358	4.551	153.93	154.08	0.15	
		M25	M25n	799.358	807.954	8.596	154.08	154.37	0.29	Marine magnetic anomaly series from M25n to M27n are rescaled from Handschu-macher *et al.* (1988)
			M25r	807.954	813.273	5.319	154.37	**154.55**	0.18	
OXFORDIAN (Tethyan)	**154.6**									Base of Kimmeridgian (as used in Tethyan ammonite stratigraphy) assigned here as base of Chron M25r
Late		M25A	M25An.1n	813.273	816.818	3.545	154.55	154.67	0.12	
			M25An.1r	816.818	818.506	1.688	154.67	154.73	0.06	
			M25An.2n	818.506	820.870	2.364	154.73	154.81	0.08	

(*cont.*)

Table 5.3 (*cont.*)

Geologic time scale				Distance (km) from Hawaiian spreading center			Age calibration (spreading-rate model)			
Period, Stage, Substage	Age of top of substage (Ma)	**Polarity chron**		Young end	Old end	*Width (km)*	Top Ma	Base Ma	*Duration (myr)*	Comments
			M25An.2r	820.870	822.558	1.688	154.81	154.86	0.06	
			M25An.3n	822.558	825.766	3.208	154.86	154.97	0.11	
			M25Ar	825.766	828.129	2.364	154.97	155.05	0.08	
		M26	M26n.1n	828.129	830.493	2.364	155.05	155.13	0.08	
			M26n.1r	830.493	832.181	1.688	155.13	155.18	0.06	
			M26n.2n	832.181	833.869	1.688	155.18	155.24	0.06	
			M26n.2r	833.869	835.558	1.688	155.24	155.30	0.06	ODP Site 761 drilled on middle of anomaly M26n yielded Ar/Ar age of 155.3 Ma (± 3.4 myr; 1-sigma)
			M26n.3n	835.558	837.246	1.688	155.30	155.35	0.06	
			M26n.3r	837.246	838.428	1.182	155.35	155.39	0.04	
			M26n.4n	838.428	842.648	4.221	155.39	155.54	0.14	
Oxfordian (Boreal)	**155.7**									Base of Kimmeridgian (as used in Boreal ammonite stratigraphy) assigned here as Chron M26r(0.2)
			M26r	842.648	847.713	5.065	155.54	155.71	0.17	
		M27	M27n	847.713	851.934	4.221	155.71	155.85	0.14	
			M27r	851.934	856.642	4.708	155.85	156.01	0.16	Marine magnetic anomaly series from M27r to M41r are rescaled from deep-tow survey of Sager *et al.* (1998) on Japanese lineations
		M28	M28n	856.642	865.256	8.614	156.01	156.29	0.29	
			M28r	865.256	869.563	4.307	156.29	156.44	0.14	
			M28An	869.563	872.417	2.855	156.44	156.53	0.10	
			M28Ar	872.417	879.128	6.711	156.53	156.76	0.23	
			M28Bn	879.128	880.581	1.452	156.76	156.81	0.05	
			M28Br	880.581	883.085	2.504	156.81	156.89	0.08	
			M28Cn	883.085	886.140	3.055	156.89	156.99	0.10	

Table 5.3 (*cont.*)

Geologic time scale				Distance (km) from Hawaiian spreading center			Age calibration (spreading-rate model)			
Period, Stage, Substage	Age of top of substage (Ma)	**Polarity chron**		Young end	Old end	*Width (km)*	Top Ma	Base Ma	*Duration (myr)*	Comments
			M28Cr	886.140	888.744	2.604	156.99	157.08	0.09	
			M28Dn	888.744	891.048	2.304	157.08	157.16	0.08	
			M28Dr	891.048	894.203	3.155	157.16	157.26	0.11	
		M29	M29n.1n	894.203	898.660	4.457	157.26	157.41	0.15	
Middle	*157.40*		M29n.1r	898.660	899.512	0.851	157.41	157.44	0.03	
			M29n.2n	899.512	901.415	1.903	157.44	157.51	0.06	
			M29r	901.415	908.176	6.761	157.51	157.73	0.23	
			M29An	908.176	909.578	1.402	157.73	157.78	0.05	
			M29Ar	909.578	911.481	1.903	157.78	157.84	0.06	
		M30	M30n	911.481	915.037	3.556	157.84	157.96	0.12	
			M30r	915.037	919.645	4.608	157.96	158.12	0.15	
			M30An	919.645	922.499	2.855	158.12	158.21	0.10	
			M30Ar	922.499	923.401	0.901	158.21	158.24	0.03	
		M31	M31n.1n	923.401	927.858	4.457	158.24	158.39	0.15	
			M31n.1r	927.858	930.212	2.354	158.39	158.47	0.08	
			M31n.2n	930.212	931.013	0.801	158.47	158.50	0.03	
			M31n.2r	931.013	932.215	1.202	158.50	158.54	0.04	
			M31n.3n	932.215	933.467	1.252	158.54	158.58	0.04	
			M31r	933.467	935.020	1.553	158.58	158.63	0.05	
		M32	M32n.1n	935.020	935.671	0.651	158.63	158.66	0.02	
			M32n.1r	935.671	936.522	0.851	158.66	158.68	0.03	
			M32n.2n	936.522	939.327	2.805	158.68	158.78	0.09	
			M32n.2r	939.327	940.829	1.502	158.78	158.83	0.05	
			M32n.3n	940.829	941.681	0.851	158.83	158.86	0.03	
			M32r	941.681	944.084	2.404	158.86	158.94	0.08	
		M33	M33n	944.084	952.598	8.514	158.94	159.22	0.29	
			M33r	952.598	956.405	3.806	159.22	159.35	0.13	
			M33An	956.405	958.658	2.254	159.35	159.43	0.08	
			M33Ar	958.658	960.662	2.003	159.43	159.49	0.07	
Early	*159.50*		M33Bn	960.662	962.615	1.953	159.49	159.56	0.07	
			M33Br	962.615	965.870	3.255	159.56	159.67	0.11	
			M33Cn.1n	965.870	966.972	1.102	159.67	159.71	0.04	
			M33Cn.1r	966.972	968.474	1.502	159.71	159.76	0.05	
			M33Cn.2n	968.474	971.980	3.506	159.76	159.87	0.12	
			M33Cr	971.980	977.840	5.860	159.87	160.07	0.20	
		M34	M34n.1n	977.840	980.294	2.454	160.07	160.15	0.08	
			M34n.1r	980.294	982.347	2.053	160.15	160.22	0.07	
			M34n.2n	982.347	983.699	1.352	160.22	160.27	0.05	
			M34n.2r	983.699	984.801	1.102	160.27	160.30	0.04	
			M34n.3n	984.801	985.702	0.901	160.30	160.33	0.03	
			M34n.3r	985.702	987.155	1.452	160.33	160.38	0.05	
			M34An	987.155	987.806	0.651	160.38	160.40	0.02	
			M34Ar	987.806	991.812	4.007	160.40	160.54	0.13	
			M34Bn.1n	991.812	994.567	2.754	160.54	160.63	0.09	
			M34Bn.1r	994.567	995.619	1.052	160.63	160.67	0.04	
			M34Bn.2n	995.619	996.320	0.701	160.67	160.69	0.02	

(*cont.*)

Table 5.3 (*cont.*)

Geologic time scale				**Distance (km) from Hawaiian spreading center**			**Age calibration** (spreading-rate model)			
Period, Stage, Substage	Age of top of substage (Ma)	**Polarity chron**		Young end	Old end	*Width (km)*	Top Ma	Base Ma	*Duration (myr)*	Comments
			M34Br	996.320	997.572	1.252	160.69	160.73	0.04	
		M35	M35n	997.572	1000.076	2.504	160.73	160.82	0.08	
			M35r	1000.076	1005.385	5.309	160.82	160.99	0.18	
		M36	M36n.1n	1005.385	1008.490	3.105	160.99	161.10	0.10	
			M36n.1r	1008.490	1010.543	2.053	161.10	161.17	0.07	
			M36An	1010.543	1011.795	1.252	161.17	**161.21**	0.04	
Middle ***JURASSIC***										
CALLOVIAN	**161.2**									Base of Oxfordian is assigned here as base of Chron M36An
Late			M36Ar	1011.795	1012.396	0.601	161.21	161.23	0.02	
			M36Bn	1012.396	1013.247	0.851	161.23	161.26	0.03	
			M36Br	1013.247	1017.304	4.057	161.26	161.39	0.14	
			M36Cn	1017.304	1019.307	2.003	161.39	161.46	0.07	
			M36Cr	1019.307	1022.863	3.556	161.46	161.58	0.12	
		M37	M37n.1n	1022.863	1029.824	6.961	161.58	161.81	0.23	
			M37n.1r	1029.824	1033.130	3.305	161.81	161.92	0.11	
			M37n.2n	1033.130	1036.636	3.506	161.92	162.04	0.12	
			M37r	1036.636	1040.542	3.906	162.04	162.17	0.13	
		M38	M38n.1n	1040.542	1043.697	3.155	162.17	162.28	0.11	
			M38n.1r	1043.697	1045.300	1.603	162.28	162.33	0.05	
			M38n.2n	1045.300	1048.054	2.754	162.33	162.43	0.09	
			M38n.2r	1048.054	1049.106	1.052	162.43	162.46	0.04	
Middle	***~162.5***		M38n.3n	1049.106	1051.410	2.304	162.46	162.54	0.08	
			M38n.3r	1051.410	1054.465	3.055	162.54	162.64	0.10	
			M38n.4n	1054.465	1061.326	6.861	162.64	162.87	0.23	
			M38n.4r	1061.326	1063.229	1.903	162.87	162.93	0.06	
			M38n.5n	1063.229	1067.436	4.207	162.93	163.08	0.14	
			M38r	1067.436	1069.940	2.504	163.08	163.16	0.08	
Early	***~163.2***	M39	M39n.1n	1069.940	1074.497	4.557	163.16	163.31	0.15	
			M39n.1r	1074.497	1078.103	3.606	163.31	163.43	0.12	
			M39n.2n	1078.103	1081.559	3.456	163.43	163.55	0.12	
			M39n.2r	1081.559	1084.263	2.704	163.55	163.64	0.09	
			M39n.3n	1084.263	1088.070	3.806	163.64	163.77	0.13	
			M39n.3r	1088.070	1090.724	2.654	163.77	163.86	0.09	
			M39n.4n	1090.724	1097.235	6.511	163.86	164.07	0.22	
			M39n.4r	1097.235	1101.291	4.057	164.07	164.21	0.14	
			M39n.5n	1101.291	1103.996	2.704	164.21	164.30	0.09	
			M39n.5r	1103.996	1107.351	3.355	164.30	164.41	0.11	
			M39n.6n	1107.351	1109.755	2.404	164.41	164.49	0.08	
			M39n.6r	1109.755	1113.161	3.406	164.49	164.61	0.11	
			M39n.7n	1113.161	1115.164	2.003	164.61	164.68	0.07	
			M39n.7r	1115.164	1117.518	2.354	164.68	164.76	0.08	

Table 5.3 (*cont.*)

Geologic time scale				Distance (km) from Hawaiian spreading center			Age calibration (spreading-rate model)			
Period, Stage, Substage	Age of top of substage (Ma)	Polarity chron		Young end	Old end	*Width (km)*	Top Ma	Base Ma	*Duration (myr)*	Comments
BATHONIAN	**164.7**									Base of Callovian is from proportional scaling of ammonite subzones, and is not yet tied to magnetic stratigraphy
Late			M39n.8n	1117.518	1120.472	2.955	164.76	164.85	0.10	
			M39r	1120.472	1123.027	2.554	164.85	164.94	0.09	
		M40	M40n.1n	1123.027	1124.880	1.853	164.94	165.00	0.06	
			M40n.1r	1124.880	1126.983	2.103	165.00	165.07	0.07	
			M40n.2n	1126.983	1130.639	3.656	165.07	165.20	0.12	
			M40n.2r	1130.639	1132.091	1.452	165.20	165.24	0.05	
			M40n.3n	1132.091	1133.444	1.352	165.24	165.29	0.05	
			M40n.3r	1133.444	1136.799	3.355	165.29	165.40	0.11	
			M40n.4n	1136.799	1138.201	1.402	165.40	165.45	0.05	
			M40r	1138.201	1141.106	2.905	165.45	165.55	0.10	
		M41	M41n	1141.106	1143.009	1.903	165.55	165.61	0.06	
Early	*~166.0*		M41r							
			Interval to ODP Site 801	1143.009	1183.075	40.065	165.61	166.95	1.34	ODP Site 801 is about 100 km beyond M41r on Japanese lineations and yielded an Ar–Ar age of 167.7 Ma (±1.4 myr; 2-sigma)
BAJOCIAN	**167.7**									Base of Bajocian is from proportional scaling of ammonite subzones, and is not yet tied to magnetic stratigraphy

[a] The Late Jurassic and Early Cretaceous M-sequence of marine magnetic anomalies is a synthetic flow profile for the Hawaiian spreading center of the Pacific. Distances are tabulated to three decimal points for preserving the relative widths of subchrons and for applying spreading-rate models, but the actual accuracy is much less.

The age model for converting the marine magnetic anomaly pattern to absolute ages is a combination of astronomical orbital tuning and a spreading-rate model derived from selected radiometric ages. The associated geological time scale framework is derived from biostratigraphic correlations to M-sequence polarity chrons. The composite biostratigraphic and geomagnetic polarity time scales for the Cretaceous and Jurassic are illustrated in Chapters 18 and 19.

lineations created by the "Japanese" spreading center, which were apparently formed at a more rapid spreading rate than the Hawaiian set. These must be rescaled to fit the Hawaiian CENT model for the composite pattern. Cande *et al.* (1978) identified a series of Jurassic magnetic anomalies in the western Pacific and named these M26 through M29. Handschumacher *et al.* (1988) modeled a complex series of magnetic anomalies recorded by an aeromagnetic survey to refine the M26–M29 pattern and to extend the M-sequence scale to "M38." The portion of their magnetic anomaly scale older than M27r was further refined by later deep-tow magnetometer surveys, and their general M25r through M27n pattern appears to be supported by magnetostratigraphic studies in British and French outcrops (e.g. Ogg and Coe, 1997). Therefore, the succession from the base of M25n to the base of M27n of Handschumacher *et al.* (1988) was proportionally scaled to be consistent with the composite CENT model of the total Hawaiian anomaly distance between these endpoints.

Sager *et al.* (1998) obtained a high-resolution deep-tow survey of the Japanese lineations older than M27n and proposed both an enhanced subdivision of M27r through M29r of Handschumacher *et al.* (1988) and extended the scale in a complex pattern to "M41." We proportionally scaled the widths of these oldest Japanese lineations observed by Sager *et al.* (1998) to fit the Hawaiian lineation scale of Channell *et al.* (1995b) by comparing the kilometers between the middle of M27r and M29r. This implies that the Japanese suite was formed at a spreading rate of approximately 2.5 times faster than the Hawaiian suite. Both the primary anomaly pattern from the deep-tow survey and a computed upward-projection to the sea surface by Sager *et al.* (1998) were transformed to "Hawaiian-distances," but only the deep-tow pattern is shown in Table 5.3. Of course, this method of projecting pre-M29r marine magnetic anomalies of the Japanese-kilometer block model onto the synthetic Hawaiian-kilometer pattern implicitly assumes that there were no relative changes in spreading rates between these ridges during the formation of this pre-M29r crust.

The synthetic profile of the full suite of M-sequence marine magnetic anomalies scaled to the relative distances in the Hawaiian lineation composite is given in Table 5.3.

5.3.3 Calibration and ages of the Middle Jurassic–Early Cretaceous geomagnetic polarity time scale

AGE CONSTRAINTS ON M-SEQUENCE SCALING

A spreading-rate model is required to transform the composite Hawaiian anomaly distances to absolute ages. This spreading-rate model is constrained by a very small set of radiometric ages and by durations of clusters of polarity chrons derived from cycle stratigraphy.

There are only three direct radiometric age constraints on the M-sequence – the top of Chron M0r, the middle of Chron M26n, and the oceanic crust older than Chron M41 – but all three have potential problems or controversial interpretations.

The top of Chron M0r is dated as 124.6 ± 0.2 Ma (Pringle and Duncan, 1995; 1-sigma; with their published Ar–Ar age of 122.8 ± 0.2 Ma from plagioclase separates recomputed to a TC sanidine monitor standard of 28.34 Ma). The dated levels are close to the top of a magnetic reversal in the basaltic edifice of MIT Guyot, for other stratigraphic evidence is most consistent with the assignment to polarity Chron M0r (e.g. Pringle *et al.*, 2003, and in prep.), as is summarized in Chapter 19.

The middle of Chron M26n is dated as 155.3 ± 3.4 Ma (Ludden, 1992; 1-sigma). This K–Ar age was obtained from a celadonite vein in basalts at ODP Site 765 drilled on magnetic anomaly M26n to the northwest of Australia. Chron M26n spans about 0.5 myr, therefore this age was arbitrarily assigned to the middle of the anomaly (or to the base of subchron M26n.2r in the block model of Handschumacher *et al.*, 1988). An additional Ar–Ar (incremental heating) analysis of basalt did not yield a plateau age, but gave a total fusion age of 155 ± 6 Ma (2-sigma), which Pálfy *et al.* (2000c, p. 930) consider "disputable." Even though each determination can be questioned, the coincidence of these two methods on different materials suggests that the 155 Ma age is acceptable for scaling the Late Jurassic M-sequence.

ODP Site 801 is about 100 km beyond anomaly M41 in the Japanese lineations on the deep-tow magnetic profile trending toward this site (Sager *et al.*, 1998), or an equivalent of about 40 km beyond M41 when projected to the composite Hawaiian M-sequence pattern. Tholeiitic basalts dated by the Ar–Ar method as 167.7 ± 1.4 Ma (2-sigma; Koppers *et al.*, 2003, who enhanced similar results from higher flows obtained by Pringle, 1992) are overlain by alkaline off-ridge basalt with an Ar–Ar age of 157.4 ± 0.5 Ma (Pringle, 1992), in turn overlain or intruded into radiolarian claystone with a disputed age assignment (see Section 18.2.2.4).

Cycle stratigraphy studies provide constraints on minimum time spans of clusters of polarity zones within the Early Cretaceous, and are reviewed in Chapter 19. Polarity zone M0r of the basal-Aptian spans 380 kyr (hence, the base of the Aptian Stage is 125.0 Ma), and the interval from the base of the Early Barremian polarity zone M3r to the base of M0r has a duration of ~4.8–5.4 myr (Herbert, 1992). The combined Valanginian and Hauterivian stages, which encompass approximately Chrons M14r to upper M5n, span a minimum of ~10.5 myr in southeastern France using modern Milankovitch-cycle periods or ~10 myr using estimated orbital periods during Early Cretaceous (Huang *et al.*, 1993, with independent corroboration by Giraud, 1995, and Giraud *et al.*, 1995).

When merged with the 125 Ma age for the base of the Aptian, the suite of cycle stratigraphy studies project an age of ~140.5 Ma or older for the base of the Valanginian Stage. This estimate cannot be reconciled with a U–Pb age of 137.1 ± 0.6 Ma (1-sigma) from a tuff from the Great Valley sequence of California with a nannofossil assemblage of latest Berriasian or earliest Valanginian (Bralower *et al.*, 1990). However, Pálfy *et al.* (2000a) suggest that many zircons from Jurassic to Cretaceous rocks in tectonically disturbed regions of California had potential Pb loss, and hence yield minimum ages.

SPREADING MODEL AND AGES OF M-SEQUENCE POLARITY CHRONS

This suite of radiometric and cycle stratigraphy constraints on the ages of the Hawaiian lineations can be fit using variable spreading rates. The ages of Site 801 and Chron M0r imply an average spreading rate for the entire M-sequence of ~28 km/myr; and the constraints from cycle stratigraphy imply that the average Early Cretaceous rates must about 25% slower than the average rate during the Middle–Late Jurassic. We opted for a minimalist model in which four extended intervals of constant-spreading rate are separated by small-amplitude jumps or ramped changes. Relative to the long-term average rate, we fit the array of constraints and estimated uncertainties by applying a factor of 1.07 to the M-sequence older than Chron M23n (mid-Kimmeridgian), a factor of 1.18 to Chrons M23n–M15n (late Berriasian) followed by a ramped slowing during the Valanginian to a factor of 0.77 for Chrons M11n–M5n (end-Hauterivian), then 0.86 for Chrons M3r–M0r (base-Aptian).

There is a continuum of other spreading-rate models to fit these few constraints, and segments of the pre-Hauterivian M-sequence can be systematically shifted to younger or older ages by over 1 myr. Future Mesozoic time scale research should have an emphasis on obtaining precise radiometric ages on M-sequence polarity chrons or on biostratigraphic levels which can be unambiguously correlated to the M-sequence pattern. Until additional constraints are available, then our age model (Table 5.3) and the associated array of projected ages on biostratigraphic datums and stage boundaries (see Chapters 18 and 19) should be considered as a schematic of the relative age relationships for Middle Jurassic–Early Cretaceous chronostratigraphy.

Most Late Jurassic (Oxfordian–Tithonian) and Early Cretaceous (Berriasian–Barremian) ammonite zones within the Tethyan faunal realm and several ammonite zones within the Boreal faunal realm have been directly calibrated to M-sequence chrons, thereby enabling age–model projections to be placed on possible stage and substage boundaries (indicated in Table 5.3, and reviewed and illustrated in Chapters 18 and 19). Many microfossil (dinoflagellate, calpionellid) and calcareous nannofossil datums have also been calibrated for this time interval.

5.4 GEOMAGNETIC POLARITY TIME SCALE FOR MIDDLE JURASSIC AND OLDER ROCKS

5.4.1 Paleozoic to Middle Jurassic

The oldest M-polarity chron from the ocean floor is of Middle Jurassic age. The geomagnetic polarity time scale for the remaining 96% of geological time has to be derived entirely by a progressive assembly and verification of the magnetostratigraphy from overlapping and coeval stratigraphic sections. The status of the geomagnetic polarity time scale for each individual period is summarized in Chapters 11–22.

Pre-Late Jurassic polarity chrons do not have a corresponding marine magnetic anomaly sequence to provide an independent nomenclature system. Some published magnetostratigraphic sections have designated the individual polarity zones by a stratigraphic numbering or lettering (upward or downward). For example, an "E" series derived from cyclic lacustrine deposits in eastern USA spans the Upper Triassic. A systematic nomenclature of polarity chrons could be developed by consensus when the completeness of these polarity sequences has been verified and unambiguously correlated to biostratigraphy. One option for a future "user-friendly" nomenclature would be a numbering of the major events from oldest to youngest within each individual stage – e.g. polarity chron "Toarcian-3n" (or "Toar.3n") would indicate the third major normal-polarity chron from the base of the Toarcian Stage. A version of this type of stage–level nomenclature was developed for a preliminary compilation of polarity patterns by Ogg (1995). However, as with the C-sequence and M-sequence, it remains somewhat subjective in designating chrons versus subchrons within a pattern that is essentially a random series.

High-resolution magnetostratigraphic studies on Paleozoic sections have mainly concentrated on compiling polarity patterns associated with major stratigraphic boundaries. Khramov and Rodionov (1981) published a composite

Paleozoic scale primarily based upon Russian outcrops, but the majority of the component magnetostratigraphic sections and their biostratigraphy have not been published in western literature. This generalized Russian scale agrees in its broad aspects with other magnetostratigraphic studies, but the details are commonly in disagreement. High-resolution studies and the predominance of reversed polarity suggest possible partial remagnetization of Paleozoic strata during the Carboniferous–Permian.

At a larger scale, some paleomagnetists, especially Russian workers, have proposed a series of Paleozoic "superzones" or "hyperzones" characterized by a dominant magnetic polarity or frequency of reversals (Irving and Pullaiah, 1976; Khramov and Rodionov, 1981; Algeo, 1996). In particular, the Late Carboniferous–Late Permian has a reversed-polarity Kiaman superzone followed by a mixed-polarity Illawarra superzone. Other suggested first-order polarity features include a "Burskan" reversed-polarity-bias superzone spanning Middle Cambrian–Middle Ordovician, and a "Nepan" normal-polarity-bias superzone spanning Late Ordovician–Late Silurian (Algeo, 1996).

5.4.2 Precambrian

Paleomagnetic reversals are observed in rocks as old as Archean (Layer *et al.*, 1996). Although there are many magnetostratigraphic studies of Precambrian basins, it is premature to attempt to construct a polarity reversal scale from them. One of the major problems of Precambrian correlation is the general absence of fossils and of dateable rocks. However, variations in the isotopes of strontium, carbon, and sulfur have been used to set up inter-continental correlations between Australia and North America for the 840 Ma to basal-Cambrian interval (Walter *et al.*, 2000). Such projects provide a basic chronological framework for future detailed magnetostratigraphic correlation and dating.

5.5 SUPERCHRONS AND POLARITY BIAS

The short subchrons within the C1r (Matuyama) reversed chron suggests that during the Matuyama polarity chron the polarity was biased to a reverse polarity with occasional short-lived normal subchrons that were uncharacteristic of the chron. This and other examples gave rise to the notion of *polarity bias* (McElhinny and McFadden, 2000) in which the field alternated between states of predominantly normal polarity followed by predominantly reversed polarity. During the normal-polarity interval the field was biased toward normal polarity with infrequent reversed episodes and conversely for the reverse-polarity interval. The apparent predominance of normal polarity in the Proterozoic of North America or of an apparent global tendency for the dominance of the reversed state for the past 1700 Ma, discussed above, supports polarity bias.

But polarity bias is at odds with the fundamental equations of dynamo theory because the equations are symmetric for the magnetic field and imply that the statistical properties of the normal and reverse polarities are identical.

It appears that the geodynamo has two states: a reversing state and a non-reversing state (McElhinny and McFadden, 2000). When the dynamo changes from the non-reversing state to the reversing state then normal and reverse chrons are produced randomly, with a gradual increase in the reversal rate over periods of tens of millions of years. As the dynamo approaches the non-reversing state again, the reversal rate decreases until reversals cease altogether. Then there follows a long interval of constant polarity – a superchron – that terminates only when the reversing state is re-established. Since superchrons are not regarded as part of the reversing state and this state is the only one for which polarity bias is being considered here, it follows that superchrons are not relevant to the question of polarity bias. Superchrons represent a different state of the geodynamo.

5.6 SUMMARY AND CONCLUSIONS

- The geomagnetic polarity time scale associated with the C- and M-sequences of marine magnetic anomalies of Late Jurassic–Neogene age is known in detail and calibrated to an array of biostratigraphic datums.
- Astronomical dating by means of Milankovitch cycles provides high-precision ages or chron durations for about half of the polarity time scale of the Cenozoic and for portions of the Cretaceous.
- Ages for the remainder of the C-sequence and for most of the M-sequence are derived from spreading-rate models for synthetic marine magnetic anomaly composites (South Atlantic for C-sequence, "Hawaiian" spreading ridge in Pacific for M-sequence).
- Many Phanerozoic polarity patterns from pre-Late Jurassic strata are calibrated to biostratigraphic scales, but most of these sequences require additional verification and complete coverage of geological stages before a systematic nomenclature can be proposed.
- Partial Precambrian polarity sequences are known from Precambrian basins; a few can be correlated across continents, but most of the Precambrian polarity sequence is poorly known and poorly dated.

6 • Radiogenic isotope geochronology

M. VILLENEUVE

The isotopic systems used for the chronometric calibration of the time scale are reviewed. Multiply concordant analyses based on U–Pb dates from the Thermal Ionization Mass Spectrometric (TIMS) method are regarded as the most robust and accurate measure of age for the Mesozoic and Paleozoic, followed by dates that imply only minor episodic Pb loss or inheritance of pre-existing material. For the Mesozoic and Cenozoic, $^{40}Ar/^{39}Ar$ mineral analyses of biotite, sanidine, and hornblende from volcanic rocks that show no evidence for thermal alteration or of excess ^{40}Ar also provide excellent material for time scale calibration.

6.1 INTRODUCTION

The chronometric calibration of stratigraphic boundaries underpins the geologic time scale. It was the discovery of radioactive decay and the recognition that its measurement in minerals gave temporal information (Holmes, 1937, 1947) that provided the impetus for the development of time scales calibrated in millions of years. Updated time scales were published as new calibration points slowly, but regularly, appeared (Holmes, 1947, 1960; Kulp, 1961; Armstrong, 1978; Harland *et al.*, 1982; Palmer, 1983).

The last quarter of a century has witnessed an explosive growth in data, an increased precision of measurement, the inter-laboratory standardization of decay constants (Steiger and Jäger, 1978) and a better understanding of the isotope systematics of analyzed materials. All have resulted in substantial changes to the database of ages used in time scale development. In particular, there is now a heavy reliance on results from the analytically precise $^{40}Ar/^{39}Ar$ and U–Pb methods at the expense of K–Ar and Rb–Sr dates, which were the mainstay of older time scales.

6.2 TYPES OF UNCERTAINTIES

Internal uncertainties are representative of a laboratory's reproducibility of analysis and are typically reported with the age. *External uncertainties* represent the reproducibility of the analysis between laboratories, the values obtained by different methods or the values obtained against absolute time. Analytical precision has steadily increased to the point that an accurate comparison of ages now requires a careful assessment of the magnitude and sources of external uncertainty, rather than the traditionally reported internal ones. In addition to inter-laboratory reproducibility, the main sources for the external uncertainties associated with a given date are (i) the uncertainties within a given isotopic system and between different isotopic systems, (ii) the calibration of standards, and (iii) the effects of geological processes.

A Geologic Time Scale 2004, eds. Felix M. Gradstein, James G. Ogg, and Alan G. Smith. Published by Cambridge University Press.

6.2.1 Decay constants, isotopic ratios, and comparison between $^{40}Ar/^{39}Ar$ and U–Pb systems

Because the decay constant (or equivalently, the half-life) determines the absolute time (years) obtained from isotopic measurements, uncertainties or errors in the constant directly affect the calculated age. For example, until 1977, two major sets of constants were in widespread use in K–Ar geochronology, referred to colloquially as "western" and "Russian," giving rise to two different sets of ages for the same material (Harland *et al.*, 1982).

In 1976, the International Union of Geological Sciences (IUGS) formed a Subcommission on Geochronology that arrived at a convention for inter-laboratory standardization of these constants (Steiger and Jäger, 1978). The recommendations were quickly and universally accepted and remain to this day a prime underpinning in any age determination. Uncertainties in these constants can be ignored when comparing ages derived *within a single isotopic system* because the uncertainty represents a systematic error that affects all ages equivalently. Thus analytical precision and accuracy become the sole determinants of age equivalence between samples when the same isotopic ratios are analyzed. However, the nature of this error changes when comparisons are made *between* isotopic systems. As highlighted by numerous authors (Mattinson, 1994a,b; Ludwig, 1998; Min *et al.*, 1998, 2000; Ludwig *et al.*,

1999; Renne *et al.*, 1998a,c), uncertainties on the accepted values need to be considered when the result is a comparison to absolute time. Furthermore, the IUGS Subcommission on Geochronology clearly stated that the "selected values are open and should be the subjects of continuing critical scrutinizing" – a process which has only recently been suggested for K–Ar decay constants (Kwon *et al.*, 2002) and more generally, for all decay constants (Begemann *et al.*, 2001).

Now that $^{40}Ar/^{39}Ar$ and U–Pb systems are effectively the only isotopic systems providing geochronologic data, intercalibration between $^{40}Ar/^{39}Ar$ and U–Pb (and ultimately absolute chronologies) becomes paramount so that all reported data represent equivalence of ages unbiased by analytical methodology (Renne *et al.*, 1998a, 1999, 2000; Min *et al.*, 2000; Villeneuve *et al.*, 2000; Begemann *et al.*, 2001).

6.2.2 Calibration of reference materials

The second source of external uncertainty concerns the calibration of reference materials in the $^{40}Ar/^{39}Ar$ and the High-Resolution–Secondary Ion Mass Spectrometry (HR–SIMS) U–Pb methods of geochronology. Because both techniques rely on comparisons with a naturally occurring standard, any bias in the apparent age of a standard propagates through to the final age determination. Some commonly used standards are homogenous at the multigrain level but exhibit substantial heterogeneities at the level of micro-scale studies. For example, MMhb-1, originally interpreted as 520.4 ± 1.4 Ma (Samson and Alexander, 1987), was demonstrated to be homogenous only at sample sizes greater than 15 mg (Baksi *et al.*, 1996). These effects may cause the final errors to be severely underestimated or the ages to be biased, depending on the number of individual standards contained in an irradiation.

6.2.3 Changes induced by geological processes

A third type of uncertainty is due to geological processes that lead to post-crystallization changes in the isotopic ratios. In an assessment of data, it is important to know whether the stated uncertainties for ages are due to analytical uncertainty alone. That is, the derived value should be reproducible within error limits *if exactly the same analytical conditions and data filtering criteria are followed*. But, as noted by Ludwig *et al.* (1999), reproducibility of a value is not itself evidence for veracity.

It is important to understand the mechanisms that disrupt isotopic ratios and the geological processes that may drive them. For example, dateable units should be selected so that thermal overprinting is absent. Such selection is as important as precision of the laboratory analysis. An example of this last is reflected by two high-precision studies of the Permian–Triassic boundary. Bowring *et al.* (1998), through U–Pb TIMS dating of eleven tuff horizons, each with multiply concordant analyses, placed the boundary at 251.4 ± 0.3 Ma; the same age was arrived at in a distant Chinese locality containing correlative boundary tuffs. However, re-analysis of many of the same units in one of Bowring's sections by Mundil *et al.* (2001a,b) suggested an age of 253 ± 0.3 Ma due to the presence of unrecognized Pb loss in the multigrain zircon data. The two dates are clearly distinct at the quoted uncertainties. Which value is correct depends to a great extent on the interpretation of the underlying isotope systematics (see also Chapter 17). To resolve this problem, samples that do not contain evidence of thermal overprinting need to be targeted.

One purpose of GTS2004 is to ensure that dates determined through different methodologies (almost exclusively U–Pb and $^{40}Ar/^{39}Ar$) have been similarly evaluated and another is to assign reasonable uncertainties to the data. In this way, overprecise data will not unduly influence the position of spline-fit curves, nor the overall errors assigned to time scale boundaries.

6.3 DATING METHODS

Below is a review of geochronometric calibration methods, with attention paid to U–Pb and K–Ar ($^{40}Ar/^{39}Ar$) and their application. All uncertainties are quoted at 1-sigma level.

6.3.1 U–Pb

Numerous articles review the U–Pb system and analytical techniques (e.g. Faure, 1986; Heaman and Parrish, 1991; Dickin, 1997). Uranium–lead decay is, in reality, two systems (^{235}U decaying to ^{207}Pb and ^{238}U decaying to ^{206}Pb). Any open-system behavior becomes evident by non-concordance of the two derived ages. Decay constants adopted by Steiger and Jäger (1977) are well determined, with uncertainties for both isotopes known to better than 0.07% (Jaffey *et al.*, 1971). As highlighted by numerous authors (Mattinson, 1994a,b; Ludwig *et al.*, 1999; Ludwig, 1998; Min *et al.*, 1998, 2000; Renne *et al.*, 1998a,c), the decay constant uncertainties are not traditionally propagated into the final age, but, in practice, should be when dates derived in more than one isotopic system are compared with one another. This also holds true for $^{207}Pb/^{206}Pb$ dates because of the explicit use of the two decay schemes in deriving the age. Although calculation of external uncertainty is complex for the U–Pb system (cf. Ludwig, 1998), for $^{206}Pb/^{238}U$

or $^{207}Pb/^{235}U$ ages, the decay constant uncertainty can be approximated as propagating in a linear fashion (as percentage error) into the final age and therefore generally represents the lowermost level of uncertainty that can be realized. For $^{207}Pb/^{206}Pb$ dates, the absolute uncertainty from decay constants is relatively unchanged at approximately ± 2.5 myr throughout the Phanerozoic (Renne *et al.*, 1998a, 2000), and rises but slowly in the Precambrian.

The U–Pb method is unique in that it readily displays post-crystallization disruption of the isotope systematics as discordance between the two derived ages. The two most notable types of disruption are inheritance of older components (especially in zircon) and post-crystallization Pb loss. If the discordance is brought about by either inheritance of material with a single age, or by episodic Pb loss, then the result can be interpreted in terms of simple mixture models and a linear regression can be carried out to arrive at the crystallization age for the mineral. Discordance and the use of $^{207}Pb/^{206}Pb$ ages results in complexities in the error analysis because of the low angle of intersection of the discordia line with concordia, which itself contains an error envelope due to decay constant uncertainty (Ludwig, 1998). As such, the most robust data consist of multiply concordant or nearly concordant data that are used to determine an age. In this way, assumptions surrounding the cause (e.g. single Pb-loss event) of post-crystallization disruption and the effects of decay constant uncertainty introduced by extrapolating data points to concordia are minimized. However, it is worth noting that there remain U–Pb determinations that use a linear regression of discordant data arrays and therefore some interpretation of the isotope systematics is required (e.g. Tucker and McKerrow, 1995; Pálfy *et al.*, 2000a). Although the resultant age picks used for time scale calibration are generally robust, it is important to remember that alternative interpretations may somewhat change the determined age or error. In general, almost all age information in the Phanerozoic (and, most certainly, in rocks younger than Mesozoic) is contained in the $^{206}Pb/^{238}U$ age alone. Thus, reliance on underlying assumptions regarding Pb loss or inheritance becomes more significant at these ages, where the parallelism of concordia with Pb loss trajectories makes it increasingly difficult to tell discordant from concordant results.

Most modern studies follow well-established techniques (Krogh, 1982a,b) that can minimize the effects of both Pb loss and inheritance in the final analyses. There has been a recent trend, especially at key time scale boundaries such as the Permian–Triassic, to produce and select only the best-quality results as exemplified by overlapping concordant analyses that can be interpreted in terms of undisturbed isotope systematics.

Material used for time scale calibration is almost exclusively from well-exposed and widespread volcanic horizons that bracket biostratigraphically calibrated sedimentary sections (Tucker and McKerrow, 1995; Bowring *et al.*, 1998; Tucker *et al.*, 1998; Mundil *et al.*, 1999, 2001c; Compston, 2000a,b; Pálfy *et al.*, 2000c). For this reason, the mineral most used for geochronological purposes is zircon because of its ubiquitous presence in these rocks, its highly refractory nature, and a crystal structure that supports trace amounts of U, but almost no Pb, upon crystallization. Corrections for any initial Pb incorporated in the crystal structure are generally minor. Diffusive Pb loss is considered slow, even at magmatic temperatures, as long as there is no radiation damage to the crystal structure. Concordance is enhanced by selection of best-quality grains by magnetic separation techniques (Krogh, 1982b) and hand picking under binocular microscope, as well as abrasion of the outer portion of crystals that are most subject to Pb loss (Krogh, 1982a).

However, the refractory nature of zircon can lead to new zircon nucleating around undissolved crystals in the magma, resulting in the presence of xenocrystic cores that may bias results to an older age. If the extent of this bias is large enough, the discordance displayed may be sufficient to indicate disruption of isotopic systematics and a linear array of data leading toward older ages. Similarly, the early crystallizing nature of zircon means that it may be subject to residence in the magma prior to final emplacement. Thus, the age may not correlate with eruption of a volcanic horizon, but rather predate it. This effect has been one of the suggested sources of the discordance between $^{40}Ar/^{39}Ar$ ages and U–Pb TIMS ages in the Fish Canyon Tuff, although recent evidence presented by Schmitz and Bowring (2001) suggests that the age difference may be sourced in errors of the ^{40}K decay constant.

Other minerals commonly dated (but not generally represented in time scale calibration studies) include monazite, titanite, and allanite. Monazite, with a sub-magmatic closure temperature between 725 (Heaman and Parrish, 1991) and 825 °C (Bingen and van Breemen, 1998), is generally restricted to peraluminous magmas. The lack of inheritance and generally high U content make monazite an ideal mineral for analytical purposes. However, the well-known presence of disequilibrium effects (Schärer, 1984; Parrish, 1990) by the incorporation of ^{230}Th (an intermediate daughter product in the $^{238}U/^{206}Pb$ decay chain) means that age determinations must be based on $^{207}Pb/^{235}U$ ages alone. Titanite, with a yet lower closure temperature of ~ 650 °C (Heaman and Parrish, 1991) generally contains low U contents and a significant proportion of common Pb in its structure. As such, high-resolution geochronology is highly dependent on the accurate correction of the latter, usually by measuring coexisting K-feldspar Pb composition and assuming identical isotopic ratios for common Pb correction within the titanite (Schmitz and Bowring, 2001).

THERMAL IONIZATION MASS SPECTROMETRIC (TIMS) METHODS

The early 1980s saw the advent of new microanalytical techniques for U–Pb zircon geochronology (Krogh, 1979, 1982a,b; Parrish, 1987; Parrish and Krogh, 1987; Roddick *et al.*, 1987). Concomitant with these developments, data reduction and error analysis techniques were significantly improved (Ludwig, 1980, 1984, 1993; Davis, 1982; Roddick, 1987, 1996). They focused work on minimizing the major sources of error with the effect of greatly improving both the efficiency and precision of analysis. Furthermore, new reference materials prepared by National Institute for Standards and Technology greatly improved standards calibration.

Inter-laboratory calibration studies of standard mineral suites are rare, but published results (Wiedenbeck *et al.*, 1995) show a high degree of agreement between laboratories. If uncertainties in decay constants and geological effects are excluded, the analytical precision for U–Pb ages derived by Thermal Ionization Mass Spectrometric (TIMS) analysis now routinely attains precision levels of less than 0.02% for rocks older than Late Paleozoic and less than 0.1% for the Phanerozoic. It should be noted that in order to attain these levels of precision in the Phanerozoic, "bulk" zircon fractions consisting of 10–50 crystals are needed, thereby increasing the odds of skewing data to older ages through inheritance or to younger ages through Pb loss (Ludwig *et al.*, 1999).

HIGH-RESOLUTION–SECONDARY ION MASS SPECTROMETRY (HR–SIMS)

Within the last 20 years the use of ion microbeam techniques has developed to the point where the analytical results may be of use in time scale calibration (Compston *et al.*, 1984). High-Resolution–Secondary Ion Mass Spectrometry (HR–SIMS, also known as SHRIMP) allows targeting of 20–30 $\times$ 2 μm portions of zircon crystals with analytical precision one to two orders of magnitude less than TIMS methodology. Because of the small amount of material analyzed, HR–SIMS geochronometry is nearly non-destructive and polished minerals, either mounted in epoxy or *in situ* within polished thin sections, provide source material for analysis. Cathodoluminescence and backscattered images can detect xenocrystic cores or metamict areas – areas whose structure has been damaged by radiation – that are unsuitable for analysis, thereby enhancing the likelihood that the material is pristine. The ability to measure trace element concentrations within zircon and monazite crystals also allows for delineation of zones marked by chemical discontinuities that may also be problem areas.

The production of U and Pb ions in the mass spectrometer by the energetic primary oxygen beam is highly dependent on mineral-specific matrix effects, with Pb ionization typically enhanced by two–five orders of magnitude relative to U. This means that derivation of the $^{206}Pb/^{238}U$ ratio, key to determining ages in the Phanerozoic, is highly dependent on the accurate measurement of $^{206}Pb/^{238}U$ in a naturally occurring standard. Furthermore, this standard must be isotopically homogenous to better than 1% on a micrometer scale, and possibly to better than 0.3% in order to meet a desired 1.0% uncertainty of an unknown (Stern and Amelin, 2001). Although high-precision measurement of this ratio in the standard can be achieved through analysis by TIMS methodologies (thereby allowing calibration of primary ratios by a technique with inherently better precision), micrometer-scale heterogeneities are evened out by the larger sample sizes required. Thus, key zircon standards have had these ratios updated; most notably SL3, which had its $^{206}Pb/^{238}U$ age shifted by 1% (Roddick and van Breemen, 1994), and SL13, which had previously undetected micrometer-scale heterogeneities ascribed to it (Compston, 2000a).

Finally, data reduction and error analysis methodologies are still undergoing review because of the relatively small number of HR–SIMS installations. All measurements of the $^{206}Pb/^{238}U$ ratio depend primarily on counting statistics, in addition to an estimate of errors related to the scatter about the Pb–U calibration line. It is uncertain whether calibration can be accurately determined using statistical error analysis on measurements (e.g. Claoué-Long *et al.*, 1995), or if inherent intra-analysis instrumental drift limits the available precision to $\sim$1%, as a study by Stern and Amelin (2003) suggests. Recently, much work has gone into finding and categorizing improved zircon standards through cross-calibration with high-precision TIMS data (Stern, 2001) or recharacterizing older standards (Compston, 2000a). Currently available HR–SIMS time scale data have uncertain status because there is a systematic difference between HR–SIMS and TIMS dates, with the HR–SIMS dates being generally younger, as highlighted in the 1.5% difference noted for the Ordovician time scale (Tucker and McKerrow, 1995; Compston, 2000a). Because of questions surrounding use of older standards (especially SL13), coupled with those surrounding the relative magnitude contributed by instrumental bias, minimum attainable uncertainty on unknowns is herein estimated at approximately ±1.0%. However, this value will rapidly diminish in the near future because of the current world-wide effort being placed on calibration of these standards that is beginning to produce better consistency between the HR–SIMS and TIMS data.

Perhaps one area where HR–SIMS has the potential to redefine time scale calibration radically is related to its ability to measure mineral grains *in situ*. Xenotime ((Y, REE)PO_4) forms as overgrowths around detrital zircon during diagenesis of siliciclastic rocks (McNaughton *et al.*, 1999; England *et al.*, 2001). Xenotime provides good material for dating because, like zircon, it has a high initial U and low common Pb content and, with a ~ 700 °C closure temperature, it is relatively resistant to later thermal resetting. Because these fragile overgrowths are usually less than 25 μm thick, HR–SIMS is the only method potentially capable of directly dating them. Dating diagenesis directly may therefore be possible in the near future. If the diagenesis is close in time to sedimentation, then the results may augment the more normal technique of dating sedimentary horizons that are bracketed by volcanic layers.

6.3.2 K–Ar and $^{40}Ar/^{39}Ar$

Good overviews of both the K–Ar method (Dalrymple and Lanphere, 1969) and the more modern $^{40}Ar/^{39}Ar$ variant (McDougall and Harrison, 1999) are available. Approximately 10% of ^{40}K decays by electron capture to the daughter product ^{40}Ar, with the remainder decaying by electron emission to ^{40}Ca, thus necessitating the use of two (in actuality, three, because of an additional minor positron decay to ^{40}Ar) decay constants. The decay constants currently accepted by the geological community represent an amalgam of a sub set of available measurements that were combined to give the values currently (Beckinsale and Gale, 1969) preferred by geochronologists. Uncertainty in the overall decay constant value of 5.543×10^{-10} a^{-1} is approximately ±0.18%, but the uncertainty associated with decay by electron capture constant of 0.572×10^{-10} a^{-1} is ±0.69%. In addition, using the same primary data set, cosmologists arrived at a value for the decay constant of ^{40}K that differs by close to 2% (Endt, 1990), although recent studies suggest a value intermediate between the two (Min *et al.*, 1998, 2000; Renne *et al.*, 1998b,c; Begemann *et al.*, 2001; Kwon *et al.*, 2002). As outlined below in the section on $^{40}Ar/^{39}Ar$ analysis, some of this built in-bias and uncertainty can be removed by calibrating externally to the K–Ar system. Derivation of ^{40}K, whether for traditional K–Ar analysis or $^{40}Ar/^{39}Ar$ is dependent on measurement of total K and the use of an accepted isotopic composition (Garner *et al.*, 1975; Steiger and Jäger, 1977). Although well calibrated, uncertainty in these ratios are traditionally not propagated into the final age.

One advantage of the K–Ar isotopic system is the ubiquity of potassium in almost all igneous rocks. Minerals such as sanidine, micas, or hornblende make dating possible for both felsic and mafic volcanic rocks. High-temperature K-feldspar (sanidine) is one of the most useful minerals because K contents of up to 10 wt. % result in production of measurable daughter Ar, well into the Cenozoic, and the ability to measure single phenocrysts with laser-based instrumentation. However, not all Ar may be degassed from sanidine (McDowell, 1983; McDougall, 1985), a prerequisite correct age determination. Muscovite, biotite, and phlogopite yield the most reliable ages, although they may incorporate excess ^{40}Ar within their crystal structure, giving rise to old dates (Roddick *et al.*, 1980). Hornblende, with a high Ca/K ratio, presents difficulty in correcting for interfering isotopes that are produced by Ca during the sample irradiation needed for $^{40}Ar/^{39}Ar$ analysis. These corrections become larger at younger ages. In addition, the low K content of hornblende, together with its common contamination by intergrowths of biotite and feldspar, limits the ability to obtain meaningful ages. The low K content also results in an increased proportion of atmospheric Ar contamination that is directly reflected in lower precision of the age.

Of these minerals, hornblende has the highest apparent closure temperatures at ~500 °C (Harrison, 1981) followed by muscovite at ~300 °C (Purdy and Jäger, 1976), and biotite at ~250 °C (Harrison *et al.*, 1985). It should be noted that these values may vary with composition and cooling rates. Hornblende, biotite, and muscovite are hydrous minerals susceptible to decrepitation – the violent break-up of the mineral – in the high-vacuum, anhydrous environment of most instruments. Decrepitation interferes with the measurement of diffusion parameters needed to model and interpret Ar concentration gradients within crystals. For these reasons, most time scale studies use unmetamorphosed volcanic rocks to minimize the effects of post-crystallization changes in isotopic ratios, the resulting mineral dates then reflect the instantaneous cooling that allows the date age to be equated with magmatic emplacement.

Empirical evidence collected over three decades demonstrates that unaltered micro- and cryptocrystalline volcanic rocks give reliable and interpretable results using either K–Ar or $^{40}Ar/^{39}Ar$ methods. Rocks consisting of, or containing, interstitial glass may give ages that are too low, possibly as a result of hydration and devitrification processes (Mankinen and Dalrymple, 1972; McDougall and Harrison, 1999). Similarly, hydrothermal alteration may preferentially target the main K-bearing phases such as plagioclase, resulting in new mineral growth and partial or complete loss of radiogenic ^{40}Ar. In general, the older the volcanic rock, the more likely that it has been exposed to thermal or hydrothermal alteration. Mineral dates, rather than whole-rock dates, are therefore preferred for rocks older than Cenozoic. The ^{40}Ar/^{39}Ar method is even more strongly recommended because of its ability to detect post-crystallization disruption of the isotope systems.

Glauconies are mixtures of the mineral glauconite with other minor clay components. Glauconies have a potentially high, though variable, K content (up to 6.5%) and form by diagenesis in marine sediments (Odin, 1982) and would be appear to be ideal for time scale calibration. However, it appears that along with its formation at temperatures as low as $\sim$100 °C, the green pellets comprising macroscopic glaucony are actually comprised of mats of small, 0.1–5 μm flakes (Smith *et al.*, 1993), resulting in Ar diffusion at very low temperatures and resultant low apparent K–Ar ages when compared to samples analyzed using higher temperature minerals (Gradstein *et al.*, 1994a). Smith *et al.* (1998) suggest that glaucony formation takes place over periods of up to 5 myr and propose that the age of the sediment is given by the oldest glauconies, typically 1–3 myr older than corresponding K–Ar ages. Argon-40/argon-39 dating requires specialized sample handling (Smith *et et al.*, 1993) because reactor-induced recoil effects on the small flakes cause ^{39}Ar (i.e. the proxy for ^{40}K) to move out of the minerals, giving apparent ages that are too old. Encapsulation in evacuated sealed glass tubes and subsequent measurement of total ^{40}Ar and ^{39}Ar can give an age equivalent to that derived from K–Ar, but no glaucony dates are used for time scale calibration in GTS2004.

Finally, there are recent reports of ^{40}Ar/^{39}Ar dating of diagenetic pyrite, although the K may reside in microscopic inclusions trapped within the pyrite crystals, rather than within the pyrite crystal structure itself (Smith *et al.*, 2001).

K–Ar METHODS

In 1989, K–Ar was the primary method used for geochronometric calibration purposes, representing over 90% of the ages used in Harland *et al.* (1990). Any systematic bias caused by the values of decay constants could generally be ignored and was further swamped by the relatively low analytical precision inherent in the method. K–Ar analysis requires separate analysis of K and ^{40}Ar, meaning that a large amount of material (typically 10–100 mg) is needed to provide splits for each analytical stream. Impure or inhomogeneous starting material can affect the final age. In addition, as sample size decreases there is increasing difficulty in precisely measuring the ever more-miniscule amounts of ^{38}Ar gas used to spike the Ar analysis.

Even recently determined K–Ar results do not surpass ±0.5% analytical precision (Renne *et al.*, 1998c; Lanphere and Baadsgaard, 2001) and neglect the need to include the uncertainties on ^{40}K decay constants. These uncertainties propagate directly into the age equation, placing a lower limit of ±0.7% on Phanerozoic dates, a value regarded as the minimum uncertainty attainable for K–Ar ages. In addition, the method is generally sensitive to the correction for contamination by atmospheric Ar. Moreover, the presence of excess ^{40}Ar is common and cannot be flagged or corrected for by the K–Ar method. Furthermore, bulk measurement of K and Ar always results in a ratio that may reflect a value modified by post-crystallization disruption of the isotopes.

Analytical indicators of post-crystallization disturbance to the K–Ar isotopic system are limited to an unreasonably high atmospheric content or unreasonably low K content. In general, only samples with radiogenic ^{40}Ar greater than 90% should be used for time scale calibration because of the dramatic increase in calculated uncertainty below that level. Potassium–argon dates should be viewed as a last resort for time scale work because ^{40}Ar/^{39}Ar dates are more stringently calibrated, far-more precise, and more readily interpretable. However, currently the standards used for ^{40}Ar/^{39}Ar dating require calibration by the K–Ar method (see below).

^{40}Ar/^{39}Ar METHODS

The development of the ^{40}Ar/^{39}Ar method (Merrihue and Turner, 1966; Mitchell, 1968; Dalrymple and Lanphere, 1971) has led to greater flexibility of analysis, decreased sample size, and better indicators of isotopic disturbance. In this method, some ^{39}K is transformed to ^{39}Ar by bombardment with fast neutrons in a nuclear reactor (for detailed description of methodology see McDougall and Harrison, 1999). By comparing with a standard placed in the same neutron flux, the ^{40}Ar*/^{40}K ratio in the sample can be found and a date determined. Separate analyses of K and Ar are unnecessary, thereby dispensing with potential inhomogeneity between sample splits, decreasing needed sample amounts, and utilizing the ability of mass spectrometers to measure isotopic ratios. Theoretical models for the distribution and movement of Ar within minerals (Turner *et al.*, 1966; Dodson, 1973; Lovera *et al.*, 1989) have allowed for the determination of dates on partially disturbed samples, although in some cases the methods are contentious (Parsons *et al.*, 1999). The step-heating method, using a temperature-calibrated furnace, represents application of diffusion modeling to isotopic ratios within mineral grains, but poor sensitivity and resolution of first-generation mass spectrometers inhibited the ultimate analytical precision of many of the dates produced.

The development of a new generation of noble gas mass spectrometers in the late 1980s has led to a two–three order of magnitude decrease in sample size and hence the application of laser microbeam sampling techniques. Infrared (e.g. CO_2) lasers excel at qualitatively mimicking traditional step-heating analysis and can simultaneously heat entire grains or groups of grains, but at the expense of spatial resolution. Ultraviolet (UV) lasers (e.g. wavelength-quadrupled Nd–YAG lasers) have poor thermal control but give spatial resolution down to tens of micrometers, allowing direct measurement of isotopic gradients within individual crystals. Key to the efficacy of both laser types is their ability to couple with most silicate minerals (Kelley, 1995).

Although use of $^{40}Ar/^{39}Ar$ analytical techniques now allows for measurement of isotope ratios to less than 0.1%, there is still an ultimate reliance on calibration to a natural mineral standard with a precisely known $^{40}Ar^*/^{40}K$ ratio (Berger, 1975; Renne *et al.*, 1998c). Part of the difficulty in assessing $^{40}Ar/^{39}Ar$ standards comes about because, unlike SHRIMP, the initial calibration of these standards requires analysis by the K–Ar method, which is inherently less precise than the method relying on the standard. As noted above, even the best K–Ar analyses cannot achieve precision levels much below 0.5% (e.g. Renne *et al.*, 1998c). This difficulty in calibrating standards results in determinations of conflicting apparent ages for standards. Fish Canyon Tuff (FCT) is the main exemplar of this problem, with apparent K–Ar ages that differ by over 2% (cf. summary in Dazé *et al.*, 2003). The result is that an age determination on a sample of unknown age is biased relative to the apparent $^{40}Ar^*/^{40}K$ chosen for the standard.

Thus, in order to compare relative ages of two unknowns one must ensure the equivalence of the apparent age between the reference flux monitors. However, this process is further compounded by the proliferation of standards commonly used for $^{40}Ar/^{39}Ar$ analysis. Fortunately, a number of studies (Roddick, 1983; Baksi *et al.*, 1996; Renne *et al.*, 1998c) have used the ability to measure $^{39}Ar/^{40}Ar$ ratios of standards precisely relative to one another at a less than 0.1% level to provide tight integration. Ignoring grain-to-grain reproducibility, this allows one monitor to be substituted for another using standard intercalibration factors, although uncertainty introduced by the use of secondary or tertiary standards needs to be propagated into the age of the unknown (Karner *et al.*, 1995; Renne *et al.*, 1998c). Fortunately, very few standards remain that are more than one step removed from the primary standard. Table 6.1 lists the apparent age of standards used to intercalibrate $^{40}Ar/^{39}Ar$ data in this publication.

Table 6.1 *Preferred apparent ages for standards used in* $^{40}Ar/^{39}Ar$ *geochronology*

Standard name	Apparent age[a] (Ma)	Reference
FCT-SAN, FCT-3	28.02	Renne *et al.* (1994, 1998b), Villeneuve *et al.* (2000), Baksi *et al.* (1996)
MMhb-1	523.1 ± 2.6	Renne *et al.* (1998b)[c]
Hb3gr[b]	1072 ± 1	Roddick (1983), Turner *et al.* (1971)
TCR	28.34 ± 0.16	Renne *et al.* (1998b)[d]
GA-1550	98.79 ± 0.1	Renne *et al.* (1998b)[e]
Alder Ck.	$1.194 \pm .007$	Renne *et al.* (1998b)
GHC-305	105.2 ± 0.7	Renne *et al.* (1998b)
LP-6	128.4 ± 0.3	Baksi *et al.* (1996)
Bern 4 Mu	18.56 ± 0.07	Baksi *et al.* (1996)
Bern 4 Bi	17.25 ± 0.06	Baksi *et al.* (1996)

[a] All ages are referenced to Fish Canyon Tuff sanidine (FCT-SAN) = 28.02 Ma.
[b] Hb3gr based upon primary calibration of (Turner *et al.*, 1971) and data reported in Roddick (1983).
[c] Standard is non-reproducible for laser-scale work.
[d] Baksi *et al.* (1996) get 28.07 Ma when corrected to FC = 28.02 Ma.
[e] Baksi *et al.* (1996) get 98.04 Ma when corrected to FC = 28.02 Ma.

It is worth noting that sample homogeneity is as important for $^{40}Ar/^{39}Ar$ analysis as it is for HR–SIMS. Although most monitors listed in Table 6.1 have been found to be relatively homogenous at individual grain levels, one of the most commonly used flux monitors, MMhb-1 (Samson and Alexander, 1987), proves to have poor reproducibility when used for laser-scale work. At least three studies (Baksi *et al.*, 1996; Renne *et al.*, 1998c; McDougall and Harrison, 1999) have found an inherent 0.5% grain-to-grain uncertainty. In general, the J-factor, a measure of neutron flux and the value used to restore measured $^{40}Ar/^{39}Ar$ values to the needed $^{40}Ar^*/^{40}K$ ratio, is calculated by reference to a best-fit curve. Some studies rely on regression techniques to define the reference line, and effectively apply a standard mean error to scattered data, thereby incorrectly diminishing J-factor uncertainty below the level warranted by the natural variation present in MMhb-1. However, this standard remains in popular usage, although it is clear that 0.5% represents the minimum attainable uncertainty for any unknown relying on MMhb-1 as a standard. Even when the J-factor has been correctly assessed, it should be treated as an uncertainty applied after averaging of analytical data. Therefore, when replicate analyses of co-irradiated samples is carried out, J-factor uncertainty should be applied to the derived data after averaging of the analyses. In certain cases, it appears that J-factor uncertainty is applied to individual analyses and then results combined, resulting in overprecise determinations of age uncertainty due to the averaging effect.

Reference of $^{40}Ar/^{39}Ar$ data to a common standard assures that relative ages can be assigned within the K–Ar isotopic system. However, the decay constant uncertainties would dominate derived ages when making comparisons to absolute ages (Min *et al.*, 1998, 2000; Renne *et al.*, 1998b,c; Villeneuve *et al.*, 2000; Begemann *et al.*, 2001; Kwon *et al.*, 2002). This remains true if primary calibration of the monitor is based upon K–Ar analysis, a method that is self-limiting because of the large analytical uncertainty inherent in the methodology. In this case, overall age precision can be substantially degraded, as exemplified by the increased uncertainty of dates from the end-Cretaceous (K–T) boundary to the ±1.0% level when uncertainties of decay constants and calibration of the flux monitor are propagated into the determined age (Renne *et al.*, 1998c).

An alternative approach depends on arriving at an apparent age for the standard based upon, for example, U–Pb dates. The $^{40}Ar^{*}/^{40}K$ of the standard is then calculated by use of IUGS decay constants (Villeneuve *et al.*, 2000), resulting in an apparent age that is shifted to incorporate some of the systematic bias introduced by the potential use of incorrect decay constant values. Although imperfect, the method minimizes the uncertainties in decay constants and K–Ar calibration. For this reason, we have chosen to use FCT sanidine (FCT-SAN) = 28.02 Ma as our primary standard.

This value was chosen because it has been consistently arrived at (within error) by separate calibrations resulting in ages of 28.04 ± 0.23 Ma (Renne *et al.*, 1997) and 28.05 Ma (Min *et al.*, 2000) against absolute time, 28.03 ± 0.09 Ma (Renne *et al.*, 1994) and 28.15 ± 0.19 Ma (Hilgen *et al.*, 1997) by orbital tuning, and 27.98 ± 0.15 Ma by measurement against $^{207}Pb/^{235}U$ (Villeneuve *et al.*, 2000). Furthermore, FCT-SAN is well represented in $^{40}Ar/^{39}Ar$ intercalibration studies, thereby allowing secondary standards to be corrected. In this way it is anticipated that accurate and precise intercalibration of $^{40}Ar/^{39}Ar$, U–Pb, and absolute time can be accomplished. For example, while completing this chapter, new single-grained sanidine datings from the Mdelilla Basin in Morocco suggest an astronomically derived age of 28.24 ± 00.1 Ma for the FCT-SAN (Section 21.5.1), potentially adding further precision.

In general, even external calibration results in uncertainties ranging from 0.3 to 0.8%, providing a lower limit on the attainable precision of $^{40}Ar/^{39}Ar$ ages. Coupled with intercalibration factors and limitation on derivation of the J-factor, it is estimated that a minimum uncertainty of ±0.3% is applied to $^{40}Ar/^{39}Ar$ results (Roddick, 1988; Scaillet, 2000) in the absence of specialized analytical protocols. For samples using MMhb-1 as flux monitor, higher uncertainty limits are likely warranted. Table 6.2 lists the lower limits of precision estimates typically attainable on ages, organized by methodology, as discussed in this and previous sections.

Table 6.2 *Estimate of typically attainable lower limits on precision of ages by methodology*

Isotopic system	Precision limit[a]	Limiting factors
TIMS		
$^{207}Pb/^{206}Pb$	±2.5 Ma	Decay constant uncertainty
$^{206}Pb/^{238}U$	±0.1%	Decay constant, analytical precision
HR–SIMS		
$^{207}Pb/^{206}Pb$	less than ±0.3%	Counting statistics, dependent on age
$^{206}Pb/^{238}U$	±1%	Analytical precision
K–Ar	±0.7%	Analytical methodology, decay constant
$^{40}Ar/^{39}Ar$	±0.3%	External calibration, dependent on flux monitor characteristics

[a] Estimates based on criteria outlined in text. Specialized sample handling and analytical protocols may warrant lower limits and are ascertained on a case-by-case basis. All uncertainties listed to 1-sigma.

6.3.3 Other methods

Few other methods can attain the accuracy or precision of the $^{40}Ar/^{39}Ar$ and U–Pb TIMS methods and, for this reason, key chronostratic calibrations now depend almost exclusively upon them, with minor infill from K–Ar. Rubidium–Strontium (Rb–Sr) decay was once widespread in its use for geochronological purposes, but recognition of the mobility of both Rb and Sr in the presence of aqueous fluids or thermal disturbances has led to the method falling into disfavor as a precision chronometric tool. Although it is possible to date high-Rb phases such as micas or K-feldspar, these minerals can often be more efficiently and accurately dated by $^{40}Ar/^{39}Ar$. Samarium–Neodymium (Sm–Nd) decay is of limited applicability because of relatively low distribution coefficients between the parent and daughter atoms during crustal processes, resulting in relatively little control on generation of isochrons.

One isotopic system that displays great potential is Rhenium–Osmium (Re–Os). Work on measurement of ages of sulfide minerals, notably molybdenite (Selby *et al.*, 2000), shows that high-precision (±0.2%) results can be achieved on samples consisting of a few grains. However, development of techniques for analysis of Re–Os ages from organic-rich fine-grained strata has the potential of providing direct ages of sedimentation. Deposition of black shales in reducing environments leads to concentration of both Re and Os into these rocks, almost all derived from seawater at the time of deposition (Ravizza and Turekian, 1989; Creaser *et al.*, 2002). Although initial studies suggested a link between hydrocarbon maturation level (Cohen *et al.*, 1999) and degree of age concordance and precision, recent work by Creaser *et al.* (2002) suggests that using samples with a consistent $^{187}Os/^{188}Os$ ratio may yield more precise ages. Currently, analytical precision is limited to around 1–2%, making results of limited use in time scale calibration. However, continued advances in both analytical techniques and the understanding of the geochemistry of Re and Os hold much promise for application of this technique in the future.

6.4 SUMMARY AND CONCLUSIONS

As outlined above, quoted uncertainties can be used as a proxy for analytical quality, but do not necessarily reflect where external uncertainty begins to outweigh analytical uncertainty. This time scale uses the following approaches to ensure that quoted ages approximate not only the absolute time reference frame, but also uncertainties surrounding the ages.

U–Pb TIMS results consisting of multiply concordant analyses (e.g. Tucker and McKerrow, 1995; Bowring and Erwin, 1998; Bowring *et al.*, 1998; Landing *et al.*, 1998; Pálfy *et al.*, 2000c; Mundil *et al.*, 2001c) are viewed as the most robust and accurate measure of age for the Mesozoic and Paleozoic. This is followed by linearly discordant data arrays that can be readily interpreted in terms of minor episodic Pb loss or inheritance of pre-existing material with uniform age. For the Mesozoic and Cenozoic, $^{40}Ar/^{39}Ar$ mineral analyses of biotite, sanidine, and hornblende from volcanic rocks are viewed as generally robust, especially if step-wise heating was done to check on the absence of thermal alteration and of excess ^{40}Ar (Roddick *et al.*, 1980). Analysis of holocrystalline volcanic rock by this method is anticipated to be relatively robust, but is more prone to disruption of isotopic systematics. Glaucony is generally not suitable for analysis by $^{40}Ar/^{39}Ar$ methodology without undertaking special analytical and data reduction procedures (Smith *et al.*, 1993) because it is prone to resetting. Finally, because of its inability to detect subtle disruption in isotopic systematics, the K–Ar method is used only if no other analyses are available and the samples have appropriate K content and low atmospheric contamination.

In order to equate $^{40}Ar/^{39}Ar$ ages with U–Pb TIMS and absolute chronologies, all fluence monitor corrections are leveled to FCT sanidine (FCT-SAN) = 28.02 Ma. This value was chosen because of the consistency of this age in external calibration (Renne *et al.*, 1994, 1997; Villeneuve *et al.*, 2000) and the independence from decay constant and first principles calibration uncertainities that degrade precision (Renne *et al.*, 1998c; Villeneuve *et al.*, 2000). All other apparent ages of monitors derive from high-precision intercalibration studies, as listed in Table 6.1. As outlined above and listed in Table 6.2, estimates of minimum errors have been assigned. These error levels are approximations based upon what could reasonably be expected during routine analysis and through propagation of external uncertainties. If the authors detail specialized protocols of analysis or error handling, their quoted error limits may be substituted.

Disparate results were not simply averaged. From a statistical standpoint, such an approach is unwarranted, unless results pass Student's *t*-test, indicating a high probability that they represent the same population.

7 • Strontium isotope stratigraphy

J. M. MCARTHUR AND R. J. HOWARTH

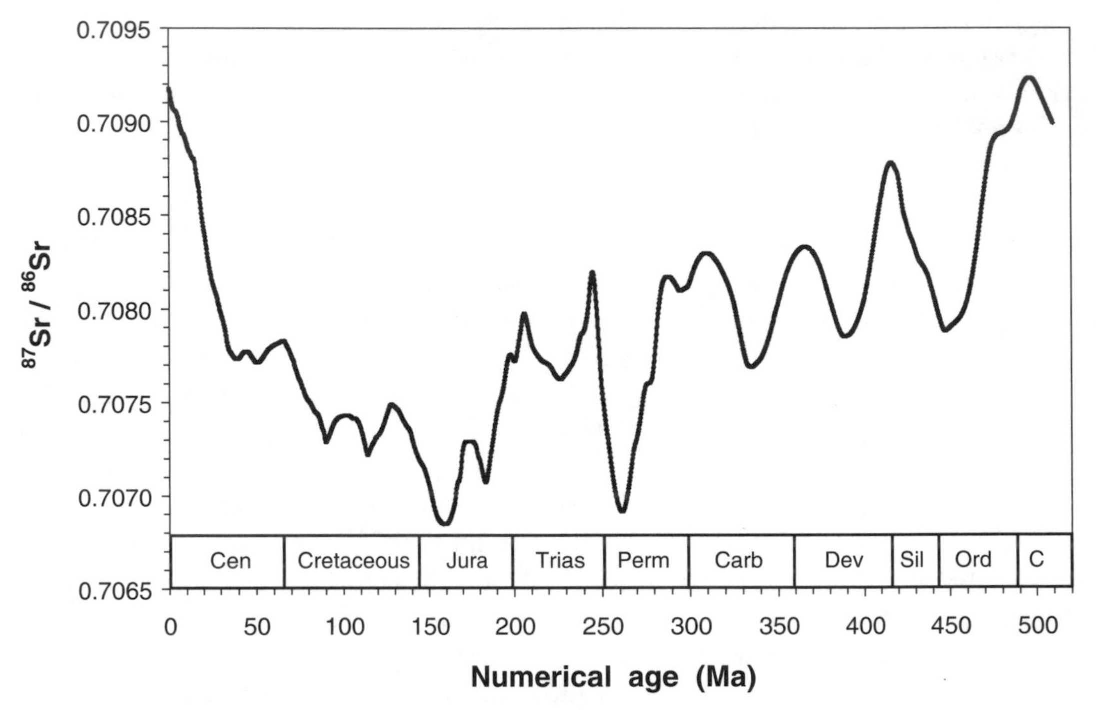

Figure 7.1 Variation of $^{87}Sr/^{86}Sr$ through Phanerozoic time. LOWESS fit to data sources in Table7.1.

The $^{87}Sr/^{86}Sr$ value of Sr dissolved in the world's oceans has varied though time, which allows one to date and correlate sediments. This variation and its stratigraphic resolution is discussed and graphically displayed.

7.1 INTRODUCTION

The ability to date and correlate sediments using Sr isotopes relies on the fact that the $^{87}Sr/^{86}Sr$ value of Sr dissolved in the world's oceans has varied though time. In Fig. 7.1, we show this variation, plotted according to the time scale presented in this volume. More detail is given in Fig. 7.2, on which we plot both the curve of $^{87}Sr/^{86}Sr$ through time and the data used to derive it. Comparison of the measured $^{87}Sr/^{86}Sr$ of Sr in a marine mineral with a detailed curve of $^{87}Sr/^{86}Sr$ through time can yield a numerical age for the mineral. Alternatively, $^{87}Sr/^{86}Sr$ can be used to correlate between stratigraphic sections and sequences by comparison of the $^{87}Sr/^{86}Sr$ values in minerals from each (Fig. 7.3). Such correlation does not require a detailed knowledge of the trend through time of $^{87}Sr/^{86}Sr$, but it is useful to know the general trend in order to avoid possible confusion in correlation near turning points on the Sr curve.

Strontium isotope stratigraphy (SIS) can be used to estimate the duration of stratigraphic gaps (Miller *et al.*, 1988), estimate the duration of biozones (McArthur *et al.*, 1993, 2000, 2004) and stages (Weedon and Jenkyns, 1999), and to distinguish marine from non-marine environments (Schmitz *et al.*, 1991; Poyato-Ariza *et al.*, 1998). The degree to which such things can be accomplished rests, in part, on how well the trend in marine $^{87}Sr/^{86}Sr$ through time can be defined, and it is this issue that concerns us here. For a more detailed account of SIS, the reader is referred to reviews by McArthur (1994) and Veizer *et al.* (1997, 1999).

A Geologic Time Scale 2004, eds. Felix M. Gradstein, James G. Ogg, and Alan G. Smith. Published by Cambridge University Press.

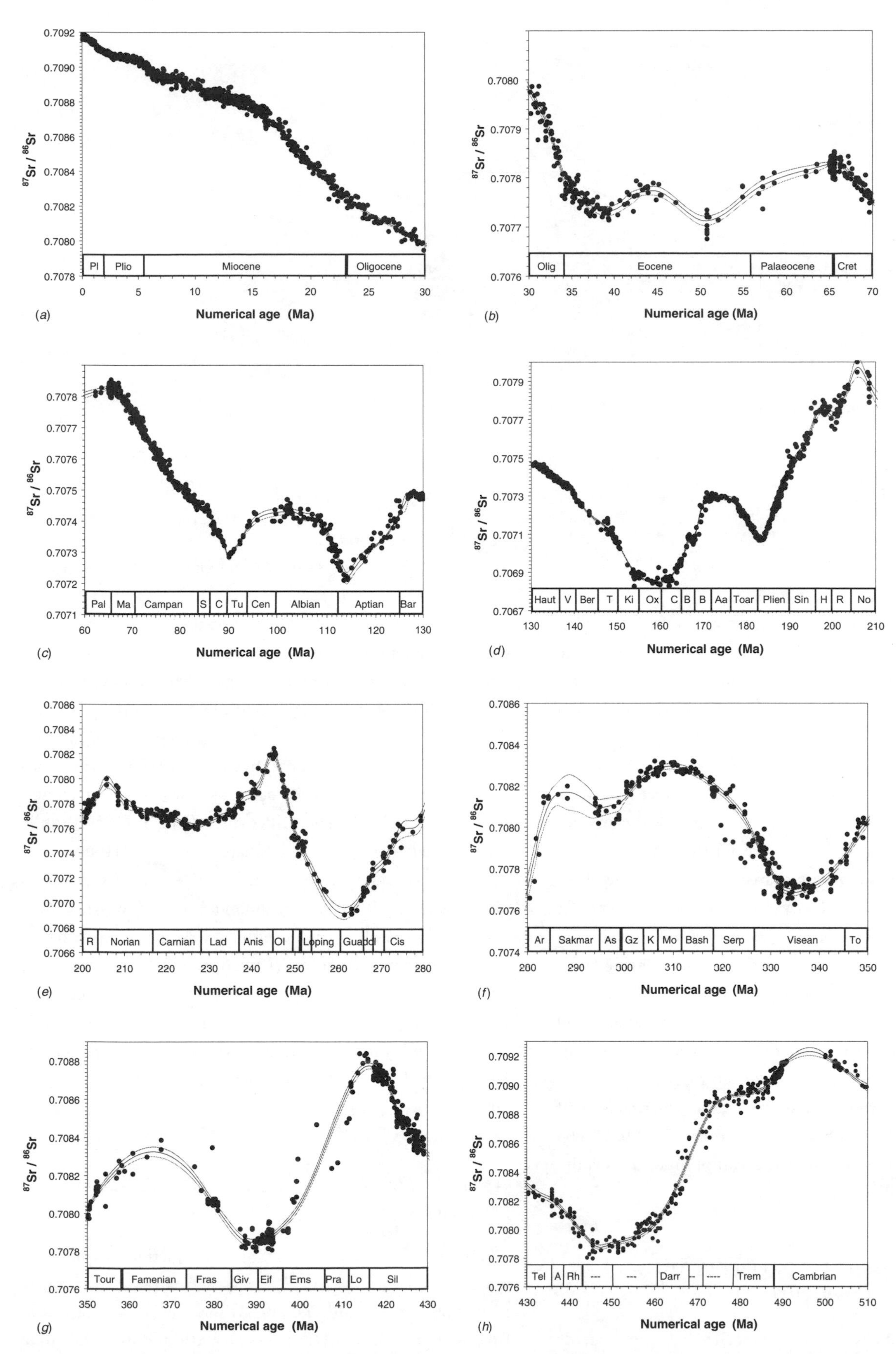

Figure 7.2 Details of the variation of $^{87}Sr/^{86}Sr$ through time given in Fig. 7.1, showing width of 95% confidence intervals. See text for a discussion of its parts.

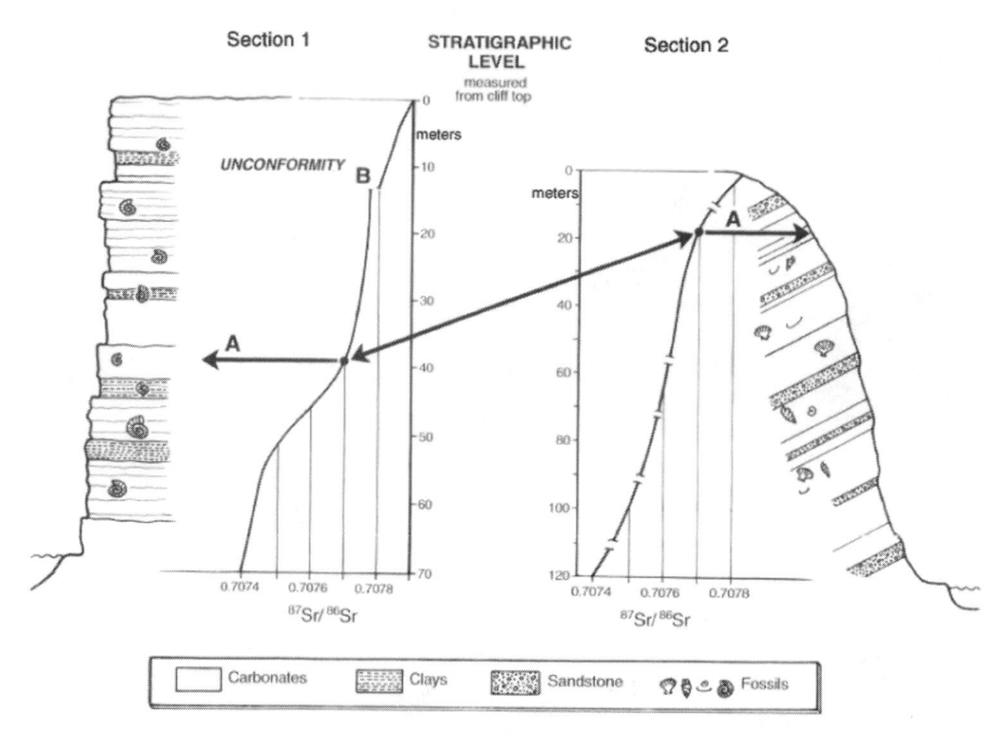

Figure 7.3 Correlation with $^{87}Sr/^{86}Sr$. Values of $^{87}Sr/^{86}Sr$ are matched between $^{87}Sr/^{86}Sr$ profiles constructed for independent sections.

The method works only for marine minerals. Practitioners assume that the oceans are homogenous with respect to $^{87}Sr/^{86}Sr$ and always were so. Uniformity of $^{87}Sr/^{86}Sr$ is expected for two reasons. First, because the residence time of Sr in the oceans ($\approx 10^6$ years) is far longer than the time it takes currents to mix the oceans ($\approx 10^3$ years), so the oceans are thoroughly mixed on time scales that are short relative to the rates of gain and loss of Sr. Second, because the amount of Sr in the sea (7.6 μg/l) is much greater than the amount in rivers (variable, but two orders of magnitude less), the effect of riverine dilution is small; seawater maintains a $^{87}Sr/^{86}Sr$ value that is characteristic of the open ocean until it is diluted to salinities well below those supportive of fully marine fauna (Andersson *et al.*, 1992). Tests of the homogeneity of $^{87}Sr/^{86}Sr$ in modern open oceans and some restricted seas (DePaolo and Ingram, 1985; Andersson *et al.*, 1992; Paytan *et al.*, 1993) confirm that it appears to be homogenous at an analytical precision of ±0.000 020. Since those studies were done, the precision of measurement of $^{87}Sr/^{86}Sr$ has improved to around ±0.000 003 for replicate determinations, so the assumption of uniformity now requires re-evaluation.

7.2 MATERIALS FOR STRONTIUM ISOTOPE STRATIGRAPHY

Of the materials that have been used for SIS, belemnite guards (Jones *et al.*, 1994a,b; McArthur *et al.*, 2000) and brachiopod shells (Veizer *et al.*, 1999) have proven useful, since they resist diagenetic alteration better than other forms of biogenic calcite. Early diagenetic marine carbonate cements have made an important, if volumetrically minor, contribution to SIS calibration in the lower Phanerozoic (Carpenter *et al.*, 1991). Foraminiferal calcite, largely from DSDP/ODP sites, has yielded the curve for the Neogene (see works of Farrell and others, Hodell and others, Miller and others; Table 7.1), while acid-leached, nannofossil-carbonate ooze (McArthur *et al.*, 1993), inoceramids (Bralower *et al.*, 1997), atoll carbonates (Jenkyns *et al.*, 1995), and ammonoid aragonite (McArthur *et al.*, 1994) have all yielded useful data. Attempts to use barite have met with mixed success (Paytan *et al.*, 1993; Martin *et al.*, 1995; Mearon *et al.*, 2003). The use of conodonts seems to work for samples with a color-alteration index of around 1, which implies minimal alteration (Martin and Macdougall, 1995; Ruppel *et al.*, 1996; Ebneth *et al.*, 2001; Korte *et al.*, 2003).

7.3 *A GEOLOGIC TIME SCALE 2004* (GTS2004) DATABASE

The standard curve of $^{87}Sr/^{86}Sr$ as a function of time presented here (Figs. 7.1 and 7.2, and available in a supplementary tabulation from the authors: j.mcarthur@ucl.ac.uk) is updated from that given in McArthur *et al.* (2001) and uses 3875 data pairs from the sources listed in Table 7.1. This table also gives

Table 7.1 *Sources of data used for the LOWESS fit*

Author	Normalizer ($\times 10^6$)	Age range (Ma)	
Azmy *et al.* (1999)	30	417	443
Banner and Kaufman (1994)	−3	337	342
Barrera *et al.* (1997): Site 463; new age model	13	67.7	74.5
Bertram *et al.* (1992): conodonts	11	420	435
Bralower *et al.* (1997): inoceramids	0	94.7	116.3
Brand and Brenckle (2001): some	0	319	319
Bruckschen *et al.* (1999)	30	325	359
Callomon and Dietl (2000)	0	164.2	164.4
Carpenter *et al.* (1991)	3	379	380
Clemens *et al.* (1993, 1995)	−6	0.0	0.2
Cummins and Elderfield (1994): brachiopods	11	327	331
Denison *et al.* (1993)	102	45.6	65.3
Denison *et al.* (1994)	102	257	360
Denison *et al.* (1997)	102	364	438
Denison *et al.* (1998)	102	445	510
DePaolo and Ingram (1985): Palaeogene	−59	38.1	65.65
Diener *et al.* (1996): brachiopods	30	378	400
Ebneth *et al.* (2001): Texas and Australia	33	487.5	491
M. Engkilde, *pers. comm.* (1998)	0	144.2	175.9
Farrell *et al.* (1995): pruned 2.8–3.5 Ma	−9	0.0	7.0
Henderson *et al.* (1994)	17	0.0	0.37
Hodell *et al.* (1991): Holes 588 and 588A	18	7.3	18.3
Hodell and Woodruff (1994): new age model	18	10.9	23.3
Jenkyns *et al.* (1995)	−12	99.7	125.2
Jones *et al.* (1994a,b): some	22	100.8	199.7
Koepnick *et al.* (1990)	102	201	251
Korte *et al.* (2003)	25	200	252
Martin and Macdougall (1995)	−12	248	295
Martin *et al.* (1999): <13.8 Ma	22	5.0	13.8
McArthur *et al.* (1993)	0	71.1	89.1
McArthur *et al.* (1993)	0	69.4	84.3
McArthur *et al.* (1994)	0	71.4	99.1
McArthur *et al.* (1998)	0	65.4	68.1
McArthur *et al.* (2000)	0	177.1	189.6
McArthur and Kennedy (unpub. data)	0	96.0	107.8
McArthur and Morton (2000)	0	170.8	174.2
McArthur *et al.* (in review)	0	2.6	3.6
McArthur *et al.* (2004)	0	124.2	133.4
McArthur and Janssen (unpub. data)	0	133.6	138.4
Mead and Hodell (1995)	18	18.6	45.3
Miller *et al.* (1988)	−14	22.9	33.7
Miller *et al.* (1991)	−14	9.1	24.1
Montañez *et al.* (1996)	−6	500	505
Oslick *et al.* (1994): >16 Ma	−14	16.0	25.0
Qing *et al.* (1998): <463 Ma	5	420	489
Ruppel *et al.* (1996)	−7	417	442
Sugarman *et al.* (1995)	−14	65.7	72.4
Zachos *et al.* (1992, 1999)	0	23.2	42.2

the amount added to, or subtracted from, the literature data in order to correct for apparent inter-laboratory bias in measurement of $^{87}Sr/^{86}Sr$. Such bias is assumed to represent systematic error and to be correctable by such a normalization process. As replication of $^{87}Sr/^{86}Sr$ measurement can give mean values precise to ±0.000 003 (Jones *et al.*, 1994a; McArthur *et al.*, 2001), inter-laboratory bias must be quantified to this precision if SIS is to realize its full potential.

We correct data to a common value of 0.710 248 for standard reference material NIST-987 (formerly known as SRM-987) or a value of 0.709 175 for EN-1, a modern *Tridachna* clam from Enewetak Atoll (prepared by the USGS). Some older work is normalized to a $SrCO_3$ standard, known as "E and A," that was prepared by the Eimer and Amend company (New York; now owned by Fisher Chemical). The $^{87}Sr/^{86}Sr$ value of E and A is 0.708 022 ± 4 (2 s.e., $n = 34$) relative to a value for NIST-987 of 0.710 248 (Jones *et al.*, 1994a). In a few cases, our normalizer is based on evaluation of inter-laboratory bias that is independent of the published source of data, so it may be different from that given in the source.

7.3.1 Numerical ages

The revised SIS calibration curve given here (Figs. 7.1 and 7.2) uses the GTS2004 time scale of this volume. Where original data were reported to other time scales, the original ages have been converted to the current time scale using the formulae of Wei (1994) and the nearest pair of numerically dated stratigraphic tie points. In some instances, where local or regional stratigraphy has advanced since the publication of a source, we have revised the age models used in original publications.

The calibration curve shown in Figs. 7.1 and 7.2 is based on measurement of $^{87}Sr/^{86}Sr$ in samples dated by biostratigraphy, magnetostratigraphy, and astrochronology (mostly the first two). The difficulty of assigning numerical ages to sedimentary rocks by the first two methods is well known. Users of the calibration curve, and the equivalent look-up tables derived from it that enable rapid conversion of $^{87}Sr/^{86}Sr$ to age and vice versa (McArthur *et al.*, 2001), must recognize that the original numerical ages on which the curve is based may include uncertainties derived from interpolation, extrapolation, and indirect stratigraphic correlations and may suffer from problems of boundary recognition (both bio- and magnetostratigraphic), diachroneity, and assumptions concerning sedimentation rate, all of which contribute uncertainly to the age models used to generate the calibration line. Furthermore, age models are ultimately based (mostly) on radiometric dates and are as accurate as those dates. Interpolation of ages between tie points, however, may be more precise, although necessarily systematically inaccurate.

7.3.2 Fitting the database

We used the statistical non-parametric regression method LOWESS (LOcally WEighted Scatterplot Smoother of Cleveland, 1979, 1981; Chambers *et al.*, 1983; Thisted, 1988; Cleveland *et al.*, 1992) to obtain a best-fit curve for the $^{87}Sr/^{86}Sr$ data as a function of time. Details of the fitting procedure are given in Howarth and McArthur (1997). Because of the complex shape of the fit, and the very uneven density of data points through time, the curve was optimized by being fitted in 29 overlapping local segments. These were then joined using splines at segment junctions. To obtain a table for predicting age from $^{87}Sr/^{86}Sr$, and the lower and upper confidence limits on the age, we used inverse interpolation of the fitted curve of $^{87}Sr/^{86}Sr$ and its 95% confidence intervals (CIs) as a function of age. The complete table is available from j.mcarthur@ucl.ac.uk.

7.3.3 The quality of the fit

CONFIDENCE LIMITS ON THE LOWESS FIT

In addition to the best-fit curve of estimated $^{87}Sr/^{86}Sr$ as a function of age, the LOWESS fitting process also provides a two-sided, 95% CI, on the estimates of age. These CIs are included in Fig. 7.2, but are best seen in Fig. 7.4 as a half-width interval plotted against time. The width of the CI varies with numerical age, and is dependent on both the density and spread of the calibration data. For substantial segments of the Mesozoic, values approach ±0.000 005 and are seldom more than ±0.000 010. Where data are abundant and samples well preserved, e.g. 0–7 Ma, the half-width CI is around ±0.000 003. Where data are few, e.g. most of the Permian, the uncertainty is much greater. Well-preserved samples become rarer with increasing age so the uncertainty envelope increases with age; nevertheless, achieving a precision of ±0.000 015 for the entire Paleozoic is not an unrealistic goal.

Assuming that the half-widths of the upper and lower confidence intervals are approximately equal, and that U = (upper age CI − estimated age) and L = (estimated age − lower age CI), then the overall uncertainty on an age derived from the curve can be computed by combining the uncertainties on the measurement and the fitted curve as follows:

$$s_{\text{total}} = \left(s_m^2 + s_c^2\right)^{1/2}, \quad (7.1)$$

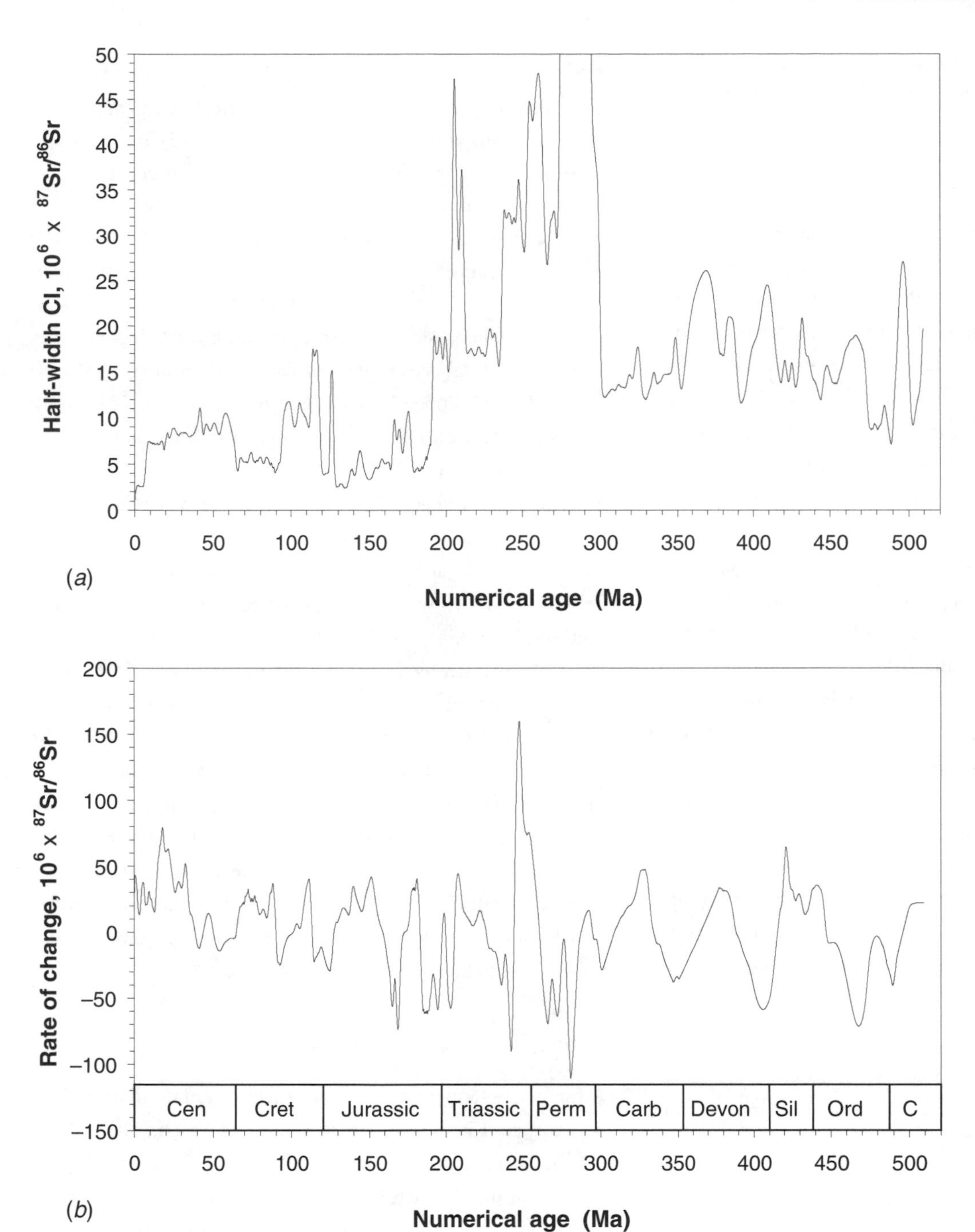

Figure 7.4 (*a*) Half-width of the 95% confidence intervals on the LOWESS fit, and (*b*) rate of change with time of $^{87}Sr/^{86}Sr$.

where s_m is the standard deviation of the estimated numerical age of the sample and $s_c = ((L + U)/2)/1.96$. If L and U are different enough, then it may be preferable to use upper and lower bounds for s_{total} by replacing s_c with L/1.96 and U/1.96, respectively.

Inter-laboratory bias introduces an additional source of uncertainty, since few laboratories have reported precise estimates of the $^{87}Sr/^{86}Sr$ value of their standards, i.e. a comparison of mean $^{87}Sr/^{86}Sr$ in NIST-987 and EN-1, each given to a precision of less than ±0.000 005. The uncertainty from this source can be as high as 0.000 020. As a consequence, inter-laboratory bias may limit the quality of dates and correlations determined using SIS.

CONFIDENCE LIMITS ON MEASURED $^{87}Sr/^{86}Sr$

The uncertainty with which the mean (m) $^{87}Sr/^{86}Sr$ of a sample is known, from n independent determinations of $^{87}Sr/^{86}Sr$, may be quantified if one assumes that the measurement errors are normally distributed and so a two-sided confidence interval applies:

$$m \pm t_{1-\alpha/2,\ n-1}(s/n^{1/2}), \quad (7.2)$$

where s is the standard deviation of n observed $^{87}Sr/^{86}Sr$ values, and $t_{1-\alpha/2,n-1}$ is the $100(1-\alpha/2)$th percentile of Student's t-statistic with $(n-1)$ degrees of freedom; α is the risk (specified as a proportion) that the true (but unknown) value of $^{87}Sr/^{86}Sr$ in the mineral, which is estimated by m, will fall outside the specified confidence interval. Thus α is commonly set to 0.05 (5%) in order to obtain two-sided 95% confidence limits on m. The t-statistic is used for this purpose rather than the $100(1-\alpha/2)$th percentile of the cumulative normal distribution in order to correct for the fact that the number of replicate determinations of $^{87}Sr/^{86}Sr$ is finite. Increasing n decreases the uncertainty in m. For example, the multipliers for two-sided 95% confidence limits when $n = 2, 3, 4, 5$, and 10 are 12.71, 4.30, 3.18, 2.78, and 2.26, respectively.

It may be possible to obtain only a single determination of $^{87}Sr/^{86}Sr$ (x) for a given mineral sample. If, for some reason, there exists a *prior* estimate of the expected value of $^{87}Sr/^{86}Sr$ (a), e.g. from measurements previously made on presumed similar material, or the ratio has been estimated from the $^{87}Sr/^{86}Sr$ curve and a knowledge of the sample's stratigraphic position, then, assuming x is the centre of a normal distribution, Blachman and Machol (1987) showed that a two-sided $100(1-\alpha)$% confidence interval on x is given by:

$$x \pm (1 + 0.484/\alpha)|x - a|. \quad (7.3)$$

If α is 0.05, the multiplier equals 9.68. Less-conservative bounds are obtained by inverting the prediction interval for a single future observation. This gives:

$$x \pm z_{1-\alpha/2}(1 + 1/n_0)^{1/2}s_0. \quad (7.4)$$

In this case, s_0 is a *prior* estimate of the standard deviation of the distribution (assumed normal) from which x is drawn, e.g. the pooled standard deviation based on n_0 sets of previous determinations of similar samples, and $z_{1-\alpha/2}$ is the $100(1-\alpha/2)$th percentile of the cumulative normal distribution. If α is 0.05, the multiplier equals 1.96.

The best analytical precision on $^{87}Sr/^{86}Sr$ that has so far been obtained by repeated replicate measurements of $^{87}Sr/^{86}Sr$ over a period of time is ±0.000 003 (2 s.e.; Jones *et al.*, 1994a; McArthur *et al.*, 2001). Figure 7.4 shows that little of the global Sr curve has yet been defined to this degree of precision.

NUMERICAL RESOLUTION

The uncertainty of an estimated numerical age obtained using the calibration curve (Figs. 7.1 and 7.2) depends on: (i) the width of the 95% CI on the calibration curve, (ii) the uncertainty with which the $^{87}Sr/^{86}Sr$ of a sample is known, and (iii) the slope of the curve. Given that the best-defined parts of the calibration curve have half-width CIs no better than ±0.000 003, and that this is also the best-attainable precision on measurement of $^{87}Sr/^{86}Sr$, application of (i) above gives a minimum total uncertainty in dating of around 0.000 004. Given that the slope of the calibration curve (Fig. 7.4) rarely exceeds a value of 0.000 060 per myr, it follows that the precision in dating with $^{87}Sr/^{86}Sr$ will not be better than about ±0.1 myr and will generally be much worse. Correlation with $^{87}Sr/^{86}Sr$ avoids the uncertainty involved in assigning numerical ages and the accuracy with which it can be accomplished depends upon: (i) the sedimentation rate of the sequences, or sections, being correlated; (ii) the uncertainty with which the $^{87}Sr/^{86}Sr$ of a sample is known; and (iii) the slope of the curve.

Under optimum conditions, e.g. with well-preserved material and where the rate of change with time of marine $^{87}Sr/^{86}Sr$ is steep, the precision with which SIS can date and/or correlate marine strata can surpass foraminiferal biostratigraphy in the Cenozoic and ammonite biostratigraphy in the Mesozoic. The utility and accuracy of SIS declines as the target rocks get older, since the method relies on analysis of well-preserved samples, mostly biogenic calcite, and these become less common and more likely to be altered as age increases.

RUBIDIUM CONTAMINATION

A further potential problem arises from the fact that samples may contain Rb, the radioactive isotope of which, ^{87}Rb, decays to ^{87}Sr, thereby altering the $^{87}Sr/^{86}Sr$ of even a perfectly preserved sample. The Rb^+ ion is too large, and its marine abundance too low, for it to be found in worrying amounts in biogenic calcite, but the larger cation site in aragonite will accommodate Rb more easily. Rubidium should therefore be monitored in all samples, if the Sr/Rb weight (ppm) ratio is >8000, samples of Phanerozoic age will have altered their $^{87}Sr/^{86}Sr$ by <0.000 003. As a rule of thumb, concentrations of Rb above 0.1 ppm may require a correction to their $^{87}Sr/^{86}Sr$. A table for making such Rb corrections can be found in McArthur (1994).

7.4 COMMENTS ON THE LOWESS FIT

Some details of the GTS2004 database and fit require comment.

Pliocene to now For the period from 0 to 7 Ma, we rely mostly on the data of Farrell *et al.* (1995) excepting between 2.8 and 3.4 Ma. In this interval, these authors' data are too high by up to 0.000 020 and have been pruned and supplemented by $^{87}Sr/^{86}Sr$ data from the astronomically calibrated Pliocene type section at Punta Piccola, Sicily, (McArthur *et al.*, unpub.)

Early Miocene In the LOWESS fit of McArthur *et al.* (2001) the high scatter of data between 22 and 24 Ma was attributed to the effects of diagenetic alteration of the material analyzed. We now believe that explanation to be incorrect and that the scatter results from an uncertain age model for DSDP Site 588C (Hodell *et al.*, 1991), so we no longer use data for 588C; we retain data for Sites 588 and 588A after updating the age models to this time scale. For Site 289 (Hodell and Woodruff, 1994), we use a revised age model that includes breaks in the sequence between 522 and 544 meters below seafloor (mbsf).

Paleogene The $^{87}Sr/^{86}Sr$ curve for the Paleogene shows sufficient slope for it to be potentially useful for dating (Fig. 7.2*b*). From the K–P boundary (65.5 Ma) value of 0.707 83, $^{87}Sr/^{86}Sr$ declines to 0.707 72 in the Ypresian (51Ma) before rising sharply to a maximum of 0.707 78 in the early Lutetian (46 Ma) and then declining again to a second minimum of 0.707 73 in the earliest Bartonian (40 Ma). Thereafter, the ratio increases steeply until modern times. For the Paleocene, the rate of decrease in $^{87}Sr/^{86}Sr$ of around 0.000 08 per myr should give a resolution in dating no better than 0.5 myr, and then only if both curve and $^{87}Sr/^{86}Sr$ measurement of the sample achieve the best-attainable precision. Given that this quality has been achieved in the late Neogene (Farrell *et al.*, 1995), Late Cretaceous (McArthur *et al.*, 1993), early Jurassic (Pliensbachian–Toarcian, McArthur *et al.*, 2000) and Hauterivian times (McArthur *et al.*, 2004), it seems possible to do so.

Maastrichtian For the Late Maastrichtian interval, we use the data of Sugarman *et al.* (1995) and Barrera *et al.* (1997), for DSDP Site 463, the latter after recalibration to the age model of Li and Keller (1999).

Aptian–Albian Of Bralower *et al.*s' (1997) data, we use only that for inoceramids. We have adjusted the Albian boundary ages of Bralower *et al.* (1997) to those in this volume, but retain his apportionment of time between them. The scaling given in this volume distorts the Sr isotope curve in an unreasonable way (cf. Fig. 7.5*a*,*b*), increasing its slope in the basal-Albian,

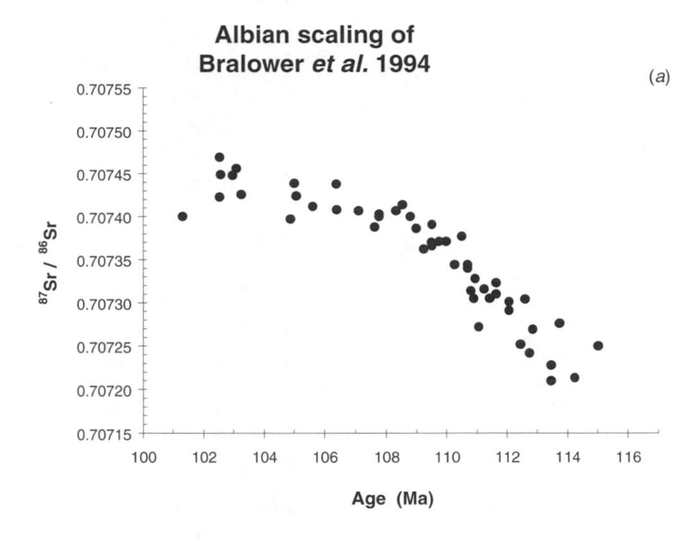

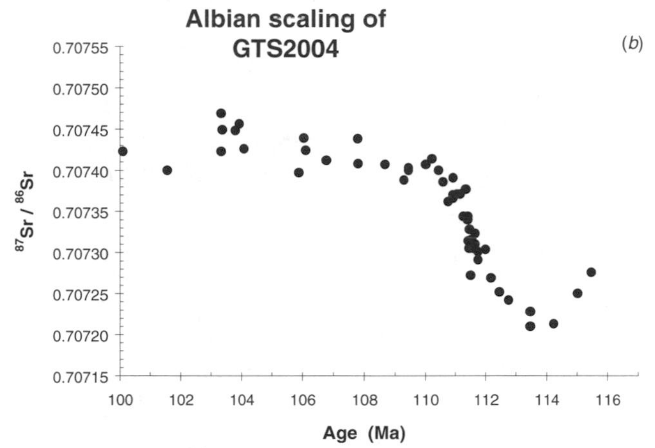

Figure 7.5 Trend in $^{87}Sr/^{86}Sr$ in the Albian–Aptian: data of Bralower *et al.* (1997) for inoceramids plotted against (*a*) the biozonation given by the authors and (*b*) that given in this volume. The age scaling of Bralower *et al.* (1997) is used for the construction of the LOWESS fit.

introducing a break-in-slope at the Aptian–Albian boundary, and stretching data into a plateau above 110 Ma.

Jurassic The data of Jones *et al.* (1994a,b) for the Toarcian, Berriasian, Valanginian, Hauterivian, and Barremian stages are replaced by data in McArthur *et al.* (2000), McArthur *et al.* (2004), and McArthur and Janssen (unpub. data). For the Valanginian and Berriasian, these data have been assigned numerical ages based on equal zone duration, excepting that the uppermost two Valanginian zones (of *Criosarasinella furcillata* and *Neocomites peregrinus*) are allotted a duration that is one half that of other zones. For the Hauterivian and Barremian time, ages are assigned on the basis of a polynomial model that assigns a linear increase of $^{87}Sr/^{86}Sr$ with time through the Hauterivian and lowermost Barremian, and a maxima and

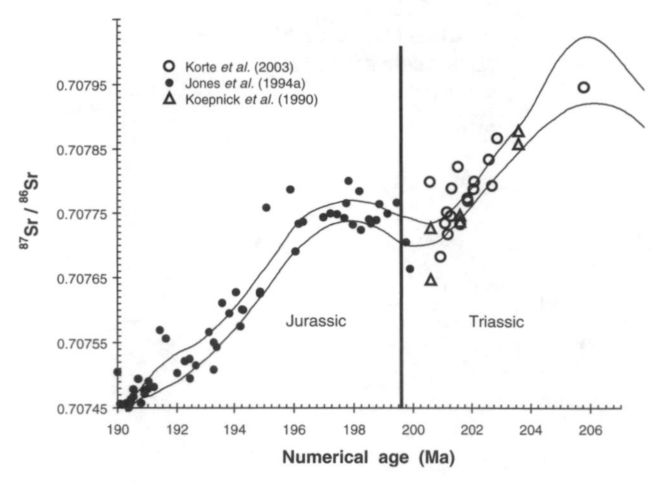

Figure 7.6 Trend in $^{87}Sr/^{86}Sr$ at the Triassic–Jurassic boundary and the accompanying LOWESS fit, showing the 95% confidence intervals on the mean line.

downturn in $^{87}Sr/^{86}Sr$ through the *fissicostatum, elegans*, and *denkmanni* ammonite zones (McArthur *et al.*, 2004).

Triassic Two Rhaetian data are from Jones *et al.* (1994a). The rest are from Koepnick *et al.* (1990) and Korte *et al.* (2003). The peaks in the earliest and latest Triassic are shown in both the second and third sets of data, but as the second data set is based on analysis of whole rocks, and the last (around the peaks) is based on conodonts, which do not preserve their isotope ratios well, the amplitudes of the peaks may be enhanced by artifacts of diagenesis. At the Triassic–Jurassic boundary there is a possible mismatch (Figs. 7.2*d* and 7.6) between the trend of the UK data of Jones *et al.* (1994a), which is calibrated with ammonites, and the trend of the data of Korte *et al.* (2003) and Koepnick *et al.* (1990), which is calibrated with conodonts. The putative mismatch may result from problems of integrating these different biostratigraphic schemes; nevertheless, the two Rhaetian samples of Jones *et al.* (1994a) fall on the Triassic trend of the other authors, so it may be real and we have honored the data in making a fit through this interval (Fig. 7.6). The steepness of the rise in $^{87}Sr/^{86}Sr$ in the very earliest Triassic, and the apparent break-in-slope of the $^{87}Sr/^{86}Sr$ curve at the Permian–Triassic boundary (Fig. 7.2*e*), might be revealing an undue compression of the time scale in the basal Triassic.

Permian The Ochoan data of Denison *et al.* (1994) and the data for the latest Permian given by Martin *et al.* (1995) differ by up to 3.5 myr around the Permian–Triassic boundary. We include both data sets, despite the decrease in precision this causes in the fit, because the differences are certainly caused by problems of age assignment and correlation. As with some Carboniferous data, $^{87}Sr/^{86}Sr$ might here be used to correlate the different sections used by these authors, rather than be composited to form a global curve.

Carboniferous The Carboniferous data rely heavily on that of Bruckschen *et al.* (1999) but those data show a large spread, especially in the Serphukovian and Visean Stages; even after extremes flyers are ignored, values of $^{87}Sr/^{86}Sr$ range from 0.707 637 to 0.707 805 around 332 Ma. Bruckschen *et al.*s' (1999) data from Germany group more tightly than do their data from Belgium, and are mostly higher (by about 0.000 070). The high spread of the data increases the width of the confidence interval in this part of the Carboniferous. The $^{87}Sr/^{86}Sr$ data might be better used to refine correlation, particularly between the USA and Europe, than be used to construct a global standard curve. Finally, much of the data of Denison *et al.* (1994) for the late Carboniferous is some 0.000 100 higher in $^{87}Sr/^{86}Sr$ than data given in Bruckschen *et al.* (1999), so we use only the latter in our fit from 304 to 315 Ma.

Ordovician The trend in $^{87}Sr/^{86}Sr$ across the Cambrian–Ordovician boundary differs markedly between data sets (Fig. 7.7). The least slope is seen in the data of Gao and Land (1991) for the Arbuckle Limestone of Oklahoma. The differences may be caused by incompatible local age models. It is important to determine the trend because different trends necessitate different rates of decline in $^{87}Sr/^{86}Sr$ in the mid Ordovician (*sensu lato*), and the steeper ones having been

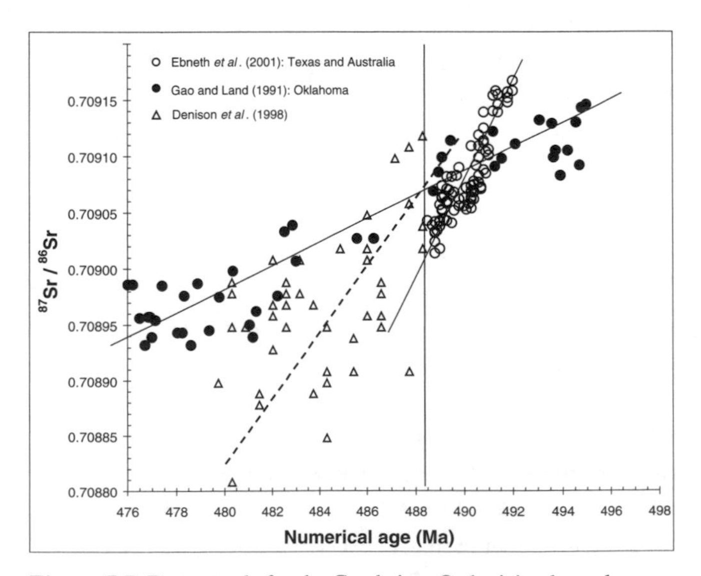

Figure 7.7 Data trends for the Cambrian–Ordovician boundary, showing the differences between authors.

interpreted as reflecting major geologic events (Shields *et al.*, 2003). The data of Qing *et al.* (1998) show a step around the Late–Middle Ordovician boundary of 0.708 041 to 0.708 698 over <1 myr, which must be a stratigraphic artifact. A lower, but still steep, decrease of $^{87}Sr/^{86}Sr$ starts in the late Darriwilian in the data of Shields *et al.* (2003); the decrease is less steep again for the data of Denison *et al.* (1998), probably because it starts in the early Darriwilian (equivalent). Where unusually steep rates of change of $^{87}Sr/^{86}Sr$ with time have been noted before, they have diminished with improved correlation, improved age models, or identification of altered samples. We have chosen to use Ebneth *et al.* (2001) for the Cambrian–Ordovician boundary interval and continue the Early and Middle Ordovician trend using the data of Denison *et al.* (1998), because that data relate to a few well-studied localities in the USA, and result in the least-steep decline through the Ordovician. The data of Shields *et al.* (2003) and Qing *et al.* (1998) supplement Denison *et al.* (1998) in the Late Ordovician.

Data gaps Finally, Fig. 7.2 reveals a paucity of reliable data for many intervals of time (the late Albian to Turonian, most of the Kimmeridgian and Berriasian, the Bajocian and Bathonian, much of the Permian and the Devonian, some of the Ordovician, and most of the Cambrian). This lack is reflected in the large (>0.000 015) half-width of the confidence interval on the mean for the LOWESS fit (Fig. 7.4); to reduce this uncertainty substantially will require some three–five accurate and precise $^{87}Sr/^{86}Sr$ values per biozone.

8 • Geomathematics

F. P. AGTERBERG

The inputs for the calculation of a numerical time scale are a set of radiometric dates with variable uncertainty in both time (in myr) and stratigraphic position (in biozones). For the Paleozoic and Mesozoic, these selected input dates are irregularly distributed with respect to a biostratigraphic scale derived from graphical correlation, constrained optimization, or successive biozonal units; and for the Cenozoic, the dates are correlated to relative seafloor distances of marine magnetic anomalies. Spline fitting combined with Ripley's Maximum Likelihood fitting of a Functional Relationship produces a linear time scale with error bars on the boundary ages and durations of geologic stages. These methods were applied to Ordovician–Silurian, Devonian, Carboniferous–Permian, the Permian–Triassic boundary, Late Cretaceous, and Paleogene date sets.

8.1 HISTORY AND OVERVIEW

8.1.1 Statistical estimation of chronostratigraphic boundary ages (with error bars)

The starting point for construction of numerical time scales, as described in this chapter, is a data set of ages, measured in millions of years, with 2-sigma error bars, for samples positioned along a relative stratigraphic scale of which the unit is approximately proportional to time (also measured in millions of years). Spline-curve fitting is used to relate the observed ages to their stratigraphic position. During this process the ages are weighted according to their variances based on the lengths of their error bars. A chi-square test can be used for identifying and reducing the weights of the relatively few outliers exhibiting error bars that are much narrower than expected on the basis of most ages in the data set.

Stratigraphic uncertainty is incorporated in the weights assigned to the observed ages during spline-curve fitting. In the final stage of analysis, Ripley's Maximum Likelihood fitting of a Functional Relationship (MLFR) algorithm is used for error estimation, resulting in 2-sigma error bars for the estimated chronostratigraphic boundary ages and stage durations. The method is applied to Ordovician–Silurian, Devonian, Carboniferous–Permian, Late Cretaceous, and Paleogene data sets.

In this chapter, previous methods used to construct time scales are briefly reviewed. The approach adopted by GTS2004 places more emphasis on the relative stratigraphic position of the few samples for which precise age dates are available. It is based on methods, Sections 8.1.3–8.1.4, of straight-line construction applied to more or less homogeneous data sets previously developed for Paleozoic periods. Application of the method to dating specific boundaries is addressed in Sections 8.2–8.3. The chapter ends with a summary of results (Section 8.4). Appendix 3 provides previously undocumented mathematical aspects of the approach.

The weights assigned to the age dates are inversely proportional to their variances. The latter are proportional to the squares of the lengths of the 2-sigma error bars. Consequently, an age date with an error bar that would be 10 times smaller would receive 100 times as much weight during the statistical analysis. This is one of the main reasons for testing all ages in the same data set for homogeneity. This aspect is emphasized in Section 8.2.

The Late Cretaceous and Paleogene applications primarily use $^{40}Ar/^{39}Ar$ age dates for which it is important to distinguish between external and internal uncertainty. For example, the Late Cretaceous data set uses $^{40}Ar/^{39}Ar$ ages with published errors that do not consider uncertainties associated with measurement of the K–Ar age of the fluence monitor or errors related to the determination of decay constants. Although the data set is homogeneous from a statistical point of view, uncertainties of both the slope and the intercept of the final best-fitting straight line have been increased to account for external uncertainties. This type of adjustment and others are discussed in more detail in Section 8.3.

Data tables with zonal assignments and age dates for each of the periods on which the calculations were carried out are presented in Chapters 12–17, 19, and 20.

A Geologic Time Scale 2004, eds. Felix M. Gradstein, James G. Ogg, and Alan G. Smith. Published by Cambridge University Press.

8.1.2 History

Geological time scales are constructed by combining stratigraphic information with radiometric dates and their standard deviations. The stratigraphic record to be used includes litho-, bio-, chrono-, cyclo-, and magnetostratigraphy. Statistical methods should embody concepts and data available for the periods considered. For the construction of several previous time scales, ages of stage boundaries were estimated by application of the chronogram method (Harland *et al.*, 1982, 1990), or maximum likelihood (Gradstein *et al.*, 1994a, 1995), to a world-wide database of chronostratigraphically classified dates. These methods, which can also be applied to more closely spaced zone boundaries (Pálfy *et al.*, 2000a), resulted in age estimates accompanied by approximate 95% confidence intervals. A final time scale was obtained by calibration using graphical and curve-fitting methods including cubic smoothing splines.

New challenges for construction of the GTS2004 numerical time scale include:

1. use of new age determinations that can be an order of magnitude more precise than earlier dates,
2. consideration of information on external versus internal uncertainty of age determinations especially for $^{40}Ar/^{39}Ar$ dates, and
3. astronomical calibration of parts of the time scale by cycle-tuning

Odin (1994) discussed three separate approaches to numerical time scale construction: statistical, geochronological, and graphical methods. Gradstein *et al.* (1994a, 1995) used all three approaches in a step-wise procedure involving maximum likelihood, use of stratigraphically constrained dates, and recalibration by curve fitting.

The chronogram method used in Harland *et al.* (1982, 1990) is suitable for estimation of the age of chronostratigraphic boundaries from a radiometric database when most rock samples used for age determination are subject to significant relative uncertainty. Inconsistencies in the vicinity of chronostratigraphic boundaries then can be ascribed to imprecision of the age determination method.

Cox and Dalrymple (1967) originally developed an approach for estimating the age of Cenozoic chron boundaries from inconsistent K–Ar age determinations of basaltic rocks. Harland *et al.* (1982, 1990) adopted this method in their calculations of ages of stage boundaries for the 1982 and 1990 time scales. The basic principle of this approach is as follows.

Assuming a hypothetical trial age for an observed chronostratigraphic boundary, rock samples from above this boundary should be younger, and those below it should be older. An inconsistent date is either an older date for a rock sample known to be younger than the trial date, or a younger date for a sample known to be older. The difference between each inconsistent date and the trial age can be standardized by dividing it by the standard deviation of the inconsistent date. Thus relatively imprecise dates receive less weight than more precise dates.

The underlying assumptions are that: (1) the rock samples are uniformly distributed along the time axis, and (2) the error of each date satisfies a normal (Gaussian) error distribution with standard deviation equal to that of the age determination method used.

Standardized differences between inconsistent dates and trial age can be squared and the sum of squares (written as E^2) determined for all inconsistent dates corresponding to the same trial age. Chronograms constructed by Harland *et al.* (1982) were basket-shaped plots of E^2 against different trial ages spaced at narrow time intervals. The optimum choice of age was at the trial age where E^2 was a minimum.

Agterberg (1988) made the following improvement to this method. In addition to inconsistent dates, there are generally more consistent dates for any trial age selected for a chronostratigraphic boundary. The statistical maximum likelihood method can be used to combine consistent with inconsistent dates resulting in an improved estimate of the age of the chronostratigraphic boundary considered. Each standardized difference with respect to a trial age was interpreted as the fractile of the normal distribution in standard form, and transformed into its corresponding probability. Summation of the logarithmically transformed probabilities then yields the log-likelihood value of the trial date. In this type of calculation, inconsistent dates receive more weight than consistent dates. Consequently, the improvement resulting from using consistent dates, in addition to the inconsistent dates, is relatively minor. The log-likelihood function is beehive-shaped. For examples, see Gradstein *et al.* (1995).

A general disadvantage of chronogram methods is that the relative stratigraphic position of the sample is generalized with respect to stage boundaries that are relatively far apart in time. The relative stratigraphic position of one sample with respect to another within the same stage is not considered. A better approach is to incorporate relative stratigraphic positions of fewer samples for which precise age determinations are available.

8.1.3 Spline-curve fitting with consideration of stratigraphic uncertainty

The objective of the statistical analysis described in this chapter is to combine age determinations with stratigraphic information in order to estimate the ages of chronostratigraphic boundaries, together with their 2-sigma error bars. Additionally, the approach can be used for calculating error bars of stage or zone durations.

The first stage of the approach consists of fitting a cubic smoothing spline curve according to the method previously described in detail for the Mesozoic time scale in Gradstein *et al.* (1994a, 1995) and Agterberg (1988, 1994). Age determinations are plotted in the vertical direction (along the y-axis) against relative stratigraphic position (x-axis). Each age determination is weighted according to the inverse of its variance corresponding to the published 2-sigma (or 1-sigma) error bar. If stratigraphic uncertainty is incorporated, this variance becomes $s_t^2(y) = s^2(x) + s^2(y)$ instead of $s^2(y)$.

A cubic-smoothing spline $f(x)$ is fully determined by n pairs of values (x_i, y_i), the standard deviations of the dates $s(y_i)$, and a smoothing factor (SF) representing the square root of the average value of the squares of scaled residuals $r_i = (y_i - f(x_i))/s(y_i)$. If all $s(y_i)$ values are unbiased, SF ≈ 1, or SF is equal to a value slightly less than 1 (cf. Agterberg, 1994, p. 874). If SF significantly exceeds 1, this suggests that some or all of the variances used are too small (under-reported). Thus the spline-fitting method provides an independent method of assessing mutual consistency and average precision of published 2-sigma error bars.

The method of "leave-off-one" cross-validation can be used to determine the optimum smoothing factor. In this method, all observed dates y_i, between the oldest and youngest one, are successively left out from spline fitting with pre-selected trial values of SF. The result is $(n - 2)$ spline curves for each SF tried. The cross-validation value for any SF is the sum of squares of deviations between y_i and estimated values on the $(n - 2)$ spline curves with the same x_i values as y_i. The best SF has the smallest cross-validation value.

It is noted that even if the cross-validation pattern shows a well-developed minimum at a value not close to 1, adoption of the optimum SF value instead of SF = 1 generally constitutes only a minor improvement of the spline curve. In Paleozoic applications (Section 8.2), the optimum SF is consistently greater than 1. Setting SF = 1 would result in slightly more curved spline curves.

Relative stratigraphic position along the x-axis is according to a continuous scale that is the same for all age determinations used. For the Late Cretaceous, bentonite ages originally determined by Obradovich (1993) are used. These occur within ammonite zones of the US Western Interior sedimentary sequence. The position of each sample is plotted along a scale derived by numbering all ammonite zones (with or without bentonites). The x_i values represent positions of bentonites within the ammonite zones that contain them. For example, ammonite zone No. 25 contains bentonite date No. 7, therefore $x_7 = 25$; date No. 15 is reported to occur at the base of zone No. 47, therefore $x_{15} = 47.5$.

Composite standard zonal scales are used for relative stratigraphic position in the Ordovician–Silurian and Carboniferous–Permian data sets. Relative positions of Ordovician and Silurian ages are determined by means of a composite stratigraphic standard scale derived by Cooper and Sadler (cf. Chapters 12 and 13) by application of the CONOP method to graptolite zones. For the Devonian, the optimized zonal scale of M. House (cf. Chapter 14), representing relative stratigraphic ranges of ammonites and conodont taxa within the Devonian, is used. Similar methods are used to obtain dates for other geologic periods.

To some extent, the scale initially used for relative stratigraphic position determines the shape of the final spline curve. A relative stratigraphic scale should be used that is as close as possible to the numerical geologic time scale (in millions of years) except for a linear transformation. Relative stratigraphic scales used in the past include scales based on sediment accumulation corrected for differences in rates of sedimentation, the hypothesis of equal duration of stages (Harland *et al.*, 1982), and the hypothesis of equal duration of biozones (Kent and Gradstein, 1986; Harland *et al.*, 1990; Gradstein *et al.*, 1995).

Unless its unit is consistently proportional to geologic time measured in millions of years, the numerical time scale resulting from spline fitting is not linearly related to the initial relative time scale. It is, however, linearly related to geologic time in millions of years. This allows for gradual changes over time in the original, hypothetical process on the basis of which the initial relative time scale is constructed. For example, deviations from a straight line on a fitted spline curve may represent corrections of changes in sedimentation rate or rate of evolutional change.

Relative stratigraphic position has its own uncertainty. For the Late Cretaceous, the stratigraphic error bar is relatively narrow because it is generally equal to the unit representing the interval between the base and the top of the ammonite zone containing one of the bentonite layers sampled. Because stratigraphic uncertainty is relatively little, statistical treatment of the Late Cretaceous data set is easier than in Paleozoic applications. Stratigraphic uncertainty can also be neglected for the Paleogene data set, where stratigraphic position will be set equal to distance from spreading center in the South Atlantic (Section 8.3.2).

In Paleozoic applications, the stratigraphic error bars can be relatively wide. Stratigraphic uncertainty then adds to the uncertainty expressed by the 2-sigma error bars of the age determinations. A strategy to cope with this additional source of uncertainty will be discussed later.

8.1.4 General comments on straight-line fitting

McKerrow *et al.* (1985) described the following type of method to construct a numerical time scale for the Ordovician, Silurian, and Devonian. Use was made of an iterative construction involving a sequence of diagrams wherein the isotopic age of the sample was plotted along the x-axis and its stratigraphic age along the y-axis. They stated:

> Most graphs are constructed with definite numerical scales along both the x and y axes; this is not the case with fig. 1, where only the horizontal (x) axis is numerical. The vertical (y) axis is a stratigraphic time scale, showing periods, series, stages and zones; the precise duration of each of these time divisions is unknown. In fact the whole object of this documentation is to determine, as far as possible with the evidence available, what estimates can be given on the duration of these stratigraphic divisions. Thus in the course of preparing this figure, we have constructed a series of graphs, each with slightly differing vertical scales, until we obtained a scale which allowed a straight line to pass through almost all the rectangles representing the analytical errors (2σ) and the stratigraphic uncertainties in the data we use.

In a later paper on the Ordovician time scale, Cooper (1999b) used 14 analytically reliable and stratigraphically controlled high-resolution TIMS U–Pb zircon dates and a single Sm–Nd date. Adopting a modified version of the McKerrow method, Cooper plotted these Ordovician dates along a relative time scale that was then re-proportioned as necessary to achieve a good fit with a straight line obtained by linear regression. This method of relative shortening and lengthening of parts of the Ordovician time scale was based mainly on a comparison of sediment accumulation rates in widely different regions and, to some extent, on empirical re-proportioning.

Agterberg (2002) subjected Cooper's (1999b) data to spline fitting and found that the optimum smoothing factor corresponds to a straight-line fit. He then used Ripley's MLFR method (cf. Appendix 3) to fit a straight line in which stratigraphic uncertainty was considered as well. Cooper's original stratigraphic error bars were converted into standard deviations as follows.

Each stratigraphic standard deviation $s(x)$ was set 15% larger than one-quarter of the error bar shown in Cooper's (1999b) Fig. 1. This is because the rectangular frequency distribution model better represents this type of stratigraphic uncertainty than the Gaussian frequency distribution model employed for the measurement errors of the dates.

If the length of a stratigraphic error bar is written as q, the variance of a rectangular frequency density distribution (with base q) is $q^2/12$ instead of $q^2/16$ for the Gaussian distribution. This translates into a standard deviation $s(x) = 1.15/(q/4)$. The stratigraphic error bars to be used for analysis in this documentation were treated in the same manner.

Spline fitting provides estimated $f(x)$ values that are independent of origin and scale of the relative time scale along the x-axis. However, when stratigraphic standard deviations $s(x_i)$ are to be combined with $s(y_i)$ values, care should be taken that both $s(x_i)$ and $s(y_i)$ are expressed in millions of years. For $s(x_i)$ values, a good approximation is obtained by setting the interval between oldest and youngest observed ages along the relative scale equal to the difference between these two ages in millions of years.

If the original relative time scale is not linearly related to the numerical time scale, spline fitting provides an objective method for shortening or lengthening of parts of the relative time scale. This is because a straight-line pattern is obtained when the original dates y_i are plotted against the $f(x_i)$ values on the best-fitting spline curve. Writing $Y = f(x)$, the result, generally, is a plot on which the points (y_i, Y_i) scatter around a straight line that, to a good approximation, slopes at 45° and passes through the origin. The unit along both axes on this plot is age in millions of years.

This approach can be taken one step further: the scatter of the points about the best-fitting straight line in the final plot can be used to calculate 2-sigma error bars on the $f(x)$ value (in millions of years) of any x including the chronostratigraphic boundaries of interest. Suppose that the straight line on the final plot is written in the form $Y' = a + bx$. The intercept a (≈ 0) and slope b (≈ 1) can be regarded as random variables with expected values that can be estimated, together with their variances $s^2(a)$ and $s^2(b)$, and their covariance $s(a, b)$. Because x is not a random variable, Y' is a random variable with variance $s^2(Y') = s^2(a) + x^2 \times s^2(b) + 2x \times s(a, b)$.

The preceding equation results in the familiar hyperbolic 95% confidence interval around a best-fitting straight line $Y' = a + bx$. It is an application of the general expression for the variance of a linear function of random variables (see, for example, Hald, 1952, p. 118).

Some remarks may be appropriate regarding notation. The subscripts i (for observations) and k (for chronostratigraphic boundaries) are omitted when this does not create misunderstanding. Final estimates of the ages of chronostratigraphic boundaries may be the end product of an iterative process involving the fitting of several, successive curves or straight lines. The final estimates to be used satisfy $Y = f(x)$, representing values on the last spline curve fitted. The values $Y' = a + bx$ then fall on a straight line obtained after setting $x = Y$ for the final supplementary (MLFR) straight-line fitting. The purpose of this MLFR straight-line fitting is not to improve Y, because $Y' \approx Y$, but to estimate $s^2(Y') \approx s^2(Y)$, which cannot be estimated otherwise. As a rule, the x-axis on a plot is horizontal (and the y-axis is vertical), but best-fit Y values become x values during supplementary straight-line fitting. Final 95% confidence intervals on estimated values are usually referred to as 2-sigma error bars.

8.2 PALEOZOIC APPLICATIONS

8.2.1 Straight-line fitting for Ordovician and Silurian ages

The initial relative stratigraphic scale of the Ordovician–Silurian is the CONOP composite standard as calculated by Cooper and Sadler (Chapters 12 and 13) for graptolite zones. A plot of the 22 ages in this data set against the CONOP scale gives a nearly linear pattern except for possible curvature near the base of the Ordovician.

Table 8.1 shows results of zero-stratigraphic-error spline-curve fitting with chi-square testing of scaled residuals. Individual scaled residuals are either positive or negative and should be approximately distributed as Z values (from the "normal" Gaussian frequency distribution). Their squares, then, are chi-square distributed with one degree of freedom. The sum of squares of several scaled residuals is also approximately distributed as chi-square but with a larger number of degrees of freedom. This method will be discussed in more detail in Section 8.2.2.

There are two small probabilities in the last column of Table 8.1. However, because the initial pattern is close to a straight line, it was decided to repeat the analysis incorporating stratigraphic uncertainty from the beginning. Chi-square testing results are shown in Table 8.2. Again there are two small probabilities in the last column.

The first relatively large chi-square value (14.92 in row 3 of Table 8.2) is for a 426.8 ± 0.85 Ma $^{40}Ar/^{39}Ar$ date discussed

Table 8.1 *Preliminary Ordovician–Silurian spline-curve values (Ma) compared to 1-sigma age determination errors*[a]

E	O	$s(y)$	O−E	Chi-square	Prob.
417.92	417.6	0.5	−0.32	0.41	0.52
423.24	421.0	1.00	−2.24	5.02	0.03
423.28	426.8	0.85	3.52	17.16	0.00
431.38	430.1	1.20	−1.28	1.13	0.29
439.10	436.2	2.50	−2.90	1.35	0.25
439.96	438.7	1.05	−1.26	1.45	0.23
445.25	445.7	1.20	0.45	0.14	0.71
452.18	454.1	1.10	1.92	3.04	0.08
453.41	453.1	0.65	−0.31	0.23	0.63
454.40	454.5	0.25	0.10	0.15	0.69
455.24	457.4	1.10	2.16	3.84	0.05
455.24	454.8	0.85	−0.44	0.27	0.60
456.97	456.9	0.90	−0.07	0.01	0.94
456.98	455.0	1.50	−1.98	1.74	0.19
462.11	460.4	1.10	−1.71	2.41	0.12
464.72	464.6	0.90	−0.12	0.02	0.90
467.03	464.0	1.00	−3.03	9.20	0.00
467.04	465.7	1.05	−1.34	1.62	0.20
467.91	469.0	2.00	1.09	0.30	0.59
481.19	483.0	0.50	1.81	13.05	0.00
489.55	489.0	0.30	−0.55	3.40	0.07
490.33	491.0	0.50	0.67	1.77	0.18

[a] E values on zero stratigraphic error spline curve; O, Observed values; $s(y)$ = 1-sigma error; chi-square = $(O-E)^2/s^2(y)$; Prob., probability.

Table 8.2 *Ordovician–Silurian: re-calculation of spline curve using $s(t)$ instead of $s(y)$*[a]

E	$s(x)$	O	$s(y)$	$s(t)$	Chi-square	Prob.
417.97	0.42	417.6	0.5	0.66	0.32	0.58
423.20	0.28	421.0	1.00	1.04	4.49	0.03
423.24	0.35	426.8	0.85	0.92	14.92	0.00
431.61	1.13	430.1	1.20	1.65	0.83	0.36
439.42	0.28	436.2	2.50	2.52	1.64	0.20
440.27	0.28	438.7	1.05	1.09	2.09	0.15
445.41	0.28	445.7	1.20	1.23	0.05	0.82
452.15	0.57	454.1	1.10	1.24	2.48	0.12
453.37	0.28	453.1	0.65	0.71	0.14	0.71
454.35	0.28	454.5	0.25	0.38	0.16	0.69
455.19	0.28	457.4	1.10	1.14	3.80	0.05
455.19	0.28	454.8	0.85	0.90	0.19	0.67
456.91	0.64	456.9	0.90	1.10	0.00	1.00
456.92	0.28	455.0	1.50	1.53	1.57	0.21
461.94	0.42	460.4	1.10	1.18	1.71	0.19
464.41	0.28	464.6	0.90	0.94	0.04	0.84
466.55	0.57	464.0	1.00	1.15	4.93	0.03
466.55	0.28	465.7	1.05	1.09	0.61	0.43
467.35	0.28	469.0	2.00	2.02	0.67	0.41
479.83	0.99	483.0	0.50	1.11	8.14	0.00
490.07	1.13	489.0	0.30	1.17	0.84	0.36
491.10	1.20	491.0	0.50	1.30	0.01	0.94

[a] E values on re-calculated spline curve; $s(x)$, stratigraphic uncertainty; $s^2(t) = s^2(x) + s^2(y)$; Prob., probability.

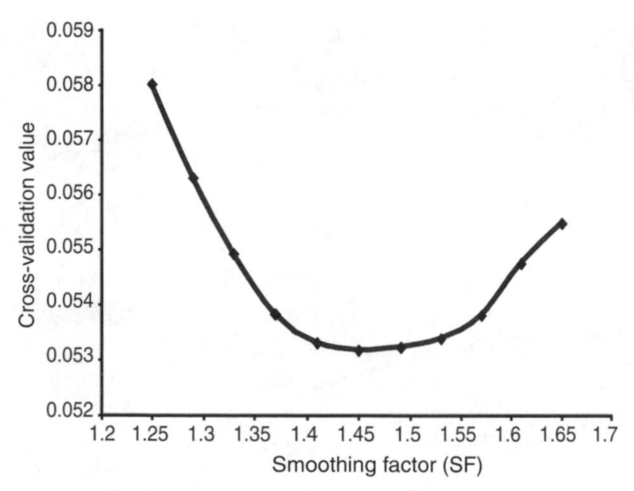

Figure 8.1 Ordovician–Silurian spline-curve cross-validation.

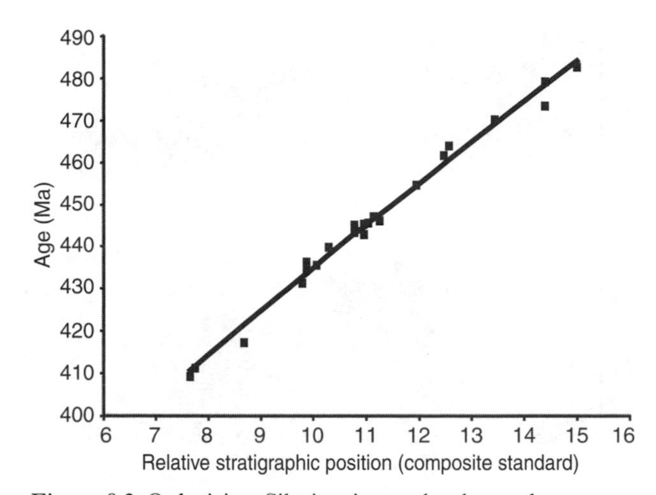

Figure 8.2 Ordovician–Silurian time scale: observed versus estimated values.

Table 8.3 *Ordovician–Silurian: calculation of 95% confidence interval (2-sigma)*[a]

x (Spline)	y (Obs.)	$s(t)$	Chi-square	Prob.	Y^1 (Cal.)	2-sigma
415.97	417.6	0.66	6.95	0.01	415.87	0.99
421.75	421.0	1.04	0.41	0.52	421.66	0.87
421.79	426.8	5.29	0.93	0.34	421.71	0.87
431.16	430.1	1.65	0.37	0.54	431.10	0.68
439.63	436.2	2.52	1.81	0.18	439.59	0.54
440.52	438.7	1.09	2.68	0.10	440.48	0.53
445.78	445.7	1.23	0.00	0.97	445.75	0.47
452.51	454.1	1.24	1.67	0.1962	452.50	0.44
453.72	453.1	0.71	0.74	0.3886	453.71	0.44
454.69	454.5	0.38	0.26	0.6127	454.69	0.44
455.53	457.4	1.14	2.72	0.0991	455.53	0.44
455.53	454.8	0.90	0.66	0.4171	455.53	0.44
457.23	456.9	1.10	0.09	0.7591	457.24	0.45
457.24	455.0	1.53	2.17	0.1409	457.25	0.45
462.18	460.4	1.18	2.31	0.1285	462.19	0.49
464.55	464.6	0.94	0.00	0.9735	464.57	0.52
466.57	464.0	1.15	5.09	0.0240	466.59	0.55
466.57	465.7	1.09	0.68	0.4102	466.60	0.55
467.31	469.0	2.02	0.67	0.4121	467.34	0.56
478.78	483.0	1.11	14.01	0.0002	478.84	0.77
488.40	489.0	1.17	0.19	0.6632	488.49	0.97
489.37	491.0	1.30	1.40	0.2369	489.46	0.99

[a] x (Spline) on curve of Fig. 8.8; $s^2(t) = s^2(x) + s^2(y)$; Prob., probability.

in Section 8.2.2. Replacing its $s(t)$ value by 0.53 myr (instead of 0.92 myr) after setting its probability equal to 0.5, and refitting the spline curve produced the results shown in Figs. 8.1 and 8.2.

The smoothest possible spline curve is a straight line. The cross-validation value does not show any increase toward larger smoothing factors (SF) if the best-fitting spline curve is a straight line. Although the minimum cross-validation value in Fig. 8.1 is not as sharply defined as those in other applications (e.g. Fig. 8.5, see later), it is possible to select an optimum smoothing factor of about 1.45 that corresponds to a curved spline rather than a straight line.

Table 8.3 shows results based on the spline curve of Fig. 8.2. The third chi-square value (14.01) from the bottom of Table 8.3

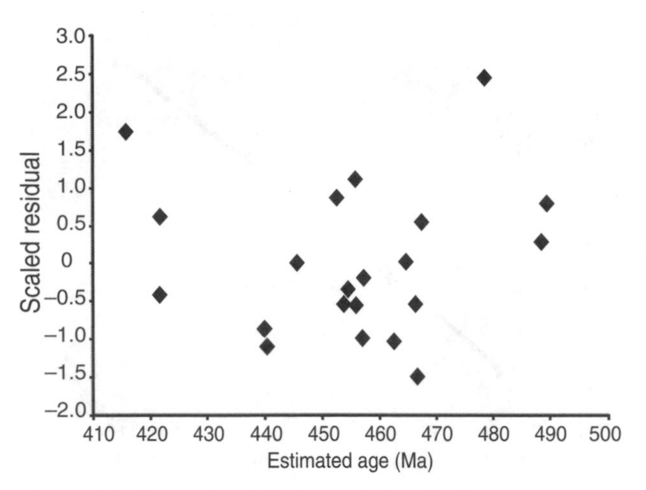

Figure 8.3 Ordovician–Silurian scaled residuals.

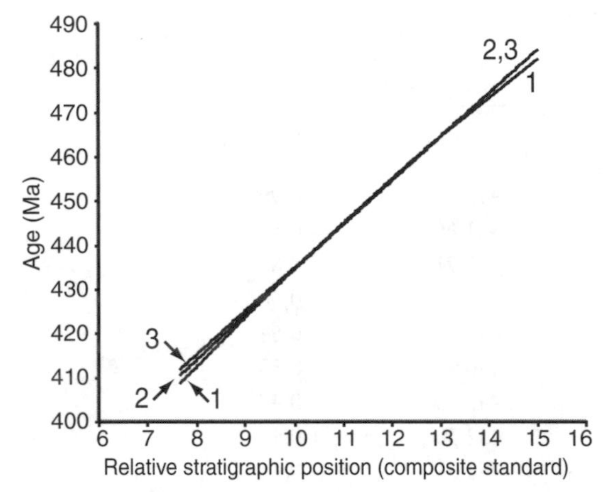

Figure 8.4 Comparison of Ordovician–Silurian spline curves: 1, based on original data; 2, as shown in Fig. 8.2; 3, alternative model (see text).

is too large for the 483 ± 1 Ma age, which could be an outlier. Dividing the scaled residuals by their standard deviation (1.51) produces the plot of Fig. 8.3 on which this age is the only Z value outside the (−1.96, 1.96) 95% confidence interval.

Inspection of the stratigraphic error bar of 483 ± 1 Ma shows that it is asymmetrical, extending downward along the CONOP composite standard scale. The current approach is based on the assumption that all error bars are symmetrical. It is not possible to account for the asymmetry of this error bar exactly, but moving its position toward a higher age along the relative stratigraphic scale would be equivalent to reducing its scaled residual value.

From the preceding analysis it follows that the spline curve of Fig. 8.2 is probably the best fit. As an experiment, Fig. 8.4 shows three spline curves obtained after enlarging the $s(t)$ value of none, one, and both of the two ages with the large chi-square values in Table 8.2, respectively. The preferred curve of Fig. 8.4

Table 8.4 *Final estimates*[a] *of the age of Ordovician and Silurian stage boundaries and durations of stages*

Period	Stage	Base (Ma)	Duration (myr)
Devonian		416.0 ± 2.8	
Silurian			
	Pridolian	418.7 ± 2.7	2.7 ± 0.1
	Ludfordian	421.3 ± 2.6	2.5 ± 0.1
	Gorstian	422.9 ± 2.5	1.7 ± 0.1
	Homerian	426.2 ± 2.4	3.3 ± 0.1
	Scheinwoodian	428.2 ± 2.3	2.0 ± 0.1
	Telychian	436.0 ± 1.9	7.9 ± 0.2
	Aeronian	439.0 ± 1.8	2.9 ± 0.1
	Rhuddanian	443.7 ± 1.5	4.7 ± 0.1
Ordovician			
	Stage 6	450.2 ± 1.6	6.5 ± 0.2
	Stage 5	460.9 ± 1.6	10.6 ± 0.3
	Stage 4 (Darriwilian)	468.1 ± 1.6	7.3 ± 0.2
	Stage 3	471.8 ± 1.6	3.7 ± 0.1
	Stage 2	478.6 ± 1.7	6.8 ± 0.2
	Stage 1 (Tremadocian)	488.3 ± 1.7	9.7 ± 0.2
Cambrian			

[a] Estimates of uncertainty in 2-sigma.

(Series 2) is close to being a straight line except near the base of the Ordovician where it shows a slight bend occupying a position between the other two splines.

Differences between Y' (from the MLFR line) and spline-curve values (column 1) are negligibly small in Table 8.3. The 2-sigma values in the last column show the typical pattern of a 95% confidence belt for a best-fitting straight line that is narrowest at the center of the cluster of data points. At the top of the Silurian (416.0 Ma), 2-sigma is 0.99 myr. In the "sample-point distribution adjustment" (cf. Section 8.3.1), this relatively large value, corresponding to the youngest stage boundary, is accepted as a conservative estimate of 2-sigma for the entire data set.

Final estimates of the ages of the Ordovician and Silurian stage boundaries and durations are given in Table 8.4. *Note: This was computed before the Ordovician subcommission revised the definition of Stage 6 (see Chapter 12).*

The method used to estimate 2-sigma values (95% confidence intervals) for stage boundaries and durations will be discussed in more detail in Sections 8.2.2 and 8.3.1. Summarizing, it can be said that the 2-sigma value of a stage base is equal to the product of the age estimate (millions of years) and an age uncertainty factor that amounts to 0.00345 for the Ordovician–Silurian data set. The corresponding duration uncertainty factor is 0.0249. A comparison of uncertainty factors for all applications will be given in Table 8.16.

Ordovician 2-sigma values were calculated from the uncertainty factors for the Ordovician–Silurian data set. To provide a smooth transition from Ordovician to Devonian uncertainties, Silurian 2-sigma values were increased toward the base of the Devonian by linear interpolation between base-Rhuddanian (the basal Silurian Stage) and base-Lochkovian (the basal Devonian stage), a process described as "ramping."

8.2.2 Straight-line fitting for Devonian ages

Table 8.5 shows results of fitting a zero-stratigraphic error line to the 14 Devonian dates used. The values in column 1 fall on a spline curve similar to the one previously obtained for the Ordovician–Silurian. The last two values in column 1 are equal because they correspond to samples from Gorstian strata of the *N. nilssoni–L. scanicus* conodont zones. These samples have the same position along the initial relative stratigraphic scale by House (see Chapter 14) although their dates are different.

As shown in column 4 of Table 8.5, the differences between observed and estimated dates are relatively large for these last two values. Division by their $s(y)$ values and squaring them gives the approximate chi-square values in column 5. Each chi-square value has a single degree of freedom and can be converted into a probability to test the hypothesis that it is not greater than can be expected on the basis of the $s(y)$ value used for scaling the residual.

Table 8.5 *Preliminary Devonian spline-curve values compared to 1-sigma age determination errors*[a]

E	O	$s(y)$	O−E	Chi-square	Prob.	New $s(y)$
359.32	358.3	2.10	−1.02	0.23	0.63	2.10
359.32	360.4	2.80	1.08	0.15	0.70	2.80
359.69	361.0	2.05	1.31	0.41	0.52	2.05
363.85	363.6	0.80	−0.25	0.10	0.76	0.80
380.61	381.1	0.80	0.49	0.37	0.54	0.80
390.22	391.4	0.90	1.18	1.73	0.19	0.90
390.22	390.0	0.25	−0.22	0.76	0.38	0.25
406.43	408.3	0.95	1.87	3.86	0.05	0.95
410.35	409.9	3.30	−0.45	0.02	0.89	3.30
415.07	413.4	3.30	−1.67	0.26	0.61	3.30
417.89	417.6	0.95	−0.29	0.10	0.76	0.95
421.22	420.2	1.95	−1.02	0.27	0.60	1.95
424.45	420.7	1.10	−3.75	11.62	0.00	5.56
424.45	426.8	0.85	2.35	7.65	0.01	3.49

[a] E-values on zero-stratigraphic error spline curve; O, observed values; $s(y)$ = 1-sigma error; chi-square = $(O-E)^2/s^2(y)$; probability that s = 1-sigma; new $s(y)$ for $P < 0.01$ only; Prob., probability.

As can be seen from column 6, these probabilities for the last two dates are very small and it is likely that the $s(y)$ values are too small. Replacement of probabilities that are too small by 0.5 results in the revised $s(y)$ values in the last column of Table 8.5. Setting the probability equal to 50% is equivalent to replacing the chi-square value by 0.4549. This is the same as adopting a new Z value of 0.674, because chi-square with a single degree of freedom is Z^2.

The details of replacing the $s(y)$ value are illustrated in the following example for the bottom row of Table 8.5. The original chi-square value (7.65) was divided by the new chi-square value (0.4549). This yields 16.82. Multiplication of the square root of this result (4.10) by the original $s(y)$ value yields the new $s(y)$ value of 4.10×0.85 (i.e. 3.49).

The bottom row is for 426.8 ± 1.7 Ma, representing the only $^{40}Ar/^{39}Ar$ date in the Devonian data set. Originally it was reported as 423.7 ± 1.7 Ma (Kunk *et al.*, 1985) and its new value used here was the result of J-factor conversion. The other Gorstian date (420.7 ± 2.2 Ma) is an average age taken from the K–Ar and Rb-Sr dates of Wyborn *et al.* (1982). These two are the only non-zircon dates in the Devonian data set and it is likely that their precision was overstated. For the $^{40}Ar/^{39}Ar$ date, this is because external uncertainty was not considered (cf. discussion in Section 8.3.1 on the nature of Late Cretaceous $^{40}Ar/^{39}Ar$ dates).

Column 2 of Table 8.6 shows stratigraphic standard deviations $s(x)$ of the Devonian dates based on measurements along the House scale. The $s(x)$ values of the two Gorstian dates are not abnormally large. In fact, they are smaller than most other $s(x)$ values. This indicates that the large chi-square values of Table 8.5 cannot be due to stratigraphic uncertainty.

For the preceding reasons, the new $s(y)$ values derived in Table 8.5 were accepted for fitting a new spline curve with $s(t)$ replacing $s(y)$. The recalculated spline-curve values in the first column of Table 8.6 differ significantly from those in column 1, Table 8.5. The $s(t)$ values that satisfy $s^2(t) = s^2(x) + s^2(y)$ can be used for a new chi-square test as shown in Table 8.6.

The probabilities in column 6 of Table 8.6 indicate that the seventh Devonian date listed (390 ± 0.5 Ma) has a 2-sigma value that is under-reported. This is the monazite age of Roden *et al.* (1990). Its $s(t)$ value was increased to 3.09 after replacing its probability of 0.0018 by 0.5.

It should be kept in mind that the preceding chi-square test does not provide a totally objective rule for deciding whether or not a published 2-sigma bar is too narrow. Sample size also has to be considered: $n = 14$ dates were used for the Devonian.

Table 8.6 *Devonian: re-calculation of spline curve using $s(t)$ instead of $s(y)$*[a]

Y	$s(x)$	y	$s(y)$	$s(t)$	Chi-square	Prob.
358.31	1.55	358.3	2.10	2.62	0.00	1.00
358.31	0.77	360.4	2.80	2.91	0.52	0.47
358.71	1.55	361.0	2.05	2.58	0.79	0.37
362.89	0.77	363.6	0.80	1.12	0.41	0.52
378.13	1.86	381.1	0.80	2.04	2.12	0.15
392.09	0.46	391.4	0.90	1.01	0.46	0.50
392.09	0.62	390.0	0.25	0.67	9.76	0.00
401.94	2.94	408.3	0.95	3.11	4.18	0.04
405.20	2.79	409.9	3.30	4.33	1.18	0.28
410.75	0.77	413.4	3.30	3.39	0.61	0.43
414.87	0.62	417.6	0.95	1.14	5.76	0.02
419.75	0.93	420.2	1.95	2.16	0.04	0.84
424.26	0.62	420.7	5.56	5.59	0.40	0.52
424.26	0.62	426.8	3.49	3.54	0.52	0.47

[a] Y values on re-calculated spline curve; $s(x)$ stratigraphic uncertainty; $s^2(t) = s^2(t) + s^2(y)$; Prob., probability.

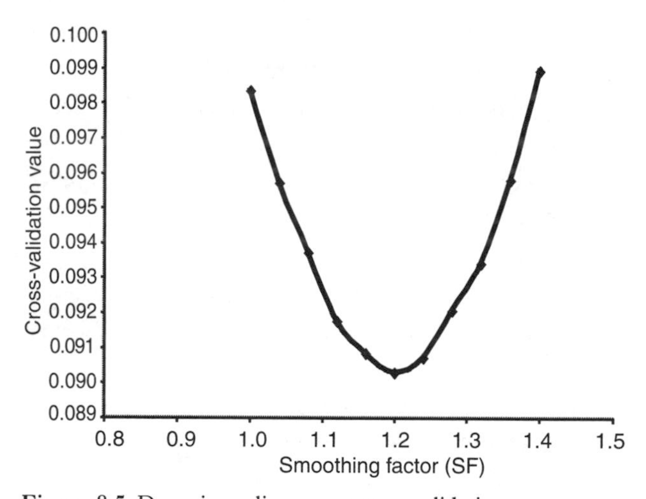

Figure 8.5 Devonian spline-curve cross-validation.

One can ask questions such as the following. What are the possibilities that probabilities less than 0.0007 or 0.0057 occur when $n = 14$ as in Table 8.5? These secondary probabilities are $1 - (1 - 0.0007)^{14} = 0.0098$ and 0.077, respectively. The corresponding probability for the seventh date of Table 8.3 is 0.025. These estimates are not as small as those for single dates listed in Tables 8.5 and 8.6. Because dating methods have improved significantly during the past ten years, previously accepted dates may have large errors that preclude their use in the time scale.

Results of final Devonian spline-curve fitting are shown in Figs. 8.5 and 8.6. The values in the first column of Table 8.5 fall on the spline curve of Fig. 8.6. The MLFR method was

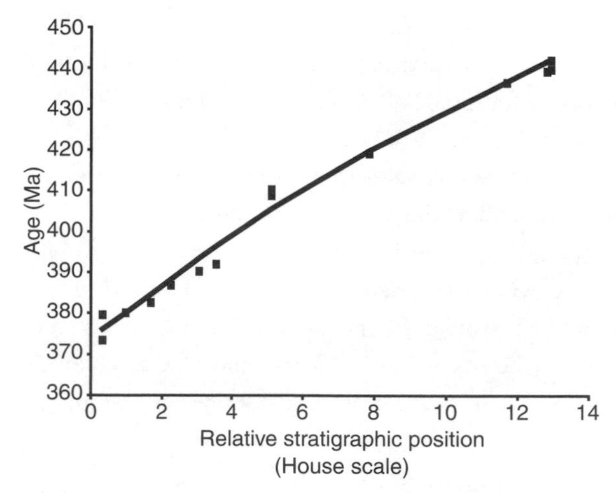

Figure 8.6 Devonian time scale: observed versus estimated values.

used to fit the corresponding straight line. This gave the Y' and 2-sigma error bars based on $s(Y')$ shown in columns 6 and 7.

The 2-sigma values of Table 8.7 and the scaled residuals were adjusted for sample-point distribution and also for average square of scaled residuals. This was achieved as follows: the sum of squares of residuals standardized with respect to $s(t_i)$ would exceed its expected value if age determination or stratigraphic error were systematically underestimated. For this reason, final scaled residuals (last column of Table 8.7) were computed by artificially setting their average square equal to 1. This procedure reduces the values in the preceding column by about 29% in absolute value.

The preceding correction of setting the average square of scaled residuals equal to unity results in slightly wider error bars. In the Paleozoic applications described in this chapter, scaled residuals $r_i = (y_i - Y_i)/s(t)$ are on average greater than expected even after eliminating the effects of outliers. This suggests that $s(t)$ values may slightly overestimate true precision by the same factor slightly greater than one.

AVERAGE SQUARE OF SCALED RESIDUAL ADJUSTMENT

The 2-sigma values in Table 8.7 were based on the variance with equation $s^2(Y') = s^2(a) + x^2 \times s^2(b) + 2x \times s(a, b)$. There is the implied assumption that all $s(t)$ values are unbiased. If the average square of scaled residuals is slightly greater than 1, it is reasonable to assume that all $s(t)$ values are slightly less than the values that should have been used. This second adjustment amounts to multiplying all initial 2-sigma values by 1.504. It was applied in addition to the sample-point distribution adjustment (cf. Sections 8.2.1 and 8.3.1).

Table 8.7 *Devonian: calculation of 95% confidence interval (2-sigma) and final scaled residuals (FSR)*[a]

X (Spline)	$s(x)$	Y (Obs.)	$s(y)$	$s(t)$	$Y^!$ (Cal.)	2-sigma	$(O-E)/s(t)$	FSR
358.10	1.55	358.3	2.10	2.62	357.83	1.98	0.18	0.14
358.10	0.77	360.4	2.80	2.91	357.83	1.98	0.88	0.68
358.55	1.55	361.0	2.05	2.58	358.29	1.96	1.05	0.81
363.31	0.77	363.6	0.80	1.12	363.09	1.77	0.46	0.35
380.29	1.86	381.1	0.80	2.04	380.20	1.26	0.44	0.34
394.53	0.46	391.4	0.90	1.01	394.56	1.17	−3.11	−2.41
394.53	0.62	390.0	3.03	3.09	394.56	1.17	−1.47	−1.14
403.81	2.94	408.3	0.95	3.11	403.91	1.33	1.41	1.09
406.81	2.79	409.9	3.30	4.33	406.94	1.41	0.68	0.53
411.88	0.77	413.4	3.30	3.39	412.05	1.56	0.40	0.31
415.64	0.62	417.6	0.95	1.14	415.83	1.69	1.55	1.20
420.08	0.93	420.2	1.95	2.16	420.31	1.86	−0.05	−0.04
424.18	0.62	420.7	5.56	5.59	424.44	2.02	−0.67	−0.52
424.18	0.62	426.8	3.49	3.54	424.44	2.02	0.67	0.51

[a] X (Spline) as on curve of Fig. 8.5; $s^2(t) = s^2(x) + s^2(y)$; O = Y(Observed), E = $Y^!$(Calculated).

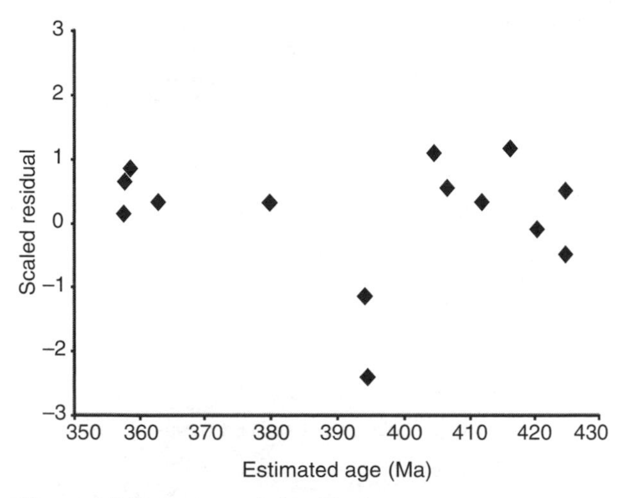

Figure 8.7 Devonian scaled residuals.

Table 8.8 *Final estimates*[a] *for the age of Devonian stage boundaries and durations of stages*

Period	Stage	Base (Ma)	Duration (myr)
Carboniferous		359.2 ± 2.5	
Devonian			
	Famennian	374.5 ± 2.6	15.3 ± 0.6
	Frasnian	385.3 ± 2.6	10.8 ± 0.4
	Givetian	391.8 ± 2.7	6.5 ± 0.3
	Eifelian	397.5 ± 2.7	5.7 ± 0.2
	Emsian	407.0 ± 2.8	9.5 ± 0.4
	Pragian	411.2 ± 2.8	4.2 ± 0.2
	Lochkovian	416.7 ± 2.9	4.8 ± 0.2
Silurian			

[a] Estimates of uncertainty in 2-sigma.

Figure 8.7 shows the final scaled residuals also listed in the last column of Table 8.7. Only one of the 14 Z values falls outside the ±1.96 interval. This value is −2.41. Because the size of the Devonian data set is relatively small ($n = 14$), it is good to keep in mind that the Z test is approximate. Use of a slightly wider 95% confidence interval of ±2.18 based on Student's t-test for 12 degrees of freedom is more appropriate (cf. Table 8.16, Section 8.4).

Final estimates of Devonian stage boundaries and durations are given in Table 8.8. However, it is better to replace the base-Devonian estimate of 416.7 ± 2.9 Ma by 416.0 ± 2.8 Ma, resulting from statistical analysis of the Ordovician–Silurian data set, in the previous section. This replacement has already been made in calculation of the duration of the Lochkovian.

8.2.3 Carboniferous–Permian

CONOP relative stratigraphic positions of 26 Carboniferous–Permian dates were used to construct Figs. 8.8 and 8.9 according to the method described in Sections 8.2.1 and 8.2.2. Permian uncertainties were based on Carboniferous–Permian MLFR results, but ramping was used in the Carboniferous to provide a gradual transition between uncertainties for base-Asselian and the larger Devonian uncertainties. Final results are shown in Table 8.9.

The estimate of 251.0 ± 0.4 Ma for the Permian–Triassic boundary is derived in Section 8.2.4.

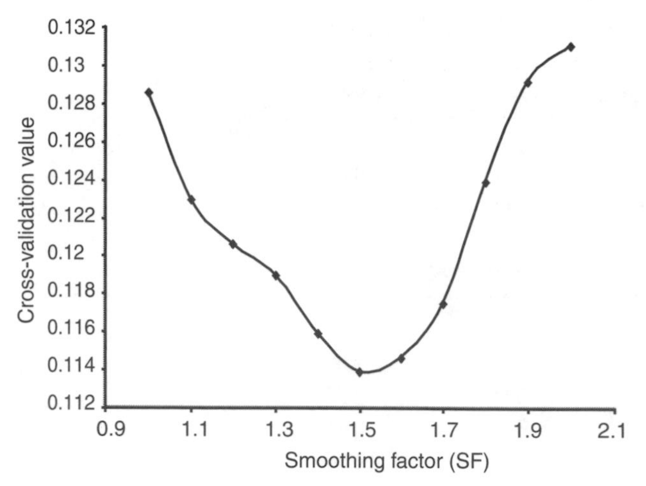

Figure 8.8 Carboniferous–Permian spline-curve cross-validation.

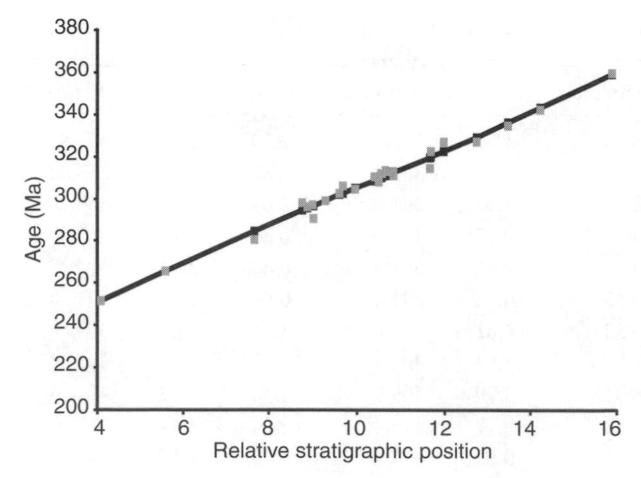

Figure 8.9 Carboniferous–Permian time scale: observed (gray squares) versus estimated values.

Table 8.9 *Final estimates[a] for the age of the Carboniferous and Permian stage boundaries and durations of stages*

Period	Stage	(Ma) Base (Ma)	Duration (myr)
Triassic		251.0 ± 0.4	
Permian			
	Changhsingian	253.8 ± 0.7	2.8 ± 0.1
	Wuchiapingian	260.4 ± 0.7	6.6 ± 0.1
	Capitanian	265.8 ± 0.7	5.4 ± 0.1
	Wordian	268.0 ± 0.7	2.2 ± 0.1
	Roadian	270.6 ± 0.7	2.5 ± 0.1
	Kungurian	275.6 ± 0.7	5.0 ± 0.1
	Artinskian	284.4 ± 0.7	8.8 ± 0.2
	Sakmarian	294.6 ± 0.8	10.2 ± 0.2
	Asselian	299.0 ± 0.8	4.4 ± 0.1
Carboniferous			
	Gzhelian	303.9 ± 0.9	4.9 ± 0.1
	Kasimovian	306.5 ± 1.0	2.6 ± 0.1
	Moskovian	311.7 ± 1.1	5.2 ± 0.1
	Bashkirian	318.1 ± 1.3	6.4 ± 0.2
	Serpukhovian	326.4 ± 1.6	8.4 ± 0.2
	Visean	345.3 ± 2.1	18.9 ± 0.7
	Tournaisian	359.2 ± 2.5	13.9 ± 0.6
Devonian			

[a] Estimates of uncertainty in 2-sigma.

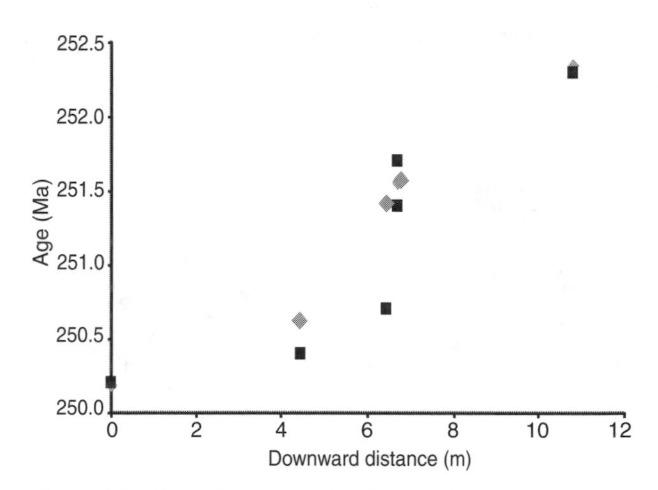

Figure 8.10 Preliminary Permian–Triassic boundary spline curve based on complete data set: observed (squares) versus estimated (diamonds) values.

8.2.4 Permian–Triassic boundary

A single spline curve was fitted to TIMS dates at the GSSP sections at Meishan, China, and the correlative locality near Heshan, China. Initially, use was made of all eight dates listed in Table 17.2 (under column "Used in Ch. 8"). This includes five dates in the immediate vicinity of the boundary. These are the four dates at 6.70 m (18 cm below the boundary at 6.62 m) and the 250.7 Ma date at 6.44 m (8 cm above the boundary). The preliminary results are shown in Figs. 8.10 and 8.11.

The spline curve derived from all eight dates is S-shaped and indicates significant decrease in sedimentation rate in the immediate vicinity of the Permian–Triassic boundary. Figure 8.10 is a comparison of input and output values. It shows that the point at 6.44 m is quite far below its calculated spline value. Figure 8.11 is an analysis of the scaled residuals (two points with 6.70 m, 251.6 coincide). Because these scaled residuals should all behave like Z values, it can be concluded that the point 8 cm above the boundary is clearly anomalous.

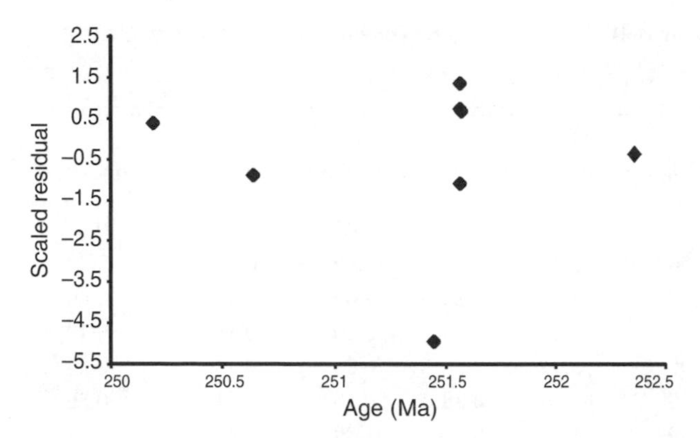

Figure 8.11 Permian–Triassic boundary: scaled residuals for Fig. 8.10.

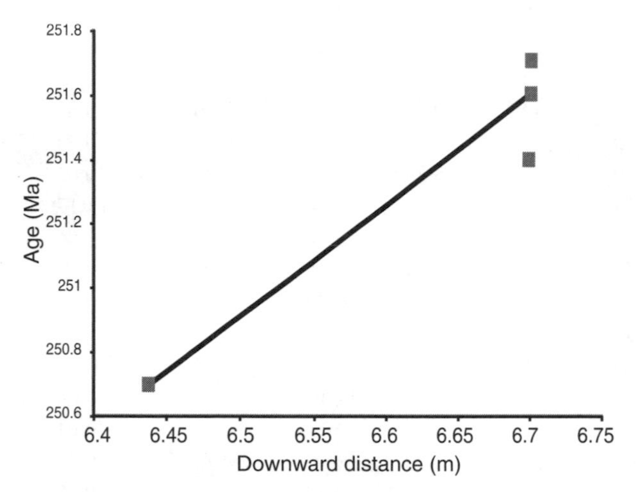

Figure 8.12 Permian–Triassic boundary: two ash layers only – input–output comparison.

A way out of this dilemma is to use the five dates in the immediate vicinity of the boundary (from two ash layers) only. The best-fitting spline then is the straight line shown in Fig. 8.12. The scaled residuals for this solution behave properly (see Fig. 8.13).

The estimate based on Figs. 8.10 and 8.11 was 251.5 Ma. The new estimate based on Figs. 8.12 and 8.13 is 251.0 ± 0.4 Ma. This new estimate would imply that the rate of sedimentation remained constant between 8 cm above and 18 cm below the boundary, respectively. In Fig. 8.10 this sedimentation rate corresponds to a straight line connecting the point at 6.44 m with the cluster of four dates at 6.70 m. The slope of this line is very steep indicating that the sedimentation rate at the boundary was much less than at the first two input points and at the last input point.

The MLFR line, corresponding to the straight line of Fig. 8.12, has intercept $a = 0.020\,566$, with $s(a) = 42.83$,

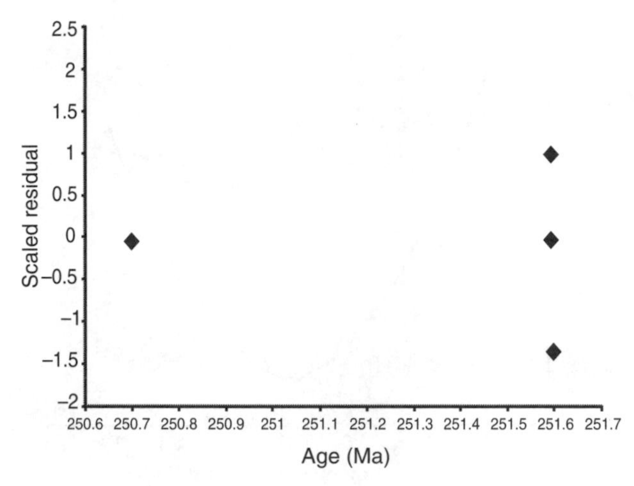

Figure 8.13 Permian–Triassic boundary: two ash layers only – scaled residuals.

and slope $b = 0.999\,918$, with $s(b) = 0.1703$. From these statistics (and the covariance), a 2-sigma uncertainty estimate of ±0.1994 myr was derived. However, this estimate was enlarged to ±0.4 myr for the following reason.

On average, the China P–T dates have a 1-sigma error of 0.0537%. This is less than the 1-sigma error of 0.1% listed in Table 6.1. For this reason, the TIMS dates at the P–T boundary were multiplied by approximately 1.86 to account for limiting decay-constant effect. This gives a final P–T age estimate of 251.0 ± 0.4 Ma.

It is noted that use of 0.1% (1-sigma) external uncertainty for U–Pb determinations at the Permian–Triassic boundary compares favorably to the 1% (1-sigma) external uncertainty used for $^{40}Ar/^{39}Ar$ determinations near the Cretaceous–Paleogene boundary in Sections 8.3.1 and 8.3.2.

8.3 LATE CRETACEOUS AND PALEOGENE APPLICATIONS

8.3.1 Late Cretaceous

As mentioned previously, application of the preceding method to the Late Cretaceous data set is simpler than to Paleozoic data sets because stratigraphic error bars are negligibly small for the former. Also, this data set is homogeneous, in that it does not present significant problems with outliers, as encountered in Paleozoic applications.

The 29 $^{40}Ar/^{39}Ar$ bentonite dates of Obradovich (1993; see Table 17.2, column "Used in Ch. 8") were re-calibrated with the new age of the Taylor Creek sanidine (TCs) monitor standard set equal to 28.34 Ma (Renne *et al.*, 1998c) instead of the 28.32 Ma age used by Obradovich. Because the new age is

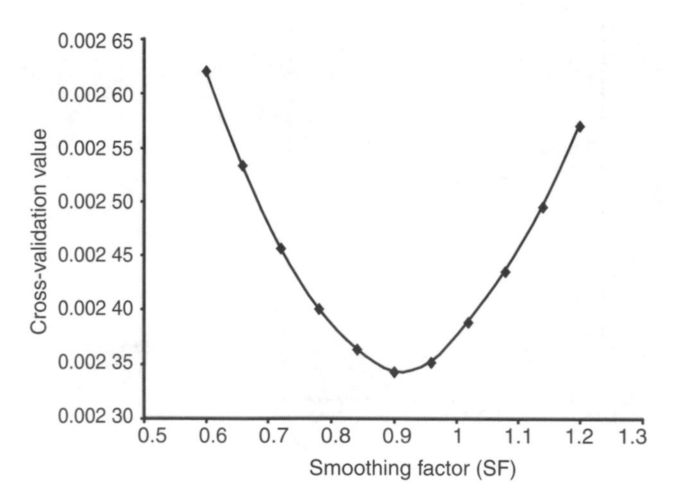

Figure 8.14 Late Cretaceous spline-curve cross-validation.

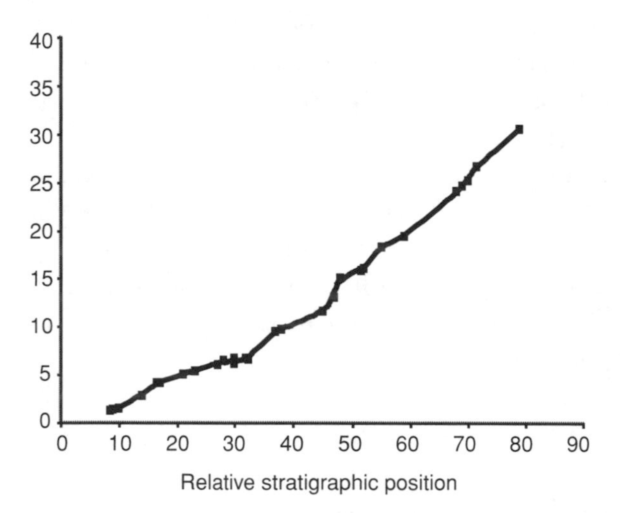

Figure 8.15 Late Cretaceous time scale: observed versus estimated values.

Table 8.10 *Late Cretaceous: calculation of 95% confidence interval (2-sigma)*[a]

X (Spline)	Y (Obs.)	$s(Y)$	Y(Cal.)	2-sigma	(O−E)/s
69.46	69.47	0.19	69.45	0.20	0.04
73.50	73.40	0.20	73.49	0.16	−0.49
74.67	74.81	0.23	74.67	0.15	0.60
75.40	75.42	0.20	75.40	0.15	0.10
76.06	75.94	0.36	76.06	0.14	−0.34
80.50	80.60	0.28	80.50	0.11	0.36
81.89	81.77	0.17	81.89	0.10	−0.73
83.79	83.97	0.22	83.79	0.09	0.82
83.99	84.15	0.20	83.99	0.09	0.79
85.33	84.94	0.14	85.33	0.09	−2.80
86.42	86.98	0.20	86.42	0.09	2.87
88.26	88.40	0.30	88.26	0.09	0.46
90.31	90.27	0.36	90.31	0.10	−0.11
90.69	90.57	0.23	90.69	0.10	−0.53
93.15	93.46	0.32	93.15	0.11	0.99
93.30	93.31	0.28	93.31	0.11	0.02
93.59	93.36	0.20	93.60	0.12	−1.17
93.59	93.85	0.25	93.60	0.12	1.04
93.59	93.65	0.29	93.60	0.12	0.19
93.78	93.55	0.45	93.78	0.12	−0.52
93.93	93.97	0.36	93.93	0.12	0.12
94.66	94.70	0.31	94.67	0.12	0.13
94.98	95.00	0.27	94.99	0.13	0.06
95.84	95.85	0.31	95.84	0.13	0.04
96.01	95.93	0.23	96.02	0.13	−0.38
97.12	97.24	0.35	97.13	0.14	0.33
98.53	98.59	0.21	98.54	0.16	0.29
98.74	98.61	0.35	98.74	0.16	−0.36
98.83	98.81	0.30	98.84	0.16	−0.07

[a] X(Spline) on curve of Fig. 8.2; $O = Y$(Observed), $E = Y$(Calculated).

nearly equal to the old age, J-factor conversion change is minor. The corresponding error bars are not changed at this stage of the analysis.

Cross-validation (Fig. 8.14) shows that SF $= 0.9$ is a good choice for the smoothing factor. The resulting spline curve (Fig. 8.15) provides an excellent fit to the data. The observed y_i values, with their standard deviations $s(y_i)$, are shown in Table 8.10. The spline-curve values are given in the first column of Table 8.10. These were used as x_i values for fitting the $Y' = a + bx$ line, whose values are shown in the fourth column.

The main purpose of regressing the observed dates on the spline-curve values is to calculate the 2-sigma values of the estimated spline values. These are given in column 5 of Table 8.10. They range from 0.09 to 0.20 myr. A second purpose is to calculate 2-sigma error bars for duration estimates. Each duration estimate is equal to b times the difference between the chronostratigraphic boundaries selected. Consequently, its uncertainty is proportional to duration multiplied by $s(b)$.

Column 6 in Table 8.10 shows differences between observed and estimated values divided by $s(y_i)$. The sum of squares of these values is 23.5. Theoretically, this sum is distributed as chi-square with $(n - 2 =)$ 27 degrees of freedom and an expected value equal to 27. The difference between 23.5 and 27 is not statistically significant. It is equivalent to the optimum smoothing factor SF $= 0.9$ being somewhat less than 1.

Because the sum of squared residuals is theoretically distributed as chi-square with $(n - 2)$ degrees of freedom, the standard deviation of scaled residuals $s(r_i)$ will be estimated by taking the square root of their sum divided by $(n - 2)$ instead of n. Consequently, $n \times \text{SF}^2$ is equivalent to $(n - 2) \times s^2(r_i)$.

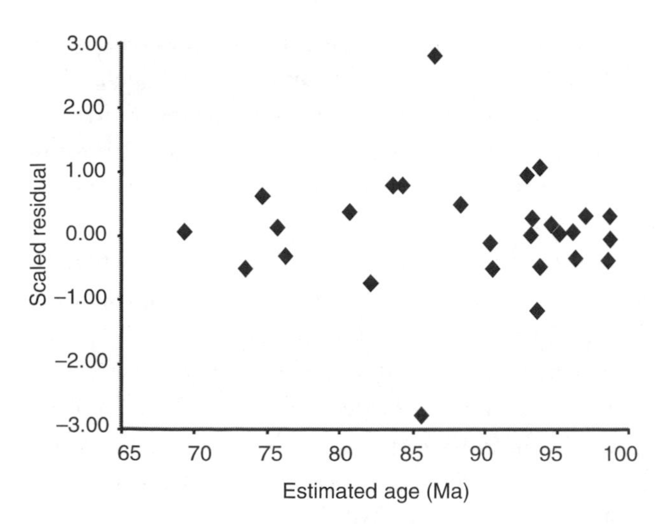

Figure 8.16 Late Cretaceous scaled residuals.

Figure 8.16 is a plot of the scaled residuals in the last column of Table 8.10. Ideally, the scaled residuals are approximately distributed as Z-values, with 68% of them contained within the interval $(-1, 1)$, and 95% within $(-1.96, 1.96)$. This condition is roughly satisfied in Fig. 8.16.

On the whole, the original error bars of Obradovich (1993) are unbiased. However, two scaled residuals (−2.80 and 2.87) have greater absolute values than normal. They are for the observed dates of 84.94 and 86.98 Ma, respectively. These two dates are probably less and greater than expected. Because their expected values are nearly equal to one another with residuals of opposite signs, the effect of this anomaly on the spline curve is negligibly small and further adjustment is not required.

Except for the Cretaceous–Paleogene boundary[1] estimate, which is included for reference purposes, the stage boundary ages, Table 8.11, were obtained using the spline curve of Fig. 8.15. This suite assumed that the current placement of these boundaries within the Western Interior ammonite zonation is equivalent to their proposed global definition based on other fossil groups (see Chapter 19). However, only the base-Turonian GSSP coincides exactly with a Western Interior ammonite level.

The 2-sigma error bar estimates are wider than those listed in Table 8.10 for the original bentonite ages. There are two reasons for this: (1) adjustment related to age distribution of samples in data set, and (2) adjustment to account for external uncertainties associated with the $^{40}Ar/^{39}Ar$ age determination method. Only the first of these adjustments had been previously applied to Paleozoic data sets. The second adjustment is replaced by an error bar correction for relatively few individual dates.

[1] For the Cretaceous–Paleogene boundary, an estimate based on the Paleogene spline curve was adopted.

Table 8.11 *Final age estimates[a] for Late Cretaceous stage boundaries as commonly assigned in the ammonite zonation of the US Western Interior and durations of stages*

Period	Stage	Base (Ma)	Duration (myr)
Paleogene		65.5 ± 0.3[b]	
Cretaceous			
	Maastrichtian	70.6 ± 0.6	5.1 ± 0.3
	Campanian	83.5 ± 0.7	13.0 ± 0.7
	Santonian	85.8 ± 0.7	2.3 ± 0.1
	Coniacian	89.1 ± 0.8	3.2 ± 0.2
	Turonian	93.5 ± 0.8	4.5 ± 0.3

[a] Estimate of uncertainty in 2-sigma.
[b] For the Paleogene–Cretaceous boundary, an estimate based on the Paleogene spline curve was used.

SAMPLE-POINT DISTRIBUTION ADJUSTMENT

The 2-sigma values in column 5 of Table 8.10 show a distinct pattern of narrowing toward the center of the cluster of observation points (cf. Fig. 8.16). This pattern is artificial in that it reflects the fact that there are no data points outside the range between the youngest and oldest samples used. The following procedure was adopted to correct for this.

The youngest stratigraphic boundary in the data set generally has an age that is slightly older than the age of the youngest sample used. For the Late Cretaceous, this is the Campanian–Maastrichtian Stage boundary of 71.0 Ma. The preliminary 2-sigma error bar for this value is 0.19 myr, i.e. slightly less than the largest value of 0.20 myr listed for the youngest date in Table 8.10. This value provides a conservative estimate of the single initial 2-sigma value to be used for the entire Late Cretaceous data set.

Preliminary 2-sigma values for the Late Cretaceous stage boundaries were set equal to 0.19 myr by multiplication of the stage boundary age by a factor of $0.1881/70.796 = 0.002\,657$ representing the ratio of 2-sigma and the age of the Campanian–Maastrichtian Stage boundary.

ADJUSTMENT TO ACCOUNT FOR EXTERNAL VARIABILITY OF LATE CRETACEOUS $^{40}Ar/^{39}Ar$ DATES

A source of uncertainty associated with $^{40}Ar/^{39}Ar$ dates, including those of Obradovich (1993), is that, traditionally, the error quoted for $^{40}Ar/^{39}Ar$ only includes internal uncertainties related to the measurement of $^{40}Ar^*/^{39}Ar_K$ and the J-factor. As discussed by McDougall and Harrison (1999), it is increasingly widely recognized that the quoted error should also include external errors associated with measurement of the K–Ar age of the fluence monitor and errors related to the determination of the decay constants. This becomes essential when ages from different systems are used together for time scale construction.

If internal uncertainty alone is considered, errors on different samples may be mutually consistent, but all dates would reflect the same uncertainty associated with decay constants and fluence monitor. In other words, although all Late Cretaceous observations and their standard deviations are fully explained by a single straight-line pattern, it is likely that the slope and intercept of this line are biased because of external errors.

Karner and Renne (1998) and Renne *et al.* (1998c) have discussed these questions and derived expressions that take the external errors into account. As an example of back-calculating with external errors, Renne *et al.* (1998c) computed that the mean age of 65.16 ± 0.08 Ma reported by Swisher *et al.* (1993) for an ash layer probably coincident with the Cretaceous–Paleogene boundary becomes 65.46 ± 1.26 Ma. This estimated decrease in precision is rather large. Consequently, the full 2-sigma error for the $^{40}Ar/^{39}Ar$ age of the Cretaceous–Paleogene boundary would be 2%.

A second problem that has been identified only relatively recently is that the decay constants themselves may be biased. For example, by calibrating with respect to high-precision reference ages based on the U–Pb system, Kwon *et al.* (2002) estimated that the decay constant of ^{40}K is $\lambda = 5.476 \pm 0.034 \times 10^{-10}/yr$. This estimate falls between the value of $5.543 \pm 0.020 \times 10^{-10}/yr$ currently used by geochronologists (Steiger and Jäger, 1977) and $5.428 \pm 0.068 \times 10^{-10}/yr$ of nuclear physicists (Endt and van der Leun, 1973).

The preceding values of λ are accompanied by relatively narrow 2-sigma 95% confidence intervals suggesting that one or more of these estimates are biased. Villeneuve *et al.* (2000) have demonstrated that due to a buffering effect, changes in $^{40}Ar/^{39}Ar$ ages that would result from revising the ^{40}K decay constant are relatively small.

Unfortunately, it is not possible at this time to revise the 2-sigma error bars on all dates published in the early 1990s in order to incorporate the external uncertainty. Possible bias of the decay constants is unknown at present and cannot be considered at all.

An estimate of the average 2-sigma error bar on Obradovich's (1993) bentonite dates is 0.61%. According to Renne (pers. comm.), it is prudent to associate a 2-sigma error of 2% with these Ar–Ar dates (also see preceding discussion). This would incorporate external precision uncertainties as well as a possible lack of accuracy of the decay constants. Adoption of the 2% error is equivalent to multiplying all 2-sigma error bar estimates by a factor of 3.2628.

The second type of adjustment discussed in Section 8.3 need not be made for data sets that almost exclusively consist of U–Pb zircon dates. If one or relatively few $^{40}Ar/^{39}Ar$ dates are considered together with a number of high-precision unbiased U–Pb dates, it is possible to subject the $^{40}Ar/^{39}Ar$ dates to J-factor conversion first, and then to test whether or not their scaled residuals are statistically too large. If the published error bar of an $^{40}Ar/^{39}Ar$ date is too large because external uncertainty was not considered, it can be adjusted to a median or average value. This method is developed in Section 8.3.2.

TOTAL ADJUSTMENT AND 2-SIGMA ERROR BAR OF STAGE DURATION

According to adjustment 1, all estimated ages are to be multiplied by (0.1881/70.796) 0.002 657 first. As a result of adjustment 2, the 2-sigma error bars of the estimated values are to be multiplied by 3.2628. Thus total adjustment of an estimated error bar consists of multiplying the age by ($0.002\,657 \times 3.2628$) 0.008 67. This procedure resulted in the error bars on Late Cretaceous stages reported at the beginning of Section 8.3.

For duration ages, 2-sigma error bars are obtained by multiplying the duration (in millions of years) by 0.033 43, representing twice the standard deviation $s(b)$, i.e. 0.005 123 multiplied by 3.2628 for adjustment 2 and two other, relatively minor, adjustments (duration uncertainty corrections 1 and 2 in Table 8.16 later). It can be argued that the equivalent of adjustment 1 should be applied to duration uncertainty as well. For this reason, stage duration uncertainties were multiplied by the age of the oldest stage boundary (93.6 Ma) divided by the age of the youngest stage boundary (70.6 Ma) in the data set. Duration uncertainty correction 1, therefore, consists of multiplication by 1.326 (cf. Table 8.16 later).

It is noted that duration uncertainty correction 1 is on the conservative side (slightly too large) for younger stage boundaries in the data set.

Finally, it can be argued that stage duration is the difference between two point estimates; in contrast to a stage boundary age, which is a single point estimate. For single point estimates, the 95% confidence belt is controlled by Student's t-distribution (± 2.052 for the Late Cretaceous data set). For

Table 8.12 *Linear relation between Campanian and Maastrichtian $^{87}Sr/^{86}Sr$ ratios and age*

Code No.	Obradovich's biozone name	Spline age of base (Ma)	Duration (spline)	Mean $^{87}Sr/^{86}Sr$	Sr–Sr age estimate	Duration (Sr–Sr)
1	*B. clinolobatus*	69.7		0.707 729	69.7	
2	*B. grandis*	70.1	0.4	0.707 757	70.2	0.5
3	*B. baculus*	70.6	0.5	0.707 739	70.7	0.5
4	*B. eliasi*	71.0	0.5	0.707 734	71.2	0.5
5	*B. jenseni*	71.6	0.5	0.707 728	71.7	0.5
6	*B. reesidei*	72.1	0.6	0.707 697	72.2	0.5
7	*B. cuneatus*	72.8	0.6	0.707 688	72.6	0.4
8	*B. compressus*	73.5	0.7	0.707 679	73.1	0.5
9	*D. chenyennense*	74.3	0.8		73.9	0.8
10	*E. jenneyi*	75.0	0.8	0.707 637	74.8	0.9

durations, the entire best-fitting straight line can be used with the 95% confidence belt controlled by $(2F)^{0.5}$, where $F(2, n-2)$ is the F-distribution with 2 and $(n-2)$ degrees of freedom (± 2.590 for the Late Cretaceous data set). Duration uncertainty correction 2, therefore, consists of multiplication by 1.264 (cf. Table 8.16). The combined effect of these two minor adjustments consists of multiplication of the initial estimate of the duration uncertainty factor by 1.674. Thus the duration uncertainty factor for the Late Cretaceous becomes 0.0559.

The resulting Late Cretaceous duration uncertainties are shown in Table 8.11. Statistics for the Late Cretaceous are shown in Table 8.16 for comparison with similar estimates obtained for other data sets. It should be kept in mind that all age and duration uncertainty factors were derived using the same types of adjustments. In comparison with the Late Cretaceous, total adjustments of age and duration uncertainties for Paleozoic data sets (Sections 8.2.1–8.2.3) are relatively small because adjustment 2 (3.2628 for the Late Cretaceous) was not required.

COMPARISON TO $^{87}Sr/^{86}Sr$ CURVE

The $^{87}Sr/^{86}Sr$ ratio changes approximately linearly with respect to time for segments of the geological time scale. This is illustrated in the Table 8.12 following tabulation for parts of the Maastrichtian and Campanian.

The biozones are as in Obradovich (1993); zone base ages are as on the spline curve of Fig. 8.15, and spline durations are equal to differences between successive zone base ages. Mean $^{87}Sr/^{86}Sr$ and Sr–Sr age estimates are from McArthur (pers. comm.). Because of difficulties associated with curve fitting of successive approximately linear Sr–Sr patterns that can show reversals in time, our Late Cretaceous time scale remains based on the spline curve of Fig. 8.15. The relative geological time scale in Fig. 8.15 represents the initial assumption of equal duration of the Obradovich biozones. Table 8.12 shows that the modified duration estimates derived from the spline curve are comparable to the durations estimated by McArthur for this part of the Late Cretaceous record.

8.3.2 Paleogene time scale

This section contains brief documentation on methods used to derive the estimated ages of Paleogene stage boundaries and durations illustrated in Table 8.15.

Paleogene spline fitting was performed on 18 original dates (see Table 20.2). The 18 estimated values are shown in the Spline-1 column of Table 8.13. They are relatively close to the original dates shown in the next column. The largest difference is 1.12 myr for Chron C31n (base). All differences were subjected to the chi-square test. The probability for C31n (base) is 0.000 43, indicating that the original date (69.01 Ma) is probably an anomalous outlier. This conclusion is confirmed when sample size ($n = 18$) is considered. Then the probability amounts to 0.0039 that 69.01 Ma is compatible with the other 17 dates.

Rejection of this date results in 17 dates with estimated values shown in the Spline-2 column of Table 8.13. Differences between original dates are shown in the next column and are less than 0.5 myr. Division by the standard deviation (1-sigma values) yields the Z values in the last column, which are approximately normally distributed with unit variance indicating homogeneity of the reduced data set.

An optimum smoothing factor (SF) could not be determined by cross-validation and a default value of SF = 1 was

Table 8.13 *Paleogene spline-curve fitting*[a]

Polarity Chron	Spline-1	Original	1-sigma	Chi-square	Probability	Spline-2	Difference	Z-value	BFSL	2-sigma
C6An.1r (base)	20.34	20.336	0.02	0.0009	0.9761	20.33	0.00	0.21	20.33	0.0310
C6Cn.2n (base)	23.03	23.03	0.02	0.0036	0.9522	23.04	−0.01	−0.41	23.04	0.0291
C9n (base)	28.04	28.10	0.15	0.1505	0.6980	27.83	0.27	1.83	27.83	0.0297
approx. C13r.9	33.87	33.70	0.20	0.7319	0.3923	33.89	−0.19	−0.93	33.89	0.0370
C15n (base)	35.13	35.20	0.135	0.3021	0.5826	35.04	0.16	1.17	35.04	0.0390
approx. C21n.67	45.64	45.60	0.19	0.0482	0.8263	45.94	−0.34	−1.81	45.95	0.0619
C24n.1n (base)	52.81	52.80	0.15	0.0081	0.9283	53.00	−0.20	−1.36	53.01	0.0786
approx C24r.5	55.11	55.07	0.25	0.0260	0.8719	55.23	−0.16	−0.63	55.23	0.0840
C27n (top)	61.72	61.77	0.15	0.1076	0.7429	61.65	0.12	0.80	61.65	0.0999
C28n (top)	63.18	63.25	0.15	0.2110	0.6460	63.10	0.15	0.97	63.11	0.1036
C29n (top)	64.52	64.53	0.15	0.0035	0.9527	64.43	0.10	0.65	64.44	0.1069
C29n (base)	65.24	65.20	0.15	0.0602	0.8062	65.12	0.08	0.55	65.12	0.1086
C29r.56	65.59	65.50	0.15	0.3885	0.5331	65.45	0.05	0.34	65.45	0.1094
C30n (top)	66.05	65.88	0.15	1.2935	0.2554	65.86	0.02	0.13	65.87	0.1105
C31n (base)	68.13	69.01	0.25	12.3876	0.0004					
C32n (top)	70.94	70.44	0.325	2.3291	0.1270	70.96	−0.52	−1.60	70.97	0.1233
C33n (base)	79.34	79.34	0.25	0.0002	0.9888	79.54	−0.20	−0.81	79.55	0.1452
C33r (base)	84.40	84.40	0.25	0.0000	0.9968	84.35	0.05	0.21	84.36	0.1575

[a] Spline-1 is based on all 18 dates. Chi-square test indicates that original value for Chron C31n(base) is anomalous. Spline-2 is based on remaining 17 dates. All Z values are within (−1.96, 1.96) 95% confidence interval. BFSL, best-fitting straight line is followed by its 2-sigma value. Dates are given in Ma.

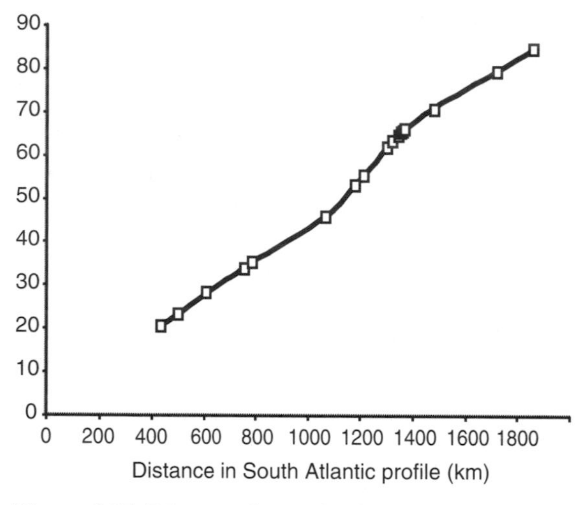

Figure 8.17 Paleogene time scale: observed versus estimated values.

used. Rejection of the 69.01 ± 0.5 Ma age for the base of Late Maastrichtian polarity Chron C31n (Hicks *et al.*, 1999) can also be justified because it was derived by projecting a constant sedimentation rate derived from $^{40}Ar/^{39}Ar$ ages in marine shale upward by 100 m across a sharp contact into the nearshore clastic-rich facies containing the polarity reversal (Röhl *et al.*, 2001, p. 179).

Figure 8.17 shows the estimated Spline-2 curve in comparison with original dates. The degree of correspondence is excellent. Table 8.14 shows estimated Spline-2 ages of Paleogene (and Late Cretaceous) stage boundaries. The Cretaceous–Paleogene boundary (Chron C29r.56) was rounded upward to 65.5 Ma in the table at the beginning of this section.

Table 8.14 also shows 2-sigma values for estimated stage boundary ages and stage durations derived by the method previously used on Obradovich's (1993) $^{40}Ar/^{39}Ar$ ages for Late Cretaceous Western Interior ammonite zones. Weighted linear regression of original dates on Spline-2 values gave a straight line with slope (1.000 12) approximately equal to one and intercept (−0.0028 Ma) approximately equal to zero. Estimated values on this straight line are shown in the BFSL column of Table 8.13.

The standard deviations and covariance of slope and intercept of BFSL were used to estimate its 95% confidence belt (2-sigma column in Table 8.13). Contrary to confidence belts previously derived for Paleozoic and Late Cretaceous data sets, 2-sigma has its minimum not near the center but near the top of the data set. This is because small standard deviations were assigned to the two Neogene stages used (cf. 1-sigma column in Table 8.13).

The 2-sigma value in the last column of Table 8.13 is ±0.1094 myr at the Maastrichtian–Danian boundary (C29r.56) and this value was used for sample point distribution adjustment.

The Paleogene $^{40}Ar/^{39}Ar$ ages used have standard deviations ranging from 0.135 to 0.325 myr (1-sigma column, Table 8.13). Re-calculated as age percentage values, they range

Table 8.14 *Other Paleogene spline curves*[a]

Polarity chron	Spline-2	Original	OSD	Spline-2A	Spline-2B
C6An.1r (base)	20.33	20.336	<0.02	20.33	20.33
C6Cn.2n (base)	23.04	23.03	<0.02	23.04	23.04
C9n (base)	27.83 ± 0.11	28.10	0.15	27.82 ± 0.01	27.83 ± 0.15
approx. C13r.9	33.89 ± 0.13	33.70	0.20	33.88 ± 0.02	33.89 ± 0.18
C15n (base)	35.04 ± 0.14	35.20	0.135	35.04 ± 0.02	35.05 ± 0.19
approx. C21n.67	45.94 ± 0.18	45.60	0.19	45.96 ± 0.02	45.94 ± 0.24
C24n.1n (base)	53.00 ± 0.21	52.80	0.15	53.05 ± 0.03	52.97 ± 0.28
approx C24r.5	55.23 ± 0.22	55.07	0.25	55.29 ± 0.03	55.17 ± 0.29
C27n (top)	61.65 ± 0.24	61.77	0.01	61.77 ± 0.03	61.52 ± 0.33
C28n (top)	63.10 ± 0.25	63.25	0.05	63.22 ± 0.03	62.97 ± 0.33
C29n (top)	64.43 ± 0.25	64.53	0.02	64.52 ± 0.03	64.30 ± 0.34
C29n (base)	65.12 ± 0.26	65.20	0.02	65.18 ± 0.03	64.99 ± 0.35
C29r.56	65.45 ± 0.26	65.50	0.05	65.49 ± 0.03	65.33 ± 0.35
C30n (top)	65.86 ± 0.26	65.88	0.01	65.89 ± 0.03	65.74 ± 0.35
C32n (top)	70.96 ± 0.28	70.44	0.325	70.85 ± 0.04	70.89 ± 0.38
C33n (base)	79.54 ± 0.31	79.34	0.25	79.52 ± 0.04	79.53 ± 0.42
C33r (base)	84.35 ± 0.33	84.40	0.25	84.36 ± 0.04	84.35 ± 0.45

[a] Spline-2 and original dates as in Table 8.8. Spline-2A is based on original standard deviations (OSD); Spline-2B is based on 1-sigma = 0.25 in Paleocene.

from 0.28 to 0.59%, with an average standard deviation of 0.43%. Adopting the larger 1% standard estimate (equivalent to Renne's 2% estimate of 2-sigma at the Cretaceous–Paleogene boundary) results in a further correction factor of 2.34.

Final age and duration uncertainty factors become 0.003 92 and 0.005 73, respectively. The resulting Paleogene stage boundary 2-sigma errors range from 0.13 to 0.26 myr (see Table 8.15) and all Paleogene stage duration uncertainty estimates are less than 0.05 myr.

The Paleogene time scale spans the transition from the precise Neogene time scale to the Cretaceous and older time scales that are largely based on age determinations. From the latest Maastrichtian to the Paleocene–Eocene boundary, the Paleogene time scale is based on cyclo-magnetostratigraphy tied to the 65.5 Ma age estimate for the Cretaceous–Paleogene boundary. Above this Milankovitch cycle-based interval, the Paleogene time scale is largely based on $^{40}Ar/^{39}Ar$ age determinations.

In the preceding analysis it was assumed that the ±0.1 myr uncertainty widely used for the Cretaceous–Paleogene boundary is too low because external uncertainties associated with $^{40}Ar/^{39}Ar$ ages were not considered. For this reason, ±0.1 myr was replaced by ±0.3 myr for uncertainty at Cretaceous–Paleogene and stage boundaries within the associated cycle-based interval. This correction ensures compatibility of Cretaceous–Paleogene uncertainty with uncertainties of other Paleogene and Late Cretaceous $^{40}Ar/^{39}Ar$ ages (1-sigma column, Table 8.13).

Table 8.15 *Final estimates*[a] *of the age of Paleogene stage boundaries and duration of stages*

Period	Stage	Base (Ma)	Duration (myr)
Neogene			
	Aquitanian	23.0 ± 0.0	
Paleogene			
	Chattian	28.4 ± 0.1	5.4 ± 0.0
	Rupelian	33.9 ± 0.1	5.4 ± 0.0
	Priabonian	37.2 ± 0.1	3.3 ± 0.0
	Bartonian	40.4 ± 0.2	3.2 ± 0.0
	Lutetian	48.6 ± 0.2	8.2 ± 0.1
	Ypresian	55.8 ± 0.2	7.2 ± 0.1
	Thanetian	58.7 ± 0.2	2.9 ± 0.0
	Selandian	61.7 ± 0.2	3.0 ± 0.0
	Danian	65.5 ± 0.3	3.7 ± 0.0
Cretaceous			

[a] Estimates of uncertainty in 2-sigma.

In comparison with Late Cretaceous 1-sigma values listed in Table 8.13, the ±0.3 myr relative uncertainty adopted for Cretaceous–Paleogene is slightly less. As a percentage value (0.46% of 65.5 Ma), it is nearly equal to the 0.43% average used in Section 8.3.1 as a basis for making the further correction to 2% (2-sigma) $^{40}Ar/^{39}Ar$ age uncertainty at Cretaceous–Paleogene boundary.

Also because ^{40}K decay constants and their uncertainties will probably be revised, it is better to overestimate than to underestimate time scale uncertainties associated with $^{40}Ar/^{39}Ar$ ages (cf. Late Cretaceous time scale documentation).

It is interesting to compare the new Cretaceous–Paleogene boundary age estimate of 65.5 ± 0.3 Ma with earlier estimates in Gradstein *et al.* (1995). The recommended value was 65.0 ± 0.1 Ma, but other (maximum likelihood) estimates reported in this publication were 66.1 ± 0.4 Ma (high- and low-temperature dates) and 66.0 ± 0.7 Ma (high-temperature dates only). The 2-sigma confidence interval of the new estimate shows overlap with those of the maximum likelihood estimates but not with 65.0 ± 0.1 Ma. The latter 2-sigma confidence interval was much too narrow, mainly because external $^{40}Ar/^{39}Ar$ age uncertainties were not considered before 1995.

Two experiments have been performed to assess the effect of choosing ± 0.3 myr further instead of the ± 0.1 myr relative uncertainty at Cretaceous–Paleogene boundary, with results shown in Table 8.14. No changes were made in the choice of smoothing factor (SF = 1), sample point distribution adjustment, and further correction to 2% (2-sigma) uncertainty of the Cretaceous–Paleogene $^{40}Ar/^{39}Ar$ age.

First the preceding analysis was repeated using ± 0.1 myr and other published uncertainties from latest Maastrichtian to the Paleocene–Eocene boundary (OSD column, Table 8.14). The results of this first experiment are shown in the Spline-2A column of Table 8.14. Estimated stage boundary ages are nearly equal to those in the Original column and relatively close to those in the Spline-2 column. However, Spline-2A uncertainties are about ten times less than Spline-2 uncertainties. It is likely that the Spline-2A uncertainties shown in Table 8.14 are too small by an order of magnitude.

In the second experiment (Spline-2B, Table 8.14), relative Cretaceous–Paleogene uncertainty was enlarged to ± 0.5 myr (instead of ± 0.3 myr for Spline-2) from latest Maastrichtian to the Paleocene–Eocene boundary. This resulted in moderate magnification of deviations between Original and Spline-2 age estimates (e.g. the Spline-2B Cretaceous–Paleogene boundary estimate becomes 65.3 Ma instead of 65.5 Ma). Spline-2B uncertainties are about 1.4 times greater than Spline-2 uncertainties.

The two additional experiments indicate that the change in relative Cretaceous–Paleogene boundary uncertainty from ± 0.1 to ± 0.3 myr has a much greater effect on final stage boundary uncertainties than the change from ± 0.3 to ± 0.5 myr. These supplementary results help to justify the validity of the results reported in the Paleogene time scale at the beginning of this book.

8.4 CONCLUDING REMARKS

Table 8.16 shows summary statistics for final spline curves and MLFR lines used to construct numerical geological time scales for the five data sets. The MLFR lines all have slopes close to 1 and intercepts close to 0 Ma. From $Y' \approx Y$ it follows that $s^2(Y') \approx s^2(Y)$ can indeed be used to construct 2-sigma error bars on the stage boundary ages and stage duration estimates.

A possible disadvantage of spline-curve fitting is that age dates included in the data set are weighted according to their variance. Thus the size of the error bars constitutes important input for obtaining valid results. Here the technique used checked that individual ages do not carry too much weight in comparison with their neighbors and other ages in the data sets. After reducing the influence of relatively few outliers, the data sets were shown to be approximately homogeneous.

The 2-sigma error bars of stage durations are narrower than those of the stage boundaries themselves. This is because the 2-sigma bars of durations are proportional to the duration itself, with twice the error of the slope of the MLFR line as a proportionality constant. The stage boundary estimates are subject to additional uncertainty related to their position along the time scale.

The error bars of successive stage boundaries are positively correlated. In this they differ from the 2-sigma error bars of observed ages in the data sets that are statistically independent. This positive correlation can be illustrated as follows: duration of the Gorstian is only 1.7 ± 0.1 myr; age estimates of its top and base are 421.3 ± 2.6 and 422.9 ± 2.5 Ma, respectively. It is not permissible to select an older age from within the confidence interval for the top and combine it with a younger age from within the confidence interval for the base.

The 2-sigma error bar of a duration estimate approaches zero when duration decreases. Thus reporting on the 95% confidence intervals of durations helps in understanding the nature of the positive correlation between the limits of the 95% confidence intervals of successive age estimates.

A final comment concerns application of the sample point distribution adjustment to age and duration uncertainty factors. Because these factors are applied to estimated age, there is a relatively small linear increase in each data set that is proportional to age. In general, an increase in uncertainty with age would be in accordance with the facts that age determination methods tend to have constant proportional errors, while stratigraphic uncertainty may also increase with age. However,

Table 8.16 *Summary statistics*[a]

	Ordovician–Silurian	Devonian	Carboniferous–Permian	Late Cretaceous	Paleogene
Youngest stage boundary (Ma)	416	359.2	251	70.6	23
Oldest stage boundary (Ma)	488.3	416	359.2	93.6	65.5
Number of ages in data set	22	14	26	29	17
Smoothing factor	1.45	1.2	1.565	0.9	1
Intercept	−0.143	−0.729	−0.0026	−0.041	−0.0028
SD (intercept)	0.648	2.121	0.0340	0.445	0.03326
Slope	1.0026	1.008	1.00023	1.00048	1.00012
SD (slope)	0.0116	0.0227	0.00280	0.00512	0.00122
Student's t-factor	2.086	2.179	2.064	2.052	2.131
Age uncertainty factor	0.00345	0.00685	0.00261	0.00867	0.00392
Initial duration uncertainty	0.01675	0.0272	0.0115	0.0334	0.00573
Duration uncertainty correction 1	1.174	1.158	1.431	1.326	2.848
Duration uncertainty correction 2	1.267	1.279	1.264	1.262	1.273
Duration uncertainty factor	0.0249	0.0403	0.0208	0.0559	0.00730

[a] SD, standard deviation. See Section 8.3.1 for explanation of "duration uncertainty factor" corrections.

it can be argued that possible proportional uncertainties were already considered in the spline-MLFR approach.

The final straight line used for each data set has the equation $Y' = a + bx$. However, it is also possible to fit a straight line with equation $Y' = bx$. This means that the best-fitting line is forced to go through an "origin" with $Y' = x = 0$. This approach can be used not only in ordinary least squares but also in least squares with errors in both x and y, as in Ripley's MLFR approach.

Advantages of fitting $Y' = bx$ instead of $Y' = a + bx$ would be that both uncertainty factors could be controlled by $s(b)$ alone. This would imply that they are proportional to estimated age. Unfortunately, it does not seem feasible to define a fixed origin where there would be zero uncertainty. The Present (0 Ma) is too far removed from the ages in the data sets. However, the following experiment makes use of the fact that the age of the Permian–Triassic boundary is assumed to be known with precision (±0.4 Ma).

Suppose that the origin for the Devonian data set is defined to occur at the base of the Triassic (251.0 Ma). This assumption results in an MLFR-based Devonian age uncertainty factor estimate of 0.0043. This is only slightly less than the 0.00685 uncertainty factor, previously used for all Devonian stage boundaries (cf. Table 8.16). The experiment seems to confirm that the estimate based on fitting $Y' = a + bx$ is reasonable although on the conservative side.

Unfortunately, the new model has a single parameter only and it forces the duration uncertainty factor to be approximately equal to the age uncertainty factor. The Devonian duration uncertainty factor actually used (0.0403) is almost ten times larger than 0.0043 (cf. Table 8.16). It would seem that the estimate for the Devonian duration uncertainty factor resulting from fitting $Y' = bx$ is much too small. This is because the $Y' = a + bx$ model is more flexible than the $Y' = bx$ model, allowing separate estimation of the two uncertainty factors.

Part III • Geologic periods

9 • The Precambrian: the Archean and Proterozoic Eons

L. J. ROBB, A. H. KNOLL, K. A. PLUMB, G. A. SHIELDS, H. STRAUSS, AND J. VEIZER

Archean and Proterozoic time scales are currently defined chronometrically, with subdivisions into eras and periods being defined and allocated boundaries in terms of a round number of millions of years before present. Isotope stratigraphy is increasingly used to identify tectonic, chemical, and biological changes. The Neoproterozoic Era is characterized by at least two, and possibly four, severe and extensive glaciogenic events; for this era, chronostratigraphic subdivisions following established Phanerozoic practices are possible.

9.1 INTRODUCTION

The "Precambrian" is not a formal stratigraphic term and simply refers to all rocks that formed prior to the beginning of the Cambrian Period. The task of establishing a rigorously defined and globally acceptable time scale for the Precambrian is an exceedingly difficult, and often frustrating, exercise. The reason for this is related to the fact that studying the Earth becomes increasingly difficult and uncertain the further one goes back in geological time.

The lack of a diverse and well-preserved fossil record, the generally decreasing volume of supracrustal rocks, and increasing degree of metamorphism and tectonic disturbance, as well as the uncertainties in the configuration and assembly of the continents, all contribute to making the establishment of a chronostratigraphic time scale beyond the Phanerozoic Eon problematical.

The Phanerozoic Eon broadly coincides with the most recent supercontinent cycle – a relatively well-understood sequence of geological events during which Pangea was assembled and dispersed. It is also the time period when multicellular life underwent enormous diversification and proliferation. Accordingly, the geologic time scale for the Phanerozoic Eon is defined in terms of a globally correlative, chronostratigraphic methodology that is rigorously constrained using biostratigraphy as well as a host of supporting techniques such as isotope, sequence and magnetostratigraphy. The detailed description of the Phanerozoic time scale is dealt with in subsequent chapters.

By contrast, the Archean and Proterozoic time scales are currently defined chronometrically, with subdivisions into eras and periods being defined and allocated boundaries in terms of a round number of millions of years before present (Ma BP, or simply Ma). Given the difficulties mentioned above, as well as the incomplete state of current knowledge, this scheme is widely considered to be the most appropriate solution for the definition of a Precambrian time scale. The scheme may not be a lasting solution as the defined units lack the geological context and definition that is required if they are to be recognized by the intrinsic features of their geologic history rather than simply by numerical dates. (One suggestion for an alternative in subdividing Precambrian time is presented by W. Bleeker in Chapter 10.)

The present chapter will, nevertheless, describe the existing chronometric time scales for the Archean and Proterozoic Eons, as well as some of the advances made over the past decade toward providing an isotopic and paleobiological context to this time scale. In addition, consideration will be given to the significant advances that have been made toward establishing a chronostratigraphic framework for the final few hundred million years of the Proterozoic Eon.

The highly distinctive biological and environmental characteristics of the Neoproterozoic Era, between 1000 Ma (arbitrarily defined chronometrically) and 542 Ma (the base of the Cambrian Period and chronostratigraphically defined) have resulted in considerable advances in the development of a time scale for this interval. The Neoproterozoic represents a transition zone between the rigorous chronostratigraphic detail of the Phanerozoic Eon and the geochronometric subdivisions of the Precambrian time scale and, as such, special attention is given to this interval. It will be a long time before such developments can be applied to the whole of the Proterozoic, or to the Archean, and the present chapter should, therefore, be regarded as an interim report on a challenging topic.

A Geologic Time Scale 2004, eds. Felix M. Gradstein, James G. Ogg, and Alan G. Smith. Published by Cambridge University Press.

9.2 HISTORY AND RECOMMENDED SUBDIVISION

The Subcommission on Precambrian Stratigraphy (SPS) of the International Commission on Stratigraphy (ICS) worked for more than two decades to develop a proposal for the subdivision and nomenclature of the Precambrian. This process was accompanied by published reports of progress and by calls for discussion and comment. It culminated in a comprehensive review and preliminary set of proposals by Plumb and James (1986).

Comments received subsequent to this publication were reviewed at a meeting of the SPS in 1988, and the proposed time scale refined into a final recommendation. Subsequent postal ballots, first by SPS and then by ICS during early 1989, achieved in excess of the required majority of 60%, and the proposal was submitted to the International Union of Geological Sciences (IUGS) for ratification at the 28th International Geological Congress. Discussion was deferred to a later meeting in early 1990, at which time IUGS ratified the proposal as the recommended international time scale for the chronometric subdivision and nomenclature of Precambrian time and of the Proterozoic Eon (Fig. 9.1).

These definitions were subsequently formalized in Plumb (1991). The SPS has since addressed the subdivision of the Archean Eon. The ninth meeting of SPS in 1991 unanimously agreed to a provisional proposal for a four-fold subdivision of the Archean into eras (Fig. 9.1). This was formally accepted by full ballot of the subcommission following the tenth meeting in 1995. While, at the time of writing, this scheme has still to be formally submitted to and ratified by ICS/IUGS, the provisional units are increasingly appearing in the literature.

The majority of written comment received by the SPS has favored the principle of chronometric subdivisions, with most of the dissension related to the ages selected for particular boundaries. Some of the debate is also related to the disparity in techniques and reliability of age determinations available during the initial work of the SPS in the 1970s and 80s. With the proliferation of more accurate and precise age data (mainly U–Pb isotope zircon dates), there is no doubt that future workers will be well placed to reconsider the efficacy of these subdivisions and perhaps to provide intervals that are either defined in a different way or better constrained with respect to the major events of Archean and Proterozoic Earth history.

Among the many approaches to subdivision that have been considered (Plumb and James, 1986) is the concept of "geons," which are defined as 100-million-year intervals counting backward from the present (Hofmann, 1990, 1999). In this scheme for example, Geon 18 is the interval from 1800.0 to 1899.9 Ma, and Geon 34 is the interval from 3400.0 to 3499.9 Ma. There are many advantages in the simplicity of such a scheme and especially so for the protracted Precambrian, which lacks the definitive chronostratigraphic rigor of the Phanerozoic Eon.

Eon	Era	Period	Age (Ma)
Ph	Paleozoic	Cambrian	542
Proterozoic	Neoproterozoic	Ediacaran	630
		Cryogenian	850
		Tonian	1000
	Mesoproterozoic	Stenian	1200
		Ectasian	1400
		Calymmian	1600
	Paleoproterozoic	Statherian	1800
		Orosirian	2050
		Rhyacian	2300
		Siderian	2500
Archean	Neoarchean		2800
	Mesoarchean		3200
	Paleoarchean		3600
	Eoarchean	Lower limit is not defined	

Figure 9.1 Formal subdivision of the Precambrian time scale. The Ediacaran Period and its GSSP in the Flinders Range, South Australia was ratified during 2004.

It could be argued, however, that such a scheme is little different from simply describing the ages of rocks by direct isotopic dating, and would detract from the objective of attaining a compatible chronostratigraphic scale for the Precambrian. Consideration of these issues will have to come from the work of future Precambrian stratigraphers.

The simplest and most direct way of describing the age of an individual rock body is by reference to its isotopic age. The need for a time scale in the Precambrian is for consistent international classification and comparison, and for describing rock units for which direct isotopic age data are not available.

The Precambrian time scale is presently divided into the chronometric subdivisions illustrated in Fig. 9.1, where time boundaries have been selected to delimit, as far as possible, the principal cycles of sedimentation, orogeny, and magmatism. The boundaries are defined in years before present, without specific reference to any rock bodies.

The Precambrian is formally divided into an older *Archean* Eon and a younger *Proterozoic* Eon. The boundary between the two eons is defined to be at 2500 Ma. The end of the Proterozoic coincides with the beginning of the Cambrian Period, the GSSP of which has been formally defined at Fortune Head, Newfoundland, Canada (see Chapter 11). This decision was ratified by the ICS and the IUGS at the Kyoto IGC (Brasier *et al.*, 1994).

9.2.1 The Archean Eon

The Archean Eon is divided into four eras named, in conformity with nomenclature of the Proterozoic and Phanerozoic, *Eoarchean, Paleoarchean, Mesoarchean,* and *Neoarchean*.

The Eoarchean has no lower limit and an upper limit at 3600 Ma. This is followed in time by the Paleoarchean (ranging from 3600 to 3200 Ma), the Mesoarchean (ranging from 3200 to 2800 Ma), and the Neoarchean (ranging from 2800 to 2500 Ma). None of the eras in the Archean Eon have been further subdivided into periods. The boundary between the Archean and Proterozoic is defined as 2500 Ma. Descriptive terms such as "Hadean" and "Swazian" are no longer recommended in the present chronometric scheme.

The Archean has no defined lower boundary because primitive geological terranes are still being discovered and the ages of the oldest known rocks and minerals on Earth are continually being pushed back in time. The oldest dated mineral on Earth at present is a detrital zircon extracted from sample W74, a metaconglomerate from the Jack Hills area of Western Australia; this zircon grain yielded a U–Pb isotope date of 4408 ± 8 Ma and the sample itself contained several other zircons with ages between 4100 and 4300 Ma (Wilde *et al.*, 2001). Although the conglomerate itself is much younger, the existence of these grains clearly and unequivocally points to the presence of continental (i.e. granitic *sensu lato*) crust dating back to within about 150 million years of the Earth's formation at about 4560 Ma.

The oldest known rock yet dated is the Acasta orthogneiss from the Slave Craton of Canada, which has yielded a U–Pb zircon age of 4031 ± 3 Ma (Bowring and Williams, 1999).

The oldest, well-mapped, and geographically contiguous segment of Archean crust is the Itsaq gneiss complex (previously referred to as the Amîtsoq gneisses) and the Isua greenstone belt of southwest Greenland. The supracrustal rocks of the Isua belt have yielded U–Pb zircon ages that range from 3806 ± 2 (felsic metavolcanic rock; Compston *et al.*, 1986) to 3707 ± 6 Ma (banded iron formation; Nutman *et al.*, 1997).

The diverse and complex Itsaq orthogneiss suite yields a variety of U–Pb zircon ages ranging between ~3900 and 3600 Ma (Nutman *et al.*, 1996, 1999). The oldest orthogneiss yet dated from the Itsaq gneisses is 3872 ± 10 Ma. This very old age, together with that of the Acasta gneiss, is, however, controversial because of the difficulties that exist in unambiguously and objectively distinguishing xenocrystic from *in situ* zircons (Kamber *et al.*, 2001). These authors suggest, on the basis of regionally compiled whole rock Rb–Sr, Sm–Nd, and Pb–Pb isochrons, that the majority of the Itsaq gneiss complex formed at around 3655 ± 45 Ma, whereas the Acasta gneiss may be even younger at 3371 ± 59 Ma. The fact that the late, heavy meteorite bombardment affecting the Earth–Moon system, which peaked at around 3900 Ma (Cohen *et al.*, 2000), will have comprehensively destroyed most of the pre-existing terrestrial crust makes it unlikely that large, contiguous tracts of pre-3900 Ma Archean crust will ever be found.

The oldest contiguous section of the Itsaq gneiss complex, about which there seems to be no contention regarding the age, is a little deformed tonalite that has yielded a U–Pb zircon age of 3818 ± 8 Ma (Nutman *et al.*, 1999). Rocks of extreme age (i.e. >3900 Ma) clearly did exist, but they appear to be either very sparsely preserved or disaggregated and now seen only as xenocrysts or detritus.

9.2.2 The Proterozoic Eon

The Proterozoic Eon is subdivided into three eras named, in conformity with nomenclature of the Phanerozoic and from oldest to youngest, the *Paleoproterozoic, Mesoproterozoic*, and *Neoproterozoic*. These eras have boundaries placed at 2500,

1600, 1000 Ma, and the base of the Cambrian (Fig. 9.1). Each era of the Proterozoic is subdivided into periods as below.

The Paleoproterozoic is subdivided into four periods named, from oldest to youngest, the *Siderian, Rhyacian, Orosirian,* and *Statherian*, with their boundaries placed at 2500, 2300, 2050, 1800, and 1600 Ma.

The Mesoproterozoic is subdivided into three periods named, from oldest to youngest, *Calymmian, Ectasian*, and *Stenian*, with their boundaries placed at 1600, 1400, 1200, and 1000 Ma, respectively.

The Neoproterozoic is divided into three periods. The lower two, with the boundary between them placed at 850 Ma, arc, from oldest to youngest, referred to as *Tonian* and *Cryogenian*. In late 2003, the ICS Subcommission on the Terminal Proterozoic System, after a decade of discussions and deliberations and increasingly focused ballots, approved the name for the third Period as *Ediacaran*. It replaces the provisional name *Neoproterozoic III*. The ICS subcommission in late 2003 also approved by voting that the Ediacaran Period has a GGSP at the base of the Nuccaleena Formation cap carbonate, immediately above the Elatina diamictite in the Enorama Creek section, Flinders Ranges, South Australia. Pending accurate age determination, its boundary with the Cryogenian is provisionally taken as 630 Ma.

9.3 NOMENCLATURE OF THE SUBDIVISIONS

9.3.1 Eons

The names of the Precambrian Eons are derived from the original terms Archeozoic and Proterozoic, which refer, respectively, to "early" and "primitive" (animal) life. Although this definition is now no longer considered appropriate, the currently used terms *Archean* and *Proterozoic* are now entrenched in the literature and widely accepted.

9.3.2 Eras

A 1988 meeting of the SPS considered and accepted a written suggestion from H. J. Hofmann (pers. comm., 1987) that the three Proterozoic Eras be called *Paleoproterozoic*, *Mesoproterozoic*, and *Neoproterozoic*. The terms are simple and understandable, have grammatically correct Greek roots and will, therefore, transcribe into other languages with minimal translation. They are analogous to equivalent units in the Phanerozoic and are now widely used in the literature.

9.3.3 Periods

Following from the historical development of Phanerozoic chronostratigraphy, many workers have attributed a fundamental significance to stratigraphic nomenclature at the period level, although the rationale for this is not immediately clear given the varied derivation of terms. Some, such as Cambrian, Silurian, and Devonian, have a clear geographic connotation, while others such as Carboniferous and Triassic are derived from processes or events. All, however, are associated with well-understood physical and biological events, and period names convey substantial history and connotation. Classically, type sections for periods or systems were considered to characterize the global characteristics of a given interval of time. Realization that geological processes were diverse and diachronous challenged this notion and led to the development of the *boundary stratotype* principle, which became the focus of activity for most subcommissions and working groups of ICS.

Since the boundary stratotype principle has not been applied to the Precambrian (except, to a certain extent, in the Neoproterozoic), the SPS has indicated that nomenclature for Proterozoic time units should simply comprise a set of convenient labels designed to be as unambiguous as possible. No fundamental significance should be attached to any terms and the nomenclature should be unique and not duplicate names already in use, such as established or obsolete local chronostratigraphic units. Many different schemes were considered (Plumb and James, 1986) and the nomenclature finally selected for the periods of the Proterozoic Eon comprised Greek-based conceptual names that reflect, but do not define, geological history. The names relate to geological processes that are typical, *but not necessarily diagnostic*, of the interval they describe. Although the scheme could incorrectly exert an influence on decisions as to where to place particular rock bodies or geological events in the time scale, the SPS has stressed that the nomenclature simply comprises a set of convenient labels for those who require such subdivisions.

It is anticipated that increased usage of the stratigraphic nomenclature for the Proterozoic Eon will accompany improved understanding and resolution of Precambrian geological evolution. An explanation of the derivation of the names for periods in the Proterozoic is provided in Table 9.1.

9.4 THE NEOPROTEROZOIC

Archean and earlier Proterozoic time will not soon yield to chronostratigraphic subdivision, but there is reason for cautious optimism that, in tandem, geochronology and

Table 9.1 *Explanation of nomenclature used at the period level in the Proterozoic Eon*

Period name	Derivation and geological process	
Cryogenian	Cryos = ice, genesis = birth	*"Global glaciation"*
	Glacial deposits, which typify the late Proterozoic, are most abundant during this interval	
Tonian	Tonas = stretch	
	Further expansion of major platform cover (e.g. Upper Riphean, Russia; Qingbaikou, China; basins of Northwest Africa), following final cratonization of polymetamorphic mobile belts, below	
Stenian	Stenos = narrow	*"Narrow belts of intense metamorphism and deformation"*
	Narrow polymetamorphic belts, characteristic of the mid Proterozoic, separated the abundant platforms and were orogenically active at about this time (e.g. Grenville, Central Australia)	
Ectasian	Ectsis = extension	*"Continued expansion of platform covers"*
	Platforms continue to be prominent components of most shields	
Calymmian	Calymma = cover	*"Platform covers"*
	Characterized by expansion of existing platform covers, or by new platforms on recently cratonized basement (e.g. Riphean of Russia)	
Statherian	Statheros = stable, firm	*"Stabilization of cratons; cratonization"*
	This period is characterized on most continents by either new platforms (e.g. North China, North Australia) or final cratonization of fold belts (e.g. Baltic Shield, North America)	
Orosirian	Orosira = mountain range	*"Global orogenic period"*
	The interval between 1900 and 1850 Ma was an episode of orogeny on virtually all continents	
Rhyacian	Rhyax = stream of lava	*"Injection of layered complexes"*
	The Bushveld Complex (and similar layered intrusions) is an outstanding event of this time; the age of the Bushveld seems unlikely to change dramatically	
Siderian	Sideros = iron	*"Banded iron formations"* (BIF)
	The earliest Proterozoic is widely recognized for an abundance of BIF, which peaked just after the Archean–Proterozoic boundary	

stratigraphy will lead to redefined Neoproterozoic periods that reflect established Phanerozoic practices. Four considerations prompt this optimism.

1. The morphological complexity and evolutionary turnover rates of eukaryotic organisms both increased through Neoproterozoic time, providing a basis for at least broad biostratigraphic correlations.
2. Strong secular variations in the C, Sr, and (less well established) S isotopic compositions of Neoproterozoic seawater offer an independent avenue for the characterization and correlation of sedimentary rocks.
3. The extraordinary biological and geochemical events of Neoproterozoic Earth history are linked, at least in part, to global ice ages that punctuated the era.
4. Well-dated ash beds have begun to place strong constraints on the timing of these events (see Knoll, 2001, for a recent review).

As discussed below, many uncertainties attend stratigraphic interpretations of Neoproterozoic paleontology, isotopes, and climate history. There is no consensus on the number of Neoproterozoic ice ages or the correlation of individual tillites. Neither do we know whether reliable assemblage zones can be recognized among Ediacaran fauna. Such uncertainties, however, present agendas for research not insurmountable obstacles. Already, the international stratigraphic community is nearing ratification of a chronostratigraphically defined period immediately before the Cambrian. The initial boundary of this period, with the name Ediacaran, is defined by a global stratotype section and point placed at the base of the cap carbonate above a tillite judged to record the younger of two globally extensive Neoproterozoic ice ages. Ratification of this GSSP in the Flinders Range, South Australia, by ICS and IUGS took place in 2004. Fossils and isotopes provide means for the subdivision of this period (Knoll and Walter, 1992; Knoll, 2001).

9.5 ISOTOPE STRATIGRAPHY IN THE PRECAMBRIAN

Temporal trends in the isotopic compositions (Sr, C, O, and S) of marine sedimentary rocks have been used to hypothesize

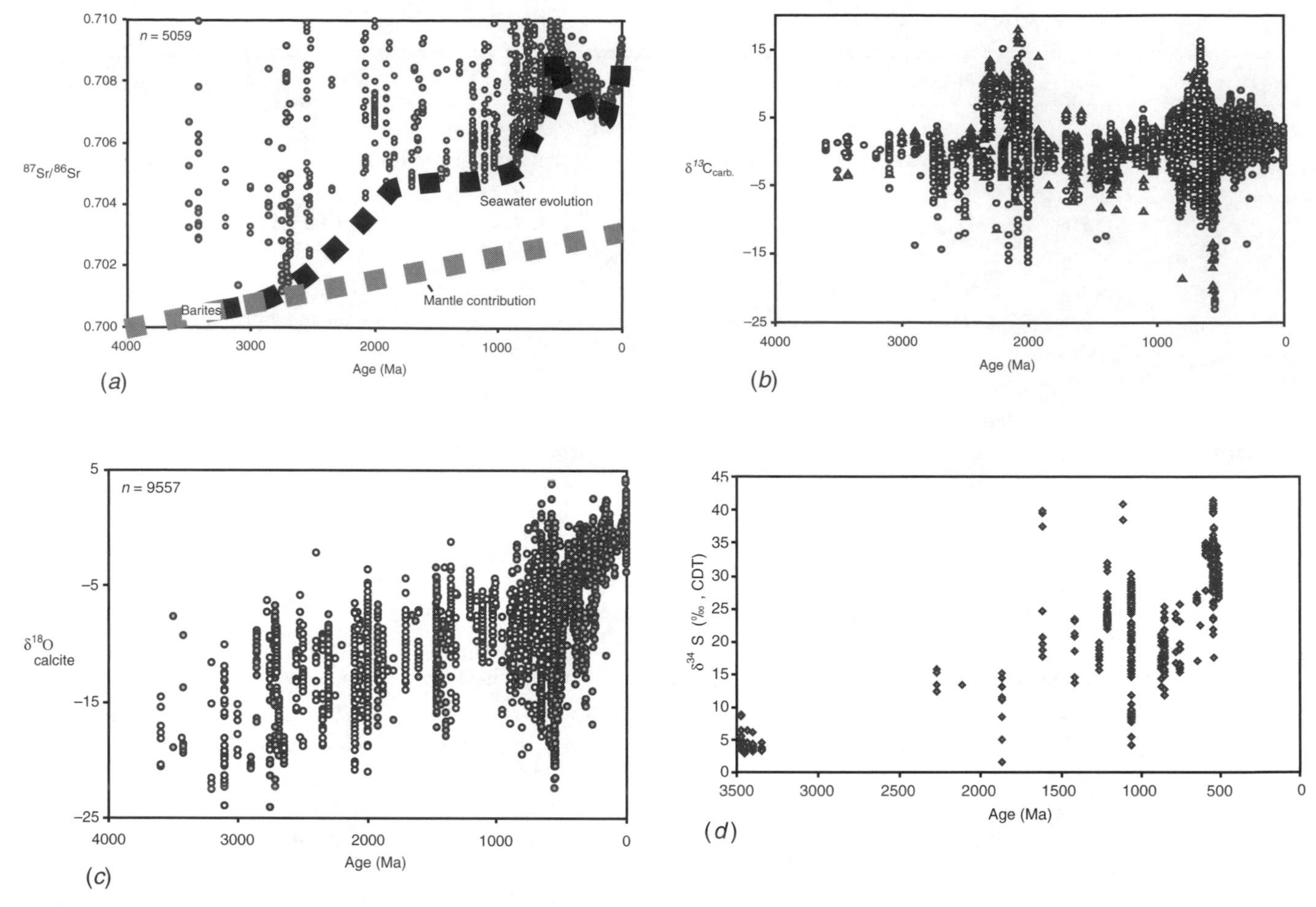

Figure 9.2 Strontium (*a*), carbon (*b*), oxygen (*c*), and sulfur (*d*) isotope geochemistry of marine carbonate and evaporite minerals through Earth history (modified after Shields and Veizer, 2002; Strauss, 2001a). Phanerozoic data are from Veizer *et al.* (1999). In Fig. 9.2(*a–c*), the filled symbols represent samples that are relatively well constrained in time (better than ±50 myr); in Fig. 9.2(*b*), triangles represent dolostone and circles represent limestone. In Fig. 9.1(*d*), open symbols represent analyses of trace sulfate.

tectonic, chemical, and biological change during the Precambrian. Fluctuations in seawater $^{87}Sr/^{86}Sr$, which can be retained in well-preserved marine carbonate rocks, reflect changes in the relative contributions of the continental versus mantle chemical reservoirs to ocean composition (Veizer, 1989). By contrast, the stable isotopes of C and S have generally been used to recognize changes in the biogeochemical cycling of these elements, which may be related to tectonic events, biological innovation, and the oxidation state at the Earth's surface (Schidlowski *et al.*, 1975; DesMarais *et al.*, 1992; Canfield and Teske, 1996; Farquhar *et al.*, 2000; Goddéris and Veizer, 2000). Recently, attention has also been paid to the potential of Sr and C isotopes in the global stratigraphic correlation of Paleoproterozoic (Melezhik *et al.*, 1999) and Neoproterozoic rock successions (Kaufman and Knoll, 1995; Shields, 1999; Walter *et al.*, 2000).

9.5.1 Strontium isotope stratigraphy

Strontium isotope ratios have the greatest potential for global stratigraphic correlation in the Precambrian because they represent a global signal, whereas absolute amplitudes of C and O isotopic excursions tend to vary. A compilation of all published $^{87}Sr/^{86}Sr$ data for the Precambrian and Cambrian is shown in Fig. 9.2*a*. These data were obtained from marine carbonate samples from 654 lithological units, the distribution of which is strongly skewed toward the period between 750 and 500 Ma. All data have been normalized to a NBS 987 standard value of 0.710 25. The considerable spread within sample populations implies that post-depositional alteration has significantly affected carbonate $^{87}Sr/^{86}Sr$, which has served in almost all cases to increase $^{87}Sr/^{86}Sr$. For this reason the lower part of the $^{87}Sr/^{86}Sr$ band is likely to represent a maximum constraint on seawater $^{87}Sr/^{86}Sr$. These "least altered" $^{87}Sr/^{86}Sr$ ratios reveal a deflection away from mantle-like $^{87}Sr/^{86}Sr$ before ~2.5 Ga to more radiogenic $^{87}Sr/^{86}Sr$ after ~2.5 Ga (Veizer and Compston, 1976). This switch is consistent with a change from a "mantle"-buffered to a "river"-buffered global ocean around this time, and is likely to result from a combination of: (1) decreasing heat flux from the mantle, and (2) intensified formation of continental crust (Veizer *et al.*, 1982).

A second major increase in $^{87}Sr/^{86}Sr$ from 0.7052 to 0.7092 took place between ~1000 and 500 Ma, implying steadily increasing continental influence on ocean chemistry during this time. This is consistent with elevated rates of tectonic uplift and erosion of highly radiogenic crust, possibly related to the birth, break-up, and dispersal of the supercontinent Rodinia (Meert and McPowell, 2001).

9.5.2 Carbon isotope stratigraphy

The $\delta^{13}C$ values reflect changes in the biogeochemical redox cycling of carbon, while long-term trends (>100 Ma) are likely to reflect real shifts in the proportion of carbonate versus organic carbon burial (Schidlowski, 1993). Published Precambrian $\delta^{13}C$ data are relatively numerous with over 10 000 samples measured from 561 distinct lithological units.

The compilation shown in Fig. 9.2*b* confirms that marine bicarbonate $\delta^{13}C$ remained close to 0% during much of Precambrian time (Schidlowski *et al.*, 1975). Two prolonged intervals of anomalously high (>10‰) and variable (up to 20‰) $\delta^{13}C$ can be identified: mid Paleoproterozoic (2.3–1.9 Ga) and mid Neoproterozoic (0.8–0.6 Ga). The extent to which high marine carbonate $\delta^{13}C$ for either period represents anomalous global seawater bicarbonate $\delta^{13}C$ or local isotopic enrichment has not been established (Melezhik *et al.*, 1999). However, the observation that comparable $\delta^{13}C$ trends can be found in many sedimentary successions, thought on independent grounds to be of the same age, makes a purely local interpretation unlikely, at least for the Neoproterozoic. Mean $\delta^{13}C$ values reveal a sustained increase at these times, which is consistent with a real increase in organic carbon burial and storage rates. Atmospheric oxygen concentrations are likely to have risen as a consequence (DesMarais *et al.*, 1992).

9.5.3 Oxygen isotope stratigraphy

Despite considerable scatter due to post-depositional alteration, primary variation, and analytical inconsistencies, calcite $\delta^{18}O$ values are generally depleted throughout the Precambrian relative to most of the Phanerozoic (Fig. 9.2*c*). Mean $\delta^{18}O$ values from both calcite and dolomite increase in parallel through the Precambrian with a roughly constant isotopic discrimination that probably reflects differences in their equilibrium isotopic fractionations during precipitation from seawater (Land, 1980). The low $\delta^{18}O$ values of most Precambrian carbonates are consistent with the well-documented increase in marine calcite $\delta^{18}O$ during the Phanerozoic, which has been interpreted as resulting from a tectonically controlled, first-order increase in seawater $\delta^{18}O$ (Veizer *et al.*, 2000). The extent to which seawater $\delta^{18}O$ can have changed over geological history is still a matter of controversy (Muehlenbachs, 1998; Goddéris and Veizer, 2000; Wallmann, 2001).

9.5.4 Sulfur isotope stratigraphy

The sulfur isotope record for Precambrian seawater sulfate, based on the analysis of anhydrite, gypsum, barite, and trace amounts of sulfate in carbonates, cherts, and phosphorites (e.g. Strauss, 1993, 2001a,b; Shields *et al.*, 1999; Kah *et al.*, 2001; Strauss *et al.*, 2001) is much more fragmentary than the carbonate-based isotope records for strontium, carbon, and oxygen. This is a consequence of a rather limited number of preserved evaporite deposits. Accordingly, the results of some 500 measurements for only 26 stratigraphic units are distributed unevenly throughout the first 4 Ga of Earth's evolution, and provide poor temporal resolution (Fig. 9.2*d*).

The early Archean sulfate sulfur isotope record is based on the analyses of barites from Australia, southern Africa, and India, which are generally taken to reflect the seawater signature. The $\delta^{34}S$ values display a range between +2.7 and +8.7‰, averaging $+4.0 \pm 1.1$‰ ($n = 67$). The temporal evolution throughout the Archean and Proterozoic displays a first-order increase in $\delta^{34}S$ toward an average value of $+32.1 \pm 3.7$‰ ($n = 134$) for the terminal Neoproterozoic. This rise is likely not linear, but poor time resolution prohibits detailed evaluation.

The very positive sulfur isotope values of terminal Neoproterozoic age have been recorded from numerous evaporite deposits around the world. Stratigraphically, these values occur in units above the late Proterozoic episode of global glaciation and continue into the Cambrian. Despite the fact that these units can be only loosely correlated, extremely positive $\delta^{34}S$ values appear to reflect a globally representative signature for seawater sulfate.

The temporal evolution in the sulfur isotopic composition of seawater sulfate throughout Earth's history reflects changes in the (bio)geochemical cycling of sulfur.

By far the largest isotope effect accompanies the bacterial reduction of sulfate, preferentially utilizing the lighter ^{32}S isotope (Kaplan and Rittenberg, 1964; Canfield, 2001; Detmers *et al.*, 2001). As a consequence of mass balance, the remaining sulfate pool, now present as evaporitic sulfate in the rock record, is enriched in ^{34}S. Thus, the observed increase toward more positive $\delta^{34}S$ values through time could reflect the growing importance of a biologically controlled sulfur cycle. Due to the poor time resolution, however, no conclusive evidence exists as far as the timing is concerned.

Based on the sulfur isotopic composition of sedimentary pyrite (reflecting the biologically mediated product), unequivocal evidence suggests the late Archean–early Paleoproterozoic as the transition period from a mantle-dominated to a biologically controlled sulfur cycle (Canfield and Raiswell, 1999; Goddéris and Veizer, 2000; Strauss, 2001b), despite a few earlier occurrences of biogenic pyrite that reveal substantial isotopic fractionation (Shen *et al.*, 2001).

9.5.5 The role of isotope stratigraphy in global correlation

Phanerozoic events and trends can be identified and correlated using a stratigraphic framework based primarily on biostratigraphy with only relatively minor contributions from other stratigraphic methods. However, no such framework exists for most of the Precambrian, which makes global stratigraphic correlation considerably less straightforward. Recently, several competing isotope-based "global stratigraphic calibration schemes" have been proposed, which facilitate global correlation, at least in the Neoproterozoic. These schemes integrate relevant radiometric ages with a variously weighted combination of sequence, litho-, bio-, and chemostratigraphy (e.g. Kaufman and Knoll, 1995; Shields, 1999; Walter *et al.*, 2000). It is important to remember that none of these schemes will turn out to be correct in their entirety but must be treated as working hypotheses that can be repeatedly tested and improved upon in an iterative fashion by future research. In this regard, they differ little from biostratigraphic correlations of Phanerozoic rocks.

Isotope stratigraphy forms a vital component of these calibration schemes due to the rapidly changing $^{87}Sr/^{86}Sr$ ratios and the extreme $\delta^{13}C$ values recorded in Neoproterozoic marine carbonates (Fig. 9.3). However, care must be taken that isotope-based correlation avoids circular reasoning by combining as much supporting evidence as possible. In order to lessen the degree of circularity, some researchers choose to illustrate their correlation hypotheses as "stacked sections." In this approach, stratigraphic sections are shown alongside each other together with potentially useful correlation criteria, e.g. U–Pb age constraints, major sequence boundaries, glaciogenic units, marker fossils, $\delta^{13}C$ excursions, and least-altered $^{87}Sr/^{86}Sr$ ratios. One such example is given in Fig. 9.4. The stacked section approach is advantageous because the sequence of appearance of correlative features within each of these sections is known from the local stratigraphy, which normally cannot be refuted. Therefore, the repetition of the same sequence of features in entirely unrelated sections raises a strong argument for the global importance of these features as well as for the mutual correlation of these sections. Section stacking allows more complete isotopic and biotic evolutionary records to be established (Fig. 9.3), which can then assist in the recognition of hiatuses.

Two aspects of Fig. 9.3 are unusual and require additional comment. First, $^{87}Sr/^{86}Sr$ features are shown as "chemozones." This was first recommended by McArthur (1994) and circumvents the problem of not knowing the precise Neoproterozoic $^{87}Sr/^{86}Sr$ curve. The $^{87}Sr/^{86}Sr$ chemozones also help in the differentiation of otherwise indistinguishable $\delta^{13}C$ excursions (Shields, 1999). Second, $\delta^{13}C$ features in Fig. 9.3 have been assigned the names of the lithological units in which they are best expressed, e.g. the Keele $\delta^{13}C$ peak (*Ke*) or the Yudoma $\delta^{13}C$ negative excursion (*Ya*). Chronological ordering (e.g. 1, 2, 3; A, B, C; i, ii, iii) is best avoided because this approach necessitates the continual re-invention of new schemes as stratigraphic records become more complete, which can lead to considerable confusion (Hancock, 2000).

9.6 BIOSTRATIGRAPHY IN THE NEOPROTEROZOIC

Fossils are abundant in Phanerozoic rocks and very much less so in latest Archean and Proterozoic strata. Surprisingly perhaps, stromatolites are just as common in late Archean as in Proterozoic strata. The Proterozoic Eon presents a sliding scale of biostratigraphic opportunity. Cyanobacteria are particularly widespread in Proterozoic cherts and shales, but their record strongly suggests early diversification followed by the long persistence of little varying lineages. In consequence, cyanobacterial fossils tend to provide better indicators of environment than age. Nonetheless, to the extent that the nature of coastal marine environments have shifted through the Proterozoic Eon, the stratigraphic distributions of cyanobacterial fossils should reflect this change. For example, the marine precipitates found commonly in Paleo- and Mesoproterozoic carbonates deposited on tidal flats largely disappear by

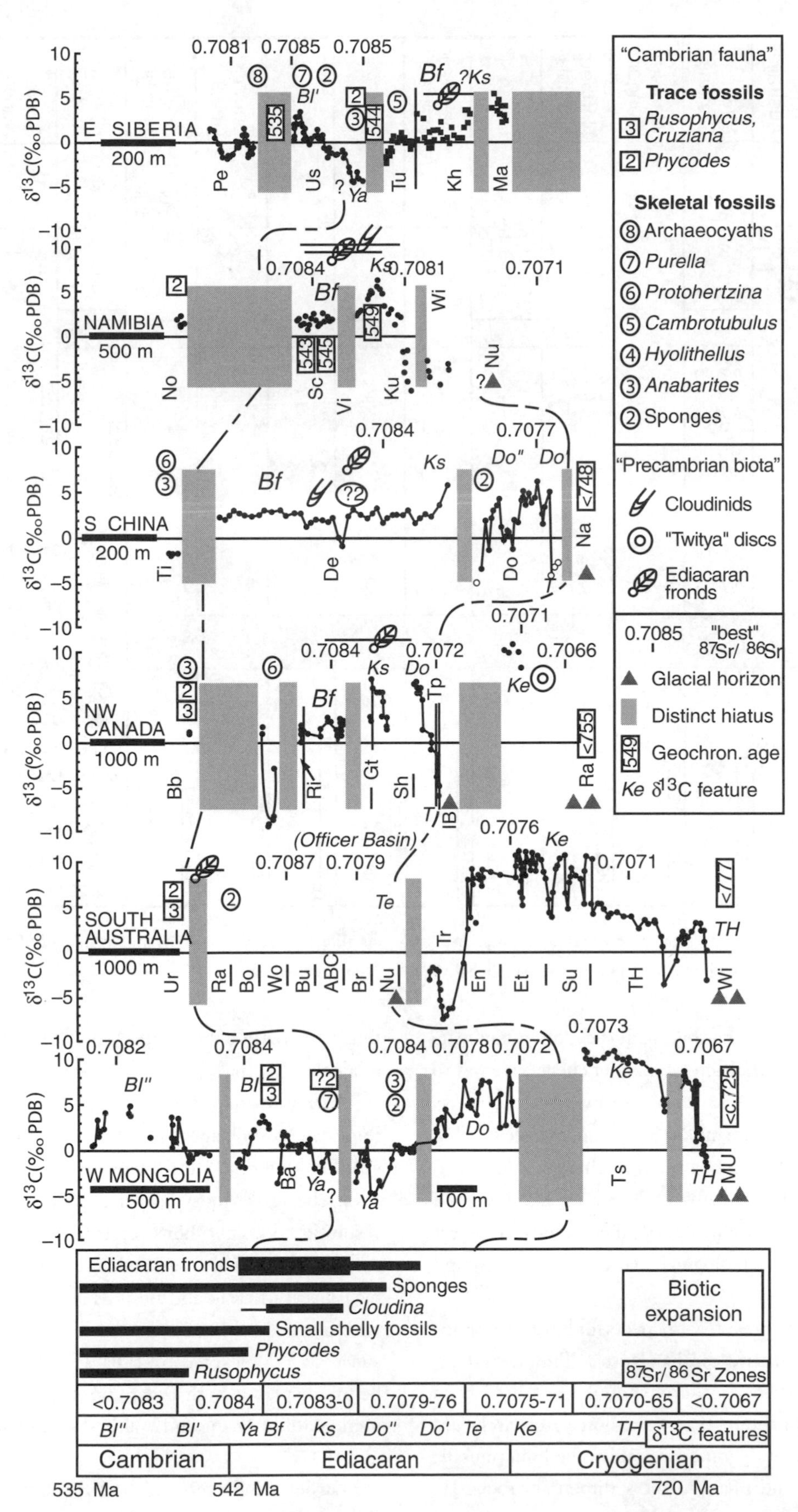

Figure 9.3 Attempted stratigraphic correlation between key Neoproterozoic–Cambrian successions world-wide (modified after Shields, 1999). Note non-linear age scale, which exaggerates the importance of the latest Neoproterozoic. Resultant order of animal fossil first appearances is given at base together with chemical zonation.

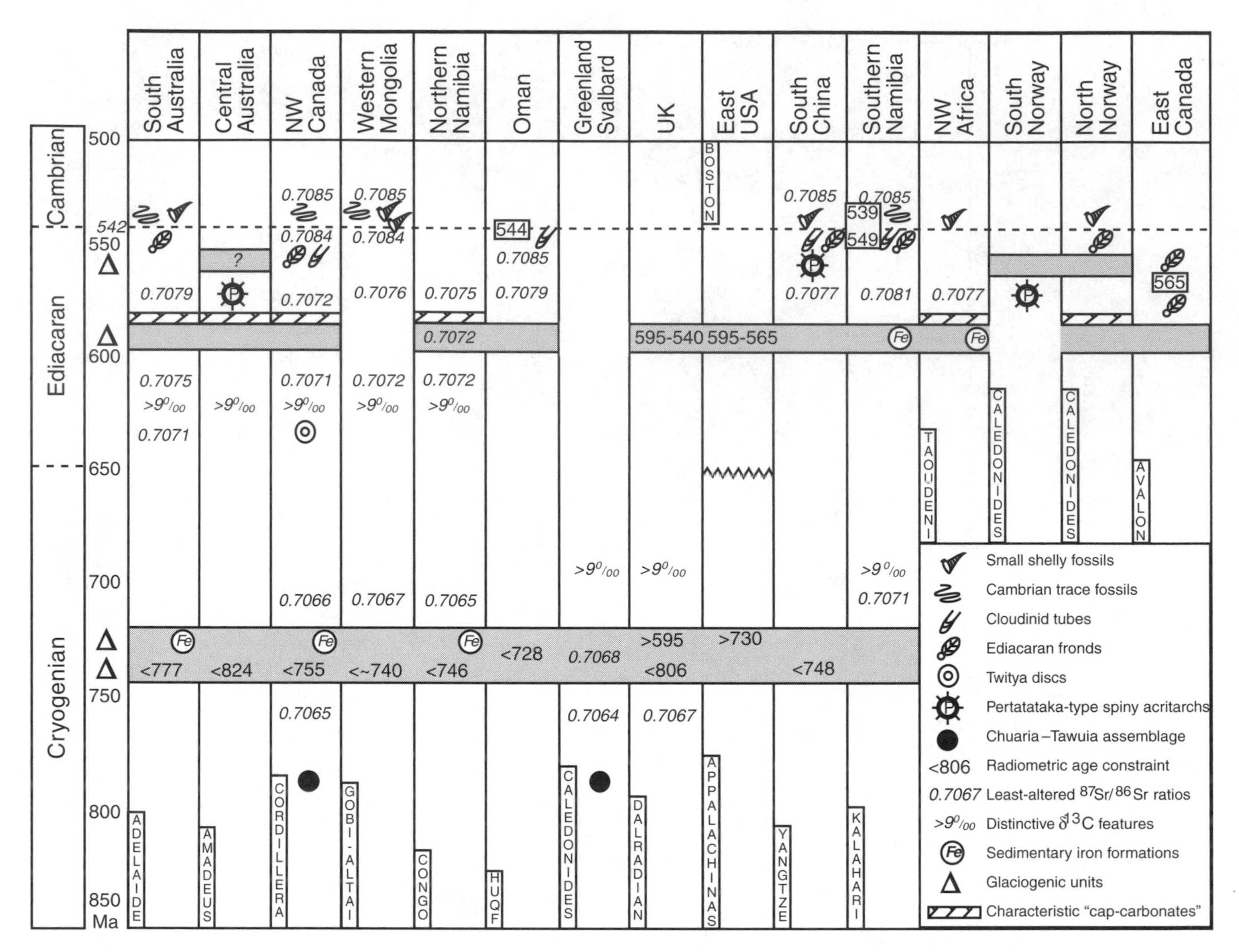

Figure 9.4 Age constraints and correlation criteria are consistent with, but do not prove the existence of, two, possibly three, globally significant glacial episodes (shown shaded) during the mid–late Neoproterozoic.

the Neoproterozoic. Mirroring this change, the abundant mat-building entophysalid cyanobacteria in older cherts give way to cyanobacteria such as *Polybessurus* whose living counterparts exist on unlithified carbonate muds (Knoll and Sergeev, 1995). Stromatolites also show broad patterns of change in form and microfabric through the Proterozoic Eon and, like observed stratigraphic patterns in cyanobacteria, these appear to reflect environmental as much as biological changes (Grotzinger and Knoll, 1999).

Eukaryotic fossils offer better prospects for biostratigraphy. Beginning with pioneering research in Russia (Timofeev, 1966) and Sweden (Vidal, 1976), acritarch populations have been used to correlate Neoproterozoic successions. Acritarchs are generally spheroidal to polygonal, organic-walled microfossils found in Proterozoic and Phanerozoic sedimentary rocks. By definition, their biological relationships are uncertain – if you know it is a green alga, you don't place it among the acritarchs – but most are thought to represent the vegetative walls or spores of algae.

Acritarchs have been used to divide the Early Cambrian into five assemblages or zones, demonstrating a biostratigraphic potential equivalent to that of contemporaneous invertebrates (Moczydlowska, 1991). This biostratigraphic resolution does not, however, extend downward into Proterozoic rocks. Acritarch assemblages from successions marked by Ediacaran casts and moulds are characteristically simple, with few taxa of mostly unornamented spheroids. In contrast, successions bracketed by Ediacaran fossils above and tillites below display a high diversity of large, flamboyantly ornamented forms (Zang and Walter, 1992; Zhang *et al.*, 1998), seen globally.

Earlier Neoproterozoic rocks contain distinctive microfossils, including ornamented acritarchs and vase-shaped microfossils (now known to be related to lobose and filose amoebae;

Porter *et al.*, 2003). These fossils are distinctively Neoproterozoic in aspect, but the degree to which finer-scale zonation can be justified remains uncertain (reviewed by Knoll, 1996). Until recently, paleontologists commonly assumed that older Proterozoic rocks contained only simple acritarchs, but distinctively ornamented forms have now been documented in Mesoproterozoic rocks from China and Australia (Xiao *et al.*, 1997; Javaux *et al.*, 2001). Thus, biostratigraphic correlation may eventually prove feasible for the second half of the Proterozoic Eon.

Eukaryotic macrofossils are less common in Proterozoic rocks, but these, too, have time distributions that suggest potential biostratigraphic utility. The distinctive macrofossils (algae?) *Horodyskia* and *Grypania* have been found in several Paleo- to Mesoproterozoic successions, but not younger rocks. In contrast, carbonaceous compressions such as *Tawuia* and *Longfengshania* have been identified in earlier Neoproterozoic shales, but not in older or younger rocks. Conspicuously branching algae and flanged tubules possibly produced by cnidarian-grade animals are limited to terminal Proterozoic shales (Hofmann, 1994; Steiner, 1994; Xiao *et al.*, 2001).

Pre-Ediacaran animals have been claimed from time to time, but a continuous record of abundant, widespread, distinctive, and, therefore, stratigraphically useful animal fossils begins only in the terminal Proterozoic period. Centimeter-scale compressions of spheroidal, sac-like fossils occur beneath the younger tillite in northwestern Canada (Hofmann *et al.*, 1990); these may be metazoans, but most could alternatively have algal origins. Microscopic animal remains, including embryos and small cnidarian-like colonies occur in phosphatic rocks just above the younger tillite in South China (Xiao and Knoll, 2000; Xiao *et al.*, 2000). Canonical Ediacaran fossils, in turn, occur in Newfoundland sandstones dated, from intercalated ash beds, at ~575 Ma. The Newfoundland assemblage includes complex frondose structures, but not distinctively bilaterian body or trace fossils. In contrast, ~555 Ma rocks from the White Sea contain both body impressions and trackways of bilaterian animals (Martin *et al.*, 2000).

As a group, Ediacaran fossils clearly define a discrete interval of latest Proterozoic time. Assemblage zones may allow subdivision of this interval – for example, the distinctive frondose fossil *Swartpuntia* has been found near the Proterozoic–Cambrian boundary on several continents – but for now, this possibility remains conjectural. Microbial reefs in latest Proterozoic (<549 Ma) carbonates also contain calcified macrofossils (*Cloudina* and *Namacalathus*) of demonstrated biostratigraphic value (Grant, 1990; Grotzinger *et al.*, 2000).

In summary, fossils document evolutionary and environmental change through Proterozoic time, and eukaryotic fossils document accelerating evolution through the Neoproterozoic Era. Thus, in tandem with isotopic chemostratigraphy, paleontology provides tools for the correlation of Proterozoic and, especially, Neoproterozoic sedimentary rocks. These are likely to remain fruitful areas of research in the short term and will undoubtedly lead to much improved global stratigraphic correlations in the Precambrian.

9.7 NEOPROTEROZOIC ICE AGES AND CHRONOMETRIC CONSTRAINTS

Evidence for continental glaciation is found in mid and late Neoproterozoic successions world-wide. This evidence usually comprises glaciomarine (or glaciolacustrine) diamictites (mixtites) that contain exotic, occasionally striated and faceted "dropstones" embedded within finer grained, often finely stratified sediment. More rarely, striated or grooved pavements, roches moutonnées, eskers, and tunnel-valleys, as well as other periglacial features, such as ice-wedge pseudomorphs and wind-blown loess, imply peripheral terrestrial glaciation. Some of the Neoproterozoic glaciations appear to have been unusually severe and extensive, with currently indisputable evidence from Australia of grounded ice at equatorial latitudes during the Marinoan glaciation (Sohl *et al.*, 1999).

At present it is unclear how many glaciations occurred during the Neoproterozoic Era or whether all these glaciogenic deposits represent times of global glaciation (Evans, 2000). Fig. 9.4 illustrates how global stratigraphic correlation criteria are consistent with, but not yet proof of, two major phases of glaciation, one mid Neoproterozoic phase at ~730 ± 15 Ma and a younger phase at ~580 ± 10 Ma. The timing of the later Neoproterozoic glacial episode is poorly constrained in most regions of the world except eastern North America.

At present, there is no evidence for any early Neoproterozoic, i.e. pre-750 Ma glaciation, which had been envisaged previously (Eyles and Young, 1994). In particular, recent geochronological evidence (Rainaud *et al.*, 2001) is inconsistent with a pre-Sturtian glaciation in Africa. Current bio- and chemostratigraphic constraints would allow for at least regional glaciation during the very latest Neoproterozoic, sometime between 570 and 550 Ma (Fig. 9.4).

The earlier mid Neoproterozoic glacial episode commonly comprises two major diamictite–mudstone sequences, which have been interpreted to represent two or more glacial advance–retreat cycles (Eyles and Young, 1994), although this particular interpretation is no longer widely accepted. Sedimentary iron formations are frequently associated with, but are not unique to, this mid Neoproterozoic episode of glaciation.

By contrast, late Neoproterozoic glaciogenic units are commonly capped by a characteristically deformed, generally <5 m thick, dolostone unit of enigmatic origin. This "cap carbonate" is never found overlying clearly metazoan fossils, such as small shelly fossils, or Ediacara-type complex fronds, and might become a robust criterion for global stratigraphic correlation (Kennedy *et al.*, 1998).

It seems likely that there were at least two distinct episodes of global glaciation during the mid–late Neoproterozoic: from 745 to 725 Ma and from 590 to >550 Ma, each episode of which seems likely to have comprised more than one cycle of glacial advance and retreat.

At present, we are unable to date these glaciations worldwide. Most age constraints are ambiguous, while some glaciogenic deposits may represent only local glaciation. For these reasons, the currently widespread use of the terms "Varangerian" ("Varangian"), "Marinoan," or "Sturtian" is not recommended unless direct correlation with the sections of origin of these names has been seriously attempted. In their place, we advise use of the terms "upper" or "late Proterozoic" and "mid Neoproterozoic," respectively.

9.8 SUMMARY

The science of biostratigraphy is applied essentially to rocks of the Phanerozoic Eon. As illustrated in the ensuing chapters of this book, a Phanerozoic time scale that is "biostratigraphically conceived and chronostratigraphically defined" (Knoll, 2001) is now reasonably well established on a global scale.

By contrast, more than seven-eighths of Earth history, between ~4500 and 500 Ma, is subdivided into segments of geological time that are more arbitrarily defined in terms of geochronometry. These segments occur at the eon level, where the Archean Eon is defined as applying to any rock older than 2500 Ma and the Proterozoic Eon as applying to any rock whose age is younger than 2500 Ma but older than 542 Ma (the age of the presently defined GSSP for the base of the Cambrian Period; see Chapter 11).

Further division occurs at the era level, and these are defined by adding the prefixes Eo (>3600 Ma), Paleo (3600–3200 Ma), Meso (3200–2800 Ma), and Neo (2800–2500 Ma) to the term Archean and, likewise, the prefixes Paleo (2500–1600 Ma), Meso (1600–1000 Ma), and Neo (1000–542 Ma) to the term Proterozoic (Fig. 9.1).

Only the Proterozoic Eon is currently further subdivided into periods and these subdivisions are also shown in Table 9.1. The periods of the Proterozoic Eon, although still arbitrarily allocated, have been defined in terms of, and are chosen to coincide with, major geological processes characterizing this time interval.

Despite the major differences that exist in the application of stratigraphic principles in the Precambrian compared to the Phanerozoic Eon, steady progress is being made in the definition of a time scale in older successions, and particularly in the Proterozoic Eon.

The Neoproterozoic Era is characterized by at least two, and possibly four, severe and extensive glaciogenic events, the legacies of which have been to deposit highly characteristic diamictic and chemical sediments that may provide the basis for well-dated marker horizons on a global scale.

The recognition of a Neoproterozoic supercontinent, Rodinia, provides an additional framework in terms of which sequence stratigraphy, magmatism, and orogenic cycles are being used, and will be utilized in the future, to facilitate global correlations. The definition of GSSPs older than the base of the Cambrian has commenced in the Neoproterozoic Era and may then work progressively back in geological time as the pattern of Precambrian crustal evolution is unraveled.

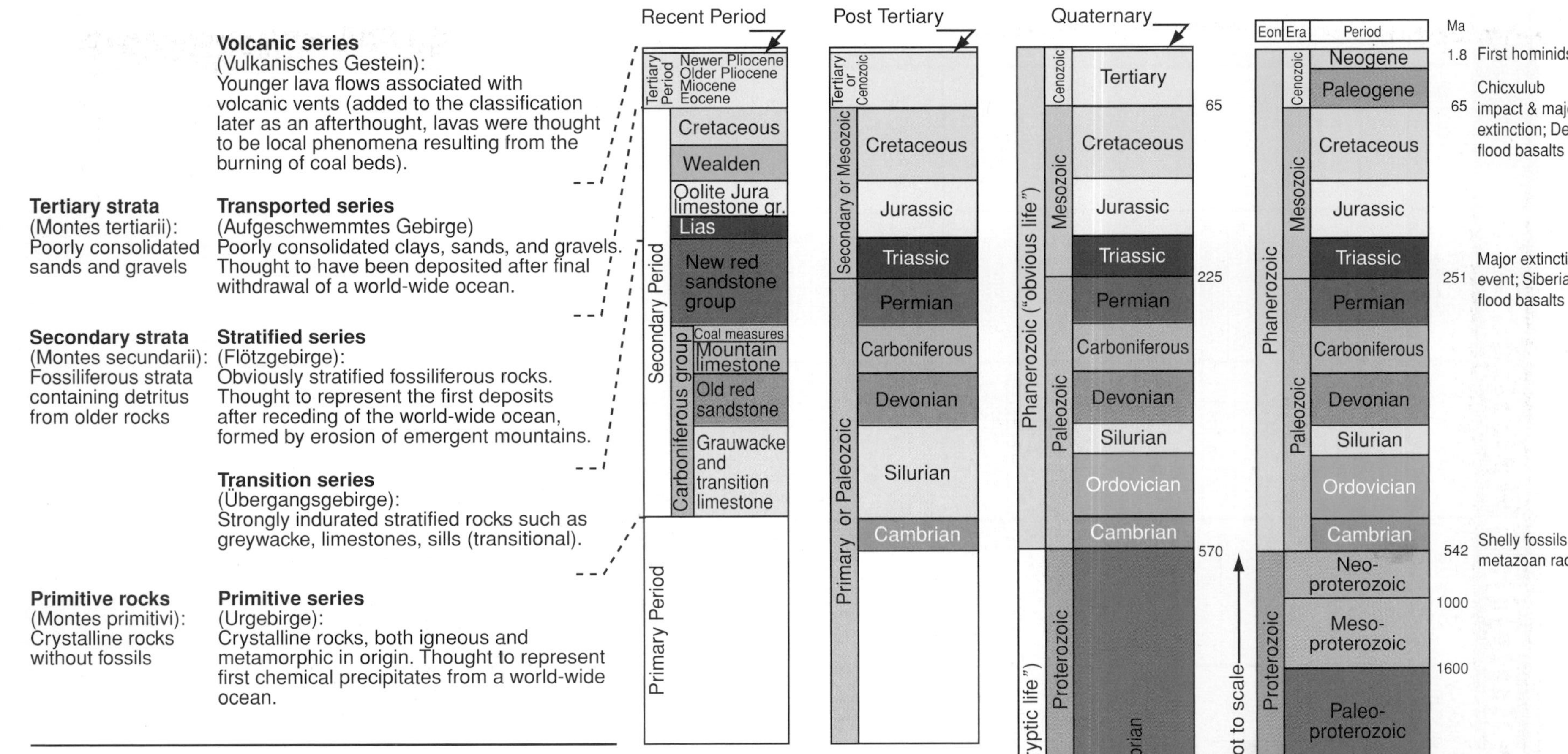

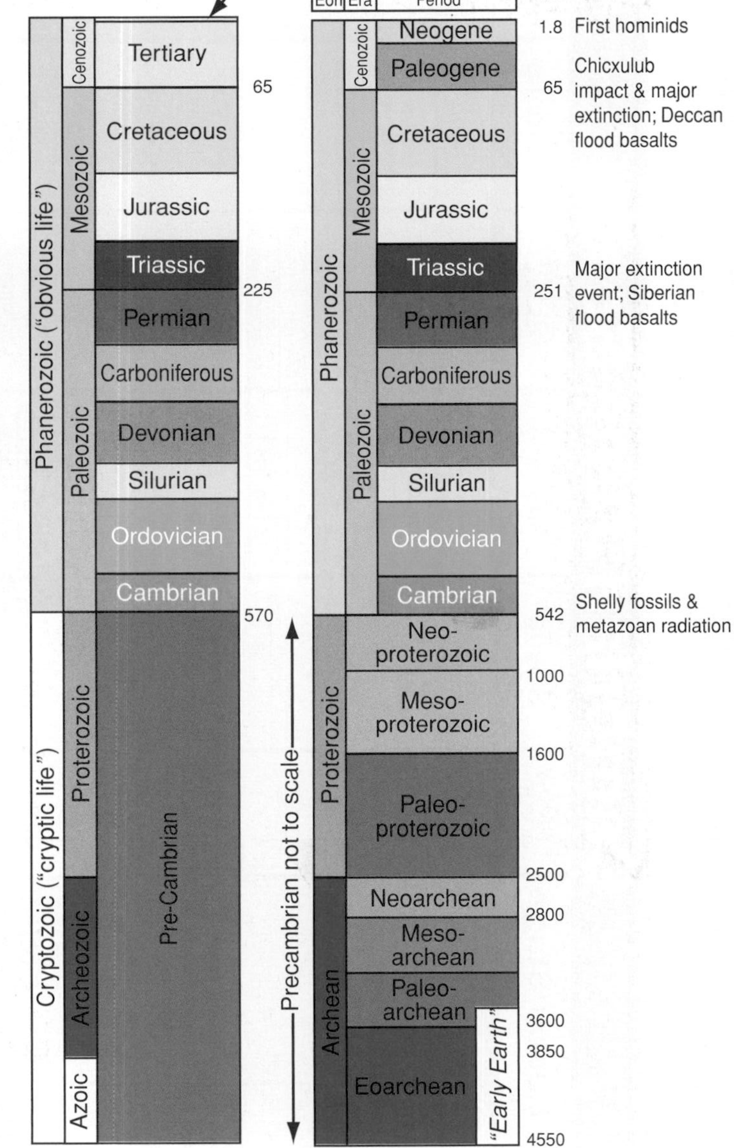

The geological time scale is as much an artifact of history (i.e. the growth of geology as a science), as it is a logical construct. Its origins go back to Nicholas Steno (Niels Stensen, 1638–1686), who was the first to formulate the "law of superposition" clearly. Nearly a century later, Giovanni Arduino (1714–1795) ordered rocks in one of the first crude stratigraphic schemes. These early attempts to classify rocks in terms of their origin and crude stratigraphy were based largely on rock types.

Similarly, in Abraham G. Werner's (1749–1817) stratigraphic scheme, a young intrusive granite rock would have been part of his basal unit, the "Urgebirge", as magmatic processes and cross-cutting relationships were not yet properly understood. Werner considered crystalline rocks to be precipitates from seawater. He thus was the leading proponent of the "neptunist" school of thought. This interpretation was later challenged by the "plutonist" school of thought, championed by James Hutton (1729–1797), which interpreted crystalline rocks, correctly, to have formed by crystallization of cooling magmas.

Cambrian Time Scale

AGE (Ma): 490, 495, 500, 505, 510

Epoch/Stage

- Ordovician
- 488.3 ± 1.7
- Furongian
  - 6th stage
  - Paibian
- 501 ± 2.0
- Middle
  - 4th stage
  - 3rd stage
  - 2nd stage
  - 1st stage
- 513 ± 2.0
- Early

Polarity chron

No data: major polar wander

Siberia

- Rhabdinopora flabelliformis
- Dolgeuloma
- Kaninia
- ?
- Kujandaspis
- ?
- Amorphella–Yurakia
- Faciura–Garbiella
- Maspakites–"Idahoia" Raashellina
- Pedinocephalina–Toxotis
- Acro.granulosa–Kold. prolixa
- L.laevigata–Oidalag.trispinifer
- A. limbataeformis
- Anopolenus henrici–Corynexochus perforatus
- Tomagnostus fissus
- Ptychagnostus gibbus
- Kounamkites
- Oryctocara–Schistocephalus antiquus
- Anabaraspis splendens
- Lermontovia grandis
- Bergeroniellus ketemensis
- Bergeroniellus ornata

Australia

- Cordylodus lindstromi–Rhabdino? scitulum
- C. prolindstromi
- Cordylodus proavus
- Mictosaukia perplexa
- Neoagn. quasibilobus–S. nomas
- Sinosaukia impages
- Rh. clarki maximus–Rh. papilio
- Rhaptagn. bifax–N.denticulatus
- Rh. clarki prolatus–Caz. secatrix
- R. c. patulus/ C.squamosa–H. lilyensis
- Peichiashania tertia–P. quarta
- Peichiash. secunda–Pro. glabella
- Wentsuia iota–Rhaptagn. apsis
- Irvingella tropica
- Stigmatoa diloma
- Erixanium sentum
- Proceratopyge cryptica
- Glyptagnostus reticulatus
- Glyptagnostus stolidotus
- Achmarhachis quasivespa
- Erediaspis eretes
- Lejopyge laevigata
- Goniagnostus nathorsti
- Ptychagnostus punctuosus
- Euagnostus opimus
- Acidusus atavus
- Ptychagnostus gibbus
- Xystridura templetonenesis–Redlichia chinensis
- Pararaia janeae
- ?
- Pararaia bunyerooensis

South China

- Hysterolenus – Onychopyge
- Leiostr. constrictum–Shen. brevica
- Mictosaukia striata–Fatocephalus
- Archaeul. taoyuanense–Leioagn. cf. bexelli
- Lot.punctatus–Hedinaspis regalis
- Probinacunaspis nasalis–Peichiashania hunanensis
- Eolotagnostus decoratus – Kaolishaniella
- Rhaptagnostus ciliensis
- Onchonot. cf. kuruktagensis
- Agnost. clavata–Irving. angustilimbata
- Corynexochus plumula – Sinop. cf. kiangshanensis
- Innitagnostus inexpectans – Proceratopyge protracta
- Glyptagnostus reticulatus
- Glyptagnostus stolidotus
- Linguagnostis reconditus
- Proagnostus bulbus
- Lejopyge laevigata
- Goniagnostus nathorsti
- Ptychagnostus punctuosus
- Acidusus atavus
- Ptychagnostus gibbus
- Oryctocephalus indicus
- Bathynotus
- Protoryctocephalus
- Arthricocephalites – Changaspis
- ?
- Arthricocephalus

Laurentia

- Symphysurina bulbosa
- S. brevispicata M. depressa
- Eurekia apopsis
- Saukiella serotina
- Saukiella junia
- Saukiella pyrene–Rasettia maga
- Ellipso–cephaloides
- Idahoia
- Taenicephalus
- Elvinia
- Dunderbergia
- Aphelaspis
- Crepicephalus
- Cedaria
- Lejopyge laevigata
- Ptychagnostus punctuosus
- Acidusus atavus
- Ptychagnostus gibbus
- Pentagnostus praecurrens
- Peronopsis bonnerensis
- Albertella
- "Plagiura–Poliella"
- Bonnia – Olenellus

Bioevents

- Cordylodus proavus (Co)
- Irvingella + Agnostotes (Tr)
- Glyptagnostus reticulatus (Tr)
- Glyptagnostus stolidotus (Tr)
- Linguagnostis reconditus (Tr)
- Lejopyge laevigata
- Ptychagnostus punctuosus (Tr)
- Acidusus atavus (Tr)
- Ptychagnostus gibbus (Tr)
- Oryctocephalus indicus (Tr)
- Protolenus–Hamatolenus–Cobboldites–Oryctocara (Tr)

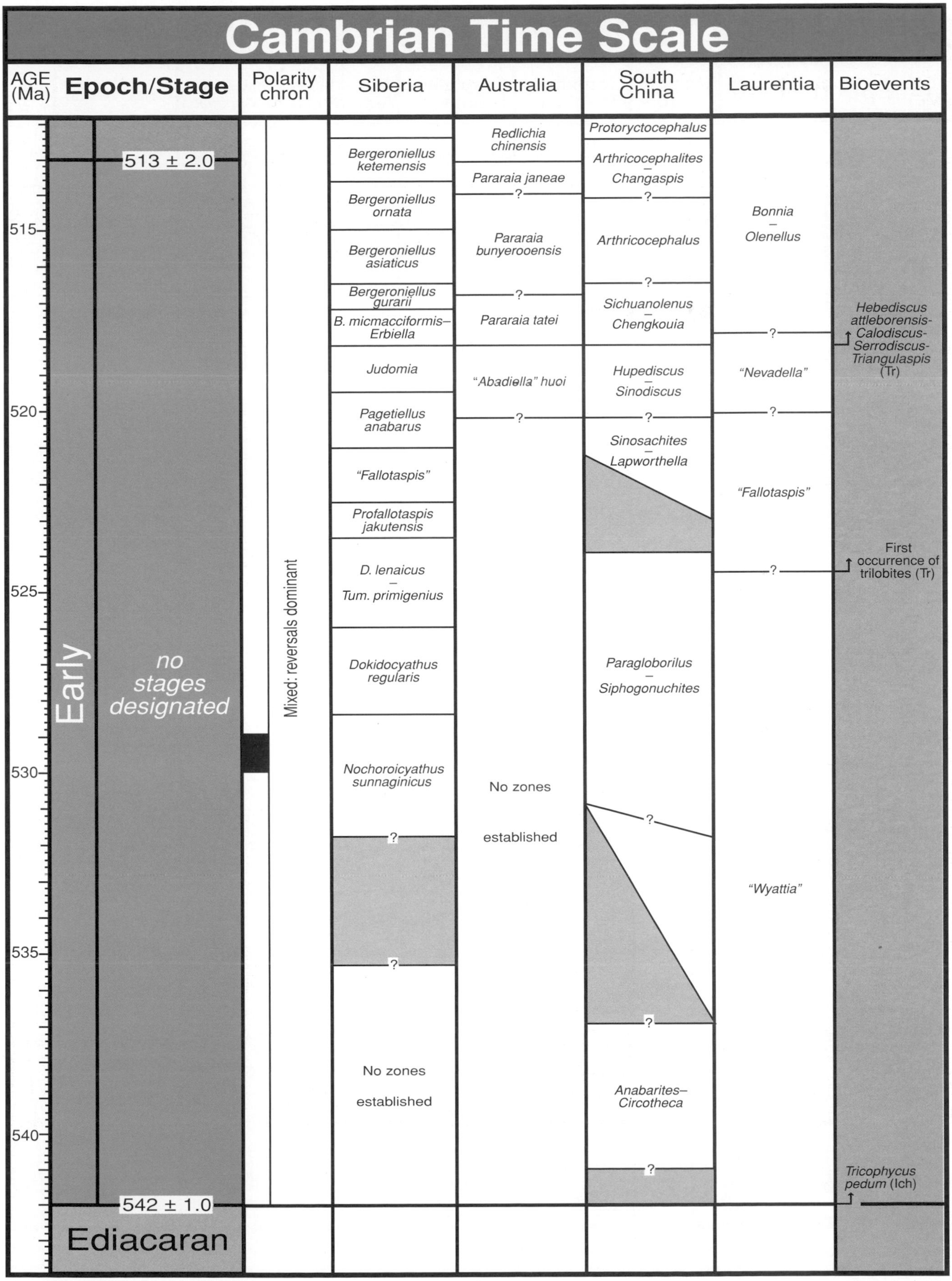

Cambrian Time Scale

AGE (Ma)	Epoch/Stage	Polarity chron	Siberia	Australia	South China	Laurentia	Bioevents

AGE (Ma): 515, 520, 525, 530, 535, 540

Epoch/Stage: 513 ± 2.0; Early; no stages designated; 542 ± 1.0; Ediacaran

Polarity chron: Mixed: reversals dominant

Siberia:
- *Bergeroniellus ketemensis*
- *Bergeroniellus ornata*
- *Bergeroniellus asiaticus*
- *Bergeroniellus gurarii*
- *B. micmacciformis–Erbiella*
- *Judomia*
- *Pagetiellus anabarus*
- *"Fallotaspis"*
- *Profallotaspis jakutensis*
- *D. lenaicus – Tum. primigenius*
- *Dokidocyathus regularis*
- *Nochoroicyathus sunnaginicus*
- ?
- ?
- No zones established

Australia:
- *Redlichia chinensis*
- *Pararaia janeae*
- ?
- *Pararaia bunyerooensis*
- ?
- *Pararaia tatei*
- *"Abadiella" huoi*
- ?
- No zones established

South China:
- *Protoryctocephalus*
- *Arthricocephalites – Changaspis*
- ?
- *Arthricocephalus*
- ?
- *Sichuanolenus – Chengkouia*
- *Hupediscus – Sinodiscus*
- ?
- *Sinosachites – Lapworthella*
- *Paragloborilus – Siphogonuchites*
- ?
- ?
- *Anabarites–Circotheca*
- ?

Laurentia:
- *Bonnia – Olenellus*
- ?
- *"Nevadella"*
- ?
- *"Fallotaspis"*
- ?
- *"Wyattia"*

Bioevents:
- *Hebediscus attleborensis-Calodiscus-Serrodiscus-Triangulaspis* (Tr)
- First occurrence of trilobites (Tr)
- *Tricophycus pedum* (Ich)

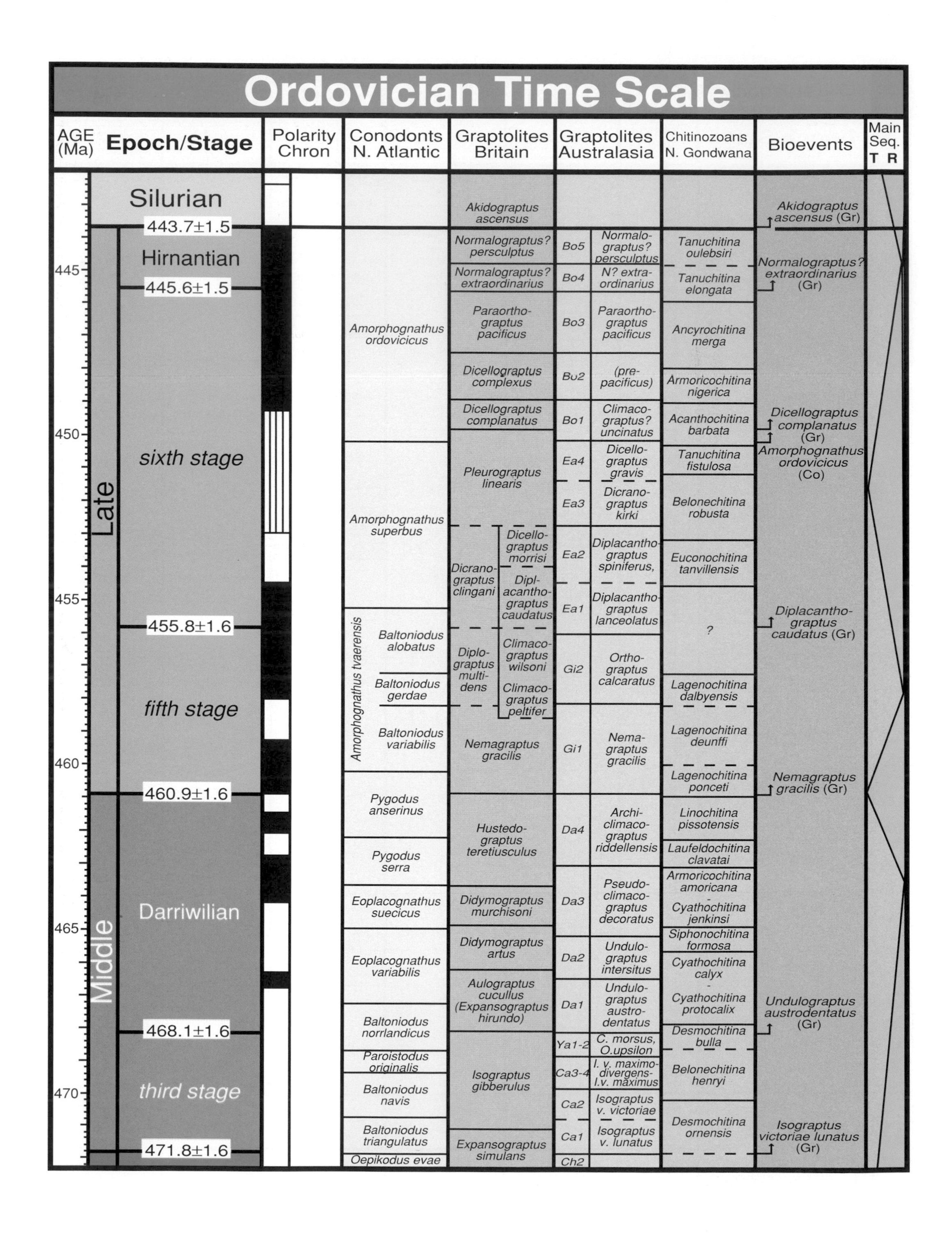

Ordovician Time Scale
AGE (Ma)
Epoch/Stage
Polarity Chron
Conodonts N. Atlantic
Graptolites Britain
Graptolites Australasia
Chitinozoans N. Gondwana
Bioevents
Main Seq.
T R
445
450
455
460
465
470
Silurian
443.7±1.5
Hirnantian
445.6±1.5
sixth stage
Late
455.8±1.6
fifth stage
460.9±1.6
Darriwilian
Middle
468.1±1.6
third stage
471.8±1.6
Amorphognathus ordovicicus
Amorphognathus superbus
Amorphognathus tvaerensis
Baltoniodus alobatus
Baltoniodus gerdae
Baltoniodus variabilis
Pygodus anserinus
Pygodus serra
Eoplacognathus suecicus
Eoplacognathus variabilis
Baltoniodus norrlandicus
Paroistodus originalis
Baltoniodus navis
Baltoniodus triangulatus
Oepikodus evae
Akidograptus ascensus
Normalograptus? persculptus
Normalograptus? extraordinarius
Paraorthograptus pacificus
Dicellograptus complexus
Dicellograptus complanatus
Pleurograptus linearis
Dicranograptus clingani
Dicellograptus morrisi
Diplacanthograptus caudatus
Diplograptus multidens
Climacograptus wilsoni
Climacograptus peltifer
Nemagraptus gracilis
Hustedograptus teretiusculus
Didymograptus murchisoni
Didymograptus artus
Aulograptus cucullus (Expansograptus hirundo)
Isograptus gibberulus
Expansograptus simulans
Bo5
Normalograptus? persculptus
Bo4
N? extraordinarius
Bo3
Paraorthograptus pacificus
Bo2
(pre-pacificus)
Bo1
Climacograptus? uncinatus
Ea4
Dicellograptus gravis
Ea3
Dicranograptus kirki
Ea2
Diplacanthograptus spiniferus,
Ea1
Diplacanthograptus lanceolatus
Gi2
Orthograptus calcaratus
Gi1
Nemagraptus gracilis
Da4
Archiclimacograptus riddellensis
Da3
Pseudoclimacograptus decoratus
Da2
Undulograptus intersitus
Da1
Undulograptus austrodentatus
Ya1-2
C. morsus, O.upsilon
Ca3-4
I. v. maximodivergens-I.v. maximus
Ca2
Isograptus v. victoriae
Ca1
Isograptus v. lunatus
Ch2
Tanuchitina oulebsiri
Tanuchitina elongata
Ancyrochitina merga
Armoricochitina nigerica
Acanthochitina barbata
Tanuchitina fistulosa
Belonechitina robusta
Euconochitina tanvillensis
?
Lagenochitina dalbyensis
Lagenochitina deunffi
Lagenochitina ponceti
Linochitina pissotensis
Laufeldochitina clavatai
Armoricochitina armoricana - Cyathochitina jenkinsi
Siphonochitina formosa
Cyathochitina calyx - Cyathochitina protocalix
Desmochitina bulla
Belonechitina henryi
Desmochitina ornensis
Akidograptus ascensus (Gr)
Normalograptus? extraordinarius (Gr)
Dicellograptus complanatus (Gr)
Amorphognathus ordovicicus (Co)
Diplacanthograptus caudatus (Gr)
Nemagraptus gracilis (Gr)
Undulograptus austrodentatus (Gr)
Isograptus victoriae lunatus (Gr)

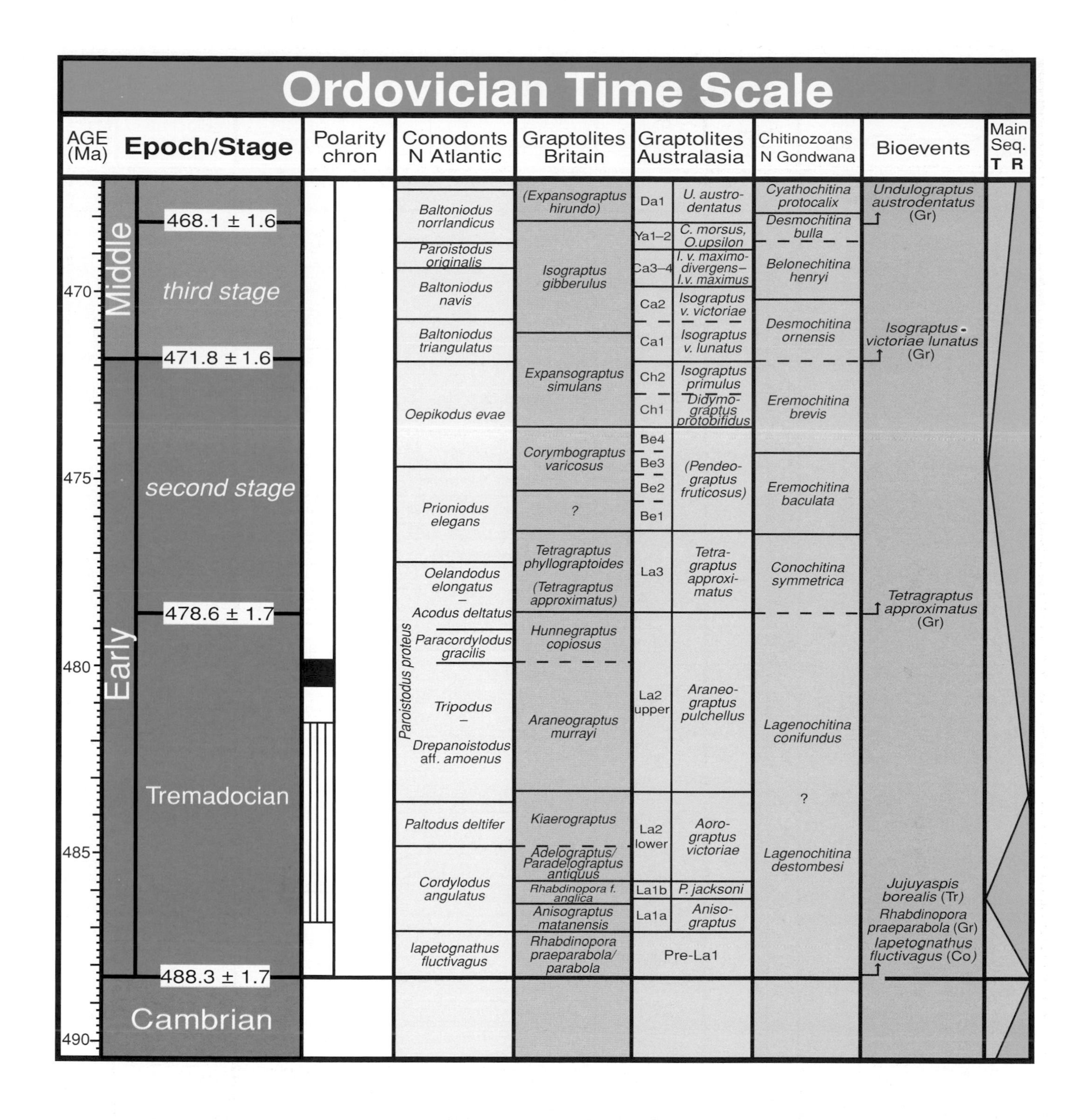

Ordovician Time Scale
AGE (Ma)
Epoch/Stage
Polarity chron
Conodonts N Atlantic
Graptolites Britain
Graptolites Australasia
Chitinozoans N Gondwana
Bioevents
Main Seq.
T R
Middle
Early
third stage
second stage
Tremadocian
Cambrian
468.1 ± 1.6
471.8 ± 1.6
478.6 ± 1.7
488.3 ± 1.7
470
475
480
485
490
Baltoniodus norrlandicus
Paroistodus originalis
Baltoniodus navis
Baltoniodus triangulatus
Oepikodus evae
Prioniodus elegans
Oelandodus elongatus
–
Acodus deltatus
Paracordylodus gracilis
Paroistodus proteus
Tripodus
–
Drepanoistodus aff. amoenus
Paltodus deltifer
Cordylodus angulatus
Iapetognathus fluctivagus
(Expansograptus hirundo)
Isograptus gibberulus
Expansograptus simulans
Corymbograptus varicosus
?
Tetragraptus phyllograptoides
(Tetragraptus approximatus)
Hunnegraptus copiosus
Araneograptus murrayi
Kiaerograptus
Adelograptus/ Paradelograptus antiquus
Rhabdinopora f. anglica
Anisograptus matanensis
Rhabdinopora praeparabola/ parabola
Da1
Ya1–2
Ca3–4
Ca2
Ca1
Ch2
Ch1
Be4
Be3
Be2
Be1
La3
La2 upper
La2 lower
La1b
La1a
Pre-La1
U. austro-dentatus
C. morsus, O.upsilon
I. v. maximo-divergens–I.v. maximus
Isograptus v. victoriae
Isograptus v. lunatus
Isograptus primulus
Didymo-graptus protobifidus
(Pendeo-graptus fruticosus)
Tetra-graptus approxi-matus
Araneo-graptus pulchellus
Aoro-graptus victoriae
P. jacksoni
Aniso-graptus
Cyathochitina protocalix
Desmochitina bulla
Belonechitina henryi
Desmochitina ornensis
Eremochitina brevis
Eremochitina baculata
Conochitina symmetrica
Lagenochitina conifundus
?
Lagenochitina destombesi
Undulograptus austrodentatus (Gr)
Isograptus victoriae lunatus (Gr)
Tetragraptus approximatus (Gr)
Jujuyaspis borealis (Tr)
Rhabdinopora praeparabola (Gr)
Iapetognathus fluctivagus (Co)

Silurian Time Scale

AGE (Ma)	Epoch/Stage	Polarity chron	Baltic regional stages	Graptolites	Conodonts	Biostratigraphic events	Main Seq. T R

AGE (Ma): 420, 425, 430, 435, 440

Epoch/Stage:

- Devonian
- 416.0 ± 2.8
- Pridoli
- 418.7 ± 2.7
- Ludlow: Ludfordian
- 421.3 ± 2.6
- Ludlow: Gorstian
- 422.9 ± 2.5
- Wenlock: Homerian
- 426.2 ± 2.4
- Wenlock: Sheinwoodian
- 428.2 ± 2.3
- Llandovery: Telychian
- 436.0 ± 1.9
- Llandovery: Aeronian
- 439.0 ± 1.8
- Llandovery: Rhuddanian
- 443.7 ± 1.5
- Ordovician

Polarity chron:

- Ludlow–Pridoli mixed interval
- Wenlock normal interval
- Llandovery mixed interval
- No data
- Lland. mixed

Baltic regional stages:

- Ohesaare
- Kaugetuma
- Kuressaare
- Paadla
- Rootsikula
- Jaagarahu
- Jaani
- Adavere
- Raikkula
- Juuru

Graptolites:

- *Monograptus uniformis*
- *Monograptus transgrediens* / *Monograptus bouceki-perneri*
- *Monograptus lochkovensis* / *Monograptus branikensis*
- *Monogr. parultimus–ultimus*
- *Monograptus formosus*
- *Neocucullograptus kozlowskii* / *B. cornuatus–P. podoliensis*
- *Saetograptus leintwardinensis–Saetograptus linearis*
- *Lobograptus scanicus*
- *Neodiversograptus nilssoni*
- *Colonograptus ludensis*
- *Colono. praedeubeli–deubeli*
- *Pristio. parvus–Gotho. nassa*
- *Cyrtograptus lundgreni*
- *Cyrtograptus perneri* / *Cyrtograptus rigidus*
- *Monogr. beloph.-anten.* / *Monogr. riccartonensis*
- *Cyrtograptus murchisoni*
- *Cyrtograptus centrifugus*
- *Cyrtograptus insectus*
- *Cyrtograptus lapworthi*
- *Oktavites spiralis*
- *Monoclimacis crenulata–Monoclimacis griestoniensis*
- *Monograptus crispus*
- *Spirograptus turriculatus*
- *Spirograptus guerichi*
- *Stimulograptus sedgwickii*
- *Lituigraptus convolutus*
- *Monograptus argenteus*
- *Demirastrites pectinatus–Demirastrites triangulatus*
- *Coronograptus cyphus*
- *Orthograptus vesiculosus*
- *Parakidograptus acuminatus*
- *Akidograptus ascensus*

Conodonts:

- *Icriodus woschmidt–postwoschmidti*
- *Oulodus elegans detortus*
- *Ozarkodina remscheidensis* Interval Zone
- *Ozarkodina crispa*
- *Ozarkodina snajdri*
- Interval Zone
- *Polygnathoides siluricus*
- *Ancoradella ploeckensis*
- Not zoned
- *K. stauros*
- *Ozarkodina bohemica*
- *Ozarkodina sagitta sagitta*
- *Ozarkodina sagitta rhenana*
- *Kockelella ranuliformis*
- *Pterospathodus amorphognathoides*
- *Pterospathodus celloni*
- *Pterospathodus tenuis – Distomodus staurognathoides*
- *Distomodus kentuckyensis*
- *Rexroadus nathani*

Biostratigraphic events:

- *Mongraptus uniformis* (Gr)
- *Monograptus parultimus* (Gr)
- *Saetograptus leintwardinensis* (Gr)
- *Neodiversogr. nilssoni* (Gr)
- *Cyrtograptus lundgreni* (Gr)
- *Margachitina margaritana* (Ch)
- *Eocelia intermedia* (Br)
- *Monograptus austerus sequens* (Gr)
- *Akidograptus ascensus* (Gr)

Main Seq.: T R

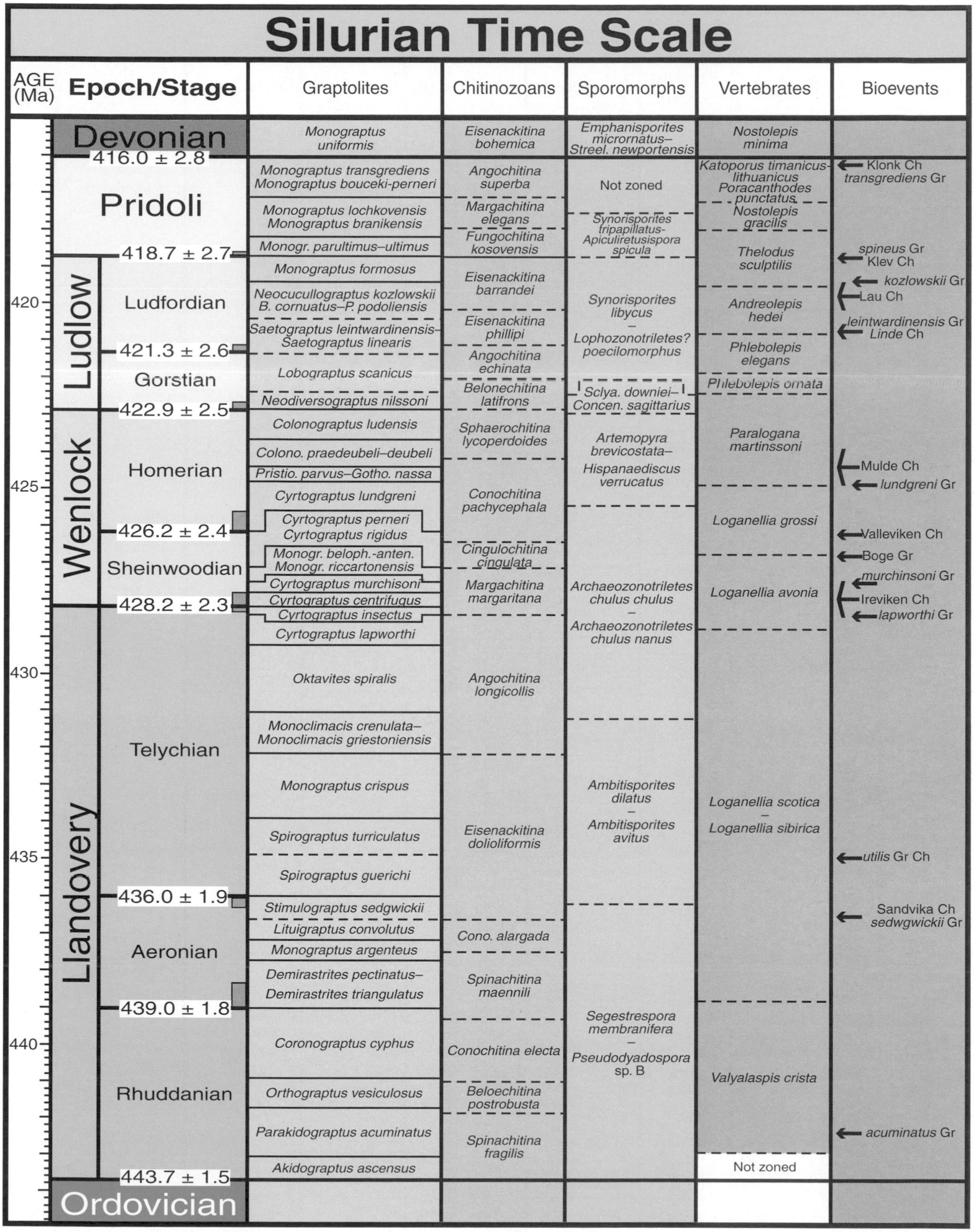

Silurian Time Scale

AGE (Ma) | Epoch/Stage | Graptolites | Chitinozoans | Sporomorphs | Vertebrates | Bioevents

Age axis: 420, 425, 430, 435, 440

Epoch/Stage

- Devonian
- 416.0 ± 2.8
- Pridoli
- 418.7 ± 2.7
- Ludlow: Ludfordian
- 421.3 ± 2.6
- Ludlow: Gorstian
- 422.9 ± 2.5
- Wenlock: Homerian
- 426.2 ± 2.4
- Wenlock: Sheinwoodian
- 428.2 ± 2.3
- Llandovery: Telychian
- 436.0 ± 1.9
- Llandovery: Aeronian
- 439.0 ± 1.8
- Llandovery: Rhuddanian
- 443.7 ± 1.5
- Ordovician

Graptolites

- *Monograptus uniformis*
- *Monograptus transgrediens* *Monograptus bouceki-perneri*
- *Monograptus lochkovensis* *Monograptus branikensis*
- *Monogr. parultimus–ultimus*
- *Monograptus formosus*
- *Neocucullograptus kozlowskii* *B. cornuatus–P. podoliensis*
- *Saetograptus leintwardinensis–Saetograptus linearis*
- *Lobograptus scanicus*
- *Neodiversograptus nilssoni*
- *Colonograptus ludensis*
- *Colono. praedeubeli–deubeli*
- *Pristio. parvus–Gotho. nassa*
- *Cyrtograptus lundgreni*
- *Cyrtograptus perneri* *Cyrtograptus rigidus*
- *Monogr. beloph.-anten.* *Monogr. riccartonensis*
- *Cyrtograptus murchisoni*
- *Cyrtograptus centrifugus*
- *Cyrtograptus insectus*
- *Cyrtograptus lapworthi*
- *Oktavites spiralis*
- *Monoclimacis crenulata–Monoclimacis griestoniensis*
- *Monograptus crispus*
- *Spirograptus turriculatus*
- *Spirograptus guerichi*
- *Stimulograptus sedgwickii*
- *Lituigraptus convolutus*
- *Monograptus argenteus*
- *Demirastrites pectinatus–Demirastrites triangulatus*
- *Coronograptus cyphus*
- *Orthograptus vesiculosus*
- *Parakidograptus acuminatus*
- *Akidograptus ascensus*

Chitinozoans

- *Eisenackitina bohemica*
- *Angochitina superba*
- *Margachitina elegans*
- *Fungochitina kosovensis*
- *Eisenackitina barrandei*
- *Eisenackitina phillipi*
- *Angochitina echinata*
- *Belonechitina latifrons*
- *Sphaerochitina lycoperdoides*
- *Conochitina pachycephala*
- *Cingulochitina cingulata*
- *Margachitina margaritana*
- *Angochitina longicollis*
- *Eisenackitina dolioliformis*
- *Cono. alargada*
- *Spinachitina maennili*
- *Conochitina electa*
- *Beloechitina postrobusta*
- *Spinachitina fragilis*

Sporomorphs

- *Emphanisporites micrornatus–Streel. newportensis*
- Not zoned
- *Synorisporites tripapillatus-Apiculiretusispora spicula*
- *Synorisporites libycus* – *Lophozonotriletes? poecilomorphus*
- *Sclya. downiei–Concen. sagittarius*
- *Artemopyra brevicostata–Hispanaediscus verrucatus*
- *Archaeozonotriletes chulus chulus* – *Archaeozonotriletes chulus nanus*
- *Ambitisporites dilatus* – *Ambitisporites avitus*
- *Segestrespora membranifera* – *Pseudodyadospora* sp. B

Vertebrates

- *Nostolepis minima*
- *Katoporus timanicus-lithuanicus* *Poracanthodes punctatus*
- *Nostolepis gracilis*
- *Thelodus sculptilis*
- *Andreolepis hedei*
- *Phlebolepis elegans*
- *Phlebolepis ornata*
- *Paralogana martinssoni*
- *Loganellia grossi*
- *Loganellia avonia*
- *Loganellia scotica* – *Loganellia sibirica*
- *Valyalaspis crista*
- Not zoned

Bioevents

- ← Klonk Ch; *transgrediens* Gr
- ← *spineus* Gr; Klev Ch
- ← *kozlowskii* Gr; Lau Ch
- ← *leintwardinensis* Gr; *Linde* Ch
- ← Mulde Ch; *lundgreni* Gr
- ← Valleviken Ch
- ← Boge Gr
- ← *murchinsoni* Gr; Ireviken Ch; *lapworthi* Gr
- ← *utilis* Gr Ch
- ← Sandvika Ch; *sedwgwickii* Gr
- ← *acuminatus* Gr

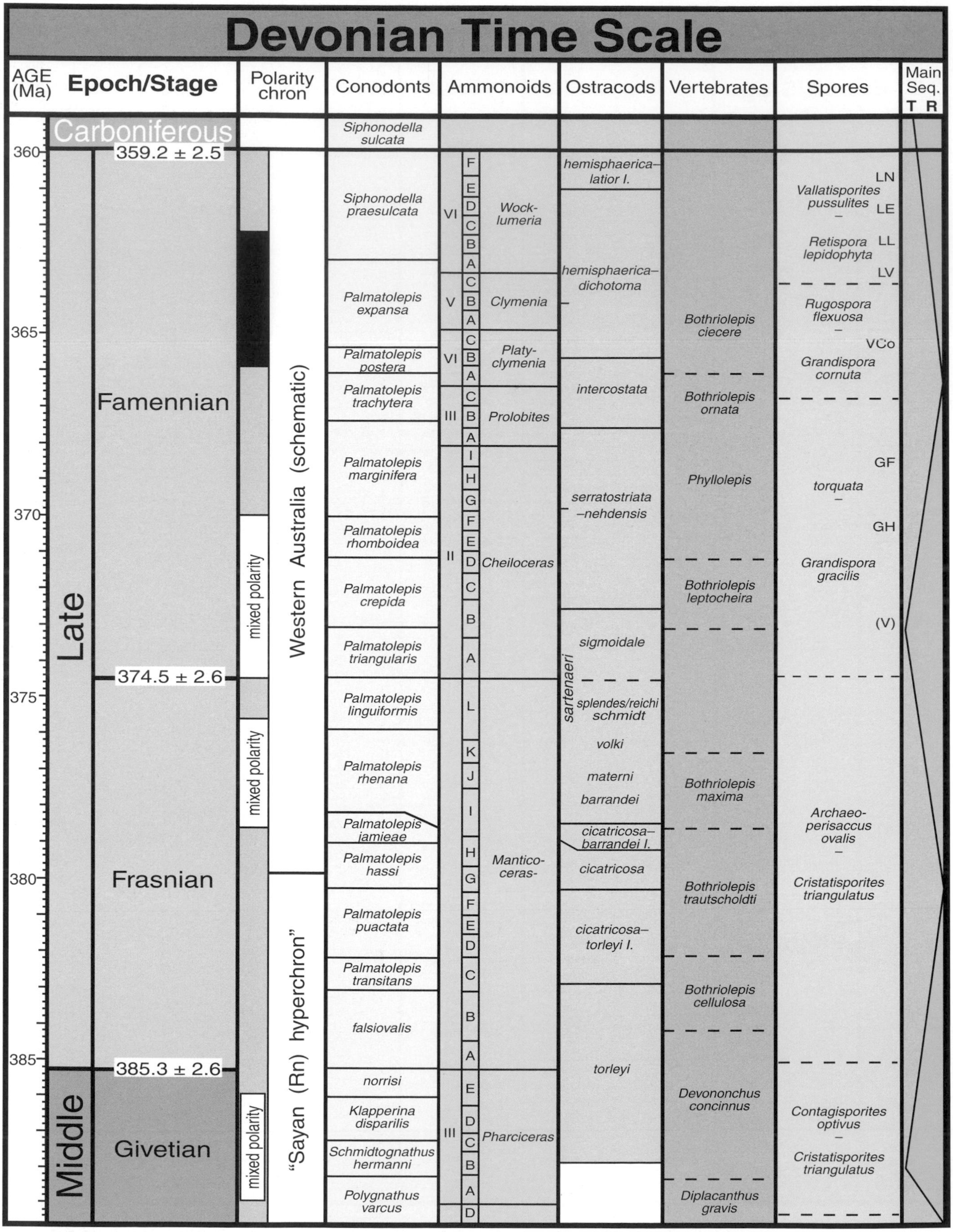

Devonian Time Scale

AGE (Ma): 360, 365, 370, 375, 380, 385

Epoch/Stage

- Carboniferous
- 359.2 ± 2.5
- Late
  - Famennian
  - 374.5 ± 2.6
  - Frasnian
- 385.3 ± 2.6
- Middle
  - Givetian

Polarity chron

- Western Australia (schematic)
- mixed polarity
- mixed polarity
- "Sayan (Rn) hyperchron"
- mixed polarity

Conodonts

- *Siphonodella sulcata*
- *Siphonodella praesulcata*
- *Palmatolepis expansa*
- *Palmatolepis postera*
- *Palmatolepis trachytera*
- *Palmatolepis marginifera*
- *Palmatolepis rhomboidea*
- *Palmatolepis crepida*
- *Palmatolepis triangularis*
- *Palmatolepis linguiformis*
- *Palmatolepis rhenana*
- *Palmatolepis jamieae*
- *Palmatolepis hassi*
- *Palmatolepis puactata*
- *Palmatolepis transitans*
- *falsiovalis*
- *norrisi*
- *Klapperina disparilis*
- *Schmidtognathus hermanni*
- *Polygnathus varcus*

Ammonoids

- VI (F, E, D, C, B, A) — *Wocklumeria*
- V (C, B, A) — *Clymenia*
- VI (C, B, A) — *Platyclymenia*
- III (C, B, A) — *Prolobites*
- II (I, H, G, F, E, D, C, B, A) — *Cheiloceras*
- (L, K, J, I, H, G, F, E, D, C, B, A) — *Manticoceras-*
- III (E, D, C, B, A) — *Pharciceras*
- D

Ostracods

- *hemisphaerica–latior l.*
- *hemisphaerica–dichotoma*
- *intercostata*
- *serratostriata–nehdensis*
- *sigmoidale*
- *sartenaeri*
- *splendes/reichi schmidt*
- *volki*
- *materni*
- *barrandei*
- *cicatricosa–barrandei l.*
- *cicatricosa*
- *cicatricosa–torleyi l.*
- *torleyi*

Vertebrates

- *Bothriolepis ciecere*
- *Bothriolepis ornata*
- *Phyllolepis*
- *Bothriolepis leptocheira*
- *Bothriolepis maxima*
- *Bothriolepis trautscholdti*
- *Bothriolepis cellulosa*
- *Devononchus concinnus*
- *Diplacanthus gravis*

Spores

- LN
- *Vallatisporites pussulites* – LE
- *Retispora lepidophyta* LL
- LV
- *Rugospora flexuosa* –
- VCo
- *Grandispora cornuta*
- GF
- *torquata* –
- GH
- *Grandispora gracilis*
- (V)
- *Archaeoperisaccus ovalis* –
- *Cristatisporites triangulatus*
- *Contagisporites optivus* –
- *Cristatisporites triangulatus*

Main Seq.

T R

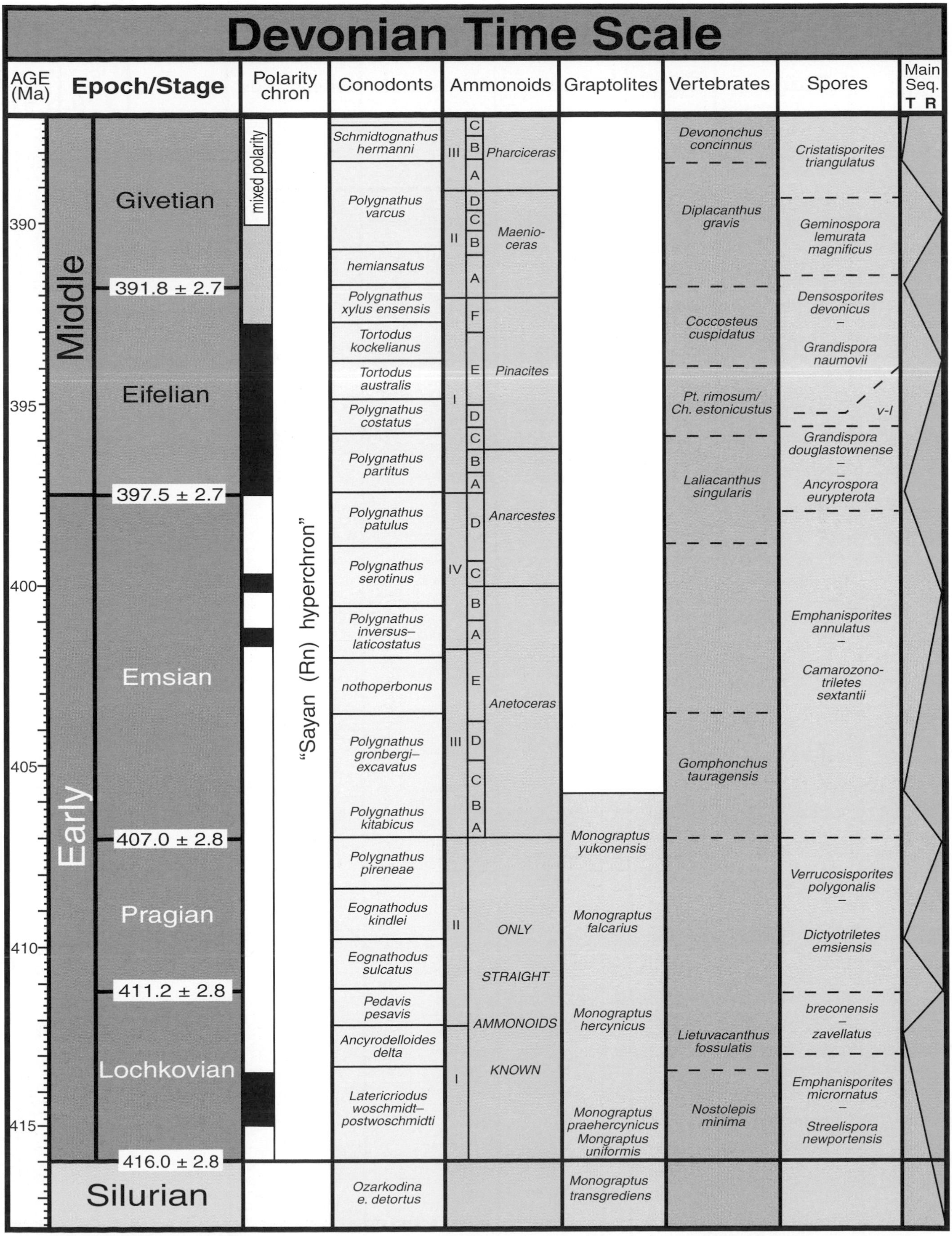

Devonian Time Scale

AGE (Ma) | **Epoch/Stage** | **Polarity chron** | **Conodonts** | **Ammonoids** | **Graptolites** | **Vertebrates** | **Spores** | **Main Seq. T R**

AGE (Ma): 390, 395, 400, 405, 410, 415

Epoch/Stage:

- Middle
  - Givetian
  - 391.8 ± 2.7
  - Eifelian
  - 397.5 ± 2.7
- Early
  - Emsian
  - 407.0 ± 2.8
  - Pragian
  - 411.2 ± 2.8
  - Lochkovian
  - 416.0 ± 2.8
- Silurian

Polarity chron: mixed polarity; "Sayan (Rn) hyperchron"

Conodonts:

- *Schmidtognathus hermanni*
- *Polygnathus varcus*
- *hemiansatus*
- *Polygnathus xylus ensensis*
- *Tortodus kockelianus*
- *Tortodus australis*
- *Polygnathus costatus*
- *Polygnathus partitus*
- *Polygnathus patulus*
- *Polygnathus serotinus*
- *Polygnathus inversus–laticostatus*
- *nothoperbonus*
- *Polygnathus gronbergi–excavatus*
- *Polygnathus kitabicus*
- *Polygnathus pireneae*
- *Eognathodus kindlei*
- *Eognathodus sulcatus*
- *Pedavis pesavis*
- *Ancyrodelloides delta*
- *Latericriodus woschmidt–postwoschmidti*
- *Ozarkodina e. detortus*

Ammonoids:

- III: C, B, A — *Pharciceras*
- II: D, C, B, A — *Maenio-ceras*
- I: F, E, D, C, B, A — *Pinacites*; *Anarcestes*
- IV: D, C, B, A — *Anarcestes*; *Anetoceras*
- III: E, D, C, B, A — *Anetoceras*
- II, I — ONLY STRAIGHT AMMONOIDS KNOWN

Graptolites:

- *Monograptus yukonensis*
- *Monograptus falcarius*
- *Monograptus hercynicus*
- *Monograptus praehercynicus*, *Mongraptus uniformis*
- *Monograptus transgrediens*

Vertebrates:

- *Devononchus concinnus*
- *Diplacanthus gravis*
- *Coccosteus cuspidatus*
- *Pt. rimosum/ Ch. estonicustus*
- *Laliacanthus singularis*
- *Gomphonchus tauragensis*
- *Lietuvacanthus fossulatis*
- *Nostolepis minima*

Spores:

- *Cristatisporites triangulatus*
- *Geminospora lemurata magnificus*
- *Densosporites devonicus* – *Grandispora naumovii*
- *v-I*
- *Grandispora douglastownense* – *Ancyrospora eurypterota*
- *Emphanisporites annulatus* – *Camarozonotriletes sextantii*
- *Verrucosisporites polygonalis* – *Dictyotriletes emsiensis*
- *breconensis* – *zavellatus*
- *Emphanisporites micrornatus* – *Streelispora newportensis*

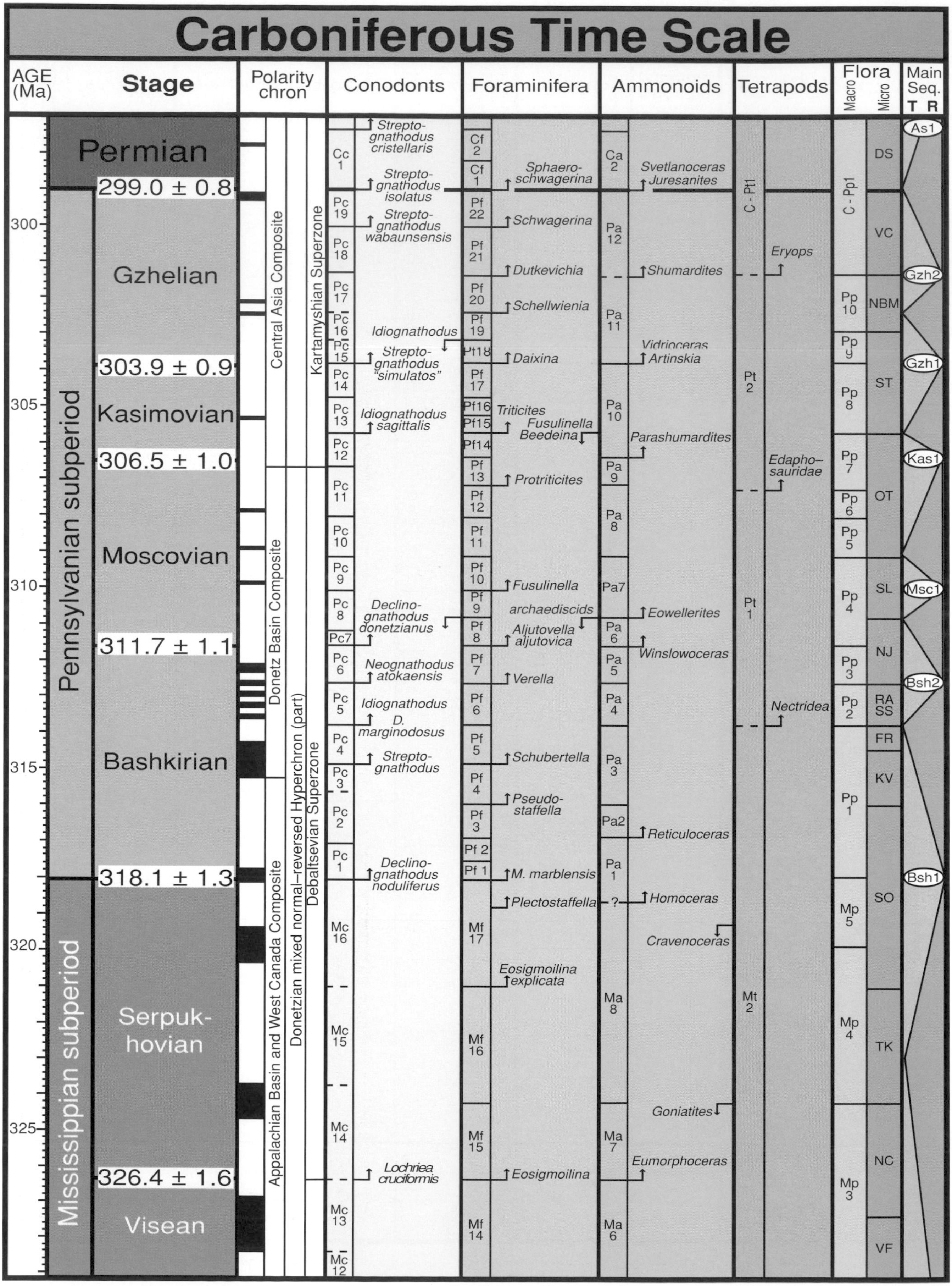

Carboniferous Time Scale

AGE (Ma) | Stage | Polarity chron | Conodonts | Foraminifera | Ammonoids | Tetrapods | Flora (Macro, Micro) | Main Seq. T R

AGE (Ma): 300, 305, 310, 315, 320, 325

Stage:
Permian
299.0 ± 0.8
Pennsylvanian subperiod
Gzhelian
303.9 ± 0.9
Kasimovian
306.5 ± 1.0
Moscovian
311.7 ± 1.1
Bashkirian
318.1 ± 1.3
Mississippian subperiod
Serpuk-hovian
326.4 ± 1.6
Visean

Polarity chron:
Central Asia Composite
Kartamyshian Superzone
Donetz Basin Composite
Appalachian Basin and West Canada Composite
Donetzian mixed normal–reversed Hyperchron (part)
Debaltsevian Superzone

Conodonts:
Cc 1 – Strepto-gnathodus cristellaris
Strepto-gnathodus isolatus
Pc 19 – Strepto-gnathodus wabaunsensis
Pc 18
Pc 17
Pc 16 – Idiognathodus
Pc 15 – Strepto-gnathodus "simulatos"
Pc 14
Pc 13 – Idiognathodus sagittalis
Pc 12
Pc 11
Pc 10
Pc 9
Pc 8 – Declino-gnathodus donetzianus
Pc7
Pc 6 – Neognathodus atokaensis
Pc 5 – Idiognathodus D. marginodosus
Pc 4 – Strepto-gnathodus
Pc 3
Pc 2
Pc 1 – Declino-gnathodus noduliferus
Mc 16
Mc 15
Mc 14 – Lochriea cruciformis
Mc 13
Mc 12

Foraminifera:
Cf 2
Cf 1 – Sphaero-schwagerina
Pf 22 – Schwagerina
Pf 21 – Dutkevichia
Pf 20 – Schellwienia
Pf 19
Pf18 – Daixina
Pf 17
Pf16 – Triticites
Pf15 – Fusulinella, Beedeina
Pf14
Pf 13 – Protriticites
Pf 12
Pf 11
Pf 10 – Fusulinella
Pf 9 – archaediscids
Pf 8 – Aljutovella aljutovica
Pf 7 – Verella
Pf 6
Pf 5 – Schubertella
Pf 4 – Pseudo-staffella
Pf 3
Pf 2
Pf 1 – M. marblensis
Plectostaffella
Mf 17 – Eosigmoilina explicata
Mf 16
Mf 15 – Eosigmoilina
Mf 14

Ammonoids:
Ca 2 – Svetlanoceras, Juresanites
Pa 12 – Shumardites
Pa 11 – Vidrioceras, Artinskia
Pa 10 – Parashumardites
Pa 9
Pa 8
Pa7 – Eowellerites
Pa 6 – Winslowoceras
Pa 5
Pa 4
Pa 3
Pa2 – Reticuloceras
Pa 1
? – Homoceras
Cravenoceras
Ma 8
Goniatites
Ma 7 – Eumorphoceras
Ma 6

Tetrapods:
C - Pt1 – Eryops
Pt 2 – Edapho-sauridae
Pt 1 – Nectridea
Mt 2

Flora – Macro:
C - Pp1
Pp 10
Pp 9
Pp 8
Pp 7
Pp 6
Pp 5
Pp 4
Pp 3
Pp 2
Pp 1
Mp 5
Mp 4
Mp 3

Flora – Micro:
DS, VC, NBM, ST, OT, SL, NJ, RA SS, FR, KV, SO, TK, NC, VF

Main Seq. T R:
As1, Gzh2, Gzh1, Kas1, Msc1, Bsh2, Bsh1

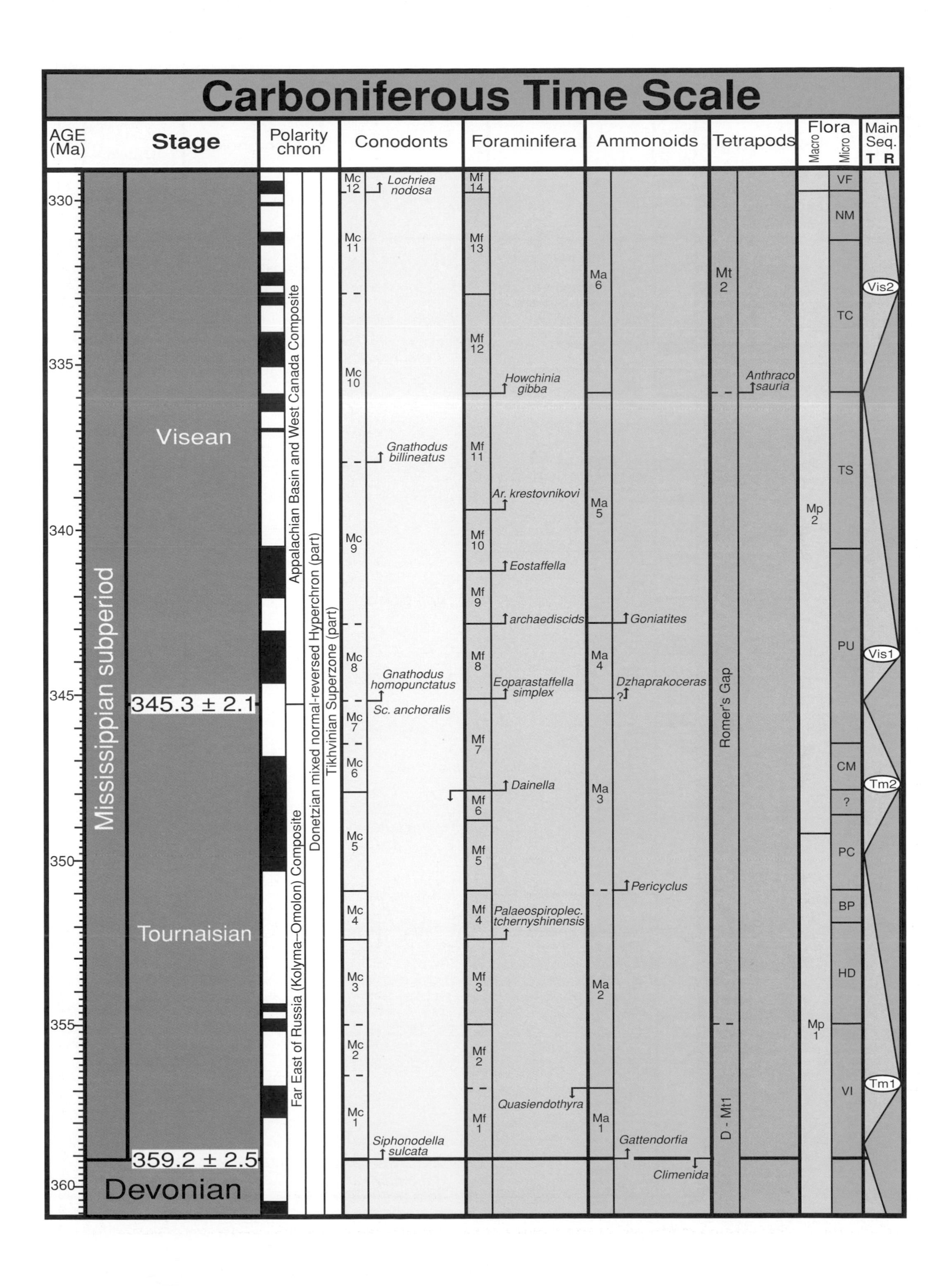

Carboniferous Time Scale
AGE (Ma)
Stage
Polarity chron
Conodonts
Foraminifera
Ammonoids
Tetrapods
Flora
Macro
Micro
Main Seq.
T R
330
335
340
345
350
355
360
Mississippian subperiod
Visean
Tournaisian
Devonian
345.3 ± 2.1
359.2 ± 2.5
Appalachian Basin and West Canada Composite
Far East of Russia (Kolyma–Omolon) Composite
Donetzian mixed normal-reversed Hyperchron (part)
Tikhvinian Superzone (part)
Mc 12
Mc 11
Mc 10
Mc 9
Mc 8
Mc 7
Mc 6
Mc 5
Mc 4
Mc 3
Mc 2
Mc 1
Lochriea nodosa
Gnathodus billineatus
Gnathodus homopunctatus
Sc. anchoralis
Siphonodella sulcata
Mf 14
Mf 13
Mf 12
Mf 11
Mf 10
Mf 9
Mf 8
Mf 7
Mf 6
Mf 5
Mf 4
Mf 3
Mf 2
Mf 1
Howchinia gibba
Ar. krestovnikovi
Eostaffella
archaediscids
Eoparastaffella simplex
Dainella
Palaeospiroplec. tchernyshinensis
Quasiendothyra
Ma 6
Ma 5
Ma 4
Ma 3
Ma 2
Ma 1
Goniatites
Dzhaprakoceras
Pericyclus
Gattendorfia
Climenida
Mt 2
Anthracosauria
Romer's Gap
D - Mt1
Mp 2
Mp 1
VF
NM
TC
TS
PU
CM
?
PC
BP
HD
VI
Vis2
Vis1
Tm2
Tm1

Permian Time Scale

AGE (Ma)	Epoch	Stage	Conodonts	Fusulinids	Ammonoids	Vertebrates
250	Triassic		*Hindeodus parvus*		*Otoceras*	
251.0 ± 0.4	Lopingian	Changhsingian	*Clarkina meishanensis*, *yini*, *changxingensis*, *subcarinata*, *wangi*	*Paleofusulina*	*Pseudotirolites*, *Paratirolites*, *Iranites*	*endothiodontid/dicynodontid empire*
253.8 ± 0.7 (255)	Lopingian	Wuchiapingian	*Clarkina orientalis*, *transcaucasica*, *guangyuanensis*, *leveni*, *asymetrica*, *dukouensis*, *postbitteri*	*Colaniella*	*Araxoceras*, *Anderssonoceras*, *Roadoceras–Doulingoceras*	*endothiodontid/dicynodontid empire*
260.4 ± 0.7 (260)	Guadalupian	Capitanian	*Jinogondolella altudaensis*, *postserrata*	*Lepidolina*, *Yabeina*, *Polydiexodina*, *Neoschwagerina magaritae*	*Timorites*	*tapinocephalid empire*
265.8 ± 0.7 (265)	Guadalupian	Wordian	*Jinogondolella aserrata*	*Neoschwagerina*		*tapinocephalid empire*
268 ± 0.7	Guadalupian	Roadian	*Jinogondolella nankingensis*	*Cancellina cutalensis*	*Waagenoceras*, *Demarezites*	*tapinocephalid empire*
270.6 ± 0.7 (270)	Cisuralian	Kungurian	*Neostreptognathodus sulcoplicatus*, *prayi*, *pnevi*	*Misellina*	*Pseudovidrioceras*, *Propinacoceras*	
275.6 ± 0.7 (275)	Cisuralian	Artinskian	*Neostreptognathodus pequopensis*	*Chalaroschwagerina*, *Pamirina*	*Uraloceras*	

Polarity chron | Main Seq. T R

Permian Time Scale

AGE (Ma)	Epoch/Stage		Polarity chron	Conodonts		Fusulinids	Ammonoids	Vertebrates	Main Seq. T R
280	Cisuralian	Artinskian		*Sweetognathus*	*whitei*	*Chalaroschwagerina* *Pamirina* *Parafusulina*	*Uraloceras* *Aktubinskia–Artinskia*	*edaphosaurid empire*	
284.4 ± 0.7									
285		Sakmarian			*binodosus*	*Pseudofusulina*	*Sakmarites*		
290					*merrilli*	*Schwagerina vernueili* *Schwagerina moelleri*			
294.6 ± 0.8									
295		Asselian		*Streptognathodus*	*barskovi* *postfusus* *fusus* *constrictus* *isolatus*	*Sphaeroschwagerina gigas* *Pseudoschwagerina robusta* *Schwagerina nux* *Sphaeroschwagerina fusiformis* *Sphaeroschwagerina vulgaris*	*Svetlanoceras*		
299 ± 0.8									
300	Carboniferous				*binodosus*	*Ultradaixina bosbytauensis*	*Shumardites–Emilites*		

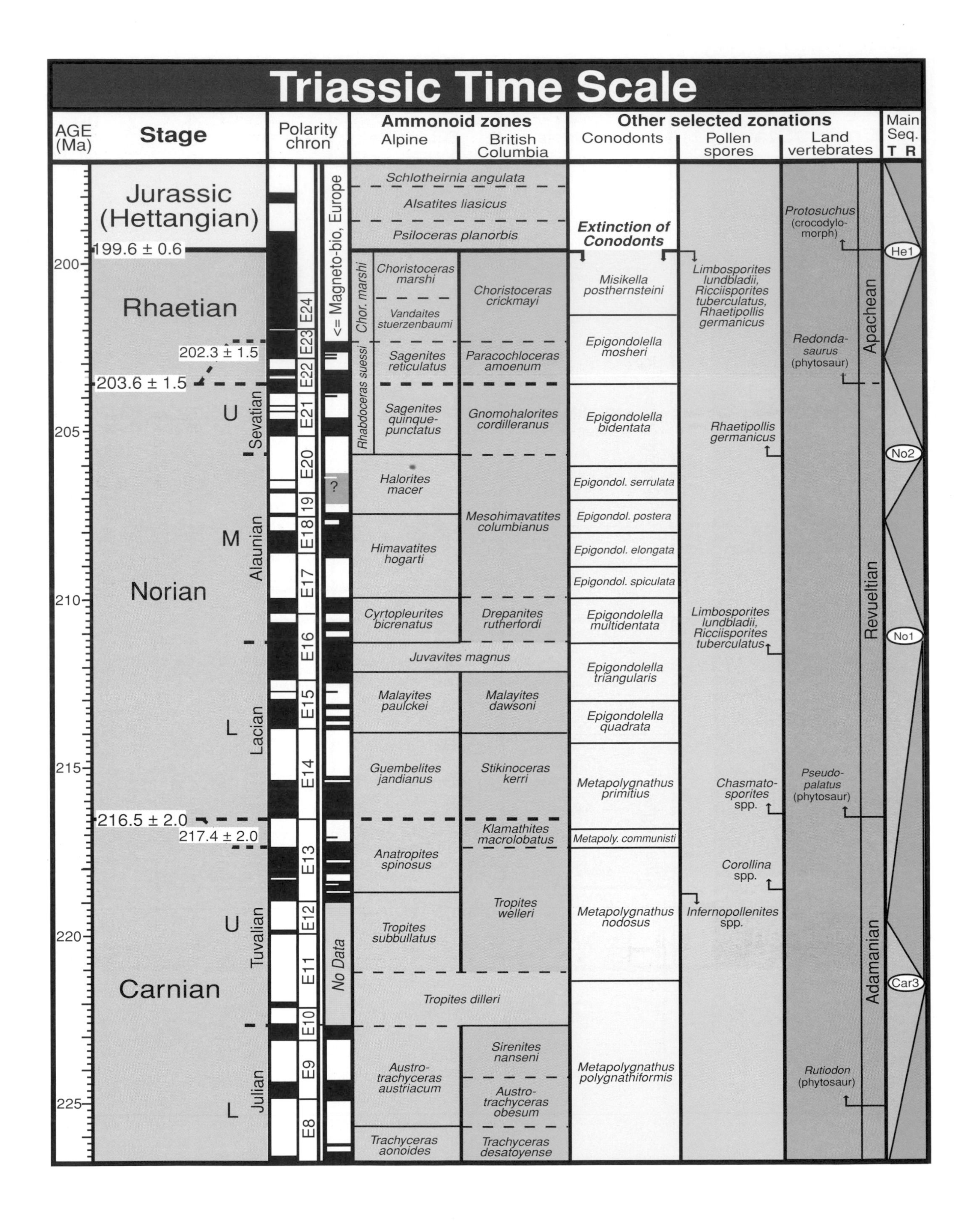

Triassic Time Scale
AGE (Ma)
Stage
Polarity chron
Ammonoid zones
Alpine
British Columbia
Other selected zonations
Conodonts
Pollen spores
Land vertebrates
Main Seq.
T R
Jurassic (Hettangian)
199.6 ± 0.6
200
Rhaetian
202.3 ± 1.5
203.6 ± 1.5
U
Sevatian
205
M
Alaunian
Norian
210
L
Lacian
215
216.5 ± 2.0
217.4 ± 2.0
U
Tuvalian
220
Carnian
L
Julian
225
E24
E23
E22
E21
E20
E19
E18
E17
E16
E15
E14
E13
E12
E11
E10
E9
E8
<= Magneto-bio, Europe
?
No Data
Schlotheirnia angulata
Alsatites liasicus
Psiloceras planorbis
Chor. marshi
Rhabdoceras suessi
Choristoceras marshi
Vandaites stuerzenbaumi
Sagenites reticulatus
Sagenites quinque-punctatus
Halorites macer
Himavatites hogarti
Cyrtopleurites bicrenatus
Juvavites magnus
Malayites paulckei
Guembelites jandianus
Anatropites spinosus
Tropites subbullatus
Tropites dilleri
Austro-trachyceras austriacum
Trachyceras aonoides
Choristoceras crickmayi
Paracochloceras amoenum
Gnomohalorites cordilleranus
Mesohimavatites columbianus
Drepanites rutherfordi
Malayites dawsoni
Stikinoceras kerri
Klamathites macrolobatus
Tropites welleri
Sirenites nanseni
Austro-trachyceras obesum
Trachyceras desatoyense
Extinction of Conodonts
Misikella posthernsteini
Epigondolella mosheri
Epigondolella bidentata
Epigondol. serrulata
Epigondol. postera
Epigondol. elongata
Epigondol. spiculata
Epigondolella multidentata
Epigondolella triangularis
Epigondolella quadrata
Metapolygnathus primitius
Metapoly. communisti
Metapolygnathus nodosus
Metapolygnathus polygnathiformis
Limbosporites lundbladii, Ricciisporites tuberculatus, Rhaetipollis germanicus
Rhaetipollis germanicus
Limbosporites lundbladii, Ricciisporites tuberculatus
Chasmato-sporites spp.
Corollina spp.
Infernopollenites spp.
Protosuchus (crocodylo-morph)
Redonda-saurus (phytosaur)
Pseudo-palatus (phytosaur)
Rutiodon (phytosaur)
Apachean
Revueltian
Adamanian
He1
No2
No1
Car3

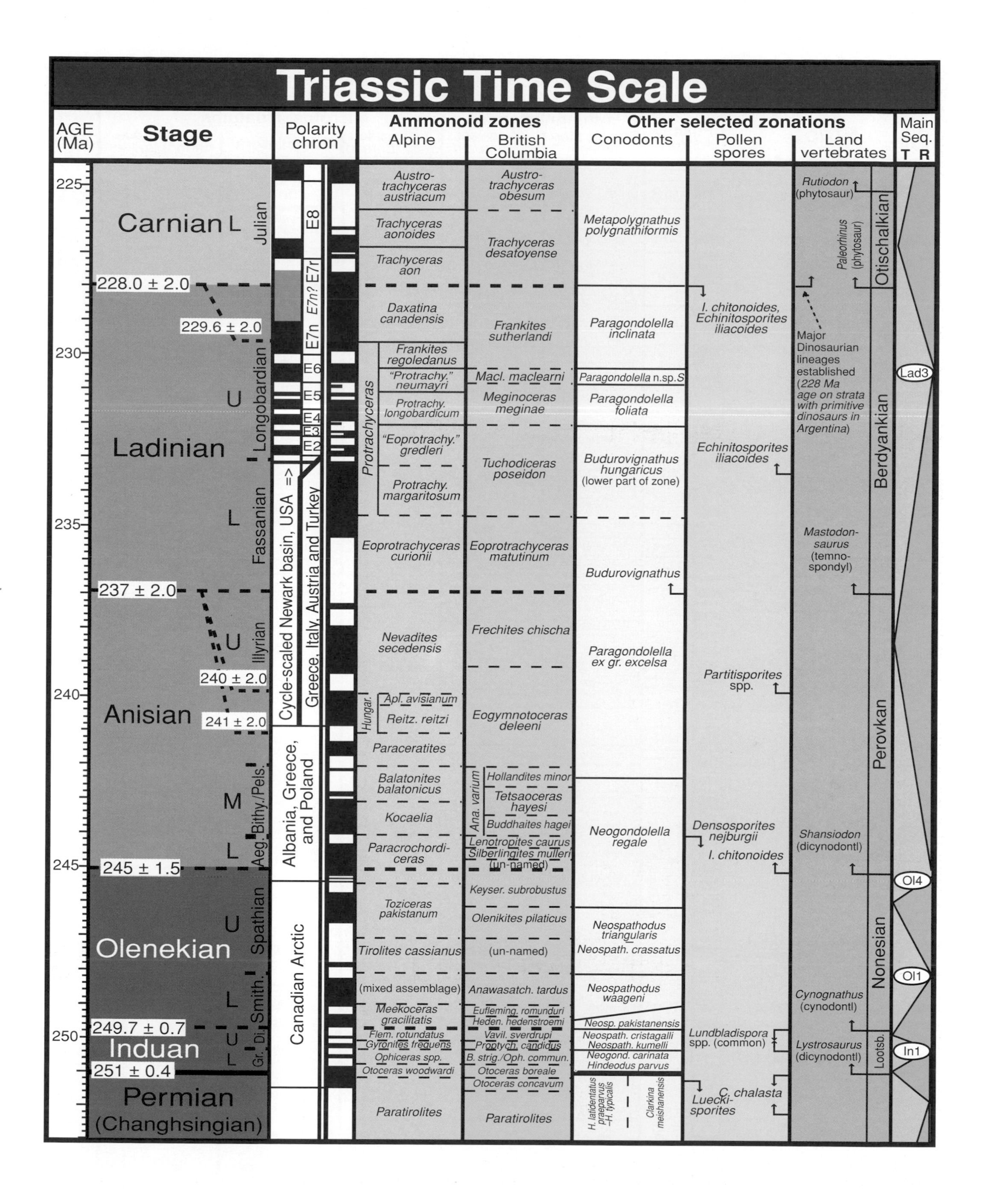
Triassic Time Scale
AGE (Ma)
Stage
Polarity chron
Ammonoid zones
Alpine
British Columbia
Other selected zonations
Conodonts
Pollen spores
Land vertebrates
Main Seq.
T R
225
230
235
240
245
250
Carnian
L
Julian
228.0 ± 2.0
229.6 ± 2.0
Ladinian
U
Longobardian
L
Fassanian
237 ± 2.0
Anisian
U
Illyrian
240 ± 2.0
241 ± 2.0
M
Bithy./Pels.
L
Aeg.
245 ± 1.5
Olenekian
U
Spathian
L
Smith.
249.7 ± 0.7
Induan
U
Di.
L
Gr.
251 ± 0.4
Permian
(Changhsingian)
E8
E7r
E7n?
E7n
E6
E5
E4
E3
E2
Cycle-scaled Newark basin, USA =>
Greece, Italy, Austria and Turkey
Albania, Greece, and Poland
Canadian Arctic
Austro-trachyceras austriacum
Trachyceras aonoides
Trachyceras aon
Daxatina canadensis
Frankites regoledanus
"Protrachy." neumayri
Protrachy. longobardicum
"Eoprotrachy." gredleri
Protrachy. margaritosum
Protrachyceras
Eoprotrachyceras curionii
Nevadites secedensis
Hungar.
Apl. avisianum
Reitz. reitzi
Paraceratites
Balatonites balatonicus
Kocaelia
Paracrochordi-ceras
Toziceras pakistanum
Tirolites cassianus
(mixed assemblage)
Meekoceras gracilitatis
Flem. rotundatus
Gyronites frequens
Ophiceras spp.
Otoceras woodwardi
Paratirolites
Austro-trachyceras obesum
Trachyceras desatoyense
Frankites sutherlandi
Macl. maclearni
Meginoceras meginae
Tuchodiceras poseidon
Eoprotrachyceras matutinum
Frechites chischa
Eogymnotoceras deleeni
Ana. varium
Hollandites minor
Tetsaoceras hayesi
Buddhaites hagei
Lenotropites caurus
Silberlingites mulleri
(un-named)
Keyser. subrobustus
Olenikites pilaticus
(un-named)
Anawasatch. tardus
Eufleming. romunduri
Heden. hedenstroemi
Vavil. sverdrupi
Proptych. candidus
B. strig./Oph. commun.
Otoceras boreale
Otoceras concavum
Paratirolites
Metapolygnathus polygnathiformis
Paragondolella inclinata
Paragondolella n.sp.S
Paragondolella foliata
Budurovignathus hungaricus (lower part of zone)
Budurovignathus
Paragondolella ex gr. excelsa
Neogondolella regale
Neospathodus triangularis
Neospath. crassatus
Neospathodus waageni
Neosp. pakistanensis
Neospath. cristagalli
Neospath. kumelli
Neogond. carinata
Hindeodus parvus
H. latidentatus praeparvus -H. typicalis
Clarkina meishanensis
I. chitonoides, Echinitosporites iliacoides
Echinitosporites iliacoides
Partitisporites spp.
Densosporites nejburgii
I. chitonoides
Lundbladispora spp. (common)
C. chalasta
Lueckisporites
Rutiodon (phytosaur)
Paleorhinus (phytosaur)
Major Dinosaurian lineages established (228 Ma age on strata with primitive dinosaurs in Argentina)
Mastodon-saurus (temno-spondyl)
Shansiodon (dicynodontl)
Cynognathus (cynodontl)
Lystrosaurus (dicynodontl)
Otischalkian
Berdyankian
Perovkan
Nonesian
Lootsb.
Lad3
Ol4
Ol1
In1

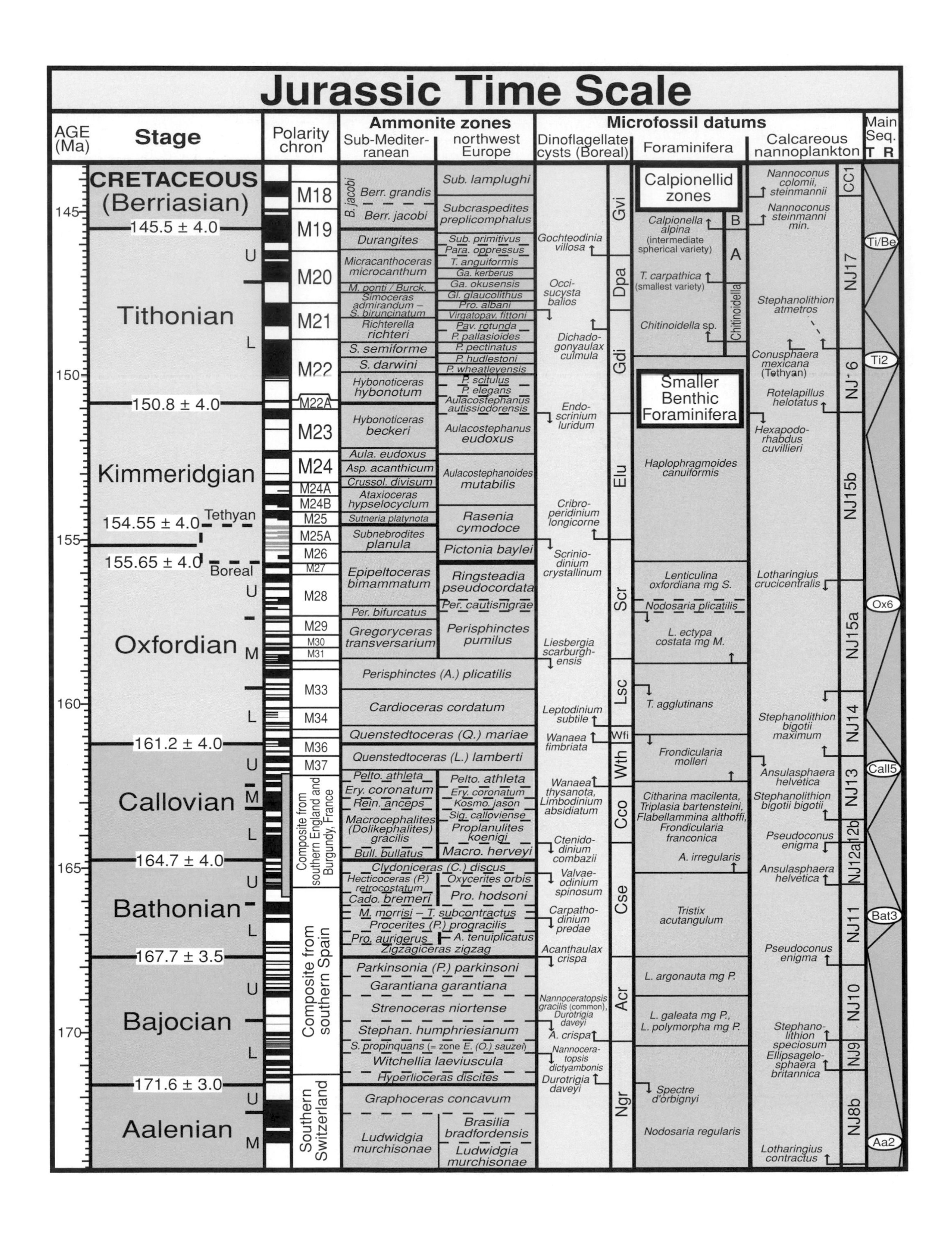

Jurassic Time Scale
AGE (Ma)
Stage
Polarity chron
Ammonite zones
Sub-Mediterranean
northwest Europe
Microfossil datums
Dinoflagellate cysts (Boreal)
Foraminifera
Calcareous nannoplankton
Main Seq. T R
CRETACEOUS (Berriasian)
145.5 ± 4.0
Tithonian
150.8 ± 4.0
Kimmeridgian
154.55 ± 4.0
Tethyan
155.65 ± 4.0
Boreal
Oxfordian
161.2 ± 4.0
Callovian
164.7 ± 4.0
Bathonian
167.7 ± 3.5
Bajocian
171.6 ± 3.0
Aalenian
145
150
155
160
165
170
U
L
M
M18
M19
M20
M21
M22
M22A
M23
M24
M24A
M24B
M25
M25A
M26
M27
M28
M29
M30
M31
M33
M34
M36
M37
Composite from southern England and Burgundy, France
Composite from southern Spain
Southern Switzerland
B. jacobi
Berr. grandis
Berr. jacobi
Durangites
Micracanthoceras microcanthum
M. ponti / Burck.
Simoceras admirandum – S. biruncinatum
Richterella richteri
S. semiforme
S. darwini
Hybonoticeras hybonotum
Hybonoticeras beckeri
Aula. eudoxus
Asp. acanthicum
Crussol. divisum
Ataxioceras hypselocyclum
Sutneria platynota
Subnebrodites planula
Epipeltoceras bimammatum
Per. bifurcatus
Gregoryceras transversarium
Sub. lamplughi
Subcraspedites preplicomphalus
Sub. primitivus
Para. oppressus
T. anguiformis
Ga. kerberus
Ga. okusensis
Gl. glaucolithus
Pro. albani
Virgatopav. fittoni
Pav. rotunda
P. pallasioides
P. pectinatus
P. hudlestoni
P. wheatleyensis
P. scitulus
P. elegans
Aulacostephanus autissiodorensis
Aulacostephanus eudoxus
Aulacostephanoides mutabilis
Rasenia cymodoce
Pictonia baylei
Ringsteadia pseudocordata
Per. cautisnigrae
Perisphinctes pumilus
Perisphinctes (A.) plicatilis
Cardioceras cordatum
Quenstedtoceras (Q.) mariae
Quenstedtoceras (L.) lamberti
Pelto. athleta
Ery. coronatum
Rein. anceps
Macrocephalites (Dolikephalites) gracilis
Bull. bullatus
Pelto. athleta
Ery. coronatum
Kosmo. jason
Sig. calloviense
Proplanulites koenigi
Macro. herveyi
Clydoniceras (C.) discus
Hecticoceras (P.) retrocostatum
Cado. bremeri
Oxycerites orbis
Pro. hodsoni
M. morrisi – T. subcontractus
Procerites (P.) progracilis
Pro. aurigerus
A. tenuiplicatus
Zigzagiceras zigzag
Parkinsonia (P.) parkinsoni
Garantiana garantiana
Strenoceras niortense
Stephan. humphriesianum
S. propinquans (= zone E. (O.) sauzei)
Witchellia laeviuscula
Hyperlioceras discites
Graphoceras concavum
Ludwidgia murchisonae
Brasilia bradfordensis
Ludwigia murchisonae
Gochteodinia villosa
Occisucysta balios
Dichadogonyaulax culmula
Endoscrinium luridum
Cribroperidinium longicorne
Scriniodinium crystallinum
Liesbergia scarburghensis
Leptodinium subtile
Wanaea fimbriata
Wanaea thysanota, Limbodinium absidiatum
Ctenidodinium combazii
Valvaeodinium spinosum
Carpathodinium predae
Acanthaulax crispa
Nannoceratopsis gracilis (common), Durotrigia daveyi
A. crispa
Nannoceratopsis dictyambonis
Durotrigia daveyi
Gvi
Dpa
Gdi
Elu
Scr
Lsc
Wfi
Wth
Cco
Cse
Acr
Ngr
Calpionellid zones
Calpionella alpina (intermediate spherical variety)
B
A
T. carpathica (smallest variety)
Chitinoidella sp.
Chitinoidella
Smaller Benthic Foraminifera
Haplophragmoides canuiformis
Lenticulina oxfordiana mg S.
Nodosaria plicatilis
L. ectypa costata mg M.
T. agglutinans
Frondicularia molleri
Citharina macilenta, Triplasia bartensteini, Flabellammina althoffi, Frondicularia franconica
A. irregularis
Tristix acutangulum
L. argonauta mg P.
L. galeata mg P., L. polymorpha mg P.
Spectre d'orbignyi
Nodosaria regularis
Nannoconus colomii, steinmannii
Nannoconus steinmanni min.
Stephanolithion atmetros
Conusphaera mexicana (Tethyan)
Rotelapillus helotatus
Hexapodorhabdus cuvillieri
Lotharingius crucicentralis
Stephanolithion bigotii maximum
Ansulasphaera helvetica
Stephanolithion bigotii bigotii
Pseudoconus enigma
Ansulasphaera helvetica
Pseudoconus enigma
Stephanolithion speciosum
Ellipsagelosphaera britannica
Lotharingius contractus
CC1
NJ17
NJ 6
NJ15b
NJ15a
NJ14
NJ13
NJ12b
NJ12a
NJ11
NJ10
NJ9
NJ8b
Ti/Be
Ti2
Ox6
Call5
Bat3
Aa2

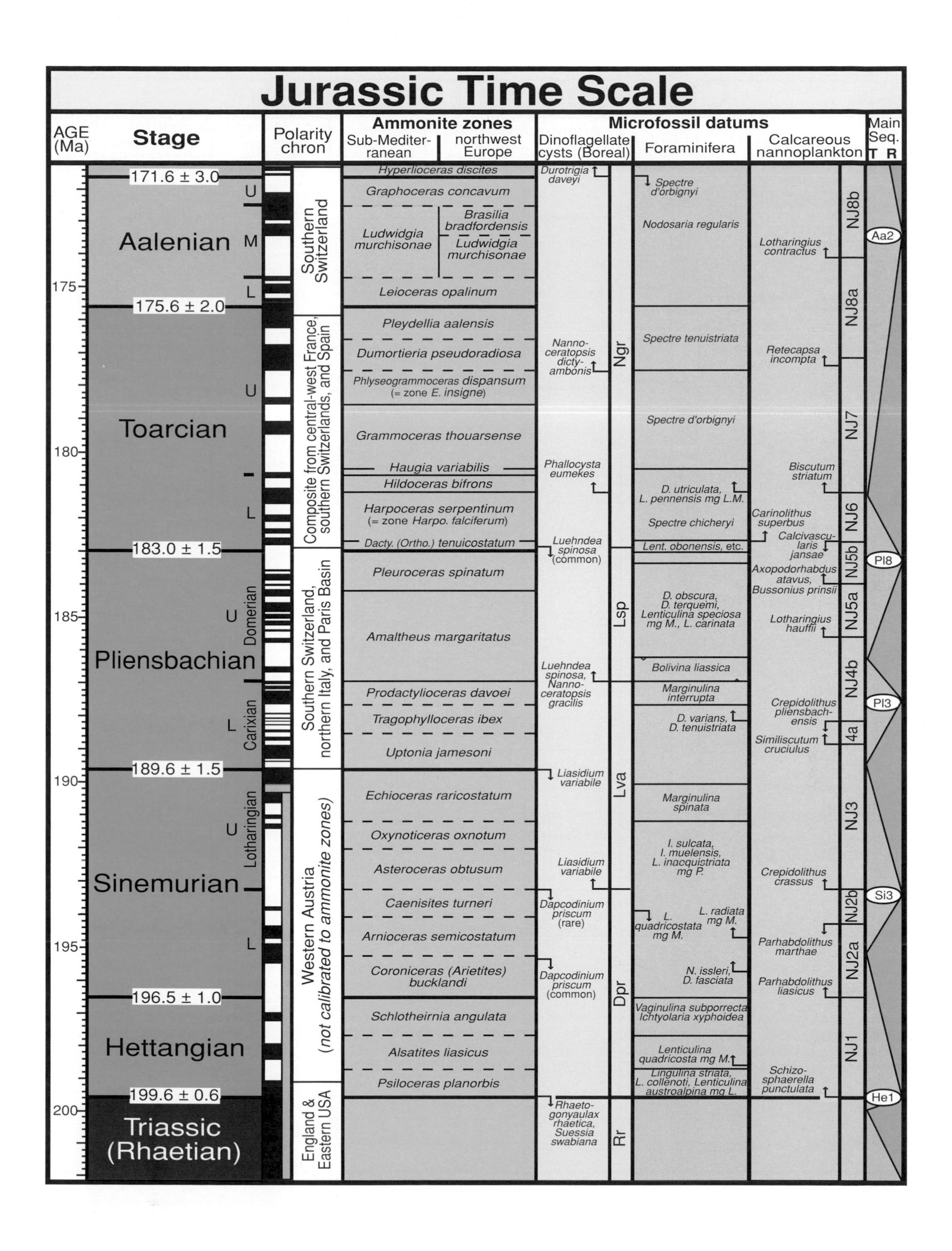

Jurassic Time Scale
AGE (Ma)
Stage
Polarity chron
Ammonite zones
Sub-Mediterranean
northwest Europe
Microfossil datums
Dinoflagellate cysts (Boreal)
Foraminifera
Calcareous nannoplankton
Main Seq.
T R
171.6 ± 3.0
175.6 ± 2.0
183.0 ± 1.5
189.6 ± 1.5
196.5 ± 1.0
199.6 ± 0.6
175
180
185
190
195
200
Aalenian
U
M
L
Toarcian
U
L
Pliensbachian
Domerian
U
Carixian
L
Sinemurian
Lotharingian
U
L
Hettangian
Triassic (Rhaetian)
Southern Switzerland
Composite from central-west France, southern Switzerlands, and Spain
Southern Switzerland, northern Italy, and Paris Basin
Western Austria (not calibrated to ammonite zones)
England & Eastern USA
Hyperlioceras discites
Graphoceras concavum
Ludwidgia murchisonae
Brasilia bradfordensis
Ludwidgia murchisonae
Leioceras opalinum
Pleydellia aalensis
Dumortieria pseudoradiosa
Phlyseogrammoceras dispansum (= zone E. insigne)
Grammoceras thouarsense
Haugia variabilis
Hildoceras bifrons
Harpoceras serpentinum (= zone Harpo. falciferum)
Dacty. (Ortho.) tenuicostatum
Pleuroceras spinatum
Amaltheus margaritatus
Prodactylioceras davoei
Tragophylloceras ibex
Uptonia jamesoni
Echioceras raricostatum
Oxynoticeras oxnotum
Asteroceras obtusum
Caenisites turneri
Arnioceras semicostatum
Coroniceras (Arietites) bucklandi
Schlotheirnia angulata
Alsatites liasicus
Psiloceras planorbis
Durotrigia daveyi
Nannoceratopsis dictyambonis
Phallocysta eumekes
Luehndea spinosa (common)
Luehndea spinosa, Nannoceratopsis gracilis
Liasidium variabile
Liasidium variabile
Dapcodinium priscum (rare)
Dapcodinium priscum (common)
Rhaetogonyaulax rhaetica, Suessia swabiana
Ngr
Lsp
Lva
Dpr
Rr
Spectre d'orbignyi
Nodosaria regularis
Spectre tenuistriata
Spectre d'orbignyi
D. utriculata, L. pennensis mg L.M.
Spectre chicheryi
Lent. obonensis, etc.
D. obscura, D. terquemi, Lenticulina speciosa mg M., L. carinata
Bolivina liassica
Marginulina interrupta
D. varians, D. tenuistriata
Marginulina spinata
I. sulcata, I. muelensis, L. inaequistriata mg P.
L. quadricostata mg M.
L. radiata mg M.
N. issleri, D. fasciata
Vaginulina subporrecta Ichtyolaria xyphoidea
Lenticulina quadricosta mg M.
Lingulina striata, L. collenoti, Lenticulina austroalpina mg L.
Lotharingius contractus
Retecapsa incompta
Biscutum striatum
Carinolithus superbus
Calcivascularis jansae
Axopodorhabdus atavus, Bussonius prinsii
Lotharingius hauffii
Crepidolithus pliensbachensis
Similiscutum cruciulus
Crepidolithus crassus
Parhabdolithus marthae
Parhabdolithus liasicus
Schizosphaerella punctulata
NJ8b
NJ8a
NJ7
NJ6
NJ5b
NJ5a
NJ4b
4a
NJ3
NJ2b
NJ2a
NJ1
Aa2
Pl8
Pl3
Si3
He1

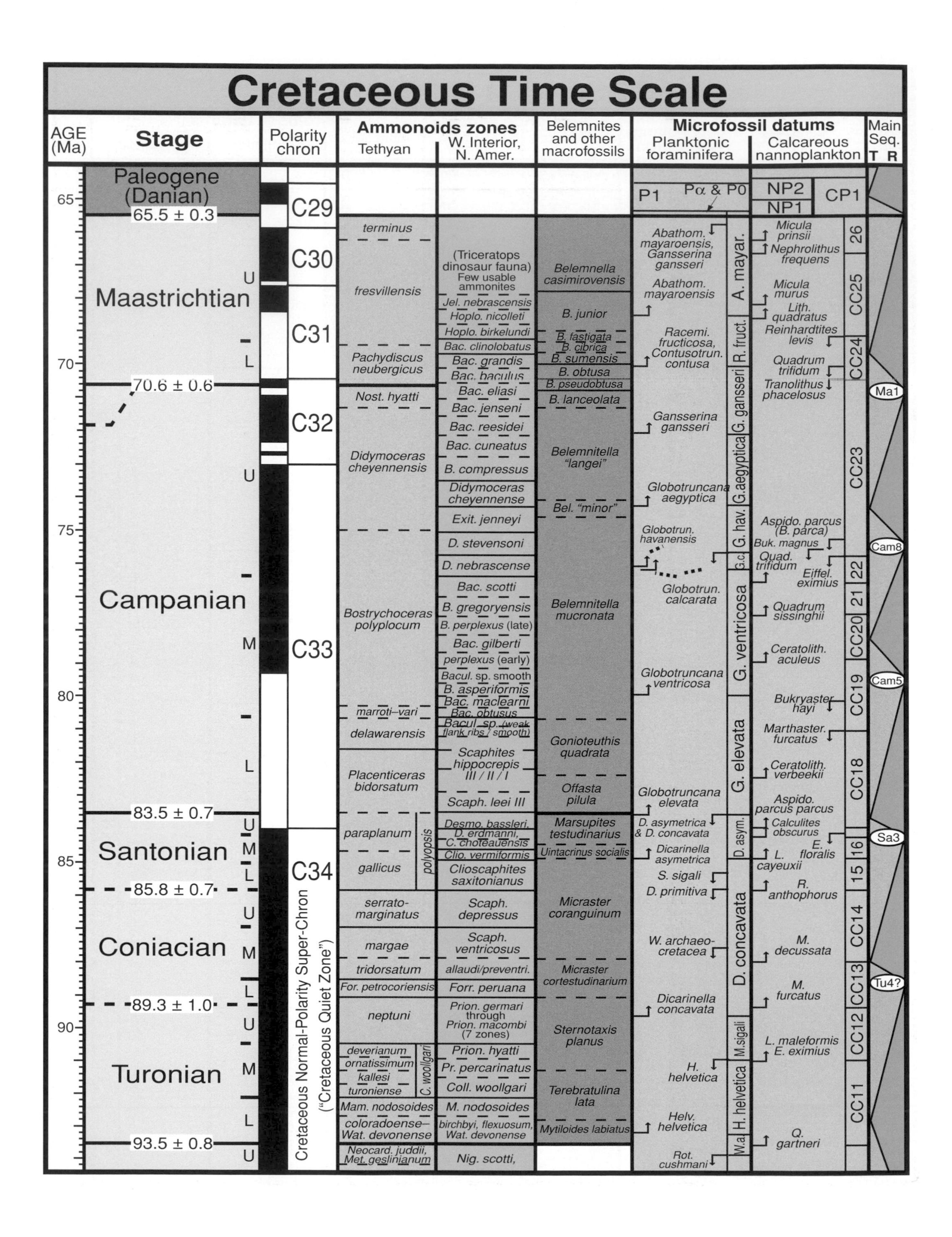

Cretaceous Time Scale
AGE (Ma)
Stage
Polarity chron
Ammonoids zones
Tethyan
W. Interior, N. Amer.
Belemnites and other macrofossils
Microfossil datums
Planktonic foraminifera
Calcareous nannoplankton
Main Seq.
T R
Paleogene (Danian)
65.5 ± 0.3
Maastrichtian
70.6 ± 0.6
Campanian
83.5 ± 0.7
Santonian
85.8 ± 0.7
Coniacian
89.3 ± 1.0
Turonian
93.5 ± 0.8
U
L
M
65
70
75
80
85
90
C29
C30
C31
C32
C33
C34
Cretaceous Normal-Polarity Super-Chron
("Cretaceous Quiet Zone")
terminus
fresvillensis
Pachydiscus neubergicus
Nost. hyatti
Didymoceras cheyennensis
Bostrychoceras polyplocum
marroti–vari
delawarensis
Placenticeras bidorsatum
paraplanum
gallicus
polyopsis
serrato-marginatus
margae
tridorsatum
For. petrocoriensis
neptuni
deverianum
ornatissimum
kallesi
turoniense
C. woollgari
Mam. nodosoides
coloradoense–Wat. devonense
Neocard. juddii, Met. geslinianum
(Triceratops dinosaur fauna) Few usable ammonites
Jel. nebrascensis
Hoplo. nicolleti
Hoplo. birkelundi
Bac. clinolobatus
Bac. grandis
Bac. baculus
Bac. eliasi
Bac. jenseni
Bac. reesidei
Bac. cuneatus
B. compressus
Didymoceras cheyennense
Exit. jenneyi
D. stevensoni
D. nebrascense
Bac. scotti
B. gregoryensis
B. perplexus (late)
Bac. gilberti
perplexus (early)
Bacul. sp. smooth
B. asperiformis
Bac. maclearni
Bac. obtusus
Bacul. sp. (weak flank ribs / smooth)
Scaphites hippocrepis III / II / I
Scaph. leei III
Desmo. bassleri,
D. erdmanni,
C. choteauensis
Clio. vermiformis
Clioscaphites saxitonianus
Scaph. depressus
Scaph. ventricosus
allaudi/preventri.
Forr. peruana
Prion. germari through Prion. macombi (7 zones)
Prion. hyatti
Pr. percarinatus
Coll. woollgari
M. nodosoides
birchbyi, flexuosum, Wat. devonense
Nig. scotti,
Belemnella casimirovensis
B. junior
B. fastigata
B. cibrica
B. sumensis
B. obtusa
B. pseudobtusa
B. lanceolata
Belemnitella "langei"
Bel. "minor"
Belemnitella mucronata
Gonioteuthis quadrata
Offasta pilula
Marsupites testudinarius
Uintacrinus socialis
Micraster coranguinum
Micraster cortestudinarium
Sternotaxis planus
Terebratulina lata
Mytiloides labiatus
P1
Pα & P0
Abathom. mayaroensis, Gansserina gansseri
Abathom. mayaroensis
Racemi. fructicosa, Contusotrun. contusa
Gansserina gansseri
Globotruncana aegyptica
Globotrun. havanensis
Globotrun. calcarata
Globotruncana ventricosa
Globotruncana elevata
D. asymetrica & D. concavata
Dicarinella asymetrica
S. sigali
D. primitiva
W. archaeo-cretacea
Dicarinella concavata
H. helvetica
Helv. helvetica
Rot. cushmani
A. mayar.
R. fruct.
G. gansseri
G. aegyptica
G. hav.
G.c
G. ventricosa
G. elevata
D. asym.
D. concavata
M. sigali
H. helvetica
W.a
NP2
NP1
CP1
Micula prinsii
Nephrolithus frequens
Micula murus
Lith. quadratus
Reinhardtites levis
Quadrum trifidum
Tranolithus phacelosus
Aspido. parcus (B. parca)
Buk. magnus
Quad. trifidum
Eiffel. eximius
Quadrum sissinghii
Ceratolith. aculeus
Bukryaster hayi
Marthaster. furcatus
Ceratolith. verbeekii
Aspido. parcus parcus
Calculites obscurus
L. cayeuxii
E. floralis
R. anthophorus
M. decussata
M. furcatus
L. maleformis
E. eximius
Q. gartneri
26
CC25
CC24
CC23
22
21
CC20
CC19
CC18
16
15
CC14
CC13
CC12
CC11
Ma1
Cam8
Cam5
Sa3
Tu4?

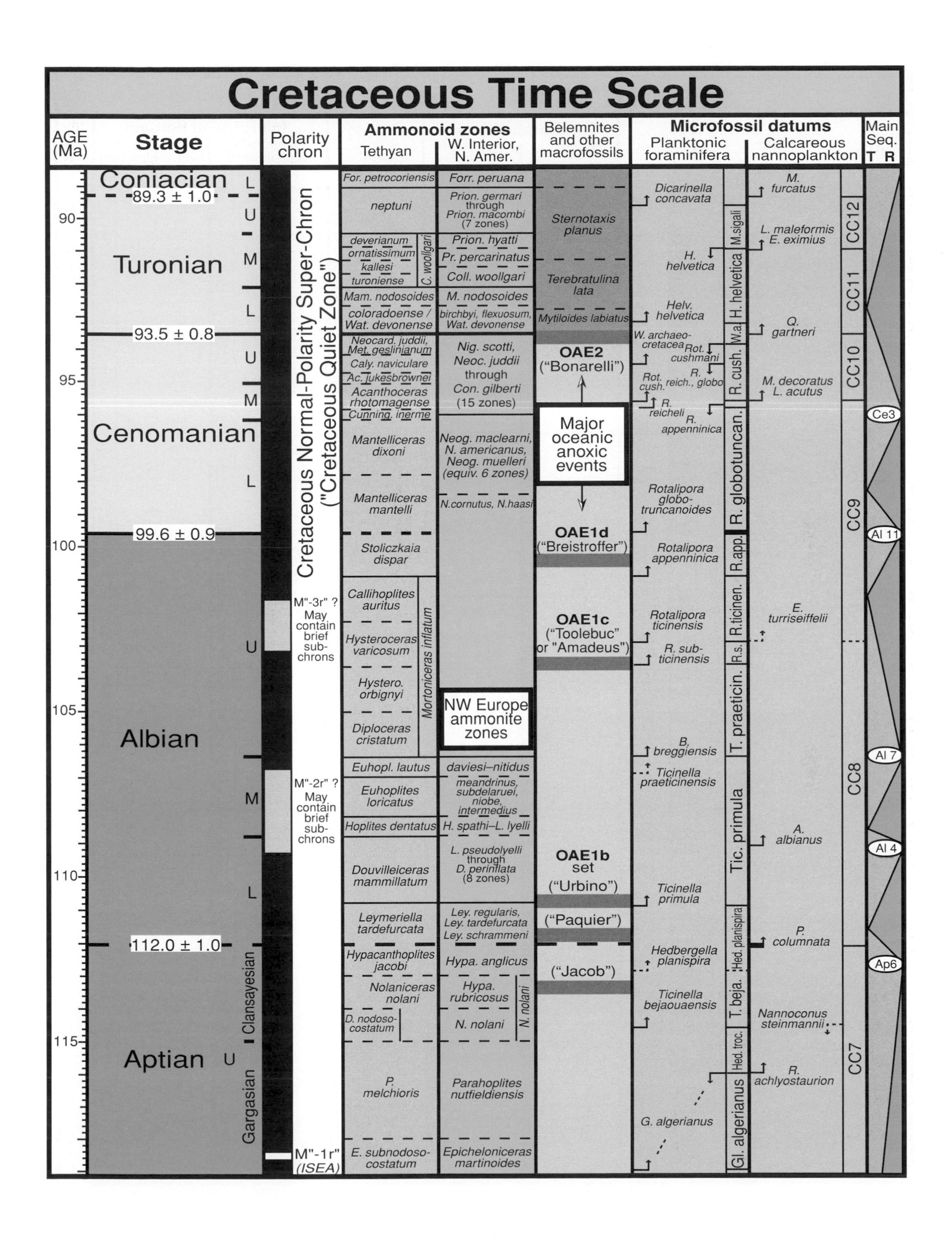

Cretaceous Time Scale
AGE (Ma)
Stage
Polarity chron
Ammonoid zones
Tethyan
W. Interior, N. Amer.
Belemnites and other macrofossils
Microfossil datums
Planktonic foraminifera
Calcareous nannoplankton
Main Seq. T R
90
95
100
105
110
115
Coniacian
L
89.3 ± 1.0
U
Turonian
M
L
93.5 ± 0.8
U
M
Cenomanian
L
99.6 ± 0.9
Albian
U
M
L
112.0 ± 1.0
Aptian
U
Clansayesian
Gargasian
Cretaceous Normal-Polarity Super-Chron ("Cretaceous Quiet Zone")
M"-3r" ? May contain brief sub-chrons
M"-2r" ? May contain brief sub-chrons
M"-1r" (ISEA)
For. petrocoriensis
neptuni
deverianum
ornatissimum
kallesi
turoniense
C. woollgari
Mam. nodosoides
coloradoense / Wat. devonense
Neocard. juddii, Met. geslinianum
Caly. naviculare
Ac. jukesbrownei
Acanthoceras rhotomagense
Cunning. inerme
Mantelliceras dixoni
Mantelliceras mantelli
Stoliczkaia dispar
Callihoplites auritus
Hysteroceras varicosum
Hystero. orbignyi
Diploceras cristatum
Mortoniceras inflatum
Euhopl. lautus
Euhoplites loricatus
Hoplites dentatus
Douvilleiceras mammillatum
Leymeriella tardefurcata
Hypacanthoplites jacobi
Nolaniceras nolani
D. nodoso-costatum
P. melchioris
E. subnodoso-costatum
Forr. peruana
Prion. germari through Prion. macombi (7 zones)
Prion. hyatti
Pr. percarinatus
Coll. woollgari
M. nodosoides
birchbyi, flexuosum, Wat. devonense
Nig. scotti, Neoc. juddii through Con. gilberti (15 zones)
Neog. maclearni, N. americanus, Neog. muelleri (equiv. 6 zones)
N.cornutus, N.haasi
NW Europe ammonite zones
daviesi–nitidus
meandrinus, subdelaruei, niobe, intermedius
H. spathi–L. lyelli
L. pseudolyelli through D. perinflata (8 zones)
Ley. regularis, Ley. tardefurcata, Ley. schrammeni
Hypa. anglicus
Hypa. rubricosus
N. nolani
N. nolani
Parahoplites nutfieldiensis
Epicheloniceras martinoides
Sternotaxis planus
Terebratulina lata
Mytiloides labiatus
OAE2 ("Bonarelli")
Major oceanic anoxic events
OAE1d ("Breistroffer")
OAE1c ("Toolebuc" or "Amadeus")
OAE1b set
("Urbino")
("Paquier")
("Jacob")
Dicarinella concavata
H. helvetica
Helv. helvetica
W. archaeo-cretacea
Rot. cushmani
R. reich., globo
Rot. cush.
R. reicheli
R. appenninica
Rotalipora globo-truncanoides
Rotalipora appenninica
Rotalipora ticinensis
R. sub-ticinensis
B. breggiensis
Ticinella praeticinensis
Ticinella primula
Hedbergella planispira
Ticinella bejaouaensis
G. algerianus
M.sigali
H. helvetica
W.a.
R. cush.
R. globotuncan.
R.app.
R.ticinen.
R.s.
T. praeticin.
Tic. primula
Hed. planispira
T. beja.
Hed. troc.
Gl. algerianus
M. furcatus
L. maleformis
E. eximius
Q. gartneri
M. decoratus
L. acutus
E. turriseiffelii
A. albianus
P. columnata
Nannoconus steinmannii
R. achlyostaurion
CC12
CC11
CC10
CC9
CC8
CC7
Ce3
Al 11
Al 7
Al 4
Ap6

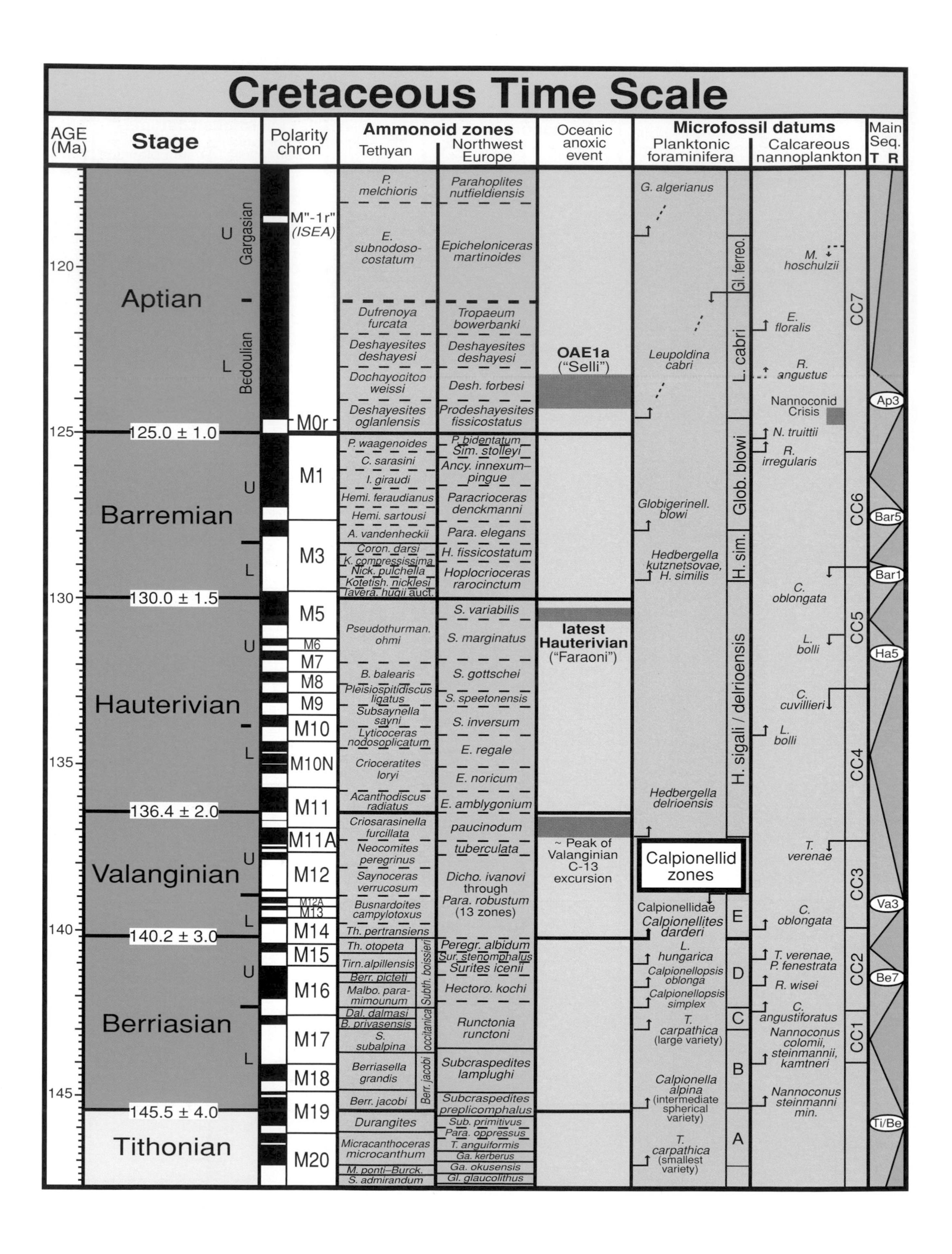

Cretaceous Time Scale
AGE (Ma)
Stage
Polarity chron
Ammonoid zones
Tethyan
Northwest Europe
Oceanic anoxic event
Microfossil datums
Planktonic foraminifera
Calcareous nannoplankton
Main Seq.
T R
120
125
130
135
140
145
Aptian
Gargasian
Bedoulian
U
L
Barremian
Hauterivian
Valanginian
Berriasian
Tithonian
125.0 ± 1.0
130.0 ± 1.5
136.4 ± 2.0
140.2 ± 3.0
145.5 ± 4.0
M"-1r"
(ISEA)
M0r
M1
M3
M5
M6
M7
M8
M9
M10
M10N
M11
M11A
M12
M12A
M13
M14
M15
M16
M17
M18
M19
M20
P. melchioris
E. subnodoso-costatum
Dufrenoya furcata
Deshayesites deshayesi
Dochayocitos weissi
Deshayesites oglanlensis
P. waagenoides
C. sarasini
I. giraudi
Hemi. feraudianus
Hemi. sartousi
A. vandenheckii
Coron. darsi
K. compressissima
Nick. pulchella
Kotetish. nicklesi
Tavera. hugii auct.
Pseudothurman. ohmi
B. balearis
Pleisiospitidiscus ligatus
Subsaynella sayni
Lyticoceras nodosoplicatum
Crioceratites loryi
Acanthodiscus radiatus
Criosarasinella furcillata
Neocomites peregrinus
Saynoceras verrucosum
Busnardoites campylotoxus
Th. pertransiens
Th. otopeta
Tirn. alpillensis
Berr. picteti
Malbo. para-mimounum
Dal. dalmasi
B. privasensis
S. subalpina
Berriasella grandis
Berr. jacobi
Subth. boissieri
occitanica
Berr. jacobi
Durangites
Micracanthoceras microcanthum
M. ponti–Burck.
S. admirandum
Parahoplites nutfieldiensis
Epicheloniceras martinoides
Tropaeum bowerbanki
Deshayesites deshayesi
Desh. forbesi
Prodeshayesites fissicostatus
P. bidentatum
Sim. stolleyi
Ancy. innexum–pingue
Paracrioceras denckmanni
Para. elegans
H. fissicostatum
Hoplocrioceras rarocinctum
S. variabilis
S. marginatus
S. gottschei
S. speetonensis
S. inversum
E. regale
E. noricum
E. amblygonium
paucinodum
tuberculata
Dicho. ivanovi through Para. robustum (13 zones)
Peregr. albidum
Sur. stenomphalus
Surites icenii
Hectoro. kochi
Runctonia runctoni
Subcraspedites lamplughi
Subcraspedites preplicomphalus
Sub. primitivus
Para. oppressus
T. anguiformis
Ga. kerberus
Ga. okusensis
Gl. glaucolithus
OAE1a
("Selli")
latest Hauterivian
("Faraoni")
~ Peak of Valanginian C-13 excursion
G. algerianus
Leupoldina cabri
Globigerinell. blowi
Hedbergella kutznetsovae, H. similis
Hedbergella delrioensis
Calpionellid zones
Calpionellidae
Calpionellites darderi
L. hungarica
Calpionellopsis oblonga
Calpionellopsis simplex
T. carpathica (large variety)
Calpionella alpina (intermediate spherical variety)
T. carpathica (smallest variety)
Gl. ferreo.
L. cabri
Glob. blowi
H. sim.
H. sigali / delrioensis
E
D
C
B
A
M. hoschulzii
E. floralis
R. angustus
Nannoconid Crisis
N. truittii
R. irregularis
C. oblongata
L. bolli
C. cuvillieri
L. bolli
T. verenae
C. oblongata
T. verenae, P. fenestrata
R. wisei
C. angustiforatus
Nannoconus colomii, steinmannii, kamtneri
Nannoconus steinmanni min.
CC7
CC6
CC5
CC4
CC3
CC2
CC1
Ap3
Bar5
Bar1
Ha5
Va3
Be7
Ti/Be

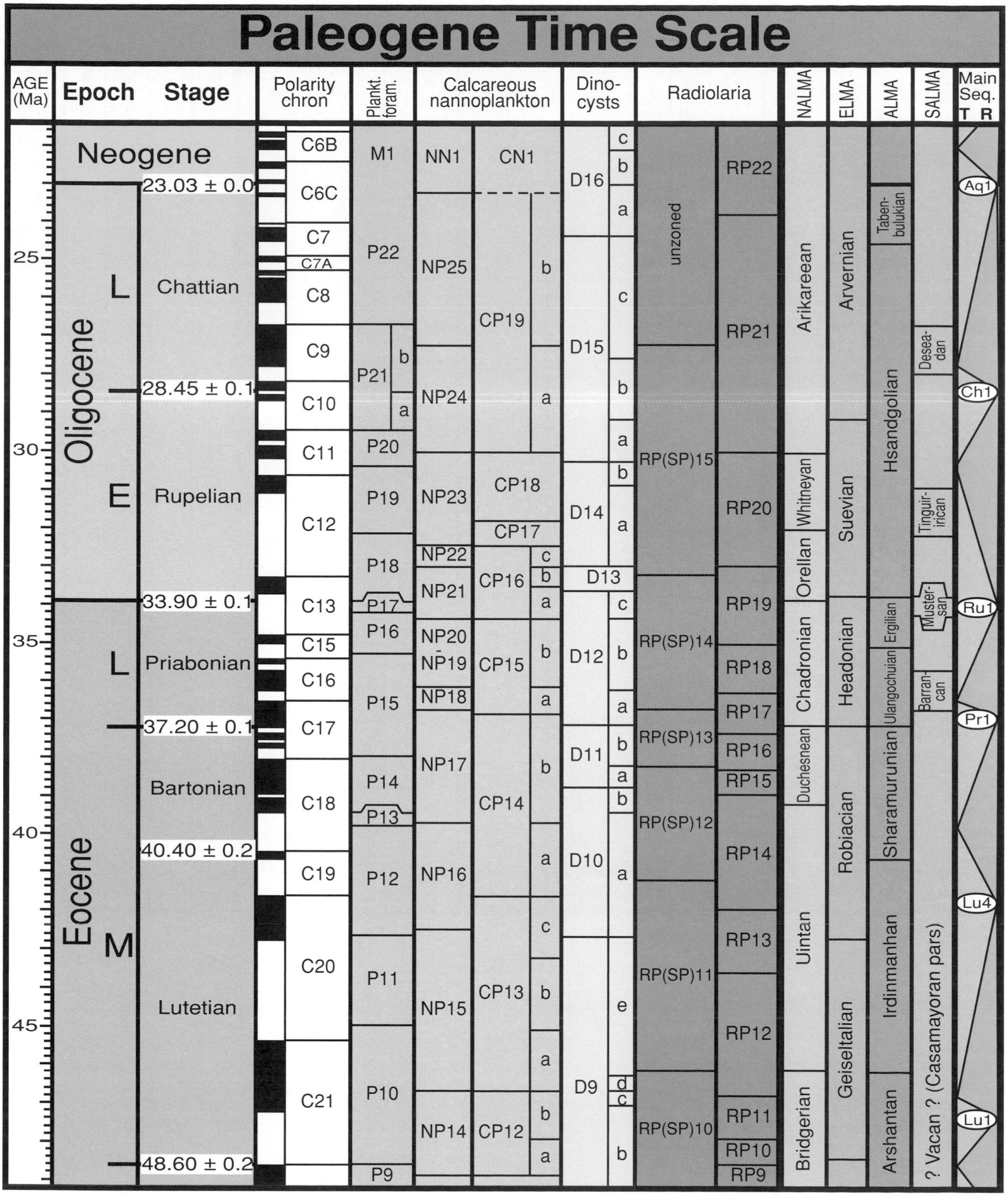

Paleogene Time Scale

AGE (Ma): 25, 30, 35, 40, 45

Epoch / Stage: Neogene; Oligocene – L: Chattian; E: Rupelian; Eocene – L: Priabonian; Bartonian; M: Lutetian

Boundary ages: 23.03 ± 0.0; 28.45 ± 0.1; 33.90 ± 0.1; 37.20 ± 0.1; 40.40 ± 0.2; 48.60 ± 0.2

Polarity chron: C6B, C6C, C7, C7A, C8, C9, C10, C11, C12, C13, C15, C16, C17, C18, C19, C20, C21

Plankt. foram.: M1, P22, P21 (b, a), P20, P19, P18, P17, P16, P15, P14, P13, P12, P11, P10, P9

Calcareous nannoplankton: NN1, NP25, NP24, NP23, NP22, NP21, NP20, NP19, NP18, NP17, NP16, NP15, NP14; CN1, CP19 (b, a), CP18, CP17, CP16 (c, b, a), CP15 (b, a), CP14 (b, a), CP13 (c, b, a), CP12 (b, a)

Dino-cysts: D16 (c, b, a), D15 (c, b, a), D14 (b, a), D13, D12 (c, b, a), D11 (b, a), D10 (b, a), D9 (e, d, c, b)

Radiolaria: unzoned, RP(SP)15, RP(SP)14, RP(SP)13, RP(SP)12, RP(SP)11, RP(SP)10; RP22, RP21, RP20, RP19, RP18, RP17, RP16, RP15, RP14, RP13, RP12, RP11, RP10, RP9

NALMA: Arikareean, Whitneyan, Orellan, Chadronian, Duchesnean, Uintan, Bridgerian

ELMA: Arvernian, Suevian, Headonian, Robiacian, Geiseltalian

ALMA: Taben-bulukian, Hsandgolian, Ergilian, Ulangochuian, Sharamurunian, Irdinmanhan, Arshantan

SALMA: Desea-dan, Tinguir-irican, Muster-san, Barran-can, ? Vacan ? (Casamayoran pars)

Main Seq. T R: Aq1, Ch1, Ru1, Pr1, Lu4, Lu1

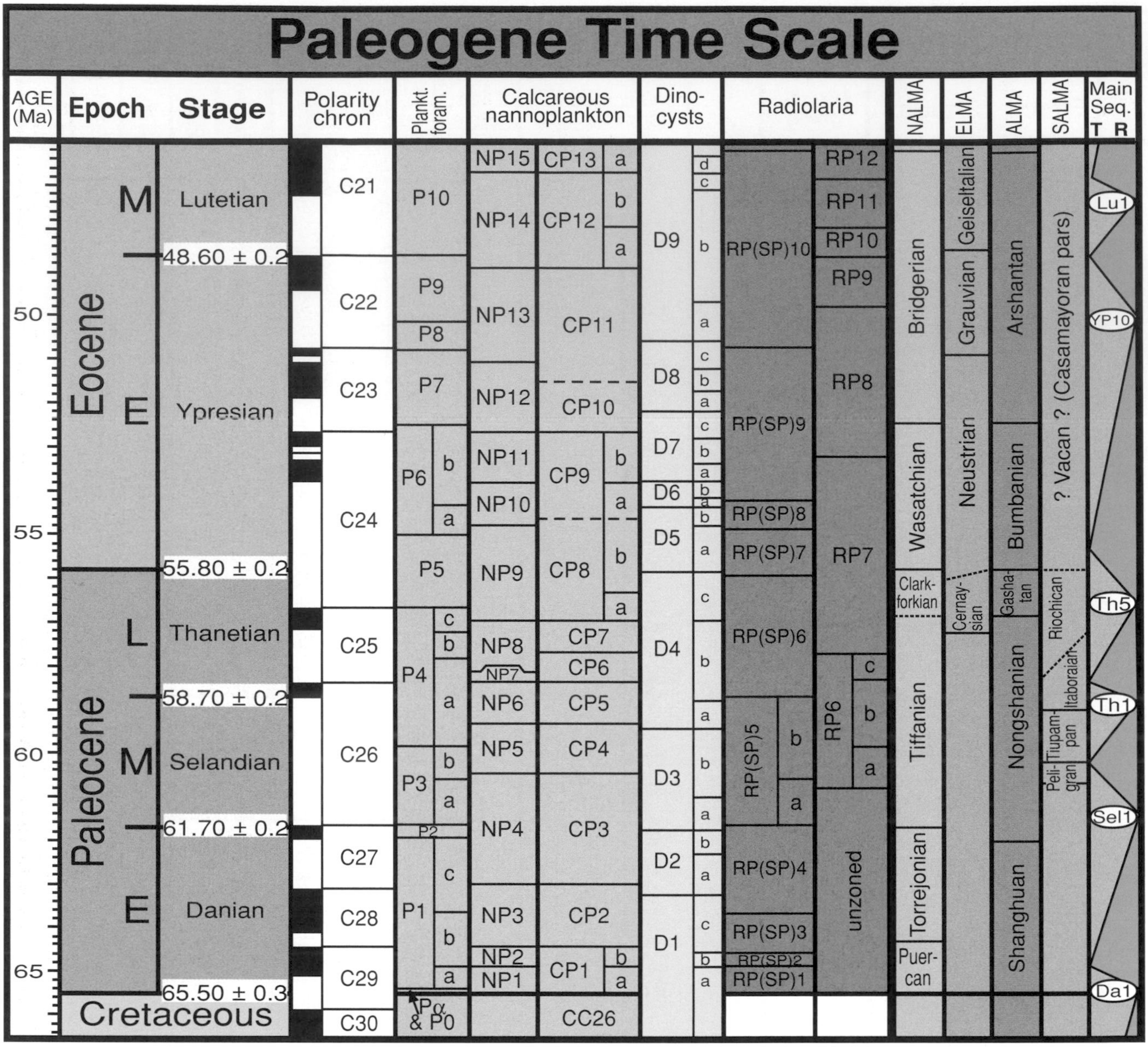

Paleogene Time Scale

AGE (Ma) | Epoch | Stage | Polarity chron | Plankt. foram. | Calcareous nannoplankton | Dino-cysts | Radiolaria | NALMA | ELMA | ALMA | SALMA | Main Seq. T R

AGE (Ma): 50, 55, 60, 65

Epoch: Eocene (M, E); Paleocene (L, M, E); Cretaceous

Stage: Lutetian, 48.60 ± 0.2, Ypresian, 55.80 ± 0.2, Thanetian, 58.70 ± 0.2, Selandian, 61.70 ± 0.2, Danian, 65.50 ± 0.3

Polarity chron: C21, C22, C23, C24, C25, C26, C27, C28, C29, C30

Plankt. foram.: P10, P9, P8, P7, P6 (b, a), P5, P4 (c, b, a), P3 (b, a), P2, P1 (c, b, a), Pα & P0

Calcareous nannoplankton: NP15, NP14, NP13, NP12, NP11, NP10, NP9, NP8, NP7, NP6, NP5, NP4, NP3, NP2, NP1; CP13 (a), CP12 (b, a), CP11, CP10, CP9 (b, a), CP8 (b, a), CP7, CP6, CP5, CP4, CP3, CP2, CP1 (b, a), CC26

Dino-cysts: D9 (d, c, b, a), D8 (c, b, a), D7 (c, b, a), D6 (b, a), D5 (b, a), D4 (c, b, a), D3 (b, a), D2 (b, a), D1 (c, b, a)

Radiolaria: RP(SP)10, RP(SP)9, RP(SP)8, RP(SP)7, RP(SP)6, RP(SP)5 (b, a), RP(SP)4, RP(SP)3, RP(SP)2, RP(SP)1; RP12, RP11, RP10, RP9, RP8, RP7, RP6 (c, b, a), unzoned

NALMA: Bridgerian, Wasatchian, Clark-forkian, Tiffanian, Torrejonian, Puer-can

ELMA: Geiseltalian, Grauvian, Neustrian, Cernay-sian

ALMA: Arshantan, Bumbanian, Gasha-tan, Nongshanian, Shanghuan

SALMA: ? Vacan ? (Casamayoran pars), Riochican, Itaboraian, Tiupam-pan, Peli-gran

Main Seq. T R: Lu1, YP10, Th5, Th1, Sel1, Da1

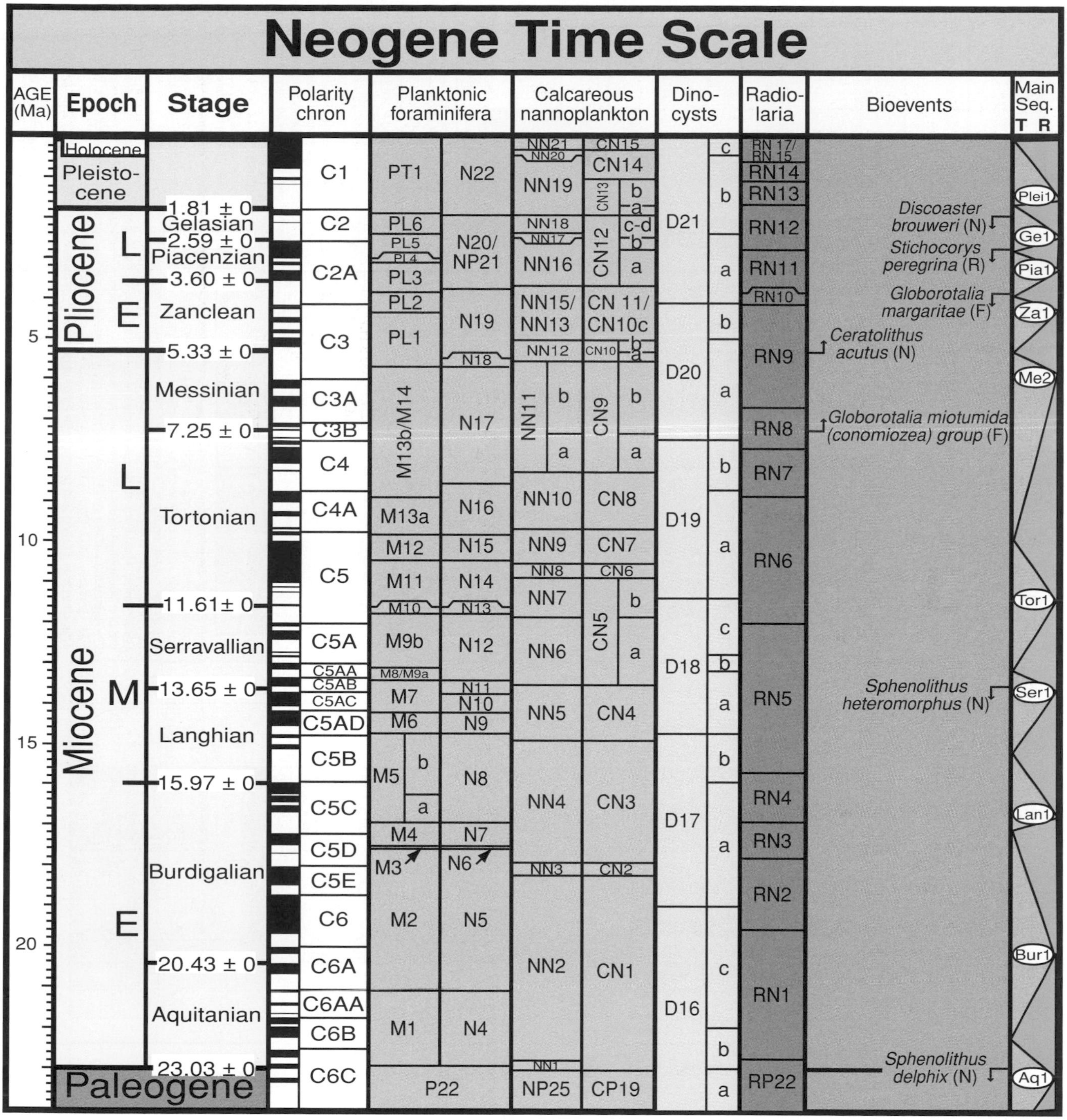

Neogene Time Scale

AGE (Ma)	Epoch	Stage	Polarity chron	Planktonic foraminifera	Calcareous nannoplankton	Dino-cysts	Radio-laria	Bioevents	Main Seq. T R
5, 10, 15, 20	Holocene; Pleistocene; Pliocene (L, E); Miocene (L, M, E); Paleogene	1.81 ± 0; Gelasian; 2.59 ± 0; Piacenzian; 3.60 ± 0; Zanclean; 5.33 ± 0; Messinian; 7.25 ± 0; Tortonian; 11.61± 0; Serravallian; 13.65 ± 0; Langhian; 15.97 ± 0; Burdigalian; 20.43 ± 0; Aquitanian; 23.03 ± 0	C1; C2; C2A; C3; C3A; C3B; C4; C4A; C5; C5A; C5AA; C5AB; C5AC; C5AD; C5B; C5C; C5D; C5E; C6; C6A; C6AA; C6B; C6C	PT1 / N22; PL6, PL5, PL4, PL3 / N20/NP21; PL2, PL1 / N19; N18; M13b/M14 / N17; M13a / N16; M12 / N15; M11 / N14; M10 / N13; M9b / N12; M8/M9a; M7 / N11, N10; M6 / N9; M5 (b, a) / N8; M4 / N7; M3 / N6; M2 / N5; M1 / N4; P22	NN21 / CN15; NN20 / CN14; NN19 / CN13 (b, a); NN18, NN17, NN16 / CN12 (c-d, b, a); NN15/NN13 / CN 11/CN10c; NN12 / CN10 (b, a); NN11 (b, a) / CN9 (b, a); NN10 / CN8; NN9 / CN7; NN8 / CN6; NN7, NN6 / CN5 (b, a); NN5 / CN4; NN4 / CN3; NN3 / CN2; NN2 / CN1; NN1; NP25 / CP19	D21 (c, b, a); D20 (b, a); D19 (b, a); D18 (c, b, a); D17 (b, a); D16 (c, b, a)	RN 17/RN 15; RN14; RN13; RN12; RN11; RN10; RN9; RN8; RN7; RN6; RN5; RN4; RN3; RN2; RN1; RP22	*Discoaster brouweri* (N); *Stichocorys peregrina* (R); *Globorotalia margaritae* (F); *Ceratolithus acutus* (N); *Globorotalia miotumida (conomiozea) group* (F); *Sphenolithus heteromorphus* (N); *Sphenolithus delphix* (N)	Plei1; Ge1; Pia1; Za1; Me2; Tor1; Ser1; Lan1; Bur1; Aq1

INTERNATIONAL STRATIGRAPHIC CHART

International Commission on Stratigraphy

Eonothem Eon	Erathem Era	System Period	Series Epoch	Stage Age	Age Ma	GSSP
Phanerozoic	Cenozoic	Quaternary*	Holocene		0.0115	
			Pleistocene	Upper	0.126	
				Middle	0.781	
				Lower	1.806	GSSP
		Neogene	Pliocene	Gelasian	2.588	GSSP
				Piacenzian	3.600	GSSP
				Zanclean	5.332	GSSP
			Miocene	Messinian	7.246	GSSP
				Tortonian	11.608	GSSP
				Serravallian	13.65	
				Langhian	15.97	
				Burdigalian	20.43	
				Aquitanian	23.03	GSSP
		Paleogene	Oligocene	Chattian	28.4 ±0.1	
				Rupelian	33.9 ±0.1	GSSP
			Eocene	Priabonian	37.2 ±0.1	
				Bartonian	40.4 ±0.2	
				Lutetian	48.6 ±0.2	
				Ypresian	55.8 ±0.2	GSSP
			Paleocene	Thanetian	58.7 ±0.2	
				Selandian	61.7 ±0.2	
				Danian	65.5 ±0.3	GSSP
	Mesozoic	Cretaceous	Upper	Maastrichtian	70.6 ±0.6	GSSP
				Campanian	83.5 ±0.7	
				Santonian	85.8 ±0.7	
				Coniacian	89.3 ±1.0	
				Turonian	93.5 ±0.8	GSSP
				Cenomanian	99.6 ±0.9	GSSP
			Lower	Albian	112.0 ±1.0	
				Aptian	125.0 ±1.0	
				Barremian	130.0 ±1.5	
				Hauterivian	136.4 ±2.0	
				Valanginian	140.2 ±3.0	
				Berriasian	145.5 ±4.0	

* proposed by INQUA

Eonothem Eon	Erathem Era	System Period	Series Epoch	Stage Age	Age Ma	GSSP
					145.5 ±4.0	
Phanerozoic	Mesozoic	Jurassic	Upper	Tithonian	150.8 ±4.0	
				Kimmeridgian	155.7 ±4.0	
				Oxfordian	161.2 ±4.0	
			Middle	Callovian	164.7 ±4.0	
				Bathonian	167.7 ±3.5	
				Bajocian	171.6 ±3.0	GSSP
				Aalenian	175.6 ±2.0	GSSP
			Lower	Toarcian	183.0 ±1.5	
				Pliensbachian	189.6 ±1.5	GSSP
				Sinemurian	196.5 ±1.0	GSSP
				Hettangian	199.6 ±0.6	
		Triassic	Upper	Rhaetian	203.6 ±1.5	
				Norian	216.5 ±2.0	
				Carnian	228.0 ±2.0	
			Middle	Ladinian	237.0 ±2.0	
				Anisian	245.0 ±1.5	
			Lower	Olenekian	249.7 ±0.7	
				Induan	251.0 ±0.4	GSSP
	Paleozoic	Permian	Lopingian	Changhsingian	253.8 ±0.7	
				Wuchiapingian	260.4 ±0.7	GSSP
			Guadalupian	Capitanian	265.8 ±0.7	GSSP
				Wordian	268.0 ±0.7	GSSP
				Roadian	270.6 ±0.7	GSSP
			Cisuralian	Kungurian	275.6 ±0.7	
				Artinskian	284.4 ±0.7	
				Sakmarian	294.6 ±0.8	
				Asselian	299.0 ±0.8	GSSP
		Carboniferous	Pennsylvanian Upper	Gzhelian	303.9 ±0.9	
				Kasimovian	306.5 ±1.0	
			Pennsylvanian Middle	Moscovian	311.7 ±1.1	
			Pennsylvanian Lower	Bashkirian	318.1 ±1.3	GSSP
			Mississippian Upper	Serpukhovian	326.4 ±1.6	
			Mississippian Middle	Visean	345.3 ±2.1	
			Mississippian Lower	Tournaisian	359.2 ±2.5	GSSP

Eonothem Eon	Erathem Era	System Period	Series Epoch	Stage Age	Age Ma	GSSP
					359.2 ±2.5	
Phanerozoic	Paleozoic	Devonian	Upper	Famennian	374.5 ±2.6	GSSP
				Frasnian	385.3 ±2.6	GSSP
			Middle	Givetian	391.8 ±2.7	GSSP
				Eifelian	397.5 ±2.7	GSSP
			Lower	Emsian	407.0 ±2.8	GSSP
				Pragian	411.2 ±2.8	GSSP
				Lochkovian	416.0 ±2.8	GSSP
		Silurian	Pridoli		418.7 ±2.7	GSSP
			Ludlow	Ludfordian	421.3 ±2.6	GSSP
				Gorstian	422.9 ±2.5	GSSP
			Wenlock	Homerian	426.2 ±2.4	GSSP
				Sheinwoodian	428.2 ±2.3	GSSP
			Llandovery	Telychian	436.0 ±1.9	GSSP
				Aeronian	439.0 ±1.8	GSSP
				Rhuddanian	443.7 ±1.5	GSSP
		Ordovician	Upper	Hirnantian	445.6 ±1.5	
					455.8 ±1.6	
					460.9 ±1.6	GSSP
			Middle	Darriwilian	468.1 ±1.6	GSSP
					471.8 ±1.6	
			Lower		478.6 ±1.7	GSSP
				Tremadocian	488.3 ±1.7	GSSP
		Cambrian	Furongian			
				Paibian	501.0 ±2.0	GSSP
			Middle			
					513.0 ±2.0	
			Lower			
					542.0 ±1.0	GSSP

	Eonothem Eon	Erathem Era	System Period	Age Ma	GSSP GSSA
				542	
Precambrian	Proterozoic	Neoproterozoic	Ediacaran	~630	GSSP
			Cryogenian	850	GSSA
			Tonian	1000	GSSA
		Mesoproterozoic	Stenian	1200	GSSA
			Ectasian	1400	GSSA
			Calymmian	1600	GSSA
		Paleoproterozoic	Statherian	1800	GSSA
			Orosirian	2050	GSSA
			Rhyacian	2300	GSSA
			Siderian	2500	GSSA
	Archean	Neoarchean		2800	GSSA
		Mesoarchean		3200	GSSA
		Paleoarchean		3600	GSSA
		Eoarchean	Lower limit is not defined		

Subdivisions of the global geologic record are formally defined by their lower boundary. Each unit of the Phanerozoic interval (~542 Ma to Present) and the base of the Ediacaran is defined by a Global Standard Section and Point (GSSP) at its base, whereas the Precambrian Interval is formally subdivided by absolute age, Global Standard Stratigraphic Age (GSSA).

This chart gives an overview of the international chronostratigraphic units,their rank, their names and formal status. These units are approved by the International Commission on Stratigraphy (ICS) and ratified by the International Union of Geological Sciences (IUGS).

The Guidelines of the ICS (Remane et al., 1996, Episodes, 19: 77-81) regulate the selection and definition of the international units of geologic time. Many GSSP's actually have a 'golden' spike (GSSP) and Stage and/or System name plaque mounted at the boundary level in the boundary stratotype section, whereas a GSSA is an abstract age without reference to a specific level in a rock section on Earth. Updated descriptions of each GSSP and GSSA are posted on the ICS website (*www.stratigraphy.org*).

Some stages within the Ordovician and Cambrian will be formally named upon international agreement on their GSSP limits. Most intra-stage boundaries (e.g., Middle and Upper Aptian) are not formally defined. Numerical ages of the unit boundaries in the Phanerozoic are subject to revision. Colors are according to the Commission for the Geological Map of the World (*www.cgmw.org*). The listed numerical ages are from 'A Geologic Time Scale 2004', by F.M. Gradstein, J.G. Ogg, A.G. Smith, et al. (2004; Cambridge University Press).

This chart was drafted and printed with funding generously provided for the GTS Project 2004 by ExxonMobil, Statoil Norway, ChevronTexaco and BP. The chart was produced by Gabi Ogg.

10 • Toward a "natural" Precambrian time scale

W. BLEEKER

It is proposed that Precambrian time should be subdivided into eons and eras that reflect natural stages in planetary evolution rather than being subdivided by a scheme based on numerical ages. The six eons can be briefly characterized as:

1. "Accretion and differentiation:" a time span of planet formation, growth, and differentiation up to the Moon-forming giant impact event.
2. Hadean (Cloud, 1972): an extended time span of intense bombardment and its consequences, but no preserved supracrustals.
3. Archean: an episode of increasing crustal record from the oldest supracrustals of Isua to the onset of giant iron formation deposition in the Hamersley Basin, likely related to increasing oxygenation of the atmosphere.
4. "Transition:" a time span with deposition of giant iron formations up to the first bona fide continental red beds.
5. Proterozoic: the time span of a nearly modern plate-tectonic Earth but without metazoan life.
6. Phanerozoic: Earth characterized by metazoan life forms of increasing complexity and diversity.

10.1 INTRODUCTION

The evolution of the geological time scale since the eighteenth century reflects in many ways the evolution of the geological sciences as a whole. On the one hand, the evolving time scale provides the essential nomenclature to classify, analyze, and communicate Earth history, while on the other it closely reflects the overall intellectual framework in which Earth history, as recorded in the rock record, is viewed.

Contrary to historical practice, however, and against the specific critique of many leading scholars in the field (e.g. Cloud, 1987; Crook, 1989; Nisbet, 1991), the Subcommission on Precambrian Stratigraphy chose a purely numerical basis of absolute ages for subdividing over 4000 million years of geological history (Plumb and James, 1986; Plumb, 1991; see also Chapter 9). The ages assigned to the boundaries of Proterozoic periods are not uniformly distributed, but were theoretically chosen to delimit principal cycles of sedimentation and tectonics (Table 9.1). The resulting time scale is "convenient" in terms of round numbers, but is divorced from key events in the stratigraphic record (Fig. 9.1).

10.2 CURRENT PRECAMBRIAN SUBDIVISIONS AND PROBLEMS

The term "Precambrian" is an informal stratigraphic term that encompasses all geologic time prior to the Cambrian Period. It traces its origin through its older hyphenated form "pre-Cambrian" to the early efforts by Sedgwick and Murchison to divide the fossiliferous "grauwacke" strata of Wales into correlatable systems (see Hallam, 1992a, pp. 65–86, for an engaging historical overview). With the Cambrian System becoming established in the mid-nineteenth century, older non-fossiliferous strata, schists, and crystalline rocks were loosely referred to as "pre-Cambrian." In modern usage, the Precambrian spans the time from the birth of planet Earth at about 4560 Ma to the biostratigraphically defined onset of the Cambrian Period, presently dated at 542 ± 1 Ma (see Chapter 11). It is formally subdivided into the Proterozoic and Archean Eons, with the base of Archean presently undefined (Fig. 9.1).

This Precambrian time scale, while innovative in design, has a few major problems.

First, a purely chronometic definition, such as the 2500 Ma definition of the Archean–Proterozoic boundary, is not, and cannot be, located precisely in the stratigraphic record. Systematic errors due to uncertainties in decay constants (Ludwig, 2000; Begemann *et al.*, 2001; see also Chapter 6), together with errors in analytical measurements, mean that the uncertainties in dates are about ±6.5 million years at ~2500 Ma and ±10 million years at 4000 Ma. Definition of boundaries in terms of arbitrary, round, absolute ages, although superficially appealing, is therefore naïve. Correlation of such boundaries between distant sections, on the basis of even our best geochronometers (U–Pb ages on single zircons), can be no better than ±5–10 million years (relative to absolute ages), even if all other sources

A Geologic Time Scale 2004, eds. Felix M. Gradstein, James G. Ogg, and Alan G. Smith. Published by Cambridge University Press.

of uncertainty (e.g. analytical scatter, Pb loss, or cryptic inheritance) are negligible. In principle, this fundamental uncertainty could be reduced by defining boundaries explicitly in terms of $^{207}Pb/^{206}Pb$ zircon ages or isotopic ratios, rather than absolute age numbers, but this would result in a time scale that is even less transparent. Furthermore, it would not solve the problem of intercalibration between different geochronometers, most of which suffer from greater decay constant uncertainties then the U–Pb system (see Chapter 6). As new values of the decay constants become available, rocks that had previously been assigned to the Archean might become Proterozoic or vice versa.

In fairness to many years of significant efforts by the Subcommission on Precambrian Stratigraphy, a positive outcome of their Precambrian time scale (Fig. 10.1) has been stabilization, at least temporarily and albeit artificially, of the debate on the age and significance of the Archean–Proterozoic boundary. This has facilitated, to some extent, the recent literature on the late Archean.

Second, the boundaries within the Precambrian scale are defined by a completely different method to the Phanerozoic time scale, in which boundaries are based on GSSPs in stratigraphic sections. The one exception is the ratification in 2004 of a GSSP in Australia for the youngest subdivision of the Proterozoic, at the base of the Ediacaran (see Chapter 9). Its upper boundary is defined by the GSSP for the base of the Cambrian in Newfoundland (see Chapter 11).

Third, the formal or proposed subdivisions (e.g. "Mesoarchean," by Plumb, 1991, and Lumbers and Card, 1991) of the current Precambrian time scale are either not being used or are used inconsistently. Subdivision names, chosen to avoid reference to particular sections (Plumb and James, 1986), are also a factor in their non-usage. For example, a search of the Georef reference database for such terms as the Ectasian or Calymmian periods yielded zero results. Due to current interest in Neoproterozoic glaciations, the Cryogenian faired better, but was only used in 14 recent papers. In effect, Precambrian stratigraphers are simply ignoring the formal terminology for subdivisions.

Fourth, the present time scale is incomplete, leaving the lower boundary of the Archean undefined. Hence, the popularity of the informal term "early Earth," loosely defined as Earth's first gigayear, to encompass the earliest episodes of our planet's history.

10.3 A "NATURAL" PRECAMBRIAN TIME SCALE

Thus a completely different view of how to subdivide the Precambrian is that the subdivisions should, like their Phanerozoic counterparts, be defined in terms of the extant rock record, i.e. by GSSPs and that these GSSPs should be placed so as to focus attention on what appear to be key stages in Precambrian history. These "key events" or "key transitions" in the stratigraphic record would establish a "natural" time scale for the evolution of planet Earth (Bleeker, 2002). Ideally, these "key events" should be observable globally, but at a minimum should be significant in one well-preserved section. In the preferred case where a specific key event or transition can be recognized in multiple sections world-wide, *that* section which shows the most detailed development of the stratigraphic interval of interest should be selected as the basis for defining a boundary, complete with a "golden spike" at the horizon or rock unit that defines the boundary. These type sections and their golden spikes will provide the physical standards for absolute age calibration, and against which other global sections can be compared.

To stimulate debate, a proposal for an improved Precambrian time scale is presented below. It attempts to provide a "natural" time scale for the Precambrian in which eons and eras, and their boundaries, are defined in terms of key events and transitions in Earth evolution. To some extent such a development was anticipated by proponents of the present chronometric scheme, which they state (see Chapter 9): "may not be a lasting solution as the defined units lack the geological context and definition required if they are to be recognized by the intrinsic features of their geologic history rather than simply numerical dates." Given the rapid advances in planetary science, geochronology, and Precambrian geology, an improved "natural" time scale for the Precambrian appears timely.

A tentative proposal for a "natural" subdivision of the first half of Earth history is given in Fig. 10.2. Parts of it are "conservative" in that it maintains present terminology at the eon and era rank, but it is built around first-order benchmarks in the extant stratigraphic record.

The first-order benchmarks are:

1. An early eon of "Accretion and Differentiation" (perhaps to be termed "Genesis"?).
2. The interval from the Moon-forming catastrophic impact to the end of heavy bombardment (a redefined Hadean).
3. A redefined Archean whose base is the end of the heavy bombardment.
4. A presently unnamed "Transition" Eon, whose base marks the onset of giant banded iron formations.
5. The Proterozoic, whose base is taken as the first continental red beds and an extended Paleozoic, whose base is the basal Ediacaran, ushering in Ediacaran metazoans.

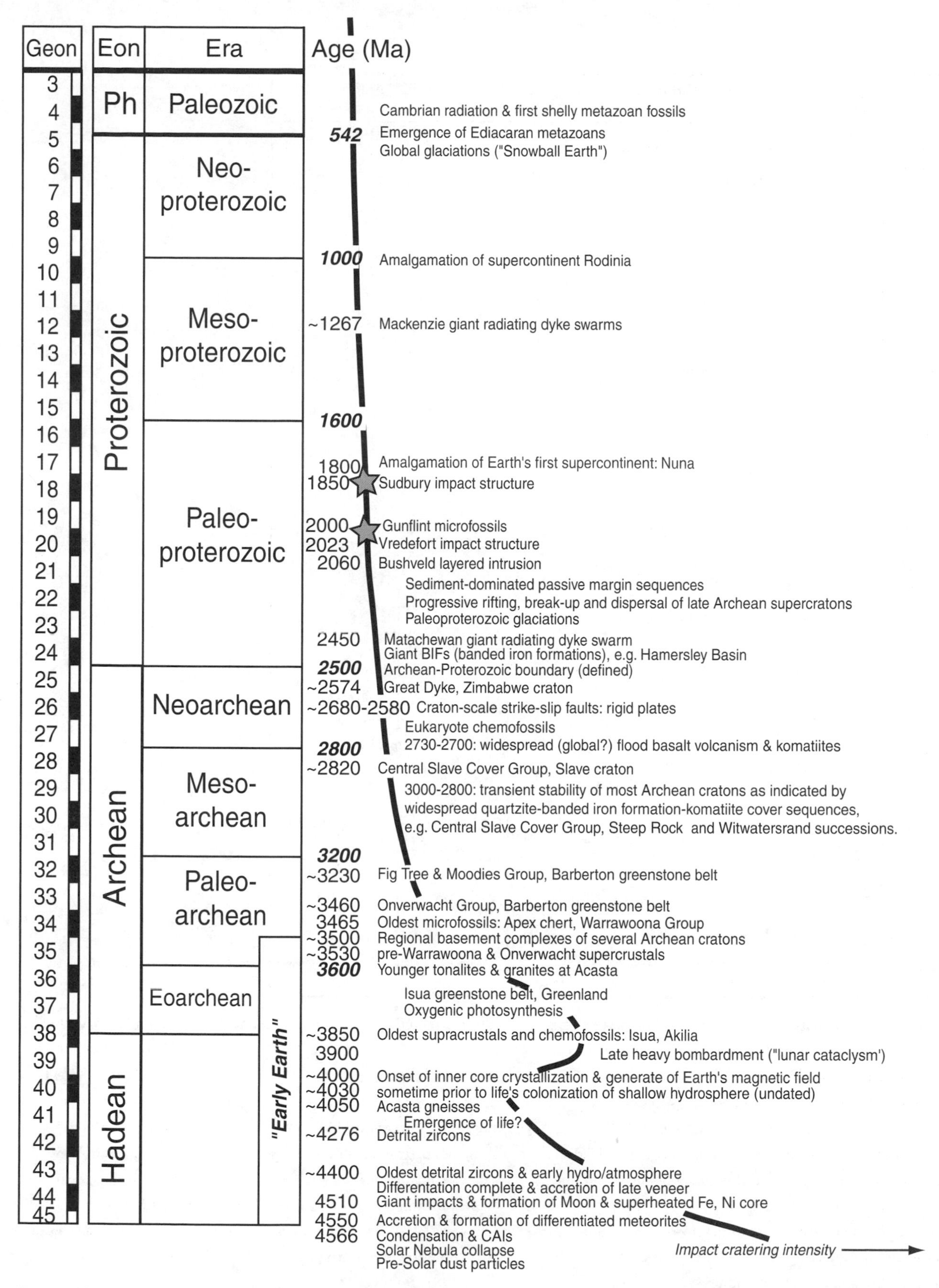

Figure 10.1 Formal subdivisions of the Precambrian annotated with key events in Earth's evolution. Geon scale from Hofmann (1990, 1991) provides a quick chronometric shorthand notation. The interval highlighted "early Earth" is an informal designation commonly used for Earth's first gigayear from the time of accretion to ~3.5 Ga. Exponentially decreasing impact intensity (curve on right) is schematic and includes the "late heavy bombardment" episode. Stars indicate Sudbury and Vredefort inpact structures with diameters >50 km.

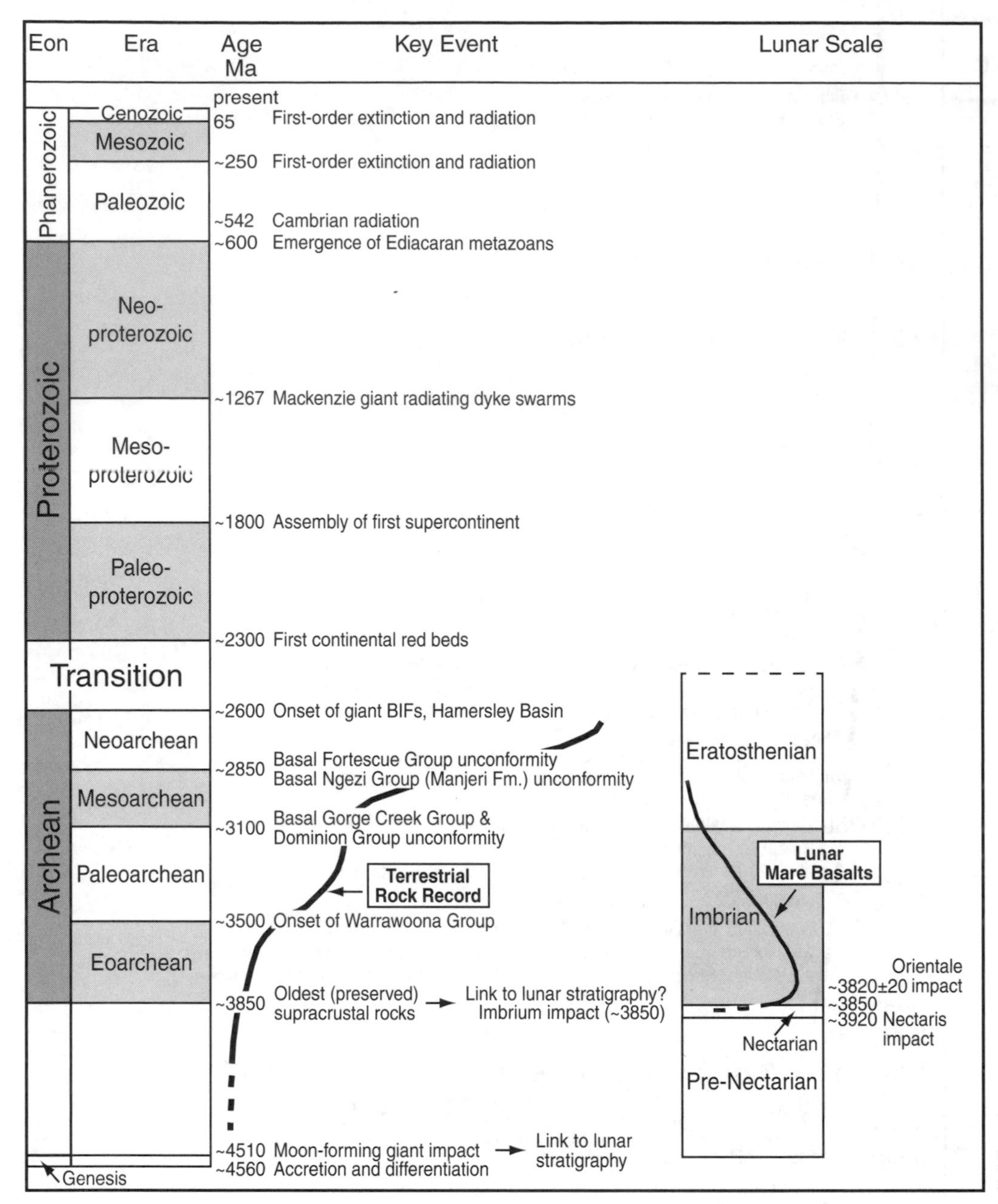

Figure 10.2 Proposal for a "natural" Precambrian time scale. Earth history is divided into six eons, with boundaries defined by what can be considered first-order "watersheds" in the evolution of our planet. The six eons can be briefly characterized as follows: "Accretion and Differentiation," planet formation, growth and differentiation up to the Moon-forming giant impact event; Hadean (Cloud, 1972), intense bombardment and its consequences, but no preserved supracrustals; Archean, increasing crustal record from the oldest supracrustals of Isua to the onset of giant iron formation deposition in the Hamersley Basin, likely related to increasing oxygenation of the atmosphere; "Transition," starting with deposition of giant iron formations up to the first bona fide continental red beds; Proterozoic, a nearly modern plate-tectonic Earth but without metazoan life; and the Phanerozoic, characterized by metazoan life forms of increasing complexity and diversity. The latter definition of the Phanerozoic (e.g. Cloud, 1988) would involve moving its lower boundary to encompass an Ediacaran Period. Some of the boundaries are currently poorly calibrated in absolute time, whereas the onset of the Archean should "float" with the oldest preserved supracrustal rocks, a distinction currently held by 3820–3850 Ma rocks of the Isua greenstone belt. Comparison to the lunar time scale is shown (e.g. Guest and Greeley, 1977; Murray *et al.*, 1981; Spudis, 1999).

10.3.1 The "Accretion and Differentiation" or "Genesis" Eon

Earth's first eon (perhaps more provocatively called "Genesis") spanned the time from planetary formation and early differentiation to the giant impact event that led to the formation of the Moon.

The "birth" of planet Earth and other planetary bodies of the inner solar system is a somewhat fuzzy concept, depending on what stage in the accretionary growth history is selected as the threshold for planet formation. Here, the 4550–4560 Ma age of the oldest differentiated meteorites (Allègre *et al.*, 1995; see also Patterson, 1956), including basaltic achondrites, is chosen as the birth date of planet Earth.

The Moon-forming event, likely involving a Mars-size planetary body (Hartmann and Davis, 1975; Cameron and Ward, 1976; Canup and Asphaug, 2001), is thought to have occurred at ~4510 Ma (Lee *et al.*, 1997). Dynamical modeling studies of the catastrophic impact event (e.g. Canup and Asphaug, 2001) suggest that the material blasted into orbit, derived in part from the impactor and in part from the Earth's mantle, and re-aggregated in a relatively short time to form the Moon. Therefore, the end of Earth's first eon of about 45 myr duration at the same time marks the onset of the lunar time scale (Fig. 10.2). Clearly, no true GSSP can be defined on Earth for the base or the top of the "accretion and differentiation" eon. Both are boundaries that will have to be defined by age, but an age linked to geological evolution rather than to an arbitrary number.

10.3.2 The Hadean Eon

The term "Hadean" was initially defined by Cloud (1972, 1988), with the Earth–Moon system clearly in mind. The Hadean Eon should be re-defined to span the time from the Moon-forming catastrophic impact to the end of heavy bombardment. The former is likely to have "re-set" the Earth in fundamental ways: e.g. excavation of part of the mantle to form the Moon; impact erosion and partial vaporization of an early crust; impact erosion of an early atmosphere; and extensive melting to form a temporary magma ocean. The nature and timing (~4.0–3.8 Ga) of a discrete "late heavy bombardment phase" is still the subject of considerable debate, but in practical terms the end of this intense bombardment may well correlate with the first preservation of terrestrial supracrustal rocks, a distinction currently held by the ~3850–3820 Ma clastic and volcanic rocks of the Isua greenstone belt (Nutman *et al.*, 1993, 1997).Thus the Hadean Eon is characterized by intense bombardment, a sparse record of ancient infracrustal rocks (e.g. Acasta gneisses; Stern and Bleeker, 1998; Bowring and Williams, 1999), some detrital zircons (Wilde *et al.*, 2001), but no preserved supracrustal rocks. From a historical perspective, the Hadean (Cloud, 1972) has priority over Harland's Priscoan (Harland *et al.*, 1982). The Hadean Eon lasted about 700 myr.

10.3.3 The Archean and "Transition" Eons

The base of the Archean can be defined by the first (preserved) occurrence of supracrustal rocks in the rock record. Hence, a GSSP should be placed at the ~3850–3820 Ma clastic and volcanic rocks in the Isua greenstone belt.

The Archean Eon is characterized by the protracted, >1 billion year evolution from granite–greenstone belts to less-deformed sequences. The transition in tectonic style was once considered too synchronous on all cratons, but it is now realized that the Archean–Proterozoic "boundary" is a transition in tectonic styles that is diachronous: in some cratons it took place as early as 3.1 Ga, whereas in others it occurred as late as 2.5 Ga or even later (e.g. Windley, 1984; Blake and Groves, 1987; Cloud, 1987; Nisbet, 1991; Bleeker, 2003b).

Interestingly, this broad transition in tectonic behavior overlaps changes in the atmosphere, in particular, a rise in oxygen partial pressure (e.g. see summary in Windley, 1995, and references therein), in turn marked by the incoming of new types of sediment such as the giant banded iron formations at about 2600 Ma and the first continental red beds at about 2300 Ma (Fig. 10.2). A "Transition" Eon is therefore proposed between the Archean and Proterozoic proper (see also Cloud, 1987) that encapsulates this episode of planetary evolution events. This has the added advantage of shortening the Paleoproterozoic, which is currently 900 myr in duration.

Thus a GSSP for the lower boundary of the "Transition" Eon could be placed to include the bulk of the giant banded iron formations, such as in the Hamersley Basin, Western Australia, which has a remarkably complete stratigraphic record (e.g. Trendall *et al.*, 1998; Pickard, 2002). Its approximate age would be ~2600 Ma.

The Archean Eon can be subdivided into the Eoarchean, Paleoarchean, Mesoarchean, and Neoarchean Eras (Fig. 10.2). The base of the Eoarchean is the base of the Archean itself. Archean cratons contain the first successions of mature quartz sandstones that lie unconformably on underlying greenstones and crystalline basement. The unconformities at the base of these successions can be traced over large areas. Each succession represents a significant tectonic event of unknown origin, but viewed as a whole, they provide important breaks in the Archean stratigraphic record. Whether these breaks are of local or of global significance is not known, but, in principle, GSSPs for the Mesoarchean and Neoarchean Eras could be placed so as to include the bulk of an appropriate quartz sandstone succession. For example, the base of the Mesoarchean could be located so as to include the Gorge Creek Group, dated at ~3.22–3.05 Ga (Buick *et al.*, 2002; van Kranendonk *et al.*, 2002), which overlies the basalt-dominated Warrawoona Group throughout much of the East Pilbara granite–greenstone terrain (van Kranendonk *et al.*, 2002).

Do these unconformities have global significance and, if so, how diachronous are they? The answers to these questions are not known. However, provisional GSSPs could be placed close to the base of a quartz arenite succession where a magnetic chron boundary was also present. This would have the advantage of defining the boundary in rock and also of providing a unique correlation marker. As knowledge progressed, the GSSP might be moved elsewhere and the origin "golden spike" would have evolved into a rusty iron spike. The GSSPs cannot be located at an unconformity without introducing the problem of "missing time." Alternatively, geochemical markers might be capable of subdividing the Archean Eon or Archean eras.

10.3.4 The Proterozoic Eon

Conventionally, it might be argued that a suitable lower boundary for the base of the Proterozoic could be defined by the first craton-scale fracturing and dyke intrusion of a typical Archean craton, such as the Great Dyke of the Zimbabwe craton of southern Africa. However, a GSSP directly associated with this event would have to be placed in a stratified sequence and no correlative sequence has yet been recognized.

Therefore, a GSSP for the base of the Proterozoic Eon (and the top of the Transition Eon) could be placed at a horizon that also reflects a stage in atmospheric evolution – the first appearance of bona fide continental red beds, such as those of the Lorraine Formation of the Huronian Supergroup, Ontario, at ~2300 Ma (Roscoe, 1973; Prasad and Roscoe, 1996).

10.3.5 The Ediacaran Period (base-Paleozoic?)

A final, and likely controversial, aspect of the proposed time scale of Fig. 10.2 is to re-define the base of the Phanerozoic (meaning "obvious life") in terms of the first appearance of metazoans – the Ediacaran fauna – rather than shelly fauna at the base of the Cambrian (Cloud, 1988; see also Nisbet, 1991). This would add the Ediacaran System to the Phanerozoic. This would mean that the base of the Ediacaran, with its GSSP in Australia, would become the boundary between the Phanerozoic and the Precambrian.

Such practice would require that the informal term "Precambrian" be replaced with the new term "Prediacaran" to designate that part of Earth history that predated an obvious (metazoan) fossil record. Such a step is not essential: the Ediacaran System could remain as the youngest period of the Neoproterozoic Era.

10.4 CONCLUSIONS

A subdivision of Precambrian time based on stages in planetary evolution appears to be a realizable goal. Adoption of such a scheme would make Precambrian geology more accessible to Earth scientists than the current classification based on the assignment of arbitrary numerical values to Precambrian eons, eras, and periods. As in the Phanerozoic Eon, boundaries would be defined by GSSPs in rock sequences, though such a definition could not be used for the onset of the terrestrial geologic time scale and the onset of the Hadean. Their boundaries need to be defined by first-order planetary events.

11 • The Cambrian Period

J. H. SHERGOLD AND R. A. COOPER

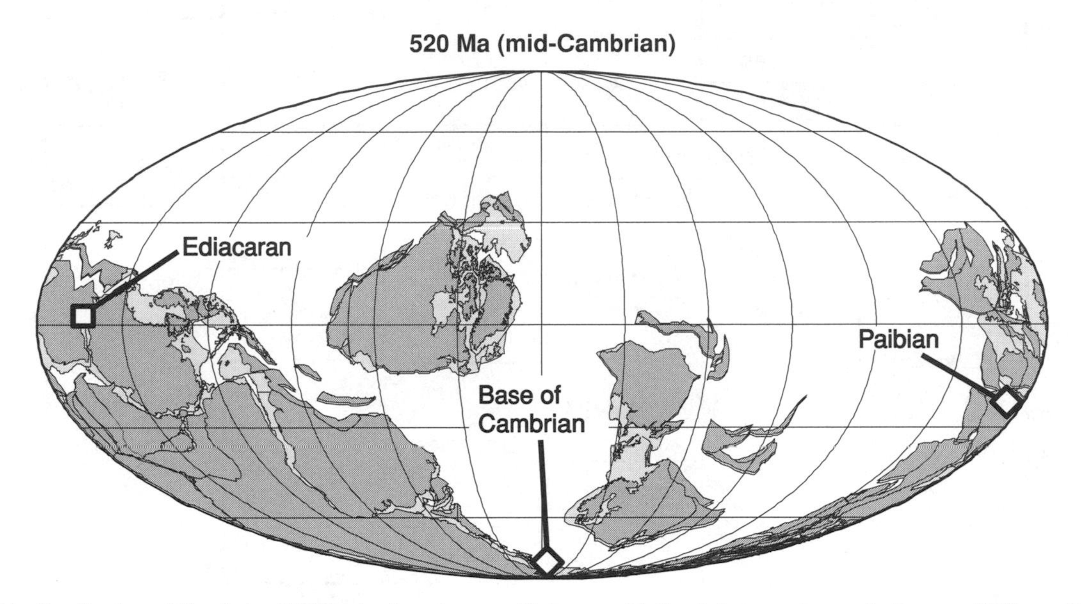

Geographic distribution of Cambrian GSSPs that have been ratified on a mid-Cambrian map (status in January 2004; see Table 2.3).

Appearance of mineralized metazoan skeletons, "explosion" in biotic diversity and disparity, occurrence of metazoan Konservatfossillagerstätten, establishment of all invertebrate phyla, strong faunal provincialism, dominance of trilobites, globally warm climate (hothouse conditions), opening of Iapetus Ocean, progressive equatorial drift, and separation of Laurentia (including Avalonia), Baltica, and Siberia characterize the Cambrian Period.

11.1 HISTORY AND SUBDIVISIONS

The Cambrian was the name given by Sedgwick (Sedgwick and Murchison, 1835) for strata exposed in North Wales. The name is taken from *Cambria*, the Roman variant of the Celtic *Cumbria*, the old name for North Wales. As used today, the name applies essentially to Sedgwick's (1852) "Lower Cambrian."

The Cambrian marks the appearance in the geological record of mineralized skeletons of multicellular animals (metazoans), and of a rapid diversification of metazoan life, commonly referred to as the "Cambrian explosion." All animal phyla appeared by the end of the Cambrian. The biostratigraphically most useful fossil group is the trilobites, which show remarkable diversification and evolutionary change, particularly in the Upper Cambrian. Inarticulate brachiopods, archaeocyathans, and acritarchs also provide good biostratigraphic control in appropriate facies.

As discussed below, international agreement on subdivision of the Cambrian lags behind most other systems. In most parts of the world, the Cambrian is divided into lower, middle, and upper parts (series) but there is no international agreement on where the boundaries should be placed and there is considerable variance in practice, particularly regarding the base of the Middle Cambrian. In this volume, we follow the Australian practice, using a three-fold subdivision with boundaries placed as in Fig. 11.1.

Cambrian global biostratigraphic markers are scarce, which is one reason why there are only two internationally ratified and formally defined Cambrian stage-level divisions (status early 2004). The problem is particularly acute in the Early Cambrian. Most continents have regional stages but the International

A Geologic Time Scale 2004, eds. Felix M. Gradstein, James G. Ogg, and Alan G. Smith. Published by Cambridge University Press.

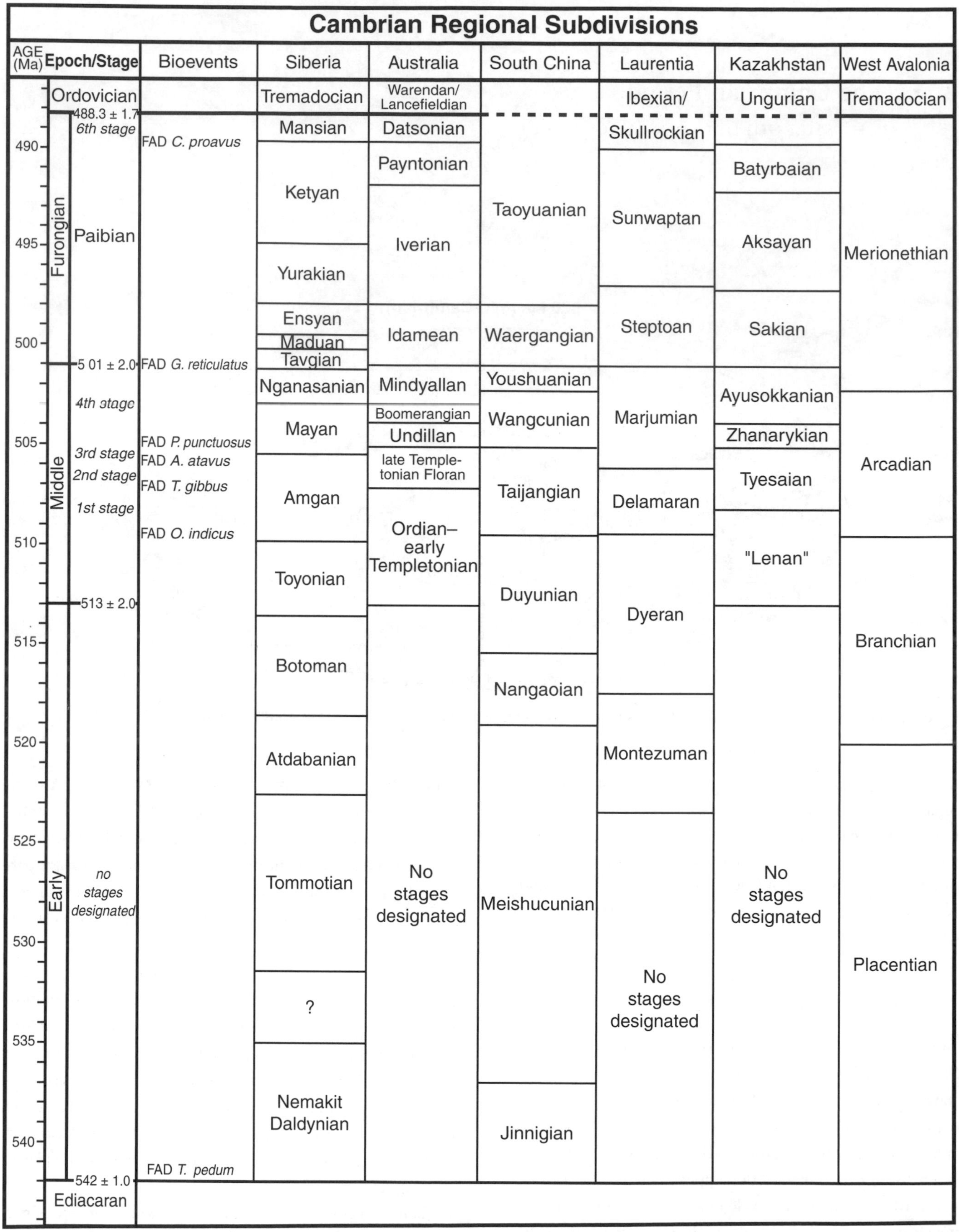

Cambrian Regional Subdivisions

AGE (Ma): 490, 495, 500, 505, 510, 515, 520, 525, 530, 535, 540

Epoch/Stage:
- Ordovician
- 488.3 ± 1.7
- Furongian: *6th stage*; Paibian
- 5 01 ± 2.0
- Middle: *4th stage*; *3rd stage*; *2nd stage*; *1st stage*
- 513 ± 2.0
- Early: *no stages designated*
- 542 ± 1.0
- Ediacaran

Bioevents:
- FAD *C. proavus*
- FAD *G. reticulatus*
- FAD *P. punctuosus*
- FAD *A. atavus*
- FAD *T. gibbus*
- FAD *O. indicus*
- FAD *T. pedum*

Siberia: Tremadocian; Mansian; Ketyan; Yurakian; Ensyan; Maduan; Tavgian; Nganasanian; Mayan; Amgan; Toyonian; Botoman; Atdabanian; Tommotian; ?; Nemakit Daldynian

Australia: Warendan/ Lancefieldian; Datsonian; Payntonian; Iverian; Idamean; Mindyallan; Boomerangian; Undillan; late Templetonian Floran; Ordian–early Templetonian; No stages designated

South China: Taoyuanian; Waergangian; Youshuanian; Wangcunian; Taijangian; Duyunian; Nangaoian; Meishucunian; Jinnigian

Laurentia: Ibexian/; Skullrockian; Sunwaptan; Steptoan; Marjumian; Delamaran; Dyeran; Montezuman; No stages designated

Kazakhstan: Ungurian; Batyrbaian; Aksayan; Sakian; Ayusokkanian; Zhanarykian; Tyesaian; "Lenan"; No stages designated

West Avalonia: Tremadocian; Merionethian; Arcadian; Branchian; Placentian

Figure 11.1 Principal stage schemes of the Cambrian (after Geyer & Shergold, 2000). The six biostratigraphic global correlation datum points are shown on the left hand side of the figure. The Early, Middle, and some of the Late Cambrian divisions have not yet been decided by the ICS; the Early and Middle Cambrian lower boundaries shown follow Australian usage. See text for caution in applying numerical scale to stage boundaries.

Subcommission on Cambrian Stratigraphy has found that no regional stage schemes are suitable for global use (International Subcommission on Cambrian Stratigraphy Newsletter, 1999–1). In the Middle and Late Cambrian, global biochronological markers, such as those based on agnostid trilobites, are present and for these series, at least, it should be possible to erect internationally acceptable stages, as is being done in the Ordovician. The only two boundaries so far defined are the base of the Cambrian System, and the base of the Paibian Stage and Furongian Series, the uppermost series of the Cambrian.

In the following discussion, therefore, the regional stage classifications of Australia and North America are outlined, as representative of two major, and widely applicable, faunal provinces, but it should be noted that Russian archaeocyathan stages are commonly used in the Lower Cambrian of Australia. In Fig. 11.2, the stages and zones used in Siberia, Kazakhstan, south China, and west Avalonia also are shown. The most promising biostratigraphic datums on which to base global stages are listed below.

11.1.1 Base of the Cambrian System and Paleozoic Erathem

The GSSP for the base of the Cambrian was the first appearance level of the ichnofossil *Phycodes pedum*, a species now commonly referred to as *Tricophycus pedum* or *Treptichnus pedum*, in the Chapel Island Formation in coastal cliffs of the 440 m thick Fortune Head section, on the Burin Peninsula of southeastern Newfoundland (Landing, 1994; Brasier *et al.*, 1994). Later work showed that the ichnofossil occurs slightly below the GSSP (Gehling *et al.*, 2001). The GSSP lies 2.4 m above the base of Member 2 in the Chapel Island Formation, just above the transition to storm-influenced facies. Member 1 of the Chapel Island Formation includes uppermost Precambrian sediments. It yields the trace fossils *Harlaniella podolica*. *Palaeopascichnus delicates* is also present, but it is widely believed not to be a trace fossil. *Harlaniella podolica* ranges into Member 2, where it is last seen 0.2 m below the GSSP.

The GSSP also defines the beginning of the Paleozoic Era. This definition recognizes the earliest record of the *activity* of complex metazoa rather than of their direct skeletal record, which comes 400 m higher in the sequence. The level is correlated with the base of the regional Nemakit–Daldynian Stage of the Olenek region, northern Siberia. Trilobites first appear some 1400 m above the boundary (base of the Branchian Series).

Unfortunately, there is no suitable magnetostratigraphy signal in the rocks of the Chapel Island Formation and, probably, no useful isotope stratigraphy either. Elsewhere, a large-magnitude, short-lived negative excursion in carbon isotopes has been equated with the boundary (Grotzinger *et al.*, 1995; Bartley *et al.*, 1998; Kimura and Watanabe, 2001).

11.1.2 Biostratigraphic datums with potential for global correlation

Geyer and Shergold (2000) listed 14 biostratigraphic levels for inter-provincial correlation. Of these, the International Subcommission on Cambrian Stratigraphy in June 2000 selected six as suitable for GSSPs for Cambrian stages, listed below. Suitable stratotype sections are now being sought.

BASE OF THE *CORDYLODUS PROAVUS* ZONE

The conodont *Cordylodus proavus* has a cosmopolitan distribution and its first appearance is used in several regions for stage base definition, for example the Skullrockian (North America), Datsonian (Australia), Xinchangian (China), and Ungurian (Kazakhstan).

BASE OF THE *GLYPTAGNOSTUS RETICULATUS* ZONE, BASE OF THE PAIBIAN STAGE, LOWERMOST FURONGIAN SERIES

This is one of the most widely recognizable trilobite horizons and one that has been suggested by many workers as suitable for definition of a global stage base. *Glyptagnostus reticulatus* is used as a zonal fossil in Siberia, Kazakhstan, the Yangtse Platform, Australia, and Laurentia. The first appearance of *G. reticulatus* marks a time of significant faunal change, that defines the base of the pterocephaliid biomere in Laurentia, the Idamean Stage in Australia, the Sakian Stage in Kazakhstan, the Kugorian Stage in Siberia, and Olenus Stage in Scandinavia.

In 2002, the base of the *G. reticulatus* Zone was formally selected by the ICS to define the base of the Paibian Stage in the lower Furongian Series. These are the first formal stage and series units in global Cambrian stratigraphy.

The GSSP for the base of the Paibian Stage and Furongian Series, which forms the uppermost series of the Cambrian, is defined at 396 m, at the lowest occurrence of the agnostoid trilobite *G. reticulatus*, in the Huaqiao Formation, Paibi Section, NW Hunan Province, South China. This level is near the

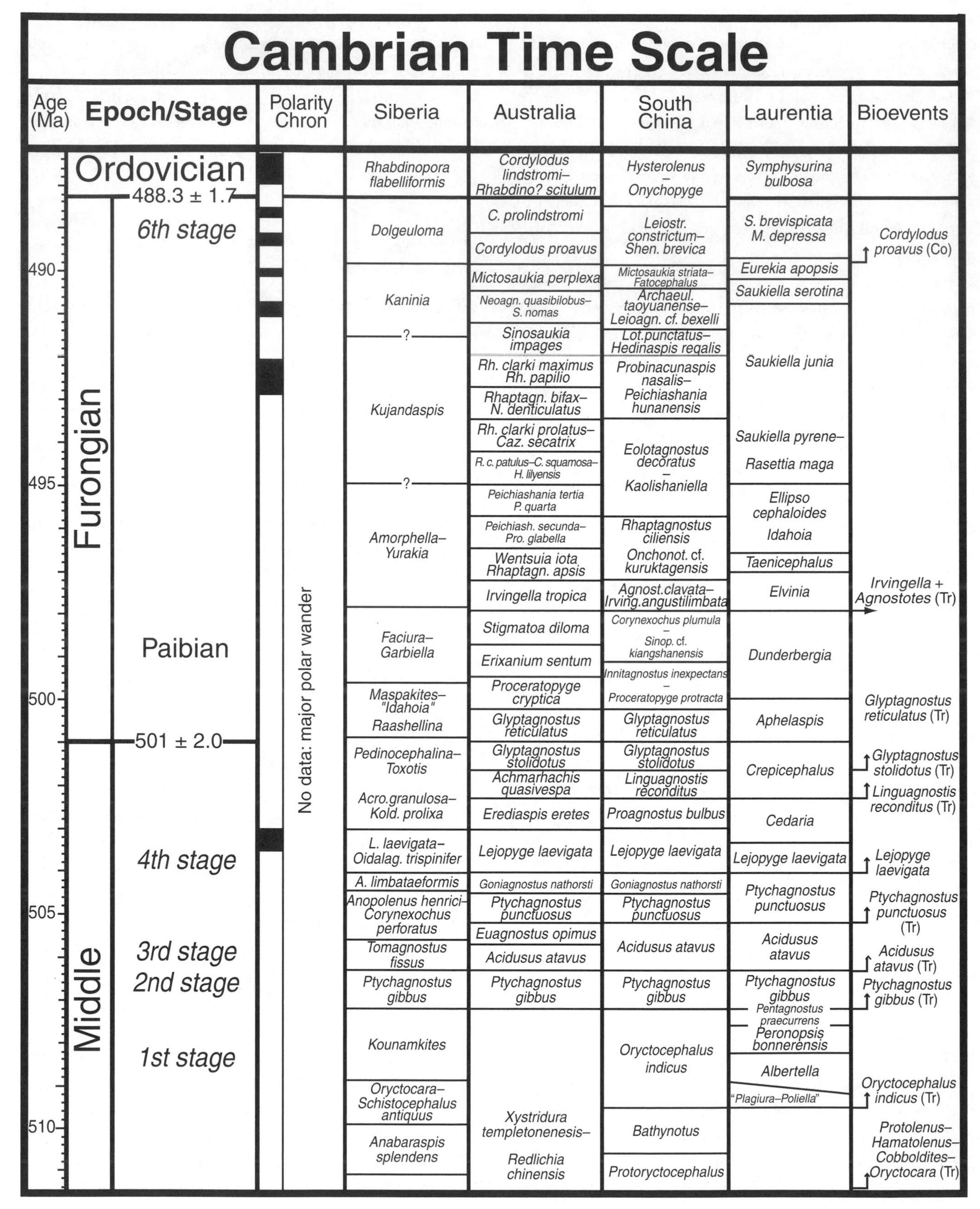

Figure 11.2 Principal biostratigraphic zonal schemes of the Cambrian. A color version of this figure can be found in the plate section.

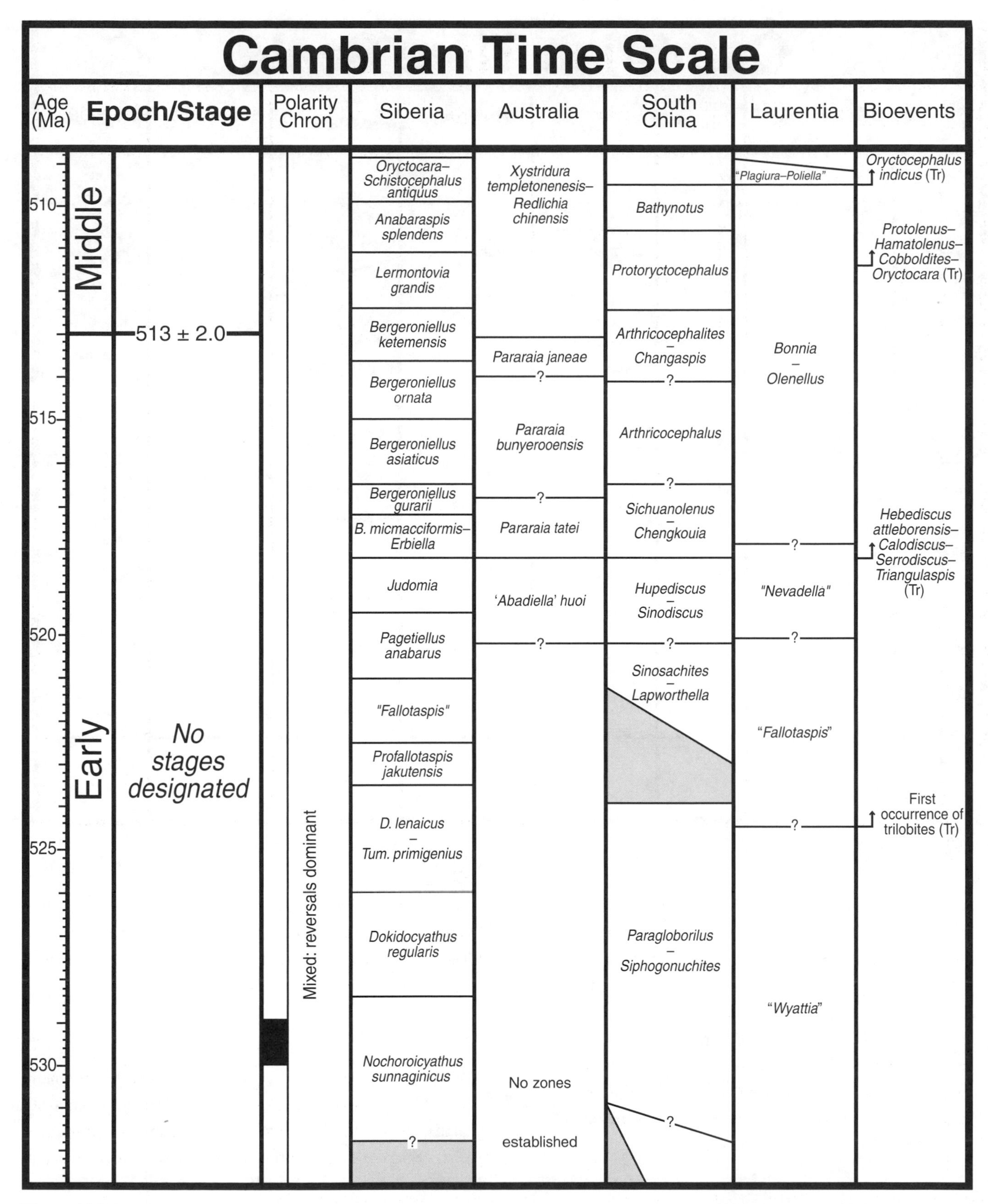

Figure 11.2 (*cont.*)

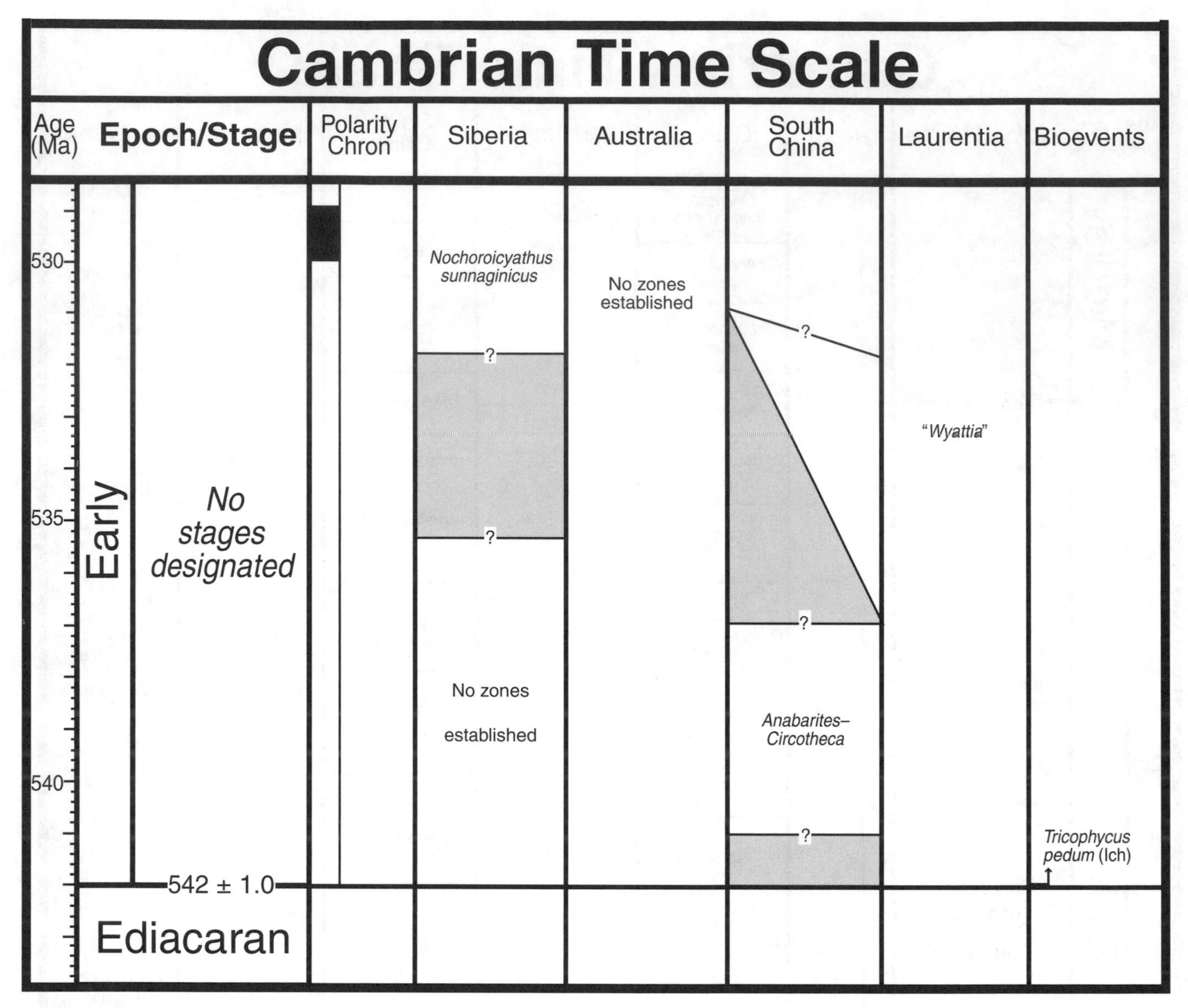

Figure 11.2 (*cont.*)

base of a large positive shift in δ^{13}C values (Fig. 11.3), referred to as the Steptoean positive carbon isotope excursion (i.e. the SPICE excursion).

The Paibian Stage is a new name, derived from Paibi, a village near the GSSP site in Hunan Province, China. It is the lower stage of the Furongian Series, which also is a new name. The name Furongian is derived from Furong, which means *lotus*, referring to Hunan, the Lotus State in China. The upper boundary of the Furongian Series is the base of the Tremadocian Series, Lower Ordovician.

BASE OF THE *PTYCHAGNOSTUS PUNCTUOSUS* ZONE

Ptychagnostus punctuosus is a widespread trilobite that is used as a zonal index species in most areas in which it is found. It is known from Australia, New Zealand, South China, northwest China, Kazakhstan, Siberian Platform, Scandinavia, Great Britain, Greenland, Canada, and the USA. Its appearance defines the base of the Undillan Stage of Australia and the base of the Zhanarykian Stage in Kazakhstan.

BASE OF THE *ACIDUSUS ATAVUS* ZONE

The trilobite *A. atavus* is known from Australia, Vietnam, China, Korea, Russia, Scandinavia, Great Britain, Greenland, Canada, and the USA. Its first appearance marks a prominent faunal change and defines the base of the Marjuman Stage in Laurentia and the base of the Floran Stage in Australia.

BASE OF THE *TRIPLAGNOSTUS GIBBUS* ZONE

Triplagnostus gibbus is a widespread trilobite used commonly as a zonal index species. It is known from Antarctica,

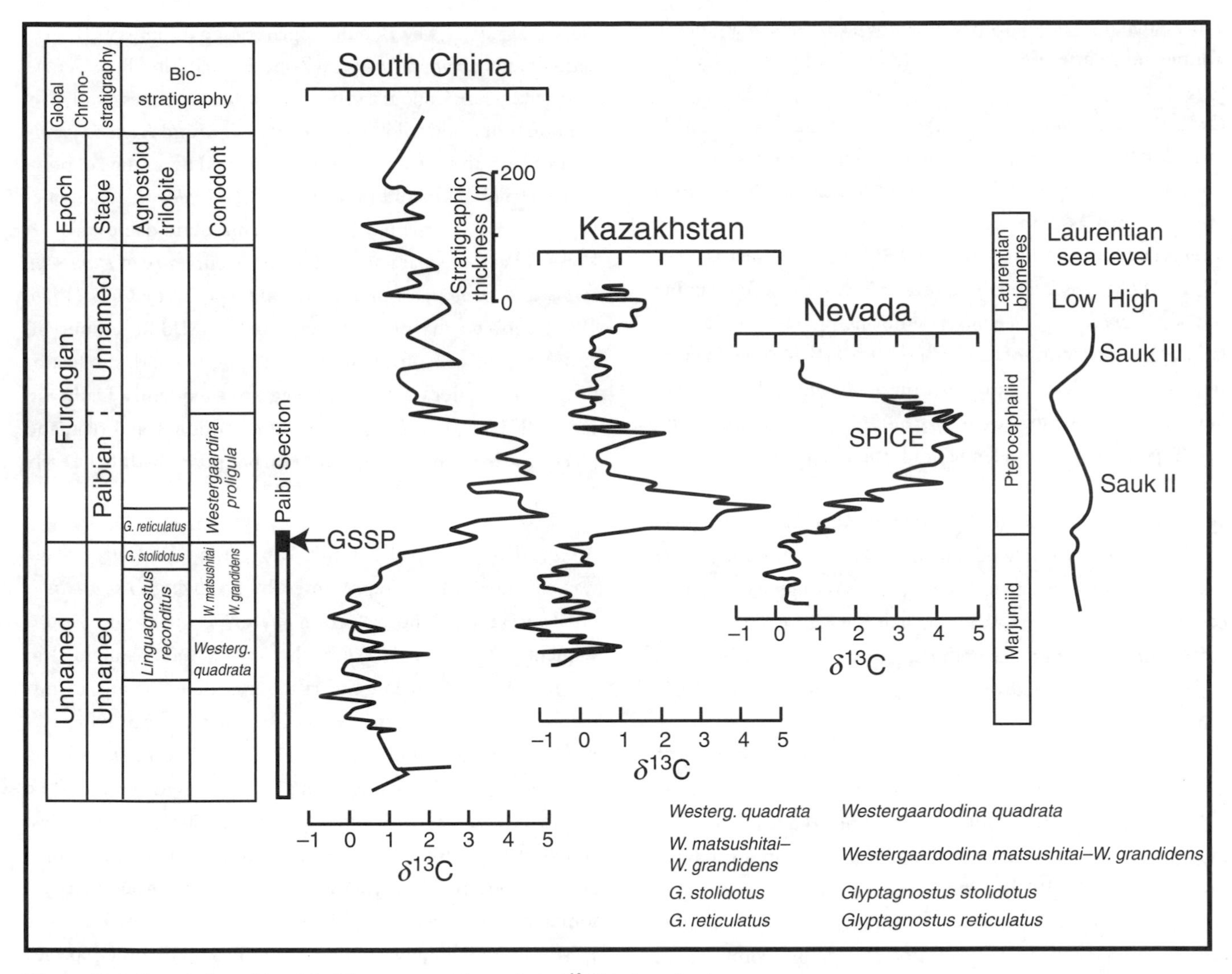

Figure 11.3 Correlation of basal-Paibian events with trends in δ^{13} C and sea level.

Australia, Kazakhstan, Russia, Poland, Scandinavia, Great Britain, Greenland, Canada, and the USA. Its first appearance defines the base of the *Paradoxides paradoxissimus* Stage in Scandinavia.

BASE OF THE *ORYCTOCEPHALUS INDICUS* ZONE

The trilobite *O. indicus* is known from Kashmir, Laurentia, Great Britain, the Yangtse Platform, and South China. It is associated with other oryctocephalids that have wider distribution, but correlation into western and central Gondwanaland would be problematic.

11.1.3 Regional Cambrian stage suites

The most intensively studied Cambrian stadial suites are those of Australia, southern China, Kazakhstan, Siberia, and North America, shown in Fig. 11.1, and Baltica.

AUSTRALIAN CAMBRIAN STAGES

Australian stages are summarized by Shergold (1995) and Young and Laurie (1996), on which the following outline is based. They are described as "biochronological" units and are defined in terms of their contained fauna (Shergold, 1995). Boundary stratotypes are therefore not designated. Ordian apart, the stages discussed below have all been erected in the Georgina Basin of western Queensland.

Pre-Ordian Stages have not yet been designated for most of the Lower Cambrian of Australia. Archaeocyathans, small shelly fossils, and trilobite correlations indicate that the Adtabanian to Toyonian Stages of the Siberian Platform and the Altay–Sayan Foldbelt of Russia can be recognized throughout southern and central Australia. Ichnocoenoses in southern Australia are thought, possibly, to represent the Tommotian and Nemakit–Daldynian (Bengtson *et al.*, 1990; Shergold in Young and Laurie, 1996).

Ordian–Lower Templetonian An Ordian Stage was originally proposed by Öpik (1968) as a time and time-rock division of the Cambrian scale characterized by the occurrence of the *Redlichia chinensis* faunal assemblage. The Templetonian Stage, "a liberal interpretation of Whitehouse's [1936] Templetonian series" (Öpik, 1968), was originally conceived by Öpik as containing the *Xystridura templetonensis* assemblage of western Queensland overlain by fauna of the *Triplagnostus gibbus* Zone. In practice, it is difficult to distinguish the *Redlichia* and *Xystridura* fauna because four species of *Xystridura*, similar eodiscoid and ptychoparioid trilobites, some bradoriid ostracodes, and chancelloriids occur in rocks of both Ordian and early Templetonian ages. Accordingly, Shergold (1995) regarded the Ordian–early Templetonian as a single stadial unit. It is retained as the earliest Middle Cambrian Stage in Australia even though it apparently correlates with the Longwangmiaoan Stage of China (Chang, 1998) and Toyonian Stage of the Siberian Platform (Zhuravlev, 1995), which are traditionally regarded as terminal Early Cambrian.

Upper Templetonian–Floran As originally defined (Öpik, 1979), the Floran Stage contained the agnostoid trilobite zones of *Acidusus atavus* and *Euagnostus opimus*. This concept was revised by Shergold (1995) to include the late Templetonian zone of *Triplagnostus gibbus*, arguing on grounds of sequence stratigraphy (Southgate and Shergold, 1991) and faunal continuity as suggested by the overlap of *A. atavus* and *T. gibbus* in western Queensland. We maintain the late Templetonian–Floran as a single unified stage. Having a global distribution, this stage is a very important datum, because, besides agnostoid trilobites, it contains oryctocephalid trilobites (Shergold, 1969) which are also significant for international correlation.

Undillan The Undillan Stage, defined by Öpik in 1979, is unrevised, and is based on the fauna of two agnostoid zones, the *Ptychagnostus punctuosus* Zone below, overlain by the *Goniagnostus nathorsti* Zone above. A third zone, based on *Doryagnostus notalibrae*, containing the overlap of 15 bizonal agnostoid species, including *P. punctuosus* and *G. nathorsti*, was recognized by Öpik (1979) in the Undilla region of the Georgina Basin. The agnostoid fauna of the Undillan Stage have cosmopolitan distribution. Agnostoids apart, other trilobites include ptychoparioids, anomocarids, mapaniids and damesellids, conocoryphids, corynexochids, nepeiids, and dolichometopids, all of widespread distribution.

Boomerangian The Boomerangian Stage (Öpik, 1979) is essentially the *Lejopyge laevigata* Zone divided into three. A *Ptychagnostus cassis* Zone at the base is overlain by zones defined by the non-agnostoid trilobites *Proampyx agra* and *Holteria arepo*. Boomerangian agnostoids are accompanied by a range of polymeroid trilobites including species of *Centropleura*, dolichometopids, olenids, mapaniids, corynexochids, and damesellids. A Zone of Passage, characterized by the occurrence of *Damesella torosa* and *Ascionepea janitrix*, was interposed by Öpik (1966, 1967) between the Boomerangian (latest Middle Cambrian) and Mindyallan (considered at that time to mark the beginning of the Upper Cambrian) Stages. Subsequently, Daily and Jago (1975) restricted this zone to the Middle Cambrian, and placed the Middle–Upper Cambrian boundary within the early Mindyallan.

Mindyallan Originally (Öpik, 1963), the Mindyallan Stage was considered to be represented by a *Glyptagnostus stolidotus* Zone (above) and a "pre-*stolidotus*" Zone (below). Subsequently (Öpik, 1966, 1967), the former was maintained in the late Mindyallan but the latter was divided into an initial Mindyallan *Erediaspis eretes* Zone and an overlying *Acmarhachis* (*Cyclagnostus*) *quasivespa* Zone. The *E. eretes* Zone contains 45 trilobites, including 18 agnostoid genera. The polymeroid trilobites belong to a wide variety of families: anomocarid, asaphiscid, catillicephalid, damesellid, leiostegid?, lonchocephalid, menomoniid, nepeiid, norwoodiid, rhyssometopid, and tricrepicephalid. The *A. quasivespa* Zone has 18 species of trilobites confined to it, but many other species range from earlier zones. Daily and Jago (1975) proposed to subdivide the *A. quasivespa* Zone into two assemblages based on the occurrence of *Leiopyge cos* Öpik and *Blackwelderia sabulosa*. Since they regarded *L. cos* to be synonymous with *L. l. armata* Westergård, a late Middle Cambrian taxon, they drew the Middle–Late Cambrian boundary between these two assemblages. Only eight species range from Öpik's *A. quasivespa* Zone into the overlying *Glyptagnostus stolidotus* Zone, which contains 75 species with partly American (asaphiscid, auritamid, catillicephalid, norwoodiid, and raymondinid) and partly Chinese (damesellid and liostracinid) relationships.

Idamean The Idamean Stage was introduced by Öpik in 1963, originally conceived as representing five successive assemblage-zones: *Glytagnostus reticulatus* with *Olenus ogilviei*, *Glypagnostus reticulatus* with *Proceratopyge nectans*, *Corynexochus plumula*, *Erixanium sentum*, and *Irvingella tropica* with *Agnostotes inconstans*. This biostratigraphic scheme was criticized by Henderson (1976, 1977) who proposed an alternative zonation in which the two zones with *Glyptagnostus* were united into a single *G. reticulatus* Zone, the *Corynexochus* Zone was renamed the *Proceratopyge cryptica* Zone, and the *Erixanium sentum* Zone was subdivided into a zone of *E. sentum* followed by a zone of *Stigmatoa diloma*. The *Irvingella tropica–Agnostotes inconstans* Zone is abbreviated to *Irvingella tropica* Zone. Henderson's scheme was adopted by Shergold (1982) and is followed here with the exception that the *Irvingella tropica* Zone be excluded from the Idamean Stage and be recognized as the initial zone of the succeeding Iverian Stage (see Shergold, 1982, for justification; Shergold, 1993).

There is a major faunal crisis at the beginning of the Idamean where no Mindyallan species survive the stadial passage (Öpik, 1966), and very few genera persist into the early Idamean. There is also a major re-organization of trilobite families as outer shelf communities dominated by agnostoids, olenids, pterocephaliids, leiostegiids, eulomids, and ceratopygids abruptly replace those of the shallow shelf Mindyallan biota. Shergold (1982) recorded a total of 69 Idamean taxa, which permit a highly resolved biochronology capable of yielding very accurate international correlations.

Iverian The Iverian Stage (Shergold, 1993) was proposed for the concept of a post-Idamean–pre-Payntonian interval in the eastern Georgina Basin, western Queensland, the only region where a probable complete sequence has so far been described (Shergold, 1972, 1975, 1980, 1982, 1993). Paleontologically, the Iverian Stage is clearly distinguished. On the basis of trilobites, it is characterized by: the occurrence of the cosmopolitan genus *Irvingella* in Australia; the diversification of the agnostoid subfamily Pseudagnostinae during which *Pseudagnostus*, *Rhaptagnostus*, and *Neoagnostus* separate, and become biostratigraphically important; diversification of the Leiostegioidea, especially the families Kaolishaniidae and Pagodiidae; the first occurrence of the Dikelocephaloidea, Remopleuridoidea, and Shumardiidae; and the separation of the true asaphids from ceratopygids. As a result, ten trilobite assemblage zones have been recognized based on successive species of *Irvingella*, *Peichiashania*, *Hapsidocare*, and *Lophosaukia* (Shergold, 1993). Over 160 trilobite taxa occur in the type area of the Iverian Stage.

Payntonian As defined by Jones *et al.* (1971), the Payntonian Stage is recognized on the basis of its trilobite assemblages (Shergold, 1975), its base lying at the point, in its type section (Black Mountain, western Queensland), where the comingled American–Asian assemblages of the Iverian are replaced by others of total Asian affinity. These are dominated by tsinaniid leiostegioidean and saukiid and ptychaspidid dikelocephaloidean and remopleuridoidean trilobites. A tripartite zonal scheme is applicable following biostratigraphic revisions suggested by Nicoll and Shergold (1992), Shergold and Nicoll (1992), and Shergold (1993). In ascending order, these zones are based on *Sinosaukia impages*, *Neoagnostus quasibilobus* with *Shergoldia nomas*, and *Mictosaukia perplexa*. These zones are fully calibrated by a comprehensive conodont biostratigraphy (Nicoll, 1990, 1991; Shergold and Nicoll, 1992). The Payntonian Stage contains a total of 30 trilobite taxa.

Datsonian The concept of the Datsonian Stage remains as defined by Jones *et al.* (1971) with its base located at the first appearance datum of the conodont *Cordylodus proavus*. Only rare trilobites, *Onychopyge* and leiostegiids, occur and these are insufficient for the establishment of a trilobite biostratigraphy. Accordingly, the Datsonian Stage is defined solely on the basis of conodonts (see discussion in Section 11.2.4 below).

NORTH AMERICAN CAMBRIAN STAGES

The development of a stadial nomenclature for Laurentia has until recently (Palmer, 1998) been complicated by the concept of the biomere ("segment of life") introduced by Palmer (1965a) and subsequently reviewed by him (1979, 1984), Stitt (1975), and Taylor (1997) among others. As originally defined, a biomere is a regional biostratigraphic unit bounded by abrupt extinction events on the shallow cratonic shelf. At these times existing evolving community complexes are replaced by low-diversity trilobite fauna dominated by simple ptychoparioid trilobites invading from the outer shelf or shelf break, which in turn rapidly evolve. Six such events have been suggested by Hollingsworth (1997), but the lower two are as yet undefined. In ascending order they are the "Olenellid," "Corynexochid," Marjumid, Pteropcephaliid, Ptychaspid, and Symphysurinid biomeres.

Ludvigsen and Westrop (1985) emphatically considered biomeres to be stages since they were based on an aggregate of trilobite zones and subzones. Palmer (1998) currently considers biomeres can be retained as subtly different units based on his (1979) modified concepts, and has subsequently (1998) proposed a sequence of stages for the Laurentian Cambrian based on trilobites, as follows. The pre-trilobite Cambrian still lacks defined stages.

Montezuman The Montezuman Stage (Palmer, 1998) is named from the Montezuma Range, Nevada. Its base is defined by characteristic fallotaspid trilobites, as in Morocco and

Siberia, which are followed by nevadiids and holmiines, and is distinguished by at least three families of Olenellina, which are different from ollenellids of the succeeding stage (see generic range charts in Palmer and Repina, 1993). The Montezuman Stage also contains the oldest Laurentian archaeocyathans.

Dyeran The Dyeran Stage (Palmer, 1998) is named from the town of Dyer, Nevada, and covers the biostratigraphic interval previously assigned to the *Olenellus* Zone, currently regarded as tripartite (Palmer and Repina, 1993). The base of the stage coincides with a major change in the olenelloid fauna following the nevadiid bearing late Montezuman. A similar change has been documented by Fritz (1992) in British Columbia. Olenellidae are characteristic.

Delamaran The Delamaran Stage (Palmer, 1998) has a boundary stratotype in the Delamar Mountains, Nevada, at the Oak Spring Summit section. The stage embraces the "Corynexochus" biomere and the *Plagiura–Poliella*, *Albertella*, and *Glossopleura* Zones in restricted shelf environments. It is characterized by ptychoparoid, corynexochid, zacanthoidid, dolichometopid, and oryctocephalid trilobites. The biostratigraphic status quo has been presented by Palmer and Halley (1979) and, in part, by Eddy and McCollum (1998).

Marjuman The Marjuman Stage (Ludvigsen and Westrop, 1985; emended Palmer, 1998) is named from Marjum Pass in the House Range, Utah. Ludvigsen and Westrop (1985) originally defined the base of the Marjuman Stage at the base of the *Acidusus atavus* Zone, but this is not a major extinction event according to Palmer (1998). Ludvigsen and Westrop (op. cit.) equated the Marjuman Stage with the Marjumiid biomere (Palmer, 1981), but this event occurs earlier on the inner shelf at the major change from trilobites of the *Glossopleura* Zone to those of the *Ehmaniella* Zone (*Proehmaniella* subzone), which has been documented by Sundberg (1994). On the open shelf this event is correlated to the base of the *Bathyuriscus–Elrathina* and *Bolaspidella* Zones, whose fauna characterize the early Marjuman.

The late Marjuman is characterized on the open shelf by cedariid trilobites, four zones of which are documented by Pratt (1992), distributed among the *Cedaria* and *Kingstonia* biofacies, regarded as bathymetrically related communities. Cedariid and crepicephalid trilobites characterize the inner shelf. The Marjuman Stage embraces agnostoid zones from *Pentagnostus praecurrens* to *Glyptagnostus stolidotus*.

Steptoean The Steptoean Stage (Ludvigsen and Westrop, 1985) was named from Steptoe Valley, in the Duck Creek Range, near McGill, eastern Nevada. The base of the stage is defined at the base of the *Aphelaspis* Zone, which also corresponds to the base of the Pterocephaliid Biomere (Palmer, 1965b). The *Aphelaspis* Zone contains *Glyptagnostus reticulatus* (Angelin), a very important species for international correlation of the Steptoean Stage. Above the *Aphelaspis* Zone, the Steptoean Stage embraces the *Dicanthopyge*, *Prehousia*, *Dunderbergia*, and lower part of the *Elvinia* Zones in restricted-shelf environments, and the *Glyptagnostus reticulatus*, *Olenaspella regularis*, *O. evansi* Zones and the *Parabolinoides calvilimbata* and *Proceratopyge rectispinata* fauna, documented by Pratt (1992), in open-shelf environments.

Late Steptoean trilobites from the *Elvinia* Zone in SE British Columbia have recently been described by Chatterton and Ludvigsen (1998). The Steptoean Stage also embraces the significant Sauk II–III discontinuity event which is widespread across Laurentia (Palmer, 1981). Palmer (1998) has suggested a parastratotype section in the Desert Range, Nevada, where the extinction event at the Marjuman–Steptoean boundary is defined at the base of the *Coosella perplexa* Zone.

Sunwaptan This stage (Ludvigsen and Westrop, 1985) is named from Sunwapta Creek, Wilcox Peak, Jasper National Park, southern Alberta. The base of the Sunwaptan Stage is taken at the *Irvingella major* subzone of the *Elvinia* Zone, which Chatterton and Ludvigsen (1998) argue should be regarded as a separate zone. This is succeeded by the *Taenicephalus* Zone, *Stigmatocephalus oweni* fauna, and *Ellipsocephaloides* Zone in the early Sunwaptan, and the *Illaenurus* Zone and most of the *Saukia* Zone in the late Sunwaptan. More than 130 trilobite taxa of Sunwaptan age have been documented in Alberta by Westrop (1986), and from the District of Mackenzie, North Western Territories, Canada, by Westrop (1995). Characteristic are dikelocephalid, ptychaspidid, parabolinoidid, saukiid, ellipsocephaloidid, illaenurid, and elviniid trilobites. The Sunwaptan Stage corresponds to the Ptychaspid Biomere (Longacre, 1970; Stitt, 1975).

Skullrockian The Skullrockian Stage (Ross *et al.*, 1997) was named from Skull Rock Pass in the House Range, Utah. It was conceived as the earliest stage of the Ibexian Series. The base of the Skullrockian is defined by conodonts at the base of the *Hirsutodontus hirsutus* Subzone of the *Cordylodus proavus* Zone. On the trilobite zonal scale this level corresponds to the base of the *Eurekia apopsis* Subzone of the *Saukia* Zone and is regarded as an independent zone by Ross *et al.* (1997). The *E. apopsis* Zone has a limited trilobite fauna, as does the overlying *Mississquoia* Zone, and the primary group for fine correlation is conodonts. The upper part of the Skullrockian Stage is Ordovician.

11.2 CAMBRIAN STRATIGRAPHY

11.2.1 Faunal provinces

The Cambrian Period is of particular biological interest in that it marks the appearance of multicellular phyla that have populated the Earth since.

Faunal provincialism was strongly developed and biostratigraphic zonal schemes generally cannot be applied beyond their provincial boundaries. This is most marked in the Early Cambrian where two major trilobite faunal provinces were developed, one characterized by Redlichiina (Redlichian Province), composed of endemic redlichiids and pandemic Ellipsocephaloidea and eodiscid fauna, the other by Olenellina (Olenellian Province), composed of endemic olenellids and pandemic ellipsocephaloids and eodiscid fauna (Chang, 1989, 1998; Palmer, 1973; Palmer and Repina, 1993).

Debrenne (1992) suggests that three archaeocyathid faunal provinces existed in the Early Cambrian: an Afro–European Province, which possibly extends to China, characterized by Anthomorphidae; an Australo–Antarctica Province characterized by Flindersicyathidae, Metacyathidae, and Syringocnemidae; and a Siberian Province characterized by all of these genera. Kruse and Shi (2000), on the other hand, applying sophisticated statistical analysis to the distribution of archaeocyathans, recognizes five provinces based on Siberia–Mongolia, Europe–Morocco, Central–East Asia, Australia–Antarctica, and North America–Koryakia. During the Middle and Late Cambrian, Chang (1989) recognizes Pacific Provinces, based on Centropleuridae, Xystriduridae, and Olenidae, respectively; and Atlantic Provinces, based on Paradoxidae and Olenidae, respectively.

Jell (1974) recognized three trilobite provinces: Columban in North and South America; Viking in Europe, maritime North America and NW Africa; and Tollchuticook in Asia, Australia, and Antarctica. Thus the trilobite fauna of Australia, China, and Kazakhstan have many elements in common, but differ markedly from those of North America (Jell, 1974; Palmer, 1973). In the latest Cambrian, conodonts provide fine biostratigraphic subdivision and substantially aid global correlation.

11.2.2 Trilobite zones

The most widely used fossil group for biostratigraphic zonation is the trilobites, the best known group of extinct arthropods. They enable fine stratigraphic subdivision and good correlation reliability in deposits of the continental shelf and platform within faunal provinces (for general statements on the occurrence of trilobites see Whittington, 1992, and Kaesler, 1997).

In the Late Cambrian, trilobite diversification and evolutionary turnover was extreme and fine zonations are established on the major paleocontinents (Jago and Haines, 1998). Through the late Middle and Late Cambrian, an interval of 18 myr (Fig. 11.2), there are 26 fossil zones, averaging 700 000 years each in duration. One group of trilobites, the agnostids, were pelagic and pandemic and enable global correlation of the Middle and Late Cambrian (Westergård, 1946; Öpik, 1979; Shergold and Laurie, 1997).

Australia, China, Russia, Scandinavia, and North America have the most complete Cambrian trilobite zonal successions. Those of Australasia and North America are shown in Fig. 11.1. Differing biostratigraphic philosophies, developed since the late eighteenth century, have resulted in different concepts of trilobite zones. In North America, for example, the concept of a zone is based on the range of a characteristic genus, which is subdivided into subzones by species or associations of species (interval-zones, Robison, 1994). A dual biostratigraphy (zones and biofacies) currently operates in Canada (Ludvigsen *et al.*, 1986). In Australia, species-zones or assemblage-zones have been most commonly applied (index fossil biostratigraphy), as in Scandinavia, Russia, and China. In Scandinavia, the Late Cambrian is minutely divided into 32 subzones on the basis of successive species of olenid trilobites (Henningsmoen, 1957).

11.2.3 Archaeocyathan zones

Regular Archaeocyatha and Radiocyatha, totalling 310 genera, are known from Lower Cambrian carbonate platform environments in 15 key regions of the Cambrian world (Kruse and Shi, 2000). They have been exploited extensively in regional biostratigraphic schemata. The most detailed archaeocyathan biostratigraphy has been developed in the Lower Cambrian of Siberia, where the Tommotian Stage embraces four successive assemblage-zones, the Atdabanian four, the Botoman three, and the Toyonian three (Debrenne and Rozanov, 1983; Zhuravlev, 1995). Archaeocyathan zones have also been established in South Australia (five), Laurentia (nine), Spain (11), and Morocco (four); (Zhuravlev, 1995). Problems associated with archaeocyathan correlation of these areas is primarily due to regional endemism. For example, Kruse and Shi (op. cit.) note that of the 240 archaeocyath species occurring in Australia and Antarctica, only 26 are shared between the two continents and only genera with wide stratigraphic distribution are common to Australia and Siberia (Zhuravlev and Gravestock, 1994).

11.2.4 Conodont zones

Protoconodonts, in particular species of *Protohertzina*, occur first in the earliest Cambrian (Nemakit–Daldynian) of the Siberian Platform where a zone of *Protohertzina–Anabarites* has been recognized at the Precambrian–Cambrian transition (Missarzhevsky, 1973, but see Bengtson *et al.*, 1990). Conodonts begin to diversify in the late Middle Cambrian, but it is not until the Late Cambrian that they are sufficiently common and differentiated to be used biostratigraphically, although Müller and Hinz (1991) dispute their utility in Scandinavia. Dong and Bergström (2001) have developed a comprehensive conodont biostratigraphy at this level in Hunan Province, China. In the latest Cambrian, they have been most intensively studied in the Great Basin (Miller, 1980, 1988, *inter alia*), western Queensland (Black Mountain; Druce and Jones, 1971; Nicoll and Shergold, 1992; Shergold and Nicoll, 1992). Faunas from these areas can be readily correlated into northeastern China and the Siberian Platform.

In Australia, from the late Iverian to the base of the earliest Ordovician (Warendan), seven successive conodont assemblages occur on the Black Mountain section. In ascending order, these are assemblage-zones based on *Teridontus nakamurai*, *Hispidodontus resimus*, *H. appressus*, *H. discretus*, *Cordylodus proavus*, *Hirsutodontus simplex*, and *Cordylodus prolindstromi* (Shergold and Nicoll, 1992). In Utah, nine subzones have been defined (Miller, 1980), based *on Proconodontus posterocostatus*, *P. muelleri*, *Eoconodontus notchpeakensis*, *Cambrooistodus minutus*, *Hirsutodontus hirsutus*, *Fryxellodontus inornatus*, *Clavohamulus elongatus*, *Hirsutodontus simplex*, and *Clavohamulus hintzei*. The conodont zones of both areas are of great assistance in correlating the provincialized (Gondwanan and Laurentian) trilobite zones.

11.2.5 Magnetostratigraphy

A magnetic polarity time scale has not been developed for the entire Cambrian, for reasons summarized by Trench (1996). Detailed studies, however, have been undertaken adjacent to the lower and upper boundaries of the system to facilitate their definitions. A rudimentary scale is available as a result of limited investigations by Kirschvink (1978a,b) and Klootwijk (1980) in central and South Australia. More detailed are studies on the Cambrian–Ordovician transition at Black Mountain in western Queensland by Ripperdan and Kirschvink (1992) and Ripperdan *et al.* (1992).

In general, Early and Middle Cambrian samples have predominantly reversed polarities with shorter periods of mixed polarity. The latest Cambrian is characterized by mainly reversed polarity with significant short intervals of normal polarity late in the *Proconodontus* and *Eoconodontus* Zones and early in the *Cordylodus proavus* and *Hirtsutodontus simplex* Zones. The initial Ordovician is dominated by periods of normal polarity. Results from Cambrian–Ordovician boundary sections in China, North America, and Kazakhstan confirm the Australian results (Kirschvink *et al.*, 1991; Apollonov *et al.*, 1992).

Based on the work of Kirschvink and Rozanov (1984) and Kirschvink *et al.* (1991), a detailed magnetostratigraphic scale is available for the Early Cambrian Tommotian and Atdabanian Stages and early archaeocyathan zones of the Siberian Platform. This, in association with carbon isotope stratigraphy, permits correlation to Morocco and South China (Kirschvink *et al.*, 1997; Kirschvink *et al.*, 1991).

11.2.6 Chemostratigraphy

A significant and stratigraphically important body of chemostratigraphic information, sometimes linked to the magnetic reversal scale, has become available over the past decade. Particularly important are carbon and strontium isotopes (Figs. 11.4 and 11.5).

Brasier (1993) and Montañez *et al.* (2000) have documented the potential of carbon isotope stratigraphy through the Cambrian Period of the Great Basin of Laurentia. A detailed reference scale has been constructed for the complete Lower Cambrian of Siberia, where eleven carbon cycles have been calibrated against the biostratigraphy by Brasier *et al.* (1994) and Brasier and Sukhov (1998). In Siberia and Mongolia, these are supplemented by $^{87}Sr/^{86}Sr$ measurements, which have permitted the recognition of a major hiatus (Fig. 11.1) in the Nemakit–Daldynian part of the Siberian sections (Derry *et al.*, 1994; Brasier *et al.*, 1996).

In Australia, carbon and strontium measurements have been obtained from the Middle Cambrian of the southern Georgina Basin, Northern Territory, by Donnelly *et al.* (1988), and have been calibrated to the agnostoid biostratigraphy there. In the Late Cambrian, the anomalous +4% shift in $\delta^{13}C$ in the Steptoean Stage, discussed by Brasier (1993), has been confirmed by Saltzman *et al.* (1998), and traced across the Great Basin to Wyoming and the upper Mississippi Valley. This event, the Steptoean Positive Carbon Isotope Excursion (SPICE), is also documented in Australia (Georgina Basin), China (Hunan), and Kazakhstan (Runnegar and Saltzman, 1998; Saltzman *et al.*, 2000; Fig. 11.3). Latest Cambrian and initial Ordovician $\delta^{13}C$ fluctuations are recorded and calibrated against the magnetostratigraphy and biostratigraphy in Australia (Black Mountain) by Ripperdan *et al.* (1992) and Saltzman *et al.* (2000). A high-resolution strontium isotope stratigraphy across the Cambrian–Ordovician boundary has been developed by Ebneth *et al.* (2001), also in Australia (Black Mountain), North America, northeastern Siberia, Baltica, China, and Kazakhstan.

Further detailed work, calibrated to the emerging global biostratigraphic framework, is required to reconstruct the isotopic evolution of seawater through all of Cambrian time. This will enable chemostratigraphy to be more widely applicable to questions of global stratigraphic correlation as well as paleoenvironmental interpretation of the Cambrian Period.

11.2.7 Cambrian evolutionary "explosion"

The "Cambrian evolutionary explosion" refers to the first metazoan evolutionary radiation, which essentially began at the beginning of the Cambrian, and during which all except one of the invertebrate phyla became established. During the Early and Middle Cambrian there was a spectacular burst in diversity (number of species and genera) and in disparity (number of distinct body plans). Brasier (1979) has given an extensive review of the Early Cambrian fossil record.

Fossil groups involved include prokaryotes and stromatolites; eukaryote protists, acritarchs, and chitinozoans; larger algae and vascular plants; Parazoa (Porifera, Chancelloriida, Radiocyatha, Archaeocyatha, Stromatoporoidea); Radiata; Bilateria (Priapulida, Sipunculida, Mollusca, Annelida, Onycophora and Tardigrada, Pogonofora, Arthropoda, Brachiopoda, Ectoprocta, Phoronida, Conodontophorida, Mitrosagophora and Tommotiida, mobergellids, Echinodermata, Hemichordata, Chaetognatha and Chordata). Brasier (1979) also comments on phyletic changes, skeletal changes, niche changes, size changes, and environment-related changes. In attempting to explain the "Cambrian explosion," Brasier (1982, 1995a) has attempted to link Cambrian "bio-events" to sea-level fluctuation and oxygen depletion, to nutrient enrichment (Brasier, 1992a,b) and, finally, to eutrophy and oligotrophy (Brasier, 1995b,c). There does appear to be a connection between chemostratigraphic cycles and biostratigraphic events (Margaritz, 1989; Brasier *et al.*, 1994).

A full review (Zhuravlev and Riding, editors, 2001) discusses the ecology of the Cambrian radiation in the context of life environments, community patterns, and dynamics, and "ecologic radiation" of major fossil groups, with important chapters on paleomagnetically and tectonically based global maps, global facies distribution supercontinental amalgamation as a trigger for the "explosion", climate change, and biotic diversity and structure.

Insights into Cambrian diversity are provided by "Konservatfossillagerstätten," deposits containing exquisitely well-preserved fossils, particularly of soft body parts. They are known globally from 35 localities if the "Orsten"-type preservation in Swedish Alum Shale, and elsewhere are considered as lagerstätten. The most spectacular lagerstätten are in the Buen Formation of Greenland, the Burgess Shale of British Columbia, the Chengjiang fauna of Maotianshan, Yunnan, and Kaili, Guizhou, China.

11.3 CAMBRIAN TIME SCALE

High-precision, biostratigraphically controlled dates in the Cambrian are extremely sparse, and ages assigned to the major boundaries have ranged widely in recent time scales (Palmer, 1983; Harland *et al.*, 1990; Tucker and McKerrow, 1995; Young and Laurie, 1996; Bowring *et al.*, 1998). We accept 11 dates as of sufficient analytical quality and biostratigraphic constraint for use in calibrating the Cambrian (Table 11.1). All are based on zircon crystals in volcanogenic rocks and are determined by the TIMS method.

Uranium–lead ages of zircons from volcanic ash beds, determined on the HR–SIMS method in the Geochronological Laboratory, Canberra, have been used for time scale calibration (Compston *et al.*, 1992; Compston and Williams, 1992; Cooper *et al.*, 1992; Perkins and Walshe, 1993). The ages have produced a time scale for the Early Paleozoic that appears to be 1–2% younger than that based on the mass spectrometric isotope dilution method (Tucker and McKerrow, 1995; Compston, 2000a,b). The cause for this systematic difference is uncertain. The standard SL13 that was used for most of the HR–SIMS dates has since been found to be inhomogeneous, but this is not thought by Compston (2000a,b) to be the cause of a systematic error. He has re-interpreted the HR–SIMS dates and also several of the TIMS dates listed in Table 11.1. In his new time scale, calibrations for the Cambrian and early Ordovician remain 1–3% younger than the TIMS scale. For the reasons discussed by Cooper and Sadler in Chapter 12, TIMS dates only are used here for calibrating the Cambrian Period.

Table 11.1 *Radioisotopic dates used for calibrating the Cambrian*[a]

No.	Sample	Locality	Formation	Comment	1-ry Biostratigraphic age	Zone/stage assignment	Bio. reliability	Reference	Age (Ma)	Type
1	Volcaniclastic sandstone	Bryn-Llyn-Fawr, N Wales	Pyritic tuffaceous ss in Dolgellau Fm.	Weighted mean $^{207}Pb/^{206}Pb$ date based on 17 concordant analyses of 22 fractions derived from 2 closely spaced volcaniclastic bands; indicates a maximum age for ash	Close to top *Acercare* Zone. Dated ash is 4 m below appearance of *Rhabdinopora* and 5 m below *R. f. parabola*. It is therefore very close to C O boundary	*Rhab. praeparabola* Zone–*Acerocare* Zone boundary	1	Landing *et al.* (1998)	Min. age 489 ± 0.6	Pb–Pb
2	Volcaniclastic sandstone	Ogof-Ddu, Criccieth, N Wales	Pyritic tuffaceous ss in Dolgellau Fm.	Four nearly concordant and 5 discordant analyses from crystal-rich volcaniclastic sandstone. Weighted mean dates of 4 concordant analyses = 490.9 ± 0.5 ($^{207}Pb/^{206}Pb$); 490.7 ± 0.7 ($^{206}Pb/^{238}U$); 490.7 ± 0.5 Ma ($^{207}Pb/^{235}U$). Suggested (maximum) age of 491 ± 1 Ma	*Peltura scarabaeoids scarabaeoids* below and *P. s. westergardi* above = Lower *Peltura scarabaeoides* Zone	Lower *P scarabaeoides* Zone	1	Davidek *et al.* (1998)	Min. age 491 ± 1	Pb–Pb, U–Pb
3	Ash bed	Taylor Nunatak, Shackleton Glacier, Antarctica	Taylor Fm.	Sample TAY-7 = 1 concordant and 1 discordant fraction (mean $^{207}Pb/^{206}Pb$ age 504.5 ± 2.9 Ma); TAY-1 = 2 concordant fractions (mean $^{207}Pb/^{206}Pb$ age of 505.7 ± 2.1 Ma); TAY-F = 1 concordant and 1 discordant fraction (mean $^{207}Pb/^{206}Pb$ age of 504.7 ± 2.2 Ma). Weighted mean of the 3 samples is 505.1 ± 1.3 Ma taken as age of the ash bed	Trilobites in carbonate bed, 1 km from dated samples. *Amphoton* cf. *oatesi, Nelsonia* cf. *schesis*, taken to indicate an Undillan, possibly late Floran, age	Undillan Stage	3	Encarnación *et al.* (1999)	505.1 ± 1.3	Pb–Pb
4	Ash bed, 3 samples from a 6.3 m interval	St John, New Brunswick	Hanford Brook Fm., Somerset St Mbr	Composite zircon ID date of 3 samples comprising a total of 39 single grain and multigrain analyses, 511 ± 1 Ma	*Protolenus* cf. *elegans* Matthew, *Ellipsocephalus* cf. *galeatus* Matthew associated in same bed. Suggests an age for the base of the Middle Cambrian no older than ∼513 Ma	*P. howleyi* Zone Late Branchian, Toyonian	1	Landing *et al.* (1998)	511 ± 1	U–Pb, Pb–Pb
5	Section Le-XI ash	S Morocco	Upper Lemdad Fm.	Zircon date. Five single grain analyses give a concordant cluster. Best estimate of age given as 517 ± 1.5 Ma	*A. gutta–pluviae* Zone, based on detailed correlation to section Le-I, 8 km away. The trilobite, *Berabichia vertumnia*, a guide to the *A. gutta–pluviae* Zone, is 21 m higher in sequence. Lower Botomian	*Antatlasia gutta–pluviae* Zone, Banian Stage, Botomian	3	Landing *et al.* (1998)	517 ± 1.5	U–Pb, Pb–Pb

6	Felsic volcanic ash	New Brunswick, Canada	Placentian Series	Upper intercept regression, based on 2 concordant analyses and 2 analyses with a "small amount of secondary Pb loss but no zircon inheritance." The U-Pb age is 530.9 ± 2.5 Ma. Moderate quality. Sample No. 6 of Tucker & McKerrow (1995)	Upper part of trace fossil zone *Rusophycus avalonensis*, Placentian series	Tommotian	2	Isachsen *et al.* (1994)	530.9 ± 2.5	U–Pb
7	Ash beds	Namibia	Nomtsas Fm.	U-Pb dates on zircons (92-N-1): 2 concordant single grain analyses and 3 discordant analyses. Detrital or inherited component present. The weighted mean $^{207}Pb/^{206}Pb$ age for the 2 concordant analyses is 539.4 ± 0.3 Ma. 'Best age' estimate given as 539.4 ± 1 Ma	*Trychophycus pedum* in lower Nomtsas Formation, which stratigraphically overlies Spitskopf Member. Regarded as a minimum age for base of Cambrian	*Trychophycus pedum*	3	Grotzinger *et al.* (1995)	539.4 ± 1	Pb–Pb
8	Ash beds	Oman	Ara Group	U-Pb dates on zircons immediately below and at the Precambrian–Cambrian boundary of, respectively, 543.2 ± 0.5 and 542.0 ± 0.3 Ma (2-sigma)	Simultaneous occurrence of an extinction of Precambrian *Namacalathus* and *Cloudina* and a negative excursion in C isotopes	Base-Cambrian	1	Amthor *et al.* (2003)	542.0 ± <0.5	U–Pb
9	Ash beds 94-N-11	Namibia	Upper Spitskopf Mbr	Dates on zircons (94-N-11): 14 fractions of from 2 to 13 grains each. Weighted mean $^{207}Pb/^{206}Pb$ age, based on 10 concordant fractions, is 543.3 ± 1. Regarded as a maximum age for basal-Cambrian boundary	*Pteridinium* and Ediacaran (dickinsonid-like) fossils 100 m higher in section. *Cloudina* present throughout Spitskoppf Mbr. Dated ash bed 'immediately below [erosional contact with] basal Cambrian' *Phycodes pedum* Zone, and base of Cambrian	Top Ediacaran, immediately below base-Cambrian	1	Grotzinger *et al.* (1995)	543.3 ± 1	Pb–Pb
10	Ash beds 91-N-1	Namibia	Lwr Spitskopf Mbr	Eight fractions (2 concordant and 6 discordant), including 4 single grains, define a near linear array. Best estimate given as 545.1 ± 1 Ma	*Cloudina* and Ediacaran fossils	Top Ediacaran	2	Grotzinger *et al.* (1995)	545.1 ± 1	U–Pb
11	Rhyolitic flows	Near Fortune Head, SE Newfld	Mooring Cove Fm.	3 concordant analyses of 10–20 crystals each, give an eruption age for rhyolite of 551.4 ± 5.9 Ma. Moderate quality. Sample no. 2 of Tucker & McKerrow (1995)	Rhyolite lies 1030 m below base *Trychophycus pedum* Zone (= base Cambrian)	Late Ediacaran	4	Tucker & McKerrow (1995)	551.4 ± 5.9	U–Pb

[a] Unless otherwise stated, all dates are high-resolution dates on zircon crystals determined by TIMS methods.
[b] Biostratigraphic precision and reliability: 1, well constrained to short zone; 2, well constrained but long zone; 3, moderately well constrained by correlation to a biostratigraphic zone; 4, poor biostratigraphic constraint.

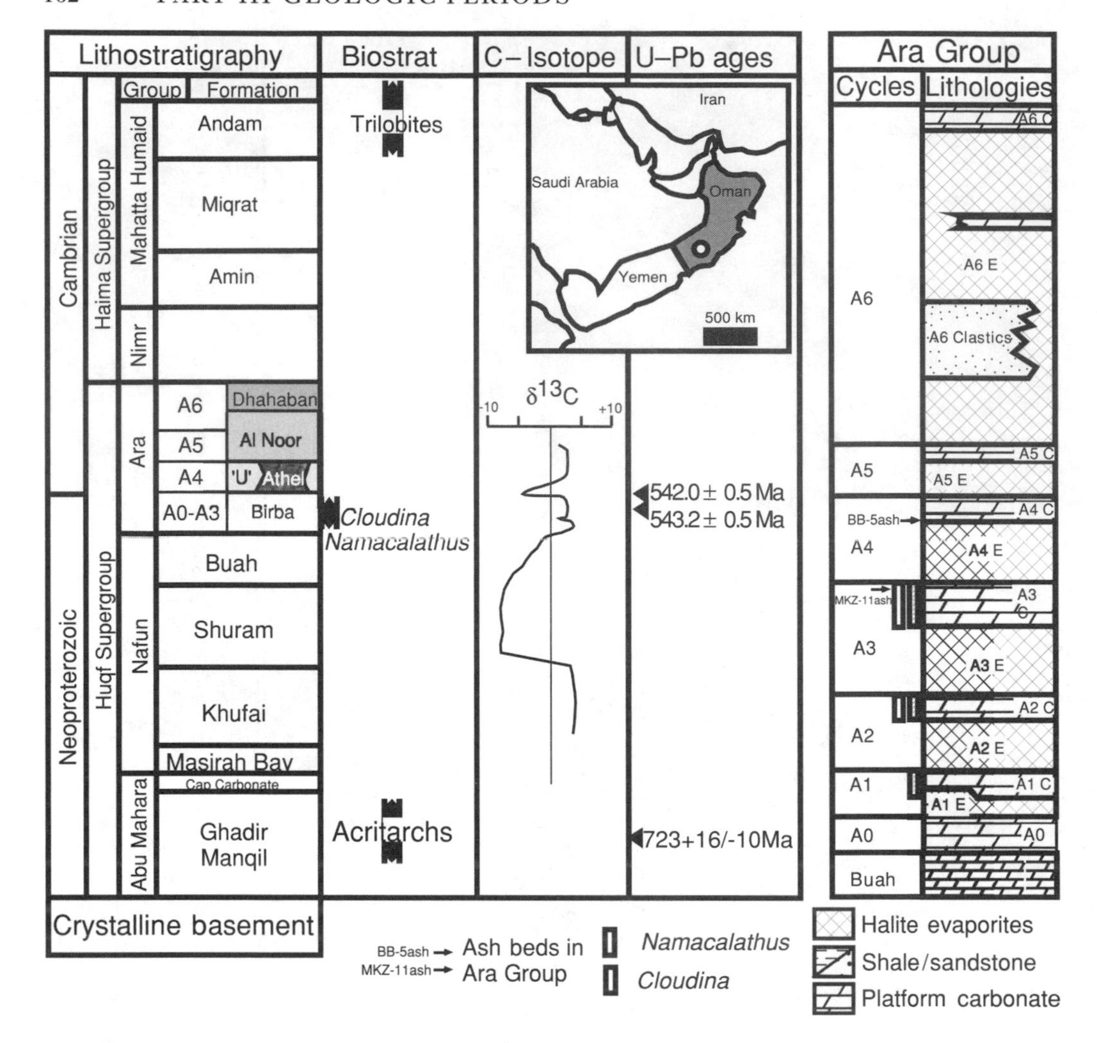

Figure 11.4 Stratigraphy of the Huqf Supergroup. Inset map shows location (unfilled circle) of the subsurface basin in the Sultanate of Oman where U–Pb dates were obtained at the Precambrian–Cambrian boundary (re-drawn after Amthor *et al.*, 2003, with permission of the authors). Chemostratigraphic and paleontologic data indicate the simultaneous occurrence of an extinction of Precambrian fossils (*Namacalathus* and *Cloudinia*) and a large-magnitude, short-lived negative excursion in carbon isotopes, which is widely equated with the boundary.

11.3.1 Age of boundaries

The maximum age of the base of the Cambrian is now reasonably well constrained. A high-quality ^{207}Pb/^{206}Pb date of 543 ± 1 Ma on volcanic ashes in the upper Spitskopf Member of the Schwarzrand Subgroup in Namibia, is assigned to the latest Ediacaran (Grotzinger *et al.*, 1995). The Spitskopf Member is overlain, with erosional contact, by the Nomtsas Formation, ^{207}Pb/^{206}Pb dated at 539.4 ± 1 Ma, the basal beds of which contain *Tricophycus pedum*. Interestingly, some elements of the globally distributed Ediacara fauna are found stratigraphically immediately above the dated ash bed in the Spitskopf Member, indicating that this fauna, characteristic of the Ediacaran, ranged into the base of the Cambrian (Grotzinger *et al.*, 1995). Similar faunal relationships have been found in South Australia (Jensen *et al.*, 1998).

The age of the base of the Cambrian is thought to be just younger than 543 Ma (Brasier *et al.*, 1994; Grotzinger *et al.*, 1995), an age consistent with other zircon dates, stratigraphically less-well constrained, from Siberia (Bowring *et al.*, 1993), and from the late Ediacaran (Grotzinger *et al.*, 1995; Tucker and McKerrow, 1995).

Recently, the Precambrian–Cambrian boundary has been identified in drill cores in Oman with tuffs on either side (Bowring *et al.*, 2003; Amthor *et al.*, 2003; Fig. 11.4). Chemostratigraphic and paleontologic data are interpreted to indicate the simultaneous occurrence of an extinction of Precambrian mineralized skeleton fossils (*Namacalathus* and

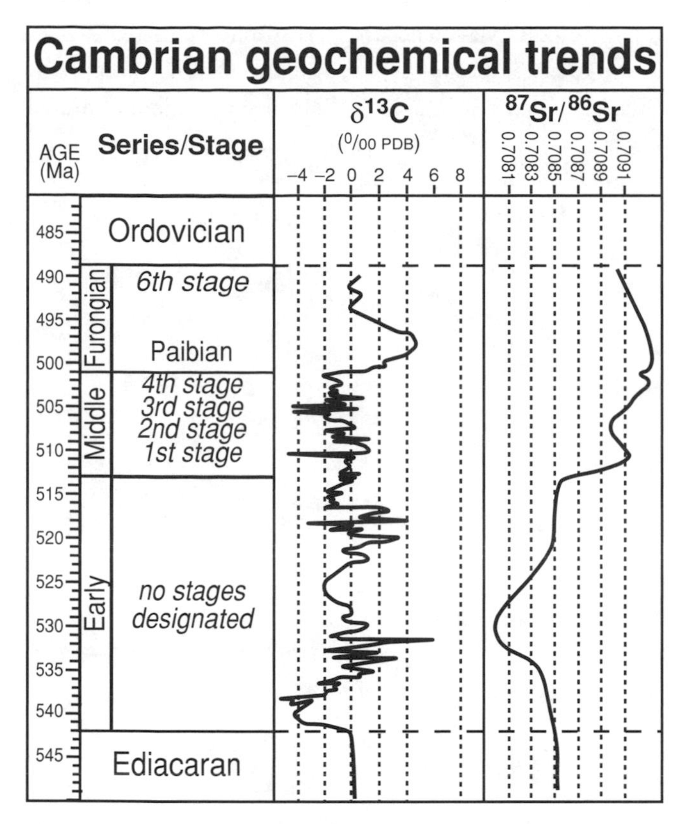

Figure 11.5 Strontium and carbon isotope trends through the Cambrian Period from published data on marine carbonate rocks and fossils. Strontium curve for the lower Paibian stage to the upper third stage is from Montañez *et al.* (2000), and for Late Cambrian and Early Cambrian is from Ebneth *et al.* (2001). The carbon-13 curve for the Late and Middle Cambrian is from Montañez *et al.* (2000) and for Early Cambrian is from Kirshvink & Raub (2003).

Cloudina) and a large-magnitude, short-lived negative excursion in carbon isotopes, which is widely equated with the boundary (Grotzinger *et al.*, 1995; Bartley *et al.*, 1998; Kimura and Watanabe, 2001). The ash bed immediately below the boundary yielded 543.2 ± 0.5 Ma, and the ash bed at the boundary 542.0 ± 0.3 Ma (2-sigma). Including external radiogenic factors the authors prefer the quoted uncertainty to be 1 myr (S. Bowring, pers. comm., 2003). The 542 ± 1 Ma date then is the best estimate for the age of the Precambrian–Cambrian boundary and the base of the Phanerozoic (Fig. 11.5).

The difference between the age of the top of the Cambrian (i.e. of the base of the Ordovician), here taken as 488.3 Ma, and the bottom, 542 ± 1 Ma, gives 54 myr as the duration of this period.

Ages of the base of the Late Cambrian and base of the Middle Cambrian are not well constrained. The Taylor Formation in Antarctica (Encarnación *et al.*, 1999) has yielded zircons with a weighted mean age of 505.1 ± 1.3 Ma on ashes interbedded with trilobite-bearing limestones assigned to the Undillan Stage. On the basis of this date, the authors estimated the age of the base of the Late Cambrian at ~500 Ma. We think 503 Ma is a better estimate based on this date. The Late Cambrian, or Upper Cambrian in this context, has a lower boundary equivalent to the base of the *Agnostus pisiformis* Zone or the base of the Mindyallan Stage of Australia. This level lies three zones below the base of the Furongian Series, the youngest division of the Cambrian. The base of the Furongian Series is therefore estimated at 501 Ma.

Ash beds associated with "upper Lower Cambrian" protolenid trilobites in southern New Brunswick (Landing *et al.*, 1998) have yielded zircons, which give a composite age based on three samples, of 511 ± 1 Ma. The beds are correlated with the middle Botoman to Toyonian Stages of Siberia. The date was taken by (Landing *et al.*, 1998) to indicate an age of 510 Ma for the base of the Middle Cambrian. This translates to 513 Ma for the base of the Middle Cambrian as applied in Australia. The Early Cambrian, therefore, occupies over half of Cambrian time.

The ages of boundaries between these levels are poorly constrained, and caution should be used when using the numerical scale in Figs. 11.1 and 11.2. In the latest Late Cambrian, a volcanic sandstone in North Wales gives a maximum age for the Lower *Peltura scarabaeoides* Zone of 491 ± 1 Ma (Davidek *et al.*, 1998). Ash beds from Morocco, taken as representing the "middle Botoman to Toyonian," help constrain the age of these stages. Five single-grain zircon analyses cluster at 517 ± 1.7 Ma (Landing *et al.*, 1998). Ash beds in New Brunswick with an age of 530 ± 2.5 Ma (Isachsen *et al.*, 1994) only weakly constrain the age of the Tommotian.

Within the Late Cambrian, which is finely zoned by trilobite biostratigraphy, stages are here proportioned according to the number of trilobite zones they contain. This method is also used for the late Middle Cambrian, where agnostoid trilobites provide reliable zonation and inter-regional correlation. The method, however, assumes a more or less constant rate of evolutionary turnover and a uniformity in paleontological practice in zonal designation, which are not only unproven, but are unlikely to be true.

Fossil diversity and abundance become increasingly rare passing downwards through the early part of the Middle Cambrian and Early Cambrian; and, as a result, the biostratigraphic framework becomes increasingly vague. In the early part of the Early Cambrian, resolution of the time scale is limited as much by lack of biostratigraphically useful fossils as by lack of radio-isotopic data. Our estimates of stage durations become correspondingly intuitive and the age of stage boundaries in the Early Cambrian shown in Figs. 11.1 and 11.2 should be regarded as highly approximate.

To summarize, the duration of the Cambrian is almost 54 myr, ranging from 542.0 to 488.3 Ma. The base of the Middle Cambrian is near 513 Ma, and that of the Furongian Series, near 501 Ma. Hence the Early Cambrian lasted 29 myr, the Middle Cambrian 12 myr, and the Furongian approximately 13 myr. Because ages for the Early–Middle and Middle Cambrian– Furongian boundaries are approximate estimates, the durations of intra-Cambrian divisions are equally tentative. More intra-Cambrian radiometric dates are urgently required.

12 • The Ordovician Period

R. A. COOPER AND P. M. SADLER

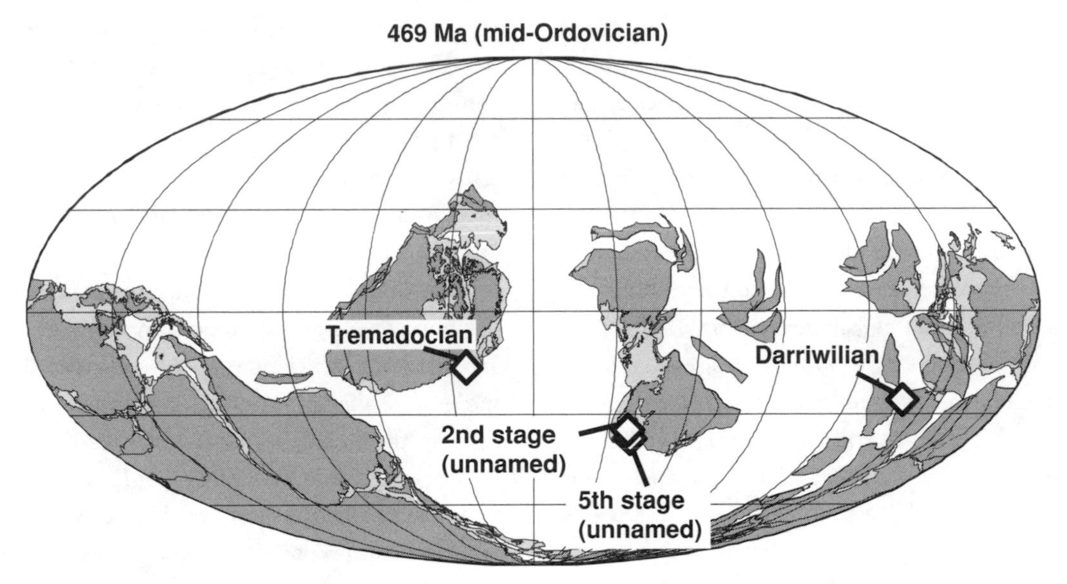

Geographic distribution of Ordovician GSSPs that have been ratified (diamonds) on a mid Ordovician map (status in January, 2004; see Table 2.3). Four of the seven Ordovician stages are not yet named, including two that have formalized GSSPs.

Rapid and sustained biotic diversification ("Ordovician radiation") to reach highest diversity levels for Paleozoic; prolonged "hot-house" climate punctuated by "ice-house" intervals and oceanic turnover; strong fluctuations in eustatic level, global glaciation, and mass extinction at end of period; appearance and evolution of pandemic planktonic graptolites and conodonts important for correlation; moderate to strong benthic faunal provincialism; re-organization and rapid migration of tectonic plates surrounding the Iapetus Ocean; migration of South Pole from North Africa to central Africa, all characterize the Ordovician period.

12.1 HISTORY AND SUBDIVISIONS

Named after the Ordovices, a northern Welsh tribe, the Ordovician was proposed as a new system by Lapworth in 1879. It was a compromise solution to the controversy over strata in North Wales that had been included by Adam Sedgwick in his Cambrian System but which were also included by Murchison as constituting the lower part of his Silurian System. Although it was initially slow to be accepted in Britain, where it was instead generally called Lower Silurian well into the twentieth century, the Ordovician was soon recognized and used elsewhere, such as in the Baltic region and Australia. The name Ordovician was officially adopted at the 1960 International Geological Congress in Copenhagen.

Black graptolite-bearing shales are widely developed in Ordovician sedimentary successions around the world. Lapworth (1879–80) described the stratigraphic distribution of British graptolites at the same time that he proposed the Ordovician System, and graptolites have played a major role in the recognition and correlation of Ordovician rocks since that time. Lapworth demonstrated as long ago as 1878, in southern Scotland, the fine biostratigraphic precision that can be achieved with this group. In the last several decades, conodonts have proved to be of similar global biostratigraphic value in the carbonate facies. In the shelly facies developed mainly on the continental shelf and platform, trilobites and brachiopods are used extensively

A Geologic Time Scale 2004, eds. Felix M. Gradstein, James G. Ogg, and Alan G. Smith. Published by Cambridge University Press.

for zonation, and coral–stromatoporoid communities enable biostratigraphic subdivision in the Late Ordovician. Chitinozoan and acritarch zonations are still being developed and both groups hold promise of providing long-range correlation with good precision.

Subdivision of the Ordovician into Upper and Lower, or Upper, Middle, and Lower, parts has been very inconsistent (Jaanusson, 1960; Webby, 1998). The International Subcommission on Ordovician Stratigraphy voted to recognize a threefold subdivision of the System (Webby, 1995), which is used here.

Because of marked faunal provincialism and facies differentiation throughout most of the Ordovician, no existing regional suite of stages or series has been found to be satisfactory in its entirety for global application. The Ordovician subcommission therefore undertook to identify the best fossil-based datums, wherever they are found, for global correlation, and to use these for definition of global chronostratigraphic (and chronologic) units (Webby, 1995, 1998). In this respect, it has deviated from the course followed by the Silurian and Devonian subcommissions, both of which have recommended the adoption of pre-existing (regional) stage or series schemes for global use.

In 1997, the Ordovician subcommission agreed to subdivide the period into three primary divisions, each to comprise two stages; and, in 2003, an uppermost seventh stage "Hirnantian" was added. It is not yet decided whether the primary subdivisions will be formally designated as Early, Middle, and Late, or will carry locality names. It is not yet decided whether they will be given the status of series, as preferred by the International Commission on Stratigraphy (ICS), or of "subperiods" (subsystems). They are referred to here as the Early, Middle, and Late Series.

During the early 1990s, the Ordovician subcommission established a number of working groups to investigate and recommend levels within the period suitable for international correlation, and therefore for defining international stages (Webby, 1995, 1998). Seven general chronostratigraphic levels have been certified as primary correlation levels for the seven international stages. They are based on the first appearance of key graptolite or conodont species. At present, four boundaries have been formally voted on and are defined by a global stratigraphic section and point (GSSP). Only three have been formally named: the Tremadocian (Stage 1), the Darriwilian (Stage 3), and the Hirmantian (Stage 7).

The seven stages are referred to here informally, as Stages 1 through 7 (Fig. 12.1).

12.1.1 Stages of the Lower Ordovician

CAMBRIAN–ORDOVICIAN BOUNDARY AND STAGE 1: THE TREMADOCIAN

The Cambrian–Ordovician boundary, as approved by the ICS in 2000, is defined by a GSSP in the Green Point section of western Newfoundland, in Bed 23 of the measured section (lower Broom Point Member, Green Point Formation; Cooper and Nowlan, 1999; Cooper *et al.*, 2001). This level coincides with the appearance of the conodont *Iapetognathus fluctivagus* (base of the *I. fluctivagus* Zone) at Green Point, and is just 4.8 m above the appearance of planktonic graptolites, which therefore can be taken as a proxy for the boundary in shale sections. This definition enables both graptolites and conodonts to be used in correlation of the boundary, and resolves a controversy extending back for at least 90 years (Henningsmoen, 1972).

The boundary also coincides with the appearance of the trilobites *Jujuyaspis borealis* and *Symphysurina bulbosa*, which are useful for correlation in carbonate successions. It lies at the peak of the largest positive excursion in the $\delta^{13}C$ curve through the boundary interval (Ripperdan and Miller, 1995; Cooper *et al.*, 2001) and during the global marine transgression that followed the Acerocare regressive event.

The name proposed for the new lowest Ordovician stage – the Tremadocian – was accepted by the International Commission on Stratigraphy in 1999. The upper and lower boundaries of the new stage almost exactly coincide with those of the British Tremadoc Series (Rushton, 1982; Fig. 12.1). The Tremadocian Stage thus encompasses the interval during which planktonic graptolites became established as a major component of the oceanic macroplankton, became widely distributed around the world, and became taxonomically diverse.

The graptolite fauna of the early part of the Tremadoc are dominated by the evolutionary complex *Rhabdinoporav* and *Anisograptus*, and that of the later part, by other anisograptids, particularly *Paradelograptus*, *Paratemnograptus*, *Kiaerograptus*, *Aorograptus*, *Araneograptus*, *Hunnegraptus*, and *Clonograptus*. Cooper (1999a) recognized nine global graptolite chronozones – in upward sequence, the zones of *Rhabdinopora praeparabola*, *R. flabelliformis parabola*, *Anisograptus matanenesis*, *R. f. anglica*, *Adelograptus*, *Paradelograptus antiquus*, *Kiaerograptus*, *Araneograptus murrayi*, and *Hunnegraptus copiosus*.

The early Tremadocian contains two widespread conodont zones: the zones of *Iapetognathus fluctivagus* and *Cordylodus angulatus*, equivalent to the graptolite zones of *R. praeparabola* to *P. antiquus*. The middle and late Tremadocian is finely subdivided into six conodont subzones (Löfgren, 1993), within the zones of *Paltodus deltifer* and *Paraiostodus proteus*,

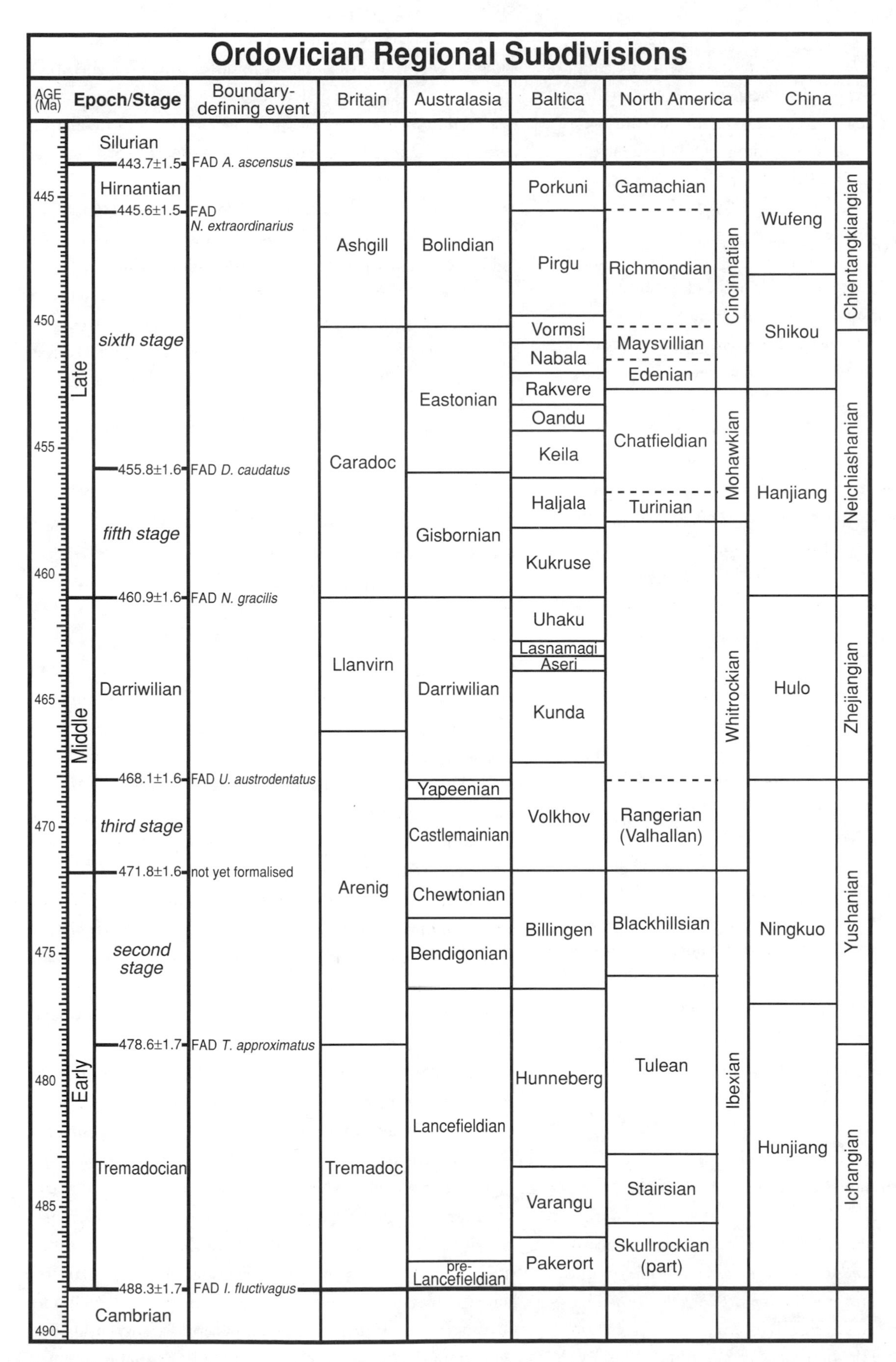

Figure 12.1 Chart showing international stages and selected regional suites of stages and series. The calibration, taken from Fig. 12.2, applies to the Australasian stage boundaries. Other regions are calibrated by correlation with the Australasian stages.

equivalent to the three graptolite zones, *Kiaerograptus, A. murrayi,* and *H. hunnebergensis.* The two sets of zones have been closely integrated by means of the conodont- and graptolite-bearing, shale–carbonate sections of southwest Sweden (Löfgren, 1993, 1996), providing a global correlation framework of high precision for the Tremadocian Stage.

STAGE 2 (UNNAMED)

The base of the second stage of the Ordovician is defined by a GSSP at the first appearance of the graptolite *Tetragraptus approximatus* in the section exposed in the Diasbrottet quarry on the northwestern slope of Mount Hunneberg, in the province of Vastergotland, southern Sweden (Maletz *et al.*, 1996). This biostratigraphic datum can be widely recognized throughout low- to middle-paleolatitude regions and has proved a distinctive and reliable marker. It is also the level adopted for base of the revised British Arenig Series (Fortey *et al.*, 1995) and thus is employed in a high-paleolatitude region.

The stage base coincides with base of the *T. approximatus* Zone of global distribution. It lies just above the base of the conodont subzone of *Oelandodus elongatus – Acodus deltatus*, the highest subzone of the zone of *P. proteus* in the Baltic (cold water) sequence (Maletz *et al.*, 1996). It lies at, or very close to, the base of the *deltatus–costatus* Zone of the midcontinent (warm water) realm. The boundary lies within the *Megistaspis (Paramegistaspis) planilimbata* trilobite zone.

During Stage 2, there is a spectacular increase in diversity and abundance of graptolites, driven by expansion of the Dichograptidae and Sigmagraptidae, and anisograptid graptolites become rare. Graptolite species diversity reaches its highest level for the Lower to Middle Ordovician (Cooper *et al.*, in press). There is also a peak in the faunal turnover rate at this time. This stage is the most finely zoned part of the Ordovician in low-paleolatitude regions such as Australasia and Laurentia.

12.1.2 Stages of the Middle Ordovician

STAGE 3 (UNNAMED)

For the base of the third stage, a level close to the base of the Australasian Castlemainian Stage and the base of the North American Whiterock Series (middle Arenig) is being considered by the Ordovician subcommission. The first appearance of the conodonts *Tripodus combsi* and *Baltoniodus triangulatus* and the first appearance of the graptolite *Isograptus victoriae lunatus* are bioevents suggested to have good international correlation value and to which to tie the boundary (Bergström, 1995). *Isograptus victoriae victoriae*, which succeeds *I. v. lunatus* in the *I. victoriae* bioseries, is also being considered. These markers are widely recognizable throughout middle- to low-paleolatitude regions, but not in high-paleolatitude regions. Finding a suitable section for the GSSP is proving difficult and further investigation is needed before specific candidate GSSPs can be proposed.

Stage 3 spans the evolutionary development of *Isograptus* (particularly the *Isograptus victoriae* and *I. caduceus* groups) and its derivatives, providing fine zonal subdivision and correlation. Two Australasian stages and six zones are represented – in upward sequence, the zones of *Isograptus victoriae lunatus, I. v. victoriae, I. v. maximus, I. v. maximodivergens* (Castlemainian stage), and *Oncograptus upsilon* and *Cardiograptus morsus* (Yapeenian stage).

STAGE 4: THE DARRIWILIAN

The GSSP for the base of the Darriwilian Stage has been placed at the first appearance of the graptolite *Undulograptus austrodentatus* in the Huangnitang section, near Changshan, Chejiang Province, southeast China (Chen and Bergström, 1995; Mitchell *et al.*, 1997). This level lies just above the first appearance of *Arieneigraptus zhejiangensis* and marks the onset of a major change in graptolite fauna, from one dominated by dichograptids and isograptids to one dominated by diplograptids and glossograptids. The rapid evolutionary radiation of the Diplograptacea, along with the appearance of several distinctive pseudisograptid and glossograptid species provide a global event that is readily recognized in many graptolite sequences around the world.

The stage base lies very close to, and immediately above, the base of the conodont zone of *M. parva*, which marks the appearance of both *M. parva* and *Baltoniodus norrlandicus*, in the Baltoscandian succession. It lies close to the appearance of the zone fossil *H. sinuosa* in the North American (midcontinent) conodont succession (Chen and Bergström, 1995).

The stage corresponds exactly to the Australasian stage after which it is named (VandenBerg and Cooper, 1992), comprising the four graptolite zones Da1 *Undulograptus austrodentatis,* Da2 *Undulograptus intersitus,* Da3 *Pseudoclimacograptus decoratus,* and Da4 *Archiclimacograptus riddellensis.* It marks the progressive increase in relative abundance of diplograptid graptolites. In the mid Darriwilian Da3 Zone, a narrow diversity peak marks a sharp, but short-lived, expansion of a diverse dichograptacean and glossograptacean assemblage, including didymograptids, sigmagraptids, sinograptids, isograptids, and glossograptids (Cooper *et al.*, in prep). The extinction event at the end of the Da3 Zone is one of the most severe within the

Ordovician graptolite succession. By the end of the Darriwilian Stage, graptolite diversity was greatly reduced.

In terms of the Baltoscandian conodont zones, the Darriwilian Stage embraces the (middle and upper) *norrlandicus* Zone, the *variabilis, suecicus,* and *serra* Zones, and the lower *anserinus* Zone. In terms of the midcontinent conodont zones, the Darriwilian Stage ranges from the *sinuosa* Zone, through *holodentata, polystrophos,* and *friendsvillensis* Zones to the lower *sweeti* Zone.

12.1.3 Stages of the Upper Ordovician

STAGE 5 (UNNAMED)

The base of Stage 5 and of the Late Ordovician Series is defined at the first appearance of the globally distributed zonal graptolite, *Nemagraptus gracilis*. The GSSP is in outcrop E14b, located on the south bank of Sularp Brook at the locality known as Fågelsång in the Province of Scania, Sweden, 1.4 m below the Fågelsång Phosphorite marker bed in the Dicellograptus Shale (Bergström *et al.*, 2000, see their Fig. 5). This level coincides with base of the *N. gracilis* graptolite zone (Finney and Bergström, 1986). *Nemagraptus gracilis* is also used to define the base of the British Caradoc series (Fortey *et al.*, 1995), the Australasian Gisbornian Stage (VandenBerg and Cooper, 1992) and the Chinese Hanjiang Series (Fig. 12.1). The base lies within the conodont zone of *Pygodus anserinus,* which has global correlation value.

During Stage 5, graptolites expand in diversity, a trend driven by proliferation of dicellograptid graptolites following an abrupt decline in the later part of the Darriwilian. Dichograptids are rare and diplograptids abundant. The stage also marks a peak in faunal turnover rate (Cooper *et al.*, in press).

STAGE 6 (UNNAMED)

In 1995, the Ordovician subcommission subdivided the Upper Ordovician Series into two stages (the fifth and sixth stages) with the boundary between them being based on the first appearance datum (FAD) of the graptolite *Dicellograptus complanatus* and/or the conodont *Amorphognathus ordovicius.* However, after 17 years of evaluating sections, no adequate stratotype section could be found for these suggested biohorizons.

Accordingly, in 2003, the subcommission voted to pursue a new strategy, which is to divide the Upper Ordovician Series into three stages with the boundaries between them placed at biohorizons with known potential for reliable global correlation and for which there exists suitable stratotype sections. The FAD of the graptolite *Diplacanthograptus caudatus* is favored as the biohorizon for defining the base of the sixth stage with candidate stratotype sections at Black Knob Ridge, Oklahoma, USA, and Hartfell Spa, southern Scotland, UK.

This revised placement of the base of the Stage 6 is much earlier than the previous level, which would have been at, or close to, the base of the British Ashgill Series and Australasian Bolindian Stage. The middle part of this Stage 6, equivalent to the Australasian zones Bol–Bo3 (*C. uncinatus–P. pacificus*), contains a rich graptolite fauna including diplograptids (*Appendispinograptus* and *Euclimacograptus*) and dicellograptids, and species diversity reaches a peak for the Ordovician (Chen *et al.*, 2000).

STAGE 7: THE HIRNANTIAN

The seventh stage (or uppermost stage of the Upper Ordovician Series) is the Hirnantian. The classical Hirnantian Stage was the uppermost subdivision of the Ashgill regional stage of Britain, and is named after the Hirnant Beds in Wales. One GSSP proposal is the base of the *Normalograptus extraordinarius–N. ojsuensis* graptolite biozone in the Wangjiawan section in China.

The Hirnantian Stage corresponds to a major climatic oscillation and sea-level excursion. Graptolites, along with many other fossil groups (see below) were drastically reduced in diversity and were almost completely wiped out during the mass extinction in graptolite zones of *Normalograptus extraordinarius* and *N. persulptus.*

The top of the stage is defined by the base of the overlying Silurian System at the base of the *Akidograptus acuminatus* Zone, marked by the first appearance of the graptolite *Akidograptus ascensus* in the Dob's Linn section of southern Scotland (Melchin and Williams, 2000).

12.2 PREVIOUS STANDARD DIVISIONS

The British series – Tremadoc, Arenig, Llanvirn, Llandeilo, Caradoc, and Ashgill – established in North Wales and England, have been those most widely used around the world, and it is likely that they will continue to be widely used in their modified form (Fortey *et al.*, 1995) until the new divisions are established and become generally accepted. The classical British series divisions that came into use at the turn of the century were applied to the British graptolite zonal succession by Elles in 1925. It is in the sense of Elles (1925) that the series have been most widely applied and correlated around the world (Skevington, 1963; Fortey *et al.*, 1995), but there has always been a problem of how to relate the graptolite-based divisions to the classical divisions which were tied to "type areas" with shelly fauna. This problem has been considerably alleviated by British stratigraphers in recent years (Fortey *et al.*, 1995, 2000).

The series have been fully reviewed and re-defined in several papers (Whittington *et al.*, 1984; Fortey *et al.*, 1991, 1995, 2000). The following summary accepts the recommendations in these works. Fortey *et al.* (1995) recommend boundary stratotype sections and levels and, for all but the Ashgill, boundaries based on a graptolite datum. They expanded the Llanvirn to include the Llandeilo of Elles, relegating the Llandeilo Series to stage status and reducing the number of series in the Ordovician to five. This classification is adopted here.

12.2.1 Tremadoc

Sedgwick (1847) introduced the stratigraphical term Tremadoc Group for trilobite-, graptolite-, and mollusc-bearing strata in Wales. Salter (1866) excluded the beds with *Rhabdinopora sociale*, referring them to the underlying Lingula flags, and Marr (1905), who introduced the term Tremadoc Series, followed Salter's classification. Most authors, however (Fearnsides, 1905, 1910; Whittington *et al.*, 1984; Fortey *et al.*, 1995), have included the *Rhabdinopora*-bearing beds in the Tremadoc, the "type area" for which is in North Wales around the town of Tremadog (Fortey *et al.*, 1995). Reviews are given in Whittington *et al.* (1984).

The base of the British Tremadoc has not been formally defined, but has been proposed in Fortey *et al.* (1991) to be at Bryn-llyn-fawr in North Wales, at the horizon of appearance of *Rhabdinopora flabelliforme, sensu lato,* about 2 m above the base of the Dol-cyn-afon Member of the Cwmhesgen Formation. This level was adopted by Fortey *et al.* (1995). The level equates to that taken as a proxy for base of the Ordovician System in shale sections (Cooper *et al.* 2001). The Tremadoc Series almost exactly corresponds in scope with the newly defined international Tremadocian Stage.

12.2.2 Arenig

Following Sedgwick's (1852) reference to rocks around the mountain Arenig Fawr as characterizing a broad lithological division of the Lower Paleozoic, overlying the Tremadoc, the Arenig Series became established as the second division in the Ordovician of Britain (Fearnsides, 1905). The base is an unconformity in the type area, and there has been much debate about its international correlation (Whittington *et al.*, 1984). Fortey *et al.* (1995) proposed that the base be taken at the base of the *Tetragraptus approximatus* Zone, thereby bringing definition of the boundary into accord with general international usage (Skevington, 1963). The best section in Britain that spans this level is Trusmador in the Lake District, but the zone base cannot yet be located with any precision. The base also accords with the base of Stage 2 of the new classification.

12.2.3 Llanvirn

The Llanvirn Group was erected by Hicks (1881) for rocks previously referred to as "Upper Arenig" and "Lower Llandeilo" in sections near Llanvirn Farm, southwest Wales. Its base has generally been taken at the base of the *Didymograptus artus* Zone (Elles, 1925) and Fortey *et al.* (1995) indicate an appropriate section for boundary stratotype in the Llanfallteg Formation in South Wales. This boundary cannot be precisely located in low paleolatitudes and has not been favored as an international correlation datum by the Ordovician subcommission. Fortey *et al.* (1995) incorporate as a stage within the Llanvirn, the entire Llandeilo Series, as it has generally been used internationally. The "historical" Llandeilo lies largely within the redefined Caradoc (see below).

12.2.4 Caradoc

The Caradoc Sandstone (Murchison, 1839), exposed adjacent to the hill Caer Caradoc, near Church Stretton, south Shropshire, is the basis of the name, Caradoc Series. The type section became accepted as the Onny River section but, unfortunately, in this region the base of the series is an unconformity (Whittington *et al.*, 1984). Consequently, there has been uncertainty about age and correlation of the base. Internationally, the base of the Caradoc Series has generally been taken at base of the graptolite zone of *Nemagraptus gracilis*, following Elles (1925) and the British Geological Survey. However, this level has been found to lie well down in the "historical" Llandeilo Series, causing much confusion. Fortey *et al.* (1995) suggest that the base be defined at the base of the *gracilis* Zone thus removing the bulk of the historical Llandeilo Series to the Caradocian. The base thus defined accords with the base of Stage 5 of the new Ordovician stage divisions proposed by the Ordovician subcommission of the ICS.

12.2.5 Ashgill

The "Ashgillian Series" was erected by Marr (1905) who subsequently (1913) designated the Cautley district of northwest Yorkshire as the "type area." The base of the Ashgill Series was proposed by Fortey *et al.* (1991) at Foggy Gill in the Cautley district, at a level marking the appearance of a number of shelly fossils, including the trilobites *Brongniartella bulbosa* and *Gravicalymene jugifera* and the brachiopods *Onniella* cf. *argentea* and *Chonetoidea* aff. *radiatula*. The absence of diagnostic graptolites and conodonts in the Foggy Gill section has led to continuing debate about correlation of this level with the graptolite and conodont successions, even within Britain (Williams and Bruton, 1983; Whittington *et al.*, 1984). Fortey *et al.* (1995) correlate the base with a level within the *Pleurograptus linearis* Zone, and close to, but just above, the base of the conodont zone of *Amorphognathus ordovicicus*. Here we follow Webby *et al.* (in prep.) and align its base with the base of the Bolindian Stage and the *A. ordovicicus* zone.

12.2.6 Australasian stages

Graptolite-based stages were established in Victoria, Australia, in the late nineteenth and early twentieth centuries (Hall, 1895; Harris and Keble, 1932; Harris and Thomas, 1938). A suite of nine stages has been used for the Ordovician of Australia and New Zealand for over 60 years (Harris and Thomas, 1938; Thomas, 1960; VandenBerg and Cooper, 1992). In upward sequence they are the Lancefieldian, Bendigonian, Chewtonian, Castlemanian, Yapeenian, Darriwilian, Gisbornian, Eastonian, and Bolindian Stages. They have proved to be widely applicable in graptolite successions around the world, particularly those representing low-paleolatitude regions (30° N–30° S), such as North America, Cordilleran South America, Greenland, and Spitsbergen. As originally defined and used, the nine stages were, in effect, groupings of graptolite zones (see below) and their lower boundaries were taken at zone boundaries (Fig. 12.2). Only one stage, the Lancefieldian, has a lower boundary formally defined by a boundary stratotype (Cooper and Stewart, 1979). In a move toward establishing the remaining stages as chronostratigraphic units, bioevents and reference sections for defining their lower boundaries, have been given by VandenBerg and Cooper (1992). One stage, the Darriwilian, has since been adopted for international use, with a lower boundary stratotype established in China (see above).

12.3 ORDOVICIAN STRATIGRAPHY

12.3.1 Biostratigraphy

The two most reliable and cosmopolitan fossil groups for correlation in the Ordovician are graptolites and conodonts. Graptolites are most abundant in shale sections, particularly those of the outer continental shelf, slope, and ocean, whereas conodonts are most abundant in carbonate sections of the shelf and platform. Together they provide a biostratigraphic correlation framework that can be applied with good precision across a wide range of facies and latitudinal zones (Bergström, 1986; Cooper, 1999b). Other fossil groups that are useful for regional and inter-regional correlation include trilobites, brachiopods, and, in Upper Ordovician carbonate facies, corals and stromatoporoids (Webby *et al.*, 2000). Chitinozoans and acritarchs appear to have good potential for global correlation (Paris, 1996).

GRAPTOLITE ZONES

Graptolites (Phylum Hemichordata) were a component of the Ordovician macroplankton. They lived at various depths in the ocean waters (Cooper *et al.*, 1991), were particularly abundant in upwelling zones along continental margins (Finney and Berry, 1997), and are found in a wide range of sedimentary facies. Most graptolite species dispersed rapidly, are widespread globally, and are of relatively short stratigraphic duration (1–3 myr). These attributes combine to make them extremely valuable fossils for zoning and correlating strata (Skevington, 1963; Bulman, 1970; Cooper and Lindholm, 1990). Together with conodonts, they are the primary fossil group for global correlation of Ordovician sequences.

The most detailed and best-established zonal scheme spanning the entire Ordovician is that of Australasia (Fig. 12.2; Harris and Thomas, 1938; VandenBerg and Cooper, 1992), widely applicable in low-paleolatitude regions ("Pacific Province," 30° N–30° S). Thirty zones, two of which are divided into subzones, are recognized, giving an average duration of 1.5 myr each. Zones are based on the stratigraphic ranges of species, most zone boundaries being tied to first appearance events. The most representative zonal scheme for middle to high paleolatitudes ("Atlantic Province") is that of southern Britain where, if the nine Tremadocian zones of Cooper (1999a) are added, 19 zones span the Tremadocian to middle Caradocian, averaging 2.4 myr each. Many species were cosmopolitan, and correlation between high- and low-paleolatitude regions is well controlled throughout most of the period. Other important graptolite zonal schemes for correlation are those of North America, Scandinavia, and China (Fig. 12.2).

CONODONT ZONES

Conodonts are tooth-like structures of primitive chordates, small eel-like animals that were predators living in shoals in

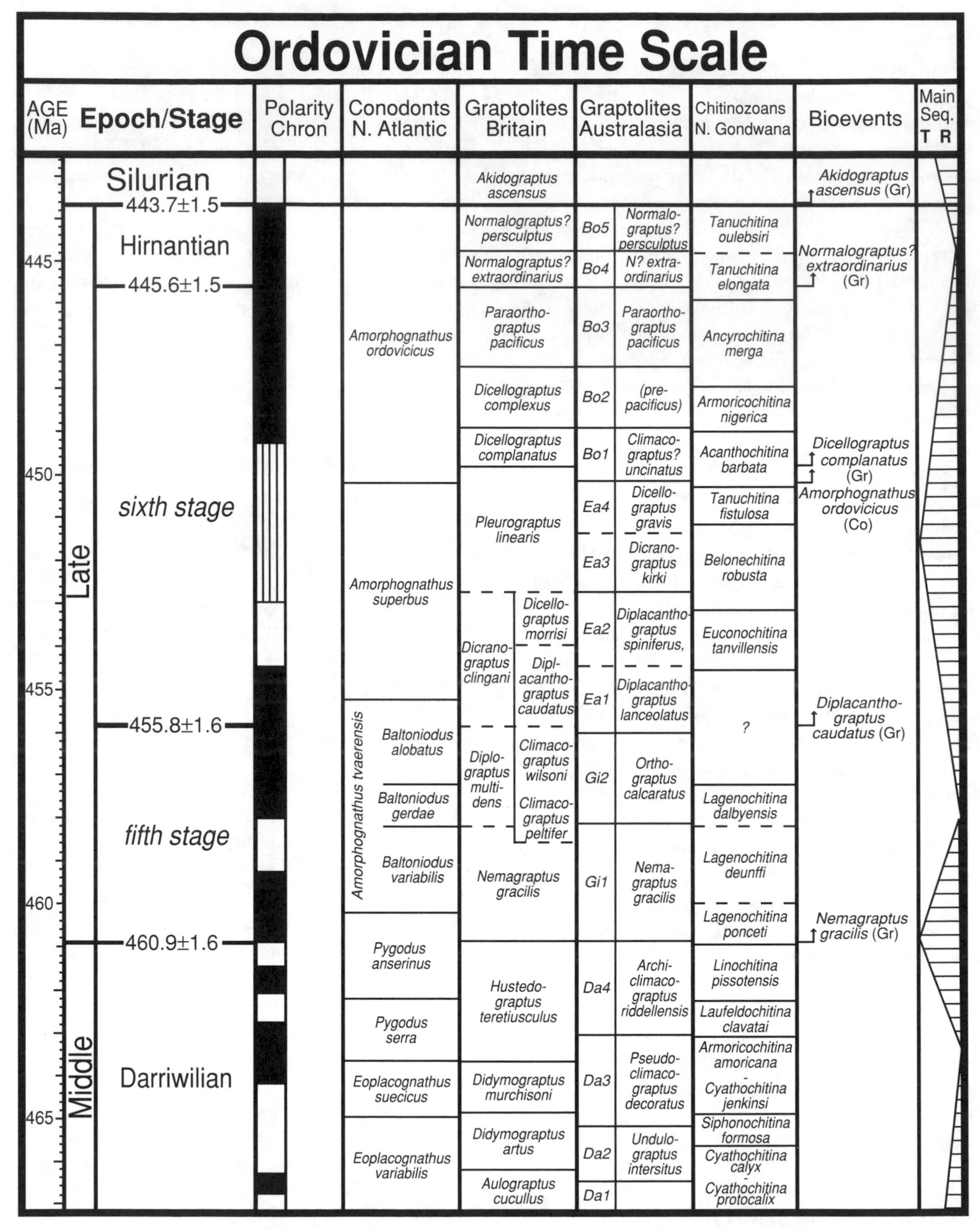

Figure 12.2 Chart showing major fossil zonal suites for the Ordovician. The Australasian graptolite zones are calibrated against the numerical time scale (see explanation in text) and other zonal suites are correlated with them. The global graptolite chronozones for the Tremadocian (Cooper, 1999a) have been added to the British zonal suite. Zones are after Löfgren (1993), Webby *et al.* (in prep.), and Bergström and Wang (1995). The magnetic polarity scale is from Idnurm *et al.* (1996). A color version of this figure is in the plate section.

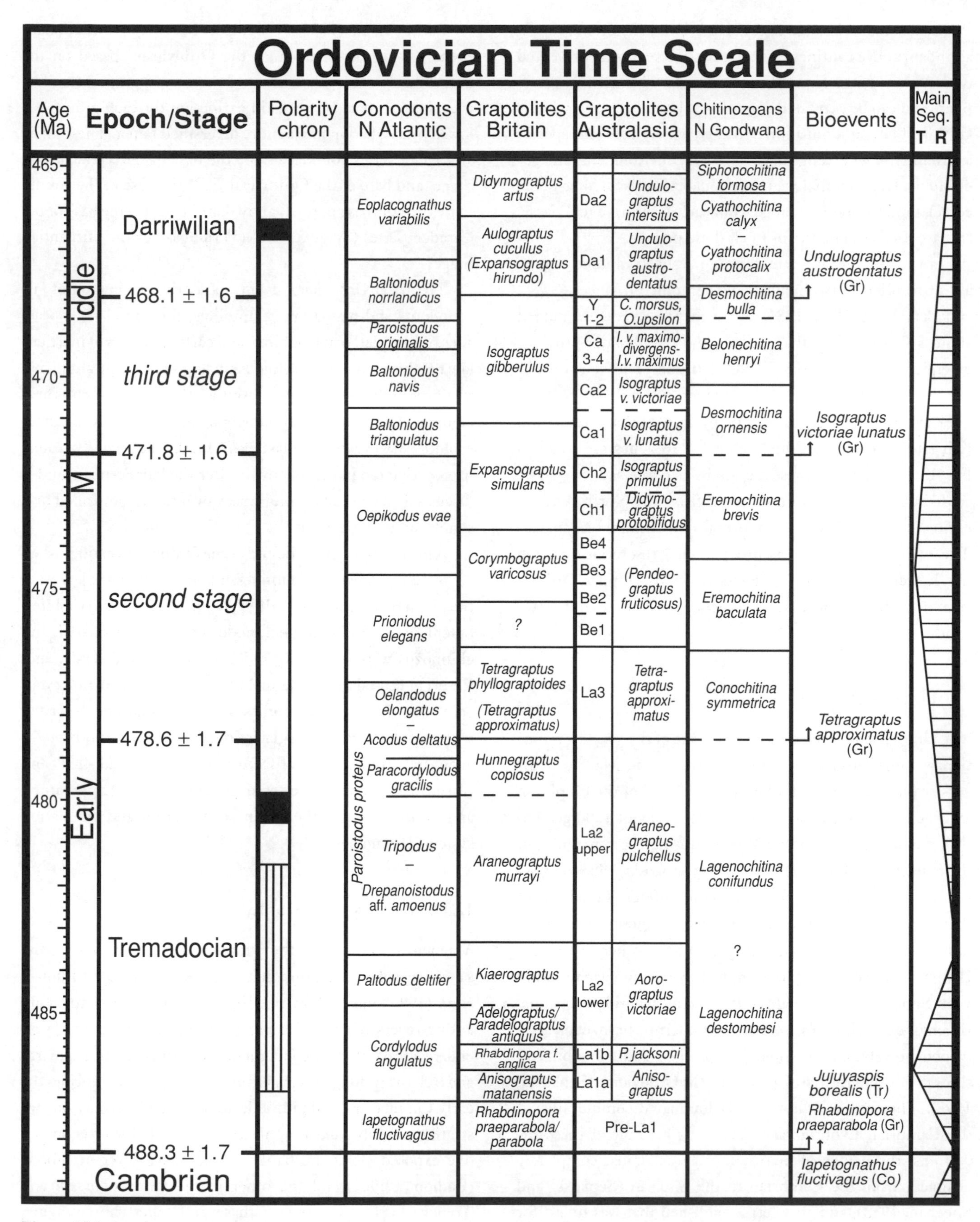

Figure 12.2 (*cont.*)

Paleozoic seas (Aldridge and Briggs, 1989). They were most abundant above continental shelves and were readily preserved in shelf carbonates. Conodonts are composed of calcium phosphate and can be extracted from the carbonate rock by acid digestion. The conodont animal roamed widely and some species are found in a wide range of sedimentary environments and geographical regions, making them valuable fossils for long-range correlation. Conodonts range from early Cambrian to Triassic and are used as zone fossils in all these periods.

In the Ordovician, conodont fauna, like graptolite fauna, are distributed in two major biogeographic provinces (Sweet and Bergström, 1976, 1984). A warm-water province ranged about 30° N and S of the Equator, and a cold-water province extended poleward from 30–40° latitude. The warm-water province, typified by the North American midcontinent region (Sweet and Bergström, 1976), contains a diverse and rich fauna that can be finely zoned – 26 zones are listed in Fig. 12.2 and some of these are subdivided into subzones. The cold-water province is best known from the North Atlantic region in which 17 zones and several subzones are recognized. The two zonal successions provide good resolution through the Lower and Middle Ordovician. In the Upper Ordovician, there are 11 warm-water zones and only three cold-water zones.

EVOLUTIONARY EVENTS

The Ordovician Period encompasses one of the greatest evolutionary radiations recorded in the Phanerozoic. Marine biodiversity increased three-fold between the end of the Cambrian and the middle Caradoc, to reach a level (about 1600 genera) that was not significantly exceeded during the remainder of the Paleozoic and the early Mesozoic (Sepkoski, 1995) or, possibly, up to the Paleocene (Alroy *et al.*, 2001). The cause of this radiation is not clear. Two spectacular bursts in diversity took place, one in the late Arenig and the other in the late Llanvirnian to early Caradocian, mainly through expansion in what Sepkoski has termed the Paleozoic Evolutionary Fauna. Brachiopods, trilobites, corals, echinoderms, bryozoans, gastropods, bivalves, ammonoids, graptolites, and conodonts all show major generic increase through the Ordovician (Sepkoski, 1995). They replace the trilobite-dominated communities of the Cambrian Evolutionary Fauna. As a result of these radiations, the nature of marine fauna was almost completely changed, from the Cambrian to the Silurian (Sepkoski and Sheehan, 1983), and a pattern established that was to last for the following 200 myr of Paleozoic time.

Barnes *et al.* (1995) recognized five major evolutionary events ("bioevents") through the Ordovician, based on the faunal histories of conodonts, trilobites, and graptolites. The events are marked by peaks in extinction for each group, followed by a radiation into a more diversified fauna. These all lie at stage or series boundaries: (1) the base of the Tremadocian Stage (and base of the Ordovician), (2) the base of the Arenig Series, (3) the base of the Darriwilian Stage, (4) the base of the Caradoc Series (Stage 5), and (5) the base of the Hirnantian Stage.

The basal-Ordovician event marks the origination of euconodonts and planktonic graptolites, the abundant appearance of new platform trilobites, and early radiation of inarticulate brachiopods and nautiloid cephalopods. Significant faunal turnover in all three groups, as well as other groups such as brachiopods and nautiloid cephalopods, mark the following three evolutionary events. The last event marks the second greatest mass extinction in the Paleozoic after the late Permian, and is discussed below. Each evolutionary event is accompanied by a significant eustatic event.

A mass extinction at the end of the Ordovician extinguished 22% of all marine animal families, making it one of the largest in the Phanerozoic (Sepkoski, 1995; Brenchley, 1989). Trilobites, brachiopods, graptolites, echinoderms, conodonts, coral, and chitinozoa were drastically reduced in generic diversity, and the event caused a major faunal turnover. The extinction event coincides with the climatic and sea-level fluctuations associated with the latest Ordovician glaciation, and was probably brought about by a combination of factors such as reduced shelf- and platform-habitable space, cold and fluctuating temperature, and perturbation of the ocean stratification and circulation systems (Brenchley, 1989).

12.3.2 Magnetostratigraphy

Magnetostratigraphy of the Ordovician is still at a reconnaissance stage of investigation. It has been summarized by Idnurm *et al.* (1996) and is shown in Fig. 12.2. Its main features are a long reversed interval corresponding to Arenig time, several reversals of the geomagnetic field during the Llanvirn, and relatively long intervals of normal polarity during the early Caradoc and Ashgill. The scale depends mainly on two stratigraphically long-ranging carbonate platform sections – one exposed along the banks of the Lena River in Siberia (Rodionov, 1966) and the other in the Baltic (Torsvik and Trench, 1991) – together with several other short-ranging, or imprecisely correlated, sections. The magnetostratigraphic

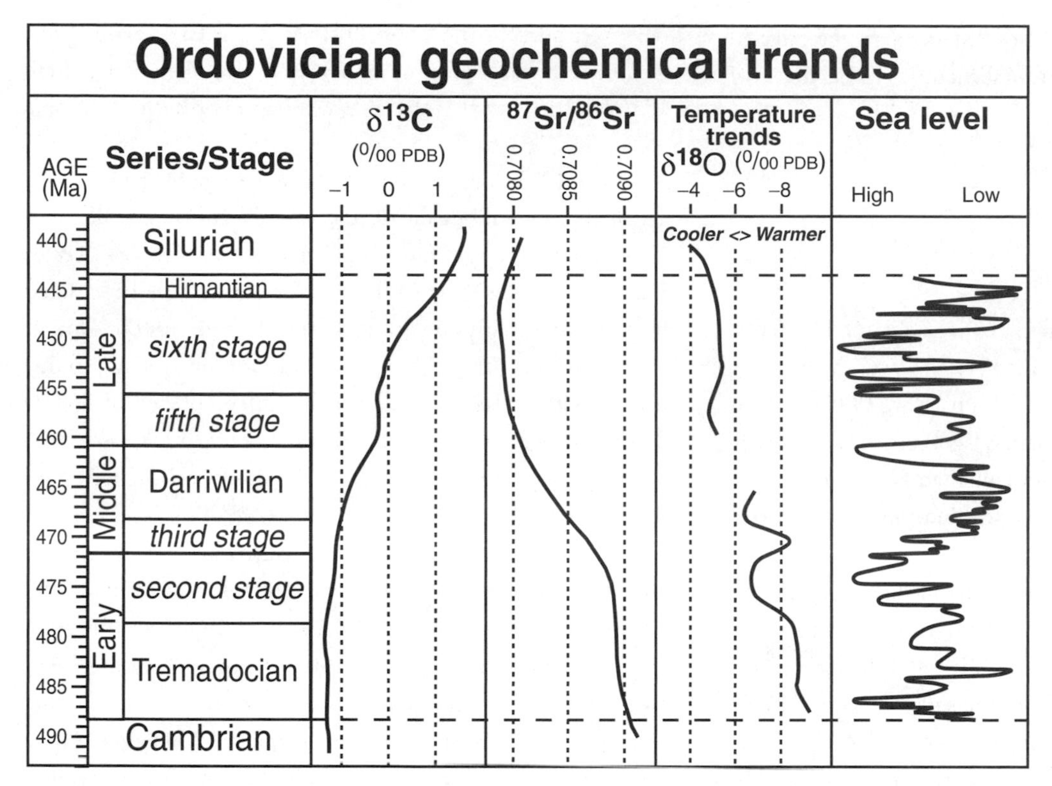

Figure 12.3 Geochemical and sea-level trends in the Ordovician Period. The schematic carbon isotope curve is a 5-myr averaging of global data (Veizer *et al.*, 1999) from Hayes *et al.* (1999; downloaded from www.nosams.whoi.edu/jmh). The $^{87}Sr/^{86}Sr$ LOWESS curve for the interval is based on the data of Qing *et al.* (1998), Denison *et al.* (1998), and Ebneth *et al.* (2001) – see Chapter 7. The oxygen isotope curves (inverted scale) are derived from a 3-myr interval averaging of global data compiled by Veizer *et al.* (1999; as downloaded from www.science.uottawa.ca/geology/isotope_data/ in January 2003). Large-scale global shifts to higher oxygen-18 values in carbonates generally are interpreted as cooler seawater or glacial episodes, but there are many other contributing factors (e.g. Veizer *et al.*, 1999, 2000; Wallmann, 2001). The Ordovician sea-level curve is from Nielsen (2003a,b).

record cannot yet be precisely tied to international stage boundaries.

12.3.3 Eustatic and climatic events

The Ordovician was a time of marked fluctuation in climate and sea level (Fig. 12.3). Eustatic curves have been proposed by several authors (e.g., Nielsen, 1992, 2003b; Nicoll *et al.*, 1992; Ross and Ross, 1995). Although concensus on a global eustatic pattern has yet to be achieved, several significant, abrupt eustatic changes are widely recognised, and include the *Acerocare*, Black Mountain, and *Ceratopyge* Regressive Events in the Tremadocian, the major transgression in the early Arenig (Stage 2), the regression at the end of the Arenig (Stage 3), the transgression in the early Caradoc (Stage 5), and the regression(s) in the late Ashgill (Stage 7). This last event was associated with glaciation in high paleolatitudes (Brenchley, 1984, 1989; Brenchley *et al.*, 1994) and a mass extinction event (Sepkoski, 1995). A distinctive cold-water shelly fauna known as the "Hirnantian fauna" can be recognized globally in this Hirnantian Stage.

12.3.4 Sr isotope stratigraphy

The $^{87}Sr/^{86}Sr$ curve for the interval (Fig. 12.3) is based on the data of Qing et al. (1998), Denison *et al.* (1998), and Ebneth *et al.* (2001). Marine $^{87}Sr/^{86}Sr$ decreases through almost the entire Ordovician, from a value of 0.709 05 at its base (Gao and Land, 1991; Ebneth *et al.*, 2001; McArthur *et al.*, 2001; Shields *et al.*, 2003) to a minimum of 0.707 88 before beginning to increase again some 2–4 myr prior to the boundary with the Silurian. The rate of change of 0.000 140 per myr between 472 and 477 Ma is the steepest known. According to Shields *et al.* (2003), this decrease and similar changes in carbon and oxygen isotopic trends mark a major reorganization of ocean chemistry and surface environment. But given that the Ordovician isotopic trends are patched together from a variety of world locations, the variable rate of change in $^{87}Sr/^{86}Sr$ through the interval might well be caused by correlation difficulties and artifacts of the time scale.

12.4 ORDOVICIAN TIME SCALE

12.4.1 Radiometric dates

Although many radiometric dates have been obtained from Ordovician rocks (e.g. Gale, 1985; Kunk *et al.*, 1985; McKerrow *et al.*, 1985; Odin, 1985; Tucker and McKerrow, 1995; Harland *et al.*, 1990; Compston and Williams, 1992) only few reach the standard of precision, analytical quality, and biostratigraphic constraint desirable for time scale calibration. An initiative started by R.J. Ross and his colleagues in North America and Britain in the late 1980s systematically undertook to sample and date British volcanic ash beds which can be biostratigraphically constrained by, or reliably correlated with, the British graptolite zones. Zircon separations from the samples were initially dated by the fission track method (Ross *et al.*, 1982), but the resulting dates had unacceptably long uncertainty intervals for refining the time scale. However, zircons from many of the samples were subsequently dated by the U–Pb method and have produced high-precision dates used in the present calibration. In addition, volcanogenic zircons have been recovered by other workers from beds in North America and Baltica, and other localities in Britain.

Tucker and McKerrow (1995) reviewed the available dates and rejected all but those based on zircon crystals and determined using the U-Pb isotope system, except for a selected number of dates by the $^{40}Ar/^{39}Ar$ and Sm–Nd methods, using other minerals. They produced a calibration for the Early Paleozoic that has been widely accepted. We follow Tucker and McKerrow in accepting only the high-resolution, biostratigraphically constrained, dates. Fourteen of the dated samples that were used by Tucker and McKerrow are used here and are shown in Table 12.1, which summarizes the isotopic data, dating method, and biostratigraphic constraints. They range in age from early Llanvirnian to earliest Silurian. Two of these are $^{40}Ar/^{39}Ar$ dates. An $^{40}Ar/^{39}Ar$ date of 454.1 ± 2.0 Ma on biotite from the Millbrig K-bentonite (Kunk *et al.*, 1985) and an $^{40}Ar/^{39}Ar$ date of 455 ± 3 Ma on biotite and sanidine from K-bentonite at Kinnekulle, Sweden, are both accepted here. However, the biostratigraphic age assignments of some samples are revised following a reassessment of the associated fauna or of its correlation.

Two dates accepted by Tucker and McKerrow are rejected here; their dated item 16, a Sm–Nd date on the Borrowdale Volcanics in the Lake District has a large error range and poor biostratigraphic constraint, and their dated item 17, from Llanwrtyd Wells in Wales, has poor biostratigraphic constraint.

Since 1995, four more dates have been published that are acceptable for calibration of the Ordovician time scale. Two K-bentonite beds in the lower member of the Los Azules Formation in the Precordillera of Argentine have been dated by the isotope dilution method (Huff *et al.*, 1997). One of these (ARG-1) has yielded three "almost concordant" multigrain fractions which give a lower intercept age of 464 ± 2 Ma. A rich graptolite fauna of early Darriwilian (Da2) Zone age is inter-bedded with the bentonites (Mitchell *et al.*, 1998).

A K-bentonite in the Chelsey Drive Group on Cape Breton Island, Nova Scotia, has given a weighted mean $^{207}Pb/^{206}Pb$ age of 483 ± 1 Ma (MSWD 0.14), based on six multigrain zircon samples (Landing *et al.*, 1997). The olenid trilobite *Peltocare rotundifrons* is found below and above the dated bentonite bed, and the conodont *Scandodus* is found 3.5 m below. These taxa suggest an age in the late Tremadocian. *Peltocare rotundifrons* is found elsewhere in the Chelsea Group associated with the graptolites *Hunnegraptus* sp. cf. *copiosus* and *Adelograptus* of *quasimodo* type, supporting the late Tremadocian assignment which is accepted here. Although the biostratigraphic constraint on this bentonite bed is not tight, the bentonite is important in lying within what is otherwise a long, undated interval in the Early Ordovician.

Two closely spaced pyritic, tuffaceous sandstone beds in the Bryn-Llyn-Fawr road section, in the Dolgellau Formation of North Wales, have given a weighted mean $^{207}Pb/^{206}Pb$ date of 489 ± 0.6 Ma based on 17 (of a total of 22) concordant zircon fractions (MSWD = 0.82, Landing *et al.*, 2000). This determination is one of the most highly concordant and precise dates available for calibration of the Cambrian and Ordovician. The age of the volcanic sandstone is inferred to be close to that of the dated zircon crystals, which, however, provide only a maximum age for the sandstone. The sandstone overlies strata with *Acerocare* Zone trilobites (Late Cambrian) and is immediately overlain by strata with planktonic graptolites, including *Rhabdinopora flabelliformis parabola*, a zonal indicator for the lowest Ordovician. The date is of high resolution and quality and the zircon-bearing bed lies very close to the Cambrian–Ordovician boundary.

Also from the Dolgellau Formation, at Ogof-Ddu near Criccieth in North Wales, zircons in a crystal-rich volcanic sandstone have yielded four "nearly concordant," and five "slightly to moderately discordant," multigrain fractions

(Davidek *et al.*, 1998). From the weighted mean $^{207}Pb/^{206}Pb$, $^{206}Pb/^{238}U$, and $^{207}Pb/^{235}U$ dates of the four nearly concordant analyses, Davidek *et al.* suggest an age of 491 ± 1 Ma for the zircons, which themselves are inferred to be of similar age to that of the enclosing volcanic sandstone. The sandstone lies 17 m below the appearance level of the early Ordovician graptolite *Rhabdinopora flabelliformis sociale* and is inter-bedded with mudstone and siltstone with calcareous nodules that contain a rich trilobite fauna representing the *Peltura scarabaeoides* Zone (Late Cambrian). Thus, this date provides a maximum age for the volcanic sandstone that underlies the Cambrian–Ordovician boundary, and is consistent with the preceding zircon date on Dolgellau volcanic sandstone.

Altogether, 22 dates are here regarded as of sufficient analytical quality and biostratigraphic constraint to be used for calibration of the Ordovician time scale (Table 12.1). They range from latest Cambrian to earliest Silurian in age. They are unevenly distributed within the Ordovician, and the Early Ordovician, in particular, is poorly calibrated. All Ordovician dated samples, except for two using the $^{40}Ar/^{39}Ar$ method, are based on zircons that are analyzed by the TIMS method.

The Ordovician chronostratigraphic scale is well constrained by isotopic dating between 453 and 457 Ma (Caradocian or approximately the fifth stage), between 461 and 468 Ma (Llanvirnian or later Darriwilian), and at 489–490 Ma (base), but elsewhere is poorly constrained. This poses a problem for interpolating stage boundaries into the linear scale of geologic time.

12.4.2 HR–SIMS (SHRIMP) dates

High-resolution-secondary ion mass spectrometry (HR–SIMS, also known as "SHRIMP") dates of zircons in Paleozoic rocks, produced in the geochronology laboratory of the Australian Geological Survey, Canberra, have been used for time scale calibration (Compston and Williams, 1992; Compston, 2000a,b). Rocks dated by the HR–SIMS method using standard SL13 appear to have produced ages that are about 1.3% younger than those based on standard QGNG (Black *et al.*, 1997). They are similarly younger than TIMS dates, even on the same samples (Tucker and McKerrow, 1995; Cooper, 1999b).

Compston (2000a,b) stated that the standard SL13 is inhomogenous at the sub-30-μm level. However, this was not thought by him to be the cause of a systematic error in HR–SIMS dates using this standard. He devised a new procedure for using the standard, and applied it to the previously published HR–SIMS analyses of Compston and Williams (1992).

Compston also adopted a new statistical method (mixing model) for separating detrital (inherited) ages from eruptive ages in zircon crystals. When applied to previous HR–SIMS analyses, some changed to younger, and others to older, ages. Many of the published TIMS dates were similarly re-interpreted, including several of those used here. Most of these were interpreted to be based on zircon populations that included an inherited component not detected by the multigrain TIMS method. For some of these re-interpreted TIMS analyses, Compston assumed a common $^{204}Pb/^{206}Pb$ ratio revised downwards by about 2% from that generally used (Stacey and Kramers, 1975). The net result was that the eruptive ages of most of the TIMS dated samples was younger than thought by the original authors.

A new time scale for the Early Paleozoic was then constructed, based on his re-interpreted HR–SIMS and TIMS dates. For the early Ordovician, this time scale is 1–3% younger than that used here. In particular, it conflicts with the two highly concordant and biostratigraphicaly well-controlled zircon TIMS dates (on the Dolgellau Formation) near the base of the Ordovician (Davidek *et al.*, 1998; Landing *et al.*, 2000). This poses a problem for time scale calibration. Clearly, Compston's ages for Ordovician zircon-bearing ashes cannot be combined with TIMS ages using conventional techniques because they are based on different assumptions. There are, in effect, two alternative calibrations for the Ordovician, one based on conventional TIMS dating, the other on HR–SIMS and re-interpreted TIMS dating. Until this problem is resolved, we use only the conventional TIMS dates in calibration.

12.4.3 Calibration of stage boundaries by composite standard optimization

Quantitative methods for interpolating period boundaries between nearby radiometric dates have been used by Harland *et al.* (1990) and, for interpolating Mesozoic stage boundaries, Gradstein *et al.* (1994a), both of whom also estimate the associated error. But where radiometric dates are sparse, as in much of the Early Paleozoic, these methods cannot be applied and some way of estimating stage durations is necessary. The compromise procedures have received little rigorous attention in the past. Generally, some arbitrary assumption is made, explicitly or implicitly.

In some previous time scales, stages are simply assumed to be of more-or-less uniform duration or scaled according to the number of biozones that they contain (Boucot, 1975; McKerrow *et al.*, 1985; Harland *et al.*, 1990). Another

Table 12.1 *List of radiometric dates used for construction of the Silurian and Ordovician time scales*

	T & M No.[a]	Sample	Locality	Formation	Comment	Primary biostratigraphic age	Zone assignment	Zonal reliability[b]	Reference	Age (Ma)	Type
1		K-bentonite	New York, USA	Kalkberg Formation	Ten small fractions of 4–12 grains each; 4 give concordant analyses; all share a $^{207}Pb/^{206}Pb$ age of ~417 Ma. Weighted mean $^{207}Pb/^{206}Pb$ age 417.6 ±1.0 Ma	Conodonts in other sections of the Kalkberg Formation	*I. woschmidti* Zone	1	Tucker *et al.* (1998)	417.6 ± 1.0	Pb–Pb
2	25	Felsic volcanic	Canberra, Australia	Laidlaw Volcanics	An average age from K–Ar (mineral) and Rb–Sr (whole rock and mineral) analyses, of 420.7 ± 2.2 Ma (SHRIMP age of 419.6 ± 5.6 Ma not used)	Interbedded with Gorstian fossiliferous sediments	*N. nilssoni–L. scanicus* Zones	1	Wyborn *et al.* (1982)	420.7 ± 2.2	K–Ar Rb–Sr
3		Bentonite	Shropshire	Middle Elton Formation	Two biotite grains with slightly "discordant" age spectra give similar total-gas $^{40}Ar/^{39}Ar$ ages (424.5 and 425 Ma). The $^{40}Ar/^{39}Ar$ weight-average plateau age of 423.7 ± 1.7 Ma is regarded as best age	Associated with graptolites indicative of *Neodiv. nilssoni* and *Lobo. scanicus* Zones, Gorstian	*N. nilssoni* and *L. scanicus* Zones	1	Kunk *et al.* (1985), Snelling (1989), Lanphere *et al.* (1977)	426.8 ± 1.7	Ar–Ar[c]
4	24	Thin volcanic ash	Welshpool, Wales	Buttington Shales	Mean $^{207}Pb/^{206}Pb$ age of "4 concordant analyses consisting of 10–20 grains per fraction analyzed"	Within the zone of *Monoclimacis crenulata*	*Monoclimacis crenulata* Zone	1	Tucker and McKerrow (1995), Tucker (USGS open file report)	430.1 ± 2.4	Pb–Pb
5	23	Ash	Esquibel Island, Alaska	Descon Formation	Ar–Ar total gas age of 436.2 ± 5 Ma at 1200 °C fusion (Kunk *et al.*, 1985)	4 m above shale with graptolites of *Coronogr. cyphus Zone* age (Churkin *et al.*, 1971)	*Coronogr. cyphus* Zone	1	Kunk *et al.* (1985), Churkin *et al.* (1971), Ross *et al.* (1982)	439.4 ± 5	Ar–Ar[c]

6	22	Ash	Dob's Linn	Birkhill Shales	Mean $^{207}Pb/^{206}Pb$ of 4 concordant analyses consisting of 13–20 grains in each fraction analyzed. Ash is 6 m above top of Ordovician – Toghill (1968), (*not 60 m as per Tucker et al., 1990, or Tucker and McKerrow, 1995*)	*Coronogr. cyphus* Zone	*Coronogr. cyphus* Zone	1	Tucker *et al.* (1990); Ross *et al.* (1982; for locality) Toghill (1968)	438.7 ± 2.1	Pb–Pb
7	21	Ash	Dob's Linn (Linn Branch)	Hartfell Shales	U–Pb age, "based on 3 concordant zircon fractions of 10–30 grains each"	Approximately 4.5 m below Ord–Sil GSSP, *Paraorthograptus pacificus* Zone	*Paraorthograptus pacificus* Zone	1	Tucker *et al.* (1990)	445.7 ± 2.4	U–Pb
8	20a	K-bentonite	Millbrig	Millbrig K-bentonite	Mean U–Pb age of 5 concordant single grain analyses, 453.1 ± 1.3 Ma (Tucker, 1992)	*Phragmodus undatus* Zone (lower)	*Phragmodus undatus* Zone (lower)	1	Bergström (1989), Huff *et al.* (1992), Tucker *et al.* (1990)	453.1 ± 1.3	U–Pb
9	20a	K-bentonite	Millbrig	Millbrig K-bentonite	$^{40}Ar/^{39}Ar$ on biotite, 451.1 ± 2.1 Ma (Kunk *et al.*, 1985)	*Phragmodus undatus* Zone (lower)	*Phragmodus undatus* Zone (lower)	1	Kunk *et al.* (1985)	457.4 ± 2.2	Ar–Ar[c]
10	20b	K-bentonite	Millbrig	Diecke K-bentonite	"Mean $^{206}Pb/^{238}U$ age of 5 concordant single grain analyses," 454.5 Ma (Tucker, 1992)	*Phragmodus undatus* Zone (lower)	*Phragmodus undatus* Zone (lower)	1	Tucker (1992), Tucker and McKerrow (1995)	454.5 ± 0.5	U–Pb
11	19	Calcareous ash	Pont-y-Ceunant, Bala, Wales	Base of Geli-grin	"Mean $^{206}Pb/^{238}U$ age of 454.8 ± 1.7" Ma (Tucker, 1992)	Rich brachiopod, trilobite, conodont fauna. "Longvillian" – above *D. multidens*, below *D. clingani*	Close to *D. miltidens–D. clingani* Zone boundary	3	Tucker *et al.* (1990), Tucker (1992)	454.8 ± 1.7	U–Pb
12	18	K-bentonite	Mossen Quarry, Kinnekulle, Sweden	Chasmops Limestone	$^{40}Ar/^{39}Ar$ dates on biotite and sanidine phenocrysts, 455.0 ± 3 Ma (Kunk and Sutter, 1984) and 454 Ma (Kunk *et al.*, 1985)	Conodonts, correlated by Bergström *et al.* (1995)	Late *D. multidens* Zone	1	Kunk and Sutter (1984), Tucker and McKerrow (1995), Bergström *et al.* (1995), Leslie and Bergström (1995)	458.3 ± 3	Ar–Ar[c]

(*cont.*)

Table 12.1 (*cont.*)

	T & M No.[a]	Sample	Locality	Formation	Comment	Primary biostratigraphic age	Zone assignment	Zonal reliability[b]	Reference	Age (Ma)	Type
13	18	K-bentonite	Mossen Quarry, Kinnekulle, Sweden	Chasmops Limestone	Mean U–Pb age of 5 concordant single grain analyses, 456.9 ± 1.8 Ma (Tucker and McKerrow, 1995)	Conodonts, correlated by Bergström *et al.* (1995)	Late *D. multidens* Zone	1	Kunk *et al.* (1988), Kunk and Sutter (1984), Tucker and McKerrow (1995), Bergström *et al.* (1995), Leslie and Bergström (1995)	456.9 ± 1.8	U–Pb
14	19	Calcareous ash	Pont-y-Ceunant, Bala, Wales	Base of Geli-grin	Mean $^{207}Pb/^{206}Pb$ age from 4 multigrain fractions of 457.4 ± 2.2 Ma; or mean $^{206}Pb/^{238}U$ age of 454.8 Ma (Tucker *et al.*, 1990)	Rich brachiopod, trilobite, conodont fauna. "Longvillian" – above *D. multidens*, below *D. clingani*	At *D. miltidens–D. clingani* Zone boundary	3	Tucker *et al.* (1990)	457.4 ± 2.2	Pb–Pb
15	15	Gritty calcareous ash	Llandrindod, central Wales	Llanvirn Series	Mean $^{207}Pb/^{206}Pb$ age of 5 concordant fractions of 1–20 grains each, give 460.4 ± 2.2 Ma	*D. murchisoni* immediately below sampled ash "considered by Elles to be close to base *G. teretiusculus* Zone"	*D. murchisoni* Zone	3	Ross *et al.* (1982), Tucker and McKerrow (1995)	460.4 ± 2.2	
16		K-bentonite	Cerro Viejo, San Juan, Argentina	Los Azules Formation	Three "almost concordant" fractions (14 grains total), 1 discordant, give a well-defined intercept age of 464 ± 2 Ma (Sample ARG-1, Huff *et al.*, 1997)	10 graptolite species listed by Mitchell *et al.* (1998)	Basal Da2	1	Huff *et al.* (1997), Mitchell *et al.* (1998)	464 ± 2	U–Pb
17	14	Indurated bentonite	Abereiddy Bay, S Wales	Lower rhyolitic tuff, Llanrian Volc Fm	Mean of 3 concordant + 1 slightly discordant $^{207}Pb/^{206}Pb$ multigrain fractions, 464.6 ± 1.8 Ma (= 76SW21 of Ross *et al.*, 1982)	Immediately overlying Cyffredin Shale is of *D. murchisoni* Zone age (Tucker and McKerrow, 1995).	*D. murchisoni* Zone	2	Tucker *et al.* (1990), Hughes *et al.* (1982)	464.6 ± 1.8	Pb–Pb
18	13	Ash flow	Arenig Fawr, Wales	Serv Formation	Mean $^{207}Pb/^{206}Pb$ age of 2 concordant, and 1 "slightly discordant," multigrain fractions	Underlying mudstone contains *D. artus* Zone graptolites	*D. artus* Zone	1	Tucker *et al.* (1990)	465.7 ± 2.1	

19	12	Rhyolite	Central New-foundland	Cutwell Group	Three small fractions give an upper intercept ($^{207}Pb/^{206}Pb$) age of 469 + 5.3 Ma	Sparse fauna of midcontinental conodonts, *Histiodella holodentata*; mid-lower *E. variabilis* Zone	Da2 Zone	2	Dunning and Krogh (1991), Stouge (pers comm.)	469 + 5 / −3	Pb–Pb
20		Volcanic sandstone	McLeod Brook, Cape Breton Is.	Chelsey Drive Group	Mean $^{206}Pb/^{238}U$–$^{207}Pb/^{235}U$ age of 8 multigrain discordant fractions define a concordant point at 483.3 + 3.9 / −2.1 Ma (0.7% discordant). Weighted mean Pb–Pb age 483 ± 1 Ma (with MSWD = 0.14) preferred	*Peltocare rotundiformis*, *Hunnegr.* cf. *copiosus*, *Adelograptus of quasimodo* type	Late Tremadocian (Hunnebergian), Late La2 Zone	2	Landing *et al.* (1997)	483 ± 1	Pb–Pb
21		Crystal-rich volcanic sandstone	Bryn-Llyn-Fawr, N Wales	Dolgellau Formation	Weighted mean $^{207}Pb/^{206}Pb$ date based on 17 concordant analyses of 22 fractions derived from 2 closely spaced volcaniclastic bands; indicates a maximum age for ash	Close to top *Acercare* Zone. Dated ash is 4 m below appearance of *Rhabdinopora* and 5 m below *R.f. parabola*. It is therefore very close to C/O boundary	*R. praeparabola* Zone–*Acercare* Zone boundary	1	Landing *et al.* (in press)	Maximum age 489 ± 0.6	Pb–Pb
22		Crystal-rich volcanic sandstone	Ogof-Ddu, Criccieth, N Wales	Dolgellau Formation	Four nearly concordant and 5 discordant analyses from crystal-rich volcaniclastic sandstone. Weighted mean dates of 4 concordant analyses = 490.9 ± 0.5 ($^{207}Pb/^{206}Pb$); 490.7 ± 0.7 ($^{206}Pb/^{238}U$); 490.7 ± 0.5 Ma ($^{207}Pb/^{235}U$). Suggested age of 491 ± 1 Ma. Maximum age only	*Peltura scarabaeoides scarabaeoides* below and *P. s. westergardi* above = Lower *Peltura scarabaeoides* Zone	Lower *P. scarabaeoides* Zone	1	Davidek *et al.* (1998)	Maximum age 491 ± 1	Pb–Pb, U–Pb

[a] T & M = Tucker and McKerrow (1995).
Items used by Tucker and McKerrow (1995) are indicated in column 2 by their item number.

[b] Biostratigraphic precision and reliability; 1, well constrained to a relatively short zone; 2, well constrained to long zone; 3, moderately well constrained to a biostratigraphic zone.
The biostratigraphic zonal assignment of dated samples is given (column 8) but the samples are located in the composite sequence (Figure 12.4) by the optimization procedure, based on the associated species (see text).

[c] $^{40}Ar/^{39}Ar$ dates have been re-calibrated with the MMhb-1 monitor standard of 523.1 Ma.

approach assumes constancy in sedimentation rate, so that stage duration can be estimated from its mean stratigraphic thickness (Churkin *et al.*, 1977). This assumption requires judicious selection of the reference section. Graphic correlation has been employed to utilize the stratigraphic thickness of one reference section in which the fossil ranges have been adjusted according to collections made elsewhere (Sweet, 1984, 1988, 1995; Kleffner, 1989; Fordham, 1992). Cooper (1999b) compared graphic correlations from different regions as a test for steadiness of depositional rate. Generally, some unspecified combination of these methods and assumptions, plus a measure of intuition, are employed (Tucker and McKerrow, 1995).

As discussed in Chapters 1, 3, and 8, a number of quantitative, or semi-quantitative, methods of biostratigraphic correlation can be adapted to derive *relative* time scales – that is, scales that estimate the relative proportions of stages without recourse to radiometric dates. These are, of course, uncalibrated in millions of years. The *con*strained *op*timization (CONOP) correlation technique, developed by Kemple *et al.* (1995) and Sadler (1999) uses evolutionary programming techniques to find a composite range chart with optimal fit to all the field observations. Thus, it resembles a multidimensional graphic correlation in the sense that it considers all the local stratigraphic sections. It differs, however, in treating all sections simultaneously. A closer analogy exists between CONOP and algorithms that search for the most parsimonious cladogram. For time scale work, CONOP requires a well-studied, pandemic fossil group with good biostratigraphic utility. Dated tuff beds that can be associated with these fossils may be included in the optimization process. For building Ordovician (and Silurian) time scales, the excellent state of graptolite biostratigraphy and the review of radiometric data by Tucker and McKerrow (1995) render the CONOP method suitable. The principles and methodology of applying constrained optimization to time scale development are summarized by Sadler and Cooper (2004, see also Chapter 3), who use the graptolite successions of the Ordovician and Silurian to develop and demonstrate the method.

Over 230 measured stratigraphic sections in graptolitic shales from around the world and containing 1400 species have been compiled in a database that spans the latest Cambrian–Early Devonian (Sadler and Cooper, 2004). The Ordovician part includes 119 sections, containing 669 taxa with ranges wholly or partly in the Ordovician (Fig. 12.4). These provide more than 1400 range-end events (first and last appearance events) and other marker horizons (tuff beds) to build a global composite sequence of events in the Ordovician, and for global correlation of sections. With the exception of the very basal portion, every biostratigraphic event level in the Ordovician composite is spanned by at least 10, and up to 30, measured sections (Fig. 12.4). Because planktonic graptolites do not range below the base of the Ordovician, eight trilobite and seven conodont species that are present in graptolite-bearing sections in the boundary interval and which range down into the Late Cambrian have been included in the database. These help compensate for the lack of graptolites.

In the global composite (Table 12.2), the order and spacing of events provides a proxy for a time scale. The order and spacing are determined separately, because the spacing requires simplifying assumptions that need not compromise determination of the optimal order. First, the optimal order of events is established by minimizing misfit in sequence between the composite and all of the individual sections. Misfit is determined primarily by the net number of event horizons through which the observed range ends must be extended in order to make all the observed range charts fit the same composite sequence. The composite, at this stage, is only an ordinal sequence of events, and is based purely on the sequences of events observed in the measured sections.

Next, the spaces between events in the optimal composite sequence are scaled by the following procedure. The observed fossil ranges in the individual sections are extended as necessary to match the composite sequence and missing taxa are added. The total thickness of each section is then rescaled according to the number of events that it spans in the composite

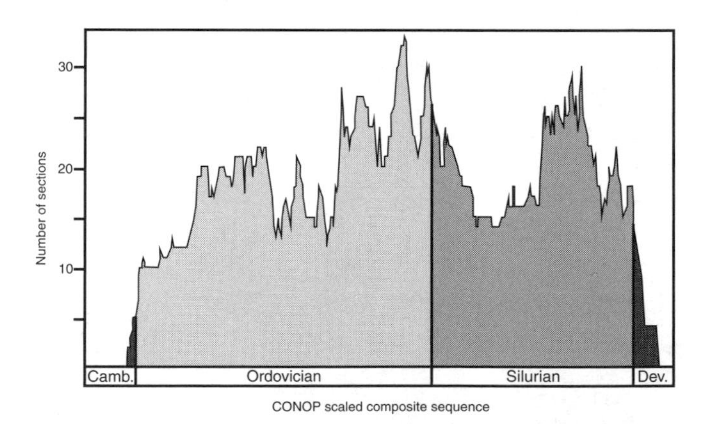

Figure 12.4 Stratigraphic section frequency through the Ordovician–Silurian composite sequence. The graph shows the number of sections for each level in the CONOP ordinal sequence used to build the Ordovician–Silurian time scale. Note that the ordinal sequence is not a time scale. (From Sadler and Cooper, 2004.)

Table 12.2 *Ordovician to early Devonian composite zonation[a] (from Sadler and Cooper, 2004)*

	Stage	Australasian Zone	Zone or zonal group	Confidence grading[b]	Proxy species in composite (FA unless otherwise specified)	Level in composite
DEVONIAN	Lochkovian		*Monograptus yukonensis*	4	*M. yukonensis*	3440.09
	Lochkovian		*Monograptus hercinicus hercinicus*	4	*M. h. hercinicus*	3379.30
	Base of Devonian		*Monograptus uniformis*	4	*M. uniformis*	3346.15
SILURIAN	Pridoli		*Monograptus perneri–transgrediens*	4	*M. perneri–transgrediens*	3317.54
	Pridoli		*Monograptus bouceki*	4	*M. bouceki*	3314.37
	Pridoli		*Monograptus lochkovensis*	4	*M. lochkovensis*	3284.20
	Pridoli		*Monogr. lochkovensis–branikensis*	4	*M. lochkov branikensis*	3283.06
	Pridoli		*Monograptus ultimus*	4	*M. ultimus*	3260.01
	Pridoli		*Monograptus parultimus*	4	*M. parultimus*	3255.50
	Ludfordian		*Monograptus formosus*	4	*Monograptus hamulosus*	3236.79
	Ludfordian		*Neocucullograptus kozlowskii*	0	uncertain	
	Ludfordian		*B. cornuatus–P. podoliensis*	1	*Monograptus helicoides*	3198.57
	Ludfordian		*Saetogr. lientwardensis–linearis*	1	*Saetogr. leintwardensis*	3171.35
	Gorstian		*Lobograptus scanicus*	2	*Colonograptus roemeri*	3133.97
	Gorstian		*Neodiversograptus nilssoni*	3	*Neodiversogr. nilssoni*	3116.40
	Homeran		*Colonograptus ludensis*	4	*Colonograptus ludensis*	3102.62
	Homeran		*Colonogr. praedeubeli–deubeli*	4	*Lobograptus sherrardae*	3080.43
	Homeran		*Pristiogr. parvus–Gothogr. nassa*	4	*Gothograptus nassa*	3071.26
	Homeran		*Cyrtograptus lundgreni*	5	*Cyrt. lundgreni*	3021.36
	Sheinwoodian		*Cyrtograptus perneri*	4	*Cyrt. perneri*	3007.78
	Sheinwoodian		*Cyrtograptus rigidus*	3	*Prist. dubius pseudolatus*	2988.69
	Sheinwoodian		*Monogr. belophorus–antennularis*	3	*M. antennularus*	2982.32
	Sheinwoodian		*Monograptus riccartonensis*	4	*M. riccartonensis*	2974.60
	Sheinwoodian		*Cyrtograptus murchisoni*	4	*Cyrt. murchisoni*	2964.37
	Sheinwoodian		*Cyrtograptus centrifugus*	4	*C. centrifugus*	2946.44
	Sheinwoodian		*Crytograptus insectus*	4	*Monograptus shottoni*	2942.17
	Telychian		*Cyrtograptus lapworthi*	2–3	*Torquigraptus pregracillis*	2912.85
	Telychian		*Oktavites spiralis*	3	*Oktavites falax*	2854.39
	Telychian		*Monogr. griestoniensis–crenulata*	3	*M. griestonensis*	2821.20
	Telychian		*Monograptus crispus*	4	*Monograptus galaensis*	2797.45
	Telychian		*Spirograptus turriculatus*	1	*Pseudopl. obesus obesus*	2754.00
	Telychian		*Spirograptus guerichi*	4	*Monograptus turriculatus*	2683.77
	Aeronian		*Stimulograptus sedgwickii*	2	*Rastrites longispinus*	2663.86
	Aeronian		*Lituigraptus convolutus*	4	*L. convolutus*	2643.07
	Aeronian		*Monograptus argenteus–leptotheca*	3	*Pristiograptus gregarius*	2628.21
	Aeronian		*Demirastr. triangulatus–pectinatus*	4	*Rastrites longispinus*	2588.83
	Rhuddinian		*Coronograptus cyphus*	5	*C. cyphus*	2537.47
	Rhuddinian		*Orthograptus vesiculosis*	4	*Dimorphogr. elongatus*	2507.25
	Rhuddinian		*Akidograptus acuminatus*	5	*A. acuminatus praecedens*	2480.21
	Base of Silurian		*Akidograptus ascensus*	5	*A. ascensus*	2437.74

(*cont.*)

Table 12.2 (*cont.*)

	Stage	Australasian Zone	Zone or zonal group	Confidence grading[b]	Proxy species in composite (FA unless otherwise specified)	Level in composite
ORDOVICIAN	Bolindian	Bo5	*Normalograptus persculptus*	5	*Normalo. persculptus*	2403.18
	Bolindian	Bo4	*Normalograptus extraodinarius*	3	*Diplograptus improvisus*	2376.14
	Bolindian	Bo3	*Paraorthograptus pacificus*	4	*Paraorthogr. pacificus*	2313.68
	Bolindian	Bo2	pre-*pacificus*	3	*Sinoretiograptus mirabilis*	2272.82
	Bolindian	Bo1	*Climacograptus? uncinatus*	3	*Dicellogr. complanatus*	2227.40
	Eastonian	Ea4	*Dicellograptus gravis*	1–2	*Leptograptus capillaris*	2160.00
	Eastonian	Ea3	*Dicranograptus kirki*	3	*Neurogr. margarit.* LA	2145.15
	Eastonian	Ea2	*Diplacanthogr. spiniferus*	2	*Pseudo scharenbergi* LA	2083.82
	Eastonian	Ea1	*Diplacanthpgr. lanceolatus*	3	*Orthograptus pageanus*	2025.69
	Gisbornian	Gi2	*Orthograptus calcaratus*	3	*Corynoides calicularis*	1967.60
	Gisbornian	Gi1	*Nemagraptus gracilis*	5	*Nemagraptus gracilis*	1889.43
	Darriwilian	Da4a&b	*Pterograptus elegans*	3	*Kalpinograptus ovatus*	1821.35
	Darriwilian	Da3	*Pseudoclimacogr. decoratus*	4	*Holmograptus spinosus*	1755.67
	Darriwilian	Da2	*Undulograptus intersitus*	3	*Holmograptus callotheca*	1714.87
	Darriwilian	Da1	*Undulograptus austrodentatus*	4	*Undulogr. formosus*	1661.35
	Yapeenian	Ya1–2	*Oncograptus upsilon*	5	*Pseudisogr. manubriatus*	1638.36
	Castlemainian	Ca3–4	*Isograptus victoriae maximus*	4	*I. v. maximus*	1604.62
	Castlemainian	Ca2	*Isograptus victoriae victoriae*	4	*I. v. victoriae*	1585.90
	Castlemainian	Ca1	*Isograptus victoriae lunatus*	3	*Tetra. reclinatus reclinatus*	1546.57
	Chewtonian	Ch1–2	*Didymograptellus protobifidus*	5	*Didymograptus bidens*	1490.17
	Bendigonian	Be1–4	*Pendeograptus fruticosus*	4	*Trichograptus* sp.	1405.14
	Lancefieldian	La3	*Tetragraptus approximatus*	5	*T. phyllograptoides*	1335.29
	Lancefieldian	La2b	*Araneograptus murrayi*	4	*A. murrayi*	1186.49
	Lancefieldian	La2a	*Adelograptus victoriae*	5	*Temno. magnificus*	1117.75
	Lancefieldian	La1b	*Psigraptus*	3	*P. jacksoni*	1105.25
	Lancefieldian	La1a	*Anisograptus*	3	*A.ruedmanni*	1068.95
	Lancefieldian	pre-La1	*Rhabd. flab. praeparabola–parabola*	3	*R. f. praeparabola*	1037.51
	Base of Ordovician		*Iapetognathus fluctivagus*	4	*Iapetognathus fluctivagus*	1032.68

[a] The columns are the zones (or zonal groupings) that were calibrated by the CONOP method, proxy species that were used for defining zone boundaries in the composite, a confidence rating on how well defined the zonal boundary is in the composite, and level in composite.

[b] Levels of confidence on zone boundary placements: 1, very low; 2, low; 3, moderate; 4, high; 5, very high.

sequence (assumes that net biologic change is a more reasonable guide to relative duration, *in the long term*, than raw stratigraphic thickness). Finally, the spacing of every pair of adjacent events in the composite is determined from the *average* of the *re-scaled* spacings of events in the sections (assumes that relative thickness is a reasonable guide to relative duration *in the short term*).

The scaling of the composite is therefore derived from all of the sections and it is the ratio of the thicknesses between events that is used, not the absolute thickness. The influence of aberrant sections, incomplete preservation, and non-uniform depositional rates is thus minimized. Where evolutionary change is rapid, many range-end events fall at the same horizon in measured sections; these "zero spacings" are included in the

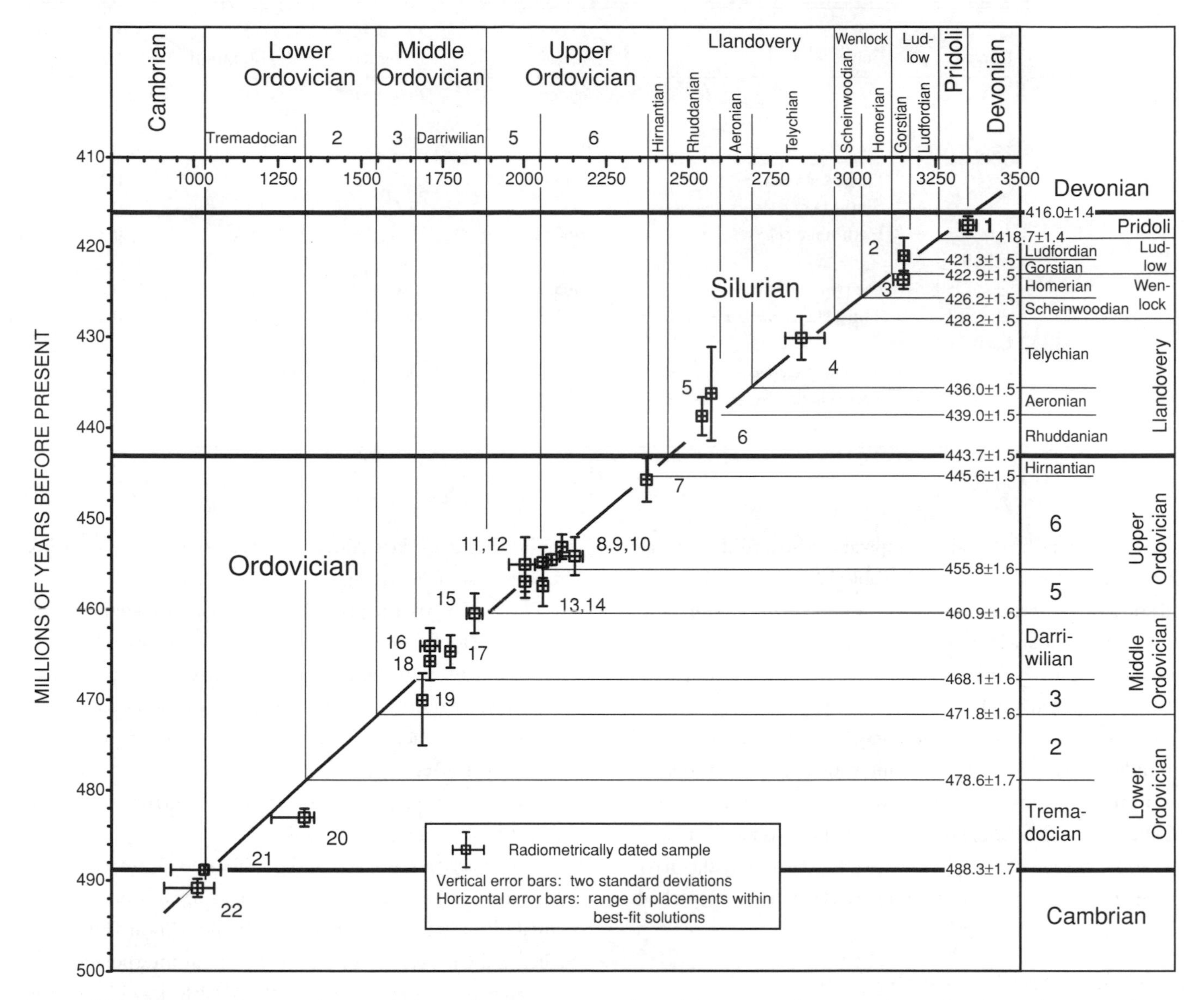

Figure 12.5 Graph showing the principle of constructing the Ordovician and Silurian time scale, using a plot of radiometric age dates for Late Cambrian through Silurian against the CONOP ordinal sequence of global graptolite events, which is compiled in Table 12.2. The numbers of the dates correspond to those listed in Table 12.1. Vertical error bars represent 2 standard deviations; horizontal error bars depict range of placement within best-fit solution. Although the best-fit line shown in this graph is a regression line with very slight curvature ($R^2 = 0.99$), the final calculations that yielded the ages on Ordovician and Silurian stage boundaries, error estimates, and stage duration estimates employ cubic-spline fitting. The spline fit closely follows the best-fit line shown. Ripley's MLFR algorithm for maximum likelihood fitting of a functional relationship was employed for 2-sigma error bars on the age of stage boundaries and duration of stages (see text). In order to obtain age constraints for the Cambrian–Ordovician, Ordovician–Silurian, and Silurian–Devonian boundaries, critical age dates below and above these boundaries were also included in the calculations.

averaging process and prevent high diversity from being misinterpreted in terms of long time intervals. The procedure is more vulnerable to intervals of extraordinarily low diversity. The graptolite clade survived long after the close of the Ordovician Period, but had very low diversity at the start of the Ordovician. We should anticipate that the composite spacings are least reliable near the base of the period.

Table 12.3 *Ages and durations of Ordovician stages*

Period	Epoch	Stage	Age of base (Ma)	Est. ± myr (2-sigma)	Duration	Est. ± myr (2-sigma)
Silurian			**443.7**	1.5		
Ordovician						
	Late					
		Hirnantian Stage	445.6	1.5	1.9	0.1
		Sixth stage *(not yet named)*	455.8	1.6	10.2	0.3
		Fifth stage *(not yet named)*	460.9	1.6	5.1	0.2
	Middle					
		Darriwilian Stage	468.1	1.6	7.3	0.2
		Third stage *(not yet named)*	471.8	1.6	3.7	0.1
	Early					
		Second stage *(not yet named)*	478.6	1.7	6.8	0.1
		Tremadocian Stage, base of Ordovician	**488.3**	1.7	9.7	0.2

Events in the final, scaled composite are spaced along a scale of arbitrary "composite units" (Table 12.2). The scaled composite is itself a proxy time scale. Graptolite zonal boundaries and stage boundaries can be located in the scaled composite to produce a relative time scale for the Ordovician. Uncertainty estimates are provided by relaxed-fit intervals, which give the range of positions within the composite at which the dated event lies, in successive runs during which "best-fit" is relaxed by a small percentage (Sadler and Cooper, 2004). For the geologic time scale in this book, uncertainty estimates on stage duration and age of stage boundaries are derived by the maximum likelihood fitting of a function relation (MLFR) method.

12.4.4 Age of stage boundaries

Because radiometrically dated beds were included in the optimizing process, the composite relative time scale already contains the necessary ingredients for calibration to time. The dated levels allow correlation of the composite with a linear time scale. An advanced bivariate fit provides a natural test of the spacing procedure and allows interpolation of the ages of zone and stage boundaries according to their position on the best-fit line.

The best fit for the two-way plot of 22 radiometric ages along the abscissa, and the composite with zones and stages along the ordinate (Fig. 12.5) was calculated with a cubic-spline fitting method that combines stratigraphic uncertainty estimates with the 2-sigma error bars of the radiometric data. The smoothing factor of 1.452 for the cubic spline was calculated with cross-validation, and the 2-sigma error on the estimated ages for the Ordovician stages with Ripley's MLFR procedure (see Chapter 8).

The Ripley MLFR procedure to assist with spline fitting worked well for the Ordovician–Silurian data set in Table 12.1, and the 2-sigma value for the base of the Ordovician is 1.7 myr; the error value decreases stratigraphically upward because it is assumed that for each period the relative 2-sigma error remains proportional to age rather than constant. It is a reasonable assumption that for any age determination, errors tend to be proportional to age.

A comparison was made of the calibrated Ordovician scale with a best-fitting interpolation using a second-order polynomial ($R^2 = 0.9965$), which is mildly non-linear (Fig. 12.5; Sadler and Cooper, 2004). The polynomial interpolation results fall within the error limits of the MLFR fit, with the ages of bases as follows: Silurian, 443.7 Ma; Hirnantian, 445.6 Ma; Stage 6, 455.8 Ma; Stage 5, 460.5 Ma; Darriwilian,468.1 Ma; Stage 3, 472 Ma; Stage 2, 479.3 Ma; and Tremadocian, 490 Ma.

The calculated durations and ages with estimates of uncertainty (2-sigma) of the Ordovician stages are given in Table 12.3. Note that rounding procedures can result in discrepancy between stage durations and ages of boundaries.

The base of the Ordovician in the calibrated CONOP scale, using cubic-spline fitting and equal weight on all dates, is 488.3 ± 1.7 Ma. This age overlaps with that of the high-resolution zircon date from the Dolgellau Formation in Wales (maximum age is 489 ± 1.6 Ma, Landing *et al.* 2000). The dated bed is biostratigraphically well constrained, lying close to the bio-stratigraphic base of the Tremadocian in Wales. Our age for the base of the Ordovician also accords well with a second high-quality date (491 ± 1 Ma) on latest Cambrian (*Acerocare* Zone) in North Wales (Davidek *et al.*, 1998). This age is younger than that given by Tucker and McKerrow (1995, 495 Ma) and Harland *et al.* (1990, 510 Ma) and considerably older than that of Compston (2000a, ~477 Ma).

The duration of the Ordovician is 44.6 myr. The base of the Middle Ordovician is 471.8 Ma and the base of the Late Ordovician is 460.9 Ma. The Early Ordovician Series lasted for 16.5 myr, the Middle Ordovician for 11 myr, and the Late Ordovician for 17.1 myr.

13 • The Silurian Period

M. J. MELCHIN, R. A. COOPER, AND P. M. SADLER

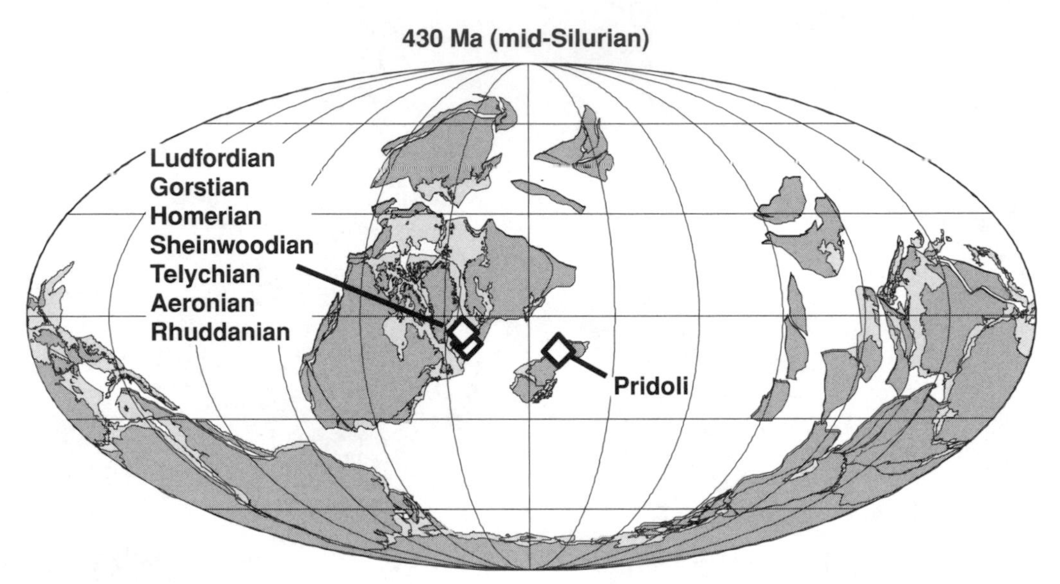

Geographic distribution of Silurian GSSPs. All eight Silurian GSSPs are defined. Only the Pridoli GSSP in the Czech Republic lies outside the United Kingdom. Of the remaining seven, one is in Scotland (the Rhuddanian GSSP) and the others are in a small area in England or adjacent parts of central Wales. No attempt has been made to distinguish between these six GSSPs on the map.

Rapid recovery in biodiversity after end-Ordovician extinction event, sustained greenhouse climate punctuated by short ice-house events, strong eustatic sea-level fluctuations and oceanic turnover, associated with extinction of moderate scale, colonization of land, general convergence of continental plates and low levels of faunal provincialism, closure of Iapetus Ocean, narrowing of Rheic Ocean, migration of South Pole over South American and southern African Gondwana.

13.1 HISTORY AND SUBDIVISIONS

The Silurian System was erected by Murchison (1839) and named after the Silures, a Welsh borderland tribe. As originally conceived, the Silurian embraced rocks that were claimed as Cambrian by Sedgwick, leading to a protracted debate. The disputed rocks were separated out as the new Ordovician System by Lapworth in 1879 but the debate lingered and Lower Silurian was used in Britain in parallel with Ordovician for many years, while Gotlandian was used in parallel with Upper Silurian in some regions. Eventually, the name Silurian was officially adopted in its restricted (Upper Silurian) sense by the IGC in Copenhagen in 1960 (Sorgenfrei, 1964). The rather complex nomenclatural history of definition and subdivision of the Silurian Period has been reviewed by Whittard (1961), Cocks *et al.* (1971), and Holland (1989).

As with the Ordovician, black shales are widely developed in Silurian sedimentary successions around the world and graptolites have proved to be valuable fossils for correlation. However, unlike the Ordovician, it was a time of relatively low faunal provincialism. Over 30 successive graptolite zones are recognized widely around the world providing a subdivision and correlation framework of extraordinary precision. Based on the present scale and generalized graptolite zonation there are 36 globally recognizable zones within a span of slightly less that 24 myr. Although variable in duration, the zones represent an average of 0.65 myr, significantly less for post-Llandovery time. Conodonts have proved to be of considerable global biostratigraphic value in shallow-water carbonate facies. A rich and diverse fauna are present in the shelly facies where trilobites and brachiopods are used extensively for

A Geologic Time Scale 2004, eds. Felix M. Gradstein, James G. Ogg, and Alan G. Smith. Published by Cambridge University Press.

zonation, and coral–stromatoporoid communities enable local biostratigraphic subdivision and correlation. Chitinozoan and acritarch zonations have been developed for several regions and are proving to be useful for correlation in many circumstances. Vertebrate microfossil and radiolarian zonations are also being developed for the Silurian.

The Silurian comprises four series, the Llandovery, Wenlock, Ludlow, and Pridoli in upward sequence, informally grouped into lower and upper Silurian subsystems. All series and their constituent stages (Fig. 13.1) have designated lower boundary Global Stratotype Sections and Points, or GSSPs (Bassett, 1985; Cocks, 1985).

13.1.1 Llandovery Series

Named from the type area in Dyfed – formerly Pembroke – in Southern Wales, the Llandovery Series comprises three stages, approved by the ICS (Bassett, 1985), the Rhuddanian, Aeronian, and Telychian stages. The Aeronian Stage is approximately, but not exactly, equivalent to the previously employed Idwian and Fronian stages.

RHUDDANIAN

Although the stage is named for the Cefn-Rhuddan Farm in the Llandovery area, its lower boundary GSSP is at Dob's Linn in the Southern Uplands of Scotland, defined at a point 1.6 m above the base of the Birkhill Shale in the Linn Brach Trench section. This point has been regarded as coincident with the local base of the *Parakidograptus acuminatus* Zone (Cocks, 1985). Re-sampling and systematic revisions have shown, however, that *Parakidograptus acuminatus* has its first occurrence datum 1.6 m above this level and that the succession can be readily subdivided, both at this section and globally, into a lower *Akidograptus ascensus* Zone and an upper *Parakidograptus acuminatus* Zone (Melchin and Williams, 2000). Melchin and Williams (2000) therefore proposed that the base-*Akidograptus ascensus* Zone, marked by the first occurrences of *A. ascensus* and *Parakidograptus praematurus* (the latter was identified by Williams (1983) as *P. acuminatus sensu lato*), be regarded as the biostratigraphic horizon that marks the base of the Silurian. Thus re-defined, the Rhuddanian Stage spans the *Akidograptus ascensus* to *Coronograptus cyphus* Zones.

AERONIAN

The Aeronian Stage is named for the Cemcoed-Aeron Farm in the Llandovery area. The GSSP is located in the Trefawr Formation, in the Trefawr track section 500 m north of the Cemcoed-Aeron Farm, between Locality 71 and 72 of Cocks *et al.* (1984). The GSSP is just below the level of occurrence of *Monograptus austerus sequens,* which indicates the *Monograptus triangulatus* Zone (Bassett, 1985; Cocks, 1989). However, *M. austerus sequens* has previously been reported from only one other locality, also in Wales (Sudbury, 1958; Hutt, 1974), where its level of first occurrence is within but significantly higher than the base of the *M. triangulatus* Zone. In addition, at the stratotype section, the *M. triangulatus* Zone is represented by only a single fossil locality. Thus, although the GSSP can be shown to occur between levels representing the *C. cyphus* and *M. triangulatus* Zones, it cannot be shown to correlate precisely with the boundary between those zones.

The Aeronian Stage is generally regarded as extending through the *Stimulograptus sedgwickii* Zone, although in parts of Wales this zone can be subdivided into a lower *Stimulograptus sedgwickii* Zone and an upper *Stimulograptus halli* Zone (Loydell, 1991). In this case, the latter is regarded as the uppermost graptolite zone of the Aeronian Stage.

TELYCHIAN

The Telychian Stage is named for the Pen-lan-Telych Farm. The GSSP is located in an abandoned quarry that forms part of the Cefn-Cerig Road section, at locality 162 of Cocks *et al.* (1984) and Cocks (1989), approximately 31 m below the top of the Wormwood Formation. Biostratigraphically, it is marked at a level above the highest occurrence of the brachiopod *Eocoelia intermedia* and below the first appearance of *Eocoelia curtisi* (Bassett, 1985). This was regarded as corresponding with the base of the *Spirograptus turriculatus* Zone and was considered to be supported by the occurrence of *Paradiversograptus runcinatus* in the beds above the stratotype level, although not at the stratotype section.

Loydell *et al.* (1993) revised the species of the genus *Spirograptus* and found that the stratigraphically lower specimens previously assigned to *S. turriculatus* and *S. turriculatus minor* actually belonged to a distinct, new species, *S. guerichi*. Accordingly, those authors found that the strata that had previously been assigned to the lower part of the *S. turriculatus* Zone (the *minor* Zone of some authors) could be regarded as belonging to a globally correlative *S. guerichi* Zone. This zone is now regarded as the lowest graptolite zone of the Telychian. However, it should be noted that *Paradiversograptus runcinatus*, the only identifiable graptolite found in the lowest Telchian beds in the vicinity of the stratotype, is known to have its first occurrence in upper Aeronian strata elsewhere in Wales (Loydell, 1991), although it reaches its acme in the lower *S. guerichi* Zone. In addition, Doyle *et al.* (1991) showed that

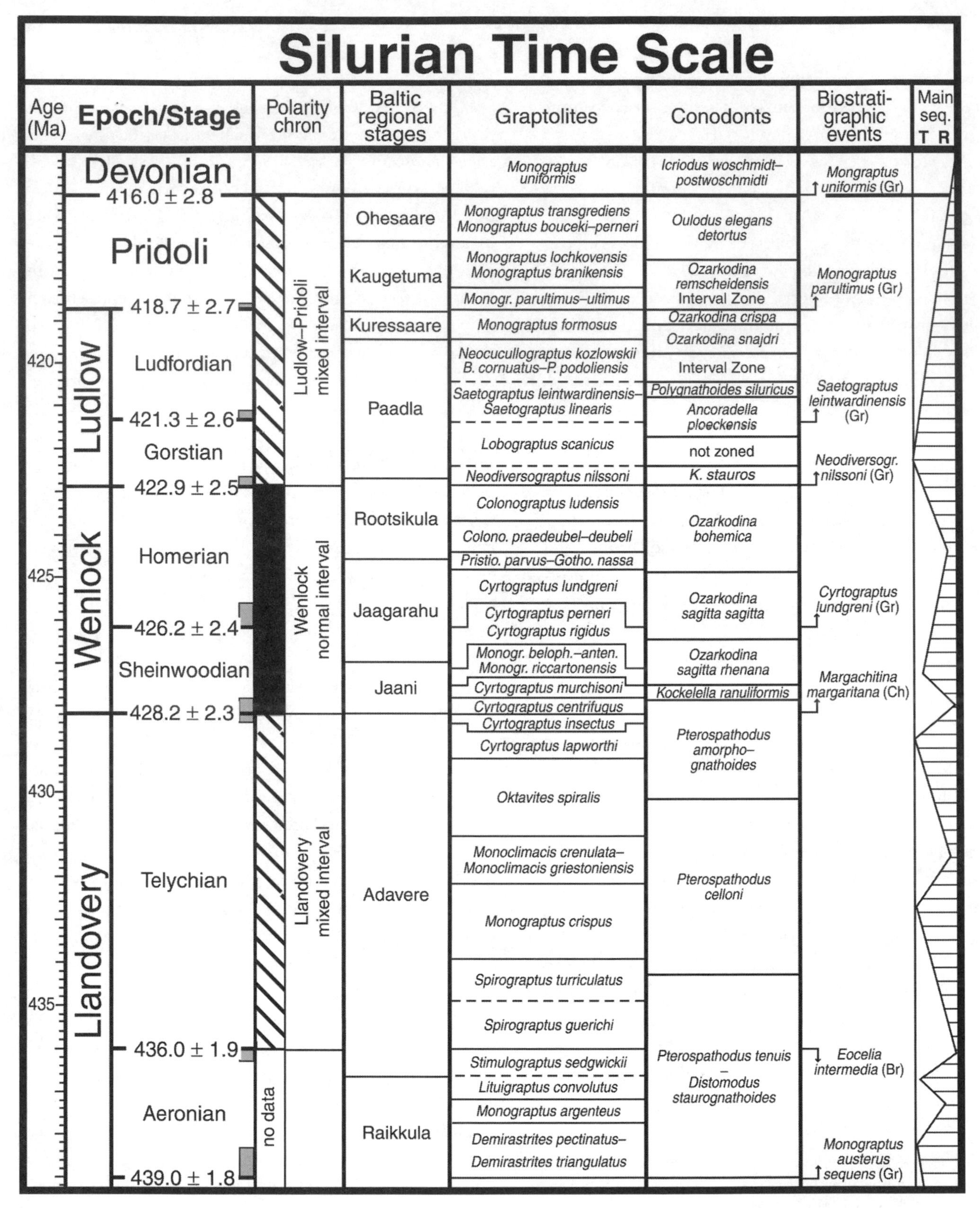

Figure 13.1 Silurian time scale, geomagnetic polarity scheme, Baltic regional stages, graptolite and conodont zonal schemes, biostratigraphic events used to mark stage boundaries at the GSSPs, and main transgressive–regressive sequences. International stages and epochs from Bassett (1985); Baltic regional stages from Bassett *et al.* (1989); generalized graptolite zonation based on Koren' *et al.* (1996), modified herein; generalized conodont zonation based on Subcommission on Silurian Stratigraphy (1995); main sequences generalized from Ross and Ross (1996), Johnson *et al.* (1998), and Loydell (1998). Shaded boxes at stage boundaries indicate interval of uncertainty in correlation between stratotype points and the graptolite zonation (see text for explanation). Gr, graptolite; Br, brachiopod; Ch, Chitinozoa. (Reproduced in colour plate section.)

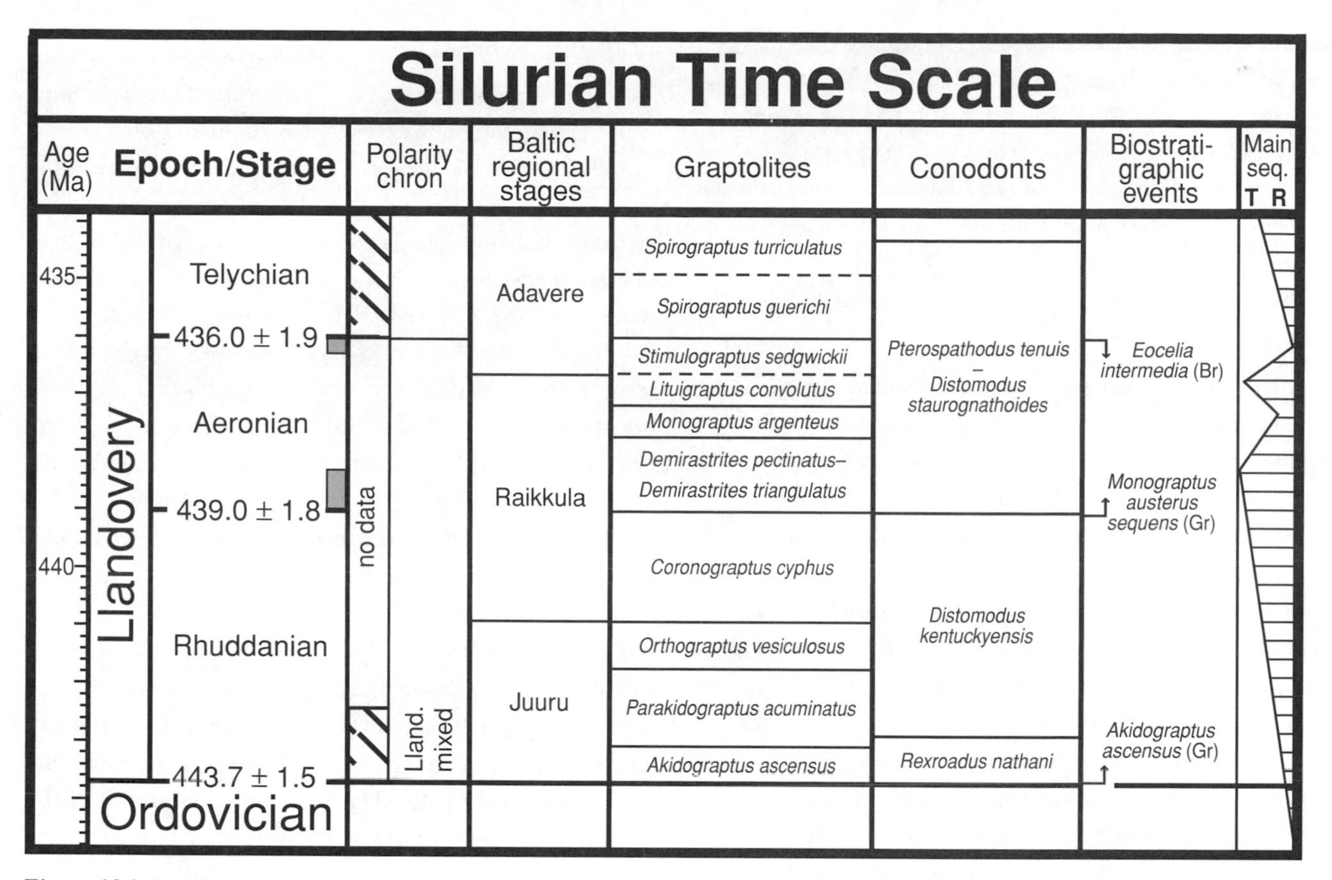

Figure 13.1 (*cont.*)

elsewhere in Britain, the last occurrence of *Eocoelia intermedia* occurs within the upper part of the *Stimulograptus sedgwickii* Zone rather than at the base of the *S. guerichi* Zone. Thus, there remains uncertainly regarding the precise correlation of the stratotype point with the graptolite zonation, but it appears to occur within the upper part of the *S. sedgwickii* Zone.

Standard graptolite zonations have previously shown the Telychian extending upward through the *Monoclimacis crenulata* Zone, overlain by the *Cyrtograptus centrifugus* Zone. However, later work, especially in Bohemia (Štorch, 1994) and Wales (Loydell and Cave, 1996) demonstrated that the "standard" British zonation was incomplete and that between the *M. crenulata* and *C. centrifugus* zones there occur strata readily assignable to the *Oktavites spiralis*, *Stomatograptus grandis–Cyrtograptus lapworthi*, and *Cyrtograptus insectus* Zones. Therefore, the Telychian is now regarded as extending from the *S. guerichi* Zone to the *C. insectus* Zone.

13.1.2 Wenlock Series

The Wenlock Series is named for the type area, Wenlock Edge, in the Welsh borderlands of England. It has been divided into two stages, the lower Sheinwoodian Stage and the upper Homerian Stage.

SHEINWOODIAN

The type locality for the Sheinwoodian Stage occurs in Hughley Brook, 200 m southeast of Leasowes Farm and 500 m northeast of Hughley Church. The GSSP is the base of the Buildwas Formation at this locality, as described by Bassett *et al.* (1975) and Bassett (1989). The stratotype point occurs within the *Pterospathodus amorphognathoides* conodont zone, between the base of acritarch zone 5 and the last occurrence of *P. amorphognathoides* (Mabillard and Aldridge, 1985). This level is considered to be approximately correlative with the base of the *Cyrtograptus centrifugus* graptolite zone although no graptolites are known to occur in the boundary interval at the stratotype section.

The occurrence of *Monoclimacis* aff. *vomerina* and *Pristiograptus watneyae* higher in the Buildwas Formation, together with species indicative of the *M. greistoniensis* and *M. crenulata* zones in the underlying Purple Shales Formation, were regarded as indicating that the stratotype point was near the *centrifugus–crenulata* zonal boundary. However, as noted above, it has since been demonstrated that three graptolite zones can be identified between these zones. In addition, some chitinozoan and conodont data indicate that the base of the Buildwas Formation may correlate with a level at or near the base of the *Cyrtograptus insectus* Zone (Jeppsson, 1997; Mullins, 2000), although Loydell *et al.* (2003) and Mullins and Aldridge (in press) provided evidence that the GSSP is more closely correlative with the base of the *Cyrtograptus murchisoni* Zone. At the present time, we must regard the correlation of the basal-Wenlock stratotype point with the graptolite zonation as somewhat uncertain.

HOMERIAN

The stratotype locality for the base of the Homerian Stage is in the north bank of a small stream that flows into a tributary of Sheinton Brook in Whitwell Coppice, which is 500 m north of the hamlet of Homer. The GSSP is within the Apedale Member of the Coalbrookdale Formation, at the point of first appearance of a graptolite fauna containing *Cyrtograptus lundgreni*. Underlying strata contain graptolites of the *C. ellesae* Zone (Bassett *et al.*, 1975; Bassett, 1989). Although the fauna of the *C. ellesae* and *C. lundgreni* Zones seem to be stratigraphically and taxonomically distinct in this and some other regions, recent work in Wales has shown a succession where the first occurrence of *C. lundgreni* is below that of *C. ellesae* (Zalasiewicz *et al.*, 1998). This suggests the possibility that the ranges of the zonal index taxa may be incomplete at the stratotype section and that the stratotype point for the Homerian may be within the *C. lundgreni* Zone rather than at its base.

13.1.3 Ludlow Series

The Ludlow Series is named for the type area near the town of Ludlow, in Shropshire, UK. It has been divided into two stages, the lower Gorstian Stage and the upper Ludfordian Stage.

GORSTIAN

The stratotype locality for the base of the Gorstian Stage is the disused Pitch Coppice Quarry, 4.5 km westsouthwest of the town of Ludlow. The GSSP is the base of the Lower Elton Formation where it overlies the Much Wenlock Limestone Formation (Holland *et al.*, 1963; Lawson and White, 1989). Graptolites questionably assigned to *Neodiversograptus nilssoni* and *Saetograptus varians* collected immediately above the base of the Lower Elton Formation indicate that this unit occurs within the *N. nilssoni* Zone. However, the absence of graptolites from other parts of the Homerian–Gorstian interval in the type area make it impossible to correlate the stratotype point precisely with the base of that zone. Lawson and White (1989) note that neither the shelly fossils nor the conodonts are useful in providing a more refined biostratigraphic definition of the stratotype point.

LUDFORDIAN

The locality for the GSSP of the Ludfordian Stage is at Sunnyhill Quarry, approximately 2.5 km southwest of the town of Ludlow. The level coincides with the contact between the Upper Bringewood Formation and the Lower Leintwardine Formation (Lawson and White, 1989). The graptolite *Saetograptus leintwardinensis leintwardinensis* occurs in the basal beds of the Lower Leintwardine Formation, although apparently not at the stratotype locality, and it becomes common higher in the formation. The underlying Upper Bringewood Formation is devoid of identifiable graptolites, although the Lower Bringewood Formation contains graptolites indicative of the *tumiscens–incipiens* Zones. Thus, the stratotype point is considered to approximate the base of the *leintwardinensis* Zone (Lawson and White, 1989), although it may occur within the lower part of that zone.

The stratotype level is also marked by the disappearance of a number of distinctive brachiopod taxa as well as changes in relative abundances among others. No distinctive conodont taxa appear at the stratotype point. However, there are significant changes in the palynological assemblages at or near the formational contact (Lawson and White, 1989).

13.1.4 Pridoli Series

The Pridoli Series is named for the Pridoli area, near Prague, Bohemia, Czech Republic. This series has not been subdivided into stages. The GSSP is in bed 96 of the Pozary section in the Daleje Valley, near Reporyjie, Prague (Kříz, 1989), approximately 2 m above the base of the Pozary Formation, marked by the first appearance of *Monograptus parultimus*. Graptolites are absent from the immediately underlying strata, however, so it is possible that the stratotype point lies within the lower part of that zone. A number of other fossil groups are common in the type area of the Pridoli besides the graptolites, but only chitinozoans show potential for more detailed biostratigraphic correlation in this region. The base of the *Fungochitina kosovensis* chitinozoan zone occurs approximately 20 cm above the stratotype point.

13.1.5 Other important stage classifications

Although a number of regions of the world have had a regional series and stage classification, the majority of these have fallen out of usage since the global standard series and stage stratotypes were defined. The generally high degree of faunal cosmopolitanism has greatly facilitated the global usage of the standard time scale. However, there remain significant intervals in which correlation between the graptolite bio-stratigraphic scale and that of conodonts and carbonate shelf facies are imprecise. As a result, in the East Baltic region, which has a long history of detailed faunal and stratigraphic study in mainly carbonate strata, workers continue to refer to a scale of regional stages (East Baltic regional stages, Fig. 13.1) based on faunal and facies changes in that area (Bassett *et al.*, 1989).

13.2 SILURIAN STRATIGRAPHY

Much of the sedimentary record of the Silurian in basinal and continental margin settings is represented by graptolite-bearing mudrocks. In the paleo-tropical regions, epicontinental settings are dominated by carbonate successions, commonly with well-developed reefs. In later Silurian time, large evaporite basins developed in some epeiric basins. Glacial deposits, mainly of Llandovery age, have been recognized in Brazil (Caputo, 1998).

The Silurian Period was generally a time of relative convergence of continental land masses and narrowing and closing of ocean basins (Cocks, 2000). One of the results of this is the generally low degree of provincialism seen in marine fauna. However, another result of this convergence is tectonic uplift in several orogenic belts and a number of areas that were dominated by marine sedimentation through much of Ordovician and Early Silurian time become sites of continental sedimentation or non-deposition by the end of the Silurian.

As in the Ordovician, the two fossil groups that have been most widely used in Silurian biostratigraphic correlation are the graptolites and conodonts. Reference to the GSSPs for the series and stages is mainly by relation to the graptolite zones. A number of shelly fossil groups, especially brachiopods, have proven to be useful for regional correlation. In addition, palynomorphs, especially Chitinozoa, are proving to be increasingly useful in Silurian biostratigraphy.

13.2.1 Biostratigraphy

GRAPTOLITE ZONES

Graptolites (Phylum Hemichordata) were a component of the Silurian macroplankton. As in the Ordovician they lived at various depths in the ocean waters (e.g. Underwood, 1993), were particularly abundant in upwelling zones along continental margins (Finney and Berry, 1997), and are found in a wide range of sedimentary facies. Most graptolite species dispersed rapidly, are geographically widespread, and are of relatively short stratigraphic duration (0.5–4 myr). These attributes combine to make them extremely valuable fossils for zonation and correlation of strata. Together with conodonts, they are the primary fossil group for global correlation of Silurian sequences and the biostratigraphic levels that are used to define the Silurian GSSPs are all based on graptolite zones.

The most established zonal scheme as the "standard" for Silurian graptolite biostratigraphy (Harland *et al.*, 1990) has traditionally been based on the British zonation (e.g. Rickards, 1976), except for the Pridoli zonation, which is based on the succession in Bohemia (e.g. Kříz *et al.*, 1986). However, recent efforts have been undertaken to establish a globally recognizable, standard zonation, based on widely recognizable episodes of faunal change rather than the succession of any particular region.

A first step toward this was the publication of the "generalized graptolite zonal sequence" (Subcommission on Silurian Stratigraphy, 1995; Koren' *et al.*, 1996), which was assembled for the purpose of a coordinated study of global paleogeography, but was also used for a study of patterns of global diversity and survivorship in Silurian graptolites (Melchin *et al.*, 1998). In the course of the latter study it was found that a number of the generalized zones recognized by Koren' *et al.* (1996) could be readily subdivided and still recognized in several different paleogeographic regions of the world. Therefore, the generalized graptolite zonation presented here represents a refinement of the one published by Koren' *et al.* (1996). The criterion used in this zonation was that the zonal base should be recognizable in at least three different paleogeographic regions of the world. In some instances, the "thicknesses" of the intervals of two or more successive zones were too thin to be depicted as distinct units on Fig. 13.1. In such cases the zones are grouped together as one unit.

Figure 13.1 shows the remarkable precision in correlation obtainable using Silurian graptolites. The 31 graptolite zonal divisions of the Silurian span 26 myr and thus average 830 000 years each in duration. In the Wenlock–Pridoli interval, this precision is even better – 680 000 years each. At the regional level, where more zones are recognized, the precision is better again.

CONODONT ZONES

Conodonts are tooth-like structures of primitive chordates that were free-swimming in various marine environments. Although they were most abundant in subtidal, carbonate shelf environments they have also been found in nearshore and deeper marine sediments (Aldridge and Briggs, 1989). Conodonts are composed of calcium phosphate and can be extracted from the carbonate rock by acid digestion. Many conodont species are found in a wide range of sedimentary environments and geographical regions, making them valuable fossils for long-range correlation. Patterns of provincialism in Silurian conodonts appear to be less-well defined than for the Ordovician, although there are still significant faunal distinctions between shallow- and deeper-water fauna as well as inter-regional differences.

High-resolution conodont zonations are available for some intervals of the Silurian for particular regions (especially the Baltic region, e.g. Jeppsson, 1998). A globally recognizable and generalized zonation for Silurian conodonts was proposed by the Subcommission on Silurian Stratigraphy (1995) based on globally recognizable biostratigraphic horizons (Fig. 13.1). High-resolution conodont zonations are available for some intervals of the Silurian for particular regions, especially in Europe (e.g. Jeppsson, 1997; Corradini and Serpagli, 1999), and some of those zonal levels are recognizable in other parts of the world.

CHITINOZOAN ZONES

Chitinozoa are organic-walled microfossils of unknown biological affinities, although many accept the hypothesis that they were the planktonic egg capsules of some metazoan. They occur in a variety of marine facies and many species were geographically widespread and relatively short lived. The Subcommission on Silurian Stratigraphy (1995) and Verniers *et al.* (1995) proposed a global biozonation for Silurian Chitinozoa, based on correlation of well-known successions in Laurentia, Avalonia, Baltica, and Gondwana (Fig. 13.2). Global biozonal levels are defined by well-established taxa whose first appearances are regarded as synchronous in two or more paleogeographically distinct regions. Many of these biozonal levels have been defined or recognized in direct reference to GSSPs.

OTHER ZONAL GROUPS

The same efforts of the Subcommission on Silurian Stratigraphy (1995) that produced the generalized zonations for graptolites, conodonts, and chitinozoans, also yielded Silurian zonations for spores and vertebrates (Fig. 13.2). The former, which has been slightly revised by Burgess and Richardson (1995), is particularly important in that it provides the possibility for biostratigraphic correlation in terrestrial strata and between the terrestrial and marine realm. The vertebrate zonation, based mainly on disarticulated remains (ichthyoliths), was also published by Märss *et al.* (1995).

BIOEVENTS

The Silurian Period has a well-documented record of bioevents. The best known is the biotic recovery in the Llandovery that follows the Late Ordovician mass extinction event (e.g. Sheehan, 2001). However, a number of extinction events of varying intensity have been described among graptolites, conodonts, and some other fossil groups (see Melchin *et al.*, 1998; Jeppsson, 1998; Fig. 13.2). Some of those extinction events in the Lower Silurian correlate well with the glacial episodes described by Caputo (1998). Many of them also correlate with excursions in the carbon and oxygen isotope chemostratigraphic record. Jeppsson (1998) related all of these bioevents to changes in climate and oceanic state.

13.2.2 Physical stratigraphy

MAGNETOSTRATIGRAPHY

Understanding of the magnetostratigraphic scale for the Silurian is still in a very preliminary state and is based on incomplete data from only a few localities. Based on the presently available data it appears that much of the Silurian is characterized by a mixed polarity, with a predominantly normal phase through much of the Wenlock (Fig. 13.1; Trench *et al.*, 1993).

CHEMOSTRATIGRAPHY

Chemostratigraphic analyses are now available at fairly high levels of stratigraphic resolution for all or most of the Silurian Period for the following isotopic systems: $\delta^{18}O$, $\delta^{13}C$, and Sr (Fig. 13.3).

Oxygen isotope stratigraphy A number of recently published oxygen isotope ($\delta^{18}O$) curves for the Silurian show very similar trends (Samtleben *et al.*, 1996; Wenzel and Joachimski, 1996; Bickert *et al.*, 1997b; Azmy *et al.*, 1998; Heath *et al.*, 1998). There is a slight general trend toward reduction in $\delta^{18}O$ values through Silurian time, possibly as a result of overall global warming (see "temperature trends" in Fig. 13.3, after Azmy *et al.*, 1998). Superimposed on this trend are a number of significant positive excursions of approximately 1–2.5% magnitude. Those events in the early and late Aeronian and earliest Wenlock, can be related to episodes of continental glaciation that have been identified in South America (Caputo, 1998). Those events in the late Homerian and Ludfordian also correspond to an episode of eustatic fall, but not with any known glaciation. Samtleben *et al.* (2000) suggested that these are related changes from times of predominantly humid to more arid climates as indicated by facies changes in carbonate platforms. As a result of the fact that most of the $\delta^{18}O$ data are from Baltica, especially from the latest Llandovery upward, the global nature and synchroneity of these excursions cannot be tested with confidence.

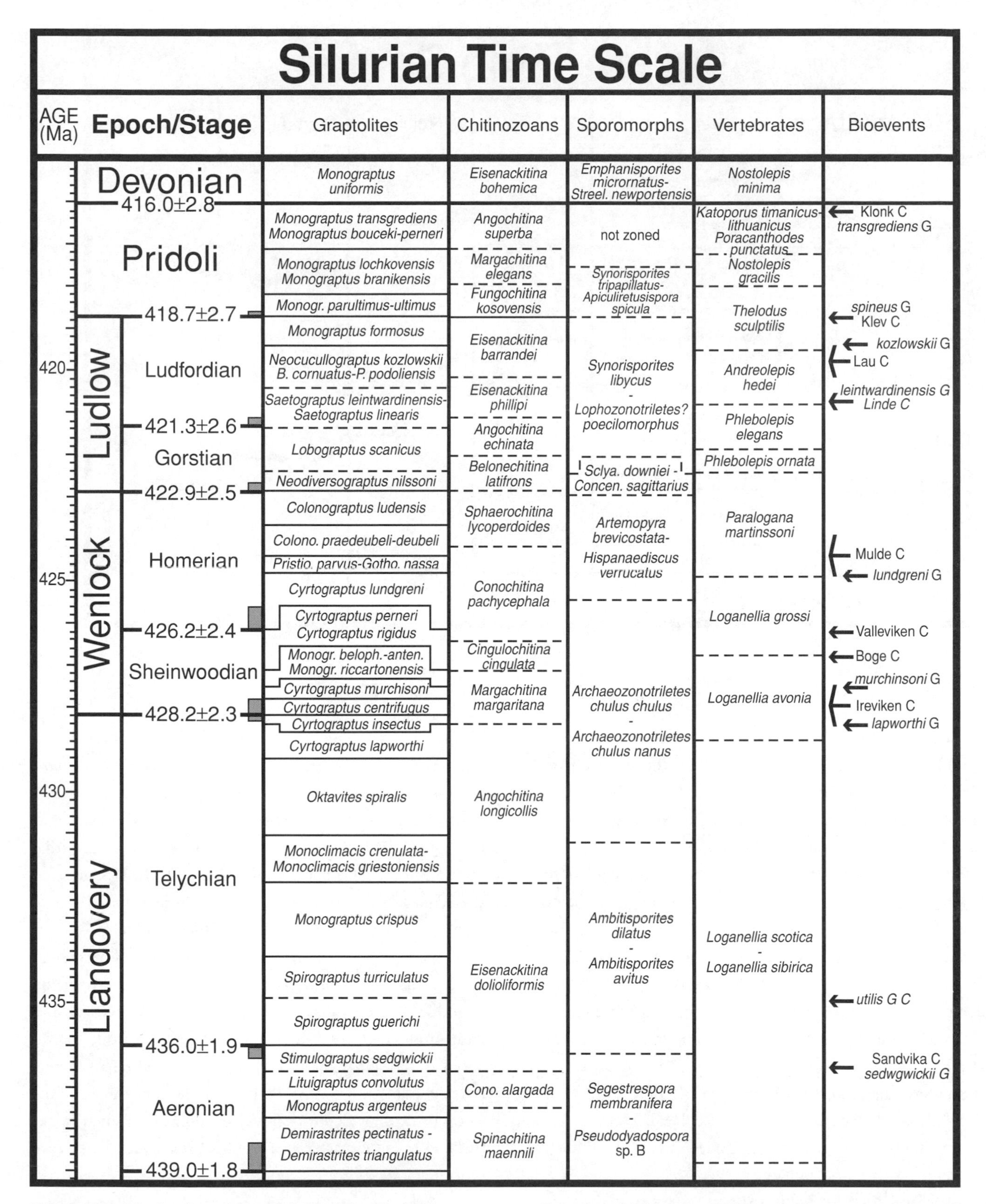

Figure 13.2 Silurian time scale, graptolite (from Fig. 13.1), chitinozoan, sporomorph, and vertebrate zonal schemes, and important bioevents recognized in graptolites (Gr) and conodonts (Ch). Chitinozoan zonation from Verniers *et al.* (1995); spore zonation from Subcommission on Silurian Stratigraphy (1995) and Burgess and Richardson (1995); vertebrate zonation from Märss *et al.* (1995); graptolite bioevents from Melchin *et al.* (1998); and conodont bioevents from Jeppsson (1998). A color version of this figure is in the plate section.

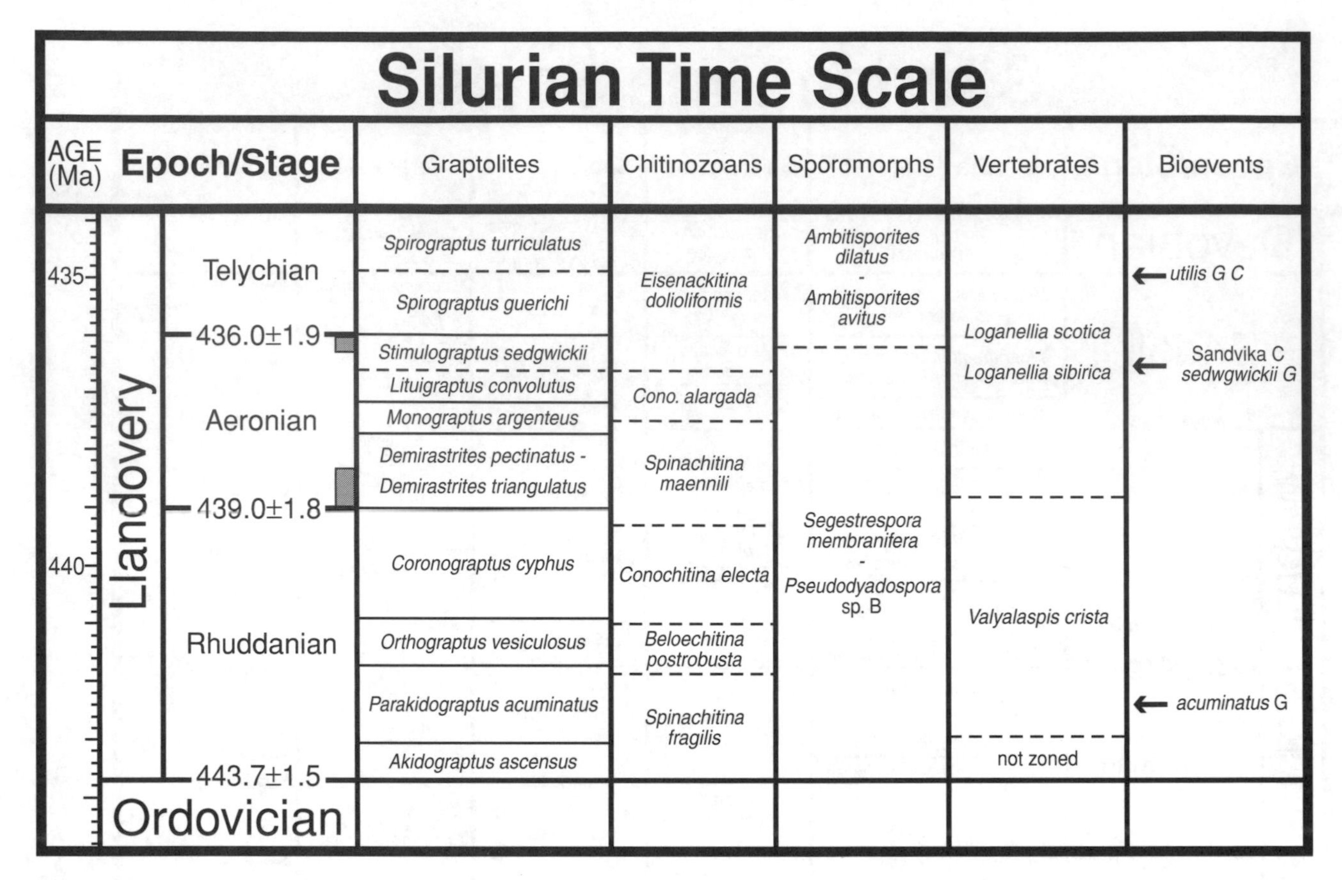

Figure 13.2 (*cont.*)

Carbon isotope stratigraphy The carbon isotope curve (δ^{13}C, Figure 13.3, after Azmy *et al.*, 1998), like the δ^{18}O curve, displays a general, slight tendency toward declining values through the Silurian. This trend is punctuated by several significant, positive excursions. Episodes of δ^{13}C values that are 2–6% above the background, which are prolonged enough to span more than one graptolite zone, are seen in the early Sheinwoodian, late Homerian, and early–mid Ludfordian (see also Kaljo *et al.*, 1998). High-resolution records (Kaljo *et al.*, 1998; Heath *et al.*, 1998; Melchin and Holmden, 2000) also show smaller-scale positive excursions of 1–3% in the early and late Aeronian. In addition, the general trends and main excursions in the carbon isotope record can be seen clearly in δ^{13}C analyses of unaltered brachiopods (Azmy *et al.*, 1998; Heath *et al.*, 1998), whole-rock carbonates (Kaljo *et al.*, 1998; Melchin and Holmden, 2000), and organic matter (Melchin and Holmden, 2000; Zimmerman *et al.*, 2000).

As with the δ^{18}O record, the positive excursions in the δ^{13}C curve appear to coincide with episodes of eustatic lowstand. In the Llandovery and early Wenlock, these are reported to be episodes of continental glaciation (Caputo, 1998). In the later Silurian there are no reported glacial deposits that coincide with the times of sea-level fall and the causes of these events are not known. However, the fact that the later events show similar isotopic trends to the earlier episodes and also show similar changes in faunal extinction rates and diversity patterns (e.g. Melchin *et al.*, 1998; Jeppsson, 1998) indicate that they may also be glaciogenic in origin, despite the lack of physical evidence in the form of known glacial deposits. Many of the studies attempting to relate glaciation and positive δ^{13}C excursions have focused on the high-magnitude event associated with the Late Ordovician extinction. Models to explain this and later events have focused either on changes in biological productivity and depositional rates of organic matter associated with changing climates and ocean circulation patterns (e.g. Armstrong, 1996; Jeppson, 1998), or changes in the balance of rates of deposition and erosion of carbonates, silicates and organic matter as a result of sea-level change (e.g. Kump *et al.*, 1999).

Strontium isotope stratigraphy The most recently published Silurian strontium isotope curves (^{87}Sr/^{86}Sr, Fig. 13.3; Ruppel *et al.*, 1998; Azmy *et al.*, 1999) both show the same secular increase in the ^{87}Sr/^{86}Sr ratio through the Silurian, from

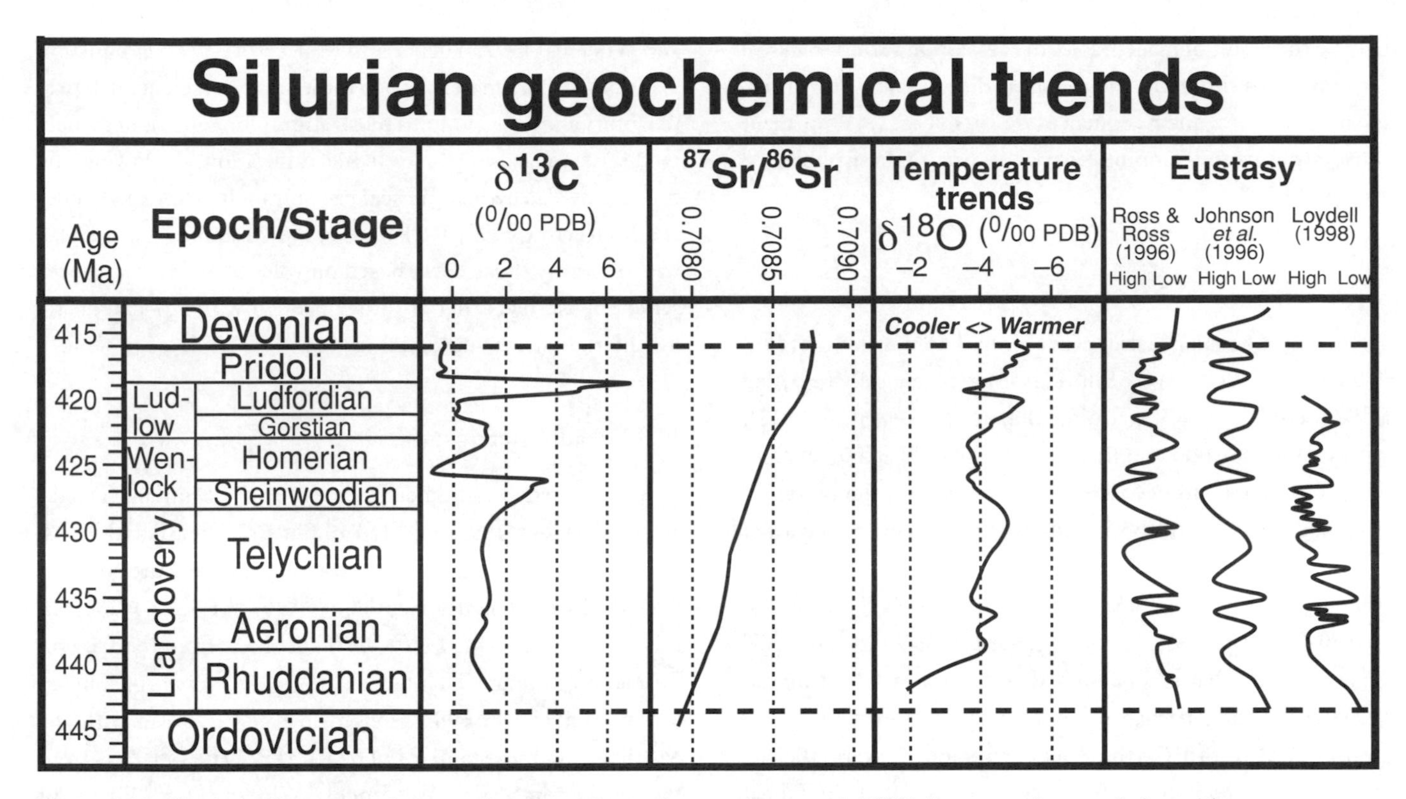

Figure 13.3 Silurian stable isotope chemostratigraphy and eustatic trends. Oxygen and carbon stable isotope curves are from Azmy *et al.* (1998) and the strontium curve is from Ruppel *et al.* (1998) and Azmy *et al.* (1999). Eustasy scales are relative, with sources as indicated.

0.707 95 at its base to 0.708 76 at its top. This increase through the interval has been attributed to an increase in riverine influx of radiogenic Sr due to climatic warming. In the lower Rhuddanian and Gorstian, the rate of increase in $^{87}Sr/^{86}Sr$ is slightly higher than elsewhere, but it is not clear if this higher rate is real or is an artifact of the time scale. The data of Ruppel *et al.* (1998) show a local decrease in $^{87}Sr/^{86}Sr$ in the *staurognathoides* Zone. The minimum probably can be ascribed to stratigraphic reasons, rather than a change in marine $^{87}Sr/^{86}Sr$, as these samples were from a separate locality to most of the others. In addition, there may be regional differences in diagenetic influences. Both Ruppel and Azmy note the potential for future use of the Sr isotope signal for stratigraphic correlation in the Silurian.

EUSTASY

There have been several different approaches to the problem of estimating eustatic changes for the Silurian Period. Johnson (1996) summarized a series of earlier papers in which the sea-level histories of individual regions have been reconstructed mainly based on the use of benthic assemblages and sedimentary structures. Sea-level curves for each region are then correlated and compared to identify global signals. Johnson *et al.* (1998; see Fig. 13.3) has added to this the study of submergence of paleotopographic features as measures of absolute sea-level change. Ross and Ross (1996; see Fig. 13.3) have also employed biofacies data, but have incorporated some lithofacies information, although their curve is based almost entirely on sections in Laurentia. In contrast, Loydell (1998; see Fig. 13.3) has defined episodes of sea-level rise and fall based on identification of transgressive and regressive systems in deeper-water, graptolitic shales, in which transgressive systems are recognized as being generally condensed and organic-rich with diverse graptolite fauna, whereas regressive systems are thicker and less organic-rich with more depauperate graptolite assemblages. This method has some advantages in that the graptolite biostratigraphy provides a more precise and globally correlatable temporal control. In addition, it is also based mainly on deeper-water successions, which are less susceptible to truncation by subaerial exposure and erosion. On the other hand, Loydell's method provides no means of deriving quantitative estimates of magnitudes of sea-level change. Due to the fact that the eustatic curves derived by Johnson *et al.* (1998) and Ross and Ross (1996) are based mainly on data from carbonate platform facies, whereas Loydell's curve is based on information from basinal shale successions, some of the apparent asynchroneity between them may be the result of inaccuracies in correlation rather than real differences in timing of the events as discerned by the different methods. The main sequences shown in Fig. 13.1 represent those eustatic trends common to two or three of these published curves.

CLIMATIC EVENTS

Following the major glacial event of the Late Ordovician (e.g. Brenchley *et al.*, 1994) the Silurian Period is generally regarded as a time of gradually warming climate (Frakes *et al.*, 1992), with smaller episodes of glacial advance in the Llandovery and early Wenlock (Caputo, 1998). As noted above, the existence of episodes of eustatic low, positive excursions in the record of $\delta^{18}O$ and $\delta^{13}C$ and significant faunal change in the later Silurian (e.g. Late Homerian), which in many ways resemble the glacially related events of the Late Ordovician and Early Silurian, indicate that it is possible that there were later Silurian glacial episodes that have yet to be recognized in the physical stratigraphic record. On the other hand, some workers (Bickert *et al.*, 1997b; Jeppsson, 1998) have suggested that the faunal and isotopic shifts of the Silurian can be related to changes in climate and oceanic state from more humid episodes of higher productivity and mainly estuarine circulation in epicontinental seas, to more arid states of lower productivity and mainly anti-estuarine circulation. These models are largely based on studies centered around the Baltic carbonate platform region.

VOLCANISM AND K-BENTONITE STRATIGRAPHY

Volcanic ash beds or K-bentonites have been widely reported throughout the Silurian, particularly from Europe and eastern North America (Bergström *et al.*, 1998a). Geochemical studies of these bentonites suggest that those distributed over Laurentia, Avalonia, and Baltica can be attributed to at least three different volcanic centres within the closing Iapetus and Rheic ocean basins (Bergström *et al.*, 1997; Huff *et al.*, 2000). However, some of these individual K-bentonite units have been shown by geochemical fingerprinting to be geographically very widespread and serve as excellent marker beds for high-resolution, regional correlations (e.g. Bergström *et al.*, 1998b; Batchelor and Evans, 2000).

13.3 SILURIAN TIME SCALE

As with the Ordovician, many isotopically dated rocks have been used for calibrating the Silurian time scale (Gale, 1985; Kunk *et al.*, 1985; McKerrow *et al.*, 1985; Odin, 1985; Snelling, 1985; Harland *et al.*, 1990; Tucker *et al.*, 1990; Compston and Williams, 1992; Tucker and McKerrow, 1995; Compston, 2000a,b). Unfortunately, few of these are of sufficient analytical reliability and precision and biostratigraphic constraint to meet modern standards for time scale calibration. Following the practice by Paleozoic time scale geochronologists (e.g. Tucker and McKerrow, 1995; Tucker *et al.*, 1998, Landing *et al.*, 2000), we use mainly those dates based on volcanogenic zircons preserved in ash beds that are inter-bedded with age-diagnostic fossiliferous strata and dated using the U–Pb decay system.

13.3.1 Radiometric dates

Of the published radiometric dates for the Silurian, Tucker and McKerrow (1995) rejected all but six. Their dated items, numbered 22, 23, 24, and 25 are all accepted here. Date No. 22 is from the Birkhill Shales, Scotland ($^{207}Pb/^{206}Pb$ age of 438.7 ± 2.1 Ma, Tucker *et al.*, 1990). Date No. 23 is from the Descon Formation, Esquibel Island, Canada ($^{40}Ar/^{39}Ar$ age of 436.2 ± 5 Ma, Kunk *et al.*, 1985; revised to 439.4 ± 5 Ma with the MMhb-1 monitor standard of 523.1 Ma). Date No. 24 is from the Buttington Shales, Welshpool, Wales ($^{207}Pb/^{206}Pb$ age of 430.1 ± 2.4 Ma, Tucker, 1991 *in* Tucker and McKerrow, 1995), and date No. 25 is from the Laidlaw Volcanics in Canberra, Australia (combined K–Ar and Rb–Sr age of 420.7 ± 2.2 Ma, Wyborn *et al.*, 1982). In addition, their dated item No. 21, from the Late Ordovician (Ashgill) Hartfell Shales, Scotland (U–Pb age of 445.7 ± 2.4 Ma, Tucker *et al.*, 1990) is used for constraining the age of the base of the Silurian.

On the other hand, the following dated items of Tucker and McKerrow are not used here: Sample No. 26, from the Upper Whitcliffe Formation, Ludlow, UK (420.2 ± 3.9 Ma, Tucker, 1991 *in* Tucker and McKerrow, 1995) because it lacks a precise link with graptolite biostratigraphy and has large analytical error; No. 27, from the Glencoe Volcanics, Scotland (421 ± 4 Ma, Thirwall, 1988), because of poor biostratigraphic constraint; and No. 28, from the Arbuthnot Group, Scotland (Rb–Sr biotite age of 411.9 ± 1.8 Ma and Sm–Nd garnet age of 411.9 ± 1.9 Ma, Thirwall, 1988), because of poor biostratigraphic constraint.

In addition to the above, we use the following two published dates. First, Kunk *et al.* (1985) report an $^{40}Ar/^{39}Ar$ weighted-average plateau age of 423.7 ± 1.7 Ma (revised to 426.8 ± 1.7 using MMhb-1 of 523.1 Ma) for a bentonite bed in the Middle Elton Formation, Hopedale, Shropshire, UK. The sample was collected by Ross *et al.* (1982, sample 76Sh25) who state that graptolites present indicate the *Neodiversograptus nilssoni* to *Lobograptus scanicus* Zones. The age is outside the dating error of the 420.7 ± 2.2 Ma date for the Laidlaw volcanics of the same biostratigraphic range, but extreme values of the dates can be accommodated in the top and bottom of the zonal interval. Hence, the dates contribute to the best-fit interpolation. Second, Tucker *et al.* (1998) report a weighted mean $^{207}Pb/^{206}Pb$ age of 417.6 ± 1.0 Ma based on nine analyses of zircons in a K-bentonite from the Kalkberg Formation (Helderberg Group), Cherry Valley, New York. The Kalkberg Formation, and most of the lower Helderberg Group, contains the zonal conodont *Icriodus woschmidti*, indicating a level that can be correlated with the *M. uniformis* graptolite zone. This date, near the base of the Early Devonian, helps constrain the age of the upper limit of the Silurian.

Thus we have a total of seven radiometric dates for calibration of the Silurian stages. Considering that only five of these are from rocks biostratigraphically dated as Silurian, it is clear that further radiometric dating is urgently needed to improve calibration. Details of the dated samples used here are given in Table 12.1. The four U–Pb dates are determined using the thermal ionization mass spectrometric (TIMS) method.

As discussed in Chapters 6, 12, and 14, zircon ages determined with the high-resolution–secondary ion mass spectrometry (HR-SIMS) method in the Canberra Geochronological Laboratory (Compston and Williams, 1992; Compston, 2000a,b), using standard SL13, are systematically different from conventional TIMS dates, even on the same samples. Until resolved, HR-SIMS dates are not employed in Silurian time scale construction.

13.3.2 Methods to estimate relative duration of zones and stages

Because there are so few high-quality radiometric dates available for calibration of the Silurian, the quantitative methods of interpolating stage boundaries used in other parts of the time scale by Harland *et al.* (1990) and Gradstein *et al.* (1994a) cannot be applied. As with the Ordovician, various proxy time scales have been devised by previous workers, in which the stage durations are estimated by some other method. Almost all of them rely on assumptions about either evolutionary or depositional rate.

The chronogram (minimized misfit) method of boundary calibration developed by Harland *et al.* (1990) is a means of estimating the age of stage boundaries from radiometric dates. Twenty-four radiometric dates in Silurian rocks were used, spanning a wide range of dating methods. Almost none of these dates are accepted here as of sufficient analytical reliability and biostratigraphic control. It is not surprising, therefore, that Harland *et al.* (1990) found that most chronograms for Silurian stage boundaries were "either poor or meaningless and [stage durations] must be estimated using chron interpolation." For chrons, they used graptolite zones. Thus, the number of graptolite zones present was a primary guide to the duration of the stages. This rule, however, was not consistently applied. The method assumes that graptolite zones are uniform in duration, which in turn assumes a constancy in evolutionary rate and in paleontological practice (Fig. 13.4*a*). Similar assumptions about steady rates of evolution in Silurian time scale construction are employed, in whole or in part, by Gale (1985), McKerrow *et al.* (1985), and Tucker and McKerrow (1995). In Fig. 13.4*c*, we have proportioned the stages according to the number of graptolite zones they contain (see Fig. 13.1) against the GTS2004 time scale, which demonstrates that the "proportional biozone" interpolation method does, indeed, approximate the relative durations of Silurian stages.

In an alternative approach, the thicknesses of fossiliferous stratigraphic sections through Silurian zones and stages are used to estimate their duration. Kleffner (1989, 1995) developed a composite section through the Silurian (Fig. 13.4*b*), using the graphic correlation method of Shaw (1964). Forty-two conodont- and graptolite-bearing, carbonate and carbonate–shale sections, mostly in North America but also from Europe, were used, and the Cellon section in Austria was selected as the standard reference section. In this exercise, the stratigraphic ranges of species in the Cellon section are extended to match the maximum ranges observed among the correlated sections, and species not present in the Cellon section are similarly composited into it based on their maximum ranges in other sections. The Cellon section thus becomes a composite, scaled, range chart of all compared sections. The conodont and graptolite zone boundaries recognized in the composite section are regarded as defining chronozones because they are based on global rather than local stratigraphic ranges. The composite section was then regarded by Kleffner as a proxy time scale. It is graduated in "composite time units," which are derived from the original stratigraphic thickness units of the Cellon section. Zone and stage boundaries are located within it, based on the composite stratigraphic ranges of the defining species. However, calibration of this time scale has proved problematic; the duration of the late Ludlow and Pridoli epochs appears to be disproportionately long when plotted against the radiometric dates, particularly that from the Kalkberg Formation (Fordham, 1992, 1998; Fig. 13.4*b*). Fordham (1998) illustrates how dependent the graphic correlation time scale is on choice of the initial standard reference section, the Cellon section in Austria, where the section for this time interval is unusually thick. Some method of minimizing, or normalizing for, thickness bias is therefore needed, and composite sections derived by graphic correlation, without a test for steadiness of depositional rate,

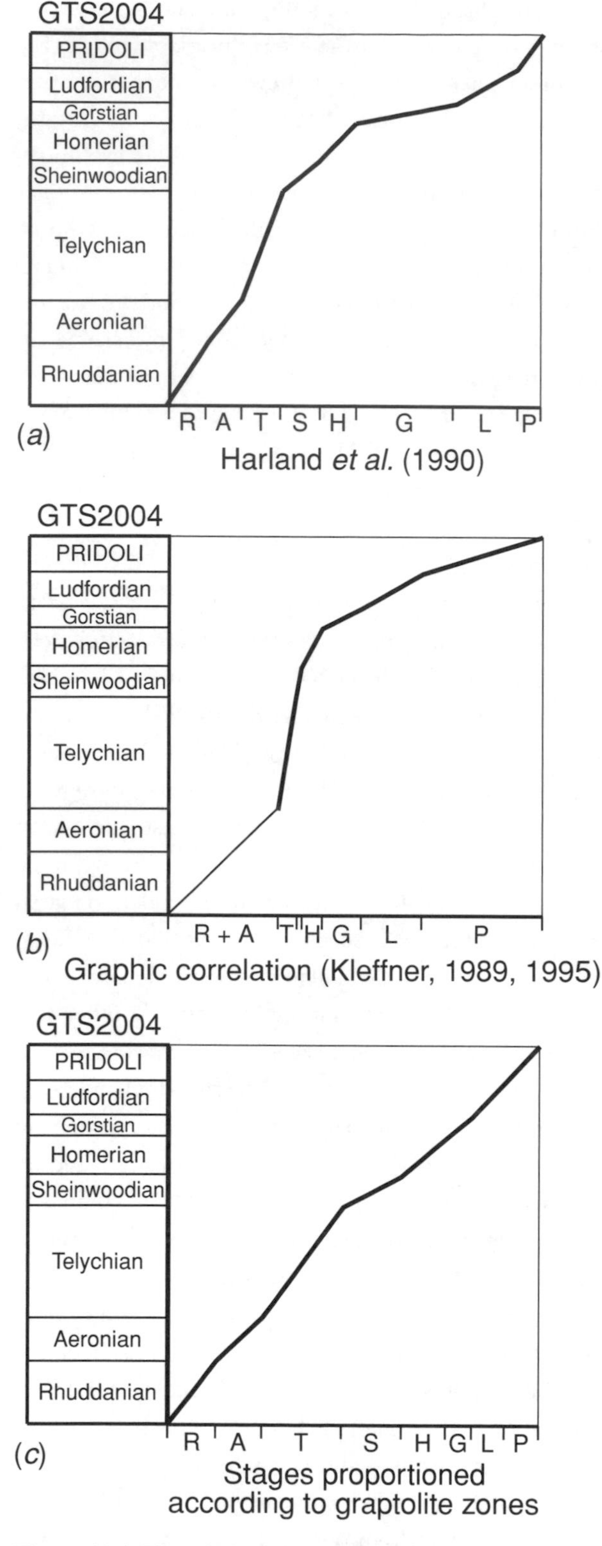

Figure 13.4 The relative duration of Silurian stages in time scales produced by three methods compared with the GTS2004 scale. (*a*) Harland *et al.* (1990) proportioned stages according to relative numbers of graptolite zones in the zonation at that time. (*b*) Kleffner (1989, 1995) used a composite section from graphical correlation. (*c*) Stages proportioned according to their component graptolite zones as shown in the zonation of Fig. 13.1.

such as that used for the Ordovician (Cooper, 1992, 1999b) can be unreliable.

13.3.3 Calibration of stage boundaries by composite standard optimization

As in the Ordovician, Silurian deep-water shales have the prerequisite for a high-resolution time scale, i.e. rich successions of graptolite fauna, inter-bedded ash layers with dateable zircon crystals, and minimally interrupted accumulation. The computer optimization method of Kemple *et al.* (1995) and Sadler (1999), using program CONOP, can be used to combine graptolite ranges in many measured sections into a scaled composite. The method and results are described in more detail in Chapters 3 and 12, and are summarized here. The graptolite successions in over 200 deep-water stratigraphic sections from around the world have thus been used to derive a global, scaled, composite section that ranges from late Cambrian to early Devonian (Sadler and Cooper, 2004; Table 12.2). In the Late Cambrian and basal-Ordovician, where graptolites are sparse or absent, the record is supplemented with conodonts and trilobites. Nearly 1200 taxa provide over 2300 biostratigraphic range-end events that have been used to build the composite. The database includes 115 graptolite-bearing sections that range wholly, or in part, in the Silurian, containing 550 graptolite species, and provide the basis for the Silurian time scale (Fig. 13.1).

The method used for developing a relative (proxy) time scale for the Silurian is the same as that used for the Ordovician. It utilizes both evolutionary rates and stratigraphic thickness, but has built-in normalizing and minimizing procedures to reduce bias, and also a test for linearity of the scale. All sections have been treated simultaneously, avoiding the bias introduced by choice of the initial standard reference section as in graphic correlation. The result is an optimized, scaled, global composite sequence of events. The order and spacing of events in the scaled composite serve as a proxy time scale (Table 12.2). Stage boundaries are readily located in the composite, as most are tied to graptolite first-appearance events.

13.3.4 Age of stage boundaries

The seven radiometric-dated ash beds discussed above, in combination with the CONOP composite to scale the eight Silurian stages, allow calculation of the Silurian time scale; the seven dates are part of the data set of 22 presented in Table 12.1 and also used to calculate the Ordovician time scale. The best-fit line for the two-way plot of radiometric ages along the abscissa and the zones and stages along the ordinate (see Fig. 12.5) was

Table 13.1 *Ages and durations of Silurian stages*

Period	Epoch	Stage	Age of base (Ma)	Est. ± myr (2-sigma)	Duration	Est. ± myr (2-sigma)
Devonian			**416.0**	2.8		
Silurian						
	Pridoli (not subdivided into stages)		418.7	2.7	2.7	0.1
	Ludlow					
		Ludfordian	421.3	2.6	2.5	0.1
		Gorstian	422.9	2.5	1.7	0.1
	Wenlock					
		Homerian	426.2	2.4	3.3	0.1
		Sheinwoodian	428.2	2.3	2.0	0.1
	Llandovery					
		Telychian	436.0	1.9	7.8	0.2
		Aeronian	439.0	1.8	3.0	0.1
		Rhuddanian	**443.7**	1.5	4.7	0.1
Ordovician						

calculated with a cubic-spline-fitting method that combines stratigraphic uncertainty estimates with 2-sigma radiometric data error bars. The smoothing factor for the cubic spline of 1.452 was calculated with cross-validation, and the 2-sigma error on the estimated ages for the Silurian stages with Ripley's MLFR procedure (see Chapters 8 and 12).

Error bars of 2 myr were too narrow according to the chi-square test of residuals. One of these was the error bar on the revised Kunk *et al.* (1985) Ar-Ar date of 426.8 ± 1.7 Ma; the "average" error bar for the whole graph was used for it instead. The other error bar with a chi-square probability of less than 0.01 was item No. 20 in Table 12.1 from Landing *et al.* (1997) of 483 ± 1 Ma. Because its stratigraphic error bar is not symmetric, and reaches the best-fit line, it was retained.

The Ripley MLFR procedure to assist with spline fitting worked well for the Ordovician–Silurian data set, and the 2-sigma value for the base-Silurian is 1.5 myr; the error value decreases stratigraphically upward because it is assumed that for each period the relative 2-sigma error remains constant (rather than the error bar width itself). This is a simple assumption along the line that for any age determination method, errors tend to be proportional to age, as also used for other periods.

To obtain some idea of the sensitivity of the results to these simplifying assumptions, the Silurian time scale was also calculated via a second-order polynomial, which is close to linear (R = 0.9965i) (Sadler and Cooper, 2004). Its results fall within the error limits of the mathematically more advanced fit.

The calculated duration and ages, with estimates of uncertainty (2-sigma), of the Silurian stages are given in Table 13.1. Note that results of rounding procedures can yield a discrepancy between durations and ages of boundaries.

The age of the base of the Silurian Period in the calibrated CONOP scale, using cubic-spline fitting, is 443.7 ± 1.5 Ma and for the top of the Silurian is 416.0 ± 1.4 Ma. The duration of the Silurian in the GTS2004 time scale is 27.7 myr. These ages differ from those in Harland *et al.* (1990), which placed the base-Silurian near 439 Ma and the top near 408.5 Ma, and from Compston (2000a), who assigned dates of ~438.7 Ma (base) and 417.6 Ma (top).

Ages for the bases of the Silurian epochs are: Llandovery, 443.7 Ma; Wenlock, 428.2 Ma; Ludlow 422.9, Ma; and Pridoli, 418.7 Ma.

Compared to stages in the Ordovician and Devonian Periods, all but the Rhuddanian and Telychian are considerably shorter in duration. Although the combination of accurate radiometrics and a detailed composite standard yielded ages with small uncertainties, the Silurian time scale needs more radiometric age dates to strengthen the interpolations. At present, there is less than one radiometric date per stage on average, with only one date in the long Telychian and none at all in the Sheinwoodian.

The Silurian time scale, shown in Figs. 13.1 and 13.2 calibrates stage and zone boundaries with a precision not previously achieved. The generalized graptolite zones used here are based on Koren' *et al.* (1996). Generalized conodont and chitinozoan zonal schemes are correlated with the graptolite zones, along with spore and vertebrate zones. Of the few regional stage schemes still used, that of the Baltic is most useful.

14 • The Devonian Period

M. R. HOUSE† AND F. M. GRADSTEIN

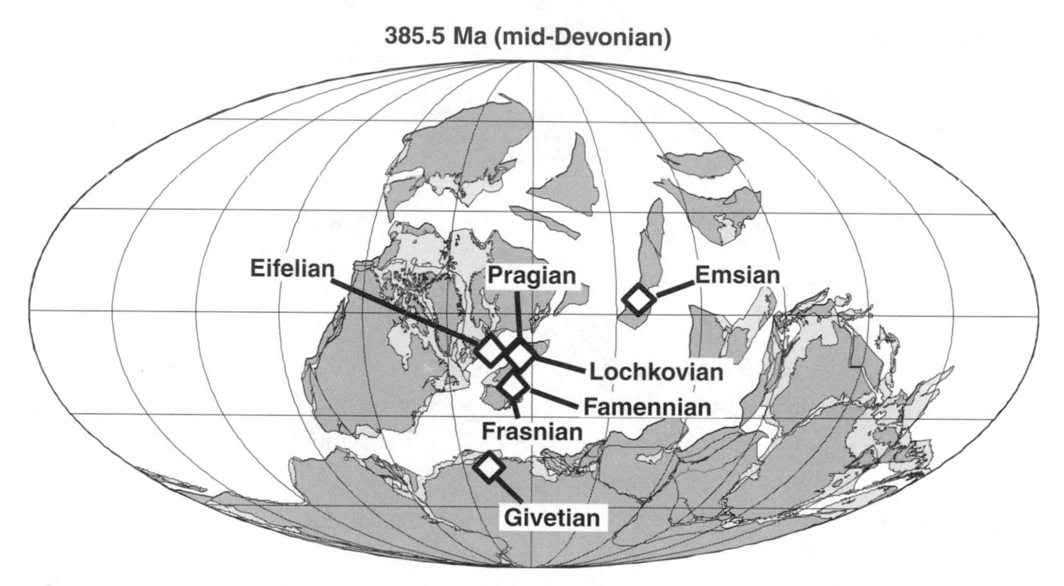

Geographic distribution of Devonian GSSPs. All seven Devonian stages are defined by GSSPs.

The Devonian Period was a time of exceptionally high sea-level stand and inferred widespread equable climates, but glaciations occur during the Late Devonian of the south polar areas of Gondwana. Most present-day continental areas and shelves were grouped in one hemisphere. Following the tectonic events of the Caledonian orogeny of Laurasia, many "Old Red Sandstone" terrestrial deposits formed. This is the time of greatest carbonate production and the greatest diversity of marine fauna in the Paleozoic. Vascular plants and forests became established and, before the end of the period, land tetrapods appeared.

14.1 HISTORY AND SUBDIVISIONS

Until recently there was considerable confusion in the use of the term Devonian because the boundaries were used differently in various parts of the world. A broad, international review of the Devonian Period is given in *Devonian Geology of the World* (McMillan *et al.*, 1988) and in House (1991).

A Geologic Time Scale 2004, eds. Felix M. Gradstein, James G. Ogg, and Alan G. Smith. Published by Cambridge University Press.

The Devonian System was established by Sedgwick and Murchison (1839) when it was recognized through the then unpublished work of Lonsdale (1840) that marine rocks in southwest England were the equivalent of terrestrial Old Red Sandstone deposits in Wales, the north of England, and Scotland: an early recognition of facies change. Murchison's definition of the boundary between the Silurian System and the Old Red Sandstone in Wales and the Welsh Borders has some ambiguities, but general opinion is that the Ludlow Bone Bed was very close to the intention. However, other boundaries were used over the next century (White, 1950). The result was that there was no clear definition of the boundary in what may be called the type area, and no consistent practice among British geologists. All this was little help internationally.

Following the detailed work on British graptolites by Ellis and Wood (1901–18) it was recognized that graptolites were last present in the late Ludlow below the Ludlow Bone Bed. Therefore, the extinction of graptolites was considered to be the major guide to the position of the base of the Devonian elsewhere in the world. It was not until 1960 that it became clear from evidence outside the British Isles that graptolites continued long after the time equivalent of the Ludlow Bone Bed. This evidence came from the work of R. Thorsteinson (Geological Survey of Canada) on Ellesmere Island and the association of monograptids with Emsian (late Early Devonian) rocks in Europe. Thus the base of the Devonian, as defined by the extinction of graptolites, especially in continental Europe and Asia, belonged to levels well above the Ludlow Bone Bed. This definition raised the Silurian–Devonian boundary to near the *Psammosteus* Limestone (Richardson *et al.*, 1981).

As to the Devonian–Carboniferous boundary, it could be argued that Sedgwick and Murchison (1840) placed the top of the Devonian at a fairly unambiguous boundary in north Devon, but faunal and floral studies were not then precise enough for accurate correlation. As a result, stratigraphic levels were taken that were subsequently demonstrated to be inaccurate, or other names were used for strata where there was some uncertainty of assignment. In the latter category were names such as Kinderhookian, in North America, and Etroungt and Strunian, in continental Europe.

In 1960, a committee was established by the International Union of Geological Sciences (IUGS) to make recommendations on the position of the Silurian–Devonian boundary, which led to its recommendations being accepted at the International Geological Congress (IGC) in Montreal in 1972. Martinsson (1977) summarizes the procedure and conclusions. A separate working party considered the Devonian–Carboniferous boundary.

Recommendations of GSSPs for all boundaries of period, series, and stage divisions for the Devonian were completed by the Subcommission on Devonian Stratigraphy (SDS) and ratified by IUGS by 1996, and by the following year accounts of all decisions had been published. Summary accounts have been published by Bultynck (2000a,b). The SDS is now considering substage definitions.

The Devonian is divided into the Lower, Middle, and Upper Series, The Lower Devonian is divided into the Lochkovian, Pragian, and Emsian Stages, the Middle into the Eifelian and Givetian Stages, and the Upper into the Frasnian and Famennian Stages.

The standard international chronostratigraphic divisions for the Devonian are given in Fig. 14.1. There are also many widely used terms as local and regional stages, e.g. for the Lower Devonian, the Gedinnian and Siegenian. For the Emsian, a subdivision into two is currently under consideration by the SDS: the Czech terms Zlichovian and Dalejan are presently local terms. It was the miscorrelation of the Dalejan deepening event with the Eifelian deepening event that led to many complications in the definition of the Lower–Middle Devonian boundary. The term Couvinian is now a regional term only. The Givetian also may come to be formally subdivided because the base of the Upper Devonian, and Frasnian, is drawn at a level above what was formerly regarded by many as early Frasnian. The German Upper Devonian Stufen, Adorf, Nehden, Hemberg, Dasberg, and Wocklum are now regional terms, but a formal subdivision of the Famennian is likely. The Strunian falls within the late Famennian of the revised terminology. It is important to note that the newly defined boundaries refer to stage names which may have been used differently in the past.

14.1.1 Lower Devonian Series

The GSSP for the Silurian–Devonian boundary, basal-Lower Devonian Series, and basal-Lochkovian Stage is at Klonk in the Czech Republic as documented in Martinsson (1977), in which D. J. McLaren recounts the scientific and political problems in reaching a decision on its placement. Important faunal characters used in the initial definition were the entry of the graptolite *Monograptus uniformis* and the *Warburgella rugosa* group of trilobites.

LOCHKOVIAN

The basal-Devonian and basal-Lochkovian GSSP is situated southwest of Prague, in the Czech Republic, in the Paleozoic area known as the Barrandium, where the Klonk section near Suchomasty is a natural 34 m cliff section embracing the latest Silurian (Pridoli) and the early Lochkovian. The sequence comprises rhythmically deposited allochthonous limestones with autochthonous intervening shales. The GSSP is within Bed 20, a 7–10 cm unit "immediately below the sudden and abundant occurrence of *M. uniformis* and *M. uniformis angustidens* in the upper part of that bed" (Martinsson, 1977, p. 21; Jaeger, 1977). Chlupáč and Hladil (2000) review the stratigraphy of the type section and summarize the detailed subsequent work on the faunal and floral sequence at the GSSP. Conodont and graptolite fauna have enabled this stage to be recognized in most parts of the world. Richardson and McGregor (1986) and Richardson *et al.* (2000) have provided spore evidence for correlation in other areas.

PRAGIAN

The base of this stage is defined by the GSSP at Velka Chuchle, near Prague, Czech Republic (Chlupáč and Oliver 1989; reviewed by Chlupáč, 2000). A primary correlation marker for

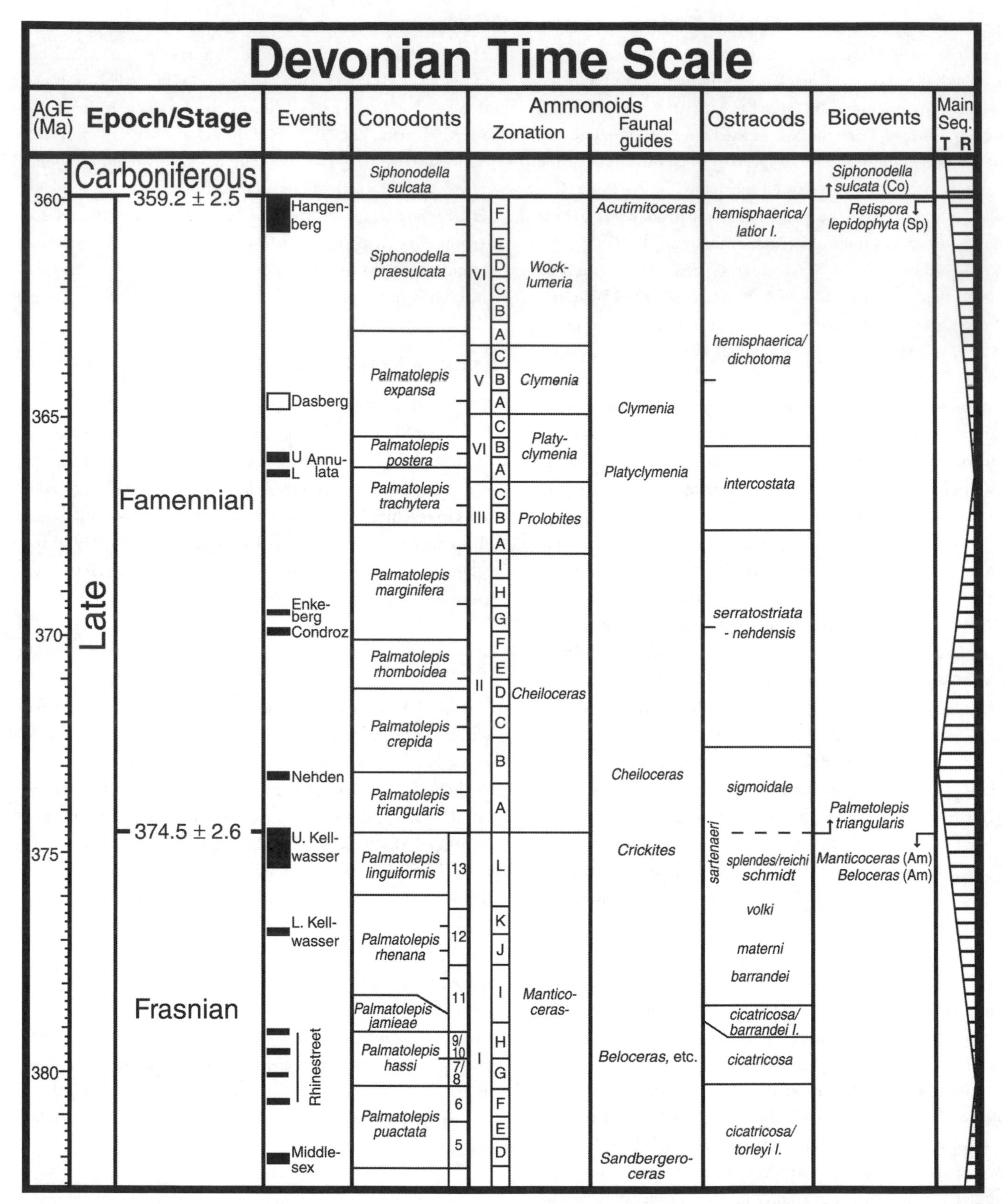

Figure 14.1 Epoch and stage boundaries and major marine biostratigraphic zonations for the Devonian Period with significant "event" levels (e.g. widespread anoxic facies or important outcrops) and principal eustatic trends. Biostratigraphic scales include conodonts, ammonoid zonation with selected guide taxa, graptolites, and ostracods. The combined conodont and ammonite scales were developed by M. R. House for scaling Devonian stages (see Section 14.4). Details of the scales are given in the text and in House (2002) and Bultynck (2000a,b). A color version of parts of this figure is in the plate section.

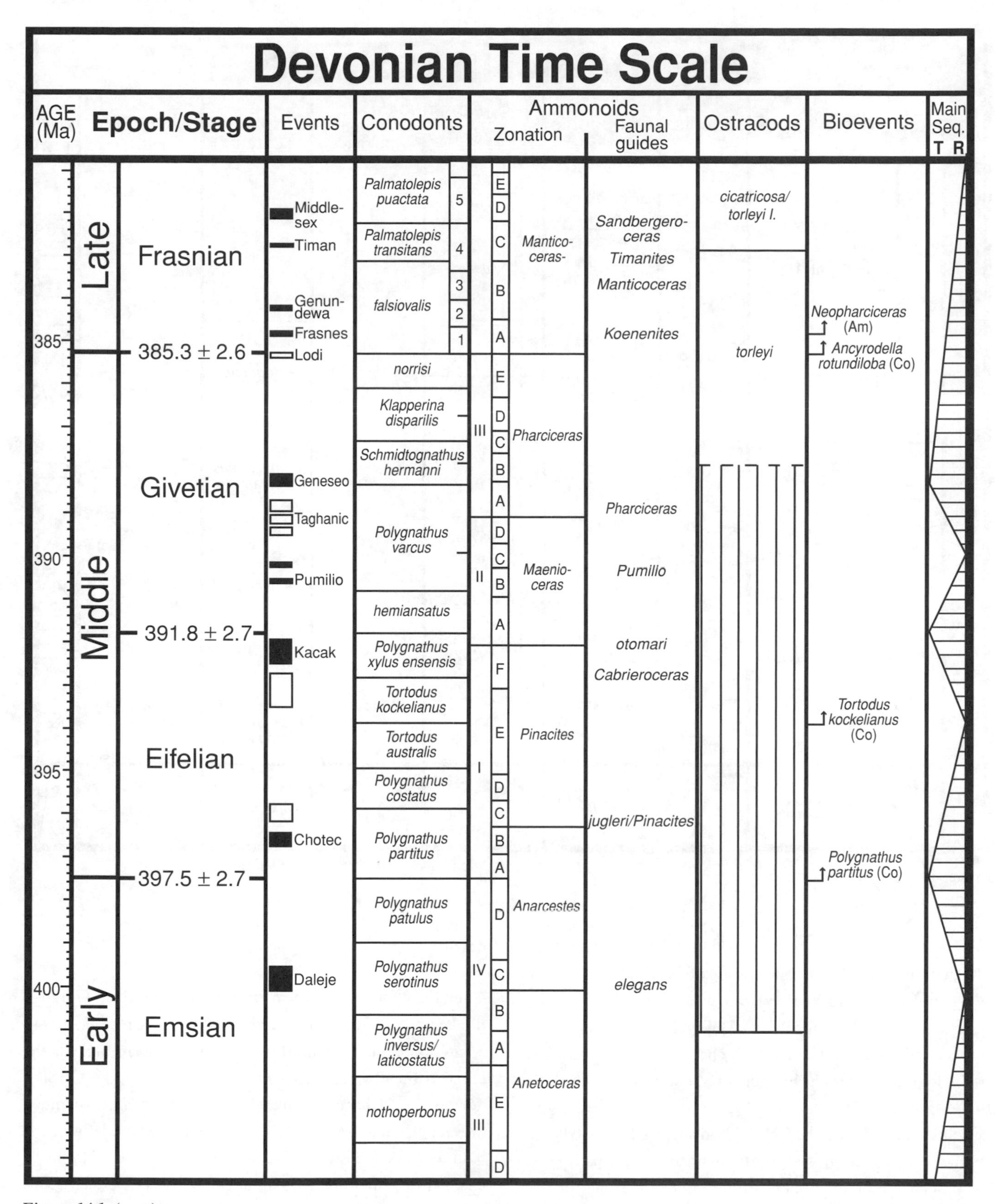

Figure 14.1 (*cont.*)

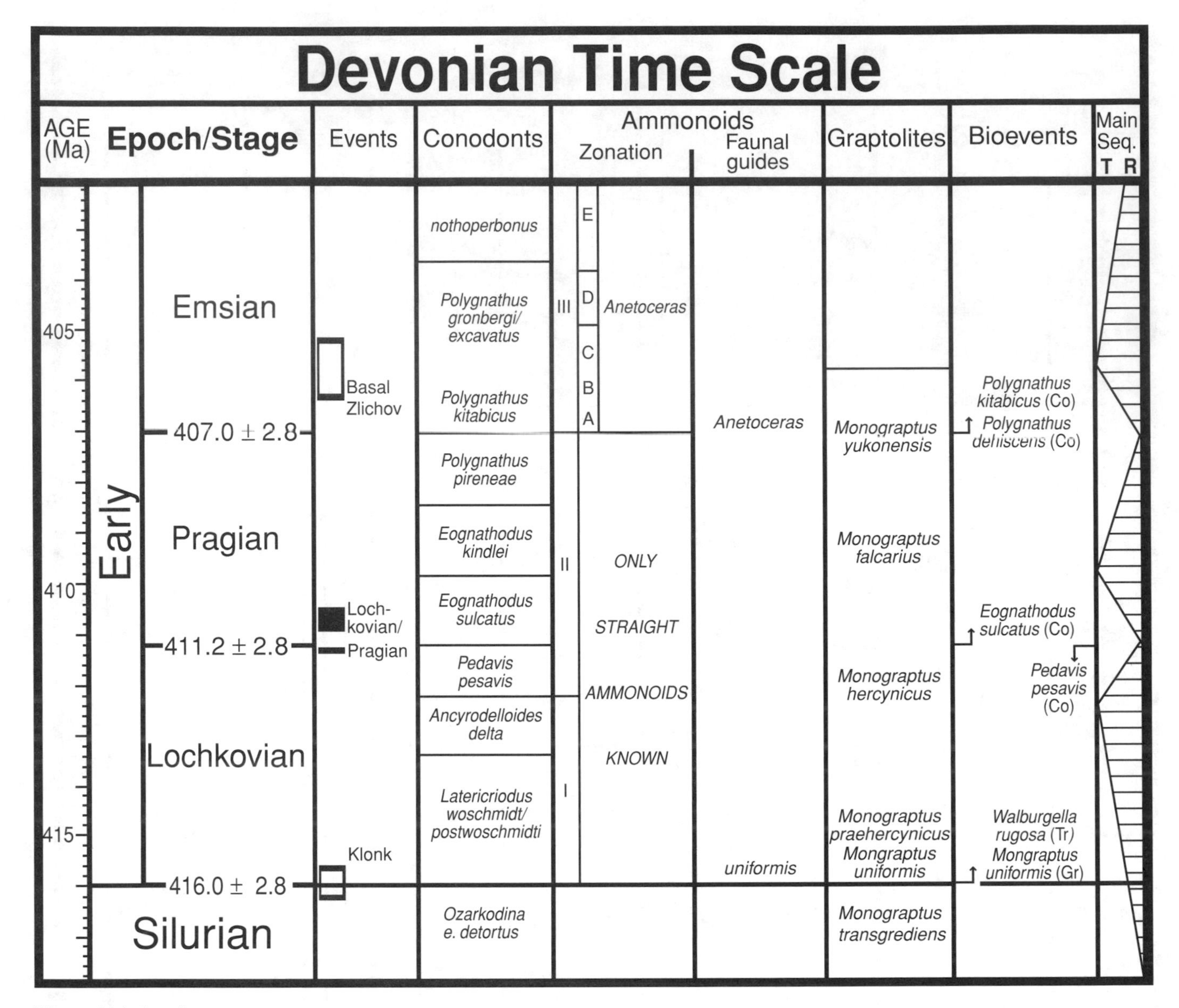

Figure 14.1 (*cont.*)

this boundary is the entry of the conodont *Eognathodus sulcatus*. Other forms which are important include the dacryoconarids *Nowakia sororcula* and *Now. arcuaria*, which enter shortly above the boundary at a level formerly taken as the base of the stage, the usage of which has been slightly changed. Chitinozoans are useful for recognizing the boundary interval and provide a link to the spore zonation shown in Fig. 14.1. Ranges of other invertebrate taxa around the GSSP are given by Chlupáč (2000). As the spore zonation is currently defined (Steemans, 1989), the base of the Pragian falls within a spore zone such that it cannot be accurately defined using palynology, but, nevertheless, fairly accurate placing has been achieved in several areas elsewhere (Richardson *et al.*, 2000).

EMSIAN

The GSSP for the basal-Emsian is in the Zinzalban Gorge of the Kitab National Park in Uzbechistan (Yolkin *et al.*, 1998). A key conodont marking the base of the Emsian Stage is the entry of the *Polygnathus dehiscens*. However, in the GSSP a new and somewhat controversial species, *P. kitabicus*, not much different from *P. dehiscens*, has been established for defining the boundary point. This level is distinctly below the level formerly taken as the top of the Pragian in Bohemia. The Czech terms Zlichovian and Dalejan are currently used informally.

A major paleoecological changeover in pelagic areas is shown by the reduction and loss of the graptolites and the uniserial monograptids become extinct within the early Emsian. Shortly above the boundary, coiled ammonoids then enter and become a dominant group in marine facies until their extinction at the close of the Cretaceous. The Daleje Shale, within the Emsian, marks a transgressive pulse that is widely recognized globally and gave rise to the term Daleje Event. In the past, this level in many areas was erroneously assigned to the Eifelian.

The Emsian Stage appears to have a relatively long duration, and the SDS is discussing a potential formal subdivision into substages.

14.1.2 Middle Devonian Series

EIFELIAN

The base of the Middle Devonian Series and of the Eifelian Stage is drawn at a GSSP at Wetteldorf Richtschnitt in the Prüm Syncline of the Eifel District of Germany (Ziegler and Werner, 1982; Ziegler, 2000). The recommendation of the SDS was ratified at meetings of the IUGS held with the ICS in Moscow in 1984. A primary correlation marker for this boundary is the junction of the *patulus* and *partitus* conodont zones, which lies just below the anoxic pulse of the Chote Event. The GSSP is in a trench and the locality is protected by a building erected by the Senckenbergische Naturforschende Gesellschaft in 1990. As this boundary is in the Eifel Hills, its definition involved minor change from historical usage of the term Eifelian in the area, but the upper boundary, discussed below, had been defined in several ways.

GIVETIAN

The base of the Givetian Stage is drawn at a GSSP in southern Morocco at Jebel Mech Irdane ("Hill of the Little Mouse") in the Tafilalt area of the Anti Atlas, 12 km southwest of Rissani (Walliser *et al.*, 1996; Walliser, 2000). A primary correlation marker for this boundary is the base of the conodont *hemiansatus* Zone, corresponding to the upper part of the former *ensensis* Zone. The ammonoid *Maenioceras* Zones commence a little below the boundary, and the spore *Geminospora lemurata* slightly above it.

The SDS's recommendation was ratified at an IUGS meeting in London in 1994. The original Givet Limestone, and Assisse de Givet, in the Ardennes gave the name to the stage, which had been defined in several different ways, based either on local neritic characters, or on inferred correlation with fauna in pelagic areas. The lower boundary of the present stage is not too far below the base of the Givet Limestone and the entry of the classical Givetian brachiopod *Stringocephalus* (Bultynck and Hollevoet, 1999). In Morocco, the boundary falls in the upper part of the anoxic pulse of the Kačák Event.

14.1.3 Upper Devonian Series

FRASNIAN

There have been considerable historic differences in the definitions of the base of the Upper Devonian internationally. These disputes are now rendered irrelevant by the selection of a GSSP for the base of the Upper Devonian Series and of the Frasnian Stage at Puech de la Suque, near Cessenon in the Montagne Noire, France (Klapper *et al.*, 1987; House *et al.*, 2000). The section is overturned and the GSSP falls within Bed 42 of the succession. The initial guide to the boundary was the conodont, *Ancyrodella rotundiloba* early morph Klapper, which in the type area for the naming of the stage (Frasnes, Belgium) almost coincides with the base of the Frasnes Group. An important advantage of this conodont taxon is that it occurs in the pelagic as well as in the neritic facies. But there has been subsequent dispute over taxonomy of this group, which has been summarized by Klapper (2000a). However, the boundary is defined as a point in a rock succession (the GSSP) and not by a zone fossil. The GSSP level corresponds to the base of Montagne Noire zone MN 1 (Klapper, 1989), or of the *soluta* Zone (Yudina, 1995), or falls within the *falsiovalis* Zone (Ziegler and Sandberg, 1990) according to the terminology followed. The entry of the goniatite genus *Neopharciceras* occurs immediately above the GSSP in Bed 43 (Korn, in House *et al.*, 2000), and this provides a useful correlation with Asian successions.

FAMENNIAN

In the past, the base of the Famennian Stage has been placed at different levels, reflecting, in part, problems of correlation between faunal groups. The new definition for the base of the Famennian proposed by the SDS was ratified by the IUGS in 1993. The GSSP for the base of the Famennian is very close to the base of the Famennian as formerly used in the Famenne area in Belgium (Bultynck and Martin, 1995). The GSSP is above the Upper Coumiac Quarry, near Cessenon, Montagne Noire, France (Klapper *et al.*, 1993; House, 2000). The boundary falls at the junction of the *linguiformis* and Lower *triangularis* conodont zones (Klapper, 2000a). The goniatites *Manticoceras* and

Beloceras become extinct at the boundary: these are the last representatives of the Gephuroceratidae and Beloceratidae. The genus *Cheiloceras* enters significantly higher. At the GSSP section, both anoxic pulses of the Lower and Upper Kellwasser Events are present. The GSSP level is immediately above the latter, a level known to be one of major extinctions in other regions.

BASE OF THE CARBONIFEROUS

Following recommendations of a special working party on the boundary, set up by ICS, the IUGS, in 1989, accepted a GSSP for the Devonian–Carboniferous boundary at La Serre, near Clermont l' Herault, Montagne Noire, France (Paproth *et al.*, 1991; Feist *et al.*, 2000). The guide to the base is the conodont *Siphonodella sulcata*, but there has been some dispute on the definition of this form.

The Devonian–Carboniferous boundary is well-defined palynologically by the last appearance of *Retispora lepidophyta*, which disappears just beneath the first occurrence of *Siphonodella sulcata* in the Hasselbachtal auxiliary section in Germany (Higgs *et al.*, 1993). Major extinctions of the ammonoid goniatites and clymenids, and of trilobites, occur below the boundary, which falls within the anoxic pulses of the Hangenberg Event.

14.2 DEVONIAN STRATIGRAPHY

14.2.1 Biostratigraphy

There are very refined zonations in pelagic facies using, especially, ammonoids, conodonts, ostracods, dacryoconarids, and, for the lower Devonian, monograptid graptolites. For neritic facies, brachiopods, trilobites, and ostracods are important. Chitinozoans occur both in the pelagic and in the neritic facies. In the terrestrial facies, spores, macroplants, and fish are especially useful. In some facies, spores and acritarchs are found in all these regimes. Many problems still remain on the correlation of the refined pelagic zonations with those of neritic and terrestrial facies. A listing of zones and their general correlation is given in Figs.14.1 and 14.2 and elaborated in Bultynck (2000b).

CONODONT ZONATIONS

The tooth-like microfossils of calcium phosphate can be etched by dissolution from many marine lithologies. Fundamental studies by Bischoff and Ziegler (1957) and Ziegler (1962) have led to a very refined zonation of considerable value in Devonian correlation. Terminology for the late Devonian has been revised by Ziegler and Sandberg (1990). A Frasnian zonation based on different criteria has been published by Klapper (1989, 2000b) and its correlation with the Ziegler and Sandberg revision, based on the Adorf section, is given in Klapper and Becker (1999). The current standard zonation is shown on Fig. 14.1.

AMMONOID ZONATIONS

Coiled ammonoids appear in the late Early Devonian. The zonation compiled by Wedekind in 1917, established that goniatites and clymenids provide an important correlation tool in marine, and especially pelagic, facies. Much further precision has been added: for the Lower Devonian by Becker and House (1994), Ruan (1996), and Klug (2001); for the Middle Devonian by Becker and House (1994); for the Upper Devonian (Frasnian) by Becker *et al.* (1993); and for the Famennian by the work of Schindewolf (1937), Becker (1993a), and others. The current situation has been summarized by Becker and House (2000), who have extended the original zonal scheme of Wedekind, using Latin numbering, into the Middle and Lower Devonian. The zonation is shown on Fig. 14.1.

OSTRACOD ZONATION

The development of pelagic ostracoda in the Devonian has contributed a very detailed biostratigraphic tool (Fig. 14.1). Major initial studies were conducted by Rabien (1954) and Blumenstengel (1965). Benthic ostracods are excellent indicators of ecotypes, but a global biozonation has not been established. The biostratigraphic role of ostracods in the Devonian is reviewed by Groos-Uffenorde *et al.* (2000).

DACRYCONARID ZONATION

The use of planktonic tentaculitids for stratigraphy has resulted in the recognition of their importance for biostratigraphy. Major contributions result from the work of Alberti (1993, 2000).

SPORE AND ACRITARCH ZONATIONS

Several basic spore zonations have been proposed. A scheme by Streel *et al.* (1987) used lettered abbreviations, while that of Richardson and McGregor (1986) used a more standard form (Fig. 14.2). For the Lower Devonian the work of Steemans

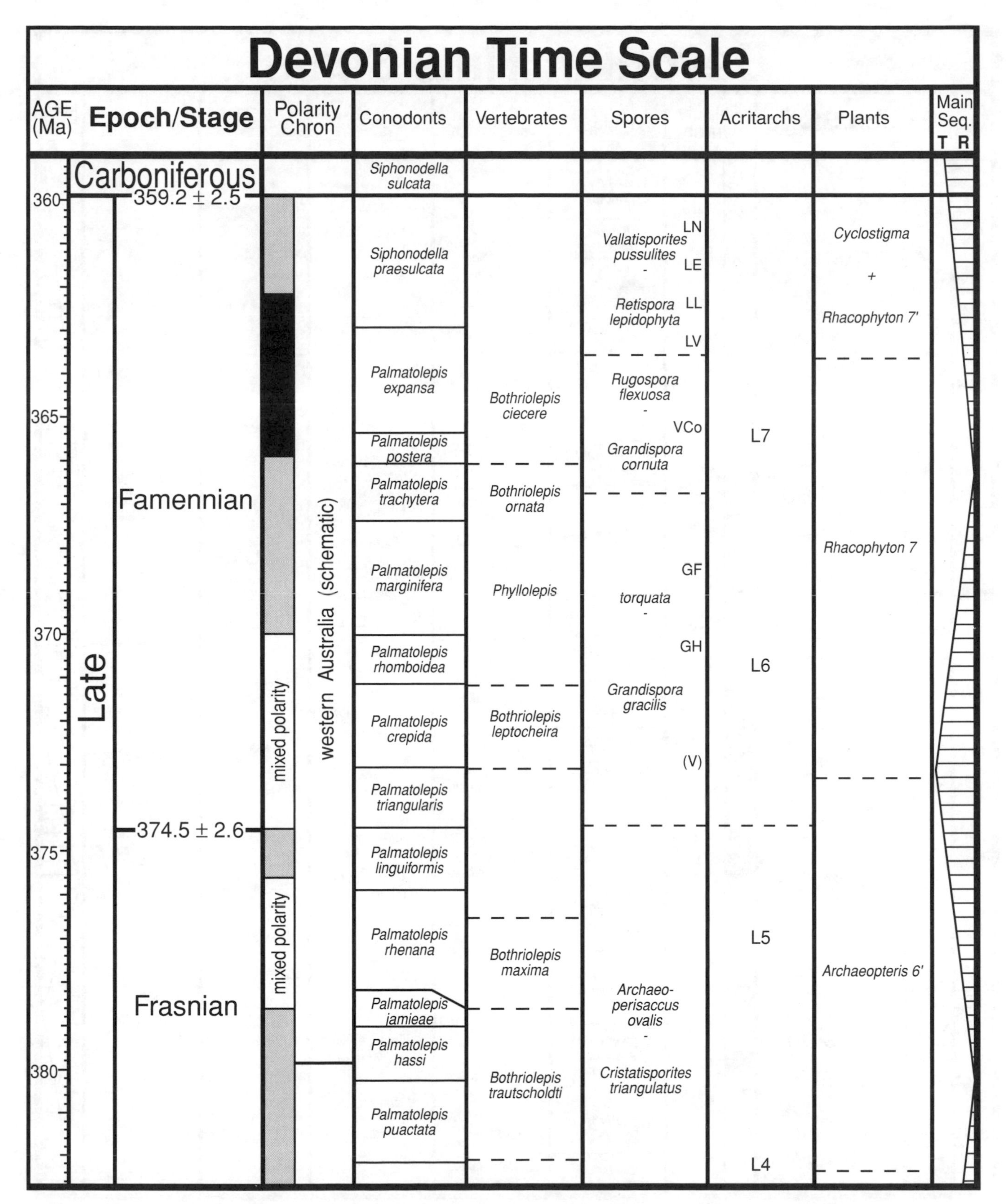

Figure 14.2 Devonian terrestrial facies zonations for vertebrates, spores, and plants, and the marine zonation of acritarchs. Biostratigraphic zonation of conodonts is shown for correlation. The geomagnetic polarity scale and principal eustatic trends are generalized. A color version of parts of this figure is in the plate section.

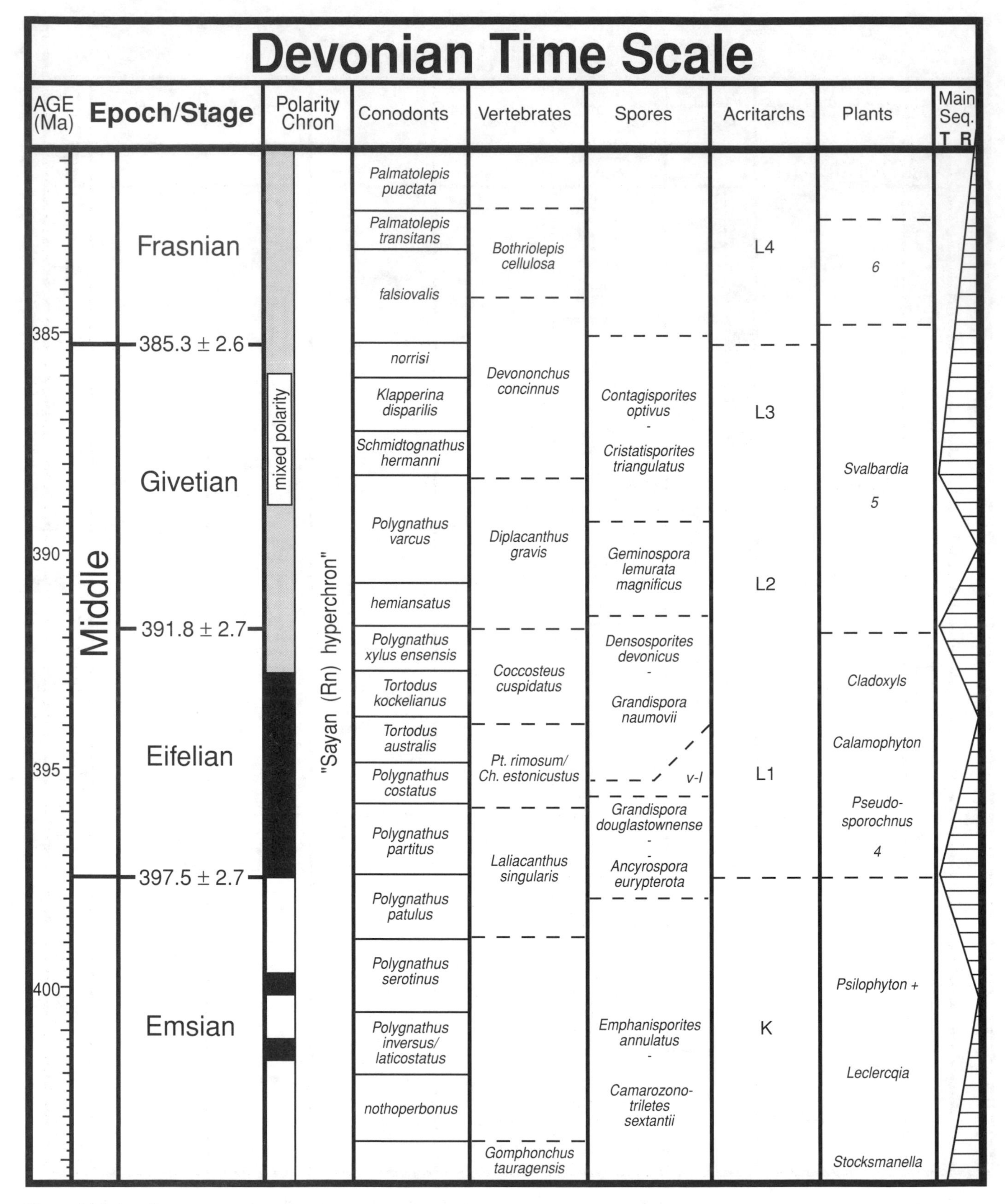

Figure 14.2 (*cont.*)

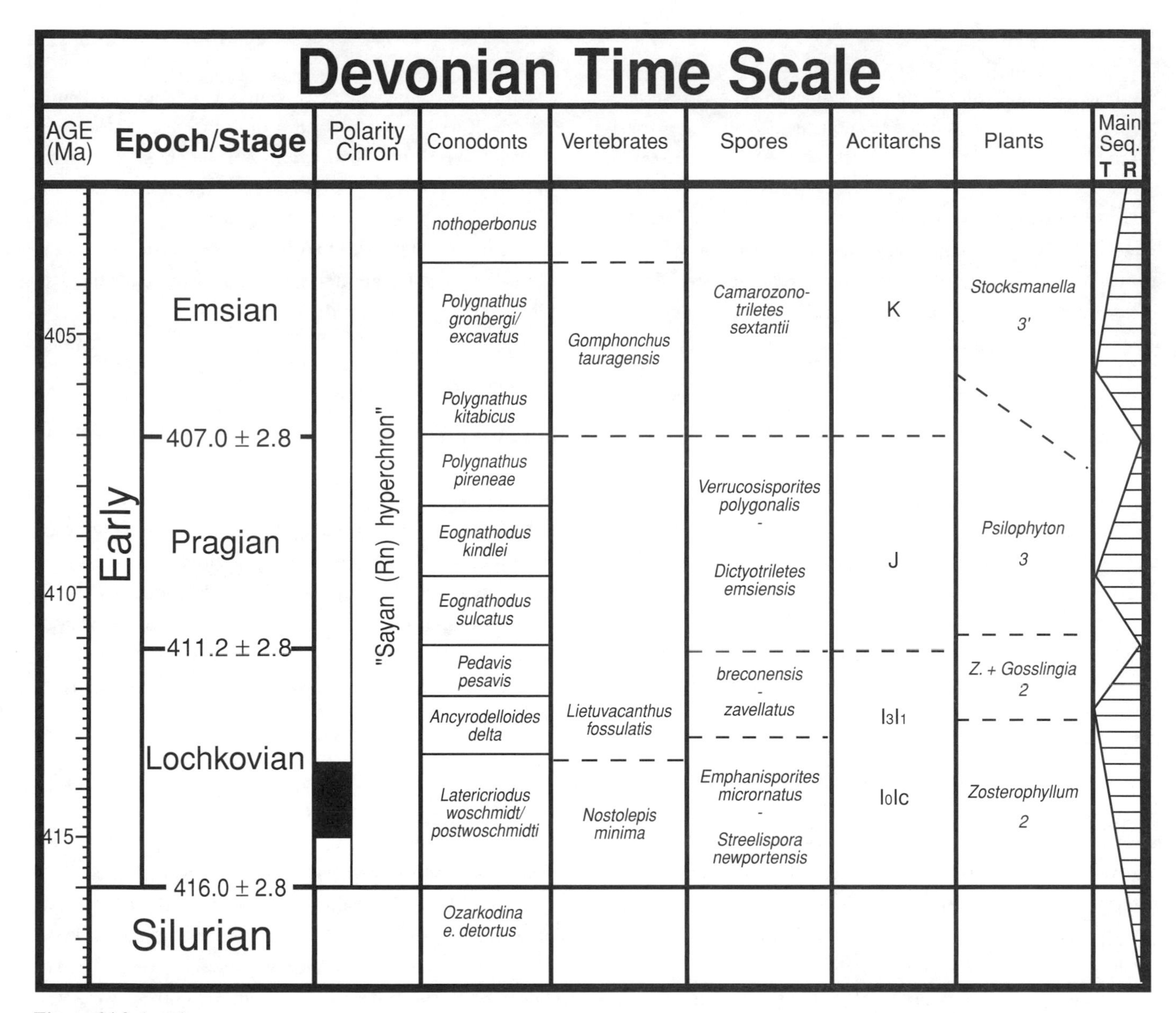

Figure 14.2 (*cont.*)

(1989) is fundamental. An analysis of spore records in relation to the conodont zonation of eastern Europe is given by Avkhimovitch *et al.* (1993). Megaspore zonation in relation to GSSP boundaries has been treated in detail by Streel *et al.* (2000a) and Mark-Kurik *et al.* (1999); the Frasnian–Famennian boundary is also reviewed by Streel *et al.* (2000b).

Acritarchs are organic-walled microfossils of unknown and probably varied biological affinities, but their stratigraphical use, especially in the Middle and Upper Devonian has been increasing rapidly. At present, the current zonation falls behind that of some other groups in resolving power. Reviews of zonations are given by Molyneux *et al.* (1996) and Le Hérissé *et al.* (2000).

PLANT MEGAFOSSIL ZONATION

Vascular plants, which begin well before the Devonian, rise in dominance during the period and forests occur around the Middle–Upper Devonian boundary. A broad division of the Devonian into seven zones was suggested by Banks (1980) and this has been further refined by Edwards *et al.* (2000), whose zonation, extending Banks' numbers, is shown on Fig. 14.2.

VERTEBRATE ZONATIONS

The Devonian is marked by the evolution of tetrapods, vertebrates with limbs and digits, which are only of very broad stratigraphical use due to the rarity of specimens and relative isolation of most known genera. The earliest body fossils are known from the late Frasnian of eastern Europe and Scotland (Clack, 2002). During the Devonian these vertebrates are in many ways fish-like and aquatic, fully terrestrial tetrapods not being known before the earliest Carboniferous (Tournaisian).

In terrestrial facies, fish are helpful in age determination (Fig. 14.2). Several zonations have developed using various fish groups and teeth and other micro-vertebrate remains. Major zonations include those for thelodonts and heterostracons and for placoderm and acanthodian fish, and are of importance for the difficult problem of marine to non-marine correlation. There has been an especial growth in recent years in the study of microvertebrates.

14.2.2 Physical stratigraphy

EXTINCTION AND ANOXIA EVENT STRATIGRAPHY

Several time-specific facies have been recognized in the Devonian and often coincide with extinction events. Many of these are associated with dark limestone or shale episodes and are interpreted as anoxic events. The events are generally named after type sections showing characteristic facies development (Fig. 14.1). General reviews of these episodes include those by Walliser (1996) and House (1985, 2002). The Kačák event is reviewed by Budil (1995); the Taghanic Event by Aboussalem and Becker (2001); the Kellwasser (Frasnian–Famennian Event) by Buggisch (1991), Schindler (1993), and Racki and House (2002); and the Hangenberg Event by Caplin and Bustin (1999) and Streel *et al.* (2000b).

Although much emphasized, the Kellwasser Event (Frasnian–Famennian boundary) may not be the greatest extinction event at invertebrate family level in the Devonian. The Taghanic, Hangenberg, and, perhaps, Kačák Events are almost similar in their extent, and much depends on how the extinctions are defined (House, 2002).

About 73 marine invertebrate families become extinct in the Famennian, many at the close of the Hangenberg Event, but how many families actually become extinct during the anoxic part for the Hangenberg Event is not yet documented. The Hangenberg Event is an especially important extinction for ammonoids. Second in importance at the stage level are the ~71 family extinctions in the Givetian, of which many are during the Taghanic Event. The Taghanic Event is associated with the break-up of the widespread carbonate platforms and associated loss of biota. Next in numerical importance are the 63 families reported lost at the close of the Frasnian in association with the Upper Kellwasser Event (Frasnian–Famennian boundary) Event.

The Lower and Middle Devonian extinctions at family level are all less than 50 families in each stage. These data, considered in relation to extinctions as a percentage of the total known in each stage, give 16.3% for the Famennian, 14.4% for the Givetian, and 14.9% for the Frasnian (data from House, 2002). Unfortunately, the Devonian radiometric time scale, and the data records, are not yet precise enough to elucidate extinction rates in any detail.

CYCLOSTRATIGRAPHY

Milankovitch orbital cyclicity has been applied for determining times scales in the Devonian only for the Famennian (Bai *et al.*, 1995) and Givetian (House, 1995). In neither case has detailed harmonic analysis been applied to sections, but the presumption has been made that dominant cycles may be precession or eccentricity cycles. The inadequacies of the radiometric scale make checking of this difficult. The main result has been to demonstrate that conodont zones are not of equal duration, yet that presumption has been used to attempt refined subdivision of the radiometric scale.

GEOCHEMISTRY

Much precise chemical analysis has been done (Fig. 14.3), especially over the problematic events, but work has concentrated on condensed pelagic successions. In the 1970s, iridium anomalies were sought on the supposition that, at least, the Frasnian–Famennian boundary extinctions were related to bolide impact. Anomalies found are not so tightly constrained. Wider European studies commenced with the work of Buggisch (1972, 1991).

Carbon isotope data show that global $\delta^{13}C$ positive excursions may be associated with the late Frasnian Kellwasser Event (Joachimski and Buggisch, 1993; Joachimski *et al.*, 2002). Similar positive excursions have been identified for other events, for example for the Hangenberg Event (Schönlaub *et al.*, 1992; Caplin and Bustin, 1999).

Cerium excursions have been noted (Girard and Lecuyer, 2002). Wider-scale work has been done on the Chinese Upper Devonian (Bai *et al.*, 1995). Anomalies are identified for many elements in Australia for many of the Devonian anoxic–extinction events (Talent *et al.*, 1993).

The $^{87}Sr/^{86}Sr$ of marine Sr through the interval (Carpenter *et al.*, 1991; Diener *et al.*, 1996; Denison *et al.*, 1997) shows a broad trough, centered on the earliest Givetian, that reaches

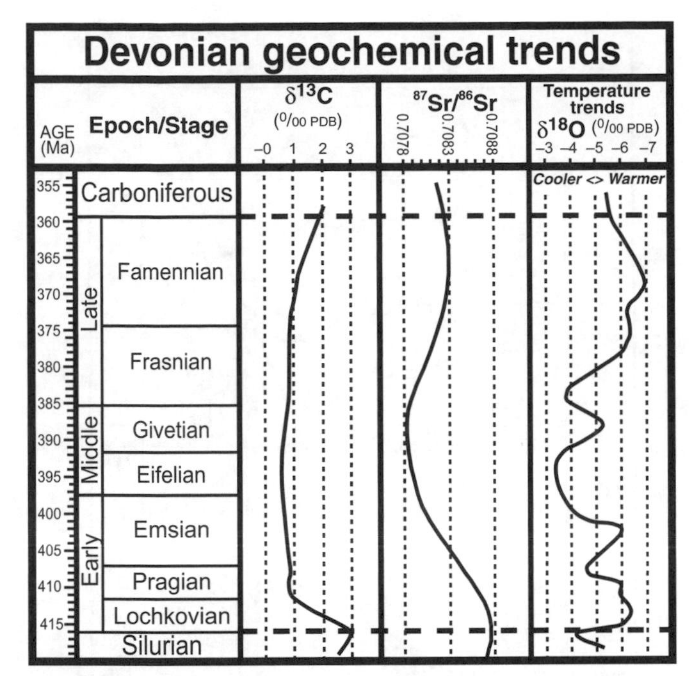

Figure 14.3 Geochemical trends in the Devonian Period. The schematic carbon isotope curve is a 5-myr averaging of global data (Veizer *et al.*, 1999) from Hayes *et al.* (1999; downloaded from www.nosams.whoi.edu/jmh). The ^{87}Sr/^{86}Sr LOWESS curve for the interval is based on the data of Denison *et al.* (1997) and Carpenter *et al.* (1991) – see Chapter 7. The oxygen isotope curves (inverted scale) are derived from a 3-myr interval averaging of global data compiled by Veizer *et al.* (1999; as downloaded from www.science.uottawa.ca/geology/isotope_data/ in Jan 2003). Large-scale global shifts to higher oxygen-18 values in carbonates are generally interpreted as cooler seawater or glacial episodes, but there are many other contributing factors (e.g., Veizer *et al.*, 1999, 2000; Wallman, 2001).

0.7078 at its minima between an early Devonian high of 0.7088 and an early Carboniferous high of 0.7083. This range is large enough to offer much potential for using Sr isotope stratigraphy in the interval.

SEA-LEVEL CHANGES

In a broad sense, sea-level changes in the Devonian have been commented upon at least since the collative work of French in the nineteenth century. A more systematic comparison between New York (USA) and Europe was published by House (1983) and this was improved using detailed conodont evidence for the Laurussian area by Johnson *et al.* (1985).

Interpretations for Asia are reviewed in House and Ziegler (1997). The terminology introduced by Johnson *et al.* (1985) has been widely adopted. More detailed work, tied to precise biostratigraphy, for example for the Frasnian of Australia (Becker *et al.*, 1993; Becker and House, 2000), has indicated that there are considerable local variations.

MAGNETIC SUSCEPTIBILITY

The usage of "magnetic susceptibility event and cyclostratigraphy" signatures for the Devonian has been advanced through the work of Ellwood, Crick, and others (Crick *et al.*, 1997; Ellwood *et al.*, 2000, 2001). Changes in the magnetic susceptibility of Devonian sediments are generally related to low and high stands of sea level. These, in turn, reflect climatic regimes, but their relation to any orbitally forced parameters still has to be demonstrated. Detailed sampling (for example, Lochkovian, 216 samples; Pragian, 553 samples; Zlichovian, 1200 samples) provides a precise history, the resolution of which, while not time specific, gives a seismograph-like record with potentially much more resolution than is possible by biostratigraphic means. The importance of this work has been demonstrated especially in relation to GSSPs and the events (relative sea-level changes?) often associated with them. The Silurian–Devonian boundary interval has been analyzed by Crick *et al.* (2001), and the Lower Devonian generally by Ellwood *et al.* (2001). Work on the Eifelian–Givetian boundary and associated Kačák Event (Crick *et al.*, 1997) has established correlations between North Africa, southern France, and the Czech Republic, and led to the proposal for establishing a "magnetostratotype" (Crick *et al.*, 2000). The Frasnian–Famennian boundary and Kellwasser Event have also been analyzed by Crick *et al.* (2000).

14.3 DEVONIAN TIME SCALE

14.3.1 Previous scales

Devonian radiometric time scales are reviewed by Tucker *et al.* (1998), Compston (2000b), and Williams *et al.* (2000a). Noticeable in reviews of the Devonian time scale are the limited number of good-quality and chronstratigraphically fixed dates within the period. As a result, various assumptions have been made such as equal duration of stages or of zones, particularly for conodonts; even though the available cyclostratigraphic data show these assumptions should be used with caution. Older time scales that assumed constant sedimentation rates, or relied on the subjective views of specialists on probable durations are now generally discarded. The fact that conodont and ammonoid zones are especially well discriminated in the Famennian may have led to the overemphasis of the length of that stage, apparent in most compilations. A review of Devonian time scale building and its results is given in Williams

Table 14.1 *Comparison of age and duration of the Devonian Period in selected time scales*

	Base (Ma)	Top (Ma)	Duration (myr)
Harland *et al.* (1990)	408.5	362.5	46
Young & Laurie (1996)	410	354	44
Tucker *et al.* (1998)	418	362	56
Compston (2000b) (1)	411.6	359.6	52
Compston (2000b) (2)	418	362	56
Williams *et al.* (2000a)	418	362	56
Menning *et al.* (2000)	–	353.7	–
GTS 2004	416	359.2	56.8

et al. (2000a), who essentially plot new dates, particularly HR–SIMS dates, against the TIMS ages time scale of Tucker *et al.* (1998). A comparison of the age of the base and/or top of the Devonian Period with the corresponding duration from selected time scales published during the 1990s is given in Table 14.1.

Although a substantial number of key dates were published after Harland *et al.* (1990) and Young and Laurie (1996) presented their fairly limited evidence and interpolations, the Devonian time scale is still not well established. A critical problem is the discrepancy between U–Pb zircon dates for key Devonian levels using TIMS and HR–SIMS (SHRIMP) dates as discussed by Tucker *et al.* (1998), Compston (2000b), and Williams *et al.* (2000a). The discrepancy is well illustrated in plots of the radiometric age dates for the Devonian superimposed on the time scale of Tucker *et al.* (1998) along the *y*-axis (Fig. 14.4). The black boxes that constrain time line *a* are TIMS ages of Tucker *et al.* (1998); the unfilled boxes that constrain time line *b* are the SHRIMP age dates of Compston (2000b). Time line *b* on average gives 1.3% younger ages for Devonian stages than time line *a*.

On a more detailed level, there is an extraordinary discrepancy between two estimates for the duration of the Fammenian – 5.1 and 14.5 myr – in two separate scales given by Compston (2000b, p. 1144). Hence, there is an urgent need for samples from the same localities and horizons to be tested at several laboratories. For example, for the Devonian–Carboniferous boundary, the Hasselbachtal date of around 353 Ma (Claoué-Long *et al.*, 1993) is at variance with other dates. A critical look at the original 36 zircon determinations by Compston (2000b) suggests the bentonite could be ~1% older, and that reprocessing would be desirable. Unfortunately, R. T. Becker (pers. comm., 2002), who collected the original samples, writes that slumping in the valley there has restricted further

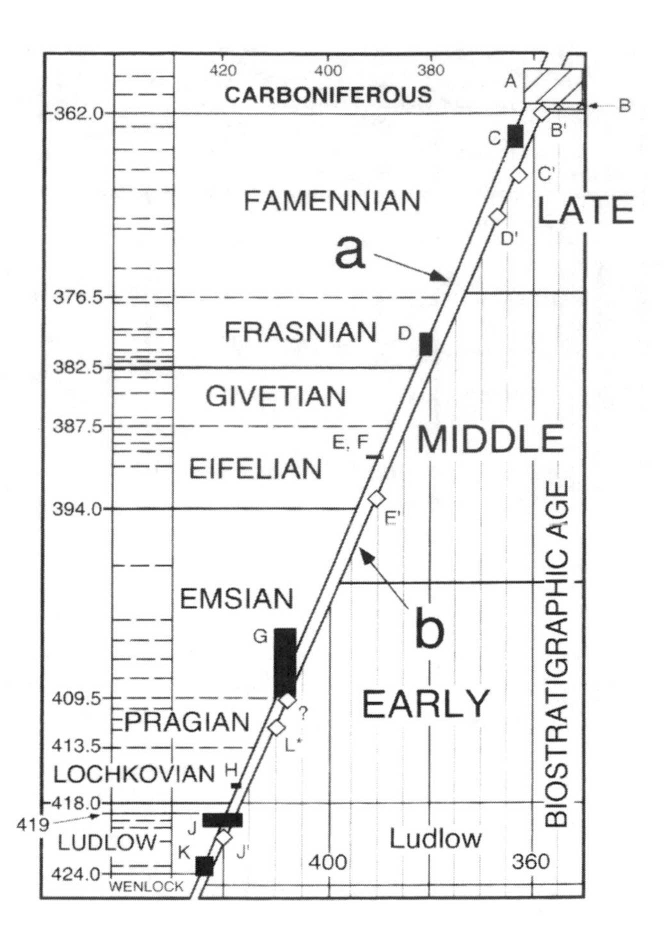

Figure 14.4 Plot of radiometric age dates for the Devonian superimposed on the time scale of Tucker *et al.* (1998) along the *y*-axis. The black boxes that constrain time line *a* are the isotope dilution ages of Tucker *et al.* (1998); the unfilled boxes that constrain time line *b* are the HR-SIMS dates calibrated with SL13 zircons of Compston (2000b). Reproduced with the permission of E. A. Williams and the Geological Society of London.

collecting. Recent U–Pb measurements on Hasselbachtal zircons extracted from large samples collected prior to slumping were unsuccessful due to the weak radiogenic signal in the very small zircons (B. Kaufmann, pers. comm., 2003).

In addition to the problems posed by accurate laboratory procedures, by standards, and by constants in radiometric age dating, there is the matter of critical analysis of the biostratigraphic assignments of the sample levels. For example, the Australian Phanerozoic time scale (Young and Laurie 1996) was analyzed by Klapper (2000b), who found basic flaws in the biostratigraphic assumptions in relation to the series boundary data points. The response (Young, 2000) shows little sign that the criticism has been understood. Next is the matter of the Los Frailes Stockwork, Rio Tinto (date B on Fig. 14.4) where an age just above the Devonian–Carboniferous boundary is plotted although there is no exact biostratigraphic placement. Similarly,

Williams *et al.* (2000b) show a U–Pb age for the Horses Glen Lower Tuffs of 387.5 ± 0.2 Ma. The unusually low analytical error strongly contrasts with a stratigraphic range spanning late Frasnian and possibly well into Famennian time; thus, the date does not contribute to the Devonian time scale.

14.3.2 Radiometric data

The detailed and high-resolution conodont–ammonoid zonation for the Devonian, with over 35 zones (Fig. 14.1), is in stark contrast to the handful of radiometric dates employed in Devonian time scale building. With a duration of the Devonian Period close to 60 myr, the zones on average give a resolution better than 2 myr, whereas on average there is less than one radiometric date per 10 myr. Worse, there are no dates for the Givetian and single dates only for the Frasnian and Famennian. TIMS or $^{40}Ar/^{39}Ar$ age dates directly assigned to conodont–ammonoid levels are needed to constrain those three stages.

The Devonian time scale presented here must be considered tentative. We have made some simplifying assumptions by using "normalized" HR–SIMS dates, and incorporated a relative scaling of stages according to component conodont–ammonoid zones.

Table 14.2 lists Devonian radiometric dates, with emphasis on available TIMS ages, that were judged suitable for time scale construction. In order to provide constraints on the age of the base and top of the Devonian, three Late Silurian and three earliest Carboniferous age dates are also included. The Late Silurian dates were calibrated to the CONOP biostratigraphic scheme (see Chapters 12 and 13). The Rb–Sr isochron date of 361.0 ± 4.1 Ma in the lower *S. sulcata* Zone of early Tournaisian is allowed with low weight in the fitting method.

The $^{40}Ar/^{39}Ar$ date in Kunk *et al.* (1985) in a bentonite in the *N. nilssoni* and *L. scanicus* conodont zones of Gorstian age was recalculated with the MMhb-1 monitor standard of 523.1 Ma to a new value of 426.8 ± 1.7 Ma. The U–Pb HR–SIMS dates of Jagodzinski and Black (1999; Items 8 and 9 in Table 14.2) were not calibrated by these authors with the SL13 monitor standard but with the QGNG standard. This practice increases the ages by 1.3% and slightly increases the standard error, and brings them more in line with nearby TIMS ages (Table 14.2). Two HR–SIMS age dates calibrated with the SL13 standard and assigned to the *Siphonella sulcata* conodont zone, the basal zone of the Tournaisian Stage, immediately above the Devonian–Carboniferous boundary (Items 1 and 2 in Table 14.2) were increased by 1.3% to yield ages of 358.3 ± 4.2 and 360.4 ± 5.6 Ma. As mentioned earlier, Compston (2000b) also indicates that a ~1% age increase is desirable. The ages are in good agreement with a TIMS date of 363.6 ± 1.6 Ma for an Upper Famennian level equivalent to the upper *Pa. expansa* conodont zone. Williams *et al.* (2000a) used the original HR–SIMS dates in the *S. sulcata* zone to make a case that the Carboniferous–Devonian boundary is more likely to be near 362 Ma rather than near 354 Ma or even younger in age.

A new $^{207}Pb/^{206}Pb$ date for the Hunsrück Slate (Kirnbauer and Reischmann, 2001) gives an age of 388.7 ± 1.2 Ma for early Emsian strata. This is much younger than can reasonably be expected from the preferred regression used, and hence is rejected. Given the discrepancy, one might consider that the biostratigraphic age is not well constrained; for example, the supposed early Emsian age of the index trilobite *Chotecops ferdinandi* may be much longer (D. Bruton, pers. comm., 2002).

14.3.3 Age of stage boundaries

The selected age dates for the Devonian Period in Table 14.2 were plotted against the relative zone–stage scale of Fig. 14.1 (the 'House scale') with approximate radiometric and stratigraphic errors (Fig. 14.5). The relative scale in Fig. 14.1 'juggles' the high-resolution ammonoid and conodont (sub) zonations such that shortest segments are fairly close to equal duration. The latter is not unlike the abandoned chron concept of GTS89 (see Chapter 1). Note that the relative position of Items 10–13 in Fig. 14.5, plotted as boxes relative to earliest Devonian–latest Silurian zones, is relatively unimportant. In the final analysis, the age obtained from the Devonian data of 416.7 ± 2.9 Ma for the top of the Silurian was overruled by 416.0 ± 1.4 Ma, resulting from the better constrained data for the Ordovician–Silurian boundary (see Chapters 12 and 13).

The best-fit line for the two-way plot in Fig. 14.5 of radiometric ages against the zones and stages was calculated with a cubic-spline-fitting method that combines stratigraphic uncertainty estimates with the 2-sigma error bars of the radiometric data. This is achieved with Ripley's MLFR procedure (see Chapter 8); a smoothing factor of about 1.4 was calculated with cross-validation. The chi-square test of residuals indicated that the 0.5-myr error bar on the 390.0 Ma monazite date (No. 6) was too narrow, and an 'average' error bar was used instead. This, together with the large error bars on dates 8 and 9 reduces the duration of the Emsian. A straight-line fit, as produced by Tucker *et al.* (1998), would substantially increase the duration of the Emsian Stage.

Table 14.2 *Selected isotopic dates for the Devonian time scale*

No.	Sample	Locality	Formation	Comment	Zone and age	Biostratigraphic reliability[a]	Reference	Age (Ma)	Type
1	K-bentonite, 1 cm	Hasselbachtal, Ruhr Basin, Nordrhein–Westfalen, Germany	Hangenberg Limestone	Original HR–SIMS (SHRIMP) age on zircons in ash bed 79 of 353.7 ± 4.2 Ma, calibrated with SL13, was corrected with 1.3% age to yield 358.3 Ma[b] Ash bed 79 is 0.35 m above the base of the *S. sulcata* Zone	FAD of *Siphonodella sulcata* in the *S. sulcata* Zone, Lower Tournaisian, and immediately below ammonoid *Acumitoceras antecedens*	1	Claoué-Long *et al.* (1993), Roberts *et al.* (1995b)	358.3 ± 4.2	U–Pb
2	Pumice tuff	Glenbawn Section, Australia	Kingsfield Formation	Original HR–SIMS (SHRIMP) age on zircons of 355.8 ± 5.6 Ma, calibrated with SL13, was corrected with 1.3% age to yield 360.4 Ma[b]	*S. sulcata* (? to lower *duplicata*) Zone, Lower Tournaisian	1	Claoué-Long *et al.* (1993), Roberts *et al.* (1995b)	360.4 ± 5.6	U–Pb
3	Carbonate and shale	Nanbiancun Section, China		Rb–Sr isochron method	Eight samples in carbonate and shale lower *S. sulcata* Zone, Lower Tournaisian	1	Yu (1988)	361.0 ± 4.1	Rb-Sr
4	Pumice tuff	Caldera complex, New Brunswick, Canada	Carrow Fm and Bailey Rock Rhyolite	Weighted mean ages on zircons of 363.3 ± 2.2 Ma (below spore-bearing bed) and 363.4 ± 1.8 Ma (just above spore-bearing bed) combine to 363.6 ± 1.6 Ma	Spore-bearing horizon between the two dated rhyolites in the *pusillites–lepidophyta* spore zone (FA2d), equivalent to the upper *Pa. expansa* conodont zone, upper Famennian	1	Tucker *et al.* (1998)	363.6 ± 1.6	U–Pb
5	K-bentonite	Little War Gap, East Tennessee, USA	Chattanooga Shale	Nine analyses on single grain or small-fraction zircons give a weighted average date of 381.1 ± 1.3 Ma. Two zircon analyses were rejected as being contaminated due to trace inheritance	*Pa. hassi* conodont zones, Frasnian. This age in Belpre Ash is correlated to Centre Hill Ash, Frasnian	1	Tucker *et al.* (1998), Over (1999)	381.1 ± 1.6	U–Pb
6	K-bentonite	Whyteville, Virginia, USA	Tioga Zone with 7 ashes	Four concordant zircon analyses in one ash bed give 391.4 ± 1.8 Ma. Data agrees with the concordant monazite age of 390.0 ± 0.5 Ma for the same ash at Union County, PA, USA	All Tioga ashes are between conodont-bearing strata of the *Po. c. costatus* zone, middle Eifelian	1	Tucker *et al.* (1998; zircon age), Roden *et al.* (1990; monazite age)	391.4 ± 1.8 (zircon age) 390.0 ± 0.5 (monazite age)	U–Pb

7	K-bentonite	Sprout Brook, Cherry Valley, NY, USA	Esopus Fm.	Four concordant zircon analyses in the lower ash; 5 analyses in higher ash are internally discordant and rejected	Indirect correlation with brachiopods to the conodont zones *Po. dehiscens*, *Po. gronbergi*, and lower *Po. inversus*, lower Emsian	2	Tucker *et al.* (1998)	408.3 ± 1.9	U–Pb
8	Tuff	Limekilns	Winburn	U–Pb dates on 25 zircons, one outlier date deleted; HR–SIMS (SHRIMP) age was calibrated to standard QGNG which gives older age than SL13[c]	*Nowakia acuaria*, Pragian index in basal shale below age sample; sample age estimate Pragian–early Emsian	2	Jagodzinski and Black (1999)	409.9 ± 6.6	U–Pb
9	Volcaniclastic	Type section	Turondale	27 zircon dates calibrated against the QGNG standard in the HR–SIMS (SHRIMP) method that is older than the SL13 standard[c]	Replaced brachiopod fauna; its most likely extrapolated age is late Lochkovian	2	Jagodzinski and Black (1999)	413.4 ± 6.6	U–Pb
10	K-bentonite	Kalkberg, NY, USA	Kalkberg	10 small fractions of 4–12 zircon grains each give 4 concordant ages; weighted mean average age is 417.6 ± 1.0 Ma	*I. woschmidti* conodont zone in another section of Kalkberg Fm gives early Lochkovian age	1	Tucker *et al.* (1998)	417.6 ± 1.9	U–Pb
11	Tuff	Ludlow, UK	Upper Whitcliffe	1 cm clay-rich tuff has isotope dilution age of 420.2 ± 3.9 Ma, using 4 concordant zircon analyses, each of 5–20 grains	From regional correlations considered close to Ludlow–Pridoli boundary; rejected by Melchin *et al.* (this study) because of poor biostratigraphic constraints, but allowed here with low weight in the fitting method	3	Tucker and McKerrow (1995), Tucker *et al.* (1998)	420.2 ± 3.9	U–Pb
12	Felsic volcanics	Canberra, Australia	Laidlaw	Average age from K–Ar (mineral) and Rb–Sr (whole rock and mineral) of 420.7 ± 2.2 Ma. Three families of older and younger HR–SIMS (SHRIMP) ages on zircons rejected due to inheritance contamination (Compston, 2000b).	Interbedded with Gorstian fossiliferous strata of the *N. nilssoni–L. scanicus* conodont zones	1	Wyborn *et al.* (1982)	420.7 ± 2.2 (95% limit)	K–Ar and Rb–Sr

(*cont.*)

Table 14.2 (*cont.*)

No.	Sample	Locality	Formation	Comment	Zone and age	Biostratigraphic reliability[a]	Reference	Age (Ma)	Type
13	Bentonite	Shropshire, UK	Middle Elton	Two biotite grains with slightly "discordant" age spectra give similar total-gas $^{40}Ar/^{39}Ar$ ages (424.5 and 425 Ma). The $^{40}Ar/^{39}Ar$ weight average plateau age of 423.7 ± 1.7 Ma is regarded as best age[d]	*N. nilssoni* and *L. scanicus* conodont zones, Gorstian	1	Kunk *et al.* (1985)	426.8 ± 1.7 (2-sigma)	Ar–Ar

[a] $^{40}Ar/^{39}Ar$ and HR–SIMS U–Pb ages have been adjusted for internal errors (see Notes *b–d* following and Chapter 6); measurement errors are 95% (2-sigma).

[b] Tucker & McKerrow (1995) and Cooper *et al.* (this study) found HR-SIMS age dates to be consistently younger than TIMS age dates for the same zircon suites by about 1.3% when plotting Ordovician, Silurian, or Devonian age dates. Apparently, previously unrecognized heterogeneity in the zircon Sri Lanka 13 (SL13) standard affects the accuracy of many HR–SIMS dates. Both Williams *et al.* (2000) and Compston (2000b) and our statistical analysis of dates around the Devonian–Carboniferous boundary (see this chapter) favor the HR–SIMS dates Nos. 1 and 2 to be shifted 1–2% (see also discussions in Chapters 12 and 14).

[c] Instead of common standard SL13, monitor standard QGNG was utilized; this practice shifts the age date from 406 to 409.9 Ma and slightly increases the standard error. The authors argue that, because the older age agrees well with the isotope dilution ages used to build the Devonian time scale, this age can be incorporated in the Tucker and McKerrow (1995) data set; a practice followed here. The QGNG standard requires further testing (Compston, 2000b).

[d] Age calibrated with MMhb-1 monitor standard of 523.1 Ma; Kunk *et al.* (1985) used an age of 519.4 Ma for this standard.

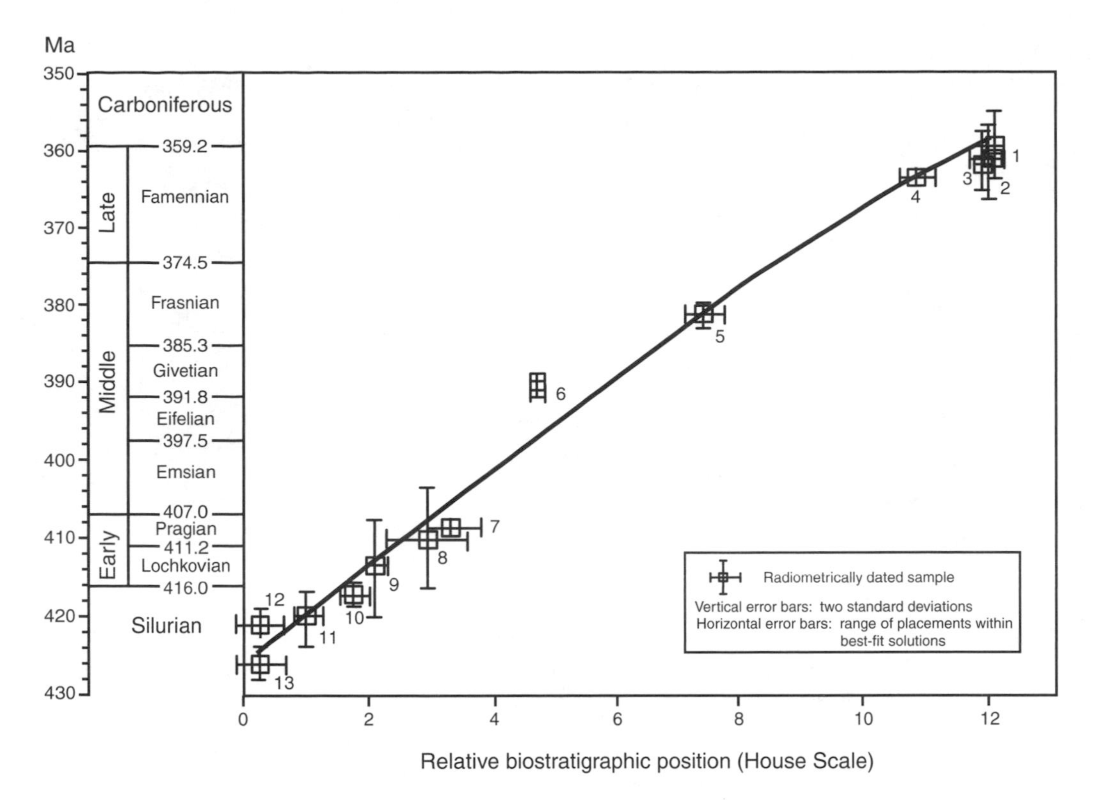

Figure 14.5 Construction of the Devonian time scale. Radiometric age dates for Devonian in Table 14.1 are plotted against the "House" relative zonal scale (shown in Figure 14.1). The "House scale" is a merger of high-resolution ammonoid and conodont (sub) zones into equal subzonal units. The best-fit line of the x/y data is a spline that is minimally smoothed. In order to obtain constraints for the ages of the Silurian–Devonian and Devonian–Carboniferous boundaries, critical age dates just below and above these boundaries were included in the calculations. For details see Chapters 8 and 14.

To get some idea of the sensitivity of the scaling to these simplifying assumptions, the Devonian time scale was also calculated using the relative scale of Tucker *et al.* (1998) where the stages are proportioned to an empirical scheme of graphic correlation and/or biostratigraphic intuition. The *pesavis* conodont zone is now placed in the Early Pragian. There is no significant difference between the ages in the "House" and the Tucker scales, but this might have been different had there been more detailed age dates to match the much higher resolution in the zonal schemes. For this study we prefer the House scale because it reduces the duration of the Emsian and Lochkovian stages in line with biostratigraphic reasoning and intuition. The scale also increases the duration of the Pragian Stage, which agrees with the analysis of cyclicity in the limestones in the classical Devonian sections of the Barrandian (Czech Republic) that suggest that the Pragian is not much shorter than the underlying Lochkovian (Chlupáč, 2000).

In a recent study, B. Kaufmann (pers. comm., 2003; see Section 14.4) obtained new U–Pb zircon ages from two classical Devonian localities in Germany. The first is from Steinbruch Schmidt (Kellerwald) where a 3 cm thick bentonite up in the *rhenana* conodont zone, of late Frasnian age yielded a U–Pb age of 376.0 ± 1.5 Ma. A second U–Pb age date of 398.6 ± 1.7 Ma is given in the Wetteldorf Section, the GSSP for the Lower–Middle Devonian boundary. Although preliminary and awaiting full documentation, both dates agree well with the results obtained here. Kaufmann makes the observation (an opinion echoed by M. Villeneuve) that it is an enigma that the 391.4 ± 1.8 Ma zircon date and the 390.0 ± 0.5 Ma monazite date of the Tioga Ash (see Table 14.2) do not fit the spline curve of Fig. 14.5. The new U–Pb date of 398.6 ± 1.7 Ma agrees well with the cubic-spline curve.

The calculated duration and ages with estimates of uncertainty (2-sigma) of the Devonian stages are given in Table 14.3. The total duration of the Devonian is 56.8 myr, which matches estimates of 56 myr by Tucker *et al.* (1998), Williams *et al.* (2000a), and Compston (2000b). Estimation of uncertainty limits on the stage boundaries may be improved with more detailed information on the stratigraphic position of the radiometric age dates. An improved scaling of the Devonian Period will require a detailed composite zonal standard as was calculated for the Ordovician and Silurian, and, especially, additional radiometric age dates at precise stratigraphic levels.

Table 14.3 *Ages and durations of Devonian stages*

Period	Epoch	Stage	Age of base (Ma)	Est. ± myr (2-sigma)	Duration	Est. ± myr (2-sigma)
Carboniferous			**359.2**	2.5		
Devonian						
	Late					
		Famennian	374.5	2.6	15.3	0.6
		Frasnian	385.3	2.6	10.8	0.4
	Middle					
		Givetian	391.8	2.7	6.5	0.3
		Eifelian	397.5	2.7	5.7	0.2
	Early					
		Emsian	407.0	2.8	9.5	0.4
		Pragian	411.2	2.8	4.2	0.2
		Lochkovian, (base of Devonian)	**416.0**	2.8	4.8	0.2
Silurian						

14.4 APPENDIX

B. Kaufmann (University of Tübingen, Germany) kindly provided some details of new datings, that reached the editors during completion and proofing of this chapter; the new dates, once peer-reviewed, are intended for publication. With their tight stratigraphic and linear age constraints, the preliminary new dates indicate the Devonian–Carboniferous boundary to be near 360 Ma, and the Pragian–Emsian boundary to be slightly below 410 Ma.

ITEM A

Sample: K-bentonite of bed 36
Locality: Steinbruch Schmidt, Germany
Comment: 24 single-zircon analyses yielded 17 concordant $^{206}Pb/^{238}U$ results that form on elongate cluster
Zone and stage: Middle part of Late *rhenana* conodont zone, Upper Frasnian
Age: 376.0±1.5 Ma

ITEM B

Sample: K-bentonite
Locality: Wetteldorf Section,Germany
Formation: Heisdorf Formation
Comment: Five concordant U–Pb TIMS analyses on single zircons
Zone and stage: Uppermost *Patulus* Zone 13 m above Lower–Middle Devonian boundary (GSSP)
Age: 398.6 ± 1.7 Ma

ITEM C

Sample: Hans-Platte volcaniclastic layer
Locality: Eschenbach quarry near Bundenbach, Germany
Formation: Hunsrück Slate
Comment: Ten analyses of single zircons; five concordant $^{206}Pb/^{238}U$ results
Zone and stage: Upper *Nowakia zlichovensis* dacryoconarid zone = middle–upper *excavatus* conodont zone, Lower Emsian
Age: 410.6 ± 0.9 Ma

ITEM D

Sample: K-bentonites of bed 79
Locality: Hasselbachtal, Germany
Formation: Hangenberg Limestone
Comment: Single zircons; three concordant points form a tight cluster with a $^{206}Pb/^{238}U$ concordia age
Zone and stage: *S. sulcata* conodont zone, 43 cm above Devonian–Carboniferous boundary, Lower Tournaisian
Age 360.5 ± 1.0 Ma

ITEM E

Sample: K-bentonites of bed 70
Locality: Hasselbachtal,Germany
Formation: Hangenberg Limestone
Comment: Ten single zircons; five concordant points form a tight cluster with a $^{206}Pb/^{238}U$ concordia age

Zone and stage: Lower *duplicata* conodont zone, 100 cm above Devonian–Carboniferous boundary, Lower Tournaisian

Age: 358.7 ± 0.7 Ma

New conodont biostratigraphy (*Icriodus curvicandata* and *I. celtibericus*) of the Carlisle stratigraphic unit, above the basal-Esopus Formation, suggests the 408.3 ± 1.9 Ma age of the Esopus Formation (Item 7 in Table 13.2) is now constrained to the lower *excavatus* Zone, lowermost Emsian. This will better constrain that interval than shown in the approximate fit of Fig. 14.5.

Kaufmann also kindly provided stratigraphic details of the isotopic ages of the Keel–Enach and Horses Glen tuffs (Williams *et al.*, 2000b). These dates were not included in the Devonian scale due to considerable stratigraphic age uncertainty. Combined single-zircon grain U–Pb dates in two Keel–Enach tuffs yielded 384.9 ± 0.4 Ma, with a biostratigraphic assignment of Tco Oppel miospore zone, correlated to an interval from the base of the *hermanni–cristatus* to approximately the top of the *falsiovalis* conodont zones, late Givetian–earliest Frasnian. The Horses Glen tuffs yielded a multigrain zircon U–Pb age of 378.5 ± 0.2 Ma. The Moll's Gap quarry microflora in an outcrop in Old Red Sandstone *below* the tuffs is assigned to lower miospore zone IV, which correlates to the upper *hassi–linguiformis* conodont zones, mid–late Frasnian. Hence, the date itself could range stratigraphically higher into the Famennian. A visual inspection of these two dates relative to the spline curve in Fig. 14.5 suggests that the latter fits the time scale trend but the former is stratigraphically too young.

15 • The Carboniferous Period

V. DAVYDOV, B. R. WARDLAW, AND F. M. GRADSTEIN

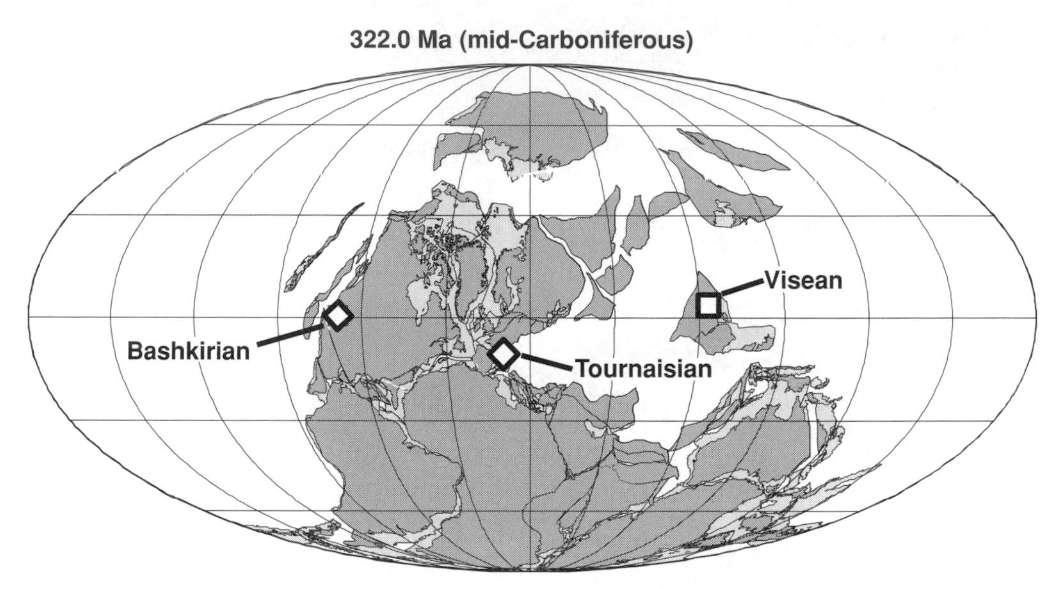

Geographic distribution of Carboniferous GSSPs that have been ratified (diamonds) or are candidates (squares) on a mid-Carboniferous map (status in January, 2004; see Table 2.3). Only the GSSPs for the Bashkirian (base of the Pennsylvanian) and Tournaisian (base of the Mississippian) have been formalized.

The supercontinent Pangea formed. Major changes in ocean circulation; biogeographic differentiation; high bio-provincialism; diversification of land plants and increased continental weathering rates and storage of organic carbon as coal; drawdown of atmospheric CO_2 and significant cooling, major glaciation, and sharp sea-level fluctuations; cyclic marine sequences; appearance of reptiles (with amniotic egg reproduction) and occupation of new (dry-land) niches; extinction or decreasing role of early Paleozoic biota such as stromatoporids, tabulate corals, trilobites, ostracods, heavily armored marine fish; appearance or very rapid diversification of foraminifera, ammonoids, freshwater pelecpods, gastropods, sharks, ray-finned fishes, and wingless insects. Late Carboniferous–Early Permian Kiaman Superchron is the longest known period of predominantly reversed polarity.

15.1 HISTORY AND SUBDIVISIONS

Because of climatic variability, the Carboniferous was a time of incredible diversification and abundant terrestrial biota. It signifies Earth's first episode of massive coal formation. The commercial production of coal led to the early development of Carboniferous stratigraphic classifications in three major regions: western Europe, eastern Europe, and North America.

Indeed, the name Carboniferous is derived from the Italian *Carbonarium* (charcoal producer) or Latin *carbo* (charcoal) and *ferous* (i.e. bearing). The term Carboniferous was first used in an adjectival sense by Kirwan (1799) and as an informal term for the section heading to describe "Coal-measures or Carboniferous Strata" by Farey in 1811 (Ramsbottom, 1984). Later it became a common term for coal-producing sediments in Great Britain and western Europe. Four stratigraphic units, in ascending order, were included:

1. the Old Red Sandstone, later assigned to the Devonian;
2. the Mountain, or Carboniferous Limestone, first listed by William Phillips in 1818;
3. the Millstone Grit, proposed by Whitehurst in 1778; and
4. the Coal Measures, proposed by Farey in 1807.

A Geologic Time Scale 2004, eds. Felix M. Gradstein, James G. Ogg, and Alan G. Smith. Published by Cambridge University Press.

Conybeare and Phillips in 1822 constituted these units as the Carboniferous or Medial Order, and Phillips in 1835 as the Carboniferous System (Ramsbottom, 1984).

Even though the Carboniferous was one of the first-established geological periods, it is one of the most complicated and confusing in terms of stratigraphic classification and correlation. The reasons for this confusion are:

1. the assembly of the supercontinent Pangea by the collision of Laurussia with Gondwana, the Variscan, and Hercynian orogenies, and the subsequent nearly complete separation of tropical and subtropical shelves by this supercontinent; and
2. the Gondwana superglaciation and the consequent drastic climatic changes and sea-level fluctuations, and significant biogeographic differentiation.

The current subdivision of the Carboniferous Period as favored by the Subcommission on Carboniferous Stratigraphy of the ICS is shown in Fig. 15.1, and selected regional stages are illustrated in Fig. 15.2. As of January 2004, only the major division of the Carboniferous Period into the Mississippian and Pennsylvanian sub-periods has been officially ratified by the IUGS.

15.1.1 Evolution of traditional European and Russian subdivisions

In western Europe, historically, the Carboniferous System was divided into the Lower Series (marine Mountain Limestone) and the Upper Series (predominantly terrestrial Millstone Grit and Coal Measures. Between 1841 and 1845, Murchison and others, in collaboration with Russian geologists and based on Russian sections, divided the Carboniferous into Lower, Middle, and Upper Stages, which were later elevated to Series (Moeller, 1878, 1880). In terms of recent chronostratigraphy, Moeller's Upper Series included Upper Pennsylvanian and most of the Cisuralian (Lower Permian) Series, excluding the Kungurian Stage, and therefore all three proposed series were approximately equal. Subsequently, the Cisuralian portion of the succession was excluded from Moeller's Upper Series. The remaining Lower, Middle, and revised Upper Series then constituted a three-fold subdivision of the Carboniferous that became the tradition in Russia and surrounding territories in Eastern Europe and Asia.

Munier-Chalmas and de Lapparent (1893) named these three Carboniferous subdivisions the Dinantian (Mountain Limestone), Moscovian, and Uralian Stages. The Westphalian (Millstone Grit and Coal Measures combined) and Stephanian Stages were also established by these geologists in terrestrial successions of western Europe as equivalents to the Moscovian and Uralian in the marine facies. This was the first attempt to build a dual marine–terrestrial classification for the Carboniferous, which is still advocated by some geologists (Wagner and Winkler Prins, 1997), but is not accepted by the International Commission on Stratigraphy.

In the nineteenth century the majority of Russian geologists accepted a two-fold subdivision of the Carboniferous. Nikitin (1890) proposed Serpukhovian (type locality near Serpukhov City) and placed it at the highest position in the Lower Carboniferous. He proposed to divide the Upper Carboniferous into the Moscovian and Gzhelian stages with type localities around Moscow (Moscovian) and a series of exposures near the villages of Gzhel, Pavlovo-Posad, and Noginsk (Gzhelian), along the Klyazma River and in the Oksko–Tsna Swell (Nikitin, 1890, pp. 77–78). The Moscovian originally included all the present-day Kasimovian, and the Gzhelian included the rest of the Carboniferous and the Asselian Stage (Lower Permian) of the modern scale. The Kasimovian (initially known as the *Tiguliferina* Horizon) was separated from the Moscovian by Ivanov (1926) and was named by Dan'shin (1947). Teodorovich (1949) suggested this unit should be a chronostratigraphic stage and regarded the Gzhelian (in the modern sense) to be the uppermost stage of the Carboniferous.

Ruzhenzev (1945) proposed the Orenburgian Stage, equal to the "*Pseudofusulina*" Horizon of Rauser-Chernousova (1937), in the southern Urals, as the latest Carboniferous stage. Because of a miscorrelation to the Russian Platform, the Orenburgian was thought to be the equivalent of the lower Asselian (Pnev *et al.*, 1975; Harland *et al.*, 1990) and was removed from the Russian stratigraphic scale. However, this miscorrelation was recognized and the Orenburgian was returned to the Carboniferous once again (Davydov and Popov, 1986), where it is generally merged with the Gzhelian Stage. It has been suggested that the Orenburgian should be the terminal stage of the Carboniferous above the Gzhelian (Ivanova and Rosovskaya, 1967; Davydov, 2001).

Development of Carboniferous stratigraphy during the nineteenth and early twentieth centuries in western Europe was summarized in the Heerlen Congresses in 1927 and 1935 (Heerlen Classification), in which two-fold division of the Carboniferous was formalized and the Lower Carboniferous was replaced by the Dinantian. The latter was divided into two stages: the Tournaisian (proposed by Koninck in 1872) and the Visean (proposed by Dupont in 1883). The Upper Carboniferous (later named Silesian) was also divided into two stages: the Namurian (name proposed by Purves in 1883 for

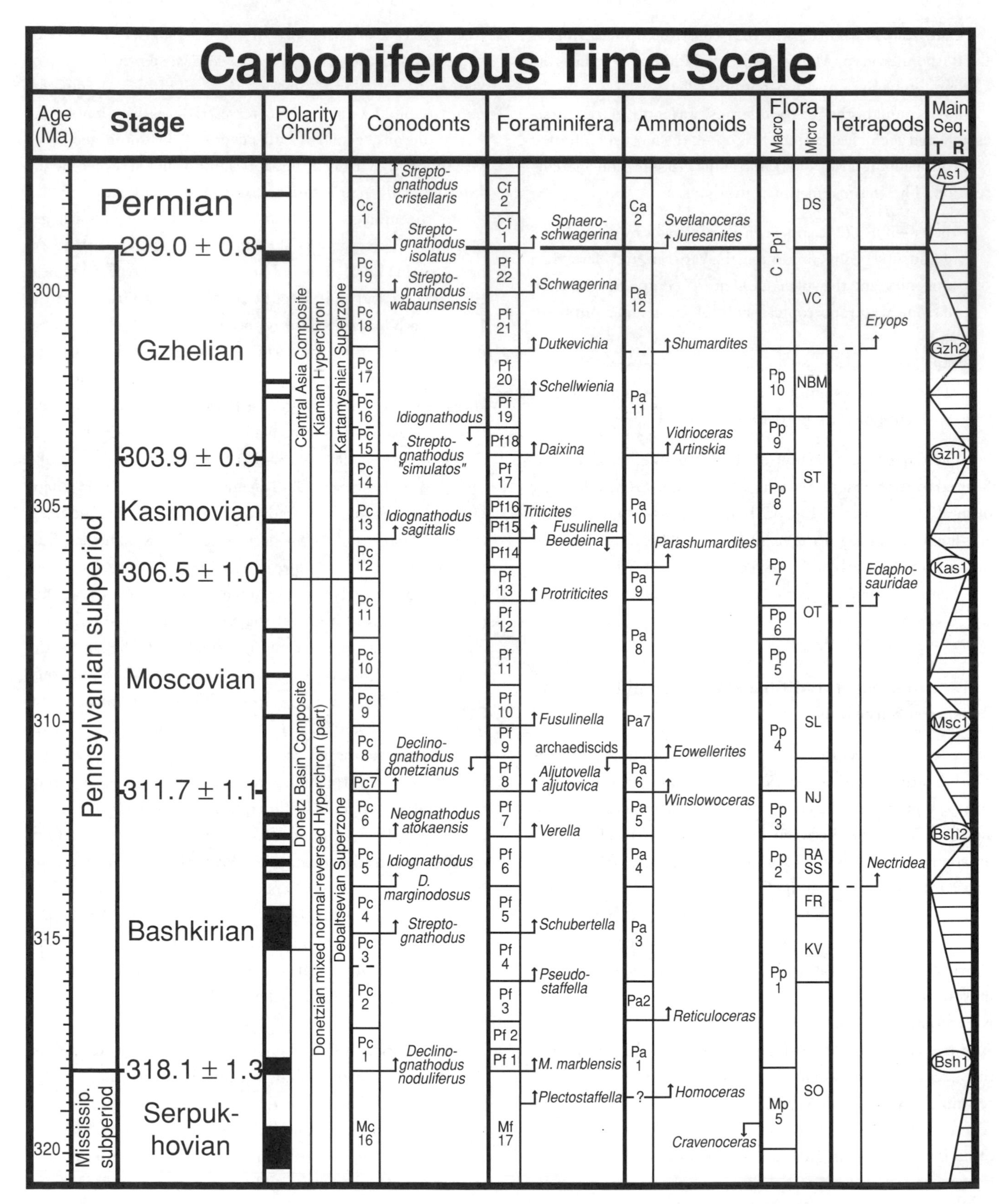

Figure 15.1 Subperiod and stage boundaries for the Carboniferous Period, with pelagic zonations of conodonts, foraminifera and ammonoids, and with terrestrial facies zonations of vertebrates, spores, and plants. The geomagnetic polarity scale with Russian-based nomenclature of super- and hyperchrons (see text) and the principal eustatic trends are also displayed. A color version of this figure is in the plate section.

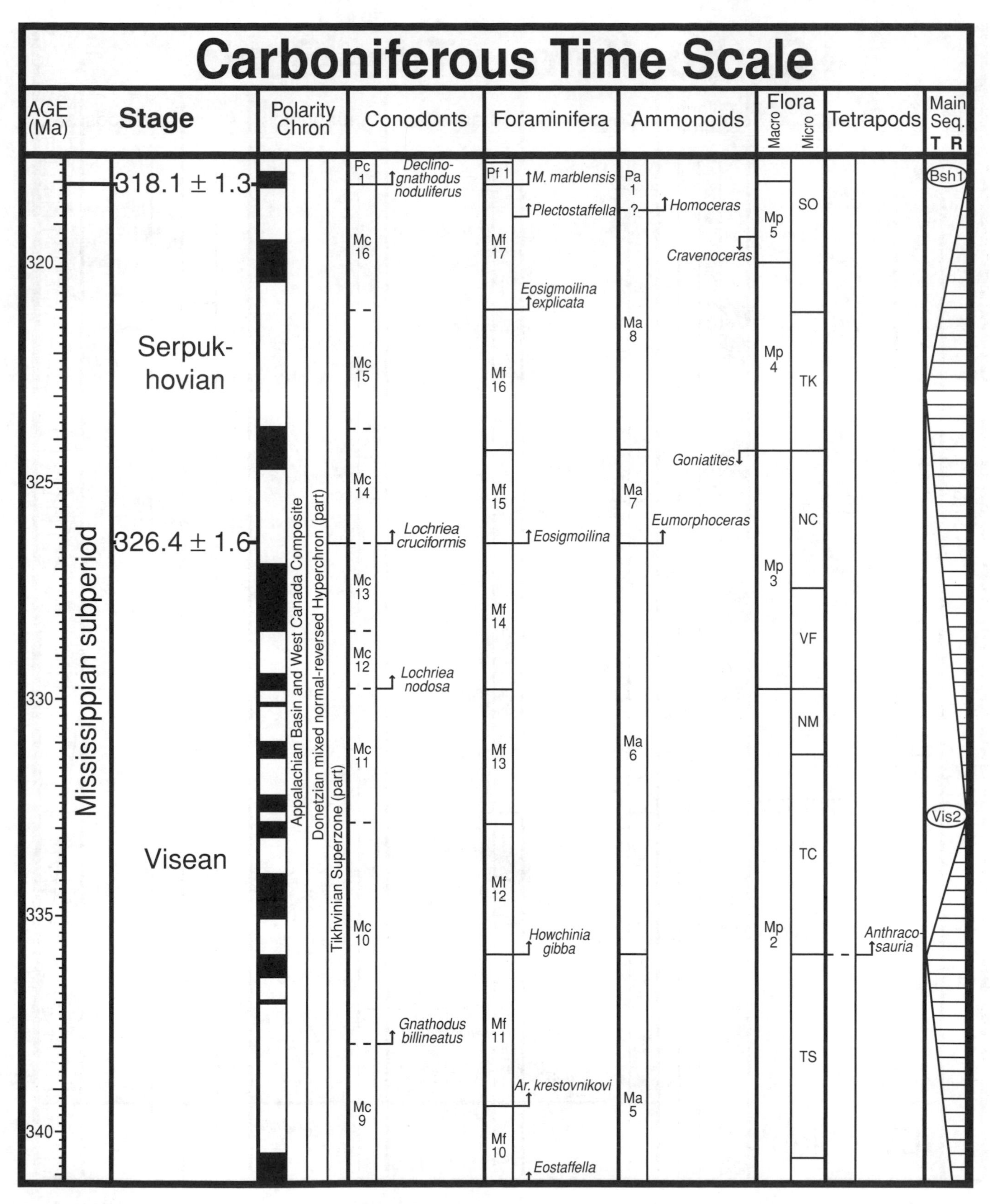

Figure 15.1 (*cont.*)

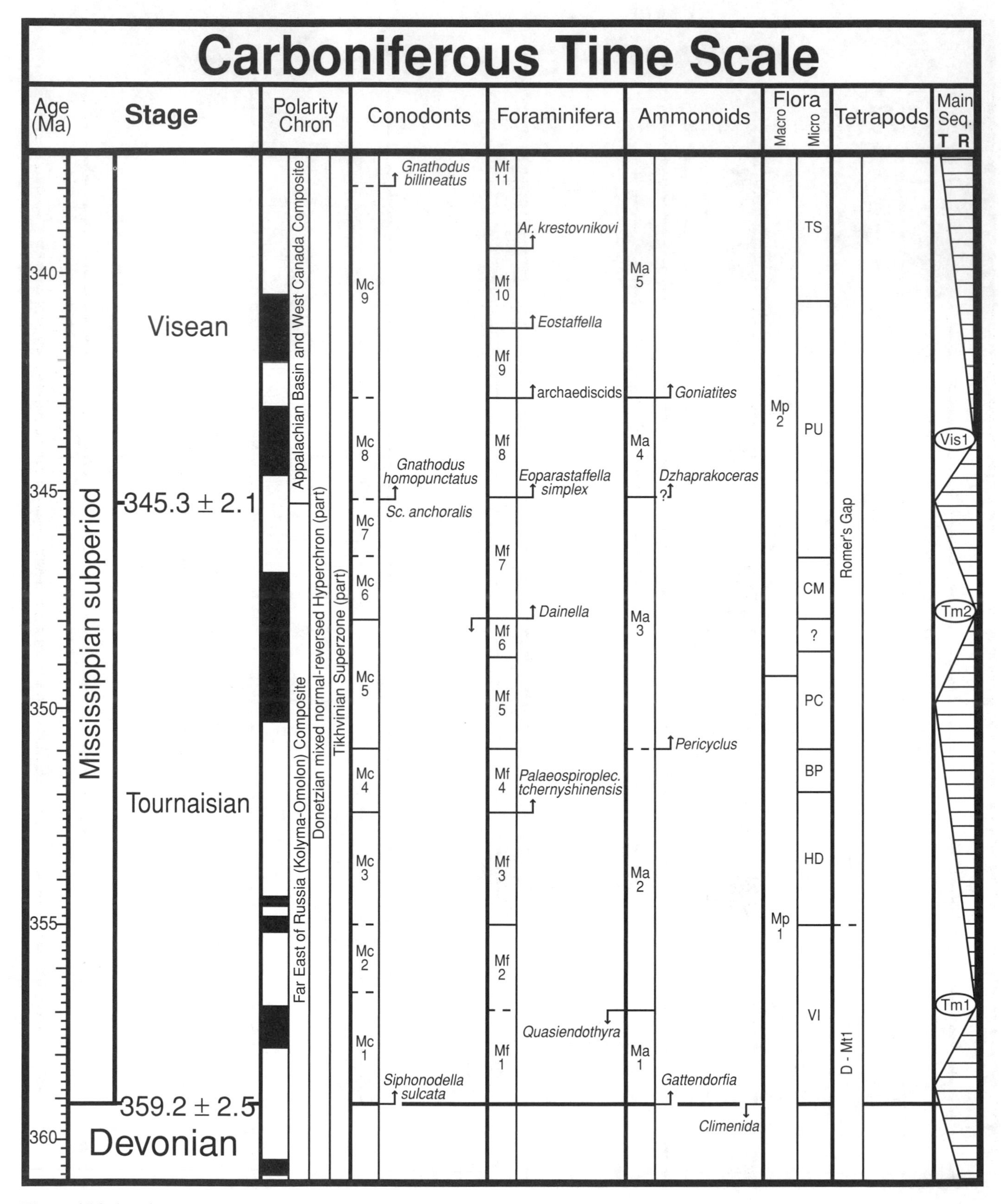

Figure 15.1 (*cont.*)

Carboniferous Regional Subdivisions

AGE (Ma)	Subperiod	Stage	Boundary age	Boundary-defining event	Russian Platform	Western Europe		North America	South China
		Permian			Sjuranian	Autunian	Lebach	Bursumian	Mapingian
300	Pennsylvanian subperiod	Gzhelian	299 ± 0.8	FAD *S. isolatus*	Melekhovian		Kuzel		Xiaodushanian
					Noginian			Virgilian	
					Pavlovoposadian	Stephanian	Stephanian C		
					Rusavkian				
		Kasimovian	303.9 ± 0.9		Dorogovilovian		Stephanian B	Missourian	
305					Khamovnichean		A Barruelian		
					Krevyakian		Cantabrian		
		Moscovian	306.5 ± 1.0		Peskovian	Westphalian	D Asturian	Desmoinesian	Dalanian
					Myachkovian				
					Podolskian				
					Kashirian				
310					Tsninian		C Bolsovian	Atokan	
					Vereian				
		Bashkirian	311.7 ± 1.1		Melekesian		B Duckmantian		
					Cheremshanian		A Langsettian		Huashibanian
					Prikarnian	Namurian	Yeadonian	Morrowan	
315					Severokeltmenian		Marsdenian		
					Krasnopolyanian		Kinderscoutian		
					Voznesenian		Alportian		Luosuan
							Chokierian		
	Mississippian subperiod	Serpukhovian	318.1 ± 1.3	FAD *D. noduliferus*	Zapaltyubian		Arnsbergian	Chesterian	Dewunian
320					Protvian				
					Steshevian				
325					Tarussian		Pendleian		Shangsian
		Visean	326.4 ± 1.6	FAD *L. cruciformis*	Venevian	Brigantian			
330					Mikhailovian	Asbian			
					Aleksian	Holkerian		Meramecian	Jiusian
335					Tulian				
340					Bobrikovian	Arundian		Osagean	
					Radaevian	Chadian			
345		Tournaisian	345.3 ± 2.1		Kosvinian	Courceyan	Ivorian		Tangbagouan
					Kizolovian				
350					Cherepetsian		Hastarian	Kinderhookian	
					Karakubian				
					Upinian				
355					Malevian				
					Gumerovian				
360		Devonian	359.2 ± 2.5	FAD *S. sulcata*	Ziganian		Famennian	Chatauquan	Gelaohean

Figure 15.2 Correlation of regional Carboniferous stages and other stratigraphic subdivisions.

the equivalent of the Millstone Grit), and the Westphalian, each with three divisions (A, B, and C; Ramsbottom, 1984).

Further biostratigraphic studies in marine and terrestrial successions in western Europe, particularly of goniatites and plants, resulted in significant refinement of the regional stratigraphic scale into 16 stages from the Tournaisian through Westphalian (George *et al.*, 1976). Details of the history of the Carboniferous classification in western Europe can be found in George *et al.* (1976), Ramsbottom (1984), and Wagner and Winkler Prins (1997).

15.1.2 Evolution of traditional North American subdivisions

In the USA, the Carboniferous was subdivided into the Mississippian, or lower Carboniferous, proposed by Alexander Winchell in 1870, and the Pennsylvanian, or upper Carboniferous, proposed by J. J. Stevenson in 1888, each of which was put forward as an independent system by Williams (1891). Predominantly marine rocks of early Carboniferous age were assigned to the Mississippian, with type localities in the upper Mississippi Valley. Coal-productive beds in the state of Pennsylvania were termed Pennsylvanian and considered as stratigraphic equivalents of the western European "Coal Measures." The Pennsylvanian included the Pottsville Conglomerate, Lower Productive Coal Measure, and Upper Barren Coal Measures. The type localities for the Pennsylvanian stages occur in marine cyclic sequences in the North American mid-continent basin in Arkansas (Morrowan), Oklahoma (Atokan), central Iowa (Desmoinesian), along the Missouri River in Iowa and Missouri (Missourian), and in east–central Kansas (Virgilian). More details on the establishment of the Pennsylvanian stages can be found in Heckel (1999).

15.1.3 Status of the international scale

The first attempt to build a global Carboniferous scale that integrated the various chronostratigraphic classifications was made during the Eighth International Congress on Carboniferous Stratigraphy and Geology in Moscow (Bouroz *et al.*, 1978). Two subsystems of the Carboniferous, the Mississippian and the Pennsylvanian, were subdivided into two series to successfully merge the usage of recognizing two systems in North America with the three-fold divisions in the former USSR (Wagner and Winkler Prins, 1994). Tournaisian, Visean, Serpukhovian, Bashkirian, Moscovian, Kasimovian, and Gzhelian were proposed as global Carboniferous stages (Fig. 15.1). Bouroz *et al.*'s proposal significantly advanced the development of international Carboniferous stratigraphy. Within a few years several working groups were founded (Brenckle and Manger, 1990) and two major Carboniferous boundaries – the base of the Carboniferous Period and the mid-Carboniferous (or Mississippian–Pennsylvanian) boundary – have been accepted as GSSPs (Paproth *et al.*, 1991; Lane *et al.*, 1999). The top of the Carboniferous is established by the GSSP for the base of the Permian (Davydov *et al.*, 1998).

15.1.4 Mississippian subsystem (Lower Carboniferous)

After several glaciation events during the Late Devonian (Famennian) with corresponding extinction events in marine and terrestrial biota (e.g. the Kellwasser and Hangenberg Events), the beginning of the Carboniferous (early–mid Tournaisian) was probably free of ice sheets (Walliser, 1984b; Joachimski and Buggisch, 1993; Streel *et al.*, 2000b). However, except for late Visean–early Serpukhovian (Chesterian), most of the rest of the Carboniferous was a time of extensive glaciation (Bruckschen *et al.*, 1999; Mii *et al.*, 1999, 2001) accompanied by high-frequency sea-level fluctuations and corresponding global transgressive–regressive sequences (Ramsbottom, 1973, 1981; Veevers and Powell, 1987).

During the Mississippian, there was generally unobstructed marine communication between paleo-Tethys and Panthalassan shelves. Therefore, Mississippian marine fauna are generally world-wide in their distribution, and latitudinal differences are more strongly developed than are longitudinal differences (Ross and Ross, 1988).

DEVONIAN–CARBONIFEROUS BOUNDARY AND THE TOURNAISIAN STAGE

The Devonian–Carboniferous boundary is a time of global regression with a major sequence boundary located slightly above it (Paproth *et al.*, 1991).

The GSSP for the base of the Tournaisian Stage, Mississippian sub-period, and Carboniferous Period has been defined in the La Serre section, Montagne Noire, southern France (Paproth *et al.*, 1991). The section lies on the southern slope of the La Serre hill, 2.5 km southwest of the village of Cabriéres, near the "classic" base of the *Gattendorfia* ammonoid zone.

The boundary is defined in Bed 89 at the FAD of the conodont *Siphonodella sulcata*, within the evolutionary lineage of *S. praesulcata–S. sulcata*. The transitional boundary beds in the section consist of platy and nodular cephalopod-bearing and oolitic limestone, and contain a mixture of pelagic and near-shore biota that show signs of minor transport before lithification, but no signs of stratigraphic reworking (Feist and Flajs, 1987). Abundant conodonts with shallow-marine invertebrate assemblages (diverse brachiopods, trilobites, corals, and rare ostracods, cephalopods, foraminifera, algae, and microproblematica) make La Serre section unique. Foraminifera, algae, and microproblematica are found only in the Carboniferous part of the section, and miospores and acritarches are almost completely absent in the transitional beds. The distribution of the latter group of fossils is well established in the Auxiliary Stratotype sections (ASS) at Hasselbachtal (Sauerland area, Germany) and Nanbiancun (southern China; Becker and Paproth, 1993; Wang, 1993). The last appearance datum (LAD) of the miospore *Retispora lepidophyta* (*Retispora lepidophyta–Verrucosisporites nitidus* Biozone [LN]) almost coincides with the FAD of the conodont *Siphonodella sulcata*.

In the ammonoid zonation, the base of the Tournaisian coincides with the first entry of *Gattendorfia* and *Gattenpleura* (Kullmann *et al.*, 1990). The major extinctions of the ammonoid goniatites and clymenids occurred during the Hangenberg Event of the latest Famennian extinction crisis (Streel *et al.*, 2000b). The FAD of the foraminiferal species *Chernyshinella glomiformis* and *Tournayella beata* characterizes the base of Tournaisian in its type area in the Franco–Belgian basin (Conil *et al.*, 1990).

VISEAN

Historically, the Visean Stage, like the Tournaisian, was established in southern Belgium. The base of the Visean was first officially defined in 1967 by the International Congress on Carboniferous Stratigraphy and Geology in Sheffield at the lowest *marbre noir* (i.e. first black limestone) intercalation in the Leffe facies at the Bastion Section in the Dinant Basin (George *et al.*, 1976; Hance *et al.*, 1997). This boundary coincided with the first occurrence of the foraminiferal genus *Eoparastaffella*, and occurs less than 1 m below the FAD of the conodont *Gnathodus homopunctatus* (Conil *et al.*, 1990).

Because of historical priority and achievement of stratigraphic stability in international stratigraphic practice, the Tournaisian–Visean boundary working group decided to keep the base of the Visean as close as possible to the classical level (Paproth, 1996). A significant turnover in pelagic fossils (either conodonts or ammonoids) does not occur near the classical Tournaisian–Visean boundary, so it was proposed to define the boundary within the foraminiferal phylogenetic lineage of *Eoparastaffella*. Simple morphological parameters, identifying two successive morphotypes (types 1 and 2, corresponding, respectively, to *E. rotunda typical* and *E. simplex typical*) can be used by non-foraminiferal specialists to recognize this evolution and identify the base of the Visean. This proposal has been accepted as the most appropriate definition of the classic base of the Visean (Work, 2002). The proposed definition is the first benthic marker within the entire Paleozoic. A proposed GSSP section for the base of the Visean Stage is the Penchong section, Guangxi, South China (Devuyst *et al.*, 2003).

The disappearance of deeper-water 'Tournaisian' conodonts and the entry of shallow-water 'Visean' foraminifera in the classic Franco–Belgian basin appears to be an ecological event related to a significant sea-level drop in late Tournaisian–early Visean time. Hence, this faunal change may be of questionable chronostratigraphic significance (Hance *et al.*, 1997).

SERPUKHOVIAN

The Serpukhovian Stage was proposed by Nikitin (1890) as the terminal stage of the lower Carboniferous "Series" and was named after Serpukhov City, where the stage is exposed in a series of quarries and localities along the Oka River. The Serpukhovian is perhaps the shortest chronostratigraphic unit in the Mississippian. In the type area of the Moscow syncline, it is represented by an unconformably bounded sequence of open- to restricted-marine shallow carbonate and mixed carbonate–siliciclastic rocks.

The Serpukhovian in the Russian Platform and surrounding regions is informally divided into lower and upper substages and into five horizons (regional stages): Tarusian (Tarusky), Steshevian (Steshevsky), Protvinian (Protvinsky), Zapaltubinian (Zapaltubinsky), and Voskresenian (Voskresensky), the last two being almost completely absent in the type area (Makhlina *et al.*, 1993).

More complete successions of the upper Serpukhovian and its transition to the Bashkirian are present in the Donetz Basin, in the northern portion of Timan–Pechora, and in the Urals (Aizenverg *et al.*, 1983; Kulagina *et al.*, 1992; Kossovaya *et al.*, 2001).

The lower boundary of the Serpukhovian still remains uncertain. The working group of the Subcommission on the Carboniferous of the ICS, responsible for defining the GSSP close to the Visean–Serpukhovian (Namurian) boundary, was only recently established. The Visean–Serpukhovian transition coincides with a major Gondwanan glaciation (Mii *et al.*, 2001) and sharp climatic changes coupled with severe marine biotic endemism. In the type area in the Moscow Basin, the Visean–Serpukhovian transition deposits were restricted, and connections with open basins in the east (Urals), south

(Precaspian), and north (Timan–Pechora) were short lived and/or limited. Benthic fauna in the Russian Platform are diachronous in their distribution (Makhlina *et al.*, 1993) and, therefore, were strongly controlled by ecological rather than by evolutionary factors.

The lower boundary of the Serpukhovian in the Moscow Basin has been referred to the base of the foraminiferal zone *Pseudoendothyra globosa–Neoarchaediscus parvus*. The latter corresponds with the Cf7 foraminifera Zone of Conil *et al.* (1990) in Franco–Belgian basins. Gibshman (2001) has recently re-studied the foraminifera in the type locality of the Serpukhovian Zaborie quarry, near Serpukhov City, and placed the lower boundary of Serpukhovian between the latest Visean *Eostaffella tenebrosa* Zone and the earliest Serpukhovian *Neoarchaediscus parvus* Zone. This boundary is purportedly coincident with a eustatic maximum flooding surface. Similarly, in the Timan–Pechora region, the Visean–Serpukhovian boundary is supposedly coincident with a maximum flooding surface (Kossovaya *et al.*, 2001).

The base of the Serpukhovian Stage closely coincides with the base of Namurian and, therefore, the Pendleian Stage of Great Britain. It is an important chronostratigraphic boundary recognized in western Europe as a widespread marine flooding surface (Ramsbottom, 1977).

15.1.5 Pennsylvanian subsystem (Upper Carboniferous)

MISSISSIPPIAN–PENNSYLVANIAN BOUNDARY AND THE BASHKIRIAN STAGE

The beginning of the Pennsylvanian coincides with a significant mid-Carboniferous glaciation and consequent sea-level drop forming a sequence boundary in many sections. Veevers and Powell (1987) proposed that this boundary marks the beginning of a major Gondwanan glaciation and climate cooling. However, the glaciation likely started earlier in the Serpukhovian and reached its maximum near the mid Carboniferous (Popp *et al.*, 1986; Gonzales, 1990; Grossman *et al.*, 1993; Mii *et al.*, 2001).

The Bashkirian Stage was established in the early 1930s (Semikhatova, 1934) in the mountains of Bashkiria (currently Bashkortostan Republic) in the southern Urals, Russia. Several sections have been suggested as a stratotype for the Bashkirian, but since the mid 1970s, the Askyn section, one of the best exposed and studied sections in the type area, has been accepted as a hypostratotype for the stage (Semikhatova *et al.*, 1979; Sinitsyna and Synitsyn, 1987; Nemirovskaya and Alekseev, 1995; Kulagina *et al.*, 2001). The basal beds of the Bashkirian as it was defined, were originally characterized by the appearance of the foraminiferal species *Pseudostaffella antique,* i.e. the base of the Bashkirian within the chronostratigraphic succession was definitely higher than its modern position.

The GSSP for the beginning of the Pennsylvanian and of the Bashkirian Stage is defined within the transition from *Gnathodus girty simplex* to *Declinognathodus noduliferus s.l.*, and has been fixed in Arrow Canyon, in the Great Basin, Nevada, USA (Lane *et al.*, 1999). The type section is located approximately 75 km northwest of Las Vegas, on the east side of a strike valley immediately north of Arrow Canyon gorge. The section begins at the top of the Battleship Wash Formation, goes up through the overlying Indian Springs Formation, and into the lower part of the Bird Spring Formation, where the GSSP boundary horizon is located. During Carboniferous time, Arrow Canyon was situated near the paleoequator in a pericratonic tropical to subtropical seaway that extended from southern California through western Canada and into Alaska (Lane *et al.*, 1999). The succession containing this the Mid-Carboniferous Boundary interval (Bird Spring Formation) comprises numerous high-order transgressive–regressive sequences driven by glacio-eustatic fluctuations caused by ongoing glaciation in Gondwana. Transgressive bioclastic and mixed bioclastic–siliciclastic limestones and fine sandstone sequences are separated by regressive mudstone, oolitic, and brecciated limestone intervals with paleosols.

The Mid-Carboniferous Boundary GSSP is coincident with the base of the *noduliferus–primus* conodont zone of Baesemann and Lane (1985). Beds below the Mid-Carboniferous Boundary contain an abundant archediscacean foraminiferal assemblage that is dominated by the eosigmoilines *Eosigmoilina robertsoni* and *Brenckleina rugosa*, and these species disappear slightly above the Mid-Carboniferous Boundary (samples 62 and 63 at Arrow Canyon). The foraminifer *Globivalvulina bulloides* is a useful marker to approximate the boundary, although in Japan, Arctic Alaska, the Pyrenees, and the Donetz Basin it appears slightly below the boundary. The foraminifera *Millerella pressa* and *M. marblensis* are also informal markers for the boundary, although primitive *Milerella* appears in the late Mississippian (Rauser-Chernousova *et al.*, 1996; Brenckle *et al.*, 1997).

In the eastern hemisphere (Arctic, Pyrenees, Russian Platform, Urals, Donetz Basin, Central Asia, Japan), the base of the Bashkirian is marked by the occurrence of foraminiferal species *Plectostaffella varvariensis*, *P. jakhensis*, *P. posochovae,* and *P. bogdanovkensis*; however, the first primitive representatives of the genus appear slightly below the

Mid-Carboniferous Boundary. The base of the Bashkirian in the Urals and surrounding regions coincides with the boundary between the *Monotaxinoides transitorius* and *Plectostaffella varvariensis* (or *Plectostaffella posokhovae*) foraminiferal zones (Kulagina *et al.*, 1992).

In the ammonoid zonation, the Mid-Carboniferous Boundary is identified at the base of the *Homoceras* Zone or *Isohomoceras subglobosum* Zone of Great Britain, Nevada, and Central Asia (Ramsbottom, 1977; Ruzhenzev and Bogoslovskaya, 1978; Nikolaeva, 1995). However, it is most likely that the first *Isohomoceras subglobosum* in Nevada and Central Asia occurs slightly earlier, in the latest Mississippian (Serpukhoviani; Nemirovskaya and Nigmadganov, 1994; Titus *et al.*, 1997).

MOSCOVIAN

The classic sedimentary sequence of shallow-marine bioclastic limestone intercalated with colorful clays outcropping in and around the Moscow area gave rise to the Moscovian Stage (Nikitin, 1890). Nikitin proposed the quarries near the village of Myachkovo to be the type section of the stage. However, as mentioned earlier, the original Moscovian unit also included strata at locations around the villages of Khamovniki and Dorogomilovo, now within Moscow, and near the villages of Voskresensk and Yauza, east of Moscow. These sediments are now referred to the early–late Kasimovian Stage in terms of modern chronostratigraphy. Fortunately, the sequence at Myachkovo quarry belongs to the late Moscovian, keeping the Moscovian Stage valid.

Ivanov (1926) was the first to recognize the significant difference between fauna in the limestone of Khamovniki, Dorogomilovo, Voskresensk, and Yauza and fauna from Moscovian strata. He named these different sequences the "*Teguliferina*" Horizon (i.e. a regional stage in Russian sense) after an attractive and common brachiopod. Dan'shin (1947) named this horizon Kasimovian, as these sediments are well exposed near the city of Kasimov in the Oksk–Tsna Basin.

Ivanova and Khvorova (1955) and Makhlina *et al.* (1979) provided detailed sedimentology and stratigraphy of Moscovian through Gzhelian Stages in the Moscow Basin. Four horizons with chronostratigraphic meaning in the Moscovian Stage were established by Ivanov (1926), in ascending order: Vereisky, Kashirsky, Podol'sky, and Myachkovsky. Later these units were proposed to be substages or stages (Ivanova and Khvorova, 1955), and widely used in this sense in equatorial and subequatorial belts of the eastern hemisphere. However, these units were not recognized elsewhere.

In the type area of the Moscow Basin, strata of the Moscovian Stage unconformably overlie limestones assigned to the Mississippian, with Bashkirian strata missing. The lowermost Moscovian (Vereian Horizon) is made up of three formations: Shatskaya, Aljutovskaya, and Ordynskaya (Ivanova and Khvorova, 1955). The lower formation is a siliciclastic and largely terrestrial sequence with no marine fossils recovered. Recently, the Shatskaya Formation has been excluded from the Moscovian and the proposed beginning of the Moscovian Stage placed at the base of the Aljutovskaya Formation. In the type section of the Aljutovskaya Formation, 3 m above its base, a characteristic fusulinid species *Aljutovella aljutovica* and the conodont species *Declinognathodus donetzianus* were found (Makhlina *et al.*, 2001). *Aljutovella aljutovica* is an index species of thc foraminiferal zone commonly accepted to define the base of the Moscovian in the former Soviet Union. The *Aljutovella aljutovica* fusulinid zone correlates widely through the entire northern and eastern margins of Pangea (Solovieva, 1986). However, because of significant provincialism of fusulinid assemblages between eastern and western hemispheres, only conodonts can be used to define the base of the Moscovian Stage in the global chronostratigraphic scale. Nemirovskaya (1999) proposed the *Declinognathodus donetzianus* and/or *Idiognathoides postsulcatus* conodont species as indicative of the base of the Moscovian Stage. These species are widely distributed geographically (in the Moscow Basin, Urals, northwestern Europe, Spain, Canadian Arctic, Alaska, North American midcontinent, and Japan) and their evolutionary position is clearly recognized (Nemirovskaya, 1999; Goreva and Alekseev, 2001). Both species occur in the basal beds of the Moscovian Stage in its type area, and hence can be used for defining the traditional Bashkirian–Moscovian boundary.

Another conodont, *Neognathodus atokaensis,* has a wide geographical distribution, and formally could be a good candidate for the index species to define the base of the Moscovian. This species is ecologically dependent, has a high rate of morphological variability, and is not considered by conodont workers to be a single species (Nemirovskaya, 1999).

It is most likely that the base of the Moscovian in North America approximates with the first appearance of the genus *Profusulinella* (Moore and Thompson, 1949; Groves *et al.*, 1999).

In the ammonoid zonation, the Bashkirian–Moscovian boundary coincides with either the base of the *Winslowoceras–Diaboloceras* Zone (a zone based on genera, rather than species) on the Russian Platform and Urals (Ruzhenzev and Bogoslovskaya, 1978), or with the *Eowellerites* Zone in western Europe and North America (Ramsbottom and Saunders,

1985). It is most likely that the base of the *Eowellerites* Zone is slightly older than the base of the *Winslowoceras–Diaboloceras* Zone.

KASIMOVIAN

The Kasimovian Stage was originally included in the Moscovian (Nikitin, 1890). It was the last stage in the Pennsylvanian Series to be established in the Moscow Basin (Teodorovich, 1949). Ivanov (1926) was the first to recognize the independence of the "*Teguliferina*" Horizon (Kasimovian Stage) from the Moscovian, and considered it to be an "Upper Carboniferous" unit (in the sense of the three-fold Russian classification of the Carboniferous) as opposed to the "Middle Carboniferous" Moscovian Stage. He placed the base of his "*Teguliferina*" Horizon at the limestone conglomerate in the base of the local "sharsha" lithostratigraphic unit.

Macrofossil assemblages of the *Teguliferina* Horizon in the type region, comprising mostly brachiopods and pelecypods, have only local correlative potential. A more widely recognizable definition of the Kasimovian Stage was proposed by Rauser-Chernousova and Reitlinger (1954), with the first occurrence of the fusulinid species *Obsoletes obsoletes* and *Protriticites pseudomontiparus.*

Although these events are only found in the "sharsha" unit of the upper Suvorovskya Formation of the lowermost Kasimovian (i.e. a few meters above the original boundary proposed by Ivanov, 1926), the base of the Kasimovian has conventionally been placed at the conglomerate marking the base of the local "garnasha" lithostratigraphic unit of the lower upper Suvorovskya Formation, i.e. slightly below its original position.

The fusulinid definition proposed by Rauser-Chernousova and Reitlinger (1954) for the Kasimovian has been accepted for years in the shallow-marine sections in Eurasia. After the evolution of *Obsoletes* and *Protriticites* became better understood (Kireeva, 1964; Davydov, 1990; Remizova, 1992), it was suggested that the Moscovian–Kasimovian boundary be placed between the latest Moscovian *Praeobsoletes burkemensis* and *Protriticites ovatus* fusulinid zone and the early Kasimovian *Obsoletes obsoletes* and *Protriticites pseudomontiparus* fusulinid zone. In the Moscow Basin, the boundary was placed 1.0–1.5 m below the base of the "garnasha" lithostratigraphic unit, at the base of the local "lyska" lithostratigraphic unit within the upper Peskovskaya Formation, which traditionally has been included in the uppermost Moscovian (Davydov, 1997). This boundary coincides with a eustatic sea-level low, and, consequently, with disconformities represented in many sections around the world (Davydov, 1997, 1999; Ehrenberg *et al.*, 1998; Davydov and Krainer, 1999; Leven and Davydov, 2001). It should be noted that in the North American midcontinent this boundary occurs two major cycles below the previously proposed upper Pennsylvanian major sequence boundary at the base of Hepler Shales of the Pleasanton Group (Ross and Ross, 1988; Heckel *et al.*, 1998).

Although the working group of the ICS Subcommision for the Carboniferous to establish the GSSP for the Moscovian–Kasimovian boundary has been active for more than ten years and has gained significant insight into boundary correlations (Villa, 2001), no decision regarding definition of this boundary has been forthcoming. The conodont species *Streptognathodus subexcelsus,* that characterized the basal beds of the Kasimovian Stage in the Moscow Basin, has been described just recently (Goreva and Alekseev, 2001), and its correlation potential is not clear. *Idiognathodus saggitalis* and "*Idiognathodus*" *nodocarinatus,* proposed as potential index taxa for a GSSP of the Moscovian–Kasimovian boundary (Barrick and Lambert, 1999; Alekseev *et al.*, 2002), appear well above the existing boundary, and would cause a redefinition of both the Moscovian and Kasimovian Stages, and most likely the replacement of the Kasimovian Stage with the Missourian and, correspondingly, the Gzhelian Stage with the Virgilian.

An alternative definition of the GSSP within the evolutionary chronocline of primitive and advanced fusulinids belonging in the genus *Protriticites* has been proposed recently (Davydov *et al.*, 1999b; Davydov, 2002). These fauna are widely distributed throughout the Tethyan and Boreal biogeographic provinces and have now also been found in the North American Great Basin within mid to upper Desmoinesian Strata (Wahlman *et al.*, 1997; Davydov *et al.*, 1999b).

Although usage of benthic markers to define a GSSP could be considered a disadvantage, it is successfully used for the base of the Visean (Sevastopulo *et al.*, 2002). The proposed fusulinid definition will keep the Moscovian–Kasimovian boundary at its classical level and will provide stability of the generally accepted Pennsylvanian scale.

In the ammonite zonation, the Moscovian–Kasimovian boundary is set at the base of the *Dunbarites–Parashumardites* Zone of Ruzhenzev (1955). However, this zone is different from the *Parashumardites* Zone of Ramsbottom and Saunders (1985), which characterizes the base of the Missourian. The *Dunbarites–Parashumardites* Zone occurs at the base of the Wewoka Formation of the upper Desmoinesian (Ruzhenzev, 1974). The base of the *Dunbarites–Parashumardites* Zone most probably correlates with the *Wewokites* Subzone of the *Wellerites* Zone of Boardman *et al.* (1994). A long-standing suggestion from ammonoid study that the base of the

Kasimovian Stage lies within the upper Desmoinesian (Ruzhenzev, 1974) is now supported by evidence from fusulinids (Davydov, 1997; Davydov *et al.*, 1999b; Wahlman *et al.*, 1997) and conodonts (Heckel *et al.*, 1998).

GZHELIAN

The Gzhelian Stage, proposed in the late nineteenth century (Nikitin, 1890), became widely used only in the middle of the twentieth century, in connection with the fusulinid zonation established in the Moscow Basin (Rauser-Chernousova, 1941) and elsewhere in the Boreal and Tethyan faunal provinces. The name of the stage comes from the small village of Gzhel', famous for porcelain teapots and earthenware, produced from the lower Gzhelian clay, which is still mined around this area. A series of localities proposed by Nikitin as types of the stage near the villages of Rusavkino, Amerevo, Pavlovo-Posad and Noginsk, became type sections for a regional division (i.e. horizons) of the stage (Rosovskaya, 1950; Makhlina *et al.*, 1979). The top of the Gzhelian has changed position through time and has stabilized only recently when the GSSP for the base of the Permian was ratified (Davydov *et al.*, 1998).

The base of the stage exposed near the village of Gzhel' is an unconformity (Makhlina *et al.*, 1979), and is interpreted to be a second-order sequence boundary (Briand *et al.*, 1998). The traditional lower boundary of the Gzhelian Stage is the first appearance of the fusulinid species *Rauserites rossicus* and *R. stuckenbergi* (Rauser-Chernousova, 1941; Rosovskaya, 1950). However, in the type location near Gzhel', only *Rauserites rossicus* has been recovered. It is likely that *R. stuckenbergi* appears slightly higher in the section (Rauser-Chernousova, 1958; Davydov, 1986). The first appearance of the fusulinids *Daixina*, *Jigulites*, and *Rugosofusulina* has also been proposed as an operational index for the lower boundary of the Gzhelian Stage (Rosovskaya, 1975; Rauser-Chernousova and Schegolev, 1979; Davydov, 1990). These genera are absent from the Americas, except for the Canadian Arctic.

The ICS working group for the Moscovian–Kasimovian boundary is attempting to define the Kasimovian–Gzhelian boundary. The first appearance of *Streptognathodus zethus*, supposedly found in the lowest Gzhelian (Rechitian Horizon) in the Moscow Basin and in the Little Pawnee Shale Member of the Haskell Formation in the North American midcontinent (Heckel *et al.*, 1998; Barrick and Heckel, 2000), was indicated to be close to the Kasimovian–Gzhelian boundary (Heckel *et al.*, 1998). However, the FAD of *S. zethus* is middle Kasimovian (Chernykh, 2002) and cannot be used to define the base of the Gzhelian Stage. Villa (2001) suggested that either the first appearance of *S. zethus* or the first appearance of more advanced *Idiognathodus simulator* (*sensu stricto*) could be used to mark the Kasimovian–Gzhelian boundary, and the latter option is now being actively pursed by the working/task group.

On the Russian Platform, in the Donetz Basin, southern Urals, and Central Asia, the base of the Gzhelian coincides with a low sea-level sequence boundary (Alekseev *et al.*, 1996; Briand *et al.*, 1998; Davydov *et al.*, 1999b; Leven and Davydov, 2001). On the North American midcontinent, the Haskell Limestone is interpreted to represent the transition to a highstand system tract-maximum flooding surface (Heckel, 1989; Ritter, 1995). Therefore, these two levels cannot be isochronous. This evidence suggests that a proposed definition of the base of Gzhelian and its correlation with the re-defined base of Virgilian (at the Haskel Formation) requires further consideration.

The ammonoid definition of the Gzhelian and its base is not certain. The lower boundary of the *Shumardites–Vidrioceras* ammonoid zone has been conventionally placed at the base of the Gzhelian Stage (Bogoslovskaya *et al.*, 1999). However, the genus *Shumardites*, in the southern Urals, ranges from the upper Gzhelian to the top of the Carboniferous (Popov *et al.*, 1985; Davydov, 2001) and has a similar stratigraphic distribution in North America (Boardman *et al.*, 1994). Several ammonoid genera, which were considered to be "typical" Orenburgian–Virgilian (i.e. upper Gzhelian) were found recently in lower Gzhelian, Kasimovian, and, even, Moscovian strata (Popov, 1992; Boardman *et al.*, 1994).

15.2 CARBONIFEROUS STRATIGRAPHY

15.2.1 Biostratigraphy

BIOSTRATIGRAPHIC ZONATIONS OF AMMONOIDS, FORAMINIFERA, AND CONODONTS

Brachiopods were the first fossil group used to calibrate the Carboniferous stratigraphic sequence: for the Mississippian in Western Europe (Delepine, 1911) and the Pennsylvanian in Eastern Europe (Nikitin, 1890). Although brachiopods are still used within certain regions (Carter, 1990; Legrand-Blain, 1990; Poletaev and Lazarev, 1994; etc.), their correlation potential is considered to be more local.

Since the beginning of the twentieth century, the ammonoid successions in the Mississippian and early Pennsylvanian of western Europe (Wedekind, 1918; Bisat, 1924, 1928; Schmidt, 1925; Ramsbottom and Saunders, 1985) and in the entire Carboniferous in eastern Europe (Ruzhenzev, 1965;

Ruzhenzev and Bogoslovskaya, 1978) have served as chronostratigraphic standards in inter-regional and global correlation (Fig. 15.1).

Together with the ammonoid zonations, the foraminiferal zonation established in the Mississippian (Lipina and Reitlinger, 1970; Mamet and Skipp, 1970; Conil *et al.*, 1977) and Pennsylvanian (Rauser-Chernousova, 1941, 1949; Rosovskaya, 1950; Solovieva, 1977; Ross and Ross, 1988) is the most practical inter-regional biostratigraphic standard. However, because of significant provincialism in both groups of fossils, particularly at the time of assembly of Pangea, and the beginning of the Carboniferous Gondwana glaciation, different standard zonations are used for eastern and western hemispheres (i.e. Eurasia and surrounding areas in the east versus America in the west).

The conodont succession, although least studied and also to some degree provincial, over the last three decades has become the most reliable tool for the calibration and geochronological boundary definition within the Carboniferous (Fig. 15.1). The zonations established in the Mississippian and Pennsylvanian in North America and Europe (Higgins, 1975; Sandberg *et al.*, 1978; Dunn, 1970; Lane *et al.*, 1971, 1980; Barskov and Alekseev, 1975; Barskov *et al.*, 1980) are used world-wide and are actively refined (Nemirovskaya and Alekseev, 1995; Skompski *et al.*, 1995; Heckel *et al.*, 1998; Nemirovskaya, 1999; Lambert *et al.*, 2001; Chernykh, 2002).

OTHER MARINE MICRO- AND MACROFAUNA (CORALS, RADIOLARIA, OSTRACODS)

The majority of the other biostratigraphical events are calibrated relative to ammonoid, conodont, and foraminiferal scales. Although a global coral zonation has been proposed (Sando, 1990), Mississippian corals have generally only regional biostratigraphic significance (Poty, 1985; Kossovaya, 1996; Bamber and Fedorowski, 1998), but great biogeographic and paleoenvironmental utility.

Radiolaria as a pelagic microfossil group have high biostratigraphic potential (Nazarov and Ormiston, 1985; Holdsworth and Murchey, 1988; Gourmelon, 1987; Nazarov, 1988; Braun and Schmidt-Effing, 1993; Won, 1998; Chuvashov *et al.*, 1999). However, their zonation in the Carboniferous is still poorly developed and has been utilized, generally, only at regional levels.

Ostracods are widely distributed in the Carboniferous and in several regions were utilized for detailed regional biostratigraphy (Chizhova, 1977; Gorak, 1977; Crasquin, 1985; Abushik *et al.*, 1990). However, the chronostratigraphic value of ostracods is limited by paleoecological, facies, and paleoclimatic factors.

Crinoids, because of the rare occurrence of complete specimens, are not commonly used in Carboniferous biostratigraphy. However, an alternative classification of crinoid columnals and stems and biostratigraphic zonation at the stadial level has recently been proposed (Stukalina, 1988). Because of the ontogenetical approach of this classification, it can potentially be used for detailed biostratigraphy over large areas. Crinoid stems occur in a wide variety of facies and environmental conditions.

PLANTS AND PALYNOLOGY

Plants were one of the first fossil groups utilized for dating Carboniferous coal basins. The Carboniferous was a time of considerable evolutionary activity in the creation of megataxa and their provincial differentiations. Zonal floristic successions have been proposed in several regions along the equatorial climatic belt and were unified by Wagner (1984) as 16 floral zones for the entire Carboniferous. Floral successions in the higher latitudes, such as the Angara Province in the northern hemisphere (Meyen *et al.*, 1996) and the Gondwana Province in the southern hemisphere (Archangelsky *et al.*, 1995) are much less developed.

Palynological assemblages provide one of the commonest dating methods for Carboniferous coal fields and hydrocarbon deposits. Although the palynological spore assemblages are different from region to region, they are successfully used in several basins (Inosova *et al.*, 1976; Clayton *et al.*, 1977; Byvsheva *et al.*, 1979; Owens, 1984; Higgs *et al.*, 1988; Byvsheva and Umnova, 1993; Peppers, 1996).

15.2.2 Physical stratigraphy

MAGNETOSTRATIGRAPHY

The Carboniferous polarity scale is not yet well known, but current views, based on composite sections, are summarized in Fig. 15.1. The most significant general feature of the Carboniferous magnetic field is a long late Carboniferous interval of reverse polarity that continues into the early Permian, forming the Kiaman chron of predominantly reverse polarity.

The Mississippian and lower Pennsylvanian in terms of magnetostratigraphic classification developed in the former Soviet Union belong to the Donetzian mixed normal–reversal megazone with two superzones: Tikhvinian and Debaltzevian (Khramov and Rodionov, 1981). The former includes the stratigraphic interval from the upper Devonian (mid Fransian) up to the base of the Serpukhovian. The Debaltzevian Superzone is equal to the Serpukhovian–Moscovian stratigraphic interval. The Tournaisian magneto-zonal portion of the scale is based on data from the far east of Russia and Belgian classical sections (Kolesov, 1984, 2001).

The Visean–Serpukhovian–Bashkirian part of the magneto-zonal scale is based on sections from northern and central Appalachia Pennsylvania, New Brunswick, Nova Scotia; (DiVenere and Opdyke, 1991a,b), Donetz Basin, and Moscow Basin (Khramov *et al.*, 1974; Khramov, 2000). Post-Bashkirian magnetostratigraphy is based on sections from Australia, Donetz Basin, southern Urals, North Caucasus, and Central Asia (Irving and Parry, 1963; Khramov and Davydov, 1984, 1993; Davydov and Khramov, 1991; Opdyke *et al.*, 2000).

The late Pennsylvanian belongs to the Permo–Carboniferous Reversed Hyperchron interval, or "Kiaman," as originally described in Australia by Irving and Parry (1963). The Kiaman Hyperchron is recognized as a very long period of reversed polarity of Earth's magnetic field spanning approximately 50–60 Ma, and has been considered one of the important Paleozoic magnetostratigraphic markers. The age of the base of the Kiaman in terms of the chronological scale has not been resolved. In Australia, the base of the Kiaman was originally defined immediately above the Paterson Volcanics beds dated as 309–310 Ma (Palmer *et al.*, 1985): i.e. as it was thought to be close to the base of Westphalian of the western European stratigraphic scale, or to the base of Atokan in terms of North American scale. However the age of the Paterson Volcanics has been reinterpreted recently as 328.5 ± 1.4 Ma (Claoué-Long *et al.*, 1995), which would place the base of the Kiaman in the type area within the Visean (Opdyke and Channell, 1996).

The youngest pre-Kiaman normal polarity zone age has been found within approximately Westphalian A Strata (i.e. mid Atokan and latest Bashkirian) in Nova Scotia and in the Northern Appalachians (DiVenere and Opdyke, 1991a). Similarly, a normal polarity zone radiometrically constrained within 317 and 313 Ma has been found recently in eastern New South Wales, Australia (Opdyke *et al.*, 2000), essentially at the same level as in the Northern Appalachians.

Khramov (1987) found in Donetz Basin three normal polarity zones within the lower–middle Moscovian. Opdyke *et al.* (1993) re-studied Bashkirian–Moscovian sequences in Donetz Basin but did not confirm the proposed magnetozones as Moscovian. However, Khramov (2000) suggested that the zones in the Bashkirian–Moscovian in Donetz Basin are present in the siliciclastic rocks (siltstones and sandstones), which were not studied by Opdyke *et al.* (1993).

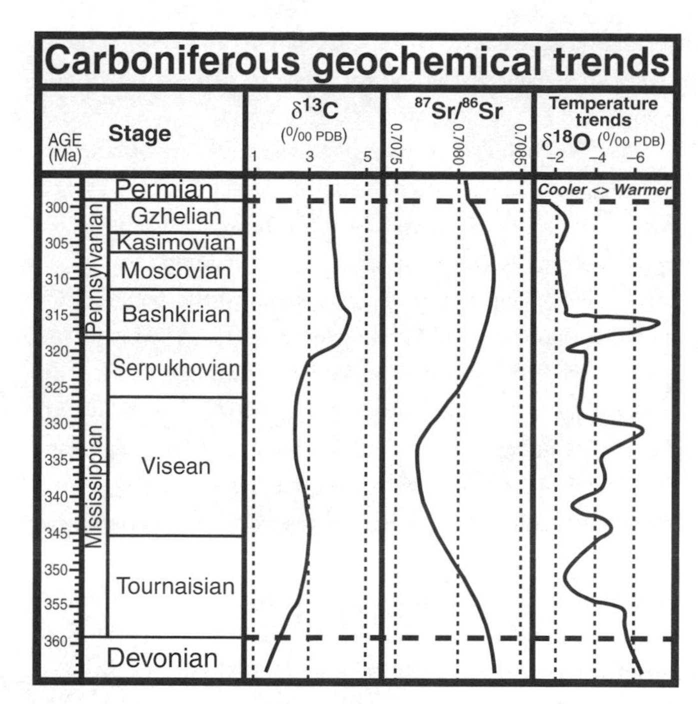

Figure 15.3 Geochemical trends during the Carboniferous Period. The schematic carbon isotope curve is a 5-myr averaging of global data (Veizer *et al.*, 1999) from Hayes *et al.* (1999; downloaded from www.nosams.whoi.edu/jmh). The $^{87}Sr/^{86}Sr$ LOWESS curve for the interval is based on the data of Denison *et al.* (1994) and other sources – see Chapter 7. The oxygen isotope curves (inverted scale) are derived from a 3-myr interval averaging of global data compiled by Veizer *et al.* (1999; as downloaded from www.science.uottawa.ca/geology/isotope_data/ in January 2003). Large-scale global shifts to higher oxygen-18 values in carbonates generally are interpreted as cooler seawater or glacial episodes, but there are many other contributing factors (e.g. Veizer *et al.*, 1999, 2000; Wallmann, 2001).

A single, but pronounced normal polarity zone within the lower Kiaman Hyperzone, which was originally found in Donetz Basin (Khramov, 1963) was reported in southwest USA, Germany, Caucasus, southern Urals, and Central Asia (Peterson and Nairn, 1971; Dachroth, 1976; Khramov and Davydov, 1984). This zone has been named the "*Kartamyshian*" (Davydov and Khramov, 1991). In the southern Urals and Central Asia, the *Kartamyshian* Zone has been precisely constrained within the uppermost Carboniferous fusulinid zone *Ultradaixiana sbytauensis–Schwagerina robusta* and, therefore, is an important magnetostratigraphic marker for separation of the Carboniferous and Permian Periods.

CHEMICAL STRATIGRAPHY: STABLE ISOTOPES OF Sr, O, AND C

In the Carboniferous, chemostratigraphy has only recently become useful (Fig. 15.3). The degree of lithification in many Palaeozoic rocks makes alteration of original chemical signals a significant problem (Grossman *et al.*, 1996).

For $^{87}Sr/^{86}Sr$ stratigraphy, the most useful fossils are brachiopods with thick, large, non-luminiscent shells (Bruckschen *et al.*, 1999), but whole rocks have also proven useful in establishing trends (Denison *et al.*, 1994). Conodonts, which alter easily during burial diagenesis, have little potential for chemostratigraphy unless they have a CAI of 1 (Martin and Macdougall, 1995; Ruppel *et al.*, 1996). The range of $^{87}Sr/^{86}Sr$ in the Carboniferous is large (from 0.707 67 to 0.708 31; Fig. 15.3) and defines a broad minimum between highs approximately coincident with the boundaries of the period. The poor quality of the $^{87}Sr/^{86}Sr$ curve for the period owes much to the difficulty of correlating sections used in curve construction.

Reconnaissance curves for $\delta^{13}C$ and $\delta^{18}O$ have been produced by Veizer *et al.* (1999; Fig. 15.3). Refined curves of these isotopes precisely tied to the chronostratigraphic scale have been published (Mii *et al.*, 1999; Bruckschen *et al.*, 1999; Grossman *et al.*, 2002; Saltzman, 2003).

Two major, possibly global, peaks in $\delta^{13}C$ distribution were found in the Carboniferous. A brief positive excursion of $\delta^{13}C$ up to +7% corresponds to the middle Tournaisian (upper Kinderkhokian) *S. isosticha* conodont zone (lower Mc5 Zone) in southeast Idaho, southeast Nevada, northeastern Utah, northern Iowa, and in the Dinant Basin, western Europe (Bruckschen and Veizer, 1997; Mii *et al.*, 1999; Saltzman, 2002). A large $\delta^{13}C$ shift, of approximately +6% has been documented across the Mississippian–Pennsylvanian transition in the paleo-Tethyan regions (Popp *et al.*, 1986; Bruckschen *et al.* 1999; Mii *et al.*, 1999; Grossman *et al.*, 2002; Saltzman, 2003). In North America, this shift is late Chesterian and early Morrowan (Serpukhovian–Bashkirian) and is reduced by +1.5–2.5%, which most probably reflects changes in ocean circulation patterns associated with the closing of the equatorial seaway. Based on the timing of the $\delta^{13}C$ divergence between North America and Europe, the isolation of Paleo-Tethys began in the late Serpukhovian. Similarly, the onset of the provincialism of benthic fauna (western Euramerica versus Paleo-Tethys) is coincident with this paleotectonic event (equatorial seaway closure). All these data support a scenario in which the closure of a subequatorial oceanic gateway during the assembly of Pangea altered the oceanic distribution of nutrients and led to enhanced poleward transport of heat and moisture. This change marks the transition from a cool, moisture-starved Gondwana to the ice-house world of the Pennsylvanian and Early Permian (Saltzman, 2003).

It should be noted that $\delta^{13}C$ shifts preceded major coal burial. The middle Tournaisian positive $\delta^{13}C$ excursion is 5–6 myr earlier than late Tournaisian and early Visean coal burial in the Moscow Basin, East Siberia, and Urals; the large $\delta^{13}C$ shift across the Mississippian–Pennsylvanian transition also appears approximately 5–6 myr earlier than the major coal burial in western Europe, Donetz Basin, the Appalachian Basin, and northern China. If this is a general rule, then it might be expected that for the Permian coal-forming episodes, a $\delta^{13}C$ shift in the Artinskian–Kungurian would have preceded the burial of Kungurian–Roadian coal and a similar shift in Capitanian time would have preceded Wuchiapingian coal burial.

SEQUENCE STRATIGRAPHY

Sequence stratigraphy is widely applicable to the Carboniferous cyclic sequences because they are generally considered to be glacio-eustatic and therefore global in distribution (Ramsbottom, 1977; Caputo and Crowell, 1985; Veevers and Powell, 1987). Ross and Ross (1987, 1988) proposed a global framework of transgressive–regressive depositional cycles in the marine shallow-water cratonic shelves of the Carboniferous–Permian tropical and subtropical regions.

It is generally accepted by most biostratigraphers and paleontologists that sea-level fluctuations are one of the major factors that drive the evolution of marine organisms. Current theory suggests that generally all zonal and/or stadial benthic assemblages are bounded by unconformities (Ross and Ross, 1994, 1995b, 1996). Based on this assumption we conclude that sea-level lowstands, represented in the sedimentary record as sequence boundaries, coincide with major evolutionary extinctions of shallow-water fauna and that the origination of species is closely associated with the subsequent transgression. As applied to foraminiferal zonation in the Russian Platform and Urals regions, this model suggests that in most cases foraminiferal zones are bounded by eustatic sequences although some species may extend through several depositional sequences. Another important insight derived from this model is that many series and stage boundaries appear associated with major and long hiatuses during times of lowered sea levels (Ross and Ross, 1995a, p. 220).

Many geologists expect sequence boundaries to correspond with system, series, and stage boundaries and foraminiferal zonal subdivisions even when sedimentological evidence is

Table 15.1 *Comparison of age estimates and durations of Carboniferous sub-periods from selected time scales published in recent years*

	Mississippian			Pennsylvanian	
	Base (Ma)	*Top (Ma)*	*Duration (myr)*	*Top (Ma)*	*Duration (myr)*
Harland *et al.* (1990)	362.5	322.8	39.7	290	32.8
Young and Laurie (1996)	354	314	40	298	16
Tucker *et al.* (1998)	362				
Compston (2000a)	359.6				
Compston (2000b)	362				
Ross and Ross (1988)	360	320	40	286	34
Jones (1995)	356	317	39	300	17
Menning *et al.* (2000)[a]	354	312	42	292	20
Menning *et al.* (2000)[b]	354	320	34	292	28
Menning and Hendrich (2002)	358	320	38	296	24
Heckel (2002)[c]		320		290	30
GTS2004	359.2	318.1	41.1	299.0	19.1

[a] Data from Ar–Ar analyses only.
[b] Data from U–Pb, Ar–K, and Rb–Sr analyses only.
[c] Includes sedimentation rate estimates.

lacking (Izart *et al.*, 1998, 1999; Stemmerik *et al.*, 1995; Samuelsberg and Pickard, 1999). That this is not always the case is shown by the Carboniferous–Permian boundary (base of the Asselian Stage). The global cyclic deposition model suggests that the base of the Asselian is a major sequence boundary. The boundary that has been established in the southern Urals is suggested to be synchronous with the Roca Shales, near the base of the Grenola Limestone in the midcontinent of North America, and with the base of Neal Ranch Formation, in West Texas (Ross and Ross, 1985, 1988). But recent conodont data suggest that the base of the Grenola Limestone correlates with the uppermost Grey Limestone of the Gaptank Formation, West Texas, and with the uppermost Asselian, southern Urals (Wardlaw and Davydov, 2000), and therefore differs from the Carboniferous–Permian boundary in the southern Urals by at least 3–5 myr.

15.3 CARBONIFEROUS TIME SCALE

15.3.1 Previous scales

There is very little agreement among recent reviews of the Carboniferous time scale in estimates of the boundary of the two subsystems (Table 15.1), although new data suggest that at least the base and top of the Carboniferous may now achieve some measure of numerical age stability, as explained below.

As with the Devonian, there is a limited number of good-quality and chronostratigraphically precise dates within the period. As a result, various assumptions have been made about the duration of stages or of zones. For example, GTS89 in Harland *et al.* (1990) assigned equal duration to 50 biochrons from the foraminiferal zones in the Donetz Basin.

15.3.2 Radiometric data

The detailed and high-resolution conodont–ammonoid zonation for the Carboniferous, with over 35–40 zones (Fig. 15.1) is in contrast to the 21 radiometric dates employed in Carboniferous time scale building. With a duration of the Carboniferous Period close to 60 myr, the average resolution of a zone is equivalent to 1.5–1.0 myr, whereas the radiometric resolution is much less. Worse, some stages have very few dates. It is important that the $^{40}Ar/^{39}Ar$ age dates assigned to well-constrained western European Namurian and Westphalian stages be correlated with the standard marine stages of the Moscow syncline defined by conodonts, foraminifera, and ammonoids. Correlation is much more difficult for the overlying continental Stephanian and may have to rely heavily on numerical dating.

The Carboniferous time scale presented here must be considered tentative. The normalization and interpolation methods and relative scaling of stages with conodont and ammonoid zones are explained below.

Table 15.2 lists Carboniferous radiometric dates, with emphasis on available $^{40}Ar/^{39}Ar$, HR–SIMS, and TIMS ages, that were judged suitable for the Carboniferous time scale. For most of the Mississippian, spanning 30–35 myr, there are few age dates. Unlike the Ordovician through Devonian, where detailed comparison of HR–SIMS and TIMS ages was performed (see Chapters 12–14, and Fig. 14.4), there is no systematic comparison between these different types of dates that may assist, or at least lead to, an empirical calibration. Hence, the HR–SIMS dates are used with caution.

Table 15.2 *Selected isotopic radiometric age dates for the Carboniferous time scale using TIMS, HR–SIMS, and $^{40}Ar/^{39}Ar$ methods*

No.	Sample	Locality	Formation	Comment	Zone and age	Biostratigraphic reliability	References	Age (Ma)	Type	Range of the horizontal error bars for each age date, in Standard Composite Units (SCU)
1	Zircon	Usolka section, southern Urals region, Russia	Tuffaceous marls	ID-TIMS date of 299.0 ± 1.0 Ma (2-sigma) for the youngest zircon population	Four samples are precisely constrained within Asselian–Gzhelian transition: first sample 1.0 m below base of the Permian; second, 0.4 m below base; third, 0.05 m above base of the Permian; fourth, 1.5 m above the base. Carboniferous–Permian boundary precisely defined in the basis of conodonts	1	Ramezani *et al.* (2003); Chuvashov *et al.* (1990a,b); Davydov *et al.* (2002)	299.0 ± 1.0 (2-sigma)	U–Pb	1914–1921 (1917)
2	Paleosol calcite	Sacramento Mts.	Paleosol calcite from cycle 50, Holder Fm	34 U–Pb dates on paleo-soil calcite give 306.0 ± 2.6 Ma (2-sigma)	From regional correlations considered to be within lower Virgilian; although unknown biostratigraphic constraints it is allowed here with low weight in the fitting method	3	Rasbury *et al.* (1998)	306.0 ± 2.6 (2-sigma)	U–Pb	1837–1889 (1890)
3	Sanidine	Baden-Baden, SW Germany	Tuff Baden-Baden, possibly upper Stephanian C, but stratigraphic position uncertain	Ar–Ar plateau age of sanidines from coal tonsteins, proposed age 300.3 ± 0.6 Ma (1-sigma). Re-calibration with MMhb-1 of 523.1 yields an age of 302.51 ± 0.6 Ma (1-sigma)	From inter-regional correlations considered to be within lower to middle Gzhelian; although of poor biostratigraphic constraint it is allowed here with low weight in the fitting method	3	Hess and Lippolt (1986); Davydov (1990); Burger *et al.* (1997)	302.51 ± 0.6 (1-sigma)	Ar–Ar	1841–1907 (1820)

4	Sanidine	Saar, SW Germany	Tonstein 0, Dilsburger Fm of middle Stephanian A	Ar–Ar plateau age of sanidines from coal tonsteins of 302.0 ± 0.6 Ma (1-sigma). Re-calibration with MMhb-1 of 523.1 yields an age of 304.22 ± 0.6 Ma (1-sigma)	From inter-regional correlations considered to be within lower Kasimovian; although of poor biostratigraphic constraint it is allowed here with low weight in the fitting method	2		Burger *et al.* (1997); Davydov (1992)	304.22 ± 0.6 (2-sigma)	Ar–Ar	1809–1841 (1842)
5	Sanidine	Donetz Basin, 3 km north of Lutugino, Ukraine	C_2^6 (L) Formation, l_3^1coal tonstein located between Limestone L_4 and L_5	Ar–Ar plateau age of sanidines from coal tonsteins of 305.5 ± 1.5 Ma (1-sigma), calibrated against MMhb-1 of 520.4 ± 1.7 Ma and BMus (laboratory muscovite standard of 326.2 ± 1.9 Ma); re-calibration with MMhb-1 of 523.1 yields an age of 307.75 ± 1.5 Ma (1-sigma)	Presence of *Neognathodus atokaensis* below Limestone L_4 and appearance of *Neognathodus medadultimus* and *N. colombiensis* above Limestone L_6 allow to correlation of Limestone L_4–L_6 interval with middle–upper Kashirian (upper lower Moscovian) of Moscow Basin	1		Hess *et al.* (1999); Nemirovskaya (1999); Nemirovskaya *et al.* (1999); Makhlina *et al.* (2001)	307.75 ± 1.5 (1-sigma)	Ar–Ar	1745–1757
6	Sanidine	Saar, SW Germany	Tonstein 1, Heiligenwalder Fm of lower Westphalian D	Ar–Ar plateau age of sanidines from coal tonsteins of 308.0 ± 1.8 Ma (1-sigma) was originally calibrated with P-207 muscovite (82.6 ± 1.0 Ma), and with MMhb-1 of 520.4 ± 1.7 Ma; re-calibration with MMhb-1 of 523.1 yields an age of 310.26 ± 1.8 Ma (2-sigma)	From inter-regional correlations considered to be close to lower–upper Moscovian boundary; although of poor biostratigraphic constraint it is allowed here with low weight in the fitting method	2		Burger *et al.* (1997); Nemirovskaya (1999)	310.26 ± 1.8 (2-sigma)	Ar–Ar	1745–1783

(*cont.*)

Table 15.2 (*cont.*)

No.	Sample	Locality	Formation	Comment	Zone and age	Biostratigraphic reliability	References	Age (Ma)	Type	Range of the horizontal error bars for each age date, in Standard Composite Units (SCU)
7	Sanidine	Saar, SW Germany	Tonstein 3, Sulzbacher Fm of middle Westphalian C	Ar–Ar plateau age of sanidines from coal tonsteins of 309.7 ± 2.0 Ma (1-sigma), was originally calibrated with P-207 muscovite (82.6 ± 1.0 Ma) and with MMhb-1 of 520.4 ± 1.7 Ma; re-calibration with MMhb-1 of 523.1 yields 312.0 ± 2.0 Ma (1-sigma)	From inter-regional correlations considered to be close to lower–upper Moscovian boundary; although of poor biostratigraphic constraint it is allowed here with low weight in the fitting method	2	Burger *et al.* (1997); Nemirovskaya (1999)	312.0 ± 2.0 (1-sigma)	Ar–Ar	1727–1757
8	Sanidine	Ruhr Basin, Nordrhein–Westfalen, Germany	Hagen 1; Dorstener Fm of Westphalian C	Ar–Ar plateau age of sanidines from coal tonsteins of 309.5 ± 1.6 Ma (1-sigma), calibrated with MMhb-1 of 520.4 ± 1.7 Ma; re-calibration with MMhb-1 of 523.1 yields 311.78 ± 1.6 Ma (1-sigma)	No specifics on biostratigraphy available; lower Westphalian C; from inter-regional correlation considered to be within lower Moscovian; although of poor biostratigraphic constraint it is allowed here with low weight in the fitting method	2	Burger *et al.* (1997); Nemirovskaya (1999)	311.78 ± 1.6 (1-sigma)	Ar–Ar	1722–1757
9	Sanidine	Ruhr Basin, Nordrhein–Westfalen, Germany	Hagen 4 tonsteins; Dorstener Fm of Westphalian C	Ar–Ar plateau age of sanidines from coal tonsteins of 310.0 ± 1.0 Ma (1-sigma), calibrated with MMhb-1 of 520.4 ± 1.7 Ma; re-calibration with MMhb-1 of 523.1 yields 313.0 ± 1.0 Ma (1-sigma)	No specifics on biostratigraphy available; lower Westphalian C; from inter-regional correlation considered to be within lower Moscovian; although of poor biostratigraphic constraint it is allowed here with low weight in the fitting method	2	Burger *et al.* (1997); Nemirovskaya (1999)	313.0 ± 1.0 (1-sigma)	Ar–Ar	1715–1735

10	Sanidine	Seven samples collected along a 300 km traverse in the central part of the Appalachian Basin	Fire Clay Coal bed of Breathitt Fm	Ar–Ar plateau age of sanidines from coal tonsteins of 310.9 ± 0.8 Ma (1-sigma), calibrated with MMhb-1 hornblende and FCT-3 sanidine; re-calibration with MMhb-1 of 523.1 yields 313.18 ± 0.8 Ma (1-sigma)	Upper part of *Diaboloceras nermeiri–Bisatoceras micrompalus* ammonoid zone, middle Atokan (i.e. early Moscovian) age	1	Kunk and Rice (1994)	313.18 ± 0.8 (1-sigma)	Ar–Ar	1715–1739
11	Zircon from tonstein	Ruhr Basin, Nordrhein–Westfalen, Germany	Z1 Tonstein, Horster Fm of Westphalian B	40 U–Pb HR–SIMS dates on 37 zircon grains give 311.0 ± 3.4 Ma (2-sigma), calibrated to standard zircon SL13; 6 zircon grains have a variable and young (Permian and Cretaceous) apparent age	Z1 tonstein occurs immediately below Agir marine band (base of Westphalian C) with conodonts *Declinagnathodus marginadosus*, *Idiognathodus aljutovensis*, and *I. tuberculatus*. Uppermost Westphalian B (i.e. uppermost Bashkirian)	1	Claoué-Long *et al.* (1995); Nemirovskaya (1999); Menning *et al.* (2003)	311.0 ± 3.4 (2-sigma)	U–Pb	1698–1709
12	Sanidine from tonstein	Ruhr Basin, Nordrhein–Westfalen, Germany	Z1 Tonstein, Horster Fm of Westphalian B	Ar–Ar plateau age of sanidines from coal tonsteins of 310.7 ± 1.3 Ma (1-sigma), calibrated with MMhb-1 of 520.4 ± 1.7 Ma; re-calibration with MMhb-1 of 523.1 Ma yields 313.0 ± 1.3 Ma (1-sigma)	Z1 tonstein occurs immediately below Agir marine band (base of Westphalian C) with conodonts *Declinagnathodus marginadosus*, *Idiognathodus aljutovensis*, and *I. tuberculatus*. Uppermost Westphalian B (i.e. uppermost Bashkirian)	1	Burger *et al.* (1997); Nemirovskaya (1999); Menning *et al.* (2001)	313.0 ± 1.3 (1-sigma)	Ar–Ar	1698–1709
13	Zircon from K-bentonite	Harewood Borehole, North Yorkshire, UK	Arnsbergian Stage of England	U–Pb HR–SIMS dates on zircon grains give 314.4 ± 4.6 and 314.5 ± 4.6 Ma (2-sigma), calibrated to standard zircon SL13	Upper part of *Euromorphoceras yatsae* (E2a3) Marine Band and upper part of *Cravenoceratoides nitidus* (E2b2) Marine Band (i.e. upper Arnsbergian and/or uppermost Serpukhovian)	1	Riley *et al.* (1994)	314.45 ± 4.6 (2-sigma)	U–Pb	1573–1599

(cont.)

Table 15.2 (*cont.*)

No.	Sample	Locality	Formation	Comment	Zone and age	Biostratigraphic reliability	References	Age (Ma)	Type	Range of the horizontal error bars for each age date, in Standard Composite Units (SCU)
14	Sanidine from tonstein	Upper Silesian Basin, Ostrava, Czech Republic	Ostrava Fm, Poruba Member; Samples COT479-1 and COT 479-2	Ar–Ar plateau age of sanidines from coal tonsteins of 319.9 ± 1.7 Ma (1-sigma), was originally calibrated with P-207 muscovite (82.6 ± 1.0 Ma), and with MMhb-1 of 520.4 ± 1.7 Ma; re-calibration with MMhb-1 of 523.1 yields an age of 322.3 ± 1.7 Ma (1-sigma)	Middle Arnsbergian boundary of Namurian A (i.e. within upper Serpukhovian)	2	Hess and Lippolt (1986); Menning *et al.* (1997)	322.3 ± 1.7 (1-sigma)	Ar–Ar	1567–1592 (1583)
15	Sanidine from tonstein	Upper Silesian Basin, Ostrava, Czech Republic	Ostrava Fm, Jaklovec Member; Samples COT365-1, COT335-1, and COT335-2	Ar–Ar plateau analysis of sanidines from coal tonsteins of 325.4 ± 2.5, 323.7 ± 1.7, and 324.8 ± 1.2 Ma (all 2-sigma); was originally calibrated with P-207 muscovite (82.6 ± 1.0 Ma), and with MMhb-1 of 520.4 ± 1.7 Ma; re-calibration with MMhb-1 of 523.1 yields an age of 327.0 ± 2.2 Ma (2-sigma)	Lower Arnsbergian of lower Namurian A (i.e. within lower Serpukhovian)	2	Hess and Lippolt (1986); Menning *et al.* (2000)	327.0 ± 2.2 (2-sigma)	Ar–Ar	1525–1570 (1542)
16	Zircons	Rheinisches Schieferge-birge, Germany	Medebach Formation, Lowermost Brigantian	U–Pb TIMS dates of 326.8 ± 0.98 Ma (2-sigma)	The bentonite layer has been found within *G. crenistria* ammonoid zone of lowermost Brigantian of uppermost Visean	1	Trapp and Kaufmann (2002)	326.8 ± 0.98 (2-sigma)	U–Pb	1429–1442 (1436)

17	Zircons	Harz, Germany	Lerbach Formation, middle Holkerian	TIMS dates of 334.25 ± 0.95 Ma (2-sigma) and 334.0 ± 0.56 Ma (2-sigma)	Two bentonite layers were found in *Gnathodus texanus–Gnathodus praebillineatus* conodont zones of middle Holkerian, lower Visean	1		Trapp and Kaufmann (2002)	334.25 ± 0.95 (2-sigma)	U–Pb	1329–1336 (1332)
18	Ignimbrite	Rouchel Block, Eastern Australia	Curra Keith Tongue, basal unit of Native Dog Ignimbrite Member of Isismurra Fm	56 U–Pb HR–SIMS dates on zircon grains calibrated to standard zircon SL13; five grains have younger ages significantly below normal distribution line (NDL) and are interpreted to have lost Pb subsequent to crystallization; some grains have older ages significantly above the NDL and are identified as xenocrysts	Ignimbrite lies between the top of local brachiopod zone *S. burlingtonensis* and the base of *O. austarlis* local brachiopod zone and correlated to a level closely coincident with the Tournaisian–Visean boundary	1		Roberts *et al.* (1995a,b)	342.1 ± 4.2 (2-sigma)	U–Pb	1182–1247 (1228)
19	Zircons	One of the sections is Hasselbach-tal, Germany		Geomathematical–statistical treatment of latest Silurian through earliest Carboniferous dates in Table 14.2 calculated the Devonian time scale, and yielded an age of 359.2 ± 2.5 Ma for the Devonian–Carboniferous boundary (see Chapters 8 and 14)	FAD of *Siphonella sulcata* defines the D/C boundary, immediately below the ammonoid *Acumitoceras antecedens*	1		See Chapters 8 and 14	359.2 ± 2.5 (2-sigma)	U–Pb	992–999

Position of stage boundaries in the composite scale (CU) of Figs. 15.4 and 15.5
Base of Asselian (Permian) – 1933 CU
Base of Gzhelian – 1838 CU
Base of Kasimovian – 1796 CU
Base of Moscovian – 1712 CU
Base of Bashkirian – 1607 CU
Base of Serpukhovian – 1472 CU
Base of Visean – 1187 CU
Base of Tournaisian – 992 CU
1 CU = 64.681 K

The first HR–SIMS is from the base of the Carboniferous at Hasselbachtal, Germany, and the second from the Curra Keith Tongue, Australia. Although biostratigraphic calibration of both dates is excellent, the analytical error of the applied technique is relatively high (up to 3.2 myr, 2-sigma). Two new TIMS dates from the Visean and Serpukhovian (dates No. 16 and 17 in Table 15.2) of the marine sequences of Germany have recently been published (Trapp and Kaufmann, 2002). More TIMS dates for northern Pangea are necessary in order to build a reliable Carboniferous time scale. Variscan volcanic activity in western and eastern Europe provide good potential for that purpose.

Radiometric data for the Serpukhovian–Gzhelian interval, spanning the younger part of Carboniferous time, lasting approximately 25–35 myr, depending on the date for the base of the Permian, mostly come from paralic successions of western Europe with marine horizons becoming scarce upwards. Biostratigraphic correlation of these successions with the marine sections of eastern Europe continues to be a problem. The recent discovery of numerous Pennsylvanian tuffs containing high-quality zircons in the southern Urals provides an exceptional potential to build a reliable Late Carboniferous time scale.

The $^{40}Ar/^{39}Ar$ dates cited for the Carboniferous in Table 15.2, which were calibrated against MMhb-1 at 520.4 Ma, were re-calibrated to an MMhb-1 of 523.1 Ma, making those dates 0.74% older.

For the Devonian–Carboniferous boundary, the age of 359.2 ± 2.5 Ma, derived from interpolation of the Devonian time scale, was used, as explained in detail in Chapters 8 and 14. For the Carboniferous–Permian boundary, Ramezani *et al.* (2003) report new dates from ash beds in the southern Urals. One ash layer in the Usolka section lies 0.6 m above the Carboniferous–Permian boundary and contains the conodont *Streptognathodus isolatus*, the index species of the base of the Permian. Preliminary results from conventional U–Pb TIMS analyses of zircons from this and other ash beds suggest an age of 299 ± 1 Ma for the Carboniferous–Permian boundary.

15.3.3 Carboniferous and Permian composite standard

In order to integrate and calibrate zonal successions of foraminifera, conodonts, and ammonoids relative to each other, and to construct the Carboniferous and Permian chronostratigraphic scale, graphic correlation was undertaken (Figs. 15.4 and 15.5). As explained in some detail in Chapter 3, graphic correlation produces a composite standard that when calibrated in linear time yields the Carboniferous and Permian time scale.

The composite standard (CS) for the Carboniferous includes data from most of the type and key sections of the Serpukhovian through Gzhelian in the Russian Platform and Urals, and also from well-studied regions such as Donetz Basin, Central Asia, Timan–Pechora, Spitsbergen, and the subsurface of Barents Sea. In total, over 40 of the most complete sections were composited. Three major fossil groups were involved in the analysis: foraminifera (~70% of the total data), conodonts (~25% of the total data), and ammonoid species and genera (~5% of the total data). Over 4000 taxa (predominantly species) were used. Uppermost Devonian–Tournaisian and Visean data came mostly from Central Asia, Urals, and Donetz Basin (12 sections). This part of the CS is less well developed and is not calibrated against the stratotype sections of the Tournaisian or Visean. Stratotypes of the Serpukhovian, Bashkirian, Moscovian Kasimovian, and Gzhelian were used as primary sections to build the entire CS. Most of the sections may include potential defects, such as changes in sedimentation rate, discontinuities, or an incomplete fossil record. Hence, the resulting scale was adjusted by adding several more complete sections from the Urals, Donetz Basin, Cantabrian Mountains, and Central Asia. A second adjustment was done with intuitive scaling of stages primarily based on evolution of foraminifera. Integration of zonal successions in the Tournaisian and Visean is less well developed, and the zonal scale there came from the assumption of equal duration of conodont zones. Next, the conodont zones were calibrated into the foraminiferal and ammonoid zonation.

The composite standard (CS) for the Permian includes data from most of the stratotype and key sections of the Cisuralian Series of the Russian Platform and Urals, the Arctic, Central Asia, and Donetz Basin; the Guadalupian Series includes data from the Guadalupian Mountains, West Texas, Central Asia, and South China; while the Lopingian Series includes data from South China, Iran, and the Carnian Alps. Over 20 of the most complete sections were composited. As for the Carboniferous, the same fossil groups were used for the composite section: foraminifera (~50% of the total data), conodonts (~45% of the total data), and ammonoid species and genera (~5% of the total data) totaling over 2500 taxa (predominantly species). The Cisuralian is well characterized by all three fossil groups; the Guadalupian is well characterized by conodonts and foraminifera and only poorly by ammonoids; the Lopingian is basically characterized by conodonts and ammonoids because foraminiferal data sets are poor. Because there are no

System	Stage	c	f	a
Triassic (Induan)		Tc1		Ta1
Permian	Changhsingian	Lc7, Lc6	Lf4, Lf3	La3
Permian	Wuchiapingian	Lc5, Lc4, Lc3, Lc2, Lc1	Lf2, Lf1	La2, La1
Permian	Capitanian	Gc4, Gc3	Gf6, Gf5	Ga4
Permian	Wordian	Gc2	Gf4, Gf3	Ga3, Ga2
Permian	Roadian	Gc1	Gf2, Gf1	Ga1
Permian	Kungurian	Cc12, Cc11, Cc10	Cf13, Cf12	Ca6, Ca5
Permian	Artinskian	Cc9, Cc8	Cf11, Cf10, Cf9	Ca4, Ca3
Permian	Sakmarian	Cc7, Cc6	Cf8, Cf7, Cf6	Ca2, Ca1
Permian	Asselian	Cc5, Cc4, Cc3, Cc2, Cc1	Cf5, Cf4, Cf3, Cf2, Cf1	Ca1

Scale: 1950, 2000, 2050, 2100, 2150, 2200, 2250, 2300, 2350, 2400, 2450, 2500, 2550, 2600, 2650

System	Stage	Substage	c	f	a
Carboniferous	Gzhelian 7	Melekhovian	Pc19	Pf22	Pa12
		Noginian	Pc18	Pf21	Pa12
		Pavlovoposadian	Pc17	Pf20	Pa11
		Rusavkian	Pc16, Pc15	Pf19, Pf18	Pa11
	Kasimovian 6	Dorogovilovian	Pc14	Pf17	Pa10
		Khamovnichean	Pc13	Pf16, Pf15	Pa10
		Krevyakian	Pc12	Pf14	Pa10
	Moscovian 5	Peskovian	Pc11	Pf13	Pa9
		Myachkovian	Pc11	Pf12	Pa8
		Podolskian	Pc10	Pf11	Pa8
		Kashirian	Pc9	Pf10	Pa7
		Tsninian	Pc8	Pf9	Pa7
		Vereian	Pc8, Pc7	Pf8	Pa6
	Bashkirian 4	Melekesian	Pc6	Pf7	Pa5
		Cheremshanian	Pc5	Pf6	Pa4
		Prikamian	Pc4	Pf5	Pa3
		Severokeltmian	Pc3	Pf4	Pa3
		Krasnopolyanian	Pc2	Pf3	Pa2
		Voznesenian	Pc1	Pf2, Pf1	Pa1
	Serpukhovian 3	Zapaltjubian	Mc16	Mf17	Ma8
		Protvian	Mc16, Mc15	Mf16	Ma8
		Steshevian	Mc15	Mf16	Ma8
		Tarussian	Mc14	Mf15	Ma7
	Visean 2	Venevian	Mc13, Mc12	Mf14	Ma6
		Mikhalovian	Mc11	Mf13	Ma6
		Aleksian	Mc10	Mf12	Ma6
		Tulian	Mc10	Mf11	Ma5
		Bobrikovian	Mc9	Mf10, Mf9	Ma5
		Radaevian	Mc8	Mf8	Ma4
	Tournaisian 1	Kosvinian	Mc7, Mc6	Mf7	
		Kizelovian	Mc5	Mf6	Ma3
		Cherepetian	Mc5	Mf5	Ma3
		Karakubian	Mc4	Mf4	Ma2
		Upinian	Mc3	Mf3	Ma2
		Malevian	Mc2	Mf2	Ma2
		Gumerovian	Mc1	Mf1	Ma1
Devonian		Ziganian			

Scale: 1000, 1050, 1100, 1150, 1200, 1250, 1300, 1350, 1400, 1450, 1500, 1550, 1600, 1650, 1700, 1750, 1800, 1850, 1900

Figure 15.4 Relative zonal scale used to build the Carboniferous time scale shown in Figure 15.5. The scale is build by means of a composite standard technique using graphic correlation. For details see text.

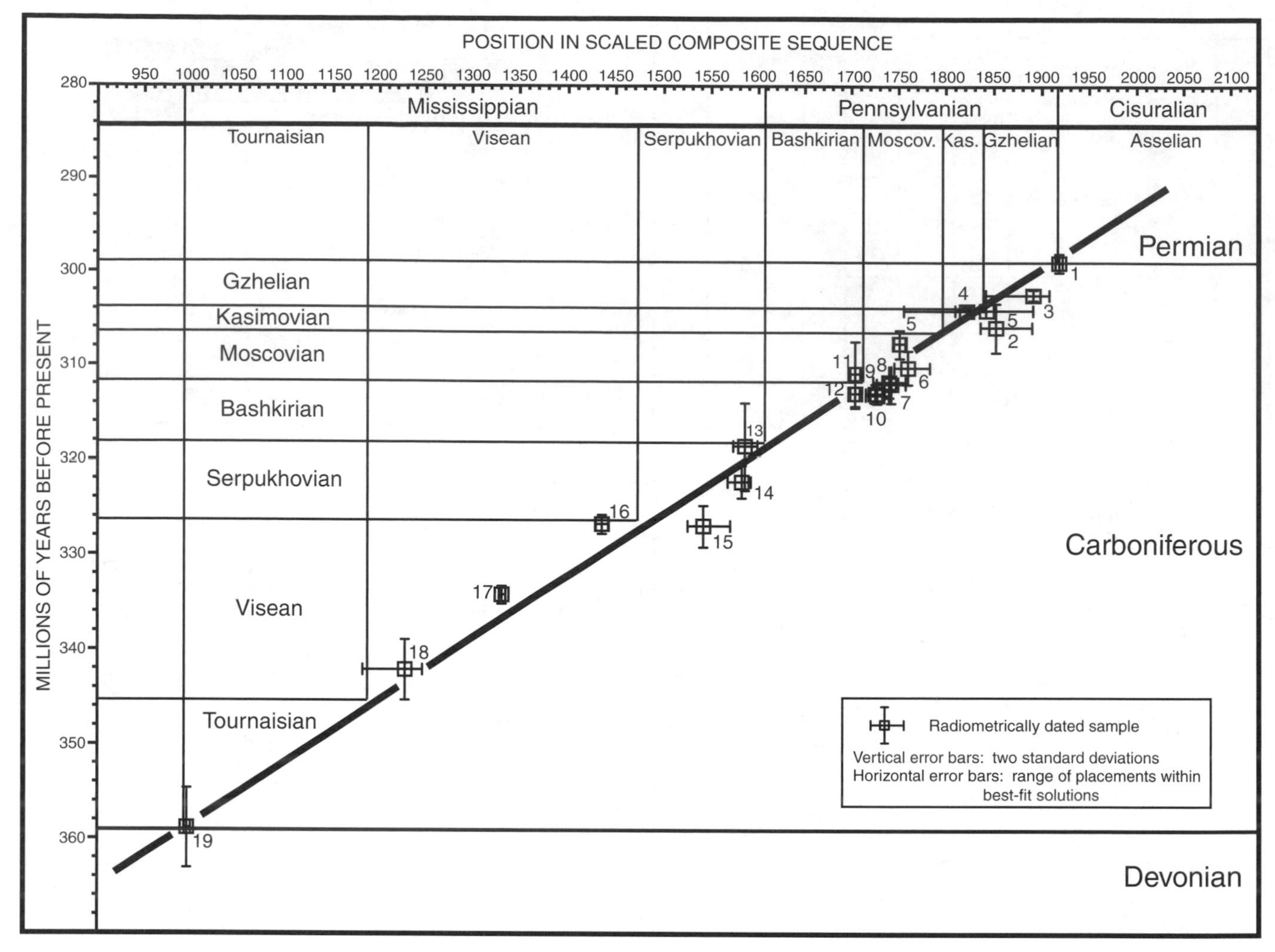

Figure 15.5 Construction of the Carboniferous time scale. Radiometric age dates for the Carboniferous are given in Table 15.1 and are plotted against the relative zonal scale shown in Figure 15.4. The best-fit line of the data is a cubic spline, minimally smoothed. For details see text.

direct data between the Guadalupian and Lopingian shallow-shelf foraminiferal succession for Tethys and the deeper-water conodont succession of the newly established international scale, these two biotic sequences were calibrated approximately against each other.

The GraphCor software, developed by Hood (1977), has been used in the compositing process. The process of compositing was repeated several times, until fossil ranges stabilized. The number of iterations varied from section to section, but was generally between two and four. Although a mature CS is considered to be time-significant, it is only an idealized stratigraphic section, and a proxy for a chronostratigraphic scale. Comparison of the CS to sections yields lines of correlation (LOCs) that are a function of the rock accumulation rates, not linear time. That is, LOCs depict how rock accumulation rates in one section differ relative to those in the CS. In order for a CS to become a stratigraphic time scale, it must be calibrated by radiometric dates tied to biostratigraphic boundaries. Unfortunately, almost none of the radiometric dates, with a few exceptions from the Urals and South China, were directly incorporated in the CS.

A qualitative approach to LOC positioning was therefore used. First and last appearance datums (FADs and LADs) of all taxa were considered individually. Species ranges were evaluated according to their relative biostratigraphic importance. Index taxa defining foraminiferal and/or conodont zones and short-ranging species were given a greater weight in defining the LOC position, while long-ranging species provided supplementary information. FADs and LADs of taxa were treated differently. For instance, the FAD of *Raphconilia* spp. is an excellent marker for the base of the Gzhelian throughout northern Pangea. However, the genus ranges at least

Table 15.3 *Ages and durations of Carboniferous stages*

Period	Subperiod	Stage	Age of base (Ma)	Est. ± myr (2-sigma)	Duration	Est. ± myr (2-sigma)
Permian			**299.0**	0.8		
Carboniferous						
	Pennsylvanian					
		Gzhelian	303.9	0.9	4.9	0.1
		Kasimovian	306.5	1.0	2.6	0.0
		Moscovian	311.7	1.1	5.2	0.1
		Bashkirian, base Pennsylvanian	318.1	1.3	6.4	0.2
	Mississippian					
		Serpukhovian	326.4	1.6	8.4	0.2
		Visean	345.3	2.1	18.9	0.7
		Tournaisian, base Mississippian, (base Carboniferous)	**359.2**	2.5	13.9	0.8
Devonian						

through the Cisuralian, and the component species have stratigraphic ranges exceeding one or more stages. The base of each foraminiferal and/or conodont zone was included into the database and treated as if it were a taxon; this was done for two reasons. First, bases of these zones were used in the correlation process in order to provide relative time indicators on the graphs. Second, data points defining zonal bases were critical in controlling LOC positions. The position of a zonal base reflects the summarized knowledge of a particular fossil assemblage. Zonal bases represent additional data points on graphs and help to establish the most reasonable LOC pattern even if index species were not recovered. This approach decreases the influence of local factors – paleoenvironment, climate, facies, migration, etc. – in establishing the most reasonable correlation pattern. It is important to mention that zonal bases alone cannot be successfully used in graphical correlation analysis because only a very general correlation pattern can be inferred from such data. Unlike multiple taxonomic ranges, whose FADs and LADs represent continuous succession of biostratigraphic events, zonal bases define only the presence of some parts of the corresponding biozones. For instance, breaks in stratigraphic sections can be confidently inferred only if a zone is totally missing. That is, the use of zonal bases alone does not allow interpretation of sedimentation patterns within zones and amounts to little more than traditional biostratigraphy. In addition to species ranges, generic ranges were employed in order to increase the number of taxa utilized in the correlation. The range of a genus, defined by the first and last appearance of its stratigraphically oldest and youngest species, respectively, can be entered into the database as an independent taxon, and thus used in the correlation process. Stratigraphic ranges of genera are considerably longer than those of component species and are less useful in determining a LOC position. Nevertheless, they provide additional data points on the graph and may be helpful in correlating sections with inadequate sampling.

The quality of the Carboniferous–Permian CS is variable. The scale is shown in Fig. 15.4, with the composite standard units (CU) needed for time scale calibration inserted in Table 15.2. The CS and the age dates were analyzed mathematically to derive the Carboniferous and Permian time scale.

15.3.4 Age of stage boundaries

The Carboniferous and Permian time scales were calculated simultaneously using the composite standard as explained above and the radiometric data in Tables 15.2 and 16.1. The selected radiometric age dates for the two periods were plotted against the relative zone–stage scale of Fig. 15.4 with radiometric and stratigraphic errors shown approximately. The best-fit line for the two-way plot in Fig. 15.5 of radiometric ages against the zones and stages was calculated with a cubic spline that combines stratigraphic uncertainty estimates with the 2-sigma error bars of the radiometric data. This is achieved with the Ripley MLFR procedure (See Chapter 8). A generalized Devonian age uncertainty factor of 0.006 85 was also adopted for the Carboniferous and Permian spline, with the exception of base-Triassic and base-Permian, which have their own uncertainty factors derived from superior radiometric and stratigraphic information.

Based on two successive chi-square tests, following the procedure as used for the Ordovician–Devonian, several ages were adjusted. The Rasbury age of 298.0 ± 1.4 (Ma No. 9 in Table 15.2) was given slightly higher CSU values to fit the spline better. Note its stratigraphic position is tentative and subject to larger than average uncertainty. An external error of 1% was applied for the HR–SIMS dates, as explained in Chapter 6, and one $^{40}Ar/^{39}Ar$ 1-sigma value was enlarged (from 1.1 to 6.5 myr; No. 15 in Table 15.2). The final spline curve is nearly a straight line: its smoothing value was calculated with the cross-validation technique.

The calculated duration and ages with estimates of uncertainty (2-sigma) of the Carboniferous stages are given in Table 15.3. Uncertainty in stage duration is less than uncertainty in the age of stage boundaries.

The total duration of the Carboniferous is just over 60 myr: the Mississippian (Early Carboniferous) lasted 41 myr and the Pennsylvanian (Late Carboniferous) 19 myr.

16 • The Permian Period

B. R. WARDLAW, V. DAVYDOV, AND F. M. GRADSTEIN

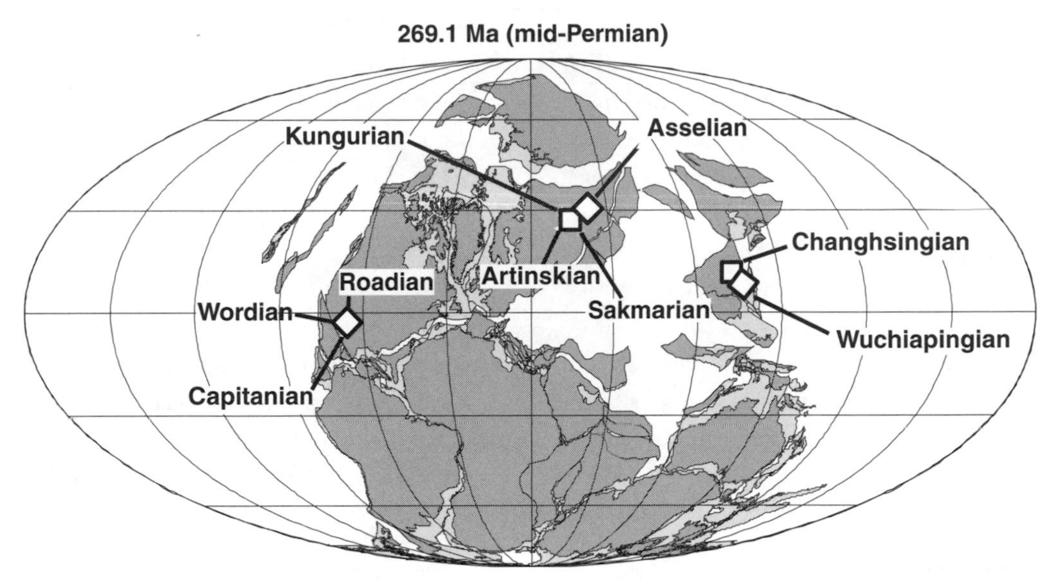

Geographic distribution of Permian GSSPs that have been ratified (diamonds) or are candidates (squares) on a mid-Permian map (status in January, 2004; see Table 2.3). GSSPs for the Changhsingian and Early Permian are not yet formalized.

Pangea moves north. Ice-house to greenhouse (humid to arid) climate transition; dramatic reduction of coal swamps and amphibian habitat; some spore-bearing plants extinct; major evaporites; changes in internal and external carbonate invertebrate skeletons; major diversification of fusulinid foraminifera, ammonoids, bryozoans, and brachiopods, then major end-Permian extinction of fusulinid foraminifera, trilobites, rugose and tabulate corals, blastoids, acanthodians, placoderms, and pelycosaurs; dramatic reduction of bryozoans, brachiopods, ammonoids, sharks, bony fish, crinoids, eurypterids, ostracodes, and echinoderms.

16.1 HISTORY AND SUBDIVISIONS

In 1841, after a tour of Imperial Russia, R. I. Murchison, in collaboration with Russian geologists, named the Permian System to take in the "vast series of beds of marls, schists, limestones, sandstones and conglomerates" that surmounted the Carboniferous System throughout a great arc stretching from the Volga eastwards to the Urals and from the Sea of Archangel to the southern steppes of Orenburg. He named it for the ancient kingdom of Permia in the centre of that territory, and the city of Perm that lies on the flanks of the Urals. In 1845 he included rocks now known as Kungurian–Tatarian in age, and for a time the underlying strata (Artinskian, etc.) were known as Permo-Carboniferous, i.e. intermediate between Carboniferous and Permian (Dunbar, 1940).

As early as 1822 (e.g. Conybeare and Phillips, 1822), the Magnesian Limestone and New Red Sandstone of England were well known, as were the equivalent German Rotliegendes and Zechstein (a traditional miner's name) with its valuable Kupferschiefer. However, all these rocks lacked richly fossiliferous strata, were difficult to correlate, and unsuitable to justify the erection of a new system in western Europe.

In North America, J. Mancou in 1853 recognized Permian rocks in a large area from the Mississippi to the Rio Colorado and noted two divisions analogous to those in western Europe. He accordingly suggested the name Dyassic as more suitable

A Geologic Time Scale 2004, eds. Felix M. Gradstein, James G. Ogg, and Alan G. Smith. Published by Cambridge University Press.

than Permian and proposed a combined Dyas and Trias as a major period (Zittel, 1901). For further historical details on the history of the Permian Period we refer to GTS89 (pp. 46 and 47). In Germany, the term Dyas is now being recommended for the combined Rotliegendes and Zechstein (German Stratigraphic Commission, 2002).

Permian biostratigraphy has been greatly refined over the last two decades, especially through a detailed understanding of the distribution of conodonts in relation to ammonoids and fusulinids.

The Permian divides itself naturally into three series (Fig. 16.1). In the classic area of the southern Urals, the Upper Carboniferous and Lower Permian (now Cisuralian) are well represented by marine deposits and abundant biota. This marine dominance disappears in the Kungurian, and the Middle Permian (now Guadalupian) and Upper Permian (now Lopingian) are dominated by terrestrial–marginal-marine deposits. The Guadalupian deposits of West Texas are dominated by diversified and well-studied marine fossil assemblages, and the deposits are the subject of seminal studies in sequence stratigraphy. China, Iran, and the Trans-Caucasian region are exemplary for their Upper Permian deposits and biota.

16.1.1 The Cisuralian Series: Lower Permian

The base of the Permian was originally defined in the Ural Mountains of Russia to coincide with strata marking the initiation of evaporite deposition (Murchison, 1841), now recognized as the Kungurian Stage. Since 1841, the base has been lowered repeatedly to include a succession of fauna with post-Carboniferous affinities. Karpinsky (1874) identified clastic successions that Murchison had included in the British Millstone Grit as being younger, transitional between Carboniferous and Permian, and termed them the Artinskian Series. His subsequent classic study of the abundant ammonoid fauna (Karpinsky, 1889) led him to add the interval to the Permian. Further study, especially of ammonoids, led Ruzhenzev (1936) to recognize the Sakmarian as an independent lower subdivision of the Artinskian. In turn, he subdivided the Sakmarian, and referred the lower interval to the Asselian Stage (Ruzhenzev, 1954). The base of the Asselian and of the Permian period was defined by the appearance of the ammonoid families Paragastrioceratidae, Metalegoceratidae, and Popanoceratidae, concurrent with the first inflated fusulinaceans referable to "*Schwagerina*" (i.e. *Sphaeroschwagerina*).

The base of the Asselian Stage, defined by reference to both ammonoids and fusulinaceans, received progressively greater recognition and eventually official Russian status (Resolutions of the Interdepartmental Stratigraphic Committee of Russia and its Permanent Commissions – Resolution on Carboniferous/Permian Boundary. St. Petersburg, 1992, pp. 52–56) following Ruzhenzev's original proposal.

The Cisuralian was proposed by Waterhouse (1982) to comprise the Asselian, Sakmarian, and Artinskian Stages. The Kungurian was included in the Cisuralian (Jin *et al.*, 1997), so that it corresponded to the Lower Permian as recognized in Russia (Likharew, 1966; Kotlyar and Stepanov, 1984) and corresponded better to the Rotliegendes of Harland *et al.* (1990). The Uralian Series, named by de Lapparent in 1900, interpreted by Gerasimov (1937) to include pre-Kungurian stages of the Lower Permian, and utilized by Jin *et al.* (1994), has been abandoned because the name caused too much confusion from a history of varied usage.

ASSELIAN

The GSSP for the beginning of the Permian Period and of the Asselian Stage is located at Aidaralash Creek, Atobe region, northern Kazakhstan (Davydov *et al.*, 1998). The section is approximately 50 km southeast of the city of Atobe. A stone and concrete marker with a plaque has been erected at the Aidaralash section, marking the exact location of the GSSP, and the boundary between the Carboniferous and Permian Periods.

The strata of Late Paleozoic age at Aidaralash Creek were deposited on a narrow, shallow-marine shelf that formed the western boundary of the orogenic zone to the east. The fluvial–deltaic conglomerate–sandstone successions grade upwards into transgressive, marginal marine sequences (beach and upper shore face) that, in turn, grade upwards into massive mudstone–siltstone and fine sandstone beds with ammonoids, conodonts, and radiolaria, interpreted as maximum flooding units. The maximum flooding zone is overlain by a regressive sequence (progressively, offshore to shoreface to delta front), which in turn is capped by an unconformity with an overlying conglomerate. The critical GSSP interval is completely within a maximum flooding unit; free of disconformities.

The position of the GSSP is at the first occurrence of the conodont *Streptognathodus isolatus*, which developed from an advanced morphotype in the *S. wabaunsensis* chronocline. This is located 27 m above the base of Bed 19, Airdaralash Creek (Davydov *et al.*, 1998).

The first occurrences of *Streptognathodus invaginatus* and *S. nodulinearis*, also morphotypes of the "*wabaunsensis*" morphocline, nearly coincide with the first occurrence of *S. isolatus* in many sections, and can be used as accessory indicators for the boundary.

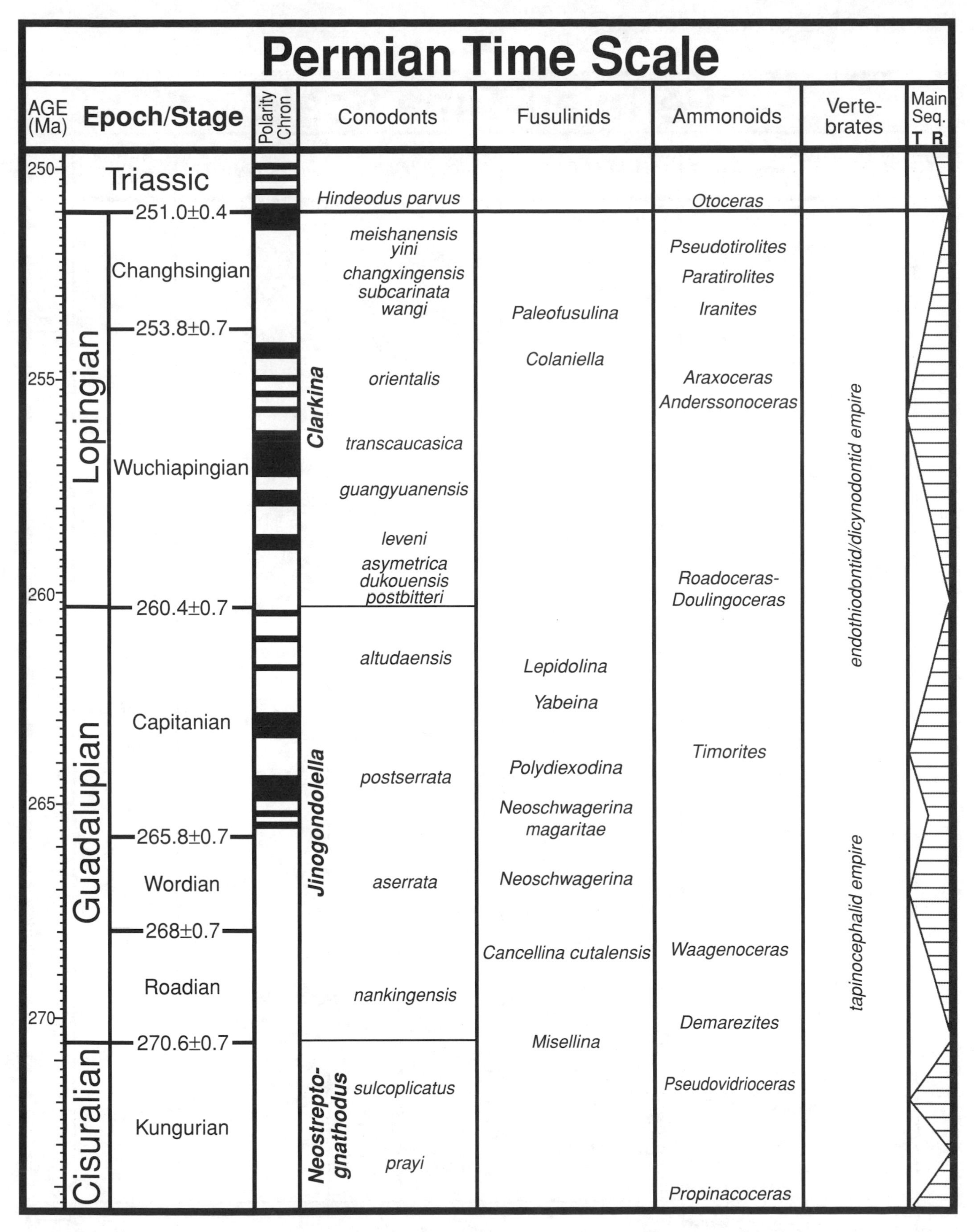

Figure 16.1 Permian time scale and magnetic polarity, conodont, fusulinid, ammonoid, vertebrate zonations, and general transgressive–regressive sequences. Magnetic polarity scale compiled from Irving (1963), Khramov, (1963), Peterson & Nairn (1971), Khramov & Davydov (1984), Steiner (1988), Haag & Heller (1991), Heller *et al.* (1995), Opdyke & Channell (1996), and Opdyke *et al.*, (2000). A color version of this figure is in the color plate section.

Permian Time Scale

AGE (Ma)	Epoch/Stage		Polarity Chron	Conodonts		Fusulinids	Ammonoids	Vertebrates	Main Seq. T R
	Cisuralian	275.6 ± 0.7			*Neostreptognathodus pequopensis*				
280		Artinskian		*Sweetognathus*	*whitei*	*Chalaroschwagerina* *Pamirina* *Parafusulina*	*Uraloceras*		
							Aktubinskia–Artinskia		
285		284.4 ± 0.7			*binodosus*		*Sakmarites*	*edaphosaurid empire*	
290		Sakmarian				*Pseudofusulina*			
					merrilli	*Schwagerina vernueili* *Schwagerina moelleri*			
295		294.6 ± 0.8		*Streptognathodus*	*barskovi*	*Sphaeroschwagerina gigas*	*Svetlanoceras*		
		Asselian			*postfusus*	*Pseudoschwagerina robusta*			
					fusus	*Schwagerina nux*			
					constrictus	*Sphaeroschwagerina fusiformis*			
					isolatus	*Sphaeroschwagerina vulgaris*			
300	Carboniferous	299 ± 0.8			*binodosus*	*Ultradaixina bosbytauensis*	*Shumardites–Emilites*		

Figure 16.1 (*cont.*)

The GSSP is 6.3 m below the traditional fusulinid boundary, i.e. the base of the *Sphaeroschwagerina vulgaris aktjubensis–S. fusiformis* Zone (Davydov *et al.*, 1998; Figs. 16.2 and 16.3). The latter can be widely correlated with Spitsbergen, the Russian Platform, Urals, Central Asia, China, and Japan, and is of practical value in identifying the boundary between the Orenburgian and overlying Asselian Stages.

The traditional ammonoid boundary, 26.8 m above the GSSP, includes the termination of the *Prouddenites–Uddenites* lineage at the top of Bed 19, and the introduction of the Permian taxa *Svetlanoceras primore* and *Prostacheoceras principale* in Bed 20 (Davydov *et al.*, 1998; Figs. 16.2 and 16.3). The evolution from *Artinskia irinae* to *A. kazakhstanica* may be a chronocline that crosses the Carboniferous–Permian boundary. A problem with the ammonoid taxa is that they are relatively rare, and many taxa may be endemic to the Urals.

Utilization of magnetostratigraphy to assist with recognition and correlation of the Carboniferous–Permian boundary is difficult because it is in the Kiaman Long Reversed-Polarity Chron (see the geologic time scale chart insert of this book). However, Davydov *et al.* (1998) cite reports that show that most of the *Ultradaixina bosbytauensi–Schwagerina robusta* fusulinid zone, just below the Carboniferous–Permian boundary in Aidaralash is characterized by normal polarity. That same stratigraphic polarity relationship is also known elsewhere from the southern Urals, and the northern Caucasus and Donetz Basin, and possibly correlates with the normally polarized magnetic zone in the Manebach Formation of the Thuringian Forest (Menning, 1987).

The conodont succession observed at Airdaralash is displayed in several sections in the southern Urals, especially the basinal reference section at Usolka. It is also displayed in the Red Eagle cyclothem of the midcontinent of the USA (Boardman *et al.*, 1998), in the West Texas regional stratotype, Wolfcamp Hills (Wardlaw and Davydov, 2000), and in China (Wang, 2000), as well as in many other intervening localities, and, therefore, serves as an excellent boundary definition.

SAKMARIAN

The proposed boundary for the Sakmarian Stage is very near the level originally proposed by Ruzhenzev (1950b) in the same section, Kondurovsky, Orenburg Province, Russia. A conodont succession exhibiting the evolutionary change from *Sweetognathus expansus* to *S. merrilli* at 115 m above the base (uppermost Bed 11 of Chuvashov *et al.*, 1993a,b) is the proposed definition. The boundary originally proposed by Ruzhenzev (1950b) was at the base of Bed 11, at an unconformable formation break and based on the change in fusulinaceans fauna, with *Schwagerina (Pseudofusulina) moelleri* occurring above the break. The actual introduction of the *S. moelleri* group occurs in Beds 6–12 (Wardlaw *et al.*, 1999), with traditional *S. moelleri* occurring in Bed 12, just a few meters above the first occurrence of *S. merrilli*. *Schwagerina merrilli* is widespread and its FAD is well constrained throughout Kansas in the upper part of the Eiss Limestone of the Bader Limestone.

ARTINSKIAN

The Artinskian Stage was proposed by Karpinsky in 1874, for the sandstone of the Kashkabash Mountain on the right bank of the Ufa River, near the village of Arty, as the stratotype. Karpinsky (1891) studied abundant and diverse ammonoids in several exposures and small quarries along the Ufa River. The taxonomically diverse ammonoid assemblage from the Arty area was distinctly more advanced than the Sakmarian one in terms of cephalopod evolution and this stimulated Karpinsky (1874) to define two belts with ammonoids: the lower, at Sakmara River, and the upper, at Ufa River.

The Sakmarian–Artinskian boundary deposits are well represented in the Dal'ny Tulkus section, a counterpart of the Usolka section. The upper part of the Sakmarian Stage (Beds 28–31) at the Usolka River and Bed 18 at the Dal'ny Tulkus section are composed of dark-colored marl, argillite, and carbonate mudstone, or less commonly, detrital limestone with fusulinids, radiolaria, rare ammonoids, and bivalves. The upper part of the Sakmarian includes fusulinids characteristic of the Sterlitamakian Horizon including *Pseudofusulina longa*, *P. fortissima*, *P. plicatissima*, *P. urdalensis*, and *P. urdalensis abnormis*.

The best GSSP section appears to be the Dal'ny Tulkus section, in Russia, but a point cannot yet be defined precisely; however, the definition will be placed at the FAD of *Sweetognathus whitei,* derived from *S. binodosus*. The succession of *S. binodosus* to *S. whitei* can also be recognized in the lower Great Bear Cape Formation, southwest Ellesmere Island (Henderson, 1988; Beauchamp and Henderson, 1994; Mei *et al.*, 2002) and in the Schroyer–Florence Limestones of the Chase Group, Kansas (Wardlaw *et al.*, 2003).

KUNGURIAN

The stratotype of the Kungurian Stage was not defined when the stage itself was established (Stuckenberg, 1890). Sometime later, the carbonate–sulphate section exposed along the Sylva River, upstream of the town of Kungur, was arbitrarily accepted for the stratotype. In line with the new position of the Kungurian lower boundary at the base of the Sarana Horizon (Chuvashov *et al.*, 1999), the stratotype section in this area consists of:

1. the Sarana Horizon, including the reefal limestone Sylva Formation, and its lateral equivalent the Shurtan Formation, composed of marls and clayey limestone;
2. the Filippovskian Horizon; and
3. the Iren' Horizon.

A disadvantage of the section is the poor fossil content of the limey Kamai Formation underlying the Sarana Horizon; it contains only small benthic foraminifera, bryozoans, and brachiopods inappropriate for age determination. Nevertheless, many features indicate that the formation corresponds to the Sarana Horizon.

The Shurtan Formation and the lateral Sylva bioclastic limestone facies yield conodonts of the *Neostreptognathodus pnevi* Zone. However, another section of Artinskian–Kungurian boundary deposits, located near the Mechetlino settlement on the Yuryuzan' River, has good fauna both below and above the boundary interval, and has been selected as the probable stratotype of the Kungurian lower boundary.

The Cisuralian is divided into stages based on the FADs of specific species from two lineages of conodonts. The first, the base of the Permian, and for that matter the zones of the Asselian, is based on the widespread occurrence of *Streptognathodus* species, as is most of the upper Pennsylvanian. The members of the genus become progressively rarer and less widespread after the beginning of the Sakmarian. The Sakmarian through Kungurian stages are based on a lineage of *Sweetognathus* and its derived daughter *Neostreptognathodus*.

The Asselian through Artinskian of the Uralian foredeep can also be recognized by a conodont succession of *Mesogondolella* species. These species are very common in the Urals, but are rare elsewhere in the world.

16.1.2 The Guadalupian Series: Middle Permian

The Guadalupian was first proposed by Girty at the turn of the last century for the spectacular fossils found in the Guadalupe and Glass Mountains of West Texas. These fauna have been well documented and represent an unprecedented display in an exhumed, well-preserved backreef, reef, and basin. The West Texas depositional basins represent a tropical North American faunal suite, well removed from the more typical tropical Tethyan fauna of Asia and Europe. The Middle Permian was a time of strong provincialism and presents some complexities for correlation. The formal establishment of the Guadalupian and its constituent stages is based on the evolution of a single genus of conodont, *Jinogondolella*. The genus has a limited distribution, though it is common in West Texas and South China.

ROADIAN

The GSSP for the base of the Roadian Stage, Guadalupian Epoch, Middle Permian, is in Stratotype Canyon, Guadalupe Mountains National Park, Texas, USA. The marker horizon is the first evolutionary appearance of the conodont *Jingondolella nankingensis* from its ancestor *Mesogondolella idahoensis*, at 42.7 m above the base of the black, thin-bedded limestones of the Cutoff Formation, and 29 cm below a prominent shale band in the upper part of the El Centro Member (Glenister *et al.*, 1999). This member consists of skeletal carbonate mudstone with one shale bed, deposited in a basinal setting, proximal to the slope. In terms of magnetostratigraphy, the Cutoff Formation indicates reversed polarity, and may fall in the Kiaman reversed superchron. The GSSP of the Roadian Stage was ratified in 2001.

WORDIAN

The GSSP for the beginning of the Wordian Stage in the Guadalupian Epoch is located in Guadalupe Pass, Texas; a short distance from Stratotype Canyon. The marker horizon for this stage is the first evolutionary appearance of *Jinogondolella aserrata* from its ancestor *J. nankingensis* at 7.6 m above the base of the Getaway Ledge outcrop section in Guadalupe Pass, Guadalupe Mountains National Park, Texas, USA. This level is just below the top of the Getaway Limestone Member of the Cherry Canyon Formation, a succession of skeletal carbonate mudstone in a base of slope depositional setting (Glenister *et al.*, 1999). Like the Roadian sediments in the type area, the Wordian Stage limestones of the Guadalupian National Park also display reversed polarity. The GSSP of the Wordian Stage was ratified in early 2001.

CAPITANIAN

Like the GSSPs for the Roadian and Wordian Stages in the Middle Permian, the GSSP for the Capitanian Stage was also selected in the Guadalupian National Park. The marker horizon for the Capitanian Stage is the first evolutionary appearance of the conodont *Jinogondolella postserrata* within the lineage *nanginkensis–aserrata–postserrata*. This level is at 4.5 m in the outcrop section at Nipple Hill, in the upper Pinery Limestone Member of the Bell Canyon Formation (Glenister *et al.*, 1999). The GSSP is in monotonous, pelagic carbonates, representing a lower slope depositional setting. The level of the date is 37.2 m below the base of the *J. postserrata* entry (Bowring *et al.*, 1998). Few samples in the Pinery Limestone and overlying Lamar Limestone of the Bell Canyon Formation display normal polarity, indicative of the approximate position of the Illawara Reversal. The GSSP of the Capitanian Stage was ratified in early 2001.

The abundant and well-preserved conodont fauna of West Texas show that the genus of *Jinogondolella* and its species evolved through short-lived transitional morphotypes, generally through a mosaic of paedomorphogenesis (retention of juvenile characters in later and later growth stages). The first species of the genus, *J. nankingensis*, is also the marker for the Guadalupian and its basal stage, the Roadian. The species is abundant in West Texas and South China but occurs rarely in several other sites (i.e. Canadian Arctic, Pamirs); however, its distribution along the western coast of Pangea represents a geographical cline from the tropical Delaware Basin (West Texas) to the upwelling-influenced Phosphoria Basin (Idaho) to temperate Canadian Arctic and exhibits overlap with several genera, especially within the Phosphoria Basin where it is abundant, and provides excellent correlation globally. The cline is confirmed to be geographic (nearly synchronous) by the co-occurrence and coincident range with *Neostreptognathodus newelli* in the Delaware and Phosphoria Basins.

The Illawarra geomagnetic reversal is an important tie point for the Guadalupian and the base of the Capitanian. The Illawarra reversal is well known from the lower part of the Tatarian in the Volga region of Russia (Gialanella *et al.*, 1997). It has also been documented from Pakistan. Haag and Heller (1991) show that normal polarity starts at the base of the Wargal Formation in the Nammal Gorge, Salt Range, which is the base of the Illawarra reversal (F. Heller, pers. comm; 1998). Peterson and Nairn (1971) record a reversal in West Texas–New Mexico that has been interpreted with additional study by Menning (2000) to occur just below the Pinery Limestone Member of the Bell Canyon Formation.

The ammonoid genus *Waagenoceras* has long been associated with the Guadalupian, and, in particular, the Wordian. The distribution of the conodont *Jinogondolella*, which characterizes the Guadalupian, is common in the tropical zone of South China and West Texas and the upwelling area of the Phosphoria Basin. Its rarer appearances around the margin of the Tethyan tropical zone and temperate zones bordering the Tethys and the northern margin of Pangea links it to other faunal provinces, making the Guadalupian an appropriate standard for the Middle Permian.

16.1.3 The Lopingian Series: Upper Permian

The Lopingian (Huang, 1932), Dxhulfian (Furnish, 1973), Transcaucasian, and Yichangian (Waterhouse, 1982) have been proposed for the uppermost Permian series. Of these, the Lopingian appears to be the first formally designated series name to be based on relatively complete marine sequences. The Lopingian Series comprises two stages: the Wuchiapingian and the Changhsingian.

The Tatarian of the traditional Volga region of Russia is mostly a continental deposit and corresponds largely to the upper Guadalupian; it does not serve in a comprehensive subdivision of the upper Permian.

The upper boundary of the Permian (i.e. the base of the Triassic) in the original type area, the Buntsandstein of Germany, and in the Urals is non-marine and unsuitable for world-wide correlation. The functional definition for the base of the Triassic has been the base of the ammonoid *Otoceras* Zone of the Himalayas (Griesbach, 1880). The first appearance of the conodont *Hindeodus parvus* is more widespread than *Otoceras* and provides a precise basis for base-Triassic (Chapter 17).

WUCHIAPINGIAN

The boundary between the Guadalupian and Lopingian Series and the base of the Wuchiapingian Stage was historically designated to coincide with a global regression, i.e. with the boundary surface between the Middle and the Upper Absaroka Megasequences. Extensive surveys of marine sections demonstrate that few sections can be considered to be continuous across the Guadalupian–Lopingian boundary. Sections with a complete succession of open-marine fauna are particularly rare. Guadalupian–Lopingian boundary successions were reported from Abadeh and Jolfa in Central Iran, from southwestern USA, and the Salt Range. The Laibin Syncline in Guangxi Province, China, is unique among these sections in that it contains a complete and inter-regionally correlatable succession of open-marine conodont zones and other diverse fossils. The GSSP for the Lopingian Series coincides with the first occurrence of *Clarkina postbitteri postbitteri* within an evolutionary lineage from *C. postbitteri hongshuiensis* to *C. dukouensis* at the base of Bed 6k of the Penglaitan section. The Tieqiao (Rail-Bridge) section on the western slope of the syncline is proposed as a supplementary reference section

(Jin *et al.*, 2001). The Wuchiapingian GSSP was ratified in 2004.

CHANGHSINGIAN

Initially, the GSSP for the Changhsingian Stage was formally recommended as the horizon between the *Clarkina orientalis* and the *Clarkina subcarinata* Zones, which was located at the base of Bed 2, the base of the Changxing Limestone in Section D at Meishan, Changxing County, Zhejiang Province, China (Zhao *et al.*, 1981). Further research suggests that the base of the Changhsingian Stage should be defined within the *Clarkina longicuspidata–Clarkina wangi* lineage based on a better understanding of conodont taxonomy and evolution. It is proposed that the GSSP for the Changhsingian now be defined by the first occurrence of *C. wangi* within Bed 4 in Section D at Meishan, Changxing County, China.

The change in the proposed boundary is slight, from the base of Bed 2 to within Bed 4. The basal part of the Changxing Limestone, and, therefore, the boundary interval of the Wuchiapingian–Changhsingian, is marked by the occurrence of advanced forms of *Palaeofusulina*, and the ammonoid families Tapashanitidae and Pseudotirolitidae, which still mark the lower Changhsingian in the new definition.

The Dzhulfian and Dorashamian Stages of Transcaucasia correspond, respectively, to the Wuchiapingian and Changhsingian. However, the successions in the basal part of the Dzhulfian Stage and the top portion of the Dorashamian Stage are not as well developed in their type areas as corresponding intervals in the standard succession of South China (Iranian–Chinese Research Group, 1995).

16.2 REGIONAL CORRELATIONS

16.2.1 Russian Platform

Increased research on the Permian of Russia (e.g. Chuvashov and Nairn, 1993; Esaulova *et al.*, 1998; Chuvashov *et al.*, 2002) has led to significant changes from the traditional stratigraphic scheme for the Russian platform, discussed earlier. The scheme was largely based on the distribution of ammonoids. For the Lower Permian, i.e. Cisuralian, the much more common fusulinids and conodonts now provide a refined zonation. The traditional "Upper" Permian units of Russia have much less common ammonoids, fusulinids, and conodonts than the Cisuralian, but detailed work has greatly improved biostratigraphic correlation (Figs. 16.2 and 16.3).

The Kungurian is sparsely fossiliferous at best. To improve its correlatability, the upper horizon of the traditional Artinskian, the Saranian, the last unit with a well-developed, fully marine fauna, was taken as a reliable horizon for worldwide correlation. Further, the conodont succession from *Neostreptognathodus pequopensis* to *Neostreptognathodus pnevi* was taken as a reliable evolutionary event to establish a base for the Kungurian (Chuvashov *et al.*, 2002). This boundary, discussed under the Kungurian of the Cisuralian in Section 16.4, corresponds to a *Neostreptognathodus* evolutionary event in the regional stratotype of the Leonardian, which indicates that the newly revised Kungurian and the Leonardian are basically equivalent.

The Ufimian has been abandoned by the All Russian Stratigraphic Commission because the Ufimian represents terrestrial and marginal-marine facies of the upper Kungurian (the Solikamian, lower Ufimian) and the lower Kazanian (the Sheshmian, upper Ufimian). However, there is no known section that shows Sokian (lower Kazanian) lying on Irenian (upper Kungurian), so problems still exist in interpreting this boundary interval, even if it is only in perception.

The Tatarian, basically a series of stacked soils in its type area, is difficult to correlate. It does contain the Illawarra Geomagnetic Reversal in its lower part, which ties that part to the upper Wordian–Capitanian interval. How young the upper Tatarian is, is an open question. Sequence stratigraphy suggests that the Tatarian and Capitanian are roughly equivalent. There is a sharp changeover in fossil species at the Tatarian–Vetluzhian boundary suggesting a significant unconformity (Fig. 16.3).

16.2.2 Germanic Basin

The Germanic Basin is only briefly dealt with here. Zechstein 1 contains a fairly good marine fauna that includes *Merrillina divergens* (Fig. 16.2). This occurs above the Illawarra Reversal, which is within the upper part of the Rotleigendes. Both tie Zechstein 1 to the Capitanian. How much time is reflected in the remaining Zechstein units is unknown; the evaporites could represent very short depositional intervals between long hiatuses or a very short interval similar to that of the Ochoan of West Texas.

16.2.3 Pamirs

The fusulinids and, to a lesser extent, the ammonoids (Fig. 16.4) are well known from the Pamirs. The fusulinid zonation serves as the standard for the Permian of the Tethyan. Conodonts appear to indicate some correlation potential and need significant investigation.

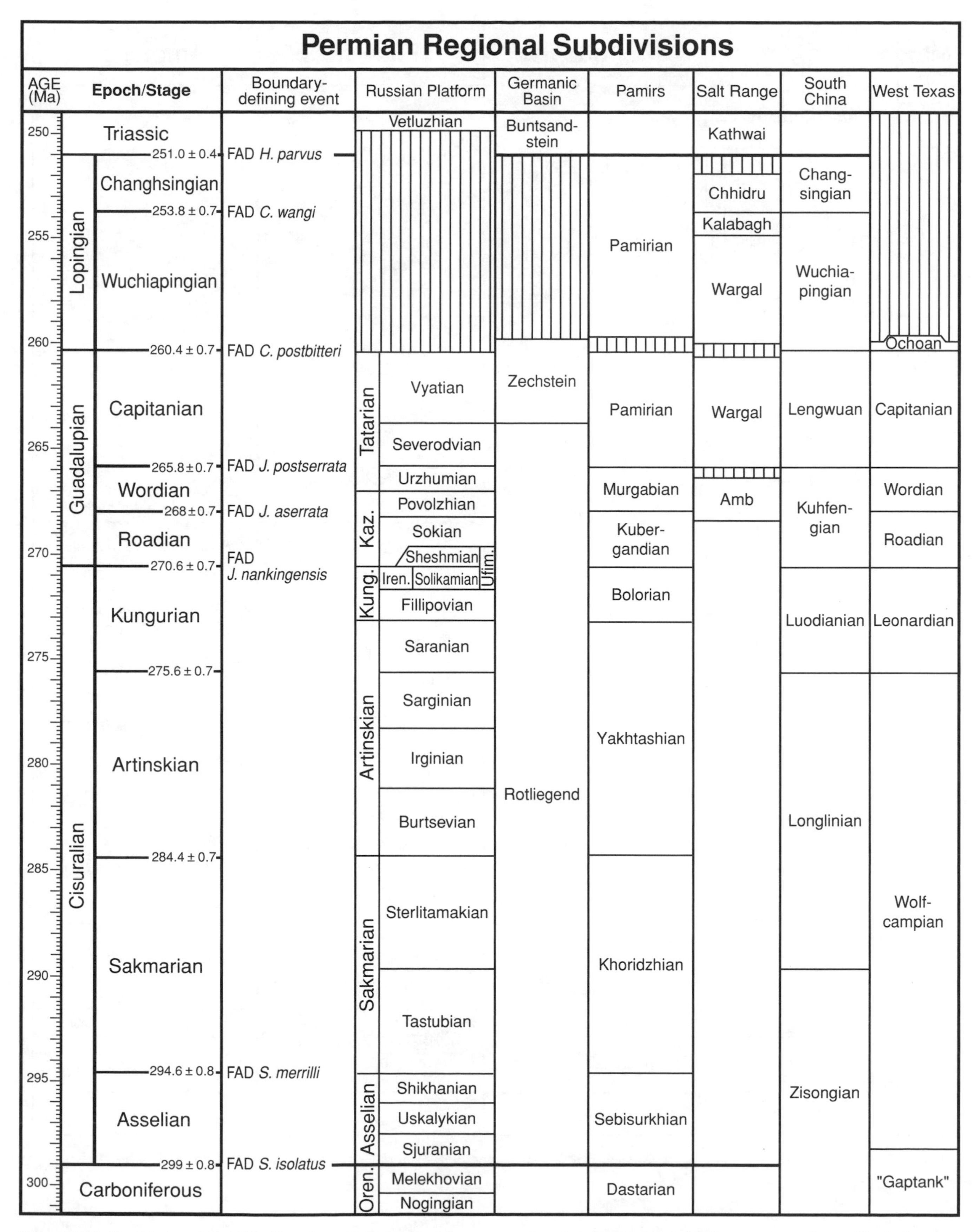

Figure 16.2 Correlation chart of the Permian with international subdivisions and boundary-defining events (left columns) and selected regional stage and substage nomenclature. Vertical pattern indicates a widespread regional hiatus.

Ma	Russian Platform		Conodonts	Fusulinids	Ammonoids
		Vetluzhian			
250					
260	Tatarian	Vyatian			
		Severodvian			
		Urzhumian			
	Kaz.	Povolzhian	*Merrillina divergens* *Kamagnathus volgensis* *Kamagnathus khalimbadzhae*		*Sverdrupites amundseni* *Sverdrupites harkeri*
		Sokian			
270	Uf.	Sheshmian			
	Kung. / Uf.	Irenian / Solikamian	*Neostreptognathodus pnevi*		*Tumaroceras zavadovskyi* *Epijuresanites musaltini* *Uraloceras tschuvaschovi*
	Kung.	Fillipovian			
	Artinskian	Saranian		*Parafusulina solidissima* *Parafusulina jenkinsi*	*Propinacoceras aktjubensis* *Neocrimites fredericksi*
		Sarginian	*Neostreptognathodus pequopensis*		
280		Irginian		*Parafusulina lutugini* *Pseudofusulina solida*	*Popanoceras annae* *Popanoceras tschernowi*
		Burtsevian	*Sweetognathus whitei*	*Pseudofusulina pedisequa* *Pseudofusulina concavutas*	
	Sakmarian	Sterlitamakian	*Mesogondolella bisselli*	*Pseudofusulina urdalensis* *Pseudofusulina plicatissima*	*Preshumardites sakmarae* *Synartinskia principalis*
290		Tastubian	*Mesogondolella visibilis*	*Schwagerina verneuili*	
			Mesogondolella lata		
			Mesogondolella uralensis	*Schwagerina moelleri*	*Svetlanoceras strigosum*
	Asselian	Shikhanian	*M. pseudostriata* *St. postfusus–M. striata*	*Sphaeroschwagerina gigas*	
		Uskalykian	*St. fusus–M. simulata* *St. constrictus–M. adentata*	*Pseudoschwagerina robusta* *Schwagerina nux*	*Svetlanoceras serpentinum*
		Sjuranian	*St. cristellaris* *St. isolatus*	*Sphaeroschwagerina fusiformis* *Sphaeroschw. vulgaris aktjubensis*	*Svetlanoceras primore*
300	Oren.	Melekhovian	*Streptognathodus wabaunsensis*	*Utradaixina bosbytauensis*	*Shumardites aktjubensis*
		Noginian	*St. elongatus* *St. ruzhenzevi*	*Daixina sokensis* *Daixina ruzhenzevi*	*Shumardites confessus*
	Carboniferous				

Figure 16.3 Regional correlation chart for the Russian Platform with conodont, fusulinid, and ammonoid zonations.

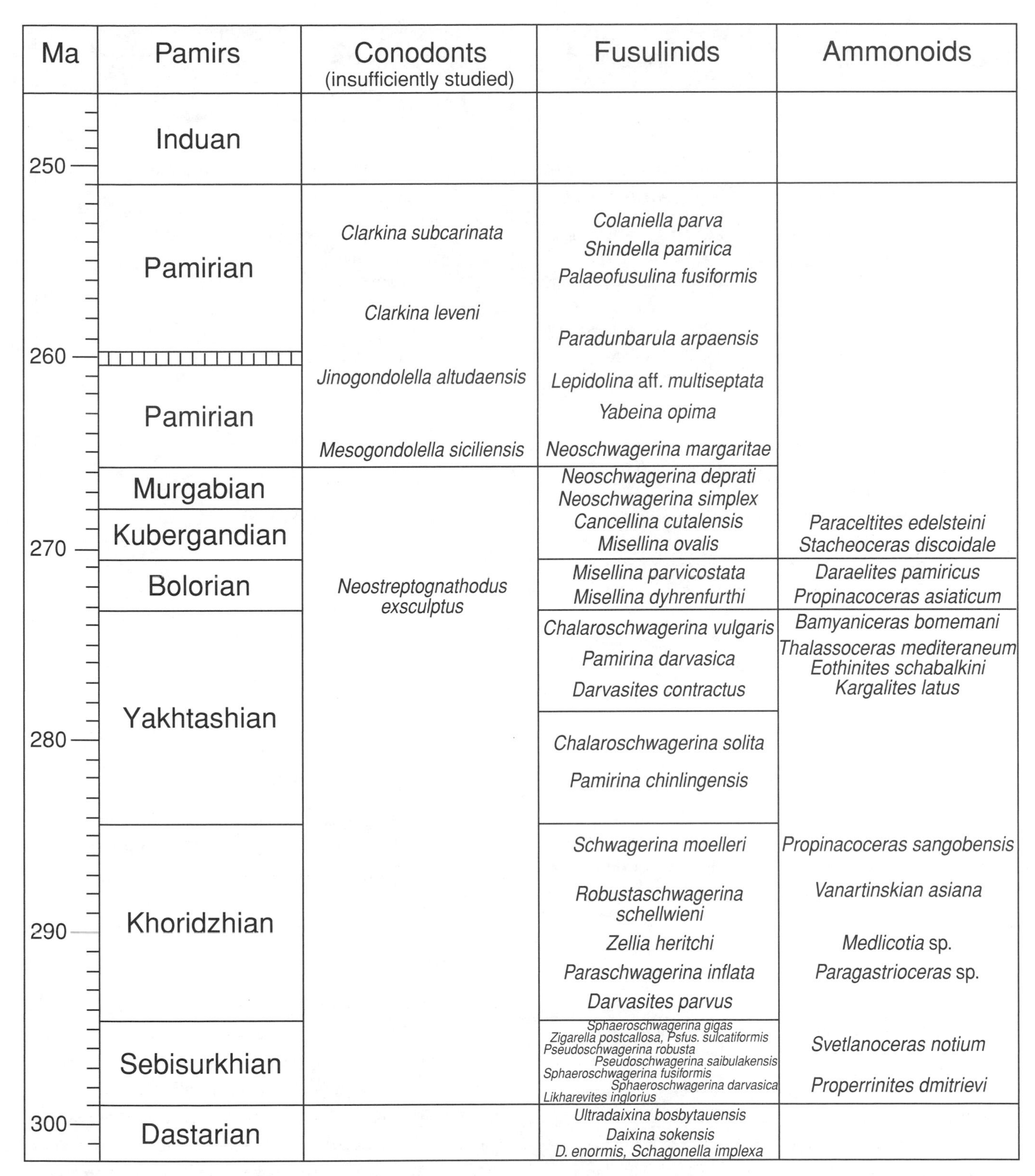

Figure 16.4 Regional correlation chart for the Pamirs with conodont, fusulinid, and ammonoid zonations.

16.2.4 Salt Range

The Salt Range (Fig. 16.5) does not serve as a regional standard, but has inter-bedded temperate and Tethyan fauna. Of major importance is the overlapping of ranges of *Merrillina divergens*, *Neoschwagerina margaritae*, and the Illawarra Reversal within the lower part of the Wargal Limestone, below a significant unconformity within that formation, indicating

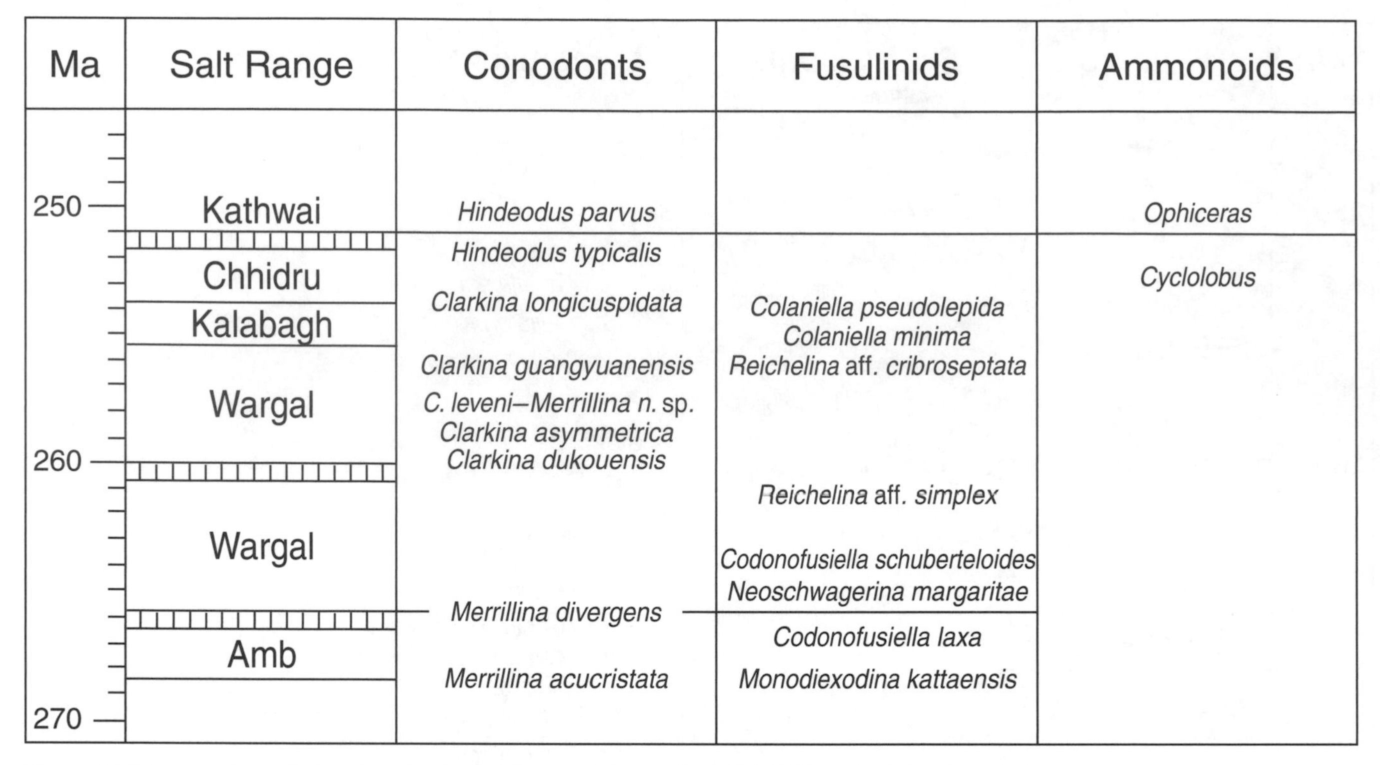

Figure 16.5 Regional correlation chart for the Salt Range with conodont, fusulinid, and ammonoid zonations.

that all are at least part Capitanian. Above the unconformity occur common *Iranognathus* and a succession of *Clarkina* species.

16.2.5 South China

The regional stages developed for South China (Fig. 16.6) are very much defined by their fossil constituents. From Sheng and Jin (1994) the following can be discerned:

Zisongian – is based on the fusulinid *Pseudoschwagerina uddeni–P. texana* Zone and the *Sphaeroschwagerina* zone.

Longlinian – is based on the biostratigraphic sequence between the last *Pseudoschwagerina* and the first *Misellina* (not an easy concept to establish as a stage).

Luodianian – is based on the fusulinid *Misellina* Zone.

Xiangboan – is based on the fusulinid *Cancellina* Zone.

Kuhfengian – is based on the first occurrence of the conodont *Jinogondolella nankingensis* (the same definition of the Roadian).

Lengwuan – includes the ranges of the conodonts *Jinogondolella shannoni* and *J. xuanhanensis* and is based on the first occurrence of *J. postserrata* (the same definition of the Capitanian).

Wuchiapingian – is based on the first occurrence of the conodont genus *Clarkina*.

Changhsingian – is based on the first occurrence of the conodont *Clarkina subcarinata* (*sensu latu*) near the base of the Changxing Limestone.

Both the Wuchiapingian and Changhsingian definitions have been modified and proposed as potential stratotypes and are dealt with under the Lopingian (Upper Permian).

16.2.6 West Texas

Both the Wolfcampian and Ochoan (Fig. 16.7), one based on a sequence of delta front conglomerates, sands, and silts, and one based on basin-filling evaporites, pose problems for correlation. Conodonts from scattered units in the Wolfcampian suggest that the top bed of the Grey Limestone Member of the Gaptank Formation contains *Streptognathodus isolatus* Zone conodonts and, therefore, signifies the base of the Permian. The overlying Neal Ranch Formation has conodont and fusulinid fauna in scattered limestones, and contains *Streptognathodus isolatus* and *S. barskovi* (Wardlaw and Davydov, 2000). The upper part of the Neal Ranch Formation and the lower part of the overlying Lenox Hills Formation yield only sparse fauna. *Sweetognathus whitei* is present along with common fusulinids in the upper

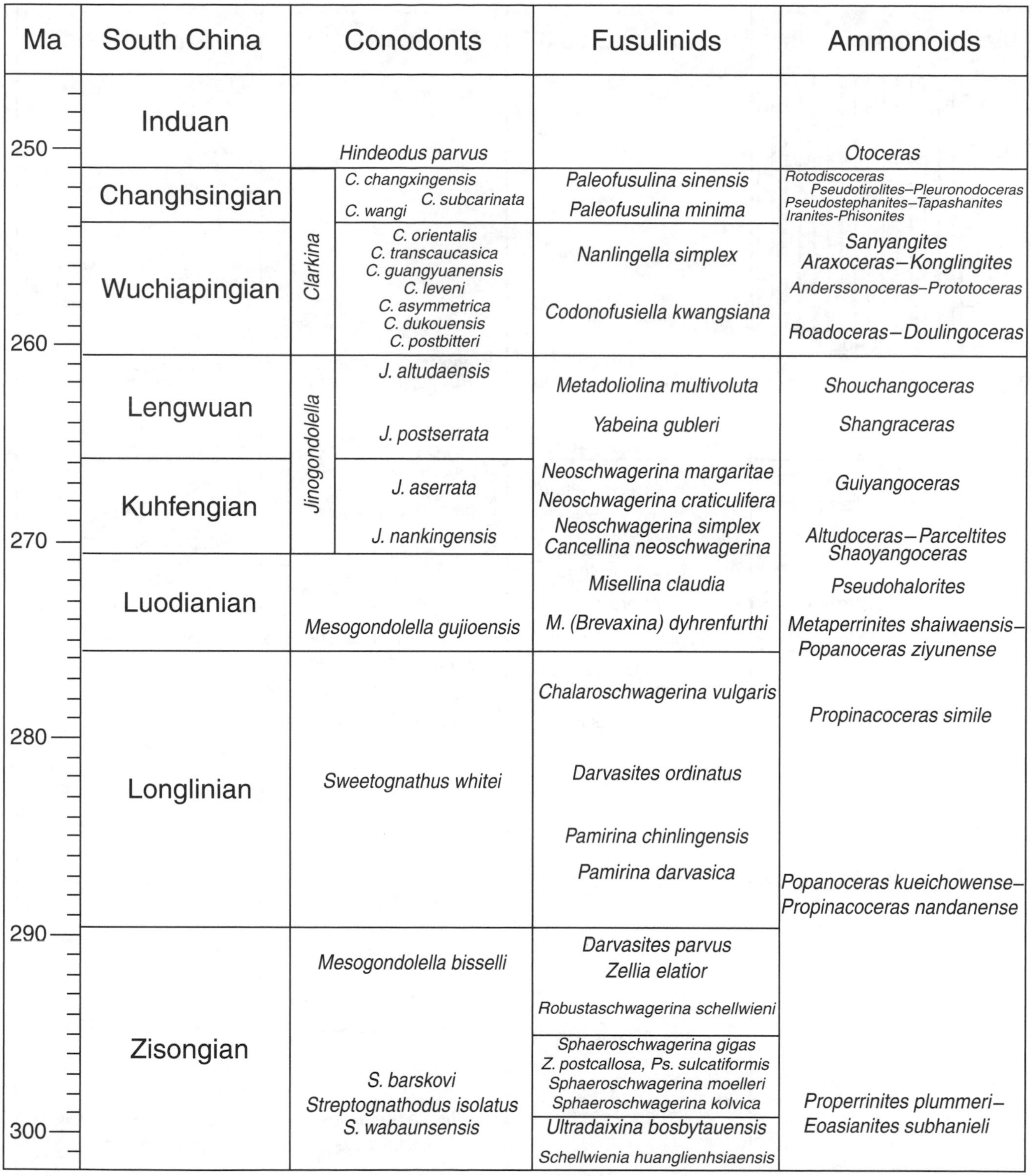

Ma	South China	Conodonts	Fusulinids	Ammonoids
250	Induan	*Hindeodus parvus*		*Otoceras*
	Changhsingian	*Clarkina*: *C. changxingensis*, *C. subcarinata*, *C. wangi*	*Paleofusulina sinensis*, *Paleofusulina minima*	*Rotodiscoceras*, *Pseudotirolites–Pleuronodoceras*, *Pseudostephanites–Tapashanites*, *Iranites-Phisonites*
260	Wuchiapingian	*Clarkina*: *C. orientalis*, *C. transcaucasica*, *C. guangyuanensis*, *C. leveni*, *C. asymmetrica*, *C. dukouensis*, *C. postbitteri*	*Nanlingella simplex*, *Codonofusiella kwangsiana*	*Sanyangites*, *Araxoceras–Konglingites*, *Anderssonoceras–Prototoceras*, *Roadoceras–Doulingoceras*
	Lengwuan	*Jinogondolella*: *J. altudaensis*, *J. postserrata*	*Metadoliolina multivoluta*, *Yabeina gubleri*	*Shouchangoceras*, *Shangraceras*
270	Kuhfengian	*Jinogondolella*: *J. aserrata*, *J. nankingensis*	*Neoschwagerina margaritae*, *Neoschwagerina craticulifera*, *Neoschwagerina simplex*, *Cancellina neoschwagerina*	*Guiyangoceras*, *Altudoceras–Parceltites*, *Shaoyangoceras*
	Luodianian	*Mesogondolella gujioensis*	*Misellina claudia*, *M. (Brevaxina) dyhrenfurthi*	*Pseudohalorites*, *Metaperrinites shaiwaensis–Popanoceras ziyunense*
280	Longlinian	*Sweetognathus whitei*	*Chalaroschwagerina vulgaris*, *Darvasites ordinatus*, *Pamirina chinlingensis*, *Pamirina darvasica*	*Propinacoceras simile*, *Popanoceras kueichowense–Propinacoceras nandanense*
290–300	Zisongian	*Mesogondolella bisselli*, *S. barskovi*, *Streptognathodus isolatus*, *S. wabaunsensis*	*Darvasites parvus*, *Zellia elatior*, *Robustaschwagerina schellwieni*, *Sphaeroschwagerina gigas*, *Z. postcallosa*, *Ps. sulcatiformis*, *Sphaeroschwagerina moelleri*, *Sphaeroschwagerina kolvica*, *Ultradaixina bosbytauensis*, *Schellwienia huanglienhsiaensis*	*Properrinites plummeri–Eoasianites subhanieli*

Figure 16.6 Regional correlation chart for South China with conodont, fusulinid, and ammonoid zonations.

part of the Lenox Hills Formation. The Skinner Ranch and Cathedral Mountain Formations (Leonardian) yield an abundance of fauna. Similarly, the Road Canyon and Word Formations yield abundant and diverse fauna. Slope and basinal equivalents of the Vidrio Formation and Capitan Limestone also yield excellent fauna.

In the base of the Skinner Ranch Formation, below the first occurrence of common *Neostreptognathodus "exsculptus"*

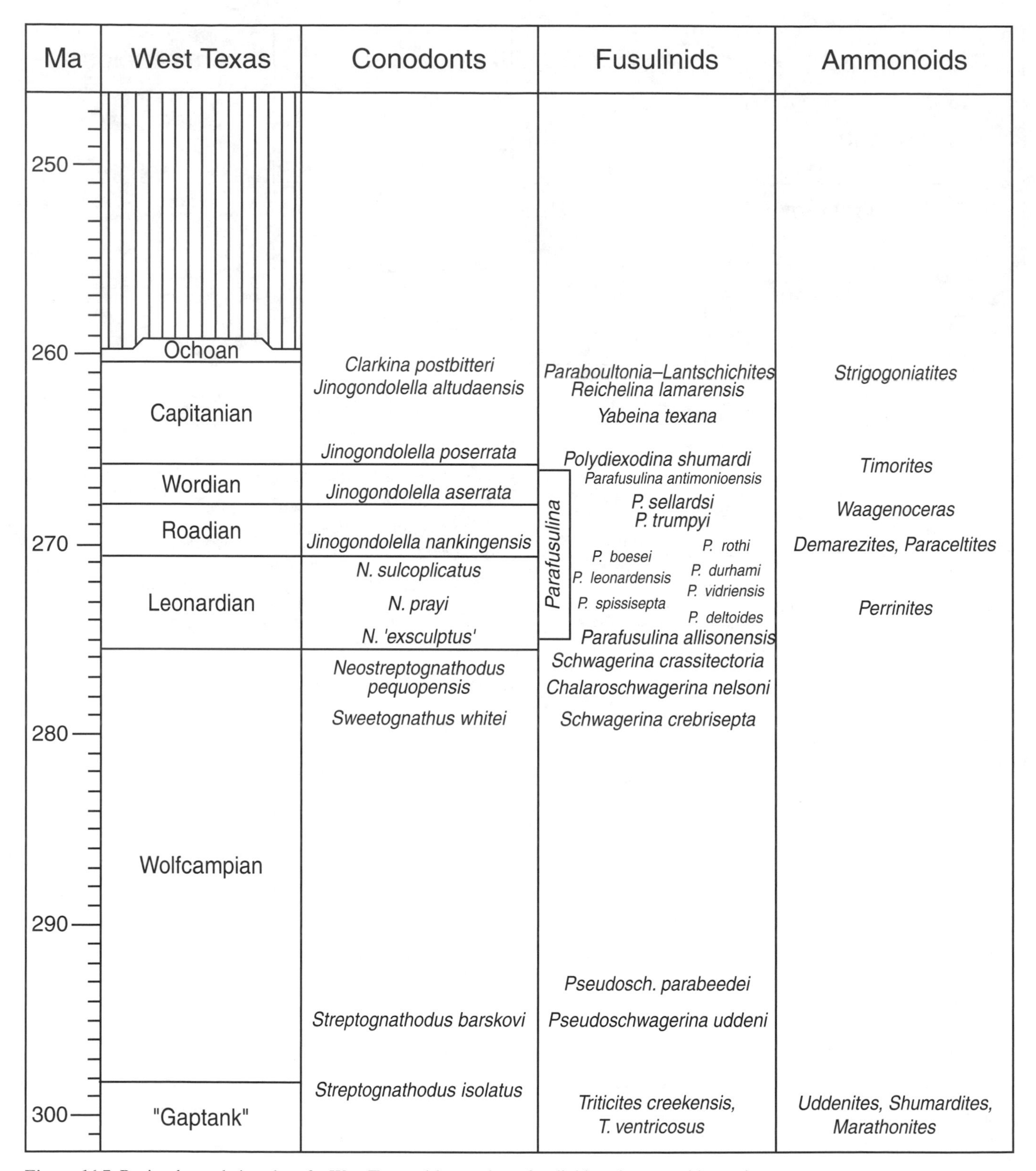

Figure 16.7 Regional correlation chart for West Texas with conodont, fusulinid, and ammonoid zonations.

are sparse fauna of *N. pequopensis.* In the Tansill Formation (the shelf equivalent of the Lamar Limestone Member of the Bell Canyon Formation), there are sparse fauna dominated by a species of *Sweetina* (Croft, 1978). Virtually the same species is found overlying the evaporites of the Ochoan stage, within the Rustler Formation (above the Castille and Salado Formations), implying that the deposition of the basin-filling evaporites occurred in less than one conodont zone.

16.3 PERMIAN STRATIGRAPHY

16.3.1 Biostratigraphy

CONODONT ZONATION

Since conodont biostratigraphy is critical to this construction of a Permian time scale, it is briefly reviewed here (Fig. 16.1).

The Lower Permian (Cisuralian Series) conodont zonation is derived from a variety of published and unpublished sources. The Asselian and Sakmarian zonation, based on the succession of *Streptognathodus* species, is from Chernykh *et al.* (1997), Boardman *et al.* (1998), and Wardlaw *et al.* (1999). This succession is well represented in Kansas and the southern Ural Mountains of Russia.

The Artinskian zonation is based on species of *Sweetognathus*, *Streptognathodus*, and *Neostreptognathodus* and reflects the major changeover in the forms dominating shelf fauna during this interval. It is largely based on unpublished material from the southern Urals, Russia, and Kazhastan, and the Great Basin, USA.

The Kungurian zonation is based on the succession of *Neostreptognathodus* species modified from Wardlaw and Grant (1987) from West Texas, USA.

The Middle Permian (Guadalupian Series) conodont zonation is from Wardlaw and Lambert (1999) and Wardlaw (2000), except that the rapid succession of upper Guadalupian *Jinogondolella* (*altudaensis, prexuanhanensis, xuanhanensis,* and *crofti*) are all overlapped by *J. altudaensis* and considered as subzone indicators of that zone. This succession is well represented in West Texas and South China.

The Upper Permian (Lopingian Series) conodont zonation of successive *Clarkina* species from Mei *et al.* (1994a, 1998a), as modified by Wardlaw and Mei (1998), reflects the current compromise (Henderson *et al.*, 2001) for definition for the base of the Lopingian based on the redefined *C. postbitteri postbitteri*. Also, even though the classic definition of the Changhsingian has been the first appearance of *C. subcarinata*, a dramatic changeover in fauna occurs within Bed 4 in the Changxing Limestone from one dominated by *C. orientalis* and *longicuspidata* (Wuchiapingian) to one dominated by *C. wangi* (Mei *et al.*, 2001), which marks a boundary closer to the base of the formation and more acceptable to non-conodont workers. This succession is well represented in both South China and the Dzhulfa area of Iran and Transcaucasia.

FUSULINACEAN ZONES

In the southern Urals, sequence boundaries coincide with the bases of several fusulinacean zones. Lowstands correspond to significant fusulinacean extinction. The base of the Asselian (i.e. the base of the Permian), the base of the Sakmarian, and some fusulinacean zones coincide with highstands. Therefore, fusulinacean speciation appears to be associated with both highstands and lowstands. Sea-level lowstands may have been very stressful for global fusulinacean assemblages and may have been a catalyst for both speciation and extinction. Highstands also may have created environmental opportunities and appear to be more closely associated with fusulinacean speciation than extinction. Sequence boundaries located within fusulinacean zones, perhaps reflect local tectonism or local climatic changes.

16.3.2 Physical stratigraphy

THE ILLAWARRA GEOMAGNETIC POLARITY REVERSAL

As mentioned earlier, the Illawarra geomagnetic reversal is an important tie point for the Guadalupian series and the base of the Capitanian Stage. The reversal is near the top of the *J. aserrata* Zone in West Texas and near the top of the *M. praedivergens* Zone in the Salt Range.

In the Guadalupe Mountains, a tuff at or about the projected position of the Illawarra, within the top of the range of *J. aserrata*, just below the first occurrence of *J. postserrata*, the indicator for the Capitanian, yields a date of 265.3 Ma.

GEOCHEMISTRY

The seawater $^{87}Sr/^{86}Sr$ curve (Fig. 16.8) indicates a minimum value for all of Phanerozoic time occurred in the Capitanian. The driving mechanism of the late Permian variations appears to have been climate change rather than tectonic; Pangea was assembled and fairly stable through this interval. The period of decreasing seawater $^{87}Sr/^{86}Sr$ for the Early–Middle Permian is associated with a waning ice age, as well as high continental aridity and low external runoff attributed to the huge Pangean landmass. Though not apparent in the oxygen-18 curve, there is faunal evidence for significant cooling after the Roadian in the Sverdrup Basin, continuing for most of the rest of the Permian (Henderson, 2002) and cooling in the middle to late Wuchiapingian in the Salt Range, indicating bipolarity in cooling at this time.

Marine extinctions began in a step-wise function following the Guadalupian (Jin, 1994; Zhou *et al.*, 1996). This suggests that global cooling and the begining of mass extinction coincided with the increase in $^{87}Sr/^{86}Sr$. This further suggests an amelioration of the climate with the cooling event, which led

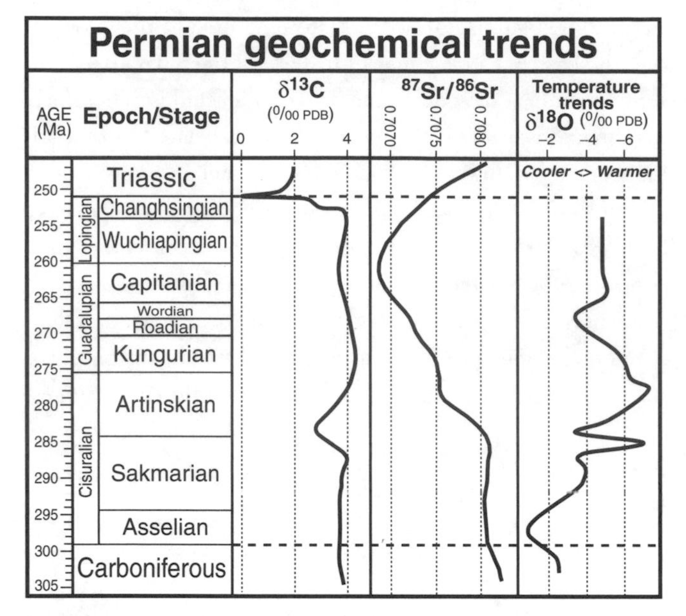

Figure 16.8 Geochemical trends during the Permian Period. The schematic carbon isotope curve is a 5-myr averaging of global data (Veizer *et al.*, 1999) from Hayes *et al.* (1999; downloaded from www.nosams.whoi.edu/jmh). The $^{87}Sr/^{86}Sr$ LOWESS curve for the interval is based on the data of Denison *et al.* (1994) and Martin & Macdougall (1995) – see Chapter 7. The oxygen isotope curves (inverted scale) are derived from a 3-myr interval averaging of global data compiled by Veizer *et al.* (1999; as downloaded from www.science.uottawa.ca/geology/isotope_data/, January 2003). Large-scale global shifts to higher oxygen-18 values in carbonates are generally interpreted as cooler seawater or glacial episodes, but there are many other contributing factors (e.g. Veizer *et al.*, 1999; Wallman, 2001).

to a decrease in continentality, an increase in precipitation, and an overall declining sea level in the Upper Permian, with more area exposed to erosional processes.

There is a significant carbon excursion just below the Permian–Triassic boundary. This excursion is intimately related to the final extinction event of the Permian, but it is unclear just how or what it completely reflects (Erwin, 1993, 1995). However, it does indicate a sharp decrease in productivity and burial of organic carbon at the very end of the Permian.

16.4 PERMIAN TIME SCALE

16.4.1 Permian composite standard

In order to integrate and calibrate zonal successions of foraminifera, conodonts, and ammonoids, and to construct a Permian composite biostratigraphic scale the composite standard technique of graphic correlation was undertaken. Graphic correlation produces a composite standard that when interpreted chronostratigraphically, and calibrated with selected radiometric age dates, yields a linear time scale.

The composite standard technique was applied to a joint Carboniferous–Permian data set that is discussed in Chapter 15 (Figs. 15.4 and 15.5).

16.4.2 Radiometric age dates

Although precise age constraints are in place for the base and top of the Permian, the Permian time scale is among the least internally constrained in the Phanerozoic. Table 16.1 lists eleven stratigraphic levels with radiometric dates, and only two (items 4 and 5) that are not in the uppermost and lowermost stages of the Permian. What is required are many more stratigraphically precise and analytically accurate radiometric dates. Until such time, the intra-Permian scale is heavily dependent on relative zonal scaling and is approximate at best.

Chuvashov *et al.* (1996) report one biostratigraphically well-constrained U–Pb HR–SIMS date of 290.6 ± 3.0 Ma in the Usolka section of the southern Urals for the base of the *constrictus* Zone (and corresponding local fusulinid zone) in lower Middle Asselian. This is one conodont zone above the Carboniferous–Permian boundary.

Ramezani *et al.* (2003) have collected volcanic ash beds within mid-ramp carbonate as well as offshore mixed carbonate–siliciclastic successions in three southern Urals sections: Usolka, Dal'ny Tulkus road cut, and Dal'ny Tulkus quarry. One ash layer in the Usolka section is 0.6 m above the Carboniferous–Permian boundary. It contains numerous zircons and the conodont *Streptognathodus isolatus* – the index species of the base of the Permian. Preliminary results from TIMS analyses of the U–Pb ratios in zircons from this ash, and others stratigraphically just above and below the Carboniferous–Permian boundary, suggest an age of 299 Ma for the boundary, to which we assign an uncertainty of 1 myr. Continuing paleontologic and geochronologic studies in the southern Urals should provide a robust data set for precise calibration of the Upper Carboniferous through Cisuralian interval of the time scale.

By comparison with the observed trend in the Ordovician through Devonian (Chapters 12–14) that HR–SIMS dates fall up to 1.3% behind TIMS dates for the same interval, the Chuvashov *et al.* (1996) date of 290.6 ± 3.0 Ma might even be as old as 297.4 Ma. Thus, this date might not disagree with the new age estimate for the Carboniferous–Permian boundary near 299 Ma.

Table 16.1 *Selected U–Pb radiometric dates for the Permian time scale*

No.	Sample	Locality	Formation	Comment	Zone and age	Biostratigraphic reliability	Reference	Age (Ma)	Type	Range and centre point of the horizontal error bars for each date in Composite Standard Units (CSU)[a]
1	Zircons	Meishan section, the GSSP of the base of the Triassic; Changhsing, Zhejiang Province, S. China	Grayish ash, illite–montmorillonite clay, white bentonite; Changhsing and Chinglung Fms	Set of TIMS zircon dates obtained at MIT geochronological laboratory. See Chapter 17 and Table 17.1	Closely sampled across the Permian–Triassic boundary section; *Clarkina changsingensis* conodont zone and into basal range of *H. parvus*	1	Bowring *et al.* (1998), *Jin et al.* (2000)	251.0 ± 0.2 (2-sigma); see Chapter 19 and Table 19.1	U–Pb	2637–2657 (2647)
2	Zircons	Matan section, near Heshan, Guangxi Province, S. China	Pyroclastic deposits, Talung Formation	Set of TIMS zircon dates obtained at MIT geochronological laboratory. See Chapter 17 and Table 17.1	Closely sampled across the Permian–Triassic boundary section; *Clarkina changsingensis* conodont zone and into basal range of *H. parvus*		Bowring *et al.* (1998), Jin *et al.* (2000)		U–Pb	
3	Zircons	Penglaitan section, banks of Hong Shui River near town of Labin, S. China	Yellow bentonite	TIMS dates were obtained at MIT geochronological laboratory	Ammonoids *Rotodiscoceras* and *Pleuronodoceras* are found beneath this unit, confirming a late Changhsingian age	1	Bowring *et al.* (1998), Jin *et al.* (2000)	253.4 ± 0.2 (2-sigma)	U–Pb	2626–2650 (2629)
4	Zircons	Ash bed occurs between the Hegler and Pinery Limestone Members of the Bell Canyon Formation; Nipple Hill, Guadalupe Mountains National Park, Texas	Grayish-green bentonite	TIMS dates were obtained at MIT geochronological laboratory	Sample collected 20 m below the base of the *Jinogondolella postserrata* conodont zone that defines the base of the Capitanian stage of the Guadalupian Series	1	Bowring *et al.* (1998)	265.3 ± 0.2 (2-sigma)	U–Pb	2435–2445 (2438)

(*cont.*)

Table 16.1 (*cont.*)

No.	Sample	Locality	Formation	Comment	Zone and age	Biostratigraphic reliability	Reference	Age (Ma)	Type	Range and centre point of the horizontal error bars for each date in Composite Standard Units (CSU)[a]
5	Zircons	Belaya River section and Sim section, southern Urals region, Russia	Bentonite	HR–SIMS monitor standard age not available. The dates were obtained at the Australian Geological Organization, hence it is assumed here that the HR–SIMS age was calibrated to standard zircon SL13. The U–Pb date of the zircon grains is 280.3 ± 2.5 Ma (2-sigma)	These two samples are precisely constrained within late Sakmarian based on ammonoids and fusulinids	1	Roberts *et al.* (1996), Chuvashov *et al.* (1996)	280.3 ± 2.5 (2-sigma)	U–Pb	2132–2197 (2148)
6	Zircons	Usolka section, southern Urals region, Russia	Tuffaceous marls	HR–SIMS monitor standard age not listed. The U–Pb zircon dates of 290.0 ± 3.2 and 290.3 ± 3.2 Ma (all 2-sigma) were obtained at the Australian Geological Organization, hence it is assumed here that the HR–SIMS age was calibrated to standard zircon SL13. The average age is 290.15 ± 2.26 Ma (2-sigma)	These samples are precisely constrained within early middle Asselian based on conodonts, ammonoids, and fusulinids	1	Roberts *et al.* (1996), Chuvashov *et al.* (1996), Davydov *et al.* (2002)	290.15 ± 2.26 (2-sigma)	U–Pb	1949–1968 (1958)

7	Zircons	Lodeve–Becken, Frankreich Germany	Tuff, Lower–Middle Viala Formation, Lower Rotliegendes	TIMS date of 295.53 ± 0.54 Ma (2-sigma) for the youngest zircon population	Viala Formation from the regional correlation is placed in the uppermost Lower Rotliegendes. Schneider and Roscher (2002) correlate this formation with Nikitovskaya and Slavyansjkaya, Formations of Donetz Basin, assigned a middle to late Asselian age (Davydov, 1990).	3	Trapp & Kaufmann (2002)	295.53 ± 0.54 (2-sigma)	U–Pb	1950–2000 (1975)
8	Zircons	Saar, SW Germany	Pappelberg tuff horizon of Jeckenbach unit of Meisenheim Fm of Lebach Group of Lower Rotliegend	34 U–Pb HR–SIMS dates on zircon grains give 297.0 ± 3.2 Ma (1-sigma) calibrated to standard zircon SL13 and AS3	Based on plants and considered to be within the Carboniferous–Permian transition (uppermost Gzhelian–lowermost Asselian); although of poor biostratigraphic constraint, it is allowed here with low weight in the fitting method	3	Königer *et al.* (2002), Davydov (1992)	297.0 ± 3.2 (1-sigma)	U–Th–Pb	1940–2027 (1960)
9	Paleosol calcite	Subsurface, UNOCAL well, Central Basin Platform, West Texas	Paleosol calcite from cycle 0, UNOCAL core	7 U–Pb TIMS dates on paleosol calcite give 298.0 ± 1.4 Ma (2-sigma)	Biostratigraphic constraints of these samples provided from unpublished data and from regional correlations considered to be within lower Wolfcampian and somewhere above the Permian base; although poor biostratigraphic constraint, it is allowed here with low weight in the fitting method	3	Rasbury *et al.* (1998)	298.0 ± 1.4 (2-sigma)	U–Pb	1956–2044 (1992)

(cont.)

Table 16.1 (*cont.*)

No.	Sample	Locality	Formation	Comment	Zone and age	Biostratigraphic reliability	Reference	Age (Ma)	Type	Range and centre point of the horizontal error bars for each date in Composite Standard Units (CSU)[a]
10	Zircon	Usolka section, southern Urals region, Russia	Tuffaceous marls	ID-TIMS date of 299.0 ± 1.0 Ma (2-sigma) for the youngest zircon population	Four samples are precisely constrained within Asselian–Gzhelian transition: first sample 1.0 m below base of the Permian; second, 0.4 m below base; third, 0.05 m above base of the Permian; and fourth, 1.5 m above the base. Carboniferous–Permian boundary precisely defined in the basis of conodonts, ammonoids, and fusulinids		Ramezani *et al.* (2003), Chuvashov *et al.* (1990a), Davydov *et al.* (2002)	299.0 ± 1.0 (2-sigma)	U–Pb	1914–1921 (1917)

[a] Position of stage boundaries in composite standard scale units (CSU)

Base of Induan (Triassic), 2654 CSU
Base of Changhsingian, 2616 CSU
Base of Wuchiapingian, 2516 CSU
Base of Capitanian, 2434 CSU
Base of Wordian, 2400 CSU
Base of Roadian, 2361 CSU
Base of Kungurian, 2284 CSU
Base of Artinskian, 2148 CSU
Base of Sakmarian, 1987 CSU
Base of Asselian (Permian), 1916 CSU

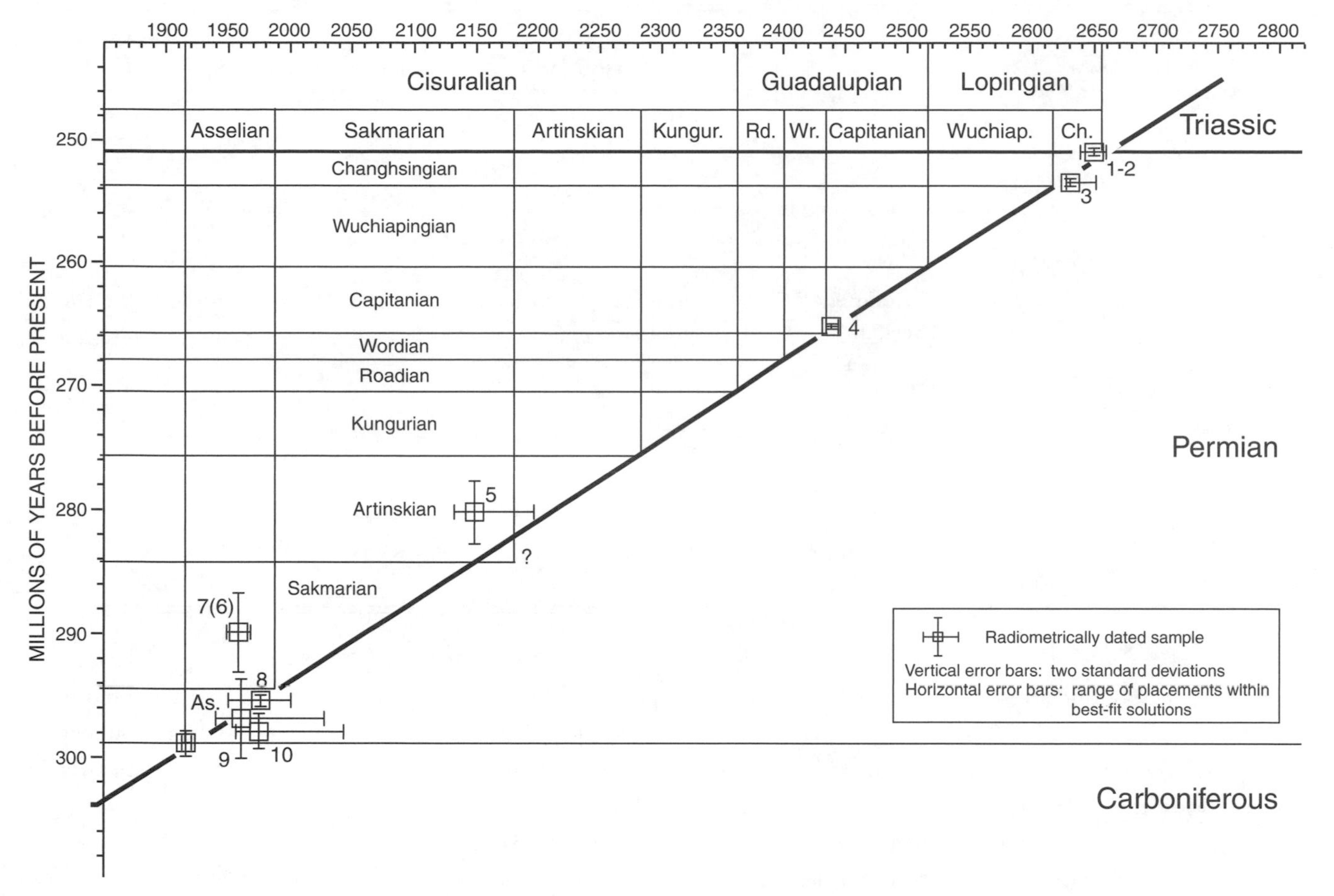

Figure 16.9 Construction of the Permian time scale. Radiometric age dates for the Permian in Table 16.1 are plotted against the relative zonal scale of Figure 15.4. For details see text.

Chuvashov *et al.* (1996) also report less well-constrained HR–SIMS ages for the uppermost Sakmarian (280.3 ± 2.4 Ma) and the lowermost Artinskian (280.3 ± 2.6 Ma). We project an age of 283 Ma for the Sakmarian–Artinskian boundary as defined by conodonts, which falls close to the margin of error for both dates.

Bowring *et al.* (1998) discuss several important TIMS dates, of which a few are well constrained within the proposed stratotypes for the Middle and Late Permian. In particular, an age of 265.3 ± 0.2 Ma from just below the recently approved GSSP for the base of the Capitanian coincides with the estimated age of 265 Ma by Menning (1995) for the Illawarra reversal. Menning (*in* Glenister *et al.*, 1999) places the Illawarra reversal within this important section between the isotopically dated horizon and the conodont-defined base of the Capitanian.

Uranium–lead dating of zircons in a volcanic ash in the lower unit of the Bell Canyon Formation, below the Pinery Limestone Member in the Capitanian type section, yields a date of 265.3 ± 0.2 Ma, which provides an estimate for the base of the Capitanian Stage.

Bowring *et al.* (1998) report several TIMS dates from the Meishan section in China, including the lower boundary of the Changhsingian Stage. A date of 253.4 ± 0.2 Ma is derived from Bed 7, above the first occurrence of *Clarkina wangi* (top of Bed 4) and immediately below the first occurrence of *Clarkina subcarinata* (*sensu strictu*) (Bed 8). Extrapolation for the sediment rate for the *C. subcarinata* and *C. changxingensis* Zones places the age of the base of the *wangi* Zone and the base of the Changhsingian Stage at about 254 Ma.

Computation of the age of the Permian–Triassic boundary, based on two series of consecutive TIMS age dates of Bowring *et al.* (1998) at the GSSP sections at Meishan and a correlative locality near Heshan in China is reported in Table 17.1. The final estimate for the age of the boundary is 251.0 ± 0.2 Ma.

Table 16.2 *Ages and durations of Permian stages*

Period	*Epoch*	Stage	Age of base (Ma)	Est. ± myr (2-sigma)	Duration	Est. ± myr (2-sigma)
Triassic			**251.0**	0.4		
Permian						
	Lopingian					
		Changhsingian	253.8	0.7	2.8	0.1
		Wuchiapingian	260.4	0.7	6.6	0.1
	Guadalupian					
		Capitanian	265.8	0.7	5.4	0.1
		Wordian	268.0	0.7	2.2	0.0
		Roadian	270.6	0.7	2.5	0.1
	Cisuralian					
		Kungurian	275.6	0.7	5.0	0.1
		Artinskian	284.4	0.7	8.8	0.2
		Sakmarian	294.6	0.8	10.2	0.2
		Asselian, (base of Permian)	**299.0**	0.8	4.4	0.1
Carboniferous						

16.4.3 Age of stage boundaries

This study indicates that the Permian lasted from 299 to 251 Ma, with the Sakmarian and Artinskian being the longest stages (Table 16.2). Calculated uncertainty estimates of the age of stage boundaries and duration of stages, to a considerable extent, is dependent on the stratigraphic precision of the Permian composite standard. Changes to zone–stage and zone–age date calibrations will produce shifts in interpolated ages of stage boundaries. The calculated error bars will decrease with more TIMS type radiometric dates; at least one or two per stage is a desirable goal. Uncertainty in stage duration is always less than uncertainty on the age of stage boundaries. More details on methodology, plotted in Fig. 16.9, are given in Chapter 15.

With the base of the Permian at 299 Ma, at the base of the *isolatus* conodont zone, and the top of the Permian, at the base of the *parvus* conodont zone at 251.0 Ma, the Permian lasted 52 myr. The Cisuralian and Early Permian lasted 28.4 myr, the Guadalupian and Mid Permian 10.2 myr, and the Lopingian and Late Permian 10.4 myr. Permian conodont zones appear to range in age from 0.7 to 3.0 myr.

17 • The Triassic Period

J. G. OGG

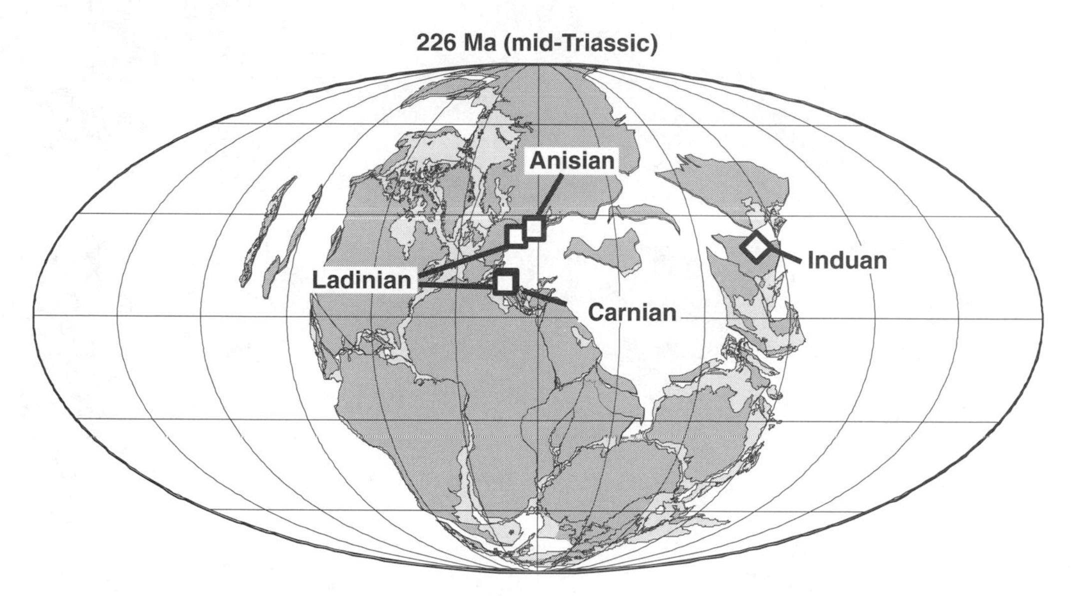

Geographic distribution of Triassic GSSPs that have been ratified (diamonds) or are candidates (squares) on a mid-Triassic map (status in January 2004; see Table 2.3). Only the base-Triassic has a formalized GSSP.

The Mesozoic begins with a gradual recovery of plant and animal life after the end-Permian mass extinction. Ammonites and conodonts are the main correlation tools for marine deposits. The Pangea supercontinent has no known glacial episodes. The modulation of its monsoonal climate by Milankovitch cycles left sedimentary signatures useful for high-resolution scaling. Dinosaurs begin to dominate the terrestrial ecosystems in latest Triassic. A relatively quiet period in Earth history.

17.1 HISTORY AND SUBDIVISIONS

The *Trias* of Friedrich August von Alberti (1834), director of the salt mines in the Württemberg district of southern Germany, united a trio of formations widespread in southern Germany – a lower Buntsandstein (colored sandstone), Muschelkalk (clam limestone), and an upper Keuper (nonmarine reddish beds). These continental and shallow-marine formations were difficult to correlate beyond Germany; therefore, most of the traditional stages (Anisian, Ladinian, Carnian, Norian, Rhaetian) were named from ammonoid-rich successions of the Northern Calcareous Alps of Austria. However, the stratigraphy of these Austrian tectonic slices proved unsuitable for establishing formal boundary stratotypes, or even deducing the sequential order of the stages (Tozer, 1984). For example, the Norian was originally considered to underlie the Carnian Stage, but after a convolute scientific–political debate (reviewed in Tozer, 1984), the Norian was established as the younger stage. Over 50 different stage names have been proposed for subdividing the Triassic (tabulated in Tozer, 1984).

In 1992, the Subcommission on Triassic Stratigraphy (International Commission on Stratigraphy) adopted a suite of seven standard Triassic stages. However, the general lack of unambiguous historical precedents for placement of Triassic stage boundaries has retarded establishment of formal GSSPs. (Table 17.1). Substages with geographic names are commonly used by European stratigraphers, whereas North American stratigraphers prefer a generic lower/middle/upper nomenclature (e.g. Fassanian substage versus Lower Ladinian).

A Geologic Time Scale 2004, eds. Felix M. Gradstein, James G. Ogg, and Alan G. Smith. Published by Cambridge University Press.

Table 17.1 *Triassic stage definitions (GSSP assignments or status) and informal division into substages*

Stage	Stage boundary *Substage base*	Global Boundary Stratotype Section and Point					
		Status	Location and point	Primary markers	Other correlations	Comments	General references (GSSP, correlations)
Upper Triassic							
Rhaetian	*Upper*						
	Rhaetian/Norian	*Guide event is undecided*	Key sections in Austria, Canada (NW Moresby Island, Queen Charlotte Islands, British Columbia), and Turkey	Near lowest occurrence of ammonoid *Cochloceras*, conodonts *Misikella* spp. and *Epigondolella mosheri*, and radiolarian *Proparvicingula moniliformis*		New task group being organized for this boundary. Non-marine auxiliary GSSP sections have been identified. No substages anticipated	
Norian	*Upper*						
	Middle	Informal European usage					
	Norian/Carnian	*Guide event is undecided*	Candidates include Canada (Black Bear Ridge in NE B.C. or NW Moresby Island, Queen Charlotte Islands, British Columbia); Sicily (Pizzo Mondello); Slovakia (Silicka Brezova section), Turkey (Antalya Taurus), and Oman	Base of *Klamathites macrolobatus* or *Stikinoceras kerri* ammonoid zones and the *Metapolygnathus communisti* or *M. primitius* conodont zones	Halobia bivalves, especially the FAD of *H. austriaca*, may provide a means of base-Carnian correlation in Tethyan strata that lack ammonoids	British Columbia section has superb fossil succession, but Sicily section has magnetostratigraphic record that permits correlation to non-marine Newark successions in eastern USA. Decision anticipated in 2004. A non-marine auxiliary GSSP candidate is at Petrified Forest Park, New Mexico, USA	Krystyn and Gallet (2002), Krystyn *et al.* (2002)

Carnian	*Upper*	Informal European usage					
	Middle						
	Carnian/Ladinian	*Guide event is undecided*	Candidate section at Prati di Stuores, Dolomites, northern Italy. Important reference sections in Spiti (India) and New Pass, Nevada (USA)	Near first occurrence of the ammonoids *Daxatina* or *Trachyceras,* and of the conodont *Metapolygnathus polygnathiformis*		Better fossiliferous successions are present in condensed sections in Spiti in the Himalayas, although access is not so good. Decision on this boundary will probably be delayed beyond 2003	
Middle Triassic							
Ladinian	*Upper*						
	Middle	Informal European usage					
	Ladinian/Anisian	*Guide event is undecided*	Leading candidates are Bagolino (Italy) and Felsoons (Hungary). Important reference sections in the Humboldt Range, Nevada (USA)	Alternate levels include base of *Reitzi, Secedensis*, or *Curionii* ammonite zone; or near first occurrence of the conodont genus *Budurovignathus*		Hungarian section is nicely prepared and has radiometric dates and ammonoid stratigraphy, but the preferred boundary level is not favored by some workers	Muttoni *et al.* (1996), Pálfy and Vörös (1998)
Anisian	*Upper*	Informal European usage				Upper substage of Anisian is sometimes called "Illyrian"	
	Middle	Informal European usage				Middle Anisian substage subdivided informally into a lower "Bithynian" and an upper "Pelsonian"	
	Anisian/Olenekian	Proposed 2002	Candidate section probable at Desli Caira, Dobrogea, Romania; significant sections in Guizhou Province (China)	Ammonite, near lowest occurrences of genera *Japonites, Paradanubites*, and *Paracrochordiceras*; and of the conodont *Chiosella timorensis*	Proposed level is slightly below the base of a normal-polarity magnetic zone	Lower substage of Anisian is sometimes called "Aegean"	Muttoni *et al.* (1995, 1998), Orchard and Tozer (1997a,b)

(*cont.*)

Table 17.1 (*cont.*)

Stage	Stage boundary *Substage base*	Global boundary stratotype section and point: Status	Location and point	Primary markers	Other correlations	Comments	General references (GSSP, correlations)
Lower Triassic							
Olenekian	*Upper*	Informal "Spathian"/ "Smithian" substage boundary	Originally defined (base of *Olenikites pilaticus* ammonite zone of Tozer) at Cape Stallworthy of Axel Heiberg Island, Canadian Arctic, at approximately 25 m above base of Lower Shale Member of the Blaa Mountain Formation	Ammonite, lowest occurrence of *Tirolites* and *Columbites* species	*Olenikites pilanticus* (informal boundary marker in Canadian Arctic) resembles a member of *Columbites* fauna of Idaho (USA). The *Columbites* beds include *Tirolites illyricus*, indicating at least an approximate correlation with the *Tirolites cassianus* ammonite zone of the Mediterranean region	Spathian and Smithian are the informal name for the lower and upper substages of the Olenekian Stage	Tozer (1967)
	Olenekian/Induan	*Guide event is undecided*	Candidates include Russia (South Primorye, near Vladivostock) or China (either in Anhui or Guizhou Provinces). Important sections also in Spiti	Near lowest occurrence of *Hedenstroemia* or *Meekoceras gracilitatis* ammonites, and of the conodont *Neospathodus waageni*	Magnetostratigraphy across the Induan–Olenekian transition at South Primorye yielded only normal polarity (Zakharov *et al.*, 2000); but is reported as near the top of a normal-polarity zone in China. Carbon isotopes may have a broad positive peak at boundary interval, then a major negative excursion in lower Olenekian (Chinese Lower Triassic Working Group, 2002)		Kiparisova and Popov (1964), Zakharov *et al.* (2000, 2002)

Induan	*Upper*	Informal "Dienerian"/ "Griesbachian" substage boundary	Originally defined south of Diener Creek, which drains into Otto Fiord on Ellesmere Island, Canadian Arctic (80.95° N, 88.83° W) at approximately 120 m above base of Blind Fiord Formation	Ammonite, lowest occurrence of abundant *Proptychites* species (base of *Proptychites candidus* zone of Tozer) and Gyronitidae family	Boundary between *Otoceras* and "*Meekoceras*" beds in Himalayas, and between *Ophiceras connectens* bed and the Lower Ceratite Limestone in the Salt Range (Pakistan)	Griesbachian and Dienerian are the informal name for the lower and upper substages of the Induan Stage; but former "lower Griesbachian" is now within uppermost Permian	Tozer (1967)
	Triassic/Permian (=base of Mesozoic Era)	Ratified 2001	Meishan Section D, about 2 km SE of Meishan town, 0.5 km N of Baoqing village (119.72° E, 31.07° N), Changxing County, Zhejiang Province, southern China. Middle of Bed 27 (base of 27c), about 15 cm above base of Yinkeng Formation	Conodont, lowest occurrence of *Hindeodus parvus* within phylomorphogenetic lineage *H. typicalis–H. latidentatus praeparvus–H. parvus–H. postparvus*	Termination of major negative carbon isotope excursion. About 1 myr after peak of Late Permian extinctions. Base of *Otoceras woodwardi* ammonite zone (Tethyan) and top of *Otoceras boreale* Zone (Boreal) are projected just above this level. End of fungi spike are projected just below. Fusulinid foraminifera become extinct in latest Permian. In continental settings, this boundary is close to the disappearance of typical Permian *Dicynodon* tetrapods after a interval of co-occurrence with "Triassic" *Lystrosaurus*	Meishan section is very compact – basal Triassic *H. parvus* Zone spans only 8 cm, and two brief late Permian conodont zones (but not latest Permian) are absent. A non-marine auxiliary section may be selected in the future, with candidates from China and South Africa being documented	Yin *et al* (1996, 2001), Yin (1996), Kozur (1998)

Important sources for this summary of Triassic subdivisions, biostratigraphic zonations, and correlation of individual stages include the *Albertiana* newsletters of the Subcommission on Triassic Stratigraphy (ICS) and Tozer (1967, 1984).

17.1.1 End-Permian ecological catastrophes and the base of the Mesozoic

The Paleozoic Era terminated in a complex environmental catastrophe and mass extinction of life. This sharp evolutionary division led J. Phillips (1840, 1841), a British paleontologist, to introduce Mesozoic (*middle animal life*, with Triassic at the base) between the Paleozoic (*old animal life*, ending with the Permian) and Kainozoic (*recent animal life*, after the Cretaceous). The latest Permian to earliest Triassic events include pronounced negative carbon isotope and strontium isotope anomalies, a positive sulfur isotope excursion, immense subaerial volcanism covering Siberia, widespread anoxic oceanic conditions, a major sea-level regression and exposure of shelves followed by a major transgression, and the progressive disappearance of up to 80% of marine genera (see reviews by Holser and Magaritz, 1987; Erwin, 1993; Kozur, 1998; Hallam and Wignall, 1999; Erwin *et al.*, 2002).

Even though many of the ecological features resemble the aftermath of the asteroid impact that concluded the Mesozoic, the end-Paleozoic strata have not yielded unambiguous signatures of a bolide catastrophe. A common hypothesis is that release of aerosols and/or greenhouse gases associated with the enormous Siberian continental flood basalts and oceanic feedbacks dramatically altered global climate. A combination of enhanced greenhouse warming punctuated by cold volcanic winter episodes with a progressive transgression of low-oxygen seas into shallow-water environments precipitated a progression of environmental and ecological feedbacks. Ecological recovery was delayed until the later half of the Early Triassic, and up to half of the genera that seemed to have disappeared at the Paleozoic–Mesozoic boundary interval re-emerged (Lazarus taxa; e.g. see reviews by Kozur, 1998; Erwin *et al.*, 2002).

REDEFINING THE PALEOZOIC–MESOZOIC BOUNDARY

The mass disappearance of Paleozoic fauna and flora, coupled with the widespread occurrence of a major regression–transgression unconformity in most regions, led to a dilemma. It was easy to recognize the bleak final act of the Permian, but how should the beginning of the Mesozoic be defined? The base of the Bundsandstein in south west Germany defined the original Trias concept (Alberti, 1834), but it is a diachronous boundary within continental beds, now assigned to the upper Permian. Similarly, the base of the Werfen Group (base of Tesero Oolite) in the Italian Alps is a diachronous facies boundary. Ammonoids are the common biostratigraphic tool throughout the Mesozoic, and the *Otoceras* ammonoid genus was long considered to be the first "Triassic" form. Therefore, Griesbach (1880) assigned the Triassic base to the base of the *Otoceras woodwardi* Zone in the Himalayan region, but this species is only known from the Perigondwana paleomargin of eastern Tethys (e.g. Iran to Nepal). The first occurrence of *Otoceras* species in the Arctic realm (*Otoceras concavum* Zone) was used by Tozer (1967, 1986, 1994a) for a Boreal marker of the base of the Triassic, but is now known to appear significantly prior to *Otoceras woodwardi* in the Tethyan realm (Krystyn and Orchard, 1996). The progressive evolution of the conodont *Hindeodus* genera through the Permian–Triassic boundary interval provided global correlation markers with no obvious facies dependence; however, conodont biostratigraphy requires special processing and identification experience. Non-biological correlation markers, such as carbon isotope excursions or magnetic polarity changes are conclusive when preserved (e.g. Newell, 1994), but can suffer from diagenetic overprints.

In 2000, the Triassic subcommission chose the first occurrence of the conodont *Hindeodus parvus* (equivalent to *Isarcicella parva* of some earlier conodont studies) within the evolutionary lineage *Hindeodus typicalis–H. latidentatus praeparvus–H. parvus–H. postparvus* as the primary correlation marker for the base of the Mesozoic Era and Triassic Period. This biostratigraphic event is the first cosmopolitan correlation level associated with the initial stages of recovery following the end-Permian mass extinctions and environmental changes. Global correlations indicate that this conodont species appears just after the carbon isotope minimum and end of a widespread spike in marine fungi abundance – events considered as possible proxies of minimum biological abundance in both marine and terrestrial settings (H. Kozur pers. comm., 2001). This level is slightly lower than the base of the *Otoceras woodwardi* ammonoid zone of the Himalayas. The revised definition assigns the *Otoceras concavum* and lowermost portion of *Otoceras boreale* ammonoid zones of the Arctic (the lower part of the "Griesbachian" substage of Tozer, 1967) into the Permian (Orchard and Tozer, 1997 a,b). In continental settings, the conodont event is close to the disappearance of typical Permian *Dicynodon* tetrapods after an interval of co-occurrence with "Triassic" dicynodont *Lystrosaurus* (Kozur, 1998).

However, the choice of this conodont to serve as the primary marker for the base of the Triassic implies that former traditional concepts of the Permian–Triassic boundary, such as the disappearance of typical Permian marine fauna, rapid facies changes, extensive volcanism, and isotope anomalies are now assigned to the latest Permian.

PALEOZOIC–MESOZOIC BOUNDARY STRATOTYPE (BASE OF THE TRIASSIC)

The GSSP for the base of the Triassic is at Meishan, Zhejiang Province, southern China, where it coincides with the first occurrence of conodont *Hindeodus parvus* (Yin *et al.*, 2001).

Approximately 18 cm below the GSSP is the former "boundary clay" bentonite (Bed 25), which has yielded high-resolution $^{40}Ar/^{39}Ar$ and U–Pb ages coinciding with the main phase of Siberian flood basalts at approximately 251 Ma (Renne *et al.*, 1995; Bowring, 1998; Metcalfe *et al.*, 1999; Erwin *et al.*, 2002; Kamo *et al.*, 2003). This "boundary clay" is now placed approximately two conodont zones below the new base of the Triassic. Another ash clay approximately 8 cm above the GSSP (Bed 28) has yielded zircon U–Pb ages approximately 0.7 myr younger than the Bed 25 bentonite (Bowring, 1998; Metcalfe *et al.*, 1999; Section 17.3.1).

Other important reference sections for the events across the Permian–Triassic boundary are located in the Dolomites of Italy (e.g. Broglio Loriga and Cassinis, 1992; Wignall and Hallam, 1992), in the Canadian Arctic (e.g. Tozer, 1967), in the Salt Ranges of Pakistan (e.g. Baud *et al.*, 1996), and the Dzhulfa section of Armenia. However, older literature about these sections commonly used placements for the "Permian–Triassic boundary" that are not coincident with the GSSP level in Meishan (e.g. see review in Kozur, 1998).

17.1.2 Lower Triassic

A multitude of stage and substage nomenclatures have been applied to the Lower Triassic interval. The Triassic subcommission adopted the current subdivision into a lower Induan Stage and an upper Olenekian Stage in 1992. The Induan and Olenekian stages of Kiparisova and Popov (1956, revised in 1964) were named after exposures in the Indus river basin in the Hindustan region of Asia and in the lower reaches of the Olenek River basin of Arctic Siberia, respectively.

A suite of four ammonoid-zoned substages is widely used. In an imaginative procedural twist, these Griesbachian, Dienerian, Smithian, and Spathian substages are named after exposures along associated small creeks on Ellesmere and Axel Heiberg Islands in the Canadian Arctic, which in turn were named after important Triassic paleontologists (Tozer, 1965).

INDUAN

The Induan Stage is informally divided into two substages. The lower substage, Griesbachian, is named after Griesbach Creek on northwest Axel Heiberg Island. The Dienerian substage is named after Diener Creek of northwest Ellesmere Island. The Griesbachian–Dienerian boundary is marked by the appearance of Gyronitidae ammonoids. This substage boundary is recognized in Canada and the Himalayas as the boundary between *Otoceras* and *Meekoceras* ammonoid-bearing beds of Diener (1912) and in the Salt Range of Pakistan at the base of the Lower Ceratite Limestone (Tozer, 1967).

The redefinition of the Permian–Triassic boundary implies that the lower portion of the original Griesbachian of Tozer (1965, 1967) is now assigned to the uppermost Permian.

OLENEKIAN

History, definition, and boundary stratotype candidates The Olenekian Stage was defined in Arctic Siberia, whereas the Induan Stage was defined in the Hindustan region of Pakistan–India. Neither region has fossiliferous strata spanning their mutual boundary – the Induan in the Olenek River basin is marginal marine to lagoonal, and ammonoids in the transitional interval in the Hindustan region are rare or absent (Zakharov, 1994). The lower Olenekian is marked by the appearance of a diverse ammonoid assemblage of *Hedenstroemia*, *Meekoceras*, *Juvenites*, *Pseudoprospingites*, *Arctoceras*, *Flemingites*, and *Euflemingites*. A sea-level regression caused a scarcity of age-diagnostic conodonts and bivalves during the latest Induan to earliest Olenekian, but the transition seems to be within the lower portion of the *Neospathodus pakistanensis* conodont zone (Zakharov, 1994; Paull, 1997: Orchard and Tozer, 1997a,b).

Proposed biostratigraphic definitions of the stage boundary are the highest occurrence of the ammonoid *Gyronites subdharmus* and the lowest occurrence of the representatives of the *Meekoceras* or *Hedenstroemia* ammonoid genera (Zahkarov *et al.*, 2000, 2002) or the lowest occurrence of the conodont *Neospathodus waageni*. Two candidate GSSPs for this transition are in the South Primorye region of southeast coastal Siberia near Vladivostok – Tri Kamnya Cape to Orel cliff section and the Abrek Bay section (Zahkarov *et al.*, 2000, 2002). A third candidate is a roadside outcrop near Chaohu city in the Anhui Province of eastern China (Jinnan *et al.*, 2001; Chinese Lower Triassic Working Group, 2002).

Smithian and Spathian substages The two informal substages of the Olenekian Stage are named after Smith and Spath creeks on Ellesmere Island of the Canadian Arctic. The base of the Smithian substage was originally defined as the base of a broad *Euflemingites romunderi* ammonoid zone (Tozer, 1965, 1967), then the biostratigraphy was revised to add a *Hedenstroemia hedenstroemi* ammonoid zone (e.g. Orchard and Tozer, 1997a,b).

The Spathian substage is characterized by *Tirolites*, *Columbites*, *Subcolumbites*, *Prohungarites*, and *Keyserlingites* ammonoid genera. The Smithian–Spathian boundary was placed at the base of the *Olenekites pilaticus* ammonoid zone, but there appears to be a missing biostratigraphic interval in the type region (Tozer, 1967; Orchard and Tozer, 1997a,b), An alternate proposed nomenclature of a lower "Ayaxian" and an upper "Russian" substage (Zakharov, 1994) would have the same limits as the current Smithian and Spathian subdivisions.

17.1.3 Middle Triassic

ANISIAN

History, definition, and boundary stratotype candidates The Anisian Stage was named after limestone formations near the Enns (= Anisus) River at Grossreifling, Austria (Waagen and Diener, 1895). The original Anisian stratotype lacks ammonoids in the lower portion, and the lower limit was later clarified in the Mediterreanean region (Assereto, 1974). The appearance of a number of ammonoid genera, like *Aegeiceras*, *Japonites*, *Paracrochordiceras*, and *Paradanubites*, may be used to define the base of the Anisian (e.g. Gaetani, 1994). This level is slightly preceded by the lowest occurrence of the conodont *Chiosella timorensis* and by the base of a normal-polarity magnetic zone (Muttoni *et al.*, 1995, 1998). The conodont *Chiosella timorensis* provides a correlation to North American stratigraphy (Orchard and Tozer, 1997a,b).

The leading candidates for the base-Anisian GSSP are Desli Caira in Dobrogea, Romania, and in the Guizhou Province of China (M. Orchard, pers. comm., 2002).

Anisian substages The Anisian Stage has three to four informal substages. The Lower Anisian (also called "Aegean" or Egean) was originally defined in beds with *Paracrochordiceras* ammonoids at Mount Marathovouno on Chios Island (Greece) in the Aegean Sea by Assereto (1974). The Middle Anisian is sometimes subdivided into two substages: a lower "Bithynian," named after the Kokaeli Peninsula (Bithynia) of Turkey by Assereto (1974), and an upper "Pelsonian," from the Latin name for the region around Lake Balaton in Hungary (Pia, 1930) spanning the *Balatonites balatonicus* ammonoid zone (Assereto, 1974). The Upper Anisian is also called "Illyrian" after the Latin term for Bosnia (Pia, 1930).

LADINIAN

History, definition, and boundary stratotype candidates The Ladinian Stage arose after a heated semantic argument of "Was ist norisch?" (Bittner, 1892), when it was realized that most of the strata that had defined a "pre-Carnian" Norian Stage (Mojsisovics, 1869) were actually deposited *after* the Carnian (Mojsisovics, 1893). This debate and the emergence of the Ladinian Stage split the Vienna geological establishment (vividly reviewed by Tozer, 1984). The Ladinian, after the Ladini inhabitants of the Dolomites region of northern Italy, encompassed the Wengen and Buchenstein beds (Bittner, 1892).

Major revision and even partial inversion of the upper Triassic stratigraphy, coupled with uncertainties about correlation potentials and definition of ammonoid zones, contributed to a delay in assigning the basal limit of the Ladinian Stage. The base of the Ladinian is still a topic of intense debate (e.g. Gaetani, 1993; Brack and Rieber, 1994, 1996; Mietto and Manfrin, 1995; Pálfy and Vörös, 1998).

There are four contenders for the primary correlation criteria to assign the Ladinian–Anisian boundary. In sequential order upwards:

1. Lowest occurrence of representatives of the ammonoid genus *Kellnerites* (defines the base of *Reitziites reitzi* ammonoid zone of Vörös *et al.*, 1996), which coincides with a widespread ammonoid and radiolarian turnover (e.g. Vörös *et al.*, 1996; Pálfy and Vörös, 1998).
2. Lowest occurrence of the *Nevadites* ammonoid genus (defines base of *Nevadites secedensis* Zone), which corresponds to a major radiation in ammonoids.
3. Lowest occurrence of *Eoprotrachyceras* ammonoid genus (defines the base of *Eoprotrachyceras curionii* Zone in Italy–Hungary and the base of *Protrachyceras subasperum* ammonoid zone in Nevada), which is the only ammonoid marker whose identity has remained stable (e.g. Brack and Rieber, 1994, 1996).
4. Lowest occurrence of the *Budurovignathus* conodont genus, which is found in most sections (e.g. Muttoni *et al.*, 1996; Orchard and Tozer, 1997a,b).

At a working group meeting in 2002, two other proposals arose:

5. First occurrence of *Reitziites reitzi* ammonoid (which does not correspond to the base of the "*R. reitzi*" Zone).
6. The first occurrence of *Aplococeras avisianum* ammonoid (N. Preto, pers. comm., December 2002).

All the lowest occurrences of ammonoid genera may have only local chronostratigraphic precision, because none of the western Tethys ammonoid species, with the possible exception of *A. avisianum*, are recorded in either Nevada or the Canadian Rocky Mountains. Also, well-established Jurassic examples suggest that first appearances of genera in Europe and in North America may be offset by as much as one ammonoid zone (Pálfy and Vörös, 1998, p. 25).

Two potential GSSP localities are: (a) the historically important Felsöörs section in the Balaton Highland of Hungary; and (b) the section of Bagolino in the Lombardian Alps, west of the Dolomites region of northern Italy. The former section only adequately exhibits the lowest correlation criteria (base of *Kellnerites*, e.g. Vörös *et al.*, 1996; Kovács *et al.*, 1994); whereas the Bagolino section is the most complete one for the entire boundary interval, yields radiometric ages, and can be directly tied to nearby magnetostratigraphy (e.g. Brack *et al.*, 1995; Muttoni *et al.*, 1996).

Important reference sections for Ladinian ammonoid–conodont correlation to North America are found in the Humboldt Range, Nevada, USA.

Ladinian substages von Mojsisovics (1893) divided the Ladinian into two substages: the Lower, or Fassanian (named after Val di Fassa in northern Italy, where it was equated to the Buchenstein Beds and Marmolada Limestone), and Upper, or Longobardian (named after the Langobard people of northern Italy, and spanning the Wengen Beds). The substage boundary is approximately at the base of the "*Eoprotrachyceras*" *gredleri* ammonoid zone in the Alpine zonation or the base of *Meginoceras meginae* ammonoid zone in the Canadian zonation.

17.1.4 Upper Triassic

The Upper Triassic consists of three stages – Carnian, Norian, and Rhaetian – that were originally defined by characteristic ammonoids (Mojsisovics, 1869). However, these units were originally recognized in different locations in the northern Alps of Austria with uncertain stratigraphic relationships. Indeed, until 1892, Norian units were considered to underlie the Carnian, and it was only after a major geological controversy was the name "Norian" applied to the same units after recognition that they were younger than Carnian (reviewed in Tozer, 1984). No formal definitions have yet been agreed for these stage boundaries.

CARNIAN

History, definition, and boundary stratotype candidates The Carnian Stage, either named after localities in the Kärnten (Carinthia) region of Austria or after the nearby Carnian Alps, was originally applied to Hallsatt Limestone beds bearing ammonoids of *Trachyceras* and *Tropites* (Mojsisovics, 1869, p. 127). The first occurrence of ammonoid *Trachyceras* (= base of *Trachyceras aon* Zone in Tethys or *Trachyceras desatoyense* in Canada) was the traditional base, although it appears that a *Trachyceras* datum would be asynchronous and not cosmopolitan (e.g. Mietto and Manfrin, 1999). Mojsisovics (1893) included the St. Cassian beds of northern Italy in a revised Carnian subdivision, therefore the level with first occurrence of the cosmopolitan ammonoid *Daxatina* at the Prati di Stuores type locality in the Dolomites has been proposed for the base-Carnian GSSP (Broglio Loriga *et al.*, 1998). The *Daxatina* appearance would imply lowering of the base of the Carnian Stage to the middle of the "Ladinian" *Frankites regoledanus* ammonoid zone. This high-sedimentation-rate Prati di Stuores section has a magnetostratigraphy that can be correlated to a more extensive section in Austria, but the conodont succession is very diluted. Two other potential GSSPs with multiple biostratigraphic successions (status from 2002 *Annual Report of Subcommission on Triassic Stratigraphy*) are in Spiti, Himalaya, northwest India (Balini *et al.*, 1998, 2001), and New Pass section, Nevada, USA.

Carnian substages Mojsisovics (1893) subdivided the Carnian into three substages (Cordevolian, Julian, and Tuvalian) corresponding to three ammonoid zones, but later stratigraphers combined their lower two substages. The boundary between the Lower Carnian (or "Julian," named after the Julian Alps region of southern Austria) and the Upper Carnian (or "Tuvalian," named after the Tuval mountains, the Roman term for the region between Berchtesgaden and Hallein near Salzburg, Austria) is traditionally placed at the first occurrence of *Tropites* ammonoids (i.e. the base of the *Tropites subbullatus* ammonoid Zone of Tethys and *Tropites dilleri* Zone of Canada). The ammonoid change at this substage boundary is more significant than at the base of the Carnian or at the base of the overlying Norian (Tozer, 1984).

NORIAN

History, revised definition, and boundary stratotype candidates Norian derives its name from the Roman province south of the Danube (Mojsisovics, 1869), and the stratigraphic extent of strata given this name had a contorted history (reviewed in Tozer, 1984). Ammonoid successions in Nevada and British Columbia led to a proposal that the base of the Norian be assigned to the base of the *Stikinoceras kerri* ammonoid zone, overlying the *Klamathites macrolobatus* Zone (Silberling and Tozer, 1968). This level is approximately coeval with a Tethyan placement between the *Anatropites* and *Guembelites jandianus* ammonoid zones (Krystyn, 1980; Orchard *et al.*, 2000). However, this ammonoid-based level does not correspond to an unequivocal microfossil signal, whereas the base of the *K. macrolobatus* ammonoid zone is coincident with the first occurrences of conodont *Metapolygnathus communisti* and some radiolarian species (Orchard *et al.*, 2000). A potential GSSP for this lowered base of the Norian (*M. communista* datum) is at Black Bear Ridge on Williston Lake, northeast British Columbia (Orchard *et al.*, 2001), although it did not yield magnetostratigraphy. Other candidate sections are Pizzo Mondello in Sicily (Muttoni *et al.*, 2001) or Silicka Brezova section in Slovakia, both of which have yielded magnetic polarity patterns and conodonts (e.g. Krystyn and Gallet, 2002; Krystyn *et al.*, 2002).

Norian substages The Norian is traditionally subdivided into three substages, following Mojsisovics (1893). The boundary between the lower Norian (or "Lacian," after the Roman name for the Salzkammergut region of the northern Austrian Alps) and middle Norian (or "Alaunian," named for the Alauns, who lived in the Hallein region of Austria during Roman times) is the base of the Tethyan *Cyrtopleurites bicrenatus* ammonoid zone. The base of the upper Norian (or "Sevatian," after the Celtic tribe who lived between the Inn and Enns Rivers of Austria) is generally assigned as the base of the North American *Gnomohalorites cordilleranus* ammonoid zone or the Tethyan *Sagenites quinquepunctatus* ammonoid zone, but there is no consistent usage (Kozur, 1998).

RHAETIAN

The Rhaetian, the first Triassic stage to be established, has led a vague existence. Carl Wilhelm Riter von Gümbel (1861) defined the Rhaetian as equivalent to strata with the *Rhaetavicula contorta* ammonoid zone, with the Kössen Beds of Austria regarded as typical. The acceptance of the Norian stage (1895) and later biostratigraphic investigations of the Kössen Beds suggested significant overlap, and the Rhaetian was eliminated in some Triassic time scales (e.g. Zapfe, 1974; Palmer, 1983; Tozer, 1984). In 1991, the Subcommission on Triassic Stratigraphy decided to retain the Rhaetian as an independent stage, but there is no agreement on its extent (e.g. Krystyn, 1990; Tozer, 1990; Kozur, 1999). Potential biostratigraphic levels for the base of the Rhaetian include the highest occurrence of *Monotis* bivalves or the lowest occurrence of ammonoid *Cochloceras*, of conodont *Epigondolella mosheri*, or a *Misikella* species, or of radiolarian *Proparvicingula moniliformis*.

17.2 TRIASSIC STRATIGRAPHY

The ammonoid successions of the Alps and Canada have historically served as global primary standards for the Triassic (reviews in Tozer, 1967, 1984). Biostratigraphic, magnetostratigraphic, chemostratigraphic, and other events are typically calibrated to these standard ammonoid zones. An extensive compilation and inter-correlation of Triassic stratigraphy of European basins was coordinated by Hardenbol *et al.* (1998) and is partially summarized in Fig. 17.1.

17.2.1 Marine macrofossils

Ammonoids dominate the historical zonation of the Triassic, but conodonts have become a major tool for global correlation. Thin-shelled bivalves (e.g. *Daonella*, *Halobia*, etc.)

Figure 17.1 Triassic time scale with selected biostratigraphic zonations, magnetic polarity chrons, and major depositional sequences. Potential definitions of stage boundaries are indicated by dashed lines, with our selected assignment for the main Phanerozoic time scale drawn as the horizontal; but the final decisions will be made by the International Commission on Stratigraphy after 2003 (see text for full discussion and ICS site at www.stratigraphy.org for current status). Uncertainties on computed ages of stage boundaries are the estimated 95% confidence limits (see text). Details on sources, calibrations, and scaling of ammonoid scales are given in the text. Datums in the pollen-spore column were selected from a compilation by Hochuli *et al.* (in Hardenbol *et al.*, 1998). Land vertebrate scale of global assemblage zones is from Lucas (1999). Major flooding or regressive trends of depositional sequences as recognized in Alpine and Boreal regions are labeled at the sequence boundary immediately preceding the maximum lowstand of the respective third-order sequence (generalized from Hardenbol *et al.*, 1998). A color version of this figure is in the plate section.

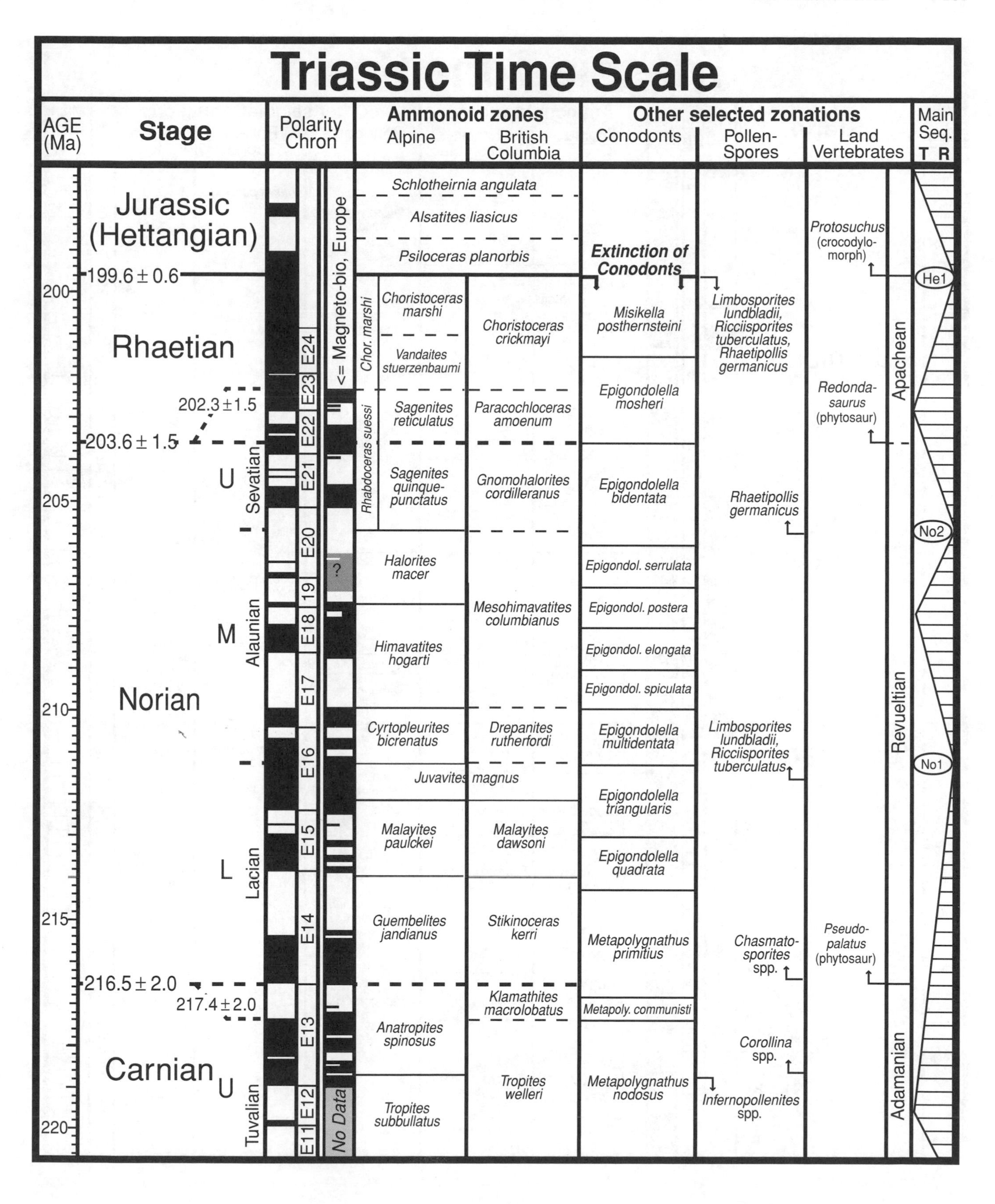

Triassic Time Scale
AGE (Ma)
Stage
Polarity Chron
Ammonoid zones
Alpine
British Columbia
Other selected zonations
Conodonts
Pollen-Spores
Land Vertebrates
Main Seq. T R
Jurassic (Hettangian)
199.6 ± 0.6
Rhaetian
202.3 ± 1.5
203.6 ± 1.5
Sevatian
U
Norian
Alaunian
M
Lacian
L
216.5 ± 2.0
217.4 ± 2.0
Carnian
Tuvalian
U
200
205
210
215
220
E24
E23
E22
E21
E20
19
E18
E17
E16
E15
E14
E13
E12
E11
<= Magneto-bio, Europe
?
No Data
Schlotheirnia angulata
Alsatites liasicus
Psiloceras planorbis
Chor. marshi
Rhabdoceras suessi
Choristoceras marshi
Vandaites stuerzenbaumi
Sagenites reticulatus
Sagenites quinque-punctatus
Halorites macer
Himavatites hogarti
Cyrtopleurites bicrenatus
Juvavites magnus
Malayites paulckei
Guembelites jandianus
Anatropites spinosus
Tropites subbullatus
Choristoceras crickmayi
Paracochloceras amoenum
Gnomohalorites cordilleranus
Mesohimavatites columbianus
Drepanites rutherfordi
Malayites dawsoni
Stikinoceras kerri
Klamathites macrolobatus
Tropites welleri
Extinction of Conodonts
Misikella posthernsteini
Epigondolella mosheri
Epigondolella bidentata
Epigondol. serrulata
Epigondol. postera
Epigondol. elongata
Epigondol. spiculata
Epigondolella multidentata
Epigondolella triangularis
Epigondolella quadrata
Metapolygnathus primitius
Metapoly. communisti
Metapolygnathus nodosus
Limbosporites lundbladii, Ricciisporites tuberculatus, Rhaetipollis germanicus
Rhaetipollis germanicus
Limbosporites lundbladii, Ricciisporites tuberculatus
Chasmato-sporites spp.
Corollina spp.
Infernopollenites spp.
Protosuchus (crocodylo-morph)
Redonda-saurus (phytosaur)
Pseudo-palatus (phytosaur)
Apachean
Revueltian
Adamanian
He1
No2
No1

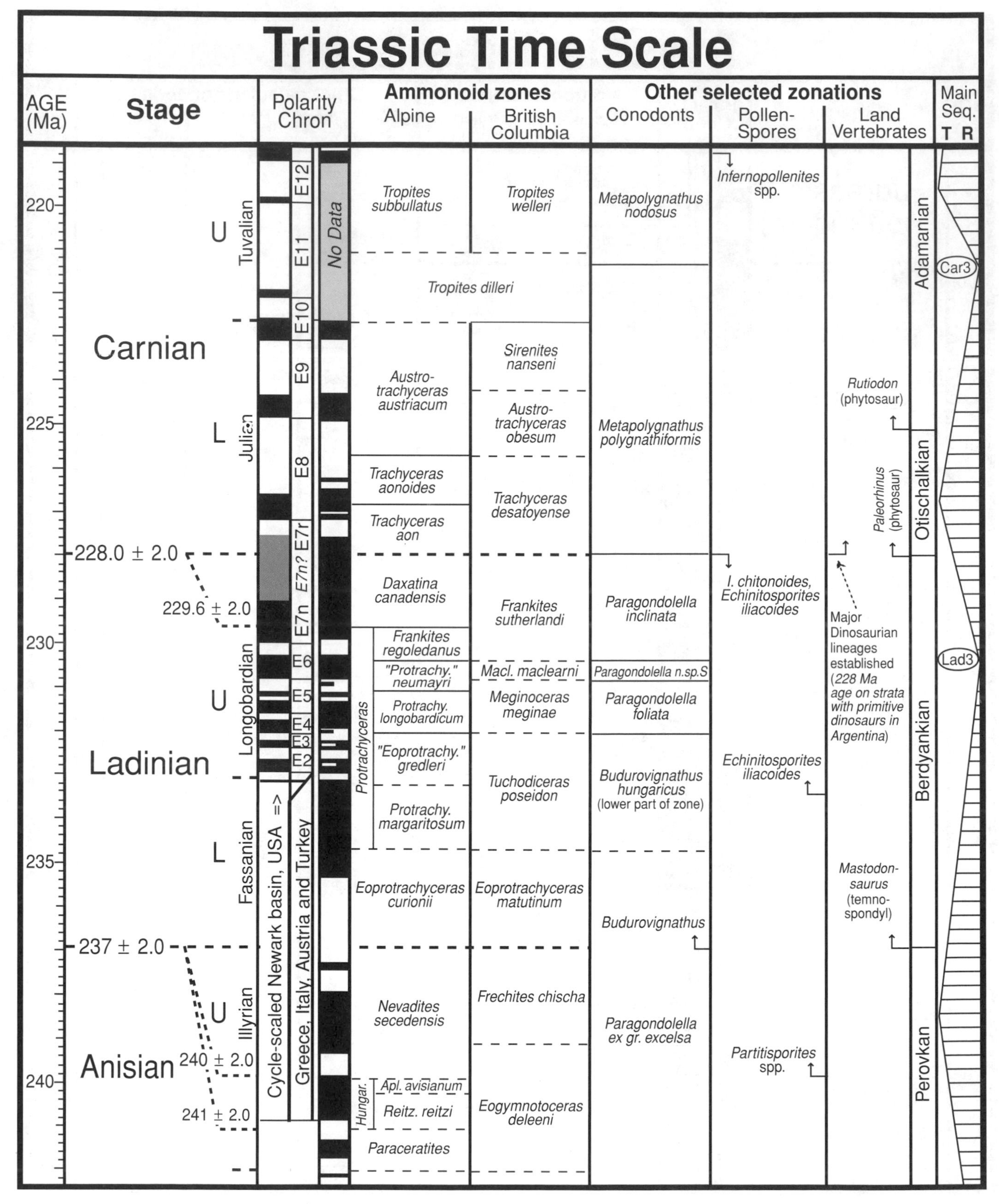

Figure 17.1 (*cont.*)

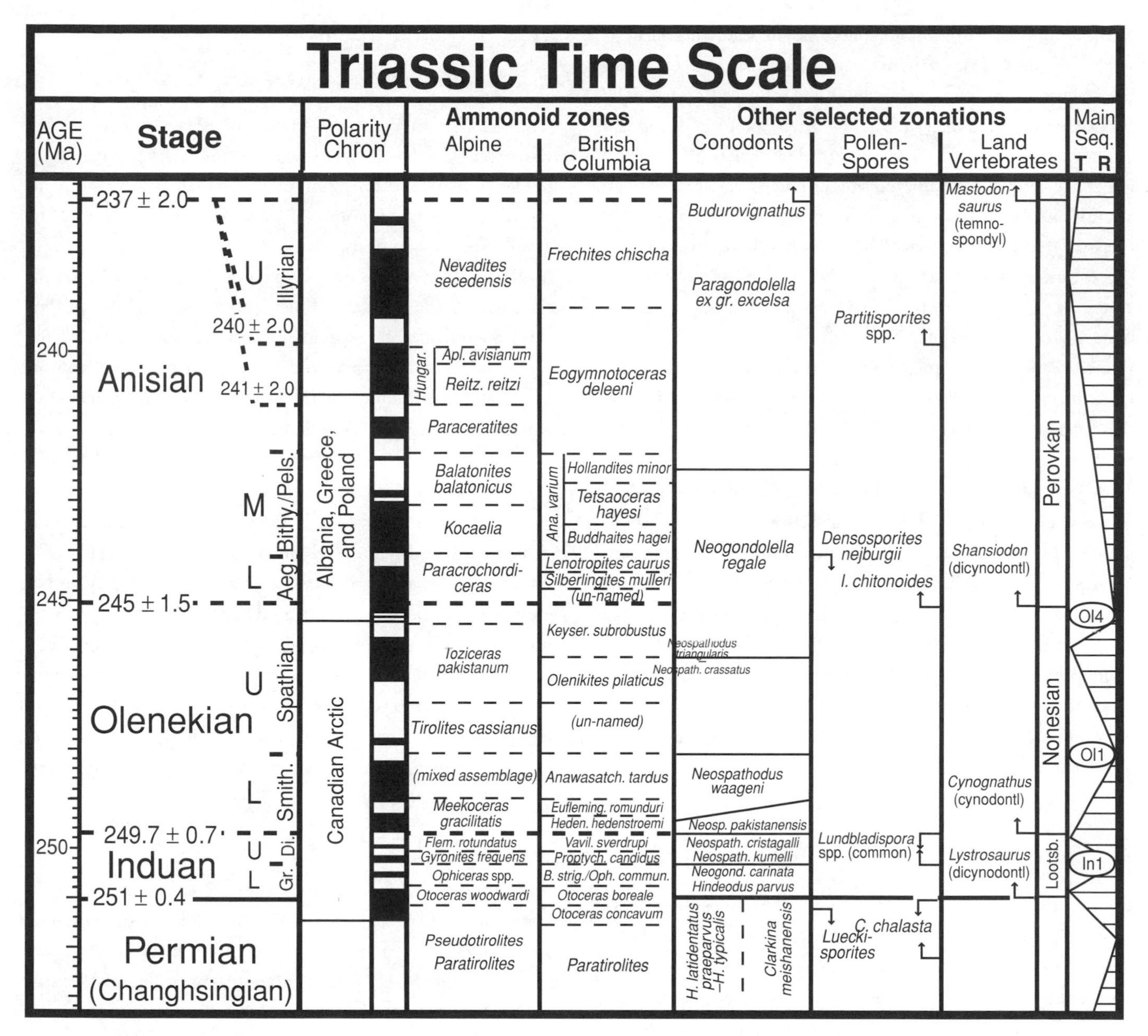

Figure 17.1 (*cont.*)

provide important regional markers. During much of the Triassic, the sedimentary record across the Pangea supercontinent was dominated by terrestrial deposits, therefore widespread tetrapod and plant remains are important for global correlation.

AMMONOIDS

Ammonoids experienced their maximum variety during the Triassic, reaching a peak in the middle Norian before being decimated at the end of the Triassic. Only a single major genera, *Phylloceras*, seems to have survived into the basal-Jurassic to enable the rapidly evolving ammonites to reconquer the Jurassic seas.

Despite their historical importance in subdividing the Triassic, there is not yet a standardized ammonoid zonation (or nomenclature) for Alpine regions. For example, Mietto and Manfrin (1995) proposed a generalized standard for the Middle Triassic of the Tethyan realm that utilized first appearances of widespread genera to define zones and major species to define subzones. But this zonal scheme was immediately rejected by some Alpine workers (e.g. critiques by Brack and Rieber, 1996, and Vörös *et al.*, 1996). A Triassic succession and nomenclature for Canada was compiled by Tozer (1967) with later modifications (e.g. Tozer, 1994b).

The ammonoid scale in Fig. 17.1 displays a hybrid selection, rather than an international standard. These calibrations of the component ammonoid zones to each other and the scaling to absolute time involved various sources and methods. The Alpine ammonoid scale is modified from a compilation by van Veen and Gianolla (1998). The inter-calibrated Canadian ammonoid and conodont scales are modified from Orchard and Tozer (1997b). Estimated ages of Early Triassic Canadian ammonoid zones are derived from graphic-correlation scaling of associated conodont datums (Sweet and Bergström, 1986). Age estimates for Anisian and for Rhaetian Tethyan ammonoid zones are derived by proportional scaling to number of component subzones. Ages of Ladinian through Norian Tethyan ammonoid zones are assigned through bio-magnetostratigraphic calibration to the Milankovitch-scaled magnetic polarity time scale derived from the Newark basin of eastern USA (see Section 17.2.4), and their correspondingly more precise absolute age assignments are indicated by solid lines in Fig. 17.1. Suggested correlations of Arctic and Tethyan ammonoid zones are mainly from van Veen (1998) and Ginolla, de Zanche and Mietto (1998), and the relatively higher-precision correlations levels are shown as solid lines. The extensive biostratigraphic chart series by Hardenbol *et al.* (1998) contains detailed information and correlation on ammonoid subzones and other biostratigraphic events, but there is no consensus on aspects of the nomenclature and inter-correlations. Ammonoid biozone names and associated assigned ages are summarized in Table 17.3.

CONODONTS

Conodonts are distinctive variations in the phosphatic jaws of an enigmatic pelagic swimmer and enable widespread correlation of Triassic strata. After surviving the end-Permian mass extinctions, the conodonts mysteriously vanished at the end of the Triassic. Compilations of the calibrations among conodont zones and ammonoids have been accomplished for several realms, including Canada (e.g. Orchard and Tozer, 1997a,b), Tethys region (e.g. Muttoni *et al.*, 1998; Krystyn *et al.*, 2002), and European basins (Vrielynck, 1998).

BIVALVES

A succession of thin-shelled clams (pectinacean pterioid bivalves) were largely restricted to deeper-water settings and tend to occur in high densities on certain bedding planes (Hallam, 1981a). These genera (*Claria*, *Enteropleura*, *Daonella*, *Halobia*, *Monotis*, etc.), which have no modern counterparts, are valuable for global Triassic correlations. Their Jurassic and Cretaceous relatives (*Bositra*, *Buchia*, inoceramids) occupied the same settings.

17.2.2 Terrestrial macrofossils

TETRAPODS AND DINOSAUR STRATIGRAPHY

The Triassic non-marine strata of Pangea was divided into eight "land-vertebrate faunachrons" (LVFs) by Lucas (1998a,b, 1999), indicating an inter-continental correlation potential equivalent to ammonoid-based stage-level units (Fig. 17.1). The first appearance of the dicynodont *Lystrosaurus* (a plant-eating, semi-aquatic "mammal-like" reptile) marks the base of the Triassic. The Lower and Middle Triassic can be subdivided by a progressive succession of first occurrences of other "mammal-like" reptiles of cynodont and dicynodont types followed by the temnospondyl *Mastodonsaurus* amphibian. The Late Triassic has four LVFs defined by progressive first occurrences of crocodile-like phytosaur genera.

The earliest dinosaurs are known from Carnian strata across Pangea, and represent a small-sized, yet diverse, component of late Carnian tetrapod fauna (Heckert and Lucas, 1999). The major dinosaur lineages were already established by the beginning of the Carnian, as indicated by a 227.8 ± 0.3 Ma date on the Ischigualasto assemblage of Argentina. Dinosaurs became dominant land reptiles in the Norian, and evolved their characteristic large size. It is uncertain whether this major turnover in terrigenous ecosystems was restricted to non-marine settings or was also a major marine extinction event (e.g. Benton, 1993).

PLANTS, POLLEN, AND SPORES

Spores and pollen are important for correlation of marine and terrestrial strata. Selected compilations of Triassic palynology and plant ecosystem evolution are given in Traverse (1988) and Wing and Sues (1992). In contrast to the end-Permian mass extinctions of shallow-marine fauna, terrestrial plants experienced only a minor decline in diversity. Widespread marine clastics of lower Triassic (Olenekian to lower Induan) record a uniquely cosmopolitan "acritarch spike" assemblage of lycopsid spores, small acanthomorph acritarchs, and *Lunatisporites* coniferalean pollen (Balme and Foster, 1996).

Palynoflora zonations for Triassic strata have been compiled for the Alpine and Germanic region (e.g. Brugman, 1986; Hochuli, 1998), Australia (Helby *et al.*, 1987), southwest USA (Litwin *et al.*, 1991; Cornet, 1993), Newark Basin of eastern USA (e.g. Cornet, 1977; Cornet and Olsen, 1985), and the Arctic (Van Veen *et al.*, 1998).

Palynoflora correlations are particularly important in resolving the succession of ecosystem and volcanic events associated with the latest Triassic mass extinctions and the transition to the Jurassic recovery, although aspects of the precision of inter-continental correlations and the calibration to marine-based chronostratigraphy remain unresolved (e.g. Morbey, 1975; Fisher and Dunay, 1981; Cornet and Olsen, 1985; Hallam, 1990; Fowell and Olsen, 1993, 1995; van Veen, 1995, and references there in). In particular, Newark rift basins in eastern USA display a transition from diverse assemblages of monsaccate and bisaccate pollen to palynoflora containing 60–90% *Corollina meyeriana* spores that occur slightly below flood basalts dated at approximately 202 Ma and was assigned as a regional marker for the Triassic–Jurassic boundary (e.g. Cornet, 1977, pp. 175–184; Cornet and Olsen, 1985; Fowell and Olsen, 1993). However, similar palynological changes are recorded near the base of the typical Rhaetian of Europe (e.g. Schuurman, 1979; Orbell, 1983; van Veen, 1995, and written comm., January 1994). This Newark palynology episode and associated volcanics may correspond to the onset of the series of mass extinctions during the latest Norian and Rhaetian that precede the marine Triassic–Jurassic boundary as defined by the first occurrence of Hettangian ammonite genus *Psiloceras* at 199.6 ± 0.3 Ma (e.g. Pálfy *et al.*, 2000b).

17.2.3 Marine microfossils

Except for radiolaria, marine microfossil biostratigraphy has not yet been developed as a widespread correlation tool within the Triassic. In contrast to Permian and Jurassic syntheses, the benthic foraminifera stratigraphy of the Triassic has not been compiled on a global scale. A stratigraphic summary of larger benthic foraminifera of the Tethyan realm is illustrated by Peybernes (1998a,b).

Calcareous nannofossils are only known from Carnian and younger strata, and the first "real coccoliths" appear in the Norian (von Salis, 1998). Records of dinoflagellates are also rare, and the oldest representative of this group may be middle Triassic (Hochuli, 1998).

Radiolarian datums and zonations have been compiled from alpine exposures (e.g. De Wever, 1982, 1998; Kozur and Mostler, 1994), Japan (e.g. Sugiyama, 1997) and western North America (e.g. Blome, 1984; Carter, 1993; Carter and Orchard, 2000).

17.2.4 Physical stratigraphy

MAGNETOSTRATIGRAPHY

Throughout the Triassic, there are about two–four magnetic polarity zones per million years with no evident polarity bias or extended intervals of constant polarity. A concentrated effort by paleomagnetists during the 1990s led to a composite magnetic polarity scale that has been calibrated to ammonoid and conodont zones, and partially to sequence stratigraphy and orbital cycles (Fig. 17.1). The reference Early Triassic magnetic polarity time scale derived from Canadian Arctic sections that included substage stratotypes (Ogg and Steiner, 1991) has been correlated to sections in the Italian Dolomites, Spitsbergen, Poland Buntsandstein, China, southwest USA, and Iran (e.g. Steiner *et al.*, 1988, 1993; Graziano and Ogg, 1994; Heller *et al.*, 1995; Hounslow *et al.*, 1996; Nawrocki, 1997; Gallet *et al.*, 2000; Scholger *et al.*, 2000). A composite Anisian through lowermost Carnian polarity pattern has been progressively resolved from studies in Greece, Italy, Albania, Poland, and Turkey (e.g. summaries by Gallet *et al.*, 1998; Muttoni *et al.*, 1998; Nawrocki and Szulc, 2000), although some details from these discontinuous sections are uncertain. Except for a small interval in the middle Carnian, the Carnian through lower Rhaetian polarity pattern has been calibrated to biostratigraphy from sections in Turkey, Italy, and Austria (Gallet *et al.*, 1992; Muttoni *et al.*, 2001; Krystyn *et al.*, 2002).

The Triassic magnetic polarity scale shown in Fig. 17.1 (right-hand column of polarity scales) is a spliced composite of other compilations: (a) Lower Triassic from Ogg and Steiner (1991), with compensation for sedimentation rate variations by Graziano and Ogg (1994); (b) uppermost Spathian through lower Ladinian from Muttoni *et al.* (1998) and Nawrocki and Szulc (2000); (c) upper Ladinian through lowermost Carnian from Gallet *et al.* (1998); (d) lower Carnian from Gallet *et al.* (1992); and (e) upper Carnian to lower Rhaetian from Krystyn *et al.* (2002).

During the late Triassic, a series of rift valleys along the western margin of the future Central Atlantic accumulated very thick successions of lacustrine deposits that recorded climatic responses to Milankovitch orbital cycles. Drilling of these Newark Basin strata has yielded a complete 30-myr cycle-scaled pattern of the magnetic reversal history during the late Triassic (Kent *et al.*, 1995). The uppermost polarity chron E23 of the Newark polarity sequence is overlain by the Orange Mountain basalts, part of the onset of a regional Central Atlantic Magmatic Province dated at 201.5 ± 1 Ma (e.g. Sutter, 1988; Dunning and Hodych, 1990). Therefore, Kent and Olsen (1999; and on-line version, 2002) assign an age of 202 Ma to the top of E23 and tuned the cyclic stratigraphy using the 404-kyr eccentricity cycle and a 1.75-myr long modulating cycle to project the ages of the late Triassic polarity pattern (Fig. 17.1). According to palynology, the base of the Norian is within Newark polarity chron E13 (Olsen *et al.*, 1996).

Correlation between the cycle-scaled Newark polarity pattern and the composites of biostratigraphic–calibrated portions of the Ladinian through Rhaetian is difficult. The correlation has been hindered by variable sedimentation and tectonic distortions of several sections, taxonomic disagreements on biostratigraphic age assignments and absence of unambiguous polarity pattern (Muttoni *et al.*, 2001; Krystyn *et al.*, 2002). The potential correlation illustrated in Fig. 17.1 was initially suggested during a combined scaling process: (a) using a 202 Ma age for the top of Newark polarity chron E23 and assigning an age of 199.6 Ma to the top of the Triassic (Pálfy *et al.*, 2000b), (b) scaling relative durations of Ladinian through Rhaetian ammonoid zones according to the number of component subzones, (c) plotting the composite polarity patterns of the Tethyan region within each ammonoid zone as displayed in published compilations, and (d) constraining the base of the Norian to be near Newark polarity chron E13. Several potential correlations were suggested visually by the relative alignment of intervals dominated by normal or by reversed polarity (e.g. normal-polarity dominance of Newark chrons E15 through E17n compared to lower Norian pattern, and reverse-polarity dominance of middle Carnian compared to polarity bias in upper Lower Carnian) and the apparent frequency of reversals (e.g. rapid reversals during Newark polarity chrons E2–E6 compared to the middle Ladinian pattern). Based on these visual matches and age constraints, the relative placements of ammonoid zones were adjusted while retaining the within-zone polarity scaling as tentatively suggested by previous compilations.

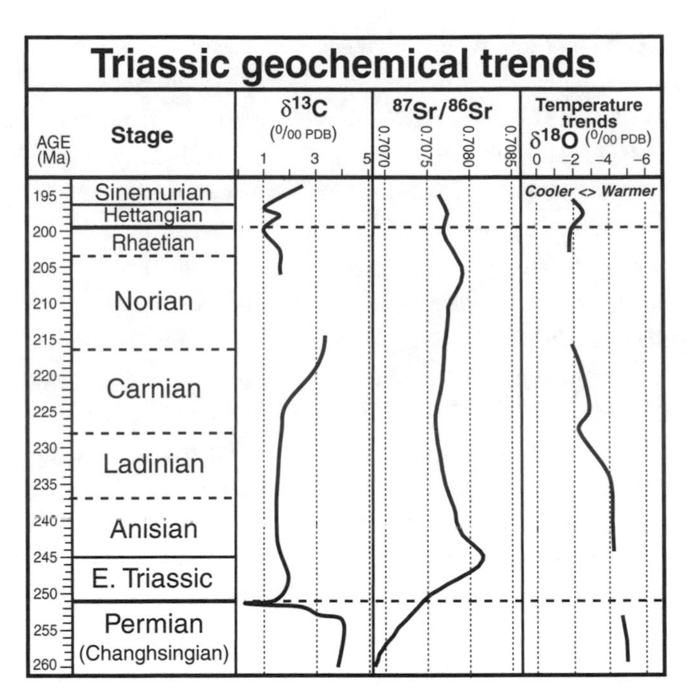

Figure 17.2 Triassic trends and excursions in carbon and oxygen stable isotopes and in marine $^{87}Sr/^{86}Sr$ ratio from calcareous fossils and bulk carbonate. The schematic carbon isotope curve is a 5-myr averaging of data (Veizer *et al.*, 1999) from Hayes *et al.* (1999; downloaded from www.nosams.whoi.edu/jmh), with details on the end-Permian and end-Triassic excursions generalized from Erwin *et al.* (2002) and Jenkyns *et al.* (2002), respectively. The strontium isotope curve is based on a LOWESS fit to data of Koepnick *et al.* (1990), Korte *et al.* (2003), and other sources – see text and also Chapter 7. The oxygen isotope curve (inverted scale) is derived from a 3-myr interval averaging of global data compiled by Veizer *et al.* (1999; as downloaded from www.science.uottawa.ca/geology/isotope_data/ in January 2003). Large-scale global shifts to higher oxygen-18 values in carbonates are generally interpreted as cooler seawater or glacial episodes, but there are many other contributing factors (e.g. Veizer *et al.*, 1999, 2000; Wallmann *et al.*, 2001). Gaps in curves indicate absence of isotope data in these compilations.

CHEMICAL STRATIGRAPHY

Recognition of global excursions in carbon and strontium isotopes during the Triassic is relatively less developed than in adjacent geological periods owing to a dearth of extended sections with high-resolution biostratigraphy and unambiguous global correlation, especially within the Middle and Upper Triassic. Only the Permian–Triassic boundary interval has a well-documented suite of chemical signatures with global correlation potential.

Stable isotopes of carbon and oxygen The end-Permian mass extinctions are coincident with an abrupt negative excursion in carbon isotopes in both marine and terrestrial settings. This excursion may have been caused by a combination of decreased marine productivity and an influx of light carbon from volcanic, soil-carbon or methane sources (e.g. Holser and Magaritz, 1987; Baud *et al.*, 1989; Krull and Retallack, 2000; Yin *et al.*, 2001; Erwin *et al.*, 2002; Sephton *et al.*, 2002a). The base of the Triassic is just after this carbon isotope minimum, but the carbon isotope values never return to the relatively heavy $\partial^{13}C$ of +4 per mil that characterizes the Late Permian (Fig.17.2).

Carbon isotopes generally remain in the +1.5 to +2.0 per mil range throughout the majority of the Triassic except for a possible broad $\partial^{13}C$ peak to +3 per mil in upper Carnian and lower Norian (Veizer *et al.*, 1999; Hayes *et al.*, 1999). A brief positive $\partial^{13}C$ peak observed in organic carbon components at the Norian–Rhaetian boundary has been tentatively attributed to widespread oceanic stagnation coincident with extinction of deep-water invertebrate fauna (Sephton *et al.*, 2002b). The late Rhaetian (end-Triassic) mass extinctions coincide with a negative carbon isotope excursion, which, like the end-Permian event, may be linked to widespread volcanism, oceanic productivity collapse, and release of methane (e.g. Pálfy *et al.*, 2001; Ward *et al.*, 2001; Hesselbo *et al.*, 2002). However, the interpretation of these carbon isotope records may be distorted by facies variations in some sections (Morante and Hallam, 1996).

Oxygen isotope trends and other paleoenvironmental evidence indicate a general progressive cooling from Late Permian through the Triassic. The global oxygen-18 pattern suggests a total cooling of about 4 °C in tropical seas (Veizer *et al.*, 2000; Wallmann, 2001).

Strontium isotope ratios The curve of marine $^{87}Sr/^{86}Sr$ through the Triassic (Fig. 17.2) is a broad trough bounded by twin peaks in the earliest and latest Triassic. Values of $^{87}Sr/^{86}Sr$ show a sharp rise through the latest Permian into an earliest Triassic maxima (0.708 22) and a major decline from the very latest Triassic maxima (0.707 95) through the earliest Jurassic (Koepnick *et al.*, 1990; Korte *et al.*, 2003).

The major rise in marine $^{87}Sr/^{86}Sr$ during the late Permian and earliest Triassic may be a long-term response to increased continental weathering following the glacial climates of the early Permian (Martin and Macdougall, 1995). The abrupt initiation of the end-Triassic downturn in marine $^{87}Sr/^{86}Sr$ coincided with major flood basalt outpouring along the future Central Atlantic seaway, and the prolonged decline in $^{87}Sr/^{86}Sr$ continued to the end of the Pliensbachian (Jones *et al.*, 1994a; McArthur *et al.*, 2001; Jenkyns *et al.*, 2002). The extra-ordinary double-peaked shape of the Triassic $^{87}Sr/^{86}Sr$ curve is shown by the data of both Koepnick *et al.* (1990) and Korte *et al.* (2003); in the former it is defined by analysis of whole rocks and in the latter by analysis of conodonts, which are not robust to alteration. It therefore remains to be seen whether the twin peaks are real. If they are, the rate of change of $^{87}Sr/^{86}Sr$ through the interval offers some promise for global correlation at high-resolution, especially in the regions of the maxima, where $^{87}Sr/^{86}Sr$ changes rapidly with time.

CYCLE STRATIGRAPHY

Late Triassic lacustrine sediments and Middle Triassic carbonate platforms have been classic locations in recognition of Milankovitch orbital-climate cycles and developing cyclostratigraphy techniques and applications.

The thick Newark group of lacustrine sediments accumulated in a series of closed rift basins in eastern North America during the initial phases of the break-up of Pangea. During the late Ladinian through earliest Jurassic, environments within these tropical basins oscillated between semi-stagnant deep lakes and arid playas as the intensity of monsoonal rains responded to Earth's precession modulated by short-term (100 kyr) and long-term (400 kyr) eccentricity cycles. Spectral analysis of sediment facies successions in a series of deep-drilling cores enabled compilation of a cycle-scaled record and magnetic polarity time scale that is unprecedented in its 30-myr temporal span (e.g. Kent *et al.*, 1995; Olsen *et al.*, 1996; Kent and Olsen, 1999). As noted in the Triassic magnetostratigraphy discussion, the exact calibration of this polarity pattern to marine-based Triassic chronostratigraphy is controversial, but this remarkable cyclic deposit will be an important component of any late Triassic time scale.

The Latemar massif in the Italian Dolomites was an atoll-like feature with a core of flat-lying Anisian and Ladinian platform carbonates. Oscillations in sea level created over 500 thin depositional cycles (Goldhammer *et al.*, 1987). Stacking patterns and spectral analysis indicate that the sea-level oscillations were driven by precession modulated by short-term (100 kyr) eccentricity, implying that the Latemar deposit spans approximately 10 myr (Goldhammer *et al.*, 1990; Hinnov and Goldhammer, 1991). In contrast, U–Pb ages from coeval tuff-bearing basinal deposits appear to constrain the Latemar platform to span only 2–4 myr (e.g. Brack *et al.*, 1996, 1997; Mundil *et al.*, 1996; Hardie and Hinnov, 1997). Possible solutions to this disparity include a very high rate of platform construction (200 m/myr) that recorded sub-Milankovitch sea-level oscillations or reworked zircons in the tuffs.

Studies of similar facies oscillations in upper Triassic platform carbonates in the Austrian Alps played an important role in developing fundamental concepts of cyclostratigraphy (e.g. Fischer, 1964), but the reality of regular cyclicity in these same deposits has also been questioned (e.g. Satterley, 1996). Interbedded marls and limestones of shallow-marine origin spanning the Permian–Triassic boundary interval in these Austrian Alps display cycles with ratios matching Milankovitch eccentricity, obliquity, and precession periodicity, and which indicate that the latest Permian extinction and negative carbon isotope spike spanned less than 30 kyr (Rampino *et al.*, 2000, 2002).

SEQUENCE STRATIGRAPHY

Triassic sea-level trends have been compiled for Boreal basins (e.g. Embry, 1988; Mørk *et al.*, 1989; Skjold *et al.*, 1998), the classic Germanic Trias (e.g. Aigner and Bachmann, 1992; Geluk and Röhling, 1997), the Dolomites and Italian Alps (e.g. De Zanche *et al.*, 1993; Gaetani *et al.*, 1998; Gianolla *et al.*, 1998), and other regions. Many of these sea-level trends appear to correlate on an inter-basin to global scale (e.g. Haq *et al.*, 1988; Hallam, 1992b; Embry, 1997; Gianolla and Jacquin,

1998). Some common features are schematically illustrated in Fig. 17.1, based on a graphical summary and systematic numbering system derived from several studies (Jacquin and Vail, 1998).

The Triassic sea-level trend is dominated by a single cycle – a progressive transgression that began in the latest Permian and peaks in the Anisian–Ladinian boundary interval, followed by a regression to the Late Norian (sequence boundary "No2" on Fig. 17.1). A second major transgression begins in the Rhaetian that peaks in the late Jurassic. The initiation of each major transgressive episode coincides with a major mass extinction (end-Permian and end-Triassic) and widespread anoxic environments on the shelves (e.g. Hallam and Wignall, 1999). Superimposed on these main cycles are at least four second-order facies cycles (major sequence boundaries Ol4 in uppermost Olenekian, Lad3 in uppermost Ladinian, Car3 in mid-Carnian, and No2 in uppermost Norian), and up to 23 third-order sequences (Gianolla and Jacquin, 1998).

OTHER MAJOR STRATIGRAPHIC EVENTS

The main pulses of the voluminous Siberian Trap flood basalts are approximately coeval with the latest Permian mass extinctions, and the waning stages of this volcanic activity may have continued into the earliest Triassic (e.g. Renne *et al.*, 1995; Erwin *et al.*, 2002, and references therein).

Two major events punctuated the Late Triassic – one of the largest flood basalt episodes during the Phanerozoic and a major impact. The Central Atlantic Magmatic Province, with ages centered on 200 Ma (just before the Triassic–Jurassic boundary as recognized in the marine realm), has a total extrusive volume approaching 4 million cubic kilometers (e.g. Marzoli *et al.*, 1999; Hames *et al.*, 2000). These extensive volcanics have mostly normal magnetic polarity, and were probably mainly emplaced within less than 2 myr during polarity chrons E23n–E24n of the Newark Basin geomagnetic polarity time scale (Olsen and Kent, 2001). The waves of mass extinctions during the latest Triassic may have been partially linked to the environmental side-effects of these continental flood basalts (e.g. Olsen, 1999).

The 100-km diameter Manicouagan impact structure of Quebec has a U–Pb age of 214 ± 1 Ma (Hodych and Dunning, 1992). This impact event and associated environmental catastrophe may have contributed to the large-scale turnover of continental tetrapods at the Carnian–Norian boundary interval, during which dinosaurs attained dominance over competing families of carnivorous, the condontians and mammal-like reptiles (e.g. Benton, 1986, 1993). It is ironic that both the beginning and the end of the reign of the dinosaurs may have been delimited by major impacts. However, there has not yet been global recognition of the ejecta or other evidence of the Manicouagan impact in the stratigraphic record.

17.3 TRIASSIC TIME SCALE

Only four clusters of radiometric ages constrain the Triassic time scale – base of Triassic at 251 ±0.4 Ma, mid-Middle Triassic (base of *Nevadites secedensis* ammonoid zone) at 240 ± 1 Ma, volcanics at the top of Newark polarity chron E23 at 202 Ma ± 1 Ma, and the Triassic–Jurassic boundary at 199.6 ± 0.3 Ma (2-sigma; Table 17.2). Different dating methods yield generally consistent results for each interval, but there are disagreements on the precise ages. Most Triassic stage boundaries and the relative ages of biostratigraphic zones within stages must be interpolated using scaling from cycle stratigraphy, graphical correlation, or other assumptions. The age estimates and derivation procedures for the selected primary biostratigraphic scales are summarized in Table 17.3, and the stage boundaries and durations are summarized in Table 17.4.

17.3.1 Base-Triassic to lower Ladinian

The Permian–Triassic boundary is 251.0 ± 0.4 Ma (95% confidence). We derived this age using a cubic-spline interpolation with a monotonic fit on the stratigraphic set of eight TIMS ages on zircons in tuff beds across the Permian–Triassic boundary at the GSSP section of Meishan and at a coeval locality in China (Bowring *et al.*, 1998; see Table 17.2 under column 'Used in Ch. 8)'. Zircons from a bentonite (Bed 25) located 18 cm below the base-Triassic GSSP (Bed 27c) at Meishan, China, has yielded consistent U–Pb ages by HR–SIMS (251.1 ± 3.6 Ma; Claoué-Long *et al.*, 1991, 1995) and TIMS (251.4 ± 0.3 Ma; Bowring *et al.*, 1998), which is about 1 myr older than $^{40}Ar/^{39}Ar$ ages from sanidine feldspar in the same bed (249.9 ± 0.15 Ma, or ±1.5 myr if uncertainties in standards are incorporated) (Renne *et al.*, 1995). Zircons from a bentonite located 8 cm above the GSSP in this section yield a U–Pb (TIMS method) age of 250.7 ± 0.3 Ma (Bowring *et al.*, 1998).

An alternate selection of the zircon data from 8 cm above the GSSP could imply an older age of 252.5 ± 0.3 Ma for this level, thereby extrapolating an age of ~253 Ma for base-Triassic (Mundil *et al.*, 2001c). These authors concluded that scatter beyond analytical uncertainty due to Pb loss or subtle reworking of zircons creates inhomogenous zircon populations. However, the 251 Ma age for the base-Triassic is consistent with zircon-derived ages obtained from independent Permian–Triassic

Table 17.2 *Selected radiometric ages that constrain the Mesozoic time scale [see Footnote a]*

STAGE	Primary control on GTS2004	Dated material	Formation and location	Zone and age	Comments (age, reliability, reviews, etc.)	Primary and review references	Revised age (Ma)	Uncertainty (myr) & type (if not 1-sigma)		Method, calib.
Cretaceous-Paleogene Boundary Interval										
	secondary guide	Bentonite biotite and sanidine	Z-Coal, Hell Creek, Montana	Just above K-P boundary.	Rb-Sr and U-Pb are only 1.5% younger than K-Ar on same bed by OB93 (given below), but error bars do not overlap. This was noted by Odin-Obradovich in NDS82.	Baadsgaard & Lerbekmo, in NDS82.	63.6	0.2		Rb-Sr
	secondary guide	Zircon from bentonite	Z-Coal, Hell Creek, Montana	Just above K-P boundary.		Baadsgaard & Lerbekmo, in NDS82.	63.6	0.4		U-Pb
	x	Bentonite biotite	K-P Boundary, Z-Coal (Hell Creek, Montana), Nevis Coal (Red Deer Valley, Alberta), Ferris Coal (Frenchman River, Saskatchewan)	Just above K-P boundary.	Source publication has U-Pb, Rb-Sr and K-Ar (sanidine) on 3 separate deposits, and effectively replaces the earlier Baadsgaard & Lerbekmo dates in NDS82.	Baadsgaard *et al.*, 1988	65.4	1.1		K-Ar
	secondary guide	*Weighted mean*			*Weighted mean is from following suite of Baadsgaard et al., 1988.*	Baadsgaard *et al.*, 1988	*64*	*1.2*	*95%*	
		Sanidine	Z-coal, Hell Creek, Montana	Z-coal bentonite.		"	64.6	1.0	95%	K-Ar
		Biotite				"	63.7	0.6	95%	Rb-Sr
		Zircon				"	63.9	0.7	95%	U-Pb
		Sanidine	Red Deer Aalley, Alberta	Nevis coal bentonite.		"	64.8	1.4	95%	K-Ar
		Biotite				"	63.9	0.6	95%	Rb-Sr
		Zircon				"	64.3	0.7	95%	U-Pb
		Sanidine	Frenchman River, Saskatchewan	Ferris coal bentonite.		"	65.8	1.2	95%	K-Ar
		Biotite				"	64.5	0.4	95%	Rb-Sr
		Zircon				"	64.4	0.7	95%	U-Pb
	x	Sanidine	Red Desert			Obradovich *et al.* 1986.	65.5	1.0		Ar-Ar
	x	Bentonite sanidine	Denver Fm, Golden, Colorado			Obradovich, 1984; OB93.	65.85	0.8		Ar-Ar (1)
	x	Sanidine	Z-coal, Hell Creek, Montana	Danian or Maastrichtian.		Obradovich, 1984; OB93.	66.05	0.9		Ar-Ar (1)
	x	Bentonite sanidine	Z-coal, Hell Creek, Montana	Z-coal bentonite.		OB93	65.40	0.31	95%	Ar-Ar (1), Laser fusion
	x	Bentonite sanidine				OB93	65.48	0.21	95%	Ar-Ar (1), Laser fusion

(*cont.*)

Table 17.2 (*cont.*)

STAGE	Primary control on GTS2004	Dated material	Formation and location	Zone and age	Comments (age, reliability, reviews, etc.)	Primary and review references	Revised age (Ma)	Uncertainty (myr) & type (if not 1-sigma)		Method, calib.
	x	Tektite	Beloc, Haiti	Haitian tektites.	K/P boundary weighted mean of tektite-sanidine set is 65.4 Ma (+/− 0.1 m.y. 95% confidence) OB93, pg. 394. However, comparing to his table, this mean seems to exclude the tektite ages.	OB93	65.02	0.29	95%	Ar-Ar (1), Laser fusion
	x	Tektite				OB93	65.24	0.45	95%	Ar-Ar (1), Laser fusion
CRETACEOUS										
Maastrichtian										
	x	Bentonite sanidine	Kneehills Tuff, Alberta	Dinosaur zone of *Triceratops*, base. Probably corresponds to Late Maastrichtian.		OB93	66.85	1.1	95%	Ar-Ar (1), Laser fusion
	x	Bentonite sanidine	Red Bird section, Wyoming	*Baculites clinolobatus* ammonite zone, top (therefore, upper limit of Early Maastrichtian).		OB93	69.47	0.37	95%	Ar-Ar (1), Laser fusion
Campanian										
	x	Bentonite sanidine	Bearpaw Shale, Montana	*Baculites compressus* ammonite zone (second zone below *B. reesidei*, second zone above *E. jenneyi*).		OB93	73.40	0.39	95%	Ar-Ar (1), Laser fusion
	x	Bentonite sanidine	upper Mancos Shale, Colorado	*Exiteloceras jenneyi* ammonite zone.		OB93	74.81	0.45	95%	Ar-Ar (1), stepwise heating
	x	Bentonite sanidine	Foreman, Arkansas	Within foraminifer zone of *Globotruncanita calcarata* (ammonite zone not known).		OB93	75.42	0.39	95%	Ar-Ar (1), Laser fusion
	x	Bentonite sanidine	Lewis Shale, New Mexico	*Baculites scotti* ammonite zone (zone below *D. nebrascense* Zone).		OB93	75.94	0.72	95%	Ar-Ar (1), Laser fusion
	x	Bentonite sanidine	Ardmore Bentonite, Wyoming	*Baculites obtusus* ammonite zone.		OB93	80.60	0.55	95%	Ar-Ar (1), Laser fusion
	x	Bentonite sanidine	Ardmore Bentonite, Elk Basin, Wyoming	*Baculites obtusus* ammonite zone.		Hicks *et al.* (1993)	80.77	0.55	95%	Ar-Ar (1), Laser fusion
	x	Bentonite sanidine	Eagle Sandstone, Montana	*Scaphites hippocrepis II* ammonite zone, lower part (third zone above Campanian/Santonian boundary in North American Western Interior zonation).		OB93	81.77	0.34	95%	Ar-Ar (1), Laser fusion

Santonian										
	x	Bentonite sanidine	Telegraph Creek Fm., Montana	*Desmoscaphites bassleri* ammonite zone (uppermost zone of Santonian in North American Western Interior zonation).		OB93	83.97	0.43	95%	Ar–Ar (1), Laser fusion
	x	Bentonite sanidine	Tombigbee Sand Mbr. of Eutaw Fm., Mississippi	20m below *Boehmoceras* fauna of "Upper Santonian"		OB93	84.15	0.40	95%	Ar–Ar (1), Laser fusion
	x	Bentonite sanidine	Austin Chalk, Texas	Top of *Cladoceramus undulatoplicatus* inoceramus bivalve zone (Early Santonian).	Top of *C. undulatoplicatus* is potential Middle/Lower Santonian marker.	OB93	84.94	0.28	95%	Ar–Ar (1), stepwise heating
Coniacian										
	x	Bentonite sanidine	Cody Shale, Montana	*Scaphites depressus–Protexanites bourgeoisianus* ammonite zone (uppermost Coniacian in N. American Western Interior zonation).		OB93	86.98	0.39	95%	Ar–Ar (1), Laser fusion
	x	Bentonite sanidine	Marias River Shale, Montana	*Forresteria alluaudi–Scaphites preventricosus* ammonite zone.		OB93	88.40	0.60	95%	Ar–Ar (1), Laser fusion
Turonian										
	x	Bentonite sanidine	Mancos Shale, New Mexico	Top of *Prionocyclus macombi* ammonite zone (middle of Upper Turonian in North American Western Interior zonation).		OB93	90.27	0.72	95%	Ar–Ar (1), Laser fusion
	x	Bentonite sanidine	Ferron Sandstone, Utah	*Prionocyclus hyatti* ammonite zone (zone below *P. macombi* Zone).		OB93	90.57	0.45	95%	Ar–Ar (1), Laser fusion
	x	Bentonite sanidine	Mancos Shale, Arizona (correlated to bed PCB-17 at top of lower Bridge Creek Limestone Mbr, Pueblo, Colorado)	Base of *Vascoceras (Greenhornoceras) birchbyi* ammonite zone (middle Middle Turonian in North American Western Interior zonation).		OB93	93.47	0.63	95%	Ar–Ar (1), Laser fusion
	x	Bentonite sanidine	Mancos Shale, New Mexico (correlated to bed PBC-17 at Pueblo)	*Pseudaspidoceras flexuosum* ammonite zone (zone below *V. birchbyi* Zone).		OB93	93.32	0.55	95%	Ar–Ar (1), Laser fusion
Cenomanian										
	x	Bentonite sanidine	Greenhorn Limestone, Nebraska	Upper third of *Neocardioceras juddii* ammonite zone (one zone below Cenomanian/Turonian boundary in North American Western Interior zonation).	This suite of three ages by OB93 in the *N. juddii* ammonite zone places a narrow constraint on the Cenomanian/Turonian boundary.	OB93	93.37	0.40	95%	Ar–Ar (1), Laser fusion
	x	Bentonite sanidine	San Juan County, New Mexico	*N. juddii* ammonite zone.		OB93	93.85	0.49	95%	Ar–Ar (1), Laser fusion
	x	Bentonite sanidine	San Juan County, New Mexico	*N. juddii* ammonite zone.		OB93	93.66	0.58	95%	Ar–Ar (1), Laser fusion

(*cont.*)

Table 17.2 (*cont.*)

STAGE	Primary control on GTS2004	Dated material	Formation and location	Zone and age	Comments (age, reliability, reviews, etc.)	Primary and review references	Revised age (Ma)	Uncertainty (myr) & type (if not 1-sigma)	Method, calib.
	x	Bentonite sanidine	Mancos Shale, Arizona (correlated to bed PBC-4 at Pueblo)	*Euomphaloceras septemseriatum* ammonite zone (two zones below *N. juddii* Zone).		OB93	93.56	0.89 95%	Ar-Ar (1), Laser fusion
	x	Bentonite sanidine	Mancos Shale, Arizona (correlated to bed PBC-1 at Pueblo)	*Vascoceras diartianum* portion of *Sciponoceras gracile* ammonite zone (below *E. septemseriatum* Zone).		OB93	93.97	0.72 95%	Ar-Ar (1), Laser fusion
	x	Bentonite sanidine	Frontier Fm., Wyoming	*Dunveganoceras pondi* portion of *S. gracile* ammonite zone.		OB93	94.70	0.61 95%	Ar-Ar (1), Laser fusion
	x	Bentonite sanidine	Frontier Fm, Wyoming	*Acanthoceras amphibolum* ammonite zone.		OB93	95.00	0.53 95%	Ar-Ar (1), Laser fusion
	x	Bentonite sanidine	Graneros Shale, Colorado	*Conlinoceras tarrantense* (=*Conlinoceras gilberti*) ammonite zone.		OB93	95.85	0.61 95%	Ar-Ar (1), Laser fusion
	x	Bentonite sanidine	Frontier Fm, Wyoming	27m below *C. gilberti* zone.		OB93	95.93	0.45 95%	Ar-Ar (1), Laser fusion
	x	Bentonite sanidine	Casper, Wyoming	Top of Mowry Shale.		OB93	97.24	0.69 95%	Ar-Ar (1), Laser fusion
	x	Bentonite sanidine	Colorado Shale, Montana	*Neogastroplites cornutus* ammonite zone (zone above *N. hassi* Zone).		OB93	98.59	0.41 95%	Ar-Ar (1), Laser fusion
	x	Bentonite sanidine	Thermopolis Shale, Wyoming	*Neogastroplites haasi* ammonite zone (basal zone of Cenomanian in North American Western Interior ammonite zonation).	Obradovich *et al.*, 2002: "Entire Neogastroplites succession in the Western Interior would most likely be of late *M. saxbii* and/or *M. dixoni* ages" [of European zonation.]	OB93	98.61	0.70 95%	Ar-Ar (1), Laser fusion
	x	Bentonite sanidine	Thermopolis Shale, Wyoming	*N. haasi* zone?		OB93	98.81	0.59 95%	Ar-Ar (1), Laser fusion
	x	Sanidine from tuff	middle part of Yezo Group, Hokkaido, Japan	*Mantelliceras saxbii* ammonite subzone.		Obradovich *et al.*, 2002	99.05	0.38 95%	Ar-Ar (1), Laser fusion
	x	Sanidine from tuff	middle part of Yezo Group, Hokkaido, Japan	*Graysonites woodridgei* ammonite zone (equivalent to European subzone of *Neostlingoceras carcitanense*).		Obradovich *et al.*, 2002	99.23	0.37 95%	Ar-Ar (1), Laser fusion
Albian									
	secondary guide	Bentonite sanidine	Hulcross Fm, British Columbia	*Pseudopulchellia pattoni* zone of North American Western Interior. Correlated to mid-Middle Albian, *Euhoplites loricatus* zone of Europe.		OB93	107.2	0.3 95%	Ar-Ar (1), Laser fusion

Aptian										
	secondary guide	Bentonite sanidine	Sarstedt, near Hannover	*Parahoplites nutfieldiensis* ammonite zone.		OB93	114.1	1.3	95%	Ar–Ar (1), Laser fusion
	secondary guide	Sanidine	Fuller's Earth, Sandgate Beds, Petteson Court, Redhill, Surrey, England	*Parahoplites nutfieldiensis* ammonite zone.	Large K–Ar range on sanidines from redeposited volcaniclastics (secondary bentonites) (Jeans, 1977, evaluated as NDS147). *[See Footnote b]*	Jeans *et al.*, 1982; OB93.	112	1		Ar–Ar (3), stepwise heating
	x	Plagioclase separates from basalt	MIT guyot, western Pacific	MIT seamount basalt at reversed-upward-to-normal polarity transition; overlain by Lower Aptian marine sediments. Interpreted as top of Chron M0r.	Paleomagnetics were originally interpreted as upper polarity zone M1r, but is re-assigned as top of Chron M0r (see text of M-sequence calibration).	Pringle & Duncan, 1995; Pringle *et al.* 2003.	124.6	0.2		Ar–Ar (2); stepwise heating
	derived product			Barremian–Aptian boundary. Projected age of base of polarity chron M0r (defines base of Aptian) based on dating of reversed-polarity basalt on MIT Guyot. Chron M0r is 0.38 myr in duration.		ODP Leg 144	125.0	0.5		
Barremian										
	secondary guide			Barremian or Hauterivian. Resolution Guyot polarity zone, interpreted as chron M5r, hence uppermost Hauterivian.	Chron M3r may be an alternative assignment?	ODP Leg 145	127.6	0.2		
Hauterivian Valanginian Berriasian										
	secondary guide	Zircon from tuff	Grindstone Creek tuff, Great Valley sequence, California	latest Berriasian or lower Valanginian. Tuff in mudstone is assigned to *Buchia uncitoides* and *B. pacifica* bivalve zones and NK2A nannofossil zone. Nannofossil taxa suggests correlation to interval of polarity chrons M15 to M16 of latest Berriasian.	This 137 Ma age is difficult to reconcile with E.Cret. cycle stratigraphy constraints and the 125 Ma age for base-Aptian. Zircons may have some Pb loss, as is "common for Jurassic from this region" (see Pálfy *et al.* 2000c's comment on Josephine and Galicia formations in California).	Bralower *et al.* 1990.	137.1	0.6	U–Pb	
JURASSIC *Tithonian*										
	secondary guide	Bentonite sanidine	Brushy Basin Member 6, upper Morrison Formation	Kimmeridgian to Tithonian. Set of 6 stratigraphic intervals in upper part of Brushy Basin member above a change in clay mineralogy, considered to be a regional hiatus. Ostracod and charophyte assemblages suggest Kimmeridgian in lower part to Tithonian in upper part.	The suite of 6 ages from the Brushy Basin Member are stratigraphically consistent, but there is no reliable correlation to marine biostratigraphy. *[See Footnote 7]*	Kowallis *et al.*, 1998; Pálfy *et al.*, 2000c.	148.1	1	95%	Ar–Ar, Laser fusion
	secondary guide	Bentonite sanidine	Brushy Basin Member 5, upper Morrison Formation	"	"	Kowallis *et al.*, 1998; Pálfy *et al.*, 2000c.	149	.8	95%	Ar–Ar, Laser fusion

(*cont.*)

Table 17.2 (*cont.*)

STAGE	Primary control on GTS2004	Dated material	Formation and location	Zone and age	Comments (age, reliability, reviews, etc.)	Primary and review references	Revised age (Ma)	Uncertainty (myr) & type (if not 1-sigma)		Method, calib.
	secondary guide	Bentonite sanidine	Brushy Basin Member 4, upper Morrison Formation	"	"	Kowallis *et al.*, 1998; Pálfy *et al.*, 2000c.	149.3	1	95%	Ar-Ar, Laser fusion
	secondary guide	Bentonite sanidine	Brushy Basin Member 3, upper Morrison Formation	"	"	Kowallis *et al.*, 1998; Pálfy *et al.*, 2000c.	149.3	1.1	95%	Ar-Ar, Laser fusion
	secondary guide	Bentonite sanidine	Brushy Basin Member 2, upper Morrison Formation	"	"	Kowallis *et al.*, 1998; Pálfy *et al.*, 2000c.	150.2	1	95%	Ar-Ar, Laser fusion
	secondary guide	Bentonite sanidine	Brushy Basin Member 1, upper Morrison Formation	"	"	Kowallis *et al.*, 1998; Pálfy *et al.*, 2000c.	150.3	.5	95%	Ar-Ar, Laser fusion
Kimmeridgian										
	secondary guide	Bentonite sanidine	Tidwell Member, lower Morrison Formation, Utah	Late Oxfordian or Early Kimmeridgian. Morrison Formation is above a regional unconformity on Stump Formation, which has middle Oxfordian carioceratid ammonites. Lower Morrison could be Early Kimmeridgian in age.	"	Kowallis *et al.*, 1998; Pálfy *et al.*, 2000c.	154.9	1	95%	Ar-Ar, Laser fusion
	x	Celadonite vein in oceanic basalt	ODP Site 765, Argo Abyssal Plain, off NW Australia	Oxf/Kimm boundary interval. Magnetic anomaly M26.	Constrains M-sequence magnetic polarity time scale through assumption of constant Pacific spreading rate to anomaly M0r (base Aptian). *[See Footnote c]*	Ludden, 1992; Pálfy *et al.*, 2000c.	155.3	3.4	63%	K-Ar
Oxfordian *Callovian*										
	secondary guide	Zircon from tuff layer	Chacay Melehué 2, Neuquén Basin, Argentina	Near boundary of the regional *bodenbenderi* and *proximum* zones, that is approximately equated to the Boreal *calloviense* Zone or Mediterranean *gracilis* Zone of late Early Callovian (Riccardi et al., 1991).	"Published" only in newsletter. Pálfy (pers. commun., 2001) expressed concerns about this age constraint.	Odin, G.S., *et al.*, 1992; Pálfy *et al.*, 2000c.	160.5	0.3	95%	U-Pb
Bathonian										
	secondary guide	Zircon from tuff layer	Chacay Melehué 1, Neuquén Basin, Argentina	Near boundary of the regional *steinmanni* and *vergarensis* ammonite zones, that is equated to the Bathonian-Callovian boundary (Riccardi *et al.*, 1991).	"Published" only in newsletter. Pálfy (pers. commun., 2001) expressed concerns about this age constraint.	Odin *et al.*, 1992; Pálfy *et al.*, 2000c.	161	0.5	95%	U-Pb

	secondary guide	Zircons from reworked volcanic ash	Copper River, Ashman Formation, NW British Columbia	Ash is in ammonite-rich strata yielding Late Bathonian to Early Callovian assemblages.		Pálfy *et al.*, 2000b; Pálfy *et al.*, 2000c.	162.6	+2.9/−7.0	95%	U-Pb
	x	Oceanic thoeiitic basalt	ODP Site 801, western Pacific	Lower basalt is probably Bajocian or Bathonian. *[See Footnote d]*	ODP Site 801 is about 100 km beyond anomaly M41 on the Sager *et al.* (1998) deep-tow magnetic profile trending toward this site. *[See Footnote d]*	Pringle 1992; SEPM95; Pálfy *et al.*, 2000c.	167.7	1.4	95%	Ar-Ar (4)
Bajocian										
	secondary guide	Bentonite sanidine	Gunlock, Carmel Formation, Utah, USA	Bivalve correlation of Carmel Formation to ammonite-bearing Twin Creek Formation indicates an age of late Early to early Late Bajocian (according to Imlay, 1967, 1980).	Age is slightly older than previously determinations using K-Ar and Rb-Sr methods. *[See Footnote 7]*	Kowallis *et al.*, 1993; Pálfy *et al.*, 2000c.	166.3	0.8	95%	Ar-Ar
Aalenian										
	secondary guide	Zircons from felsic volcanic unit	Treaty Ridge, Salmon River Formation, Iskut River map area, British Columbia	Late Aalenian to Early Bajocian. Underlying mudstone is Upper Aalenian (*Erycitoides* cf. *howelli* ammonite zone of western North America zonation), and overlying siltstone is Lower Bajocian.		Friedman & Anderson, unpub. data, 1997; Pálfy *et al.*, 2000c.	177.3	0.8	95%	U-Pb
	secondary guide	Rhyolite	Eskay rhyolite east, Eskay Creek gold mine, Iskut River map area, NW British Columbia	Late Toarcian or Aalenian. Flow-banded rhyolite is overlain by Upper Aalenian strata (*Erycitoides* cf. *howelli* ammonite zone of western North America zonation).		Childe, 1996; Pálfy *et al.*, 2000c.	174.1	+4.5/−1.1	95%	U-Pb
	secondary guide	Rhyolite	Eskay rhyolite west, Eskay Creek gold mine, Iskut River map area, NW British Columbia	Late Toarcian or Aalenian. Same rhyolite unit as PSM35, and considered similar in age.		Childe, 1996; Pálfy *et al.*, 2000c.	175.1	4.7	95%	U-Pb
Toarcian										
	x	Zircons from dacite flow	Julian Lake dacite, Salmon River Formation, Iskut River area, NW British Columbia	Submarine flow overlain by volcanic sandstones of uppermost Toarcian (*Yakounensis* ammonite zone of western North America).	Several hundred meters upsection, another dacite flow yielded an age of 172.3 Ma (±1.0), but lacks adequate biostratigraphic constraints.	Mortensen & Lewis, unpub. data, 1996; Pálfy *et al.*, 2000c.	178	1	95%	U-Pb
	x	Zircons from volcanic ash	Yakoun River, Queen Charlotte Islands, British Columbia	Volcanic ash layer near base of *Crassicosta* ammonite zone (late Middle Toarcian) at type section of this regional western North American zone.		Pálfy *et al.*, 1997; Pálfy *et al.*, 2000c.	181.4	1.2	95%	U-Pb
	x	Zircons from felsic crystal tuff	Mount Brock volcanics within Hazelton Group, Mount Brock, British Columbia	Early to early Middle Toarcian (*Kanense* to *Planulata* ammonite zones of western North America).		Pálfy *et al.*, 2000b; Pálfy *et al.*, 2000c.	180.4	+11.2/−0.4	95%	U-Pb
	x	Zircons from quartz monzonite	McEwan Creek pluton, Spatsizi River map area, NW British Columbia	Intrudes volcanics with intercalated sediments of Early Toarcian. Pluton is perhaps co-magmatic with volcanism.		Evenchick & McNicoll, 1993; Pálfy *et al.*, 2000c.	183.2	0.7	95%	U-Pb

(*cont.*)

Table 17.2 (*cont.*)

STAGE	Primary control on GTS2004	Dated material	Formation and location	Zone and age	Comments (age, reliability, reviews, etc.)	Primary and review references	Revised age (Ma)	Uncertainty (myr) & type (if not 1-sigma)		Method, calib.
Pliensbachian										
	x	Dacite tuff	Whitehorse, Nordenskiold volcanics, Yukon	Late Pliensbachian (*Kunae* ammonite zone of western North America).		Hart, 1997; Pálfy *et al.*, 2000c.	184.1	+5.8/−1.6	95%	U-Pb
	x	Zircon from rhyolite tuff	Skinhead Lake, NW British Columbia	Late Pliensbachian (*Kunae* ammonite zone of western North America).		Pálfy *et al.*, 2000b; Pálfy *et al.*, 2000c.	184.7	0.9	95%	U-Pb
	x	Zircon from crystal tuff	Atlin Lake (Copper Island), Laberge Group, NW British Columbia	Late Pliensbachian (*Kunae* ammonite zone of western North America).		Johannson & McNicoll, 1997; Pálfy *et al.*, 2000c.	185.8	0.7	95%	U-Pb
	x	Zircon from tuff	Atlin Lake (East shore), Laberge Group, NW British Columbia	Early Pliensbachian (*Whiteavesi* ammonite zone of western North America).		Mihalynuk & Gabites, unpub. data, 1996; Pálfy *et al.*, 2000c.	187.5	1	95%	U-Pb
Sinemurian										
	x	Zircon from dacite tuff	Telkwa Range 1, Telkwa Formation, British Columbia	Late Sinemurian with ammonite assemblages characteristic of *Plesechioceras? harbledownense* ammonite zone of western North America.	Appendix text in Pálfy *et al.* (2000c, item 17–18) seems to cite this age as 194.0 Ma (+9.1, −1.8).	Pálfy *et al.*, 2000b; Pálfy *et al.*, 2000c.	192.8	+6.4/−0.6	95%	U-Pb
	x	Zircons from andesite flow	Ashman Ridge 2, Telkwa Formation, British Columbia	Andesite flow is immediately below sandstone with early Late Sinemurian ammonites (*Varians* assemblage).		Pálfy *et al.*, 2000b; Pálfy *et al.*, 2000c.	192	Min. age		U-Pb
Hettangian										
	x	Zircon from tuff	Puale Bay 3, Talkeetna Formation, Alaska Peninsula	Middle to upper Hettangian (*Franziceras* or *Pseudoaetomoceras* ammonite zone, western North America zonation).		Pálfy *et al.*, 1999; Pálfy *et al.*, 2000c.	197.8	+1.1/−0.4	95%	U-Pb
	x	Zircon from tuff	Puale Bay 2, Talkeetna Formation, Alaska Peninsula	Middle to upper Hettangian (*Franziceras* or *Pseudoaetomoceras* ammonite zone, western North America zonation).		Pálfy *et al.*, 1999; Pálfy *et al.*, 2000c.	197.8	1	95%	U-Pb
TRIASSIC *Rhaetian*										
	x	Zircon from tuff	Kunga Island, British Columbia	Latest Rhaetian. Tuff layer interbedded in marine sediments of uppermost Triassic, immediately below the Triassic-Jurassic boundary defined by conodonts, radiolarians and ammonites.		Pálfy *et al.*, 2000a; Pálfy *et al.*, 2000c.	199.6	0.4	95%	U-Pb

	secondary guide	Zircon and badde-leyite from basalt	Palisades sill, Newark Basin, eastern U.S.	Rhaetian to earliest Hettangian. *[See Footnote e]*		Dunning & Hodych, 1990; Pálfy *et al.*, 2000c.	200.9	1	95%	U-Pb
	secondary guide	Zircon from basalt	Gettysburg sill, Newark Basin, eastern U.S.	*[Same as above]*.		Dunning & Hodych, *1990*; Pálfy *et al*., 2000c.	201.3	1	95%	U-Pb
	secondary guide	Zircon from basalt	North Mountain basalt, Fundy Basin	Rhaetian to earliest Hettangian. *[See Footnote e]*		Hodych & Dunning, 1992; Pálfy *et al.*, 2000c.	201.7	+1.4/−1.1	95%	U-Pb
Norian *Carnian* *Ladinian*										
	secondary guide		Predazzo volcanics, Dolomites, Italy	Ladinian or younger. Pálfy *et al.* (2003) "tentatively assign to the *Regoledanus* ammonite zone" (highest zone of Ladinian).		Brack *et al.*, 1997, Geology, 25: 471-472.	237.3	+0.4/−1.0		U-Pb
	x	Zircon	Upper Bandekalke, below onset of late Ladinian pillow basalts, 49 m level in Seceda section, NW Dolomites, N Italy	Upper Ladinian. Strata has *Daonella lommeli* bivalves (just above highest *Daon. pichleri*) and corresponds to "*Protrachyceras archelaus*" ammonite zone [*probably Protrachyceras longobardicum* ammonite subzone in other Ladinian schemes].		Mundil *et al.*, 1996; Brack *et al.*, 1996.	238.0	+0.4/−0.7	95%	U-Pb
	x	Zircon	upper Buchenstein Beds, 85.5 m level in Bagolino section, eastern Lombardian Alps, N Italy.	Middle Ladinian. Strata has *Daonella pichleri* and *D. indica* bivalves and is assigned to lowermost "*Protrachyceras archelaus*" amonite zone *[probably Protrachyceras longobardicum* ammonite subzone in other Ladinian schemes].		Mundil *et al.*, 1996; Brack *et al.*, 1996.	237.9	+1.0/ − 0.7	95%	U-Pb
	x	Zircon	middle Buchenstein Beds, 72.2 m level in Bagolino section, eastern Lombardian Alps, N Italy.	Lower Ladinian. Above are beds with *Eoprotrachyceras margaritosum* ammonites, and below ones with *Arpadites* and *Protrachyceras,* therefore probably in *Protrachyceras gredleri* ammonite zone. [Pálfy *et al.*, 2003, suggest it is uppermost portion of *P. gredleri* zone.].	"statistically indistinguishable from the dates" of the underlying four subzones of the Reitzi Zone [Pálfy *et al.*, 2003].	Mundil *et al.*, 1996; Brack *et al.*, 1996; Pálfy *et al.*, 2003.	238.8	+0.5/−0.2	95%	U-Pb
	x	Zircons from tuff	Neptunian dyke into Tagyon Dolomite, Litér dolomite quarry, Balaton Highlands, Hungary	Redeposited tuff in neptunian dyke that contains ammonite assemblage assigned to upper *Protrachyceras gredleri* ammonite zone.		Pálfy *et al.*, 2003.	238.7	0.6	95%	U-Pb (TIMS)
Anisian/Ladinian Boundary interval										
	secondary guide	Sanidine from bentonite	Tuff in bituminous shales (Grenzbitumen horizon), Bed 71, Monte San Giorgio, Lake Lugano, Switzerland	*Daonella* lamellibrach species are 150 cm below lowest dated sample. Assigned to lower *Nevadites secendensis* ammonite zone by Brack *et al.*, 1996.	High sanidine (water-clear) ages from Ar-Ar plateau agree with conventional K-Ar. 9 tuff samples, 5 outcrops.	Hellmann & Lippolt, 1979, 1981; Hellmann, in NDS82.	233	2.5		Ar-Ar stepwise heating, plus K-Ar

(*cont.*)

Table 17.2 (*cont.*)

STAGE	Primary control on GTS2004	Dated material	Formation and location	Zone and age	Comments (age, reliability, reviews, etc.)	Primary and review references	Revised age (Ma)	Uncertainty (myr) & type (if not 1-sigma)		Method, calib.
	x	Zircon	Tuff in bituminous shales (Grenzbitumen horizon), Bed 71, Miniera Val Porina, 820m elevation, Monte San Giorgio, Lake Lugano, Switzerland	Lower *Nevadites secedensis* ammonite zone. Same level as NDS196 bentonite with Ar-Ar age.		Mundil *et al.*, 1996; Brack *et al.*, 1996.	241.2	0.8	95%	U-Pb
	x	Zircon	Tc tuffs at base of Knollenkalke, lower Buchenstein Beds, 9.35 m level in Seceda section, NW Dolomites, N Italy	Lower *Nevadites secedensis* ammonite zone.		Mundil *et al.*, 1996; Brack *et al.*, 1996.	241.2	+0.8/−0.6	95%	U-Pb
	x	Zircons from tuff	Vászoly Formation, Forrás Hill, Felsöörs, Balaton Highlands, Hungary	Upper part of *R. reitzi* subzone, third of four subzones within *Reitziites reitzii* ammonite zone.	*[See Footnote f]*	Pálfy *et al.*, 2003; L. Hinnov, pers.comm., 2002.	240.4	0.4	95%	U-Pb (TIMS)
	x	Zircons from tuff	Vászoly Formation, Forrás Hill, Felsöörs, Balaton Highlands, Hungary	Near base of *R. reitzi* subzone, third of four subzones within *Reitziites reitzii* ammonite zone.		Pálfy *et al.*, 2003, J. Geol. Soc. London, 160:271–284.	240.5	0.5	95%	U-Pb (TIMS)
	x	Zircons from tuff	Vászoly Formation, Forrás Hill, Felsöörs, Balaton Highlands, Hungary	Near base of *Hyparpadites liepoldti* subzone, second of four subzones within *Reitziites reitzii* ammonite zone.		Pálfy *et al.*, 2003.	241.2	0.4	95%	U-Pb (TIMS)
	x	Zircons from tuff	Vászoly Formation, Forrás Hill, Felsöörs, Balaton Highlands, Hungary	Middle part of *Kellnerites felsoeoersensis* subzone, first (lowest) of four subzones within *Reitziites reitzii* ammonite zone. "Would correlate to Kellnerites beds at Bagolino, or to the lower Contrin Fm. at Seceda" (L. Hinnov, pers. comm. 2002).		Pálfy *et al.*, 2003.	241.1	0.5	95%	U-Pb (TIMS)
Anisian										
Early Triassic										
Permian/Triassic Boundary interval										
	x	Zircon	Bed 36 (6.7 m above Bed 25), Meishan quarry D, SE China	Lower Triassic. Within *Claraia* bivalve and *H. parvus* conodont zone (lower Triassic). 6.7 m above P-T boundary.	Mundil *et al.* (2001) suggest "zircon population is characterized by non-analytical scatter, possibly mainly from xenocrystic contamination".	Bowring *et al.*, 1998; Mundil *et al.*, 2001.	250.2	0.2	95%	U-Pb TIMS
	x	Zircon	Bed 33 (2.25 m above Bed 25), Meishan quarry D, SE China	Basal Triassic. Within *Claraia* bivalve and *H. parvus* conodont zone (lower Triassic). 2.25 m above P-T boundary.	Mundil*et al.* (2001) suggest "set appears to be slightly biased by Pb loss".	Bowring *et al.*, 1998; Mundil *et al.*, 2001.	250.4	0.5	95%	U-Pb TIMS

	x	Zircon	Bed 28, Meishan quarry Z, SE China	Bed 28 is 8 cm above Permian-Triassic GSSP.	*[See Footnote g]*	Bowring *et al.*, 1998; Metcalfe & Mundil, 2001.	250.7	0.3	95%	U-Pb TIMS
	secondary guide	Zircon	Bed 28, Meishan quarry Z, SE China	Basal Triassic. Bed 28 is 8 cm above Permian-Triassic GSSP.		Mundil *et al.*, 2001; Metcalfe & Mundil, 2001.	252.5	0.3		U-Pb TIMS
PERMIAN *top of Permian*	x	Zircon	Bed 25 volcanic clay, Meishan quarry D (GSSP), SE China	Uppermost Permian. Bed 25, a 5-cm volcanic clay, is the approximate culmination of the late-Permian mass extinction. This former "boundary clay" level, is 18 cm below GSSP Bed 27c at Meishan D section, corresponding to first appearance of conodont *Hindeodus parvus.*		Bowring *et al.*, 1998; Metcalfe & Mundil, 2001.	251.4	0.3	95%	U-Pb TIMS
	secondary guide	Zircon	Bed 25 volcanic clay, Meishan (GSSP), SE China	"	Re-evaluation of zircon U-Pb data suggests complex behaviors, and an age can not be reliably assigned other than greater than 254 Ma.	Mundil *et al.*, 2001; Metcalfe & Mundil, 2001.	> 254			U-Pb TIMS
	secondary guide	Zircon	Bed 25 volcanic clay, Meishan (GSSP), SE China	"	Mundil *et al.* (2001) consider that this age is "clearly too young". They suggest either "ash is older than about 257 Ma (the apparent age of the two oldest concordant grains)" or "if these two grains are xenocrysts, that the ash is slightly older than 254 Ma".	Mundil *et al.*, 2001.	249.0	0.8	95%	U-Pb TIMS
	secondary guide	Zircon	Bed 25 volcanic clay, Meishan (GSSP), SE China	"		Claoué-Long *et al.*, 1991; 1995; Metcalfe & Mundil, 2001.	251.1	3.6	95%	U-Pb, HR-SIMS
	secondary guide	Sanidine	Bed 25 volcanic clay, Meishan (GSSP), SE China	"		Z. Zichao, in Odin, 1992.	250	3		Rb-Sr
	secondary guide	Sanidine	Bed 25 volcanic clay, Meishan (GSSP), SE China	"	Internal error. Estimated external error that "incorporated uncertainty in the standard's age" is 1.52 myr (Renne *et al.*, 1995).	Renne *et al.* 1995; Metcalfe & Mundil, 2001.	249.9	0.15	95%	Ar-Ar (5)
	secondary guide	Sanidine	Bentonite in Shangsi section, Sichuan province, China	Approx. 5 cm below former paleontologically-defined boundary. If similar stratigraphic-thickness trends as Meishan GSSP, then this is probably 20 cm below revised base-Triassic definition.	Internal error. Estimated external error that "incorporated uncertainty in the standard's age" is 1.13 myr (Renne *et al.*, 1995).	Renne *et al.* 1995.	250.0	0.36	95%	Ar-Ar (5)
	x	Zircon	Graded beds of silicic pyroclstic rocks, Talung Formation, 2 km from town of Heshan, Guangxi Province, S. China	Corresponds to upper part of Changhsing Fm. Just below base of Triassic. Approximately correlative with Bed 25 at Meishan.	When combined with the Meishan dates, the many independent Heshan samples indicate that base of Triassic is near 251 Ma (see Chapter 8).	Bowring *et al.*, 1998.	251.6	0.1	95%	U-Pb TIMS

(*cont.*)

Table 17.2 (*cont.*)

STAGE	Primary control on GTS2004	Dated material	Formation and location	Zone and age	Comments (age, reliability, reviews, etc.)	Primary and review references	Revised age (Ma)	Uncertainty (myr) & type (if not 1-sigma)		Method, calib.
	x	Zircon	Graded beds of silicic pyroclstic rocks, Talung Formation, 2 km from town of Heshan, Guangxi Province, S. China	Same level as Bowring H-Matan96-7 (just below base of Triassic).		Bowring *et al.*, 1998.	251.7	0.2	95%	U-Pb TIMS
	x	Zircon	Graded beds of silicic pyroclstic rocks, Talung Formation, 2 km from town of Heshan, Guangxi Province, S. China	Below Bowring H-Matan96-7 and 96-6, but in same pyroclastic sequence.		Bowring *et al.*, 1998.	251.6	0.1	95%	U-Pb TIMS
	secondary guide	Zircon and Badellyite	Noril'sk-1 gabroic intrusion	Siberian Traps. Age is from Noril'sk-1 gabbroic intrusion, which cuts lower third of the Traps.		Kamo *et al.* 1996; Bowring *et al.*, 1998.	251.2	0.3	95%	U-Pb
	secondary guide	Biotite	Noril'sk-1 gabroic intrusion	Siberian Traps. Age is from Noril'sk-1 gabbroic intrusion, which cuts lower third of the Traps.	Ar-Ar on bulk hornblende sample yielded isochron of 249.3 $\pm$1.6 Ma.	Renne, 1995.	250.1	1.5	95%	Ar-Ar
	secondary guide	Hornblende, Whole Rock	Siberian Traps	Siberian Traps.	Increased uncertainty on recalculated age "incorporated uncertainty in the standard's age" (Renne *et al.*, 1995). Internal error was only 0.30 myr.	Renne & Basu, 1991; Renne 1995.	250	1.6	95%	Ar-Ar (6)
Changhsingian										
	x	Zircon	Tuff in Bed 20, mid-Beishan Member, Meishan quarry Z, SE China	4.3 m below base-Triassic GSSP. Upper Changhsingian.	Mundil *et al.* (2001) conclude this age "is statistically robust"	Bowring *et al.*, 1998; Mundil *et al.*, 2001.	252.3	0.3	95%	U-Pb TIMS
	x	Zircon	9 m-thick pyroclastic sequence, Penglaitan, banks of Hong Shui River, town of Laibin, S. China	Ammonoids *Rotodiscoceras* and *Pleuronodoceras* indicate within late Changhsingian stage, "but its exact location relative to the conodont zonation is uncertain".		Bowring *et al.*, 1998.	252.4	0.2	95%	U-Pb TIMS
	x	Zircon	Clay within Bed 15 at Meishan section, SE China	17.3 m below base-Triassic GSSP.	Age "is extremely difficult to reconcile with the results [*by Mundil et al.*] from the other analyzed horizons .. ca. 255 Ma would be expected".	Mundil *et al.*, 2001.	252.0	0.4	95%	U-Pb TIMS

x	Zircon	9-cm-thick tuffaceous sandstone (Bed 7) near the base of the Baoqing Member (lower unit of Changhsing Fm.), Meishan quarry D, SE China	Basal part of *Clarkina subcarinata* conodont zone. First occurrence of conodont *Clarkina wangi* is at top of Bed 4. Immediately below dated-sample is the first occurrence of *Clarkina subcarinata* (sensu strictu) (Bed 8). "Extrapolation for the sediment rate for the *subcarinata* and *changxingensis* zones places the age of the base of the *C. wangi* Zone and the base of the Changhsingian at about 254 Ma." (see chapter on Permian).	Mundil *et al.* (2001) suggest zircons affected by "significant amounts of Pb loss", and an age of 254.7± 0.2 Ma "may be considered as a minimum age for this horizon".	Bowring *et al.*, 1998; Mundil *et al.*, 2001.	253.4	0.2	95%	U-Pb TIMS
x	Zircon	Ash layer between Bed 3 and Bed 4, near base of Baoqing Member, lower unit of Changhsing Fm, Meishan, SE China	Basal part of *Clarkina subcarinata* conodont zone. 41 m below base-Triassic GSSP, and 7 m above *Clarkina orientalis* zone of uppermost Wuchiapingian stage.	"Age of the oldest crystal (257.2 ± 0.7 Ma) is the best estimate of the minimum age of this unit".	Mundil *et al.*, 2001.	257.2	0.7	95%	U-Pb TIMS

Footnotes:

(a) The main requirements for this selection of Mesozoic ages used for constructing an absolute time scale were utilization of Ar-Ar or U-Pb methods, adequate stratigraphic control (to within a substage), and low uncertainties. Radiometric ages from intrusions, Rb-Sr or K-Ar methods, and material that publications have indicated may have significant Pb loss or other problems were generally omitted. A secondary suite that aided in indicating approximate ages for some geologic stages, but did not have adequate stratigraphic or radiometric control, is also listed. In Cretaceous entries, reference "OB93" is Obradovich (1993).

(b) *[Aptian]* Jeans *et al.* (1982) later concluded from sanidine separations that the Fuller Earth ages represent variable mixtures of 300 Ma sanidines into these re-deposited ash layers. Therefore, Jeans *et al.* (1982) selected the youngest date as most reliable. Fuller's Earth seams are "local deposits . . . carried into shallow marine environments by small rivers, draining areas rich in freshly fallen ash." (NDS147, by Kreuzer, in NDS82). OB93 cites Fuller's Earth as reworked ash fall "so some concern is expressed about the result".

(c) *[Kimm. Site 765]* An additional Ar/Ar (incremental heating) analysis of basalt did not yield a plateau age, but gave a total fusion age of 155 Ma ± 6 (2-sigma), which Pálfy *et al.* (2000c, p. 930) consider 'disputable'.

(d) *[Bathonian ODP Site 801]* Revised age of 166.8 Ma is adjusted for Milankovitch-calibrated Ar-Ar monitor standard, following Larson & Erba (1999). The 167.7 Ma age is based on new basalt analyses by Koppers *et al.* (2004). Incorporation of external errors would increase the uncertainty. The dated basalts are overlain by alkaline off-ridge basalt with Ar-Ar age of 157.4 Ma (+/− 0.5), which is overlain (or, perhaps, intruded into) radiolarian claystones with disputed calibration to geological stages (Bajocian to Oxfordian, see discussion of radiolarian stratigraphy in Jurassic chapter).

(e) *[Rhaetian]* Palisades and Gettysburg sills are considered to be feeders for regional basalt that immediately overlies a turnover in palynology and vertebrate assemblages that was tentatively considered equivalent to Triassic-Jurassic boundary. Base of North Mountain basalt flows are 20 m above a similar turnover. (See discussion in Triassic chapter.)

(f) *[Anisian/Ladinian boundary]* L. Hinnov (pers. comm. 2002) comments that the set of U/Pb ages from Hungary and N. Italy would imply an implausible rapid rate of carbonate platform growth of the Latemar buildup in the Italian Dolomites; indeed, the fastest in the geological record. Alternatively, there is a significant amount of inherited older zircons within the series of regional tuffs.

(g) *[Permian/Triassic boundary, China]* Metcalfe & Mundil (2001) conclude "isotopic ages reported by Bowring *et al.* (1998) for individual layers from Meishan appear to be underestimated by as much as 1%." However, the suite is fully consistent, agrees with dating in Heshan, and implies the base of Triassic is older than 250.7 Ma.

Notes in Methods & calibration column:

(1) TCR monitor age of 28.32 Ma converted to 28.34 Ma.
(2) TCR monitor age of 27.92 Ma. Recomputed for 28.34 Ma.
(3) Steiger & Jaeger (1977) decay constants.
(4) Recalibrated to FCT-3 biotite (28.04 ± 0.18 Ma) by Koppers *et al.*, 2004.
(5) FCT = 28.03 Ma. Renne *et al.*, 1998 suggest increasing this uncertainty on basal-Triassic Ar/Ar ages to ± 4.6 myr to compensate for possible Ar/Ar systematic errors.
(6) FCT monitor age of 27.84 Ma converted to 28.03 Ma.
(7) *[Late Jurassic]* Pálfy *et al.* (2000c) suggested increasing all Ar/Ar uncertainties by an additional 2.5 myr to indicate 'external errors' due to calibration of radiometric decay constants and other factors.

Table 17.3 *Triassic time scale for Alpine and/or Tethyan ammonite zones*[a]

Stage, *substage*	Ammonite zone (Alpine or Tethyan)	Age of base of zone (Ma)	Comments
Hettangian (top)		**196.5**	Cycle-scaled linear Sr isotope trend
	Schlotheimia angulata	197.7	
	Alsatites liasicus	198.8	
	Psiloceras planorbis	199.6	
TRIASSIC (*Rhaetian*) (*top*)		**199.6**	Radiometric age control
	Choristoceras marshi	201.0	Rhaetian zones are given equal duration
	Vandaites stuerzenbaumi	202.3	Considered subzone of larger *C. marshi* Zone in some schemes
	Sagenites reticulatus	**203.6**	Formerly considered Norian. Tentative magnetostratigraphy correlation to middle of Chron E22 of Newark polarity pattern
Norian			Ages of Norian, Carnian, and uppermost Ladinian ammonite zones (Alpine/Tethyan) are from tentative magnetostratigraphy correlations to the Newark polarity pattern that is scaled by Milankovitch cycles (see Fig. 17.1). Potential chron assignments are listed
Sevatian			
	Sagenites quenquepunctatus	205.7	mid-Chron E20r
Alaunian			
	Halorites macer	207.5	base Chron E19r
	Himavatites hogarti	210.0	base Chron E17r
	Cyrtopleurites bicrenatus	211.3	mid-Chron E16n
Lacian			
	Juvavites magnus	212.2	lower Chron E16n
	Malayites	214.0	base Chron E15n
	Guembelites jandianus	**216.5**	base Chron E14n
Carnian			
Tuvalian			
	Anatropiites	218.7	lower Chron E13n
	Tropites subbullatus	221.0	uncertain (no magnetostratigraphy)
	Tropites dilleri	222.7	uncertain (no magnetostratigraphy)
Julian			
	Austrotrachyceras austriacum	225.7	mid-Chron E8r
	Trachyceras aonoides	226.8	*T. aonoides* and *T. aon* are combined into a general *Trachyceras* Zone in some zonal schemes. Mid-Chron E8n
	Trachyceras aon	**228.0**	upper Chron E7n
Ladinian			
Longobardian			
	Daxatina cfr. canadensis	229.6	lower Chron E7n
	Protrachyceras (Frankites regoledanus s.z.)	230.4	upper Chron E6n
	(*"Protrachyceras" neumayri s.z.*)	231.1	lower Chron E5r
	(*Protrachyceras longobardicum s.z.*)	232.1	upper Chron E3r
	(*"Eoprotrachyceras" gredleri s.z.*)	233.2	Just below base of Newark pattern
	(*Protrachyceras margaritosum s.z.*)	234.7	
Fassanian			
	Eoprotrachyceras curionii	237.0	Radiometric age (±1.5 myr)
Anisian			
Illyrian			
	Nevadites	240.0	Radiometric age (±1 myr)

Table 17.3 (*cont.*)

Stage, *substage*	Ammonite zone (Alpine or Tethyan)	Age of base of zone (Ma)	Comments
	Hungarites (Aplococeras avisianum s.z.)	240.3	
	(Reitziites reitzi s.z.)	241.0	Radiometric age (±1 myr)
	Paraceratites	242.1	
Pelsonian			Lower Anisian zones were proportionally scaled according to component subzones
	Balatonites	243.1	
Bithynian			
	Kocaelia	244.1	
Aegean			
	Paracrochordiceras	245.2	
Olenekian			Olenekian and Induan zones were scaled relative to published potential correlations to conodont zones. The composite conodont scale for Early Triassic (Sweet and Bergstöm, 1986) was linearly scaled between base-Triassic and base-Anisian
Spathian			
	[gap zone]	245.5	
	Toziceras pakistanum	247.1	
	Tirolites cassianus	248.1	
Smithian			
	Wasatchites spiniger + A. pluriformis + A. prahlada	249.0	
	Meekoceras gracilitatis	249.7	
Induan			
Dienerian			
	Flemingites rohilla	250.1	
	Gyronites frequens	250.3	
Griesbachian			
	Ophiceras connectens + Ophiceras tibeticum	250.8	
	Otoceras woodwardi (upper)	251.0	
Late Permian		251.0	Radiometric age on GSSP
	Otoceras woodwardi (lower)	251.2	Continued same scaling into uppermost Permian
	Otoceras concavum	251.7	
	Paratirolites		

[a] The method and computation/interpolation details incorporated for the assignment of each age for primary ammonite zones and associated stage boundaries are briefly noted. The ages assigned to other Triassic events were generally relative to this ammonite scale.

boundary sections (Bowring *et al.*, 1998; Table 17.2). As pointed out by Kamo *et al.* (2003), a significantly older age interpretation for the Permian–Triassic boundary would causally separate the end of the Permian mass extinction event and the Siberian flood-volcanic rocks that largely extruded between 251.4 and 251.1 Ma. Therefore, Kamo *et al.* (2003) also favor an age for the Permian–Triassic boundary near 251 Ma.

The base of the Middle Triassic *Nevadites secedensis* ammonoid zone is assigned as 240 ± 1 Ma. Uranium–lead ages from a sequence of tuffs within the underlying *Reitziites reitzi* ammonite zone in Hungary (equivalent to the *Hungarites* ammonoid zone of Mietto and Manfrin (1995)) range downward from 240.4 ± 0.4 to 241.1 ± 0.5 Ma (Pálfy *et al.*, 2002a). However, this age succession is slightly younger than a pair of 241.2 ± 0.8 Ma (95% confidence limit) ages from the lower *Nevadites secedensis* Zone in Italy and Switzerland (Mundil *et al.*, 1996; Brack *et al.*, 1996). One possibility is that some of the dated zircons are recycled older material, therefore a generalized age

Table 17.4 *Ages and durations of Triassic stages [see footnote]*

Boundary	Stage	Age (Ma)	Estimated uncertainty on age (2-sigma)	Duration (myr)	Estimated uncertainty on duration (2-sigma)	Status; primary marker	Calibration and comments
	JURASSIC (Hettangian)						
Top of TRIASSIC (base of Hettangian)		**199.6**	*0.6*			Not defined, but traditionally is the first occurrence of the smooth *planorbis* group within the ammonite genus *Psiloceras*.	U-Pb age on tuff near candidate GSSP level is 199.6 ± 0.3 Ma(2-sigma); increased to ±0.6 (2-sigma) to incorporate potential external errors.
	Rhaetian			**4.0**	*1.0*		
Base of Rhaetian		**203.6**	*1.5*			Not defined. Assigned here as base of *Sagenites reticulatus* ammonoid zone and first occurrence of conodont *Epigondolla mosheri* (assumed to be coincident)	Age of stage boundary estimated according to matching magnetostratigraphy to cycle-scaled magnetic polarity chrons of Newark series. Estimated placement is lower part of Newark Chron E22n.
	Norian			**13.2**	*0.5*		Duration and associated uncertainty assumes that the magnetostratigraphic correlation of stage boundary to Newark polarity scale is accurate.
Base of Norian		**216.5**	*2.0*			Not defined. Assigned here as bases of *Klamathites macrolobatus* and *Stikinoceras kerri* ammonoid zones (assumed to be coincident)	Age of stage boundary estimated according to matching magnetostratigraphy to cycle-scaled magnetic polarity chrons of Newark series. Estimated placement is base of Newark Chron E14n.
	Carnian			**11.5**	*0.5*		Duration and associated uncertainty assumes that the magnetostratigraphic correlation of stage boundary to Newark polarity scale is accurate.
Base of Carnian		**228.0**	*2.0*			Not defined. Assigned here as first occurrences of *Trachyceras* ammonite genus and conodont *Metapolygnathus polygnathiformis*.	Age of stage boundary estimated according to matching magnetostratigraphy to cycle-scaled magnetic polarity chrons of Newark series. Estimated placement is upper Newark Chron E7n.
	Ladinian			**9.0**	*0.5*		Duration and associated uncertainty assumes that the magnetostratigraphic correlation of stage boundary to Newark polarity scale is accurate.
Base of Ladinian		**237.0**	*1.5*			Not defined. Assigned here as base of *Eoprotrachyceras curionii* ammonite zone, and near first occurrence of *Budurovignathus* conodont genus.	Age of stage boundary estimated by assigning equal ammonite subzones between base of *Nevadites secedensis* zone (U-Pb age of 240 ± 1 Ma) through top of *Protrachyceras margaritosum* ammonoid zone, which has a projected age of 233 ± 1 Ma from cycle-scaled magnetic polarity chrons.
	Anisian			**8.0**	*1.5*		
Base of Anisian		**245.0**	*1.5*			Not defined. Assigned here as first occurrence of *Paracrochordiceras* ammonite genus.	Age estimated from proportional scaling of early Anisian (below 240 ± 1 Ma U-Pb age at base of *Nevadites secedensis* ammonoid zone) relative to Early Triassic (basal age of 251.0 ± 0.4 Ma (95% confidence limit)) according to component ammonite subzones
	Olenekian			**4.7**	*1.0*		Olenekian and Induan scaling within Early Triassic is from the composite standard of conodont zones derived from graphic correlation of several sections by Sweet & Bergström (1986).
Base of Olenekian		**249.7**	*0.7*			Not defined. Assigned here as slightly above first occurrence	Age derived from relative position on composite standard. Relatively low

Table 17.4 (*cont.*)

Boundary	Stage	Age (Ma)	Estimated uncertainty on age (2-sigma)	Duration (myr)	Estimated uncertainty on duration (2-sigma)	Status; primary marker	Calibration and comments
			0.7			of conodont *Neosp. pakistanensis*, or approximately at base of *Meekoceras gracilitatis* ammonite zone.	uncertainty due to precision of base-Triassic.
	Induan			**1.3**	*0.3*		
Base of TRIASSIC (base of Induan)		**251.0**	*0.4*			GSSP, lowest occurrence of conodont *Hindeodus parvus*.	Bracketing U-Pb ages on tuffs near GSSP level.
	PERMIAN						

Only the base of the Triassic has a ratified international definition, therefore the computed ages are for the selected potential definitions (as indicated). The estimates age of other potential markers are shown in Fig. 17.1. See text for derivation of uncertainty estimates.

of 240 ± 1 Ma (95% confidence) was assigned to the base of the *N. secedensis* ammonoid zone.

Ages derived by both the $^{40}Ar/^{39}Ar$ and K–Ar methods on feldspars from the same lower *N. secedensis* tuff in Switzerland yielded 233 ± 5 Ma (Hellman and Lippolt, 1981; 95% confidence limit), which had been a key control age in earlier Triassic time scales (e.g. Harland *et al.*, 1990; Gradstein *et al.*, 1994b, 1995). The cause of the discrepancy among ages in the upper Anisian–lower Ladinian interval obtained by different U–Pb studies and the $^{40}Ar/^{39}Ar$ versus U–Pb results is not resolved (e.g. Hardie and Hinnov, 1997; Brack *et al.*, 1997).

There are no other significant well-constrained radiometric ages between the base-Triassic and the lower-Middle Triassic. Therefore, the ages of intervening datums were extrapolated by utilizing a composite standard from graphical correlation of Lower Triassic sections (Sweet and Bergström, 1986) and by proportional scaling of ammonite zones according to subzones.

The age of the base of the Anisian (245 Ma) was proportionally assigned between the 251 and 240 Ma control ages according to the relative number of ammonoid subzones in the Early Triassic (17 subzonal units) and in the pre-*Nevadites secedensis* Anisian subzonal count (15 subzonal units). Subzone counts were from van Veen (compiled for Gradstein *et al.*, 1995), and ammonite zones that had no subzonal divisions were arbitrarily assigned a 1.5 "subzonal" equivalent. Pre-*Nevadites secedensis* ammonoid zones in the Anisian were proportionally scaled according to their relative number of subzones.

Substage and ammonoid zone boundaries within the Early Triassic were proportionally assigned according to the relative spacing of major conodont datums computed from global graphic correlation (Sweet and Bergstöm, 1986). The calibration of Canadian ammonoid zones to these conodont datums was from Orchard and Tozer (1997a,b). The apparent compact Induan Stage (only 1.3 myr duration) compared to the overlying Olenekian Stage (4.7 myr duration) reflects the relative thicknesses of strata in sections used for the graphic correlation of conodont datums, even though the Induan and Olenekian stages have approximately equal numbers of ammonoid subzones. Perhaps the Induan experienced a relatively rapid evolution of ammonoids and corresponding zones of shorter duration following the near-extinction of ammonoids at the end of the Permian.

17.3.2 Upper Ladinian to base-Jurassic

Ages for Upper Ladinian through Norian substages and Tethyan ammonite zones were assigned from their apparent magnetostratigraphic correlations to the cycle-scaled Newark Basin magnetic polarity time scale, as previously explained in the Triassic magnetostratigraphy section. The volcanic flows that cap the main Newark Basin lacustrine succession (top of polarity chron E23) are part of the extensive Central Atlantic Magmatic Province (CAMP) that erupted at 202 ± 1 Ma (Olsen and Kent, 2001; Olsen *et al.*, 2003), therefore the ages of the Newark polarity pattern are relative to this control age (Fig. 17.1). This magnetostratigraphy correlation extends downward to the base of the *Eoprotrachyceras gredleri* ammonoid zone, and ages of individual ammonoid zones were read from the associated absolute age scale.

Ages for the ammonite zones in the uppermost Anisian and Lower Ladinian (base of *Nevadites secedensis* at 240 Ma through top of *Protrachyceras margaritosum* ammonoid zone at 233 Ma) were computed according to the relative number of component subzones. Rhaetian ammonoid zones were given equal durations.

The Jurassic–Triassic boundary is tightly constrained by a U–Pb age of 199.6 ± 0.3 Ma (95% confidence limits) from zircons in a tuff directly below the Triassic–Jurassic boundary in British Columbia (Pálfy *et al.*, 2000b).

17.3.3 Estimated uncertainties on stage ages and durations

In contrast to the Paleozoic stages, in which ages of stages are interpolated by spline fitting of multisection composite standards derived from constrained optimization (CONOP) or graphical correlation techniques to an array of radiometric dates, the Mesozoic time scales have integrated several techniques with varying degrees of precision. Therefore, the uncertainties on the calculated ages of stages vary greatly among intervals. In the following discussion of Triassic ages, all uncertainties are 95% confidence (2-sigma) limits.

The Early Triassic ages were scaled between a precise age of 251.0 ± 0.4 Ma for the base-Triassic and a less precise age of 240 ± 1 Ma for the base of the *Nevadites secedensis* ammonoid zone. Therefore, the base-Anisian age of 245 Ma extrapolated from proportional scaling to ammonite subzones will have a larger uncertainty (estimated as 1.5 myr) than either of these equal-distant constraints. The duration of the relatively short-duration Induan Stage of the basal-Triassic is tightly constrained by the composite standard, so would have an uncertainty (estimated as 0.7 myr) closer to the 0.4-myr uncertainty for the base-Triassic.

The Late Triassic is governed by the potential magnetostratigraphic correlations to the Newark magnetic polarity pattern, which is precisely scaled by Milankovitch cycles. The assigned ages for this Newark reference scale are tied to the 202 Ma for the overlying basalts, which Olsen and Kent (2001) suggested had an uncertainty of ±1 myr, therefore a conservative uncertainty of ±1.5 myr (for base-Rhaetian) to ±2 myr was assigned for the set of derived ages. However, this is a systematic uncertainty, and the durations of the stages have a much lower uncertainty (assigned as 0.5 myr; which implicitly assumes that the suite of magnetostratigraphic correlations are correct). The top of the Triassic is taken to be 199.6 ± 0.6 myr, after adjustment of the published uncertainty (±0.3 myr) to partially incorporate potential systematic errors (see Chapter 6), therefore the uncertainty on the duration of the Rhaetian Stage is relatively large.

An increase in the accuracy of the Triassic time scale requires a verified magnetostratigraphic correlation within the upper Triassic to the Newark magnetic reference pattern. At least one high-precision radiometric age with tight biostratigraphic constraints is required within the Norian–Carnian, and an independent set of radiometric ages are needed within the lower Triassic through Ladinian that are outside of the Italian–Hungary region. The studied section in the latter region may have re-worked zircons hampering detailed age dates. Application of Milankovitch cycle scaling to lower Triassic deposits is highly desirable.

17.3.4 Triassic time scale summary

The Triassic Period spans 51.4 myr, between 251.0 ± 0.4 and 199.6 ± 0.6 Ma. The Early, Middle, and Late Triassic Epochs have quite unequal durations of 6, 17, and 28.4 myr, respectively, which is a rather odd division of time for a tripartite period. Stages within the Triassic range from one of the shortest in the Phanerozoic (Induan Stage of 1.3-myr duration), to two of the longest (Carnian of 11.5-myr and Norian of 13-myr durations).

18 • The Jurassic Period

J. G. OGG

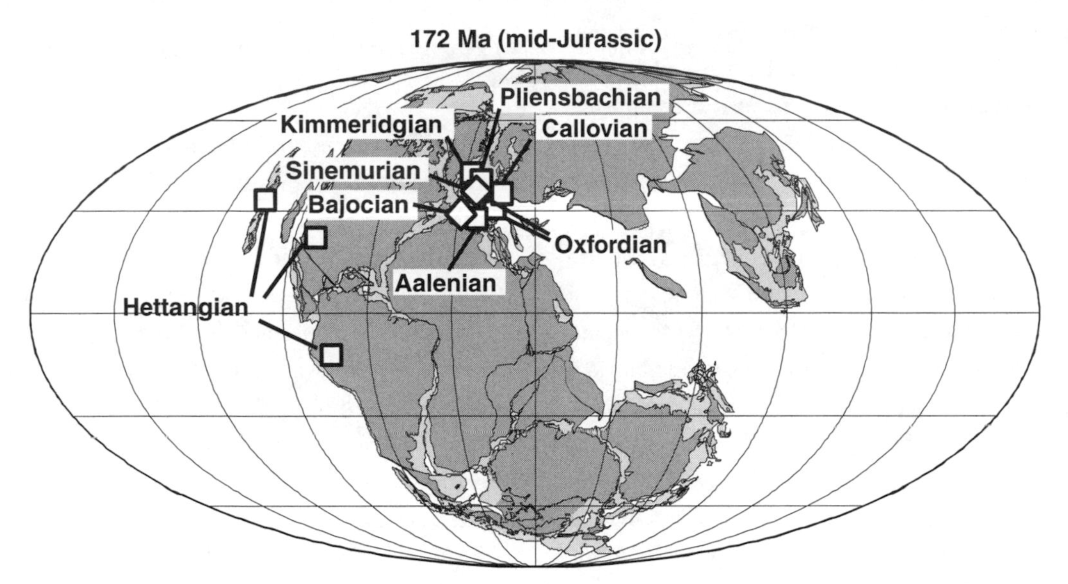

Geographic distribution of Jurassic GSSPs that have been ratified (diamonds) or are candidates (squares) on a mid-Jurassic map (status in January 2004; see Table 2.3). Overlaps in Europe have obscured some GSSPs, and not all candidate sections are indicated (see Table 18.1 for more extensive listing). GSSPs for the base-Jurassic, Late Jurassic stages, and some Middle Jurassic stages are undefined. The projection center is at 30° E to place the center of the continents in the center of the map.

Dinosaurs dominated the land surface. Ammonites are the main fossils for correlating marine deposits. Pangea supercontinent began to break up, and at the end of the Middle Jurassic the Central Atlantic was born. Organic-rich sediments in several locations eventually became the source rocks helping to fuel modern civilization.

18.1 HISTORY AND SUBDIVISIONS

18.1.1 Overview of the Jurassic

The term "Jura Kalkstein" was applied by Alexander von Humboldt (1799) to a series of carbonate shelf deposits exposed in the mountainous Jura region of northernmost Switzerland, and he first recognized that these strata were distinct from the German Muschelkalk (middle Triassic), although he erroneously considered his unit to be older. Alexander Brongniart (1829) coined the term "Terrains Jurassiques" when correlating the "Jura Kalkstein" to the Lower Oolite Series (now assigned to Middle Jurassic) of the British succession. Leopold von Buch (1839) established a three-fold subdivision for the Jurassic. The basic framework of von Buch has been retained as the three Jurassic series, although the nomenclature has varied (Black–Brown–White, Lias–Dogger–Malm, and currently Lower–Middle–Upper).

The immense wealth of fossils, particularly ammonites, in the Jurassic strata of Britain, France, Germany, and Switzerland was a magnet for innovative geologists, and modern concepts of biostratigraphy, chronostratigraphy, correlation, and paleogeography grew out of their studies. Alcide d'Orbigny (1842–51, 1852), a French paleontologist, grouped the Jurassic ammonite and other fossil assemblages of France and England into ten main divisions, which he termed "étages" (stages). Seven of d'Orbigny's stages are used today, but none of them has retained its original stratigraphic range. Simultaneously,

A Geologic Time Scale 2004, eds. Felix M. Gradstein, James G. Ogg, and Alan G. Smith. Published by Cambridge University Press.

Quenstedt (1848) subdivided each of the three Jurassic series of von Buch of the Swabian Alb of southwestern Germany into six lithostratigraphic subdivisions, which he characterized by ammonites and other fossils and denoted by Greek letters (*alpha–zeta*; Geyer and Gwinner, 1979). Alfred Oppel (1856–8), Quenstedt's pupil, subdivided the Jurassic stages into biostratigraphic zones, was the first to correlate Jurassic units successfully among England, France, and southwestern Germany, and modified d'Orbigny's stage framework.

Ammonites have provided a high-resolution correlation and subdivision of Jurassic strata throughout the globe (e.g. Arkell, 1956). The bases of nearly all Jurassic stages and substages are traditionally assigned to the base of ammonite zones in marginal-marine sections in western Europe (e.g. Oppel, 1856–8), and this philosophy was formalized at the Colloque du Jurassique à Luxembourg 1962 (Maubeuge, 1964; see also Morton, 1974), where the majority of the current suite of eleven Jurassic stages were defined in terms of component ammonite zones. Therefore, the process of assigning bases of Jurassic stages at GSSPs continues this historical practice, in which the GSSP placement is commonly locked into recognizing or defining the basal ammonite horizon of the lowest component zone. However, much of the historical subdivision of the Jurassic was limited to shallow-marine deposits of the northwest European region (England to southwest Germany), therefore, establishing reliable high-resolution correlation to tropical (Tethyan), Pacific, deep-sea, continental, and other settings has commonly remained tenuous. In particular, this difficulty in global correlation has frustrated efforts to define with GSSPs both the base and the top of the Jurassic and the bases of the Kimmeridgian and Tithonian stages.

Detailed reviews of the history, subdivisions, biostratigraphic zonations, and correlation of individual Jurassic stages are compiled in several sources, including Arkell (1933, 1956), Cope *et al.* (1980a,b), Harland *et al.* (1982, 1990), Krymholts *et al.* (1982, 1988), Burger (1995), and Groupe Français d'Etude du Jurassique (1997), and our brief summaries have been distilled from their narratives.

18.1.2 Lower Jurassic

A marine transgression in northwest Europe during the latest Triassic and earliest Jurassic resulted in widespread clay-rich calcareous deposits. These distinctive strata in southwest Germany were called the Black Jurassic (schwarzen Jura) by von Buch (1839), and were called Lias in southern England by Conybeare and Phillips (1822). The base of the historical Hettangian Stage is the initial influx of ammonites into southern England during the early stages of the transgression. This series was subdivided into three stages (Sinemurian, Liasian, and Toarcian) by d'Orbigny (1842–51, 1852), then Oppel (1856–8) replaced the Liasian with the Pliensbachian Stage and Renevier (1864) separated the lower Sinemurian as a distinct Hettangian Stage. Widespread hiatuses or condensation horizons mark the bases of the classical Sinemurian, Pliensbachian, and Toarcian stages.

TRIASSIC–JURASSIC BOUNDARY AND THE HETTANGIAN STAGE

The original Sinemurian Stage of d'Orbigny (1842–51, 1852) extended to the base of the Jurassic. Indeed, the Lower Jurassic tentatively included the Rhaetian (Bonebed of southwest Germany, portions of Penarth Beds in England, Rhätische Gruppe of German and Austrian Alps, etc.), which is now assigned to the uppermost Triassic. Overlying this basal unit, Oppel (1856–8) assigned the base of his Jurassic to the lowest ammonite assemblage which is characterized by the *planorbis* species, and referred to characteristic coastal sections in southern England including Lyme Regis in Dorset and Watchet in Somerset.

Renevier (1864) proposed the Hettangian Stage to encompass the *Psiloceras planorbis* and *Schlotheimia angulatus* ammonite zones as interpreted by Oppel. The stage was named after a quarry near the village of Hettange-Grande in Lorraine (northeastern France), 22 km south of Luxembourg, although the strata in this locality are primarily sandstone with no fossils in the lowermost part.

The latest Triassic and Triassic–Jurassic boundary interval span one of the five most significant mass extinctions of the Phanerozoic, including termination of conodonts and major declines of ammonites and bivalves (e.g. Hallam, 1996; Bloos, 1999). This progressive decline, coupled with the low-diversity survivor fauna and transgressive facies migration during the early Hettangian, has greatly limited the choice of markers for defining the base of the Jurassic. The base of the Hettangian is traditionally assigned to the first occurrence of the smooth *planorbis* group within the ammonite genus *Psiloceras*, which are ubiquitous from the eastern Pacific and Tethys to the European Boreal province. Ammonite diversity was very low in late Rhaetian time (*Choristoceras marshi* Zone), and the Hettangian genus *Psiloceras* must be derived from the Triassic genera of the family Discophyllitidae, which lives mainly in the open sea (von Hillebrandt, 1997). The Triassic–Jurassic boundary, as recognized in the marine realm, is within the earliest stages of a transgression following a major sequence boundary (He1 in

Fig. 18.1) and eustatic lowstand (Hesselbo and Jenkyns, 1998; Hallam and Wignall, 1999).

The age of the Triassic–Jurassic boundary is constrained by a U–Pb zircon age of 199.6 ± 0.3 Ma on a tuff layer in the uppermost Rhaetian (top of Triassic) on Kunga Island (Pálfy *et al.*, 2000a). A floral turnover and peak in tetrapod extinction in eastern North America, that had been considered to coincide with the Triassic–Jurassic boundary (e.g. Fowell and Olsen, 1993), has an age no younger than 200.6 Ma; therefore, this continental level appears to represent part of the progressive loss of diversity within the latest Triassic (Pálfy *et al.*, 2000a,b). Olsen *et al.* (2003) favor an age of 202 Ma for this "continental Triassic–Jurassic boundary," based on the average of ages from the overlying basalts in the Newark Basin successions.

There are four main candidates for the placement of the base-Jurassic GSSP (Warrington, 1999, 2003; Table 18.1): (1) Chilingote, Peru, on the west side of the Utcubamba Valley; (2) southeast shore of Kunga Island, Queen Charlotte Islands, British Columbia, Canada; (3) Muller or New York Canyon area, Gabbs Valley Range, Nevada; and (4) St. Audrie's Bay, Somerset, England. Only the Peru and Nevada sections contain ammonite assemblages of both the uppermost Rhaetian and lowermost Hettangian; but St. Audrie's Bay has magnetostratigraphy and Kunga Island has dated tuff layers.

There is no accepted grouping into substages of the three Hettangian ammonite zones (*Psiloceras planorbis*, *Alsatites liasicus*, and *Schlotheimia angulata*).

SINEMURIAN

History, definition, and boundary stratotype The Sinemurian Stage was named by d'Orbigny (1842–51, 1852) after the town of Semur-en-Auxois (Sinemurum Briennense castrum in Latin) in the Cote d'Or department of eastern central France. After the establishment of the Hettangian Stage removed the lower ammonite zones (Renevier, 1864), the base of the Sinemurian was traditionally assigned to the proliferation of the Arietitidae ammonite group, particularly the lowest occurrence of the early genera *Vermiceras* and *Metophioceras* (base of *Metophioceras conybeari* subzone of the *Arietites bucklandi* Zone. However, the stage boundary was never defined by a generally accepted species or assemblage (Sinemurian Boundary Working Group, 2000). In addition, a gap exists between the Hettangian and Sinemurian throughout most of northwest Europe.

Only in rapidly subsiding troughs in western Britain was sedimentation continuous across the boundary interval. Therefore, the boundary GSSP was placed in inter-bedded limestone and claystone at coastal exposures near East Quantoxhead, Somerset, England (Page *et al.*, 2000; Sinemurian Boundary Working Group, 2000; Bloos and Page, 2002; Table 18.1). The GSSP is at the lowest occurrence of arietitid ammonite genera *Vermiceras* and *Metophioceras*. This level is just below the highest occurrence of the ammonite genera *Schlotheimia* that is characteristic of the uppermost Hettangian. This turnover of ammonite genera is a global event that marks the boundary interval (Sinemurian Boundary Working Group, 2000; Bloos and Page, 2002).

Sinemurian substages The Sinemurian has two substages. The Colloque du Jurassique à Luxembourg 1962 (Maubeuge, 1964) assigned the base of an upper stage, called Lotharingian (named by Haug, 1910, after the Lorraine region of France), to the base of the *Caenisites turneri* ammonite zone. However, current usage follows Oppel (1856–8) in assigning the base of the Lotharingian substage at the base of the overlying *Asteroceras*

Figure 18.1 Jurassic time scale with selected biostratigraphic zonations, magnetic polarity chrons, and major depositional sequences. The primary absolute-age stratigraphic scales are the ammonite zonations of northwest Europe for Hettangian through Bajocian and of the sub-Mediterranean province for the Bathonian stage (modified from J. Thierry in Hardenbol *et al.*, 1998, pp. 776–777), and the M-sequence magnetic polarity chrons for Callovian through Tithonian stages. Ages of stage boundaries and other stratigraphic events are from their direct calibration to the primary stratigraphic scale (e.g. magnetostratigraphic correlation of proposed basal-Tithonian ammonite zone boundary) or extrapolated from published correlation estimates (e.g. Mesozoic chronostratigraphy charts of Hardenbol *et al.*, 1998) – see text for details. Uncertainty estimates for stage boundaries are given at 95% confidence limits. Dashed lines denote relatively uncertain calibrations of other biostratigraphic events to the primary scale, or intervals in which ammonite zones have been arbitrarily scaled proportional to the relative number of subzones. Most subzonal units are omitted, and only a generalized ammonite stratigraphy is given for some intervals (see biostratigraphic chart series in Hardenbol *et al.*, 1998, for full listing and correlation web). Ammonite biozone names and associated assigned ages are summarized in Table 18.2. Major flooding or regressive trends of depositional sequences of northwest Europe are labeled at the sequence boundary immediately preceding the maximum lowstand of the respective third-order sequence (extracted from Hardenbol *et al.*, 1998). A color version of this figure is in the plate section.

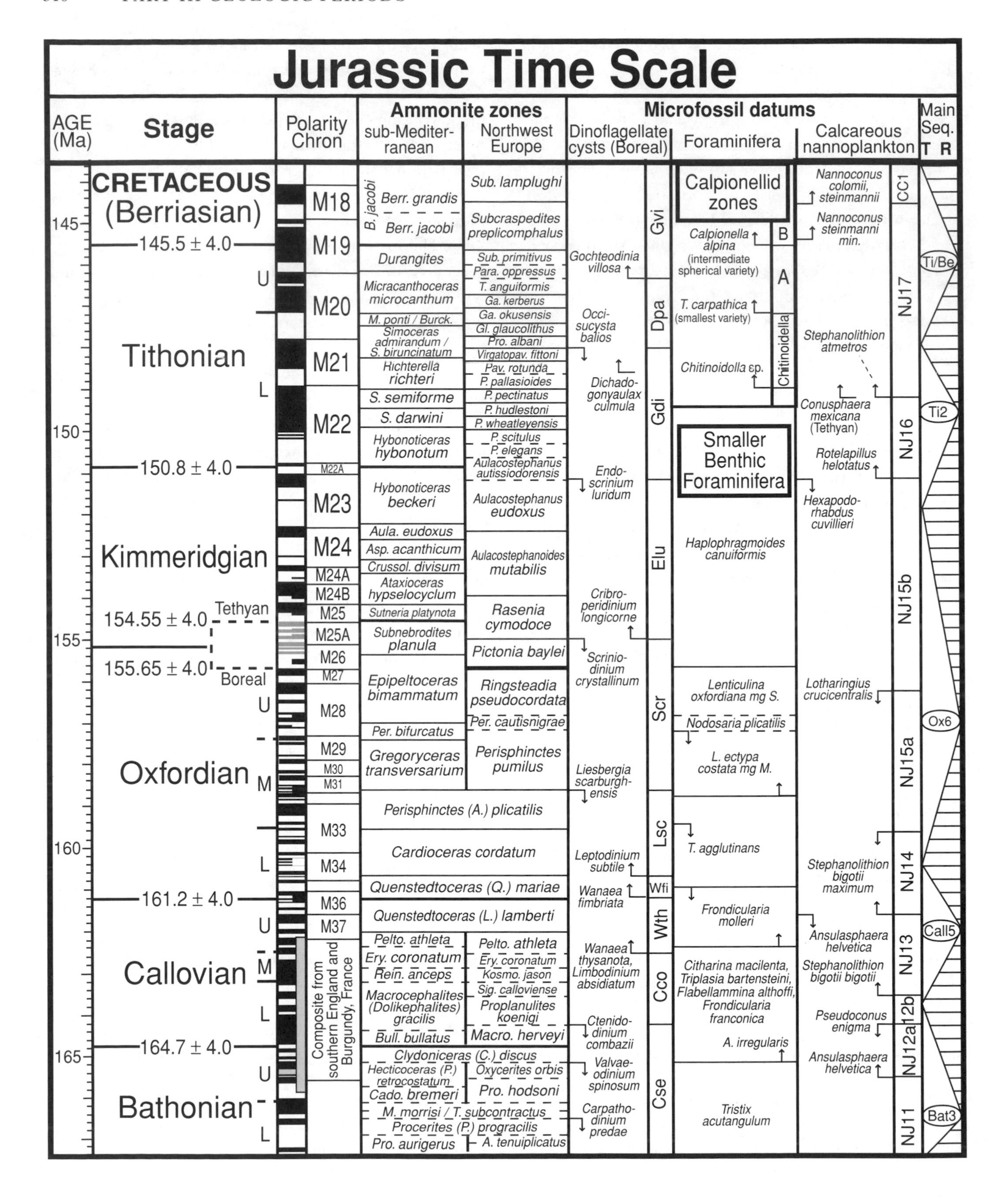
Jurassic Time Scale
AGE (Ma)
Stage
Polarity Chron
Ammonite zones
sub-Mediterranean
Northwest Europe
Microfossil datums
Dinoflagellate cysts (Boreal)
Foraminifera
Calcareous nannoplankton
Main Seq.
T R
CRETACEOUS (Berriasian)
145.5 ± 4.0
Tithonian
U
L
150.8 ± 4.0
Kimmeridgian
154.55 ± 4.0
Tethyan
155.65 ± 4.0
Boreal
Oxfordian
M
161.2 ± 4.0
Callovian
164.7 ± 4.0
Bathonian
145
150
155
160
165
M18
M19
M20
M21
M22
M22A
M23
M24
M24A
M24B
M25
M25A
M26
M27
M28
M29
M30
M31
M33
M34
M36
M37
Composite from southern England and Burgundy, France
B. jacobi
Berr. grandis
Berr. jacobi
Durangites
Micracanthoceras microcanthum
M. ponti / Burck.
Simoceras admirandum / S. biruncinatum
Richterella richteri
S. semiforme
S. darwini
Hybonoticeras hybonotum
Hybonoticeras beckeri
Aula. eudoxus
Asp. acanthicum
Crussol. divisum
Ataxioceras hypselocyclum
Sutneria platynota
Subnebrodites planula
Epipeltoceras bimammatum
Per. bifurcatus
Gregoryceras transversarium
Perisphinctes (A.) plicatilis
Cardioceras cordatum
Quenstedtoceras (Q.) mariae
Quenstedtoceras (L.) lamberti
Pelto. athleta
Ery. coronatum
Rein. anceps
Macrocephalites (Dolikephalites) gracilis
Bull. bullatus
Clydoniceras (C.) discus
Hecticoceras (P.) retrocostatum
Cado. bremeri
M. morrisi / T. subcontractus
Procerites (P.) progracilis
Pro. aurigerus
Sub. lamplughi
Subcraspedites preplicomphalus
Sub. primitivus
Para. oppressus
T. anguiformis
Ga. kerberus
Ga. okusensis
Gl. glaucolithus
Pro. albani
Virgatopav. fittoni
Pav. rotunda
P. pallasioides
P. pectinatus
P. hudlestoni
P. wheatleyensis
P. scitulus
P. elegans
Aulacostephanus autissiodorensis
Aulacostephanus eudoxus
Aulacostephanoides mutabilis
Rasenia cymodoce
Pictonia baylei
Ringsteadia pseudocordata
Per. cautisnigrae
Perisphinctes pumilus
Pelto. athleta
Ery. coronatum
Kosmo. jason
Sig. calloviense
Proplanulites koenigi
Macro. herveyi
Oxycerites orbis
Pro. hodsoni
A. tenuiplicatus
Gochteodinia villosa
Occisucysta balios
Dichadogonyaulax culmula
Endoscrinium luridum
Cribroperidinium longicorne
Scriniodinium crystallinum
Liesbergia scarburghensis
Leptodinium subtile
Wanaea fimbriata
Wanaea thysanota, Limbodinium absidiatum
Ctenidodinium combazii
Valvaeodinium spinosum
Carpathodinium predae
Gvi
Dpa
Gdi
Elu
Scr
Lsc
Wfi
Wth
Cco
Cse
Calpionellid zones
Calpionella alpina (intermediate spherical variety)
B
A
T. carpathica (smallest variety)
Chitinoidolla sp.
Chitinoidella
Smaller Benthic Foraminifera
Haplophragmoides canuiformis
Lenticulina oxfordiana mg S.
Nodosaria plicatilis
L. ectypa costata mg M.
T. agglutinans
Frondicularia molleri
Citharina macilenta, Triplasia bartensteini, Flabellammina althoffi, Frondicularia franconica
A. irregularis
Tristix acutangulum
Nannoconus colomii, steinmannii
Nannoconus steinmanni min.
Stephanolithion atmetros
Conusphaera mexicana (Tethyan)
Rotelapillus helotatus
Hexapodorhabdus cuvillieri
Lotharingius crucicentralis
Stephanolithion bigotii maximum
Ansulasphaera helvetica
Stephanolithion bigotii bigotii
Pseudoconus enigma
Ansulasphaera helvetica
CC1
NJ17
NJ16
NJ15b
NJ15a
NJ14
NJ13
NJ12b
NJ12a
NJ11
Ti/Be
Ti2
Ox6
Call5
Bat3

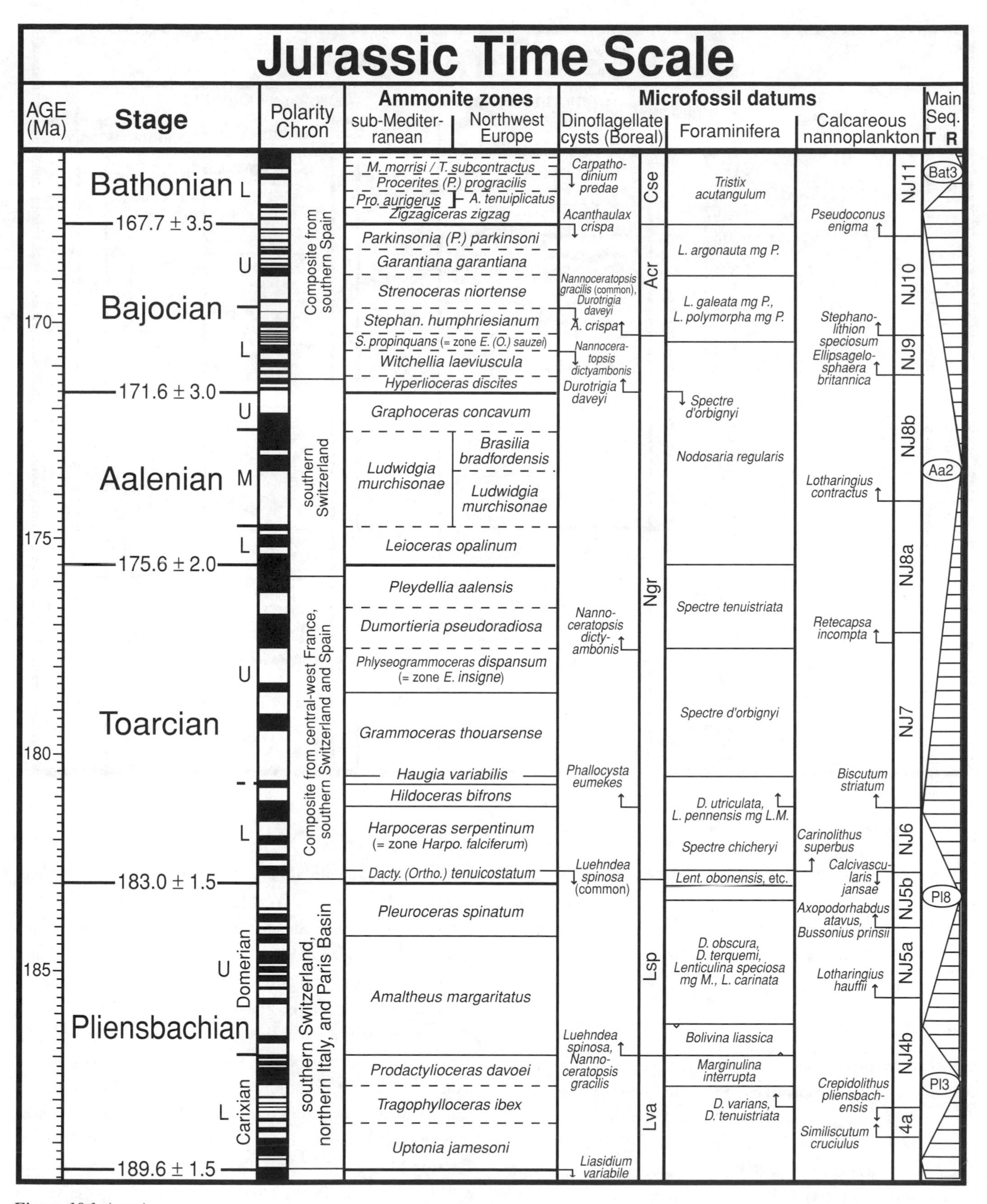

Figure 18.1 (*cont.*)

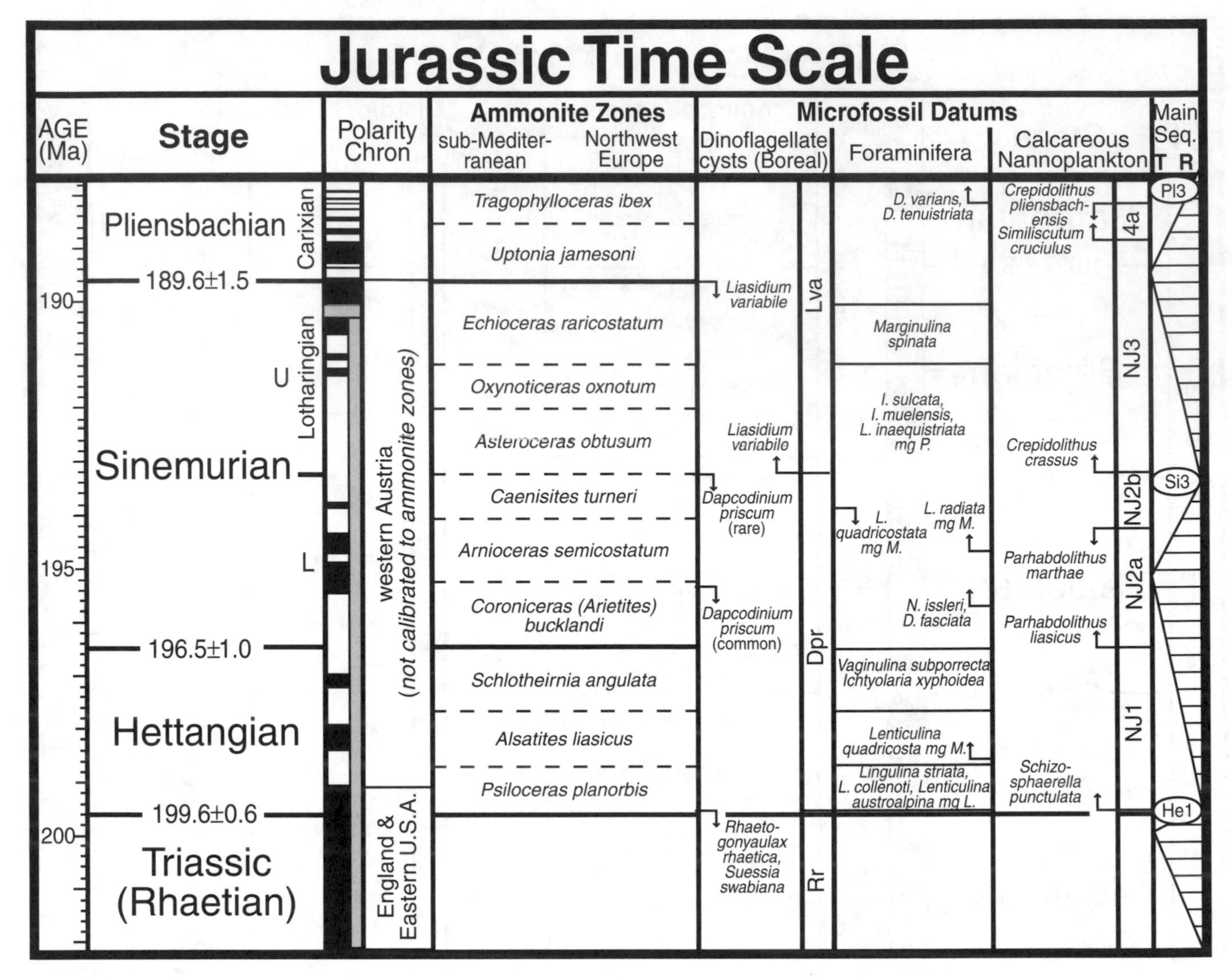

Figure 18.1 (*cont.*)

obtusum Zone (e.g. Krymholts *et al.*, 1982–8; Groupe Français d'Etude du Jurassique, 1997). The lower substage does not have a secondary name, and there is no recommendation for a potential GSSP for the substage boundary.

PLIENSBACHIAN

History, definition, and boundary stratotype The Pliensbachian Stage was proposed by Oppel (1856–8) to replace the Liasian stage of d'Orbigny, which lacked a type locality for its base. The stage was named after the outcrops along the Pliensbach stream near the village of Pliensbach (Geppingen, 35 km southeast of Stuttgart) in the Baden–Württemberg district of Germany. Even though this section lies unconformably on the underlying Sinemurian, the lowest ammonite subzone (*Phricodoceras taylori* subzone of the *Uptonia jamesoni* Zone) in this section is traditionally used as the base of the Pliensbachian Stage (e.g. Dean *et al.*, 1961; Meister, 1999a,b).

At this level, the Psiloceratoidea ammonites, which dominated the Hettangian and Sinemurian, disappear and the Eoderoceratoidea superfamily diversifies and dominates the northeast European fauna of the Pliensbachian Stage (Meister *et al.*, 2003). This faunal event occurs globally, but a stratigraphic gap between the Pliensbachian and Sinemurian sequences is a common feature. Of 27 regions considered by the Pliensbachian boundary working group, only a single candidate in Yorkshire, England, was satisfactory for a potential GSSP (Meister, 1999a,b; Meister *et al.*, 2003). At the clay-rich coastal section of Wine Haven at Robin Hood's Bay, Yorkshire, the GSSP coincides with the lowest ammonite occurrences of *Bifericeras donovani* species and of *Apoderoceras* genera (Table 18.1). A section at Aselfingen in the Baden–Württemberg district of

Table 18.1 *Jurassic stage definitions (GSSP assignments or status) and informal division into substages. Stages are ordered stratigraphically (youngest at top)*

Stage	Stage boundary *Substage base*	Global Boundary Stratotype Section and Point: Status	Location and point	Primary markers	Other correlations	Comments	References (GSSP, correlations)
Upper Jurassic *Tithonian*	*Upper*	Informal usage within sub-Mediterranean province (Tethyan faunal domain)		Base of *Micracanthoceras microcanthum* ammonite zone.	At this level there is a major turnover in ammonite assemblages and calpionellid microfossils become important in the biostratigraphic correlation of pelagic limestone. Calibrated to base of normal-polarity Chron M20n.		Groupe Français d'Étude du Jurassique, 1997.
	Tithonian/ Kimmeridgian	Candidate sections	Southeastern France – Crussol mountain on the Rhône river just west of Valence, and a quarry at Canjuers – or in Swabia region of southern Germany.	Ammonite, simultaneous lowest occurrence of the ammonites *Hybonoticeras* aff *hybonotum* and *Glochiceras lithographicum* (base of *H. hybonotum* Zone), immediately followed by the lowest occurrence of the *Gravesia* genera.	Base of normal-polarity Chron M22An.	Traditional placement of Kimmeridgian-Tithonian boundary in sub-boreal realm is at base of *Pectinatites (Virgatosphinctoides) elegans* ammonite zone, but this datum is known to be younger than the sub-Mediterranean definition of the boundary.	Atrops, 1982, 1994
Kimmeridgian	*Upper*	Informal usage within sub-Boreal province (Boreal faunal domain)		Base of *Aulacostephanoides mutabilis* ammonite zone.	Probably near the base of normal-polarity Chron M24Bn.	In the sub-Mediterranean realm, the base of an Upper Kimmeridgian substage is typically assigned to the base of the *Aspidoceras acanthicum* Zone. Magnetostratigraphy suggests that this level (lower part of Chron M24r) is about 0.9 myr younger than the sub-Boreal assignment (Ogg, Coe & Atrops, in preparation).	Groupe Français d'Étude du Jurassique, 1997.
	Kimmeridgian/ Oxfordian (traditional Tethyan placement)	Candidate sections	Crussol mountain on the Rhône river just west of Valence, and Châteauneuf d'Oze in the Haute Provence district	Ammonite, base of *Sutneria platynota* Zone.	Just above base of reversed-polarity Chron M25r.	Crussol candidate yielded magnetostratigraphy. Châteauneuf d'Oze candidate did not yield a magnetostratigraphy, but appears more suitable for microfossil biostratigraphy (dinoflagellate cysts) and chemostratigraphy. Magnetostratigraphy, sequence stratigraphy and ammonite constraints indicate that the base of the *S. platynota* zone approximately correlates with the middle of the *Rasenia cymodoce* Zone of the sub-Boreal province (Boreal faunal realm) (Ogg & Coe, in preparation).	Melendez & Atrops, 1999; Atrops, 1994.

(*cont.*)

Table 18.1 (*cont.*)

		Global Boundary Stratotype Section and Point					
Stage	Stage boundary *Substage base*	Status	Location and point	Primary markers	Other correlations	Comments	References (GSSP, correlations)
	Kimmeridgian/ Oxfordian (traditional Boreal placement)	Candidate section	Staffin Bay (Isle of Skye, northwest Scotland)	Ammonite, concident bases of *Pictonia baylei* Zone (Subboreal province) and *Amoeboceras bauhini* Zone (Boreal province)	The base of the *P. baylei* zone is commonly a minor hiatus at a maximum flooding surface (Coe, 1995). In the Staffin Bay section, the Kimmeridgian–Oxfordian boundary is just above the base of a reversed-polarity zone assigned to lower part of polarity Chron M26r (Ogg & Coe, 1997 and in preparation).	Magnetostratigraphy, sequence stratigraphy and ammonite constraints indicate that the base of the *P. baylei* zone approximately correlates with the base of the *Taramelliceras hauffianum* Subzone of uppermost *Epipeltoceras bimammatum* Zone of the sub-Mediterranean province (Tethyan faunal realm) (Ogg & Coe, in preparation).	Melendez & Atrops, 1999; Wright, 1973, 1989; Riding & Thomas, 1997; Wierzbowski, 2002, 2003.
Oxfordian	*Upper*	Informal usage within sub-Boreal province (Boreal faunal domain)		Base of *Perisphinctes cautisnigrae* ammonite zone.		In the sub-Mediterranean province (Tethyan domain), the base of an Upper Oxfordian substage is commonly assigned to the base of the *Perisphinctes (Dichotomoceras) bifurcatus* Zone, and this level is probably slightly older than the Boreal placement. Base of *P. bifucatus* is within a normal-polarity zone that is correlated to magnetic Chron M29n.	Groupe Français d'Étude du Jurassique, 1997.
	Middle	Informal European usage		Base of *Perisphinctes (Arisphinctes) plicatilis* ammonite zone.	Base of *P. plicatilis* Zone is within a brief normal-polarity magnetic zone that is correlated to marine magnetic anomaly M33b(n) (Ogg & Coe, 1997 and in preparation).		Groupe Français d'Étude du Jurassique, 1997.
	Oxfordian/ Callovian	Site of GSSP is undecided.	Leading candidates in Tethyan realm are Thuoux and Savournon near Serres (Provence, Chaines Subalpin) in southeast France. A candidate in Boreal realm is a coastal section (Ham Cliff) near Weymouth (Dorset) in southern England.	Ammonite, *Brightia thuouxensis* Horizon at base of the *Cardioceras scarburgense* Subzone (*Quenstedtoceras mariae* Zone).	Boundary interval is contact of range zone of *Quenstedtoceras mariae* to underlying range zone of *Q. lamberti*. The French candidate section has dinoflagellate markers, but no other macrofossils or microfossils. The boundary in England coincides with a maximum flooding surface (Coe, 1995) and is within a brief normal-polarity magnetic zone that is correlated to M-sequence marine magnetic anomaly M36An (Coe & Ogg, in preparation).	Candidate GSSP sections in France did not preserve a primary magnetostratigraphy, and other faunal groups or chemostratigraphy are not documented.	Melendez, 1999; Fortwengler & Marchand, 1994.

Middle Jurassic							
Callovian	*Upper*	Informal European usage		Base of *Peltoceras (Peltoceras) athleta* ammonite zone*(Kosmoceras (Lobokosmoceras) phaeinum* Subzone).		In the sub-Mediterranean province, the Middle/Upper Callovian boundary is placed at the base of the *Peltoceras (Peltoceras) athleta* Zone*(Hecticoceras (Orbignyiceras) trezeense* Subzone), or approximately a subzone higher than in the sub-Boreal province.	Groupe Français d'Étude du Jurassique, 1997.
	Middle	Informal usage within sub-Boreal province (Boreal faunal domain)		Base of *Kosmoceras (Zugokosmoceras) jason* ammonite zone (at base of *Kosmoceras (Zugokosmoceras) medea* Subzone).	Lower/Middle Callovian substage boundary coincides with a moderate sequence boundary (Call3 of Hardenbol *et al.*, 1998).	In the sub-Mediterranean province, the Lower/Middle Callovian boundary is placed at the base of the *Reineckeia anceps* ammonite zone (*Reineckeia stuebeli* Subzone), which is considered approximately coeval with the sub-Boreal substage boundary placement.	Groupe Français d'Étude du Jurassique, 1997.
	Callovian/ Bathonian	Candidate section	Excavated section of Macrocephalen-Oolith formation in forest preserve "Quellgebiet des Roschbachs" in the upper Eyach valley, about 1 km west of Pfeffingen village in the Albstadt district of the Swabian Alb (30 km south of Tübingen, southwest Germany). *[See footnote a]*	Ammonite, lowest occurrence of the genus *Kepplerites* (Kosmoceratidae), which defines the *Kepplerites (Kepplerites) keppleri* horizon at base of *K. keppleri* Subzone of *Macrocephalites herveyi* Zone in the sub-Boreal province (Great Britain to southwest Germany).	In the sub-Mediterranean province (southern Paris Basin to north Africa and Italy, the basal Callovian zone is the *Bullatimorphites (Kheraiceras) bullatus* Zone defined by the range of the index species. A major latest Bathonian sequence boundary (Bat5 of Hardenbol *et al.*, 1998) is widespread in lower *C. discus* Subzone, and a minor sequence boundary (Call0) coincides with the Bathonian-Callovian boundary level.	The Macrocephalen-Oolith formation, a condensed facies of iron-oolite-bearing clay to marly limestone, is easily eroded, therefore the complete "Roschbachs" section is only exposed by escavation, then reburied after sampling to prevent removal of its rich ammonite fauna by amateur fossil collectors. The flat-lying section is similar to the profile of Macrocephalen-Ooliths diagrammed by Dietl (1994, fig. 4), but the relative thicknesses are different. *[See footnote a]*	Dietl, 1994; Callomon, 1999; Callomon & Dietl, 2000
Bathonian	*Upper*	Informal usage within sub-Mediterranean province (Tethyan faunal domain)		Base of *Hecticoceras (Prohecticoceras) retrocostatum* ammonite zone.	Approximately coincides with minor sequence boundary (Bat4 of Hardenbol *et al.*, 1998) in NW European basins.	In the Northwest European province (Boreal domain), a substage boundary is commonly assigned to the base of the *Procerites (Procerites) hodsoni* Zone, which is a significantly older level. This level is just above a major sequence boundary (Bat3 of Hardenbol *et al.*, 1998, see Figure 12.1) in NW European basins.	Groupe Français d'Étude du Jurassique, 1997.
	Middle	Informal European usage		Base of *Procentes progracilis* ammonite zone.			

(*cont.*)

Table 18.1 (*cont.*)

Stage	Stage boundary *Substage base*	Global Boundary Stratotype Section and Point: Status	Location and point	Primary markers	Other correlations	Comments	References (GSSP, correlations)
	Bathonian/ Bajocian	Proposed in 1988, but may not be suitable for GSSP.	Ravin du Bès-Bas Auran near Digne, 4 km west of Barrême, Basses-Alpes, southeast France. Proposed GSSP was base of Bed 23 of Sturani (1967) in section of interbedded limestone and marl. Another candidate GSSP is at Cabo Mondego, Portugal.	Ammonite, base of the *Zigzagiceras zigzag* Zone (base of *Parkinsonia (Gonolkites) convergens* Subzone) as marked by the lowest occurrence of *Parkinsonia (G.) convergens*, *P. (P.) pachypleura* and *Morphoceras parvum*.	Just prior to the peak of a major transgression trend in NW Europe	Strata at proposed GSSP did not preserve a primary magnetostratigraphy, are barren of dinoflagellate cysts, and the uppermost Bajocian (*Parkinsonia (Parkinsonia) bomfordi Subzone*) may be absent. A nearby auxiliary GSSP section of La Palud near Castellane places the *P. bomfordi*–*P. convergens* subzone boundary at Bed 44 (rather than Bed 39, as placed in Innocenti *et al.*, 1988) (Mangold, 1999).	Mangold, 1999; Innocenti *et al.*, 1988.
Bajocian	*Upper*	Informal European usage		Base of *Strenoceras (Strenoceras) niortense* ammonite zone.	Major turnover of ammonite genera occurs at this level as *Teloceras* disappear, and Perisphinctaceae, *Leptosphinctes*, *Strenoceras* and *Garantiana (Orthogarantiana)* appear with some overlap.		Krymholts *et al.*, 1982/1988; Groupe Français d'Étude du Jurassique, 1997.
	Bajocian/ Aalenian	Ratified 1996	Cabo Mondego, Portugal (Murtinheira coastal section at the foot of Cabo Mondego cliff, southwest of the village of Murtinheira, 40 km west of Coimbra and 7 km north of Figueira da Foz). GSSP is base of Bed AB11 (section of Henriques *et al.*, 1988, which corresponds to Bed M337 of Henriques *et al.*, 1994) at 77.8 m level as measured from the base of the coastal section in rhythmic alternations of gray limestone and marl.	Ammonite, lowest occurrence of the genus *Hyperlioceras (Toxolioceras)*, which defines the base of the *Hyperlioceras discites* Zone.	The GSSP is just below the lowest occurrences of calcareous nannofossils *Watznaueria communis* and *W. fossacincta*. The GSSP at Capo Mondego coincides with the boundary between a reversed-polarity zone in the uppermost Aalenian to a normal-polarity zone spanning the lowermost *H. discites* Zone (Henriques *et al.*, 1994), which is consistent with a composite magnetic pattern derived from other Aalenian–Bajocian studies (e.g., compilation by Ogg, 1995, shown in Figure 12.1). Boundary is near a major sequence boundary (Bj1 of Hardenbol *et al.*, 1998) in NW European basins.	Auxiliary Stratotype Point is at Bearreraig Bay, about 10 km north of Portree on the eastern coast of the Isle of Skye in western Scotland. The boundary level is at the base of Bed U10 in the lower Udairn Shale Member, 12.4 m above the base of the section as revised by Morton (in Pavia *et al.*, 1995). Base of the Bajocian in this section is marked by radiation of gonyaulacacean dinoflagellate cysts, is within the NJ8b nannofossil Subzone, and is just above the lowest occurrence of inoceramid bivalve *Mytiloceramus polyplocus*.	Henriques *et al.*, 1994; Pavia & Enay, 1997; enhanced CD-ROM from M.H. Henriques (hhenriq@cygnus.ci.uc.pt)
Aalenian	*Upper*	Informal European usage		Base of *Graphoceras concavum* ammonite zone.	Transgression above major lowstand (Aa2) in NW European basins		

	Middle	Informal European usage		Base of *Ludwigia murchisonae* ammonite zone.	Just above a major sequence boundary (Aa1 of Hardenbol *et al.*, 1998) in NW European basins.		
	Aalenian/ Toarcian	Ratified 2000	Fuentelsalz at Nuevalos, Spain, in central sector of the Castelian Branch of the Iberian Range, about 170 km ENE of Madrid and 30 km north of Molina de Aragon). GSSP is base of calcareous Bed 107 within an expanded uppermost Toarcian-lowermost Aalenian succession of flat-lying rhythmic alternations of marl and limestone.	Ammonite, lowest occurrence of genus *Leioceras* (base of *Leioceras opalinum* Zone), which evolved from *Pleydellia*.	Evolution of the ammonite Subfamily Grammoceratinae and Leioceratinae. Boundary interval is within a normal-polarity magnetozone which, with the underlying reversed-polarity magnetozone in the lower part of the *Pleydellia aalensis* zone (uppermost zone of Toarcian). Diversity changes are recorded by brachiopods, bivalves, benthic foraminifera, ostracods and calcareous nannofossils in the Fuentelsaz section; although the most significant faunal events generally take place in the uppermost Toarcian before the boundary (Goy *et al.*, 1996). The boundary interval is the lower part of a minor trangressive systems tract (Hardenbol *et al.*, 1998).	Wittnau at Freiburg in south Germany was the other main GSSP candidate (Ohmert, 1996).	Goy *et al.*, 1994, 1996; Cresta *et al.*, 2001.
Lower Jurassic							
Toarcian	*Upper*	Informal European usage		Base of *Grammoceras thouarsense* ammonite zone.		An alternate two-fold subdivision of Toarcian places an Upper/Lower substage boundary at the base of the *Haugia variabilis* ammonite zone.	Groupe Français d'Étude du Jurassique, 1997
	Middle	Informal European usage		Base of *Hildoceras bifrons* ammonite zone.	Major maximum flooding surface in NW European basins.		Groupe Français d'Étude du Jurassique, 1997
	Toarcian/ Pliensbachian	Location and global correlation debated.	Main candidate profile for the GSSP is Ponte da Trovã-Cruz dos Remedios secton at Peniche, Portugal. Another candidate is at Aït Moussa in the Middle Atlas of Morocco.	Ammonite, lowest occurrence of a diversified *Eodactylites* fauna (Simplex horizon, sensu Goy *et al.*, 1997) with the association *Paltarpites-Tiltoniceras-Eodactylites*, which correlates with the northwest European *Palterpites paltus* horizon/subzone.	Toarcian-Pliensbachian boundary interval is marked by a massive surge of Dactylioceratide (*Eodactylites*) and Hildoceratide ammonite families of Tethyan origin and extinction of Boreal amaltheid family. Base of Toarcian is an important maximum flooding surface above a major sequence boundary (Pl8) and minimum in seawater strontium-isotope ratios.	Widespread condensation or gaps at the base of the Toarcian strata (*Dactylioceras tenuicostatum* ammonite zone) necessitated selection of candidate GSSPs in the Mediterranean region where gaps in the succession are less pronounced	Elmi, 1999, 2003.
Pliensbachian	*Upper (Carixian substage)*	Informal European usage		Base of *Amaltheus margaritatus* ammonite zone.	Appearance of the *Amaltheus* ammonite genera (typically *Amaltheus stokesi*).	The Domerian is the informal name for the lower substage of Pliensbachian.	Colloque du Jurassique à Luxembourg 1962 (1964)

(*cont.*)

Table 18.1 (*cont.*)

Stage	Stage boundary *Substage base*	Global Boundary Stratotype Section and Point: Status	Location and point	Primary markers	Other correlations	Comments	References (GSSP, correlations)
	Pliensbachian/ Sinemurian	ICS voting 2003	Wine Haven section of claystone, Robin Hood's Bay, Yorkshire, England. GSSP is at base of clay Bed 73b, 6 cm above a thin calcareous nodule Bed 72 of Hesselbo and Jenkyns (1995).	Ammonite, lowest occurrences of *Bifericeras donovani* and of genera *Apoderoceras* and *Gleviceras*.	Lowest occurrences of ammonites *Apoderoceras nodogigas, A. leckenbyi, Tetraspidoceras quadrarmatum,* and *P. taylori*. Uppermost Sinemurian has the disappearance of the Echioceratidae ammonite family. Dinoflagellate cysts are absent, microfauna studies have not been published, and magnetostratigraphy is unavailable. The potential boundary level displays a seawater 87Sr/86Sr ratio of 0.707425 and oxygen isotopes from belemnites suggest a local seawater temperature drop of about 5 C (Hesselbo *et al.*, 2000; Meister, 2001). This level is just below a maximum flooding surface in British sections (Hesselbo & Jenkyns, 1998).	Of 27 regions considered by the Pliensbachian boundary working group, only a single candidate in Yorkshire, England, appeared to be satisfactory for a potential GSSP. A section at Aselfingen in the historical area in southwest Germany is condensed limestone and clay with rare ammonites, but allows calibration of secondary markers in ostracods and dinocysts (Meister, 1999a). Boundary corresponds to Strontium 87/86 ratio of 0.707425 ± 0.000021.	Meister, 1999a,b, 2001; Hesselbo *et al.*, 2000, Meister *et al.*, 2003.
Sinemurian	*Upper (Lotharingian substage)*	Informal European usage		Base of *Asteroceras obtusum* ammonite zone.	Just above a major sequence boundary (Si3) in NW European basins.	Lower substage of Sinemurian does not have a secondary informal name.	Krymholts *et al.*, 1982/1988; Groupe Français d'Étude du Jurassique, 1997
	Sinemurian/ Hettangian	Ratified 2000	East Quantoxhead section of interbedded limestone and claystone at the coastal exposures 500m north of court house of village of Quantock's Head, 6 km east of Watchet, southern coast of the Bristol Channel, West Somerset, England. Within a bituminous shale between calcareous claystone Beds 145 and 146 of Whittaker and Green (1983).	Ammonite, lowest occurrence of arietitid genera *Metophioceras s. str.,* and *Vermiceras* (base of *Metophioceras conybearoides* subzone of *Coroniceras (Arietites) bucklandi* ammonite zone). Just below the highest occurrence of the genera *Schlotheimia* that is characteristic of the uppermost Hettangian.	Sinemurian–Hettangian boundary is within a transgressive episode following a latest Hettangian lowstand in British sections (Hesselbo & Jenkyns, 1998). Foraminifer *Lingulina tenera* plex (latest Hettangian) and appearance of *Planularia inaequistriata* and the *Frondicularia terquemi* plexus group (basal Sinemurian). There are no conspicuous changes in ostracods, palynology, pelecypods or brachiopods across the boundary interval, and magnetostratigraphy was not successful.	This turnover of ammonite genera is a world-wide event that marks the boundary interval (Sinemurian Boundary Working Group, 2000).	Page *et al.*, 2000; Sinemurian Boundary Working Group, 2000; Bloos & Page, 2002.

Hettangian	*No recommended substages*						
	Hettangian/Rhaetian (= base of Jurassic)	Debated criteria and location	Four main candidates: (1) Chilingote, Peru on the west side of the Utcubamba Valley (Hillebrandt, 1997, 1994), (2) southeast shore of Kunga Island, Queen Charlotte Islands, British Columbia, Canada (Tipper *et al.*, 1994; Carter *et al.*, 1998), (3) New York Canyon area, Gabbs Valley Range, Nevada (Guex *et al.*, 1997; Guex, 1995), and (4) St. Audrie's By, Somerset, England (Warrington *et al.*, 1994; Page and Bloos, 1998).	Ammonite, first occurrence of the smooth *planorbis* group within the ammonite genus *Psiloceras*.	Extinction of conodonts. Radiolarian assemblages (top of the latest Triassic *Globolaxtorum tozeri* zone and base of earliest Hettangian *Conoptum merum* zone) (e.g., Carter *et al.*, 1998). Earliest stage of transgression after major eustatic lowstand.	The *Planorbis* ammonite group is described under several names *(planorbis, tilmanni, pacificum, calliphyllum,* etc.), which probably characterize local morphologic varieties of a single species or closely related coeval species (Guex *et al.*, 1997). Peru and Nevada sections contain ammonite assemblages of both the uppermost Rhaetian and lowermost Hettangian. Only the uppermost Triassic of the Somerset section has a published magnetic polarity stratigraphy (Briden and Daniels, 1999), but an expanded bio-magnetostratigraphy for that section is forthcoming (cited by Warrington and Bloos, 2001). Age of the Triassic–Jurassic boundary is constrained by a U-Pb zircon age of 199.6 ± 0.3 Ma on a tuff layer in the uppermost Rhaetian (top of Triassic) at the Kunga Island (British Columbia) section (Pálfy *et al.*, 2000a).	Bloos, 1999; Warrington, 1999; Page & Bloos, 1998; Pálfy *et al.*, 2000a.

Footnotes:

[a] *[Callovian/Bathonian candidate GSSP]* Most of the *K. keppleri* Subzone (basal Callovian) is encompassed within 70 cm and overlies a 8-cm-thick bed of uppermost Bathonian (the *hockstetteri* horizon (var. *hockstetteri* of *Clydoniceras discus)* of the *C. discus* Subzone. The boundary stratigraphic interval is bounded by unconformities – the basal Callovian *keppleri* Subzone is overlain by the *Kepplerities (Gowericeras) gowerianus* Subzone of the *Proplanulites koenigi* Zone implying omission of the two upper subzones of the *M. herveyi* Zone, and the uppermost Bathonian *hockstetteri* Horizon overlies the *Hecticoceras (Prohecticoceras) blanazense* Subzone of the *Oxycerites orbis* Zone implying omission of the majority of the *C. discus* Zone. Preliminary magnetostratigraphy of the candidate GSSP section at Roschbachs (Ogg and Dietl, unpublished) suggests significant omission surfaces where polarity zone and ammonite biozone boundaries coincided, including the Bathonian–Callovian contact, although the main polarity pattern may provide a useful secondary correlation tool for the boundary interval into other provinces.

Germany comprises condensed limestone and clay with rare ammonites, but allows calibration of secondary markers in ostracods and dinocysts (Meister, 1999a,b).

Pliensbachian substages The Pliensbachian has two substages. The lower substage of Carixian was named by Lang (1913) after Carixia, the Latin name for Charmouth, France. The upper stage was named Domerian by Bonarelli (1894, 1895) after the type section in the Medolo formation at Monte Domaro in the Lombardian Alps of northern Italy.

The Colloque du Jurassique à Luxembourg 1962 (1964) assigned the boundary between the Carixian and Domerian substages to the base of the *Amaltheus margaritatus* ammonite zone, at the appearance of the *Amaltheus* genera (typically *Amaltheus stokesi*). This level is just below the sequence boundary ("Pl5" of Hardenbol *et al.*, 1998) in British sections (Hesselbo and Jenkyns, 1998; Fig. 18.1).

TOARCIAN

History, definition, and boundary stratotype candidates The Toarcian Stage was defined by d'Orbigny (1842–51, 1852) at the Vrines quarry, 2 km northwest of the village of Thouars (Toarcium in Latin) in the Deux-Sévres region of west-central France. The thin-bedded succession of blue-gray marl and clayey limestone spans the entire Toarcian with 27 ammonite horizons grouped into eight ammonite zones (Gabilly, 1976).

The Toarcian–Pliensbachian boundary interval marks a major extinction event in western Europe among rhynchonellid brachiopods, ostracod fauna, benthic foraminifera, and bivalves, and turnover in ammonites and belemnites, but the extinction event appears to be a phenomenon of regional, not global, extent (Hallam, 1986). The base of the Toarcian is marked by a massive surge of Dactylioceratide (*Eodactylites*) ammonites and extinction of the Boreal amaltheid family. Seawater strontium isotope ratios, which had been declining since the Hettangian, reach a minimum in the latest Pliensbachian.

The base of the Toarcian strata at Thouars, and throughout northwest Europe, is an important maximum flooding surface and associated condensation or gaps above a major sequence boundary (Pl8 in Fig. 18.1) in the *Dactylioceras tenuicostatum* ammonite zone. This widespread hiatus necessitates selection of candidate GSSPs in the Mediterranean region where gaps in the succession are less pronounced (Elmi, 1999). The primary marker of the Toarcian GSSP will be the lowest occurrence of a diversified *Eodactylites* ammonite fauna (Simplex horizon, *sensu* Goy *et al.*, 1996; Table 18.1). The best profile currently available is in Peniche, Portugal, where the Tethyan *Eodactylites* fauna is succeeded by an "English" *Orthodactylites* succession (Elmi, 2003).

Toarcian substages There is no agreement on the number of substages of the Toarcian. A binary subdivision following that by Oppel (1856–8) places a substage boundary at the base of the *Haugia variabilis* ammonite zone, at the appearance of abundant Phymatoceratinae group of ammonites, particularly the *Haugia* genus (e.g. Krymholts *et al.*, 1982–8; Burger, 1995). An alternate three-substage division (e.g. Groupe Français d'Etude du Jurassique, 1997) groups the *Haugia variabilis* and underlying *Hildoceras bifrons* Zones into a Middle Toarcian, and places the limit of an Upper Toarcian at the base of the *Grammoceras thouarsense* Zone. There are no recommendations for a potential GSSP(s) for the substage boundary(s).

18.1.3 Middle Jurassic

The black clays that are typical of the Early Jurassic (*Schwarzer Jura*) are overlain in southwestern Germany by strata containing clayey sandstone and brown-weathering ferruginous oolite. Therefore, these strata were grouped as the Brown Jurassic (*Brauner Jura*) by von Buch (1839), and this lithologic change has been retained in the assignment of the base of the Middle Jurassic (the base of the Aalenian). This Middle Jurassic interval is characterized by shallow-marine carbonates and siliciclastics in southern England, which comprised the Lower Oolite group of Conybeare and Phillips (1822) or the expanded "Bathonian Stage" of d'Omalius d'Halloy (1868, p. 470). The lower portion (Lower Oolite or Dogger strata) of this "Bathonian" of southern England was classified as a separate Bajocian Stage by d'Orbigny (1842–51, 1852). In turn, Mayer-Eymar (1864) separated the lower portion of d'Orbigny's "Bajocian" into a distinct Aalenian Stage.

The upper limit of the Middle Jurassic or Dogger of Oppel (1856–8) was placed at the base of the Kellaway Rock of England, hence at the base of the associated Callovian Stage of d'Orbigny (1842–51, 1852). The Colloque du Jurassique à Luxembourg 1962 (Maubeuge, 1964) reassigned the Callovian Stage into the Middle Jurassic series as preferred by Arkell (1956).

The bases of the Aalenian and Bajocian stages (and probably soon the Bathonian) have been marked by GSSPs in expanded sections of rhythmic alternations of limestone and marl. The placement of a base for the Callovian Stage has been hindered by a ubiquitous condensation or hiatuses in strata of northwest Europe.

AALENIAN

History, definition, and boundary stratotype The Aalenian Stage was proposed by C. Mayer-Eymar (1864) for the lowest part of the "Braunjura" in the vicinity of Aalen at the northeastern margin of the Swabian Alb (southwestern Germany) where iron ore was mined from the associated ferruginous oolite sandstones (Dietl and Etzold, 1977; Rieber, 1984). His lithologic-based definition truncated the Bajocian Stage of d'Orbigny (1842–51, 1852) at the base of the *Sonninia sowerbyi* ammonite zone.

The biostratigraphic recognition of the base of the Middle Jurassic was traditionally assigned to the evolution of the ammonite subfamily Grammoceratinae and Leioceratinae, in particular the first occurrence of species of the genus *Leioceras*, which evolved from *Pleydellia*. The Aalenian GSSP in the Fuentelsaz section in Spain corresponds to this ammonite marker (Goy *et al.*, 1994, 1996; Cresta *et al.*, 2001; Table 18.1). This section of alternating marl and limestone yielded a magnetostratigraphy that could be correlated to a composite magnetic pattern derived from other sections in Europe. A secondary reference section for the base of Aalenian is at Wittnau, near Freiburg, south Germany (Ohmert, 1996).

Aalenian substages The four ammonite zones of the Aalenian are grouped into three substages: the Lower Aalenian is equivalent to the *Leioceras opalinum* Zone, the Middle Aalenian comprises the *Ludwigia murchisonae* and *Brasilia bradfordensis* Zones, and the Upper Aalenian is the *Graphoceras concavum* Zone.

BAJOCIAN

History, definition, and boundary stratotype The Bajocian Stage was named by d'Orbigny (1842–51, 1852) after the town of Bayeux, Normandy (Bajoce in Latin). The abandoned quarries from which the stage was first described are now overgrown, and the nearby coastal cliff section of Les Hachettes indicates that most of the lower Bajocian is a hiatus and erosional surface, and the upper Bajocian is largely condensed in a 15-cm-thick layer (Rioult, 1964). Ammonite lists of d'Orbigny indicate that he erroneously assigned species of the upper Toarcian to the lower Bajocian and vice versa. This confusion was one reason why Mayer-Eymar (1864) distinguished the Aalenian Stage for the deposits between the Toarcian and Bajocian.

The Colloque du Jurassique à Luxembourg 1962 (Maubeuge, 1964) defined the Bajocian Stage to begin at the base of the *Sonninia sowerbyi* ammonite zone and to extend to the top of the *Parkinsonia parkinsoni* Zone. However, the holotype of the *Sonninia sowerby* index species was later discovered to be a nucleus of a large Sonniniidae (*Papilliceras*) from the overlying *Otoites sauzei* Zone (Westermann and Riccardi, 1972). Therefore, the basal ammonite zone of the Bajocian was redefined as the *Hyperlioceras discites* Zone, with the zonal base marked by the lowest occurrence of the ammonite genus *Hyperlioceras* (*Toxolioceras*), which evolved from *Graphoceras* (both in ammonite family Graphoceratidae).

Two sections recorded this ammonite datum with supplementary biostratigraphic and magnetostratigraphic data: Murtinheira at Cabo Mondego, Portugal (selected for the GSSP), and Bearreraig Bay on the Isle of Skye, Scotland (selected as an auxiliary stratotype point), (Pavia and Enay, 1997). The GSSP at coastal Cabo Mondego (Table 18.1) comprises rhythmic alternations of gray limestone and marl (Henriques, 1992; Henriques *et al.*, 1994), and was ratified in 1996 (Pavia and Enay, 1997).

Bajocian substages The base of the Upper Bajocian is the base of the *Strenoceras (Strenoceras) niortense* ammonite zone. (In older literature, the base was assigned as the base of the "*Strenoceras*" *subfurcatum* Zone, until it was recognized by Dietl (1981) that the holotype of the index species belongs to *Garantiana* and had originated from the overlying zone, therefore this zone became invalid.) A major turnover of ammonite genera occurs at this level (Table 18.1).

BATHONIAN

History, definition, and proposed boundary stratotype The former Bathonian Stage of d'Omalius d'Halloy (1843) was named after the town of Bath in southern England, where strata characterized by oolitic limestone are exposed in a number of quarries, but these are incomplete and lack adequate characterization by ammonites (Torrens, 1965). The lower half of the originally "Bathonian" exposed in Normandy was reclassified as the Bajocian Stage in the system of d'Orbigny (1842–51, 1852), but he did not specify a revised stratotype for the shortened Bathonian, nor provide an unambiguous lower boundary. Indeed, d'Orbigny's description suggests that he included the present "Lower Bathonian" substage within his Bajocian (Rioult, 1964). A century of confusion ended when the base of the Bathonian Stage was defined by the Colloque du Jurassique à Luxembourg 1962 (Maubeuge, 1964) as the base of the *Zigzagiceras zigzag* ammonite zone.

The basal-Bathonian is well developed in southeastern France. A GSSP was suggested within inter-bedded limestone and marl at Ravin du Bès-Bas Auran, near Digne, Basses-Alpes (Innocenti *et al.*, 1988; Table 18.1). However, the strata do not preserve a primary magnetostratigraphy and are barren of dinoflagellate cysts (Mangold, 1999). In addition, the uppermost Bajocian (*Parkinsonia (Parkinsonia) bomfordi* subzone) may be absent, indicating that the proposed GSSP level is a hiatus (Dietl, 1995, as reported by Mangold, 1997). Another GSSP candidate is Cabo Mondego, Portugal, the same section that defines the base of the underlying Bajocian, but its uppermost Bajocian zone is poorly preserved (Fernández-López, 2003).

Bathonian substages The Bathonian is generally divided into three substages, with the base of the Middle Bathonian placed at the base of the *Procentes progracilis* ammonite zone.

A divergence of ammonite assemblages in the upper Middle Bathonian has resulted in different bases of an Upper Bathonian substage in each province. In the sub-Mediterranean province (Tethyan domain), a Middle–Upper Bathonian boundary is assigned to the base of the *Hecticoceras (Prohecticoceras) retrocostatum* Zone. In the northwest European province (Boreal domain), a substage boundary is commonly assigned to the base of the *Procerites (Procerites) hodsoni* Zone, which is a significantly older level (Groupe Français d'Etude du Jurassique, 1997).

CALLOVIAN

History, definition, and proposed boundary stratotype The Callovian Stage was named by d'Orbigny (1842–51, 1852) after the village of Kelloway, Wiltshire, England, 3 km northeast of Chippenham. The "Kelloways Stone" contains abundant cephalopods, including *Ammonites calloviensis* (*Sigaloceras calloviensis* in current taxonomy), and d'Orbigny considered "*Calloviens*" to be a derivative of Kelloway. Oppel (1856–8) placed the base of his "Kelloway gruppe" at the base of the *Macrocephalites macrocephalus* Zone, or at the lithologic contact of the Upper Cornbrash with the underlying Forest Marble Formation (currently the upper part of the *Clydoniceras discus* subzone of uppermost Bathonian). At this contact, ammonites of the genus *Macrocephalites* replace *Clydoniceras*, but much of the upper Cornbrash is condensed and/or a deposit "representing but a fraction of the time-intervals involved" (Cope *et al.*, 1980a).

Callomon (1964, 1999) noted that the base of the *Macrocephalites macrocephalus* subzone in "standard chronostratigraphy" was initially defined by Arkell (1956) as the base of Bed 4, at the Sutton Bingham section, near Yeovil, Somerset, England; therefore, this served as the de facto GSSP for the base of the Callovian Stage. However, the lowest occurrence of *Macrocephalites* genera was later discovered to be in strata equivalent to the Upper Bathonian (Dietl, 1981; Dietl and Callomon, 1988), therefore the "standard" *Macrocephalites macrocephalus* Zone was abandoned, and the base of the Callovian was assigned to the lowest occurrence of the genus *Kepplerites* (Kosmoceratidae), which defined a basal horizon of *Kepplerites* (*Kepplerites*) *keppleri* (base of *K. keppleri* subzone of *Macrocephalites herveyi* Zone) in the sub-Boreal province (UK to southwest Germany). The uppermost Bathonian is the *hockstetteri* horizon (var. *hockstetteri* of *Clydoniceras discus*) of the *C. discus* subzone, *C. discus* Zone.

A continuous transition between the uppermost Bathonian and basal-Callovian is rarely preserved. A proposed GSSP with an apparently complete boundary at the resolution level of ammonite successions is in the Albstadt district of the Swabian Alb, southwest Germany (Dietl, 1994; Callomon and Dietl, 2000; Table 18.1). The Macrocephalen–Oolith Formation (Unit ε of the Brown Jura facies) is a condensed facies of iron-oolite-bearing clay to marly limestone, and the compact Bathonian–Callovian boundary interval is bounded by unconformities and may contain minor hiatuses (Table 18.1). Therefore, this suggested GSSP has not yet been adopted by the International Stratigraphic Commission.

In the sub-Mediterranean province (southern Paris Basin to north Africa and Italy), the basal-Callovian zone is the *Bullatimorphites* (*Kheraiceras*) *bullatus* Zone defined by the range of the index species (Groupe Français d'Etude du Jurassique, 1997). Strong ammonite biogeographic differences required these two regions to have distinct and poorly correlated zonations until the middle of the Callovian.

Callovian substages The Callovian Stage is generally divided into three substages. The substage boundaries correspond to two important changes in ammonite fauna, but ammonite provincialism and utilization of different faunal successions led to different placements within each realm that do not necessarily coincide (Groupe Français d'Etude du Jurassique, 1997).

In the sub-Boreal province, the Lower–Middle Callovian boundary is placed at the base of the *Kosmoceras* (*Zugokosmoceras*) *jason* Zone (base of *Kosmoceras* (*Zugokosmoceras*) *medea* subzone), above the *Sigaloceras* (*Sigaloceras*) *calloviense* Zone (*Sigaloceras* (*Catasigaloceras*) *enodatum* subzone). In the sub-Mediterranean province, the Lower–Middle Callovian boundary is placed at the base of the *Reineckeia anceps* Zone (*Reineckeia stuebeli* subzone), above the *Macrocephalites* (*Dolikephalites*) *gracilis* Zone (*Indosphinctes patina* subzone). These two levels are considered approximately coeval.

The Middle–Upper Callovian boundary in the sub-Boreal province is assigned to the base of the *Peltoceras* (*Peltoceras*) *athleta* Zone (*Kosmoceras* (*Lobokosmoceras*) *phaeinum* subzone), above the *Erymnoceras coronatum* Zone (*Kosmoceras* (*Zugokosmoceras*) *grossouvrei* subzone). The Middle–Upper Callovian boundary in the sub-Mediterranean province is assigned to the base of the *Peltoceras* (*Peltoceras*) *athleta* Zone (*Hecticoceras* (*Orbignyiceras*) *trezeense* subzone), above the *Erymnoceras coronatum* Zone (*Rehmannia* (*Loczyceras*) *rota* subzone), or approximately a subzone higher than in the sub-Boreal province (Groupe Français d'Etude du Jurassique, 1997).

18.1.4 Upper Jurassic

The brownish-weathering deposits of the Middle Jurassic (*Brauner Jura*) in southwestern Germany are overlain by units dominated by calcareous claystone and limestone. Therefore, these carbonates were grouped as the White Jurassic (*Weisser Jura*) by von Buch (1839), and the base of the current Upper Jurassic (the base of the Oxfordian) coincides approximately with this lithologic change. This Upper Jurassic interval, or former "Malm," is approximately equivalent to the Middle and Upper Oolite group of Conybeare and Phillips (1822) in England. Both the White Jurassic of southwest Germany and the English strata undergo a shallowing upward in the latest Jurassic, and are erosionally truncated or are overlain by non-marine deposits. Southern England provided the reference sections when d'Orbigny (1842–1851, 1852) subdivided the Upper Jurassic into four stages (Oxfordian, Corallian, Kimmeridgian, and Portlandian), and designated the base of the Cretaceous as the Purbeck Stage, followed by the Neocomian Stage. Oppel (1856–1858) eliminated d'Orbigny's Corallian and Portlandian Stages, and extended the Kimmeridgian to the base of the Purbeckian (also considered to be Cretaceous). Oppel left an interval "unassigned" between his Oxfordian and Kimmeridgian groups (his *Diceras arietina* Zone, approximately equivalent to the Upper Calcareous Grit Formation of Dorset, England). Later, Oppel (1865) created a new uppermost Jurassic stage, the Tithonian, in the Mediterranean region that encompassed the upper part of his previous "Kimmeridgian group" and extended to the base of the Neocomian Stage. However, Oppel did not specify the limits or reference sections for the Tithonian Stage concept. The situation was further distorted when the "Berriasian" Stage of Coquand (1871) came into common use to designate the lowermost Cretaceous, even though it overlapped with the original concept of the Tithonian Stage.

The combination of (1) the shuffling of Upper Jurassic stage nomenclature coupled with imprecise definitions, (2) a pronounced faunal provincialism during the majority of the Upper Jurassic that precluded precise correlation even within northwest Europe, and (3) widespread hiatuses in the reference sections resulted in a proliferation of regional stage and substage nomenclature. Finally, after a century of debate, the Colloque du Jurassique à Luxembourg 1962 (1964) voted to "return to the original sense of this stage [Oxfordian] as defined by A. d'Orbigny and given precision by W. J. Arkell" and to discontinue regional usage of a "Purbeckian" Stage, because it was primarily a local facies. However, the controversy over other Upper Jurassic stage definitions or the placement of the Jurassic–Cretaceous boundary led the Collogue to "refer the question back for consultation among interested specialists." During the 1980s and 1990s, the International Subcommission on Jurassic Stratigraphy established that the Upper Jurassic consists of the Oxfordian, Kimmeridgian, and Tithonian stages. Through a fortunate episode of biogeography, an inter-regional biostratigraphic definition of the base of the Oxfordian is well established with ammonites. However, it has proven difficult to correlate potential definitions for the bases of the Kimmeridgian and Tithonian stages, and long-held traditions of regional equivalence have proven to be erroneous.

OXFORDIAN

History, definition, and boundary stratotype candidates The Oxfordian Stage of d'Orbigny (1842–51, 1852) was named after the town of Oxford, Oxfordshire, England, with reference to the Oxford Clay Formation, and was overlain by his "Coralline Stage." Oppel (1856–8) incorporated the majority of the "Corallian Stage" into his expanded Oxfordian group. Oppel assigned the base of his "Oxfordian Stage" to both the contact between the Oxford Clay Formation and the underlying Kelloway Rock in Yorkshire (now considered to be approximately the Lower–Middle Callovian boundary) and to the top of the *Peltoceras athleta* ammonite zone (now considered to be middle of the Upper Callovian). He also left "unassigned" a suite of strata between his "Oxfordian" and overlying "Kimmeridgian Stage."

Ammonites across the Callovian–Oxfordian boundary interval were studied by Arkell (1939, 1946), who placed the boundary at the contact of the range zones of *Vertumniceras mariae* (now placed in the *Quenstedtoceras* genus) above *Quenstedtoceras lamberti*, or the base of the Oxford Clay Formation in Yorkshire. This is consistent with the historical usage in southwest Germany, where the Upper or "White" Jurassic begins just above the Lamberti Knollen bed. The Colloque du Jurassique à Luxembourg 1962 (Maubeuge, 1964) selected Arkell's biostratigraphic definition for the base of the Oxfordian Stage. The Colloque also assigned the upper limit of the Oxfordian as the top of the *Ringsteadia pseudocordata* ammonite zone in the Boreal realm.

The ammonite succession across the *Q. lamberti–Q. mariae* interval has been studied in expanded dark clay sections in southeast France (e.g. Fortwengler and Marchand, 1994) and two complementary sections were recommended as basal-Oxfordian GSSPs (Melendez, 1999; Table 18.1). The basal-Oxfordian was proposed as the *Brightia thuouxensis* Horizon at base of the *Cardioceras scarburgense* subzone (*Quenstedtoceras mariae* Zone), above the uppermost Callovian *Cardioceras paucicostatum* Horizon. However, except for dinoflagellates, these French sections have not yet proved suitable for other stratigraphic correlation methods, therefore the GSSP proposal was suspended (e.g. Melendez, 2002, 2003). A coastal section near Weymouth (Dorset, southern England) is in a facies suitable for magnetostratigraphy, where the base of the Oxfordian is correlated to a brief normal-polarity marine magnetic anomaly M36An (Coe and Ogg, unpublished), and is being considered as a reference section for the Callovian–Oxfordian boundary in the Boreal domain (Melendez, 1999; Table 18.1).

Oxfordian substages Traditionally, the base of the Middle Oxfordian substage is placed at base of the *Perisphinctes (Arisphinctes) plicatilis* Zone.

Beginning with the Middle Oxfordian, faunal differentiation in separate basins became more pronounced and has inhibited standardization and correlation of regional ammonite zones. In addition, even though regional zonal nomenclatures have commonly remained constant, the assigned boundaries of biostratigraphic units have undergone re-definition (e.g. Glowniak, 1997; Groupe Français d'Etude du Jurassique, 1997).

In the sub-Mediterranean province (Tethyan domain), the base of an Upper Oxfordian substage is commonly assigned to the base of the *Perisphinctes (Dichotomoceras) bifurcatus* Zone. In the sub-Boreal province (Boreal domain), this substage boundary is assigned to the base of the *Perisphinctes cautisnigrae* Zone. These two levels may be approximately synchronous (Groupe Français d'Etude du Jurassique, 1997).

KIMMERIDGIAN

History, revised definition, and boundary stratotype candidates The Kimmeridgian Stage was named by d'Orbigny (1842–51, 1852) after the coastal village of Kimmeridge in Dorset, England, where the spectacular cliffs of dark gray Kimmeridge Clay expose a continuous record of that interval. Oppel (1856–8) expanded the Kimmeridgian downward by incorporating a portion of d'Orbigny's former Corallian Stage, but, rather than assign a boundary between the Oxfordian and Kimmeridgian, he left the intervening Upper Calcareous Grit Formation "unassigned." Oppel initially indicated that the Kimmeridgian "group" would continue upward to the base of the Purbeck (his base of the Cretaceous), but later, Oppel (1865) inserted a Tithonian Stage as the uppermost Jurassic stage. Therefore, neither boundary of the Kimmeridgian Stage was adequately defined.

The Oxfordian–Kimmeridgian boundary was defined by Salfeld (1914) after studying the Perisphinctidae ammonite succession from the boundary interval. He proposed that the boundary should be placed between the *Ringsteadia anglica* Zone (now called *Ringsteadia pseudocordata* Zone) in the uppermost Oxfordian and the appearance of *Pictonia* at the base of the Kimmeridgian. The Colloque du Jurassique à Luxembourg 1962 (Maubeuge, 1964) fixed the base of the Kimmeridgian as the base of the *Pictonia baylei* Zone. However, due to faunal provincialism that began in the middle Oxfordian, the ammonite zonation of England (Boreal domain) could not be correlated to the sub-Mediterranean province (Tethyan domain). The Colloque du Jurassique à Luxembourg 1962 (Maubeuge, 1964) indicated that this level was equivalent to the base of the *Sutneria platynota* Zone of the sub-Mediterranean province.

However, this presumed equivalence was later demonstrated to be incorrect from comparisons of dinoflagellate cyst assemblages (Brenner, 1988; Melendez and Atrops, 1999) and rare incursions of ammonites from the Boreal domain into the sub-Mediterranean successions in Poland and the Swabian Alb (Atrops *et al.*, 1993; Matyja and Wierzbowski, 1997; Schweigert and Callomon, 1997). The chain of logic from ammonite assemblages is (1) ammonite *Pictonia densicostata* (SALFELD) occurs in the lower *Pictonia baylei* Zone, which is the basal zone of the "Kimmeridgian" Stage as currently used in sub-Boreal province, in Dorset; (2) *Pictonia densicostata* occurs with *Amoeboceras bauhini* (OPPEL) in South Ferriby in eastern England and on the Isle of Skye in northwest Scotland; and (3) *Amoeboceras bauhini* occurs in the *Taramelliceras hauffianum* subzone of the *Epipeltoceras bimammatum* Zone (middle of the "upper Oxfordian" as currently used in the sub-Mediterranean province) in southwest Germany and in Poland. Assuming a narrow and synchronous range of *A. bauhini* among these localities, then the current placement of the base of the "Kimmeridgian" Stage in Britain is significantly younger (approximately one-and-a-half ammonite zones, or about 1 myr) than the base of the "Kimmeridgian" Stage as used in the sub-Mediterranean province (Fig. 18.1). This biostratigraphic conclusion is supported by comparing magnetostratigraphy and sequence stratigraphy patterns of the UK with different regions in the sub-Mediterranean province (Ogg and Coe, 1997, and in prep.).

This temporal offset has created a dilemma in selecting a GSSP for the base of the Kimmeridgian Stage, because an initial choice must be made between faunal provinces and the corresponding definition of the Oxfordian–Kimmeridgian boundary (reviewed in Atrops, 1999; Melendez and Atrops, 1999; and Wierzbowski, 2001).

In the Boreal realm, the leading candidate for a GSSP is the coincident base of the *Pictonia baylei* Zone (sub-Boreal province) and *Amoeboceras bauhini* Zone (Boreal province) in a succession of medium-gray clays at Staffin Bay (Isle of Skye, northwest Scotland) containing abundant ammonites and dinoflagellate cysts (Wright, 1973, 1989; Riding and Thomas, 1997; Melendez and Atrops, 1999; Wierzbowski, 2002, 2003). At this Staffin Bay section, the base of the *P. baylei* Zone is just above the base of a reversed-polarity zone assigned to lower Chron M26r (Ogg and Coe, 1997, and in prep.).

In the sub-Mediterranean province, southwestern France has two main candidates for a GSSP at the base of the *S. platynota* Zone in inter-bedded limestone and marl: Crussol mountain on the Rhône river just west of Valence (Atrops, 1982, 1994) and Châteauneuf d'Oze in the Haute Provence district (Atrops, 1994; Melendez and Atrops, 1999). The base of the *S. platynota* Zone at Crussol is just above the base of reversed-polarity zone M25r (Ogg and Atrops, in prep.). Châteauneuf d'Oze did not yield a magnetostratigraphy, but appears more suitable for microfossil biostratigraphy (dinoflagellate cysts) and chemostratigraphy.

Kimmeridgian substage The dichotomy in defining the base of the Kimmeridgian Stage among paleogeographic faunal realms precludes standardizing substages. Traditional usage places a base of an Upper Kimmeridgian substage in the sub-Boreal realm at the base of the ammonite *Aulacostephanoides mutabilis* Zone, whereas in the sub-Mediterranean realm it is typically assigned to the base of the *Aspidoceras acanthicum* Zone. Magnetostratigraphy suggests that the sub-Mediterreanean substage base is about 0.9 myr younger than the sub-Boreal assignment (Ogg *et al.*, in prep.).

TITHONIAN

History and revised definition In an enlightened departure from the stratotype concept, Oppel (1865) defined the Tithonian Stage solely on biostratigraphy. In mythology, Tithon is the spouse of Eos (Aurora), goddess of dawn, and Oppel used this name in a poetic allusion to the dawn of the Cretaceous. He referenced characteristic Tithonian sections in western Europe from Poland to Austria.

The base of Oppel's Tithonian was placed at the top of the Kimmeridgian *Aulacostephanus eudoxus* ammonite zone, which can be recognized in both the sub-Boreal and sub-Mediterranean realms. Later, Neumayr (1873) established the *Hybonoticeras beckeri* Zone above the *A. eudoxus* in the sub-Mediterranean realm and also assigned it to the Kimmeridgian Stage.

Neumayr's revised placement of the Tithonian–Kimmeridgian boundary corresponded closely with the boundary between the "Portlandian" and Kimmeridgian stages as initially assigned by Alcide d'Orbigny (1842–51, 1852), who had assigned "des *Ammonites giganteus* et *Irius*" as Portlandian index fossils. However, d'Orbigny did not visit England, and he inadvertently combined fossil assemblages from outcrops at Bologna in Italy with a name derived from a "type section" on the Isle of Portland in England. The "*Ammonite irius*" is one representative of the *Gravesia* genera, which have a lowest occurrence in the basal *Hybonoticeras hybonotum* Zone of the revised Tithonian. Accordingly, the *Gravesia gravesiana* ammonite zone was assigned as the basal zone of the British Portlandian (Tithonian) Stage by Salfeld (1913). Following Salfeld's oral presentation to the Geological Society of London, it was noted that the chronostratigraphic term "Kimmeridgian" only partially encompassed the "Kimmeridgian Clay" Formation, therefore it was recommended that Salfeld "should invent a dual nomenclature – one for the stratigraphical and another for the zoological sequence" and replace 'Kimmeridgian Stage" with a new name (in Salfeld, 1913). Unfortunately, this enlightened recommendation was not pursued, and a confusing equivalence of a "Kimmeridgian" Stage with the "Kimmeridge Clay" Formation and associated lifting of the base of d'Orbigny's "Portlandian" Stage became common usage in England, but a lower Tithonian–Kimmeridgian boundary was used elsewhere in Europe. The Kimmeridge Clay Formation was arbitrarily subdivided into a lower and upper member at the approximate Kimmeridgian–Tithonian boundary level at the lowest occurrence of *Gravesia gravesiana* at the "Maple Ledge" bed (reviewed in Cox and Gallois, 1981).

The "Tithonian" was formally adopted as the name of the uppermost stage of the Jurassic by a vote of the International Commission on Stratigraphy in September 1990.

The Second Colloquium on the Jurassic System, held in Luxembourg in 1967, recommended that the top of the Kimmeridgian be assigned to the base of the *Gravesia gravesiana* Zone (Anonymous, 1970). However, Cope (1967) subdivided the lowermost Tithonian portion of the Upper Kimmeridge Clay into several ammonite zones based on successive species of his reconstituted *Pectinatites* genera and abandoned Salfeld's two *Gravesia* Zones. Cope raised the upper limit of the uppermost Kimmeridgian *Aulacostephanus autissiodorensis* Zone to the base of his new *Pectinatites* (*Virgatosphinctoides*) *elegans* Zone, thereby effectively lifting the associated biostratigraphic division between Lower and Upper Kimmeridge Clay. Cox and Gallois (1981) note that the top of the international Kimmeridgian Stage now falls within the middle of Cope's expanded *A. autissiodorensis* Zone in the sub-Boreal realm, therefore, they suggest reinstating a truncated *Gravesia gravesiana* Zone below the *P.* (*V.*) *elegans* Zone.

Boundary stratotype candidates The Tithonian–Kimmeridgian boundary interval in the Tethyan faunal realm is marked by the simultaneous lowest occurrence of the ammonites *Hybonoticeras* aff. *hybonotum* and *Glochiceras lithographicum* (the base of the *H. hybonotum* Zone), immediately followed by the lowest occurrence of the *Gravesia* genera. Candidate sections in southeast France for the GSSP include thick pelagic limestone outcrops at Crussol mountain on the Rhône river just west of Valence (Atrops, 1982, 1994) and at a quarry at Canjuers (Var district, southeast; France; Atrops *et al.*, in prep.). The base of *H. hybonotum* Zone is at the base of normal-polarity Chron M22An at Crussol (Ogg and Atrops, in prep.). The Canjuers quarry did not yield a magnetostratigraphy, but its ammonite succession is better established. Other potential GSSP sections are in the Swabian region of southern Germany.

Tithonian substages In the Tethyan faunal domain, the base of an Upper Tithonian substage is traditionally assigned to a major turnover in ammonite assemblages at the base of the *Micracanthoceras microcanthum* Zone. This level is approximately where calpionellid microfossils become important in the biostratigraphic correlation of pelagic limestone and is at the base of normal-polarity Chron M20n.

In the Boreal realm, the boundary between Lower and Middle "Volgian" regional substages assigned at the base of the *Dorsoplanites panderi* ammonite zone may be approximately coeval. However, in the sub-Boreal faunal province in Britain, the base of the "Portlandian" regional stage is traditionally placed at the base of the *Progalbanites albani* ammonite zone, at the base of the Portland Sand Formation, which may be significantly higher.

Bolonian–Portlandian and Volgian regional stages of Europe The century of controversy over the subdivision and nomenclature for the uppermost Jurassic and lowermost Cretaceous stages, coupled with markedly distinct ammonite assemblages in different regions of Europe, led to extensive usage of regional stages (see Fig. 19.2).

Though the Portlandian was used in Russia, the Volgian Stage was later established in western Russia by Nikitin (1881), capped by a Ryazanian horizon (Bogoslovsky, 1897), and extended downward so as to equate with the Tithonian (Resolution, 1955; reviewed in Krymholts *et al.*, 1982–8). The Volgian is zoned by continuous ammonite assemblages (but generally containing several stratigraphic breaks) that are extensively distributed in the Boreal faunal realm, therefore it became another widely used standard in the northern high latitudes outside of Britain. In 1996, the Russian Interdepartmental Stratigraphic Committee resolved to equate the Lower and Middle Volgian (*Ilowaiskya klimovi* through *Epivirgatites nikitini* ammonite zones) to the Tithonian Stage, assign the Upper Volgian (*Kachpurites fulgens* through *Craspedites nodiger* ammonite zones) to the lowermost Cretaceous, and use only Tithonian and Berriasian as chronostratigraphic units in the Russian geological time scale (Rostovtsev and Prozorowsky, 1997).

The "Bolonian" and "Portlandian" have been promoted as "secondary standard stages" of a Tithonian "superstage" for usage in English and French regional geology, especially in Dorset (e.g. Cope 1993, 1995, 2003; Taylor *et al.*, 2001). The "Bolonian" is equivalent to the upper Kimmeridgian Clay Formation between the Tithonian–Kimmeridgian boundary and the base of the Portland Sand, or *Pectinatites* (*Virgatosphinctoides*) *elegans* through *Virgatopaviovia fittoni* ammonite zones. The overlying "Portlandian" is traditionally equivalent to the Portland Group in Dorset, or *Progalbanites albani* through the *Titanites anguiformis* ammonite zones.

18.2 JURASSIC STRATIGRAPHY

The ammonite successions of Europe have historically served as the global primary standard for the Jurassic. Biostratigraphic, magnetostratigraphic, chemostratigraphic, and other events are typically calibrated to these standard European ammonite zones. However, especially during the middle Oxfordian through Tithonian, different faunal assemblages occur in the various paleogeographic realms, such as Boreal (Arctic and northernmost Europe), sub-Boreal (northern Europe), sub-Mediterranean (southern Europe), and Tethyan (southernmost Europe and margins of the former Tethys seaway). An extensive compilation and inter-correlation of Jurassic stratigraphy of European basins was coordinated by Hardenbol *et al.* (1998), and Fig. 18.1 is a summary of a portion of their comprehensive chart series. These European basins contain the majority of the proposed GSSP sites and alternative sites for the chronostratigraphic framework of the Jurassic.

18.2.1 Macrofossil zonations

Ammonites dominate the historical zonation of the Jurassic. Brachiopod and bivalve assemblages provide important regional markers. Terrestrial biostratigraphy of dinosaurs and palynology has a less-precise calibration to the marine stratigraphy.

AMMONITES

Alfred Oppel (1856–8) developed the concept of a biostratigraphic zone, and used ammonites to define two-thirds of his 33 Jurassic zones. Jurassic ammonite zonations have undergone constant revision since Oppel, and the Jurassic is currently subdivided into 70–80 zones and typically has 160–170 subzones in each faunal realm. Reviews of the development, definitions, and inter-correlation of European ammonite zonations are presented in Thierry (in Hardenbol *et al.*, 1998, pp. 776–777 plus correlation charts), Krymholts *et al.* (1982–8), and Groupe Français d'Etude du Jurassique (1997). Correlation of the regional ammonite zones of western North America to the northwest European standard is summarized in Pálfy *et al.* (2000a).

In contrast to standard biostratigraphic usage and most Cretaceous ammonite zones, the "index" or "name" species of some Jurassic ammonite zones is not always an indicator of the definition of those zones. The biostratigraphic range of an index species, such as *Aulacostephanoides mutabilis*, can be entirely independent of the limits of the designated "*mutabilis* Zone" (Callomon, 1985, 1995). Partly to alleviate the resulting confusion, several British workers have advocated designating "standard chronozones" with regional equivalents of GSSPs to replace the former associated ammonite zones (e.g. see tables in Cox, 1990). It was believed that the evolutionary rates of the ammonites were relatively fast that so there was no practical difference between an ammonite-based biostratigraphy (zonation) and chronostratigraphy. Therefore, a standard Eudoxus Chronozone (capitalized, non-italics, no genus designation) of the Kimmeridgian in southern England has a base defined at bed E1 at a quarry near Westbury, Wiltshire (Birkelund *et al.*, 1983), and continues to the base of the Autissiodorensis Chronozone, which is assigned as the top of Flats Stone Band bed at beach exposures of the Kimmeridge Clay near Kimmeridge village, Dorset (Cox and Gallois, 1981; Cox, 1990). However, the base of the "Eudoxus" ammonite assemblage zone, as independently used in France has been assigned to the lowest occurrence of *Orthaspidoceras orthocera*, which is significantly lower than the base of the Eudoxus Chronozone of England (Ogg *et al.*, in prep.), and neither is delimited by the observed range (biozone) of *Aulacostephanus eudoxus*. Both systems – the nomenclature of Jurassic ammonite assemblage zones and the regional designation of standard chronozones – have defenders (mainly Jurassic ammonite specialists) and critics. For clarity in the chart in Fig. 18.1 (see also Table 18.2), we have included the genera of the ammonite "index" species, but with caution that these are not always the "guide" species of the named zone.

OTHER MARINE MACROFAUNA

Brachiopod zonations for northwest Europe and for the northern part of the Tethyan province provide important markers within individual basins and approach the resolution of ammonite zones in some stages of the Jurassic (e.g. Alméras *et al.*, 1997; Laurin, 1998). The correlation potential of brachiopods and the slower-evolving successions of bivalves and gastropods are compromised by their benthic habits which can be reflected in ecological-facies associations and provincialism (reviewed in Cope *et al.*, 1980b). Belemnite zones within the Jurassic can provide correlation to the stage or substage level (e.g. Combemorel, 1998).

Ostracodes are small (0.2–1.5 mm) crustaceans with bivalved, calcified shells, which are a major constituent of shallow-marine and brackish benthic fauna. Ostracode datums can approach the resolution of ammonite zones, especially within portions of the Lower and Middle Jurassic (e.g. see reviews by Cox, 1990; Colin, 1998).

DINOSAURS AND OTHER VERTEBRATES

Dinosaurs are the most famous Jurassic fauna. This summary of major trends is from Lucas (1997). Dinosaurs dominated the land herbivores and carnivores during the Early and Middle

Table 18.2 *Jurassic time scale for ammonite zones*[a]

Stage	Zone: Boreal (or Cosmopolitan for early Jurassic)	Method	Calibration	Age (base of zone) (Ma)	Comments	Zone: Tethyan (sub-Mediterranean)	Method	Calibration	Age (base of zone) (Ma)	Comments
Berriasian						*Berriasian*				
	Subcraspedites (Volgidiscus) lamplughi	magstrat	M18n.2 (±0.5)	144.5		*Berriasella jacobi*	magstrat	M19n.2n.55 (±0.05)	144.5	
JURASSIC (Tithonian) *(top)*				**145.5**	Base of *B. jacobi* Zone	**JURASSIC (Tithonian)** *(top)*			**145.5**	
	Subcraspedites (S.) preplicomphalus	magstrat	M19n.3 (±0.2)	145.6						
	Subcraspedites (Swinnertonia) primitivus	magstrat	M19r.8 (±0.5, hiatus)	146.0						
	Paracraspedites oppressus	magstrat	M20n.1n.5 (±0.5, est.)	146.3	Base Oppressus = Base of Lulworth Beds (Purbeck) in Dorset; but usefulness of this zone is disputed	*Durangites*	magstrat	M19r.1 (±0.2)	146.1	Base *Calpionellid* Zone A2 is base of *Durangites* Zone
	Titanites anguiformis	magstrat	M20n.2n.7 (±0.2)	146.7	Anguiformis = Portland Freestone Member in Dorset					
	Galbanites (Kerberites) kerberus	magstrat	M20n.2n.2 (±0.2)	147.0	Kerberites = Upper Cherty Beds in Dorset					
	Galbanites okusensis	magstrat	M20r.6 (±0.1)	147.4	*Okusensis–Glaucolithus* = West Weare "sandstone" (dolomite) and overlying Lower Cherty Member in Dorset					
	Glaucolithites glaucolithus	magstrat	M20r.1 (±0.1)	147.7						

Progalbanites albani	magstrat	M21n.7 (±0.1)	148.0	Albani = basal-Portland Sand = Black Noire and Exogyra Bed in Dorset, hiatus is common	*Micracanthoceras microcanthum* (lower subzone is *Simplisphinctes* and upper is *Paraulacosphinctes transitorius*)	magstrat	base M20n (±0.1)	147.2	Base *Calpionellid* Zone A1 is just above base of *Microcanthum* Zone
Virgatopavlovia fittoni (top of Kimmeridge Clay formation)	magstrat	M21n.3 (± 0.1)	148.3	Base of *V. fittoni* = top of Kimmeridge Clay formation in Dorset	*Micracanthoceras ponti–Burckhardticeras*	magstrat	M20r.5 (±0.1)	147.5	
Pavlovia rotunda	magstrat	M21r.8 (±0.3, est.)	148.6						
Pavlovia pallasioides	magstrat	M22n.95 (±0.1)	149.0		*Simoceras admirandum–Simoceras biruncinatum*	magstrat	M21n.4 (±0.1)	148.2	Base *Chitinoidella* is near top of *Admirandum* Zone
Pectinatites (P.) pectinatus	magstrat	M22n.7 (±0.1)	149.3		*Richterella richteri*	magstrat	M22n.95 (±0.05)	149.0	
Pectinatites (Arkellites) hudlestoni	magstrat	M22n.45 (±0.1)	149.6		*Semiformiceras semiforme* (= zone of *Haploceras (Volanites) verruciferum*)	magstrat	M22n.6 (±0.1)	149.4	
Pectinatites (Virgatosphinctoides) wheatleyensis	magstrat	M22n.2 (±0.1)	150.0		*Semiformiceras darwini* (= zone of *Virgatosimoceras albertinum*)	magstrat	M22n.25 (±0.05)	149.9	
Pectinatites (Virgatosphinctoides) scitulus	magstrat	M22r.8 (±0.2, est)	150.3						
Pectinatites (Virgatosphinctoides) elegans	magstrat	M22r.2 (±0.2, est)	150.6	Projects to middle of M22r	*Hybonoticeras hybonotum*	magstrat	base M22An (±0.1)	150.8	
Kimmeridgian (top)			**150.8**		*Kimmeridgian (top)*			**150.8**	
Aulacostephanus autissiodorensis	magstrat (for zone base)	M23n.5 (±0.5, est.)	151.2		*Hybonoticeras beckeri*	magstrat	M23r.2r.1 (±0.05)	152.2	
Aulacostephanus eudoxus	magstrat	M24n.5	152.4		*Aulacostephanus eudoxus*	magstrat	M24r.1r.8 (±0.1)	152.6	
					Aspidoceras acanthicum	magstrat	M24r.2r.6 (±0.1)	153.1	

(*cont.*)

Table 18.2 (*cont.*)

Stage	Zone: Boreal (or Cosmopolitan for early Jurassic)	Method	Calibration	Age (base of zone) (Ma)	Comments	Zone: Tethyan (sub-Mediterranean)	Method	Calibration	Age (base of zone) (Ma)	Comments
	Aulacostephanoides mutabilis	magstrat	base M24Bn (±0.3) (estimate)	153.9		*Crussoliceras divisum*	magstrat	M24Ar.7 (±0.1)	153.4	
						Ataxioceras (A.) hypselocyclum	magstrat	M25n.8 (±0.1)	154.1	
						Sutneria platynota	magstrat	M25r.1 (±0.1)	154.5	
	Rasenia cymodoce	magstrat	base M25An.3n (±0.1)	155.0		*Oxfordian* (sub-Mediterranean usage) *(top)*			**154.5**	
	Pictonia baylei	magstrat	M26r.2 (±0.2)	155.7	Marine anomaly model is uncertain for M26 subchrons	*Subnebrodites planula*	magstrat	base M26n.3n (±0.1)	155.4	Marine anomaly model is uncertain for M26 subchrons
Oxfordian (sub-Boreal usage) *(top)*				**155.7**						
	Ringsteadia pseudocordata	magstrat	base M28Bn (estimate)	156.8		*Epipeltoceras bimammatum*	magstrat	base M28Cn (±0.1)	157.0	
	Perisphinctes cautisnigrae	magstrat	base M28Dn (estimate)	157.2		*Perisphinctes (Dichotomoceras) bifurcatus*	magstrat	M29n.1n.1 (±0.2)	157.4	
	Perisphinctes pumilus	magstrat	M31r.5 (±0.2)	158.6	Assumed to be same as base of *Transversarium*	*Gregoryceras transversarium*	magstrat	M31r.5 (±0.2)	158.6	
	Perisphinctes (Arisphinctes) plicatilis	magstrat	M33Bn.3 (±0.2)	159.5		*Perisphinctes (Arisphinctes) plicatilis*			159.5	
	Cardioceras (C.) cordatum	magstrat	M34Bn.1r.3 (±0.3)	160.6		*Cardioceras (C.) cordatum*			160.6	
	Quenstedtoceras (Q.) mariae	magstrat	base M36An (±0.3)	161.2		*Quenstedtoceras (Q.) mariae*			161.2	

Callovian (top)		Callovian = 18 subzones	**161.2**		*Callovian (top)*	161.2
Quenstedtoceras (Lamberticeras) lamberti	magstrat	M37.2n.3 (±0.2)	**162.0**	Lamberti (2 subzones) base set by magstrat. Rest of Callovian zones proportioned to subzones (16 subzones)	*Quenstedtoceras (Lamberticeras) lamberti*	162.0
Peltoceras (P.) athleta	equal subzones		162.5	*Athleta–Coronatum* zonal boundary in Boreal zonation is placed one subzone lower than in Tethyan	*Peltoceras (P.) athleta*	162.3
Erymnoceras coronatum	equal subzones		162.9		*Erymnoceras coronatum*	162.9
Kosmoceras (Zugokosmoceras) jason	equal subzones		163.2		*Reineckeia anceps*	163.2
Sigaloceras (S.) calloviense	equal subzones		163.5			
Proplanulites koenigi	equal subzones		164.2		*Macrocephalites (Dolikephalites) gracilis*	164.4
Macrocephalites herveyi	equal subzones		164.7		*Bullatimorphites (Kheraiceras) bullatus*	164.7
Bathonian (top)		Bathonian = 15 subzones	**164.7**	Bajo–Bath–Callov zones are proportionally scaled according to their component subzones	*Bathonian (top)*	
Clydoniceras (C.) discus	equal subzones		165.1		*Clydoniceras (C.) discus*	165.1
Oxycerites orbis	equal subzones		165.5		*Hecticoceras (Prohecticoceras) retrocostatum*	165.7
Procerites (P.) hodsoni	equal subzones		166.1		*Cadomites (C.) bremeri*	166.1
Morrisiceras (M.) morrisi	equal subzones		166.3		*Morrisiceras (M.) morrisi*	166.3
Tulites (T.) subcontractus	equal subzones		166.5		*Tulites (T.) subcontractus*	166.5
Procerites (P.) progracilis	equal subzones		166.9		*Procerites (P.) progracilis*	166.9

(*cont.*)

Table 18.2 (*cont.*)

Stage	Zone: Boreal (or Cosmopolitan for early Jurassic)	Method	Calibration	Age (base of zone) (Ma)	Comments	Zone: Tethyan (sub-Mediterranean)	Method	Calibration	Age (base of zone) (Ma)	Comments
	Asphinctites tenuiplicatus	equal subzones		167.1		*Procerites (Siemiradzkia) aurigerus*			167.3	
	Zigzagiceras (Z.) zigzag	equal subzones		167.7		*Zigzagiceras (Z.) zigzag*			167.7	
Bajocian (top)			Bajocian = 20 subzones	**167.7**		*Bajocian (top)*				
	Parkinsonia (P.) parkinsoni	equal subzones		168.3		Tethyan zones are the same as Boreal zones from Hettangian through Bajocian				
	Garantiana (G.) garantiana	equal subzones		168.9						
	Strenoceras (S.) niortense	equal subzones		169.6						
	Stephanoceras (S.) humphriesianum	equal subzones		170.2						
	Sonninia propinquans (= former zone *of Emileia (Otoites) sauzei*)	equal subzones		170.6						
	Witchellia laeviuscula	equal subzones		171.2						
	Hyperlioceras (H.) discites	equal subzones		171.6						
Aalenian (top)			Total duration of Aalenian set by cycle stratigraphy as 4 myr	**171.6**						

Table 18.2 *(cont.)*

Stage	Zone: Boreal (or Cosmopolitan for early Jurassic)	Method	Calibration	Age (base of zone) (Ma)	Comments
	Graphoceras concavum	equal subzones		172.5	
	Brazilia bradfordensis	equal subzones		173.4	
	Ludwigia murchisonae	equal subzones		174.7	
	Leioceras opalinum	equal subzones		175.6	
Toarcian (top)			Duration of Toarcian is set by cycle stratigraphy	**175.6**	
	Pleydellia aalensis	equal subzones		176.6	
	Dumortieria pseudoradiosa	equal subzones		177.6	
	Phlyseogrammoceras dispansum	equal subzones		178.5	
	Grammoceras thouarsense	equal subzones	Control on upper Toarcian equal-subzone scale	**180.5**	
	Haugia variabilis	Linear Sr isotope trend		180.7	
	Hildoceras bifrons	Linear Sr isotope trend		181.2	
	Harpoceras serpentinum (= former zone of *Harpoceras falciferum*)	Linear Sr isotope trend		182.7	
	Dactylioceras (Orthodactylites) tenuicostatum	Linear Sr isotope trend		183.0	
Pliensbachian (top)			Cycle-scaled linear Sr isotope trend	**183.0**	
	Pleuroceras spinatum	Linear Sr isotope trend		184.2	
	Amaltheus margaritatus	Linear Sr isotope trend	Control on lower Pliensbachian equal-subzone scale	**187.0**	
	Prodactylioceras davoei	equal subzones		187.7	
	Tragophylloceras ibex	equal subzones		188.5	
	Uptonia jamesoni	equal subzones		189.6	
Sinemurian (top)			Cycle-scaled linear Sr isotope trend	**189.6**	
	Echioceras raricostatum	equal subzones		191.2	
	Oxynoticeras oxnotum	equal subzones		192.0	
	Asteroceras obtusum	equal subzones		193.3	
	Caenisites tuneri	equal subzones		194.1	
	Arnioceras semicostatum	equal subzones		195.3	
	Coroniceras (Arietites) bucklandi	equal subzones		196.5	
Hettangian (top)			Cycle-scaled linear Sr isotope trend	**196.5**	
	Schlotheimia angulata	equal subzones		197.7	
	Alsatites liasicus	equal subzones		198.8	
	Psiloceras planorbis	equal subzones		199.6	
TRIASSIC (Rhaetian) *(top)*			Radiometric age	**199.6**	

[a] Methodology for deriving each age for primary ammonite zones and associated stage boundaries are given with summary of computational details. For ammonite zones that are scaled using relative numbers of subzones, these subzonal counts are from compilation by J. Thierry (in Gradstein *et al.*, 1994a, 1995, and illustrated in Hardenbol *et al.*, 1998).

Jurassic, but left only a sketchy record. In contrast, rich fossil deposits of the Late Jurassic document the evolution of the largest land animals that ever lived. Plant eaters included huge sauropods, large stegosaurids, and moderate-sized ornithopids (iguanodontids and hypsilophodontids). The kings of the carnivores were allosaurid theropods.

18.2.2 Microfossil zonations

Major microfossil biostratigraphic zonations for the Jurassic incorporate dinoflagellate cysts, calcareous nannoplankton, benthic foraminifera, calpionellids, and radiolaria.

DINOFLAGELLATE CYSTS

Organic-walled cysts of dinoflagellates are an important correlation tool for the North Sea, and the datums are correlated directly to ammonite zones of the Boreal realm (e.g. Woollam and Riding, 1983; Riding and Ioannides, 1996). A few selected markers and associated dinoflagellate zones for the Boreal realm are shown in Fig. 18.1 (extracted from Ioannides *et al.*, 1998). Several of these markers also occur in the Tethyan realm, but the ranges and correlation to ammonite zones are not as well established (Habib and Drugg, 1983; Ioannides *et al.*, 1998). Independent dinocyst zonations have been developed for the Jurassic of Australia (Helby *et al.*, 1987), for the Upper Jurassic of New Zealand (Wilson, 1984), and for other basins.

CALCAREOUS NANNOFOSSILS

The beginning of the major transgression during the late Triassic coincides with the earliest known calcareous nannofossils. The major radiation of Jurassic placolith coccoliths (plates from coccolithophorid algae) took place during the late Sinemurian to Pliensbachian (reviewed in de Kaenel *et al.*, 1996). A major re-organization of Tethyan nannofossil assemblages took place in the late Tithonian, followed by the initiation of the nannofossil-rich limestone that characterizes the pelagic realm in the Cretaceous. Jurassic nannofossil zonations and markers in the Boreal–sub-Boreal realm are calibrated to ammonite zones in northwestern Europe (e.g. Bown *et al.*, 1988; Bown, 1998), and the generalized scale in Fig. 18.1 is modified from the compilation by von Salis *et al.* (1998). Nannofossil datums in the Tethyan–sub-Mediterranean realm are partially calibrated to ammonite zones (de Kaenel *et al.*, 1996) and to latest Jurassic magnetic polarity zones (e.g. Bralower *et al.*, 1989).

FORAMINIFERA AND CALPIONELLIDS

Planktonic foraminifera did not evolve until the Middle Jurassic and are localized in occurrence, therefore Jurassic foraminifera assemblages are dominated by calcareous and agglutinated benthic forms. Compilations of foraminifer zonations and events are available for the British and North Sea region (e.g. Copestake *et al.*, 1989), for larger benthic foraminifera in the Tethyan domain (Peybernes, 1998b), and generalized for smaller benthic foraminifera in European basins (Ruget and Nicollin, 1998). The benthic foraminifer zonation in Fig. 18.1 is generalized from the later compilation.

Calpionellids are vase-shaped pelagic microfossils of uncertain origin, which appeared in the late Tithonian and continued until the middle of the Early Cretaceous (Remane, 1985). They provide important correlation markers, especially in pelagic carbonates of the Tethyan–Atlantic seaway (reviewed by Remane, 1998).

RADIOLARIA

Siliceous radiolaria are a major component of Jurassic pelagic sediments deposited under high-productivity conditions, but their tests are rarely preserved jointly with aragonitic ammonite shells. Detailed radiolarian zonations for the Middle and Late Jurassic have been developed for the western margin of North America (e.g. Pessagno *et al.*, 1993), for Japan (Matsuoka and Yao, 1986; Matsuoka, 1992), for the former Tethyan seaways exposed in Europe (Baumgartner, 1987; INTERRAD Jurassic–Cretaceous Working Group, 1995; De Wever, 1998), and for the North Sea (Dyer and Copestake, 1989; Dyer, 1998). These zonations can be partially correlated to each other (e.g. Pessagno and Meyerhoff Hull, 1996).

However, calibration of the radiolarian assemblages to standard geological stages and reference ammonite scales of Europe has been challenging and controversial. An example is the divergent correlations for the radiolarian assemblage overlying basalt at Ocean Drilling Program (ODP) Site 801, which provided a key age control on the Callovian–Oxfordian portion of the marine magnetic anomaly M-sequence and global spreading rates. The basal sediment assemblages were originally interpreted as "late Bathonian–early Callovian" (Shipboard Scientific Party, 1990; Matsuoka, 1992; reviewed in Ogg *et al.*, 1992), interpreted as equivalent to the "middle Oxfordian" of western North America by Pessagno and Meyerhoff Hull (1996), and assigned to "Bajocian" in the zonal calibrations developed by the INTERRAD Jurassic–Cretaceous Working Group (1995) for the Mediterranean region (Bartolini *et al.*,

2000; Bartolini and Larson, 2001). Possible contributing factors to this divergence are diachroneity of radiolarian or ammonite datums and ranges among basins, errors in taxonomy assignments, imprecise correlation of radiolarian markers to regional ammonite stratigraphy, and miscorrelation of ammonite assemblages among paleogeographic provinces (Pessagno and Meyerhoff Hull, 1996).

18.2.3 Physical stratigraphy

MAGNETOSTRATIGRAPHY

The M-sequence of marine magnetic anomalies is the template for calibrating magnetostratigraphy from Upper Jurassic fossiliferous sections. Several high-resolution studies of Kimmeridgian and Tithonian strata have correlated ammonites, calpionellids, and calcareous nannofossils from the sub-Mediterranean faunal province and DSDP cores from the central Atlantic to polarity chrons M25 through M18 (e.g. Bralower *et al.*, 1989: syntheses in Ogg, 1988; Ogg *et al.*, 1991a; Ogg and Atrops, in prep.). Calibration of Boreal sections to this magnetic time scale has been partially achieved for the Tithonian (Ogg *et al.*, 1994) and the middle Kimmeridgian (Ogg and Coe, in prep.). (Based upon magnetostratigraphy of coastal sections along the French side of the English Channel (Ogg and Wimbledon, in prep.), the reversed-polarity interval of the *Progalbanites albani* ammonite zone at the base of the Portland Sand is now assigned as Chron M20r, instead of M21r as interpreted by Ogg *et al.* (1994).) The suite of calibrations shown in Fig. 18.1 is scaled to a variable spreading model of the Pacific M-sequence during the Kimmeridgian through lower Cretaceous (see Chapters 5 and tables therein). These calibrations constrain the relative duration of each ammonite zone within the Kimmeridgian and Tithonian stages.

The oldest magnetic anomaly that is documented in all ocean basins is M25, which has been correlated to the base of the *Sutneria platynota* Zone, which is the base of the Kimmeridgian as traditionally assigned in the sub-Mediterranean province (Ogg *et al.*, 1984; Ogg and Atrops, in prep.). Magnetic profiles over pre-M26 oceanic crust in the Pacific (Handschumacher *et al.*, 1988) have been supported and extended by deep-tow surveys (Sager *et al.*, 1998), thereby indicating a possible set of between 50 and 100 polarity chrons within the Callovian and Oxfordian stages. In contrast to the well-resolved major magnetic anomalies younger than M25, it is uncertain how many of the modeled short-duration, low-amplitude polarity intervals in the deep-tow data are paleomagnetic intensity fluctuations rather than actual geomagnetic reversals. An array of magnetostratigraphic studies in Oxfordian strata with sub-Mediterranean ammonite zonation in Spain and Poland yielded reversal patterns that were consistent with the extended Pacific model (e.g. Steiner *et al.*, 1986; Juárez *et al.*, 1994, 1995; Ogg and Gutowski, 1996). The magnetostratigraphy of a suite of overlapping sections in Great Britain with sub-Boreal and Boreal ammonite zones, coupled with a revised correlation of the Oxfordian–Kimmeridgian boundary interval between the sub-Mediterranean and Boreal provinces, enabled construction of a complete Oxfordian magnetic polarity time scale for the combined Boreal and Tethyan faunal realms and correlation to the main features of the Pacific magnetic anomaly pattern from M25 through M36 (Ogg and Coe, 1997; Ogg and Coe, in prep.). This calibrated M-sequence scale for the Oxfordian is illustrated in Fig. 18.1 with ammonite zones of both faunal realms scaled to a spreading-rate model for M25 through M41 (see Chapter 5). Minor divergences of the magnetostratigraphic pattern (right-hand scale) with the modeled deep-tow signature (left-hand scale) are generally consistent with uncertainties in the marine magnetic model (Sager *et al.*, 1998).

The magnetostratigraphy of the Upper Callovian in European sections is also consistent with the simple marine magnetic anomaly pattern of M36 through M38. However, the marine magnetic profiles of the presumed Callovian portion of the "Jurassic Quiet Zone" older than M38 display a very low-amplitude short-wavelength oscillation of magnetic intensity of uncertain origin (Sager *et al.*, 1998). Magnetographic investigations of Lower and Middle Callovian strata in England, France, and Poland (e.g. Ogg *et al.*, 1991b; Belkaaloul *et al.*, 1995, 1997; Ogg with Garcia, Coe, and Dietl, unpublished) are generally dominated by normal polarity with only a few correlative reversed-polarity intervals, mainly within the *Macrocephalites (Dolikephalites) gracilis* and *Clydoniceras discus* ammonite zones. Therefore, it remains to be demonstrated that the modeled short-wavelength fluctuations over Pacific crust older than Late Callovian are a reliable indicator of the history of the geomagnetic field. Figure 18.1 displays the tentative common polarity pattern derived from French and English sections scaled to equal-duration ammonite subzones.

The Early and Middle Jurassic magnetic polarity patterns are primarily compiled from ammonite-bearing sections in Europe. A suite of Bathonian and Bajocian sections in Spain displays a rapidly changing magnetic polarity pattern (Steiner *et al.*, 1987), but this has not yet been fully verified elsewhere.

The Aalenian, Toarcian, and Pliensbachian polarity pattern is primarily derived from a detailed study in southern Switzerland (Horner, 1983; Horner and Heller, 1983). This pattern has been consistent with later magnetostratigraphy work across the base-Bajocian GSSP at Cabo Mondego, Portugal (Henriques *et al.*, 1994; Pavia and Enay, 1997), the base-Aalenian GSSP at Fuentelsaz, Spain (Goy *et al.*, 1996; Cresta, 1999), and the relatively condensed type Toarcian section at Thouars, France (Galbrun *et al.*, 1988, using the inter-province ammonite zone correlation in Hardenbol *et al.*, 1998). The Pliensbachian portion has been enhanced and recalibrated using boreholes in the Paris Basin (Moreau *et al.*, 2002; Fig. 18.1).

The Hettangian and Sinemurian stages have not yet yielded a verified magnetostratigraphy. The Sinemurian and Hettangian appear to be dominated by reversed polarity (Steiner and Ogg, 1988; Gallet *et al.*, 1993), with perhaps several minor normal-polarity zones (Yang *et al.*, 1996), and the basal-Hettangian may be dominated by normal polarity (Kent *et al.*, 1995; Yang *et al.*, 1996; Posen *et al.*, 1998; Kent and Olsen, 1999). Only a possible schematic pattern, not correlated to biostratigraphy, is shown in Fig. 18.1.

Polarity chrons prior to the Callovian do not have a corresponding marine magnetic anomaly sequence to provide an independent nomenclature or scaling system. A compilation by Ogg (1995) suggested using abbreviations derived from the corresponding ammonite zones or stage. However, until the polarity pattern spanning each stage has been adequately verified, a standardized nomenclature system is not possible.

CHEMICAL STRATIGRAPHY

A comprehensive review of Jurassic chemostratigraphy trends and excursions is compiled by Jenkyns *et al.* (2002). Only the major features are summarized below.

Stable isotopes of carbon Two major negative excursions in carbon isotopes, accompanied by an abundance of organic-carbon-rich sediments, are recognized in the Jurassic in the Toarcian and Oxfordian, and lesser anomalies occur in other intervals (reviewed in Jenkyns *et al.*, 2002; Fig. 18.2).

Lower Toarcian pelagic limestone in the upper *Harpoceras serpentinus* ammonite zone (also known as upper *Harpoceras falciferum* ammonite zone) display a major positive carbon-13 excursion (e.g. Jenkyns and Clayton, 1986, 1997). High-resolution isotope stratigraphy (McArthur *et al.*, 2000) suggests that the positive excursion consists of multiple oscillations, but identification of global trends may be compromised by superimposed local and basinal effects (Jiménez *et al.*, 1996). This carbon isotope excursion in carbonates follows a widespread occurrence of organic-carbon-rich marine deposits, such as the Posidonienschiefer of Germany and the Jet Rock of England, that are associated with a major transgression during the upper *Dactylioceras tenuicostatum* and lower *H. serpentinus* Zones (e.g. Hallam, 1981b; Jenkyns, 1988). These organic-rich sediments are associated with pronounced negative carbon-13 excursions in both organic and carbonate components, which is one of the largest of the Phanerozoic (e.g. Hesselbo *et al.*, 2000b, 2003; Schouten *et al.*, 2000). This early Toarcian "oceanic anoxic event" at approximately 183 Ma persisted for only about 0.5 myr (McArthur *et al.*, 2000). The

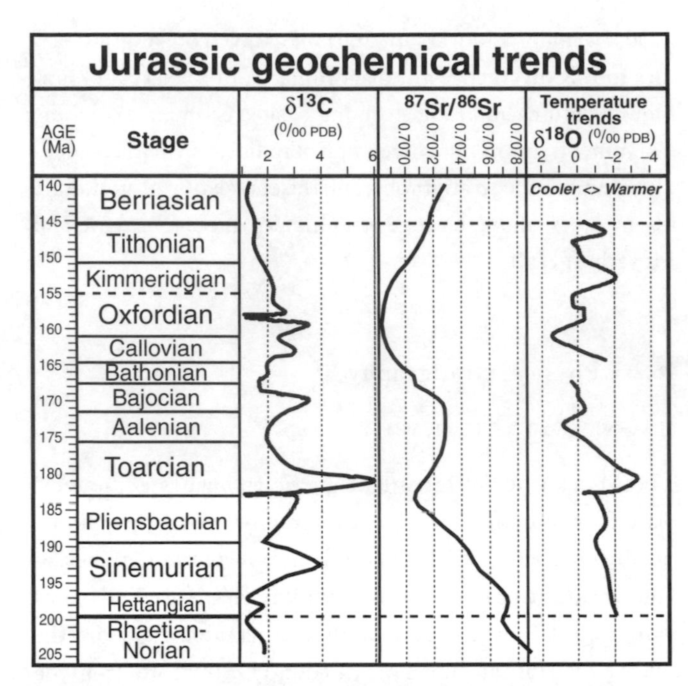

Figure 18.2 Smoothed trends and excursions in carbon and oxygen stable isotopes and in marine $^{87}Sr/^{86}Sr$ ratio during the Jurassic. The schematic carbon isotope curve is generalized from the compilations in Bartolini *et al.* (1996) and Jenkyns *et al.* (2002), with details on the complex Toarcian and the Oxfordian excursions from McArthur *et al.* (2000) and Padden *et al.* (2001), respectively. The strontium isotope curve is a LOWESS fit to data from several sources – see text and Chapter 7. The schematic oxygen isotope curve is mainly from belemnites from Europe, but the main trends are also observed in New Zealand (late Jurassic) and Canada (middle Jurassic) (from Jenkyns *et al.*, 2002). Overall global shifts to higher oxygen-18 values in carbonates are generally interpreted as cooler seawater, but there are many other contributing factors. The data for these schematic carbon and oxygen isotope curves have significant scatter, and should be considered as only general indications of the trends.

event seems to have been coincident with the eruption of the Karoo–Ferrar flood basalts across South Africa and Antarctica (Kerr, 2000; Pálfy and Smith, 2000) and a methane release from undersea destabilization of frozen methane hydrates may have contributed to this excursion (Hesselbo *et al.*, 2000b; Cohen *et al.*, 2004).

Middle Oxfordian pelagic sediments indicate a broad peak in carbon-13 (Jenkyns, 1996) that contains a pair of brief (50 kyr) major negative excursions within the upper *Gregoryceras transversarium* ammonite zone (Padden *et al.*, 2001, 2002). Other major features are a negative excursion in carbon-13 near the Sinemurian–Pliensbachian boundary, and generally low values in the Bathonian–Bajocian boundary interval and Tithonian (Jenkyns *et al.*, 2002). Excursions in this general trend in carbon isotope are reported within the Triassic–Jurassic boundary interval (e.g. Pálfy *et al.*, 2001; Hesselbo *et al.*, 2002; see Chapter 17), and in the lower Bajocian and upper Bathonian to lower Callovian pelagic sediments in Italy (Bartolini *et al.*, 1996, 1999; Morettini *et al.*, 2002). However, these postulated excursions have not yet been calibrated to ammonite zones. The Jurassic–Cretaceous boundary interval lacks any carbon isotope excursions to aid in global correlation of the system boundary (Weissert and Channell, 1989).

Stable isotopes of oxygen and temperature trends The Jurassic is generally considered as an interval of sustained warmth without any documented glacial deposits. Oxygen isotope records of oceanic temperature trends are patchy and heavily biased toward records from Europe and Russia (e.g. Veizer *et al.*, 2000; Jenkyns *et al.*, 2002). These suggest an overall warm period (lighter oxygen-18 values) from Hettangian to Toarcian, cooler temperatures during Aalenian through Oxfordian (but with potential anomalies), and moderate temperatures during the Kimmeridgian and Tithonian (Fig. 18.2). Except for the implied "cold snap" during the Callovian–Oxfordian from the isotope compilation, these general temperature trends are consistent with other paleoclimate indicators (Jenkyns *et al.*, 2002).

Strontium and osmium isotope ratios The curve of marine $^{87}Sr/^{86}Sr$ through the Jurassic is a broad valley centered on the early Oxfordian, with a subordinate trough in latest Pliensbachian time (Fig. 18.2). Both minima broadly coincide with major carbon isotope excursions. Except within the Aalenian and the Oxfordian, the rapidly changing ratios enable global correlation at high resolution (McArthur *et al.*, 2001).

At the end of the Triassic, an abrupt downturn in marine $^{87}Sr/^{86}Sr$ from the latest Triassic peak (0.707 95) coincided with major flood basalt outpouring of the "Central Atlantic Magmatic Province" along the future Central Atlantic seaway (McArthur *et al.*, 2001). The strontium isotope ratio continued a steady decline through the Early Jurassic to a trough (0.707 08) in the latest Pliensbachian (Jones *et al.*, 1994a; McArthur *et al.*, 2000; Jones and Jenkyns, 2001). By assuming that this decrease was linear through the Hettangian, Sinemurian, and Pliensbachian stages, and scaling the slope with cycle stratigraphy in the lower Pliensbachian, Weedon and Jenkyns (1999) estimated that the minimum duration of these three stages was 2.86, 7.62, and 6.67 myr, respectively, for a total of 17.15 myr. A similar calculation using an expanded Lower Jurassic database (McArthur *et al.*, 2001) yields 3.10, 6.90, and 6.60 myr, respectively, for a total of 16.90 myr.

Strontium isotope ratios progressively rose during the Toarcian to a sustained plateau (0.707 30) through the Aalenian. If the Pliensbachian fall and Toarcian rise are assumed to be linear segments, then a high-resolution time scale can be constructed for scaling ammonite subzones and carbon isotope excursions (McArthur *et al.*, 2000). Strontium isotope ratios again decreased through the Bajocian to middle Callovian, with a shoulder spanning the Bajocian–Bathonian boundary (Jones *et al.*, 1994b; M. Engkilde, unpublished) that may be real or may be an artifact of differential rates of sedimentation and the imposed time scale.

During the early Oxfordian, marine $^{87}Sr/^{86}Sr$ reached its lowest ratio (0.706 86) throughout the entire Phanerozoic (McArthur *et al.*, 2001). This pronounced episode may indicate a major pulse of seafloor hydrothermal activity (Jones *et al.*, 1994b; Jones and Jenkyns, 2001), which is supported by interpretations of other geochemical, deep-sea sediment, and spreading-rate evidence (e.g. Ogg *et al.*, 1992). From the middle Oxfordian, the strontium isotope ratio began a long-term increase that peaked in the *P. elegans* ammonite zone of the Barremian Stage of Early Cretaceous.

As with Sr, the $^{187}Os/^{188}Os$ value of marine Os reflects competing fluxes of Os from the input of ^{187}Os-rich fluids, from continental weathering, and of non-radiogenic ^{188}Os-rich fluids from hydrothermal alteration of oceanic crust and from dissolution of extraterrestrial material from meteorites, but with a shorter residence time (about 40 kyr). Jurassic $^{187}Os/^{188}Os$ ratios have been derived from organic-rich shales in southern England (Cohen *et al.*, 1999). The extremely non-radiogenic $^{187}Os/^{188}Os$ ratio of 0.15 in the basal-Hettangian (*Psiloceras planorbis* ammonite zone) is similar to the anomalous end-Cretaceous excursion produced by the ^{188}Os-rich bolide impact, but is interpreted as hydrothermal activity associated with the eruption of the Central Atlantic Magmatic Province (Cohen *et al.*, 1999; Cohen and Coe, 2002). In contrast to the descending strontium trend, the $^{187}Os/^{188}Os$ ratio progressively rises through the Hettangian and Sinemurian, and attains a ratio of 0.8 in the lower Toarcian. The average upper Kimmeridgian and lower Tithonian ratio is about 0.6.

CYCLE STRATIGRAPHY AND ESTIMATES OF STAGE DURATIONS

Period oscillations in composition or physical properties within several marine successions in the Jurassic have been interpreted as Milankovitch orbital-climate cycles. Sudden shifts in the relative dominance of obliquity (37–38 kyr in Jurassic) and precession (20 kyr) within Jurassic cyclostratigraphy sections in England and Italy (30°–35° N paleolatitude) can be associated with times of significant global environmental change (e.g. Hinnov and Park, 1999; Weedon *et al.*, 1999).

Hettangian and Sinemurian Obliquity-dominated cyclicity in the Blue Lias Formation of southern England yields minimum durations of 1.29 myr for the Hettangian Stage and 0.34 myr for the *Arietites bucklandi* ammonite zone of lowest Sinemurian, but these estimates incorporate known stratigraphic breaks (Weedon *et al.*, 1999). Assuming that marine $^{87}Sr/^{86}Sr$ had a linear decrease through this interval identical to the precession-cycle-scaled trend of the lower Pliensbachian (−0.000 042 per myr), this yields minimum durations of 2.86 myr for the Hettangian Stage and 7.62 myr for the total Sinemurian Stage (Weedon and Jenkyns, 1999), or 3.10 and 6.90 myr using the LOWESS-fit Sr-curve of McArthur *et al.* (2001).

Pliensbachian Precession-dominated cyclicity in the Belemnite Marls of southern England, combined with linear strontium isotope trends and cyclostratigraphic data from Robin Hood's Bay, northeast England (van Buchem *et al.*, 1994), and Breggia Gorge, southern Switzerland (Weedon, 1989), indicate a minimum duration of 6.67 myr for the Pliensbachian Stage (Weedon and Jenkyns, 1999). This is consistent with the minimum estimate of 5 myr derived from precession-dominated strata in northern Italy (Hinnov and Park, 1999).

Toarcian and Aalenian Obliquity-dominated cyclic carbonates of the Sogno Formation in northern Italy yielded a combined duration of 11.37 ± 0.05 myr for the Toarcian and Aalenian stages (Hinnov and Park, 1999; Hinnov *et al.*, 1999). A mid-Toarcian date of 181.4 ± 0.6 Ma (1-sigma; Pálfy *et al.*, 1997) is projected to fall 132 obliquity cycles (4.9 myr) below the Aalenian–Bajocian boundary, implying an age for the boundary of 176.5 ± 0.6 Ma (Hinnov and Park, 1999). This age estimate for the Aalenian–Bajocian boundary is identical to an earlier statistical fit of combined radiometric ages and biozones by Gradstein *et al.* (1995).

Kimmeridgian and Tithonian Lithologic and magnetic-susceptibility variations within the Kimmeridge Clay Formation of southern England appear to be associated with obliquity with perhaps minor contributions from precession (e.g. Waterhouse, 1995; Weedon *et al.*, 1999). Tuning all susceptibility peaks to a fixed 38-kyr obliquity cycle implies that the *Aulacostephanus autissiodorensis* ammonite zone of the uppermost Kimmeridgian Stage spans 1.35 myr and the overlying *Pectinatites* (*Virgatosphinctoides*) *elegans* Zone of basal-Tithonian spans 0.55 myr (Weedon *et al.*, 1999). However, if this obliquity tuning is applied to all overlying susceptibility peaks, then the upper portion of the Kimmeridgian Clay (*P. elegans* through *Virgatopaviovia fittoni* ammonite zones) spans a minimum of ~4.3 myr (Weedon, 2001; Weedon *et al.*, 2004). If verified, this cycle tuning would imply that the "Upper Kimmeridgian" regional stage of classical usage in England coincides with the entire Tithonian Stage of the sub-Mediterranean (Tethyan) faunal realm. This equivalency has not been incorporated into our Jurassic time scale (Fig. 18.1), pending satisfactory consistency with other correlation constraints.

SEQUENCE STRATIGRAPHY

Jurassic sea-level trends have been compiled for different basins (e.g. Partington *et al.*, 1993; Coe, 1996; Sahagian *et al.*, 1996; Gygi *et al.*, 1998; Hesselbo and Jenkyns, 1998) and on a global scale (e.g. Hallam, 1978, 1981b, 1988, 2001; Haq *et al.*, 1988; Hardenbol *et al.*, 1998).

The main Jurassic sea-level trend is a progressive transgression from the latest Triassic until the late Kimmeridgian. A major regressive trend begins in the Tithonian and reaches a minimum in the late Berriasian.

Superimposed on these main cycles are several major sequences (especially a lesser regressive trend during middle Toarcian and Aalenian). The larger-scale deepening and shallowing trends compiled by Hardenbol *et al.* (1998) are summarized in Fig. 18.1. Assignments of small-scale sequences depend on interpretation models for the response of sediment facies (other than obvious relative sea-level falls or flooding surfaces) to relative sea-level changes, therefore interpretations vary among stratigraphers for assigning small-scale sequences within a given region (e.g. comparison charts within Hesselbo and Jenkyns, 1998; Newell, 2000; and Taylor *et al.*, 2001).

OTHER MAJOR STRATIGRAPHIC EVENTS

Large igneous provinces The Central Atlantic Magmatic Province has ages centered on 200 Ma (Olsen *et al.*, 2003), and probably peaked just before the Triassic–Jurassic boundary as recognized in the marine realm (Pálfy *et al.*, 2000b), and may be the largest known flood basalt (e.g. Marzoli *et al.*, 1999; Hames *et al.*, 2000). (This major volcanic outpouring is summarized in Chapter 17.)

The majority of the Karoo flood basalts in South Africa and the Farrar volcanics in Antarctica erupted at 183 ± 2 Ma (Pálfy and Smith, 2000). This large igneous province coincides with major geochemical anomalies, organic-rich strata, and faunal extinctions within the earliest Toarcian (e.g. Jones and Jenkyns, 2001; Wignall, 2001; Cohen and Coe, 2002; Pálfy *et al.*, 2002b).

Large impact events The Puchezh–Katunki crater in Russia, with an apparent diameter of 80 km, has an age reported as $\sim 167 \pm 3$ Ma, implying a large impact within the Bathonian–Bajocian interval.

A modest iridium anomaly has been reported from near the palynology-defined Triassic–Jurassic boundary in eastern USA, and associated features, such as a fern spike and apparent suddenness of the terrestrial extinctions suggest a possible impact relationship (Olsen *et al.*, 2003).

18.3 JURASSIC TIME SCALE

The time scale for the Jurassic Period and the scaling of events within each geological stage combines many types of stratigraphic information: a selection of key radiometric dates, constraints on durations from cycle stratigraphy, applying linear trends to strontium isotopic variation within certain intervals, proportional scaling of some ammonite zone successions according to their subzonal numbers, and applying the M-sequence magnetic polarity time scale derived from estimates of Pacific seafloor-spreading rates. Table 18.2 summarizes the scaling methods and assumptions for each stage and each component ammonite zone.

18.3.1 Selection of radiometric ages

Suites of high-precision U–Pb and $^{40}Ar/^{39}Ar$ ages from ammonite-zoned strata have largely rendered obsolete Jurassic time scales published before 1996, which had incorporated ages from less precise K–Ar and Rb–Sr methods. Pálfy (1995) critiqued the databases and methodology of previous Jurassic time scales, and his team began a systematic effort to establish a detailed Early and Middle Jurassic age array from U–Pb analysis of zircons from volcaniclastics inter-bedded with ammonite-zoned strata in British Columbia and Alaska (Pálfy *et al.*, 1997, 1999, 2000b,c; Pálfy and Smith, 2000). A series of studies have calibrated North American ammonite zones or specific ammonite datums to the standard northwest European zones and associated definition of geological stages (summarized in Pálfy *et al.*, 2000a). This effort required re-evaluation of biostratigraphic material (e.g. Pálfy *et al.*, 1997, observed that there were approximately 25% erroneous identifications of Toarcian ammonites at the species level in other reference biostratigraphic sections).

Pálfy *et al.* (2000a) compiled a database of 55 latest Triassic through Tithonian ages that were derived solely from U–Pb and $^{40}Ar/^{39}Ar$ methods, of which only 12 were from publications of pre-1995 vintage. Their detailed analysis of the stratigraphic control and radiometric behavior of each item (Appendix 1 in Pálfy *et al.*, 2000a) is partially summarized in Table 17.1.

Ten other U–Pb and $^{40}Ar/^{39}Ar$ ages from the Jurassic that had been used in Harland *et al.* (1990), and hence also incorporated into Gradstein *et al.* (1994a, 1995), did not meet the more rigorous standards for radiometric behavior or stratigraphic control of Pálfy *et al.* (2000a, their Appendix 1). They also excluded all previous high-temperature-mineral ages derived by the K–Ar (and Rb–Sr) method, which are considered less reliable and do not easily allow detection of a geochemical error, in which loss of radiogenic daughter isotopes produces a younger apparent age (e.g. see comparisons in Pálfy, 1995). Unexpectedly, after rejecting all K–Ar ages in favor of a select U–Pb and $^{40}Ar/^{39}Ar$ database, their statistical estimates of all Jurassic stage boundaries (Pálfy *et al.*, 2000a) are systematically shifted to younger ages relative to the previous all-inclusive fit by Gradstein *et al.* (1994a, 1995). This result suggests that some K–Ar ages may provide useful age approximations, albeit with greater analytical uncertainties, of stratigraphic units.

With these caveats, our selected ages for constructing the Jurassic time scale are summarized in Table 17.2. This suite was selected from a larger compilation of published ages on the basis of both stratigraphic and radiometric precision.

18.3.2 Hettangian through Aalenian

The Early Jurassic time scale integrates radiometric ages with cycle stratigraphy, linear trends in strontium isotope ratios, and relative numbers of ammonite subzones. In most cases, this integrated stratigraphy yields estimates for the stage boundaries that are close to a statistical fit of U–Pb age dates by Pálfy *et al.* (2000a), but our procedure also yields a high-resolution scaling of component ammonite zones within some stages.

The *base of the Jurassic* (*base of the Hettangian*) is tightly constrained by a U–Pb age of *199.6* ± 0.3 Ma (95% confidence limits) from zircons in a tuff directly below the Triassic–Jurassic boundary in British Columbia (Pálfy *et al.*, 2000b). This is currently the only well-documented U–Pb zircon age with narrow 95% confidence limits (2-sigma) that directly constrains a stage boundary within the Jurassic! The Hettangian spans 3.1 myr according to the cycle-scaled Sr trend (Weedon and Jenkyns, 1999, applied to LOWESS fit of McArthur *et al.*, 2001). Therefore, the *base of the Sinemurian* is at *196.5 Ma*. The three Hettangian ammonite zones are proportionally scaled relative to subzones (Table 18.2).

The Sinemurian spans 6.90 myr according to the cycle-scaled Sr trend (Weedon and Jenkyns, 1999, applied to LOWESS fit of McArthur *et al.*, 2001). Therefore, the *base of the Pliensbachian* is at *189.6 Ma*. The six Sinemurian ammonite zones are proportionally scaled relative to subzones.

The Pliensbachian spans 6.60 myr according to the cycle-scaled Sr trend (Weedon and Jenkyns, 1999, applied to LOWESS fit of McArthur *et al.*, 2001). Therefore, the *base of the Toarcian* is at *183.0 Ma*. This is only slightly younger, but within the narrow error bars, than the 183.6 Ma (+1.7/−1.1 myr) estimate from a statistical fit of U–Pb age dates (Pálfy *et al.*, 2000a). Cycle stratigraphy implies that the lower Pliensbachian spans 2.65 myr; and we have proportionally scaled the durations of its three ammonite zones relative to their subzones. The two ammonite zones of the upper Pliensbachian (spanning 3.95 myr) are proportions scaled according to their placement along a linear Sr trend (McArthur *et al.*, 2000; Table 18.2): *Amaltheus margaritatus* = 2.75 myr, *Pleuroceras spinatum* = 1.20 myr.

Zircons from strata of British Columbia that are equivalent to the lower part of the *Haugia variabilis* ammonite zone of the upper Middle Toarcian yield a U–Pb age of 181.4 ± 1.2 Ma (95% confidence limits; Pálfy *et al.*, 1997). The younger limit (~180.5 Ma) of the standard deviation on this age was used to scale the four ammonite zones (14 subzones) of the Early and Middle Toarcian according to their relative proportions along a linearly increasing Sr curve (McArthur *et al.*, 2000; Table 18.2): *Dactylioceras tenuicostatum* = 0.3 myr, *Harpoceras falciferum* = 1.5 myr, *Hidoceras bifrons* = 0.5 myr, and *Haugia variabilis* = 0.2 myr.

The combined Toarcian and Aalenian spans 11.37 ± 0.05 myr according to obliquity-cycle stratigraphy (Hinnov and Park, 1999; Hinnov *et al.*, 1999), which projects the Aalenian–Bajocian boundary as ~171.6 Ma. This assignment is younger, but within the large uncertainty of the estimated 174.0 Ma (+1.2/−7.9 myr) from a statistical fit of U–Pb age dates (Pálfy *et al.*, 2000a). Cycle stratigraphy in pelagic strata from a former seamount at Bugarone, central Italy, suggests that the Aalenian spans only 4 myr (L. Hinnov, pers. comm., 2001). Therefore, the *base of the Aalenian* is estimated as *175.6 Ma*. Durations of the four Late Toarcian and the four Aalenian ammonite zones are proportionally scaled relative to equal subzones.

(Note: Pálfy *et al.* (2000a) estimated the base of the Aalenian as 178.0 Ma (+1.0/−1.5 myr), but this estimate was mainly constrained by a U–Pb date from a possible late Aalenian volcanic unit in British Columbia of 177.3 ± 0.8 Ma (95% confidence limits), of which only one of the four samples yielded a concordant analysis.)

18.3.3 Bajocian through Callovian

The Callovian, Bathonian, and Bajocian stages are poorly constrained by radiometric dates. The Toarcian–Aalenian cycle stratigraphy implies that the *base of the Bajocian* is at *171.6 Ma*. The M-sequence calibration of the ammonite zones within the Oxfordian Stage, coupled with the age model for the corresponding magnetic anomalies in the Pacific (see Chapter 5) imply that the base of the Oxfordian is approximately 161.1 Ma. Therefore, the Callovian, Bathonian, and Bajocian stages span 10.5 myr.

Durations of the seven Bajocian, eight Bathonian, and seven Callovian ammonite zones are scaled proportional to their subzones in the selected standard zonation (e.g. see diagrams in Hardenbol *et al.*, 1998; Table 18.2). Using this simplistic assumption of equal average subzones, the Bajocian (20 subzones) spans 3.9 myr. The *base of the Bathonian* is therefore estimated as *167.7 Ma* (which is within the estimate of 166.0 Ma (+3.8/−5.6 myr) by Pálfy *et al.*, 2000a), and this stage spans 3.0 myr (15 subzones).

The *base of the Callovian* is therefore assigned as *164.7 Ma*, and the Callovian (18 subzones) spans 3.5 myr. Pálfy *et al.*

(2000a) assigned a much younger age of 160.4 Ma (+1.1/−0.5 myr) to the base of the Callovian, which was largely derived from a reported U–Pb date of 160.5 ± 0.3 Ma (95% confidence limits) from the equivalent of the Bathonian–Callovian boundary in Argentina (Odin *et al.*, 1992). However, it is difficult to reconcile this Bathonian–Callovian boundary age with the age model for the Oxfordian through Callovian M-sequence magnetic anomalies in the Pacific, which is constrained by an Ar–Ar age on oceanic crust at Pacific ODP Site 801.

18.3.4 Oxfordian through Tithonian

The Late Jurassic time scale is based on the magnetostratigraphic correlation of Tethyan and Boreal ammonite zones to the M-sequence polarity time scale. The radiometric age constraints and associated assignment of absolute ages to the M-sequence magnetic anomaly pattern in the Pacific is explained in detail in Chapter 5.

There are no high-precision ages obtained directly on volcanogenic units within middle Callovian through Tithonian marine sediments. Volcanic ash horizons within the continental facies of the upper Morrison Formation of western USA have several $^{40}Ar/^{39}Ar$ ages spanning the 148–151 Ma interval (Kowallis *et al.*, 1998), and palynology, ostracode, and charophyte assemblages are interpreted as Kimmeridgian to early Tithonian (Litwin *et al.*, 1998; Schudack *et al.*, 1998). The Morrison Formation has several magnetostratigraphic studies (e.g. Steiner *et al.*, 1994), but unambiguous correlation and associated age control on the M-sequence has not yet been possible (F. Peterson, pers. comm., 2001).

The ages of the Oxfordian through Tithonian stages are determined by the calibration of their magnetostratigraphy. The *base of the Oxfordian* (base of polarity Chron M36An; Ogg and Coe, 1997, and in prep.) is at *161.2 Ma*. The *base of the Kimmeridgian* as currently assigned in *Boreal* realm ammonite stratigraphy (lower part of polarity Chron M26r; Ogg and Coe, 1997, and in prep.) is at *155.65 Ma*, whereas its base as currently assigned in *Tethyan* (sub-Mediterranean) stratigraphy (base of polarity Chron M25r; Ogg *et al.*, in prep.) is at *154.55 Ma*. This base-Kimmeridgian age agrees with an independent statistical estimate of 154.7 Ma (+3.8/−3.3) for the Oxfordian–Kimmeridgian boundary derived from a radiometric age database by Pálfy *et al.* (2000c).

The *base of the Tithonian* (base of polarity Chron M22An; Ogg and Atrops, in prep.) is at *150.8 Ma*. Oxfordian, Kimmeridgian, and Tithonian ammonite zones are scaled according to their calibration to the magnetic polarity time scale (Table 18.2). The *Jurassic–Cretaceous boundary* or the base of the Berriasian, which is assigned here as the base of *Berriasella jacobi* ammonite zone (middle of polarity Chron M19n.2n), is at *145.5 Ma*.

18.3.5 Estimated uncertainties on stage ages and durations

A variety of methods were applied to estimate Jurassic stage boundaries, and each of these has different degrees of imprecision. In general, owing to a dearth of verified high-precision radiometric ages between the Aalenian and the Albian, the Jurassic age assignments are the least accurate for any portion of the Phanerozoic time scale. In the following discussion, the uncertainties are expressed as 2-sigma (95% confidence limits).

The Early Jurassic scale merged high-precision radiometric ages, Milankovitch cycle tuning, and linear strontium isotope segments, therefore the precision of the interpolated stage ages are approximately the same as the radiometric constraints – from ±0.6 myr for base-Hettangian (after increasing the cited ±0.3 myr uncertainty on the U–Pb age to incorporate potential systematic errors) increasing to ±1.5 myr for base-Toarcian. The early part of the Middle Jurassic is scaled by cycle stratigraphy relative to the base-Toarcian age, but these duration estimates have not yet been verified in independent sections, therefore a conservative uncertainty of ±2 myr was applied to the base-Aalenian age and ±3 myr for the base-Bajocian age.

The choice of spreading-rate model for the synthetic profile of the M-sequence magnetic anomaly lineations, coupled with the uncertainties on the two constraining ages on Chron M26n (155 ± 6 Ma) and Site 801 (167.7 ± 1.4 Ma; see Chapter 5), implies that the magnetostratigraphic calibration of the Oxfordian through Tithonian stage boundaries have a high degree of uncertainty (estimated as ±4 myr), which is the highest uncertainty on the age of any stage boundary within the Phanerozoic (see Chapter 23). The choice of a spreading model will also affect the durations by expanding or contracting the reference scale. We conservatively estimate this uncertainty on durations as being ±35% for a realistic range of spreading models (Table 18.3).

The base-Bathonian and base-Callovian are proportionally scaled according to their component ammonite subzones between the calculated ages for the base-Bajocian (from cycle stratigraphy) and base-Oxfordian (from the spreading model), and therefore have equivalent or greater uncertainties.

Table 18.3 *Ages and duration of Jurassic stages (see footnote)*

Boundary	Stage	Age (Ma)	*Estimated uncertainty on age (2-sigma)*	Duration (myr)	*Estimated uncertainty on duration (2-sigma)*	Status; primary marker	Calibration and comments
	CRETACEOUS (Berriasian)						
Top of JURASSIC (base of Berriasian)		**145.5**	*4.0*			Not defined. Base of Cretaceous (base of Berriasian) placed here as the leading candidate of the base of *Berriasella jacobi* ammonite zone or base of Calpionellid Zone B; which is calibrated as Chron M19n.2n(0.55)	Chron ages and duration of Tithonian through Oxfordian stages depend upon selected Pacific spreading rate model. Not well constrained. A slower spreading rate would yield a longer duration.
	Tithonian			**5.3**	*1.8*		Uncertainties on Tithonian through Oxfordian durations are assigned as 35% of duration, which is probable range of realistic spreading models.
Base of Tithonian		**150.8**	*4.0*			Not defined, but candidate marker is the lowest occurrence of ammonite *Hybonoticeras aff hybonotum*, which is at base of Chron M22An.	Computed from M-sequence spreading model.
	Kimmeridgian			**3.8** *or* **4.9**	*~1.5 myr*		Duration of Kimmeridgian depends on eventual decision for Kimm/Oxf boundary definition.
Base of Kimmeridgian (Tethyan)		**154.6**	*4.0*			Traditional usage is base of *Sutneria platynota* Zone, which is just above base of Chron M25r.	Computed from M-sequence spreading model.
Base of Kimmeridgian (Boreal)		**155.7**	*4.0*			Traditional usage is base of *Pictonia baylei* Zone, which is lower part of polarity Chron M26r.	Computed from M-sequence spreading model.
	Oxfordian			**5.5** *or* **6.6**	*~2 myr*		Duration of Oxfordian depends on eventual decision for Kimm/Oxf boundary definition.
Base of Oxfordian		**161.2**	*4.0*			Not defined, but candidate marker is *Quenstedtoceras mariae* ammonite zone, which is correlated to base of Chron M36An.	Computed from M-sequence spreading model.
	Callovian			**3.5**	*1.0*		Equal subzones in Bajo-Bath-Callov (53 subzones span 10.5 myr). Callovian has 18 subzones.
Base of Callovian		**164.7**	*4.0*			Not defined, but candidate marker is base of *Macrocephalites herveyi* ammonite zone	Scaled according to equal-subzones from computed ages for base-Bajocian and base-Oxfordian.
	Bathonian			**3.0**	*1.0*		Equal subzones in Bajo-Bath-Callov (53 subzones span 10.5 myr). Bathonian has 15 subzones.
Base of Bathonian		**167.7**	*3.5*			Not defined, but candidate marker is base of *Zigzagiceras zigzag* ammonite zone.	Scaled according to equal-subzones from computed ages for base-Bajocian and base-Oxfordian.
	Bajocian			**3.9**	*1.0*		Equal subzones in Bajo-Bath-Callov (53 subzones span 10.5 myr). Bathonian has 20 subzones.
Base of Bajocian		**171.6**	*3.0*			GSSP; lowest occurrence of the ammonite genus *Hyperlioceras (Toxolioceras)*.	Age derived from cycle-scaling from mid-Toarcian U-Pb dates.

Table 18.3 *(cont.)*

Boundary	Stage	Age (Ma)	Estimated uncertainty on age (2-sigma)	Duration (myr)	Estimated uncertainty on duration (2-sigma)	Status; primary marker	Calibration and comments
	Aalenian			**4.0**	*1.0*		Total duration of Aalenian is estimated as 4.0 myr from cycle stratigraphy.
Base of Aalenian		**175.6**	*2.0*			GSSP; lowest occurrence of the ammonite genus *Leioceras.*	Age derived from cycle-scaling from mid-Toarcian U-Pb dates.
	Toarcian			**7.4**	*1.0*		Total duration of Toarcian is estimated as ~7.4 myr from cycle stratigraphy.
Base of Toarcian		**183.0**	*1.5*			Not defined, but candidate marker is lowest occurrence of the diversified *Eodactylites* ammonite fauna.	Age derived from cycle-scaling from basal-Jurassic and mid-Toarcian U-Pb dates.
	Pliensbachian			**6.6**	*0.8*		Total duration of Pliensbachian is estimated as 6.6 myr from assumed linear Sr trend, scaled by cycle stratigraphy.
Base of Pliensbachian		**189.6**	*1.5*			GSSP pending; lowest occurrence of the ammonite species *Bifericeras donovani* and *Apoderoceras* genera.	Age derived from cycle-scaling from basal-Jurassic and mid-Toarcian U-Pb dates.
	Sinemurian			**6.9**	*0.8*		Total duration of Sinemurian is estimated as 6.9 myr from assumed linear Sr trend, scaled by cycle stratigraphy.
Base of Sinemurian		**196.5**	*1.0*			GSSP; lowest occurrence of the ammonite genera *Vermiceras* and *Metophioceras.*	Age derived from cycle-scaling from basal-Jurassic U-Pb date.
	Hettangian			**3.1**	*0.4*		Total duration of Hettangian is estimated as 4.1 myr from assumed linear Sr trend, scaled by cycle stratigraphy.
Base of JURASSIC (base of Hettangian)		**199.6**	*0.6*			Not defined, but traditionally assigned to the first occurrence of the smooth *planorbis* group within the ammonite genus *Psiloceras.*	U-Pb age on tuff near candidate GSSP level is 199.6 ± 0.3 Ma(2-sigma); increased to ± 0.6 (2-sigma) to incorporate potential external errors.
	TRIASSIC (Rhaetian)						

For stages that lack a ratified definition (as of Nov 2004), the computed age is for the indicated primary marker. See text for discussion of derivation of uncertainty estimates.

18.3.6 Summary of the Jurassic time scale

The Jurassic spanned 54 myr, between 199.6 ± 0.6 and 145.5 ± 4.0 Ma. The Early, Middle, and Late Jurassic segments lasted 24, 14.4, and 15.7 myr, respectively. Ages for the Middle and Late Jurassic segments are relatively uncertain. An improved time scale for the Jurassic would require additional high-precision radiometric ages with reliable stratigraphic control, especially a data set that can be used to calibrate the Late Jurassic magnetic polarity scale, and application of Milankovitch cycle stratigraphy to upper Jurassic deposits.

19 • The Cretaceous Period

J. G. OGG, F. P. AGTERBERG, AND F. M. GRADSTEIN

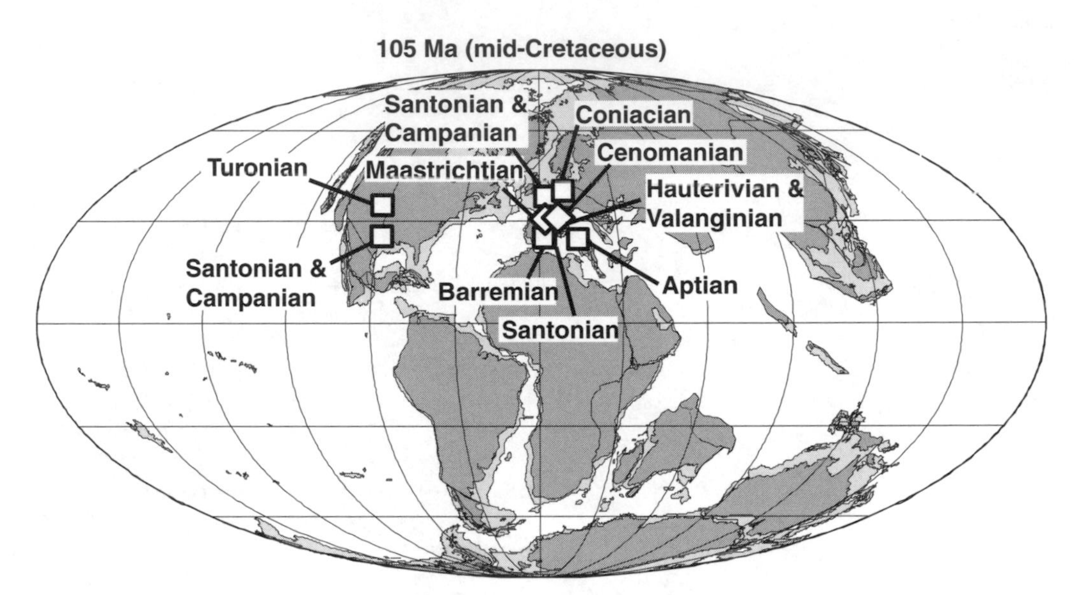

Geographic distribution of Cretaceous GSSPs that have been ratified (diamonds) or are candidates (squares) on a mid-Cretaceous (105 Ma) map (status in January, 2004; see Table 2.3). Not all candidates are shown. GSSPs for the base-Cretaceous and entire Lower Cretaceous are not yet formalized. The projection center is at 30° W.

The Pangea supercontinent fractures into the modern drifting continents. An explosion of calcareous nannoplankton and foraminifera in the warm seas creates massive chalk deposits. A surge in undersea volcanic activity and spreading-ridge formation enhances super-greenhouse conditions in the middle–Late Cretaceous. Angiosperm plants bloom on the dinosaur-dominated land during the Late Cretaceous. The Cretaceous dramatically ends with an asteroid impact.

19.1 HISTORY AND SUBDIVISIONS

19.1.1 Overview of the Cretaceous

Chalk (*creta* in Latin) characterizes a major unit of strata around the Paris Basin and extends across much of Europe. The "Terrain Crétacé" was established by d'Omalius d'Halloy (1822) and defined by him in 1823 to include "the formation of the chalk, with its tufas, its sands and its clays."

A Geologic Time Scale 2004, eds. Felix M. Gradstein, James G. Ogg, and Alan G. Smith. Published by Cambridge University Press.

William Smith had already mapped four stratigraphic units between the "lower clay" (Eocene) and "Portland Stone" (Jurassic), which were grouped by Conybeare and Phillips (1822) into two divisions: the Chalk and the formations below. This two-fold division, adopted in England and France at an early stage, has persisted as the two Cretaceous series and epochs.

Alcide d'Orbigny (1840) grouped the Cretaceous fossil assemblages of France into five divisions, which he termed "étages" (stages): Neocomian, Aptian, Albian, Turonian, and Senonian. He later added Urgonian (between the Neocomian and Aptian) and Cenomanian (between the Albian and Turonian). The broad subdivision "Neocomian," coined by Thurmann in 1836 for strata in the vicinity of Neuchâtel, Switzerland, has been replaced by the three stages of Berriasian, Valanginian, and Hauterivian following the recommendation of Barbier, Debelmas, and Thieuloy in the Colloque sur le Crétacé inférieur (Barbier and Thieuloy, 1965). A "Barremian" Stage and an expanded Aptian Stage have replaced the "Urgonian" Stage. The broad Senonian (originally defined by d'Orbigny to

Table 19.1 *Historical stratotypes of Cretaceous stages*

Stage	Regional	Location	Author (year)	Traditional base
Maastrichtian		Maastricht, southern Netherlands	Dumont (1849)	Lowest occurrence of belemnite *Belemnella lanceolata*
Campanian		Grande Champagne hillside near Aubeterre-sur-Dronne, northern Aquitaine, France	Coquand (1857a)	Lowest occurrence of ammonite *Placenticeras bidorsatum*
Santonian		Saintes, Cognac region, France	Coquand (1857a)	Lowest occurrence of inoceramid bivalve *Cladoceramus undulatoplicatus*
Coniacian		Richemont Seminary near Cognac, France	Coquand (1857a)	Lowest occurrence of ammonite *Forresteria (Harleites) petrocoriensis*
Turonian		Tours region, Touraine, France	d'Orbigny (1842, 1847)	Lowest occurrence of ammonite *Mammites nodosoides*
Cenomanian		Le Mans town, Sarthe region, northern France	d'Orbigny (1847)	Lowest occurrence of ammonite *Mantelliceras* genus
Albian		Aube region, northeast France	d'Orbigny (1842)	Lowest occurrence of NW Europe ammonite *Leymeriella schrammeni*
Aptian		Apt village, Vaucluse, Basse-Alpes region, southeast France	d'Orbigny (1840)	Lowest occurrence of NW Europe ammonite *Prodeshayesites*
Barremian		Barrême village, Basse-Alpes region, southeast France	Coquand (1861)	Lowest occurrence of ammonite *Spitidiscus hugii*
Hauterivian		Hauterive area, Neuchâtel, northwest Switzerland	Renevier (1874)	Lowest occurrence of ammonite *Acanthodiscus* genus
Valanginian		Seyon Gorge near Valangin, Neuchâtel, northwest Switzerland	Desor (1854); Desor and Gressly (1859)	Lowest occurrence of ammonite "*Thurmanniceras*" *otopeta*
Berriasian		Berrias village, Ardèche, southeast France	Coquand (1871)	Lowest occurrence of ammonite *Berriasella jacobi*
	Ryazanian			Lowest occurrence of ammonite *Ructonia ructoni*
Tithonian		None	Oppel (1865)	

encompass all strata from the section at Villedieu through the section at Maastricht) has been divided into four stages: Coniacian, Santonian, Campanian, and Maastrichtian. The uppermost chalks of the Danish coast, which had traditionally been included within the Cretaceous, are now placed as the Danian Stage of the lowermost Cenozoic. The termination of the Cretaceous is now defined by a major mass extinction and impact-generated iridium anomaly horizon at the base of this Danian Stage.

The boundaries of these twelve Cretaceous stages were primarily defined by ammonoids in France and the Netherlands (Birkelund *et al.*, 1984; Kennedy, 1984; Table 19.1). Refined recognition or proposed definitions of the basal boundaries of these stages at the Second International Symposium on Cretaceous Stage Boundaries (Rawson *et al.*, 1996b) have encompassed other global criteria, including geomagnetic reversals, carbon isotope excursions, and microfossil datums (Table 19.2). A key problem in Cretaceous chronostratigraphy is the difficulty in correlating biostratigraphic events and associated stage boundary definitions from the cluster of historical stratotype regions within the Paris Basin, Jura, and southeast France into other paleogeographic and paleoceanographic realms.

Historical usage, coupled with the expertise of members of the various boundary working groups, has dictated a preferential selection of boundary stratotypes for most stages and substages within western European basins, and the accompanying considerations for global correlation have generally not addressed the circum-Pacific regions.

19.1.2 Lower Cretaceous

The subdivisions of the Berriasian through Aptian stages of the Lower Cretaceous were originally derived from exposures in southeast France and adjacent northwest Switzerland, where the paleogeographic Vocontian trough and adjacent margins preserved a nearly continuous record, often in limestone–marl successions (Table 19.1). The fauna and microflora thriving

Table 19.2 *Cretaceous stage definitions (GSSP assignments or status) and informal division into substages. Stages are ordered stratigraphically (youngest at top)*

Stage	Stage boundary *Substage base*	Global Boundary Stratotype Section and Point: Status	Location and point	Primary markers	Other correlations	Comments	References (GSSP, correlations)
Upper Cretaceous							
Maastrichtian	*Upper*	Under discussion	Potentially at Zumaya, northern Spain.	Ammonite, *Pachydiscus fresvillensis* first occurrence.	Approximately coeval with belemnite *Belemnitella junior* lowest occurrence.		Odin *et al.*, 1996
	Maastrichtian/ Campanian	Ratified, 2001	Grande carrière quarry (abandoned), left bank of Adour River, 1.5 km north of Tercis-les-Bains village, or 8 km southeast of Dax railway station, Landes province, southwestern France. GSSP is at 115.70 m level on platform IV in pale flint facies of Les Vignes lithologic unit (beds are vertically dipping).	Ammonite, *Pachydiscus neubergicus* lowest occurrence is immediately above GSSP. GSSP level is mean of a succession of several bioevents, rather than utilizing a single primary event.	Uppermost Campanian markers include, in approximate succession, the highest occurrences (extinction) of ammonite *Nostoceras hyatti* and of dinoflagellate cyst *Corradinisphaeridium horridum*, and benthic foraminifera *Gavelinella clementiana*. Lowest Maastrichtian markers include, in approximate succession, lowest occurrence of ammonite *Diplomoceras cylindraceum*, of planktonic foraminifers *Rugoglobigerina scotti* and *Contusotruncana contusa*, and of ammonite *Hoploscaphites constrictus;* highest occurrences of dinoflagellate cysts *Raetiaedinium truncigerum, Samlandia mayii* and *S. carnarvonensis*, and highest occurrence of calcareous nannofossil *Quadrum trifidum* (at 16m above GSSP).	Magnetostratigraphy, stable isotopes, strontium isotopes and belemnites are not useful at Tercis, but it may be possible to correlate to other sections which have such stratigraphies. Boundary may be approximately coeval with beginning of magnetic polarity Chron C31r, and may correlate with base of *Belemnella obtusa* belemnite zone of NW Germany and near base of *Baculites grandis* ammonite zone of western interior of North America. However, exact equivalences are disputed (see text).	Odin, 1996; Odin *et al.*, 1996; Barchi *et al.*, 1997; Peybernès *et al.*, 1997; Burnett, 1998; Odin & Lamaurelle, 2001; Odin *et al.*, 2001.
Campanian	*Upper*	Under discussion	Under discussion	Substages of "approximately equal duration".		Informal placement in North America at lowest occurrence of ammonite *Didymoceras nebrascense.*	Hancock, Gale *et al.*, 1996
	Middle	Under discussion	Under discussion	Substages of "approximately equal duration".		Informal placement in North America at lowest occurrence of ammonite *Baculites obtusus.*	Hancock, Gale *et al.*, 1996
	Campanian/ Santonian	Working Group recom-mendation	Under discussion. Boundary stratotypes under consideration include Seaford Head (Sussex county, England), Foreness Point (Kent county, England), and Waxahachie dam-spillway (south of Dallas in north-central Texas, USA).	Crinoid, *Marsupites testudinarius* highest occurrence (extinction).	Basal-Campanian markers include the lowest occurrence of ammonite *Scaphites leei III* (uppermost of three subspecies of *S. leei* in North America), the evolutionary transition from *Gonioteuthis granulate* to *G. granulataquadrata* belemnites (used in NW Germany), the lowest occurrence of planktonic foraminifer *Globotruncana elevata,* and a small positive excursion in carbon-13. *[See footnote a for additional markers.]*	Within the early portion of reversed-polarity chron C33r.	Hancock, Gale *et al.*, 1996; Gale *et al.*, 1995; Burnett, 1998

Santonian	*Upper*	Under discussion	Under discussion	Potentially a crinoid, *Uintacrinus socialis* lowest occurrence.	Other markers near this level include the lowest occurrences of planktonic foraminifer *Globotruncanita convexa*, belemnite *Goniotheuthis granulata*, and calcareous nannofossils *Amphizygus minimus* and *Calculites obscurus*.	At or just before the beginning of reversed-polarity chron C33r.	Lamolda, Hancock, *et al.*, 1996; Montgomery *et al.*, 1998
	Middle	Under discussion	Under discussion	Potentially an Inoceramid bivalve, *Cladoceramus undulatoplicatus* highest occurrence (extinction).	Other potential markers are the lowest occurrence of inoceramid bivalve *Cordiceramus cordiformis*, and the base of the *Gonioteuthis corangiunum/westfalica* belemnite zone.	At or just before the beginning of reversed-polarity chron C33r.	Lamolda, Hancock, *et al.*, 1996
	Santonian/ Coniacian	Under-going formal vote (2003)	Cementos Portland quarry (Cantera de Margas) at Olazagutia, Navarra, Spain. Seaford Head (Sussex, England) and Ten Mile Creek (Dallas, Texas, USA) were also considered.	Inoceramid bivalve, *Cladoceramus undulatoplicatus* lowest occurrence.	Uppermost Coniacian markers include the lowest occurrences of ammonite genus *Texanites (Texanites)*, inoceramid bivalve *Magadiceramus subquadratus*, planktonic foraminifer *Sigalia deflaensis*, and calcareous nannofossils *Lithastrinus grillii* (base of nannofossil zone CC 15) and *Micula decussata* (= *Micula staurophora*). Lowermost Santonian markers include the lowest occurrences of benthic foraminifer *Stensioina polonica*, planktonic foraminifer *Dicarinella asymetrica*, and calcareous nannofossil *Lucianorhabdus cayeuxii* (base of nannofossil zone CC16).	The Olazagutia section is situated in a working quarry, therefore accessibility is not fully guaranteed, and the lack of ammonites means that correlation with other regions using macrofossils can best be based on inoceramids and echinoids. However, the micropalaeontological data includes nannofossils, foraminifers (both planktonic and benthic), and stable isotope curves have been obtained.	Lamolda, Hancock, *et al.*, 1996; Cretaceous Subcommission newsletter, Dec. 2002.
Coniacian	*Upper*	Working Group recom-mendation	Under discussion	Inoceramid bivalve, *Magadiceramus subquadratus* lowest occurrence.	Lowest occurrences of planktonic foraminifer *Sigalia carpathica* and benthic foraminifer *Stensioina exsculpta exsculpta*.	The highest occurrence of this inoceramid is near the Coniacian–Santonian stage boundary.	Kauffman *et al.*, 1996
	Middle	Working Group recom-mendation	Potentially in Lower Austin Chalk Formation near Dallas–Fort Worth (central Texas, southern USA) or in the Upper Chalk formation exposed at Seaford Head in southern England.	Inoceramid bivalve, genus *Volviceras* lowest occurrence, and specifically *Volviceramus koeneni*.	At or near the lowest occurrence of ammonite *Peroniceras (Peroniceras) tridorsatum*.		Kauffman *et al.*, 1996
	Coniacian/ Turonian	Working Group recom-mendation	Salzgitter-Salder Quarry (active quarry; 52°07.55′N, 10°19.80′E) near Salzgitter-Salder village, southwest of Hannover, Lower Saxony province, northern Germany. GSSP would be at base of Bed MK 47 near the top of the "Grauweisse Wechselfolge"	Inoceramid bivalve, *Cremnoceramus rotundatus* (*sensu* Tröger *non* Fiege) lowest occurrence.	The boundary lies between the highest occurrence of ammonite *Prionocyclus germari*, and the lowest occurrence of *Forresteria peruana* and *F. brancoi* (North America). Markers of uppermost Turonian include the lowest occurrence of calcareous nannofossil Lithastrinus septenarius (base of nannofossil zone CC13b), and an acme-zone of inoceramid bivalve *Cremnoceramus waltersdorfensis*	Significantly below the lowest occurrence of *Forresteria (Harleites) petrocoriensis*, which was the traditional marker for the base of the Coniacian in Europe. Boundary is near a minor negative carbon-isotope excursion. However, diagenesis at candidate GSSP limits microfossil and stable isotope information.	Wood *et al.*, 1994; Kauffman *et al.*, 1996; Voigt & Hilbrecht, 1997; Sikora *et al.*, pers. commun.

(*cont.*)

Table 19.2 (*cont.*)

Stage	Stage boundary *Substage base*	Global Boundary Stratotype Section and Point					References (GSSP, correlations)
		Status	Location and point	Primary markers	Other correlations	Comments	
			lithostratigraphic unit in a steeply dipping limestone and marl succession.		*waltersdorfensis*. Markers of lowermost Coniacian include the lowest occurrence of inoceramid bivalve *Cremnoceramus waltersdorfensis hannovrensis*.		
Turonian	*Upper*	Under discussion	Candidate sections are at Lengerich, Münster Basin, Westphalia, northern Germany.	Undecided. Potential datums are the lowest occurrence of the ammonite *Subprionocyclus neptuni* in the Boreal realm (e.g., Germany, England), of the ammonite *Romaniceras deverianum* in the Tethys realm (e.g., southern France, Spain), or of an inoceramid bivalve, *Inoceramus perplexus* (= *Mytiloides costellatus*).	Near the "Pewsey" carbon-isotope peak, which correlates with a transgressive surface in southern England and may provide a correlation horizon to oceanic sections.	Some of the biostratigraphic markers have discrepancies in position relative to the reference carbon-13 curves from different sections (Wiese and Kaplan, 2001).	Bengtson *et al.*, 1996; Jenkyns *et al.*, 1994; Gale, 1996; Weise & Kaplan, 2001.
	Middle	Working Group recom-mendation	Rock Canyon Anticline, east of Pueblo, Colorado, west-central USA. Suggested GSSP is at base of Bed 120 (approximately 5 meters above the GSSP defining the base of the Turonian Stage in the same sections).	Ammonite, *Collignoniceras woollgari* lowest occurrence.	Lowest occurrence of inoceramid bivalve *Mytiloides hercynicus*.		Bengtson *et al.*, 1996; Harries *et al.*, 1996
	Turonian/ Cenomanian	Ratified, 2003	Rock Canyon Anticline, west of Pueblo, Colorado, west-central USA. GSSP is base of Bed 86 in a section of limestone-marl cycles in the lower Bridge Creek Member of the Greenhorn Limestone Formation at west end of the Denver and Rio Grande Western Railroad cut near the north boundary of the Pueblo Reservoir State Recreation Area.	Ammonite, *Watinoceras devonense* lowest occurrence.	Below the boundary are the highest occurrences of planktonic foraminifer *Rotalipora* morphotypes and the lowest occurrences of inoceramid bivalves *Mytiloides hattini* and *Inoceramus pictus* and of calcareous nannofossil *Microstaurus chastius*. Above the boundary is the lowest occurrence of planktonic foraminifer *Helvetoglobotruncana helvetica* and of inoceramus bivalve *Mytiloides kosmati*.	The maximum major carbon-isotope peak associated with oceanic anoxic event 'OAE2' occurs 0.5 m above the boundary, and the overall isotopic profile enables global correlation. The boundary is near a minor sequence boundary within the middle of a major transgressive systems tract	Kennedy & Cobban, 1991; Bengtson *et al.*, 1996; Accarie *et al.*, 1996; Gale, 1996; Kennedy *et al.*, 2000; Kennedy *et al.*, in press (*Episodes*, 2004).

Cenomanian	*Upper*	Under discussion	Potentially in southern France.	Undecided.	Possibilities (from earliest to latest) are the lowest occurrence of ammonite *Acanthoceras jukesbrownei*, lowest occurrence of bivalve *Inoceramus pictus pictus*, extinction of *A. jukesbrownei*, and lowest occurrence of ammonite *Calycoceras (Proeucalycoceras) gueranger*i.		Tröger, Kennedy, *et al.*, 1996
	Middle	Working Group recom-mendation	Southerham Gray Quarry, Lewes, Sussex province, England. Suggested GSSP is couplet "B38" in a group of thin, relatively clay-rich couplets that rest on more thickly bedded marly chalks.	Ammonite, *Cunningtoniceras inerme* lowest occurrence.	Lowest occurrences of planktonic foraminifer *Rotalipora reicheli* and bivalve *Inoceramus schoendorfi*. Approximately 5 couplets (100 kyr?) above GSSP are the lowest occurrences of ammonite *Acanthoceras rhotomagense* and of bivalve *Inoceramus tenuis* and the beginning of a positive excursion in carbon-13.	Lower/Middle Cenomanian boundary interval is missing over large regions due to its coincidence with a major sequence boundary.	Tröger, Kennedy, *et al.*, 1996; Gale, 1995
	Cenomanian/ Albian	Ratified, 2001	Gully exposures descending the western flanks of Mont Risou, north of Le Chataud, 3.15 km east of Rosans, Haute-Alpes province, southeast France (5°30'43"E, 44°23'33"N). GSSP is 36 m below top of Marnes Bleues Formation, as defined by a zero datum limestone at the base of the succeeding limestone.	Planktonic foraminifer, *Rotalipora globotruncanoides* lowest occurrence.	Highest occurrence of predominately Albian ammmonites is 3m above GSSP. Lowest occurrences of Cenomanian ammonite markers of *Mantelliceras mantelli, Neostlingoceras oberlini, Hyphoplites curvatus* and *Sciponoceras roto* are 6m above GSSP. *[See footnote b for additional markers]*	Kalaat Senan region, north of Kef el Azreg in central Tunisia, was proposed as a reference section for the Albian-Cenomanian boundary interval for the Tethyan Realm. In shelf carbonates, the lowest occurrences of foraminifers *Orbitolina (Orbitolina) concava concava* and *O. (O.) conica* are commonly used as markers.	Tröger, Kennedy, *et al.*, 1996; Gale *et al.*, 1996; Kennedy *et al.*, in press (*Episodes*, 2003); Robaszynski *et al.*, 1993
Lower Cretaceous							
Albian	*Upper*	Working Group recom-mendation	Not decided. Potentially Col de Pallluel, east of Rosans, Drôme province, southeast France.	Ammonite, *Dipoloceras cristatum* lowest occurrence.	Base of the *Neohibolites oxycaudatus* belemnite zone (north Germany), and the lowest occurrences of benthic foraminifera *Citharinella pinnaeformis* s.s. and *Arenobulimina chapmani*.	Commonly coincides with a transgression above a major sequence boundary	Hart *et al.*, 1996
	Middle	Working Group recom-mendation	Not decided. Potentially within dark gray clay-rich strata forming a steep southern bank of the Marne river (Les-Côtes-Noires-de-Moëslains) about 4 km upstream from town of St-Dizier, Haute-Marne province, northern France.	Ammonite, *Lyelliceras lyelli* lowest occurrence.	Micropaleontology markers have not yet been published from this St-Dizier section, but the Lower/Middle Albian boundary interval from other sections is associated with the lowest occurrences of the benthic foraminifera *Epistominia spinulifera*, and of the calcareous nannofossils *Prediscosphaera cretacea, Dictyococcites parvidentatus, Gaarderella granulifera* and *Braarudosphaera regularis*.		Hart *et al.*, 1996; Destombes and Destombes, 1965; Owen, 1971, 1996a

(*cont.*)

Table 19.2 (*cont.*)

Stage	Stage boundary *Substage base*	Global Boundary Stratotype Section and Point					References (GSSP, correlations)
		Status	Location and point	Primary markers	Other correlations	Comments	
	Albian/ Aptian	Northern Europe (Boreal) alternative	Vörhum, about 30 km east of Hannover in northern Germany. A suggested GSSP is a 10-cm thick horizon (Bed 6b) of brown cementstone nodules within very dark sparsely shelly clay.	Ammonite, *Leymeriella (P.) schrammeni anterior* lowest occurrence.	Base of *Neohibolites strombecki* belemnite zone. Lowest occurrences of calcareous nannofossil *Prediscosphaera columnata* (subcircular form), and of benthic foraminifers *Pleurostomella obtusa, Pl. subnodosa, Arenobulimina macfadyeni, Dorothia filiformis, Vaginulina gaultina* and members of the *Gaudryina dividens* lineage.	A tuff layer lies shortly above the proposed boundary at Vöhrum, which may be suitable for radiometric dating.	Hart *et al.*, 1996; (Owen, 1979, in Hart *et al.*, 1996); Mutterlose *et al.*, 2003
		Southern Europe (Tethyan Realm) alternative	Southeast France: (1) Stream cut to the north of highway D994 at Col de Palluel, east of Rosans village, Drôme province, or at nearby Col de Pré-Guittard, or (2) Clay-rich badlands just east of the village of Tartonne, north of Barrême, Basse-Alpes province.	Undecided.	Possibilities (in approximate ascending order) are "Jacob" organic-rich shale, lowest occurrence of calcareous nannofossil *Prediscophaera columnata* (base of nannofossil zone NC8), topmost organic-rich bed of the "Killian facies", highest occurrence of ammonite *Hypacanthoplites jacobi*, lowest occurrence of ammonite *Leymeriella tardefurcata*, the "Paquier" organic-rich event (coinciding with "oceanic anoxic event" OAE 1b), and lowest occurrence of ammonite *Douvilleiceras* ex. gp. *mammillatum*.	"Paquier" (OAE 1b) organic-enrichment episode can be correlated to the upper part of the *Leymeriella acuticostata* ammonite subzone in northern Germany.	Hart *et al.*, 1996; Bréheret *et al.*, 1986; Bréheret, 1996, 1997; Kennedy *et al.*, 2000; personal communications from J. Erbacher (1997), H. Owen (1999), and J. Kennedy (2003)
Aptian	*Upper (Clasayesian)*	Informal usage		Ammonite, base of *Acanthoplites (Nolaniceras) nolani* zone.			
	Middle (Gargasian)	Informal usage		Ammonite, genus *Dufrenoya* highest occurrence (extinction) (= top of *Dufrenoya furcata* zone) in Tethyan realm; and equivalent to top of *Tropaeum bowerbanki* ammonite zone in Boreal realm.	First occurrences of ammonites *Aconoceras nisum* and *Cheloniceras (Epicheloniceras) martini.*	This level corresponds to suggested Upper/Lower two-fold subdivision in the Boreal realm.	Casey *et al.*, 1998; Ropolo *et al.*, 1998; Erba *et al.*, 1996
	Aptian/ Barremian	Working Group recom-mendation	Gorgo a Cerbara, 2 km E of Piobbico, Umbria-Marche province, central Italy. GSSP is suggested as the 893.32 m level.	Magnetic reversal, base of Chron M0r.	Just below proposed GSSP are lowest occurrences of calcareous nannofossils *Nannoconus truittii, Rucinolithus irregularis, Chiastozygus litterarius,* and *Flabellites oblongus,* and of planktonic foraminifer *Globigerinelloides blowi,* and the highest occurrence in the Tethyan realm of dinoflagellate cyst	Candidate GSSP section has cycle stratigraphy for scaling biostratigraphic and chemistratigraphic events. About 1 myr after the boundary is Oceanic Anoxic Event 'OAE1a', marked by a widespread organic-rich shale (Selli Level).	Erba *et al.*, 1996, 1999; Channell *et al.*, 1995, 2000; Aguado *et al.*, 1997; Leereveld, 1997b; Erba, 1994; Herbert, 1992

					Pseudoceratium pelliferum. Biostratigraphic markers that postdate the boundary are a "nannoconid crisis" in calcareous nannofossils and the lowest occurrence of dinoflagellate cyst *Tehamadinium tenuiceras*.		
Barremian	*Upper*	Working Group recom-mendation	Section X.KV, Barranco de Cavila near Caravaca, Murcia province, Spain. GSSP is suggested as thin limestone bed "17a" within the marl-rich section.	Ammonite, *Ancyloceras vandenheckei* lowest occurrence.	Lowest occurrence of dinoflagellate cyst *Odontochitina operculata*. Correlation to Boreal Realm is to lower part of the *Paracrioceras elegans* ammonite zone, the highest occurrence of belemnite genus *Aulacoteuthis*, and the lowest occurrence of belemnite genus *Oxyteuthis*.	Magnetostratigraphy was unsuccessful at Rio Argo, but correlation to Italian sections indicates a placement in latest part of magnetic polarity chron M3r (approximately M3r.8).	Company *et al.*, 1995; Rawson *et al.*, 1996; Wilpshaar, 1995; Leereveld, 1995, 1997b; Bartolocci *et al.*, 1992; Channell *et al.*, 1995
	Barremian/ Hauterivian	Working Group recom-mendation	Río Argos near Caravaca, Murcia province, Spain. GSSP level is not yet decided.	Ammonite, *Spitidiscus hugii – Spitidiscus vandeckii* group lowest occurrence.	Approximately midway between the highest occurrences of calcareous nannofossils *Lithraphidites bollii* and *Calcicalathina oblongata*. Just below the highest occurrence of dinoflagellate cyst *Exiguisphaera phragma* and above the lowest occurrence of dinoflagellate cyst *Druggidium rhabdoreticulatum*.	Boundary interval is within uppermost magnetic polarity zone M4n (approximately M4n.8) in Italy. Corresponds approximately to the base of the *"Hoplocrioceras" rarocinctum* ammonite zone and the lowest occurrence of the belemnite *Praeoxyteuthis pugio* in the Boreal Realm.	Company *et al.*, 1995; Rawson *et al.*, 1996; Habib & Drugg, 1983; Cecca et al., 1994; Hoedemaeker & Leereveld, 1995; Rawson *et al.*, 1996; Leereveld, 1995, 1997b; Bartolocci *et al.*, 1992; Channell *et al.*, 1995
Hauterivian	*Upper*	Working Group recom-mendation	Section near La Charce village, Drôme province, southeast France. GSSP level is not yet decided.	Calcareous nannofossil, *Cruciellipsis cuvillieri* highest occurrence (extinction).	This level is reported to coincide with the middle of the Subsaynella sayni ammonite zone of Tethyan Realm and near beginning of polarity Chron M8r. In the Boreal realm, it may be near the base of *Simbirskites (Speetoniceras) inversum* ammonite zone of England (Crux, 1989) and base of *Simbirskites staffi* ammonite zone in northwest Germany (Kemper *et al.*, 1987).		Mutterlose *et al.*, 1996; Bergen, 1994; Channell *et al.*, 1993, 1995
	Hauterivian/ Valanginian	Working Group recom-mendation	Section near La Charce village, Drôme province, southeast France. GSSP level is not yet assigned, but would probably be between Bed 188 and 189 of Reboulet (1995) (or Bed 254 in the numbering system of Bulot *et al.*, 1993).	Ammonite, genus *Acanthodiscus* lowest occurrence (especially *A. radiatus*).	Earliest Hauterivian has the lowest occurrence of the dinoflagellate cyst *Muderongia staurota* and the highest occurrence of calcareous nannofossil *Eiffelithus windii*. Probably coeval with lowest occurrence of ammonite *Endemoceras amblygonium* in Boreal Realm.	No magnetostratigraphy from La Charce section, but is probably near beginning of polarity chron M11n.	Mutterlose *et al.*, 1996; Bulot *et al.*, 1993; Reboulet, 1995; Thieuloy, 1977; Leereveld, 1995, 1997b; Bergen, 1994; Channell *et al.*, 1995
Valanginian	*Upper*	Informal European usage		Ammonite, *Saynoceras verrucosum* lowest occurrence	Onset of a major sea-level transgression		Bulot *et al.*, 1996

(*cont.*)

Table 19.2 (*cont.*)

Stage	Stage boundary *Substage base*	Global Boundary Stratotype Section and Point: Status	Location and point	Primary markers	Other correlations	Comments	References (GSSP, correlations)
	Valanginian/ Berriasian	Working Group recom-mendation	Two candidates: (1) 5 km NW of Montbrun-les-Bains, Drôme province, southeast France where suggested GSSP level is Bed 209 (a condensed, 4-cm thick, pyrite-rich bed), and (2) Barranco de Cañada Luenga, 3 km SSW of Cehegín, near Caravaca in southeast Spain.	Calpionellid, *Calpionellites darderi* lowest occurrence (base of Calpionellid Zone E); followed by the lowest occurrence of ammonite *Thurmanniceras pertransiens.*	Approximately at the base of the *Thurmanniceras pertransiens* ammonite zone and just below lowest occurrence of nannofossil *Calcicalathina oblongata. Close to lowest occurrences of* dinoflagellate cysts *Spiniferites* spp. and *Oligosphaeridium complex.* Correlation is indicated to near the base of the *Platylenticeras* ammonite zone (Germany) and the top of the *Peregrinoceras albidum* ammonite zone (England).	Coincides with rare migrations of cold-water taxa from the Boreal Realm into the Tethyan Realm. No magnetostratigraphy from candidate GSSP in France, but magnetostratographic studies from candidate GSSP in southeast Spain place the base of Calpionellid Zone E at approximately the middle of magnetic polarity Chron M14r.	Aguado *et al.*, 2000; Blanc *et al.*, 1994; Bulot *et al.*, 1996; Bulot *et al.*, 1995; Leereveld, 1995, 1997a; Bergen, 1994; Channell & Grandesso, 1987; Channell *et al.*, 1987; Ogg *et al.*, 1988.
Berriasian	*Upper*	Informal southern Europe usage		Ammonite, *Malbosiceras paramimounum* lowest occurrence (base of *Fauriella boissieri* ammonite zone).	Close to lowest occurrence of *Calpionellopsis simplex* (base of Calpionellid Zone D = *Calpionellopsis* Zone). Middle of magnetic polarity Chron M16r (M16r.5) .	Possibily within the uppermost *Runctonia runctoni* ammonite zone of the Boreal Realm.	Zakharov *et al.*, 1996; Galbrun, 1984; Ogg *et al.*, 1991.
	Berriasian/ Tithonian (= base of Cretaceous)	Under discussion (younger possibility)	La Faurie, Ravin de Dreymien (Alpes-de-Provence, France), or Section Z of Berrancode de Tollo (Río Argos near Caravaca, Murcia province, Spain).	Ammonite, *Tirnovella subalpina* lowest occurrence.	Approximately coincident with bases of *Praetollia maynci/Runctonia runctoni* ammonite zone (base of "Ryazanian" regional stage) and *Buchia okensis* buchiid zone of the Boreal Realm. Early part of polarity Chron M17r (approximately M17r.3).	Traditional base of the 'Middle Berriasian' in the Mediterranean region, but can be correlated to Boreal Realm.	Zakharov *et al.*, 1996; Hoedemaeker & Leereveld, 1995; Galbrun, 1984
	Berriasian/ Tithonian (= base of Cretaceous)	Under discussion (older possibility)	Puerto Escano section, Cordoba province, Spain.	Ammonite, *Berriasella jacobi* lowest occurrence.	Approximated by the base of Calpionellid Zone B (base *Calpionella alpina* subzone, marked by base of acme of *Calpionella subalpina*). Later part of magnetic polarity Chron M19n (approximately M19n.2n.5).	Correlation to Boreal Realm is not firmly established, but magnetostratigraphy suggests equvalence to lowermost *Subcraspedites preplicomphalus* ammonite zone.	Zakharov *et al.*, 1996, Tavera *et al.*, 1994; Hoedemaeker & Bulot, 1990; Hoedemaeker, Company, *et al.*, 1993; Ogg *et al.*, 1991

Footnotes:

[a] *[Campanian/Santonian]* Boundary is in the lower part of calcareous nannofossil zone CC17, with the first regular occurrence of *Calculites obscurus* and highest occurrence of *Lithastrinus septenarius* (= *L. moratus* sensu Varol) in uppermost Santonian. Lowest Campanian markers include, in approximate succession, the lowest occurrence of crinoid *Uintacrinus anglicus* (base of the *Offaster pilula* zone), the highest occurrences of planktonic foraminifers *Dicarinella asymetrica* and *Globotruncana concavata,* the lowest occurrence of ammonite *Menabites (Delawarella) delawarensis,* and the lowest occurrence of calcareous nannofossil *Broinsonia parca parca* (base of nannofossil zone CC18).

[b] *[Cenomanian/Albian]* Uppermost Albian markers include the highest occurrence of planktonic foraminifer *Rotalipora ticinensis,* the lowest occurrence of planktonic foraminifer *Rotalipora gandolfii,* and the lowest occurrence of calcareous nannofossil *Calculites anfractus.* Lowest Cenomanian markers include the lowest 'common' occurrence of planktonic foraminifer *Rotalipora globotruncanoides*, and lowest occurrence of bivalve *Inoceramus crippsii crippsii.* GSSP is bounded by two positive excursions in carbon-13. In many regions, the Albian-Cenomanian boundary interval is coincident with a widespread hiatus and condensation associated with a major sequence boundary.

in the tropical ocean of this Tethyan realm did not extend into the colder Boreal realm of northwest Europe and other northern regions; therefore, the Boreal equivalents of most of these Tethyan-defined stages remain uncertain.

Most primary markers under consideration for defining stage and substage boundaries of the Berriasian through Barremian are lowest or highest occurrence of species of ammonoids (Table 19.2). The proposed base of the Aptian Stage has utilized a global magnetic reversal, and a possible correlation marker for the base of the Albian Stage is a widespread organic-enrichment episode or a calcareous nannofossil.

JURASSIC–CRETACEOUS BOUNDARY AND THE BERRIASIAN STAGE

Debate over the basal boundary of the Cretaceous Period The base of the Cretaceous Period currently lacks an accepted global boundary definition, despite over a dozen international conferences and working groups dedicated to the issue since 1974 (Zakharov *et al.*, 1996).

The "Berriasian" was proposed for limestone near the village of Berrias (Ardèche province, southeast France; Coquand, 1871; reviewed in Rawson, 1983). Originally conceived as a subdivision of the Valanginian, it was subsequently often referred to as "Infra-Valanginian" until the name "Berriasian" was eventually brought back into use. The preceding Tithonian Stage of latest Jurassic was defined by Oppel (1865) to include all deposits in the Mediterranean area which lie between a restricted Kimmeridgian and "Valanginian," but no representative section was designated. To some degree, the new "Berriasian" Stage overlapped with the original concept of the Tithonian Stage, and the historical lower boundary of the Berrias stratotype (base of the *Berriasella jacobi* ammonite zone) lacks any significant faunal change. Therefore, some workers have suggested that perhaps it is more appropriate to place the Berriasian Stage in the Jurassic and assign the base of the Cretaceous to the base of the current Valanginian Stage (e.g. Rawson, 1990; Remane, 1990, cited in Zakharov *et al.*, 1996).

However, traditional usage favors an ammonoid biostratigraphic datum for defining the base of the Cretaceous as the base of the Berriasian Stage. Two lowest occurrences of new ammonoid taxa in the Mediterranean region have been proposed (Table 19.2; Zakharov *et al.*, 1996):

1. Ammonoid turnover at family level at the base of the *Berriasella jacobi* ammonite zone (e.g. Colloque sur la limite Jurassique–Crétacé, 1975; Hoedemaeker *et al.*, 1993). This horizon is approximately the base of Calpionellid Zone B and the later part of magnetic polarity Chron M19n. However, it has not been correlated with the Boreal paleogeographic realm. In a suggested boundary stratotype at the Puerto Escano section (Cordoba province, Spain), the pelagic limestone facies contains ammonoids, calpionellids, and calcareous nannoplankton (Tavera *et al.*, 1994).
2. Ammonoid turnover at the base of the *Subthurmannia subalpina* ammonite subzone (base of *Subthurmannia occitanica* Zone) in the lower portion of polarity Chron M17r. Even though this level is generally considered as the base of the "Middle Berriasian" in the Mediterranean region, it corresponds to the effective base of the ammonite zonation at the historical Berrias stratotype, "because of the virtual absence of ammonites and the frequency of intervals with reworked sediments below that zone" (Hoedemaeker *et al.*, 2003). This level also has an advantage in that it can be correlated to the Boreal realm, where it is approximately coincident with the base of the *Runctonia runctoni* ammonite zone. Potential boundary stratotypes are found in southeast France and at Rio Argos near Caravaca, Spain, where the level coincides with a depositional sequence boundary (Hoedemaeker and Leereveld, 1995).

Alternative possibilities for a primary marker for a GSSP at the base of the Cretaceous are a microfossil event (e.g. the base of Calpionellid Zone B), sequence stratigraphy (e.g. the major "Purbeckian regression" marking a major global fall in sea level), or magnetostratigraphy (e.g. the base of magnetic polarity Chron M18r was suggested by Ogg and Lowrie, 1986).

In GTS2004, we have adopted the first alternative (base of *Berriasella jacobi* ammonite zone, base of Calpionellid Zone B; later part of polarity Chron M19n) for the base of the Cretaceous period in our summary charts (e.g. Figs. 18.1, 19.1, and 19.2). Magnetostratigraphy of sections near the English Channel indicate a correlation to the lowermost portion of the *Subcraspedites preplicomphalus* ammonite zone of northwest Europe (Ogg *et al.*, 1991a, 1994; Ogg and Wimbledon, in prep.).

Upper Berriasian substage The base of the Upper Berriasian substage in the Mediterranean region is currently placed at the base of the *Malbosiceras paramimounum* ammonite subzone (the base of the *Subthurmannia boissieri* Zone; Zakharov *et al.*, 1996). This level is close to the base of Calpionellid Zone D and is in the middle of magnetic polarity zone M16r (M16r.5; Galbrun, 1984). Magnetostratigraphic correlation to southern England projects this level to the uppermost *Runctonia runctoni* ammonite zone of the Boreal realm (Ogg *et al.*, 1991a).

No boundary stratotype for this substage has yet been recommended.

Volgian, Portlandian, and Ryazanian regional stages of Europe It has been nearly impossible to correlate the Jurassic–Cretaceous boundary interval (upper Tithonian and Berriasian Stages) of the Tethyan realm into the Boreal realm, therefore independent regional units of a Volgian Stage (or Portlandian Stage in southern England) followed by a Ryazanian Stage have been traditionally used (Fig. 19.2).

The base of the regional Ryazanian Stage is traditionally assigned as the base of the *Ructonia ructoni* ammonite zone (eastern England) and at the base of the *Praetollia maynci* or *Chetaites sibiricus* ammonite zones (Russian and Siberian zonations), but these levels may not be equivalent (Hancock, 1991; Hardenbol *et al.*, 1998). The base of the Ryazanian Stage is probably near the base of the *Subthurmannia subalpina* ammonite subzone of the Berriasian Stage. The top of the Ryazanian Stage is probably close to the base of the revised definition of the Valanginian Stage.

VALANGINIAN STAGE

History, revised definition, and boundary stratotype The original type section of the Valanginian Stage included all post-Jurassic strata to the base of the "Marnes de Hauterive" in shallow-marine facies at Seyon Gorge near Valangin (Neuchâtel, Switzerland; Desor, 1854; Desor and Gressly, 1859). Because the type section is poor in ammonoids, "hypostratotype" sections at Angles (Alpes-de-Haute-Provence province) and Barret-le-Bas (Hautes-Alpes province) in southeast France have served as reference sections (Busnardo *et al.*, 1979). The base of the Valanginian was provisionally placed at the base of the *Thurmanniceras otopeta* ammonite zone (Busnardo *et al.*, 1979; Birkelund *et al.*, 1984).

Taxonomic and correlation problems with the ammonoid successions (e.g. Bulot *et al.*, 1993a) led the Valanginian Working Group (Hoedemaeker *et al.*, 2003) to recommend placing the base of the Valanginian Stage at the base of Calpionellid Zone E, defined by the first occurrence of *Calpionellites darderi* (Bulot *et al.*, 1996; Aguado *et al.*, 2000), which shifts the *T. otopeta* ammonite zone into the Berriasian Stage. This level can be traced from France to Mexico, is associated with other bioevents, and coincides with rare migrations of cold-water taxa into the Tethyan realm (reviewed in Aguado *et al.*, 2000). Magnetostratigraphic–calpionellid studies in Italy and Spain have indicated an apparent 300 000 yr variability or diachroneity of the base of Calpionellid Zone E between the middle of magnetic polarity Chron M14r to earliest part of polarity Chron M14n (Channell and Grandesso, 1987; Channell *et al.*, 1987; Ogg *et al.*, 1988). A candidate GSSP in southeast France is within alternating limestone and marl with well-preserved ammonoid and calpionellid successions, where the proposed boundary level is followed by the lowest occurrence of the ammonite *Thurmanniceras pertransiens* (Blanc *et al.*, 1994). A candidate GSSP section at Barranco de Cañada Luenga, south of Cehegín, southeast Spain, has integrated calpionellid, ammonoid, nannofossil, and magnetostratigraphy (Aguado *et al.*, 2000).

Correlation of this re-defined base of Valanginian in the Tethyan realm to the Boreal realm is not precise. Ammonoid migration episodes and nannofossil and dinoflagellate cyst comparisons suggest correspondences near the bases of the regional usages of "Valanginian" in northwest Germany and in England (Leereveld, 1995; Bulot *et al.*, 1996).

Upper Valanginian substage The base of the Upper Valanginian is traditionally placed at the base of the distinctive *Saynoceras verrucosum* ammonite zone, which can also be recognized in the west European province of the Boreal realm. Magnetostratigraphic, chemostratigraphic, and microfossil markers (e.g. Channell *et al.*, 1993) await calibration with this ammonoid datum. Potential boundary stratotypes in southeast France and the Betic Cordillera of Spain are

Figure 19.1 Cretaceous time scale with selected biostratigraphic zonations and datums, magnetic polarity chrons, and major depositional sequences. The primary absolute-age stratigraphic scales are: Cenomanian–Maastrichtian ammonite zones of the Western Interior of North America, Albian foraminifer zones, Aptian ammonite zones, and basal-Aptian through Tithonian M-sequence magnetic polarity chrons. Ages of stage boundaries and other stratigraphic events are from their direct calibration to the primary stratigraphic scale (e.g. magnetostratigraphic correlation of proposed basal-Valanginian ammonite zone boundary) or extrapolated from published correlation estimates (e.g. Mesozoic chronostratigraphy charts of Hardenbol *et al.*, 1998) – see text for details. Dashed lines denote relatively uncertain calibrations of other biostratigraphic events to the primary scale, or intervals in which ammonite zones have been arbitrarily scaled proportional to the relative number of subzones. Most subzonal units are omitted, and only a generalized ammonite stratigraphy is given for some intervals (see biostratigraphic chart series in Hardenbol *et al.*, 1998 for full listing and correlation web). Biostratigraphic ages of Aptian and Albian OAE1a through OAE1d are from Leckie *et al.* (2002). Major flooding or regressive trends of depositional sequences of north west Europe (Hardenbol *et al.*, 1998) are labeled at the maximum lowstand of the respective third-order sequence. A color version of this figure is in the plate section.

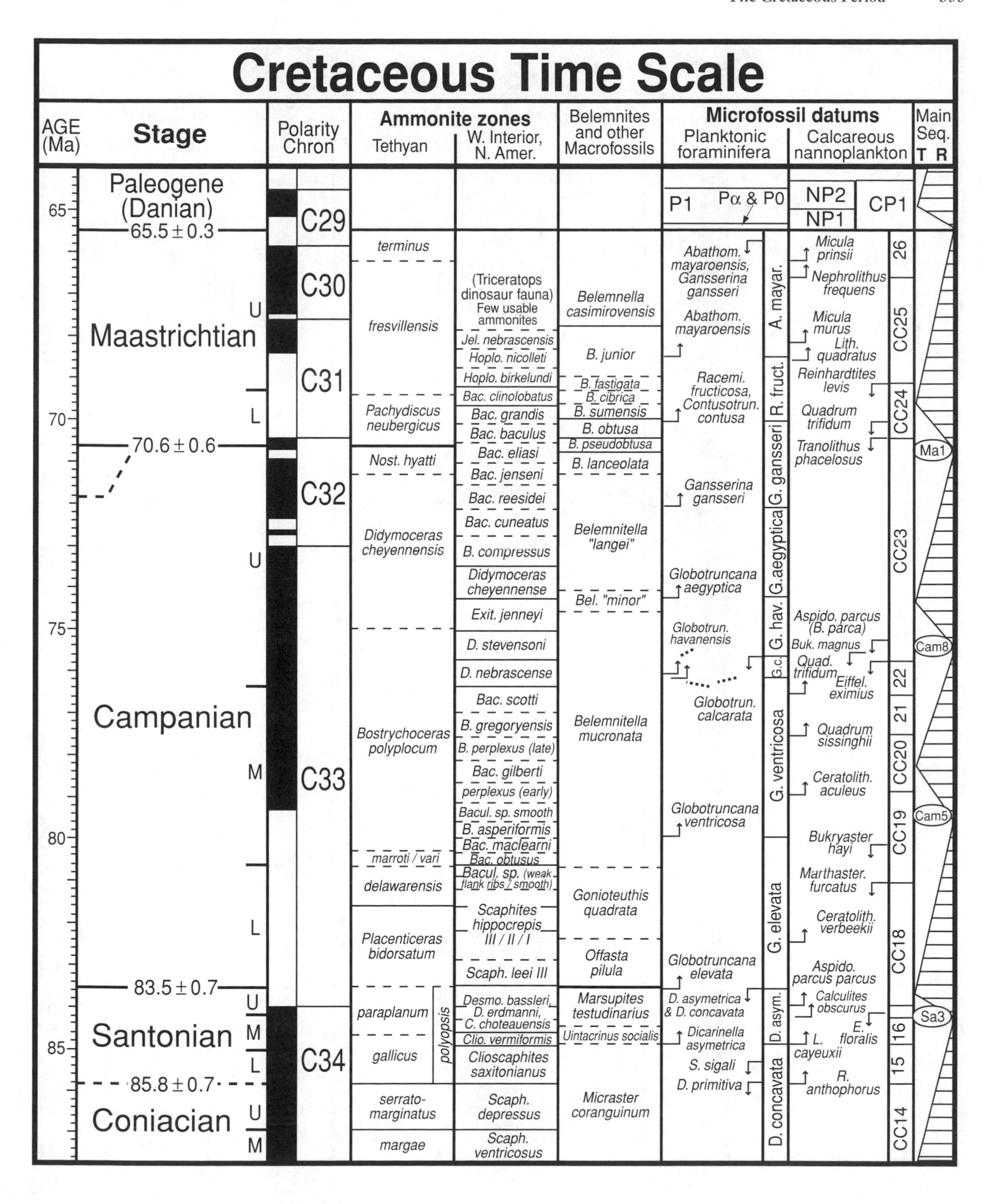
Cretaceous Time Scale
AGE (Ma)
Stage
Polarity Chron
Ammonite zones
Tethyan
W. Interior, N. Amer.
Belemnites and other Macrofossils
Microfossil datums
Planktonic foraminifera
Calcareous nannoplankton
Main Seq.
T R
Paleogene (Danian)
65.5 ± 0.3
Maastrichtian
70.6 ± 0.6
Campanian
83.5 ± 0.7
Santonian
85.8 ± 0.7
Coniacian
U
L
M
65
70
75
80
85
C29
C30
C31
C32
C33
C34
terminus
fresvillensis
Pachydiscus neubergicus
Nost. hyatti
Didymoceras cheyennensis
Bostrychoceras polyplocum
marroti / vari
delawarensis
Placenticeras bidorsatum
paraplanum
gallicus
polyopsis
serrato-marginatus
margae
(Triceratops dinosaur fauna) Few usable ammonites
Jel. nebrascensis
Hoplo. nicolleti
Hoplo. birkelundi
Bac. clinolobatus
Bac. grandis
Bac. baculus
Bac. eliasi
Bac. jenseni
Bac. reesidei
Bac. cuneatus
B. compressus
Didymoceras cheyennense
Exit. jenneyi
D. stevensoni
D. nebrascense
Bac. scotti
B. gregoryensis
B. perplexus (late)
Bac. gilberti
perplexus (early)
Bacul. sp. smooth
B. asperiformis
Bac. maclearni
Bac. obtusus
Bacul. sp. (weak flank ribs / smooth)
Scaphites hippocrepis III / II / I
Scaph. leei III
Desmo. bassleri, D. erdmanni, C. choteauensis
Clio. vermiformis
Clioscaphites saxitonianus
Scaph. depressus
Scaph. ventricosus
Belemnella casimirovensis
B. junior
B. fastigata
B. cibrica
B. sumensis
B. obtusa
B. pseudobtusa
B. lanceolata
Belemnitella "langei"
Bel. "minor"
Belemnitella mucronata
Gonioteuthis quadrata
Offasta pilula
Marsupites testudinarius
Uintacrinus socialis
Micraster coranguinum
P1
Pα & P0
Abathom. mayaroensis, Gansserina gansseri
Abathom. mayaroensis
Racemi. fructicosa, Contusotrun. contusa
Gansserina gansseri
Globotruncana aegyptica
Globotrun. havanensis
Globotrun. calcarata
Globotruncana ventricosa
Globotruncana elevata
D. asymetrica & D. concavata
Dicarinella asymetrica
S. sigali
D. primitiva
A. mayar.
R. fruct.
G. gansseri
G. aegyptica
G. hav.
G.c.
G. ventricosa
G. elevata
D. asym.
D. concavata
NP2
NP1
CP1
Micula prinsii
Nephrolithus frequens
Micula murus
Lith. quadratus
Reinhardtites levis
Quadrum trifidum
Tranolithus phacelosus
Aspido. parcus (B. parca)
Buk. magnus
Quad. trifidum
Eiffel. eximius
Quadrum sissinghii
Ceratolith. aculeus
Bukryaster hayi
Marthaster. furcatus
Ceratolith. verbeekii
Aspido. parcus parcus
Calculites obscurus
E. floralis
L. cayeuxii
R. anthophorus
26
CC25
CC24
CC23
22
21
CC20
CC19
CC18
16
15
CC14
Ma1
Cam8
Cam5
Sa3

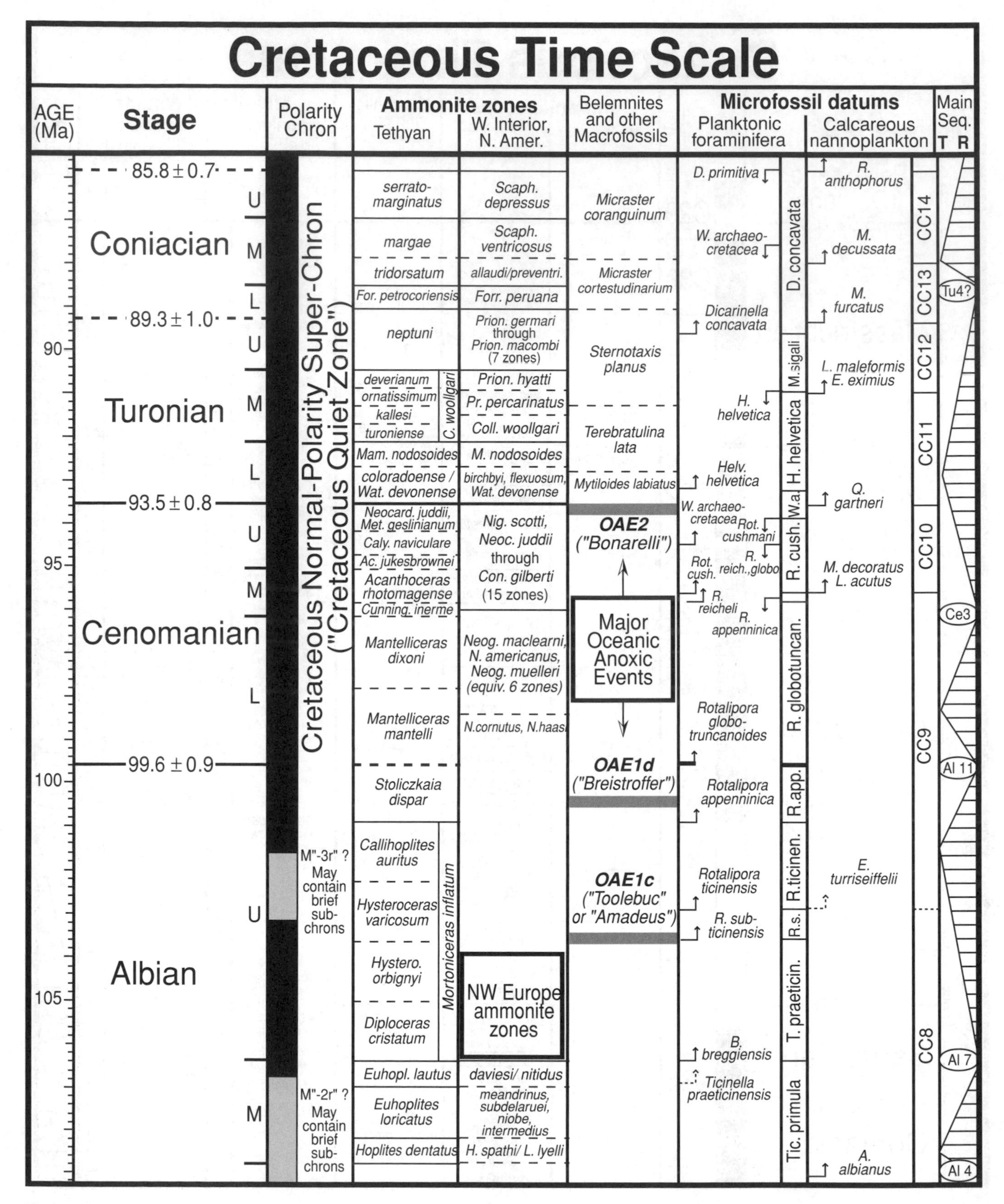

Figure 19.1 (*cont.*)

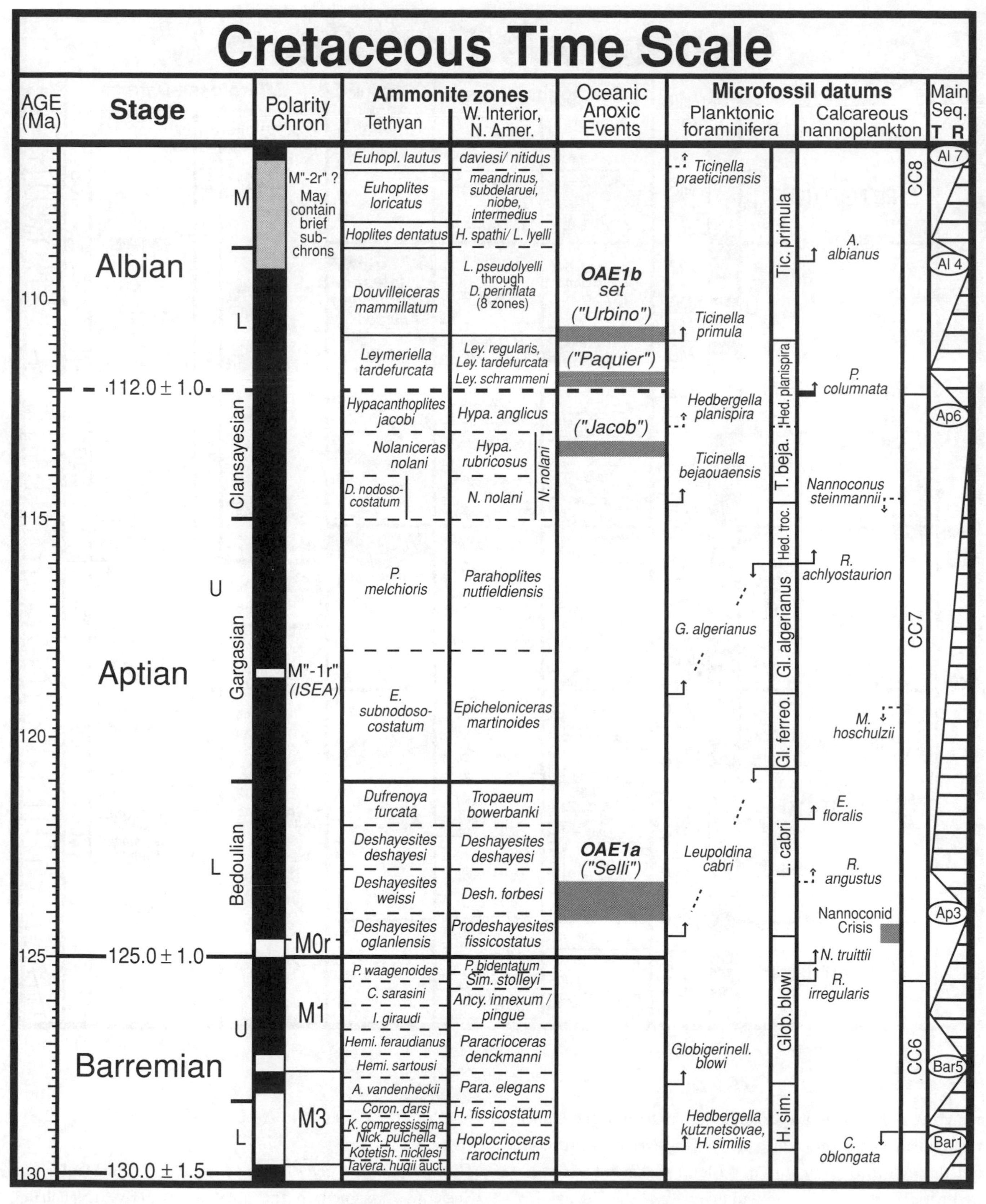

Figure 19.1 (*cont.*)

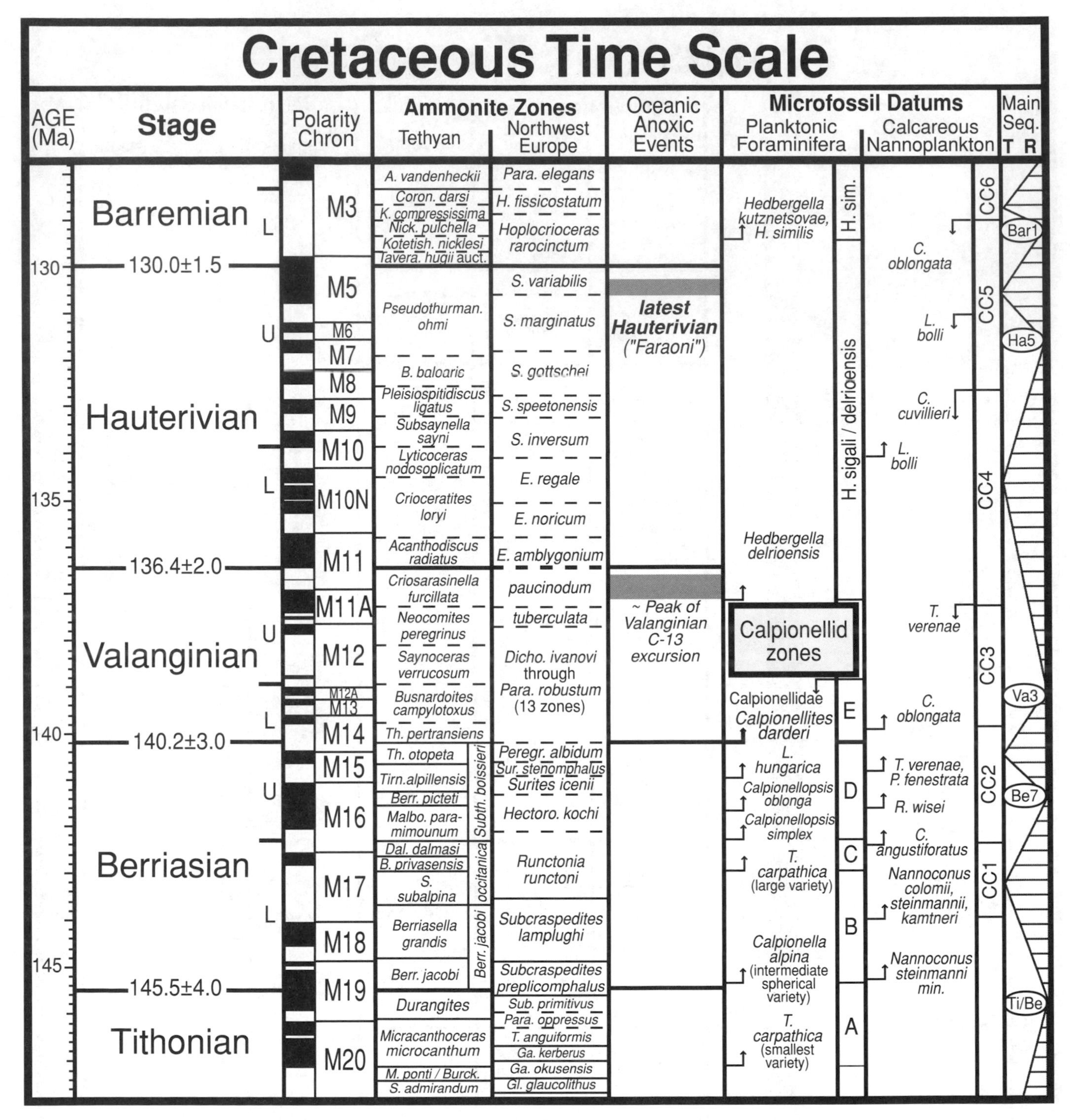

Figure 19.1 (*cont.*)

being considered (Bulot *et al.*, 1996). The base of the Upper Valanginian is the onset of a major sea-level transgression, and the following pronounced sea-level highstand is associated with the greatest ammonoid diversity of the Valanginian in the Mediterranean faunal province (Reboulet *et al.*, 1992; Hoedemaeker and Leereveld, 1995; Reboulet, 1996; Reboulet and Atrops, 1999).

HAUTERIVIAN STAGE

History, definition, and boundary stratotype The Hauterivian Stage was assigned in the area of Hauterive (Neuchâtel, Switzerland) to encompass "marnes à *Astieria*" (later transferred to Valanginian), "marnes bleues d'Hauterive," and "Pierre Jaune de Neuchâtel" (Renevier, 1874; reviewed in Rawson, 1983). The base of the Hauterivian in the Tethyan realm is traditionally recognized by the lowest occurrence of the ammonite species *Acanthodiscus radiatus*. Reboulet (1996) suggested that the base of the Radiatus Zone should be defined by the first occurrence of the ammonite genus *Acanthodiscus* (Reboulet and Atrops, 1999).

Expanded sections of ammonoid-rich, limestone–marl alternations in southeast France have served as the reference for this stage. The section at the village of La Charce in this region is the proposed global boundary stratotype for both the base of the Hauterivian Stage and base of the upper substage (Thieuloy, 1977; Mutterlose *et al.*, 1996). Inter-realm exchanges of dinoflagellate cyst and ammonoid assemblages suggest that the traditional Boreal usage of "Hauterivian" (base of *Endemoceras amblygonium* ammonite zone) is nearly coeval with this candidate GSSP level in the Tethyan realm (Leereveld, 1995; Mutterlose *et al.*, 1996). Ammonoids that are characteristic of lowest Hauterivian in Italy indicate that the base of the Hauterivian may be nearly coincident with the beginning of magnetic polarity Chron M11n (Channell *et al.*, 1995a).

Upper Hauterivian substage The traditional base of the Upper Hauterivian substage in the Tethyan realm was the highest occurrence of Neocomitinae ammonoids and the lowest occurrence of the ammonite *Subsaynella sayni*, but these events cannot be recognized in the Boreal realm. Therefore, it was recommended that the highest occurrence of the calcareous nannofossil *Cruciellipsis cuvillieri* serve as the global marker for the base of the Upper Hauterivian (Mutterlose *et al.*, 1996). The highest occurrence of *C. cuvillieri* has been reported from the middle of the *Subsaynella sayni* ammonite zone (Bergen, 1994) and is near the base of magnetic polarity zone M8r in Italy (Channell *et al.*, 1993, 1995a).

BARREMIAN STAGE

History, definition, and boundary stratotype The original Barremian Stage concept of Coquand (1861) was a vague assignment to an assemblage of belemnites and ammonoids with a type locality near Barrême (Alpes-de-Haute-Provence, southeast France), which also encompassed the current Upper Hauterivian substage. Busnardo (1965) redefined the Barremian Stage using the roadside exposures of Angles in southeast France as the type locality.

The base of the Barremian is marked by the lowest occurrence of the *Taveraidiscus hugii* auctorum (formerly called "*Spitidiscus*" *hugii*; Hoedemaeker and Reboulet, 2003), *Avramidiscus vandeckii* ammonite group, and a candidate for the global boundary stratotype is at Rio Argos, near Caravaca, Spain (Company *et al.*, 1995; Rawson *et al.*, 1996b). This boundary interval in Italy falls within uppermost magnetic polarity zone M4n (at approximately Chron M4n.8 = "M5n.8"; Bartolocci *et al.*, 1992; Channell *et al.*, 1995a). Correlation to the Boreal realm by dinoflagellate cysts indicate that this base of the Barremian Stage is approximately at the base of the *"Hoplocrioceras" rarocinctum* ammonite zone (Leereveld, 1995; Rawson *et al.*, 1996b).

Upper Barremian substage The boundary between the Lower and Upper Barremian is defined by the lowest occurrence of the ammonite *Ancyloceras vandenheckei* (Rawson *et al.*, 1996b). The proposed substage GSSP is near Caravaca, Spain (Company *et al.*, 1995). This substage boundary level is within the uppermost part of magnetic polarity zone M3r (at approximately Chron M3r.8) in Italy (Bartolocci *et al.*, 1992; Channell *et al.*, 1995a) and is correlated to the lower part of the *Paracrioceras elegans* ammonite zone of the Boreal realm (Leereveld, 1995; Rawson *et al.*, 1996b).

APTIAN STAGE

History, revised definition, and boundary stratotype The Aptian Stage was a vague designation by Alcide d'Orbigny (1840) for strata containing "Upper Neocomian" fauna and named after the village of Apt (Vaucluse province, southeast France). The French sections are poor in ammonoids, therefore the classical marker for the base of the Aptian was the lowest occurrence of the deshayesitid ammonite, *Prodeshayesites*, in northwest Europe (Rawson, 1983; Moullade *et al.*, 1998; Hoedemaeker *et al.*, 2003). However, the local lowest occurrence of this ammonite genus is commonly associated with a major progressive transgression in earliest Aptian and virtually no ammonoid-bearing section represents a continuous and complete Barremian–Aptian boundary interval (Erba *et al.*, 1996). Therefore, the proposed primary marker for the base of the Aptian Stage is the beginning of magnetic polarity Chron M0r.

The proposed global boundary stratotype in pelagic limestone at Gorgo a Cerbara, central Italy, has an integrated stratigraphy of paleomagnetics, biostratigraphy (calcareous nannofossils, planktonic foraminifera, radiolaria, dinoflagellates), carbon isotope chemostratigraphy, and cycle stratigraphy (Erba *et al.*, 1996; Erba *et al.*, 1999; Channell *et al.*, 2000). Cyclic stratigraphy implies that the lowest occurrence of a *Prodeshayesites* ammonite in uppermost polarity zone M0r occurs about 300 000 years after the boundary (Herbert, 1992; Channell *et al.*, 1995a; Erba *et al.*, 1996). An important event about 1 myr after the boundary is the oceanic anoxic event "OAE1a," marked by a widespread organic-rich shale (e.g. Selli Level in Italy, Goguel level in southeast France, Fischschiefer in northwestern Germany) and the beginning of a positive carbon-13 isotopic excursion (Weissert and Lini, 1991; Mutterlose and Böckel, 1998).

The Aptian substages The Aptian Stage has no standardized international subdivisions. In the Tethyan realm, the stage was initially subdivided, two-fold, into a lower "Bedoulian" and an upper "Gargasian" substage, which were based on sections at Cassis–La Bédoule (Bouches du Rhône province, near Marseilles) and at Gargas (near Apt), southern France. The substage boundary is the top of the *Dufrenoya furcata* ammonite zone, which is approximately equivalent to the base of the *Tropaeum bowerbanki* ammonite zone of the Boreal zonations (e.g. Ropolo *et al.*, 1998). This level corresponds to the suggested two-fold (Upper–Lower) substage boundary in the Boreal realm (e.g. Casey *et al.*, 1998). However, in the Tethyan realm, an additional uppermost substage of "Clansayesian" was added when Breistroffer (1947) moved the thin "Clansayes" horizon from the Albian into the underlying Aptian. The situation has not yet been resolved because: (a) the reference sections at Gargas (near Apt) and Clansayes are not suitable for correlation purposes (Rawson, 1983); (b) the rationale for a separate "Clansayesian" substage is questioned (e.g. Owen, 1996a; Casey *et al.*, 1998); and (c) the placement of the Bedoulian–Gargasian boundary is debated (Conte, 1994, as reported in Casey *et al.*, 1998, but see Ropolo *et al.*, 1998).

Informal subdivisions using calcareous plankton biostratigraphy commonly equate the base of the Upper Aptian to the lowest occurrences of planktonic foraminifer *Leupoldina cabri* (Tethyan faunal realm) or calcareous nannofossil *Eprolithus floralis* (Erba *et al.*, 1996). However, these biostratigraphic events occur relatively early within the Aptian Stage, at approximately 1 and 2 myr, respectively, after the beginning of the Aptian (e.g. Erba *et al.*, 1999; Larson and Erba, 1999; Premoli-Silva *et al.*, 1999).

ALBIAN STAGE

History, definition, and boundary stratotype The Albian Stage was named after the Aube region (Latin name, Alba), northeast France (d'Orbigny, 1842). Prior to 1947, the base of the Albian was assigned as the base of the *Nolaniceras nolani* ammonite zone. Breistroffer (1947) shifted this zone and the overlying *Hypacanthoplites jacobi* ammonite zone to an expanded uppermost Aptian, and placed the base of the Albian in the northwest European faunal province of the Boreal realm at the base of the *Leymeriella tardefurcata* ammonite zone, with a basal level characterized by *Leymeriella schrammeni*. However, this boundary interval in western Europe is marked by endemic ammonoid biogeography, in which the occurrences of early *Leymeriella* ammonite species appear to be restricted to northern Germany (Casey, 1996; Hart *et al.*, 1996; Mutterlose *et al.*, 2003) and the *L. schrammeni* ammonite subzone is represented by sediments in only isolated localities (e.g. the Harz foredeep of north Germany and in north Greenland; Hugh Owen, written comm., February 1999). In addition, this version of an Aptian–Albian boundary interval coincides with a widespread hiatus associated with a major sequence boundary. The base of the Albian in southeast England was placed above a transgressive condensation horizon at the lowest occurrence of ammonite *Farnhamia farnhamensis* (Casey, 1961; Hesselbo *et al.*, 1990; Hart *et al.*, 1996). However, whether the associated *F. farnhamensis* ammonite subzone is equivalent to the *L. schrammeni* subzone of northern Germany is debated (Owen, 1996a,b; Casey, 1996). As a result, definitions for the base of the Albian Stage and its component substages are currently undergoing active review (see Hancock, 2001; Owen, 2002).

An important section associated with the lowest occurrence of the earliest "anterior" form of *L. schrammeni* is at Vörhum, near Hannover, northern Germany (Hart *et al.*, 1996; Mutterlose *et al.*, 2003). However, Mutterlose *et al.* (2003) conclude that definition of "the base of the Albian on these criteria cannot be used in a global context," therefore an alternate biostratigraphic or non-biostratigraphic marker must be identified.

The Aptian–Albian boundary interval in deep-water facies of the Tethyan realm is commonly to be found in clay-rich facies punctuated by widespread organic-rich horizons and associated carbon isotope excursions. An array of options is being considered as potential definitions for the base of the Albian, including (in approximate ascending order): "Jacob" organic-rich shale, the lowest occurrence of the calcareous nannofossil

Figure 19.2 Comparison of regional ammonite zonations and associated regional stage nomenclature spanning the Jurassic–Cretaceous boundary interval (modified from a compilation by Sven Bäckström in Gradstein *et al.*, 1999). British and Tethyan ammonite zones have been independently rescaled to the GTS2004 time scale (e.g. Figs. 18.1 and 19.1), and the Russian ammonite zones are shown with their potential correlation to the British zones. However, the exact correspondence and relative ages of most of these ammonite zones are not precisely known.

Ma scale: 137, 138, 139, 140, 141, 142, 143, 144, 145, 146, 147, 148, 149, 150, 151, 152, 153, 154, 155, 156

Boreal — British

Stage	Substage	Ammonite Zone
Valanginian	Upper	pitrei
		Dichotomites spp.
	Lower	Polyptychites
		Paratolia: brancoi
		Paratolia: involutum
		Paratolia: heteropleurum
		robustum
Ryazanian	Upper	albidum
		stenomphalus
		icenii
	Lower	kochi
		runctoni
Portlandian	Upper	lamplughi
		preplicomphalus
		primitivus
	Lower	oppressus
		anguiformis
		kerberus
		okusensis
		glauconolithus
		albani
Kimmeridgian	Upper	fittoni
		rotunda
		pallasioides
		paravirgatus
		eastlecottensis
		encombensis
		reisiformis
		wheatleyensis
		smedmorensis
		scitulus
		elegans/gigas
	Lower	autissiodorensis
		eudoxus: contejeani
		eudoxus: caletanum
		eudoxus: orthocera
		mutabilis: lallierianum
		mutabilis: mutabilis
		cymodoce: chatellaionensis
		cymodoce: achilles
		cymodoce: cymodoce
		baylei: rupellense
Oxf.	Upper	pseudocordata

Boreal — Russian

Stage	Substage	Ammonite Zone
Valanginian		
Ryazanian	Upper	albidum
		tzikwinianus
		rjasenensis/spasskensis
	Lower	kochi
		Garniericeras + Riasanites
Volgian	Upper	nodiger: kaschpuricus
		nodiger: mosquensis
		subditus
		fulgens
	Middle	oppressus
		nikitini: nikitini
		nikitini: blakei
		virgatus: rosanovi
		virgatus: virgatus
		panderi: zarajskensis
		panderi: pavlovi
	Lower	pseudoscythica
		sokolovi
		klimovi
Kimmeridgian	Upper	
	Lower	
Oxf.	Upper	

Tethys — French

Stage	Substage	Ammonite Zone
Valanginian	Upper	furicillata
		peregrinus
		verrucosum
	Lower	campylotoxus
		pertransiens
Berriasian	Upper	boissieri: otopeta
		boissieri: alpillensis
		boissieri: picteti
		boissieri: paramimounum
	Middle	occitanica: dalmasi
		occitanica: privasensis
		occitanica: subalpina
	Lower	jacobi: grandis
		jacobi: jacobi
Tithonian	Upper	Durangites
		transitorius
		simplisphinctes
	Middle	peroni
		admirandum + biruncinatum
		richteri
		verruciferum
	Lower	albertinum
		hybonotum
Kimmeridgian	Upper	beckeri: setatum
		beckeri: submeula
		eudoxus: cavouri
		acanthicum: compsum
	Lower	divisum: uhlandi
		divisum: tenuicostatum
		strombecki: lothari
		strombecki: hippolytense
		platynota
Oxfordian	Upper	planula: galar/grandiplex
		planula: planula
		bimammatum: hauffianum
		bimammatum: bimammatum

Period: Cretaceous (Valanginian–Berriasian); Jurassic (Tithonian–Oxfordian)

Praediscosphaera columnata (base of nannofossil zone NC8), the top-most organic-rich bed of the "Killian facies," the highest occurrence of ammonite *H. jacobi*, the lowest occurrence of ammonite *L. tardefurcata*, the "Paquier" organic-rich event (coinciding with oceanic anoxic event OAE1b), and the lowest occurrence of ammonite *D.* ex. gp. *mammillatum* (Hart *et al.*, 1996). All of these events are represented in potential boundary stratotypes in southeast France (Kennedy *et al.*, 2000; Table 19.2).

The difficulty in resolving unambiguous global correlation near the traditional Aptian–Albian boundary has led to suggestions to use a significantly higher level – the base of the *Lyelliceras lyelli* ammonite zone, which has been traditionally assigned as the base of the Middle Albian substage (Hancock, 2001; Owen, 2002). This level would shorten the Albian by approximately 3 myr.

In GTS2004, we assigned the base of the Albian Stage to the lowest occurrence of the calcareous nannofossil *Praediscosphaera columnata* (*P. cretacea* of some earlier studies). Even though this bioevent has taxonomic problems (e.g. Mutterlose *et al.*, 2003), this nannofossil first occurrence is recorded in candidate Boreal and Tethyan boundary stratotypes near the local ammonoid-defined boundaries (e.g. Kennedy *et al.*, 2000; Mutterlose *et al.*, 2003), it is a candidate marker for global boundary correlation (e.g. Hart *et al.*, 1996; Owen, 2002) and it can be calibrated to cycle stratigraphy spanning the complete Albian Stage in Italian sections (e.g. Fiet *et al.*, 2001; Grippo *et al.*, 2004).

Albian substages The traditional base of the Middle Albian is placed at the lowest occurrence of the ammonite *Lyelliceras lyelli*. The occurrence of this ammonite in European marginal basins represents a temporary incursion of a more cosmopolitan southern form (Owen, 1996b). A proposed boundary stratotype is within dark gray clay-rich strata near St. Dizier, northern France (Hancock, 2001; Hart *et al.*, 1996).

The base of the Upper Albian is assigned to the lowest occurrence of the ammonite *Dipoloceras cristatum* (Hart *et al.*, 1996). This event commonly coincides with a transgression above a major sequence boundary (e.g. Hesselbo *et al.*, 1990; Amédro, 1992; Hardenbol *et al.*, 1998), thereby causing the Middle–Upper Albian substage boundary interval to be quite condensed and probably incomplete at key sections along the English Channel at Wissant (Pas-de-Calais province, northwest France) and at Folkstone (Kent province, UK; Hart *et al.*, 1996). Therefore, an expanded basinal clay-rich section at Col de Palluel in southeast France is being considered as a substage GSSP.

19.1.3 Upper Cretaceous

The majority of the subdivisions of the Upper Cretaceous were originally derived from exposures in eastern France and the Netherlands in marginal marine to chalk facies. The original two stages of d'Orbigny (1847) were progressively subdivided into the six current stages. Few of the classical stratotypes were suitable for placing the limits of the stages, but most are defined by GSSPs within Europe.

A diverse array of primary markers are used or are under consideration for defining stage and substage boundaries, including ammonoids, inoceramid bivalves, planktonic foraminifera, and pelagic crinoids.

CENOMANIAN STAGE

History, revised definition, and boundary stratotype In 1847, d'Orbigny replaced the lower portion of his previous Turonian Stage with a Cenomanian Stage, and assigned the type region to be the vicinity of the former Roman town of Cenomanum, now called Le Mans (Sarthe region, northern France; d'Orbigny, 1847, p. 270). The conventional ammonoid marker for the base of the Cenomanian was the lowest occurrence of the acanthoceratid genus *Mantelliceras*, which was derived from the genus *Stoliczkaia* of Upper Albian age, or the lowest occurrence of the genus *Neostingoceras*, but there are many regions where these groups are relatively rare (Hancock, 2001).

Therefore, the Cenomanian Working Group selected the lowest occurrence of the planktonic foraminifer *Rotalipora globotruncanoides* (equivalent to *R. brotzeni* of some studies) as the basal boundary criterion for the Cenomanian Stage, with the Mont Risou section in southeast France as the GSSP section (Tröger *et al.*, 1996; Gale *et al.*, 1996a; Kennedy *et al.*, in press). The GSSP was ratified in 2002. This level is slightly lower than the lowest occurrence of Cenomanian ammonoid marker of *Mantelliceras mantelli*. In many regions, the Albian–Cenomanian boundary interval is coincident with a widespread hiatus and condensation associated with a major sequence boundary (e.g. Tröger *et al.*, 1996; Hardenbol *et al.*, 1998).

Cenomanian substages The sudden entry of the ammonite genera *Cunningtoniceras* and *Acanthoceras s.s.* marks the base of the Middle Cenomanian, and replacement of *Acanthoceras* ammonites by the *Calycoceras* genus is commonly used to mark the base of the Upper Cenomanian (Hancock, 1991).

The basal stratotype for the Middle Cenomanian is defined at the lowest occurrence of the ammonite *Cunningtoniceras inerme* at the Southerham Gray Quarry, Sussex, England (Tröger *et al.*, 1996). The lowest occurrences of the ammonite *Acanthoceras rhotomagense* and the beginning of a positive carbon-13 isotope excursion are approximately five couplets (100 kyr?) higher (Gale, 1995). The Lower–Middle Cenomanian boundary interval is missing over large regions due to its coincidence with a major sequence boundary (e.g. Hardenbol *et al.*, 1998).

A marker for the base of the Upper Cenomanian has not yet been selected, but the placement will probably be near the limits or within the *Acanthoceras jukesbrownei* ammonite zone (Tröger *et al.*, 1996).

TURONIAN STAGE

History, revised definition, and boundary stratotype The Turonian Stage has suffered continual re-definition (reviewed in Bengtson *et al.*, 1996). The Turonian Stage proposed by d'Orbigny in 1842 was later divided by him (d'Orbigny, 1847) into a lower Cenomanian Stage and an upper Turonian Stage. The name is derived from the Tours or Touraine region of France (Turones and Turonia of the Romans), and d'Orbigny (1852) designated a type region lying between Saumur (on the Loire river) and Montrichard (on the Cher river). In this region, the lowest Turonian formation contains the ammonite *Mammites nodosoides*, therefore its lowest occurrence was considered the marker for the base of the Turonian Stage (e.g. Harland *et al.*, 1990). However, below this level is a world-wide oceanic anoxic event (OAE2) and associated major positive excursion in carbon-13 isotopes, and mass extinction of over half of all ammonoid and brachiopod genera (e.g. Schlanger *et al.*, 1987; Kerr, 1998).

After considering several potential placements, the Turonian Working Group assigned the base of the Turonian to the lowest occurrence of the ammonite *Watinoceras devonense*, which occurs near the termination of this global oceanic anoxic event (Bengtson *et al.*, 1996). The GSSP at Rock Canyon Anticline, east of Pueblo (Colorado, west-central USA; Kennedy and Cobban, 1991; Bengtson *et al.*, 1996; Kennedy *et al.*, 2000, in press) was ratified in 2003. The maximum major carbon isotope peak associated with the oceanic anoxic event occurs 0.5 m above the boundary. The age of the Cenomanian–Turonian boundary is well constrained by average $^{39}Ar/^{40}Ar$ ages from correlative bentonites in nearby sections of 93.25 ± 0.55 and 93.55 ± 0.4 Ma (Obradovich, 1993; Kennedy *et al.*, in press).

Turonian substages The base of the Middle Turonian is marked by the lowest occurrence of the ammonite *Collignoniceras woollgari*. The candidate for the GSSP is the base of Bed 120 in the Rock Canyon Anticline section, approximately 5 m above the GSSP defining the base of the Turonian Stage in the same section (Bengtson *et al.*, 1996; Kennedy *et al.*, 2000).

The base of the Upper Turonian is not yet formalized, but potential datums are the lowest occurrences of the ammonite *Subprionocyclus neptuni* in the Boreal realm (e.g. Germany and England), of the ammonite *Romaniceras deverianum* in the Tethyan realm (e.g. southern France and Spain), or of an inoceramid bivalve, *Inoceramus perplexus* (= *Mytiloides costellatus* in some studies) (Bengtson *et al.*, 1996; Wiese and Kaplan, 2001). Some of these biostratigraphic markers have discrepancies in position relative to the reference carbon-13 curves from different sections (Wiese and Kaplan, 2001).

CONIACIAN STAGE

History, revised definition, and boundary stratotype Coquand (1857a,b) defined the Coniacian Stage in the northern part of the Aquitaine Basin, with the type locality at Richemont Seminary near Cognac (Charente province, western France). In this region, basal-Coniacian glauconitic sands overlie Turonian rudistid-bearing limestones. The entry of the ammonoid *Forresteria (Harleites) petrocoriensis* was used to mark the base of the Coniacian Stage, but there is confusion in identifying this species and ammonoids are rare or absent in most important Coniacian sections (Hancock, 1991; Kauffman *et al.*, 1996). Therefore, the Coniacian Working Group re-defined the Coniacian Stage and its substage boundaries using lowest occurrences of widespread inoceramid bivalves.

The proposed marker for the base of the Coniacian is the lowest occurrence of inoceramid bivalve *Cremnoceramus deformis erectus* (= *C. rotundatus* (*sensu* Tröger *non* Fiege)), which is significantly below the lowest occurrence of *F.* (*H.*) *petrocoriensis* in Europe. A candidate boundary stratotype is the active Salzgitter–Salder Quarry, southwest of Hannover (Lower Saxony province, northern Germany; Kauffman *et al.*, 1996); however, extensive diagenesis limits stratigraphic usage of calcareous nannofossils, palynomorphs, and stable isotopes (Sikora *et al.*, 2003, pers. comm. 2002).

Coniacian substages The base of the Middle Coniacian is placed at the lowest occurrence of the inoceramid bivalve genus *Volviceramus*, which is at or near the lowest occurrence of ammonoid *Peroniceras (Peroniceras) tridorsatum* (Kauffman *et al.*, 1996). Potential boundary stratotypes are near Dallas–Fort Worth, Texas (southern USA) or Seaford Head, southern England.

The base of the Upper Coniacian is placed at the lowest occurrence of the inoceramid bivalve *Magadiceramus subquadratus* (Kauffman *et al.*, 1996). No stratotypes have yet been proposed for this substage boundary.

SANTONIAN STAGE

History, definition, and boundary stratotype The Santonian Stage was named after Saintes (Cognac province, southwest France) by Coquand (1857a, b), who placed the lower boundary at a glauconitic hardground capping Santonian chalk.

The lowest occurrence of the widespread inoceramid bivalve *Cladoceramus undulatoplicatus* has been selected as the marker for the base of the Santonian (Lamolda *et al.*, 1996). The candidate GSSP is a quarry at Olazagutia in the Navarra region of Spain (Table 19.2; Cretaceous Subcommission Newsletter No. 5, December 2002).

The Coniacian through lower Santonian is within the long normal-polarity Chron C34n (Lowrie and Alvarez, 1977). Magnetostratigraphic sections within the chalk succession of southern England have indicated that a reversed-polarity subzone may straddle the Coniacian–Santonian boundary (Montgomery *et al.*, 1998).

Santonian substages The Santonian has three substages, but no markers for boundary stratotypes have yet been formalized. A possible datum for the base of the Middle Santonian is the extinction of the same inoceramid bivalve, *Cladoceramus undulatoplicatus*, that marks the Coniacian–Santonian boundary (Lamolda *et al.*, 1996). A possible datum for the base of the Upper Santonian is the lowest occurrence of stemless crinoid *Uintacrinus socialis* (Lamolda *et al.*, 1996). Magnetostratigraphy of chalk successions in southern England suggested that the base of the associated *Uintacrinus socialis* Zone was at or just below the base of reversed-polarity Chron C33r (Montgomery *et al.*, 1998), but this has not yet been confirmed in other sections.

CAMPANIAN STAGE

History, revised definition, and boundary stratotype The Campanian Stage of Coquand (1857b) was named after the hillside exposures of Grande Champagne, near Aubeterre-sur-Dronne (45 km west of Périgueux, northern Aquitaine province, France), but the shallow-water limestone outcrop had no obvious base. The bulk of the type "Campanian" at Aubeterre is now classified as Maastrichtian (e.g. van Hinte, 1965; Séronie-Vivien, 1972). The base of the Campanian was placed at the lowest occurrence of ammonoid *Placenticeras bidorsatum* by de Grossouvre (1901), but this extremely rare species is not a practical marker (reviewed in Hancock *et al.*, 1996). In contrast, stemless crinoids of the *Uintacrinus* and *Marsupites* genera, with their planktonic or benthic habitat, have a near-global distribution in shelf chalks (Gale *et al.*, 1996b).

Therefore, the extinction of crinoid *Marsupites testudinarius* is the provisional boundary marker for the base of the Campanian Stage (Hancock *et al.*, 1996). A potential boundary stratotype may be in England or Texas (Table 19.2). The base of the Campanian is at or within the lower portion of reversed-polarity Chron C33r (Montgomery *et al.*, 1998).

The age of the Santonian–Campanian boundary is constrained by a $^{40}Ar/^{39}Ar$ date of 83.91 $\pm$ 0.43 Ma from the uppermost Santonian *Desmoscaphites bassleri* ammonoid zone of North America (Obradovich, 1993).

Campanian substages The Campanian will be subdivided into a Lower, a Middle, and an Upper substage of approximately equal duration, but there are as yet no formal recommendations for primary markers or boundary stratotypes for the substages (Hancock *et al.*, 1996; Odin, 2001). In the US–Canadian Western Interior, these substage bases are informally placed by Cobban (1993) as the lowest occurrences of ammonoids *Baculites obtusus* and *Didymoceras nebrascense*, respectively.

MAASTRICHTIAN STAGE

History, revised definition, and boundary stratotype The Maastrichtian Stage was introduced by André Dumont (1849) for the "Calcaire de Maastricht" with a type locality at the town of Maastricht (southern Netherlands, near the border with Belgium). The stratotype was fixed by the Comité d'étude du Maastrichtian as the section of the Tuffeau de Maastrict exposed in the ENCI company quarry at St Pietersberg on the outskirts of Maastricht, but this local detrital-carbonate facies would correspond only to part of the upper Maastrichtian in current usage (reviewed in Rawson *et al.*, 1978; Odin, 2001). A revised concept of the Maastrichtian Stage was based on belemnites in the white chalk facies. Accordingly, the base of the stage was assigned to the lowest occurrence of belemnite *Belemnella lanceolata*, with a reference section in the chalk quarry at Kronsmoor (50 km northwest of Hamburg, north Germany; e.g. Birkelund *et al.*, 1984; Schönfeld *et al.*, 1996). The lowest occurrence of ammonoid *Hoploscaphites constrictus* above this level provided a secondary marker. Comparison of strontium isotope stratigraphy and indirect correlations by ammonoids indicate that this level is approximately equivalent to the base of the *Baculites eliasi* ammonoid zone of the

US Western Interior (McArthur *et al.*, 1992; Landman and Waage, 1993).

However, the belemnite *Belemnella lanceolata* is not a useful marker into the Tethyan faunal realm, whereas the ammonoid *Pachydiscus neubergicus* has a much wider geographical distribution (reviewed in Hancock, 1991). Therefore, in a mixed decision, the Maastrichtian Working Group recommended the base of the Maastrichtian to be assigned to the lowest occurrence of ammonoid *Pachydiscus neubergicus* (Odin *et al.*, 1996).

The ratified GSSP boundary is in an abandoned quarry near the village of Tercis les Bains, southwest France, at 90 cm below a coincident lowest occurrence of *Pachydiscus neubergicus* and *Hoploscaphites constrictus* ammonoids (Odin, 1996; Odin and Lamaurelle, 2001; Odin, 2001). The GSSP level was selected as the arithmetic mean of 12 biohorizons with correlation potential, including ammonoids, dinoflagellate cysts, planktonic and benthic foraminifera, inoceramid bivalves, and calcareous nannofossils (Odin and Lamaurelle, 2001). The history, stratigraphy, paleontology, and inter-continental correlations are extensively compiled in a special volume (Odin, 2001).

It is uncertain whether the placement of the Campanian–Maastrichtian boundary at the lowest occurrence of ammonoid *Pachydiscus neubergicus* at Tercis is significantly above the "traditional" belemnite *Belemnella lanceolata* level. Correlations of the assembly of other paleontological markers at the Tercis stratotype to other regions is crucial, but yield conflicting conclusions (Odin, 2001). Independent studies of the planktonic foraminifera at Tercis had reported different assemblages and ranges (e.g. Simmons *et al.*, 1996; Peybernès *et al.*, 1997; Ward and Orr, 1997), but the biostratigraphic array has been partially reconciled (Odin, 2001). However, the paleontological successions are not always consistent among different reference sections, including those sections used for magnetostratigraphic calibration (e.g. reviews in Odin, 2001). Echinoid and dinoflagellate cyst ranges suggest that the GSSP at Tercis is essentially contemporaneous with the belemnite *Belemnella lanceolata* level (Odin and Lamaurelle, 2001). From a web of paleontological considerations, Odin (2001, pp. 775–782) proposes that the GSSP is equivalent to the basal part of the Italian *Gansserina gansseri* foraminifer zone, to near the base of the *Baculites jenseni* ammonoid zone of the US Western Interior, to the middle of magnetic polarity Chron C32n.2n, and has an approximate age of 72.5 ± 0.5 Ma.

Other interpretations of macrofossil occurrences and microfossil stratigraphy of Tercis implies that the revised base of the Maastrichtian Stage is approximately at the base of the *Belemnella obtusa* belemnite zone of northwest Germany (e.g. Hancock *et al.*, 1993; Burnett *et al.*, 1998) and has an age near 70.5 Ma. The last occurrence of the calcareous nannofossil *Quadrum trifidum* at the top of belemnite *B. obtusa* Zone in northwest Germany has a projected age of 69.9 Ma, based on strontium isotope curve calibrations to dated-bentonite sections in the US Western Interior (McArthur *et al.*, 1993; McArthur, 1994). Therefore, if this highest occurrence of *Q. trifidum* is synchronous with its position at 16 m above the GSSP at Tercis, then the estimated sediment accumulation rate of 25 m/myr in this Tercis interval indicates that the underlying GSSP level is 0.75 myr prior to this age (Odin, 2001), or at approximately 70.6 Ma. This radiometric age is within the *Baculites baculus* or overlying *Baculites grandis* ammonoid zone of the US Western Interior (McArthur *et al.*, 1993). The weakly magnetized limestones at the Tercis stratotype did not yield a magnetostratigraphy above the upper–middle Campanian, but indirect microfossil correlation to other regions can be interpreted as projecting the base-Maastrichtian GSSP to be near the beginning of reversed-polarity Chron C31r (Barchi *et al.*, 1997), which has an estimated age of approximately 70.5 Ma (Fig. 19.1). In comparison, the base of the B. *lanceolata* belemnite zone has a projected age of 71.3 Ma (Obradovich, 1993; McArthur *et al.*, 1994). While acknowledging that the situation remains controversial, this latter set of base-Maastrichtian GSSP correlations are shown preferentially in Fig. 19.1 and other summary diagrams.

Upper Maastrichtian substage The Maastrichtian will be divided into two substages, but there is no agreement on boundary criterion for the base of the Upper Maastrichtian (Odin *et al.*, 1997; Odin, 2001). One potential marker is the lowest occurrence of ammonoid *Pachydiscus fresvillensis* with a possible stratotype at Zumaya in northern Spain (Kennedy, 1984, unpublished; Odin *et al.*, 1997). The lowest occurrence of ammonoid *Hoploscaphites birkelundi* (formerly *H.* aff. *nicolleti*) is an informal marker for the base of the Upper Maastrichtian in the US Western Interior (Landman and Waage, 1993; Cobban, 1993). Alternative criteria include a magnetic polarity reversal, the extinction of rudistid reefs, or the extinction of the majority of inoceramid bivalves.

19.2 CRETACEOUS STRATIGRAPHY

An extensive compilation and inter-correlation of Cretaceous biostratigraphy of European basins and the Western Interior seaway of North America was coordinated by Hardenbol *et al.* (1998), and Fig. 19.1 is a summary of a portion of their comprehensive chart series. These European and North American basins contain the majority of the proposed GSSP sites and alternate sites for the chronostratigraphic framework of the Cretaceous.

19.2.1 Macrofossil biostratigraphy

Ammonoids dominate the historical zonation of the Cretaceous. Belemnites, inoceramid bivalves, and pelagic or benthic crinoids (e.g. *Marsupites*) provide important macrofossil horizons within the Upper Cretaceous of northwest Europe, and buchiid brachiopods are used for correlation within the Lower Cretaceous within the Boreal realm. Important microfossil biostratigraphic zonations use planktonic foraminifera, calcareous nannoplankton, dinoflagellate cysts, and calpionellids.

AMMONOIDS

Ammonites have been the traditional primary standard in different faunal realms for subdividing most of the Cretaceous Period.

Grouping of ammonoid datums into zones and subzones underwent significant revisions during the 1990s. Only a few of the Cretaceous ammonoid zones compiled by Hancock (1991) are currently used by the various Cretaceous working groups (Rawson *et al.*, 1996b; Hoedemaeker *et al.*, 2003). The relative grouping into zones also varies among regions and stages. For example, many of the high-resolution "zones" of the Western Interior of North America would be classified as "horizons" in the broader zonal schemes used in Europe. The zonal nomenclatures are currently in a state of flux as workers strive to define zones by the first occurrences of index/name species, and reorganize assignments to genera (e.g. re-defined Lower Cretaceous zones by Reboulet and Atrops (1999) and Hoedemaeker *et al.* (2003), as shown in Fig. 19.1).

Extreme faunal provincialism necessitated the establishment of different regional scales throughout most of the Cretaceous. In Fig. 19.1, selected zonal successions are displayed for the southern European portion of the Tethyan realm, for the northwest European portion of the Boreal faunal realm (Berriasian through Albian), for the Russian portion of the Boreal realm (Berriasian and Valanginian), and for the North American Western Interior (Cenomanian through Maastrichtian). These simplified scales were excerpted from compilations by Hancock, Hoedemaeker, and Thierry (Hardenbol *et al.*, 1998). The nomenclature for selected ammonoid zonations with their estimated ages are partially compiled in Table 19.3. Other major syntheses include stage-by-stage overviews by Hancock (1991) and the Second International Symposium on Cretaceous Stage Boundaries (Rawson *et al.*, 1996b), Cobban (1993) for the Western Interior of North America, Hoedemaeker *et al.* (1993) for the lower Cretaceous scale for the southern European region of the Tethyan realm, and Amédro (1992), Owen (1996a), and Casey (1996) for alternative scales for the Aptian–Albian.

OTHER MARINE MACROFAUNA

Upper Cretaceous chalk-rich successions of northwest Europe have a variety of markers and zonations based on belemnites, pelagic crinoids and other echinoderms, and inoceramids. The version of this regional scale on Fig. 19.1 is from Combemorel and Christensen (1998) and Wood *et al.* (1994).

Inoceramid bivalves provide an important tool for correlation within Late Cretaceous basins, especially within the Western Interior of North America (reviewed by D'Hondt, 1998). Rudist bivalves enable correlation of shallow-water carbonate platforms (reviewed by Masse and Philip, 1998). Ostracode datums and associated zones are correlated to Boreal and Tethyan ammonoid zones within the lower Cretaceous, and to Boreal belemnite–echinoderm zonations within the upper Cretaceous (reviewed by Colin and Babinot, 1998).

DINOSAURS AND OTHER VERTEBRATES

Dinosaurs, the most renowned group of Cretaceous vertebrates, provide only a broad biostratigraphy. The following summary of major trends is from Lucas (1997). Early Cretaceous sauropods were smaller and ornithopods (such as *Iguanodon*) were larger than their Jurassic cousins. Nodosaurid ankylosaurs replaced stegosaurs in the earliest Cretaceous. Stegosaurs, iguanodontid and hypsilophodontid ornithopods, and sauropods (except in South America) were nearly extinct by the end of the Early Cretaceous. Placental and marsupial mammals appeared near the end of the Early Cretaceous. The rapid diversification of angiosperms (flowering plants) displaced gymnosperms in the mid Cretaceous and was probably a major factor in the evolution of the suite of hadrosaurid ornithopods, ceratopsians, and ankylosaurs browsers. This suite and their tyrannosaurid and coelurosaurian theropod predators were dramatically terminated at the end of the Cretaceous.

19.2.2 Microfossil biostratigraphy

CALPIONELLIDS AND FORAMINIFERA

Calpionellids are enigmatic pelagic microfossils with distinctive vase-shaped tests in thin section. Calpionellids appeared in the Tithonian and vanished in the latest Valanginian or earliest Hauterivian (Remane, 1985), and their abundance in carbonate-rich shelf–basinal settings within the Tethyan realm enable biostratigraphic correlation prior to the evolution of diverse foraminifera. Six standard zones (Allemann *et al.*, 1971) with finer subdivisions (e.g. Remane *et al.*, 1986) provide the basic framework for inter-regional correlation (reviewed by Remane, 1998).

Planktonic foraminifera have only broad zones prior to the Aptian, then provide a series of high-resolution global markers (e.g. Bralower *et al.*, 1995, 1997; Robaszynski and Caron, 1995; Premoli Silva and Sliter, 1999; Leckie *et al.*, 2002). The selected foraminifera datums and zones in Fig. 19.1 are from Robaszynski (1998) with selected modifications (e.g. Erba *et al.*, 1999; Leckie *et al.*, 2002; composite scales by Shipboard Scientific Party, 2004).

Larger and smaller benthic foraminifera correlation to ammonoid zones is partially established (e.g. Magniez-Jannin, 1995; Arnaud-Vanneau and Bilotte, 1998).

CALCAREOUS NANNOFOSSILS

The Cretaceous was named for the immense chalk formations that blanket much of northwestern Europe. The main components of this chalk are calcareous nannofossils. Following their rapid surge in abundance at the end of the Jurassic, calcareous nannofossils remained ubiquitous throughout the Cretaceous and Cenozoic in all oceanic settings above the carbonate dissolution depth.

Calibration of major calcareous nannofossil datums to ammonoid zones or magnetic polarity zones are established for several intervals in the Tethyan and Boreal realms (e.g. Bergen, 1994; Burnett *et al.*, 1998; Erba *et al.*, 1999; Channell *et al.*, 2000). The generalized scale in Fig. 19.1 is modified from von Salis (1998), although the age placement of several datums is uncertain (J. Bergen, pers. comm.).

OTHER MICROFOSSIL GROUPS

Organic-walled cysts of dinoflagellates have been correlated directly to ammonoid zones in the Tethyan realm for Berriasian through Turonian and in the Boreal realm for Berriasian through Aptian (reviewed by Foucher and Monteil, 1998). Siliceous radiolaria (pelagic sediments) and charophytes (brackish-water algae tests) have a relatively lower resolution set of datums and zones compared to other Cretaceous microfossil groups (e.g. respective syntheses by de Wever, 1998, and Riveline, 1998).

19.2.3 Physical stratigraphy

MAGNETOSTRATIGRAPHY

Cretaceous portion of M-sequence The M-sequence of marine magnetic anomalies formed from the Late Jurassic to the earliest Aptian. Several biomagnetostratigraphic studies have correlated Early Cretaceous calpionellid, calcareous microfossil, and dinoflagellate datums to the M-sequence polarity chrons (e.g. Channell and Grandesso, 1987; Channell *et al.*, 1987, 1993, 1995b, 2000; Ogg, 1987, 1988; Ogg *et al.*, 1991a). Correlation of Tethyan ammonite zones to the M-sequence have been achieved for the Berriasian, spanning Chrons M18–M15 (Galbrun, 1984); the Berriasian–Valanginian boundary interval, spanning M15–M13 (Ogg *et al.*, 1988; Aguado *et al.*, 2000); and the Hauterivian and Barremian, spanning portions of M10N–M1 (Cecca *et al.*, 1994; Channell *et al.*, 1995a). Polarity zone M0r is a primary marker associated with the proposed GSSP at the base of the Aptian. When coupled with a spreading-rate model for the Pacific magnetic lineations within each individual stage (see Chapter 5), these correlations also constrain the relative duration of each ammonite zone within the Berriasian, Hauterivian, and Barremian stages.

Correlation of Boreal ammonite zones to the M-sequence has been indirectly achieved only for the equivalent of the Berriasian Stage in the Purbeck beds of southern England (Ogg *et al.*, 1991a, 1994).

Reported brief subchrons within the Aptian and Albian An extended normal-polarity Chron C34n or "Cretaceous Normal-Polarity Superchron" spans the early Aptian through middle Santonian. Brief reversed-polarity chrons have been reported from three intervals – middle Aptian, middle Albian, and mid-late Albian – especially within drilling cores of deep-sea sediments. Ryan *et al.* (1978) proposed a negative numbering for these three "pre-M0r" reversed-polarity events or clusters of events:

1. M"-1r" in late Aptian (Pechersky and Khramov, 1973; Jarrard, 1974; Keating and Helsley, 1978a,b,c; VandenBerg *et al.*, 1978; Hailwood, 1979; Lowrie *et al.*, 1980; VandenBerg and Wonders, 1980; Tarduno, 1990; Ogg *et al.*, 1992), which has a biostratigraphic age near the base of the *Globigerinelliodes algerianus* planktonic foraminifer zone. This subchron has also been called the "ISEA" event from an Italian outcrop sample code and has an estimated duration of less than 100 000 years (Tarduno, 1990).
2. M"-2r" set of Middle Albian events near the boundary of the *Biticinella breggiensis* and *Ticinella primula* planktonic

foraminifer zones (Jarrard, 1974; Keating and Helsley, 1978a; VandenBerg and Wonders, 1980; Tarduno, 1992; Shipboard Scientific Party, 1998; Ogg and Bardot, 2001).

3. M"-3r" set in Late Albian (Green and Brecher, 1974; Jarrard, 1974; Hailwood, 1979), which may occur at the end of the *Praediscosphaera cretacea* or within the *Eiffellithus turriseiffeli* nannoplankton zones (Tarduno, 1992).

Another reversed-polarity event, possibly near the Aptian–Albian boundary, has been reported within basalt flows with a radiometric age of 113.3 ± 1.6 Ma (Gilder *et al.*, 2003). Neither M"-2r" nor M"-3r" have been verified in outcrop sections, and further documentation is required to determine whether events represent true reversed-polarity episodes (J. E. T. Channell, pers. comm., August 1999).

Cretaceous portion of C-sequence Polarity Chrons C33r through lower C29r span the latest Santonian to the base of the Cenozoic (e.g. Alvarez *et al.*, 1977; Lowrie and Alvarez, 1977; Montgomery *et al.*, 1998). This polarity time scale has been calibrated in the Western Interior seaway of North America to regional ammonoid zones and to an array of $^{40}Ar/^{39}Ar$ dates from bentonites (e.g. Obradovich, 1993; Hicks and Obradovich, 1995; Hicks *et al.*, 1995, 1999; Lerbekmo and Braman, 2002; see tables in Chapters 5 and 20). The Paleogene–Cretaceous boundary occurs during the middle of polarity Chron C29r, and the 65.5 Ma age for this boundary (Renne *et al.*, 1998c; Hicks *et al.*, 1999) coupled with analysis of cyclic sediments (Herbert *et al.*, 1995) imply that Chron C29r begins at 65.9 Ma.

The ages on the Campanian–Maastrichtian portion of the C-sequence constrain the synthetic age–distance model for the magnetic anomalies of the South Atlantic (revised from Cande and Kent, 1992a; see Chapter 5). Spreading rates of this South Atlantic ridge remained constant at ~27 km/my during the Campanian, then decelerated smoothly through the Maastrichtian and Danian to reach a minimum of ~13 km/my during the late Paleocene (Röhl *et al.*, 2001).

CHEMICAL STRATIGRAPHY

Carbon stable isotopes and carbon-enrichment episodes At least seven significant excursions in the carbon cycle punctuate the Cretaceous stratigraphic record (Figs. 19.1 and 19.3). Most of these positive excursions (>1.5 per mil) in carbon-13 are associated with widespread organic-rich sediments and drowning of carbonate platforms, and appear to be preceded by or coincide with the eruption of major flood basalt provinces (e.g. reviews by Kerr, 1998; Weissert *et al.*, 1998; Jenkyns, 1999; Larson and Erba, 1999). The middle Cretaceous suite were originally considered to be the products of oceanic anoxic events (OAE) (e.g. Schlanger and Jenkyns, 1976), but organic-carbon preservation due to enhanced oceanic productivity may play a role (e.g. Hochuli *et al.*, 1999; Jenkyns, 1999; Leckie *et al.*, 2002). The approximate biostratigraphic ages of widespread organic-enrichment levels associated with these excursions are illustrated in Fig. 19.1. The reference nomenclature attached to the organic-rich zones varies from region to region.

1. A Late Valanginian positive carbon-13 excursion of ~2 per mil has an onset in the lower Valanginian (polarity Chron M12, *Th. campylotoxus* ammonite zone), peaks during latest Valanginian, and terminates in the early Hauterivian (*A. radiatus* ammonite zone; Lini *et al.*, 1992; Channell *et al.*, 1993; Weissert *et al.*, 1998). The Lower Valanginian (*Th. campylotoxus* Zone) onset is documented by four centimetric organic-carbon-rich layers ("Barrande" layers B1–B4) in the southeast France basin (Vocontian Basin; Reboulet, 2001; Reboulet *et al.*, 2000, 2003).
2. A latest Hauterivian (Faraoni) excursion within the *Pseudothurmannia catulloi* ammonite subzone (middle of *Ps. ohmi* Zone) is documented by a pair of organic-rich sediments in Mediterranean and Atlantic pelagic sections and coincides with a relatively minor positive carbon-13 excursion (e.g. Baudin *et al.*, 1997, 1999; Coccioni, 2003).
3. An Early Aptian excursion (OAE1a) with a double carbon-13 spike spans the *D. deshayesi* and *Tr. bowerbanki* ammonite zones. The organic-rich "Selli" event is near the base at the lower portion of the *Leopoldina cabri* foraminifer zone (e.g. Weissert and Lini, 1991; Mutterlose and Böckel, 1998; Weissert *et al.*, 1998; Hochuli *et al.*, 1999; Larson and Erba, 1999; Leckie *et al.*, 2002).
4. A multiphase Aptian–Albian boundary excursion (OAE1b) that contains at least three widespread organic-rich layers – the "Jacob," "Pacquier," and "Urbino" levels (Bréheret, 1988; Weissert and Bréheret, 1991; Bralower *et al.*, 1993; Weissert *et al.*, 1998; Leckie *et al.*, 2002).
5. A middle Albian excursion (OAE1c) with an organic-rich layer named "Amadeus" (after Mozart) or "Toolebuc" (e.g. Leckie *et al.*, 2002; Coccioni, 2003).
6. A latest Albian excursion (OAE1d); e.g. Erbacher *et al.*, 1996; Leckie *et al.*, 2002). The organic-rich layer is named "Breistroffer" or "Pialli."
7. A Cenomanian–Turonian boundary excursion (OAE2) spans the *M. geslinianum–W. devonense* ammonite zones, with the peak in uppermost Cenomanian (e.g. Schlanger

et al., 1987; Jenkyns *et al.*, 1994; Kerr, 1998; Jenkyns, 1999). The organic-rich level is named "Bonarelli" in Italy.

Oxygen stable isotopes and estimated temperature trends Oxygen isotope records of oceanic temperature trends are patchy, and some isotopic and paleontological studies have suggested that the overall warm Early Cretaceous was punctuated by cold spells during the Berriasian–Valanginian boundary interval, earliest Late Valanginian, and earliest Aptian (e.g. Weissert and Lini, 1991; Hochuli *et al.*, 1999; Miller *et al.*, 1999; Pucéat *et al.*, 2003). The general model for the late Cretaceous from isotopic records from DSDP–ODP sites, uplifted chalk, and fish tooth enamels is that there was an overall warming trend from the early Aptian to a peak in the early Turonian, followed by a gradual cooling trend through the end of the Maastrichtian (Douglas and Savin, 1975; Jenkyns *et al.*, 1994; Clarke and Jenkyns, 1999; Wilson *et al.*, 2002; Pucéat *et al.*, 2003; Fig. 19.3). Paleontologic and isotopic evidence indicate that globally averaged surface temperatures during the middle part of the Cretaceous were more than 10 °C higher than today (e.g. see reviews by Barron, 1983; Huber *et al.*, 1995, 2002; Bice *et al.*, 2002).

Strontium isotope ratios The marine $^{87}Sr/^{86}Sr$ record displays a progressive rise from the Berriasian to a maximum of 0.707 493 in the Barremian *P. elegans* ammonite zone, before decreasing to a pronounced minimum of 0.707 220 just before the Aptian–Albian boundary (Fig. 19.3). From this point it rises sharply in the Early Albian, flattens to a broad maximum through the Middle to Late Albian, before declining to a late Turonian minimum of around 0.707 275. From this minimum, it rises to 0.707 830 at (or probably a few kyr before) the end of the Cretaceous (Fig. 19.3; see details in Chapter 6).

These trends enable global correlation and dating with $^{87}Sr/^{86}Sr$ through this interval with a stratigraphic resolution that is good, except on the cusps of reversals (Aptian–Albian boundary, Late Turonian, middle Barremian (*sensu lato*), or where $^{87}Sr/^{86}Sr$ changes little with time (Middle and Late Albian).

CYCLE STRATIGRAPHY AND ESTIMATES OF STAGE DURATIONS

Analyses of astronomical Milankovitch cycles recorded as compositional or isotopic oscillations in marine successions place minimal limits on the elapsed time within associated biostratigraphic or magnetostratigraphic zones.

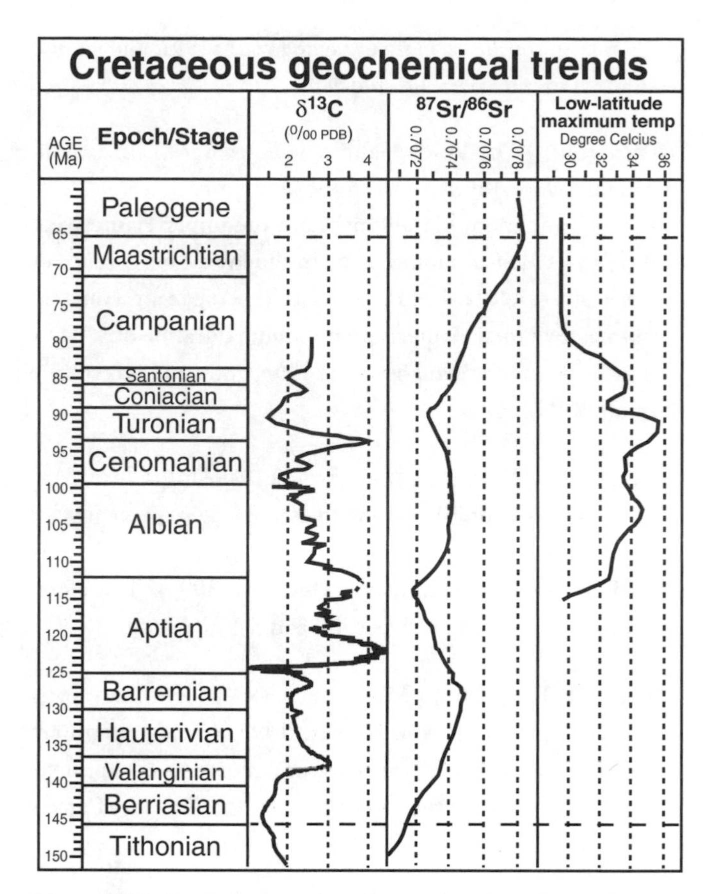

Figure 19.3 Cretaceous trends and excursions in carbon and oxygen stable isotopes and in $^{87}Sr/^{86}Sr$. The schematic carbon isotope curve is a composite derived from Weissert *et al.* (1998) for the lower Cretaceous, from Menegatti *et al.* (1998) for the lower Aptian, from Erbacher *et al.* (1996) for the upper Aptian and Albian, and from Jenkyns *et al.* (1994) for the upper Cretaceous. The Aptian–Albian reference curves are rescaled using the revised durations of foraminifer zones (Figure 19.1). The mean strontium isotope trend is a LOWESS fit to data from several sources (see text and Chapter 7). The curve of extrapolated low-latitude maximum temperatures (center of 5 °C range) is from Clarke and Jenkyns (1999).

Late Berriasian The interval from the lowest occurrence of *Calpionellopsis simplex* to the lowest occurrence of *L. hungarica* (Calpionellid Zones D1 and D2) in southeast Spain spans a minimum of 1.1–1.2 myr (Sprenger and ten Kate, 1993).

Valanginian The uppermost Berriasian *Thurmanniceras otopeta* through lower Valanginian *Busnardoites campylotoxus* ammonite zones span an estimated minimum of 3.15 myr, and the upper Valanginian *Sayno. verrucosum* ammonite zone through *Teschenites callidiscus* subzone (topmost subzone of Valanginian) span an estimated 2.75 myr. The Valanginian portion of this interval would have a minimum duration of

~5.5 myr in southeast France (Huang *et al.*, 1993; and similar result by Giraud, 1995; Giraud *et al.*, 1995).

Hauterivian The lower Hauterivian *Acanthodiscus radiatus* through *Lyticoceras nodosoplicatum* ammonite zones span an estimated minimum of 2.40 myr, and the upper Hauterivian *Subsaynella sayni* ammonite zone to the lowest occurrence of *C. emerci* (possible base of Barremian Stage) span an estimated 2.93 myr, for a total Hauterivian minimum duration of 5.3 myr in southeast France (Huang *et al.*, 1993; and similar result by Giraud, 1995).

Barremian The interval from the base of polarity Chron M3r to the base of Chron M0r spans an estimated 46.5 eccentricity bundles (estimated 106 kyr) and 274 precession beds (estimated 20 kyr) in northern Italy (Herbert, 1992). Therefore, this interval may span between 4.8 and 5.4 myr.

Aptian Polarity Chron M0r at the base of the Aptian spans 380 kyr at Gorgo a Cerbara (Herbert, 1992). Cyclostratigraphy indicates that the "Selli" carbon-enrichment layer is approximately 0.5 myr above the top of polarity Chron M0r (Larson and Erba, 1999).

Albian Cycle stratigraphy in a cored section at Piobbico (Italy) records 30.6 long-eccentricity cycles (estimated 406 kyr), hence a span of 12.4 myr between the base of nannofossil *P. columnata* (a potential marker for the Aptian–Albian boundary) and the base of planktonic foraminifer *Rotalipora globotruncanoides* (i.e. the *R. brotzeni* Zone of some earlier papers) Zone (Grippo *et al.*, 2004; which enhances an earlier study by Fiet *et al.*, 2001). This core provides an astronomical scaling for the component Albian nannofossil and planktonic foraminifer zones.

Cenomanian A complete cycle stratigraphy has been constructed for the uppermost Albian, Cenomanian, and lower Turonian in western Europe (212 precession beds) with scaling of the upper *Stoliczkaia dispar* (uppermost Albian), *Mantelliceras mantelli* through *N. juddii* (Cenomanian), and *W. devonense* through lower *C. woolgari* ammonite zones (Gale, 1995). The Cenomanian Stage spans a minimum of 4.45 myr. Additional orbital tuning indicates that the Middle and Late Cenomanian (*Cunningtoniceras inerme* through *N. juddii* ammonite zones) span 3.0 myr (Gale *et al.*, 1999), thus the entire Cenomanian may be close to 5 myr in duration.

Coniacian The Coniacian chalks in the Anglo–Paris basin display eight depositional cycles. If these correspond to a 400-kyr eccentricity rhythm, then the Coniacian Stage spans ~3.2 myr (Grant *et al.*, 1999).

SEQUENCE STRATIGRAPHY

Cretaceous marginal-marine to deep-shelf successions in Europe and North America record an abundance of basinal and regional transgressions and regressions. The common features from an extensive suite of compilations edited by de Graciansky *et al.* (1998) were assembled in a comprehensive synthesis and systematically numbered from the base of each stage (Hardenbol *et al.*, 1998). Coeval emergent horizons recorded by Aptian–Albian seamount carbonate platforms in the central Pacific Ocean imply that some of these depositional sequences reflect global eustatic sea-level oscillations (Röhl and Ogg, 1996). Details of the Cretaceous sequence stratigraphy are continually undergoing further refinement, but the major global oscillations have probably been identified. These large-scale deepening and shallowing trends are summarized in Fig. 19.1.

In a broader context, the Cretaceous strata encompass a single transgressive–regressive cycle (the "North Atlantic" cycle of Jacquin and de Graciansky, 1998). The lower boundary is a widespread unconformity of Late Berriasian age; the peak transgression occurred in the Early Turonian, and average sea levels continued to decrease into the Paleocene.

OTHER MAJOR STRATIGRAPHIC EVENTS

Large igneous provinces At least five large igneous provinces formed during the Cretaceous. Some of these seem to be associated with major distortions in the global carbon budget as reflected by excursions in carbon isotopes, widespread organic-rich shales or "OAE" episodes, and other changes in climate and oceanic chemistry (e.g. Larson and Erba, 1999; Jones and Jenkyns, 2001; Bice *et al.*, 2002). The following summary is mainly from extensive reviews by Wignall (2001) and Courtillot and Renne (2003).

1. **133 Ma**: Extrusive volcanics associated with the early stages of rifting between South America (Paraná flood basalts) and Namibia (Etendeka Traps) peaked near ~133 ± 1 Ma. The full episode may have begun at ~138 Ma and lasted about 10 myr (Stewart *et al.*, 1996), therefore it overlapped with the broad positive carbon isotope peak through the late Valanginian. Whether the eruption peak exactly coincided with the late Hauterivian "Faraoni" carbon-enrichment episode and carbon isotope excursion (schematically placed at ~131 Ma in Fig. 19.1) is

dependent upon the method used for extrapolating the ages for the Hauterivian Stage, as discussed below.

2. **122 Ma**: During the middle of the Early Aptian, the largest series of volcanic eruptions of the past quarter-billion years built the Ontong Java Plateau and Manihiki Plateau in the western equatorial Pacific. A series of deep-sea drilling legs documented that the multi-kilometer-thick series of volcanic flows forming the Ontong Java Plateau occurred during a short time span at ~122 Ma (e.g. Mahoney *et al.*, 1993), and the upper portions of this are inter-bedded with pelagic sediments of the lower portion of the *Leopoldina cabri* foraminifer zone (Mahoney *et al.*, 2001; Sikora and Bergen, 2004). A cascade of environmental effects from the eruption of the Ontong Java Plateau is the suspected culprit for the organic-rich deposits associated with the OAE1a, or Selli, episode and the onset of the large positive carbon isotope excursion of the Early Aptian (e.g. Larson, 1991; Tarduno *et al.*, 1991; Larson and Erba, 1999).
3. **118 Ma**: The Kerguelen Plateau in the southern Indian Ocean is the second largest oceanic plateau after the Ontong Java Plateau. The peak of construction of the southern and largest portion may have been simultaneous with the eruption of the formerly adjacent Rajmahal Traps of eastern India at ~118 Ma (reviewed in Wignall, 2001). This episode probably contributed to the broad carbon isotope excursion that characterizes the Aptian Stage.
4. **90 Ma**: A large Pacific oceanic plateau, the Caribbean–Columbian volcanic province, was later emplaced between the North and South American plates. This event was theorized to be a contributor to the end-Cenomanian OAE2 carbon isotope excursion (e.g. Kerr, 1998), but its apparent average age, reported as 89.5 ± 0.3 Ma, coincides instead with the Turonian–Coniacian boundary (Courtillot and Renne, 2003).
5. **65 Ma**: The Deccan traps cover a major portion of India, with a peak of eruption at ~65.5 Ma coinciding with the catastrophic termination of the Cretaceous (Courtillot and Renne, 2003).

In addition to these large igneous provinces, pulses of large-scale volcanism constructed the Shatsky Rise in the central Pacific through the mid-Early Cretaceous, and the Madagascar traps at ~88 Ma.

Large impact events Five impact craters with diameters greater than 40 km are currently documented from the Cretaceous:

1. ***Morokweng*** crater (~140 km?), South Africa, dated between 145 and 146 Ma (Koeberl *et al.*, 1997; Reimond *et al.*, 1999), coincides with the Jurassic–Cretaceous boundary interval;
2. ***Mjølnir*** crater (~40 km), Norway, at 142 ± 2.6 Ma;
3. ***Tookoonoka*** crater (~50 km), Queensland, Australia, at 128 ± 5 Ma;
4. ***Kara*** crater (~60 km), Russia, at 70.3 ± 2.2 Ma; and
5. the immense ***Chixculub*** crater (~170 km), Yucatan, Mexico, that dramatically terminated the Mesozoic Era at 65.5 Ma.

19.3 CRETACEOUS TIME SCALE

The time scale for the Cretaceous Period is derived in three segments: (i) the Berriasian through Barremian stages is determined mainly from calibration of Tethyan ammonite zones with the M-sequence of magnetic polarity chrons; (ii) the Aptian and Albian stages utilize durations of microfossil zones derived from cycle stratigraphy as a primary scaling; and (iii) the Late Cretaceous is constrained by an extensive suite of $^{40}Ar/^{39}Ar$ ages on ammonoid-rich successions of the US Western Interior.

The ages of Cretaceous stage boundaries are either yielded from their direct calibration to these primary absolute age scales (e.g. the GSSP for the base of the Turonian corresponds to the base of the *Watinoceras devonense* ammonite zone of the US Western Interior; the proposed GSSP for the base of the Aptian corresponds to the base of Chron M0r) or are assigned according to estimates of their correlation or possible placements relative to the primary scale (e.g. the yet-to-be-officially defined base of the Albian is assigned here as the first occurrence of calcareous nannofossil *Praediscosphaera columnata*, which is placed according to cycle stratigraphy relative to the radiometrically dated base of the Cenomanian).

The main stratigraphically controlled ages that are used in constructing the Cretaceous time scale are summarized in Table 17.2. This suite was selected from a larger compilation of published ages, based on both stratigraphic and radiometric precision.

19.3.1 Berriasian through Barremian

There are essentially no reliable radiometric ages with precise stratigraphic controls within the Late Jurassic or Early Cretaceous. Therefore, in a process similar to that used in the Cenozoic (e.g. Berggren *et al.*, 1995a; Chapter 20), the calibration of major biostratigraphic events to the magnetic

Table 19.3 *Selected ammonite zone scales for Cretaceous with estimated ages*[a]

Stage	Primary standard zonation	Method	Calibration of base	Basal age[b] (Ma)	Comments
	Ammonite zones of US Western Interior				
Maastrichtian				**65.50**	Radiometric age
	[Few usable ammonites]			67.88	
	Jeletzkytes nebrascensis	Early Maastrichtian (base *B. clinolobatus*) through early Cenomanian are spline fit of US Western Interior ammonite zones using recalibrated Obradovich '93 age suite		68.33	
	Hoploscaphites nicolleti			68.78	
	Hoploscaphites birkelundi			69.23	
	Baculites clinolobatus	Upper Maastrichtian zones (*J. nebrascensis* through *B. clinolobatus*) are assigned 0.45 myr durations, which is average of the lower Maastrichtian spline-scaled zones		69.68	69.42 Ma (±0.37) is within this zone
	Baculites grandis			70.11	
	Baculites baculus			70.56	Odin (2001) considered base-Maastrichtian GSSP to project to near base of *B. jenseni* ammonite zone, but another interpretation (used here) is equivalence to a level within the *B. baculus* or overlying *B. grandis* ammonite zone of the US Western Interior
Campanian				**70.60**	
	Baculites eliasi			71.04	Cobban's zonation of 1993 had estimated base-Campanian (prior to setting of GSSP) approximately at base of *B. eliasi*
	Baculites jenseni	23 Campanian ammonites zones in Western Interior scaled to fit radiometric age control		71.56	
	Baculites reesidei			72.14	
	Baculites cuneatus			72.78	
	Baculites compressus			73.50	73.35 Ma (±0.39) is within this zone
	Didymoceras cheyennense			74.28	
	Exiteloceras jenneyi	C32r/C33n boundary		75.05	74.76 Ma (±0.45) is within this zone
	Didymoceras stevensoni			75.74	
	Didymoceras nebrascense			76.38	75.89 Ma (±0.72) is within this zone
	Middle			*76.38*	
	Baculites scotti			77.00	
	Baculites gregoryensis			77.59	

Table 19.3 (*cont.*)

Stage	Primary standard zonation	Method	Calibration of base	Basal age[b] (Ma)	Comments
	Baculites perplexus (late)			78.15	
	Baculites gilberti			78.68	
	Baculites perplexus (early)			79.16	
	Baculites sp. (smooth)			79.61	
	Baculites asperiformis			80.00	
	Baculites maclearni			80.35	
	Baculites obtusus			80.64	80.54 Ma (±0.55) is within this zone
	Early			***80.64***	
	Baculites sp. (weak flank ribs)			80.91	
	Baculites sp. (smooth)			81.22	
	Scaphites hippocrepis III			81.63	
	Scaphites hippocrepis II			82.20	81.71 Ma (±0.34) is within this zone
	Scaphites hippocrepis I			82.89	
	Scaphites leei III			83.53	Base-Campanian is the extinction of crinoid *Marsupites testudinarius* (provisional boundary marker). For this scale, we have assumed equivalence with base of *S. leei III* ammonite zone of Western Interior
Santonian				**83.53**	
	Desmoscaphites bassleri			83.99	83.91 (±0.43) and 84.09 Ma (±0.40) are within this zone
	Desmoscaphites erdmanni	*D. erdmanni* through *C. vermiformis* given equal durations (pure spline method caused anomalous internal scaling where no age constraints were present)		84.31	
	Clioscaphites choteauensis			84.62	
	Clioscaphites vermiformis			84.94	
	Clioscaphites saxitonianus			85.85	84.88 Ma (±0.28) is within this zone; assigned to top Base-Santonian is lowest occurrence of the widespread inoceramid bivalve *Cladoceramus undulatoplicatus*. For this scale, we arbitrarily equated this to the base of the *C. saxitonianus* ammonite zone of Western Interior
Coniacian				**85.85**	
	Scaphites depressus–Protexanites bourgeoisianus			86.96	86.92 Ma (±0.39) is within this zone; assumed to be in lower third
	Scaphites ventricosus			87.88	
	Forresteria alluaudi–Scaphites preventricosus			88.58	88.34 Ma (±0.60) is within this zone

(*cont.*)

Table 19.3 (*cont.*)

Stage	Primary standard zonation	Method	Calibration of base	Basal age[b] (Ma)	Comments
	Forresteria peruuana			89.07	Base-Coniacian (candidate definition) is lowest occurrence of inoceramid bivalve *Cremnoceramus deformis erectus* (= *C. rotundatus* (*sensu* Tröger non Fiege)), which is significantly below the lowest occurrence of *F.* (*H.*) *petrocoriensis* in Europe. Therefore, we have arbitrarily added 0.2 myr (1/2 ammonite zone) to estimate age of base-Coniacian
Turonian				**89.27**	
	Prionocyclus germari			89.40	
	Scaphites nigricollensis			89.63	
	Scaphites whitfieldi			89.79	
	Scaphites ferronensis			89.96	
	Scaphites warreni			90.17	
	Prionocyclus macombi			90.48	90.21 Ma (±0.72) is within this zone
	Prionocyclus hyatti			90.94	90.51 Ma (±0.45) is within this zone; assigned to middle
	Prionocyclus percarinatus			91.51	
	Collignoniceras wolligari			92.13	
	Mammites nodosoides			92.70	
	Vascoceras birchbyi			93.15	93.40 Ma (±0.63) is within this zone; 92.98 Ma assigned to base
	Pseudaspidoceras flexuosum			93.41	93.25 Ma (±0.55) is within this zone; 93.33 Ma assigned to base
	Watinoceras devonense			93.55	This is same as base-Turonian (GSSP)
Cenomanian				**93.55**	
	Nigericeras scotti	*N. scotti* + *N. juddii* considered equivalent to one zone in duration; divided equally		93.60	
	Neocardioceras juddii			93.64	93.78 (±0.55), 93.59 (±0.58), and 93.49 Ma (±0.89) are within this zone
	Burroceras clydense			93.72	
	Sciponoceras gracile (*Euomphaloceras septemseriatum*)			93.85	93.49 Ma (±0.89) is within this zone
	Sciponoceras gracile (*Vascoceras diartianum*)			94.01	93.90 Ma (±0.72) is within this zone; 93.84 Ma assigned to top, 93.99 Ma to base
	Dunveganoceras conditum			94.19	
	Dunveganoceras albertense			94.38	
	Dunveganoceras problimaticum			94.57	

Table 19.3 (*cont.*)

Stage	Primary standard zonation	Method	Calibration of base	Basal age[b] (Ma)	Comments
	Calycoceras canitaurinum–Dunveganoceras pondi			94.75	94.63 Ma (±0.61) is within this zone
	Plesiacanthoceras wyomingsense			94.91	
	Acanthoceras amphibolum			95.06	94.93 Ma (±0.53) is within this zone; 95.03 Ma assigned to base
	Acanthoceras bellense			95.23	
	Acanthoceras muldoonense			95.43	
	Acanthoceras granerosense			95.68	
	Conlinoceras tarrantense–Conlinoceras gilberti			96.01	95.78 Ma (±0.61) is within *C. gilberti* Zone; 95.84 Ma assigned to base
	[blank]	3 "blank" (un-named) zones and *N. maclearni* have sub-equal durations		96.43	
	[blank]			96.89	
	[blank]			97.35	
	Neogastroplites maclearni			97.75	
	Neogastroplites americanus			98.11	
	Neogastroplites muelleri			98.41	
	Neogastroplites cornutus			98.64	98.52 Ma (±0.41) is within this zone; 98.35 Ma assigned to top
	Neogastroplites haasi			98.83	98.54 (±0.70) and 98.74 Ma (±0.59) are within this zone; 98.65 Ma assigned to top
	Base of *N. caractanense* subzone (basal subzone of *M. mantelli* Zone; basal zone of Cenomanian in Europe)	Ar–Ar of ammonite-bearing sections in Japan, plus cycle stratigraphy		99.60	The top of *N. caractanense* subzone (basal subzone of Cenomanian) is ~99.1 ± 0.4 Ma (Obradovich *et al.*, 2002) and cycle stratigraphy scaling of its duration (0.44 myr) and a slight offset to the GSSP marker below implies the base of the Cenomanian is 99.6 ± 0.4 Ma Base-Cenomanian is base of foraminifer *Rotalipora globotruncanoides*, which is 6 m lower than base of *M. mantelli* ammonite zone at the GSSP section
	Ammonite zones of European Tethyan realm				
Albian		cycle stratigraphy from base-Cenomanian		99.6	
	Stoliczkaia dispar			100.9	= base of *R. appenninica* foraminifer zone (cycle-scaled as 100.9 Ma)
	Mortoniceras (Mortoniceras) inflatum			106.4	= FAD of foraminifer *B. breggiensis* (base of *T. praeticinensis* subzone in Leckie *et al.* (2002) is cycle-scaled as 106.4 Ma
	Callihoplites auritus s. z.	Subzones of *M. inflatum* are assigned equal durations			

(*cont.*)

Table 19.3 (*cont.*)

Stage	Primary standard zonation	Method	Calibration of base	Basal age[b] (Ma)	Comments
	Hysteroceras varicosum s.z.				
	Hysteroceras orbigyni s.z.				
	Diploceras cristatum s.z.				
		Albian ammonite zones (below *M. inflatum* to top of Aptian) scaled proportionally to number of subzones			
	Euhoplites lautus			107.0	
	Euhoplites loricatus			108.2	
	Hoplites (Hoplites) dentatus			108.8	
	Douvilleiceras mammillatum			110.8	
	Leymeriella tardefurcata			112.0	Very close to FAD of *P. columnata* (subcircular) according to Mutterlose *et al.* (2003) = our base of Albian stage = 112.0 Ma from cycle-scaling
Aptian		Cycle stratigraphy from base-Cenomanian		112.0	
	Hypacanthoplites jacobi	*H. jacobi* through *N. nolani* proportionally scaled according to subzones (1 and 2, respectively)		113.0	
	Nolaniceras nolani			114.0	
	N. nolani (*Diodochoceras nodosocostatum* s.z.)			115.0	Base of *N. nolani* (hence, top of *P. nutfieldiensis* Zone) is set at 115 Ma to fit the constraint of an Ar–Ar age of 114 ± 1.3 Ma within that underlying zone. This also fits its calibration to foraminifer datums (Bellier *et al.*, 2000) on chart of Leckie *et al.* (2002; after scaling to 125 Ma at base-Aptian)
	P. melchioris				Zone is approximately equivalent to Boreal *Parahoplites nutfieldiensis* Zone
	E. subnodosocostatum			121.0	Top of foraminifer *L. cabri* (120.7 Ma in rescaling of Leckie *et al.*, 2002) is just above the top of the *D. furcata* ammonite zone at Le Bedoule (Moullade *et al.*, 1998); therefore the zonal boundary age was rounded to 121 Ma. Zone is approximately equivalent to Boreal *Cheloniceras* (*Epicheloniceras*) *martinoides* Zone
	Dufrenoyia furcata			121.9	Base of *D. furcata* Zone = base of nannofossil *E. floralis* in SE France (Moullade *et al.*, 1998), which, in turn, is estimated at 121.9 Ma in the chart in Leckie *et al.* (2002; rescaled to our base-Aptian). Zone is set equivalent to *Tropaeum* (*Tropaeum*) *bowerbanki* in this chart

Table 19.3 (*cont.*)

Stage	Primary standard zonation	Method	Calibration of base	Basal age[b] (Ma)	Comments
	Deshayesites deshayesi	*D. deshayesi* to *D. oglanlensis* (3 zones) assigned equal duration		122.9	
	Deshayesites weissi			124.0	Zone is set equivalent to *Deshayesites forbesi* in this chart
	Deshayesites oglanlensis	magstrat	Base Chron M0r	125.0	*D. tuarkyricus* Zone is used as basal-Aptian in some charts; but the *D. tuarkyricus* index species is only found in Turkmenistan; therefore, IUGS Lower Cretaceous group proposes *D. oglanlensis.* Zone is set equivalent to *Prodeshayesites fissicostatus* of Boreal in this chart
Barremian		Barremian ammonite zones of Tethyan realm are given equal durations, above (6 zones), then below (5 zones), the base-*vandenheckii* control age		125.0	
	P. waagenoides			125.6	
	C. sarasini			126.1	Formerly "Martelites" sarasini. Top 2 Barremian zones set as equivalent to Boreal "*stolleyi* " and "*bidentaum/scalare*" Zones
	I. giraudi			126.7	
	Hemihoplites feraudianus			127.2	
	Hemihoplites sartousi			127.8	'*E. barremense*' Zone is equivalent to upper-half of *A. vandenheckii* + lower-half of *H. sartousi*
	A. vandenheckii	magstrat	Upper Chron M3r	128.3	
	Coronites darsi			128.6	Zones of Lower Barremian were extensively revised (Hoedemaker *et al.*, 2003)
	Kotetishvilia compressissima			129.0	Formerly "*Pulchellia*" *compressissima*
	Nicklesia pulchella			129.3	Elevated to zone status; was a horizon
	Kotetishvilia nicklesi			129.7	Formerly "*Nicklesia*" *nicklesi*
	Spitidiscus hugii	magstrat	Chron M5n.8	130.0	Formerly "*Spitidiscus*" *hugii*
Hauterivian		Hauterivian ammonite zones are scaled according to subzones (10)		130.0	
	Pseudothurmannia ohmi			131.9	Formerly called *Pseudothurmannia angulicostata auctorum*. Three subzones (from base) are *P. ohmi*, *Pseudothurmannia catulloi*, *Pseudothurmannia picteti* (Hoedemaker *et al.*, 2003)

(*cont.*)

Table 19.3 (*cont.*)

Stage	Primary standard zonation	Method	Calibration of base	Basal age[b] (Ma)	Comments
	B. balearis			132.5	
	"*Pleisiospitidiscus ligatus*"			133.2	
	Subsaynella sayni			133.8	
	Lyticoceras nodosoplicatus			134.5	
	Crioceratites loryi			135.8	*C. loryi* Zone has two subzones (*O.* (*J.*) *jennoti* and *C. loryi*)
	Acanthodiscus radiatus	magstrat	Base Chron M11n	136.4	Contains 2 subzones
Valanginian		Valanginian ammonite zones (above *T. otopeta*) scaled proportionally to number of subzones + horizons (9)		136.4	
	Criosarasinella furcillata			137.3	Former "*Neocomites* (*Teschenites*) *pachydicranus*" Zone has been completely replaced in new zonal scheme by "Kilian Group" (2002). New upper zone has 2 subzones (*Teschenites callidiscus* and *C. furcillata*)
	Necomites peregrinus			138.1	N. peregrinus Zone has 2 subzones (*Neocomites peregrinus*, formerly in *S. verrucosum*; and *Olcostephanus* (*O.*) *nicklesi*); the former *Himantoceras trinodosum* subzone has been replaced
	Saynoceras verrucosum			138.9	*S. verrucosum* Zone now has 2 subzones (*S. verrucosum* and *Karakaschiceras pronecostatum*); (Hoedemaeker *et al.*, 2003)
	Busnardoites campylotoxus			139.8	Name "*Thurmanniceras*" *campylotoxus* has also been used for this ammonite. Divided into 2 subzones, with the upper one having two horizons
	Tirnovella pertransiens	magstrat	Chron M14r.3	140.2	
Berriasian		Berriasian ammonite zones are correlated to magnetics		140.2	
	Thurmanniceras otopeta	magstrat	Base Chron M15n	140.7	Was M15n.4 in Ogg *et al.* (1988), but revised ammonite stratigraphy by Aguado *et al.* (2000) assigns as base of M15n
	boissieri (*Tirnovella alpillensis* s.z.)	magstrat	Chron M16n.8	141.2	"*Fauriella*" and "*Berriasella*" have also been used as genera names for this species
	B. (*F.*) *boissieri* (*Berriasella picteti* s.z.)	magstrat	Chron M16n.5	141.6	Name "*Picteticeras*" *picteti* has also been used for this ammonite. See note on *Tirnovella alpillensis* s.z.

Table 19.3 (*cont.*)

Stage	Primary standard zonation	Method	Calibration of base	Basal age[b] (Ma)	Comments
	B. (*F.*) *boissieri* (*Malbosiceras paramimounum* s.z.)	magstrat	Chron M16r.5	142.3	Name "*Berriasella*" *paramimounum* has also been used for this ammonite
	Subthurmannia occitanica (*Dalmasiceras dalmasi* s.z.)	magstrat	Chron M17n.7	142.6	"*Subthurmannia* (*Strambergella*)" and "*Tirnovella*" have also been used as genera names for this species
	S. occitanica (*Berriasella privasensis* s.z.)	magstrat	Chron M17r.9	143.0	Calibration of base to magnetic polarity chrons is constrained to upper half of M17r in Berrias section
	S. occitanica (*Tirnovella subalpina* s.z.)	magstrat	Chron M17r.3	143.7	"*Neocomites*" and "*Tirnovella*" have also been used for the genus name of this species
	Berriasella jacobi (*Pseudosubplanites grandis* s.z.)	magstrat	Chron M18r.2	144.8	This zone has also been included as the upper subzone of a *Berriasella* (*Pseudosubplanites*) *exinus* Zone. See NOTE on *B. jacobi*
	B. jacobi (*B. jacobi* s.z.)	magstrat	Chron M19n.2n.55 (± 0.05)	145.5	
JURASSIC (Tithonian) (top)				**145.5**	
	Durangites	magstrat	Chron M19r.1 (± 0.2)	146.1	Base Calpionellid Zone A2 is base *Durangites* Zone

[a] The ammonite zones and component subzones for the Mediterranean faunal province of the Tethyan paleogeographic realm are those recommended by the IUGS Lower Cretaceous Ammonite Working Group (Hoedemaeker *et al.*, 2003), and their correlation to prior schemes is partially indicated in the comment column. The calculated ages for these ammonite zones were used as the primary scaling for most of the Cretaceous time scale chart (Fig. 19.1). Ages of most other ammonite zones, microfossil datums, and sea-level changes in Fig. 19.1 are derived from their relative correlations to these primary scales as shown in the charts of Hardenbol *et al.* (1998) and of Leckie *et al.* (2002).

[b] Age to base of zone.

polarity time scale is used to estimate absolute ages. In turn, for the Early Cretaceous, the M-sequence magnetic polarity time scale was constructed by incorporating key radiometric ages (the top of Chron M0r is of earliest Aptian; Chron M26n is of lower Kimmeridgian; and pre-Chron M41n is of Bajocian age) with estimated durations from cycle stratigraphy of the basal Aptian, the Barremian (~5 myr), and the combined Hauterivian–Valanginian (~10.5 myr) stages.

Ammonite zones of the Mediterranean faunal province of the Tethyan realm have been calibrated to M-sequence chrons for the Berriasian through lowest Valanginian stages, and for portions (including the proposed limits) of the Hauterivian and Barremian stages. The spreading-rate model (see Chapter 5) together with the estimated chron assignments, imply that the Valanginian Stage had a relatively short duration, which also matches its lowest number of component ammonite zones. If a set of ammonite zones lacked a direct calibration to the M-sequence, the zones in this interval were scaled according to their relative numbers of subzones (Table 19.3). The relative duration and the ages of the Berriasian through Hauterivian stages (and Kimmeridgian and Tithonian of the late Jurassic) are entirely dependent upon the selection of the Pacific spreading-rate model for the Hawaiian lineations, therefore their estimated uncertainties increase to ~2 myr for absolute ages of the stages furthest from the radiometric-age control points, and uncertainties for their durations are ~1 myr.

Ammonite zones of the sub-Boreal faunal province have been partially calibrated to M-sequence chrons only within the Berriasian. Therefore, the approximate ages of the sub-Boreal zones of Valanginian through Barremian are dashed according to potential correlations shown on the ammonite scales of Hardenbol *et al.* (1998).

Calpionellid microfossil and most calcareous nannofossil events have been directly calibrated to the M-sequence chrons.

19.3.2 Aptian and Albian

This portion of the Cretaceous time scale hinges on radiometric ages for the basal-Aptian and the base of the Cenomanian, and uses cycle stratigraphy for a major portion of the internal scaling. The most significant difference in this Cretaceous time scale compared to versions published in the 1990s (Obradovich, 1993; Gradstein *et al.*, 1994a, 1995; Larson and Erba, 1999) is the assignment of an age of 125 Ma to the base of the Aptian (base of polarity Chron M0r), rather than ~121 Ma. In addition to increasing the duration of the Aptian Stage, this results in a shift to older absolute ages for the underlying M-sequence and associated biostratigraphic and stage boundaries through the Tithonian. Therefore, an expanded explanation for this key age is appropriate.

AGE OF BASE-APTIAN

The absolute age for the base of the Aptian, which was assigned as ~121 Ma on the Gradstein *et al.* (1995) time scale, is derived from an $^{40}Ar/^{39}Ar$ radiometric date of 124.6 ± 0.3 Ma (after correcting to TCR monitor standards of Renne *et al.*, 1998c) near the top of a reversed-polarity magnetic zone recorded in the basalt edifice of MIT Guyot in the western Pacific (Pringle and Duncan, 1995; Larson and Erba, 1999). The overlying basalt is a transition into a well-developed soil that is overlain by transgressive marine sediments containing early Aptian nannofossils, then capped by a thick shallow-water carbonate platform. This reversed-polarity zone had been assigned as the uppermost portion of polarity Chron M1r of middle Barremian based on estimates of (a) the required time to form an overlying soil profile and (b) the abundance of nannoconid calcareous nannofossils within the overlying marine flooding horizon, which had been interpreted as preceding the earliest Aptian "nannoconid crisis."

However, re-evaluation of the nannofossil criteria (J. Bergen, pers. comm., 2001) and the character of the carbon isotope values from the overlying sediments at MIT Guyot are also consistent with assigning this polarity zone to polarity Chron M0r of the basal-Aptian (Pringle *et al.*, 2003). Chron M0r spans 0.4 myr (Herbert *et al.*, 1995), therefore its base, which is a proposed marker for the base of the Aptian Stage, is at approximately 125.0 Ma.

This revised age scale using 125.0 Ma for the base of the Aptian is consistent with radiometric and paleontological data from the Ontong Java Plateau. For example, the basaltic basement of ODP Site 807 on the Ontong Java Plateau, which is inter-bedded with limestone containing *Leopoldina cabri* foraminifera (Sliter and Leckie, 1993), yielded an average whole-rock Ar–Ar age of 122.4 ± 0.8 Ma (Mahoney *et al.*, 1993). ODP Leg 192 sites also documented inter-bedded late-stage volcanism of Ontong Java Plateau with *Leopoldina cabri* foraminifera (Mahoney *et al.*, 2001; Sikora and Bergen, 2004). The OAE1a, or Selli, episode of carbon-rich oceanic sediment accumulation and onset of a shift to heavier carbon isotope ratios that are characteristic of the lower to middle Aptian begins about 0.5 myr after the end of Chron M0r, and spans about 1 myr (Larson and Erba, 1999), therefore the approximate termination of OAE1a is assigned as 123 Ma. The biostratigraphic age of OAE1a is the middle of *C. litterarius* nannofossil zone and the lower part of the *Leopoldina cabri* planktonic foraminifer zone (Leckie *et al.*, 2002). This coincidence between biostratigraphic and radiometric age calibration strongly supports the concept that rapid eruption of the Ontong Java Plateau was a major cause of the conditions leading to the OAE1a carbon-rich sedimentary deposits and the associated cascading effects on the global carbon cycle and carbon isotope ratios.

Therefore, the reversed-polarity zone dated at MIT Guyot was probably M0r (e.g. Pringle *et al.*, 2003) and the base of the Aptian Stage is 125.0 Ma.

SCALING OF THE APTIAN AND ALBIAN

The Albian–Cenomanian boundary (base of planktonic foraminifer *G. truncanoides*, or just slightly older than the base of the M. *mantelli* ammonite zone) is projected at 99.6 Ma (see "Upper Cretaceous time scale" below). The Albian time scale below this base-Cenomanian control age is derived from the absolute durations of Albian planktonic foraminifer zones as scaled by cycle stratigraphy (Grippo *et al.*, 2004).

The placement of the Aptian–Albian boundary is a contentious topic. We decided to provisionally assign the base of the Albian to be the lowest occurrence of the nannofossil *Prediscophaera columnata* (equivalent to *P. cretacea* of some earlier studies) within European reference sections, despite some problems in taxonomy (Mutterlose *et al.*, 2003). This base-Albian placement has an age of 112 Ma computed from cycle stratigraphy.

The upper Aptian scale is partially constrained by an age of 114 ± 1.3 Ma from a bentonite in the upper Aptian *Parahoplites nutfieldensis* ammonite zone (Obradovich, 1993). The inter-zonal calibration of Aptian–Albian calcareous nannofossil and planktonic foraminifer zones and the relative scaling of these datums within the Aptian are primarily from Leckie *et al.* (2002), which in turn modifies the zonal scales in Bralower *et al.* (1997) by a downward extension of the range of planktonic foraminifer *Leupoldina cabri* (Erba *et al.*, 1999; Premoli Silva *et al.*, 1999).

Aptian and Albian nannofossil and planktonic foraminifer zone definitions and relative scaling of datums from Leckie *et al.* (2002), which in turn relies on Bralower *et al.* (1997) with downward extension of the range of planktonic foraminifer *Leupoldina cabri* from Erba *et al.* (1999) and Premoli-Silva *et al.* (1999). Within intervals lacking direct cycle-stratigraphic scaling of biostratigraphic events, the relative spacing of these events from Leckie *et al.* (2002) or Hardenbol *et al.* (1998) was used proportionally. Tethyan ammonite zones were calibrated to these foraminifer and nannofossil events (e.g. Moullade *et al.*, 1998; Robaszynski, 1998; Bellier *et al.*, 2000), or proportionally scaled relative to their subzones (Table 19.3). In turn, estimated relative placements of Boreal ammonite zones and major sea-level trends to the Tethyan reference scale are from the Cretaceous compilations of Hardenbol *et al.* (1998).

19.3.3 Cenomanian through Maastrichtian

A Late Cretaceous time scale for North American ammonoid zones was calibrated by Obradovich (1993) from his extensive suite of Ar–Ar dates on bentonites. He rejected all radiometric dates derived from biotites in bentonites as too young, and considered all his previous K–Ar ages on sanidines to be obsolete. Correlation of the North American ammonoid zonation and the associated radiometric ages to Upper Cretaceous European stages and zones was partially achieved through rare interchanges of ammonoid and other marine macrofauna (reviewed in Cobban, 1993) and strontium isotope curves for portions of the Campanian and Maastrichtian (e.g. McArthur *et al.*, 1993, 1994). This Late Cretaceous time scale was modified slightly by Gradstein *et al.* (1994, 1995), and we have further refined the spline fit for the component ammonoid zones (Table 19.3; and see Chapter 8).

Obradovich *et al.* (2002) report $^{40}Ar/^{39}Ar$ ages of 98.98 ± 0.38 Ma from the *Mantelliceras saxbii* ammonite subzone (the upper of the two subzones of the *Mantelliceras mantelli* Zone; the basal zone of the Cenomanian) and 99.16 ± 0.37 Ma from the *Graysonites wooldridgei* ammonite zone (equivalent to the European subzone of *Neostlingoceras carcitanense*; which is the lower of the two subzones). These ages by Obradovich *et al.* (2002) are only slightly older than his previous (Obradovich, 1993) average of 98.6 Ma from the *Neogastroplites haasi* ammonite zone, which had been provisionally considered to be the basal zone of a comparatively finer-detailed Cenomanian ammonoid zonation in the North American Western Interior. Therefore, the age for the top of the *N. carcitanense* subzone is ~99.1 Ma (±0.4 myr). In the Vocontian Basin of southeast France, there are 22 couplets within the *N. carcitanense* subzone in outcrops of apparent continuity down through the uppermost Albian (Gale, 1995). If, following Gale (1995), we assign these couplets to be precession cycles (20 kyr), then the *N. carcitanense* subzone spans at least 0.44 myr, hence it is ~99.55 Ma at the base. In the Risou GSSP, the base of *M. mantelli* is 6 m above the GSSP (base of *G. truncanoides*), but the relative spacing of uppermost Albian foraminifer datums relative to their Milankovitch-cycle-scaling in Italy (Grippo *et al.*, 2004) implies a very rapid sedimentation rate; therefore, the base of the Cenomanian is ~99.6 Ma.

The C-sequence of magnetic polarity chrons is scaled according to combined ammonoid and radiometric constraints in the US Western Interior. Other European macrofossil zones, the microfossil and nannofossil datums, and the sea-level curve are scaled relative to the estimated calibrations to either the North American ammonoid zonation (e.g. charts in Hardenbol *et al.*, 1998) or the C-sequence polarity chrons (e.g. Shipboard Scientific Party, 2004).

19.3.4 Estimated uncertainties on stage ages and durations

The Cretaceous time scale was derived in different segments with a mixture of spline fitting, cycle stratigraphy, and seafloor-spreading models. Each method has different degrees of precession. In the following discussion, the uncertainties are expressed at the 95% confidence level (2-sigma).

The Late Cretaceous uses a spline fit to combine $^{40}Ar/^{39}Ar$ ages and ammonoid zones of the US Western Interior, and the computed uncertainties have partially incorporated possible systematic offsets between $^{40}Ar/^{39}Ar$ radiometric ages and other methods (see Chapter 8). This method results in an estimated uncertainty of ±0.8 myr for the base-Turonian, and the combined radiometric and cycle-scaling for base-Cenomanian and base-Albian (or, rather our potential marker for this yet-to-be-defined level) continues this trend (see Table 19.4.).

The base-Aptian age (125.0 Ma) is tied closely to an $^{40}Ar/^{39}Ar$ radiometric age of 124.6 ± 0.6 Ma (2-sigma) for the top of Chron M0r, and we have increased the uncertainty to ±1.0 myr to incorporate partially for the external errors on this radiometric method. Cycle stratigraphy relative to this base-Aptian age was incorporated into a seafloor-spreading model, and implies a relatively low uncertainty in durations for the Barremian and Hauterivian stages. As noted in Chapter 18,

Table 19.4 *Ages and duration of Cretaceous stages (see footnote)*

Boundary	Stage	Age (Ma)	*Estimated uncertainty on age (2-sigma)*	Duration (myr)	*Estimated uncertainty on duration (2-sigma)*	Status; primary marker	Calibration, and Comments
Base of Cenozoic (base of Danian)		**65.5**	*0.3 myr*			GSSP; Iridium anomaly	Array of direct Ar-Ar ages on impact level
	Maastrichtian			**5.1**	*~0.5 myr*		Durations (and uncertainties) of Maastrichtian and Campanian depend on assumed correlation to Ar-Ar age dates of U.S. Western Interior ammonite zones.
Base of Maastrichtian		**70.6**	*0.6*			GSSP; bracketed by microfossils and regional ammonites	Odin (2001) projected the base-Maastrichtian GSSP to base of *B. jenseni* ammonite zone, but another interpretation (see text) is to a younger level, and estimated age corresponds to the base of the *B. baculus* zone of the Western Interior of North America. Ar-Ar ages bracket this ammonite datum.
	Campanian			**12.9**	*~0.7*		
Base of Campanian		**83.5**	*0.7*			Not defined; provisional boundary marker is the extinction of crinoid *Marsupites testudinarius*.	Level of proposed GSSP marker is assumed to be equivalent with the base of *S. leei III* ammonite zone of Western Interior of North America. Ar-Ar ages bracket this ammonite datum.
	Santonian			**2.3**	*0.1*		
Base of Santonian		**85.8**	*0.7*			Not defined; proposed primary marker is lowest occurrence of the widespread inoceramid bivalve *Cladoceramus undulatoplicatus*.	Level of proposed GSSP marker is assumed to be equivalent with the base of the *C. saxitonianus* ammonite zone of Western Interior of North America. Ar-Ar ages bracket this ammonite datum.
	Coniacian			**3.5**	*0.3*		
Base of Coniacian		**89.3**	*1.0*			Not defined; Base-Turonian (candidate definition) is lowest occurrence of inoceramid bivalve *Cremnoceramus deformis erectus*,	Level of proposed GSSP marker is "significantly below the lowest occurrence of *F. (H.) petrocoriensis* in Europe", which is generally shown as approximately equivalent to base of *Foreesteria peruuana* zone of Western Interior of North America. Ar-Ar ages bracket this ammonite datum in North America, and we have arbitrarily added 1/2 ammonite zone of Western Interior (0.2 myr) its age.
	Turonian			**4.2**	*0.3*		
Base of Turonian		**93.5**	*0.8*			GSSP; lowest occurrence of the ammonite *Watinoceras devonense*.	Ar-Ar ages tightly bracket this ammonite datum in North America
	Cenomanian			**6.1**	*0.3*		
Base of Cenomanian		**99.6**	*0.9*			GSSP; the lowest occurrence of the planktonic foraminifer *Rotalipora globotruncanoides*.	Ar-Ar ages bracket top of overlying ammonite subzone, then cycle stratigraphy was applied to scale GSSP datum
	Albian			**12.4**	*0.3*		Duration of Albian is from cycle stratigraphy; but depends upon definition of its base.
Base of Albian		**112.0**	*1.0*			Not defined, and controversial. Boundary shown here is one candidate; the lowest occurrence of calcareous nannofossil *Praediscosphaera columnata* (base of nannofossil zone NC8).	Cycle stratigraphy scaling from base-Cenomanian GSSP age estimate
	Aptian			**13.0**	*0.5*		

Table 19.4 (*cont.*)

Boundary Stage	Age (Ma)	*Estimated uncertainty on age (2-sigma)*	Duration (myr)	*Estimated uncertainty on duration (2-sigma)*	Status; primary marker	Calibration and comments
Base of Aptian	**125.0**	*1.0*			The proposed primary marker is the beginning of magnetic polarity Chron M0r.	Ar–Ar age at top of presumed zone M0r on Pacific guyot; and duration of Chron M0r is known from cycle stratigraphy.
Barremian			**5.0**	*0.5*		Duration of Barremian is estimated from cycle stratigraphy as ~5 myr.
Base of Barremian	**130.0**	*1.5*			Not defined, but the candidate marker is the lowest occurrence of the *Taveraidiscus hugii* auctorum, which is reported as ~Chron M5n(0.8)	Chron ages depend upon selected Pacific spreading rate model.
Hauterivian			**6.4**	*1.0*		Total duration of Hauterivian and Valanginian is estimated from cycle stratigraphy as ~10 myr.
Base of Hauterivian	**136.4**	*2.0*			Not defined, but the candidate marker is the lowest occurrence of the ammonite genus *Acanthodiscus* (especially *A. radiatus*), which is reported to be near the base of Chron M11n	Chron ages and duration of stages depend upon selected Pacific spreading rate model. Not well constrained. A faster spreading rate would yield a shorter duration, and a younger basal age.
Valanginian			**3.8**	*1.0*		Uncertainties on Valanginian through Berriasian durations are assigned as 30% of duration, which is probable range of realistic spreading models.
Base of Valanginian	**140.2**	*3.0*			Not defined, but candidate marker is the lowest occurrence of ammonite *Thurmanniceras pertransiens,* which is at ~Chron M14r(0.3)	Chron ages and duration of stages depend upon selected Pacific spreading rate model. Not well constrained. A slower spreading rate would yield a longer duration, and a older basal age.
Berriasian			**5.3**	*1.7*		
Base of Cretaceous (base of Berriasian)	**145.5**	*4.0*			Not defined. Base of Cretaceous (base of Berriasian) placed here as the leading candidate of the base of *Berriasella jacobi* ammonite zone or base of Calpionellid Zone B; which is calibrated as Chron M19n.2n(0.55)	Chron ages and duration of stages depend upon selected Pacific spreading rate model. Not well constrained.

For stages that lack a ratified definition (as of Nov 2004), the computed age is for the indicated primary marker. See text for discussion of derivation of uncertainty estimates.

this seafloor-spreading model has an estimated uncertainty of ±4 myr for Late Jurassic stage boundaries. However, the uncertainties from the spreading model should quickly converge to about ±1.5 myr for the base-Barremian.

19.3.5 Summary of Cretaceous time scale

The Cretaceous is the longest period of the Phanerozoic, spanning 80 myr from 145.5 ± 4.0 to 65.5 ± 0.3 Ma. With the Albian–Cenomanian boundary estimated at 99.6 ± 0.9 Ma, the Early Cretaceous Epoch lasted 46 myr, and the Late Cretaceous Epoch spanned 34 myr. Three of the Cretaceous stages (the Aptian, Albian, and Campanian) span over 12 myr, and are among the longest stages within the Phanerozoic.

Compared to the high precision of the Late Cretaceous time scale (with the exception of correlating the base-Maastrichtian GSSP), the Early Cretaceous and Late Jurassic are nearly devoid of verified high-precision ages that are calibrated to biostratigraphy or magnetic polarity chrons. A refinement of this portion of the Mesozoic time scale is crucial to our understanding of the rates of plate motions during the birth of the Atlantic Ocean, and should be an emphasis of future campaigns for radiometric dating and cycle tuning.

20 • The Paleogene Period

H. P. LUTERBACHER, J. R. ALI, H. BRINKHUIS, F. M. GRADSTEIN, J. J. HOOKER, S. MONECHI, J. G. OGG, J. POWELL, U. RÖHL, A. SANFILIPPO, AND B. SCHMITZ

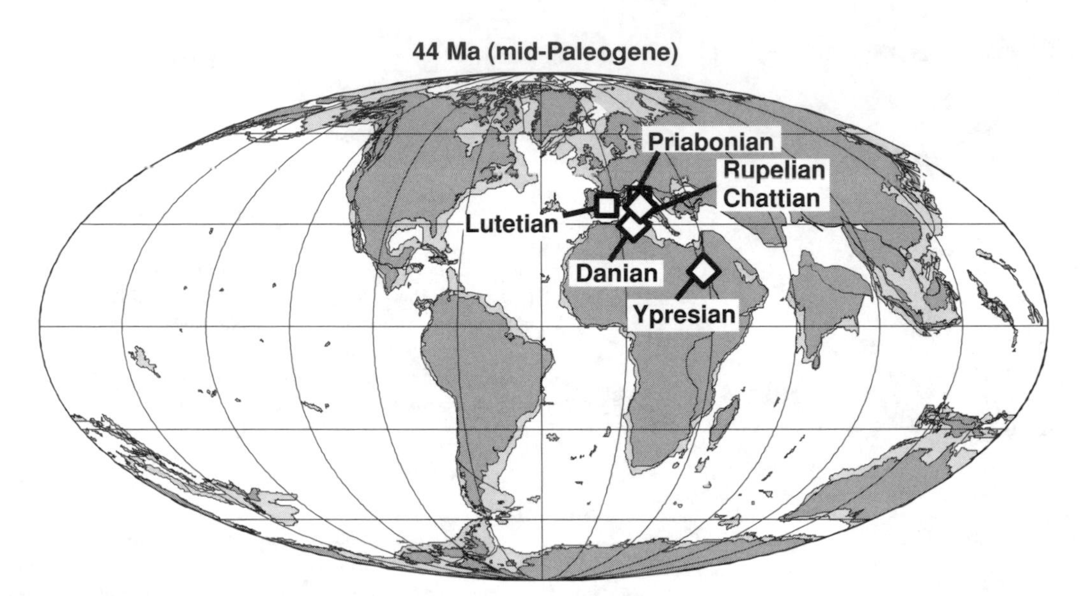

Geographic distribution of Paleogene GSSPs that have been ratified (diamonds) or are candidates (squares) on a mid-Paleogene (44 Ma) map (status in January, 2004; see Table 2.3). Not all candidate GSSPs are shown. Only the GSSPs for base-Oligocene (Rupelian), base-Eocene (Ypresian), and base-Paleocene (base-Danian) epochs have been formalized. The projection center is at 30° W.

After the global Cretaceous–Paleogene boundary impact catastrophy, planktonic foraminifera and nannofossils start new evolutionary trends, and mammals appear on Earth; global warming episode occurs at the Paleocene–Eocene boundary and significant cooling trends develop in later Paleogene in preparation for Modern life and climate. Orbital tuning of deep-marine cyclic sedimentation patterns, calibrated to the geomagnetic polarity and biostratigraphic scales, have the potential to elevate the Paleogene time scale to the level of resolution of the Neogene.

20.1 HISTORY AND SUBDIVISIONS

20.1.1 Overview of the Paleogene

The Cenozoic Era (Phillips, 1841, originally "Cainozoic," from *kainos* = new and *zoon* = animal) is subdivided into the Paleogene (*palaios* = old, *genos* = birth) and Neogene periods. The use of "Tertiary" (Arduino, 1759) and "Quaternary" (Desnoyers, 1829) is not recommended, being equally antiquated terms such as Primary and Secondary that have fallen into disuse in the twentieth century. Naumann (1866) combined in his "Paleogen Stufe," the Eocene and Oligocene, as opposed to the "Neogen Stufe" of Hörnes (1853) which included not only the Miocene and the Pliocene, but also fauna of the Pleistocene (see Chapter 21). The term "Nummulitique," which had been employed as an equivalent of Paleogene mainly by French-speaking stratigraphers such as Renevier (1873), Haug (1908–11), and Gignoux (1950), is no longer used.

The Paleogene Period is subdivided into the Paleocene, Eocene, and Oligocene epochs (Fig. 20.1). In his original subdivision of the Tertiary, Lyell (1833) introduced the term Eocene (*eos*, dawn; *kainos*, recent) for the older part of the Tertiary in the classical European Cenozoic basins in which he recognized less than 3% of extant mollusc species. Later, Beyrich (1854), working mainly in northern Germany, separated the Oligocene (*oligos*, little) from the Eocene. Finally, Schimper (1874), based on paleobotanic studies in the Paris and other basins of western

A Geologic Time Scale 2004, eds. Felix M. Gradstein, James G. Ogg, and Alan G. Smith. Published by Cambridge University Press.

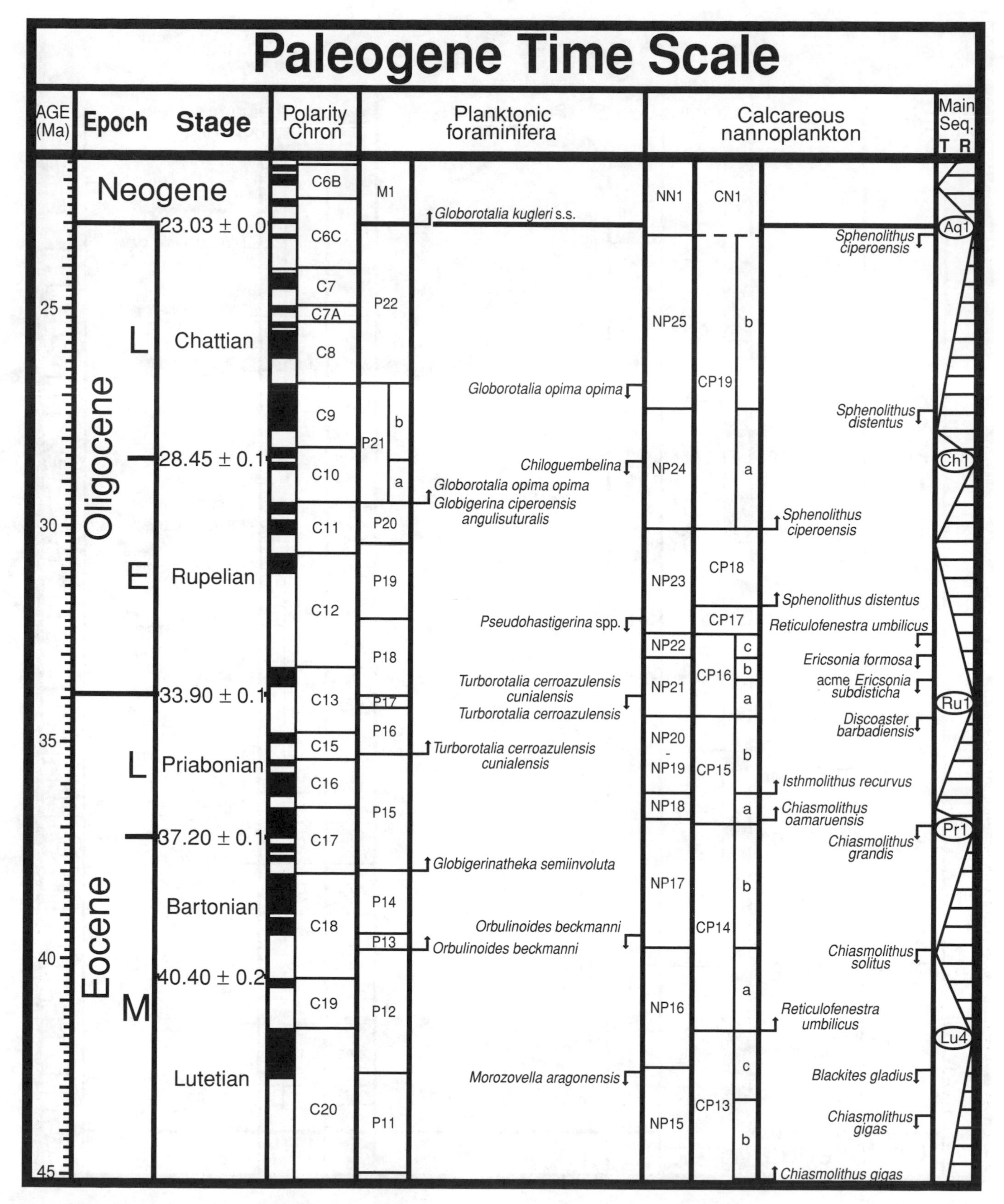

Figure 20.1 Paleogene stratigraphic subdivisions, geomagnetic polarity scale, pelagic zonations of planktonic foraminifera and calcareous nannoplankton, and main trends in eustatic sea level. A color version of this figure is in the plate section.

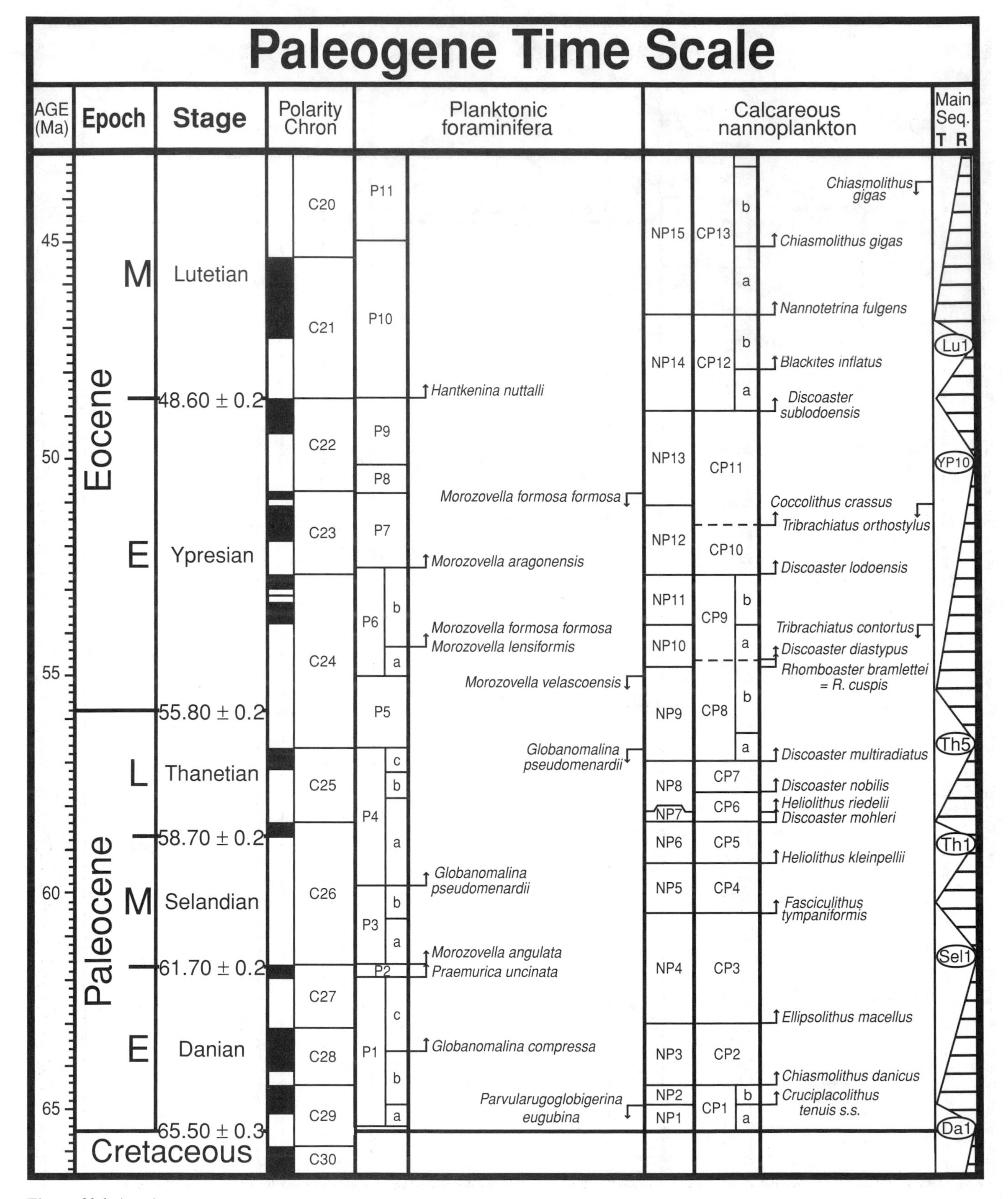

Figure 20.1 (*cont.*)

Europe, added the Paleocene (old Eocene). Whereas the Oligocene has been accepted rather smoothly, the Paleocene for a long time has been met with opposition (e.g. Mangin, 1957).

The formal assignment of seven stage names covering the three Paleogene epochs was decided by the International Subcommission on Paleogene Stratigraphy at the 1989 International Geological Congress in Washington (Jenkins and Luterbacher, 1992). As of early 2004, only the Global Stratotype Sections and Points (GSSP) defining the bases of the Paleogene epochs have been formally ratified: the base of the Danian Stage (the Cretaceous–Paleogene boundary), the base-Ypresian (the Paleocene–Eocene boundary), and the base-Rupelian (the Eocene–Oligocene boundary). Proposals for the GSSPs of the other Paleogene stages are being formulated.

20.1.2 Cretaceous–Paleogene boundary

Since the seminal paper by Alvarez *et al.* (1980) on the events at the Cretaceous–Paleogene (K–P) boundary, their causes, interrelationships, and consequences have been much debated (e.g. review by Frankel, 1999). The base-Danian GSSP has been fixed in a section at El Kef, Tunisia, at the base of the "boundary clay," the main criterion for the recognition of the K–P boundary being the increase in iridium caused by the impact of an extraterrestrial body (e.g. Hildebrand *et al.*, 1991; Frankel, 1999). It corresponds to a drastic change in the marine plankton, the extinction of the ammonites, and the demise of the dinosaurs accompanied by biotic turnover/crises in many other groups of organisms.

The El Kef GSSP section had become severely degraded over time, therefore the nearby Ain Settara or Elles sections in Tunisia are being considered by the ICS as a replacement GSSP. Its stratigraphy is equivalent in completeness.

20.1.3 Paleocene Series

DANIAN

The Danian Stage, named after its type area in Denmark, was introduced by Desor (1847) following his studies of echinids (Floris and Thomsen, 1981) within the Cerithium and the Bryozoan Limestones at Stevns Klint and Fakse on Seland. The Danian in the type area corresponds to the interval between the top of the Maastrichtian Chalk and the basal conglomerate of the Selandian. Originally considered as the youngest stage of the Cretaceous (Desor, 1847), the Danian is now included in the Paleogene (e.g. Pomerol, 1973). Its range spans the calcareous nannoplankton Zones NP1–NP4 of Martini (1971), planktonic foraminifera Zones P0–P2 (*Guembelitria cretacea* Zone to *Praemurica uncinata–Morozovella angulata* Zone) of Berggren *et al.* (1995a), and dinoflagellate cyst Viborg Zone 1 of Heilmann-Clausen (1988), which are of definite Paleogene character (e.g. Michelsen *et al.*, 1998, and references therein).

SELANDIAN

Rosenkrantz (1924) based his Selandian Stage on a succession composed of conglomerates, greensand, marls, and clays that unconformably overly the upper Danian limestones and underlying the ash-bearing Mo-Clay in Denmark (e.g. Perch-Nielsen and Hansen, 1981; Berggren, 1994). In its type area, the Selandian contains characteristic dinoflagellate cyst assemblages (Viborg 2 and 3 Zones of Heilmann-Clausen, 1985, 1988) and calcareous nannoplankton (Zones NP5–NP6 of Martini, 1971; e.g. Thomsen, 1994). The leading candidate for the base-Selandian GSSP is the deep-marine succession of the Zumaya section of northern Spain (Schmitz *et al.*, 1998) at a level close to the lowest occurrence of nannofossil *Fasciculithus tympaniformis* (base of Zone NP5). This candidate GSSP level is below the middle of polarity Chron C26r (approximately C26r(0.3)), therefore is ~1 myr younger than previous working definitions (e.g. lowest occurrence of planktonic foraminifer *Morozovella angulata* at base of Zone P3).

THANETIAN

The Thanetian is the youngest stage of the Paleocene. The name of this stage was first used by Renevier (1873), and its meaning was subsequently narrowed by Dollfus (1880). The Thanet Sands, Isle of Thanet, in southeast England constitute the type-strata. These strata contain calcareous nannoplankton Zones NP6–NP9 of Martini (1971) and dinoflagellate cyst Zones Viborg 3–6 of Heilmann-Clausen (1985, 1988) and D4b–D5a of Costa and Manum (1988). The Thanet Sands span polarity Chrons C26n–C24r (Ali and Jolley, 1996). Some stratigraphers prefer a two-fold subdivision of the Paleocene (e.g. Harland *et al.*, 1990) and include the Selandian in the Thanetian Stage.

The base of the Thanetian is recognized readily by dinoflagellate cyst and calcareous nannoplankton events (first occurrence of *Areoligera gippingensis* and last occurrence of *Palaeocystodinium bulliforme*). A candidate GSSP is in the Zumaya section in northern Spain at a level corresponding to the base of polarity Chron C26n.

20.1.4 Paleocene–Eocene boundary

The events taking place in the Paleocene–Eocene boundary interval, their inter-relations, and possible causes have been intensively studied (e.g. reviews in Aubry *et al.*, 1998, Schmitz *et al.*, 2001). The best criterion for the recognition of the Paleocene–Eocene boundary (base of the Ypresian) is the onset of a pronounced negative carbon isotope excursion (CIE, see Fig. 20.5), which allows global correlation of a wide variety of marine and terrestrial environments (e.g. Stott *et al.*, 1996). The carbon isotope excursion is the expression of the Paleocene–Eocene thermal maximum (PETM), which was possibly caused by a sudden release of methane from subsea marine clathrates (e.g. Dickens *et al.*, 1997). It coincides with major turnovers in deep-sea benthic foraminifera, calcareous nannoplankton, and mammals. The GSSP for the Paleocene–Eocene boundary in the Gabal Dababiya section near Luxor, Egypt, was ratified in 2003 (Aubry and Ouda, 2003).

20.1.5 Eocene Series

YPRESIAN

This stage was introduced by Dumont (1849) to include the shelf-facies strata lying between the terrestrial to marginally marine Landenian and the marine Brusselian. Dumont (1851) later assigned the upper more sandy part of the stage, which locally contains *Nummulites*, to the Paniselian from the Ypresian *sensu stricto*, which is typified by the Yper Clay of western Belgium (Willems *et al.*, 1981).

Biostratigraphically, the Ypresian is well restrained by its dinocyst and calcareous nannoplankton associations (e.g. Vandenberghe *et al.*, 1998; Moorkens *et al.*, 2000). In the Tethys realm, the Ilerdian (Hottinger and Schaub,1960) corresponds to an important phase in the evolution of the larger foraminifera which is not represented in northwest European basins. Magnetostratigraphic studies (Ali *et al.* 1993; Ali and Hailwood, 1995) correlate the Ypresian interval with polarity Chrons C24r–C22r.

LUTETIAN

According to its author, De Lapparent (1883), the Lutetian Stage (early-Middle Eocene) is typified by the "Calcaire grossier" of the Paris Basin. A stratotype was selected by Blondeau (1981) at St. Leu d'Esserent and St. Vaast-les-Mello, approximately 50 km north of Paris. The type-Lutetian contains typical larger foraminifera (*Nummulites laevigatus*, *Orbitolites complanatus*) and, to a lesser extent, palynomorphs and calcareous nannoplankton. The criterion selected for the recognition of the base of the Lutetian is the first appearance of *Hantkenina nuttallii*. A leading candidate for the base-Lutetian GSSP is the Fortuna section in southeastern Spain.

BARTONIAN

The late-Middle Eocene stage of Bartonian (Mayer-Eymar, 1858) is based on the Barton Beds of the Hampshire Basin, central southern England (Curry, 1981). The Barton Beds contain a rich and diversified dinoflagellate cyst assemblage including *Rhombodinium draco* and *Rhombodinium porosum*. The upper limit of the Bartonian remains poorly defined, since it corresponds to a widespread sea-level lowstand and is frequently marked by a hiatus separating it from the Priabonian.

PRIABONIAN

The Priabonian Stage (Munier Chalmas and De Lapparent, 1893) is equivalent to the Late Eocene. The historical type section in Priabona, northern Italy (Roveda, 1961; Hardenbol, 1968; Barbin, 1988) contains rare planktonic foraminifera (*Turborotalia cerroazulensis* group) assigned to zone P16 and spans calcareous nannofossil zones NP19–NP20 (Roth *et al.*, 1971). The pelagic facies of the Valle della Contessa Section, central Italy, is one of the most suitable sections for a Priabonian GSSP, and the disappearance of planktonic foraminifer *Globigerinatheka semiinvoluta* provides an important marker for correlation.

20.1.6 Eocene–Oligocene boundary

The section at Massignano, 10 km southeast of Ancona, northern Italy, has been formally ratified as the GSSP for the base of the Oligocene Series and base of the Rupelian Stage (Premoli-Silva and Jenkins, 1993). The key event marking the GSSP level is the extinction of the hantkeninid planktonic foraminifera at the top of Zone P17, which lies within nannofossil Zone NP21, and magnetic polarity Chron 13r.1. Dating of biotite grains from 19 m below the base of the GSSP with K–Ar and Ar–Ar methods suggests a numerical age of 34 Ma for the GSSP (Premoli-Silva and Jenkins, 1993). The Eocene–Oligocene boundary "interval" marks a phase of widespread cooling causing considerable faunal and floral changes (e.g. Premoli–Silva *et al.*, 1988; Vonhof *et al.*, 2000).

20.1.7 Oligocene Series

RUPELIAN

A pragmatic chronostratigraphic scheme for the standardization of Oligocene stages was provided by Hardenbol and Berggren (1978) who recognized two distinct lithostratigraphic units in northwest Europe: (i) a lower, moderately deep-marine, clayey unit, which includes the typical Rupelian rocks; and (ii) an upper, predominantly shallow-marine, sandy unit, which incorporates the type section of the Chattian.

The name Rupelian was introduced by Dumont (1849), who subdivided the unit into a lower sandy shale and an upper shale, the so-called "Argile de Rupelmonde." The latter incorporates the transgressive Boom Clay of central northern Belgium, which may be regarded as a reference for the upper Rupelian. The Boom Clay displays striking banding caused by alternating layers, with thickness on the order of tens of centimeters, caused by systematic variations in organic-carbon and carbonate content, and clay and silt, which is thought to reflect 41 and 100 Ka Milankovitch cycles (Van Echelpoel, 1994; Vandenberghe *et al.*, 1997). The Boom Clay contains a typical dinoflagellate assemblage with *Wetzeliella gochtii* (Stover and Hardenbol, 1993).

The lower Rupelian in the classical Belgium area incorporates the lowstand Ruisbroek Sands and Wintham Silts, that have a hiatus at their base due to tectonic uplift. This hiatus is considered correlative to the "Grande Coupure" in the mammal record, separating Eocene from Oligocene vertebrate fauna after the onset of significant global cooling that created new landbridges (Woodburne and Swisher, 1995). The Wintham Silts in western Belgium are assigned to the lower part of nannofossil Zone NP22, which is well above the base of the Oligocene, as evidenced from the open-marine marls at Massignano, Italy, and the marginal marine strata in the upper part of the Priabonian type section (Vandenberghe *et al.*, 1998; Brinkhuis and Visscher, 1995).

CHATTIAN

The Chattian was introduced by Fuchs (1894) with the "Kasseler Meeressande" in Hesse, Germany, as the name-giving strata. However, Goerges (1957) selected the section at Doberg, near Bünde (Westphalia, Germany), as the stratotype. This stratotype contains assemblages of the upper part of calcareous nannoplankton Zone NP24 (Martini *et al.*, 1976), miogypsinids (Drooger, 1960), and correlatable dinoflagellate cyst assemblages. The traditional northwest European Rupelian–Chattian boundary at the base of the acme of benthic foraminifer *Asterigerina guerichi* is indicative of a rapid warming trend (Van Simaeys *et al.*, in press).

Two of the best-documented sections for the boundary stratotype are at Monte Cagnero and Pieve d'Accinelli, central Italy. In addition to having good biostratigraphic and magnetostratigraphic records (Premoli-Silva *et al.*, 2000), they contain several tuffite horizons yielding radiometric dates. The candidate GSSP in the section at Monte Cagnero coincides with the planktonic foraminifer P21a–P21b zonal boundary (extinction of genus *Chiloguembelina*). In this section, the last occurrence of *Chiloguembelina cubensis* is in the lower portion of polarity Chron C9n (approximately C9n(0.25)) and coincides with the appearance of the colder-water immigrant dinoflagellate *Distatodinium biffi* in the Mediterranean (Coccioni *et al.*, in prep). However, the proposed primary marker, i.e. the extinction of *Chiloguembelina*, may be diachronous among basins and is most likely controlled by paleolatitude and paleobiogeography (Van Simaeys *et al.*, in press). As examples, in tropical Pacific Site 1218 of ODP Leg 199, the significant decrease of *C. cubensis* was observed in the upper half of polarity Chron C10r (Lyle *et al.*, 2002), and Berggren *et al.* (1995a) observed its extinction or strong reduction in the middle of polarity Chron C10n.

OLIGOCENE–MIOCENE BOUNDARY

The Paleogene–Neogene boundary (base of the Aquitanian Stage) coincides with the magnetic reversal from polarity Chron C6Cn.2r to C6Cn.2n in the section at Lemme-Carrosio, northern Italy (Steininger *et al.*, 1997a).

20.2 PALEOGENE BIOSTRATIGRAPHY

There are very refined Paleogene zonations in pelagic facies using nannofossils and planktonic foraminifera; the zonations differ between low and higher latitudes. Larger foraminifera are a key stratigraphic tool in shallow-marine tropical to temperate areas. In siliciclastic neritic facies dinoflagellates are particularly useful; information concerning Paleogene dinocyst distribution is most comprehensive for the mid latitudes of the northern hemisphere. During the Paleogene, the various continental masses had distinctive land mammals fauna, which show marked turnovers and yield detailed terrestrial zonations.

Figures 20.1 through 20.3 show the correlation of Paleogene stages, magnetostratigraphy, and standard marine zonations based on planktonic foraminifera, calcareous nannoplankton, larger foraminifera, radiolaria, and dinoflagellate cysts. The intricate terrestrial zonations for North America and Europe with mammalian fossils, and the calibration of the regional terrestrial "stages" with the Paleogene standard stages are shown in Fig. 20.4.

20.2.1 Foraminifera

PLANKTONIC FORAMINIFERA

The application of planktonic foraminifera to Paleogene stratigraphy (Fig. 20.1) has been developed mainly in two areas: the southern part of the former Soviet Union (e.g. Subbotina,1953) and the Caribbean (e.g. Bolli, 1957a,b). In the late 1950s and 1960s, planktonic foraminifera became one of the main biostratigraphic tools, and assumed even greater importance with the advent of scientific ocean drilling. Paleogene planktonic foraminifera and associated zonations have been discussed by many authors (e.g. Bolli, 1966; Blow, 1979; Bolli and Saunders, 1985; Toumarkine and Luterbacher, 1985; Berggren and Miller, 1988; Berggren *et al.*, 1995a; Olsson *et al.*, 1999). Planktonic foraminifera can be best used in open-marine deposits of the tropical and subtropical realms. Important marker species are not present in middle and higher latitudes and zonations proposed for these areas are therefore considerably less detailed (Jenkins, 1985).

Estimates of the number of planktonic foraminifera species surviving the Cretaceous–Paleogene boundary mass extinction are controversial (e.g. Olsson, 1970; Keller, 1988; Arz *et al.*, 1999). These survivors are small opportunistic species belonging to the genera *Heterohelix, Guembelitria, Globoconusa*, and *Hedbergella*. The recovery of planktonic foraminifera during the Paleocene led to a high degree of diversification, thereby allowing a detailed biostratigraphic subdivision. Several new genera develop during the early Paleocene, in particular, forms with spinose tests (*Parasubbotina*), with murica (*Praemurica*), and with smooth surfaces (*Globanomalina*). Open-marine Middle and Late Paleocene assemblages in lower latitudes are dominated by relatively large conicotruncate representatives of the genera *Morozovella, Acarinina*, and *Globanomalina*.

In many sections, the Paleocene–Eocene boundary interval in the lower part of Zone P5 (*Morozovella velascoensis* Zone) corresponds to a reduction in the morphological diversity of the planktonic foraminifera assemblages (e.g. Kelly *et al.*, 2000). The early and most of the middle Eocene is again a time of rapid evolution within the genera *Morozovella* and *Acarinina*. In addition, representatives of several new genera have their first appearance, including *Pseudohastigerina, Truncorotaloides, Globigerinatheka, Clavigerinella, Turborotalia*, and *Orbulinoides*. Several species are very characteristic and have very short stratigraphic ranges (e.g. *Orbulinoides beckmanni*).

A major change in the composition of the planktonic foraminifera assemblages took place at the end of the P15 *Truncorotaloides rohri* Zone (mid-Late Eocene) when the morozovellids and several other genera disappeared. The Eocene–Oligocene boundary corresponds to the last occurrence of representatives of the genus *Hantkenina* and of the *Turborotalia cerrazulensis* lineage. The early Oligocene is dominated by small-sized, low-diversity assemblages, but the diversity and average size of the assemblages recovered again during late Oligocene. The first massive occurrence of representatives of the genus *Globigerinoides* is observed just above the Paleogene–Neogene boundary.

LARGER (BENTHIC) FORAMINIFERA

Many of the west European centers in which the science of stratigraphy evolved during the eighteenth and nineteenth centuries are located on Paleogene strata rich in larger foraminifera. Hence, these microfossils have been paramount for the development of the Paleogene biostratigraphic timescale, as demonstrated by the now abandoned term "Nummulitique," which has been used as an equivalent of "Paleogene." Larger foraminifera continue to be a decisive stratigraphic tool in shallow-marine tropical to temperate areas (Fig. 20.2). Zones for larger foraminifera are ideally based on successions of biometric populations within phylogenetic lineages; species are essentially morphometric units. During the latter part of the twentieth century, several monographs on larger foraminifera groups have considerably increased their stratigraphic usefulness (e.g. Hottinger, 1960, 1977; Less, 1987; Schaub, 1981).

The Danian and early Selandian larger foraminifera assemblages are relatively poorly differentiated and consist mainly of a few rotaliids. They mark the start of the recovery from the extinction of virtually all larger foraminifera genera at the Cretaceous–Paleogene boundary.

Within the Thanetian, the rapid radiations of the alveolinids and orthophragminids followed by that of the nummulitids give origin to several phylogenetic lineages within each group which allow a closely spaced zonation of the younger part of the Thanetian, Ypresian, Lutetian, and early Bartonian (equivalent to the "Biarritzian" regional substage). A major change takes place within the Bartonian by the extinction of the genera *Alveolina* and *Assilina* as well as the extinction of the conspicuous large representatives of the genus *Nummulites*. The younger part of the Bartonian and the Priabonian are characterized by

Series	Stage	Zonation	Larger foraminifera	
	Aquitanian	**SBZ 24**	*Miogypsina gr. gunteri–tani*	
Oligocene	Chattian	**SBZ 23**	*Miogypsinoides, Lepidocyclinids, Nummulites bouillci*	
Oligocene	Chattian	**SBZ 22** b	*Lepidocyclinids, Nummulites vascus N. fichteli, N. bouillei*	*Cycloclypeus*
Oligocene	Rupelian	**SBZ 22** a		*Bullalveolina*
		?		
Oligocene	Rupelian	**SBZ 21**	*Nummulites vascus, N. fichteli*	
Eocene	Priabonian	**SBZ 20**	*Nummulites retiatus, Heterostegina gracilis*	
		SBZ 19	*Nummulites fabianii, N. garnieri, Discocyclina pratti minor*	
	Bartonian	**SBZ 18**	*Nummulites biedai, N. cyrenaicus*	
		SBZ 17	*Alveolina elongata, A. fragilis, A. fusiformis, Discocyclina pulcra baconica Nummulites perforatus, N. brogniarti, N. biarritzensis*	
	Lutetian	**SBZ 16**	*Nummulites herbi, N. aturicus, Assilina gigantea, Discocyclina pulcra balatonica*	
		SBZ 15	*Alveolina prorrecta, Nummulites millecaput, N. travertensis*	
		SBZ 14	*Alveolina munieri, Nummulites beneharnensis, N. boussaci, Assilina spira spira*	
		SBZ 13	*Alveolina stipes, Nummulites laevigatus, N. uranensis*	
	Cuisian	**SBZ 12**	*Alveolina violae, N. manfredi, N. campesinus, N. caupennensis, Assilina major, A. cuvillieri*	
		SBZ 11	*Alveolina cremae, A. dainellii, Nummulites praelaevigatus, N. nitidus, N. archiaci, Assilina laxispira*	
		SBZ 10	*Alveolina schwageri, A. indicatrix, Nummulites burdigalensis burdigalensis, N. planulatus, Assilina placentula, Discocyclina archiaci archiaci*	
Paleocene	Ilerdian	**SBZ 9**	*Alveolina trempina, Nummulites involutus, Assilina adrianensis*	
		SBZ 8	*Alveolina corbarica, Nummulites exilis, N. atacicus, Assilina leymeriei*	
		SBZ 7	*Alveolina moussoulensis, Nummulites praecursor, N. carcasonensis*	
		SBZ 6	*Alveolina ellipsoidalis, A. pasticillata, Nummulites minervensis*	
		SBZ 5	*Orbitolites gracilis, Alveolina vredenburgi, Nummulites gamardensis*	
	Thanetian	**SBZ 4**	*Glomalveolina levis, Nummulites catari, Assilina yvettae*	
		SBZ 3	*Glomalveolina primaeva, Fallotella alavensis, Miscellanea yvettae*	
	Selandian	**SBZ 2**	*Miscellanea globularis, Ornatononion minutus, Paralockhartia eos, Lockhartia akbari*	
	Danian	**SBZ 1**	*Bangiana hanseni, Laffitteina bibensis*	

Figure 20.2 Paleogene zonation of larger foraminifera with selected taxa.

relatively small nummulitids and other rotaliids and by orthophragminiids, which may reach considerable dimensions. Within the larger foraminifera assemblages, the Eocene–Oligocene boundary is not marked by drastic changes. The subdivisions of the Oligocene based on larger foraminifera are mainly based on the lepidocyclinids, miogypsinids, and nummulitids.

A larger foraminifera zonation of the Paleocene and Eocene of the Tethyan area has been published by Serra-Kiel *et al.* (1998) as one of the results of the IGCP 286 *Early Paleogene Benthos*. Cahuzac and Poignant (1997) proposed a similar larger foraminifera zonation for the Oligocene–Miocene of the west European basins. The "letter-stage" subdivision of the Indo-Pacific Cenozoic (Leupold and van der Vlerk, 1936; Adams, 1970; Chaproniere, 1984; Boudagher-Fadel and Banner, 1999) is based on larger foraminifera. These are also largely used in the western hemisphere (Barker and Grimsdale, 1936; Butterlin, 1988; Caudri, 1996).

The zonations proposed by Serra-Kiel *et al.* (1998) and the Oligocene part of the zonation by Cahuzac and Poignant (1997) are shown on Fig. 20.2. Correlation between the planktic zonations and those based on larger foraminifera are mainly achieved by their co-occurrence with calcareous nannoplankton and dinoflagellates (e.g. Molina *et al.*, 2000), but they are always somewhat discontinuous leaving some leeway for subjective interpretations. This is particularly true for the Paleocene and the Oligocene.

SMALLER BENTHIC FORAMINIFERA

Smaller benthic foraminifera are important for correlation of both shallow- and deep-water Paleogene sedimentary sections (Table 20.1). Although the paleoecological niches of calcareous and agglutinating benthics overlap, it is convenient to discuss each group separately.

Agglutinating foraminifera The majority of Paleogene agglutinating taxa occupy deeper marine habitats below storm wave base on fine-grained, gravity-flow-rich siliciclastic wedges off continental margins. Hence, deeper marine shales in many petroleum basins are rich in agglutinating foraminifera; the latter is particularly true for Paleogene sediments in the North Sea, offshore Norway, offshore Labrador, Beaufort–MacKenzie Delta, Carpathians (flysch sediments), and the Caribbean. In New Zealand, the restricted basin Waipawa Black Shale facies of late Teurian (late Paleocene) age contains diverse assemblages of agglutinated foraminifera with calcareous taxa either leached or originally absent due to dysaerobia (Hornibrook *et al.*, 1989). In general, fine-grained, higher-latitude (austral and boreal) sediments, poor in carbonate harbor diversified agglutinated benthic assemblages.

Table 20.1 *Selected reference literature for stratigraphic study of smaller benthic foraminifera in Paleogene sediments*

Region	Reference
New Zealand	Hornibrook *et al.* (1989)
Caribbean	Bolli *et al.* (1994)
Californian Coastal Ranges	Mallory (1959)
Mediterranean	AGIP (1982)
NW Europe, with emphasis on North Sea and UK	Jenkins and Murray (1989)
Circum N. Atlantic Margin basins (agglutinated benthics)	Gradstein *et al.* (1994b)
Global bathyal and abyssal realms (agglutinated taxa)	Kaminski and Gradstein (2004)
Atlantic bathyal and abyssal realm (calcareous benthics)	Tjalsma and Lohmann (1983)
Global bathyal and abyssal realms (calcareous benthics)	van Morkhoven *et al.* (1986)

There is no taxonomic turnover of the agglutinated fauna at the Cretaceous–Paleogene boundary, nor in the Paleogene itself, and many Paleogene taxa have originated in the Campanian. Paleoenvironmental changes in basins, often the result of shallowing due to sediment infill, basin uplift, or the return of carbonate facies truncate the stratigraphic range of various taxa. Well-known examples of such regional stratigraphic truncations include:

1. Late Paleocene disappearance of the Lizard Spring assemblage of Trinidad;
2. Middle–late Eocene disappearance of flysch-type agglutinated fauna in the Carpathians; and
3. mid-Cenozoic disappearance of various assemblages offshore Eastern Canada – along the Norwegian continental margin and in the North Sea the disappearance interval of the flysch-type assemblage tracks mid-Cenozoic basin shallowing.

There are over 150 stratigraphically useful cosmopolitan Paleogene deeper-water agglutinated benthic foraminifera, and many taxa have stratigraphic ranges that vary slightly or even markedly from basin to basin. However, correlation of wells within a basin can be accomplished with local zonations. The atlas of cosmopolitan deep-water Paleogene taxa by Kaminski and Gradstein (2004), the circum North Atlantic continental margin basins study of Gradstein *et al.* (1994b), and the New Zealand monograph by Hornibrook *et al.* (1989) contain detailed range charts and zonations. These are regions where Paleogene planktonic foraminifera are generally sporadic and only present in narrow stratigraphic intervals. For example, prominent offshore Norway interval zones based on the upper part of ranges of agglutinated taxa include the *Ammoanita ruthvenmurrayi* and *Reticulophragmium paupera* Zones in the late Paleocene, *Ammomarginulina aubertae* and *R.amplectens* s.str Zones in the late–middle Eocene and part of the late Eocene, and *Adercotryma agterbergi* in the early Oligocene.

Smaller calcareous benthic foraminifera Just as investigations of the agglutinated taxa got a boost in the late 1970s and 1980s from petroleum exploration in continental margin basins and deep-sea scientific drilling, the study of calcareous smaller benthic foraminifera also advanced significantly. Remarkably diverse bathyal–abyssal Paleogene assemblages were discovered that turned out to have stratigraphic potential, although rare or patchy single taxa distributions require the use of zonal assemblages. The assemblages at the same time track important paleoceanographic changes. The most important turnover of the calcareous benthic fauna took place during the latest Paleocene when the marine carbon reservoir experienced a major change in isotopic composition (see Section 20.3). At that time, benthic deep-sea fauna at many sites experienced dramatic reduction in diversity and composition (Thomas and Shackleton, 1996). Well known are the disappearances in bathyal–abyssal facies of *Stensioeina beccariiformis*, *Angulogavelinella avnimelechi*, various species of *Cibicidoides*, including *C. velascoensis*, and several taxa of *Aragonia*.

The genus *Turrilina* with the species *robertsi* in latest Paleocene through Eocene (planktonic foraminifer Zones P6a–P15), the species *T. alsatica* in early Oligocene (Rupelian), and the related *Rotaliatina bulimoides* in mid Eocene through early Oligocene are examples of taxa that have global correlation potential in deep neritic and bathyal environments, particularly in mid–high latitudes with shale facies. Similarly, the disappearance of *Nuttalides trumpyi* at the end of the Eocene is a useful bathyal–abyssal event, as is the appearance of the genus *Siphonina* in the late Eocene, and of *Sphaeroidina* in the early Oligocene.

20.2.2 Calcareous nannoplankton

Calcareous nannofossils form a heterogenous group of minute objects that range in size from 1 to 30 μm and are important constituents of (deeper) marine sediments. The majority of the fossils resemble the coccoliths of the exterior calcareous cover (coccosphere) of the Haptophyceae and it is therefore generally accepted that they are the fossil remains of unicellar algae. The recognition of calcareous nannofossils as a useful tool for biostratigraphic correlations is generally credited to Bramlette and co-workers (1954, 1961, 1967). Following these pioneer efforts, intensive taxonomic and biostratigraphic studies have been carried out that formed the basis for the presently existing calcareous nannofossil biozonal schemes for the Mesozoic and Cenozoic.

DEVELOPMENT OF ZONATIONS

In the middle of the last century, several authors described species important for correlation of Paleogene marine sediments in Europe and North America (Bramlette and Riedel, 1954; Deflandre and Fert, 1954; Martini, 1958, 1959a,b; Stradner, 1958, 1959a,b; Brönnimann and Stradner, 1960). In the 1960s, significant contributions to Paleogene zonations were made (Bramlette and Sullivan, 1961; Bystricka, 1963, 1965; Hay, 1964; Sullivan, 1964, 1965; Bramlette and Wilcoxon, 1967; Hay and Mohler, 1967; Hay *et al.*, 1967). But it was only in the 1970s with the onset of scientific ocean drilling that calcareous nannofossils became one of the most important biostratigraphic microfossil groups.

The two most widely used Paleogene zonal schemes are the standard zonation of Martini (1971), codified as NP, and Bukry (1973, 1975a; emended by Okada and Bukry, 1980), codified as CP (Fig. 20.1). Martini's zonation relied on studies of land sequences from largely temperate areas, whereas Bukry's zonation was developed in low-latitude oceanic sections.

High-resolution studies (e.g. Romein, 1979; Perch-Nielsen, 1985; Varol, 1989; Aubry, 1996) re-defined and subdivided these zones. For example, a four-fold subdivision was proposed for Zone NP10, and Zones NP19 and NP20 were combined because *Sphenolithus pseudoradians* at the base of Zone NP20 is an unreliable marker. Aubry and Villa (1996) emended Zone NP25 in the late Oligocene with a three-fold subdivision that accounts for the definition of the Chattian–Aquitanian boundary (Oligocene–Miocene boundary). Fornaciari and Rio (1996) proposed a two-fold subdivision of Zone NP25 for the Mediterranean region, and re-defined the top of the zone with the last common occurrence (LCO) of *Reticulofenestra bisecta*, because the zonal marker species *H. recta* extends well into the Miocene (Rio et al., 1990b). The LCO of *R. bisecta* is very close to the Oligocene–Miocene boundary (Berggren *et al.*, 1995a).

The detailed scheme of calcareous nannofossil zones and key events in Fig. 20.1 is largely based on the magnetobiostratigraphic correlations of Berggren *et al.* (1995a). The magnetobiostratigraphic calibrations include detailed data from the Contessa section (Monechi and Thierstein, 1985; Galeotti *et al.*, 2000). The top of Zone CP19b was defined by the last occurrence of *Sphenolithus ciperoensis* (emendation of Martini's zonation by Aubry and Villa, 1966; Berggren *et al.*, 2000).

Calcareous nannoplankton respond quickly to changes in the thermal structure of oceanic water masses. As a consequence, the presence and stratigraphic ranges of marker species strongly differ from low to high latitudes. Therefore, the "standard zonal schemes" are only partially applicable to sedimentary sequences of the North Sea and the sub-Antarctic South Atlantic, and modified Paleogene zonations have been proposed for these areas (Wise, 1983; van Heck and Prins, 1987; Varol, 1989, 1997; Wei and Wise, 1990a,b; Crux, 1991; Wei and Pospichal, 1991).

EVOLUTIONARY TRENDS

Following the biotic crisis at the end of the Maastrichtian, calcareous nannoplankton underwent a major diversification in the early–middle Paleocene giving rise to several key genera for biostratigraphy (e.g. *Fasciculithus, Chiasmolithus, Sphenolithus, Discoaster*, and *Helicosphaera*). A major taxonomic turnover took place at the Paleocene–Eocene boundary interval in response to the Paleocene–Eocene thermal maximum (PETM). The *Rhomboaster–Tribrachiatus* lineage and the *Discoaster* radiation characterize the zones at the Paleocene–Eocene transition (Angori and Monechi, 1996; Aubry, 1996, 1998a).

While the early Paleogene calcareous nannoplankton evolution reflects an increasing temperature trend and oligotrophic conditions, the late Paleogene calcareous nannoplankton evolution is influenced by climatic deterioration and eutrophication. A progressive decline in diversity with a low rate of evolution characterizes this interval (Aubry, 1992, 1998b). A pronounced taxonomic turnover occurred near the middle–late Eocene boundary. A sharp impoverishment in species of the genus *Discoaster* took place prior to the Eocene–Oligocene boundary with the extinction of the last two representatives of the rosette-shaped discoasters *D. barbadiensis* and *D. saipanensis* (Monechi, 1986; Nocchi *et al.*, 1988).

A large number of nannofossil extinctions took place in the early Oligocene. This decreasing diversity trend continued through the Oligocene, in particular at high latitude; it is actually reversed at low latitudes with a radiation of the genera *Sphenolithus* and *Helicosphaera*. A succession of several last occurrences characterizes the Oligocene–Miocene transition.

20.2.3 Radiolaria

Investigations contributing to a detailed Cenozoic radiolarian biostratigraphy have been carried out largely on deep-sea sediment samples in which calcareous microfossils co-occur, or on cores with magnetostratigraphy or other methods for estimating sediment age. In the case of radiolaria, this calibration is essential since they are commonly absent in stage stratotypes. For this reason, a two-step correlation via calcareous nannofossils is usually unavoidable. The latter group can then be tied to the geomagnetic polarity time scale, epoch/series boundaries, and numerical age estimates, thus allowing for accurate correlation of zonations across paleobiogeographic boundaries.

Riedel (1957) was the first to realize the potential of radiolaria for stratigraphic purposes. The originally proposed stratigraphic zonation scheme of Riedel and Sanfilippo (1970, 1971, 1978) has received only minor modifications and additions (Moore, 1971; Nigrini, 1971, 1974; Foreman, 1973; Maurasse and Glass, 1976; Saunders *et al.*, 1985) and is the generally accepted scheme for tropical areas of the world oceans (Sanfilippo *et al.*, 1985).

Code numbers for the radiolarian zonation for the tropical Pacific, Indian, and Atlantic Oceans (RP1–RP22 for the Paleogene and RN1–RN17 for the Neogene) were introduced and standardized by Sanfilippo and Nigrini (1998a). The Paleogene radiolarian zonations for the tropics (Sanfilippo and Nigrini, 1998b) and the South Pacific (Hollis, 1993, 1997; Hollis *et al.*, 1997) with calibrations to planktonic foraminifer and calcareous nannofossil zones are shown in Fig. 20.3.

A late Eocene to Pleistocene northern high-latitude zonation was developed for the Norwegian and Greenland Seas by Bjørklund (1976). Independent zonations for the Antarctic sediments have been established by Petrushevskaya (1975), Takemura (1992), and Abelmann (1990) who proposed two zones for the upper Oligocene. Takemura and Ling (1997) extended the Paleogene high-latitude Southern Ocean zonation into the late Eocene mainly based on material from ODP Legs 114 and 120.

Foreman (1973) created the first lower Paleogene low-latitude radiolarian zonation based on material from DSDP Leg 10 in the Gulf of Mexico. The lowermost upper Paleocene *Bekoma campechensis* Zone was introduced by Nishimura (1987), who subsequently (1992) subdivided this zone into three subzones. For the Paleogene record, where magnetostratigraphic data are not available for most radiolarian sequences, Sanfilippo and Nigrini (1998a) used a combination of data from previous literature and an unpublished integrated compilation chart based on data from DSDP/ODP Legs 1–135 to construct a composite chronology of radiolarian zonal boundary events tied to numerical ages.

Investigations of a nearly complete radiolarian record for the upper Lower Paleocene to upper Middle Eocene

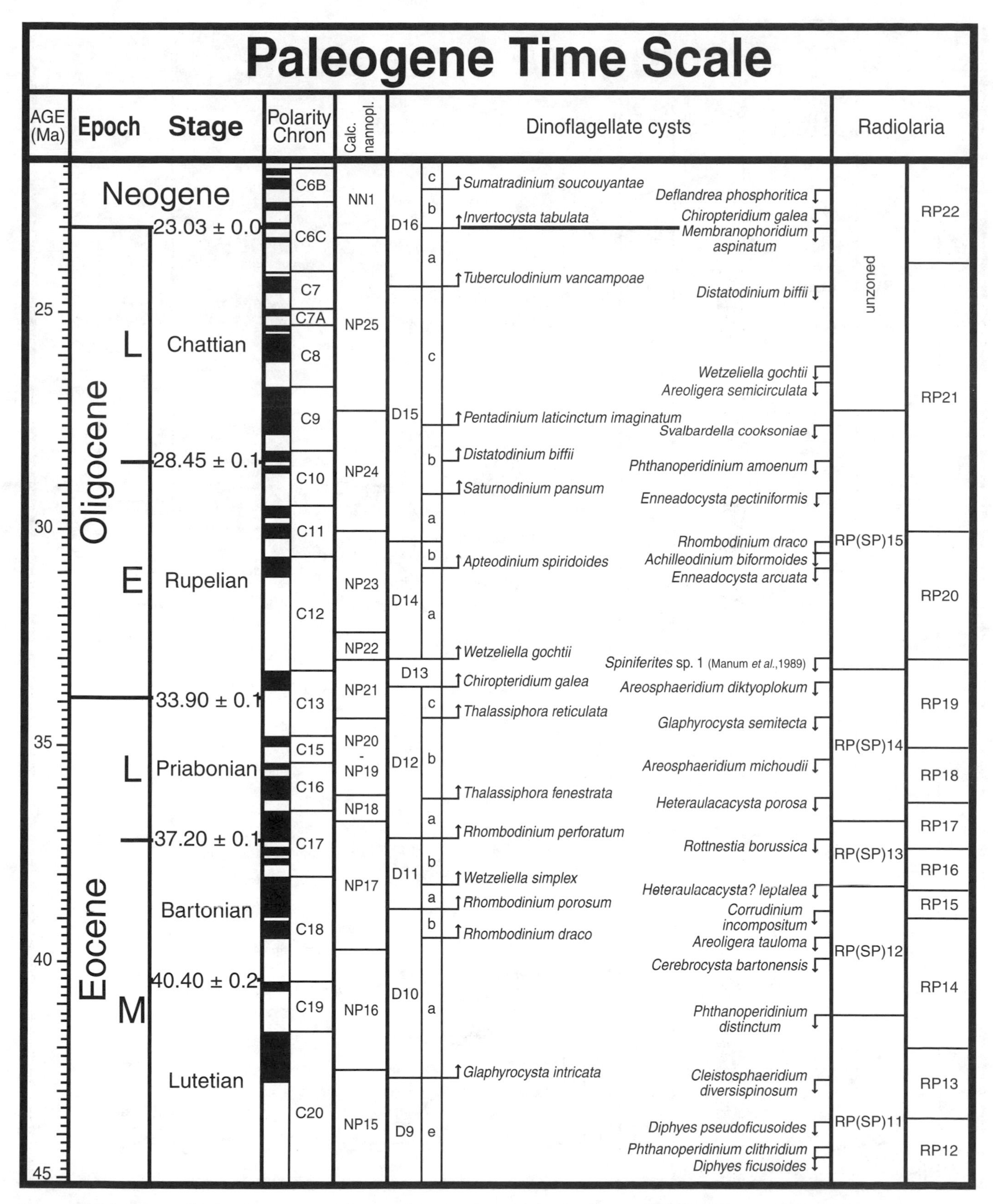

Figure 20.3 Paleogene dinoflagellate cyst zonation and datums and radiolarian zonation, with their estimated correlation to magnetostratigraphy and calcareous nannoplankton zones. Dinoflagellate stratigraphy was compiled by A. J. Powell and H. Brinkhuis. A color version of part of this figure is in the plate section.

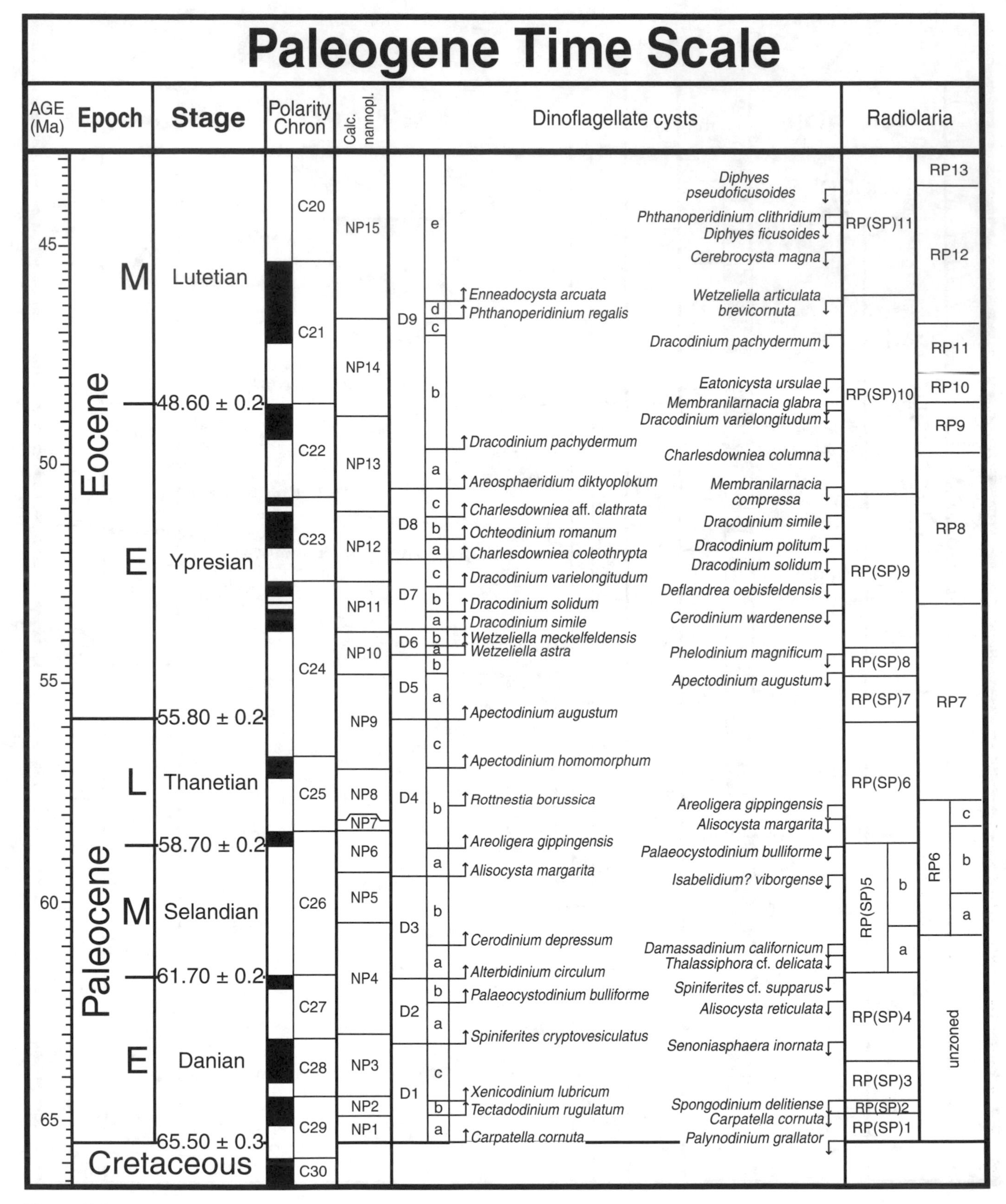

Figure 20.3 (*cont.*)

(Sanfilippo and Blome, 2001) recovered on ODP Leg 171B, western North Atlantic, indicates no major change in the composition of the radiolarian assemblage across the Paleocene–Eocene boundary. However, an abnormally large radiolarian turnover was observed near the Early–Middle Eocene boundary. Comparison of the stratigraphic ranges of species from Hole 1051A with those from the tropics indicates that a high proportion of these species have diachronous first and/or last occurrences. A correlation between the lower Paleogene standard radiolarian low-latitude and other zonal schemes presented for the Caribbean region and mid–high southern latitudes is presented in Sanfilippo and Hull (1999).

20.2.4 Dinoflagellate cysts

Dinoflagellate cysts are a group of predominantly single-celled organisms (protists) with two flagella and a eukaryotic nucleus. They form a major component of the marine plankton, with about half being autotrophic and half being phagotrophic. Others are symbionts or parasites. Although most dinoflagellates are marine, they can occur in brackish and freshwater environments, ice, snow, and wet sand. Dinoflagellates have a simple to complex life cycle, which typically includes a motile stage. Non-motile cells may be resting, temporary, vegetative, or digestion cysts, and may have organic, calcareous, or siliceous walls. Most fossil organic-walled dinoflagellate cysts (dinocysts) are considered to represent hypnozygotes, being distinguished by a restant organic wall and an excystment opening, the archeopyle (Fensome *et al.*, 1993). Almost all (fossil and extant) dinocysts are assignable to the subclass Peridiniphycidae, with a few representing the subclasses Gymnodinipgycidae and Dinophysisiphycidae (Fensome *et al.*, 1993). The bulk of Cenozoic taxa are included in the Peridiniphycidae. This subclass, the largest in the Dinophyceae, contains siliceous, calcareous, and organic-walled cysts. Here, we deal exclusively with the organic-walled group.

Applications of fossil dinocysts in global Paleogene biostratigraphy and paleoecology has been reviewed in detail in several papers including Williams and Bujak (1985), Powell (1992b), Stover *et al.* (1996), and Williams *et al.* (1998b, 2000b). These compilations indicate that the stratigraphical range of a given Paleogene dinocyst is rarely synchronous world-wide. Many authors have demonstrated climatic and environmental control on the stratigraphical distribution of taxa in the Paleogene (e.g. Brinkhuis, 1994; Wilpshaar *et al.*, 1996; Bujak and Brinkhuis, 1998; Crouch *et al.*, 2001). Several dinocyst papers also deal with the differentiation of northern and southern hemisphere assemblages and/or endemic Antarctic assemblages (e.g. Wilson, 1998; Wrenn and Hart, 1988; Truswell, 1997; Guerstein *et al.*, 2002; Brinkhuis *et al.*, 2003a,b).

Williams *et al.* (2000b) recognized the need to accommodate both latitudinal and hemispherical control of dinocyst assemblages in Paleogene distribution charts. Accordingly, these authors give ranges for low, mid, and high latitudes in both northern and southern hemispheres, and the contribution also updates data presented in Williams *et al.* (1998b).

Information concerning Paleogene dinocyst distribution is most comprehensive for the mid latitudes of the northern hemisphere. This is a reflection of the more intense study of assemblages from these regions, notably from northwest Europe (particularly the greater North Sea Basin). Although first-order calibration of dinocyst events against magnetostratigraphy is largely absent, reliable age control is possible through published studies on the biostratigraphy of type sections of the Paleogene stages. This is supplemented by large volumes of largely unpublished subsurface data. The succession of dinocyst events offshore of northwest Europe is becoming better documented in the public domain (e.g. Harland *et al.*, 1992; Mudge and Bujak, 1996; Neal, 1996; Powell *et al.*, 1996; Mangerud *et al.*, 1999).

The updated overview of Paleogene index dinocyst events for northwest Europe presented in Fig. 20.3 is supplementary to the data presented in Williams *et al.* (2000b).

20.2.5 Mammals

During the Paleogene, the various continental masses had distinctive land mammal fauna. These exhibit rapid evolution and have been much used for correlation of non-marine strata. Inter-continental correlation, however, has often proved problematic owing to endemism, except during geologically brief periods of faunal inter-change facilitated by paleogeographic features such as land bridges.

Because mammals are generally rarer as fossils than are invertebrates or microbiota, and because of the often laterally discontinuous nature of continental strata, occurrences may be in isolated exposures whose superpositional relationships are unknown. A notable exception, however, is the stratigraphically and geographically extensive sequences in western North America. Solutions to the problems of correlating isolated mammalian fauna have varied.

Series of broad biochronological–biostratigraphic units known as "land mammal ages" have been widely applied, with a separate series in each continent for North America (NALMA), Europe (ELMA), Asia (ALMA), and South America (SALMA). These can stand independently when correlation to standard marine biostratigraphies and to global chronostratigraphy–geochronology is uncertain.

Owing to endemism, smaller biostratigraphic–biochronological units vary from having continent-wide applicability to only local use. These may be conventional biozones or, commonly in Europe, reference levels (MP). Reference levels purport to order superpositionally isolated fauna according to evolutionary grade, avoiding the problem of fixing boundaries (Schmitt-Kittler, 1987). In practice, because of referral of fauna other than the reference fauna to a given reference level, a temporal range is spanned and a reference level is thus used in much the same way as a standard assemblage biozone (Aguilar *et al.*, 1997). An alternative solution to the isolation problem is the application of parsimony to concurrence (Alroy, 1992).

Calibration of mammalian biostratigraphic–biochronological systems to the geomagnetic polarity time scale is most extensively documented in North America and Europe (Aguilar *et al.*, 1997; Janis, 1998). It has been achieved either directly or via links with other biostratigraphies, through intercalation with marine strata or through co-occurrence of mammalian and other zonal indicators in paralic facies. Radiometric dating is helping parts of the central Asian, South American, and Australasian sequences (Flynn and Swisher, 1995; Aguilar *et al.*, 1997; Kay *et al.*, 1999), which are otherwise still proving difficult to correlate globally. The broadly construed Casamayoran unit was until recently thought to be early Eocene. One of its two component SALMAs, the Barrancan, has now been radiometrically dated, which shifts it to late Eocene, thereby shifting the Mustersan (Kay *et al.*, 1999). No LMA or zonal system currently exists in Africa, but magnetostratigraphic and sequence stratigraphic studies are improving calibration (Gingerich, 1992; Kappelman *et al.*, 1992; Gheerbrant *et al.*, 1998).

Three major faunal turnovers occurred in the Paleogene, at or close to epoch/series boundaries. The first, at the beginning of the Paleogene, set the scene for rapid evolutionary radiation and continental endemism following the major tetrapod extinctions. Most of this Paleocene radiation is recorded in the first one and a half million years of the epoch in North America and is represented by the Puercan NALMA. Correlation of the Puercan to continental strata elsewhere in the world is difficult.

The second major faunal turnover at the Paleocene–Eocene boundary coincides with the carbon isotope excursion (CIE). The turnover involved similar innovations in all three northern hemisphere continents, and is known as the Mammalian Dispersal Event (MDE). This marks the first appearance of many modern mammalian orders, especially primates, bats, artiodactyls, perissodactyls, and proboscideans. The CIE and MDE coincide with the beginning of the Wasatchian NALMA (Bowen *et al.*, 2001) and probably also of the Neustrian ELMA (Hooker, 1998). The end of the Wasatchian has been recently found to be older than previously thought thanks to magnetostratigraphic studies (Clyde *et al.*, 2001).

The third major faunal turnover, at the end of the Headonian, in the earliest Oligocene, was less widespread, but well represented in Europe where it is known as the Grande Coupure. It marks the extinction of many endemic European taxa and the incoming of new ones from Asia. A supposedly contemporaneous, but less-clearly dated, major turnover in central Asia at the end of the Ergilian ALMA has been termed the Mongolian Remodelling (Meng and McKenna, 1998). A less major turnover, but with some elements similar to the Grande Coupure occurs at the beginning of the Ergilian (Dashzeveg, 1993). This raises considerable doubt as to whether the Ergilian belongs in the late Eocene or early Oligocene.

Figure 20.4 shows the current state of knowledge on calibration of LMAs and relevant biozones (concurrent range zones, CRZ, and interval zones, IZ).

Figure 20.4 Mammalian zonations and biostratigraphy and events of the Paleogene. NALMA, North American Land Mammal Ages; SALMA, South American Land Mammal Ages; ALMA, Asian Land Mammal Ages; ELMA, European Land Mammal Ages; MP, European Reference Levels. The sequence of NALMAs and their subdivision follows Woodburne (1987) with calibrations to the geomagnetic polarity time scale by Butler *et al.* (1987), Prothero and Emry (1996), Williamson (1996), Clyde *et al.* (2001), and Clemens (2002). The sequence of ELMAs follows Fahlbusch (1976) with some modifications in the Eocene (see Savage and Russell, 1977; Franzen and Haubold, 1987; Hooker in Aubry *et al.*, 1998). They are divided into MP reference levels (Schmidt-Kittler, 1987; Aguilar *et al.*, 1997) and partly into biozones (Hooker, 1986, 1987, 1996). Several Paleocene localities with significant mammals are tabulated as local fauna (LF): Fontllonga 3 (Peláez-Campomanes *et. al.*, 2000); Hainin (Sigo and Marandat, in Aguilar *et al.*, 1997; Steurbaut, 1998), Menat (Gingerich, 1976), and Walbeck (references in Hooker, 1991). Calibration to the GPTS varies from direct (Engesser and Mödden and Legendre and Levêque, in Aguilar *et al.*, 1997; Hooker in Aubry *et al.*, 1998; López-Martínez and Peláez-Campomanes, 1999; Hooker and Millbank, 2001) to indirect (Franzen and Haubold, 1987; Hooker, 1986; Steurbaut, 1992; Merz *et al.*, 2000). Calibration of ALMAs and their subdivisions essentially follows Holroyd and Ciochon (1994) and Meng and McKenna (1998). That of SALMAs follows Flynn and Swisher (1995) and Kay *et al.* (1999). In the case of Europe and North America, selected first and last appearances are given. They are intended to reflect a balance between those recording important faunal changes, used in biostratigraphy and inter-continental correlation, and those highlighting inter-continental diachronism. Sources additional to the above references are: Köhler and Moyà-Solà (1999) and Stucky (1992).

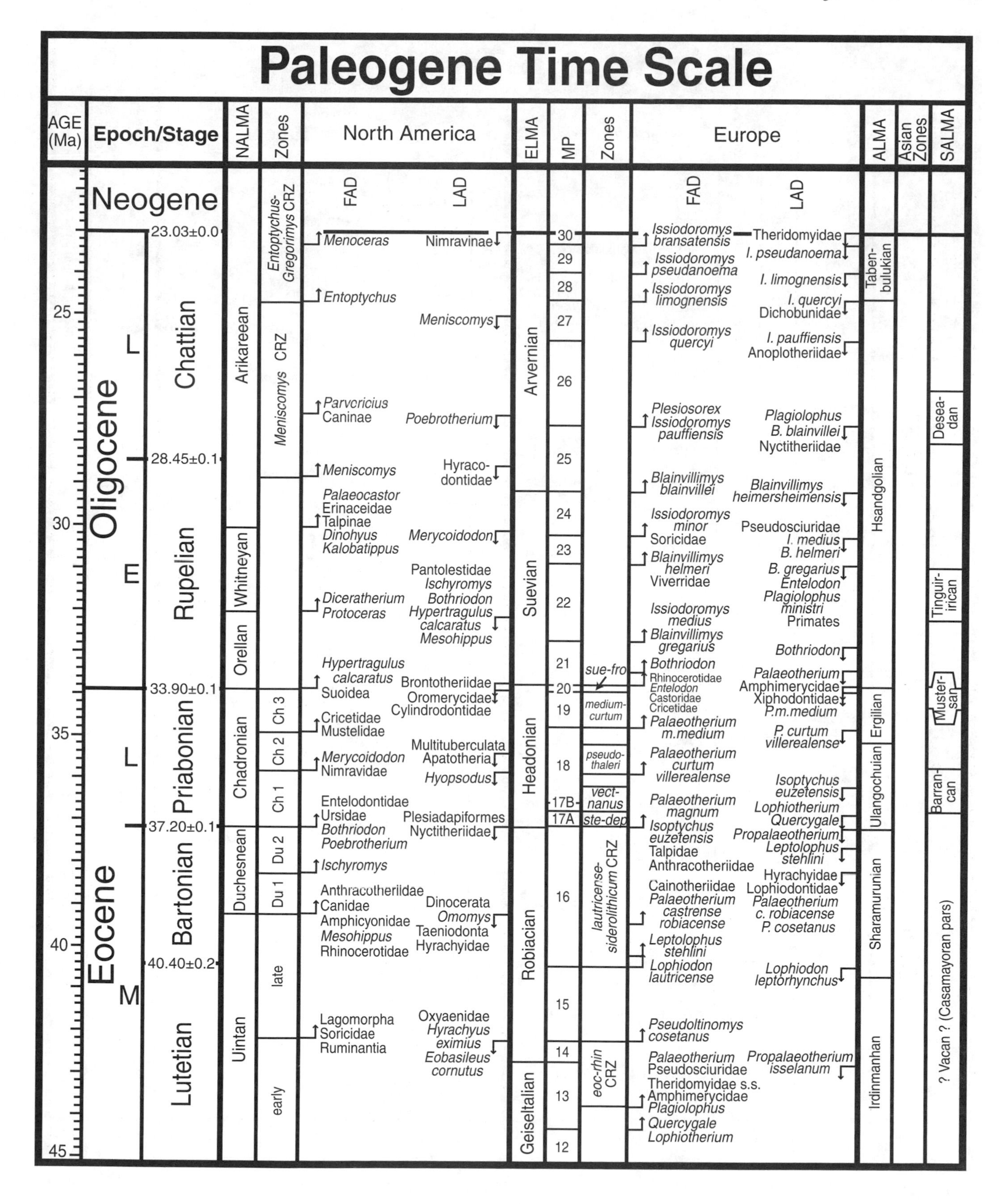

Paleogene Time Scale
AGE (Ma)
Epoch/Stage
NALMA
Zones
North America
ELMA
MP
Zones
Europe
ALMA
Asian Zones
SALMA
Neogene
23.03±0.0
25
28.45±0.1
30
33.90±0.1
35
37.20±0.1
40
40.40±0.2
45
Oligocene
L
Chattian
E
Rupelian
Eocene
L
Priabonian
Bartonian
M
Lutetian
Arikareean
Whitneyan
Orellan
Chadronian
Duchesnean
Uintan
Entoptychus-Gregorimys CRZ
Meniscomys CRZ
Ch 3
Ch 2
Ch 1
Du 2
Du 1
late
early
FAD
LAD
Menoceras
Nimravinae
Entoptychus
Meniscomys
Parvericius
Caninae
Poebrotherium
Meniscomys
Hyraco-dontidae
Palaeocastor
Erinaceidae
Talpinae
Dinohyus
Kalobatippus
Merycoidodon
Pantolestidae
Ischyromys
Diceratherium
Protoceras
Bothriodon
Hypertragulus calcaratus
Mesohippus
Hypertragulus calcaratus
Suoidea
Brontotheriidae
Oromerycidae
Cricetidae
Mustelidae
Cylindrodontidae
Multituberculata
Merycoidodon
Nimravidae
Apatotheria
Hyopsodus
Entelodontidae
Ursidae
Bothriodon
Poebrotherium
Plesiadapiformes
Nyctitheriidae
Ischyromys
Anthracotheriidae
Canidae
Amphicyonidae
Mesohippus
Rhinocerotidae
Dinocerata
Omomys
Taeniodonta
Hyrachyidae
Lagomorpha
Soricidae
Ruminantia
Oxyaenidae
Hyrachyus eximius
Eobasileus cornutus
Arvernian
Suevian
Headonian
Robiacian
Geiseltalian
30
29
28
27
26
25
24
23
22
21
20
19
18
17B
17A
16
15
14
13
12
sue-fro
medium-curtum
pseudo-thaleri
vect-nanus
ste-dep
lautricense-siderolithicum CRZ
eoc-rhin CRZ
Issiodoromys bransatensis
Theridomyidae
Issiodoromys pseudanoema
I. pseudanoema
Issiodoromys limognensis
I. limognensis
I. quercyi
Dichobunidae
Issiodoromys quercyi
I. pauffiensis
Anoplotheriidae
Plesiosorex
Issiodoromys pauffiensis
Plagiolophus
B. blainvillei
Nyctitheriidae
Blainvillimys blainvillei
Blainvillimys heimersheimensis
Issiodoromys minor
Soricidae
Pseudosciuridae
I. medius
B. helmeri
Blainvillimys helmeri
Viverridae
B. gregarius
Entelodon
Plagiolophus ministri
Primates
Issiodoromys medius
Blainvillimys gregarius
Bothriodon
Bothriodon
Rhinocerotidae
Entelodon
Castoridae
Cricetidae
Palaeotherium
Amphimerycidae
Xiphodontidae
P.m.medium
Palaeotherium m.medium
P. curtum villerealense
Palaeotherium curtum villerealense
Isoptychus euzetensis
Lophiotherium
Palaeotherium magnum
Quercygale
Isoptychus euzetensis
Propalaeotherium
Talpidae
Leptolophus stehlini
Anthracotheriidae
Hyrachyidae
Cainotheriidae
Lophiodontidae
Palaeotherium castrense robiacense
Palaeotherium c. robiacense
P. cosetanus
Leptolophus stehlini
Lophiodon lautricense
Lophiodon leptorhynchus
Pseudoltinomys cosetanus
Palaeotherium
Pseudosciuridae
Propalaeotherium isselanum
Theridomyidae s.s.
Amphimerycidae
Plagiolophus
Quercygale
Lophiotherium
Taben-bulukian
Hsandgolian
Ergilian
Ulangochuian
Sharamurunian
Irdinmanhan
Desea-dan
Tinguir-irican
Muster-san
Barran-can
? Vacan ? (Casamayoran pars)

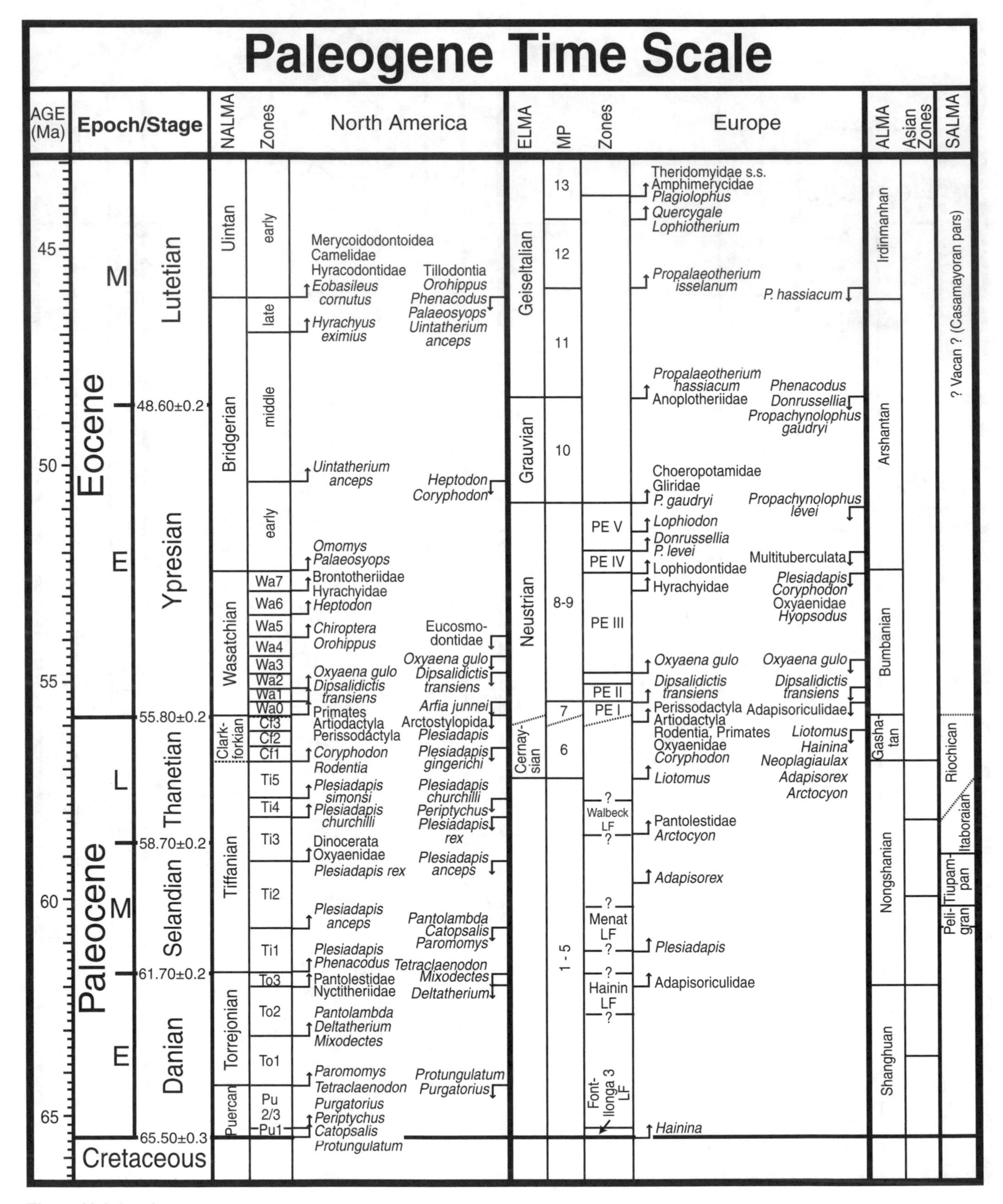

Figure 20.4 (*cont.*)

20.3 PHYSICAL STRATIGRAPHY

20.3.1 Magnetostratigraphy

Since its introduction by Heirtzler *et al.* (1968), the geomagnetic polarity time scale (GPTS) has enabled Earth's Late Cretaceous and Cenozoic history to be elucidated in considerable detail. Based upon a long magnetic anomaly profile from the South Atlantic, a linear scale was constructed back to ~75 Ma using as tie points the zero-age ridge and 3.35 Ma for the start of the Pliocene anomaly 2a. Despite the twenty-fold extrapolation, the dates of the Paleogene–Late Cretaceous anomalies have proven to be largely correct.

Thirty years after the pioneering work, the GPTS has evolved into a sophisticated dating tool as a large body of bio-, chemo-, and magnetostratigraphic data has been acquired and systematically integrated with the scale (Berggren *et al.*, 1985b; Harland *et al.*, 1990). As a tool for dating and correlation, magnetostratigraphy is particularly powerful for Paleogene rocks. The geomagnetic field reversals for this ~41-myr interval occurred on average every 0.65 myr.

The C-sequence of marine magnetic anomalies assembled and calibrated by Cande and Kent (1992a, 1995) forms the basis of the integrated magneto–biostratigraphic scale of Berggren *et al.* (1995a), and is reviewed in Chapter 5. It is constructed around a synthetic profile stitched together from several South Atlantic anomaly tracks. Specific segments of the profile are further constrained by anomaly data from other parts of the globe. For example, the floor of the northeast Pacific is used as a check on the relative spacing of the Middle Eocene through Oligocene anomalies. The scale for the interval 5.23–83.0 Ma of the CK92–CK95 scale was based on nine age-control ties, and the anomalies are linked and dated using a cubic-spline curve to smooth the effects of ocean basin spreading-rate changes. Spreading-rate changes are most apparent in the pre-45 Ma part of the curve (see Chapter 5).

As explained in the Chapter 5, in this book we use the "stratigraphic" relative placement of polarity events, as originally proposed by LaBreque, Lowrie, Channell, and others (see synthesis in Hallam *et al.*, 1985). Cande and Kent (1992a, 1995) used the inverse system, which is more convenient for calculations.

MAGNETOSTRATIGRAPHIC AGES OF THE PALEOGENE EPOCH BOUNDARIES

The first decade or so after Heirtzler *et al.* (1968) appeared saw various workers trying to refine the correlation of each of the Cenozoic–Late Cretaceous epoch and stage boundaries to the GPTS using data from marine and terrestrial sequences (e.g. LaBrecque *et al.*, 1977). Perhaps the most important contribution after this time was by Lowrie and Alvarez (1981), who synthesized a large body of magneto- and biostratigraphic data obtained by themselves and colleagues from several uplifted deep-marine Tethys sections in central Italy. Subsequently, correlation of each of the Paleogene Epoch boundaries relative to specific magnetochrons has remained essentially fixed (Harland *et al.* 1982, 1990; Berggren *et al.*, 1985b, 1995a; Haq *et al.* 1987). However, the absolute ages of the boundaries in each scale differ by as much as 3 myr, a consequence of the choice of tie point ages that were used to construct a particular GPTS.

The base of the Paleocene Epoch (Cretaceous–Paleogene boundary) is one of the most studied intervals of geological time. Many radiometric dating studies have been carried out on rocks associated with the boundary with a cluster of dates circa 65.5 Ma. This boundary is positioned at about midway in the ~0.8-myr-long polarity Chron C29r.

The base of the Eocene (base-Ypresian) is defined by the initiation of a carbon isotope excursion with a GSSP in the Gabal Dababiya section, Egypt. Although the Egyptian section does not yield magnetostratigraphic information, the carbon isotope excursion can be readily correlated globally. From cycle stratigraphy, this event at the base of the Eocene is 0.94 myr after the end of polarity Chron 25n (Norris and Röhl, 2001).

The Eocene–Oligocene boundary is in the upper part of polarity Chron C13r (C13r.86) and has an age close to 33 Ma. Note that there is no Anomaly 14 – the jump from C15n to C13r in the GPTS is an artifact of early marine magnetic studies that had mistakenly assigned this anomaly; but to avoid the confusion inherent in re-labeling the published anomaly and chron successions, the adopted convention is that the GPTS jumps from C15n to C13r.

The Oligocene–Miocene boundary corresponds to the start of Chron C6n.2n and has an age close to 23 Ma based on calibration to Milankovitch cycles (see Chapter 21).

20.3.2 Chemical stratigraphy

Chemical dating methods have contributed significantly during the last two decades to our understanding of Paleogene stratigraphy and, in particular, the major global events during the period (Fig. 20.5). Carbon and oxygen stable isotope analyses are now routinely performed in most stratigraphic studies of Paleogene sequences. Strontium isotopes and iridium are other commonly applied correlation tools. Importantly, the carbon isotopic and iridium approaches are facies independent and in

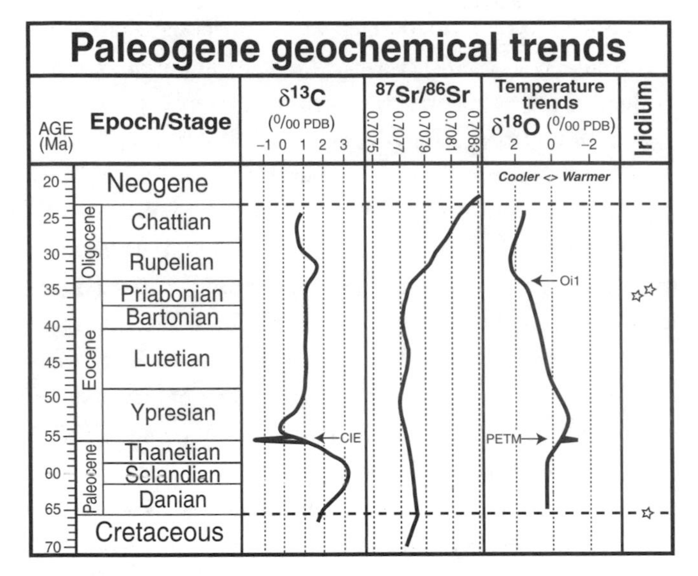

Figure 20.5 Generalized oxygen, carbon, and strontium isotope curves and impact-related iridium anomalies for the Paleogene. Oxygen and carbon isotopic curves are based on deep-sea benthic foraminifera from Zachos *et al.* (1993, 1999). The strontium isotope curve is a LOWESS fit to reference data from Miller *et al.* (1988), Denison *et al.* (1993), Mead and Hodell (1995), Schmitz *et al.* (1997), Zachos *et al.* (1999), and other sources (see Chapter 7). Iridium anomalies are from Alvarez *et al.* (1980) and Coccioni *et al.* (2000). See text for details.

the Paleogene they have been used to establish crucial high-resolution correlations between shallow- and deep-marine as well as terrestrial and marine sequences.

The Cretaceous–Paleogene boundary is marked by an iridium anomaly originating from extraterrestrial dust, spread world-wide after the impact of a major extraterrestrial body at 65 Ma (Alvarez *et al.*, 1980; Fig. 20.5). In continental sections of the US Western Interior, the iridium anomaly occurs in an impact ejecta layer, a few centimeters thick, that contains also abundant shocked quartz (Izett, 1990). The bulk of this layer was deposited during a few days, and it can be correlated world-wide with a layer in deep- and shallow-marine environments that is rich in iridium and shocked-quartz. The Cretaceous–Paleogene boundary is also marked by a 2–3% negative carbon isotopic excursion in calcite or organic matter that formed in the upper ocean water mass. This anomaly most likely reflects a dramatic decline in biological productivity associated with the impact event (Zachos and Arthur, 1986).

The onset of a 2–3% negative carbon isotopic excursion (CIE) marks the Paleocene–Eocene boundary (Kennett and Stott, 1991; Schmitz *et al.*, 2001). The isotopic anomaly developed within a few thousand years in both the deep and shallow ocean, indicating that the entire oceanic carbon reservoir experienced a rapid change in isotopic composition. It took ~200 kyr before δ^{13}C returned to its background levels. This period is named the Paleocene–Eocene thermal maximum (PETM) and is characterized by the highest global temperatures of the Cenozoic (Fig. 20.5).

The negative δ^{13}C anomaly can best be explained by outgassing of methane with low δ^{13}C from destabilized methane hydrates in seafloor sediments (Dickens *et al.*, 1997). The δ^{13}C anomaly has been used for high-resolution shelf–basin correlations (Schmitz *et al.*, 2001) and for marine–terrestrial correlations (Koch *et al.*, 1992). Atmospheric and oceanic CO_2 equilibrate isotopically, and Koch *et al.* (1992) were able to locate the Paleocene–Eocene boundary in continental sections in the Bighorn Basin by analyzing the carbon isotopes of fossil teeth of mammals that had eaten plants containing carbon from atmospheric CO_2. This study showed for the first time that the Clarkforkian–Wasatchian land-mammal turnover and the major extinction event among benthic foraminifera in the deep sea at ~55 Ma are more or less coeval, probably reflecting the same global environmental perturbation.

Another important correlation feature of the carbon isotopic record of the Paleogene is the maximum in seawater δ^{13}C for a few million years in the late Paleocene and the subsequent long-term fall in δ^{13}C across the Paleocene–Eocene transition (Zachos *et al.*, 1993; Fig. 20.5). The late Paleocene δ^{13}C maximum most likely reflects a period of increased oceanic productivity.

The marine ^{87}Sr/^{86}Sr record for the Paleogene is poorly known and so currently of no use as a stratigraphic tool before ~42 Ma (Fig. 20.5). Once a good calibration curve is constructed, the method has potential for correlation and dating in the interval 65–42 Ma, as ^{87}Sr/^{86}Sr changes by some 0.000 100 through this period. During the latest Eocene and Oligocene, ^{87}Sr/^{86}Sr increased at one of the highest rates known (about 0.000 050 per myr), thereby allowing high-resolution dating by this technique with a precision as low as 0.2 myr (McArthur *et al.*, 2001; see Chapter 7).

Oxygen isotopes measured in calcite from benthic foraminifera show a stable trend in the Paleocene (Fig. 20.5). During the PETM, δ^{18}O shows a short-term decrease associated with greenhouse warming. Minimum δ^{18}O values occur throughout the warm early Eocene followed by a gradual increase throughout the middle Eocene. A profound positive shift in δ^{18}O occurs in the early Oligocene (Oi1 Event), thereafter, δ^{18}O stabilizes through the rest of the Oligocene (Zachos *et al.*, 1993). This trend reflects the global cooling from the late Eocene into the Oligocene.

At least two small impact-related iridium anomalies have been reported from several early Late Eocene sedimentary sections (Coccioni *et al.*, 2000; Vonhof *et al.*, 2000). The impacts responsible for these two marker beds may have triggered the late Paleogene long-term global cooling.

20.3.3 Cycle stratigraphy

Orbital-climatic oscillations recorded in deep-sea sediments enabled the development of a high-resolution astronomically calibrated time scale of geomagnetic polarity reversals and late biostratigraphy for the late Cenozoic (see Chapter 21) to the base of the late Eocene (Shackleton *et al.*, 1999, 2000; Pälike *et al.*, 2001). Analyses of sedimentological and physical properties data from Paleogene sequences document the sensitivity of the ocean to Milankovitch-band forcing and the rapidity of threshold events in Paleogene paleoceanographic and paleoclimatic history. Therefore, "floating" astronomical time scales are based on the number of precession or obliquity controlled cycles multiplied by the period of the corresponding astronomical cycle at that time.

A high-resolution time scale of the Paleocene to early Eocene is constrained by the following components:

1. Milankovitch cycles of precession indicate that polarity Chron C29r in the South Atlantic (DSDP Sites 516 and 528) and in Spain spans 0.68 ± 0.04 myr. The end-Cretaceous extinction occurs in the middle of Chron C29r (Herbert *et al.*, 1995; Herbert, 1999).
2. Cycles of obliquity and precession are recorded in deep-sea sediments recovered at ODP Sites 1050 and 1051 on the Blake Nose off Florida and Site 1001 in the Caribbean Sea (Röhl *et al.*, 2001, 2003). These yield a scaling of polarity Chrons C27 through C22 (middle Danian through late Early Eocene). Combined Chrons C27n–C27r span 1.45 myr; combined Chrons C26n–C26r span 3.61 ± 0.1 myr; Chron C25r spans 1.07 myr; Chron C25n spans 0.50 ± 0.02 myr; and Chron C24r spans 2.88 ± 0.1 myr. Chron C24n is not as well delimited, but its duration is ~1.2 myr. Polarity Chron C23r spans 0.53 myr, Chron C23n is 0.74 myr, and Chron C22r spans 0.9 myr.
3. When coupled with a synthetic block model of relative widths of magnetic anomalies in the South Atlantic (Cande and Kent, 1992a, 1995), the approximate duration of polarity Chron C28n can be interpolated from this array of cycle-tuned durations of magnetic polarity chrons (Röhl *et al.*, 2000, 2001). Combined Chrons C28r–C28n span 1.3 myr.
4. The relative position of the δ^{13}C anomaly marking the Paleocene–Eocene boundary is 0.94 ± 0.02 myr after the end of Chron C25n according to precession cycles in ODP Site 1051 (Norris and Röhl, 1999; Röhl *et al.*, 2001, 2003).
5. A composite depth record from ODP Site 1052 (Leg 171B, Blake Nose) provides the basis to extend the astronomically calibrated geologic time scale into the middle Eocene and results in revised estimates for the age and duration of polarity Chrons C16 through C18 (Pälike *et al.*, 2001): from the new data, the relative duration of these chrons does not change significantly from that given by Cande and Kent (1995). Exceptions are the relative durations within Chron C16n, in which C16n.1n seems to be ~200 kyr longer and C16n.2n seems to be ~200 kyr shorter.

This array of Paleocene through earliest Eocene cycle stratigraphy estimates is consistent with radiometric ages of 65.5 ± 0.1 Ma for the end-Cretaceous extinction (Renne *et al.*, 1998c; Obradovich in Hicks *et al.*, 1999), 55.07 ± 0.5 Ma for approximately the middle of Chron C24r (Obradovich in postscript in Berggren *et al.*, 1995a) and 52.8 ± 0.3 Ma at the base of Chron C24n.1.n (Wing *et al.*, 1991; Tauxe *et al.*, 1994). The last age converts to 53.0 Ma with a change of the Ar–Ar MMhb-1 monitor from 521 to 523 Ma.

Planktonic foraminifer and calcareous nannofossil datums and zonations have been calibrated to the magnetic polarity time scale (e.g. Berggren *et al.*, 1995a), and cycle tuning constrains the associated assignment of ages. When there is international agreement on the microfossil markers for the Paleocene stage boundaries and these events are calibrated to the cycle stratigraphy of the ODP sites, then precise astronomically tuned ages can be assigned to these boundaries.

Cycle stratigraphy of intervals within the late Eocene and Oligocene Epochs are being compiled (e.g. Weedon *et al.*, 1997; Shackleton *et al.*, 1999; Pälike *et al.*, 2001) and the estimated astronomical age for the Oligocene–Miocene boundary is 23.03 Ma (see Chapter 21).

20.4 PALEOGENE TIME SCALE

20.4.1 Radiometric ages

Although the number of reliable radiometric data available for the Paleogene has increased considerably in recent years, coverage of this interval is still somewhat patchy. The radiometric ages discussed in this chapter (Table 20.2) are linked to the magnetostratigraphy scale and directly calibrate the Paleogene time scale. All are based on high-temperature radiometric age

Table 20.2 *Selected radiometric age determinations or derived ages with uncertainty limits, their calibration in polarity chrons and distances in kilometers in the South Atlantic spreading profile*

Placement in polarity chron	Age (Ma)	Uncertainty (2-sigma)	Geological age	Distance (km) in S Atlantic profile	Calibration type	Source or review references	Comments
C6An.1r (base)	20.336	<40 kyr	Aquitanian–Burdigalian boundary	434.18	Cyclo–magnetostratigraphy	See Section 21.3	
C6Cn.2n (base)	23.03	<40 kyr	Oligocene–Miocene boundary	501.55	Cyclo–magnetostratigraphy	See Section 21.3	
C9n (base)	28.10	0.30	Mid Oligocene (earliest Chattian)	607.96	K–Ar and $^{40}Ar/^{39}Ar$ ages and magnetostratigraphy in Italy	Odin *et al.* (1991), Wei (1995)	
approx. C13r.86	33.70	0.40	Eocene–Oligocene stage boundary	approx. 759.49	K–Ar and $^{40}Ar/^{39}Ar$ dating in Italy	Odin *et al.* (1991), Cande and Kent (1992a)	Not used in spline fit, in preference to more direct age-chron constraint on C15n base (35.2 Ma)
C15n (base)	35.20	0.27	Latest Eocene	791.78	K–Ar and $^{40}Ar/^{39}Ar$ ages and magnetostratigraphy in Italy	Odin *et al.* (1991), Wei (1995)	
approx. C21n.67	45.60	0.38	Middle Eocene (early Lutetian)	approx. 1071.62	$^{40}Ar/^{39}Ar$ dating of magnetostratigraphy in DSDP Hole 516F	Berggren *et al.* (1995a) (postscript)	
C24n.1n (base)	52.80	0.30	Early Eocene (mid-Ypresian)	1184.03	$^{40}Ar/^{39}Ar$ ages and magnetostratigraphy in Wyoming	Tauxe *et al.* (1994), Wei (1995)	
approx C24r.5	55.07	0.50	Just above Paleocene–Eocene boundary (earliest Ypresian)	approx. 1214.93	$^{40}Ar/^{39}Ar$ ages and magnetostratigraphy in DSDP Hole 550	Berggren *et al.* (1995a; age by Obradovich in postscript)	P–E boundary (= C-spike) is 0.94 ± 0.02 myr after the end of Chron C25n
C27n (top)	61.77	0.30	Early to Late Paleocene boundary	1303.81	Cyclo–magnetostratigraphy (Central Atlantic and Caribbean)	Röhl *et al.* (2001)	36 obliquity cycles of 41 kyr. Uncertainty is relative to K–P boundary
C28n (top)	63.25	0.30	Mid Danian	1325.71	Cyclo–magnetostratigraphy (Central Atlantic and Caribbean)	Röhl *et al.* (2001)	Interpolated by applying an intermediate spreading rate between C29(n + r) and C27

C29n (top)	64.53	0.30	Early Danian	1347.03	Cyclo–magnetostratigraphy (South Atlantic and Spain)	Herbert *et al.* (1995)	32 ± 2 precession cycles of 20.8 kyr. Uncertainty is relative to K–P boundary
C29n (base)	65.20	0.30	Earliest Danian	1358.66	Cyclo–magnetostratigraphy (South Atlantic and Spain)	Herbert *et al.* (1995)	14.5 ± 2 precession cycles of 20.8 kyr. Uncertainty is relative to K–P boundary
C29r.56	65.50	0.30	Base of Cenozoic (K–P boundary)	1364.45	$^{40}Ar/^{39}Ar$ ages and cyclostratigraphy of polarity zone C29r	Herbert *et al.* (1995), Hicks *et al.* (1999), Renne *et al.* (1998c)	Assigned position (km) in S. Atlantic. profile based on proportion position in chron, rather than attempting to adjust for slowing spreading rate. An arbitrary 0.1 myr uncertainty is assigned to this spreading-rate derived estimate
C30n (top)	65.88	0.30	Latest Maastrichtian	1371.84	Cyclo–magnetostratigraphy (South Atlantic and Spain)	Herbert *et al.* (1995)	18.5 ± 1 precession cycles of 20.8 kyr. Age and uncertainty is relative to K–T boundary
C31n (base)	69.01	0.50	Base Late Maastrichtian	1407.22	$^{40}Ar/^{39}Ar$ ages and extrapolation to polarity zone C31.n	Hicks *et al.* (1999)	
C32n (top)	70.44	0.65	Early Maastrichtian	1481.12	$^{40}Ar/^{39}Ar$ ages and magnetostratigraphy of US Western Interior	Hicks and Obradovich (1995), Hicks *et al.* (1999)	Less-precise interpolations for C31n (top) and C33n (base) by Hicks and Obradovich (1995) were not included because the large uncertainties do not constrain age–distance fit
C33n (base)	79.34	0.50	Mid Campanian	1723.76	$^{40}Ar/^{39}Ar$ ages and magnetostratigraphy of US Western Interior (Elk Basin)	Hicks *et al.* (1995)	In the magnetostratigraphy of the Elk Basin, the age of this reversal is extrapolated from underlying bentonite ages
C33r (base)	83.97	0.50	Base Late Santonian	1862.32	$^{40}Ar/^{39}Ar$ age constraints on biostratigraphy associated with magnetostratigraphy	Montgomery *et al.* (1998), Obradovich (1993)	*U. socialis* lowest occurrence in the Upper Santonian near base C33r

assignments, with the age for the FCT $^{40}Ar/^{39}Ar$ monitor at 28.02 Ma (see Chapter 6).

The Cretaceous–Paleogene boundary based on the iridium spike is well constrained by a number of radiometric data with 65.5 ± 1 Ma (Hicks *et al.*, 1999). Analysis of the radiometric dating of this boundary in Chapter 8 yields an age estimate of 65.5 ± 0.3 Ma.

There is no direct radiometric dating of the Paleocene–Eocene boundary, as defined by the $\delta^{13}C$ excursion within the lower part of C24r. A $^{40}Ar/^{39}Ar$ date from the middle of polarity Chron C24r, which is significantly above the Paleocene–Eocene boundary, is given as 55.07 ± 0.5 Ma by Obradovich (see postscript in Berggren *et al.* 1995a). The planktonic foraminifera *Morozovella velascoensis* Zone (i.e. Zone P5) is dated by Wing *et al.* (2000) in the continental sequences of Wyoming (Clarkforkian–Wasatchian boundary) as 54.98–55.2 Ma.

Argon-40/argon-39 dating and magnetostratigraphy in Wyoming at the base of polarity Chron C24n.1n of mid-Ypresian age yielded 52.80 ± 0.30 Ma (Tauxe *et al.*, 1994; Wei, 1995), while $^{40}Ar/^{39}Ar$ dating in DSDP Site 516F of approximate magnetostratigraphic level C21n(0.67) in the early Lutetian yielded 45.60 ± 0.38 Ma (Berggren *et al.*, 1995a, see postscript therein). Potassium–argon and $^{40}Ar/^{39}Ar$ dating of a level approximating C13r(0.9) near to the Eocene–Oligocene Stage boundary in Italy is dated at 33.70 ± 0.40 Ma, and earliest Chattian tuffs in Italy assigned to the base of polarity Chron C9n have a radiogenic age of 28.10 ± 0.30 Ma (Odin *et al.*, 1991; Wei, 1995; Cande and Kent, 1992a).

The age of the Oligocene–Miocene boundary as defined by its GSSP in the Carrosio–Lemme section (Steininger *et al.*, 1997a) has not been directly dated. However, the lowermost Miocene is constrained by the $^{40}Ar/^{39}Ar$ age of 21.88 ± 0.32 Ma attributed by Odin *et al.* (1997) to volcaniclastic levels close to the Raffaelo Level in the Central Apennines. Correlation of this level is somewhat difficult, but it is probably close to polarity Chron C6AA.2 and foraminifer Zone N4 of Blow (1979) and to the top of calcareous nannofossil Zone NN1 or the base of NN2.

Table 20.3 *Placement of Paleogene stage boundary definitions relative to polarity chrons and associated distance in kilometers in the South Atlantic seafloor-spreading profile (see Chapter 21 for Neogene ages)*

Stage boundary	Placement in magnetic polarity chron	Distance (km) in S Atlantic profile
Aquitanian–Burdigalian (20.336 Ma)	C6An.1r (base)	434.18
Chattian–Aquitanian (O–M boundary 23.03 Ma)	C6Cn.2n (base)	501.55
Rupelian–Chattian	C10n.1n	622.16
Priabonian–Rupelian (E–O boundary)	C13r (.86)	759.49
Bartonian–Priabonian	C17n (uppermost part)	856.19
Lutetian–Bartonian	C19n (top)	947.96
Ypresian–Lutetian	C22n (top)	1117.55
Thanetian–Ypresian (P–E boundary)	C24r.3	1222.82
Selandian–Thanetian	C26n (base)	1262.74
Danian–Selandian	C27n.8	1304.78
Maastrichtian–Danian (K–P boundary)	C29r.56	1364.45

Table 20.4 *Ages of Paleogene stage boundaries and duration of stages*

Stage name	Base (Ma)	Duration (my)
Aquitanian	23.0 ± 0.0	
Chattian	28.4 ± 0.1	5.4 ± 0.0
Rupelian	33.9 ± 0.1	5.4 ± 0.0
Priabonian	37.2 ± 0.1	3.3 ± 0.0
Bartonian	40.4 ± 0.2	3.2 ± 0.0
Lutetian	48.6 ± 0.2	8.2 ± 0.0
Ypresian	55.8 ± 0.2	7.2 ± 0.0
Thanetian	58.7 ± 0.2	2.9 ± 0.0
Selandian	61.7 ± 0.2	3.0 ± 0.0
Danian	65.5 ± 0.3	3.7 ± 0.0

20.4.2 Age and duration of stages

The Paleogene time scale is constructed from the integration of well-constrained radiometric age dates, the geomagnetic polarity time scale (GPTS), and cycle stratigraphy. A summary of key data selected to construct the time scale is given in Tables 20.2 and 20.3, with the age assignments and interval durations, rounded off to one decimal in Table 20.4. The geomathematical and statistical procedures applied to this data set are explained in detail in Chapter 8; external errors associated with Ar–Ar dating were taken into account (see Section 8.3).

The basal stage of each Paleogene epoch has been defined by GSSPs, but none of the stages within these epochs have yet been formalized. Since no time scale can be constructed using undefined units, we assigned working definitions to the remaining Paleogene stages. These assigned Paleogene stage boundaries generally correspond to the Paleogene magnetostratigraphic scale proposed by Berggren *et al.* (1995a), which was also used for the Paleocene cyclostratigraphic scale of Röhl *et al.* (2001). The Neogene cyclostratigraphic extrapolations to the base of the Miocene are explained in Chapter 21.

The Cretaceous–Paleogene boundary at polarity Chron C29r(0.56) is radiometrically dated at 65.5 ± 0.3 Ma, and the Paleogene–Neogene boundary at C6Cn.2n is astronomically estimated as 23.03 Ma. Therefore, the Paleogene Period spans 42.5 myr. This is almost 20 myr more than the Neogene Period, which spans Miocene–Recent.

The Danian–Selandian boundary is calibrated to the uppermost portion of polarity zone C27n (C27n.8), which implies an age of 61.7 ± 0.2 Ma, using a combination of cycle tuning and seafloor-spreading interpolation. The Selandian–Thanetian boundary at the base of polarity Chron C26n is interpolated to be 58.7 ± 0.2 Ma. The Paleocene–Eocene (Thanetian–Ypresian) boundary at the major organic–carbon isotope anomaly and Chron C24r.3 is 55.8 ± 0.2 Ma in age, is slightly older than the 55 Ma age on the International Stratigraphic Chart (Remane, 2000).

The duration of the Paleocene, between 65.5 ± 0.3 and 55.8 ± 0.2 Ma, is 9.6 myr. The Danian stage spans 3.7 myr, the Selandian 3.0 myr, and the Thanetian 2.9 myr.

An age of 48.6 ± 0.2 Ma at the top of Chron C22n is estimated for the boundary between Ypresian and Lutetian (the Early–Middle Eocene boundary), and an age of 40.4 ± 0.2 Ma at the top of Chron C19n is interpolated for the Lutetian–Bartonian boundary. The base of the Late Eocene Priabonian Stage is placed at the NP17–NP18 zonal boundary tied to the younger part of polarity Chron C17n. The age of its lower boundary is estimated at 37.2 ± 0.1 Ma.

With the upper limit of the Eocene Period (Priabonian–Rupelian boundary) tied to the extinction of the hantkenninid planktonic foraminifera at the top of Zone P17 within polarity Chron C13r(0.14), the best-age estimate is 33.9 ± 0.1 Ma. This is identical to an age of 34 Ma for biotites 19 m below the GSSP in the Massignano section (Premoli-Silva and Jenkins, 1993).

The duration of the Eocene is 21.9 myr, which is more than twice that of the Paleocene, with the Ypresian spanning 7.2 myr, the Lutetian 8.2 myr, the Bartonian 3.2 myr, and the Priabonian 3.3 myr. This makes the Lutetian the longest stage in the Paleogene, and the Thanetian the shortest lasting one.

The Rupelian–Chattian Stage boundary between the lower and upper series of the Oligocene corresponds to the disappearance of common Chiloguembelinid planktonic foraminifera. This boundary between planktonic foraminifer Zones P21a and P21b is in the middle of polarity Chron C10n. The boundary is assigned an age of 28.4 ± 0.1 Ma, which is consistent with an age of 28.1 ± 0.3 Ma for earliest Chattian tuffs at the top of polarity zone C9r in the Contessa Quarry section, Italy (Wei, 1995).

The base of the Aquitanian Stage and the Oligocene–Miocene boundary as defined with a GSSP in the Lemme–Carrosio section, Italy, correlates to the transition of magnetic polarity zones C6Cn.2r and C6Cn.2n with an assigned age of 23.03 Ma (see Chapter 21). The duration of the Oligocene is 10.8 myr, with the Rupelian and Chattian Stages both lasting 5.4 myr.

20.4.3 Future development of the Paleogene time scale

Improvement and consolidation of the Paleogene time scale will depend on future GSSP definitions for the remaining stage boundaries and on astronomical tuning of durations of all polarity chrons. Such tuning and calibration of the Paleogene time scale at much higher levels of resolution and precision than presently available is currently in active progress. Recently completed Ocean Drilling Program cruises (e.g. ODP Legs 198, 199, 207, 208) will provide a wealth of directly linked magneto- and biostratigraphic datums, which will need to be taken into account in possible future definitions and applications of stage boundaries. In the case of ODP Leg 199, an astronomical time scale calibration for all Oligocene polarity chrons is nearing completion (status January 2004), and provides calibrated age estimates that will extend the approach presented for the Neogene (see Chapter 21). First results confirm that the present Paleogene time scale will undergo further revisions, in contrast to the now well-calibrated Neogene. In particular, it has become clear that seafloor-spreading rates, as determined by astronomical duration calibration of magnetic anomalies, show a significantly less smooth variation than suggested by the spline fits (see also Chapter 21).

However, astronomical time scale calibrations in the Paleogene face additional theoretical uncertainties in the orbital calculations (Laskar, 1999). While some of these theoretical challenges have been resolved from geological observations, it is likely that astronomical calibrations of the geologic time scale in the earlier parts of the Cenozoic are more challenging in detail, providing additional scope for uncertainties. In the medium term, it can be predicted, though, that a complete coverage of astronomically calibrated geological markers will exist for the entire Cenozoic, and that traditional geochronological scales, astronomical calibrations, and magneto- and biostratigraphic datums will become intercorrelated.

APPENDIX

New observations by Van Simaeys (in press) and Van Simaeys *et al.* (in press) shed light on the Rupelian–Cahttian boundary, as summarized below.

The current "global" criterion for the recognition of the Rupelian–Chattian boundary, i.e. the demise of the planktonic foraminiferal genus *Chiloguembelina*, is not applicable in the North Sea Basin, home of the Rupelian and Chattian Stages stratotypes. Moreover, records from several sections (mainly ODP boreholes) indicate that the chiloguembelinid extinction is globally time-transgressive, from the Early Oligocene at high latitudes to the Late Oligocene at low latitudes. Because of the diachronous nature of the last occurrence of the genus *Chiloguembelina*, this criterion can no longer be upheld for the recognition of the Rupelian–Chattian boundary.

Detailed dinoflagellate cyst analysis enabled correlation between the restricted Oligocene North Sea Basin successions and the well-calibrated pelagic sections from central Italy. Based on the established correlations, it appears that the unconformity between the Rupelian and Chattian Stages in their stratotype area is genetically related to a 500-kyr Oligocene Glacial Maximum (OGM), and the corresponding glacio-eustatic sea-level fall. Calibrated dinoflagellate cyst events further suggest that the oldest of the time-transgressive fine glauconitic Chattian sands in the southern North Sea Basin were deposited in Chron C9n, not Chron C10n.

An important consequence of these results is that any Rupelian–Chattian GSSP should be positioned to match at least the age of the OGM, and that Chron C9r occurs below the Chattian. This would decrease the age of the Rupelian–Chattian boundary by more than 1 myr.

21 • The Neogene Period

L. LOURENS, F. HILGEN, N. J. SHACKLETON, J. LASKAR, AND D. WILSON

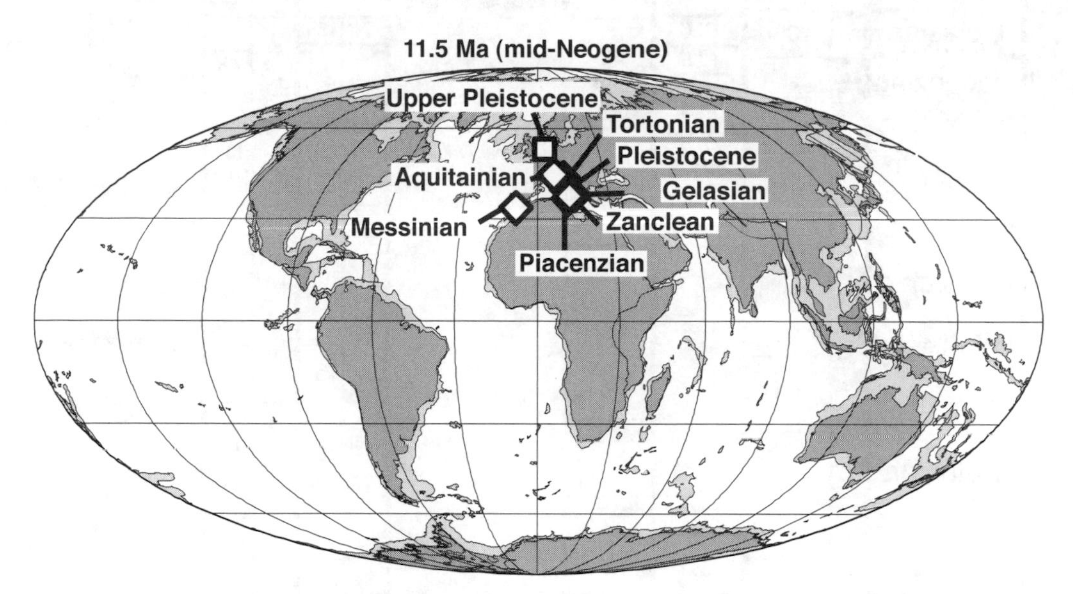

Geographic distribution of Neogene GSSPs that have been ratified (diamonds) or are candidates (squares) on a mid-Neogene (11.5 Ma) map (status in January, 2004; see Table 2.3). GSSPs within the Pleistocene Epoch and within the Middle and Early Miocene sub epochs are not yet formalized. Not all GSSP candidates are shown.

Gradual closing of Tethys and formation of more distinctive latitudinal biota belts under influence of dramatic climatic cooling phases. Marine microfossils are the backbone of Neogene chronostratigraphy; complex mammal evolution under the influence of major continental separations and climate changes; Milankovitch cyclicity in sediments and oxygen isotopes in the Atlantic and Mediterranean, assisted by the Australian–Antarctic marine magnetic polarity scale for the interval beyond 13 Ma, provide a precise and highly accurate Neogene geologic time scale.

21.1 HISTORY AND SUBDIVISIONS

The subdivision of the Neogene into its constituent stages is presently well established and internationally accepted pre-Pleistocene (Fig. 21.1). New task groups have been organized under the umbrella of the Subcommission on Quaternary Stratigraphy (SQS) to establish an international subdivision for the Pleistocene. GSSPs have been formalized for the Aquitanian (defining the Paleogene–Neogene boundary), the Tortonian and Messinian stages of the Miocene, and for the Zanclean, Piacenzian, and Gelasian stages of the Pliocene. In addition, the Pliocene–Pleistocene boundary has been defined.

21.1.1 Defining the Neogene Period

The term *Neogene* was introduced by Hörnes (1853, 1864) to differentiate the closely related molluscan fauna of the Miocene and Pliocene in the Vienna Basin from those of the Eocene (*sensu* Lyell, 1833). The original Neogene concept of Hörnes (1853, 1864) referred to the biostratigraphic division of the Tertiary (Arduino, 1760a,b) and Quaternary (Desnoyers, 1829) made by Bronn (1838), who subdivided this period (termed *Molasse Gebirge*) into three groups (termed *Molasse Gruppen*). The first Molasse Group was equated with Lyell's (1833) Eocene Epoch. The second group included Lyell's (1833) Miocene (*untere Abtheilung*) and Pliocene (*obrige Abtheilung*) Epochs, whereas the third group contained the Quaternary (*Alluvial und Quartär-Gebilde zum Theile*).

A Geologic Time Scale 2004, eds. Felix M. Gradstein, James G. Ogg, and Alan G. Smith. Published by Cambridge University Press.

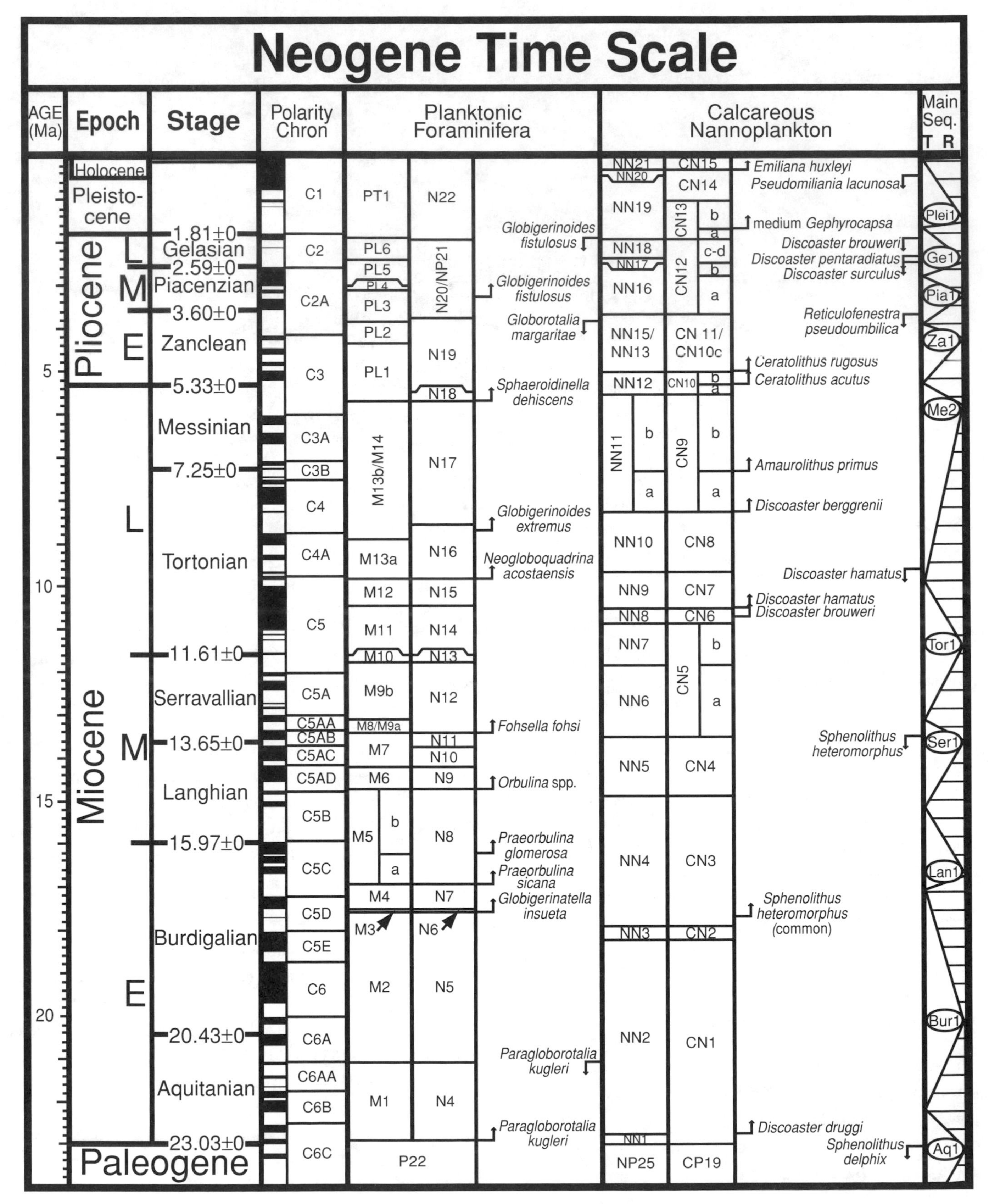

Figure 21.1 Neogene stratigraphic subdivisions, geomagnetic polarity scale, pelagic zonations, and selected datums of planktonic foraminifera and calcareous nannoplankton. Main trends in eustatic sea level are generalized from Hardenbol *et al.* (1998). A color version of this figure is in the plate section.

According to Bronn's subdivision, the so-called Löss and Diluvium deposits (*Knochen-Höhlen und der Loss*) were incorporated into the upper part of the second Molasse Group.

In 1837, Louis Agassiz presented his "Ice Age Theory" as an alternative explanation for the diluvial deposits found in large parts of Europe and North America (Agassiz, 1840), which until then were thought to be leftovers of the Biblical Flood (the "Diluvial Theory"). Soon afterwards, Forbes (1846) proposed equating the term Pleistocene (or Glacial Epoch) with the Diluvium. It took more than 25 years before Lyell formally accepted the term Pleistocene, which he had introduced in 1839 as an abbreviation for Newer Pliocene. He preferred this term above Post-Pleiocene, the term introduced by him in 1857 together with a redefinition of his Recent. Lyell (1873) recommended, therefore, that the term Pleistocene should either be abandoned or maintained; but, if retained, should be strictly synonymous with his Post-Pleiocene. Lyell (1839) mentioned that the Newer Pliocene Epoch was equivalent to the older strata of Desnoyers' Quaternary, which had been defined by a separate group of marine, lacustrine, alluvial, and volcanic rocks younger than the Tertiary in the Seine Basin: these rocks occasionally contained strange fossils (mammoth) or out-of-place bones of reindeer and Arctic birds. Thus, although Lyell never adopted the term Quaternary, it apparently encompasses both Lyell's Newer Pliocene Epoch as well as the Post-Pleiocene (i.e. Pleistocene *sensu* Forbes) and Recent Epochs (abbreviated as the Post-Tertiary Period according to Lyell, 1857).

Consequently, the scientific community was left with three different options for defining the Tertiary–Quaternary boundary at the end of the nineteenth century. The youngest option for the boundary corresponded with the base of Bronn's (1838) third Molasse Group, which should approximate the base of Lyell's (1873) Recent. The second option, in the light of Agassiz's Glacial Theory, was to equate the term Quaternary with the Ice Age and thus with the Pleistocene Epoch *sensu* Forbes (1846), although it was different from the Pleistocene in that it also included Lyell's (1857) "Recent," later named Holocene by Gervais (1867). Accordingly, the Tertiary–Quaternary boundary should correspond with the Newer Pliocene–Pleistocene boundary *sensu* Lyell (1873), whereas the Quaternary is equivalent to the term Post-Tertiary *sensu* Lyell (1857). The third and oldest option for placing the Tertiary–Quaternary boundary is at the base of Lyell's Newer Pliocene Epoch, which implies, a redefinition of Lyell's original subdivision of the Tertiary (Haug, 1908–1911).

In view of these options, current ideas about the age of the original localities should be taken into consideration. The Newer Pliocene marine strata are exposed north of Catania, along the east coast of Sicily, Southern Italy, and contain volcanic tuffs and basalts, which mark the first volcanic eruptions of Mount Etna, dated not much older than ~0.5 Ma (e.g. Tanguy *et al.*, 1997). This age estimate approximates that of the Günz glacial period (i.e. 0.6–0.55 Ma), which was dated astronomically by Köppen and Wegener (1924) using Penck and Brückner's (1909) climate curve and the insolation calculations of Milankovitch (1920). According to recent time scales, the original type sections of the Calabrian Stage, which cover the Newer Pliocene Epoch (*sensu* Gignoux, 1910, 1913), would actually fall within the middle Pleistocene, where the lower–middle Pleistocene boundary is taken at about the level of marine isotope stages 22–24 (i.e. younger than 0.9 Ma; see Rio *et al.*, 1991, and references therein). This implies that all other plausible historical definitions for the Tertiary–Quaternary boundary refer to deposits younger than 0.9 Ma.

Following a decade of study and discussion by the International Union for Quaternary Research (INQUA) Subcommission 1a on Stratigraphy (Pliocene–Pleistocene boundary) and International Geological Correlation Program Project 41 (Neogene–Quaternary boundary), a draft proposal on the choice of a boundary stratotype for the Pliocene–Pleistocene was submitted and approved by the INQUA Commission on Stratigraphy (also acting as the Subcommission on Quaternary Stratigraphy of the ICS) at the 1982 Moscow INQUA Congress. Three alternatives were discussed for placement of the Pliocene–Pleistocene boundary – one around the Brunhes–Matuyama reversal, approximating the base of Lyell's (1873) Newer Pliocene; the second around the top of the Olduvai subchron, approximately 1 myr older than Lyell's (1873) base of the Newer Pliocene; and a third option around the Gauss–Matuyama reversal, approximately 2 myr older than Lyell's (1873) base of the Newer Pliocene. The recommendations resulted in a formal proposal (cf., option 2) to define the base of the Pleistocene in the Vrica section (Calabria, Southern Italy), which was submitted to and approved by the ICS in 1983. This proposal was published two years later by Aguirre and Pasini (1985) together with the announcement by Bassett (1985) that the content of the proposal had been formally ratified by the IUGS Executive as the Global Stratotype Section and Point (GSSP) of the base-Pleistocene.

Notwithstanding formal ratification of the Pliocene–Pleistocene Boundary at Vrica, another decade and a half of heated discussions followed, ending with a formal attempt to lower the base of the Pleistocene in favor of the third option. However, such a definition ignores important chronostratigraphic concepts based on Lyell's original biostratigraphic subdivision of the Tertiary or on Desnoyers' lithostratigraphic definition of the Quaternary. And, instead, it equates the base of the Quaternary with the beginning of the "Ice Age," arbitrarily marked by the major onset of so-called northern hemisphere glaciations at 2.6 Ma, notwithstanding the fact that oldest indications for northern hemisphere glaciations in the Neogene now date back to 14 Ma (Wolf-Welling *et al.*, 1996; Thiede *et al.*, 1998). After a postal ballot within the subcommissions on Neogene Stratigraphy and Quaternary Stratigraphy (Rio *et al.*, 1998), the proposal to lower the base of the Pleistocene was rejected. In 1999, the original definition put forward by Aguirre and Pasini (1985) was reconfirmed by the ICS.

An important consequence of the historical development of the Pliocene–Pleistocene boundary is that the scientific community inherited two different notions for the Neogene. The first refers to the original concept of Hörnes (1853, 1864), according to which the Neogene includes the Miocene, Pliocene, and Pleistocene up to the Pleistocene–Holocene boundary. The second results from the choice made in London (1948) to equate the Tertiary–Quaternary with the Pliocene–Pleistocene boundary. Note, however, the confusion that arose from recommendations of the commission on the definition of the Pliocene–Pleistocene boundary. In their first recommendation:

> The commission considers that it is necessary to select a type area where the Pliocene–Pleistocene (Tertiary–Quaternary) boundary can be drawn in accordance with stratigraphic principles. [*Later on*] The Commission notes that according to evidence given this usage would place the boundary at the horizon of the first indication of climate deterioration in the Italian Neogene succession.

In other words, the Tertiary–Quaternary boundary should fall somewhere within the, apparently at that time, already generally accepted Neogene Period.

Chronostratigraphic confusion became even worse when the Tertiary – following the already obsolete Primary and Secondary – was abandoned from the IUGS-CGMW geologic time scale (e.g. Remane, 2000), resulting in an arbitrary definition of the Neogene–Quaternary Boundary in deposits originally termed (Lyell, 1833–1873) Tertiary. In the light of this decision, Berggren *et al.* (1995b) recommended abandonment of the term Quaternary from the geological time scale and extention of the Neogene Period up to the Recent, thereby including the Holocene in its definition. The Cenozoic Era is thus then divided into the Paleogene Period and the Neogene Period. We have adopted this concept in GTS2004.

21.1.2 The Miocene Series

The Miocene (Greek *meion*, "few") is the oldest epoch of the Neogene Period, and was originally proposed by Lyell (1833) as a biostratigraphic unit with more than 8% extant mollusc taxa. The Miocene has been subdivided in three subseries (Lower, Middle, and Upper).

AQUITANIAN: LOWER MIOCENE

The Aquitanian Stage (from *Aquitania*, the Latin name for Aquitaine) was introduced by Mayer-Eymar (1858) on the basis of largely lagoonal deposits in the Aquitanian Basin, France. He defined the stratotype near the village of Saucats (Gironde) in the Saint-Jean-d'Etampes valley, between La Brède and le Moulin de l'Eglise. The geographic location of the stratotype was specified and more precisely defined by Dollfus in 1909. The stratotype was ratified and confirmed at international conferences on the Mediterranean Neogene (at Aix-en-Provence, 1958; in Vienna, 1959; in Bologna, 1967).

Planktonic foraminifera from the type area place the beginning of the Aquitanian near the base of Zone N4 (of Blow, 1969) and the top in the lower part of N5 (i.e. Drooger *et al.*, 1976; Poignant *et al.*, 1997a). Calcareous nannofossils allow the recognition of Zone NN1 of Martini (1971; see Poignant *et al.*, 1997a). Strontium isotope data point to an age of close to 23 Ma for the base of the type Aquitanian according to CK92, although a continental imprint cannot totally be excluded (Cahuzac *et al.*, 1997).

Aquitanian GSSP, base-Miocene Series, base-Neogene System
Because sections in the stratotype area were not suitable for defining the boundary, other sections were investigated to select the most suitable section and boundary criterion for the lower limit of the Neogene Period, the Miocene Series, and the Aquitanian Stage. The boundary is now formally defined in middle bathyal, massive and laminated siltstones with several nodule levels in the Lemme–Carrosio section in northern Italy (Steininger *et al.*, 1997a,b). The publication by Steiniger *et al.* (1997a,b) on the GSSP pays attention to continental biostratigraphic correlations and regional chronostratigraphic equivalents; but it does not discuss the GSSP (bio)stratigraphy relative to the bordering Chattian or Aquitanian stages. The GSSP, located at the 35-m level as measured downward from the top of the section, corresponds closely with the calcareous nannofossils *Sphenolithus delphix* (its range) and *Sphenolithus capricornutus* (first occurrence, FO) and with the base of polarity Chron C6Cn.2n; the identification

of both has been confirmed through high-resolution biostratigraphic correlations to ODP Site 522 in the South Atlantic Ocean (Raffi, 1999; Shackleton *et al.*, 2000). The planktonic foraminifer *Paragloborotalia kugleri* FO is located 2 m above the boundary, and the *Globigerinoides* Acme, well known from deep-sea records outside the Mediterranean, is observed below the boundary. In addition, a diverse nannofossil (approximately base of Zone NN1) and dinoflagellate cyst record (base of *Cordosphaeridium canthaerellum* Zone) delineates the boundary interval; the *S. capricornutus* and *S. delphix* last occurrences (LOs) are pinpointed 1 and 4 m above the boundary, respectively.

The boundary corresponds approximately to oxygen isotope event Mi-1 and to the TB1.4 highstand of supercycle TB1 of Haq *et al.* (1987). Tuning of ODP Site 926 to the La2003 solution (see Section 21.7) resulted in an age for the base of the Miocene of 23.03 Ma. This age is 0.77 myr younger than in the Berggren *et al.* (1995b) time scale.

BURDIGALIAN: LOWER MIOCENE

The Burdigalian (from *Burdigala*, Latin name for the city of Bordeaux in Roman times) Stage was introduced by Depéret (1892) based on stratigraphic units in two distinct areas in France, namely near Bordeaux (lower part) and in the Rhône valley (upper part). Dollfus (1909) designated the Coquillat crag at Léognan as the Burdigalian stratotype in the Bordeaux area. Calcareous plankton studies (Poignant and Pujol, 1978; Müller and Pujol, 1979) indicate that this lower part of the Burdigalian belongs to the N5–6 Zones of Blow (1969), and to the NN2 Zone of Martini (1971). Demarcq and Carbonnel (1975) designated the St-Paul-Trois-Chateaux section as the stratotype for the (upper) Burdigalian in the Rhône valley (see Pouyet *et al.*, 1997).

Planktonic foraminifera indicate that the upper part of the Burdigalian starts in Zone N5 and ends in Zone N7. Martini (1988) placed the base of the historical stratotype in zone NN2 and the top in NN2–NN3. Strontium isotope stratigraphy provided an age range of 20.5–20.6 Ma, according to CK92, for the (lower) Burdigalian of Saucats, which constitutes the stratotypical formation (Cahuzac *et al.*, 1997) of the Burdigalian as described by Depéret (1892). No Sr isotope data are presently available for the upper Burdigalian.

Lower limit of the Burdigalian Stage No consensus exists about the criterion and hence the age for the Aquitanian–Burdigalian boundary. One option refers to the *P. kugleri* LO (N4–N5 transition) at 21.12 Ma (Berggren *et al.*, 1995a). A second option uses the *H. ampliaperta* FO (MNN2a–MNN2b transition), which corresponds to the top of the transitional interval between the Aquitanian and the Burdigalian (Fornaciari and Rio, 1996) dated at 20.43 Ma. A third option is the top of polarity Chron C6An (Berggren *et al.*, 1995a) dated at 20.04 Ma. Finally, Haq *et al.* (1987) used the *S. belemnos* FO (NN3–NN4 transition) dated at 19.03 Ma as criterion for the age of the Aquitanian–Burdigalian boundary (all ages according to our new time scale). We provisionally place the boundary coincident with the *H. ampliaperta* FO dated astronomically at 20.43 Ma at Ceara Rise, because this date lies close to the Sr estimated age (~20.0–20.1 Ma, CK92 recalibrated to our time scale) for the base of the Burdigalian in the stratotype area of Saucats, and because of the absence and presence of *H. ampliaperta* in the historical stratotypes of the Aquitanian and Burdigalian, respectively (Poignant *et al.*, 1997a,b).

LANGHIAN: MIDDLE MIOCENE

Based on marly to sandy successions exposed in (among others) the Bormida valley in the middle of the Langhe (Piedmont Basin, Italy), Pareto (1865) introduced the Langhian Stage for the middle part of the Miocene, above the now-discarded Bormidian and below the Serravallian (he considered the Tortonian as belonging to the Pliocene Series). The original concept was modified by Mayer-Eymar in 1868 who limited the term to the upper, mainly marly, part of the succession (the so-called "Pteropod Marls"). Since then, the "Pteropod Marls" of the Piedmont Basin (now known as the Cessolo Formation) have become synonymous with the Langhian and a stratotype section was selected, accordingly, near the village of Cessolo in the Bormida valley (Cita and Premoli Silva, 1960). The Langhian has long been held to be synchronous with the Burdigalian, but it is now clear that the type Burdigalian is considerably older than the type Langhian *sensu* Mayer-Eymar (1868; but not *sensu* Pareto, 1865).

Planktonic foraminiferal studies indicate that the first evolutionary appearance of *Praeorbulina glomerosa* occurs at the base of the Langhian and that the first evolutionary appearance of *Orbulina suturalis* (N8–9 boundary in the zonal scheme of Blow, 1969) is found in its upper part. Calcareous nannofossils (Fornaciari *et al.*, 1997) indicate that the base of the type Langhian predates the *Helicosphaera ampliaperta* LO and contains *Sphenolithus heteromorphus*, and hence falls within Zone NN4 of Martini (1971). The Langhian top lies within the *S. heteromorphus* range and thus belongs to Zone NN5 (Fornaciari *et al.*, 1997).

Gelati *et al.* (1993) considered the Langhian stratotype to represent a single depositional sequence, which they correlated on the basis of the planktonic foraminifera to third-order cycle 2.3, supercycle TB2, of the Global Cycle Chart of Haq *et al.* (1987).

Lower limit of the Langhian Stage, base-Middle Miocene The base of the Langhian and thus the Lower–Middle Miocene boundary is widely accepted to be approximated by the planktonic foraminifer *Praeorbulina* datum (see Rio *et al.*, 1997). It seems preferable to locate the Langhian GSSP in a position close to both Chron C5Cn and the *Praeorbulina* datum, in agreement with common and consolidated practice. However, the historical stratotype at Cessolo with terrigenous and turbiditic sediments in its lower part is less suitable for defining the GSSP. We provisionally place the Langhian GSSP so as to coincide with the top of C5Cn.1n dated astronomically at 15.974 Ma (this chapter).

SERRAVALLIAN: MIDDLE MIOCENE

The Serravallian Stage was introduced by Pareto (1865) in the same publication as that in which he introduced the Langhian. The stratigraphic position of the Serravallian in the Langhe type area is clearly defined at the lithological change from prevailing marls (of the Langhian) to dominantly sandstones (see Rio *et al.*, 1997). The Serravallian was soon abandoned in favor of the Helvetian defined by Mayer-Eymar in 1858. The validity of the Serravallian as a global chronostratigraphic unit was restored after it was realized (in the 1960s) that the Helvetian was time-equivalent to the Burdigalian. The Serravallian was proposed as the second younger division of the Middle Miocene above the Langhian and below the Tortonian at the RCMNS Congress held in Bratislava in 1975.

A stratotype section for the Serravallian was designated by Vervloet (1966) near the village of Serravalle Scrivia in agreement with the original definition of Pareto (1865). The planktonic foraminifera are poorly preserved and not very age diagnostic, although all authors agree that the Serravallian post-dates the *Orbulina universa* FO (see Rio *et al.*, 1997). Results of a high-resolution calcareous nannofossil study indicate that the Serravallian base corresponds almost perfectly with the *S. heteromorphus* LO located slightly higher in the section and marking the NN5–6 zonal boundary (see Rio *et al.*, 1997). Secondary calcareous nannofossil markers indicate that the fossiliferous lower to middle part of the type Serravallian belongs to Zone NN6 (Fornaciari *et al.*, 1996).

Lower limit of the Serravallian Stage The calcareous nannofossil *S. heteromorphus* LO is at present the most suitable guiding criterion to define the Serravallian GSSP. The *S. heteromorphus* LO event has been dated astronomically at 13.52 Ma in the tropical Atlantic (Backman and Raffi, 1997). So far, no proposal has been made for the GSSP. We provisionally place the Langhian–Serravallian boundary to coincide with the *S. heteromorphus* LO event in the Mediterranean, which has been dated astronomically at 13.654 Ma (Tremiti islands, unpubl. data).

TORTONIAN: UPPER MIOCENE

The Tortonian Stage was introduced by Mayer-Eymar (1858) as the "Blaue Mergel mit *Conus canaliculatus* and *Ancillaria glandiformis* von Tortona." A type section was designated by Gianotti (1953) in the valleys of Rio Mazzapiedi and Rio Castellania between the villages of Sant'Agatha Fossili and Castellania (see Rio *et al.*, 1997). Calcareous plankton zonal assignments for the lower part of the historical stratotype were conflicting for a long time (see Rio *et al.*, 1997) but recent studies indicate that the base of the type Tortonian corresponds almost exactly with the *N. acostaensis* FRO (first regular occurrence) in the Mediterranean, where this event has been dated astronomically at 10.57 Ma in the Monte Gibliscemi section on Sicily (Hilgen *et al.*, 2000c, this chapter). According to Hilgen *et al.* (2000c), this event post-dates the *N. acostaensis* FO in the Mediterranean by more than one million years. The base of the type Tortonian is approximated by the *D. hamatus* FO at Ceara Rise (10.55 Ma), but the same event is much younger in the Mediterranean (10.18 Ma). This strong diachroneity of the zonal marker events between low and mid latitudes complicates the assignment of the various stratigraphic levels in the Mediterranean to the standard calcareous plankton zonations.

Tortonian GSSP, base-Upper Miocene Biostratigraphic data indicate that there is a gap between the top of the type Serravallian and the base of the type Tortonian rendering all events in the interval between 11.8 and 10.55 Ma suitable for defining the boundary (Rio *et al.*, 1997; Hilgen *et al.*, 2000c). Several astronomically dated sections in the Mediterranean contain the boundary interval in a continuous deep-marine succession. The GSSP for the Serravallian–Tortonian boundary is at the midpoint of the sapropel of cycle 76 in the Monte dei Corvi section (northern Italy), a position recommended by the working group on the Serravallian and Tortonian GSSPs of the SNS. This level is dated astronomically at 11.600 Ma (Hilgen *et al.*, in press; this chapter) and coincides approximately with the calcareous nannofossil event *D. kugleri* LCO and the planktonic foraminiferal event *G. subquadratus* LCO, which are synchronous events between the mid and low latitudes. This GSSP was ratified in 2003.

MESSINIAN: UPPER MIOCENE

The Messinian Stage is named after the town of Messina on Sicily, southern Italy. It was Mayer-Eymar who in 1867–1868 introduced this name for a stage that would fill the gap between the Tortonian and Astian *sensu lato*, the latter being considered equivalent to the entire Pliocene at that time. Mayer-Eymar erected the Messinian without a detailed knowledge of the local stratigraphy of Sicily in order to compete with the Zanclean (the pre-Roman name for Messina) Stage introduced by Seguenza one year later (Seguenza, 1868). The original concept of the Messinian included diatomaceous marls at the base ("tripoli"), evaporites ("gesso"), and deep-marine marls with a diverse planktonic fauna ("trubi") on top. However, Seguenza (in 1868 and 1879) restricted the Messinian to the "tripoli" and the "gesso," considering the "trubi" to be basal-Pliocene. Owing to sedimentological and tectonic complications in the vicinity of Messina itself, Selli (1960) selected and described a neostratotype, exposed between Mt Capodarso and Mt Pasquasia (central Sicily, Italy). The section is bounded below by Tortonian marls and above by basal-Pliocene marls ("trubi"), and its microfauna has been described by d'Onofrio (1964).

Selli (1960) argued that the base of the Messinian coincides with the first marked change in the benthic microfauna, heralding the actual beginning of the Messinian salinity crisis. Since this criterion is difficult to export outside the Mediterranean, the *Globorotalia conomiozea* FO and the *Amaurolithus delicatus* FO are considered more suitable criteria to delimit the boundary (Colalongo *et al.*, 1979). Unfortunately the boundary interval is now obscured by landslides both in the neo-stratotype as well as in the Falconara section proposed by Colalongo *et al.* (1979) as an alternative boundary stratotype. Argon-40/argon-39 dating of biotites in volcanic ash layers yielded an age of 7.26 Ma for the boundary in northern Italy (Vai *et al.*, 1993), while astronomical tuning of sedimentary cycles produced an age of 7.24 Ma (Hilgen *et al.*, 1995). The section most suitable for defining the Messinian GSSP proved to be Oued Akrech, located on the Atlantic side of Morocco.

Messinian GSSP The GSSP for the lower limit of the Messinian Stage of the Late Miocene Epoch is in the Blue Marls of the Oued Akrech section, located 10 km south southeast of Rabat, Morocco, in a road-cut along a steep bluff, next to the Oued Akrech ("oued" = valley). The Neogene sediments at Oued Akrech were deposited in the Gharb Basin, which represent the westward extension and opening into the Atlantic of the Rifian Corridor from the Mediterranean to the Atlantic. High-resolution integrated stratigraphic correlations to the type area for the Messinian strata on Sicily are excellent and straightforward (Hilgen *et al.*, 2000a,b).

The basal part of the Blue Marls contains 20 color cycles, with the base of the reddish layer of cycle 15 proposed as the GSSP for the base of the Messinian Stage. The section contains an excellent and continuous faunal and magnetic polarity record across the boundary. The characteristic cycle pattern has been tied to the astronomical time scale, resulting in an astronomical age of 7.246 Ma (Hilgen *et al.*, 2000a,b; this chapter) for the Messinian GSSP. The GSSP coincides almost exactly with the first regular occurrence of the *Globorotalia miotumida (conomiozea)* group, and falls in the middle of Chron C3Br.1r of the geomagnetic polarity time scale. In terms of calcareous nannofossil biostratigraphy, the GSSP falls within Zone NN11b and Zone CN9b of Okada and Bukry (1980), with the genus *Amaurolithus* providing a series of useful events to delimit the boundary on a global scale. The *A. primus* FO predates the boundary, while the *A. delicatus* and *A. amplificus* FOs post-date the boundary. In terms of the planktonic foraminiferal biostratigraphy, the GSSP falls within the subtropical interval zone M13B, and marks the transitional Mt9–Mt10 zonal boundary of Berggren *et al.* (1995a,b). The latter coincides with the abrupt replacement of predominantly dextrally coiled *Globorotalia menardii* 5 with dominantly sinistrally coiled *G. conomiozea* group fauna. This event can be used to locate the GSSP both in the Mediterranean and in the North Atlantic.

In terms of stable isotopes, the boundary is located within the global Chron 6 carbon shift as recognized in the Oued Akrech section (Hodell *et al.*, 1994). The 1998 proposal for the Messinian GSSP was ratified in 2000 (Hilgen *et al.*, 2000a; also see SNS website, www.geo.uu.nl/SNS).

21.1.3 The Pliocene Series

In 1833, Lyell proposed the name Pliocene (Greek *pleion*, "more") for the youngest Tertiary deposits which he recognized at the time. He divided the Pliocene (more than 50% of molluscan species still living) into an "Older Pliocene" (50–90% of molluscan species still living) and a "Newer Pliocene" (90–100% of molluscan species still living) for which he subsequently introduced the name Pleistocene (Greek *pleisto*, "most"). Lyell regarded the "Sub Apennine" strata in northern

Italy in 1833 as typical for his "Older Pliocene" (i.e. the Pliocene *sensu stricto*). Since 2000, the Pliocene has been formally subdivided into three subseries: Lower, Middle, and Upper (Van Couvering *et al.*, 2000).

ZANCLEAN: LOWER PLIOCENE

The Zanclean Stage was defined by Seguenza (1868) as the early part of the Pliocene, to complement Mayer-Eymar's Astian Stage for the Upper Pliocene. The name derives from the pre-Roman name, Zanclea, for the modern town of Messina in Sicily, Italy. The classical Zanclean sediment is the white, rhythmically bedded foraminiferal ooze, then exposed in the Gravitelli valley, just northwest of Messina. These sediments belong to the Trubi Formation and locally reach a thickness of 15 m. Because the Gravitelli exposures no longer exist, Cita and Gartner (1973) proposed a Zanclean neostratotype in the coastal cliffs at Capo Rossello, along the south coast of Sicily, west of the town of Agrigento. Here the Trubi reaches a thickness in excess of 100 m.

Cita (1975a) in addition proposed to define the lower limit of the Zanclean and hence the Miocene–Pliocene boundary at the base of the Trubi marls at Capo Rossello. Because of the excellent magnetostratigraphy, Hilgen and Langereis (1988) proposed the correlative level at Eraclea Minoa as an alternative. The base of the Trubi marks the sudden catastrophic flooding of the Mediterranean at the end of the Miocene. The GSSP designation at Eraclea Minoa (see below) is unique in that it is proposed in a stratigraphically potentially incomplete sequence. Reasons for its selection are vested in the historical significance of the boundary for the traditional Pliocene, immediately above the Lago Mare strata. Since one of the aims of stratigraphy is to maintain stability and historical continuity wherever practical, the GSSP for the base of the Pliocene at Eraclea Minoa is considered highly appropriate.

Zanclean GSSP, base-Lower Pliocene, base-Pliocene Series The GSSP for the base of the Zanclean Stage, and the base of the Pliocene Series is at Eraclea Minoa, Agrigento Province, on the south coast of Sicily, Italy. The Eraclea Minoa section is the basal component of the Rossello Composite Section, which is the global reference section for the Early and Middle Pliocene part of the astronomical polarity time scale (Langereis and Hilgen, 1991; Lourens *et al.*, 1996a).

The GSSP at Eraclea Minoa is defined as the base of small-scale carbonate cycle number 1 of the Trubi marls, which locally overlie the brownish colored sandy Arenazzolo and massive gypsum of latest Messinian age with a sharp contact. The level corresponds to insolation cycle 510 with an astrochronologic age estimate of 5.332 Ma (Lourens *et al.*, 1996a; this chapter). The base of the Thvera (base of polarity Chron C3n.4n) with an astrochronological age of 5.235 Ma provides a very good approximation of the Messinian–Zanclean (Miocene–Pliocene) boundary.

Several nannofossil events are important for exporting the Zanclean GSSP outside the Mediterranean. The *Ceratolithus acutus* FO and *Discoaster quinqueramus* LO, calibrated at 5.35 and 5.58 Ma in the equatorial Atlantic (Backman and Raffi, 1997; this chapter), directly precede the level of the GSSP in the open ocean. The GSSP in turn predates the *C. rugosus* FO, calibrated at about 5.05 Ma in the western equatorial Atlantic by 285 000 years, and thus falls within standard zone NN12 of Martini (1971). The first appearances of *Globorotalia tumida* have been calibrated at 5.72 Ma, or about 385 kyr earlier than the GSSP, in tropical to warm-temperate sections outside the Mediterranean. Hence, the boundary falls within Zone PL1 of Berggren *et al.* (1995a). The proposal for the Zanclean GSSP by Van Couvering and co-workers in 1998 (see SNS website, www.geo.uu.nl/SNS) was ratified in 2000 (Van Couvering *et al.*, 2000).

PIACENZIAN: MIDDLE PLIOCENE

The Piacenzian Stage was introduced by Mayer-Eymar in 1858 (as the "Piacenzische Stufe") for the marly–clayey ("Argille Azurre") facies of the Lower Pliocene in northern Italy. Pareto (1865) adopted the name, using the French equivalent Plaisancian, and specified that the typical development is found in the hills around the village of Castell'Arquato. The name itself is derived from the city of Piacenza, Italy, and some 25–30 km from Castell'Arquato. Barbieri (1967) designated a stratotype in the Castell'Arquato section, following the indications by Pareto (1865). Initially, the stage was locally defined between the LO of *Globorotalia margaritae* and a prominent calcarenitic bed containing *Amphistegina* spp. on top of which Castell'Arquato is built. Subsequent studies (Rio *et al.*, 1988) demonstrated the presence of an hiatus at the base of the Piacenzian in the type area, and hence no GSSP could be defined there.

Piacenzian GSSP, base-Middle Pliocene The GSSP for the lower limit of the Piacenzian Stage, Middle Pliocene Series, is located in the Punta Piccola section, along the road from Porto Empedocle to Realmonte, 4 km east of Cabo Rossello, Sicily, Italy. The Rossello Composite Section, of which the Punta

Piccola section is the upper segment, is an important reference standard for the Pliocene time scale.

The Piacenzian GSSP and, hence, the Lower to Middle Pliocene boundary, is defined at the base of the beige marl of small-scale carbonate cycle 77 with an estimated astrochronological age of 3600 Ma (Lourens *et al.*, 1996a; this chapter). The primary correlation tool for the base-Piacenzian is the Gilbert–Gauss magnetic reversal of polarity Chron 2Ar–2An, which occurs immediately above the "golden spike," with an age estimate of 3.596 Ma.

As pointed out by Castradori *et al.* (1998), Pliocene biostratigraphic events may be diachronous on a global scale, which is why different fossil events are used in different regions to ascertain the base of the Piacenzian Stage. In the Mediterranean region, this level may be correlated using the temporary disappearance of the planktonic foraminifer *Globorotalia puncticulata* at 3.57 Ma, the first influx of *G. crassaformis* at 3.60 Ma, the end of the calcareous nannofossil *Discoaster pentaradiatus* acme interval at 3.61 Ma, and the *Sphenolithus* spp. LO at 3.70 Ma. The GSSP post-dates the *Reticulofenestra pseudoumbilicus* LO and thus falls within Zone NN16 of Martini (1971). The planktonic foraminifer *Hirsutella* (*Globorotalia*) *margaritae* LO, dated at 3.88 Ma, provides a good approximation for the base of the Piacenzian in open-ocean sediments from low to mid latitudes outside the Mediterranean. Hence, the GSSP coincides closely with the PL2–3 zonal boundary of Berggren *et al.* (1995a). The GSSP for the base of the Piacenzian was ratified in 1997, and published by Castradori *et al.* (1998).

GELASIAN: UPPER PLIOCENE

The Gelasian Stage is derived from the Greek name of the town of Gela (southern Sicily, Italy), close to the boundary stratotype section of Monte San Nicola, and spans the interval between the Middle Pliocene Piacenzian and the Lower Pleistocene Calabrian. In 1994, Rio and co-workers argued against the practice of extending the Piacenzian up to the Pliocene–Pleistocene boundary. They introduced a new stage, the Gelasian because none of the then existing standard Pliocene stages covered the interval between the top of the Piacenzian historical stratotype at Castell'Arquato and the Pliocene–Pleistocene boundary as defined at Vrica. The Gelasian marks a crucial period in the evolution of the Earth's ocean–climate system due to the occurrence of major obliquity-controlled northern hemisphere glacial cycles.

Gelasian GSSP, base-Upper Pliocene The sedimentary succession that includes the Gelasian GSSP at Monte San Nicola (Sicily, Italy), contains a cluster of six sapropels, which can be recognized throughout the Mediterranean, and are known as Mediterranean precession related sapropels (MPRS) or insolation-related cycles (i-cycles) 250–260 (Hilgen, 1991b; Lourens *et al.*, 1996a, 2001). The base of the homogeneous marl overlying sapropel MPRS 250, located at 62 m in the Monte San Nicola section defines the base of the Gelasian Stage, with an astronomically derived age of 2.588 Ma (this chapter). The base of the Gelasian falls within isotopic stage 103, predating the prominent glacial stage 100 by about 60 kyr (Lourens *et al.*, 1992). The Gauss–Matuyama reversal boundary has been pinpointed slightly above MPRS 250 in the Singa section (Langereis *et al.*, 1994) and provides an excellent approximation and correlation tool for the "golden spike." The *Discoaster pentaradiatus* and *D. surculus* LOs in low- and mid-latitude sites occur close to isotopic stage 99, about 80 kyr above the boundary. The GSSP thus falls in the top part of Zone NN16 of Martini (1971). The planktonic foraminifer *Globorotalia bononiensis* and *Neogloboquadrina atlantica* LOs are pinpointed at the top of isotope stage 96, both in the Mediterranean as well as in the Atlantic, i.e. some 140 kyr above the "golden spike." The GSSP falls within Zone PL5 of Berggren *et al.* (1995a).

The Gauss–Matuyama boundary and, thus, the Gelasian GSSP is approximated by the LO of the radiolaria *Stichocorys peregrina*, coincident with the base of the *Pterocanium prismaticum* Zone, and by the diatoms *Nitschia joussaea* FO at low latitudes and the *Denticulopsis kamtschatica* LO in the North Pacific mid and high latitudes. The GSSP for the base of the Gelasian was ratified in August 1996, and published by Rio *et al.* (1998).

21.1.4 The Pleistocene Series

In 1839, Lyell introduced the term Pleistocene as an abbreviation for his "Newer Pliocene" (see also Section 21.1.1 and the previous discussion on the Pliocene Series).

BASE OF THE PLEISTOCENE SERIES

The GSSP for the Pliocene–Pleistocene boundary and the base of the Pleistocene was fostered and conceived within the INQUA and ICS working group (IGCP Project 41) on the Pliocene–Pleistocene boundary. With the acceptance that the beginning of the Pleistocene no longer has a primary climatic implication, since significant climatic change cuts across time as a function of latitude, the road was paved for the acceptance of a GSSP for the Pliocene–Pleistocene boundary in the

Vrica section located in southern Italy (Aguirre and Pasini, 1985).

The Vrica section consists of bathyal, marly, and silty claystones with inter-bedded, pink–gray sapropelic marker beds. The boundary is at the base of homogeneous marls overlying the 190-cm thick sapropel bed "*e*," close to the top of Chron C2N (Olduvai). Paleontological criteria suitable for widespread correlations outside the Mediterranean are the *Discoaster brouweri* LO below the boundary and the medium-sized *Gephyrocapsa* (incl. *G. oceanica*) FO and the FCO of sinistrally coiled *Neogloboquadrina pachyderma* (Zijderveld *et al.*, 1991; Lourens *et al.*, 1996a,b, 1998; Lourens and Hilgen, 1997) directly above the boundary. The *Globigerinoides obliquus extremus* LO is located just below the boundary. In shallow-marine sections in the Mediterranean, the boundary can be traced via the arrival of the mollusc *Arctica islandica*, a northern immigrant.

The GSSP corresponds to the base of Zone MPL7 and falls within Zone PT1. It also falls within Zones NN19A and CN13. Astronomical calibration of the Vrica sapropels provides an age of 1.806 Ma for the Pliocene–Pleistocene boundary (Lourens *et al.*, 1996a,b; this chapter).

SUBDIVISION OF THE PLEISTOCENE SERIES

Working groups were organized by the ICS in 2003 under the umbrella of the SQS to clarify the state of confusion concerning the subdivision of the Pleistocene (see Chapter 22). It is planned to subdivide the Pleistocene formally into three subseries (Lower, Middle, and Upper), leaving "stages" as a general term for marine isotopic oscillations, glacial episodes, and regional stratigraphy. This is quite different from the approach proposed by the Italian Commission on Stratigraphy (2002) to establish standard chronostratigraphic subdivisions of the Pleistocene. The Italian Commission on Stratigraphy proposed using essentially the same approach as for the pre-Pleistocene part of the Cenozoic, namely (i) fixing the boundary stratotype by their lower boundary, (ii) following the principles of "base defines boundary," (iii) fixing the boundary stratotypes in marine sediments, (iv) respecting the historical usage (albeit not a true historical priority), and (v) focusing as much as possible on the best correlation potential. In their view, the stage would still be considered as the basal unit of the Pleistocene chronostratigraphic scale.

The only GSSP existing in the Pleistocene is the one defining its base in the Vrica section. No official subdivision into units has been established by the ICS, although the literature is rich in informal schemes that often enjoyed wide acceptance. In this scenario of widespread informality, there are three noteworthy exceptions: the "Calabrian" Stage and the Lower–Middle and Middle–Upper Pleistocene boundaries.

THE "CALABRIAN" STAGE: LOWER PLEISTOCENE

The Calabrian Stage was introduced by Gignoux (1910, 1913) as a substitute for Doderlein's (1870–2) Sicilian and spans Lyell's Newer Pliocene. Gignoux's Calabrian was publicized in Haug's (1908–1911) textbook, and included an important modification: the relocation of the base of the Quaternary to the base of the Calabrian. Recent studies argued that the base of most sections indicated by Gignoux as Calabrian was in fact younger than the base of Gignoux's supposedly overlying Sicilian Stage (Rio *et al.*, 1991). More precisely, taking the Lower–Middle Pleistocene boundary at about the level of isotope stage 22–24, it would actually be of Middle Pleistocene age. Thus Calabrian, unless re-defined, would be unacceptable as the base of the Lower Pleistocene. However, in agreement with the legendary recommendations adopted at the IGC in London (1948) on the definition of the Pliocene–Pleistocene boundary, the Calabrian Stage is considered by some as already defined by the GSSP in the Vrica section (e.g. Vai, 1997; Van Couvering, 1997).

THE LOWER–MIDDLE PLEISTOCENE BOUNDARY

As stated above, the SQS favors a three-fold subdivision of the Pleistocene (into Lower, Middle, and Upper). The Lower–Middle Pleistocene boundary already enjoys almost official status. According to Richmond (1996), the Brunhes–Matuyama paleomagnetic boundary was selected as the primary criterion for the adoption of this boundary by the INQUA Commission on Stratigraphy. However, no formal proposal has been put to the vote within the SQS and ICS.

Other workers (see Cita and Castradori, 1995) proposed a lower position (between the top of the Jaramillo event and the Brunhes–Matuyama boundary), and the related events (e.g. base of the *Pseudoemiliania lacunosa* nannofossil zone), as more suitable for the selection of the boundary. In this respect, the Ionian Stage was proposed for the Middle Pleistocene by Carianfi and co-workers (2001), referring to the upper part of a thick sequence of Lower and Middle Pleistocene marine silts at Montalbano Ionico in southern Italy (see Van Couvering, 1995, 1997). Preliminary studies have shown that sedimentation was continuous from "the "middle Matuyama" to the "middle Brunhes," within the framework of large-sized (>5.5 μms) *Gephyrocapsa*, small-sized *Gephyrocapsa*, and *Pseudoemiliana lacunosa* calcareous nannoplankton zones (Rio *et al.*, 1990a). The boundary stratotype of the Ionian was proposed at the level of the last warm, marine transgression event prior to the Menapian "glacial Pleistocene" at 0.9 Ma. This is MIS 25, which occurs immediately above the Jaramillo event and correlates with the base of the *P. lacunosa* Zone (Castradori, 1993). This proposal has not reached official status. A task group has been established by the SQS to establish where and how to place a GSSP for the Lower–Middle Pleistocene boundary.

THE MIDDLE–UPPER PLEISTOCENE BOUNDARY

The Middle–Upper Pleistocene boundary has generally been equated with the beginning of the last interglacial period (i.e. Eemian, Mikolino) or Marine Isotope Stage 5. The first formal usage of the boundary was probably made by Zeuner (1935, 1959), who used Köppen and Wegener's (1924) astronomical correlation of Penck and Brückner's (1909) climate curve to the insolation calculations of Milankovitch (1920). This resulted in an age of ~180 000 years BP for the Middle–Upper Pleistocene boundary. In 1962, the age of the boundary was revised by Woldstedt to 120 000 years BP. The inequality in the duration of the Pleistocene chronostratigraphic units led West (1968) to propose a lowering of the Middle–Upper Pleistocene boundary, thereby including the terrestrial Saalian and equivalent stages, but this proposal has not found wide acceptance. A working group has been established to evaluate the different options for placing the boundary (see also the SQS website at www.quaternary.stratigraphy.org.uk). A new proposal has been presented by Gibbard (2003) in which he recommends defining the Middle–Upper Pleistocene boundary (i.e. the Saalian–Eemian boundary) at 63.5 m below surface in the Amsterdam-Terminal borehole (van Leeuwen *et al.*, 2000).

21.1.5 The Holocene Series

Lyell (1857) used "Recent" for the duration of time since the appearance of "modern" (Present and Neolithic) humans. At the third International Geological Congress (IGC) in 1885 the term Holocene (Greek *holo*, "whole") introduced by Gervais in 1867 was adopted as a substitute for Lyell's (1857) "Recent."

BASE OF THE HOLOCENE SERIES

The base of the Holocene Series is informally defined at the end of the last glacial period and dated at 10 000 radiocarbon years. It was proposed at the eighth INQUA Congress in Paris in 1969 to define the Pleistocene–Holocene boundary in a varved lacustrine sequence in Sweden (see Mörner, 1976). The INQUA Holocene Commission accepted the original proposal in 1982 (Olausson, 1982), but it was not submitted to the ICS. Alternatively, it has been proposed that the boundary is defined precisely at 10 000 (^{14}C yr BP), making it the first stratigraphic boundary later than the Proterozoic to be defined chronometrically (Mangerud *et al.*, 1974). As a third alternative, it has been proposed that the GRIP ice-core should constitute the stratotype for the Last Termination, thereby encompassing the Pleistocene–Holocene boundary (Björk *et al.*, 1998). According to this classification, the age of the base of the Holocene would be assigned as 11 500 calendar years BP based on counting annual ice layers back to 14.5 kyr (Dansgaard *et al.*, 1993). Nevertheless, it is still possible that the boundary will eventually be placed in an annual-laminated lacustrine sequence, as found in western Germany (Litt *et al.*, 2001), because such sequences yield an excellent, fossil record, principally pollen and other freshwater microfossils, which facilitates regional biostratigraphic correlation.

21.2 NEOGENE STRATIGRAPHY

21.2.1 Macrofossil zonations

In contrast to Paleozoic and Mesozoic fauna (e.g. trilobites, graptolites, brachiopods, ammonites), invertebrate macrofossils are of no importance for constructing standard biozonal schemes in the Neogene. However, they do play an important role for correlations on a regional scale especially when other microfossil groups are rare or absent, such as in shallow-marine or brackish-water environments. A good example is the Para-Tethys where mollusc zonations have been erected for different basins to cope with endemic species evolution and distributions, which result from intermittently interrupted connections (Magyar *et al.*, 1999a,b). For the eastern Para-Tethys, ecozonal schemes have been developed, which employ multiple invertebrate groups at the same time as a kind of assemblage of zones to avoid the problem of endemism and strong environmental control. Other areas where invertebrate macrofossil zonal schemes have been constructed and are successfully applied include the North Sea Basin, New Zealand, and the Arctic (e.g. Thomsen and Vorren, 1986; Roe, 2001). The uncertainty in the ages of most zonal boundaries is often considerable.

VERTEBRATES

Mammals are by far the most useful organisms for dating and correlating Cenozoic terrestrial sediments, because of their rapid evolution and abundance. The general nature of the mammal record, however, is different from that of most marine fossil records. The continental record is generally less continuous, and mammal localities are typically isolated field occurrences. For example, karst localities contain very rich fauna, but lack almost any additional stratigraphic control. During the last decades an increasing amount of magnetostratigraphic and radiometric dated mammal–faunal sequences have been documented, allowing studies of faunal change at a temporal resolution as fine as 100 000 years (Daams *et al.*, 1999; Barry *et al.*, 2002). A recent development involves first-order calibrations of mammal localities to continental sections that are astronomically tuned (Abdul-Aziz, 2001).

The usefulness of detailed regional and continental mammal-based chronological systems is not always evident because of the complexity of biogeographic distributions. Provinciality and diachroneity and the problem of relative scarcity of localities causes an ongoing discussion on the philosophical and practical aspects of mammal-based biostratigraphic principles (e.g. Lindsay and Tedford, 1990; De Bruijn *et al.*, 1992; Alroy, 1998). Nevertheless, basal units have been distinguished, conforming to the strict biostratigraphic, chronostratigraphic, and geochronologic principles in international stratigraphic guides (Hedberg, 1976; North American Commission on Stratigraphic Nomenclature, 1983; Salvador, 1994), or based on a faunal approach taking specific characteristics of the total assemblage into account, which can be evolutionary-controlled morphological changes in multiple lineages, presence/absence, and abundance patterns (see Van Dam *et al.*, 2001). A faunally based system may consist of either a discrete set of reference fauna (De Bruijn *et al.*, 1992) or of a sequence of statistically derived clusters of fauna of similar composition separated by faunal breaks (Alroy, 1992, 2003). In general, the biostratigraphic method works well for local sequences, but for larger geographic areas where the recognition of biozones and their chronostratigraphic equivalents (including geologic stages) is limited, a faunal approach appears to be more useful.

Both biostratigraphic and faunal approaches have been applied to arrange the mammal record in Europe, resulting in Neogene continental stages with stratotype designations in Spain (Marks, 1971a,b) and a European Tertiary Mammal zonation (Thaler, 1965), respectively. In 1975, Mein introduced the Mammal Neogene (MN) zonation (Mein, 1975; Mein *et al.*, 1990; last update, Mein, 1999) based on a combination of features such as the presence of selected taxa and first appearances (FAs). During the International Symposium on Mammalian Stratigraphy of the European Tertiary in Munich 1975, reference localities were explicitly attached to the MN units (Fahlbusch, 1976). At present, there is still ongoing debate regarding the exact definition of the MN units, because the same MN designation has been adopted for both stratigraphically defined systems (e.g. Steininger, 1999; Agustí *et al.*, 2001) and systems strictly defined on the basis of reference fauna (De Bruijn *et al.*, 1992). Similarly, the term "Age" has been used both for units based on stratotypes and for faunal complexes (such as European land mammal ages, or ELMAs, Sen, 1997).

Discussions about European mammal-based chronological systems are very similar to and partly echo those of North American systems. A standard framework was erected by the Wood Committee (Wood *et al.*, 1941), who defined a series of North American Land Mammal "Ages" (NALMAs), which were in fact complex hybrids of local rock units and mammalian events (Prothero, 1995). From the 1950s onwards, efforts have been undertaken to re-define the boundaries of the NALMAs stratigraphically on the basis of range zones (e.g. Woodburne and Swisher, 1995). This approach has been accompanied by quantitative faunal approaches, which use calculated event and faunal sequences to delineate mammal–chronological units (Alroy, 1992, 2003). A large number of North American Neogene K–Ar dates exists (Tedford *et al.*, 1987), but age estimates of many, particularly Miocene, mammal localities can still be improved by including more $^{40}Ar/^{39}Ar$ and paleomagnetic data.

The construction of mammal-based chronological systems for other continents still lags behind those for Europe and North America. The systems of South American Land Mammal Ages (SALMAs) and Chinese Neogene Mammal Units (NMUs) are currently defined on the basis of reference fauna, with chronologies becoming more and more refined by results of numerical dating (Flynn and Swisher, 1995; Qiu *et al.*, 1999). Preliminary faunal units have also been defined for Africa (e.g. Pickford, 1981). The nature of the Australian mammalian record is patchy (Archer *et al.*, 1995), and presently no separate Neogene mammal-based scale has been created.

21.2.2 Microfossil zonations

FORAMINIFERA

Petroleum exploration in the Caribbean fueled the first serious attempts to establish a planktonic foraminiferal biostratigraphy with the purpose of differentiating between intervals in the (sub)tropical marine stratigraphic record of the Neogene (Cushman and Stainforth, 1945). These efforts culminated during the late 1960s and early 1970s in a variety of low-latitude zonal schemes (Bolli, 1966; Blow, 1969, 1979;

Postuma, 1971; Stainforth *et al.*, 1975). In particular, the N-zonal scheme of Blow (1969, 1979), which has been modified and refined by Srinivasan and Kennett (1981) after its publication (see Bolli and Saunders, 1985, for an overview), gained wide acceptance. Since the beginning of the Deep Sea Drilling Project in 1968 knowledge about distribution patterns in time and space and of taxonomy and phylogeny rapidly expanded. Syntheses are given by Kennett and Srinivasan (1983) and Berggren *et al.* (1985c, 1995a). Application of the low-latitude standard zonations proved difficult outside the tropical region and attempts were soon made afterwards to establish regional zonal schemes for the middle and high latitudes (e.g. Jenkins, 1967, 1993) and Mediterranean (Bizon and Bizon, 1972). Elaborating on these zonal schemes (Berggren 1973; Berggren *et al.*, 1995a) introduced a PL, PT, and a M-zonal scheme for (sub)tropical regions and a Mt-scheme for the temperate-transitional regions of the open ocean, where PL stands for Pliocene, PT for Pleistocene, and M for Miocene. These zonations do not represent new zonal schemes but should be regarded as "unification of existing schemes with a view of providing improved biochronologic subdivisions reflecting regional biogeographies."

Both Blow's N-zonal scheme as well as the PL, PT, and M-scheme of Berggren are accepted in this chapter as standard zonations for the (sub)tropical open ocean in the present paper (Fig. 21.1). Berggren *et al.* (1995b) revised the existing calibration of Neogene planktonic foraminiferal datum planes to the geomagnetic polarity times scale of Cande and Kent (1995) by using reliable magnetostratigraphic records of a large number of DSDP and ODP cores as well as land-based marine successions from the Mediterranean. The Mediterranean is of special importance because most of the global chronostratigraphic units of the Neogene are currently defined there and recognition of these boundaries elsewhere relies at least partly on calcareous plankton biohorizons.

For the Pliocene–Pleistocene of the Mediterranean, the MPl zonal scheme of Cita (1973, 1975b) is generally accepted. The establishment of detailed planktonic foraminiferal biostratigraphies and biochronologies for continuous and astronomically dated successions in the Miocene (e.g. Hilgen *et al.*, 1995, 2000a–c; Sierro *et al.*, 2001) resulted in the recent publication of an MMi zonation for the middle Miocene interval (Sprovieri *et al.*, 2002b), which elaborates upon earlier schemes of Iaccarino (1985) and Foresi *et al.* (1988). In this chapter, we extend the MMi codification to the top of the Miocene following the zonal scheme of Iaccarino (1985). At present, the zonation for the Mediterranean has not been completed for the older part of the Miocene. Zonal boundary positions are taken from Hilgen *et al.* (1995, 2000c) and Sierro *et al.* (2001). Detailed studies reveal that biostratigraphic resolution on a regional scale can be considerably improved by also taking into account changes in coiling direction, short-term influxes, and different morphologies (ecophenotypes; Sierro *et al.*, 2001).

CALCAREOUS NANNOFOSSILS

Calcareous nannofossils are extremely useful for Neogene stratigraphic correlations and age determination of marine outcrop and exploration well samples, including deep-sea sediments. A general background to calcareous nannofossils is presented in Chapter 20. Standard zonal schemes based on continuous successions – and not isolated samples – were developed in the beginning of the 1970s and in line with common practice we adopt these essentially low-latitude zonations of Martini (1971: NN1-21) and Bukry (1973, 1975a: CN) as standard (Fig. 21.1).

In addition, regional zonations were developed due to the limited utility of the standard low-latitude zonations in certain areas such as the mid–high latitudes in the open ocean and marginal basins such as the Mediterranean. In particular, regional zonal schemes were developed to avoid problems, such as the strong diachroneity of zonal boundary events and/or the absence or extreme rareness of primary marker species. For instance, an effective MNN zonal scheme has been established for the Mediterranean Neogene (Raffi and Rio, 1979; Rio *et al.*, 1990a; Fornaciari and Rio, 1996; Fornaciari *et al.*, 1996; Raffi *et al.*, 2004).

DIATOMS AND RADIOLARIA

Diatoms are photosynthetic, single-celled algae that flourish in many (sub)aquatic environments. Sessile living taxa exist but, together with the coccolithophorids, their free-floating taxa constitute the major part of the marine phytoplankton. The use of diatoms for biostratigraphic purposes is restricted to areas with essentially continuous preservation of biogenic silica particularly in the low-latitude and high-latitude open ocean. Modern biostratigraphic studies started in the mid 1920s from the rich diatomaceous successions exposed in California. But it has been the recovery of oceanic sediments from DSDP and ODP drilling legs that greatly facilitated Neogene diatom biostratigraphic studies and led to the currently existing zonations. The numerous zonations published document the continuous development and improvement made during the last 30 years.

For the low-latitude ocean, no unified standard zonation exists and so attention focuses mostly on the equatorial Pacific.

Following earlier studies of, for example, Burckle (1972), Barron (1985) established the zonal standard for this region based on ODP Leg 85 sites. Baldauf and Iwai (1995) published a slightly modified version based on ODP Leg 138 sites with 17 zones and 17 subzones for the last 20 myr. The excellent magnetostratigraphy of Leg 138 sites (Schneider, 1995) resulted in a very reliable and straightforward calibration of the revised diatom zonal scheme to the geomagnetic polarity time scale (GPTS) of Cande and Kent (1995) for the last 13 myr.

Regional zonations have been established for the high-latitude open ocean, in particular the different Atlantic, Indian, and Pacific sectors of the Southern Ocean. Here the strong latitudinal control on the diatom floral composition associated with the position of the polar front necessitates the construction of different zonal schemes in these regions. For instance, versions of the Neogene diatom zonation for the Atlantic sector of the Southern Ocean were recently published in two papers dedicated to the Plio–Pleistocene and Miocene biostratigraphies of ODP Leg 113 and 177 sites (Censarek and Gersonde, 2002; Zielinski and Gersonde, 2002).

The use of radiolaria for biostratigraphy, as with diatoms, is restricted to areas with sufficiently continuous preservation of biogenic silica. The potential use of radiolaria for biostratigraphic purposes was first fully realized by Riedel (1957) when he studied long deep-sea cores from the Swedish Deep-Sea Expedition. Regional zonal schemes were subsequently developed for different oceanic basins both for low latitudes as well as for mid–high latitudes. Over the years, the linkage between Neogene radiolarian biostratigraphic events and the GPTS has much improved due to the recovery of deep-sea cores with a reliable magnetostratigraphy thus allowing for a straightforward magnetostratigraphic calibration of these cores to the GPTS. Recently Sanfilippo and Nigrini (1998b) merged the existing regional zonations for the tropical Pacific, Atlantic, and Indian Oceans (Sanfilippo *et al.*, 1985; Johnson *et al.*, 1989; Moore, 1995) into a single zonal scheme and introduced formal code numbers for the radiolarian zones (RN1–17 for the Neogene) as has been done for calcareous microfossil groups (Fig. 21.2). The incorporation of radiolarian biostratigraphy in integrated zonal schemes is crucial to their application in areas where other microfossil groups are biostratigraphically of limited utility.

DINOFLAGELLATES

The biology and main types of dinoflagellates were reviewed in Chapter 20. Applications of fossil dinocysts in global Neogene biostratigraphy and paleoecology has been reviewed in detail in several papers including Williams and Bujak (1985), Powell (1992a), Stover *et al.* (1996), Mudie and Harland (1996) and Williams *et al.* (1998a; in press). These compilations indicate that the stratigraphic range of a given Neogene dinocyst is rarely synchronous world-wide. Many authors have demonstrated climatic and environmental control on the stratigraphic distribution of taxa in the Neogene (see, e.g. reviews in Mudie and Harland, 1996; Stover *et al.*, 1996; Versteegh, 1997; Rochon *et al.*, 1999).

The bulk of currently available information on Neogene dinoflagellates is derived from the northern hemisphere, and most notably from the Atlantic and Mediterranean domains. Compilations also include the US Atlantic coastal plain (de Verteuil and Norris, 1996) and southern North Sea Basin (e.g. Louwye *et al.*, 1999, 2000). Neogene southern hemisphere dinocyst studies are but few; only sporadically representatives of this fossil group have been found at all (see overviews in McMinn, 1992, 1993; Wrenn *et al.*, 1998; McMinn *et al.*, 2001). McMinn (1995) noted the apparent absence of "Neogene dinocysts" in the circum Antarctic realm, and postulated that the loss of shallow-water shelves (due to glaciation) from the Antarctic continent from the basal-Oligocene onwards contributed to this "local extinction" of the cyst-producing dinoflagellates. Rather, later studies (e.g. Wrenn *et al.*, 1998) indicate that the organic wall of the dinocysts is not resistant to the oxygen-rich waters in the Antarctic domain, and/or winnowing at depth and/or low sedimentation rates preclude these microfossils from being preserved (see also, e.g. Versteegh and Zonneveld, 2002). Southern Ocean dinocyst assemblages should be present when preservation requirements are met. The study by McMinn and co-workers, analyzing Pliocene dinocysts and diatoms from DSDP Site 594, Chatham Rise, illustrates this aspect (McMinn *et al.*, 2001). Moreover, McMinn and Wells (1997) demonstrated the potential of dinocyst analysis for paleoenvironmental reconstructions in the Quaternary in the region, using piston-cored materials from locations offshore western Tasmania. The first more comprehensive stratigraphical studies from the Neogene of the southern hemisphere are now becoming available (e.g. Wrenn *et al.*, 1998; Harland and Pudsey, 2002; Brinkhuis *et al.*, in press).

Williams *et al.* (in press) recognized the need to accommodate both latitudinal and hemispherical control of dinocyst assemblages in Neogene distribution charts. Accordingly, these authors give ranges for low, mid, and high latitudes in both northern and southern hemispheres, and the contribution also updates data presented in Williams *et al.* (1998a). An overview of index dinoflagellate cyst events of northwest Europe is summarized in Fig. 21.2.

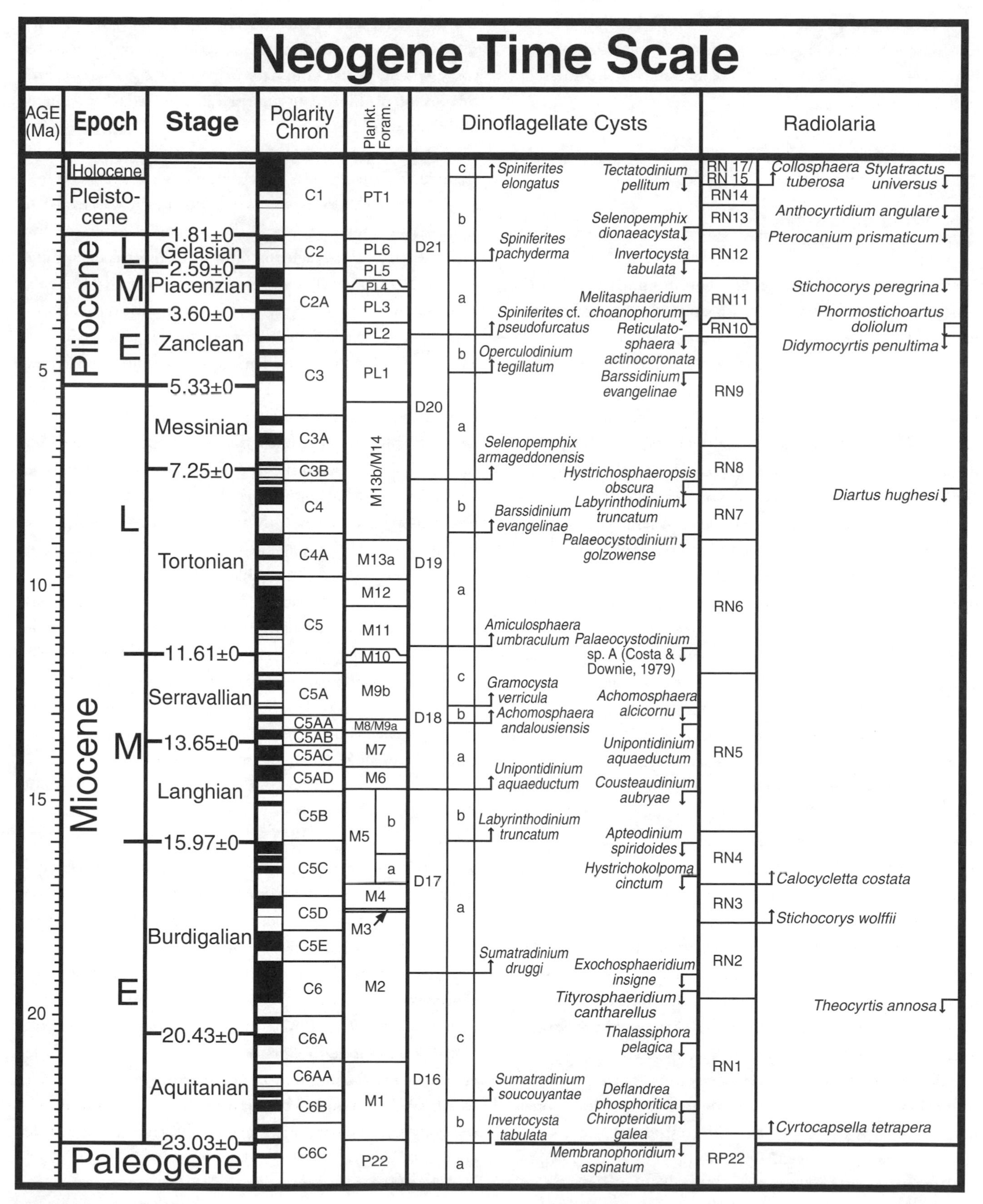

Figure 21.2 Neogene dinoflagellate cyst zonation and datums and Neogene radiolarian zonation, with estimated correlation to magnetostratigraphy and planktonic foraminifer zones. Dinoflagellate stratigraphy was compiled by A. J. Powell and H. Brinkhuis. A color version of part of this figure is in the plate section.

Dinocyst events that have been derived from astronomically dated sequences are rare, and have only regional applicability (e.g. Versteegh, 1994, 1997; Santarelli, 1997; Santarelli *et al.*, 1998).

21.2.3 Event stratigraphy

POLARITY SEQUENCES: ANOMALY PROFILES AND MAGNETOSTRATIGRAPHY

Marine magnetic anomaly profiles have been extremely successful in unraveling the reversal history of the Earth's magnetic field over the last 160 myr. Conversion of reversal distance to spreading ridge to time is achieved by the introduction of a limited number of radioisotopically controlled age calibration points and linear interpolation or cubic-spline fitting of the spreading rates between calibration points. For the last 5 myr, reversals have also been dated directly by combining polarity and radioisotopic age of volcanic samples collected throughout the world (e.g. Mankinen and Dalrymple, 1979). Directly dated polarity chrons were named after a renowned scientist in the study of the Earth's magnetic field, while subchrons were named after the locality where their existence was first proven unambiguously.

The standard nomenclature for Cenozoic marine magnetic anomalies and associated polarity chrons is based on numbering prominent positive anomalies from the ridge axis back to Anomaly 34 (Pitman *et al.*, 1968; Chapter 5 and tables therein). Fast-spreading-rate profiles from oceans other than the southern Atlantic have been studied separately, or incorporated in the synthetic South Atlantic profile of Cande and Kent (1992a), to resolve intricate details of the reversal history. However, even in such profiles, resolution is limited to 20–30 kyr and shorter events cannot usually be unambiguously resolved. Alternatively, polarity records of lava sequences can be employed to detect short polarity events especially because they can be radiometrically dated as well. Such sequences have been successfully employed to detect short polarity intervals but they do not provide a continuous record of (changes in) the Earth's magnetic field because of their episodic nature.

The details of the Neogene polarity time scale are resolved using long and continuous magnetostratigraphic records both from land-based marine and continental successions (e.g. Langereis and Hilgen, 1991; Abdul Aziz *et al.*, 2003) as well as from ODP cores for the younger half of the Neogene (e.g. Schneider, 1995). These magnetostratigraphic records are more important where cyclic successions have been directly tied to the APTS. Such records are preferable to the often rotary drilled and discontinuous records of older deep-sea cores, which, in addition, have magnetostratigraphic records that are often difficult to interpret and correlate unambiguously to the GPTS.

Numerous high-quality magnetostratigraphic records from deep-sea cores now tie the Brunhes–Matuyama boundary to marine oxygen isotope stage 19 and the succession of short reversed intervals within the Brunhes becomes increasingly clear. Reliable magnetostratigraphic records of continuous and astronomically dated successions are available for the entire Pliocene and Pleistocene as well as for the major part of the Upper Miocene. Tuned marine sections from the Mediterranean extended the APTS back to 11 Ma, and a further extension, to 13 Ma, was achieved using lacustrine successions in Spain. ODP Leg 138 sites produced an excellent magnetostratigraphy back to almost 12 Ma (Schneider, 1995), but the succession proved difficult to tune astronomically.

Major problems concerning the magnetostratigraphy lie in the interval between 13 Ma and the Oligocene–Miocene boundary. For this interval, no long and continuous magnetostratigraphic records are available except for the untuned continental section of Armantes (Krijgsman *et al.*, 1994). Nannofossil events allowed high-resolution correlation from the tuned succession at Ceara Rise to the untuned DSDP Site 522 that has a reliable magnetostratigraphy across the boundary (Shackleton *et al.*, 2000). These correlations confirm equating the Oligocene–Miocene boundary with base of polarity Chron C6Cn.2n, as suggested by Steininger *et al.* (1997a,b).

SEQUENCE STRATIGRAPHY

The modern concept of sequence stratigraphy was born out of seismic stratigraphy and started with the publication by Vail *et al.* (1977). In total, six orders of stratigraphic sequences were distinguished with durations ranging from more than 50 myr (first order) to less than 0.5 myr (fourth–sixth). Global sea-level charts were published by Vail *et al.* (1977), and in an updated version by Haq *et al.* (1987), which rely on the assumption of a global or eustatic control on sequences, especially third-order episodes. This assumption has been strongly criticized (e.g. Miall, 1992), but gained renewed support from ODP data of the New Jersey coastal plain of Late Cretaceous age (Miller *et al.*, 2003).

The youngest Neogene part of the disputed cycle chart for global sea-level fluctuations of Haq *et al.* (1987) comprises two complete second-order sequences, TB2 and TB3, plus the top of a third, TB1. The second-order sequences contain six (TB2.1–2.6) and nine (TB3.1–3.9) third-order sequences. The early Miocene according to the Haq *et al.* curve is marked by a relatively high sea-level with superimposed third-order fluctuations. Sea level started to fall after 15 Ma, culminating in a dramatic drop around 10.5 Ma which defines the TB2–3 cycle boundary. Sea level gradually rose during the late Miocene reaching its highest levels during the earliest Pliocene, followed by a gradual fall up to the Recent with superimposed high-amplitude third-order fluctuations.

Miller *et al.* (1996) correlated third-order sequence boundaries observed in seismic profiles and bore holes on the New Jersey coastal plain and continental shelf to the Oligocene and Miocene benthic isotope events of the deep-sea record. Ages of the New Jersey sequence boundaries and global $\delta^{18}O$ increases are in good agreement with one another, indicating that they both are related to falls in global sea level. Moreover, they correlate well with fluctuations in the Haq *et al.* (1987) global sea-level chart. The link between third-order sequences and sea level is strengthened by the results of high-resolution oxygen isotopc studies which suggest that the Mi1 and 5–6 events are related to a 1.2 myr obliquity cycle and thus may be related to astronomically controlled changes in Antarctic ice volume (Turco *et al.*, 2001).

Higher-order sequences have been investigated in addition to the third-order sequences. Especially the marine cyclothems of Plio–Pleistocene age in the Wanganui basin (New Zealand) have been studied in detail. The cyclothems, which correspond in duration with fifth- and sixth-order sequences but can best be regarded as parasequences according to sequence stratigraphic nomenclature, have been astronomically calibrated and correlated to the standard oxygen isotope record (Naish *et al.*, 1998). They reflect glacio-eustatic sea-level fluctuations driven by Milankovitch variations in climate and bridge the gap between sequence stratigraphy and cyclostratigraphy.

VOLCANIC ASH LAYERS AND TEPHROCHRONOLOGIES

Volcanic ash layers play an important role in establishing time stratigraphic correlations on a regional scale. A detailed tephrostratigraphy and tephrochronology has for instance been developed for the late Pleistocene–Holocene in the Mediterranean (Narcisi and Vezzoli, 1999) and North Atlantic regions (Haflidason *et al.*, 2000; Davies *et al.*, 2002), and for the middle Miocene in central Japan (Takahashi and Saito, 1997).

Several ash layers, however, can be correlated on an inter-regional or even global scale. The Toba ash of Indonesia, dated at 74 ka, has been found in the South China Sea and the Indian Ocean and its expression has also been detected in an ice core from Greenland (e.g. Buhring and Sarnthein, 2000; Schulz *et al.*, 2002). In ice cores, sulfate concentrations are employed to trace explosive volcanic activity and evidence for a number of major low-latitude eruptions has been identified in ice cores from both Greenland as well as the Antarctic. Their identification and correlation may help to solve problems related to the phasing of paleoclimate change. The tephrostratigraphy and tephrochronology for the Holocene in the North Atlantic region has been used to correlate and date paleoclimate records from peat-bogs and maares more accurately (van den Boogaard *et al.*, 2002).

TEKTITES

Tektites are extremely rare and not very important from a stratigraphic point of view but their occurrences are worth mentioning here. One of the most famous members of the tektite group are the moldavites. These tektites have been collected from several localities in Czechoslovakia, Austria, and Germany since the mid-eighteenth century (Bouska, 1989). Using detailed geological, petrological, geochemical, and geographical constraints, the numerical modeling study of Stoffler *et al.* (2002) showed that the moldavite-strewn field and the Steinheim and Ries craters in southern Germany were related and were caused by the oblique impact of a binary asteroid from a west-southwest direction. New $^{40}Ar/^{39}Ar$ dates of four tektites indicate that the moldivites are 14.5 ± 0.16 Ma (Schwarz and Lippolt, 2002).

A microtektite layer associated with the Indo Australian strew field and dated at 0.79 Ma is by far the most important one. It has a wide geographical distribution, which includes the Chinese loess plateau; the South China, Philippine, Celebes, and Sulo Seas; and parts of the Indian and Pacific Ocean, covering at least 5×10^7 km^2 (e.g. Glass and Pizzuto, 1994). The layer has been used as a time stratigraphic correlation tool and has helped to solve the problem of misleading positions of the Brunhes–Matuyama magnetic reversal boundary in loess successions (Zhou and Shackleton, 1999). Microtektites have further been reported from sedimentary deposits of a late Pliocene asteroid impact in the Southeast Pacific and from a number of deep-sea cores off the West Africa coast, the latter being associated with the Ivory Coast impact and tektite event dated at 1.07 Ma (Schneider and Kent, 1990).

21.2.4 Radiometric ages

^{40}Ar/^{39}Ar AND U–Pb

During the last 15 years, the ^{40}Ar/^{39}Ar method has replaced the conventional ^{40}K/^{40}Ar method for dating potassium-bearing minerals of Neogene age because it allows parent (first converted to ^{39}Ar by neutron irradiation) and daughter isotopes to be simultaneously measured by the same method and on the same sample aliquot. Moreover, the method is fast (20 minutes for a single age determination), is precise (with a typical standard error of 0.1%) and is reproducible. Only a small sample size is needed, and this even allows dating of single crystals. Details on the method and its analytical and external (relative to other methods) uncertainties are presented in Chapter 6.

The most widely applied secondary mineral standard when dating samples for the calibration of the geological time scale for the Neogene is the Fish Canyon Tuff (FCT) sanidine. Currently used ages for this standard in the literature (obtained via inter-calibration with primary standards) range from 27.5 (Lanphere and Baadsgaard, 2001) via 27.84 and 28.02 (Renne *et al.*, 1998c) to 28.10 Ma (Spell and McDougall, 2003). The last age was obtained by re-dating the primary standard GA-1550, but the new age for this standard is statistically indistinguishable from the previous age (Renne *et al.*, 1998c).

An alternative and independent method of determining the absolute age of the FCT sanidine is via inter-calibration with astronomical datings for magnetic reversals or for ash layers (Renne *et al.*, 1994; Hilgen *et al.*, 1997; Steenbrink *et al.*, 1999) that have been dated by the ^{40}Ar/^{39}Ar method as well. Results of the most recent and comprehensive study directed at ^{40}Ar/^{39}Ar resolution of astronomically dated volcanic ash layers in the Mediterranean Neogene revealed a significant discrepancy between the ^{40}Ar/^{39}Ar and astronomical ages, the latter being on average ~0.8% older (Kuiper, 2003).

Using the most reliable, single-crystal sanidine datings from the Melilla Basin in Morocco, an astronomically derived age of 28.24 ± 0.01 Ma is obtained for the FCT sanidine and favors the introduction of a directly astronomically dated ^{40}Ar/^{39}Ar standard. This is significantly younger than the U–Pb zircon dates of the FCT (Schmitz and Bowring, 2001). The older U–Pb age might be related to extended residence of zircons in the magma chamber (Oberli *et al.*, 2002), reflecting the complex eruptive history (Bachmann *et al.*, 2002), although this does not explain the equally old U–Pb age for titanite, which may remain open to Pb diffusive exchange until quenching upon eruption (Schmitz and Bowring, 2001). Clearly, the main advantage of using an astronomically derived age for the FCT sanidine is that problems associated with uncertainties in the decay constants and the age of the primary dating standard, which contribute mostly to the full error in ^{40}Ar/^{39}Ar dating ages, are avoided (Kuiper, 2003). This would favor the introduction of a directly astronomically dated ^{40}Ar/^{39}Ar standard.

^{14}C AND ^{230}Th/^{234}U

Over the last 20 years, it has become apparent that calculation of the radiocarbon age of a specific sample assuming a constant production of ^{14}C in atmospheric CO_2 is not valid, but that the activity of ^{14}C in atmospheric CO_2 has varied over time (e.g. Stuiver and Braziunas, 1993). The comparison between radiocarbon ages of samples in equilibrium with atmospheric CO_2 and dendrochronologically dated tree rings made it possible to construct decadal calibration data sets for the last 11 854 cal yr BP (0 yr BP = AD 1950), whereas computer modeling has provided a similar decadal calibration data set for marine samples (Stuiver *et al.*, 1998). Using ^{230}Th/^{234}U and ^{14}C measurements of corals (Bard *et al.*, 1990, 1998; Edwards *et al.*, 1993; Burr *et al.* 1998) and a floating marine varve chronology (Hughen *et al.*, 1998), the calibration curve for atmospheric, as well as marine, samples is extended to 24 000 cal BP with a spline through the measured radiocarbon ages (Stuiver *et al.*, 1998). Extending the calibration record to older periods is difficult because residual ^{14}C concentrations in samples become extremely low (Bard, 2001). Contradictory results from ^{14}C dating back to 45 000 BP using ^{230}Th/^{234}U dates of glacial varved sediments of Lake Suigetsu in Japan (Kitagawa and Van der Plicht, 1998) and a well-preserved stalagmite record of the Bahamas (Beck *et al.*, 2001) will have to be reconciled before they can be used to update and extend the INTCAL98 curve back in time (Bard, 2001).

Thorium-230/Uranium-234 dating has been successfully applied to tropical corals from sea-level highstands (Chen *et al.*, 1991; Gallup *et al.*, 1994), calcite veins (Winograd *et al.*, 1992), and bulk sediment (Henderson and Slowey, 2000) for the last ~140 000 years. Dating of older corals is more difficult partly because high-precision thorium measurements are required as errors expand rapidly as the ^{230}Th/^{238}U tends toward unity, but also due to the strong influence of subaerial diagenetic processes. To test the accuracy of the ^{230}Th/^{234}U dates, Edwards *et al.* (1997) used ^{231}Pa/^{235}U dates and confirmed the predictions of astronomical theory (e.g. Imbrie *et al.*, 1984) for the timing of sea-level change over parts of the last glacial cycle.

^{87}Sr/^{86}Sr

Strontium isotope stratigraphy provides a powerful tool to correlate and date marine sequences in the Cenozoic (Hess *et al.*,

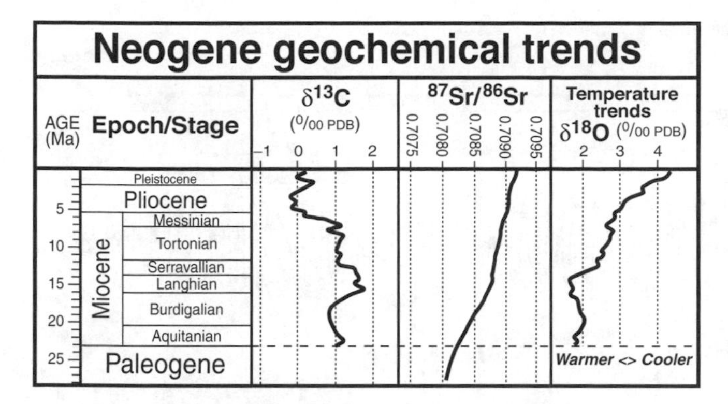

Figure 21.3 Generalized oxygen, carbon, and strontium isotope trends for the Neogene. See text for sources of oxygen and carbon curves. Strontium curve is a LOWESS fit to data from several sources (see text and Chapter 7).

1989), especially when other stratigraphic tools fail, such as at high latitudes or in carbonate platforms (e.g. Ohde and Elderfield, 1992). The method relies on matching the $^{87}Sr/^{86}Sr$ value of a sample to a standard curve of the $^{87}Sr/^{86}Sr$ of the oceans through time. Where the rate of change with time of $^{87}Sr/^{86}Sr$ is high, e.g. around 0.000 050 per myr from latest Eocene to middle Miocene times, the method has a resolution in dating as low as ±0.1 myr. At times of lower rates of change (<15 Ma), the resolution is poorer, typically around 0.5–1 myr. The large increase in seawater Sr isotope composition since the late Eocene, which is presumably related to increased chemical weathering due to mountain building of, in particular, the Himalayas and Tibetan plateau (Raymo *et al.*, 1988), ensures relative high resolution of the method in the Neogene. The Neogene calibration curve (Fig. 21.3) is based on the works of John Farrell, Ken Miller, Dave Hodell, and coworkers (see Chapter 7 for a full reference list and discussion of methodology).

21.2.5 Climate change and Milankovitch cycles

OXYGEN ISOTOPES ($\delta^{18}O$)

Stable isotopes, in particular those of oxygen, are extensively used in the Neogene to improve our understanding of paleoceanographic and paleoclimate change. In addition to unlocking past changes in the state of the ocean–climate system, stable isotopes prove to be extremely useful for global chronostratigraphic correlations on different time scales and for constructing astronomical time scales.

The global deep-sea oxygen ($\delta^{18}O$) and carbon ($\delta^{13}C$) isotope records reveal major changes in the ocean–climate system during the Neogene (Zachos *et al.*, 2001b; see Fig. 21.3). The Early to early Middle Miocene is marked by relatively light $\delta^{18}O$ values and prevailing warm climatic conditions. This warm phase culminates in the so-called mid-Miocene climatic optimum marked by a distinct $\delta^{18}O$ minimum between ~17 and 15 Ma, followed by an initial rapid and then gradual decrease in $\delta^{18}O$ in the late Middle to Late Miocene, reflecting the re-establishment of a major Antarctic ice cap. This decreasing trend is followed by an interval marked by relatively stable $\delta^{18}O$ values, albeit a step toward lighter $\delta^{18}O$ values is recognized around 5.6 Ma. The final major step toward distinctly heavier values starts around 3 Ma, reflecting the final build-up of a major northern hemisphere ice cap.

Miller *et al.* (1991) recognized a number of Oligocene (Oi1–3) and Miocene (Mi1–7) benthic foraminiferal $\delta^{18}O$ increases that were synchronous within the resolution of the applied magneto–biostratigraphic framework. These events occur superimposed on the general long-term trend described above and may reflect brief periods of enhanced Antarctic glaciation. Attempts have been made to link the events to third-order stratigraphic sequences on the New Jersey continental shelf (Miller *et al.*, 1996) and recent studies suggest that they are related to the long-period 1.2 myr obliquity cycle (Lourens and Hilgen, 1997; Turco *et al.*, 2001; Zachos *et al.*, 2001a), thus rendering support to the notion that third-order sequence boundaries might have a chronostratigraphic significance after all.

High-frequency Milankovitch variations in $\delta^{18}O$ occur superimposed on the above described long-term trends and variations throughout the Neogene. These variations mainly reflect astronomically controlled glacial cyclicity and serve as a powerful tool for high-resolution chronostratigraphic correlations. They are especially prominent from 2.8 Ma onward, but similar variations albeit of a lesser amplitude have been found throughout the entire Neogene. The labeling of individual maxima (glacials, even numbers) and minima (interglacials, odd numbers) as standard stages in late Pleistocene $\delta^{18}O$ records started with the pioneering study of Emiliani (1955) and has subsequently been extended to 3.0 Ma (e.g. Imbrie *et al.*, 1984; Raymo *et al.*, 1989) and beyond. These stages are now commonly referred to as marine isotope stages (MIS). Alternatively, Shackleton *et al.* (1995a) proposed a codification scheme for prominent and partly older $\delta^{18}O$ maxima where (downward) numbering is re-initialized at each magnetic polarity chron and subchron (see also Tiedemann *et al.*, 1994). This nomenclature has been extended to 7 Ma (Hodell *et al.*, 1994).

Similar high-frequency Milankovitch variations have been revealed in the older part of the Neogene (Zachos *et al.*, 2001b)

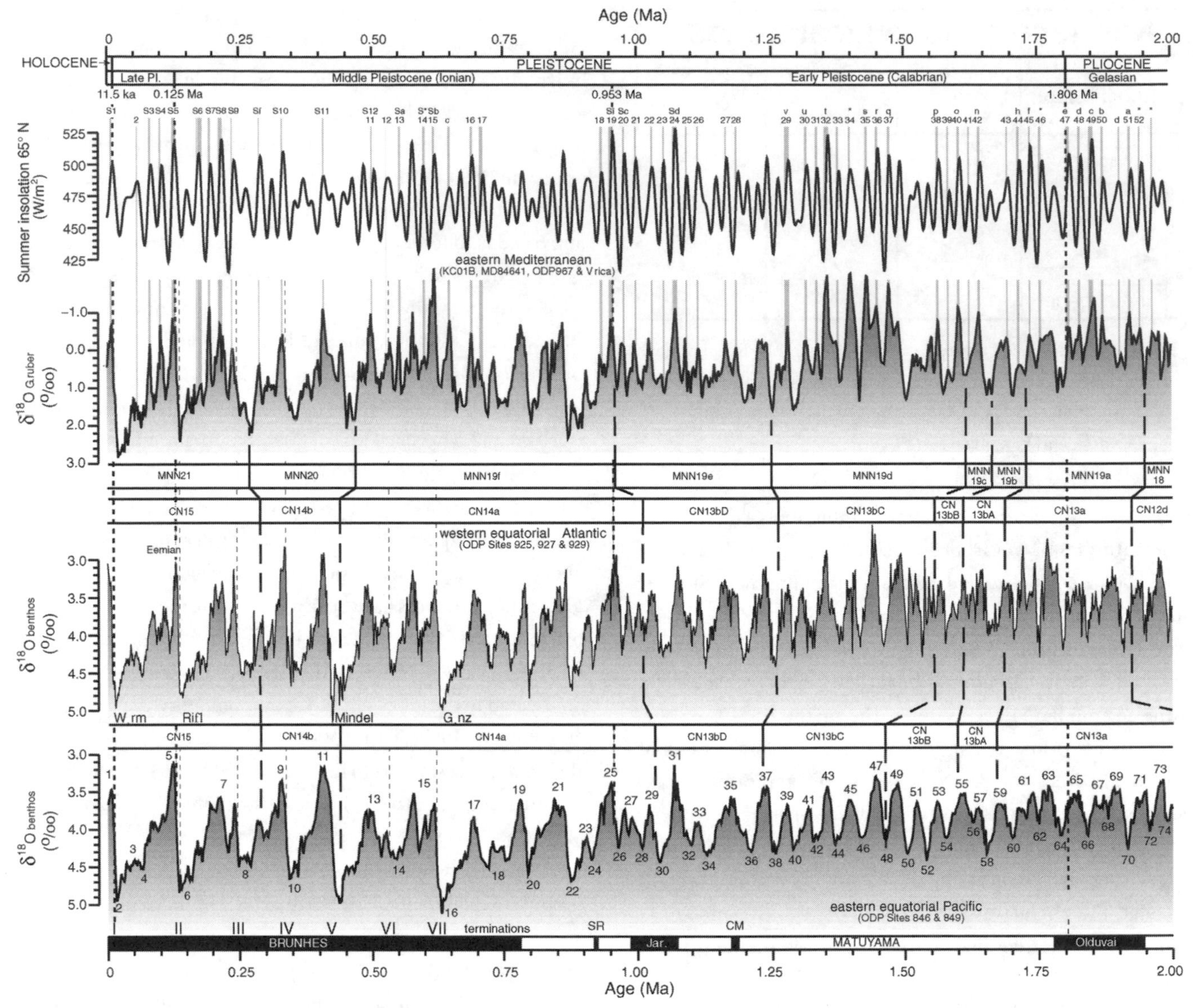

Figure 21.4 Comparison between the oxygen isotope chronologies of astronomically tuned marine sequences from the Mediterranean, western equatorial Atlantic, and eastern equatorial Pacific during the Pleistocene. The Mediterranean $\delta^{18}O$ stacked record of the planktonic foraminiferal species *Globigerinoides ruber* is based on data from ODP Site 967 (Lourens, unpublished data), MD69-KC01B (Rossignol-Strick *et al.*, 1998), MD84641 (Fontugne and Calvert, 1992), Vrica (Lourens *et al.*, 1996a,b), and Singa (Lourens *et al.*, 1992, 1996a). The age model is based on sapropel tuning to the summer insolation target curve at 65° N, using a 3-kyr time lag between maximum insolation and the correlative sapropel midpoint (upper panel). The western equatorial Atlantic and eastern equatorial Pacific benthic $\delta^{18}O$ stacked records are based on data from ODP Sites 925, 927 (Bickert *et al.*, 1997a), and 929 (Bickert *et al.*, 1997a; DeMenocal *et al.*, 1997), and ODP Sites 846 and 849 (Mix *et al.*, 1995a,b), respectively. We refrained from re-tuning the isotope records to the La2003 solution, because differences with previously adopted solutions are negligible. Magnetostratigraphy and calcareous nannofossil ages are obtained from Tables A2.1 and A2.2 in Appendix, respectively, and references therein.

and it can be anticipated that the standard oxygen isotope stratigraphy will be extended further downward to cover the entire Neogene in the future. Figs. 21.4 and 21.5 are a comparison of Pliocene and Pleistocene isotope stratigraphies of oceanic sequences from the eastern equatorial Pacific (Shackleton *et al.*, 1990, 1995a; Mix *et al.*, 1995a,b), western equatorial Atlantic (Bickert *et al.*, 1997a; Billups *et al.*, 1997; DeMenocal *et al.*, 1997; Franz and Tiedemann, 1997; Shackleton and Hall, 1997; Tiedemann and Franz, 1997) and Mediterranean (Lourens *et al.*, 1996a, unpublished data; Kroon *et al.*, 1998: re-tuned data from Fontugne and Calvert, 1992; Rossignol-Strick *et al.*, 1998; Rossignol-Strick and Paterne, 1999).

CARBON ISOTOPES (δ^{13}C)

The onset of the Neogene falls within a distinct δ^{13}C maximum which is followed by an interval with markedly lighter values before the next prominent maximum is reached between 16 and 14.5 Ma (Fig. 21.3). This maximum corresponds to the Monterey event (Flower and Kennett, 1994) and coincides approximately with the mid-Miocene climatic optimum. The carbon isotope maximum is followed by a step-wise decrease in δ^{13}C which is probably due to the expansion of C4 plants. The major second step toward lighter δ^{13}C values is termed the late Miocene carbon shift (Keigwin and Shackleton, 1980) and corresponds with a time equivalent opposite shift to heavier values in the continental record (Cerling *et al.*, 1997).

Superimposed on the general long-term trend outlined above, Woodruff and Savin (1991) distinguished seven carbon isotope maxima during the middle Miocene (CM1–7) with a potential for global correlations in the marine record. Strong covariance between the oxygen and carbon isotopic composition associated with Milankovitch cycles occurs throughout the Neogene and has been attributed to effects of climate change on ocean–atmosphere circulation, ocean productivity, and organic-carbon burial (i.e. Raymo *et al.*, 1989; Flower and Kennett, 1994; Fontugne and Calvert, 1992; Zachos *et al.*, 2001a).

SEDIMENTARY CYCLES

Apart from cyclic variations in δ^{18}O (as well as in other geochemical and faunal/floral parameters) the visual expression of Milankovitch variability as sedimentary (color) cycles has similarly been used for high-resolution correlations and astronomical calibration of the stratigraphic record. For instance, the Pliocene–Pleistocene isotope stages mentioned above are clearly discernable in northern Atlantic DSDP/ODP cores as color cycles, the glacials being dark colored due to the admixture of marine carbonate with ice-rafted debris.

In particular, marine sapropels and carbonate cycles in the Mediterranean have been used for high-resolution correlations (Figs. 21.4 and 21.5) and to construct astronomical time scales for the last 14 myr by tuning them to astronomically derived target curves (i.e. Hilgen, 1991a,b; Lourens *et al.*, 1996a; Hilgen *et al.*, 2000c). Although initially limited to the marine record of the Mediterranean, a similar approach is equally feasible for cyclic continental successions as demonstrated by Williams *et al.* (1997) and Prokopenko *et al.* (2001) for Lake Baikal and Abdul Aziz *et al.* (2003) for Orera (Spain), allowing marine–continental correlations with an unprecedented resolution. The same approach has been applied to cycles in the Chinese loess; these cycles have been astronomically tuned and correlated in detail to the standard marine isotope stratigraphy (Heslop *et al.*, 2000). Similar correlations and tuning have further been established for both marine and river terraces (Veldkamp and van den Berg, 1993), as well as for stratigraphic sequences deposited in shelf settings (Naish *et al.*, 1998; see also Section 21.2.3).

DANSGAARD–OESCHER CYCLES AND HEINRICH EVENTS

The oxygen isotope records of the ice cores demonstrate that a series of rapid warm–cold oscillations, called Dansgaard–Oeschger events, punctuated the last glaciation with an average periodicity of ~2 kyr (Oeschger *et al.*, 1984; Dansgaard *et al.*, 1993). Subsequent analyses of sea surface temperature proxy records from North Atlantic sediment cores also revealed a series of rapid temperature oscillations which closely match those in the ice cores (Bond *et al.*, 1993). These events are bundled into cooling cycles, so-called Bond cycles, lasting on average 5–10 kyr, with asymmetrical saw-tooth shapes. Each cycle culminated in massive iceberg discharge into the North Atlantic, known as Heinrich events (Heinrich, 1988; Broecker *et al.*, 1992), followed by an abrupt shift to a warmer climate.

Short-term climatic events correlative with Dansgaard–Oeschger and Heinrich events are increasingly detected in deep-sea, lake, and continental records on an almost global scale (Grimm *et al.*, 1993; Lowell *et al.*, 1995; Porter and Zhisheng, 1995; Thompson *et al.*, 1995; Behl and Kennett, 1996; Sirocko *et al.*, 1996; Chen *et al.*, 1997; Schulz *et al.*, 1998; Cacho *et al.*, 1999; Peterson *et al.*, 2000), and there is growing evidence that climate fluctuations on sub-orbital time scales are not restricted to the late Pleistocene glacial period, but that they also occur in older geological periods (e.g. Raymo *et al.*, 1998; Steenbrink *et al.*, 2003). The Heinrich events fall well within the range of the sub-Milankovitch band. Although the expression of these events is most clearly in the North Atlantic region, the ultimate cause of the climate change in this frequency band may come from a non-linear response of the climate system at low latitudes to Milankovitch forcing (Pestiaux *et al.*, 1988; Short *et al.*, 1991; Crowley *et al.*, 1992; Hagelberg *et al.*, 1994). Dansgaard–Oeschger events occur at the high-frequency end of the sub-Milankovitch band and are therefore more difficult to explain in terms of a non-linear response to Milankovitch forcing.

The origin of these short-term events is still puzzling. Proposed (competing) mechanisms invoke variations in solar activity (e.g. Stuiver *et al.*, 1991; Magny, 1993), dustiness (Overpeck *et al.*, 1996), internal oscillations in ice-sheet dynamics and stability (binge/purge model of Alley and MacAyeal, 1994), combination tones of primary orbital frequencies (McIntyre and Molfino, 1996) and variations in the atmosphere's green-house gas content and in the production rate of North Atlantic Deep Water (NADW) formation (Broecker, 1994).

Although the Dansgaard–Oeschger and Heinrich events are most clearly expressed in the less-stable glacial period, similar variations of lesser amplitude do occur in the post-glacial Holocene (O'Brien *et al.*, 1995; Alley *et al.*, 1997; Bond *et al.*, 1999) and Eemian period (Cortijo *et al.*, 1994; Keigwin *et al.*, 1994; McManus *et al.*, 1994). As a consequence, studies of the Holocene (preferably of varved records to guarantee maximum age control) and of the Eemian, as a natural analog of the Holocene, are becoming increasingly important. Part of the research now focuses on the post-glacial Holocene with the aim of corroborating potential links of climate variability in the millennial–decadal frequency band with, among others, fluctuations in solar activity as reflected in the ^{14}C record of tree rings (Stuiver and Reimer, 1993), terrestrial dust increases of Greenland ice (O'Brien *et al.*, 1995), and monsoonal driven changes in rainfall in Oman reflected in stalagmite oxygen isotope records (Neff *et al.*, 2001).

21.3 TOWARD AN ASTRONOMICALLY TUNED NEOGENE TIME SCALE (ATNTS)

21.3.1 Introduction

From the 1970s until 1994, Neogene time scales were constructed using a limited number of radioisotopic age calibration points in geomagnetic polarity sequences that were primarily derived from a seafloor anomaly profile in the South Atlantic, modified after Heirtzler *et al.* (1968). Biozonations and stage boundaries were subsequently tied to the resulting GPTS, preferably via magneto–biostratigraphic calibrations (Berggren *et al.*, 1985c). Alternatively, radioisotopic age determinations from both sides of stage boundaries were used to calculate a best-fit radioisotopic age estimate for these boundaries in a statistical way (chronogram method of Harland *et al.*, 1982, 1990; see Chapter 8).

The "standard" method to construct time scales changed drastically with the advent of the astronomical dating method to the pre-late Pleistocene. This method relies on the calibration, or tuning, of sedimentary cycles or cyclic variations in climate proxy records to target curves derived from astronomical solutions for the solar–planetary and Earth–Moon systems (for details see Chapter 4). Quasi-periodic perturbations in the shape of the Earth's orbit and the tilt of the inclination axis are caused by gravitational interactions of our planet with the Sun, the Moon, and the other planets of our Solar System. These interactions give rise to cyclic changes in the eccentricity of the Earth's orbit, with main periods of 100 000 and 413 000 years, and in the tilt (obliquity) and precession of the Earth's axis, with main periods of 41 000, and 21 000 years, respectively (Berger, 1977a). These perturbations in the Earth's orbit and rotation axis are climatically important because they affect the global, seasonal, and latitudinal distribution of the incoming solar insolation. Orbital forced climate oscillations are recorded in sedimentary archives through changes in sediment properties, fossil communities, chemical and isotopic characteristics. While Earth scientists can read these archives to reconstruct paleoclimate, astronomers have formulated models based on the mechanics of the solar–planetary system and the Earth–Moon system to compute the past variations in precession, obliquity, and eccentricity of the Earth's orbit and rotation axis. As a logical next step, sedimentary archives can be dated by matching patterns of paleoclimate variability with patterns of varying solar energy input computed from the astronomical model solutions. This astronomical tuning of the sedimentary

Figure 21.5 Comparison between the oxygen isotope chronologies of astronomically tuned marine sequences from the Mediterranean, western equatorial Atlantic, and eastern equatorial Pacific during the Pliocene. The Mediterranean $\delta^{18}O$ stacked record of the planktonic foraminiferal species *Globigerinoides ruber* is based on data from Vrica, Singa, and Rossello (Lourens *et al.*, 1992, 1996a,b). The age model is based on sapropel and carbonate cycle tuning to the summer insolation target curve at 65° N, using a 3-kyr time lag between maximum insolation and the correlative sapropel or gray-layer midpoint (left panel). The western equatorial Atlantic and eastern equatorial Pacific benthic $\delta^{18}O$ stacked records are based on data from ODP Sites 925 (Bickert *et al.*, 1997a; Billups *et al.*, 1997; Tiedemann and Franz, 1997), 926 (Shackleton and Hall, 1997; Tiedemann and Franz, 1997), 927 (Bickert *et al.*, 1997a), 928 (Franz and Tiedemann, 1997), and 929 (Bickert *et al.*, 1997a; Billups *et al.*, 1997; DeMenocal *et al.*, 1997), and ODP Sites 846 and 849 (Mix *et al.*, 1995a,b; Shackleton *et al.*, 1995b), respectively. We refrained from re-tuning the isotope records of the Mediterranean and western equatorial Pacific to the La2003 solution, because differences with the previously adopted La93 solution are very small (~1–2 kyr). Plotted are the original tuned records of ODP Sites 846 and 849 to the Ber90 solution, which still need to be re-tuned to the La2003 solution (work in progress). Magnetostratigraphy and calcareous nannofossil ages are obtained from Table A2.1 and Shackleton *et al.* (1995a), and from Table A2.2, respectively, and references therein.

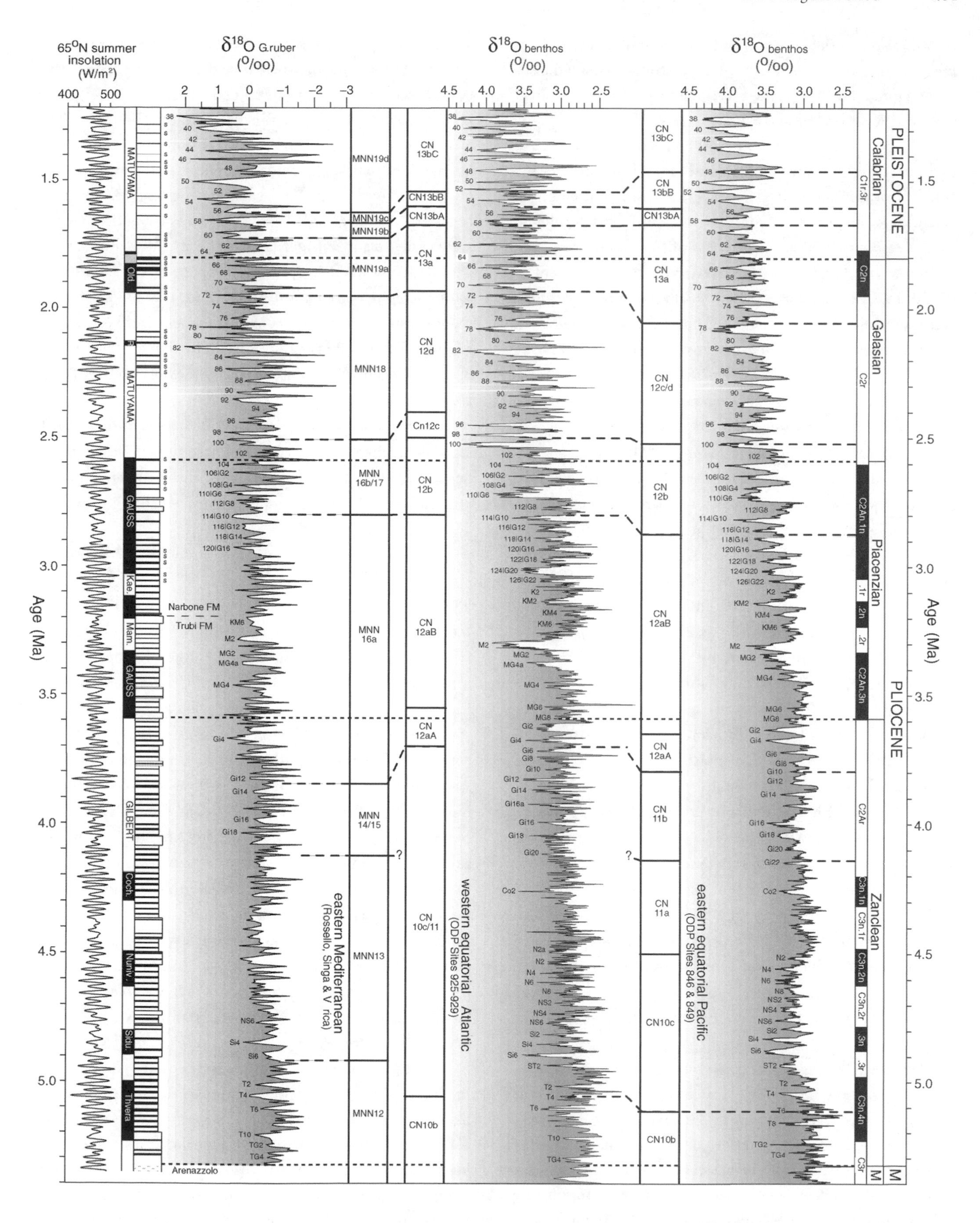
65°N summer insolation (W/m²)
δ18O G.ruber (‰)
δ18O benthos (‰)
δ18O benthos (‰)
Age (Ma)
MATUYAMA
Old.
GAUSS
Kae.
Mam.
GILBERT
Coch.
Nunv.
Sidu.
Thvera
Narbone FM
Trubi FM
Arenazzolo
MNN19d
MNN19c
MNN19b
MNN19a
MNN18
MNN 16b/17
MNN 16a
MNN 14/15
MNN13
MNN12
CN 13bC
CN13bB
CN13bA
CN 13a
CN 12d
Cn12c
CN 12b
CN 12aB
CN 12aA
CN 10c/11
CN10b
CN 12c/d
CN 11b
CN 11a
CN10c
eastern Mediterranean (Rossello, Singa & V rica)
western equatorial Atlantic (ODP Sites 925-929)
eastern equatorial Pacific (ODP Sites 846 & 849)
PLEISTOCENE
Calabrian
Gelasian
Piacenzian
PLIOCENE
Zanclean
M

record results in time scales based on measurable physical parameters that are independent from those underlying radioisotopic dating and that are tied to the Recent through a direct match with astronomical curves.

Astronomical tuning was first applied in the late Pleistocene in order to build a common high-resolution time scale for the study of orbital induced glacial cyclicity. Initial attempts to extend this time scale back in time were unsuccessful due to lack of resolution or incompleteness of the sedimentary succession. These problems were overcome with the advent of the advanced piston corer (APC) technique in ocean drilling and the drilling of multiple offset holes per site. Combined, these innovations were used to construct spliced composite sections in order to recover undisturbed and complete successions marked by high sedimentation rates. Soon afterwards, the astronomical time scale was extended to the base of the Pliocene based on ODP sites (Shackleton *et al.*, 1990) and land-based sections in the Mediterranean (Hilgen, 1991a,b), the study of the latter providing another means to overcome the problem of incompleteness of the stratigraphic record.

Acknowledging the break with tradition, Cande and Kent (1992a) incorporated an astronomically dated calibration point, the Gauss–Matuyama boundary at 2.60 Ma, to construct their geomagnetic polarity time scale (GPTS) by fitting a cubic-spline function fitted to a total of nine age calibration–anomaly distance tie points. Soon afterwards a revised GPTS (Cande and Kent, 1995) was generated with 65 Ma for the K–P boundary – instead of 66 Ma – and using the astronomical age of 5.23 Ma for the base of polarity Chron C3n.4n; ages for reversals younger than C3n.4n were set equal to the astrochronology of Shackleton *et al.* (1990) and Hilgen (1991a,b). The continuous chronology provided by the astronomical time scale implies that it is no longer necessary to interpolate between a few discrete tie points to construct a GPTS; instead, ages of reversals in the interval with an astrochronology simply become equivalent to the astrochronological values, thereby avoiding the promulgation of separate time scales (Berggren *et al.*, 1995a). Berggren *et al.* (1995b) subsequently used CK95 as the backbone to construct their revised Cenozoic time scale.

Another advantage of astrochronology is that most magnetic reversals and numerous calcareous plankton biohorizons are directly tied to this time scale via first-order calibrations. An additional advantage of the Mediterranean is that all currently defined GSSPs of Neogene stages, apart from the Aquitanian GSSP, are designated in the Mediterranean (or directly outside) in marine successions that have been astronomically dated.

But incorporating Pliocene–Pleistocene astrochronology also resulted in a kind of hybrid time scale with the younger part being based on astronomical tuning and the older pre-Pliocene part of a cubic-spline fitting function through a limited number of age calibration points in the CK92 geomagnetic polarity sequence. The hybrid character is problematical because Kuiper (2003) clearly demonstrates a consistent discrepancy in the order of ~0.8% between $^{40}Ar/^{39}Ar$ and astronomical ages despite earlier claims of mutual consistency. As a consequence, we decided to build the present Neogene time scale as much as possible on astronomical calibrations, this despite the fact that the tuning has not always been independently verified and magnetostratigraphic records are still lacking for the early and middle Miocene. We think that this is the only acceptable way to proceed in view of a strong tendency to incorporate increasingly older parts of the astrochronologic framework into the standard geological time scale. But first, we will briefly review new results which have appeared in the literature since the publication of the previous standard time scale.

21.3.2 New data

The astronomical time scale (ATS) for the Pliocene–Pleistocene has been verified and slightly modified (Lourens *et al.*, 1996a) since the publication of the previous standard time scale in 1995, by adopting the La90–93 (Laskar, 1990; Laskar *et al.*, 1993) astronomical solution instead of the Ber90 solution (Laskar, 1990; Berger and Loutre, 1991). A detailed comparison revealed that the best fit with the geological data was obtained if the La93 solution with present-day values for the dynamical ellipticity of the Earth and the tidal dissipation by the Moon, denoted $La93_{(1,1)}$, was used. This solution was subsequently employed to extend the marine-based ATS back to 13.6 Ma (Hilgen *et al.*, 1995; 2000c). An initial gap between 6.8 and 5.33 Ma has been closed by establishing an astronomical tuning for pre-evaporite and evaporite cycles of the Mediterranean Messinian which reflect the sedimentary expression of the salinity crisis (Krijgsman *et al.*, 1999; Sierro *et al.*, 2001). Good magnetostratigraphic records from the Mediterranean marine record are presently obtained back to ~10 Ma.

The tuning approach was successfully applied to the continental record and astronomical time scales have been established for Pliocene lignite-bearing successions in Greece (Van Vugt *et al.*, 1998) and Miocene lacustrine limestone containing successions in Spain (Abdul Aziz *et al.*, 2003). The Spanish record allowed the determination of astronomical ages for polarity reversals older than 10 Ma, thereby extending the Mediterranean-based ATS to ~13 Ma.

Simultaneously, Shackleton and Crowhurst (1997) developed an astronomical time scale for the interval between 5 and 14 Ma based on ODP Leg 154 Site 926 located in the equatorial Atlantic. This ODP Leg 154 time scale was extended to 34 Ma by Shackleton *et al.* (1999) and used to constrain the tidal dissipation and dynamic ellipticity parameters of the Earth over the last 25 myr (Pälike and Shackleton, 2000). The older part of this time scale includes the admittedly problematic interval between 14 and 18 Ma. Unfortunately, a reliable magnetostratigraphic record is lacking for the ODP Leg 154 sites, but detailed calcareous nannofossil event correlations to DSDP Site 522 having a reliable magnetostratigraphy showed that astronomical ages for the Oligocene–Miocene boundary interval are approximately 900 kyr younger than in most recent geological time scales (e.g. Berggren *et al.*, 1995b).

21.3.3 The La2003 solution

The orbital solutions La90–93 (Laskar, 1990; Laskar *et al.*, 1993) were obtained by a numerical integration of the averaged equations of the Solar System, including the main general relativity and lunar perturbations. The averaging process was performed using dedicated computer algebra routines. The resulting equations were huge, with about 150 000 polynomial terms, but as the short-period terms were no longer present, these equations could be integrated with a step size of 200–500 years, allowing very extensive long-term orbital computations for the Solar System.

Although solutions for the averaged equations could be improved by some new adjustments of the initial conditions and parameters, it appears that because of the improvement in computer technology, it is now possible to obtain more precise results over a few tens of millions of years using a more direct numerical integration of the gravitational equations, and this is how the new solution La2003 was obtained (Laskar *et al.*, 2003). In order to minimize the accumulation of rounding error, the numerical integration was performed with the new symplectic integrator scheme $SABA_4$ of Laskar and Robutel (2001), with a correction step for the integration of the Moon. This integrator is particularly adapted to perturbed systems where the equations of motion can be written as the sum of an integrable part (the Keplerian equations of the planets orbiting the Sun) and a small perturbation potential (here the small parameter $\in$ is of the order of the planetary masses). The error of method in the integration is then of the order of $O(\partial^8 \in) + O(\partial^2 \in^2)$, and even $O(\partial^8 \in) + O(\partial^4 \in^2)$ when a correction step is added. The step size used in the integration was $\partial = 1.826\,25$ days. The relative energy error over 50 million years is then less than 2×10^{-10}, and behaves as a random walk, testifying the absence of systematic error. The angular momentum is conserved with a relative error of less than 10^{-10}. The initial conditions of the integration were least-square adjusted to the JPL ephemeris DE406, and compared with DE406 over the full range of DE406, i.e. from -5000 to $+1000$ years from the present date (Standish, 1998). The maximum differences in the position of the Earth–Moon barycenter is less than 0.09 arcsec in longitude over the whole interval, and the difference in eccentricity is less than 10^{-8}. The variation in the longitude of the Moon is more important, as the dissipative models that are used in the two integrations are slightly different. They amount to 240 arcsec after 5000 years and to an eccentricity difference of 2×10^{-5}, but it should be noted that the perturbations due to orbital evolution of the Moon on the precession and obliquity of the Earth are very small.

The orbital model differs from La93, as it now includes all nine planets of the Solar System, including Pluto. The general relativity perturbation of the Sun is included. The Moon is treated as a separate object. The Earth–Moon system takes into account the most important coefficient (J_2) in the gravitational potential of the Earth and the Moon, and the tidal dissipation of the Earth–Moon system, as well as the influence of the precession of the Earth axis on the lunar orbit. The evolution of the Earth–Moon system and rotation of the Earth are treated in a comprehensive and coherent way, following the lines of Néron de Surgy and Laskar (1996) and Correia *et al.* (submitted). The climate friction effect, due to the change of momentum of inertia of the Earth arising from the change of ice load on the polar caps during ice ages has been estimated (Levrard and Laskar, 2003), but neglected as it was found to be too small and too uncertain to be taken into account.

Over 20 millions years, the two orbital solutions La93 and La2003 do not differ significantly. The difference increases regularly with time, amounting to about 0.02 in the eccentricity after 20 myr, about one-third of the total amplitude $\approx$ 0.063, while the difference in the orbital inclination reaches 1°, to compare to a maximum variation of 4.3°. This difference results mostly from a small difference in the main secular frequency g_6 from Jupiter and Saturn that is now 28.2450 arcsec/year instead of 28.2207 arcsec/year in the previous solution. Beyond 20 million years, the differences between the two solutions become more noticeable (these differences are described in more detail in Laskar *et al.*, 2003). The difference in obliquity over 20 myr amounts to about 2°, which means that the obliquity cycles of the two solutions become nearly out of phase after this date. It should be noted that most of the differences between the La93 and La2003 obliquity solutions result from the change in the dissipative model of the Earth–Moon system, while the single change of the orbital solution would lead to a change of only 0.6° after 20 myr.

We performed several numerical integrations of the whole system with some small variations in the initial conditions or model. In particular, we integrated two different models, one with the present accepted value for the flattening of the Sun ($\mathcal{J}_{2S} = 2 \times 10^{-7}$) and one without this effect ($\mathcal{J}_{2S} = 0$). The two solutions differ by less than 0.0036 for the eccentricity of the Earth after 30 myr and 0.015 after 40 myr. We can thus consider that the most recent solution, La2003, can be used with confidence over 30–40 million years. As in the previous solution La93, we expect that the orbital part of the La2003 solution is more accurate than the precession and obliquity solutions, as some uncertainty remains in the dissipative evolution of the Earth–Moon system.

21.3.4 ATNTS2004

To construct the Neogene time scale, we used the Pliocene–Pleistocene astronomical time scales of Lourens *et al.* (1996a), Bickert *et al.* (1997a), Tiedemann and Franz (1997), and Horng *et al.* (2002), and the Miocene astronomical time scales of Shackleton and Crowhurst (1997), Shackleton *et al.* (1999), Krijgsman *et al.* (1999), and Hilgen *et al.* (1995; 2000a–c; in press) as the starting point. Additional information about the ages of late Neogene reversal boundaries was obtained from the astronomical tuning of Miocene continental successions with a reliable magnetostratigraphy (Abdul Aziz *et al.*, 2003). The La90–93$_{(1,1)}$ solution with present-day values for dynamical ellipticity and tidal dissipation was used in all the papers mentioned above. For this chapter, the successions were retuned to the new solution La2003$_{(1,1,0)}$, again with present-day values for dynamical ellipticity and tidal dissipation, and no climate friction. As anticipated, the retuning resulted in almost negligible changes in the ages of sedimentary and geochemical (isotopes, susceptibility, color, etc.) cycles and thus of the bioevents and magnetic reversals over the last 13 myr (Table A2.1 in Appendix). In the following discussion, "(o)" and "(y)" indicate the beginning or ending, respectively, of a polarity chron.

An extra complication in constructing the present time scale for the entire Neogene is the lack of reliable magnetostratigraphic records for ODP Leg 154 sites, resulting in the lack of direct astronomical ages for reversal boundaries in the interval older than 13 Ma. Detailed nannofossil correlations to DSDP Site 522 with a reliable and well-calibrated magnetostratigraphy resulted in an age of 23.03 Ma for polarity Chron C6Cn.2n(o) and the Oligocene–Miocene boundary (Shackleton *et al.*, 2000; retuned to La2003). Detailed cyclostratigraphic correlations between Ceara Rise and DSDP Sites 521 and 521A in the southern Atlantic corroborated by biostratigraphic evidence further resulted in indirect astronomical ages of 15.160 and 15.974 Ma for polarity Chron C5Bn.2n(o) and polarity Chron C5Cn.1n(y), respectively (Shackleton *et al.*, 2001).

For the time intervals lacking orbitally dated magnetostratigraphic records, marine magnetic anomaly profiles remain the most useful data for establishing a high-precision polarity time scale. Experience gained by inspecting spreading rates for the astronomically calibrated parts of the time scale (Wilson, 1993; Krijgsman *et al.*, 1999; Abdul Aziz *et al.*, 2003) shows several useful refinements that have been developed for time scale calibration. First, noise introduced by asymmetric spreading can be nearly completely avoided by measuring total spreading distance by solving for plate rotations, and determining incremental distances by subtracting total distances (Wilson, 1993). Also, the assumption that spreading rates should change smoothly and continuously, a key aspect of Cande and Kent's calibration, is not supported by independent calibrations. Instead, rates are constant for periods of several million years, but then can experience abrupt changes. Since 13 Ma, the plate pair with the most constant motion has been Australia–Antarctica, with rates on a flowline at 98° E only varying from 65 to 69 mm/yr.

We have chosen to use the Australia–Antarctic spreading distances to close the gaps in the Miocene time scale calibration. Since this plate pair accelerated to an intermediate-spreading rate at about Chron C18 (Royer and Sandwell, 1989), only minor plate motion changes can be detected during Chrons C12r and C3An (Wilson, in prep.). A spreading rate of 67.7 mm/yr gives a good fit to the ages for polarity Chrons C5n.2n(o)–C5Ar.2n(o) from Abdul Aziz *et al.* (2003) and Chrons C5Bn.2n(base) and C5Cn.1n(top) from Shackleton *et al.* (2001). Extrapolating this rate to the beginning of the Neogene predicts an age of 23.18 Ma for polarity Chron C6Cn.2n(o). To match the age of 23.03 Ma, we introduce a minor rate change at polarity Chron C5En(o) (18.52 Ma), with a rate of 69.9 mm/yr prior to the change. The resultant ages for the reversal boundaries are added to the list of astronomically dated reversals to complete the dating of all (sub)chrons of the last 23 myr.

21.3.5 The age of the Paleogene–Neogene boundary

The astronomically derived age of the Oligocene–Miocene (Paleogene–Neogene) boundary of 23.03 Ma (Shackleton *et al.*,

2000; retuned to La2003 this chapter) in the Carrosio–Lemme section in northern Italy (Steininger *et al.*, 1997a,b) proved to be significantly different from that adopted in most recent Neogene time scales (e.g. Harland *et al.*, 1990; Berggren *et al.*, 1995b). Harland *et al.* (1990) used the chronogram method to arrive at an admittedly weakly constrained age estimate of 23.8 ± 1.0 Ma for the boundary placed at the base of polarity Chron C6Cn.2n. This age was based on 5 K–Ar ages from the Aquitanian and 19 K–Ar ages and one Rb–Sr age from the Chattian. Based on their Table 2, they were trying to satisfy an age for the boundary older than 22.00 ± 0.30, 22.50 ± 0.17, 22.60 ± 0.60, and 25.97 ± 0.58 Ma, and younger than 22.00 ± 0.55, 23.05 ± 1.14, 23.80 ± 0.27, and 26.20 ± 0.50 Ma (to list the youngest 4 out of 20 Chattian ages). Evidently, the age of 25.97 Ma should be considered an outlier since it is at nearly 4-sigma off the best fit. Discarding this one point would result in a best fit of around 22.4–22.6 Ma, or 22.8–23.2 Ma, with a decay-constant fudge of about 2%. Using current standards, the age of 23.8 Ma is unacceptable because it is derived from a suite of radiometric age determinations that are based on different and partly less-suitable minerals or even whole-rock samples (Harland *et al.*, 1990). Nevertheless it was incorporated in CK92 and CK95 as one of their age tie points.

Naish *et al.* (2001) and Wilson *et al.* (2002) argued in favor of the older option for the boundary age of 24.0 ± 0.1 Ma in their integrated stratigraphic study, including single-crystal $^{40}Ar/^{39}Ar$ age determinations of anorthoclase phenocrysts from two tephra horizons, in marine successions off Antarctica. The actual discrepancy may even increase to approximately 1.15 myr if an offset of ~0.8% between $^{40}Ar/^{39}Ar$ and astronomical ages is taken into account. Wilson *et al.* (2002) contend that the discrepancy with the astronomically calibrated ages arises from a mismatch of three 406-kyr eccentricity cycles or a 1.2 million year modulation of obliquity amplitude in the astronomical calibration of the Oligocene–Miocene time scale of Shackleton *et al.* (2000). This alternative view, however, is difficult to accept for two reasons:

1. Re-tuning of ODP Leg 154 data to La2003 not only shows a very flat phase response at the 1.2 and 170 kyr periods that modulate obliquity, but also displays a clear relationship with the ~2.4 myr component of eccentricity. This finding precludes moving the boundary age by 1.2 myr as suggested by Wilson *et al.* (2002).
2. Astronomical calibrations of sediments including a reliable magnetostratigraphy from ODP Site 1090 (K. Billups, pers. comm.) and ODP Sites 1218 and 1219 (H. Pälike, pers. comm.) by using stable isotope measurements (Billups *et al.*, 2002, and additional unpublished data) and multisensor track data (i.e. magnetic susceptibility), respectively, confirm the astronomical tuning of a large portion of the Early Miocene and latest Oligocene.

21.3.6 Incorporation of global chronostratigraphic boundaries

Incorporation of the basic chronostratigraphic units (stages) and their boundaries in the new time scale is straightforward as far as they have been formally defined. The reason for this is that they are defined in astronomically tuned sequences in the Mediterranean that have been used to build the time scale. Pliocene stage boundaries are, without exception, defined along the south coast of Sicily, Italy. They include the Gelasian, Piacenzian, and Zancleans GSSPs, which have been pinpointed in the San Nicola, Punta Piccola, and Eraclea Minoa sections and have re-tuned astrochronologic ages of 2.588, 3.600, and 5.332 Ma, respectively (Lourens *et al.*, 1996a; Castradori *et al.*, 1998; Rio *et al.*, 1998; Van Couvering *et al.*, 2000). The Eraclea Minoa section at 5.332 Ma marks the Miocene–Pliocene boundary. The Pliocene–Pleistocene boundary is formally designated at the base of the claystone overlying sapropel *e* in the Vrica section with an astronomical age of 1.806 Ma (Aguirre and Pasini, 1985; Lourens *et al.*, 1996a). Note that all mentioned ages are according to ATNTS 2004.

The GSSP of the youngest stage of the Miocene, the Messinian, has been formally defined in the Oued Akrech section located on the Atlantic side of Morocco, just outside the Mediterranean. The boundary is astronomically dated at 7.246 Ma (Hilgen *et al.*, 2000a,b) and tied via integrated stratigraphy to the Mediterranean type sections. The Tortonian GSSP at a level close to the CO of *Discoaster kugleri* in the tuned section of Monte dei Corvi in northern Italy (Hilgen *et al.*, 2002) was ratified by IUGS in 2003. A proposal for the Serravallian GSSP is anticipated mid 2004. However, it is expected that the remaining stage boundaries (Langhian and Burdigalian GSSPs) will take somewhat longer to define because suitable candidate sections in the Mediterranean that have been tuned are currently lacking.

21.3.7 Incorporation of zonal schemes

The standard low-latitude calcareous plankton zonal schemes are directly tied to the new time scale via biostratigraphic data obtained from the same sequences that have been used to construct the new time scale (Figs. 21.1 and 21.6; Tables A2.2 and

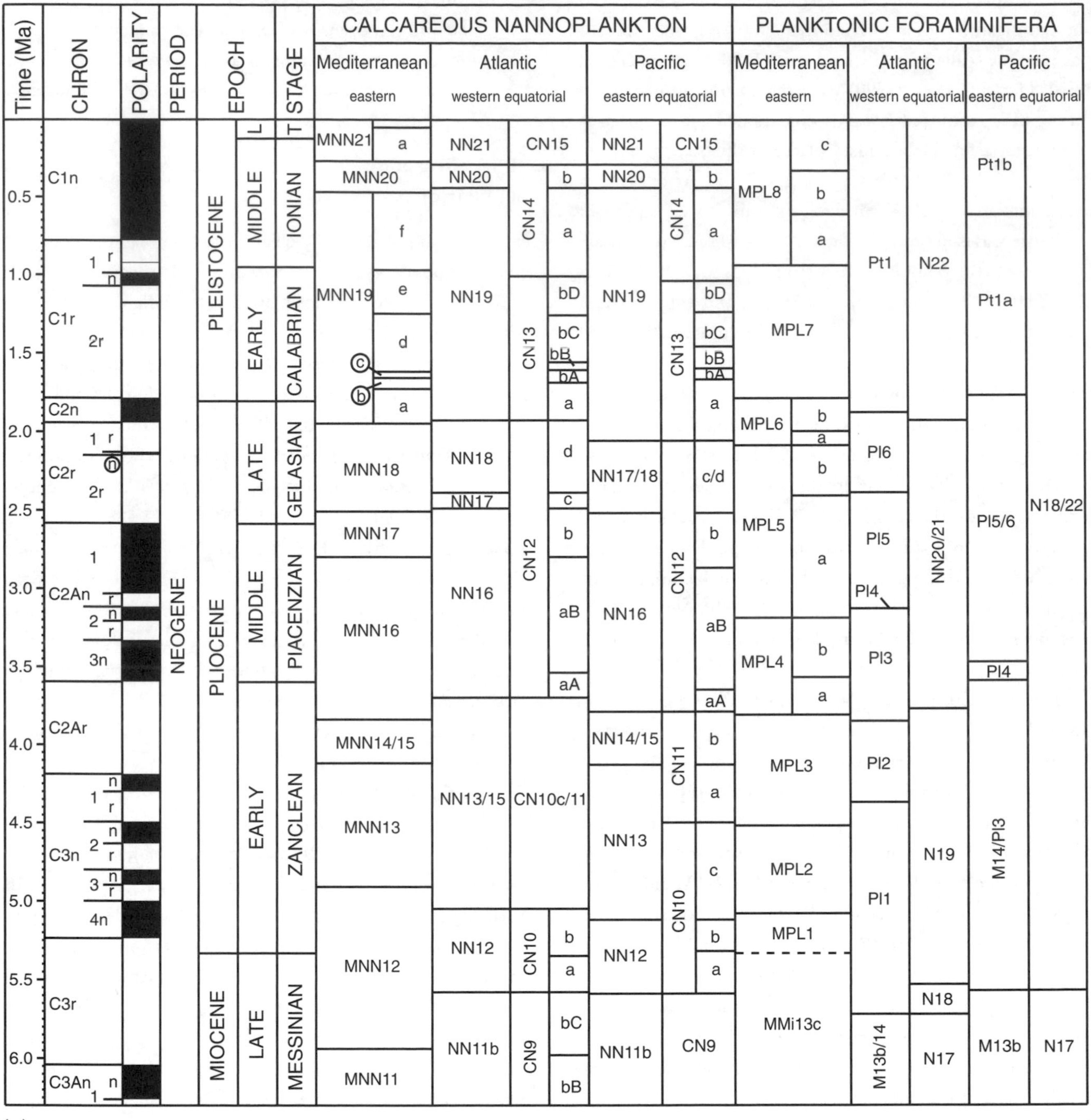

Figure 21.6 Correlation of planktonic foraminifer and calcareous nannoplankton zonal schemes for the Mediterranean, Atlantic, and Pacific. Astronomically tuned ages are based on Tables A2.1 through A2.3 in Appendix and references therein. (*a*) The Pliocene and Pleistocene. (*b*) The Middle–Late Miocene.

A2.3 in Appendix). Sample resolution is still low in the older part of the Miocene at Ceara Rise but we refrain from using biostratigraphic information from other cores because we value direct first-order calibrations to the astronomical time scale. First-order calibrations also exist for regional calcareous plankton zonations in the Mediterranean back to almost 14 Ma and for the eastern equatorial Pacific ODP Sites 677, 846, and 849 over the past ~6.0 myr (Table A2.2).

We have used revised astronomical ages for the planktonic foraminiferal bioevents (Table A2.3) used in the middle

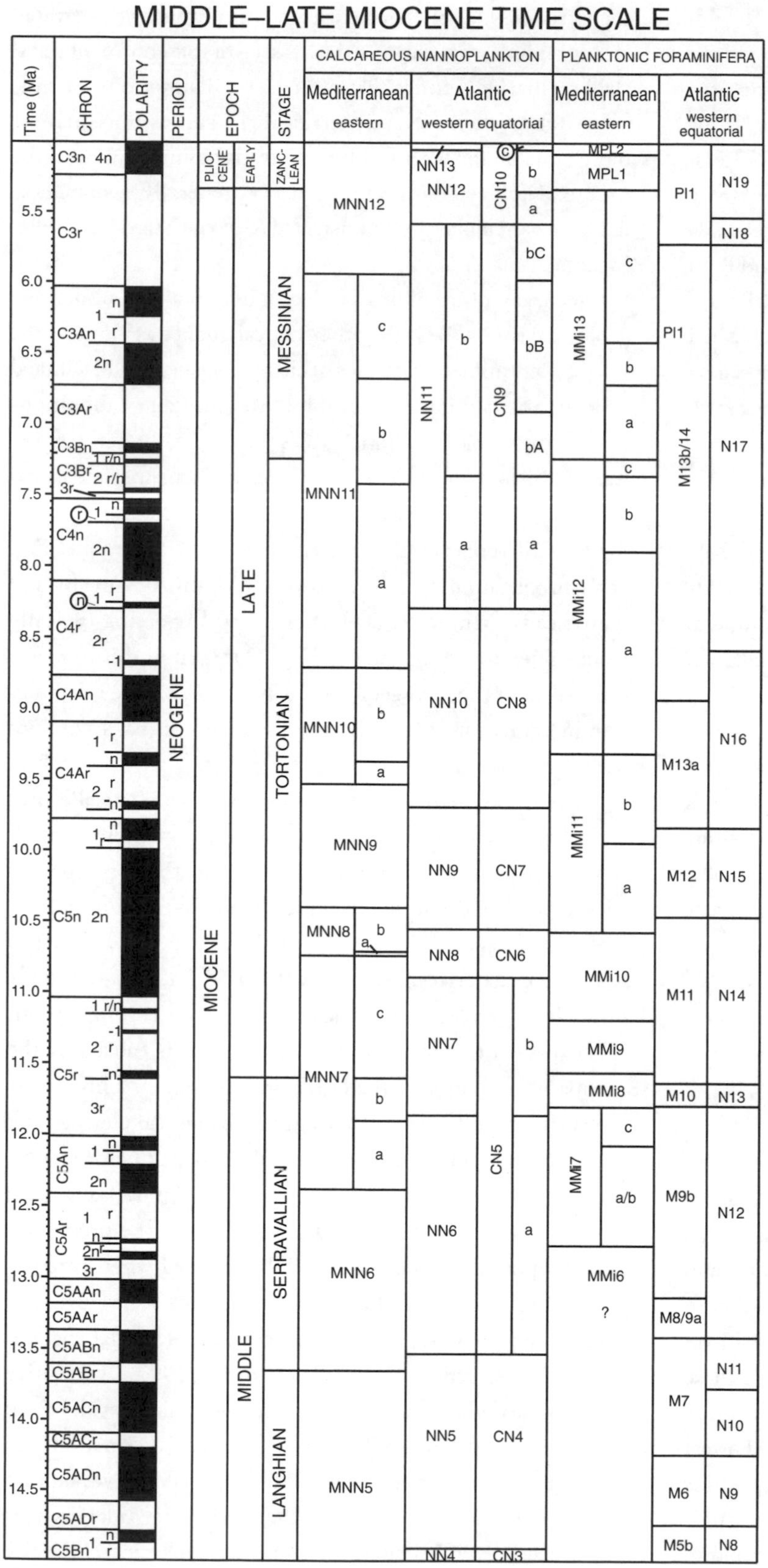

MIDDLE–LATE MIOCENE TIME SCALE

Time (Ma) | CHRON | POLARITY | PERIOD | EPOCH | STAGE | CALCAREOUS NANNOPLANKTON: Mediterranean eastern; Atlantic western equatorial | PLANKTONIC FORAMINIFERA: Mediterranean eastern; Atlantic western equatorial

Time (Ma): 5.5, 6.0, 6.5, 7.0, 7.5, 8.0, 8.5, 9.0, 9.5, 10.0, 10.5, 11.0, 11.5, 12.0, 12.5, 13.0, 13.5, 14.0, 14.5

CHRON: C3n 4n, C3r, C3An 1 n r, 2n, C3Ar, C3Bn 1 r/n, C3Br 2 r/n, 3r, C4n 1 n (r), 2n, C4r 1 r (n), 2r, -1, C4An, C4Ar 1 r n, 2 r n, C5n 1 n r, 2n, C5r 1 r/n, -1, 2 r n, 3r, C5An 1 n r, 2n, C5Ar 1 r, n, 2n r, 3r, C5AAn, C5AAr, C5ABn, C5ABr, C5ACn, C5ACr, C5ADn, C5ADr, C5Bn 1 n r

PERIOD: NEOGENE

EPOCH: PLIOCENE EARLY; MIOCENE LATE; MIDDLE

STAGE: ZANCLEAN, MESSINIAN, TORTONIAN, SERRAVALLIAN, LANGHIAN

Calcareous nannoplankton, Mediterranean eastern: MNN12; MNN11 c, b, a; MNN10 b, a; MNN9; MNN8 b, a; MNN7 c, b, a; MNN6; MNN5

Calcareous nannoplankton, Atlantic western equatorial: NN13, NN12; NN11 b, a; NN10; NN9; NN8; NN7; NN6; NN5; NN4 — CN10 (c), b, a; CN9 bC, bB, bA, a; CN8; CN7; CN6; CN5 b, a; CN4; CN3

Planktonic foraminifera, Mediterranean eastern: MPL2, MPL1; MMi13 c, b, a; MMi12 c, b, a; MMi11 b, a; MMi10; MMi9; MMi8; MMi7 c, a/b; MMi6 ?

Planktonic foraminifera, Atlantic western equatorial: Pl1; Pl1 M13b/14; M13a; M12; M11; M10; M9b; M8/9a; M7; M6; M5b — N19; N18; N17; N16; N15; N14; N13; N12; N11; N10; N9; N8

(*b*)

Figure 21.6 (*cont.*)

Miocene zonal scheme of Foresi *et al.* (2002) because the detailed tuning of Monte dei Corvi and Tremiti island (Hilgen *et al.*, 2003) is considered an improvement of the tuning proposed by Lirer *et al.* (2002) and Sprovieri *et al.* (2002a) for time equivalent sections in the Mediterranean (Fig. 21.6*b*). Moreover, we used revised astronomical ages for the planktonic foraminiferal and calcareous nannofossil bioevents of the middle and late Pleistocene zonal schemes of Cita (1976) and Raffi and Rio (1979), because a new detailed re-tuning of ODP Sites 964 and 967 and core MD69–KC01B (Lourens, unpublished data) is considered an improvement of the tuning proposed by Emeis *et al.* (2000), Kroon *et al.* (1998), Rossignol-Strick *et al.* (1998), and Langereis *et al.* (1997; Fig. 21.6*a*).

The ages for the (sub)chron reversal boundaries (Table A2.1) allow microfossil zonal schemes that are not directly tied to the astronomical time scale, but which are calibrated to CK95 to be converted to the new time scale. This approach holds for the low-, mid-, and high-latitude planktonic foraminiferal zonal schemes of Berggren *et al.* (1995a,b), and the radiolarian zonal scheme of Sanfilippo and Negrini (1998a; Figs. 21.1 and 21.2).

The incorporation of, for instance, the North American Land Mammal Ages (NALMAs) to the new time scale is problematical due to lack of fossil sites in long and continuous sections having a well-calibrated magnetostratigraphy. Ages for most NALMA boundaries are based on different radiometric dating techniques, which need to be evaluated and re-calibrated using the inter-calibration of radioisotopic and astronomical dating methods. Evidently, the chronometric consequences are largest for the Early Miocene.

21.3.8 Advantages of the new time scale

A geological time scale should preferably have a high accuracy and a high temporal and spatial resolution, requirements that are crucial to unravel cause-and-effect relationships and to determine rates of change. The systematic application of Milankovitch cycles has led to a Neogene time scale with an unprecedented resolution and accuracy thus fulfilling these requirements. The new time scale has a temporal resolution of <20 or <40 kyr depending on the period of the shortest orbital cycle used in the tuning procedure. In principle, this will eventually allow paleoclimatic and oceanographic studies in the entire Neogene with a resolution comparable to that of the late Pleistocene. Other stratigraphic tools such as high-resolution biostratigraphy, beyond the resolution of most commonly applied zonal schemes and magnetostratigraphy, including the use of short subchrons and cryptochrons, are then crucial to applying the astronomically tuned integrated stratigraphic framework in other less-favorable continental and shallow-marine settings that often lack a distinct type of Milankovitch cyclicity. Nevertheless, these successions may be more suitable for high-resolution stratigraphic studies of the Milankovitch-type of cyclicity than is generally assumed, as has been convincingly demonstrated (e.g. Van Vugt *et al.*, 1998; Steenbrink *et al.*, 2000).

Accuracy of absolute age determinations that underlie a geological time scale is the other critical quality as it allows the reliable determination of rates of change especially when linked to a high-resolution stratigraphy. Absolute age calibrations underlying geological time scales for the younger part of Earth's history (apart from the already incorporated tuned Pliocene–Pleistocene) are increasingly based on $^{40}Ar/^{39}Ar$ dating thereby replacing the conventional K–Ar technique. Full error propagation, which can be solved both analytically and numerically (Min *et al.*, 2000), for $^{40}Ar/^{39}Ar$ dating typically results in errors in the range of 2%. But even in this case, additional errors such as geological constraints for the reliability of mineral dating standards (see discussion on FCT) are not included (see Chapter 6).

By contrast, astronomically tuned ages are typically presented in three digits without error bars. The main uncertainties in astronomical dating and, hence, on the astronomical ages depend on the accuracy of the astronomical solution from which the target was derived and on the correctness of the tuning. The exact error in the presently used La2003 solution is difficult to calculate due to the complexity of the solution, but it is expected that an important uncertainty is caused by the dissipative evolution of the Earth–Moon system. This uncertainty may result from changes in the dynamical ellipticity of the Earth and/or tidal dissipation by the Moon. The La93 solution for the first time offered the possibility of modifying these two parameters. Both these parameters affect the precession and obliquity frequencies and will be reduced when entering an ice age (Laskar *et al.*, 1993). On longer time scales, dynamical ellipticity may also vary as a consequence of secular changes in mantle convection (Forte and Mitrovica, 1997). A sensitive test was carried out to estimate the error, which results from changing the dynamical ellipticity and/or tidal dissipation over the past 3 myr (Lourens *et al.*, 2001). This study revealed that the La93 solution including half the present-day tidal dissipation value or 0.9997 times the present-day dynamical ellipticity value resulted in the optimum fit between the precession and obliquity components of the insolation target curve and their related components derived from an exceptional climate proxy record of the eastern Mediterranean.

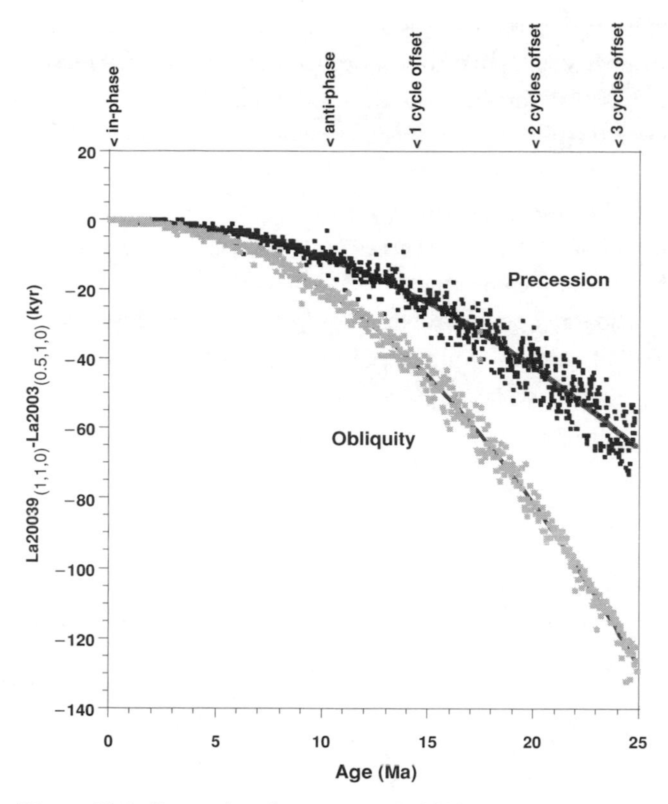

Figure 21.7 Comparison between two La2003 solutions including the present-day and half the present-day tidal dissipation value. Plotted are the differences in age between the two solutions for the correlative minima in precession and obliquity. Note that eccentricity is similar for both solutions.

To illustrate the uncertainty in the La2003 solution due to tidal dissipation we have plotted the difference (in thousands of years) of correlative minimum values in the obliquity and precession cycles between the $La2003_{(1,1,0)}$ and $La2003_{(0.5,1,0)}$ solutions for the last 25 myr (Fig. 21.7). The differences for both cycles indicate an anti-phase relationship around 10 Ma and an in-phase relationship around 15 Ma, but the precession and obliquity time series are stretched in the $La2003_{(0.5,1,0)}$ solution due to a reduced dissipitative effect and as a result contain one cycle less around 15 Ma. At ~23 Ma, both precession and obliquity cycles derived from the $La2003_{(0.5,1,0)}$ solution contain three cycles less than that obtained by the $La2003_{(1,1,0)}$ solution. Consequently the maximum uncertainty here will be ~68 kyr where proxy records are tuned to precession and ~123 kyr if tuned to obliquity. However, the additional use of amplitude modulations of precession (by eccentricity) and obliquity, and precession/obliquity interference patterns may help to constrain the correct tuning of the proxy records. It is anticipated that extension of the type of study carried out by Lourens *et al.* (2001) will solve the problem of which values for dynamical ellipticity and tidal dissipation should be incorporated in the astronomical solution.

Other errors may come from the correctness of the tuning itself. A mistuning of one cycle typically results in an error of ±20 kyr, or of ±40 kyr where obliquity is used in the tuning. However, such a tuning error may be restored if the amplitude modulation of precession (by eccentricity) or obliquity is also taken into account. Other potential tuning errors in the astronomical ages may stem from the presence of time lags between insolation forcing and climate response and registration in the stratigraphic record, and from an uncertainty in phase relations between the astronomically forced variations in the climate proxy records used for tuning and the initial insolation forcing. For the moment, we consider it less likely that phase relations used for building the astronomical time scale are incorrect, while the problem of time lags will result in a possible error of the order of one to several thousand years at a maximum.

In summary, the errors in the astronomical ages over the last 10 myr will be of the order of 0.1–0.2% and possibly even less. This is probably the case for the last 13 myr in view of the excellent fit between details in the Milankovitch cycle patterns in the Mediterranean (as confirmed in parallel sections) and the insolation target. For older time intervals, the error will remain very small, but the tuning of Ceara Rise has to be independently confirmed.

21.3.9 Summary

An astronomically tuned Neogene time scale (ATNTS2004) has been presented based on cyclic sedimentary successions from the western Equatorial Atlantic Ocean and Mediterranean. It continues the development that led Berggren *et al.* (1995a) to incorporate the Pliocene and Pleistocene astrochronology of Shackleton *et al.* (1990) and Hilgen (1991a,b) into their Neogene time scale. Construction of ATNTS2004 was made possible through: (i) technological and procedural improvements in deep-sea drilling, (ii) high-resolution studies of exposed marine sections in tectonically active areas where ancient seafloor has been rapidly uplifted, and (iii) improvements in the accuracy of astronomical solutions resulting in the La2003 numerical solution. A seafloor anomaly profile from the Australia–Antarctic plate pair was employed to complete the polarity time scale for the interval between 13 and 23 Ma due to the lack of magnetostratigraphic records for ODP Leg 154 sites. Biostratigraphic zonal schemes are either directly tied to the new time scale via first-order calibrations, such as the standard low-latitude calcareous plankton zonations, or are recalibrated to the associated polarity time scale. Formally designated chronostratigraphic boundaries (GSSPs of Neogene stages), defined in sections used to build the astronomically tuned integrated stratigraphic framework, are directly tied to the new time scale.

The new time scale results in a significantly younger age of 23.03 Ma for the Oligocene–Miocene boundary than the 23.8 Ma preferred in previous time scales; the latter age is based on radiometric age determinations that are not fully acceptable according to current standards. Intercalibration of independent astronomical and radiogenic isotopic dating methods is not yet possible, but new results (Kuiper, 2003) point to an astronomical-derived age of 28.24 ± 0.01 Ma for the Fish Canyon Tuff (FCT) sanidine and favor the introduction of an astronomically dated $^{40}Ar/^{39}Ar$ standard.

The astronomically tuned Neogene time scale, with its unprecedented accuracy (1–40 kyr) and resolution (<10 kyr), opens new perspectives for paleoclimatic and paleooceanographic studies of the entire period with a temporal resolution comparable to that of Pleistocene research (Krijgsman *et al.*, 1999; Zachos *et al.*, 2001a).

22 • The Pleistocene and Holocene Epochs

P. GIBBARD AND T. VAN KOLFSCHOTEN

This chapter focuses on the major subdivisions and events in the terrestrial sequences of the Pleistocene and Holocene, with correlations to the marine record. Current proposals for formal subdivision are outlined.

22.1 PLEISTOCENE SERIES

22.1.1 Evolution of terminology

The classification and interpretation of the youngest stratigraphic sequences, variously known as Pleistocene, Holocene, and Quaternary, have been, and still are, a matter of debate. During the first two decades of the nineteenth century, many of the sequences were attributed to the biblical flood (the "Diluvial" theory). This theory could account for unconsolidated sediments that rested unconformably on "Tertiary" rocks and capped hills, and that commonly contained exotic boulders and the remains of animals, many still extant. This origin for the "Diluvium" was accepted by most eminent geologists of the time, including Buckland and Sedgwick.

Floating ice had frequently been seen transporting exotic materials, providing an explanation for the transport of the boulders, and reinforcing the Diluvial theory. This explanation lead to adoption of the term "drift" to characterize the sediments. However, geologists working in the Alps and northern Europe had been struck by the extraordinary similarity of the "drift" deposits and their associated landforms to those being formed by modern mountain glaciers. Several observers such as Perraudin, Venetz-Sitten, and de Charpentier proposed that the glaciers had formerly been more extensive, but it was the paleontologist Agassiz who first advocated that this extension represented a time that came to be termed the Ice Age by Goethe.

After having convinced Buckland and Lyell of the validity of his Glacial Theory in 1840, Agassiz's ideas became progressively accepted. The term *Drift* became established for the widespread sands, gravels, and boulder clays thought to have been deposited by glacial ice. Meanwhile, Lyell had already proposed the term *Pleistocene* in 1839 for the post-Pliocene period closest to the present. He defined this period on the basis of its molluscan faunal content, the majority of which are still extant. However, the term Quaternary (*Quaternaire or Tertiaire récent*) had already been proposed in 1829 by Desnoyers for marine sediments in the Seine Basin (Bourdier, 1957, p. 99) – although the term had been in use from the late eighteenth century.

Both terms – Pleistocene and Quaternary – have become synonymous with the Ice Age. However, unlike the Pleistocene concept, the span of the Quaternary included Lyell's original "Recent," later named *Holocene* by the Third International Geological Congress (IGC) in London in 1885. The term *Holocene* (meaning "wholly recent") refers to the percentage of living organisms and was defined by Gervais (1867–1869) "for the post-diluvial deposits approximately corresponding to the post-glacial period" (Bourdier, 1957, p. 101). The Holocene period was originally considered to represent a fifth era or *Quinquennaire* (Parandier, 1891), but this division was deemed to be "excessive;" details are given in Bourdier (1957) and de Lumley (1976).

Because the terms Primary, Secondary, and Tertiary have been abandoned, the continued use of Quaternary is regarded by some stratigraphers as somewhat archaic. Alternative terms of *Anthropogene* (extensively used in the former USSR) or *Pleistogene* (suggested by Harland *et al.*, 1990) have been proposed, but neither found favor. Other proposals place the Holocene in the Pleistocene epoch as a stage (cf. the Flandrian: see below).

The "Quaternary" is traditionally considered to be the interval of oscillating climatic extremes (glacial and interglacial episodes) that was initiated at about 2.6 Ma, therefore encompassing the Holocene and Pleistocene epochs and Gelasian stage of late Pliocene. A formal decision on its chronostratigraphic status is pending, as advocated by ICS and INQUA (Pillans, 2004).

22.1.2 The Pliocene–Pleistocene boundary and the status of the Quaternary

In 1948 at the IGC in London, an attempt was made to identify a basal-Pleistocene boundary. The requirement that it be located in exposed marine sediment led its placement near or at the base of the Calabrian strata in southern Italy (a stage introduced by Gignoux in 1910). This horizon was close to the first indication of climatic deterioration in Italy that took place after the deposition of the Italian Neogene (Oakly, 1950). The initial Calabrian boundary was thought to be marked by the first appearance of the cold-water mollus indicators *Arctica islandica* and *Hyalinea baltica* (Sibrava, 1978), but Ruggieri and Sprovieri (1979) showed that *Hyalinea baltica* appears slightly later. Moreover, he argued for the suppression of Calabrian and its replacement by Santernian, together with a revision of the rest of the sequence. Subsequently, various sections in southern Italy competed for the position of stratotype. Haq *et al.* (1977) correlated the boundary with the top of, or slightly above, the short-lived Olduvai magnetostratigraphic event at 1.8 Ma.

The GSSP for the Pliocene–Pleistocene boundary and the beginning of the Pleistocene was placed by a joint INQUA and ICS working group (IGCP Project 41) near the top of the Olduvai subchron (1.8 Ma) and approved by ICS in 1983. The GSSP is at Vrica (39° 32′ 18.61″ N, 17° 08′ 05.79″ E), approximately 4 km south of Crotone in the Marchesato Peninsula, Calabria, southern Italy (Aquirre and Pasini, 1985; Bassett, 1985). Stratigraphic details are given in Chapter 21 (Section 21.1.4)

The decision to assign the base-Pleistocene GSSP was "*isolated from other more or less related problems, such as ... the status of the Quaternary within the chronostratigraphic scale*" (Aguirre and Pasini, 1985). Many "Quaternary" workers, especially those working with terrestrial and climatic records, now favor defining "Quaternary" as beginning significantly before the base-Pleistocene GSSP. As a result, the status (as of 2004) and chronostratigraphic rank of Quaternary has not been established, and different options for formally defining "Quaternary" are being considered (e.g. Ogg, 2004; Pillans, 2004; Pillans and Naish, 2004).

The London 1948 IUGS recommendations included the notion that the base-Pleistocene boundary should be placed at the first evidence of climatic cooling. However, the Vrica GSSP boundary level is not the first severe cold climate oscillation of the late Cenozoic. It can be argued that the first severe cold climate takes place at a stratigraphic position equivalent to the base of the Dutch terrestrial Praetiglian Stage, and some earth scientists studying Quaternary strata in northern Europe tend to begin their Pleistocene at this level. Since the record on land is highly fragmentary and difficult to correlate, eastern Europeans have their own terminology in which the Eopleistocene follows the Pliocene (cf. Section 22.2 below). These alternative sequence terminologies have been included in Fig. 22.1.

This older level corresponds to the Gauss/Matuyama magnetic epoch boundary (2.6 Ma) and the base of the Pliocene Gelasian Stage (Rio *et al.*, 1998). An equivalent level in marine sediments occurs at Monte San Nicola in Sicily and can be easily correlated with Marine Isotope Stage (MIS) 104 in the ocean sediments (see discussion in Suc *et al.*, 1997). The event is clearly defined in the marine oxygen isotope stratigraphy and coincides with the first major influx of ice-rafted debris into the middle latitude of the North Atlantic (Shackleton *et al.*, 1984; Shackleton, 1997; Partridge, 1997a). The fossil mammalian record also shows changes that are obvious near the Gauss/Matuyama reversal. Opposing views on the position for the Pliocene–Pleistocene boundary have been discussed by Van Couvering (1997) and by Partridge (1997b).

22.1.3 Division of the Pleistocene

Two major types of subdivisions have been proposed for the Pleistocene Series. A standard subdivision at stage level has been advocated by workers based on sections in elevated shallow-marine sediments in Italy (see Chapter 21 and Fig. 22.1). Earth scientists concerned with terrestrial and to a lesser extent shallow-marine sequences have adopted regional subdivision schemes. The regional schemes have found favor despite the difficulties of world-wide correlation. In these schemes, larger, subseries- (sub epoch) scale units have been adopted and are advocated here.

A quasi-formal tripartite subdivision of the Pleistocene into Lower, Middle, and Upper has been in use since the 1930s. The first usage of the terms Lower, Middle, and Upper Pleistocene was at the second International Quaternary Association (INQUA) Congress in Leningrad 1932 (Woldstedt, 1962), although they may have been used in a loose way before this time. Their first use in a formal sense in English was by Zeuner (1935, 1959) and Hopwood (1935) and was based on characteristic assemblages of vertebrate fossils in the European sequence.

The desire to make these units identifiable world-wide led the INQUA Commission on Stratigraphy/ICS Working Group on Major Subdivision of the Pleistocene (Richmond, 1996) to place the Lower–Middle boundary at the Brunhes–Matuyama magnetic reversal epoch boundary; the "Toronto Proposal" of Richmond (1996). Unfortunately, it is less easy to define the Middle–Upper boundary in the same fashion and therefore it seems expedient to consider it equivalent to the base of marine isotope stage 5 (MIS 5) following the long-established convention that the basal boundary of the Upper Pleistocene corresponds with that of the last interglacial stage (proposed at INQUA Commission on Stratigraphy working group meeting, Berlin 1995, unpublished). This proposal naturally follows from the acceptance that MIS 5, substage e, is the ocean equivalent of the terrestrial northwest European Eemian Stage interglacial (Shackleton, 1977).

The proposals of the INQUA/ICS Working Group on Major Subdivision of the Pleistocene (Richmond, 1996) can be summarized as follows:

> It was proposed that the initial Middle Pleistocene boundary be placed at the Matuyama–Brunhes magnetic polarity reversal. The reversal has not been dated directly by radiometric controls. It is significantly older than the Bishop Tuff (revised K–Ar age 738 ka; Izett, 1982), and the estimated K–Ar age of 730 ka assigned to the reversal by Mankinen & Dalrymple (1979) is too young. In Utah, the Bishop volcanic ash bed overlies a major paleosol developed in sediments that record the Matuyama–Brunhes reversal (Eardley *et al.*, 1973). The terrestrial geologic record is compatible with the astronomical age of 788 ka assigned to the reversal by Johnson (1982).

[Note: The age of this Matuyama–Brunhes reversal is estimated as 781 ka in the astronomical-tuned time scale, see Chapter 20.] The initial Late Pleistocene boundary, placed arbitrarily at the beginning of MIS 5 (at the midpoint of Termination II or the MIS 6–5 transition), is not dated directly. It was assigned provisional ages of 127 ka by CLIMAP Project members (1984) and 128 ka by SPECMAP Project members (Ruddiman and Mcintyre, 1984), based on uranium series ages of the MI substage 5e high eustatic sea-level stand. However, more recent re-evaluation of the boundary indicates that following historical precedent in northwest Europe, the Middle–Upper Pleistocene subseries boundary should correspond to the Saalian–Eemian Stage boundary rather than to the boundary in marine isotope records which is not coeval (see below). The former is positioned at the boundary stratotype of the latter at 63.5 m below surface in the Amsterdam Terminal borehole (52 E 0913: 52 22 45 N; 4 54 52 E). This parastratotype locality is also the Eemian Stage unit-stratotype (Cleveringa *et al.* 2000; van Kolfschoten and Gibbard, 2000; van Leeuwen *et al.*, 2000). Both the stage and the stage boundary are recognized on the basis of multidisciplinary biostratigraphy, the boundary being placed at the expansion of forest tree pollen above 50% of the total pollen assemblage, the standard practice in northwest Europe (Gibbard, 2003). The Saalian–Eemian Stage boundary is identified at 126 kyr in deep-sea sediment off Iberia by Sanchez-Goñi *et al.* (1999) and Shackleton *et al.* (2002).

Independently, groups of workers have advocated a subdivision based on "standard stages" comparable in scale to those defined for the Neogene, as already noted. Of particular importance is the scheme that has been developed for shallow-marine sequences in southern Italy (Fig. 22.1) and summarized in Chapter 21.

Other shallow-marine sequences, such as that from New Zealand, have also been developed. In the former USSR, and particularly in European Russia, the Pleistocene is divided into the Eopleistocene, equivalent to the Early Pleistocene subseries, and the Neopleistocene, equivalent to the Middle and Late Pleistocene subseries (Anonymous, 1982, 1984; Krasnenkov *et al.*, 1997). The most recent proposal for a revised stratigraphical scheme for the last 1 Ma in the Eastern European Plain is given by Shik *et al.* (2002).

22.2 TERRESTRIAL SEQUENCES

In contrast to the rest of the Phanerozoic, the uppermost Cenozoic has a long-established tradition of sediment sequences being divided on the basis of represented climatic changes, particularly sequences based on glacial deposits in central Europe and mid-latitude North America. This approach was adopted by early workers for terrestrial sequences because it seemed logical to divide till (glacial diamicton) sheets and non-glacial deposits or stratigraphical sequences into *glacial (Glaciation)* and *interglacial* periods, respectively (cf. West, 1968, 1977; Bowen, 1978). In other words, the divisions were fundamentally lithological. The overriding influence of climatic change on sedimentation and erosion has meant that, despite the enormous advances in knowledge during the last century and a half, climate-based classification has remained central to the subdivision of the succession. Indeed, the subdivision of the modern ocean sediment isotope stage sequence is itself based on the same basic concept (see below). It is this approach which has brought Quaternary geology so far, but at the same time causes considerable confusion to workers attempting to correlate sequences from enormously differing geographical and, thus, environmental settings. This is because of the great complexity of climatic change and the very variable effects of the changes on natural systems.

The recognition of climatic events from sediments is an inferential method and by no means straightforward. Sediments are not unambiguous indicators of contemporaneous climate, and other evidence such as fossil assemblages, characteristic sedimentary structures (including periglacial

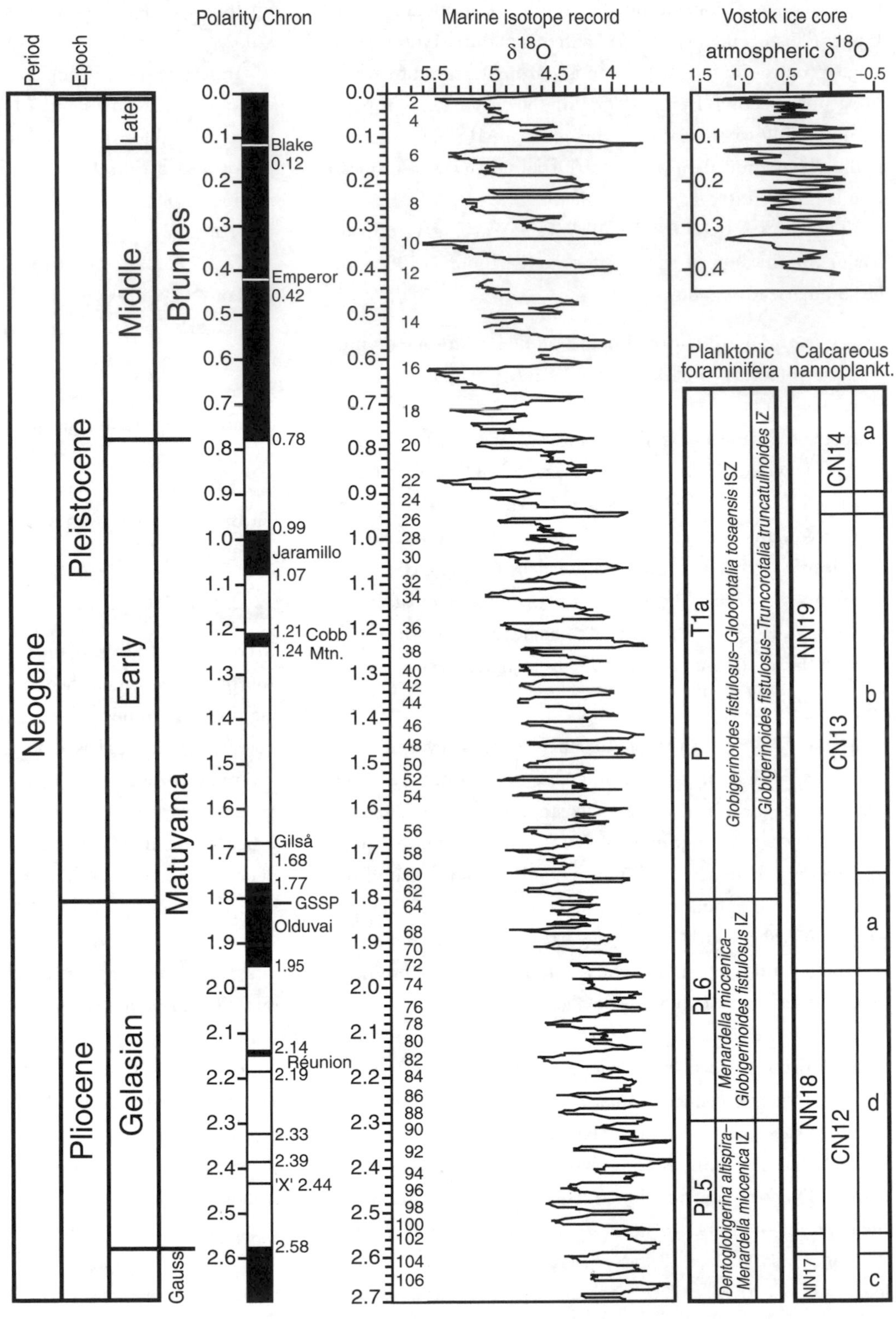

Figure 22.1 The Pleistocene-Holocene and upper Pliocene time scale. The Global Stratotype Section and Point (GSSP) for the base of the Pleistocene Epoch is indicated. The calibration of the geomagnetic polarity time scale is from oceanographic data collected and processed by S. J. Crowhurst (Delphi Project 2002) and modified from Funnell (1996). The composite marine $\delta^{18}O$ isotope sequence is from the Delphi Project (database at http://131.111.44.196 at Godwin Laboratory, University of Cambridge, UK). The micropaleontological zonations are from Berggren *et al.* (1995a). The atmospheric oxygen isotope curve from the Vostok ice coring is from Petit *et al.* (2001, Vostok Ice Core Data for 420, 000 Years, IGBP PAGES/World Data Center for Paleoclimatology Data Contribution Series #2001-076, at NOAA/NGDC Paleoclimatology Program, Boulder, CO, USA; original reference is Petit *et al.*, 1999).

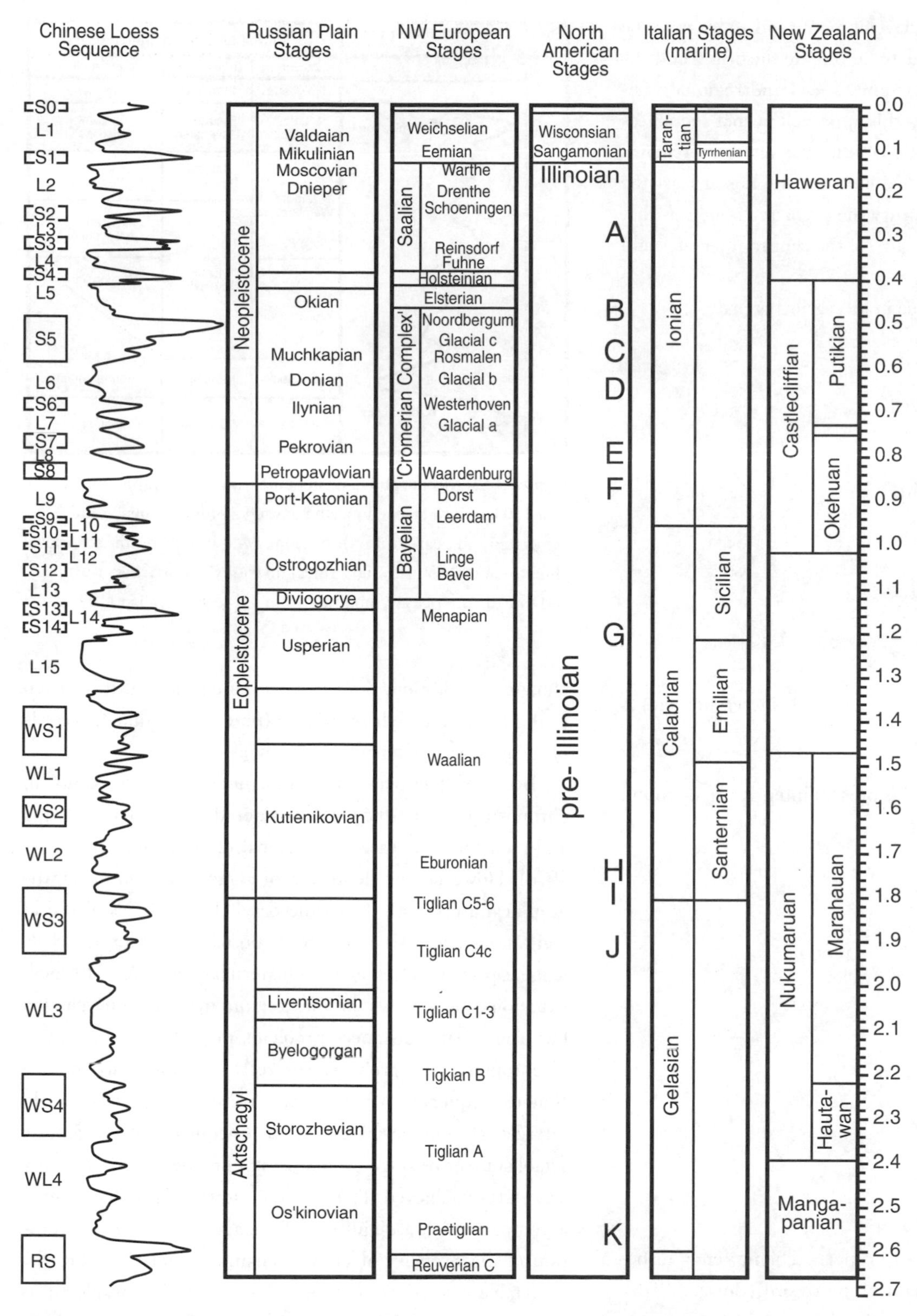

Figure 22.1 (*cont.*) This curve can be downloaded from www.ngdc.noaa.gov/paleo/icecore/antarctica/vostok/vostok_isotope.html. The Chinese loess sequence is a magnetic susceptibility signature from Luochuan (An Zhisheng *et al.*, 1990); S, soil horizon; L, loess interval; W and R, numbered older successions. The Quaternary continental successions were compiled from Zagwijn (1992), de Jong (1988), Tzedakis *et al.* (1997). European Russia succession were compiled from the Stratigraphy of the USSR: Quaternary System (Anonymous, 1982, 1984), Krasnenkov *et al.* (1997), and Shik *et al.* (2002). North America successions from Richmond (unpublished). The stage successions based on shallow-marine sequences of Italy (van Couvering, 1997) and New Zealand (Pillans, 1991) are included.

structures) or textures, soil development, and so on must be relied upon wherever possible to illuminate the origin and climatic affinities of a particular unit. Local and regional variability of climate complicates this approach in that sequences are the result of local climatic conditions, yet there remains the need to equate them with a global scale. For at least the first half of the twentieth century the preferred scale was that developed for the Alps at the turn of the century by Penck and Brückner (1909).

For the Alps, the sequence in increasing age is:

Würm Glacial (Würmian)
Riss–Würm Interglacial
Riss Glacial (Rissian)
Mindel–Riss Interglacial
Mindel Glacial (Mindelian)
Günz–Mindel Interglacial
Günz Glacial
Donau–Günz Interglacial
Donau Glacial
?Biber Glacial

For northern Europe, the sequence (with increasing age) is (see also Figs. 22.1 and 22.2):

Flandrian (i.e. Holocene), i.e. present interglacial, up to and including the present day.
Weichselian Glacial
Eemian Interglacial
Saalian Glacial Complex
Holsteinian Interglacial
Elsterian Glacial
Cromerian Complex
Bavelian Complex
Menapian Glacial
Waalian Interglacial
Eburonian Glacial
Tiglian Complex
Praetiglian Glacial

More recently, the northern European scheme tends to become replaced by the marine isotope record (Bowen, 1978). Today the burden of correlation lies in equating local, highly fragmentary, yet high-resolution terrestrial and shallow-marine sediments on the one hand, with the potentially continuous, yet comparatively lower resolution ocean isotope sequence on the other. Both are required, particularly because the ocean record sums a global situation while terrestrial sequences are dependent on local and regional conditions of climate. In addition, both need to be considered in relation to the extremely high-resolution (potentially annual) records of the ice-core sequences (see below).

Stacked Marine Isotope Record (with Marine Isotope Stages)

Terrestrial Chronostratigraphy		
N.W. Europe	British Isles	North America
Holocene	Flandrian	Holocene
Late Weichselian	Late Devensian	Late Wisconsian
Middle Weichselian	Middle Devensian	Middle Wisconsian
	Early Devensian	Early Wisconsian
Early Weichselian		Eo-Wisconsian
Eemian	Ipswichian	Sangamonian
Saalian	Wolstonian	Illinoian

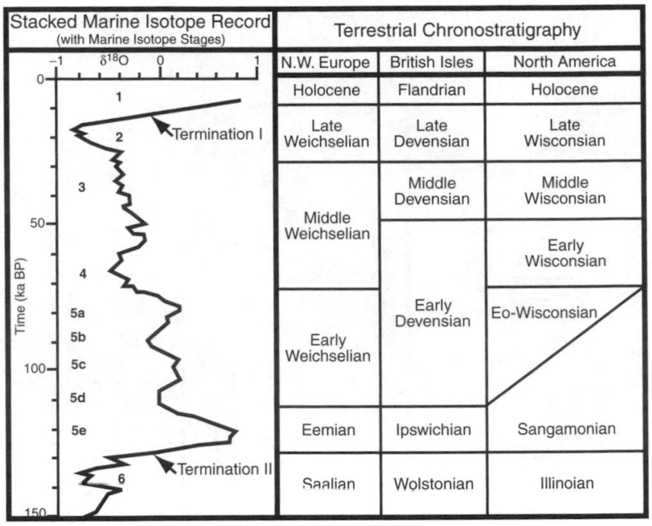

Figure 22.2 Marine and continental chronostratigraphy for the past 150 kyr. The stacked marine oxygen isotope sequence and associated stages are from Martinson *et al.* (1987), and the terrestrial climato- and chronostratigraphical divisions in northwest Europe and North America are modified from Lowe and Walker (1997).

Because stages are the fundamental working units in chronostratigraphy they are considered appropriate in scope and rank for practical intra-regional classification (Hedberg, 1976). However, the definition of stage-status chronostratigraphical units, with their time-parallel boundaries placed in continuous successions wherever possible, is a serious challenge especially in terrestrial Quaternary sequences. In these situations, boundaries in one region may be time-parallel but over greater distances problems may arise as a result of diachroneity. It is probably correct to say that only in continuous sequences which span entire interglacial–glacial–interglacial climatic cycles can an unequivocal basis for the establishment of stage events using climatic criteria be truly successfully achieved. There are the additional problems which accompany such a definition of a stage, including the question of diachroneity of climate changes themselves and the detectable responses to those changes. For example, it is well known that there are various "lag" times of geological responses to climatic stimuli. Thus, in short, climate-based units cannot be the direct equivalents of chronostratigraphical units because of the time-transgressive nature of the former. This distinction of a stage in a terrestrial sequence from that in a marine sequence should be remembered.

Before the impact of the ocean core isotope sequences, an attempt was made to formalize the climate-based stratigraphical terminology in the American Code of Stratigraphic Nomenclature (American Commission, 1961), where so-called geologic-climate units were proposed. Here a geologic-climate unit is based on an inferred widespread climatic episode defined from a subdivision of Quaternary rocks (American Commission, 1961). Several synonyms for this category of units have been suggested, the most recent being climatostratigraphical units (Mangerud *et al.*, 1974) in which an hierarchy of terms is proposed. These units are neither referred to in the standard stratigraphic codes by Hedberg (1976) nor Salvador (1994), and are not followed in New Zealand, but are included in the local Norwegian Code (Nystuen, 1986). Boundaries between geologic-climate units were to be placed at those of the stratigraphic units on which they were based.

The American Commission (1961) defines the fundamental units of the geologic-climate classification as follows:

> A *Glaciation* is a climatic episode during which extensive glaciers developed, attained a maximum extent, and receded. A *Stadial* ('Stade') is a climatic episode, representing a subdivision of a glaciation, during which a secondary advance of glaciers took place. An *Interstadial* ('Interstade') is a climatic episode within a glaciation during which a secondary recession or standstill of glaciers took place. An *Interglacial* ('Interglaciation') is an episode during which the climate was incompatible with the wide extent of glaciers that characterise a glaciation.

In Europe, following the work of Jessen and Milthers (1928), it is customary to use the terms interglacial and interstadial to define characteristic types of non-glacial climatic conditions indicated by vegetational changes (Table 22.1); interglacial to describe a temperate period with a climatic optimum at least as warm as the present interglacial (Holocene, Flandrian: see below) in the same region, and interstadial to describe a period that was either too short or too cold to allow the development of temperate deciduous forest or the equivalent of interglacial-type in the same region.

In North America, mainly in the USA, the term interglaciation is occasionally used for interglacial (cf. American Commission, 1961). Likewise, the terms stade and interstade may be used instead of stadial and interstadial, respectively (cf. American Commission, 1961). The origin of these terms is not certain but the latter almost certainly derive from the French word *stade* (m), which is unfortunate since in French *stade* means (chronostratigraphical) stage (cf. Michel *et al.*, 1997), e.g. *stade isotopique marin* = marine isotope stage.

It will be readily apparent that, although in longstanding usage, the glacially based terms are very difficult to apply outside glaciated regions, i.e. most of the world. Moreover, as Suggate and West (1969) recognized, the term Glaciation or Glacial is particularly inappropriate since modern knowledge indicates that cold rather than glacial climates have tended to characterize the periods intervening between interglacial events. They therefore proposed that the term "cold" stage (chronostratigraphy) be adopted for "glacial" or "glaciation." Likewise, they proposed the use of the term "warm" or "temperate" stage for interglacial, both being based on regional stratotypes. The local nature of these definitions indicates that they cannot necessarily be used across great distances or between different climatic provinces (Suggate and West, 1969; Suggate, 1974; West, 1968, 1977) or indeed across the terrestrial–marine facies boundary (see below). The use of mammalian biostratigraphic data, in particular the evolution of voles, offers the possibility of long-distance correlations between local assemblages. In addition, it is worth noting that the subdivision into glacial and interglacial is mainly ap plied to the Middle and Late Pleistocene.

Both interglacial and glacial, or temperate and cold stages, have been subdivided into substages and zones. This is achieved in interglacial stages using paleontological, particularly vegetational, assemblages. The cyclic pattern of interglacial vegetation that typifies all known temperate events in Europe was developed as a means of subdividing, comparing, and, therefore, characterizing temperate events by West (1968, 1977) and Turner and West (1968). In this scheme, temperate (interglacial) event sequences are subdivided into four substages: pre-temperate, early temperate, late temperate, and post-temperate. Finer-scale zonation schemes are also commonly in use throughout Europe and the former USSR (Table 22.1).

Late Middle- and Late-Pleistocene glacial stages have been divided on various bases, but in the northern hemisphere the division is based on a combination of vegetation, lithology, and occasionally pedological evidence, often resulting in an unfortunate intermixture of chrono-and climatostratigraphical terminology. The last glacial stage (Weichselian, Valdaian, Devensian, Wisconsinian) has particularly been divided into three or four substages (Early, Middle or Pleni-glacial, Late Weichselian, etc., and Late-glacial), using geochronology (^{14}C). Boundaries are defined at specific dates, especially in the last 30 ka (Table 22.1).

An independent record of Late Pleistocene and Holocene climatic changes has been derived from $\delta^{16}O/\delta^{18}O$ ratios in cores through the Greenland and Antarctic ice sheets (Johnsen *et al.*, 1972; Dansgaard *et al.*, 1993) and from other areas. In the

Table 22.1 *Examples of chronostratigraphical substage divisions of interglacial (temperate) stages and related cold (glacial) stages of the Middle and Late Pleistocene*[a,b]

	Chronostratigraphical substages[c]				Vegetational aspect	Characteristic vegetation
	Cromerian (~750ka)	Hoxnian (~350ka)	Ipswichian (~125ka)	Flandrian (post-10ka)		
Cold stage	e An	e Wo	e De		early glacial	herb-dominated
Temperate stage	Cr IV	Ho IV	Ip IV		post-temperate	birch–pine forest
	Cr III	Ho III	Ip III	Fl III	late-temperate	mixed deciduous–coniferous forest
	Cr II	Ho II	Ip II	Fl II	early-temperate	deciduous forest
	Cr I	Ho I	Ip I	Fl I	pre-temperate	birch–pine forest
Cold stage	l Be	l An	l Wo	l De	late-glacial	herb-dominated

[a] Modified after West (1968) and West & Turner (1968).
[b] For the Holocene (Flandrian), correlations with the zones of Godwin (1975) are also indicated.
[c] e, early; l, late; Be, Beestonian; An, Anglian; Wo, Wolstonian; De, Devensian.

past three decades, the drilling of cores into ice sheets in various parts of the world has revolutionized our records of detailed climatic change. Boreholes sunk particularly in Antarctica and in Greenland, and more recently into smaller ice shields in tropical mountain areas, have provided spectacularly unrivalled sequences which allow annual resolution of climatic events.

From a stratigraphical point of view, it is the recognition of patterns of a wide range of climatically controlled parameters that provide potentially very high-resolution correlation tools. Detailed patterns arise from determination of aerosol particles, dust, trace elements, spores, or pollen grains, etc., that have fallen onto the ice surface and become incorporated into the annual ice layers. They include, for example, dust from wind activity or volcanic eruptions. Trace gases such as carbon dioxide or methane can be trapped in air bubbles within ice crystals, etc. These gases can themselves be analyzed to determine their stable isotope content, particularly that of $\delta^{18}O$ and the sequences obtained can be compared to those from ocean sediments. In addition, naturally and artificially occurring radioactive isotopes present in the ice layers can be used to provide an independent chronology for dating the ice core sequences.

Most cores span the Holocene and provide an annually resolved sequence for the interglacial. However, the Vostok core in Antarctica spans a period of up to 400 kyr (Fig. 22.1; Petit *et al.*, 1999), but it is in Greenland that detailed cores from the Greenland Ice Core Project (GISP) and the Greenland Ice Sheet Project (GISP) have been obtained that provide a sequence that extends at least as far back as the Last Interglacial (Eemian). These sequences have revolutionized our understanding of patterns and rates of global climate changes, as well as the interlinking of the ocean–atmosphere–terrestrial systems (see Lowe and Walker, 1997, for a more detailed discussion).

22.3 OCEAN SEDIMENT SEQUENCES

Because the span of Quaternary time includes our own, a different order of discrimination is possible and different methods are rapidly developing. The principal development in the Pleistocene time scale depends on the regularity of the climatic cycle that was discovered around 1875 by Croll and developed, especially, by Milankovitch. This approach was not taken seriously by Quaternary geologists until Zeuner (1945), Emiliani (1955), and Evans (1971) were among those to recalculate and relate the astronomical parameters, testing, for example, 42 and 100 kyr cycles against other phenomena, such as the newly established oxygen isotope curve from the oceans. The first rigorous treatment using wide-ranging techniques was by Hays et al. (1976). Isotope studies from the bottom sediments of the world's oceans since then have indicated as many as 52 Late Cenozoic glacial ages and that the continental evidence is so incomplete as compared with the oceanic sequences that terrestrial glacial–interglacial stratigraphy in future must depend on the ocean record for chronological foundation as outlined in Chapter 21.

The marine oxygen isotope scale makes use of the fact that, when continental ice builds up as a result of global cooling and sea level is lowered, the ice is depleted in $\delta^{18}O$ relative to the ocean water, leaving the ocean water enriched in $\delta^{18}O$. The oxygen isotope composition of calcareous foraminifera and coccoliths, and of siliceous diatoms, varies in direct proportion to that of the water (cf. Shackleton and Opdyke, 1973, for discussion of the limitations of isotope stratigraphy). The

16 stages of Emiliani (1955, 1966) obtained from Caribbean and Atlantic sediment cores were extended to 22 by Shackleton and Opdyke (1973) after analysis of the V28–238 core from the equatorial Pacific. Subsequently another equatorial Pacific core, V28–239 (Shackleton and Opdyke, 1973), and an Atlantic core (Van Donk, 1976) extended the reconstruction of glacial–interglacial variability through the Pliocene–Pleistocene boundary. Later developments under the flag of the Deep Sea Drilling Program resulted in the extension of the isotope record into the Early Pleistocene and Pliocene (Shackleton and Hall, 1989; Ruddiman *et al.*, 1987; Raymo *et al.*, 1989). The sequence shown in Fig. 22.1 is a combination of measurements from cores V19–30, ODP677, and ODP846 (Crowhurst, 2002). The isotope stages recognized in core V28–238, from the eastern Pacific (Shackleton and Opdyke, 1973), are generally regarded as the "type" for the Late Pleistocene, while those defined in cores ODP 677 and 846 are those for the Middle Pleistocene and Pliocene, respectively (Shackleton and Hall, 1989; Shackleton *et al.*, 1995).

As regards nomenclature, the events differentiated in isotope sequences are termed marine isotope stages (MIS); this term is preferred by palaeoceanographers to the previously widely used oxygen isotope stages (OIS). This is because of the need to distinguish the isotope stages recognized from those identified from ice cores or speleothem sequences (Shackleton, pers. comm.). The stages are numbered from the present-day (MIS 1) backwards in time, such that cold-climate or glacial events are assigned even numbers and warm or interglacial (and interstadial) events are given odd numbers. Individual events or substages in marine isotope stages are indicated either by lower-case letters or in some cases by a decimal system, thus MIS 5 is divided into warm substages 5a, 5c, and 5e, and cold substages 5b and 5d, or 5.1, 5.3, 5.5, and 5.2 and 5.4, respectively, named from the top downwards. This apparently unconventional top-downwards nomenclature originates from Emiliani's (1955) original terminology and reflects the need to identify oscillations down cores from the ocean floor.

The biggest problem with climate-based nomenclature, like the marine isotope stratigraphy, is where the boundaries should be drawn. Ideally, the boundaries should be placed at a major climate change. However, this is problematic because of the multifactorial nature of climate. But since the events are only recognized through the responses they initiate in depositional systems and biota, a compromise must be agreed. Although there are many places at which boundaries could be drawn, in principle in ocean-sediment cores they are placed at midpoints between temperature maxima and minima. The boundary points thus defined in ocean sequences are assumed to be globally isochronous, although a drawback is that temperatures may be very locally influenced and may also show time lag. The extremely slow sedimentation rate of ocean-floor deposits and the relatively rapid mixing rate of oceanic waters argue in favour of the approach. Attempts to date these MIS boundaries are now well established (Martinson *et al.*, 1987).

22.4 LAND–SEA CORRELATION

In recent years it has become common to correlate terrestrial sequences directly with those in the oceans. This arises from the need felt to correlate local sequences to a regional or global time scale, mentioned above, occasioned by the fragmentary and highly variable nature of terrestrial sequences. The realization that more events are represented in the deep-sea, and indeed ice-core, sequences than were recognized on land, together with the growth in geochronology, has often led to the replacement of locally established terrestrial scales. Instead, direct correlations of terrestrial sequences to the global isotope scale are advanced, as advocated, for example, by Kukla (1977). The temptation to do this is understandable, but there are serious practical limitations to this approach (cf. Schlüchter, 1992).

In reality, there are very few means of directly and reliably correlating between the ocean and terrestrial sediment sequences. Direct correlation can be achieved using markers that are preserved in both rock sequences, such as magnetic reversals, radiometric dating, or tephra layers, and, rarely, fossil assemblages (particularly pollen). However, this is normally impossible over most of the record and in most geographical areas. Thus these correlations must rely totally on direct dating or less reliably on the technique of "curve matching;" a widely used approach in the Quaternary. The latter can only reliably be achieved where long, continuous terrestrial sequences are available, such as long lake profiles (e.g. Tzedakis *et al.*, 1997), but even here it is not always straightforward (e.g. Watts *et al.*, 1995) because of overprinting by local factors. Moreover, the possibility of failure to identify "leads-and-lags" in timing by the matching of curves is very real. Loess–soil sequences, such as those in China (An Zhisheng *et al.*, 1990; Fig. 22.1), have also provided very important and locally reliable correlative sequences, but they are also restricted by the need to have continuous or at least quasi-continuous sedimentation without subsequent disturbance. In discontinuous sequences, which typify land and shelf environments, correlations with ocean-basin sequences are potentially unreliable, in the absence of fossil groups distributed across the facies boundaries or potentially useful markers.

In recent years, the growth of stratigraphy recognized from short-duration, often highly characteristic, events has led to attempts to use these features as a basis for correlation. This event stratigraphy (e.g. Lowe *et al.*, 1999), typically deposition of a tephra layer or magnetic reversals, can also include geological records of other potentially significant-type events such as floods, tectonic movements, changes of sea level, climatic oscillations or rhythms, and the like. Such occurrences, often termed "sub-Milankovitch events," may be preserved in a variety of environmental settings and thus offer important potential tools for high- to very high-resolution cross-correlation.

Of particular importance are the so-called "Heinrich Layers" which represent major iceberg-rafting events in the North Atlantic Ocean (Heinrich, 1988; Bond *et al.*, 1992; Bond and Lotti, 1995). These detritus bands *can potentially* provide important lithostratigraphical markers for intercore correlation in ocean sediments and the impact of their accompanying sudden coolings (Heinrich Events) *may be* recognizable in certain sensitive terrestrial sequences (see summary in Lowe and Walker, 1997). Similarly, the so-called essentially time-parallel periods of abrupt climate change termed "terminations" (Broecker and van Donk, 1970), seen in oxygen isotope profiles, can also be recognized on land as sharp changes in pollen assemblage composition or other parameters, for example, where sufficiently long and detailed sequences are available, such as in long lake cores (cf. Tzedakis *et at.*, 1997). However, their value for correlation may be limited in high-sedimentation-rate sequences because these "terminations" are not instantaneous but have durations of several thousand years (Broecker and Henderson, 1998). These matters essentially concern questions of resolution and scale.

Of greater concern for the development of a high-resolution terrestrial stratigraphy is the precise recognition and timing of boundaries or events from the marine isotope stages on land, and indeed vice versa. Until very recently this has not been perceived as a problem since it has generally been assumed that boundaries identified using a variety of proxies on land are precisely coeval with those seen in ocean sediments. Yet we know that different proxies respond at different rates and in different ways to climate changes and these changes themselves may be time-transgressive. Moreover, changes in ocean currents, sea level, wind patterns, tectonics, and so on, may further complicate local responses reflected in coastal regions. This has been forceably demonstrated recently by work off Portugal by Sanchez-Goñi *et al.* (1999) and Shackleton *et al.* (2002) where the MIS 6/5 boundary has been shown to have not been coeval with the Saalian–Eemian Stage boundary on land, as previously assumed. The same point concerns the MIS 1–2 boundary, which pre-dates the Holocene–Pleistocene (Flandrian–Weichselian) boundary by some 2000–4000 years. Thus high-resolution land sequences and low-resolution marine sequences must be correlated with an eye to the detail since it cannot be assumed that the boundaries recognized in different situations are indeed coeval.

A different, yet equally relevant, example is the situation that occurs during MIS 3. This period was generally interpreted by ocean sediment workers as being of an "interstadial" character because it showed a decrease in $\delta^{18}O$ relative to the preceding and following MISs. Moreover, today it is known to include considerable climatic variability of a lower amplitude cyclic character (e.g. Bond cycles: Bond *et al.*, 1992; Bond and Lotti, 1995). However, on land, particularly in northwest Europe, this period is not wholly interstadial (*sensu* Jessen and Milthers, 1928; see above), but is characterized by a variable, predominantly cold, climate that is interrupted by short minor climatic ameliorations (interstadials), as evidenced by the occurrence of frozen-ground features and of biota indicating warmer and/or arid conditions (e.g. Guthrie and van Kolfschoten, in press).

Notwithstanding these problems of detail, which will no doubt be further resolved as new evidence becomes available, it is now generally possible to relate the onshore–offshore sequences fairly reliably at a coarse scale at least for the last glacial–interglacial cycle (Upper Pleistocene). This was first proposed by Woillard (1978), but is now well established (e.g. Tzedakis *et al.*, 1997). Beyond, things are very much more complicated. Witness, for example, the longstanding disagreements over the nature and duration of the northwest European Saalian Stage, already referred to above (Litt and Turner, 1993). Questions of whether the Holsteinian–Hoxnian temperate (interglacial) stage relates to MIS 9 or 11, and thus the immediately preceding Elsterian–Anglian glacial stage (cf. Zagwijn, 1992) to MIS 10 or 12 (Turner, 1998; de Beaulieu and Reille, 1995) or even MIS 8, and precisely how many interglacial-type events occur within the Saalian, leave much potential for inaccuracy that cannot be resolved by "counting-backwards" methods. In the absence of reliable dating these correlations represent little more than a matter of belief. The situation becomes significantly more difficult in the early Middle Pleistocene (Turner, 1998), in spite of the fact that there is the important marker of the Brunhes–Matuyama magnetic reversal event with which to correlate. In the Early Pleistocene, where the dominant cyclicity of the climate signal is the

42 kyr periodicity, not the 100 kyr periodicity of the later Quaternary, ocean–terrestrial sequence correlation is virtually impossible at present, except close to the major magnetic reversal boundaries.

To add further to these problems the phenomenon of delayed preservation of the magnetic signal that has been detected in some terrestrial sediments, in particular in Chinese loess (Zhou and Shackleton, 1999). This therefore suggests that it is questionable whether magnetic reversals can be used as reliable markers for inter-regional correlation in high-resolution, high sedimentation-rate sequences.

Nevertheles, dating through astronomical (and subastronomical) cycles is clearly a geochronological tool of considerable future potential, already realized in respect of the ocean and ice-core sequences (e.g. Björck *et al.*, 1998), and of singular importance to understanding rates of process operation on land once the problems of cross-facies correlation have been overcome. Perhaps the way forward should be to date fixed events – probably magnetic reversals or major climatic events – as accurately as possible, then use the astronomical cyclicity to provide a finer-scale chronology. In future, it is important that this scheme be phased-in to run in parallel and perhaps eventually to replace the fundamentally palaeontologically tuned scheme that has served stratigraphical geology so well in the past.

22.5 PLEISTOCENE–HOLOCENE BOUNDARY

In the previous edition of this book it was stated that "this boundary was thought to correspond to a climatic event around 10 000 radiocarbon years before present (BP)." At the time, the boundary was considered likely to be standardized in a varved lacustrine sequence in Sweden (cf. Mörner, 1976). It was originally proposed at the Eighth INQUA Congress in Paris in 1969 and was subsequently accepted by the INQUA Holocene Commission in 1982 (Olausson, 1982). The climatic amelioration on which this boundary is identified is well established in a variety of sediments, particularly in northern Europe and North America. In Scandinavia, it corresponds to the following boundaries: European Pollen Zones III–IV, the Younger Dryas–pre-Boreal and Late Glacial–postglacial (Mörner, 1976; Mangerud *et al.*, 1974). However, this boundary definition was not formally ratified by the ICS. If it is finally defined precisely at 10 000 (^{14}C yr BP), it would be the first stratigraphic boundary later than the Proterozoic to be defined chronometrically. This statement remains broadly accurate, albeit seen in the light of the abundant ice core evidence, now available from both hemispheres (see below), it appears highly likely that the boundary should actually occur at 11.5 kyr (ice-accumulation years: Dansgaard *et al.*, 1993). In spite of the potential accuracy of less than five years that can now be achieved in the annually laminated ice cores, it is possible that the basal boundary stratotype of the Holocene (or potentially a parastratotype) will be placed in an annually laminated lacustrine sequence in western Germany (Litt *et al.*, 2001). This is because here it can be identified to the nearest year and can be precisely radiocarbon-dated. Moreover, the sequence has yielded an excellent, easily correlated fossil record, principally pollen and other freshwater microfossils, which facilitates regional biostratigraphical correlation.

22.6 HOLOCENE SERIES

Holocene is the name for the most recent interval of Earth history and includes the present day (see Section 21.2.4). It is generally regarded as having begun 10 000 radiocarbon years, or the last 11 500 calibrated (i.e. calender) years, before present (i.e. 1950). The term "Recent" as an alternative to Holocene is invalid and should not be used. Sediments accumulating or processes operating at present should be referred to as "modern" or by similar synonyms.

The term Flandrian, derived from marine transgression sediments on the Flanders coast of Belgium (Heinzelin and Tavernier, 1957), has often been used as a synonym for Holocene (Fig. 22.2). It has been adopted by authors who consider that the last 10 000 years should have the same stage status as previous interglacial events and thus be included in the Pleistocene. In this case, the latter would thus extend to the present day (cf. West, 1968, 1977, 1979; Hyvärinen, 1978). This usage, although advocated particularly in Europe, has been losing ground in the last two decades (cf. Lowe and Walker, 1997, p. 16).

Various zonation schemes have been proposed for the Holocene (Flandrian) Epoch. The most established is that of Blytt and Sernander (cf. Lowe and Walker, 1997), which was developed for peat bogs in Scandinavia in the late nineteenth to earliest twentieth centuries. Their terminology, based on interpreted climatic changes, comprised, in chronological order, the pre-Boreal, Boreal, Atlantic, sub-Boreal, and sub-Atlantic. This scheme was refined by the Swede von Post and others, using pollen analysis throughout Europe. Today this terminology remains in use in northern Europe, although it has been largely displaced by absolute chronology, particularly ^{14}C. Dating has shown that the biostratigraphically defined zone boundaries are diachronous (cf. Godwin, 1975). An attempt to fix these boundaries to precise dates was proposed for northern Europe by Mangerud *et al.* (1974).

In prehistoric times, as well as later, climatic events have largely served to identify the divisions elaborated by modern ^{14}C, other dating techniques, tephrachronology, and dendrochronology as well as successively by archeology and human history. Using these techniques Holocene time can be divided into ultra-high-resolution divisions. For example, recent developments indicate that cyclic patterns of climate change of durations as short as 200 yr can be differentiated and potentially used for demonstrating equivalence in peat sequences.

Part IV • Summary

23 • Construction and summary of the geologic time scale

F. M. GRADSTEIN, J. G. OGG, AND A. G. SMITH

A geologic time scale (GTS2004) is presented that integrates currently available stratigraphic and geochronologic information. Key features of the new scale are outlined, how it was constructed, and how it can be improved. Major impetus to the new scale was provided through:

(a) advances in stratigraphic standardization and refinement of the International Chronostratigraphic Scale;
(b) enhanced methods of extracting linear time from the rock record, leading to numerous high-resolution ages;
(c) progress with the use of global geochemical variations, Milankovitch climate cycles, and magnetic reversals as important stratigraphic calibration tools;
(d) improved statistical techniques for extrapolating ages and associated uncertainties to the relative stratigraphic scale, using high-resolution biozonations, including composite standards, that scale stages.

23.1 CONSTRUCTION OF GTS2004

23.1.1 The components of GTS2004

The Geologic Time Scale 2004 (GTS2004) project, that commenced in 1998, has compiled integrated scales of selected components of Earth history including:

1. Formal international subdivisions of the "rock-time" chronostratigraphic scale as ratified, or being considered, by the International Commission on Stratigraphy (ICS). The brief historical review of these subdivisions shows the progress toward the goal of a full international standard for chronostratigraphy. Due to space limitations, correlations of selected regional stratigraphic scales to the international standard are only included for some periods. The choice was ours.
2. An informal proposal to subdivide Precambrian time into eons and eras that reflect natural stages in planetary evolution rather than a subdivision in arbitrary numerical ages.
3. Major biostratigraphic zonations and datums for each geologic period in the Phanerozoic. Composite zonations derived from graphical correlation or constrained optimization methods were assembled for most Paleozoic periods, and parts of the Triassic.
4. Magnetic reversal patterns throughout the Phanerozoic.
5. Major geochemical trends of strontium, carbon, and oxygen isotopes in seawater.
6. High-resolution cyclic climatic and oceanographic changes physically and chemically recorded in the sedimentary record.
7. Other significant events (large igneous provinces, impacts, etc.) which are important for global correlation or may have this future potential.
8. Radiometric dates selected for their stratigraphic importance and reliability.

This massive array of information was melded together to produce a framework for Earth geologic history scaled to linear time. The summary of the geologic time scale in Fig. 23.1 (see also Table 23.1) is a calibration of the Phanerozoic part of the International Stratigraphic Chart. Ages of chronostratigraphic boundaries and durations of stages include estimates of the 95% uncertainty (2-sigma). The Neogene portion is calibrated by astronomical cycles to within an orbital-precession oscillation (~20 kyr). Parts of the Paleocene, Cretaceous, Jurassic, and Triassic are also scaled using Milankowitch cycle durations.

We are still a considerable distance from the goal where geologic time scale calibration is achieved by precise direct astronomical tuning or radiometric age dating of all successive stage, zonal, or magnetic polarity chron boundaries. In fact, it is doubtful if the rock record on Earth harbors all the precise age information. This sparse skeleton of age control, especially prior to ~30 Ma (as of 2004), leaves considerable room for interpolation in construction of a geologic time scale. Future time scales will undoubtedly re-examine and reprocess a more expanded array of Earth history data, and will undoubtedly employ even more sophisticated means of interpolation.

A Geologic Time Scale 2004, eds. Felix M. Gradstein, James G. Ogg, and Alan G. Smith. Published by Cambridge University Press.

GEOLOGIC TIME SCALE

PHANEROZOIC

CENOZOIC

Period	Epoch		Stage	AGE (Ma)
Quaternary	Holocene			
Quaternary	Pleistocene			1.81
Neogene	Pliocene	L	Gelasian	2.59
Neogene	Pliocene	L	Piacenzian	3.60
Neogene	Pliocene	E	Zanclean	5.33
Neogene	Miocene	L	Messinian	7.25
Neogene	Miocene	L	Tortonian	11.61
Neogene	Miocene	M	Serravallian	13.65
Neogene	Miocene	M	Langhian	15.97
Neogene	Miocene	E	Burdigalian	20.43
Neogene	Miocene	E	Aquitanian	23.03
Paleogene	Oligocene	L	Chattian	28.4
Paleogene	Oligocene	E	Rupelian	33.9
Paleogene	Eocene	L	Priabonian	37.2
Paleogene	Eocene	M	Bartonian	40.4
Paleogene	Eocene	M	Lutetian	48.6
Paleogene	Eocene	E	Ypresian	55.8
Paleogene	Paleocene	L	Thanetian	58.7
Paleogene	Paleocene	M	Selandian	61.7
Paleogene	Paleocene	E	Danian	65.5

MESOZOIC

Period	Epoch	Stage	AGE (Ma)
Cretaceous	Late	Maastrichtian	65.5–70.6
Cretaceous	Late	Campanian	83.5
Cretaceous	Late	Santonian	85.8
Cretaceous	Late	Coniacian	89.3
Cretaceous	Late	Turonian	93.5
Cretaceous	Late	Cenomanian	99.6
Cretaceous	Early	Albian	112.0
Cretaceous	Early	Aptian	125.0
Cretaceous	Early	Barremian	130.0
Cretaceous	Early	Hauterivian	136.4
Cretaceous	Early	Valanginian	140.2
Cretaceous	Early	Berriasian	145.5
Jurassic	Late	Tithonian	150.8
Jurassic	Late	Kimmeridgian	155.7
Jurassic	Late	Oxfordian	161.2
Jurassic	Middle	Callovian	164.7
Jurassic	Middle	Bathonian	167.7
Jurassic	Middle	Bajocian	171.6
Jurassic	Middle	Aalenian	175.6
Jurassic	Early	Toarcian	183.0
Jurassic	Early	Pliensbachian	189.6
Jurassic	Early	Sinemurian	196.5
Jurassic	Early	Hettangian	199.6
Triassic	Late	Rhaetian	203.6
Triassic	Late	Norian	216.5
Triassic	Late	Carnian	228.0
Triassic	Middle	Ladinian	237.0
Triassic	Middle	Anisian	245.0
Triassic	Early	Olenekian	249.7
Triassic	Early	Induan	251.0

PALEOZOIC

Period	Epoch		Stage	AGE (Ma)
Permian	Lopingian		Changhsingian	251.0–253.8
Permian	Lopingian		Wuchiapingian	260.4
Permian	Guada-lupian		Capitanian	265.8
Permian	Guada-lupian		Wordian	268.0
Permian	Guada-lupian		Roadian	270.6
Permian	Cisuralian		Kungurian	275.6
Permian	Cisuralian		Artinskian	284.4
Permian	Cisuralian		Sakmarian	294.6
Permian	Cisuralian		Asselian	299.0
Carboniferous	Penn-sylvanian	Late	Gzhelian	303.9
Carboniferous	Penn-sylvanian	Late	Kasimovian	306.5
Carboniferous	Penn-sylvanian	Middle	Moscovian	311.7
Carboniferous	Penn-sylvanian	Early	Bashkirian	318.1
Carboniferous	Mississippian	Late	Serpukhovian	326.4
Carboniferous	Mississippian	Middle	Visean	345.3
Carboniferous	Mississippian	Early	Tournaisian	359.2
Devonian	Late		Famennian	374.5
Devonian	Late		Frasnian	385.3
Devonian	Middle		Givetian	391.8
Devonian	Middle		Eifelian	397.5
Devonian	Early		Emsian	407.0
Devonian	Early		Pragian	411.2
Devonian	Early		Lochkovian	416.0
Silurian	Pridoli			418.7
Silurian	Ludlow		Ludfordian	421.3
Silurian	Ludlow		Gorstian	422.9
Silurian	Wenlock		Homerian	426.2
Silurian	Wenlock		Sheinwoodian	428.2
Silurian	Llandovery		Telychian	436.0
Silurian	Llandovery		Aeronian	439.0
Silurian	Llandovery		Rhuddanian	443.7
Ordovician	Late		Hirnantian	445.6
Ordovician	Late			455.8
Ordovician	Late			460.9
Ordovician	Middle		Darriwilian	468.1
Ordovician	Middle			471.8
Ordovician	Early			478.6
Ordovician	Early		Tremadocian	488.3
Cambrian	Furongian			
Cambrian	Furongian		Paibian	501
Cambrian	Middle			513
Cambrian	Early			542.0

PRECAMBRIAN

Eon	Era	Period	AGE (Ma)
Proterozoic	Neo-proterozoic	Ediacaran	542–~630
Proterozoic	Neo-proterozoic	Cryogenian	850
Proterozoic	Neo-proterozoic	Tonian	1000
Proterozoic	Meso-proterozoic	Stenian	1200
Proterozoic	Meso-proterozoic	Ectasian	1400
Proterozoic	Meso-proterozoic	Calymmian	1600
Proterozoic	Paleo-proterozoic	Statherian	1800
Proterozoic	Paleo-proterozoic	Orosirian	2050
Proterozoic	Paleo-proterozoic	Rhyacian	2300
Proterozoic	Paleo-proterozoic	Siderian	2500
Archean	Neo-archean		2800
Archean	Meso-archean		3200
Archean	Paleo-archean		3600
Archean	Eoarchean	Lower limit is not defined	

Figure 23.1 Summary of *A Geologic Time Scale 2004*.

Table 23.1 *Summary of ages and durations of stages in GTS2004*[a]

EON, Era, System, Series, Stage	Age of Base (Ma)	Est. ± myr (2-sigma)	Comment	Duration	Est. ± myr (2-sigma)
PHANEROZOIC					
Cenozoic Era					
Neogene System					
Holocene Series					
base Holocene	11.5 ka	0.00			
Pleistocene Series					
base Upper Pleistocene subseries	0.126	0.00		0.115	0.0
base Middle Pleistocene subseries	0.781	0.00		0.655	0.0
base Pleistocene Series	1.806	0.00		1.025	0.0
Pliocene Series					
base Gelasian Stage	2.588	0.00		0.782	0.0
base Piacenzian Stage	3.600	0.00		1.01	0.0
base Zanclean Stage, base Pliocene Series	5.333	0.00		1.73	0.0
Miocene Series					
base Messinian Stage	7.248	0.00		1.92	0.0
base Tortonian Stage	11.608	0.00		4.36	0.0
base Serravallian Stage	13.65	0.00		2.04	0.0
base Langhian Stage	15.97	0.0		2.32	0.0
base Burdigalian Stage	20.43	0.0		4.46	0.0
base Aquitanian Stage, base Miocene Series, base Neogene System	23.03	0.0		2.60	0.0
Paleogene System					
Oligocene Series					
base Chattian Stage	28.4	0.1		5.4	0.0
base Rupelian Stage, base Oligocene Series	33.9	0.1		5.4	0.0
Eocene Series					
base Priabonian Stage	37.2	0.1		3.3	0.0
base Bartonian Stage	40.4	0.2		3.2	0.0
base Lutetian Stage	48.6	0.2		8.2	0.1
base Ypresian Stage, base Eocene Series	55.8	0.2		7.2	0.1
Paleocene Series					
base Thanetian Stage	58.7	0.2		2.9	0.0
base Selandian Stage	61.7	0.2		3.0	0.0
base Danian Stage, base Paleogene System, base Cenozoic	65.5	0.3		3.7	0.0
Mesozoic Era					
Cretaceous System					
Upper					
base Maastrichtian Stage	70.6	0.6	Duration uncertainty increased to reflect correlation problems to GSSP	5.1	0.5
base Campanian Stage	83.5	0.7	Duration uncertainty increased to reflect correlation problems to GSSP	12.9	0.7

(*cont.*)

Table 23.1 (*cont.*)

EON, Era, System, Series, Stage	Age of Base (Ma)	Est. ± myr (2-sigma)	Comment	Duration	Est. ± myr (2-sigma)
base Santonian Stage	85.8	0.7		2.3	0.1
base Coniacian Stage	89.3	1.0	0.2 myr added to uncertainty to account for offset to actual proposed GSSP marker	3.5	0.3
base Turonian Stage	93.5	0.8		4.2	0.3
base Cenomanian Stage	99.6	0.9		6.1	0.3
Lower					
base Albian Stage	112.0	1.0		12.4	0.3
base Aptian Stage	125.0	1.0		13.0	0.5
base Barremian Stage	130.0	1.5		5.0	0.5
base Hauterivian Stage	136.4	2.0		6.4	1.0
base Valanginian Stage	140.2	3.0		3.8	1.0
base Berriasian Stage, base Cretaceous System	145.5	4.0		5.3	1.7
Jurassic System					
Upper					
base Tithonian Stage	150.8	4.0		5.3	1.8
base Kimmeridgian Stage	155.7	4.0	Boreal placement	4.2	1.5
base Oxfordian Stage	161.2	4.0		6.2	1.5
Middle					
base Callovian Stage	164.7	4.0		3.5	1.0
base Bathonian Stage	167.7	3.5		3.0	1.0
base Bajocian Stage	171.6	3.0		3.9	1.0
base Aalenian Stage	175.6	2.0		4.0	1.0
Lower					
base Toarcian Stage	183.0	1.5		7.4	1.0
base Pliensbachian Stage	189.6	1.5		6.6	0.8
base Sinemurian Stage	196.5	1.0		6.9	0.8
base Hettangian Stage, base Jurassic System	199.6	0.6		3.1	0.5
Triassic System					
Upper					
base Rhaetian Stage	203.6	1.5		4.0	1.0
base Norian Stage	216.5	2.0		12.9	0.5
base Carnian Stage	228.0	2.0		11.5	0.5
Middle					
base Ladinian Stage	237.0	2.0		9.0	0.5
base Anisian Stage	245.0	1.5		8.0	1.5
Lower					
base Olenekian Stage	249.7	0.7		4.7	1.0
base Induan Stage, base Triassic System, base Mesozoic	251.0	0.4		1.3	0.3
Paleozoic Era					
Permian System					
Lopingian Series					
base Changhsingian Stage	253.8	0.7		2.8	0.1
base Wuchiapingian Stage	260.4	0.7		6.6	0.1

Table 23.1 (*cont.*)

EON, Era, System, Series, Stage	Age of Base (Ma)	Est. ± myr (2-sigma)	Comment	Duration	Est. ± myr (2-sigma)
Guadalupian Series					
base Capitanian Stage	265.8	0.7		5.4	0.1
base Wordian Stage	268.0	0.7		2.2	0.0
base Roadian Stage	270.6	0.7		2.6	0.1
base Kungurian Stage	275.6	0.7		5.0	0.1
base Artinskian Stage	284.4	0.7		8.8	0.2
base Sakmarian Stage	294.6	0.8		10.2	0.2
base Asselian Stage, base Permian System	299.0	0.8		4.4	0.1
Carboniferous System					
Pennsylvanian Subsystem					
base Gzhelian Stage	303.9	0.9		4.9	0.1
base Kasimovian Stage	306.5	1.0		2.6	0.0
base Moscovian Stage	311.7	1.1		5.2	0.1
base Bashkirian Stage, base Pennsylvanian Subsystem	318.1	1.3		6.4	0.2
Mississippian Subsystem					
base Serpukhovian	326.4	1.6		8.4	0.2
base Visean	345.3	2.1		18.9	0.7
base Tournaisian, base Mississippian Subsystem, base Carboniferous System	359.2	2.5		13.9	0.6
Devonian System					
Upper					
base Famennian Stage	374.5	2.6		15.3	0.6
base Frasnian Stage	385.3	2.6		10.8	0.4
Middle					
base Givetian Stage	391.8	2.7		6.5	0.3
base Eifelian Stage	397.5	2.7		5.7	0.2
Lower					
base Emsian Stage	407.0	2.8		9.5	0.4
base Pragian Stage	411.2	2.8		4.2	0.2
base Lochkovian Stage, base Devonian System	416.0	2.8		4.8	0.2
Silurian System					
Pridoli Series					
base Pridoli Series (*not subdivided in stages*)	418.7	2.7	Uncertainties "ramped" from computed base-Devonian to "low" value at base-Silurian	2.7	0.1
Ludlow Series					
base Ludfordian Stage	421.3	2.6		2.5	0.1
base Gorstian Stage	422.9	2.5		1.7	0.1
Wenlock Series					
base Homerian Stage	426.2	2.4		3.3	0.1
base Sheinwoodian Stage	428.2	2.3		2.0	0.1

(*cont.*)

Table 23.1 (*cont.*)

EON, Era, System, Series, Stage	Age of Base (Ma)	Est. ± myr (2-sigma)	Comment	Duration	Est. ± myr (2-sigma)
Llandovery Series					
base Telychian Stage	436.0	1.9		7.8	0.2
base Aeronian Stage	439.0	1.8		3.0	0.1
base Rhuddanian Stage, base Silurian System	443.7	1.5		4.7	0.1
Ordovician System					
Upper					
base Hirnantian stage	445.6	1.5		1.9	0.1
base of sixth stage *(not yet named)*	455.8	1.6		10.2	0.3
base of fifth stage *(not yet named)*	460.9	1.6		5.1	0.2
Middle					
base Darriwilian Stage	468.1	1.6		7.3	0.2
base of third stage *(not yet named)*	471.8	1.6		3.7	0.1
Lower					
base of second stage *(not yet named)*	478.6	1.7		6.8	0.1
base of Tremadocian Stage, base Ordovician System	488.3	1.7		9.7	0.2
Cambrian System					
Upper ("Furongian") Series					
upper stage(s) in Furongian	*not defined*				
base Paibian Stage, base Furongian Series	501.0	2.0	Age of boundary is "approximate estimate" (see text)		
Middle	513.0	2.0	Age of boundary is "approximate estimate" (see text)		
Lower					
base Cambrian System, base Paleozoic, base PHANEROZOIC	542.0	1.0			

[a] Uncertainties are 2-sigma (95% confidence).

23.1.2 Calibration methods to linear time used in GTS2004

The main steps involved in GTS2004 time scale construction were:

Step 1. Construct an updated global chronostratigraphic scale for the Earth's rock record.

Step 2. Identify key linear-age calibration levels for the chronostratigraphic scale using radiometric age dates, and/or apply astronomical tuning to cyclic sediment or stable isotope sequences which had biostratigraphic or magnetostratigraphic correlations.

Step 3. Interpolate the combined chronostratigraphic and chronometric scale where direct information is insufficient.

Step 4. Calculate or estimate error bars on the combined chronostratigraphic and chronometric information to obtain a geologic time scale with estimates of uncertainty on boundaries and on unit durations.

Step 5. Peer review the geologic time scale

The first step, integrating multiple types of stratigraphic information in order to construct the chronostratigraphic scale, is the most time-consuming; it summarizes and synthesizes centuries of detailed geological research. The second step, identifying which radiometric and cycle-stratigraphic studies would be used as the primary constraints for assigning linear ages, is the one that is evolving most rapidly since the last decade. Historically, time scale building went from an exercise with very few and relatively inaccurate radiometric dates, as used by Holmes (1947, 1960), to one with many dates with greatly varying analytical precision (like GTS89 or, to some extent, SEPM95). Next came studies that selected a few radiometric dates with high internal analytical precision (e.g. Cande and Kent, 1992a, 1995; Obradovich, 1993; Cooper, 1999b) or measure time relative to present using astronomical cycles (e.g. Hilgen *et al.*, 1995, 2000c; Shackleton *et al.*, 1999). This new philosophy is also adhered to in this book.

In addition to selecting radiometric ages based upon their stratigraphic control and analytical precision, we also applied the following criteria or corrections:

1. Stratigraphically constrained radiometric ages with the U–Pb method on zircons were accepted from the isotope dilution mass spectrometry (TIMS) method, but generally not from the high-resolution ion microprobe (HR-SIMS, also known as "SHRIMP") that uses the Sri Lanka (SL)13 standard. An exception is the Carboniferous Period, where there is a dearth of TIMS dates and more uncertainty.
2. $^{40}Ar/^{39}Ar$ radiometric ages were re-computed to be in accord with the revised ages for laboratory monitor standards: 523.1 ± 4.6 Ma for MMhb-1 (McClure Mountain hornblende), 28.34 ± 0.28 Ma for TCR (Taylor Creek Rhyolite sanidine) and 28.02 ± 0.28 Ma for FCT (Fish Canyon Tuff sanidine). Systematic ("external") errors and uncertainties in decay constants were partially incorporated (see Chapters 6 and 8). No glauconite dates were used.

The bases of the Paleozoic, Mesozoic, and Cenozoic are bracketed by analytically precise ages at their GSSP or primary correlation markers – 542 ± 1.0, 251.0 ± 0.4 and 65.5 ± 0.3 Ma, respectively – and there are direct age dates on base-Carboniferous, base-Permian, base-Jurassic, and base-Oligocene; but most other period or stage boundaries prior to the Neogene lack direct age control. Therefore, the third step, linear interpolation, plays a key role for most of GTS2004. This detailed and high-resolution process incorporated several techniques, depending upon the available information (Fig. 23.2):

1. A composite standard of graptolite zones spanning the latest Cambrian, Ordovician, and Silurian interval was derived from 200+ sections in oceanic and slope environment basins using the constrained optimization method (see Chapters 12 and 13). With zone thickness taken as directly proportional to zone duration, the detailed composite sequence was scaled using selected, high-precision zircon and sanidine age dates. For the Carboniferous through Permian, a composite standard of conodont, fusulinid, and ammonoid events from many classical sections was calibrated to a combination of U–Pb and $^{40}Ar/^{39}Ar$ dates with assigned external error estimates. A composite standard of conodont zones was used for Early Triassic. This procedure directly scaled all stage boundaries and biostratigraphic horizons.
2. Detailed direct ammonite-zone ages for the Late Cretaceous of the Western Interior of the USA were obtained by a cubic-spline fit of the zonal events and 25 $^{40}Ar/^{39}Ar$ dates. The base-Turonian age is directly bracketed by this $^{40}Ar/^{39}Ar$ set, and ages of other stage boundaries and stratigraphic events are estimated using calibrations to this primary scale.
3. Seafloor-spreading interpolations were done on a composite marine magnetic lineation pattern for the Late Jurassic through Early Cretaceous in the Western Pacific and for the Late Cretaceous through early Neogene in the South Atlantic Ocean. Ages of biostratigraphic events were assigned according to their calibration to these magnetic polarity time scales.
4. Astronomical tuning of cyclic sediments was used for the Neogene and Late Triassic, and portions of the Early and Middle Jurassic, the middle part of the Cretaceous, and the Paleocene. The Neogene astronomical scale is directly tied to the Present; the astronomical scale provides linear-duration constraints on polarity chrons, biostratigraphic zones, and entire stages.
5. Proportional scaling relative to component biozones or subzones. In intervals where none of the above information under Items 1–4 was available it was necessary to return to the methodology employed by past time scales. This procedure was necessary in portions of the Middle Triassic and Middle Jurassic. Devonian stages were scaled from approximate equal duration of a set of high-resolution subzones of ammonoids and conodonts, fitted to an array of high-precision dates.

The actual geomathematics employed for the above data sets (Items 1, 2, 3, and 5) constructed for the Ordovician–Silurian,

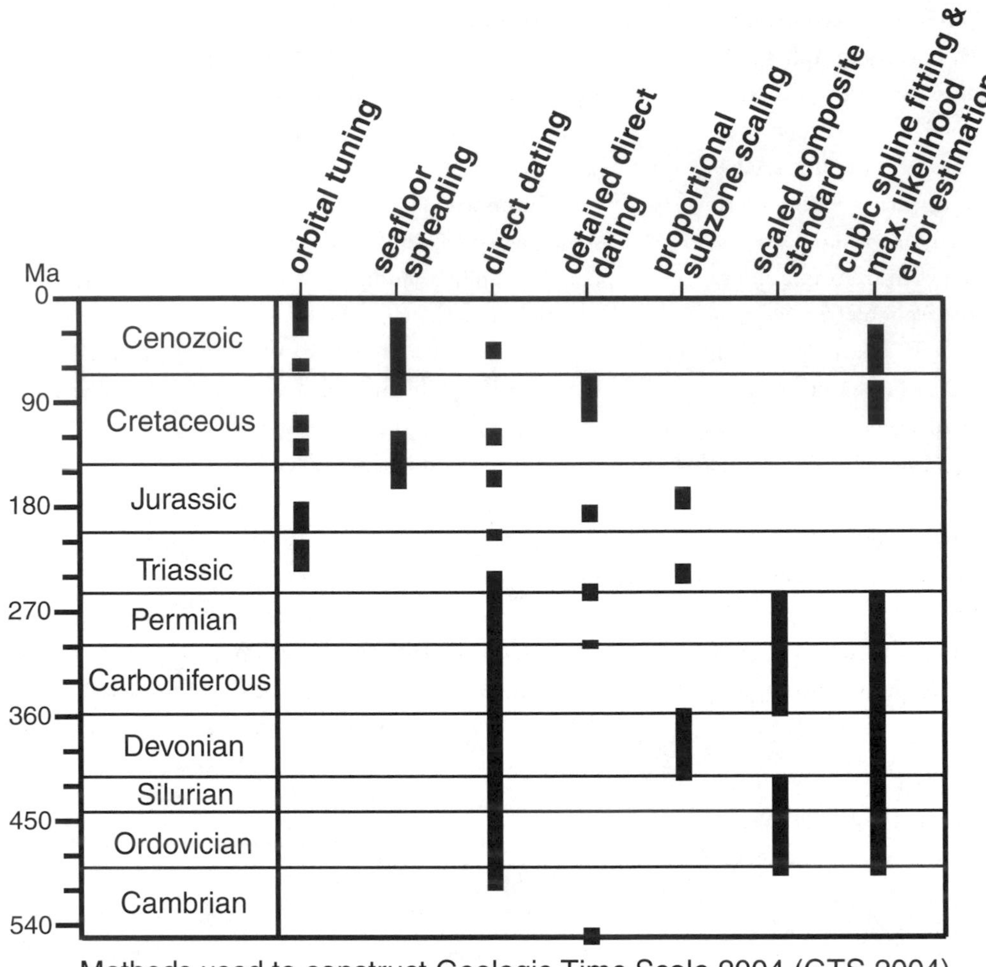

Figure 23.2 Methods used to construct *A Geologic Time Scale 2004* (GTS2004) integrate different techniques depending on the quality of data available within different intervals.

Devonian, Carboniferous–Permian, Late Cretaceous, and Paleogene involved cubic-spline curve fitting to relate the observed ages to their stratigraphic position. During this process the ages were weighted according to their variances based on the lengths of their error bars. A chi-square test was used for identifying and reducing the weights of relatively few outliers with error bars that were much narrower than could be expected on the basis of most ages in the data set.

Stratigraphic uncertainty was incorporated in the weights assigned to the observed ages during the spline-curve fitting. In the final stage of analysis, Ripley's MLFR algorithm, for maximum likelihood fitting of a functional relationship, was used for error estimation, resulting in 2-sigma (95% confidence) error bars for the estimated chronostratigraphic boundary ages and stage durations. The uncertainties on older stage boundaries generally increase owing to potential systematic errors in the different radiometric methods, rather than to the analytical precision of the laboratory measurements (Table 23.1 and Fig. 23.3). In this connection, we mention that biostratigraphic error is fossil event and fossil zone dependent, rather than age dependent.

In Mesozoic intervals that were scaled using the seafloor-spreading model or proportionally scaled using paleontological subzones, the assigned uncertainties are conservative estimates based on variability observed when applying different assumptions (see discussions in Chapters 5, 17–19). Ages and durations of Neogene stages derived from orbital tuning are considered to be accurate to within a precession cycle (~20 kyr) assuming that all cycles are correctly identified and that the theoretical astronomical tuning for progressively older deposits is precise.

23.2 FUTURE TRENDS IN GEOLOGIC TIME SCALES

The changing philosophy in time scale building has made it more important to undertake high-resolution radiometric study of critical stratigraphic boundaries and extend the

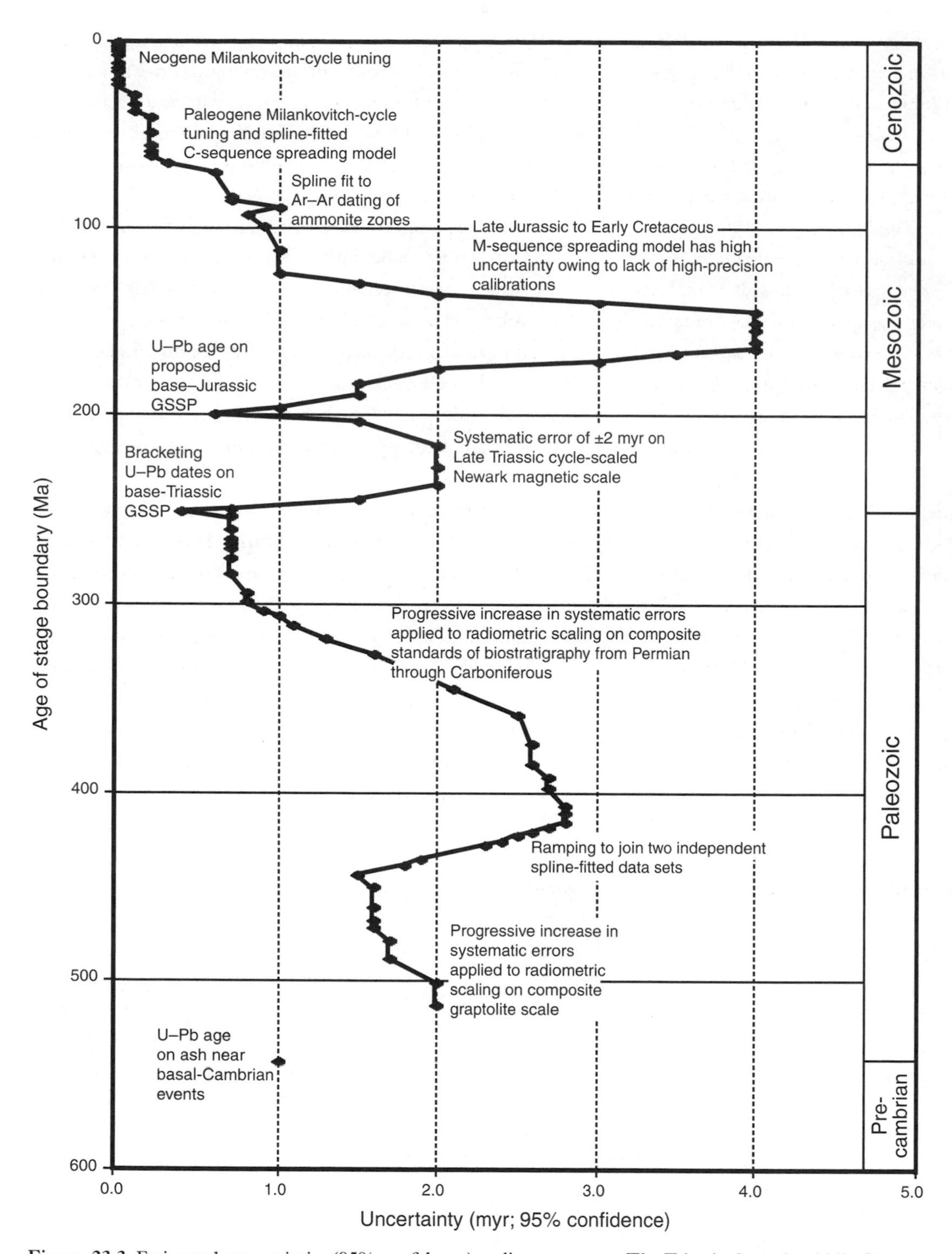

Figure 23.3 Estimated uncertainties (95% confidence) on linear ages of stage boundaries. These estimates partially incorporate potential systematic errors in radiometric methods. Orbital tuning relative to the present yields negligible uncertainties in the Neogene. The Triassic through middle Cretaceous generally has higher uncertainties owing to the dearth of precise radiometric ages and inadequate calibration of seafloor-spreading models.

astronomical tuning into progressively older sediments. Good examples are Bowring *et al.* (1998) for basal-Triassic, Amthor *et al.* (2003) for basal-Cambrian, and Hilgen *et al.* (2000c) for base-Tortonian. The philosophy is that obtaining high-precision age dating at a precisely defined stratigraphic boundary avoids stratigraphic bias and its associated uncertainty in rock and in time. In this respect, it is of vital importance that the ICS not only completes the definition of all stage boundaries, but also actively considers definition of subdivisions within the many long stages (see Chapter 2). Regional and philosophical arguments between stratigraphers should be actively resolved to reach consensus conclusions with focus on global correlation implications. Stratigraphic standardization precedes linear time calibration.

Even more refined and more time extensive scaling of zones and stages with the deterministic and probabilistic quantitative methods outlined in Chapter 3 and in Chapters 11–16 is probably feasible today, and should be pursued actively. Progress with a natural time scale for the Precambrian is also a high and challenging priority (Chapter 10), not in the least because a solar scale for all of science should soon be "over the horizon."

In the process of assembling the pieces of the new time scale, i.e. GTS2004, several decisions had to be made with respect to the global radiometric data set. This data set should be subjected to further scrutiny, both within radiometric laboratories and in the field. For example, significant discrepancies exist between U–Pb dates on the P–T boundary beds and in the middle Triassic, both of which appear to be a zircon problem, and the misalignment of HR–SIMS dating, using the SL13 standard, and TIMS dating in parts of the Paleozoic. Other decisions (Chapter 6), i.e. which $^{40}Ar/^{39}Ar$ monitor age value to use, and which decay constant, also need further study and consensus building among radiometric specialists. For example, intercalibration of independent astronomical and radioisotopic dating methods is not yet solved, but new results (Kuiper, 2003) point to an astronomically derived age of 28.24 ± 0.01 Ma for the Fish Canyon Tuff (FCT) sanidine. This precise age requires careful evaluation in the geochronologic community.

Enhanced utilization of geochemical trends and magnetic reversal patterns to resolve linear scaling of critical intervals such as the long Jurassic–Early Cretaceous and parts of the Triassic are also highly desirable. The virtual absence of reliable radiometric age dates for the long Jurassic–Cretaceous interval needs urgent correction.

In summary, improvement and consolidation of the time scale will depend on definition of the remaining stage boundaries, on astronomical tuning of durations of as many intervals as possible, on more evenly time-distributed high-resolution age dating, and on more detailed relative scaling of stages with biozones. For example, tuning and calibration of the Paleogene time scale at much higher levels of resolution and precision than are presently available will be achieved within this decade. Astronomical calibrations of the geologic time scale in its earlier parts is more challenging than in the Neogene and requires careful evaluation of uncertainties. In the medium term, it can be predicted that complete coverage of astronomically calibrated geological markers will exist for the entire Cenozoic and that traditional geochronological scales, astronomical calibrations, and magneto- and biostratigraphic datums and zonal composites will become more closely intertwined and aligned in the Mesozoic and Paleozoic.

A high-resolution geologic time scale allows more insight into the cause and effect of all physical, chemical, and biological processes that have left their enduring and wondrous mark on Earth. The order of things and the order in nature is our goal, such is the reward of our undertakings.

Appendix 1 • Recommended color coding of stages

F. M. GRADSTEIN AND J. G. OGG

Ever since time scale charts have become wallpaper items in geological offices the world over, color of stratigraphic units has been a "hot" topic. Colors stir the senses, imagination, and also indignation. In this respect, the senior editor (FMG) of this book recalls strong emotional reactions of colleagues when he innocently tried to adopt the logical, pragmatic, and easily reproducible spectral color scheme in use with the US Geological Survey for an early version of the Gradstein and Ogg (1996) wall chart, courtesy of Saga Petroleum. He had to backtrack to so-called standard colors, i.e. Harland *et al.*'s (1990) emulation of the cryptic UNESCO color scheme.

Since there is consensus among geologists that the stratigraphic color scheme should be a reproducible and practical one, we will dwell on this issue below. At the end, our choice of colors will be made clear.

There are several well-known time scale charts in circulation, e.g.

1. *A Geologic Time Scale 1989* sponsored by BP for Harland *et al.* (1990).
2. The A3 format and hand-held plasticized card of the *Geological Time Scale* by Gradstein and Ogg (1996) in *Episodes.*
3. The fifth edition of the *Geological Time Table* (Haq and van Eysinga, 1998).
4. The *Geological Time Chart 2001* compiled by Okulitch (2002) for the Canadian Geological Atlas.
5. The *Phanerozoic Timescale Wall Chart* as part of the book *An Australian Phanerozoic Time Scale* (Young and Laurie, 1996).
6. The *International Stratigraphic Chart* (Remane, 2000) sponsored by ICS, IUGS, CGMW, and UNESCO.

Unfortunately, all these charts differ in their use of color, particularly and largely so with respect to the pre-Cretaceous. The Cretaceous, in shades of green, and the Cenozoic, in shades of yellow/brownish, are more or less, but not quite, harmonized. The Gradstein and Ogg (1996) chart tried to emulate the color scheme of Harland *et al.* (1990) popular in parts of Europe, but did not quite succeed. The colors used in Remane (2000) for the divisions down to stage rank are supposed to be the same as those employed with the Commission for the Geological Map of the World (CGMW/CCGM), reason why it is instructive to learn from the discussion in Harland *et al.*'s book (1990, p. 221), which we update below with our own findings.

The Commission for the Geological Map of the World (CGMW) was conceived at the Second International Geological Congress, Bologna, 1881, and the choice of colors for (largely) time-stratigraphic purposes is discussed at length in the Congress Proceedings. Apart from color plates in the Proceedings volume, an early example of the agreed standard color scheme is presented by the *Geological Map of Europe* 1:500 000 prepared in accordance with the Congress's resolutions. It seemed appropriate to Harland *et al.* (1982) to consider adopting the 1881 scheme, but the idea was abandoned at an early stage when they found the original descriptions of the colors to be imprecise while, after the passage of nearly 100 years, the colors in the Proceedings volume and on two separate examples in the Proceedings of the European map no longer matched and no "standard" colors could therefore be determined.

The work of the CGMW has continued, mainly since the Eleventh International Geological Congress in Stockholm, 1910, with interruptions in the war years 1914–1918 and 1939–1945. The work is pursued in Paris, where it is connected with the IUGS and with UNESCO, which publishes its maps. The UNESCO *Geological World Atlas* (Choubert and Faure-Muret, 1976) represents the culmination of decades of trials and the first sheets were produced in 1974. Vojacek (1979) outlined the history of this project and provided useful background technical information about the printing of the atlas. This author also noted at that time that "with this project UNESCO also hopes to set international standards for the presentation of geological symbols and for the color designs of geological maps."

This seemed a laudable aim, and sufficient justification for adopting the UNESCO colors, but there proved to be a serious commercial obstacle to this course of action. UNESCO used no less than 35 colors (including black) in printing their atlas (Vojacek, 1979). This was impractical for the low-cost commercial printing operation envisaged by Cambridge University

Table A1.1 *Standard color scales for geologic time units (screen proportions or percentages) according to the Commission for the Geological Map of the World (CGMW) and the US Geological Survey. The CGMW colors were used for all color figures in this book. Tables of both scales (and displayed as color graphics) can be downloaded from the ICS website (www.stratigraphy.org) under "Time Scale" webpages*

Geologic unit	CODE	Comm. Geol. Map of World (Paris) color set				USGS color set			
			RGB	Exact CMYK	Rounded CMYK		RGB	CMY code	CMY
PHANEROZOIC	PH	light blue	154/217/229	40/0/1,7/0	40/0/0/0	light blue	179/226/209	301	30/0/13
Cenozoic	CZ	yellow	250/253/1	1.9/0/100/0	5/0/100/0	yellow	255/255/0	00X	0/0/100
Neogene	N	yellow hues	255/232/0	0/9/100/0	0/10/100/0	yellow hues	253/204/138	024	0/20/40
Holocene	Q2		255/251/240	0/1.5/5/0	0/0/5/0		255/255/179	003	0/0/30
Pleistocene	Q1		255/247/176	0/3/30/0	0/5/30/0		255/235/98	0A6	0/8/60
Pliocene	N2		255/255/153	0/0/40/0	0/0/40/0		254/235/172	0A3	0/8/30
Miocene	N1		255/255/0	0/0/100/0	0/0/100/0		255/222/0	01X	0/13/100
Paleogene	E	tan hues	253/154/82	0/40/60/0	0/40/60/0	orangish-brown hues	255/179/0	03X	0/30/100
Oligocene	E3		254/224/144	0/12/40/0	0/15/40/0		234/198/114	A25	8/20/50
Eocene	E2		254/209/59	0/18/75/0	0/20/75/0		234/173/67	A37	8/30/70
Paleocene	E1		235/192/109	7.6/22.5/51.8/0	10/25/50/0		235/147/1	A4X	8/40/100
Mesozoic	MZ	darker blue	103/197/202	60/0/10.2/0	60/0/10/0	dull green	127/173/81	517	50/8/70
Cretaceous	K	green hues	197/229/71	22.8/0/75/0	25/0/75/0	green hues	127/195/28	50X	50/0/100
Upper	K2		226/243/152	11.4/0/40/0	10/0/40/0		22/241/151	104	13/0/40
Lower	K1		65/170/84	75/0/75/0	75/0/75/0		179/223/127	305	30/0/50
Jurassic	J	bluish hues	226/244/224	11.4/0/10.2/0	10/0/10/0	turquoise hues	77/180/126	705	70/0/50
Upper	J3		179/226/230	30/0/3.4/0	30/0/5/0		204/235/197	202	20/0/20
Middle	J2		130/215/232	50/5/0/0	50/5/0/0		127/202/147	504	50/0/40
Lower	J1		0/160/198	100/0/0/0	100/0/0/0		102/192/146	604	60/0/40
Triassic	T	purplish hues	107/1/125	60/100/0/0	60/100/0/0	greenish-blue hues	103/195/183	602	60/0/20
Upper	T3		228/197/225	10/20/0/0	10/20/0/0		204/236/225	20A	20/0/8
Middle	T2		177/137/193	30/40/0/0	30/40/0/0		153/215/190	402	40/0/20
Lower	T1		152/57/153	40/75/0/0	40/75/0/0		103/179/159	6A3	60/8/30
Paleozoic	PZ	pale green	153/192/141	40/10/40/0	40/10/40/0	blue	128/181/213	510	50/13/0
Permian	P	orange-red hues	240/64/40	5/75/75/0	5/75/75/0	cyan hues	103/198/221	600	60/0/0
Lopingian	P3		250/154/163	0/40/20/0	0/40/20/0		179/227/238	300	30/0/0
Guadalupian	P2		240/150/82	5/40/60/0	5/40/60/0		153/216/216	40A	40/0/8
Cisuralian	P1		202/60/64	20/75/60/0	20/75/60/0		128/206/201	501	50/0/13
Carboniferous	C		103/171/160	60/12/27.6/0	60/10/30/0	light-blue	153/189/218	410	40/13/0
Pennsylvanian	C2		153/196/181	40/9/20.7/0	40/10/20/0	blue	104/159/202	620	60/20/0
Mississippian	C1		103/148/109	60/22.5/51.8/0	60/25/50/0	bluish-gray	128/145/173	531	50/30/13

Devonian	D	dark tan	203/140/55	20/40/75/0	20/40/75/0	bluish-purple hues	153/153/201	430	40/30/0
Upper	D3	grayish-green hues	204/206/169	20/12/27.6/0	20/10/30/0		203/189/220	220	20/20/0
Middle	D2		153/165/109	40/22.5/51.8/0	40/25/50/0		153/131/190	440	40/40/0
Lower	D1		153/148/73	40/30/69/0	40/30/70/0		128/125/186	540	50/40/0
Silurian	S	light cyan hues	179/226/208	30/0/13.6/0	30/0/15/0	purple hues	177/114/182	350	30/50/0
Pridoli	S4		245/251/240	3.8/0/5/0	5/0/5/0		233/199/226	A20	8/20/0
Ludlow	S3		226/244/224	11.4/0/10.2/0	10/0/10/0		202/167/209	230	20/30/0
Wenlock	S2		197/232/195	22,8/0/20.4/0	20/0/20/0		177/137/179	34A	30/40/8
Llandovery	S1		153/215/179	40/0/25.5/0	40/0/25/0		152/88/168	460	40/60/0
Ordovician	O	dark turquoise	0/146/112	100/0/60/0	100/0/60/0	pink hues	249/129/166	051	0/50/13
Upper	O3		102/192/146	60/0/40/0	60/0/40/0		251/180/189	031	0/30/13
Middle	O2		65/156/104	75/10/60/0	75/10/60/0		250/154/177	041	0/40/13
Lower	O1		1/128/85	100/10/75/0	100/10/75/0		230/125/164	A51	8/50/13
Cambrian	€	greenish-brown hues	64/133/33	75/20/100/0	75/20/100/0	orange-brown hues	251/128/95	055	0/50/50
Upper	€3		215/211/170	15.2/12/27.6/0	15/10/30/0		253/205/184	022	0/20/20
Middle	€2		182/174/109	28.5/22.5/51.8/0	30/25/50/0		232/174/151	A33	8/30/30
Lower	€1		102/169/75	60/10/75/0	60/10/75/0		231/124/114	A54	8/50/40
PRECAMBRIAN	$P_€$	red	247/67/112	0/75/30/0	0/75/30/0	brown	178/134/83	346	30/40/60
PROTEROZOIC	PR	dull orange-reddish	251/154/148	0/40/27.6/0	0/40/30/0		204/216/145	2A4	20/8/40
Neoproterozoic	NP	dark tan-orange	254/179/67	0/30/69/0	0/30/70/0	pinkish-brown hues	202/165/149	233	20/30/30
Ediacaran	NP3						234/216/188	A12	8/13/20
Cryogenian	NP2						220/171/170	132	13/30/20
Tonian	NP1						203/164/108	235	20/30/50
Mesoproterozoic	MP	light tan	253/180/105	0/30/51.8/0	0/30/50/0		221/194/136	124	13/20/40
Paleoproterozoic	PP	red	247/67/115	0/75/27.6/0	0/75/30/0		179/178/94	326	30/20/60
ARCHEAN	AR	pinkish-red	240/2/127	0/100/0/0	0/100/0/0	gray hues	153/173/172	422	40/20/20
	NA		249/155/197	0/40/3.45/0	0/40/5/0		203/205/200	211	20/13/13
	MA		234/102/177	5/60/0/0	5/60/0/0		221/194/136	322	30/20/20
	PA		244/68/159	0/75/0/0	0/75/0/0		153/151/145	433	40/30/30
	EA		218/2/127	10/100/0/0	10/100/0/0		128/144/144	533	50/30/30

Press for the Harland *et al.* 1982 and 1990 time scales, and represents a standard unlikely to be emulated by many organizations, government or private. Hence, these authors, upon kindly obtaining proof sheets of the UNESCO color scheme through the courtesy of H.J. Vojacek of Mercury-Walch in Hobart, Tasmania, simply tried to match the UNESCO legend colors as closely as possible. Their Table A6.1, on p. 222, shows the degree of equivalency in stratigraphic units.

This, then, was the situation when we (i.e. the team responsible for GTS2004) started our investigation of color, in the footsteps of all the charts mentioned earlier. Seemingly, the task would be easier now, since Remane (2000) in his Appendix 2 lists the composition of the colors of eons, eras, periods, and epochs with reference to "Gamme de Colors BRGM, 1989; for the green color: ref. Pantone 354; for the orange color: ref. Pantone 1454," followed by percent mixing tables for all units.

However, none of our professional and able drafting personnel managed to create the colors with the code provided. A letter to CGMW in mid 2001 generated a reply acknowledging code printing errors and offered some corrections. Unfortunately to no avail; from many directions reactions were received like: "cannot match colors and cannot match colors according to their [i.e. CGMW] new instructions on any normal graphics program."

Now, color mixers for electronic media come in three options: CMYG (cyan–magenta–yellow–gray), RGBG (red–green–blue–gray), or IHG (intensity–hue–grayness color wheel). The CGMW scheme, which means jumping between those schemes and the Pantone standard listed, is one not supplied with standard drafting programs available.

Another obstacle to the CGMW scheme, other than inability to match colors scientifically, is that some colors for periods are difficult to use. There are two reasons for the latter: first, the Early Paleozoic has shades of green that can be hard to distinguish from the Cretaceous, and the same is true for the Pennsylvanian and Middle–Late Jurassic. Second, it would be more logical to have a color scheme that goes from darker to lighter and is light enough in itself to allow good contrast with the names of the units in black letters. One remedy, to invert colors and have white letters on dark, intense background colors, we do not encourage.

Having gone to great lengths to investigate the UNESCO legend published by Remane (2000), we have come to the conclusion that, in the long term, it is better to have a pragmatic color scheme that can be readily used by drafting personnel and geologists the world over. Nowadays electronic reproduction is within reach of a great many scientists, and with it comes the requirement of an easy and logical color scheme. The ICS also displays its charts on the World Wide Web and encourages students and professionals to make copies of it for embedding in applications.

The choice we have made has come out in favor of two color schemes, both in RGB modes that are easily reproducible on modern personal computers. The first one matches in RGB code the color scheme according to the Commission for the Geological Map of the World (CGMW) in Paris, France (Table A1.1). The scheme runs from shades of reddish/pink for Precambrian units to greenish/brown and reddish/orange for the Paleozoic; to purple, bluish, and greenish for the Mesozoic; and to shades of yellowish/green for the Cenozoic. Fifty-nine stratigraphic units are color coded, and, for example, D-203/140/55 represents the Devonian with a color mix of 203 Red, 140 Green, and 55 Blue. Stratigraphic units are also letter coded, with D being Devonian; D1, Lower Devonian; D2, Middle Devonian; and D3, Upper Devonian. The abbreviations of the stratigraphic units, i.e. the stage, epoch, period, era, and eon notations, follow the ICS *International Stratigraphic Chart* in concert with the *World Geological Map*.

The second scheme matches the color code according to the US Geological Survey, and is known as the spectral color scheme (Table A1.1). The scheme runs from shades of gray-brown for Precambrian units to reddish/pink and bluish-gray tones in the Paleozoic, blue-greenish hues in the Mesozoic, and light brownish and yellow shades for the Cenozoic. For example, RGB153/189/218 represents the Carboniferous with a color mix of 153 Red, 189 Green, and 218 Blue.

Both color schemes can be readily downloaded from the ICS website at www.stratigraphy.org, and now provide pragmatic and satisfactory standards.

Appendix 2 • Orbital tuning calibrations and conversions for the Neogene Period

L. LOURENS, F. HILGEN, N. J. SHACKLETON, J. LASKAR, AND J. WILSON

Table A2.1 *Comparison of magnetic reversal ages based on Cande and Kent (1995) and ATNTS2004*[a]

Polarity Chron	CK'95 time scale			ATNTS2004			
Chron	Top	Base	Duration	Young	Old	Duration	Reference
C1n (Brunhes)	0.000	0.780	0.780	0	0.781	0.781	1, 2
C1r.1r (Matuyama)	0.780	0.990	0.210	0.781	0.988	0.207	1, 2
C1r.1n (Jaramillo)	0.990	1.070	0.080	0.988	1.072	0.084	1, 2
C1r.2r	1.070	1.201	0.131	1.072	1.173	0.101	1, 2
C1r.2n (Cobb Mountain)	1.201	1.211	0.010	1.173	1.185	0.012	2
C1r.3r	1.211	1.770	0.559	1.185	1.778	0.593	2
C2n (Olduvai)	1.770	1.950	0.180	1.778	1.945	0.167	2
C2r.1r	1.950	2.140	0.190	1.945	2.128	0.183	2, 3, 4, 5
C2r.1n (Reunion)	2.140	2.150	0.010	2.128	2.148	0.020	3, 4, 5
C2r.2r (Matuyama)	2.150	2.581	0.431	2.148	2.581	0.433	3, 4, 5
C2An.1n (Gauss)	2.581	3.040	0.459	2.581	3.032	0.451	3, 4, 5, 6
C2An.1r (Keana)	3.040	3.110	0.070	3.032	3.116	0.084	4, 7, 8, 9
C2An.2n	3.110	3.220	0.110	3.116	3.207	0.091	4, 7, 8, 9
C2An.2r (Mammoth)	3.220	3.330	0.110	3.207	3.330	0.123	4, 7, 8, 9
C2An.3n (Gauss)	3.330	3.580	0.250	3.330	3.596	0.266	4, 7, 8, 9
C2Ar (Gilbert)	3.580	4.180	0.600	3.596	4.187	0.591	4, 7, 8, 9
C3n.1n (Cochiti)	4.180	4.290	0.110	4.187	4.300	0.113	4, 8, 9, 10
C3n.1r	4.290	4.480	0.190	4.300	4.493	0.193	4, 8, 9, 10
C3n.2n (Nunivak)	4.480	4.620	0.140	4.493	4.631	0.138	4, 8, 9, 10
C3n.2r	4.620	4.800	0.180	4.631	4.799	0.168	4, 8, 9, 10
C3n.3n (Sidufjall)	4.800	4.890	0.090	4.799	4.896	0.097	4, 8, 9, 10
C3n.3r	4.890	4.980	0.090	4.896	4.997	0.101	4, 8, 9, 10
C3n.4n (Thvera)	4.980	5.230	0.250	4.997	5.235	0.238	4, 8, 9, 10
C3r (Gilbert)	5.230	5.894	0.664	5.235	6.033	0.798	4, 8, 9, 10, 11, 12
C3An.1n	5.894	6.137	0.243	6.033	6.252	0.219	11, 12
C3An.1r	6.137	6.269	0.132	6.252	6.436	0.184	11, 12
C3An.2n	6.269	6.567	0.298	6.436	6.733	0.297	11, 12
C3Ar	6.567	6.935	0.368	6.733	7.140	0.407	11, 12, 13
C3Bn	6.935	7.091	0.156	7.140	7.212	0.072	13
C3Br.1r	7.091	7.135	0.044	7.212	7.251	0.039	13
C3Br.1n	7.135	7.170	0.035	7.251	7.285	0.034	13
C3Br.2r	7.170	7.341	0.171	7.285	7.454	0.169	13, 14, 16
C3Br.2n	7.341	7.375	0.034	7.454	7.489	0.035	14, 16
C3Br.3r	7.375	7.432	0.057	7.489	7.528	0.039	14, 16
C4n.1n	7.432	7.562	0.130	7.528	7.642	0.114	14, 16
C4n.1r	7.562	7.650	0.088	7.642	7.695	0.053	14, 16
C4n.2n	7.650	8.072	0.422	7.695	8.108	0.413	14, 16
C4r.1r	8.072	8.225	0.153	8.108	8.254	0.146	14, 16
C4r.1n	8.225	8.257	0.032	8.254	8.300	0.046	14, 16

(*cont.*)

Table A2.1 *(cont.)*

Polarity Chron	CK'95 time scale			ATNTS2004			
Chron	Top	Base	Duration	Young	Old	Duration	Reference
C4r.2r	8.257	8.699	0.442	**8.300**	**8.769**	**0.469**	14, 16
C4r.2r-1	8.635	8.651	0.016	**8.661**	**8.699**	**0.037**	14, 16, 17
C4An	8.699	9.025	0.326	**8.769**	**9.098**	**0.329**	17
C4Ar.1r	9.025	9.230	0.205	**9.098**	**9.312**	**0.214**	14, 16, 17
C4Ar.1n	9.230	9.308	0.078	**9.312**	**9.409**	**0.097**	14, 16, 17
C4Ar.2r	9.308	9.580	0.272	**9.409**	**9.656**	**0.247**	14, 16, 17
C4Ar.2n	9.580	9.642	0.062	**9.656**	**9.717**	**0.060**	14, 16, 17
C4Ar.3r	9.642	9.740	0.098	**9.717**	**9.779**	**0.063**	14, 16, 17
C5n.1n	9.740	9.880	0.140	**9.779**	**9.934**	**0.155**	17
C5n.1r	9.880	9.920	0.040	**9.934**	**9.987**	**0.053**	17
C5n.2n	9.920	10.949	1.029	**9.987**	**11.040**	**1.053**	17, 18
C5r.1r	10.949	11.052	0.103	**11.040**	**11.118**	**0.078**	18
C5r.1n	11.052	11.099	0.047	**11.118**	**11.154**	**0.036**	18
C5r.2r	11.099	11.476	0.377	**11.154**	**11.554**	**0.400**	18
C5r.2r-1				**11.267**	**11.298**	**0.031**	18
C5r.2n	11.476	11.531	0.055	**11.554**	**11.614**	**0.060**	18
C5r.3r	11.531	11.935	0.404	**11.614**	**12.014**	**0.400**	18
C5An.1n	11.935	12.078	0.143	**12.014**	**12.116**	**0.102**	18
C5An.1r	12.078	12.184	0.106	**12.116**	**12.207**	**0.091**	18
C5An.2n	12.184	12.401	0.217	**12.207**	**12.415**	**0.208**	18
C5Ar.1r	12.401	12.678	0.277	**12.415**	**12.730**	**0.315**	18
C5Ar.1n	12.678	12.708	0.030	**12.730**	**12.765**	**0.035**	18
C5Ar.2r	12.708	12.775	0.067	**12.765**	**12.820**	**0.055**	18
C5Ar.2n	12.775	12.819	0.044	**12.820**	**12.878**	**0.058**	18
C5Ar.3r	12.819	12.991	0.172	**12.878**	*13.015*	*0.137*	18, 19
C5AAn	12.991	13.139	0.148	*13.015*	*13.183*	*0.168*	19
C5AAr	13.139	13.302	0.163	*13.183*	*13.369*	*0.186*	19
C5ABn	13.302	13.510	0.208	*13.369*	*13.605*	*0.236*	19
C5ABr	13.510	13.703	0.193	*13.605*	*13.734*	*0.129*	19
C5ACn	13.703	14.076	0.373	*13.734*	*14.095*	*0.361*	19
C5ACr	14.076	14.178	0.102	*14.095*	*14.194*	*0.099*	19
C5ADn	14.178	14.612	0.434	*14.194*	*14.581*	*0.387*	19
C5ADr	14.612	14.800	0.188	*14.581*	*14.784*	*0.203*	19
C5Bn.1n	14.800	14.888	0.088	*14.784*	*14.877*	*0.093*	19
C5Bn.1r	14.888	15.034	0.146	*14.877*	*15.032*	*0.155*	19
C5Bn.2n	15.034	15.155	0.121	*15.032*	**15.160**	*0.128*	19
C5Br	15.155	16.014	0.859	**15.160**	**15.974**	**0.814**	20
C5Cn.1n	16.014	16.293	0.279	**15.974**	*16.268*	*0.293*	19, 20
C5Cn.1r	16.293	16.327	0.034	*16.268*	*16.303*	*0.035*	19
C5Cn.2n	16.327	16.488	0.161	*16.303*	*16.472*	*0.169*	19
C5Cn.2r	16.488	16.556	0.068	*16.472*	*16.543*	*0.071*	19
C5Cn.3n	16.556	16.726	0.170	*16.543*	*16.721*	*0.178*	19
C5Cr	16.726	17.277	0.551	*16.721*	*17.235*	*0.514*	19
C5Dn	17.277	17.615	0.338	*17.235*	*17.533*	*0.298*	19
C5Dr.1r	17.615	18.281	0.666	*17.533*	*17.717*	*0.184*	19
C5Dr.1n				*17.717*	*17.740*	*0.023*	19
C5Dr.2r				*17.740*	*18.056*	*0.316*	19
C5En	18.281	18.781	0.500	*18.056*	*18.524*	*0.468*	19
C5Er	18.781	19.048	0.267	*18.524*	*18.748*	*0.224*	19
C6n	19.048	20.131	1.083	*18.748*	*19.722*	*0.974*	19

Table A2.1 *(cont.)*

Polarity Chron	CK'95 time scale			ATNTS2004			
Chron	Top	Base	Duration	Young	Old	Duration	Reference
C6r	20.131	20.518	0.387	*19.722*	*20.040*	*0.318*	19
C6An.1n	20.518	20.725	0.207	*20.040*	*20.213*	*0.173*	19
C6An.1r	20.725	20.996	0.271	*20.213*	*20.439*	*0.226*	19
C6An.2n	20.996	21.320	0.324	*20.439*	*20.709*	*0.270*	19
C6Ar	21.320	21.768	0.448	*20.709*	*21.083*	*0.374*	19
C6AAn	21.768	21.859	0.091	*21.083*	*21.159*	*0.076*	19
C6AAr.1r	21.859	22.151	0.292	*21.159*	*21.403*	*0.244*	19
C6AAr.1n	22.151	22.248	0.097	*21.403*	*21.483*	*0.080*	19
C6AAr.2r	22.248	22.459	0.211	*21.483*	*21.659*	*0.176*	19
C6AAr.2n	22.459	22.493	0.034	*21.659*	*21.688*	*0.029*	19
C6AAr.3r	22.493	22.588	0.095	*21.688*	*21.767*	*0.079*	19
C6Bn.1n	22.588	22.750	0.162	*21.767*	*21.936*	*0.169*	19
C6Bn.1r	22.750	22.804	0.054	*21.936*	*21.992*	*0.056*	19
C6Bn.2n	22.804	23.069	0.265	*21.992*	*22.268*	*0.276*	19
C6Br	23.069	23.353	0.284	*22.268*	*22.564*	*0.296*	19
C6Cn.1n	23.353	23.535	0.182	*22.564*	*22.754*	*0.190*	19
C6Cn.1r	23.535	23.677	0.142	*22.754*	*22.902*	*0.148*	19
C6Cn.2n	23.677	23.800	0.123	*22.902*	*23.030*	*0.128*	21

[a] References for Neogene age scaling of polarity chrons: 1. Shackleton *et al.* (1990), 2. Horng *et al.* (2002), 3. Hilgen (1991a), 4. Lourens *et al.* (1996a), 5. Zijderveld *et al.* (1991), 6. Langereis *et al.* (1994), 7. Zachariasse *et al.* (1989), 8. van Hoof (1993), 9. Hilgen (1991b), 10. Langereis and Hilgen (1991), 11. Krijgsman *et al.* (1999), 12. Sierro *et al.* (2001), 13. Hilgen *et al.* (2000b), 14. Hilgen *et al.* (1995), 15. Krijgsman *et al.* (1994), 16. Hilgen *et al.* (in press), 17. Abdul-Aziz (unpubl.), 18. Abdul-Aziz *et al.* (2003), 19. interpolated ages (in italics) derived from seafloor-spreading-rate history model of the Australia–Antarctic plate pair (this paper), 20. Shackleton *et al.* (2001), and 21. Shackleton *et al.* (2000).

Table A2.2 *Recalculated age estimates of calcareous nannofossil datum events according to ATNTS2004*[a–c]

Species event	Zone/subzone (transition)	ODP Legs 111 and 138		ODP Sites 925, 926		Eastern Mediterranean		Morocco	
		Age (Ma)	Reference	Age (Ma)	Reference	Age (Ma)	Reference	Age (Ma)	Reference
PLEISTOCENE–HOLOCENE	**0.0115 Ma (a)**								
Bottom acme *Emiliania huxleyi*	MNN21a–MNN21b					**0.05**	17, 18		
Bottom *Emiliania huxleyi*	CN14b–CN15, NN20–NN21, MNN20–MNN21a	**0.29**	1, 2	**0.29**	9, 10	**0.27**	17, 18		
Bottom acme small *Gephyrocapsa* spp.						**0.27**	17, 18		
Top *Pseudoemiliania lacunosa*	CN14a–CN14b, NN19–NN20, MNN19f–MNN20	**0.44**	1–5	**0.44**	9, 10	**0.47**	11, 17–19		
Top *Gephyrocapsa* sp3						**0.61**	17, 18, 20		
Top (common) *Reticulofenestra asanoi*				**0.91**	9–11	**0.90**	11, 18, 21		
Bottom *Gephyrocapsa* sp3	MNN19e–MNN19f	**1.02**	1–5			**0.97**	17, 18, 21		
Reentrance medium *Gephyrocapsa* (>4 μm)	CN13bD–CN14a	**1.04**	1–5	**1.01**	9–11	**0.96**	11, 17–19, 21		
Bottom (common) *Reticulofenestra asanoi*				**1.14**	9–11	**1.08**	18, 19, 21		
Top large *Gephyrocapsa* (>5.5 μm)	CN13bC–CN13bD, MNN19d–MNN19e	**1.24**	1–5	**1.26**	9–11	**1.25**	11, 22–24		
Top *Helicosphaera sellii*		**1.34**	1–5			**1.26**	11, 22–24		
Bottom large *Gephyrocapsa* (>5.5 μm)	CN13bB–CN13bC, MNN19c–MNN19d	**1.46**	1–5	**1.56**	9–11	**1.62**	11, 22–24		
Top *Calcidiscus macintyrei*	CN13bA–CN13bB, MNN19b–MNN19c	**1.60**	1–5	**1.61**	9–11	**1.66**	11, 22–24		
Bottom medium *Gephyrocapsa* (>4 μm)	CN13a–CN13bA, MNN19a–MNN19b	**1.67**	1–5	**1.69**	9–11	**1.73**	11, 22–24		
PLIOCENE–PLEISTOCENE									
Gelasian–E. Pleist. sub-epoch	**1.806 Ma (b)**								
Top *Discoaster brouweri*	CN12d–CN13a, NN18–NN19, MNN18–MNN19a	**2.06**	4–7	**1.93**	9, 10	**1.95**	11, 22–25		
Bottom acme *Discoaster triradiatus*				**2.14**	9, 10	**2.22**	22, 25		
Top *Discoaster pentaradiatus (quintatus)*	CN12c–CN12d, NN17–NN18, MNN17–MNN18			**2.39**	9, 10	**2.51**	20, 22, 25, 26		
Top *Discoaster surculus*	CN12b–CN12c, NN16–NN17	**2.52**	4–7	**2.49**	9, 10	**2.54**	20, 22, 25, 26		

Piacenzian–Gelasian	**2.588 Ma (c)**						
Bottom acme *Discoaster pentaradiatus (quintatus)*						**2.64**	22, 25
Subtop *Discoaster tamalis*	CN12aB–CN12b, MNN16a–MNN16b/17	**2.87**	4–7	**2.80**	10, 12	**2.80**	20, 22, 25, 26
Subtop *Discoaster asymmetricus*						**2.83**	22, 25
Second acme *Gephyrocapsa* spp.						**3.29**	22, 25
Top paracme *Discoaster pentaradiatus (quintatus)*						**3.61**	22, 25
Zanclean–Piacenzian	**3.600 Ma (d)**						
Subtop *Sphenolithus* spp.	CN12aA–CN12aB	**3.65**	4–7	**3.54**	10, 12	**3.70**	22, 25
Top *Reticulofenestra pseudoumbilica*	CN11b–CN12aA, NN14/15–NN16, MNN14/15–MNN16a	**3.79**	4–7	**3.70**	10, 12	**3.84**	22, 25
Bottom paracme *Reticulofenestra pseudoumbilica*						**3.89**	22, 25
Subbottom *Discoaster pentaradiatus*						**3.93**	22, 25
Subbottom *Discoaster tamalis*						**3.97**	22, 25
Top *C. tricorniculatus*							
Subbottom *Discoaster asymmetricus*	CN11a–CN11b, NN13–NN14/15, MNN13–MNN14/15	**4.13**	5–7			**4.12**	22, 25
Subbottom *Discoaster brouweri*						**4.12**	22, 25
Subbottom *Gephyrocapsa* spp.						**4.33**	22, 25
Top *Amaurolithus primus*	CN10c–CN11a	**4.50**	5–8				
Subbottom *Reticulofenestra pseudoumbilica*	"MNN 12–13"					**4.91**	22, 25
Subbottom *Discoaster ovata*						**4.91**	22, 25
Top *Reticulofenestra antarctica*						**4.91**	22, 25
Top *Ceratolithus acutus*		**5.04**	5–8	**5.04**	10, 12		
Cross-over *Ceratolithus acutus–C. rugosus*				**5.05**	13, 14		
Bottom *Ceratolithus rugosus*	CN10b–CN10c, NN12–NN13	**5.12**	5–8	**5.05**	13, 14		
Top *Triquetrorhabdulus rugosus*				**5.28**	13, 14		
MIOCENE–PLIOCENE							
Messinian–Zanclean	**5.332 Ma (e)**						
Bottom *Ceratolithus larrymayeri* sp1				**5.34**	13, 14		
Bottom *Ceratolithus acutus*	CN10a–CN10b	**5.32**	5–8	**5.35**	13, 14		

(*cont.*)

Table A2.2 *(cont.)*

Species event	Zone/subzone (transition)	ODP Legs 111 and 138		ODP Sites 925, 926		Eastern Mediterranean		Morocco	
		Age (Ma)	Reference	Age (Ma)	Reference	Age (Ma)	Reference	Age (Ma)	Reference
Top *Discoaster quinqueramus*	CN9bC–CN10a, NN11b–NN12	**5.59**	5–8	**5.58**	10, 13			**5.54**	38
Top *Nicklithus amplificus*	CN9bB–CN9bC, MNN11c–MNN12			**5.98**	13, 14	**5.94**	27–29	**6.00**	38
Last observed specimen *Reticulofenestra rotaria*						**5.99**	27–29		
Top *Discoaster* cf *toralus*						**6.12**	27–29		
Top common occurrence *Nicklithus amplificus*		**6.12**	5–8	**5.98**	13, 14	**6.14**	27–29		
Bottom *Discoaster* cf *toralus*						**6.25**	27–29		
Cross-over *Nicklithus amplificus*–*T. rugosus*				**6.79**	13, 14				
Bottom *Nicklithus amplificus*	CN9bA–CN9bB, MNN11b–MNN11c			**6.91**	10, 13	**6.68**	27–29		
Top *Reticulofenestra rotaria*						**6.79**	31, 32		
Bottom (common) *Helicosphaera* cf. *sellii*						**7.00**	27–32		
Bottom common *Amaurolithus delicatus*								**7.22**	39
Top paracme *Reticulofenestra pseudoumbilicus*				**7.08**	13, 14	**7.17**	28, 33–35		
Bottom common *Reticulofenestra rotaria*						**7.24**	30–35	**7.27**	39
Bottom *Reticulofenestra rotaria*						**7.41**	28, 30–35	**7.46**	39
Bottom *Amaurolithus* spp.–bottom *Amaurolithus primus*	CN9a–CN9bA, NN11a–NN11b, MNN11a–MNN11b			**7.36**	10, 13	**7.42**	28, 30–35		
Tortonian–Messinian	**7.246 Ma (f)**								
Bottom common *Discoaster surculus*				**7.79**	10, 13				
Top *Helicosphaera stalis*						**7.61**	28, 33, 35		
Bottom *Discoaster berggrenii*	CN8–CN9a, NN10–NN11a			**8.29**	10, 13				
Top *Minylitha convallis*						**8.68**	28, 33, 35		
Bottom paracme *Reticulofenestra pseudoumbilicus*	MNN10b–MNN11a			**8.79**	13, 14	**8.71**	28, 33, 35		
Top common *Minylitha convallis*						**8.88**	28, 33, 35		
Bottom common *Minylitha convallis*						**9.29**	28, 33, 35		

Bottom common *Discoaster pentaradiatus*	MNN10a–MNN10b			**9.37**	28, 33, 35
Regular *Minylitha convallis*				**9.38**	28, 33, 35
Top *Catinaster calyculus*		**9.67**	13, 14		
Top *Discoaster hamatus*	CN7–CN8, NN9–NN10, MNN9–MNN10	**9.69**	13, 14	**9.53**	28, 33, 35
Top *Catinaster coalitus*		**9.69**	13, 14		
Bottom *Minylitha convallis*		**9.75**	13, 14	**9.61**	28, 33, 35
Cross-over *Discoaster hamatus–D. neohamatus*		**9.76**	13, 14		
Cross-over *Catinaster calyculus–C. coalitus*		**10.41**	13, 14		
Bottom *Discoaster neohamatus*		**10.52**	13, 14	**9.87**	35, 36
Bottom *Discoaster hamatus*	CN6–CN7, NN8–NN9	**10.55**	13, 14	**10.18**	35, 36
Bottom *Discoaster bellus*	MNN8b–MNN9			**10.40**	35, 36
Bottom common *Helicosphaera stalis*	MNN8a–MNN8b			**10.71**	35, 36
Bottom *Discoaster brouweri*		**10.76**	13, 14	**10.73**	35, 36
Bottom *Catinaster calyculus*		**10.79**	13, 14		
Cross-over *Discoaster micros*–transition *Catinaster coalitus*		**10.85**	13, 14		
Bottom *Catinaster coalitus*	CN5b–CN6, NN7–NN8	**10.89**	13, 14	**10.73**	35, 36
Top common *Helicosphaera walbersdorfensis*	MNN7c–MNN8a			**10.74**	35, 36
Top *Coccolithus miopelagicus*		**11.02**	13, 14	**10.97**	35, 36
Top *Calcidiscus premacintyrei*		**11.21**			
Serravallian–Tortonian	**11.600 Ma (g)**				
Top common *Discoaster kugleri*	MNN7b–MNN7c	**11.58**	13, 14	**11.60**	35, 36
Bottom common *Discoaster kugleri*	CN5a–CN5b, NN6–NN7, MNN7a–MNN7b	**11.86**	14, 15	**11.90**	35, 36
Top *Coronocyclus nitescens*		**12.12**	10, 15		
Top regular *Calcidiscus premacintyrei*	MNN6–MNN7a	**12.45**	15, 16	**12.38**	36
Bottom common *Calcidiscus macintyrei*		**12.46**	15, 16		
Top *Cyclicargolithus floridanus*		**13.33**	10, 15		
Top common *Cyclicargolithus floridanus*		**13.33**	15, 16	**13.28**	36

(*cont.*)

Table A2.2 *(cont.)*

Species event	Zone/subzone (transition)	ODP Legs 111 and 138		ODP Sites 925, 926		Eastern Mediterranean		Morocco	
		Age (Ma)	Reference	Age (Ma)	Reference	Age (Ma)	Reference	Age (Ma)	Reference
Langhian–Serravallian	**13.654 Ma (h)**								
Top *Sphenolithus heteromorphus*	CN4–CN5a, NN5–NN6, MNN5b–MNN6			**13.53**	14, 15	**13.65**	37		
Top *Helicosphaera ampliaperta*	CN3–CN4, NN4–NN5			**14.91**	10, 15				
Top abundant *Discoaster deflandrei*				**15.80**	10, 15				
Burdigalian–Langhian	**15.974 Ma (i)**								
Bottom common *Sphenolithus heteromorphus*				**17.71**	10, 15				
Top common *Sphenolithus belemnos*	CN2–CN3, NN3–NN4			**17.95**	10, 15				
Top *Triquetrorhabdulus carinatus*	CN1–CN2, NN2–NN3			**18.28**	10, 15				
Bottom *Sphenolithus belemnos*				**19.03**	15				
Aquitanian–Burdigalian	**20.428 Ma (j)**								
Bottom *Helicopontosphaera ampliaperta*				**20.43**	15, 41				
Cross-over *Helicosphaera euphratis/Helicosphaera carteri*				**20.92**	15, 41				
Bottom common *Helicosphaera carteri*				**22.03**	15, 41				
Bottom *Sphenolithus disbelemnos*				**22.76**	15, 41				
Bottom *Discoaster druggi*	NN1–NN2					*22.82*	40		
OLIGOCENE–MIOCENE									
Chattian–Aquitanian	**23.030 Ma (k)**								
Top *Sphenolithus delphix*				**23.11**	10, 15				
Bottom *Sphenolithus delphix*				**23.21**	10, 15				

[a] Zonations: NN = Martini (1971), CN = Bukry (1973, 1975b), MNN = Raffi and Rio (1979), Fornaciari and Rio (1996), Fornaciari *et al.* (1996), Raffi *et al.* (2003).
[b] Bases of Neogene stages:

(a) Holocene: termination 1 in the GRIP-ice core (proposal by Björck *et al.*, 1998).
(b) Pleistocene: top sapropel e in the Vrica section, Calabria, Italy (Aguirre and Pasini, 1985; Rio *et al.*, 2000).
(c) Gelasian Stage (uppermost Pliocene): midpoint sapropel "A5" in the San Nicola section, Sicily, Italy (Rio *et al.*, 1998).
(d) Piacenzian Stage: base beige layer of $CaCO_3$ cycle 77 in the Punta Piccola section, Sicily, Italy (Castradori *et al.*, 1998).

(e) Pliocene (Zanclean Stage): base $CaCO_3$ cycle 1 in section Eraclea Minoa, Sicily, Italy (Van Couvering *et al.*, 2000).
(f) Messinian Stage (uppermost Miocene): base red layer 15 in the Oued Akrech section, Morocco (Hilgen *et al.*, 2000c).
(g) Tortonian Stage: midpoint sapropel 76 in the Monte dei Corvi section, Italy (proposal by Hilgen *et al.*, in press).
(h) Serravallian Stage: adopted age of the top *S. heteromorphus* in the Mediterranean (working group).
(i) Langhian Stage: adopted age of C5Cn.1n (top; working group).
(j) Burdigalian Stage: adopted age of bottom *H. ampliaperta* at Ceara Rise (working group).
(k) Miocene (Aquitanian Stage): C6Cn.2n (base) in the Lemme–Carrosio section, Italy (Steininger *et al.*, 1997a,b; Shackleton *et al.*, 2000).

[c] References: 1. Shackleton *et al.* (1990), 2. Shipboard Scientific Party (1988), 3. Raffi *et al.* (1993), 4. Mix *et al.* (1995a,b), 5. Raffi and Flores (1995), 6. Shackleton *et al.* (1995a), 7. Shackleton *et al.* (1995b), 8. Pälike (unpubl.), 9. Bickert *et al.* (1997a), 10. Shipboard Scientific Party (1995), 11. Raffi *et al.* (2003), 12. Tiedemann and Franz (1997), 13. Shackleton and Crowhurst (1997), 14. Backman and Raffi (1997), 15. Shackleton *et al.* (1999), 16. Turco *et al.* (2003), 17. Castradori (1993), 18. Lourens (unpubl.), 19. Sprovieri *et al.* (1998), 20. Staerker (1998), 21. Maiorano *et al.* (in press), 22. Lourens *et al.* (1996a), 23. Lourens *et al.* (1996b), 24. Lourens *et al.* (1998), 25. Driever (1988), 26. Lourens *et al.* (2001), 27. Krijgsman *et al.* (1999), 28. Raffi *et al.* (2003), 29. Hilgen and Krijgsman (1999), 30. Negri and Villa (2000), 31. Negri *et al.* (1999). 32. Krijgsman *et al.* (1997a), 33. Hilgen *et al.* (1995), 34. Hilgen *et al.* (2000a), 35. Hilgen *et al.* (in press), 36. Hilgen *et al.* (2000c), 37. Hilgen and Raffi (unpubl.), 38. Krijgsman *et al.* (unpublished), 39. Hilgen *et al.* (2000b), 40. interpolated age from Fornaciari and Rio (1996) using the position of this event (36.6 m) in relation to that of *S. delphix* LO (19.8 m) and *S. disbelemnos* FO (40.5 m) in the Poggio d'Ancona section, Italy, by using the ages derived from Ceara Rise, and 41. Fornaciari (1996).

Table A2.3 *Recalculated age estimates of planktonic foraminifer datum events according to ATNTS2004*[a,b]

Species event	Zone/subzone (transition)	ODP Legs 111 and 138		ODP Sites 925, 926		Eastern Mediterranean		Morocco	
		Age (Ma)	Reference	Age (Ma)	Reference	Age (Ma)	Reference	Age (Ma)	Reference
PLEISTOCENE–HOLOCENE	**0.0115 Ma (a)**								
Bottom *Globigerinoides ruber rosa*	"MPL8b–MPL8c"					**0.33**	15		
Top common *Neogloboquadrina* sp. (sin)	"MPL8a–MPL8b"					**0.61**	15		
Top *Globorotalia tosaensis*	Pt1a–Ptb	**0.61**	1,2						
Bottom common *Truncorotalia (Globorotalia) truncatulinoides (excelsa)*	MPL7–MPL8a					**0.94**	15,16		
Top paracme *Neogloboquadrina* sp. (sin)						**1.21**	17–19		
Top *Globoturborotalita obliquus*				**1.30**	7–9				
Bottom paracme *Neogloboquadrina* sp. (sin)						**1.37**	17–19		
Top *Neogloboquadrina acostaensis*				**1.58**	7–9				
Top *Globoturborotalita apertura*				**1.64**	7–9				
Bottom common *Neogloboquadrina* sp. (sin)	MPL6b–MPL7					**1.79**	17–21		
PLIOCENE–PLEISTOCENE									
Gelasian–E. Pleist. subepoch	**1.806 Ma (b)**								
Top *Globigerinoides fistulosus*	Pl6–Pt1a	**1.77**	3,4	**1.88**	7–9				
Bottom *Truncorotalia (Globorotalia) truncatulinoides*	N20/N21–N22, MPL6a–MPL6b			**1.93**	7–9	**2.00**	17, 20, 21		
Top *Globigerinoides extremus*				**1.98**	7–9				
Top *Menardella limbata*		**2.24**	1, 2, 5, 6						
Bottom *Pulleniatina finalis*				**2.04**	7–9				
Top *Menardella exilis*				**2.09**	7–9				
Bottom *Globoconella (Globorotalia) inflata*	MPL5b–MPL6a					**2.09**	17, 20, 21		
Reappearance *Pulleniatina*				**2.26**	7–9				
Top *Globoturborotalita woodi*				**2.30**	7–9				
Top *Menardella pertenuis*				**2.30**	7–9				
Top *Menardella limbata*				**2.39**	7–9				
Top *Menardella miocenica*	Pl5–Pl6			**2.39**	7–9				

Top *Globoconella (Globorotalia) puncticulata (bononiensis)*	MPL5a–MPL5b					**2.41**	17, 20, 21
Top *Neogloboquadrina atlantica* (sin)						**2.41**	17, 20, 21
Piacenzian–Gelasian	**2.588 Ma (c)**						
Bottom *Neogloboquadrina atlantica* (sin)						**2.72**	17,20,21
Top *Globoturborotalita decoraperta*				**2.75**	8–10		
Top *Menardella multicamerata*				**2.98**	8–10		
Top *Dentoglobigerina (Globoquadrina) altispira*	Pl4–Pl5	**3.47**	1, 2, 5, 6	**3.13**	8–10	**3.17**	17, 22, 23
Top *Sphaeroidinellopsis seminulina*	Pl3–Pl4, MPL4b–MPL5a	**3.59**	1, 2, 5, 6	**3.13**	8–10	**3.19**	17, 22, 23
Top *Globoquadrina baroemoenensis*				**3.23**	8–10		
Top *Hirsutella cibaoensis*				**3.23**	8–10		
Reappearance *Globoconella (Globorotalia) puncticulata (bononiensis)*						**3.31**	17, 22, 23
Reappearance *Truncorotalia (Globorotalia) crassaformis*						**3.35**	17, 22, 23
Disappearance *Pulleniatina*				**3.41**	8–10		
Bottom *Menardella pertenuis*				**3.52**	8–10		
Disappearance *Globoconella (Globorotalia) puncticulata*	MPL4a–MPL4b					**3.57**	17, 22, 23
Bottom *Truncorotalia (Globorotalia) crassaformis*						**3.60**	17, 22, 23
Zanclean–Piacenzian	**3.600 Ma (d)**						
Bottom *Menardella miocenica*	N19–N20/N21			**3.77**	8–10		
Top *Globorotalia plesiotumida*				**3.77**	8–10		
Top *Hirsutella (Globorotalia) margaritae*	Pl2–Pl3, MPL3–MPL4a			**3.85**	8–10	**3.81**	17, 23, 24
Top common *Hirsutella (Globorotalia) margaritae*						**3.98**	17, 23, 24
Pulleniatina sin → dex coiling change				**4.08**	8–10		
Bottom *Truncorotalia (Globorotalia) crassaformis s.l.*				**4.31**	8–10		
Top *Globoturborotalita nepenthes*	Pl1–Pl2			**4.37**	8–10		
Bottom *Menardella exilis*				**4.45**	8–10		
Bottom *Globoconella (Globorotalia) puncticulata*	MPL2–MPL3					**4.52**	17, 23, 24
Bottom *Sphaeroidinellopsis kochi*				**4.53**	8–10		

(*cont.*)

Table A2.3 *(cont.)*

Species event	Zone/subzone (transition)	ODP Legs 111 and 138		ODP Sites 925, 926		Eastern Mediterranean		Morocco	
		Age (Ma)	Reference	Age (Ma)	Reference	Age (Ma)	Reference	Age (Ma)	Reference
Bottom common *Hirsutella (Globorotalia) margaritae*	MPL1–MPL2					**5.08**	17, 23, 24		
Top acme *Sphaeroidinellopsis*						**5.21**	17, 23, 24		
Bottom acme *Sphaeroidinellopsis*						**5.30**	17, 23, 24		
Bottom *Sphaeroidinellopsis dehiscens*	N18–N19			**5.53**	8, 11				
MIOCENE–PLIOCENE									
Messinian–Zanclean	**5.332 Ma (e)**								
Bottom *Globorotalia tumida*	N17–N18, M13b/M14–Pl1	**5.57**	2, 5, 6	**5.72**	8, 11				
Bottom *Globorotalia humilis*				**5.81**	8, 11				
Bottom *Hirsutella (Globorotalia) margaritae*				**6.08**	8, 11				
2nd influx sinistral *Neogloboquadrina acostaensis* (40%)						**6.08**	25–27		
First influx sinistral *Neogloboquadrina acostaensis* (90%)						**6.12**	25–27		
Bottom *Globigerinoides conglobatus*				**6.20**	8, 11				
Influx dextral *Hirsutella (Globorotalia) scitula gr.*						**6.29**	25–27		
Neogloboquadrina acostaensis sin → dex coiling change						**6.35**	25–27	**6.37**	33
Bottom common *Turborotalita multiloba*	MMi13b–MMi13c					**6.42**	25–27		
Top *Globorotalia miotumida (conomiozea) gr.*						**6.52**	25–27		
Top *Globorotalia nicolae*	MMi13a–MMi13b					**6.72**	25–27		
Bottom *Globorotalia nicolae*						**6.83**	25–30		
Top dominant sinistral *Hirsutella (Globorotalia) scitula gr.*						**7.08**	25–28, 30		
Influx conical *Globorotalia miotumida*						**7.18**	25–28, 30	**7.22**	34
Bottom common *Globorotalia miotumida (conomiozea) gr.*	MMi12c–MMi13a					**7.24**	25–30	**7.24**	34
Top (common) *Menardella (Globorotalia) menardii* 5									

Tortonian–Messinian	**7.246 Ma (f)**						
Top paracme dominant dextral *Hirsutella (Globorotalia) scitula gr.*				**7.28**	28, 30	**7.28**	34
Bottom common *Menardella (Globorotalia) menardii* 5	MMi12b–MMi12c			**7.36**	28, 29, 30	**7.36**	
Top *Catapsydrax parvulus*				**7.45**	28, 30		
Top common *Menardella (Globorotalia) menardii* 4				**7.51**	28, 30	**7.50**	34
Bottom paracme dominant dextral *Hirsutella (Globorotalia) scitula gr.*				**7.58**	28, 30		
Highest regular occurrence *Sphaeroidinellopsis seminulina*				**7.72**	28, 30		
Bottom *Globorotalia miotumida (conomiozea) gr.*	MMi12a–MMi12b			**7.89**	28, 30		
Lowest regular occurrence *Sphaeroidinellopsis seminulina*				**7.92**	28, 30		
Frequency shift *Catapsydrax parvulus*				**8.43**	28, 30		
Bottom *Candeina nitida*		**8.44**	8, 11				
Bottom *Globorotalia plesiotumida*	N16–N17	**8.58**	8, 11				
Bottom *Globigerinoides extremus*	M13a–M13b/M14	**8.93**	11, 13				
Top common *Catapsydrax parvulus* (ls)				**8.89**	28, 30		
Influx *Globoquadrina dehiscens*				**8.99**	28, 30		
Top *Globorotalia lenguaensis*	(M13–M14?)	**8.97**	11, 13				
Top dextral *Globorotalia lenguaensis*		**9.21**	11, 13				
Lowest common *Menardella (Globorotalia) menardii* 4	MMi11b–MMi12a			**9.31**	28, 30		
Bottom dextral *Globorotalia lenguaensis*		**9.34**	11, 13				
Bottom *Hirsutella cibaoensis*		**9.44**	8, 11				
Top regular dominant dextral *Neogloboquadrina acostaensis*				**9.54**	28, 30		
Bottom *Hirsutella juanai*		**9.69**	8, 11				
Neogloboquadrina gr. sin → dex coiling change				**9.90**	31		
Top *Globorotalia partimlabiata*	MMi11a–MMi11b			**9.94**	31, 32		
Neogloboquadrina gr. dex → sin coiling change				**10.05**	31, 32		

(*cont.*)

Table A2.3 *(cont.)*

Species event	Zone/subzone (transition)	ODP Legs 111 and 138		ODP Sites 925, 926		Eastern Mediterranean		Morocco	
		Age (Ma)	Reference	Age (Ma)	Reference	Age (Ma)	Reference	Age (Ma)	Reference
Top regular small-sized *Neogloboquadrina atlantica s.s.*						**10.48**	31, 32		
Bottom regular *Neogloboquadrina acostaensis s.s.*	N15–N16, M12–M13a, MMi10–MMi11a			**9.83**	8, 11, 13	**10.57**	31, 32		
Top large sized dextral *Neogloboquadrina atlantica*						**10.84**	31, 32		
Top *Globoquadrina sp* 1						**10.89**	31, 32		
Bottom large-sized *Neogloboquadrina atlantica* (dex)						**11.12**	31, 32		
2nd influx *Neogloboquadrina* group						**11.15**	31, 32		
Top *Globorotalita challengeri*				**9.99**	11, 13				
Top *Paragloborotalia siakensis*	N14–N15, M11–M12, MMi9–MMi10			**10.46**	11, 13	**11.19**	31, 32		
Bottom *Menardella limbata*				**10.64**	8, 11				
Top *Cassigerinella chipolensis*				**10.89**	11, 13				
Bottom *Globoturborotalita apertura*				**11.18**	8, 11				
Bottom *Globorotalita challengeri*				**11.22**	11, 13				
Bottom *Globorotaloides falconarae*						**11.36**	31, 32		
Bottom regular *Globoturborotalita obliquus obliquus*				**11.25**	11, 13	**11.54**	31, 32		
Bottom *Globoturborotalita decoraperta*				**11.49**	8, 11				
Top paracme *Catapsydrax parvulus* (ls)				**11.52**	11, 13				
Top *Globigerinoides subquadratus*				**11.54**	11, 13				
Top common *Globigerinoides subquadratus*	MMi8–MMi9					**11.56**	31, 32		
Serravallian–Tortonian	**11.600 Ma (g)**								
Bottom *Globoturborotalita nepenthes*	N13–N14, M10–M11			**11.63**	11, 13				
Bottom *Neogloboquadrina gr.*						**11.78**	31, 32		
Top *Fohsella fohsi* s.l.	N12–N13, M9b–M10 MMi7c–MMi8			**11.79**	8, 11				
Bottom *Globoquadrina* sp1						**11.80**	31, 32		
Top *Clavatorella bermudezi*				**12.00**	11, 13				
Top 2nd acme *Paragloborotalia siakensis*						**12.02**	31		

Top *Paragloborotalia mayeri*	MMi7a/b–MMi7c			**12.07**	31
Bottom 2nd acme *Paragloborotalia siakensis*				**12.44**	31
Top 1st acme *Paragloborotalia siakensis*				**12.61**	31
Bottom *Paragloborotalia partimlabiata*	MMi6–MMi7a/b			**12.77**	31
Bottom (common) *Paragloborotalia mayeri*				**12.77**	31
Bottom *Globorotalia lenguaensis*		**12.84**	8, 11		
Bottom *Fohsella fohsi*		**12.85**	11, 12, 13		
Bottom *Sphaeroidinellopsis subdehiscens*		**13.02**	11, 12, 13		
Bottom *Fohsella robusta*	M8/9a–M9b	**13.13**	8, 11, 12		
Top *Cassigerinella martinezpicoi*		**13.27**	11, 12, 13		
Bottom 1st acme *Paragloborotalia siakensis*				**13.32**	31
Bottom *Fohsella "fohsi"*	N11–N12, M7–M8/9a	**13.41**	8, 11, 12		
Bottom paracme *Catapsydrax parvulus* (ls)		**13.58**	12, 13		
Langhian–Serravallian	**13.654 Ma (h)**				
Top *Globorotalia praescitula*		**13.73**	12, 13		
Bottom *Fohsella "praefohsi"*	N10–N11	**13.77**	12, 13		
Top *Fohsella peripheroronda*		**13.80**	12, 14		
Top regular *Clavatorella bermudezi*		**13.82**	12, 13		
Top *Globorotalia archeomenardii*		**13.87**	12, 14		
Bottom *Fohsella peripheroacuta*	N9–N10, M6–M7	**14.24**	12, 14		
Bottom *Globorotalia praemenardii*		**14.38**	12, 14		
Top *Praeorbulina sicana*		**14.53**	12, 14		
Top *Globigeriantella insueta s. str.*		**14.66**	12, 14		
Bottom *Orbulina universa*	N8–N9, M5b–M6	**14.74**	12, 14		
Bottom *Praeorbulina circularis*		**14.89**	12, 14		
Bottom *Clavatorella bermudezi*		**14.89**	12, 14		
Burdigalian–Langhian	**15.974 Ma (i)**				
Bottom *Globorotalia archeomenardii*		**16.26**	12, 14		
Bottom *Praeorbulina glomerosa*	M5a–M5b	**16.27**	12, 14		
Bottom *Praeorbulina sicana*	N7–N8, M4–M5a	**16.97**	12, 14		
Top *Catapsydrax dissimilis*	N6–N7, M3–M4	**17.54**	12, 14		
Bottom *Globigeriantella insueta s. str.*	N5–N6, M2–M3	**17.59**	12, 14		
Top *Globiquadrina binaiensis*		**19.09**	12, 14		
Bottom *Globigerinatella* sp.		**19.30**	12, 14		
Bottom *Globiquadrina binaiensis*		**19.30**	12, 14		

(*cont.*)

Table A2.3 *(cont.)*

Species event	Zone/subzone (transition)	ODP Legs 111 and 138		ODP Sites 925, 926		Eastern Mediterranean		Morocco	
		Age (Ma)	Reference	Age (Ma)	Reference	Age (Ma)	Reference	Age (Ma)	Reference
Aquitanian–Burdigalian	**20.428 Ma (j)**								
Top *Paragloborotalia kugleri*	N4–N5, M1–M2			**21.12**	12, 14				
Top *Paragloborotalia pseudokugleri*				**21.31**	12, 14				
Top *Globigerina ciperoensis*				**22.90**	12, 14				
Bottom *Globigerinoides trilobus s.l.*				**22.96**	12, 14				
Bottom *Paragloborotalia kugleri*	P22–N4, P22–M1			**22.96**	12, 14				
OLIGOCENE–MIOCENE									
Chattian–Aquitanian	**23.030 Ma (k)**								

[a] Zonations: N = Blow (1969), PL–PT = Berggren *et al.* (1995b), MPL = Cita (1973, 1975a,b), MMi = Sprovieri *et al.* (2002b). See Table A2.2 caption for notes on stage boundaries.

[b] References: 1. Mix *et al.* (1995a), 2. Shipboard Scientific Party (1990), 3. Shackleton *et al.* (1990), 4. Shipboard Scientific Party (1988), 5. Shackleton *et al.* (1995a), 6. Shackleton *et al.* (1995b), 7. Bickert *et al.* (1997a), 8. Chaisson and Pearson (1997), 9. Shipboard Scientific Party (1995), 10. Tiedemann and Franz (1997), 11. Shackleton and Crowhurst (1997), 12. Shackleton *et al.* (1999), 13. Turco *et al.* (2003), 14. Pearson and Chaisson (1997), 15. Lourens (unpubl.), 16. Sanvoisin *et al.* (1993), 17. Lourens *et al.* (1996a), 18. Lourens *et al.* (1996b), 19. Lourens *et al.* (1998), 20. Zijderveld *et al.* (1991), 21. Hilgen (1991a), 22. Zachariasse *et al.* (1989), 23. Hilgen (1991b), 24. Langereis and Hilgen (1991), 25. Krijgsman *et al.* (1999), 26. Hilgen and Krijgsman (1999), 27. Sierro *et al.* (2001), 28. Krijgsman *et al.* (1994), 29. Krijgsman *et al.* (1997a), 30. Hilgen *et al.* (1995), 31. Hilgen *et al.* in press, 32. Hilgen *et al.* (2000a), 33. Krijgsman *et al.* (unpubl.), 34. Hilgen *et al.* (2000b).

Appendix 3 • Geomathematics

F. P. AGTERBERG

A.3.1 MATHEMATICS OF METHODS USED FOR FINAL STRAIGHT-LINE FITTING

When there is no (or negligibly small) stratigraphic uncertainty, as for the Late Cretaceous, the line of best fit is calculated according to the following standard method. We are interested in the linear relation $v_i = \alpha + \beta x_i$, where v_i is observed as y_i (i.e. v_i + error). If $\lambda_i \cong s^2(y)$ represents variance of y_i, the problem reduces to minimizing the expression

$$Q = \sum \left[(y_i - \alpha - \beta x_i)^2 / \lambda_i \right].$$

Partial differentiation of Q with respect to α and β and setting the results equal to zero at the point where $\alpha = a$, and $\beta = b$ yields the normal equations:

$$\sum [(y_i - a - bx_i)/\lambda_i] = 0; \quad \sum [x_i(y_i - a - bx_i)/\lambda_i] = 0.$$

Press *et al.* (1992) discuss the following algorithm. Introducing weights $w_i = 1/\lambda_i$, and replacing λ_i by $s^2(y_i)$, the following sums can be computed:

$$S = \sum w_i, \quad S_x = \sum w_i x_i, \quad S_y = \sum w_i y_i,$$
$$S_{xx} = \sum w_i x_i^2, \quad \text{and} \quad S_{xy} = \sum w_i x_i y_i.$$

The normal equations become:

$$aS + bS_x = S_y; \qquad aS_x + bS_{xx} = S_{xy},$$

with solution

$$a = [S_{xx}S_y - S_x S_y]/\Delta; \qquad b = [SS_{xy} - S_x S_y]/\Delta,$$
$$\text{where } \Delta = SS_{xx} - S_x^2.$$

Furthermore, it can be shown (Press *et al.*, 1992, p. 657) that:

$$s^2(a) = S_{xx}/\Delta; \qquad s^2(b) = S/\Delta;$$
$$s(a,b) = -S_x/\Delta = -x_m s^2(b), \quad \text{where} \quad x_m = S_x/S.$$

The estimates a and b would not change if all values of $s(y_i)$ are increased by the same constant c. However, in this situation, the variances $s^2(a)$ and $s^2(b)$, and the covariance $s(a,b)$, are increased by the same factor equal to c^2. Consequently, the error bar, proportional to $s(Y')$, being the square root of the variance $s^2(Y') = s^2(a) + x^2 \cdot s^2(b) + 2x \cdot s(a,b)$, as introduced at the end of section 8.1.4, would be enlarged by the factor (i.e. by c).

For example, application of the preceding algorithm to the values in the first three columns of Table 8.10 give $a = -0.041$ and $b = 1.000\,48$, with $s(a) = 0.445$, $s(b) = 0.005\,12$, and $s(a,b) = -0.002\,27$. Estimated Y' values are shown in column 4 of Table 8.10.

Clearly, the differences between values in columns 1 and 4 of Table 8.10 are negligibly small. If the values in column 1 are written as $Y = f(x)$, then $Y' = Y$. For this reason, we will continue to use the values of column 1 for the Late Cretaceous time scale. These values fall on the spline curve of Fig. 8.15, which was used to estimate the Late Cretaceous stage boundary ages.

The preceding method of constructing error bars on estimated chronostratigraphic boundary ages can be generalized to account for stratigraphic uncertainty. Ripley and Thompson (1987) have developed a maximum likelihood fitting method for a functional relationship (MLFR) that can be adapted for use in our situation. Their method generalizes Adcock's original "major axis" method to the situation that the variances of X and Y are not equal and different for every single observation.

We now are interested in the linear relation $v_i = \alpha + \beta \cdot u_i$, where u_i and v_i are observed as x_i ($= u_i$ + error), and y_i ($= v_i$ + error), respectively. If $\kappa_i \cong s^2(x)$ and $\lambda_i \cong s^2(y)$ represent the variances of x_i and y_i, respectively, the problem reduces to minimizing the expression

$$Q = \sum \left[(x_i - u_i)^2/\kappa_i + (y_i - \alpha - \beta u_i)^2/\lambda_i \right]$$

over u_i. First minimizing over u_i and introducing the weights $w_i = 1/(\lambda_i + \beta^2 \kappa_i)$, this minimum is reached when

$$Q_{\min}(\alpha, \beta) = \sum w_i (y_i - \alpha - \beta x_i)^2.$$

The weights w_i depend on β and so does the estimate of α that satisfies

$$a = \left[\sum w_i (y_i - \beta x_i)^2 \right] \Big/ \sum w_i.$$

The slope b is found by minimizing

$$Q_{\min}(a, \beta) = \sum w_i(y_i - a - \beta x_i)^2.$$

This, in turn, yields the weights w_i and intercept a.

Ripley has written a computer algorithm in FORTRAN that provides a and b as estimates of α and β, together with their standard deviations $s(a)$ and $s(b)$. As will be discussed later in this section, in our application $a \approx 0$ and $b \approx 1$. This simplifies matters considerably because estimates $(a + bx_i \approx x_i)$ of v_i are already provided by the spline-curve values Y_k.

The covariance $s(a, b)$ can be calculated from the relation $s(a, b) = -x_m \cdot s^2(b)$, where $x_m = \sum w_i x_i / \sum w_i$ is the weighted mean x_i (cf. Fuller, 1987). From these statistics, it is possible to calculate the variance $s^2(Y_k) = s^2(a) + x^2 \cdot s^2(b) + 2x_k \cdot s(a, b)$, where the subscript k denotes a chronostratigraphic boundary of interest.

Another feature of Ripley's algorithm is that it provides a plot of the scaled residuals

$$r_i = (y_i - a - bx_i)\sqrt{w_i}(b),$$

where $w_i = 1/(\lambda_i + b^2\kappa_i)$. The sum $\sum r_i^2$ should be around $(n - 2)$ and a plot of r_i against X_i representing estimates of u_i should not show any noticeable pattern. These estimates satisfy

$$X_i = w_i[\lambda_i x_i + \kappa_i b(y_i - a)].$$

In our applications of the MLFR method, the x_i values are estimated in millions of years obtained after spline-curve fitting. Spline-curve fitting is based on the method of least squares with all errors assumed to be restricted to the dependent variable (y_i values with variances κ_i). In general, the x_i values already provide good estimates of the u_i values as well as the v_i values because $a \approx 0$ and $b \approx 1$.

The null hypotheses $\alpha = 0$ and $\beta = 1$ can be tested statistically by calculating the ratios $a/s(a)$ and $(b - 1)/s(b)$ in two Z-tests (Ripley and Thompson, 1987, p. 379). In the absence of significant bias, both ratios should be inside the $(-1.96, 1.96)$ interval with 95% probability. In our applications, a is a small negative value with $a/s(a)$ greater than -1.96, and b is slightly greater than 1 with $(b - 1)/s(b)$ less than 1.96.

If $\alpha = 0$ and $\beta = 1$, $X_i = Y_i$. Also, each weight w_i is simply the inverse of the sum of the variances of x_i and y_i, or $w_i = 1/s^2(t)$, where $s^2(t) = s^2(x) + s^2(y)$. The estimated spline-curve values are not significantly improved when the MLFR method is applied. However, the approach offers three advantages: (1) replacing $s^2(y_i)$ by $s_t^2(y_i)$ allows a new type of approximate spline-curve fitting in which stratigraphic uncertainty is considered from the beginning; (2) scaled residuals $r_i = (y_i - Y_i)/s_t(y_i)$ can be used to check whether or not any $s_t(y_i)$ value is too large; and (3) it enables us to calculate $s^2(Y_k)$ and 2-sigma error bars for estimated ages of chronostratigraphic boundaries (subscript i denotes observations and k is for chronostratigraphic boundaries).

Spline fitting with weights $w_i = 1/s_t^2(y_i)$ provides an approximate solution. If the best-fit spline is a straight line, the validity of this approximation can be verified. In all applications performed to date, setting $\beta = 1$ makes no difference to the end product. However, this approximation may become less satisfactory if the spline locally deviates strongly from a straight line. Nevertheless, it can be assumed that use of $s_t^2(y)$ generally provides better results than use of $s^2(y)$ with all stratigraphic uncertainty set equal to zero.

Bibliography

Abdul-Aziz, H., 2001, Astronomical forcing in continental sediments. *Geologica Ultraiectina* **207**: 191.

Abdul-Aziz, H., Hilgen, F. J., Krijgsman, W., and Calvo, J. P., 2003, An astronomical polarity time scale for the late middle Miocene based on cyclic continental sequences. *Journal of Geophysical Research* **108**: article 2159.

Abelmann, A., 1990, Oligocene to middle Miocene radiolarian stratigraphy of southern high latitudes from Leg 113, Sites 689 and 690, Maud Rise. In D. M. Kennett, A. Masterson, and N. J. Stewart, eds., *Proceedings of the Ocean Drilling Program, Scientific Results*, Vol. 113. College Station, TX: Ocean Drilling Program, pp. 675–708.

Absoussalem, Z. S. and Becker, R. T., 2001, *Prospects for an Upper Givetian Substage*. Berlin: Mitteilungen aus dem Museum für Naturkunde in Berlin.

Abushik, A. F., Chizhova, V. A., Guseva, E. A., and Sidaravichene, N. V., 1990, Evolution and biostratigraphy of ostracods in Paleozoic. In A. F. Abushik, ed., *Practical Guide in Microfauna of USSR: Ostracods of Paleozoic, Transactions of VSEGEI*, Vol. 4. Leningrad: Nedra, p. 356 [In Russian].

Accarie, H., Emmanuel, L., Robaszynski, F., *et al.*, 1996, La géochimie isotopique du carbone (δ^{13}C) comme outil stratigraphique; application à la limite cenomanien/turonien en Tunisie centrale. *Comptes rendus de l'Académie des Sciences, Paris, Série II, Sciences de la Terre et des Planètes* **322**: 579–86.

Adams, C. G., 1970, A reconsideration of the East Indian letter classification of the Tertiary: *Bulletin of the British Museum of Natural History, Geology Series* **19**: 87–137.

Adhémar, J., 1842, Révolutions de la Mer, Les Déluges Périodiques. Paris: Carilian-Goeury et V. Dalmort, 184 pp., 5-fold. Plates.

Agassiz, L., 1840, *Etudes sur les glaciers.* Neuchâtel.

AGIP, 1982, *Foraminiferi Padani (Terziario e Quaternario)*. San Donato Milanese: AGIP S.p.A.

Agterberg, F. P., 1988, Quality of time scales – a statistical appraisal. In D. F. Merriam, ed., *Current Trends in Geomathematics.* New York: Plenum, pp. 57–103.

1990, *Automated Stratigraphic Correlation*. Amsterdam: Elsevier.

1994, Estimation of the Mesozoic geological time scale. *Mathematical Geology* **26**: 857–76.

2002, Construction of numerical geological time scales. *Terra Nostra–Schriften der Alfred-Wegener-Stiftung*, April 2002, pp. 227–32 (Paper presented at 8th Annual IAMG Conference, Berlin, September 2002).

Agterberg, F. P. and Gradstein, F. M., 1988, Recent developments in quantitative stratigraphy. *Earth Science Reviews* **25**: 1–73.

1999, The RASC method for ranking and scaling of biostratigraphic events. *Earth Science Reviews* **46**(1–4): 1–25.

Aguado, R., Company, M., Sandoval, J., and Tavera, J. M., 1997, Biostratigraphic events at the Barremian/Aptian boundary in the Betic Cordillera, southern Spain. *Cretaceous Research* **18**: 309–29.

Aguado, R., Company, M., and Tavera, J. M., 2000, The Berriasian/Valanginian boundary in the Mediterranean region: new data from the Caravaca and Cehegin sections, SE Spain. *Cretaceous Research* **21**: 1–21.

Aguilar, J.-P., Legendre, S., and Michaux, J., 1997, *Actes du Congrès BiochroM'97*, Montpellier, Avril 14–17. Biochronologie mammalienne du Cénozoïque en Europe et domaines reliés, *Mémoires et Travaux de l'Institut de Montpellier de l'Ecole Pratique des Hautes Etudes* **21**: 1–181.

Aguirre, E. and Pasini, G., 1985, The Pliocene–Pleistocene boundary. *Episodes* **8**: 116–20.

Agusti, J., Cabrera, L., Garces, M., *et al.*, 2001, A calibrated mammal scale for the Neogene of Western Europe. State of the art. *Earth Science Reviews* **52**: 247–60.

Aigner, T. and Bachmann, G. H., 1992, Sequence-stratigraphic framework of the German Triassic. *Sedimentary Geology* **80**: 115–35.

Aizenverg, D. E., Astakhova, T. V., Berchenko, O. I., *et al.*, 1983, *Upper Serpukhovian Substage of the Donets Basin: Paleontological Characteristics.* Kiev: Naukova Dumka, pp. 1–164 [In Russian].

Alberti, F. A., von, 1834, *Beitrag zu einer Monographie des Buntern Sandsteins, Muschelkalks und Keupers und die Verbindung dieser Gebilde zu einer Formation.* Stuttgart and Tübingen: Verlag der J. G. Cottaishen Buchhandlung, 326 pp.

Alberti, G. K. B., 1993, Dacryoconarida und homoctenide Tentaculiten aus dem Unter- und Mittel Devons–I. *Courier Forschungsinstitut Senckenberg* **158**: 1–229.

2000, Planktonische Tentakuliten des Devon. IV. Dacryconarida FISHER 1962 dem Unter Devon. *Palaeontographica*, *Abteilung A* **254**: p. 1–23.

Alcock, F. J., 1934, Report of the National Committee on Stratigraphical Nomenclature. *Transactions of the Royal Society of Canada, Series III* **28**(4): 113–21.

Aldridge, R. J. and Briggs, D. E. G., 1989, A soft body of evidence: a puzzle that has confused paleontologists for more than a century has finally been solved. *Natural History* **89**: 6–9.

Alekseev, A. S., Kononova, L. I., and Nikishin, A. M., 1996, The Devonian and Carboniferous of the Moscow Syneclise (Russian Platform): stratigraphy and sea-level changes. *Tectonophysics* **268**: 149–68.

Alekseev, A. S., Goreva, N. V., Kulagina, E. I., *et al.*, 2002, *Upper Carboniferous of South Urals (Bashkiria, Russia)*. Guidebook of the field trip of the WG to establish a GSSP in the upper part of the Carboniferous System (SCCS Project 5) Moscow and Ufa, pp. 1–56.

Algeo, T. J., 1996, Geomagnetic polarity bias patterns through the Phanerozoic. *Journal of Geophysical Research* **101**: 2785–814.

Ali, J. R. and Hailwood, E. A., 1995, Magnetostratigraphy of upper Paleocene through lower middle Eocene strata of NW Europe. *Society of Economic Paleontologists and Mineralogists Special Publication*, Vol. 54. Tulsa, OK: SEPM, pp. 271–4.

1998, Magnetostratigraphic (re)calibration of the Paleocene/Eocene boundary interval in Holes 550 and 549, Goban Spur, eastern North Atlantic. *Earth and Planetary Science Letters* **161**: 201–13.

Ali, J. R. and Jolley, D. W., 1996, Chronostratigraphic framework for the Thanetian and lower Ypresian deposits of SE England. *Geological Society of London Special Publication* Vol. 101. London: Geological Society, pp. 129–44.

Ali, J. R., King, C., and Hailwood, E. A., 1993, Magnetostratigraphic calibration of early Eocene depositional sequences in the southern North Sea Basin. *Geological Society of London Special Publication,* Vol. 70. London: Geological Society, pp. 99–125.

Ali, J. R., Heilmann-Clausen, C., Thomsen, E., and Abrahamsen, N., 1994, Preliminary magnetostratigraphic results from the upper Danian and Selandian in Denmark. *Geologiska Föreningen i Stockholm Förhandlingar* **116**: 43.

Ali, J. R., Jolley, D. W., Bell, B. R., and Kelley, S. P., in press, Construction of the Paleogene time-scale: implications for timing of volcanism in the N. Atlantic. *Geological Society of London Special Publication*. London: Geological Society.

Allègre, C. J., Manhès, G., and Göpel, C., 1995, The age of the Earth. *Geochimica et Cosmochimica Acta* **59**(8): 1445–56.

Allemann, F., Catalano, R., Fares, F., and Remane, J., 1971, Standard calpionellid zonation (Upper Tithonian–Valanginian) of the Western Mediterranean province. In *II Planktonic Conference Roma 1970*. Rome: International, pp. 1337–40.

Alley, R. B. and MacAyeal, D. R., 1994, Ice-rafted debris associated with binge/purge oscillations of the Laurentide ice sheet. *Palaeoceanography* **9**: 503–11.

Alley, R. B., Mayewski, P. A., Sowers, T., *et al.*, 1997, Holocene climatic instability: a prominent, widespread event 8200 yr ago. *Geology* **25**: 483–6.

Alméras, Y., Boullier, A., and Laurin, B., 1997, Brachiopodes. In E. Cariou and P. Hantzpergue, eds., Biostratigraphie du Jurassique ouest-européen et méditerranéen: zonations parallèles et distribution des invertébrés et microfossiles, Groupe Français d'Etude du Jurassique. *Bulletin des Centres de Recherches Exploration–Production Elf–Aquitaine*, **17**: 169–95.

Alroy, J., 1992, Conjunction among taxonomic distributions and the Miocene mammalian biochronology of the great-plains. *Paleobiology* **18**: 326–43.

1998, Diachrony of mammalian appearance events: implications for biochronology. *Geology* **26**: 23–6.

2003, A quantitative North American mammalian timescale. www.nceas.ucsb.edu/~alroy/TimeScale.html

Alroy, J., Marshall, C. R., Bambach, R. K., *et al.*, 2001, Effects of sampling standardisation and estimates of Phanerozoic marine diversification. *Proceedings of the National Academy of Sciences, USA* **98**: 6261–6.

Alvarez, L. W., Alvarez, W., Asaro, E., and Michel, H. V., 1980, Extraterrestrial cause for the Cretaceous–Tertiary extinction. *Science* **208**: 1095–108.

Alvarez, W., Arthur, M. A., Fischer, A. G., *et al.*, 1977, Upper Cretaceous–Palaeocene magnetic stratigraphy at Gubbio, Italy – V. Type section for the Late Cretaceous–Palaeocene geomagnetic reversal time scale. *Geological Society of America Bulletin* **88**: 383–9.

Álvarez-Sierra, M. A., Daams, R., Lacomba, J. I., López Martínez, N., and Sacristán, M., 1987, Münchner Geowissenschaftliche Abhandlungen. *Friedrich Pfeil*. **10**: 43–8.

Amédro, F., 1992, L'Albien du bassin Anglo–Parisien: ammonites, zonation phylétique, séquences. *Bulletin des Centres de Recherches Exploration–Production Elf-Aquitaine* **16**: 187–233.

American Commission on Stratigraphic Nomenclature, 1961, Code of stratigraphic nomenclature. *Bulletin of the American Association of Petroleum Geologists* **45**: 645–60.

Amthor, J. E., Grotzinger, J. P., Schroder, S., *et al.*, 2003, Extinction of *Cloudina* and *Namacalathus* at the Precambrian boundary in Oman. *Geology* **31**(5): 431–4.

Anderson, R. Y., 1982, A long geoclimatic record from the Permian. *Journal of Geophysical Research* **87**: 7285–94.

Andersson, P. S., Wasserburg, G. J., and Ingri, J., 1992, The sources and transport of Sr and Nd isotopes in the Baltic Sea. *Earth and Planetary Science Letters* **113**: 459–72.

Angori, E. and Monechi, S., 1996, High-resolution nannofossil biostratigraphy across the Paleocene–Eocene boundary at Caravaca (southern Spain). *Israel Journal of Earth Sciences* **44**: 207–16.

Anonymous, 1882, *Congrès International de Géologie Comptes Rendus de la 2ième Session, Bologna 1881*. Bologna: Fava and Garagni.

1970, Résolution du deuxième Colloque du Jurassique à Luxembourg. In *Colloque du Jurassique à Luxembourg,* 1967. Nancy, p. 38.

1975, Colloque sur la limite Jurassique-Crétacé. *Mémoires du Bureau de Recherches Géologiques et Minières,* Vol. 86. Paris: Bureau de Recherches Géologiques et Minières, pp. 386–93 (summary).

1982, *Stratigraphy of the USSR. Quaternary System*, Vol. 1. Moscow: Nedra [In Russian].

1984, *Stratigraphy of the USSR. Quaternary System*, Vol. 2. Moscow: Nedra [In Russian].

1988, Biostratigraphy rejected for Pleistocene subdivisions. *Episodes* **11**: 228.

1992, *Resolutions of the Interdepartmental Stratigraphic Committee of Russia and its Permanent Commissions on the Carboniferous/Permian boundary (St. Petersburg)*, pp. 52–6.

1999, *International Subcommission on Cambrian Stratigraphy Newsletter*, Vol. 1.

An Zhisheng, Lui Tungsheng, Porter, S. C., *et al.*, 1990, The long-term paleomonsoon variation record by the loess–paleosol sequence in central China. *Quaternary International* **7/8**: 91–5.

Apollonov, M. K., Bekteleuov, A. K., Kirschvink, J. L., *et al.*, 1992, Paleomagnetic scale of Upper Cambrian and Lower Ordovician in Batyrbai section (Lesser Karatau, southern Kazakhstan). *Izvestiya Akademii Nauk Kazakhskoy SSR, Seriya Geologicheskaya* **4**(326): 51–7 [In Russian].

Archangelsky, S., 1990, Plant distribution in Gondwana during the late Paleozoic. In T. N. Taylor and E. L. Taylor, eds., *Antarctic Paleobiology – Its Role in the Reconstruction of Gondwana*. New York: Springer-Verlag, pp. 102–17.

Archangelsky, S., Arrondo, O. G., and Leguizamon, R. R., 1995, Floras Paleozoicas. Contribuciones a la palaeophytologia Argentina y revision y actualizacion de la obra paleobotanica de Kurtz en la Republica Argentina. Actas de la Academia Nacional de Ciencias de la Republica Argentina. *Academia Nacional de Ciencias, Argentina* **11**(1–4): 85–125.

Archer, M., Hand, S. J., and Godthelp, H., 1995, Tertiary environmental and biotic change in Australia. In E. Vrba, G. H. Denton, T. C. Partidge, and L. H. Burckle, eds., *Paleoclimate and Evolution with Emphasis on Human Origins*. New Haven: Yale University Press, Ch. 6.

Archibald, J. D. and Lofgren, D. L., 1990, Mammalian zonation near the Cretaceous–Tertiary boundary. In K. D. Rose and T. Bown, eds., *Dawn of the Age of Mammals in the Northern Part of the Rocky Mountain Interior, Geological Society of America Special Paper,* Vol. 243. Boulder, CO: Geological Society of America, pp. 31–50.

Arduino, G., 1759, *Lettera seconda sopre varie osservazioni fatti in diversi parti del territorio di Vicenza, ed altrove, appartenenti alla teoria terrestre, ed alla mineralogía*. Venezia.

1760a, In *Nuovo raccolta di opuscoli scientifici e filologici del padre abate Angiolo Calogierà (Venice)*, Vol. 6. pp. 99–180.

1760b, A letter to Sig. Cav. Antonio Valisnieri. In *Nuova raccolta di opuscoli scientifici e filologici del padre abate Angiolo Calogierà (Venice)*, Vol. 6. pp. 142–3.

Arkell, W. J., 1933, *The Jurassic System in Great Britain*. Oxford: Clarendon Press.

1939, The ammonite succession at the Woodham Brick Company's pit, Akeman Street Station, Buckinghamshire, and its bearing on the classification of the Oxford Clay. *Quarterly Journal of the Geological Society of London* **95**(2): 135–221.

1946, Standard of the European Jurassic. *Geological Society of America Bulletin* 57: 1–34.

1956, *Jurassic Geology of the World.* Edinburgh: Oliver and Boyd.

Armstrong, H. A., 1996, Biotic recovery after mass extinction: the role of climate and ocean-state in the post-glacial (Late Ordovician–Early Silurian) recovery of conodonts. In M. B. Hart, ed., *Biotic Recovery from Mass Extinction Events, Geological Society of London Special Publication,* Vol. 102. London: Geological Society, pp. 105–17.

Armstrong, R. L., 1978, Pre-Cenozoic Phanerozoic time scale – computer file of critical dates and consequences of new and in-progress decay-constant revisions. In G. V. Cohee, M. F. Glaesner, and H. D. Hedberg, eds., *Contributions to the Geologic Timescale*, Vol. 6, *Studies in Geology.* Tulsa, OK: American Association of Petroleum Geologists, pp. 73–91.

Armstrong, R. and Wilson, A. H., 2000, A SHRIMP U–Pb study of zircons from the layered sequence of the Great Dyke, Zimbabwe, and a granitoid anatectic dyke. *Earth and Planetary Science Letters* **180**(1–2): 1–12.

Arnaud-Vanneau, A. and Bilotte, M., 1998, Larger benthic foraminifera [for Cretaceous portion of Mesozoic and Cenozoic sequence chronostratigraphic framework of European basins, in Hardenbol, J. *et al.*]. In P.-C. de Graciansky, J. Hardenbol., Th. Jacquin, and P. R. Vail, eds., *Mesozoic–Cenozoic Sequence Stratigraphy of European Basins, Society of Economic Paleontologists and Mineralogists Special Publication,* Vol. 60. Tulsa, OK: SEPM, pp. 772–3, Chart 5.

Arz, J. A., Arenillas, I., Molina, E., and Dupuis, C., 1999, La extinción en masa de foraminíferos planctónicos en el límite Cretácico/Terciario (K/T) de Elles (Túnez): los efectos tafonómico y Signor-Lipps. *Revista Sociedad Geológica de España* **12**: 251–68.

Assereto, R., 1974, Aegean and Bithynian: proposal for two new Anisian substages: In H. Zapfe, ed., *Die stratigraphic der alpin–mediterraneas Trias – The Stratigraphy of the Alpine–Mediterraneas Triassic*, Vol. 2, *Schriftenreihe der Erdwissenschaftlichen Kommissionen*, Vol. 2. Vienna: Springer, pp. 23–39.

Atrops, F., 1982, La sous-famille des Ataxioceratinae (Ammonitina) dans le Kimméridgien inférieur du Sud-Est de la France. Systématique, évolution, chronostratigraphie des genres Orthosphinctes et Ataxioceras. *Documents des Laboratoires de Géologie de Lyon* **83**: 463.

1994, Upper Oxfordian to Lower Kimmeridgian ammonite successions and biostratigraphy of the Crussol and Châteauneuf d'Oze sections. In F. Atrops, D. Fortwengler, D. Marchand, and G. Melendez, eds., *IV Oxfordian and Kimmeridgian Working Group Meeting, Lyon and SE France Basin, Guide Book and Abstracts*, pp. 50–60, 106–111.

1999, Report of the Oxfordian–Kimmeridgian Boundary Working Group. *International Subcommission on Jurassic Stratigraphy Newsletter* **27**: 34–37.

Atrops, F., Gygi, R., Matyja, B. A., and Wierzbowski, A., 1993, The *Amoeboceras* faunas in the Middle Oxfordian–Lowermost Kimmeridgian, Submediterranean succession, and their correlation value. *Acta Geologica Polonica* **43**(3–4): 213–27.

Aubry, M.-P., 1983, Biostratigraphie du Paléogène de l'Europe épicontinentale du Nord-Ouest, étude fondée sur les nannofossiles calcaires. *Documents des Laboratoires de Géologie de Lyon* **89**: 317.

1992, Late Paleogene calcareous nannoplankton evolution: a tale of climatic deterioration. In D. R. Prothero and W. A. Berggren, eds., *Eocene–Oligocene Climatic and Biotic Evolution*. Princeton, NJ: Princeton University Press, pp. 272–309.

1996, Towards an Upper Paleocene–Lower Eocene high resolution stratigraphy based on calcareous nannofossil stratigraphy. *Israel Journal of Earth Sciences* **44**: 239–53.

1998a, The upper Paleocene–lower Eocene stratigraphic puzzle. *Strata* **6**: 7–9.

1998b, Early Paleogene calcareous nannoplankton evolution: a tale of climatic amelioration. In M.-P. Aubry, S. G. Lucas, and W. A. Berggren, eds., *Late Paleocene–Early Eocene Climatic and Biotic Events in the Marine and Terrestrial Records*. New York: Columbia University Press, pp. 158–203.

Aubry, M.-P. and Ouda, K., 2003, Introduction. In K. Ouda and M.-P. Aubry, eds., *The Upper Paleocene-Lower Eocene of the Upper Nile Valley*, Part 1, *Stratigraphy*. New York: Micropaleontology Press, pp. ii–iv.

Aubry, M.-P. and Villa, G., 1996, Calcareous nannoplankton stratigraphy of the Lemme–Carrosio Paleogene/Neogene global stratotype section and point. *Giornale di Geologia, ser. 3* **58**(1–2): 51–69.

Aubry, M.-P., Berggren, W. A., Kent, D. V., *et al.*, 1988, Paleogene geochronology: an integrated approach. *Paleoceanography* **3**: 707–42.

Aubry, M.-P., Lucas, S. G., and Berggren, W. A., 1998, *Late Paleocene–Early Eocene Climatic and Biotic Events in the Marine and Terrestrial Records.* New York: Columbia University Press.

Avkhimovitch, V. I., Tchibrikova, E. V., Obukhovskaya, T. G., *et al.*, 1993, Middle and Upper Devonian miospore zonation of eastern Europe. *Bulletin des Centres Recherches Exploration–Production Elf–Aquitaine* **17**: 79–147.

Azmy, K., Veizer, J., Bassett, M. G., and Copper, P., 1998, Oxygen and carbon isotopic composition of Silurian brachiopods: implications for coeval seawater and glaciations. *Geological Society of America Bulletin* **110**: 1499–512.

Azmy, K., Veizer, J., Wenzal, B., Bassett, M., and Cooper, P., 1999, Silurian strontium isotope stratigraphy. *Geological Society of America Bulletin* **111**: 475–83.

Baadsgaard, H. and Lerbekmo, J. F., 1982, The dating of bentonite beds. In G. S. Odin, ed., *Numerical Dating in Stratigraphy*, Vol. 2. Chichester: Wiley Interscience, pp. 423–40.

Baadsgaard, H., Lerbekmo, J. F., and McDougall, I., 1988, A radiometric age for the Cretaceous–Tertiary boundary based on K–Ar, Rb–Sr, and U–Pb ages of bentonites from Alberta, Saskatchewan and Montana. *Canadian Journal of Earth Sciences* **25**: 1088–97.

Bachmann, O., Dungan, M. A., and Lipman, P. W., 2002, The Fish Canyon magma body, San Juan volcanic field, Colorado: rejuvenation and eruption of an upper-crustal batholith. *Journal of Petrology* **43**: 1469–503.

Backman, J. and Raffi, I., 1997, Calibration of Miocene nannofossil events to orbitally-tuned cyclostratigraphies from Ceara Rise. In *Proceedings of the Ocean Drilling Program, Scientific Results*, Vol. 154. College Station, TX: Ocean Drilling Program, pp. 83–99.

Baesemann, J. F. and Lane, H. R., 1985, Taxonomy and evolution of the genus *Rhachistognathus Dunn* (Conodonta; Late Mississippian to early Middle Pennsylvanian). *Courier Forschungsinstitut Senckenberg*. **74**: 93–135.

Bai, S. L., 1995, Milankovitch cyclicity and time scale of the middle and upper Devonian. *International Geology Review* **37**: 1109–14.

Bai, S. L., Bai, Z. Q., Ma, X. P., Wang, D. R., and Sun, Y. L., 1995, *Devonian Events and Biostratigraphy of South China*. Peking: Peking University Press, pp. 1–303.

Baksi, A. K., Archibald, D. A., and Farrar, E., 1996, Intercalibration of $^{40}Ar/^{39}Ar$ dating standards. *Chemical Geology* **129**(3–4): 307–24.

Baldauf, J. G. and Iwai, M., 1995, Neogene diatom biostratigraphy for the eastern equatorial Pacific, Leg 138. In N. G. Pisias, L. A. Mayer, *et al.*, eds., *Proceedings of the Ocean Drilling Program, Scientific Results*, Vol. 138. College Station, TX: Ocean Drilling Program, pp. 105–128.

Balini, M., Krystyn, L., and Torti, V., 1998, In search of the Ladinian/Carnian boundary: perspectives from Spiti (Tethys Himalaya). *Albertiana* **21**: 26–32.

Balini, M., Krystyn, L., Nicora, A., and Torti, V., 2001, The Ladinian–Carnian boundary succession in Spiti (Tethys

Himalaya) and its bearing to the definition of the GSSP for the Carnian stage (Upper Triassic). *Journal of Asian Earth Sciences* **19**(3A): 3–4.

Balme, B. E. and Foster, C. B., 1996, Triassic, Explanatory Notes on Biostratigraphic Charts. Australian Geological Survey Organisation. In G. C. Young and J. R. Laurie, eds., *An Australian Phanerozoic Timescale.* Oxford: Oxford University Press, Ch. 2.7, Chart 7.

Bamber, E. W. and Fedorowski, J., 1998, Biostratigraphy and systematics of Upper Carboniferous cerioid rugose corals, Ellesmere Island, Arctic Canada. *Geological Survey of Canada Bulletin* **511**: 1–127.

Banks, H. P., 1980, Floral assemblages in the Siluro–Devonian. In D. L. Dilcher and T. N. Taylor eds., *Biostratigraphy of Fossil Plants.* Stroudsville, PA: Dowden, Hutchinson and Ross, pp. 1–24.

Banner, J. L. and Kaufman, J., 1994, The isotopic record of ocean chemistry and diagenesis preserved in non-luminescent brachiopods from Mississippian carbonate rocks, Illinois and Missouri. *Geological Society of America Bulletin.* **106**: 1074–82.

Barbier, R. and Thieuloy, J. P., 1965, Etage Valanginien. *Mémoires du Bureau de Recherches Géologiques et Minières* **34**: 79–84.

Barbieri, F., 1967, The Foraminifera in the Pliocene section Vernasca–Castell'Arquato including the "Piacenzian Stratotype." *Societa Italiana di Scienze Naturale e dd Museo Civico di Storia Naturale de Milano Mem.*, **15**: 145–63.

Barbin, V., 1988, The Eocene–Oligocene transition in shallow-water environment: the Pribonian stage type area (Vicentine, northern Italy). In I. Premoli Silva, R. Coccioni, and A. Montanari, eds., *Proceedings of the Eocene–Oligocene Boundry* Meeting, Ancona, October 1987. Ancona: Annibali/ International Union of Geological Science, pp. 163–71.

Barchi, P., Bonnemaison, M., Galbrun, B., and Renard, M., 1997, Tercis (Landes, Sud-Ouest France): point stratotype global de la limite Campanien–Maastrichtien. Résultats magnétostratigraphiques et premières données sur la nannoflore calcaire. *Bulletin de la Société Géologique de France* **168**: 133–42.

Bard, E., 2001, Extending the calibrated radiocarbon record. *Science*, **292**: 244–5.

Bard, E., Arnold, M., Hamelin, B., Tisnerat-Laborde, N., and Cabioch, G., 1998, Radiocarbon calibration by means of mass spectrometric $^{230}Th/^{234}U$ and ^{14}C ages of corals. An updated base including samples from Barbados, Mururoa and Tahiti. *Radiocarbon.* **40**: 1085–92.

Bard, E., Hamelin, B., Fairbanks, R. G., and Zinder, A., 1990, Calibration of the ^{14}C timescale over the past 30,000 years using mass spectrometric U–Th ages from Barbados corals. *Nature* **345**: p. 405–10.

Barker, R. W. and Grimsdale, T. F., 1936, A contribution to the phylogeny of the orbitoidal foraminifera with descriptions of new forms from the Eocene of Mexico. *Journal of Paleontology* **10**: 231–47.

Barnes, C. R., Fortey, R. A., and Williams, H., 1995, The pattern of global bio-events during the Ordovician Period. In O. H. Walliser, ed., *Global Events and Event Stratigraphy in the Phanerozoic.* Berlin: Springer-Verlag, pp. 139–72.

Barrell, J., 1917, Rhythms and the measurement of geological time. *Geological Society of America Bulletin* **28**: 745–904.

Barrera, E., Savin, S., Thomas, E., and Jones, C. E., 1997, Evidence for thermohaline-circulation reversals controlled by sea-level change in the latest Cretaceous. *Geology* **25**: 715–18.

Barrick, J. E. and Heckel, P. H., 2000, A provisional Conodont Zonation for Late Pennsylvanian (late Late Carboniferous) strata in Midcontinent region of North America. *Newsletter on Carboniferous Stratigraphy* **18**: 15–22.

Barrick, J. E. and Lambert, L. L., 1999, Conodont taxa proposed by Jones (1941) are valid. In P. H. Heckel, ed., *Guidebook, Fieldtrip # 8, XIV International Congress on the Carboniferous-Permian, Kansas Geological Survey Open File Report 99–27.* Lawrence, KS: Kansas Geological Survey, pp. 162–9.

Barron, E. J., 1983, A warm, equable Cretaceous: the nature of the problem. *Earth Science Review* **19**: 305–38.

Barron, J. A., 1985, Late Eocene to Holocene diatom biostratigraphy of the equatorial Pacific Ocean, Deep Sea Drilling Project Leg 85. In L. Mayer, F. Theyer, E. Thomas, *et al.*, eds., *Initial Reports of the Deep Sea Drilling Project,* Vol. 85. Washington, DC: US Govt. Printing Office, pp. 413–56.

Barry, J., Morgan, M. E., Flynn, L. J., *et al.*, 2002. Faunal and environmental change in the Late Miocene Siwaliks of Northern Pakistan. *Paleobiology* Memoir 3, supplement **28**(2): 1–72.

Barskov, I. S. and Alekseev, A. S., 1975, Conodonts of the Middle and Upper Carboniferous of Moscow Basin. *News of Academy of Sciences of USSR, Geology Series* **5**: 84–99.

Barskov, I. S., Alekseev, A. S., and Goreva, N. V., 1980, Conodonts and stratigraphic scale of the Carboniferous. *News of Academy of Sciences of USSR, Geology Series* **3**: 43–5.

Bartley, J. K., Pope, M., Knoll, A. H., Semikhatov, M. A., and Petrov, P. Y., 1998, A Vendian–Cambrian boundary succession from the northwestern margin of the Siberian Platform: stratigraphy, paleontology, chemostratigraphy and correlation. *Geological Magazine* **135**: 473–94.

Bartolini, A. and Larson, R. L., 2001, The Pacific microplate and the Pangea supercontinent in the Early to Middle Jurassic. *Geology* **29**: 735–8.

Bartolini, A., Baumgartner, P. O., and Hunziker, J., 1996, Middle and Late Jurassic carbon stable-isotope stratigraphy and radiolarite sedimentation of the Umbria-Marche Basin (Central Italy). *Eclogae Geologicae Helvetiae* **89**: 811–44.

Bartolini, A., Baumgartner, P. O., and Guex, J., 1999, Middle and Late Jurassic radiolarian palaeoecology versus carbon-isotope stratigraphy. *Palaeogeography, Palaeoclimatology, Palaeoecology* **145**: 43–60.

Bartolini, A., Larson, R., Baumgartner, P. O., and ODP Leg 185 Scientific Party, 2000, Bajocian radiolarian age of the oldest oceanic crust *in situ* (Pigafetta Basin, western Pacific, ODP Site 801, Leg 185). In *European Ocean Drilling Project Forum, Abstracts*. La Grande-Motte, France, p. 32.

Bartolocci, P., Beraldini, M., Cecca, F., *et al.*, 1992, Preliminary results on correlation between Barremian ammonites and magnetic stratigraphy in Umbria–Marche Apennines (Central Italy). *Palaeopelagos* **2**: 63–8.

Bassett, M., 1985, Towards a "common language" in stratigraphy. *Episodes* **8**: 87–92.

1989, The Wenlock Series in the type area. In C. H. Holland and M. G. Bassett, eds., *A Global Standard for the Silurian System, Geological Series*, Vol. 9. Cardiff: National Museum of Wales.

Bassett, M. G., Cocks, L. R. M., Holland, C. H., Rickards, R. B., and Warren, P. T., 1975, The type Wenlock Series. *Report of the Institute of Geological Sciences* **75**(13): 1–19.

Bassett, M. G., Kaljo, D., and Teller, L., 1989, The Baltic region. In C. H. Holland and M. G. Bassett, eds., *A Global Standard for the Silurian System, Geological Series*, Vol. 9. Cardiff: National Museum of Wales, pp. 158–70.

Bassinot, F. C., Labeyrie, L. D., Vincent, E., *et al.*, 1994, The astronomical theory of climate and the age of the Brunhes–Matuyama magnetic reversal. *Earth and Planetary Science Letters* **126**: 91–108.

Batchelor, R. A. and Evans, J., 2000, Use of strontium isotope ratios and rare earth elements in apatite microphenocrysts for characterization and correlation of Silurian metabentonites: a Scandinavian case study. *Norsk Geologisk Tidsskrift* **80**: 3–8.

Baud, A., Magaritz, M., and Holser, W. T., 1989, Permian–Triassic of the Tethys: carbon isotope studies. *Geologische Rundschau* **78**: 649–77.

Baud, A., Atudorei, V., and Sharp, Z., 1996, Late Permian and early Triassic evolution of the northern Indian margin: carbon isotope and sequence stratigraphy. *Geodinamica Acta* **9**: 57–77.

Baudin, F., Faraoni, P., Marini, A., and Pallini, G., 1997, Organic matter characterization of the "Faraoni Level" from Northern Italy (Lessini Mountains and Trento Plateau): comparison with that from Umbria–Marche Appennines. *Palaeopelagos* **7**: 41–51.

Baudin, R., Bulot, L. G., Cecca, F., *et al.*, 1999, Un équivalent du "Niveau Faraoni" dans le Bassin du Sud-Est de la France, indice possible d'un événement anoxique fini-hauterivien étendu à la téthys méditerranéenne. *Bulletin de la Société Géologique de France* **170**: 487–98.

Baumgartner, P. O., 1987, Age and genesis of Tethyan Jurassic radiolarites. *Eclogae Geologicae Helvetiae* **80**: 831–79.

Baxter, J. W. and Brenckle, P. L., 1982, Preliminary statement on Mississippian calcareous foraminiferal successions of the Midcontinent (USA) and their correlation to western Europe. *Newsletters on Stratigraphy* **11**(3): 136–53.

Bazykin, D. A. and Hinnov, L. A., 2002, Orbitally-driven depositional cyclicity of the Lower Paleozoic Aisha–Bibi seamount (Malyi Karatau, Kazakstan): integrated sedimentological and time series study. In W. G. Zempolich and H. E. Cook, eds., *Paleozoic Carbonates of the Commonwealth of Independent States (CIS): Subsurface Reservoirs and Outcrop Analogs, Society of Economic Paleontologists and Mineralogists Special Publication*, Vol. 75. Tulsa, OK: SEPM, pp. 19–41.

Beard, K. C. and Dawson, M. R., 1998, Dawn of the Age of Mammals in Asia. *Bulletin of the Carnegie Museum of Natural History* **34**: 1–348.

Beauchamp, B. and Henderson, C. M., 1994, The Lower Raanes, Great Bear Cape and Trappers Cove formations, Sverdrup Basin, Canadian Arctic: stratigraphy and conodont zonation. *Canadian Society of Petroleum Geologists Bulletin* **42**: 562–97.

Beaulieu, J. L. D. and Reille, M., 1995, Pollen records from the Velay craters: a review and correlation of the Holsteinian Interglacial with isotopic stage 11. *Mededelingen van de Rijks Geologische Dienst* **52**: 9–70.

Beck, J. W., Richards, D. A., Edwards, R. L., *et al.*, 2001, Extremely large variations of atmospheric ^{14}C concentration during the last glacial period. *Science* **292**: 2453–8.

Becker, R. T., 1993a, Stratigraphische Gliederung und Ammonoideen-Fauna im Nehdenium (Oberdevon II) von Europa und Nord-Afrika. *Courier Forschungsinstitut Senckenberg* **155**: 1–405.

1993b, Anoxia, eustatic changes, and Upper Devonian to lowermost Carboniferous global ammonoid diversity. *Systematics Association Special volume* **47**: 115–63.

Becker, R. T. and House, M. R., 1993, Sea level changes in the Upper Devonian of the Canning Basin, Western Australia. *Courier Forschungsinstitut Senckenberg* **199**: 129–46.

1994, International Devonian goniatite zonation, Emsian to Givetian, with new records from Morocco. *Courier Forschungsinstitut Senckenberg* **169**: 79–135.

2000, Devonian ammonoid zones and their correlation with established series and stage boundaries. *Courier Forschungsinstitut Senckenberg* **220**: 113–51.

Becker, R. T. and Paproth, E., 1993, Auxiliary stratotype section for the Global Stratotype Section and Point (GSSP) for the Devonian–Carboniferous boundary. *Hasselbachtal: Annales de la Société Géologique de Belgique* **115**(2): 703–6.

Becker, R. T., House, M. R., and Kirchgasser, W. T., 1993, Devonian goniatite biostratigraphy and timing of facies movements in the Frasnian of the Canning Basin, Western Australia. *Geological Society of London Special Publication*, Vol. 70. London: Geological Society, pp. 293–321.

Becker, M. L., Rasbury, E. T., Hanson, G. N., and Meyers, W. J., 2001, Refinement in the age of the Carboniferous–Permian boundary based on U–Pb dating of biostratigraphically constrained syn-sedimentary carbonates in the Appalachian region of North America. *Newsletter on Carboniferous Stratigraphy* **19**: 18–20.

Beckinsale, R. D. and Gale, N. H., 1969, A reappraisal of the decay constants and branching ratio of ^{40}K. *Earth and Planetary Science Letters* **6**: 289–94.

Begemann, F., Ludwig, K. R., Lugmair, G. W., *et al.*, 2001, Call for an improved set of decay constants for geochronological use. *Geochimica et Cosmochimica Acta* **65**(1): 111–21.

Behl, R. J. and Kennett, J. P., 1996, Brief interstadial events in the Santa Barbara basin, NE Pacific, during the past 60 kyr. *Nature* **379**: 243–6.

Belkaaloul, K. N., Aissaoui, D., Rebelle, M., and Sambet, G., 1995, Magnetostratigraphic correlations of the Jurassic carbonates from the Paris Basin; implications for petroleum exploration. In *Palaeomagnetic Applications in Hydrocarbon Exploration and Production, Geological Society Special Publication,* Vol. 98. London: Geological Society of London, pp. 173–86.

Belkaaloul, K. N., Aissaoui, D., Rebelle, M., and Sambet, G., 1997, Resolving sedimentological uncertainties using magnetostratigraphic correlation: an example from the Middle Jurassic of Burgundy, France. *Journal of Sedimentary Research* **67**: 676–85.

Bellier, J.-P., Moullade, M., and Huber, B.T., 2000, Mid-Cretaceous planktonic foraminifers from Blake Nose: revised biostratigraphic framework. In D. Kroon, R. D. Norris, and A. Klaus, eds., *Proceedings of the Ocean Drilling Program, Scientific Results*, Vol. 171B [Online: www-odp.tamu.edu/publications/171B_SR/chap_03/chapt_03.html, cited 8 Sept 2000].

Bengtson, P., with contributions from Cobban, W. A., Dodsworth, P., Gale, A. S., *et al.*, 1996, The Turonian stage and substage boundaries. *Bulletin de l'Institut Royal des Sciences Naturelles de Belgique, Sciences de la Terre*, Supplément **66**: 69–79.

Bengtson, S., Conway Morris, S., Cooper, B. J., Jell, P. A., and Runnegar, B. N., 1990, *Early Cambrian Fossils from South Australia. Association of Australasian Palaeontologists Memoir* **9**: 1–364.

Benton, M. J., 1986, The Late Triassic tetrapod extinction events. In K. Padian, ed., *The Beginning of the Age of Dinosaurs. Faunal Change Across the Triassic–Jurassic Boundary*. Cambridge: Cambridge University Press, pp. 303–20.

1993, Late Triassic extinctions and the origin of the dinosaurs. *Science* **260**: 769–70.

Bergen, J. A., 1994, Berriasian to early Aptian calcareous nannofossils from the Vocontian trough (SE France) and Deep Sea Drilling Site 534: new nannofossil taxa and a summary of low-latitude biostratigraphic events. *Journal of Nannoplankton Research* (International Nannoplankton Association Newsletter) **16**: 59–69.

Berger, A. L., 1977a, Long term variations of the Earth's orbital elements. *Celestial Mechanics* **15**: 53–74.

1977b, Support for the astronomical theory of climatic change. *Nature* **269**: 44–5.

Berger, A. L. and Loutre, M.-F., 1990, Origine des fréquences des éléments astronomiques intervenant dans le calcul de l'insolation. *Bulletin de la Classe des Sciences, 6ième Série, Académie Royale de Belgique* **1–3**: 45–106.

1991, Insolation values for the climate of the last 10 m.y. *Quaternary Science Reviews* **10**: 297–317.

1994, Astronomical forcing through geologic time. In P. DeBoer and D. G. Smith, eds., *Orbital Forcing and Cyclic Sequences. International Association of Sedimentologists Special Publication,* Vol. 19. Oxford: Blackwell Scientific, pp. 15–24.

Berger, A. L., Imbrie, J., Hays, J., Kukla, G., and Saltzman, B., 1984, *Milankovitch and Climate*, Parts 1 and 2. Dordrecht: D. Reidel.

Berger, A. L., Loutre, M.-F., and Dehant, V., 1989, Influence of the changing lunar orbit on the astronomical frequencies of pre-Quaternary insolation patterns. *Paleoceanography* **4**: 555–64.

Berger, A. L., Loutre, M.-F., and Laskar, J., 1992, Stability of the astronomical frequencies over the Earth's history for paleoclimate studies. *Science* **255**: 560–6.

Berger, A. L., Loutre, M.-F., and Tricot, C., 1993, Insolation and Earth's orbital periods. *Journal of Geophysical Research* **98**: 10 341–62.

Berger, G. T. W., 1975, ^{40}Ar/^{39}Ar step heating of thermally overprinted biotite, hornblende and potassium feldspar from Eldora, Colorado. *Earth and Planetary Science Letters* **26**: 387–408.

Berggren, W. A., 1971, *Tertiary Boundaries and Correlations.* Cambridge: Cambridge University Press, pp. 693–809.

1972, A Cenozoic time-scale: some implications for regional geology and paleobiogeography. *Lethaia* **5**: 195–215.

1973, The Pliocene time scale: calibration of planktonic foraminifera and calcareous nannoplankton zones. *Nature* **241**(5407): 391–7.

1994, In defense of the Selandian Age/Stage. *Geologiska Föreningen i Stockholm Förhandlingar* **116**: 44–6.

1998, The Cenozoic Era: Lyellian (chrono) stratigraphy and nomenclatural reform at the millennium. *Geological Society of London Special Publication*, Vol. 143. London: Geological Society, pp. 111–32.

Berggren, W. A. and Aubry, M.-P., 1996, A Late Paleocene–Early Eocene NW European and North Sea Magnetobiochronologic Correlation Network. *Geological Society of London Special Publication* Vol. 101. London: Geological Society, pp. 309–352.

1998, *The Paleocene/Eocene Epoch/Series Boundary: Chronostratigraphic Framework and Estimated Geochronology, Late Paleocene–Early Eocene Climatic and Biotic Events in the*

Marine and Terrestrial Records. New York: Columbia University Press, pp. 18–36.

Berggren, W. A. and Miller, K. G., 1988, Paleogene tropical planktonic foraminiferal biostratigraphy and magnetobiochronology. *Micropaleontology* **34**: 362–80.

Berggren, W. A. and Norris, R. D., 1997, Taxonomy and biostratigraphy of (sub) tropical Paleocene planktonic foraminifera. *Micropaleontology*, 43rd supplement **1**: 116.

Berggren, W. A., Burkle, L. H., Cita, M. B., *et al.*, 1980, Towards a Quaternary time scale. *Quaternary Research* **13**: 277–302.

Berggren, W. A., Kent, D. V., Flynn, J. J., and van Couvering, J. A., 1985a, Cenozoic geochronology. *Geological Society of America Bulletin* **96**(11): 1407–18.

Berggren, W. A., Kent, D. V., and Flynn, J. J., 1985b, Jurassic to Paleogene. Part 2. Paleogene geochronology and chronostratigraphy. In N. J. Snelling, ed., *The Chronology of the Geological Record, Geological Society of London Memoir,* Vol. 10. London: Geological Society, pp. 141–95.

Berggren, W. A., Kent, D. V., and Van Couvering, J. A., 1985c, The Neogene: Part 2, Neogene geochronology and chronostratigraphy. In N. J. Snelling, ed., *The Chronology of the Geological Record, Geological Society of London Memoir,* Vol. 10. London: Geological Society, pp. 211–50.

Berggren, W. A., Kent, D. V., Swisher, C. C., III., and Aubry, M.-P., 1995a, A revised Cenozoic geochronology and chronostratigraphy. In W. A. Berggren, D. V. Kent, H. P. Aubry, and J. Hardenbol, eds., *Geochronology, Time Scales and Global Stratigraphic Correlations: A Unified Temporal Framework for a Historical Geology, Society of Economic Paleontologists and Mineralogists Special Publication*, Vol. 54. Tulsa, OK: SEPM. pp. 129–212.

Berggren, W. A., Hilgen, F. J., Langereis, C. G., *et al.*, 1995b, Late Neogene (Pliocene–Pleistocene) chronology: new perspectives in high-resolution stratigraphy. *Geological Society of America Bulletin* **107**: 1272–87.

Berggren, W. A., Aubry, M.-P., Van Fossen, M., *et al.*, 2000, Integrated Paleocene calcareous plankton magnetobiochronology and stable isotope stratigraphy, DSDP Site 384 (NW Atlantic). *Palaeogeography, Palaeclimatology, Palaeoecology* **159**: 1–51.

Bergström, S. M., 1986, Biostratigraphic integration of Ordovician graptolite and conodont zones – a regional review. In R. B. Rickards and C. P. Hughes, eds., *Paleobiology and Biostratigraphy of Graptolites*. Oxford: Blackwell Scientific, pp. 61–78.

1989, Use of graphic correlation for assessing event-stratigraphic significance and trans-Atlantic relationships of Ordovician K-bentonites. *Proceedings of the Academy of Sciences of the Estonian SSR. Geology* **38**: 55–60.

1995, The search for global biostratigraphic reference levels in the Ordovician System: regional correlation potential of the base of the North American Whiterockian Series. In R. A. Cooper, *et al.*, eds., *Ordovician Odyssey: Short Papers for the Seventh International Symposium on the Ordovician System*, Fullerton, CA. Tulsa, OK: Society of Economic Paleontologists and Mineralogists, pp. 149–53.

Bergström, S. M. and Wang, Z.-H., 1995, Global correlation of Castlemainian to Darriwilian conodont faunas and their relations to the graptolite zone succession. In Chen Xu and S. M. Bergström, eds., *The Base of the Austrodentatus Zone as a Level for Global Subdivision of the Ordovician System*. Nanjing: Nanjing University Press, pp. 92–8.

Bergström, S. M., Huff, W. D., Kolata, D. R., and Melchin, M. J., 1997, Occurrence and significance of Silurian K-bentonite beds at Arisaig, Nova Scotia, eastern Canada. *Canadian Journal of Earth Sciences* **34**: 1630–43.

Bergström, S. M., Huff, W. D., and Kolata, D. R., 1998a, The Lower Silurian Osmundsberg K-bentonite. Part I: Stratigraphic position, distribution, and paleogeographic significance. *Geological Magazine* **135**: 1–13.

1998b, Silurian K-bentonites in North America and Europe. In J. C. Gutiérrez-Marco and I. Rábano, eds., *Proceedings of the Sixth International Graptolite Conference on the GWG (IPA) and the 1998 Field Meeting of the International Subcommission on Silurian Stratigraphy (ICS-IUGS), Instituto Technológico Geominero de España, Temas Geologico-Mineros*, Vol. 23. pp. 54–6.

Bergström, S. M., Huff, W. D., Kolata, D. R., and Bauert, H., 1995, Nomenclature, stratigraphy, chemical fingerprinting, and areal distribution of some Middle Ordovician K-bentonites in Baltoscandia. *Geologiska Forenigens Forhandlingar* **117**: 1–13.

Bergström, S. M., Finney S. C, Xu, C., *et al.*, 2000, A proposed global boundary stratotype for the base of the Upper Series of the Ordovician System: the Fågelsång section, Scania, southern Sweden. *Episodes* **23**: 102–9.

Berner, R. A. and Kothavala, Z., 2001, Geocarb III: a revised model of atmospheric CO_2 over Phanerozoic time. *American Journal of Science* **301**(2): 182–204.

Berry, W. B. N., 1987, *Growth of a Prehistoric Time Scale. Based on Organic Evolution*. Palo Alto: Blackwell Scientific, p. 202.

Bertram, C. J., Elderfield, H., Aldridge, R. J., and Conway Morris, S., 1992, $^{87}Sr/^{86}Sr$, $^{143}Nd/^{144}Nd$ and REEs in Silurian phosphatic fossils. *Earth and Planetary Science Letters* **113**: 239–49.

Beyrich, E., 1854, Über die Stellung der hessischen Tertiärbilldungen. *Berichte der Verhandlungen der königlichen: Preussischen Akademie der Wissenschaften* **1954**: 640–66.

Bice, K. L., Bawlower, T. J., Duncan, R. A., *et al.*, 2002, *Cretaceous Climate-Ocean Dynamics: Future Directions for IODP, A JOI/USSSP and NSF Sponsored Workshop* [Online: www.whoi.edu.ccod/CCOD_report.html].

Bickert, T., Curry, W. B., and Wefer, G., 1997a, Late Pliocene to Holocene (2.6–0 Ma) western equatorial Atlantic deep-water circulation: inferences from benthic stable isotopes. In N. J. Shackleton, W. B. Curry, C. Richter, and T. J. Bralower, eds., *Proceedings of the Ocean Drilling Program, Scientific Results*, Vol. 154. College Station, TX: Ocean Drilling Program, pp. 239–53.

Bickert, T., Pätzold, J., Samtleben, C., and Munnecke, A., 1997b, Paleoenvironmental changes in the Silurian indicated by stable isotopes in brachiopod shells from Gotland, Sweden. *Geochimica et Cosmochimica Acta* **61**: 2717–30.

Bickle, M. J., Martin, A., and Nisbet, E. G., 1975, Basaltic and peridotitic komatiites and stromatolites above a basal unconformity in the Belingwe greenstone belt, Rhodesia. *Earth and Planetary Science Letters* **27**(2): 155–62.

Bills, B. G., 1994, Obliquity–oblateness feedback: are climatically sensitive values of obliquity dynamically unstable? *Geophysical Research Letters* **21**: 177–80.

Billups, K., Ravelo, A. C., and Zachos, J. C., 1997, Early Pliocene deep-water circulation: stable isotope evidence for enhanced northern component deep water. In N. J. Shackleton, W. B. Curry, C. Richter, and T. J. Bralower, eds., *Proceedings of the Ocean Drilling Program, Scientific Results*, Vol. 154. College Station, TX: Ocean Drilling Program, pp. 319–30.

Billups, K., Channell, J. E. T., and Zachos, J. C., 2002, Late Oligocene to early Miocene geochronology and paleoceanography from the subantarctic South Atlantic. *Palaeoceanography* **17**(1–4): 1–11.

Bingen, B. and van Breemen, O., 1998, Tectonic regimes and terrane boundaries in the high-grade Sveconorwegian Belt of SW Norway, inferred from U–Pb zircon geochronology and geochemical signature of augen gneiss suites. *Journal of the Geological Society of London* **155**(1): 143–54.

Birkelund, T., Callomon, J. H., Clausen, C. K., Nøhr Hansen, H., and Salinas, I., 1983, The Lower Kimmeridge Clay at Westbury, Wiltshire, England. *Proceedings of the Geologists' Association* **94**: 289–309.

Birkelund, T., Hancock, J. M., Hart, M. B., *et al.*, 1984, Cretaceous stage boundaries - proposals. *Geological Society of Denmark Bulletin* **33**: 3–20.

Bisat, W. S., 1924, The Carboniferous goniatites of the north of England and their zones. *Proceedings of the Yorkshire Geological Society, new series* **20**(1): 40–124.

1928, The Carboniferous goniatite zones of England and their continental equivalents. In *1st International Congress on Carboniferous Stratigraphy and Geology*, Heerlen, June, 7–11, 1927, Vaillant-Carmanne, pp. 117–133.

Bischoff, G. and Ziegler, W., 1957, Die Conodontenchronologie des Mitteldevons und des tiefsten Oberdevons. *Abhandlungen des Hessischen Geologischen Landesamtes für Bodenforschung* **22**: 1–136.

Bittner, A., 1892, Was ist norisch? *Jahrbuch der Geologischen Reichsanstalt* **42**: 387–96.

Bizon, G. and Bizon, J. J., 1972, *Atlas des principaux foraminifères planctoniques du basin méditerranéen Oligocène à Quaternaire*. Paris Editions Technip.

Björk, S., Walker, M. J. C., Cwynar, L. C., *et al.*, 1998, An event stratigraphy for the last termination in the North Atlantic region based on the Greenland ice-core record: a proposal by the INTIMATE group. *Journal of Quaternary Science* **13**: 283–92.

Bjørklund, K. R., 1976, Radiolaria from the Norwegian Sea, Leg 38 of the Deep Sea Drilling Project. *Initial Reports of the Deep Sea Drilling Project*, Vol. 38. Washington, DC: US Govt. Printing Office, pp. 1101–68.

Blachman, N. M. and Machol, R. E., 1987, Confidence intervals based on one or more observations. *IEEE Transactions on Information Theory* **33**: 373–82.

Black, L. P., Seymour, D. B., Corbett, K. D., *et al.*, 1997, *Dating Tasmania's Oldest Geological Events, Australian Geological Survey Organisation Record 1997/15*. Canberra: Australian Geological Survey, pp. 28–57.

Blake, T. S. and Groves, D. I., 1987, Continental rifting and the Archean–Proterozoic transition. *Geology* **15**(3): 229–32.

Blanc, E., Bulot, L. G., and Paicheler, J.-C., 1994, La coupe de référence de Monbrun-les-Bains (Drôme, SE France): un stratotype potentiel pour la limite Berriasien-Valanginien. *Comptes rendus de l'Académie des Sciences, Paris, Serie II. Sciences de la Terre et des Planetes* **318**: 101–8.

Bleeker, W., 2002, Archean tectonics: a review, with illustrations from the Slave craton. In C. M. R. Fowler, C. J. Ebinger, and C. J. Hawkesworth, eds., *The Early Earth: Physical, Chemical and Biological Development, Geological Society of London, Special Publication*, Vol. 199. London: Geological Society, pp. 151–81.

2003a, Towards a "natural timescale" for 88 percent of Earth history. In *NUNA Conference 2003. New Frontiers in the Fourth Dimension: Generation, Calibration and Application of Geological Timescales*, Mont Tremblant, Quebec, March 15–18, 2003. Geological Association of Canada.

2003b, The late Archean record: a puzzle in ca. 35 pieces. *Lithos* **71**(2–4): 99–134.

Bleeker, W., Ketchum, J. W. F., and Davis, W. J., 1999a, The Central Slave Basement Complex. Part II: age and tectonic significance of high-strain zones along the basement–cover contact. *Canadian Journal of Earth Sciences* **36**(7): 1111–30.

Bleeker, W., Ketchum, J. W. F., Jackson, V. A., and Villeneuve, M. E., 1999b, The Central Slave Basement Complex. Part I: its structural topology and autochthonous cover. *Canadian Journal of Earth Sciences* **36**(7): 1083–109.

Bleil, U., 1989, The magnetostratigraphy of northwest Pacific sediments. In *Initial Reports of the Deep Sea Drilling Project*,

Vol. 86. Washington, DC: US Govt. Printing Office, pp. 441–58.

Blome, C. D., 1984, Upper Triassic radiolaria and radiolarian zonation from western North America. *Bulletin of American Paleontology* **85**: 88.

Blondeau, A., 1981, Lutetian. In C. Pomerol, ed., *Stratotypes of Paleogene Stages, Mémoire hors série 2 du Bulletin d'Information de Géologues du Bassin de Paris.* pp. 167–80.

Bloos, G., 1999, Aspekte der Wende Trias/Jura. In N. Hausche and V. Wilde, eds., *Trias – Eine ganz andere Welt*. Munich: Verlag Dr Friedrich Pfeil, pp. 43–68.

Bloos, G. and Page, K. N., 2002, Global stratotype section and point for base of Sinemurian Stage (Lower Jurassic). *Episodes* **25**: 22–8.

Blow, W. H., 1969, Late middle Eocene to Recent planktonic foraminiferal biostratigraphy. In P. Bronniman and H. H. Renz, eds., *Proceedings of the First International Conference Planktonic Microfossils 1967*, Vol. 1. Geneva: Bill/Netherlands: Leiden, pp. 199–242.

1979, *The Cainozoic Globigerinida: A Study of the Morphology, Taxonomy, Evolutionary Relationships and the Stratigraphical Distribution of some Globigerinida (mainly Globigerinacea)*, Vols. 1–3. Leiden: E. J. Brill.

Blumenstengel, H., 1965, Zur Taxionomie und Biostratigraphie verkieselter Ostracoden aus dem Thüringer Oberdevon. *Freiberger Forschungshefte* **C183**: 1–127.

Boardman, D. R., II, Work, D. M., Mapes, R. H., and Barrick, J. E., 1994, Biostratigraphy of Middle and Late Pennsylvanian (Desmoinesian–Virgilian) ammonoids. *Bulletin of Kansas Geological Survey* **232**: 1–121.

Boardman, D. R., II., Nestell, M. K., and Wardlaw, B. R., 1998, Uppermost Carboniferous and lowermost Permian deposition and conodont biostratigraphy of Kansas, USA. In Y. Jin, B. R. Wardlaw, and Y. Wang, eds., *Permian Stratigraphy, Environments and Resources*, Vol. 2, *Stratigraphy and Environments.* Palaeoworld 9, pp. 19–32.

Bogoslovskaya, M. F., Kuzina, L. F., and Leonova, T. B., 1999, Classification and stratigraphic distribution of late Paleozoic ammonoids. In A. Y. Rozanov and A. A. Shevyrev, eds., *Fossil Cephalopods*: *Recent Advances in Their Study*. Moscow: Paleontological Institute of Russian Academy of Sciences, pp. 89–124 [In Russian].

Bogoslovsky, N. L., 1897, *L'horizon de Riazan* 18. St. Petersburg: Mater. Géol. Ross.

Bolli, H. M., 1957a, The genera Globigerina and Globorotalia in the Paleocene–Lower Eocene Lizard Springs Formation of Trinidad, BWI. *Bulletin of the United States National Museum* **215**: 61–82.

1957b, Planktonic foraminifera from the Eocene Navet and San Fernando Formations of Trinidad, BWI. *Bulletin of the United States National Museum* **215**: 155–72.

1966, Zonation of Cretaceous to Pliocene marine sediments based on planktonic foraminifera. *Boletín Informativo de la Asociación Venezolana de Geología, Minería y Petróleo* **9**: 3–32.

Bolli, H. M. and Saunders, J. B., 1985, Oligocene to Holocene low latitude planktic foraminifera. In H. M. Bolli, J. B. Saunders, and K. Perch-Nielsen, eds., *Plankton Stratigraphy*. Cambridge: Cambridge University Press, pp. 155–262.

Bolli, H. M., Beckmann, J.-P., and Saunders, J. B., 1994, *Benthic Foraminiferal Stratigraphy of the South Caribbean Region.* Cambridge: Cambridge University Press.

Bonarelli, G., 1894, Contribuzione alla conoscenza del Giuralias *Lombardo. Atti R. Accad. Sci. Torino* **30**: 81–96.

1895, Fossili domeriani della Brianza. *Real. Lst. Lomb. Sci. Lett. Rendic.* (*Milano*) **28** (2): 326–47.

Bond, G. C. and Lotti, R., 1995, Iceberg discharge into the North Atlantic on millenial time scales during the Last Glaciation. *Science* **267**: 1005–10.

Bond, G. C., Kominz, M. A., and Beavan, J., 1991, Evidence for orbital forcing of Middle Cambrian peritidal cycles: Wah Wah Range, south-central Utah. In K. Franseen, W. L. Watney, C. G. S. C. Kendall, and W. Ross, eds., *Sedimentary Modeling: Computer Simulations and Methods for Improved Parameter Definition, Kansas Geological Survey Bulletin*, Vol. 233. Lawrence, KS: Kansas Geological Survey, pp. 293–317.

Bond, G., Heinrich, H., Broecker, W., *et al.*, 1992, Evidence for iceberg discharges into the North Atlantic during the last glacial period. *Nature* **360**: 245–9.

Bond, G., Broecker, W. S., Johnsen, S. J., *et al.*, 1993, Correlations between climate records from North Atlantic sediments and Greenland ice. *Nature* **365**: 143–7.

Bond, G. C., Showers, W., Elliot, M., *et al.*, 1999, The North Atlantic's 1–2 kyr climate rhythm: relation to Heinrich events, Dansgaard/Oeschger cycles and the Little Ice Age, mechanisms of global climate change at millennial time scales. *Geophysical Monograph* **112**: 35–58.

Boucot, A. J., 1975, *Evolution and Extinction Rate Controls.* Amsterdam: Elsevier.

Boudagher-Fadel, M. K. and Banner, F. T., 1999, Révision of the stratigraphic significance of the Oligocene–Miocene "letter-stages": *Revue de Micropaléontologie* **42**: 93–7.

Bourdier, F., 1957, Quaternaire. In P. Pruvost, ed., *Lexique stratigraphique international*, Vol. 1, *Europe*. Paris: Centre National de la Recherche Scientifique, pp. 99–100.

Bouroz, A., Einor, O. L., Gordon, M., Meyen, S. V., and Wagner, R. H., 1978, Proposal for an International Chronostratigraphic Classification of the Carboniferous. In S. V. Meyer, V. V. Menner, Y. A. Reytlinger, A. P. Rotay, and M. N. Solovyeva, eds., *Compte Rendu VIII International Congress on Carboniferous Stratigraphy and Geology* 1978, Vol. 1. Moscow: various publishers, pp. 36–52.

Bouska, V., 1989, Distribution of moldavites and their stratigraphic position. *Meteoritics* **24**(4): 253–4.

Bowen, D. Q., 1978, *Quaternary Geology*. Oxford: Pergamon Press.

Bowen, G. J., Koch, P. L., Gingerich, P. D., *et al.*, 2001, Refined isotope stratigraphy across the continental Paleocene–Eocene boundary at Polecat Bench in the northern Bighorn Basin. *University of Michigan Papers on Paleontology* **33**: 73–88.

Bown, P. R., 1998, *Calcareous Nannofossil Biostratigraphy*, British Micropalaeontological Society Publication Series. London: Chapman and Hall, pp. 132–99.

Bown, P. R., Cooper, M. K. E., and Lord, A. R., 1988, A calcareous nannofossil biozonation for the early to mid Mesozoic. *Newsletters on Stratigraphy* **20**: 91–114.

Bowring, S. A. and Erwin, D. H., 1998, A new look at evolutionary rates in deep time: uniting paleontology and high-precision geochronology. *GSA Today* **8**: 1–8.

Bowring, S. A. and Williams, I. S., 1999, Priscoan (4.00–4.03 Ga) orthogneisses from northwestern Canada. *Contributions to Mineralogy and Petrology* **134**: 3–16.

Bowring, S. A., Grotzinger, J. P., and Isachsen, C. E., 1993, Calibrating rates of Early Cambrian evolution. *Science* **261**(5126): 1293–98.

Bowring, S. A., Erwin, D. H., Jin, Y. G., *et al.*, 1998, U/Pb zircon geochronology and tempo of the end-Permian mass extinction. *Science* **280**(5366): 1039–45.

Bowring, S. A., Ramezani, J., and Grotzinger, J. P., 2003, High-precision U–Pb geochronology and the Cambrian–Precambrian boundary. In *NUNA Conference 2003, New Frontiers in the Fourth Dimension: Generation, Calibration and Application of Geological Timescales*, Mont Tremblant, Quebec, March 15–18, 2003. Geological Association of Canada.

Brack, P. and Rieber, H., 1994, The Anisian/Ladinian boundary: retrospective and new constraints. *Albertiana* **13**: 25–36.

1996, The new 'High-resolution Middle Triassic ammonoid standard scale' proposed by Triassic researchers from Padova: – a discussion of the Anisian/Ladinian boundary interval. *Albertiana* **17**: 42–50.

Brack, P., Rieber, H., and Mundil, R., 1995, The Anisian/Ladinian boundary interval at Bagolino (Southern Alps, Italy): I. Summary and new results on ammonoid horizons and radiometric dating. *Albertiana* **15**: 45–56.

Brack, P., Mundil, R., Oberli, F., Meier, M., and Rieber, H., 1996, Biostratigraphic and radiometric age data question the Milankovitch characteristics of the Latemar cycles (Southern Alps, Italy). *Geology* **24**: 371–5.

1997, Biostratigraphic and radiometric age data question the Milankovitch characteristics of the Latemar cycles (Southern Alps, Italy): – reply. *Geology* **25**: 471–2.

Bradley, W. H., 1929, *The Varves and Climate of the Green River Epoch, US Geological Survey Professional Paper*, Vol. 158-E. Boulder, CO: Geological Survey, pp. 87–110.

Bralower, T. J. and Mutterlose, J., 1995, Calcareous nannofossil biostratigraphy of Site 865, Allison Guyot, Central Pacific Ocean: a tropical Paleogene reference section. In *Proceedings of the Ocean Drilling Program, Scientific Results*, Vol. 143. College Station, TX: Ocean Drilling Program, pp. 31–74.

Bralower, T. J., Monechi, S., and Thierstein, H. R., 1989, Calcareous nannofossil zonation of the Jurassic–Cretaceous boundary interval and correlation with the geomagnetic polarity timescale. *Marine Micropaleontology* **14**: 153–235.

Bralower, T. J., Ludwig, K. R., Obradovich, J. D., and Jones, D. L., 1990, Berriasian (Early Cretaceous) radiometric ages from the Grindstone Creek Section, Sacramento Valley, California. *Earth and Planetary Science Letters* **98**: 62–73.

Bralower, T. J., Sliter, W. V., Arthur, M. A., *et al.*, 1993, Dysoxic/anoxic episodes in the Aptian–Albian (Early Cretaceous). In M. S. Pringle, W. W. Sager, W. V. Sliter, and S. Stein, eds., *The Mesozoic Pacific: Geology, Tectonics, and Volcanism, Geophysics Monograph*. Vol. 77. Washington, DC: American Geophysical Union, pp. 5–37.

Bralower, T. J., Leckie, R. M., Sliter, W. V., and Thiestein, H. R., 1995, An integrated Cretaceous microfossil biostratigraphy. In W. A. Berggren, D. V. Kent, and J. Hardenbol, eds., *Geochronology, Time Scales and Global Stratigraphic Correlations: A Unified Temporal Framework for a Historical Geology, Society of Economic Paleontologists and Mineralogists Special Publication,* Vol. 54. Tulsa, OK: SEPM, pp. 65–79.

Bralower, T. J., Fullagar, P. D., Paull, C. K., Dwyer, G. S., and Leckie, R. M., 1997, Mid-Cretaceous strontium-isotope stratigraphy of deep-sea sections. *Geological Society of America Bulletin* **109**: 1421–42.

Bramlette, M. N. and Riedel, W. R., 1954, Stratigraphic value of discoasters and some other microfossils related to Recent coccolithophores. *Journal of Paleontology* **28**: 385–403.

Bramlette, M. N. and Sullivan, F. R., 1961, Coccolithophorids and related nannoplankton of the early Tertiary in California. *Micropaleontology* **7**: 129–88.

Bramlette, M. N. and Wilcoxon, J. A., 1967, Middle Tertiary calcareous nannoplankton of the Cipero Section, Trinidad, W. I. *Tulane Studies in Geology and Paleontology* **5**: 93–131.

Brand, U. and Brenckle, P., 2001, Chemostratigraphy of the Mid-Carboniferous boundary global stratotype section and point (GSSP), Bird Spring Formation, Arrow canyon, Nevada, USA. *Palaeogeography, Palaeoclimatology, Palaeoecology* **165**: 321–47.

Brasier, M. D., 1979, The Cambrian radiation event. In M. R. House, ed., *The Origin of Major Invertebrate Groups, Systematics Association Special Volume.*, 12. London and New York: Academic Press, 103–59.

1982, Sea-level changes, facies changes and the late Precambrian–early Cambrian evolutionary explosion. *Precambrian Research* **17**: 105–23.

1992a, Paleoceanography and changes in the biological cycling of phosphorus across the Precambrian–Cambrian boundary. In J. H. Lipps and P. W. Signor, eds., *Origin and Early Evolution of the Metazoa*. New York: Plenum Press, pp. 483–523.

1992b, Nutrient-enriched waters and the early skeletal fossil record. *Journal of the Geological Society of London* **149**: 621–9.

1993, Towards a carbon isotope stratigraphy of the Cambrian System: potential of the Great Basin succession. In E. A. Hailwood and R. B. Kidd, eds., *High Resolution Stratigraphy, Geological Society of London Special Publication*, Vol. 70. London: Geological Society, pp. 341–50.

1995a, Fossil indicators of nutrient levels. 1: Eutrophication and climate change. In D. W. J. Bosence and P. A. Allison, eds., *Marine Palaeoenvironmental Analysis from Fossils, Geological Society of London Special Publication*, Vol. 83. London: Geological Society, pp. 113–32.

1995b, Fossil indicators of nutrient levels. 2: Evolution and extinction in relation to oligotrophy. In D. W. J. Bosence and P. A. Allison, eds., *Marine Palaeoenvironmental Analysis from Fossils, Geological Society of London Special Publication*, Vol. 83. London: Geological Society, pp. 133–50.

1995c, The basal Cambrian transition and Cambrian bio-events (from terminal Proterozoic extinctions to Cambrian biomeres). In O. H. Walliser, ed., *Global Events and Event Stratigraphy in the Phanerozoic*. Berlin: Springer, pp. 113–18.

Brasier, M. D. and Lindsay, J. F., 2001, Did supercontinental amalgamation trigger the Cambrian explosion? In A. Yu. Zhuravlev and R. Riding, eds., *The Ecology of the Cambrian Radiation*. New York: Columbia University Press, pp. 69–89.

Brasier, M. D. and Sukhov, S. S., 1998, The falling amplitude of carbon isotopic oscillations through the Lower to Middle Cambrian: northern Siberia data. *Canadian Journal of Earth Sciences* **35**: 353–73.

Brasier, M. D., Corfield, R. M., Derry, L. A., Rozanov, A. Y., and Zhuravlev, A. Y., 1994, Multiple delta ^{13}C excursions spanning the Cambrian explosion to the Botomian crisis in Siberia. *Geology* **22**: 455–8.

Brasier, M. D., Shields, G. A., Kuleshov, V. N., and Zhegallo, E. A., 1996, Integrated chemo- and biostratigraphic calibration of early animal evolution: Neoproterozoic–early Cambrian of southwest Mongolia. *Geological Magazine* **133**: 445–85.

Braun, A., 1990, Evolutionary trends and biostratigraphic potential of selected radiolarian taxa from the Early Carboniferous of Germany. *Marine Micropaleontology* **15**(3–4): 351–64.

Braun, A. and Schmidt-Effing, R., 1993, Biozonation, diagenesis and evolution of radiolarians in the Lower Carboniferous of Germany. *Marine Micropaleontology* **21**(4): 369–83.

Bréheret, J.-G., 1988, Episodes de sedimentation riche en matiere organique dans les marnes bleues d'âge Aptien et Albien de la partie pélagique du bassin vocontien. *Bulletin de la Société Géologique de France* 8(IV): 349–56.

Bréhéret, J.-G., 1997, *L'Aptien et l'Albien de la Fosse vocontienne (des bordures au bassin). Evolution de la sédimentation et enseignements sur les événements anoxiques*, Publication 25. Lille: Société Géologique du Nord, 614 pp.

Bréheret, J.-G., Caron, M., and Delamette, M., 1986, Niveaux riches en matière dans l'Albien Vocontien: Quelques caractères du paléoenvironment: Essai d'interprétation génétique. *Documents Bureau Recherches Géologiques et Minières* **110**: 141–91.

Breistroffer, M., 1947, Sur les zones d'ammonites dans l'Albien de France et d'Angleterre. *Travaux du Laboratoire de Géologie de la Faculté des Sciences de l'Université de Grenoble* **26**: 17–104.

Brenchley, P. J., 1984, Late Ordovician extinctions and their relationship to the Gondwana glaciation. In P. J. Brenchley, ed., *Fossils and Climate*. Chichester: Wiley, pp. 291–315.

1989, The Late Ordovician extinction. In S. K. Donovan, ed., *Mass Extinctions: Processes and Evidence*. London: Belhaven Press, pp. 104–32.

Brenchley, P. J., Marshall, J. D., Carden, G. A. F., *et al.*, 1994, Bathymetric and isotopic evidence for a short-lived Late Ordovician glaciation in a greenhouse period. *Geology* **22**: 295–8.

Brenckle, P. L. and Manger, W. L., 1990, Intercontinental correlation and division of the Carboniferous System: contributions from the Carboniferous Subcommission meeting CFS. *Courier Forschunginstitut Senckenberg* **130**: 1–350.

Brenckle, P. L., Marshall, F. C., Waller, S. F., and Wilhelm, M. H., 1982, Calcareous microfossils from the Mississippian Keokuk Limestone and adjacent formations, upper Mississippi Valley: their meaning for North American and intercontinental correlation. *Geologica et Palaeontologica* **15**: 47–88.

Brenckle, P. L., Baesemann, J. F., Lane, H. R., *et al.*, 1997, *Arrow Canyon, the Mid-Carboniferous Boundary Stratotype*. In P. L. Brenckle and W. R. Paqe, eds., *Paleoforams'97 guidebook, post-conference field trip to the Arrow Canyon Range, southern Nevada, USA*, Vol. 36. Special Publications, Cushman Foundation for Foraminiferal Research. Ithaca, NY: Cushman Foundation for Foraminiferal Research, pp. 13–32.

Brenner, W., 1988, Dinoflagellaten aus dem unteren Malm (Oberer Jura) von Süddeutschland; Morphologie, Ökologie, Stratigraphie. *Tübinger Mikropaläontologische Mitteilungen* **6**: 116.

Bretagnon, P., 1974, Termes à longues périodes dans le système solaire. *Astronomy and Astrophysics* **30**: 141–54.

Briand, C., Izart, A., Vaslet, D., *et al.*, 1998, Sequence stratigraphy of the Moscovian, Kasimovian and Gzhelian in the Moscow Basin. *Bulletin de la Société Géologique de France* **169**(1): 35–52.

Briden, J. C. and Daniels, B. A., 1999, Palaeomagnetic correlation of the Upper Triassic of Somerset, England, with continental Europe and eastern North America. *Journal of the Geological Society of London* **156**: 317–26.

Brinkhuis, H., 1994, Late Eocene to early Oligocene dinoflagellate cysts from the Priabonian type-area (Northeast Italy): biostratigraphy and paleoenvironmental interpretation. *Palaeogeography, Palaeoclimatology, Palaeoecology* **107**: 121–63.

Brinkhuis, H. and Visscher, H., 1995, The upper boundary of the Eocene Series: a reappraisal based on dinoflagellate cyst biostratigraphy and sequence stratigraphy. In W. A. Berggren, D. V. Kert, M.-P. Aubry, and J. Hardenbol, eds., *Geochronology, Time Scales and Global Stratigraphic Correlation, Society of Economic Paleontologists and Mineralogists Special Publication*, Vol. 54. Tulsa, OK: SEPM, pp. 295–304.

Brinkhuis, H., Munsterman, D. K., Sengers, S., *et al.*, 2003a, Late Eocene to Quaternary dinoflagellate cysts from ODP Site 1168, off western Tasmania. In Exon, *et al.*, eds, *Proceedings of the Ocean Drilling Project, Scientific Results* Vol. 189. College Station, TX: Ocean Drilling Project.

Brinkhuis, H., Sengers, S., Sluijs, A., Warnaar, J., and Williams, G. L., 2003b, Latest Cretaceous to earliest Oligocene, and Quaternary dinoflagellate cysts from ODP Site 1172, East Tasman Plateau. In N. F. Exon, J. P. Kennett, M. J. Malone, *et al.*, eds., *Proceedings of the Ocean Drilling Program, Scientific Results*, Vol. 189. College Station, TX: Ocean Drilling Program [Online: www-odp.tamu.edu/publications/189_SR/106/106.htm].

Brock, G. A., Engelbretsen, M. J., Jago, J. B., *et al.*, 2000, Palaeobiogeographical affinities of Australian Cambrian faunas. *Association of Australasian Palaeontologists Memoir* **23**: 1–61.

Broecker, W. S., 1994, Massive iceberg discharges as triggers for global climate change. *Nature* **372**: 421–4.

Broecker, W. S. and Denton, G. H., 1989, The role of ocean–atmosphere reorganizations in glacial cycles. *Geochimica et Cosmochimica Acta* **53**: 2465–501.

Broecker, W. S. and Henderson, G. M., 1998, The sequence of events surrounding Termination II and their implications for the cause of glacial–interglacial CO_2 changes. *Paleoceanography* **13**: 352–64.

Broecker, W. S. and van Donk, J., 1970, Insolation changes, ice volumes and the ^{18}O record in deep-sea cores. *Review of Geophysics and Space Physics* **8**: 169–98.

Broecker, W. S., Bond, G., Klas, M., Clark, E., and McManus, J., 1992, Origin of the northern Atlantic's Heinrich events. *Climate Dynamics* **6**: 265–73.

Broglio Loriga, C. and Cassinis, G., 1992, The Permo–Triassic boundary in the Southern Alps (Italy) and in adjacent Periadriatic regions. In W. C. Sweet, Z. Yang, J. M. Dickins, and H. Yin, eds., *Permo–Triassic Events in the Eastern Tethys. Stratigraphy, Classification, and Relations with the Western Tethys.* Cambridge: Cambridge University Press, pp. 78–97.

Broglio Loriga, C., Cirilli, S., De Zanche, V., *et al.*, 1998, A GSSP candidate for the Ladinian–Carnian boundary: the Prati di Stuores/Stuores Wiesen section (Dolomites, Italy). *Albertiana* **21**: 2–18.

Brongniart, A., 1829, *Tableau des Terrains qui Composent L'écorce du Globe ou Essai sur la Structure de la Partie Connue de la Terre.* Paris.

Bronn, H. G., 1838, *Lethea Geognostica, oder Abbildungen und Beschreibungen der für die Gebirgsformationen bezeichnensten Versteinerungen*, Vol. 2. Stuttgart. pp. 545–1356.

Brönnimann, P. and Stradner, H., 1960, Die Foraminiferen- und Discosasteriden-Zonen von Kuba und ihre interkontinentale Korrelation. *Erdölzeitschrift* **76**: 364–9.

Bruckschen, P. O. and Veizer, J., 1997, Isotope stratigraphy for the late Paleozoic greenhouse–icehouse transition: proxy signals for links between ocean chemistry, climate, and tectonics. In *Seventh Annual V. M. Goldschmidt Conference, Volume Lunar and Planetary Science Contribution 921*. Houston, TX: Lunar and Planetary Science Institute, pp. 36–37.

Bruckschen, P., Oesmann, S., and Veizer, J., 1999, Isotope stratigraphy of the European Carboniferous: proxy signals for ocean chemistry, climate and tectonics. *Chemical Geology* **161**: 127–63.

Brugman, W. A., 1986, A palynological characterization of the upper Scythian and Anisian of the Transdanubian Central Range, Hungary and the Vicentinian Alps, Italy. Unpublished Ph.D. thesis, University of Utrecht.

Budil, P., 1995, Demonstrations of the Kacák event (Middle Devonian, uppermost Eifelian) at some Barrandian localities. *Vestnik Ceského geologeckého ústavu* **70**: 1–24.

Buggisch, W., 1972, Zur geologie und Geochemie der Kellwasserkalke und ihrer begleitenden Sedimente (unteres Oberdevon). *Abhandlungen des Hessischen Geologischen Landesamtes für Bodenforschung* **62**: 1–68.

1991, The global Frasnian–Famennian 'Kellwasser'-Event. *Geologische Rundschau* **80**: 49–72.

Buhring, C. and Sarnthein, M., 2000, Toba ash layers in the South China Sea: evidence of contrasting wind directions during eruption ca. 74 ka. *Geology* **28**: 275–8.

Buick, R., Brauhart, C. W., Morant, P., *et al.*, 2002, Geochronology and stratigraphic relationships of the Sulphur Springs Group and Strelley Granite: a temporally distinct igneous province in the Archaean Pilbara Craton, Australia. *Precambrian Research* **114**(1–2): 87–120.

Bujak, J. P. and Brinkhuis, H., 1998, Global warming and dinocyst changes across the Paleocene/Eocene boundary. In M.-P. Aubry, S. G. Lucas, and W. A. Berggren, eds., *Late Paleocene–Early Eocene Climatic and Biotic Events in the Marine and Terrestrial Records*. New York: Columbia University Press, pp. 277–95.

Bukry, D., 1973, Low latitude coccolith biostratigraphic zonation. In *Initial Reports of the Deep Sea Drilling Project*, Vol. 15. Washington, DC: US Govt. Printing Office, pp. 658–77.

1975a, Coccolith and silicoflagellate stratigraphy, northwestern Pacific Ocean. In *Initial Reports of the Deep Sea Drilling Project*, Vol. 32. Washington, DC: US Govt. Printing Office, pp. 677–701.

1975b, New Miocene to Holocene stages in the oceans basins based on calcareous nannoplankton zones. In T. Saito and L. H. Burckle, eds., *Late Neogene Epoch Boundaries, Micropaleontology Press Special Paper*, New York: American Museum of Natural History Micropaleontology Press, pp. 44–60.

Bulman, O. M. B., 1970, Graptolite faunal distribution. In F. A. Middlemiss, P. F. Rawson, and G. Newall, eds., Faunal provinces in space and time, Special Issue. *Geological Journal* **4**: 47–60.

Bulot, L. G., Blanc, E., Thieuloy, J.-P., and Remane, J., 1993a, La limite Berriasien–Valanginien dans le Sud-Est de la France: données biostratigraphiques nouvelles: *Comptes rendus de l'Académie des Sciences, Paris, Série II. Sciences de la Terre et des Planetes* **317**: 387–94.

Bulot, L. G., Thieuloy, J.-P., Blanc, E., and Klein, J., 1993b, Le cadre stratigraphique du Valanginien supérieur et de l'Hauterivien du Sud-Est de la France: définition des biochronozones et caractérisation de nouveaux biohorizons. *Géologie Alpine* **68**: 13–56.

Bulot, L. G., Blanc, E., Company, M., *et al.*, 1996, The Valanginian Stage. *Bulletin de l'Institut Royal des Sciences Naturelles de Belgique, Sciences de la Terre*, Supplément, **66**: 11–18.

Bultynck, P. and Hollevoet, C., 1999, The Eifelian–Giverian boundary and Struve's Middle Devonian Great Cap in the Couvin area (Ardennes, southern Belgium). *Senckenbergiana lethaea* **79**(1): 3–11.

Bultynck, P. and Martin, F., 1995, Assessment of an old stratotype: the Frasnian/Famennian boundary at Senzeilles, Southern Belgium. *Bulletin de l'Institut Royal des Sciences Naturelles de Belgique, Sciences de la Terre* **65**: 5–34.

Bultynck, P., 2000a, Fossil groups important for boundary definition. *Courier Forschungsinstitut Senckenberg* **220**: 1–205.

2000b, Recognition of Devonian series and stage boundaries in geological areas, *Courier Forschungsinstitut Senckenberg* **225**: 1–347.

Burckle, L. H., 1972, Late Cenozoic planktonic diatom zones from the eastern equatorial Pacific. In R. Simonson, ed., *First Symposium on Recent and Fossil Marine Diatoms, Nova Hedwegia Beih.* **39**: 217–46.

Burger, D., 1995, *Jurassic. Timescales Calibration and Development Correlation Charts and Explanatory Notes*, Vol. 8, AGSO Record 1995/37. Canberra: Australian Geological Survey Organization, pp. 30.

Burger, K., Hess, J. C., and Lippolt, H. J., 1997, Tephrochronologie von Kaolin-Kohlentonstein: Mittel zur Korrelation paralischer und limnischer Ablagerungen des Oberkarbons. *Geologische Jahrbücher A* **147**: 3–39.

Burgess, N. D. and Richardson, J. B., 1995, Late Wenlock to Early Pridoli cryptospoles and miospoles from south and southwest Wales, Great Britain. *Palaeontographica* **B236**: 1–44.

Burnett, J. A., Gallagher, L. T., and Hampton, M. J., 1998, Upper Cretaceous. In P. R. Bown, ed., *Calcareous Nannofossil Biostratigraphy, British Micropalaeontological Society Publication Series*, Vol. 6. London: Chapman and Hall, pp. 132–99.

Burr, G. S., Beck, W. J., Taylor, F. W., *et al.*, 1998, A high resolution radiocarbon calibration between 11.7 and 12.4 kyr BP derived from ^{230}Th ages of corals from Espiritu Santo Island, Vanuatu. *Radiocarbon* **40**: 1093–105.

Busnardo, R., 1965, Le Stratotype du Barrémien. 1. Lithologie et macrofaune, et rapport sur l'étage Barrémien. *Mémoires du Bureau de Recherches Géologiques et Minières* **34**: 101–16, 161–9.

Busnardo, R., Thieuloy, J.-P., and Moullade, M., 1979, Hypostratotype Mesogéen de l'etage Valanginien (sud-est de la France). *Les Stratotypes Français*, Vol. 6. Éditions du Centre National de la Recherche Scientifique, p. 143.

Butler, R. F., Krause, D. W., and Gingerich, P. D., 1987, Magnetic polarity stratigraphy and biostratigraphy of middle–late Paleocene continental deposits of south-central Montana. *Journal of Geology* **95**: 647–57.

Butterlin, J., 1988, Origine et évolution des Lépidocyclines de la région caraïbe, comparaison et relations avec les lépidocyclinies des autres régions du monde. *Revue de Paléobiologie* **2**: 623–9.

Bybell, L. M. and Self-Trail, J. M., 1995, *Evolutionary, Biostratigraphic and Taxonomic Study of Calcareous Nannofossils From a Continuous P/E Boundary Section in New Jersey, US Geological Survey Professional Paper*, Vol. 1554. Boulder, CO: US Geological Survey, pp. 1–36.

Bystricka, H., 1963, Die unter-eozänen Coccolithophoriden (Flagellata) des Myjavaer Paleogens. *Geologiske Sbornik (Bratislava)* **14**: 269–81.

1965, Der stratigraphische Wert von Discoasteriden im Palaeogen der Slowakei. *Geologiske Sbornik (Bratislava)* **16**: 7–10.

Byvsheva, T. V. and Umnova, N. I., 1993, Zonal scale on spores. In M. K. Makhlina, M. V. Vdovenko, A. S. Alekseev, *et al.*, eds., *Lower Carboniferous in the Moscow Synecline and Voronezh Antecline*. Moscow: Nauka, pp. 136–41.

Byvsheva, T. V., Owens, B., and Teteriuk, V. K., 1979, Stratigraphical palynology of the Tournaisian to Stephanian deposits of the USSR and western Europe. In S. V. Meyer, V. V. Hennr, Y. A. Leytlinger, A. P. Rotay, and M. N. Solov'yena, eds., *Obshchiye problemy stratigrafii kamennougol'nykh otlozheniy*, *Compte Rendu Congrès International de Stratigraphie et de Géologie du Carbonifère*, Vol. 8. pp. 209–15.

Cacho, I., Grimalt, J. O., Pelejero, C., *et al.*, 1999, Dansgaard–Oeschger and Heinrich event imprints in Alboran Sea paleotemperatures. *Palaeoceanography* **14**(6): 698–705.

Cahuzac, B. and Poignant, A., 1997, Essai de biozonation de l'Oligo–Miocène dans les bassins européens à l'aide des grands foraminifères néritiques. *Bulletin de la Société Géologique de France* **168**: 155–69.

Cahuzac, B., Turpin, L., and Bonhomme, P., 1997, Sr isotope record in the area of the lower Miocene historical stratotypes of the Aquitanian basin. In A. Montanari, G. S. Odin, and R. Coccioni, eds., *Miocene Stratigraphy: An Integrated Approach, Developments in Paleontology and Stratigraphy*, Vol. 15. Amsterdam: Elsevier, pp. 125–47.

Callomon, J. H., 1964, Notes on the Callovian and Oxfordian Stages. In P. L. Maubeuge, ed., *Colloque du Jurassique à Luxembourg 1962*. Luxembourg: Publication de l'Institut Grand-Ducal, Section des Sciences Naturelles, Physiques et Mathématiques, pp. 269–91.

1985, Biostratigraphy, chronostratigraphy and all that – again. In O. Michelsen and A. Zeiss, eds., *International Symposium on Jurassic Stratigraphy*, Vol. 3, Erlangen, 1984. Copenhagen: Geological Survey of Denmark, pp. 611–24.

1995, Time from fossils: S. S. Buckman and Jurassic high-resolution geochronology. In M. J. Le Bas, ed., *Milestones in Geology, Geological Society of London Memoir*, Vol. 16. London: Geological Society, pp. 127–50.

1999, Report of the Bathonian–Callovian Boundary Working Group. *International Subcommission on Jurassic Stratigraphy Newsletter* **26**: 53–60.

Callomon, J. H. and Dietl, G., 2000, On the proposed basal boundary stratotype (GSSP) of the Middle Jurassic Callovian Stage. In R. L. Hall and P. L. Smith, eds., *Advances in Jurassic Research, GeoResearch Forum*, Vol. 6. Switzerland: Transtec Publications, pp. 41–54.

Cameron, A. G. W. and Ward, W. R., 1976, The origin of the Moon. *Lunar Science* **7**: 120–2.

Cande, S. C. and Kent, D. V., 1992a, A new geomagnetic polarity time scale for the Late Cretaceous and Cenozoic. *Journal of Geophysical Research* **97**: 13 917–51.

1992b, Ultrahigh resolution of marine magnetic anomaly profiles: a record of continuous paleointensity variations? *Journal of Geophysical Research* **97**: 15 075–83.

1995, Revised calibration of the geomagnetic polarity timescale for the Late Cretaceous and Cenozoic. *Journal of Geophysical Research* **100**: 6093–5.

Cande, S. C., Larson, R. L., and LaBrecque, J. L., 1978, Magnetic lineations in the Pacific Jurassic Quiet Zone. *Earth and Planetary Science Letters* **41**: 434–40.

Canfield, D. E., 2001, Isotope fractionation by natural populations of sulfate-reducing bacteria. *Geochimica et Cosmochimica Acta* **65**: 1117–24.

Canfield, D. E. and Raiswell, R., 1999, The evolution of the sulfur cycle. *American Journal of Science* **299**: 697–723.

Canfield, D. E. and Teske, A., 1996, Late Proterozoic rise in atmospheric oxygen concentration inferred from phylogenetic and sulphur-isotope studies. *Nature* **382**: 127–32.

Canup, R. M. and Asphaug, E., 2001, Origin of the Moon in a giant impact near the end of the Earth's formation. *Nature* **412**(6848): 708–12.

Caplin, M. L. and Bustin, R. M., 1999, Devonian–Carboniferous Hangenberg mass extinction event, widespread organic-rich mudrock and anoxia: causes and consequences. *Palaeogeography, Palaeoclimatology, Palaeoecology* **148**(4): 187–207.

Caputo, M. V., 1998, Ordovician–Silurian glaciations and global sea-level changes. In E. Landing and M. E. Johnson, eds., *Silurian Cycles: Linkages of Dynamic Stratigraphy with Atmospheric, Oceanic and Tectonic Changes, New York State Museum Bulletin*, Vol. 491. Albany, NY: New York State Museum, pp. 15–25.

Caputo, M. V. and Crowell, J. C., 1985, Migration of glacial centers across Gondwana during Paleozoic Era. *Geological Society of America Bulletin* **96**(8): 1020–36.

Carianfi, N., D'Alessandro, A., Girone, A., *et al.*, 2001, Pleistocene sections in the Montalbano Ionico area: the potential GSSP for Early–Middle Pleistocene in the Lucania Basin (southern Italy). *Memorie di Scienze Geologiche* **53**: 67–83.

Cariou, E. and Hantzperque, P., 1997, Biostratigraphie du Jurassique ouest-européen et méditerranéen: zonations parallèles et distribution des invertébrés et microfossiles. *Bulletin des Centres de Recherches Exploration-Production Elf-Aquitaine, Mémoire 17*. Sociele Nationale Elf-Aquitaine. p. 422.

Carpenter, S. J., Lohmann, K. C., Holden, P., *et al.*, 1991, $\delta^{18}O$ values, $^{87}Sr/^{86}Sr$ and Sr/Mg ratios of Late Devonian abiotic calcite: implications for the composition of ancient seawater. *Geochimica et Cosmochimica Acta* 55: p. 1991–2010.

Carter, C., Trexeler, J. H., and Churlein, M., 1980, Dating of graptolite zones by sedimentation rates: implications for rates of evolution. *Lethaia* **13**: 279–87.

Carter, E. S., 1993, Biochronology and paleontology of uppermost Triassic (Rhaetian) radiolarians, Queen Charlotte Islands, British Columbia, Canada. *Mémoires Géologie* (*Lausanne*) **11**: 175.

Carter, E. S. and Orchard, M. J., 2000, Intercalibrated conodont–radiolarian biostratigraphy and potential datums for the Carnian/Norian boundary within the Upper Triassic Peril Formation, Queen Charlotte Islands. *Current Research 2000-A7*, Geological Survey of Canada.

Carter, E. S., Whalen, P. A., and Guex, J., 1998, Biochronology and paleontology of Lower Jurassic (Hettangian and Sinemurian) radiolarians, Queen Charlotte Islands, British

Columbia. *Geological Survey of Canada Bulletin* **496**: 162.

Carter, J. L., 1990, Subdivision of the Lower Carboniferous in North America by means of articulate brachiopod generic ranges. In P. L. Brenkle, and W. L. Manger, eds., Intercontinental correlation and division of the Carboniferous System: contributions from the Carboniferous Subcommission meeting. *Courier Forschungsinstitut Senckenberg* **130**: 145–55.

Casey, R., 1961, The stratigraphical palaeontology of the Lower Greensand. *Palaeontology* **3**: 487–621.

1996, Lower Greensand ammonites and ammonite zonation. *Proceedings of the Geologists' Association* **107**: 69–76.

Casey, R., Boyliss, H. M., and Simpson, M. E., 1998, Observations on the lithostratigraphy and ammonite succession of the Aptian (Lower Cretaceous) Lower Greensand of Chale Bay, Isle of Wight, UK. *Cretaceous Research* **19**: 511–35.

Castradori, D., 1993, Calcareous nannofossil biostratigraphy and biochronology in eastern Mediterranean deep-sea cores. *Rivista Italiana di Paleontologia e Stratigrafia* **99**(1): 107–26.

Castradori, D., Rio, D., Hilgen, F. J., and Lourens, L. J., 1998, The Global Standard Stratotype-section and Point (GSSP) of the Piacenzian Stage (Middle Pliocene). *Episodes* **21**: 88–93.

Caudri, C. M. B., 1996, The larger foraminifera of Trinidad. *Eclogae Geologicae Helvetica* **89**: 1137–309.

Cavelier, C. and Pomerol, C., 1986, Stratigraphy of the Paleogene. *Bulletin de la Société Géologique de France* **2**(8): 255–65.

Cecca, F., Pallini, G., Erba, E., Premoli-Silva, I., and Coccioni, R., 1994, Hauterivian–Barremian chronostratigraphy based on ammonites, nannofossils, planktonic foraminifera and magnetic chrons from the Mediterranean domain. *Cretaceous Research* **15**:457–67.

Censarek, B. and Gersonde, R., 2002, Miocene diatom biostratigraphy at ODP Sites 689, 690, 1088, 1092 (Atlantic sector of the Southern Ocean). *Marine Micropaleontology* **45**: 309–56.

Cerling, T. E., Harris, J. M., MacFadden, B. J., *et al.*, 1997, Global vegetation changes through the Miocene/Pliocene boundary. *Nature* **389**: 153–8.

Chadwick, G. H., 1930, Subdivision of geologic time (abstract). *Geological Society of America Bulletin* **41**: 47.

Chaisson, W. P. and Pearson, P. N., 1997, Planktonic foraminifer biostratigraphy at Site 925: middle Miocene–Pleistocene. In N. J. Shackleton, W. B. Curry, C. Richter, and T. J. Bralower, eds., *Proceedings of the Ocean Drilling Program, Scientific Results*, Vol. 154. College Station, TX: Ocean Drilling Program, pp. 3–31.

Chambers, J. M., Cleveland, W. S., Kleiner, B., and Tukey, T. A., 1983, *Graphical Methods for Data Analysis*. Pacific Grove, CA: Wadsworth and Belmont.

Chang, W. T. Z. W., 1989, *World Cambrian Biogeography, Chinese Academy of Sciences, Developments in Geoscience*. Beijing: Science Press, pp. 20–220.

Chang, W. T., 1998, Cambrian Biogeography of the Perigondwana Faunal Realm. *Revista Española de Paleontología, no. extr. Homenaje al Prof. Gonzalo Vidal*, pp. 35–49.

Channell, J. E. T. and Grandesso, P., 1987, A revised correlation of magnetozones and calpionellid zones based on data from Italian pelagic limestone sections. *Earth and Planetary Science Letters* **85**: 222–40.

Channell, J. E. T. and Medizza, F., 1981, Upper Cretaceous and Paleogene magnetic stratigraphy and biostratigraphy from the Venetian (southern) Alps. *Earth and Planetary Science Letters* **161**: 201–13.

Channell, J. E. T., Bralower, T. J., and Grandesso, P., 1987, Biostratigraphic correlation of M-sequence polarity chrons M1 to M22 at Capriolo and Xausa (S. Alps, Italy). *Earth and Planetary Science Letters* **85**: 203–21.

Channell, J. E. T., Erba, E., and Lini, A., 1993, Magnetostratigraphic calibration of the Late Valanginian carbon isotope event in pelagic limestones from Northern Italy and Switzerland. *Earth and Planetary Science Letters* **118**: 145–66.

Channell, J. E. T., Cecca, F., and Erba, E., 1995a, Correlations of Hauterivian and Barremian (Early Cretaceous) stage boundaries to polarity chrons. *Earth and Planetary Science Letters* **134**: 125–40.

Channell, J. E. T., Erba, E., Nakanishi, M., and Tamaki, K., 1995b, Late Jurassic–Early Cretaceous time scales and oceanic magnetic anomaly block models. In W. A. Berggren, D. V. Kent, and J. Hardenbol, eds., *Geochronology, Time Scales and Global Stratigraphic Correlations. A Unified Temporal Framework for a Historical Geology. Society of Economic Paleontologists and Mineralogists Special Publication*, Vol. 54. Tulsa, OK: SEMP, pp. 51–63.

Channell, J. E. T., Erba, E., Muttoni, G., and Tremolada, F., 2000, Early Cretaceous magnetic stratigraphy in the APTICORE drill core and adjacent outcrop at Cismon (Southern Alps, Italy), and correlation to the proposed Barremian–Aptian boundary stratotype. *Geological Society of America Bulletin* **112**: 1430–43.

Chappell, J. and Shackleton, N. J., 1986, Oxygen isotopes and sea level change. *Nature* **324**: 137–40.

Chaproniere, C. G. H., 1984, Oligocene and Miocene larger Foraminiferida from Australia and New Zealand. *Bulletin of the Bureau of Mineral Resources of Australia* **188**: 1–98.

Chatterton, B. D. E. and Ludvigsen, R., 1998, Upper Steptoean (Upper Cambrian) trilobites from the McKay Group of southeastern British Columbia, Canada. *Journal of Paleontology, Supplement* **72**(2): 43.

Chen, F. H., Curran, H. A., White, B., and Wasserburg, G. J., 1991, Precise chronology of the last interglacial period. ^{234}U–^{230}Th data from fossil reefs in the Bahamas. *Geological Society of America Bulletin* **103**(1): 82–97.

Chen, F. H., Blomendal, J., Wang, J. M., Li, J. J., and Oldfield, F., 1997, High-resolution multi-proxy climate records from Chinese loess: evidence for rapid climatic changes over the last 75 kyr. *Palaeogeography, Palaeclimatology, Palaeoecology* **130**: 323–35.

Chen Xu and Bergström, S. M., 1995, *The Base of the Austrodentatus Zone as a Level for Global Subdivision of the Ordovician System.* Nanjing: Nanjing University Press, p. 117.

Chen Xu, Mitchell, C. E., Harper, D. A. T., *et al.*, 2000, Late Ordovician to earliest Silurian graptolite and brachiopod biozonation from the Yangtse region South China, with a global correlation. *Geological Magazine* **137**: 623–50.

Chernykh, V. V., 2002, Zonal scale of Kasimovian and Gzhelian stages on the bases of Streptognathodus genus. In B. I. Chuvashov and E. A. Amon, eds., *Stratigraphy and Paleogeography of Carboniferous of Eurasia.* Ekaterinburg, pp. 302–6 [In Russian].

Chernykh, V. V., Ritter, S. M., and Wardlaw, B. R., 1997, *Streptognathodus isolatus* new species (Conodonta): proposed index for the Carboniferous–Permian boundary. *Journal of Paleontology* **71**: 162–4.

Childe, F., 1996, U–Pb geochronology and Nd and Pb isotope characteristics of the Au–Ag-rich Eskay Creek volcanogenic massive sulfide deposit, British Columbia. *Economic Geology* **91**: 1209–24.

Chinese Lower Triassic Working Group, 2002, Recent studies on the Lower Triassic of Chaohu, Anhui Province, China. *Albertiana* **27**: 20–5.

Chizhova, V. A., 1977, Ostracods from Devonian–Carboniferous boundary beds of the Russian Platform. *Transactions of VNIGRI* **49**: 3–255 [In Russian].

Chlupáč, I., 1993, Geology of the Barrandian. *Senckenberg-Buch* **69**: 163.

2000, The global stratotype section and point of the lower Pragian boundary. *Courier Forschungsinstitut Senckenberg* **225**: 9–15.

Chlupáč, I. and Hladil, J., 2000, The global stratotype section and point of the Silurian–Devonian boundary. *Courier Forschungsinstitut Senckenberg* **225**: 1–7.

Chlupáč, I. and Oliver, Jr., W. A., 1989, Decision on the Lochkovian–Pragian Boundary Stratotype (Lower Devonian). *Episodes* **12**: 109–13.

Choubert, G. and Faure-Muret, A., 1976, *Geological World Atlas*. Paris: UNESCO.

Churkin, M. J., Carter, C., and Eberlein, G. D., 1971, Graptolite succession across the Ordovician–Silurian boundary in south-eastern Alaska. *Quarterly Journal of the Geological Society of London* **126**: 319–30.

Churkin, M., Carter, C., and Johnson, B. R., 1977, Subdivision of Ordovician and Silurian time scale using accumulation rates of graptolitic shale. *Geology* **5**: 452–6.

Chuvashov, B. I. and Nairn, A. E. M., 1993, *Permian System: Guides to Geological Excursions in the Uralian Type Localities, Occasional Publications ESRI*, Vol. 10. Earth Sciences and Resources Institute, p. 303.

Chuvashov, B. I., Dupina, I. G. V., Mizens, G. A., and Chernykh, V. V., 1990a, *Key-sections of the Upper Carboniferous and Lower Permian of the western slope of the Urals and Preurals.* Sverdlovsk: Uralian Branch of Russian Academy of Sciences, pp. 3–367, pls. 1–41 [In Russian].

Chuvashov, B. I., Dupina, G. V., Mizens, G. A., and Chernykh, V. V., 1990b, *Report on Upper Carboniferous and Lower Permian Sections, Western Slope of the Urals and Pre-Urals.* Sverdlovsk: Uralian Branch USSR Academy of Science, p. 368.

Chuvashov, B. I., Chernykh, V. V., Davydov, V. I., and Pnev, V. P., 1993a, Kondurovsky Section. *Permian System: Guides to Geological Excursions in the Uralian Type Localities, Occasional Publications ESRI*, Vol. 10. Earth Sciences and Resources Institute, pp. 4–31.

Chuvashov, B. I., Chernykh, V. V., and Mizens, G. A., 1993b, Zonal divisions of the boundary deposits of Carboniferous and Permian in sections of different facies in the south Urals. *Permophiles* **22**: 11–16.

Chuvashov, B. I., Foster, C. B., Mizens, G. A., Roberts, J., and Claoué-Long, J. C., 1996, Radiometric (SHRIMP) dates for some biostratigraphic horizons and event levels from the Russian and Eastern Australian Upper Carboniferous and Permian. *Permophiles* **28**: 29–36.

Chuvashov, B. I., Amon, E. O., Karidrua, M., and Prust, Z. N., 1999, Late Paleozoic radiolarians from the polyfacies formations of Uralian Foreland. *Stratigrafiya, Geologicheskaya Korrelyatsiya* **7**(1): 41–55 [In Russian and English].

Chuvashov, B. I., Chernykh, V. V., and Ivanova, R. M., 2000, Carbonate and pre-flysch formations in the Krasnousol'sky section – new data on lithology and biostratigraphy. In B. I. Chuvashov, ed., *Stratigraphy and Paleontology of the Urals.* pp. 18–24 [In Russian].

Chuvashov, B. I., Chernykh, V. V., and Bogoslovskaya, M. F., 2002, Biostratigraphic characteristic of stage stratotypes of the Permian System. *Stratigraphy and Geological Correlation* **10**(4): 317–33.

Ciaranfi, N., *et al.*, 1997, Preface, the new Pleistocene. In J. van Couvering, ed., *The Pleistocene Boundary and the Beginning of the Quaternary.* Cambridge, Cambridge University Press, pp. ii–xvii.

Cita, M. B., 1973, Pliocene stratigraphy and chronostratigraphy. In W. B. F. Ryan, K. J. Hsu, *et al.*, eds., *Initial Reports of the Deep Sea Drilling Project,* Vol. 13. Washington, DC: US Govt. Printing Office, pp. 1343–79.

1975a, The Miocene/Pliocene boundary: history and definition. In T. Saito and L. Burckle, eds., *Late Neogene Epoch Boundaries, Micropaleontology Special Publication*, Vol. 1. pp. 1–30.

1975b, Studi sul Pliocene e gli strati di passaggio dal Miocene al Pliocene. VII. Planktonic foraminiferal zonation of the Mediterranean Pliocene deep sea record. *Rivista Italiana di Paleontologia e Stratigrafia* **81**: 527–44.

1976, Planktonic foraminiferal biostratigraphy of the Mediterranean Neogene. In Y. Takayanagi and T. Saito, eds., *Progress in Micropaleontology: Selected Papers in Honor of Professor Kiyoshi Asan.* New York: Micropaleontology Press, pp. 47–68.

Cita, M. B. and Castradori, D., 1995, Workshop on marine sections from the Gulf of Taranto (southern Italy) usable as potential stratotypes for the GSSP of the Lower, Middle and Upper Pleistocene (Bari, Italy, September 29 to October 4, 1994). *Il Quaternario* **7**(2): 677–92.

Cita, M. B. and Gartner, S., 1973, Studi sul Pliocene e gli strati di passaggio dal Miocene al Pliocene, IV. The stratotype Zanclean foraminiferal and nannofossil biostratigraphy. *Rivista Italiana di Paleontologia e Stratigrafia* **79**: 503–58.

Cita, M. B. and Premoli-Silva, I., 1960, Pelagic foraminifera from the type Langhian. *Proceedings International Paleontological Union*, Vol. XXII. Nordon: International, Paleontological, Union, pp. 39–50.

Clack, J. A., 2002, *Gaining Ground: The Origin and Evolution of Tetrapods.* Indiana University Press, 381 pp.

Claoué-Long, J. C., Zhang, Z., Ma, G., and Du, S., 1991, The age of the Permian–Triassic boundary. *Earth and Planetary Science Letters* **105**(1–3): 182–90.

Claoué-Long, J. C., Jones, P. J., and Roberts, J., 1993, The age of the Devonian–Carboniferous boundary. *Annales de la Société Géologique de Belgique* **115**: 531–49.

Claoué-Long, J. C., Compston, W., Roberts, J., and Fanning, C. M., 1995, Two Carboniferous ages: a comparison between SHRIMP zircon dating with conventional zircon ages and $^{40}Ar/^{39}Ar$ analysis. In W. A. Berggren, D. V. Kent, M.-P. Aubry, and J. Hardenbol, eds., *Geochronology, Time Scales and Global Stratigraphic Correlations: A Unified Temporal Framework for a Historical Geology, Special Publication, Society of Economic Paleontologists and Mineralogists*, Vol. 54. Tulsa, OK: SEPM, pp. 3–21.

Clarke, L. J. and Jenkyns, H. C., 1999, New oxygen evidence for long-term Cretaceous climatic change in the Southern Hemisphere. *Geology* **27**: 699–702.

Clayton, G., Coquel, R., Doubinger, J., *et al.*, 1977, Carboniferous miospores of western Europe. Rapport de 12 Commission Internationale de Microflore du Paleozoic Working Group on Carboniferous Stratigraphic Palynology. *Mededelingen vans Rijks Geologische Dienst* **29**: 1–11.

Clayton, G., Higgs, K. T., and Keegan, J. B., 1978, The Dinantian palynostratigraphy of the British Isles Series/Source. *Palynology* **2**: 215–16.

Clemens, S. C., Farrell, J. W., and Gromet, L. P., 1993, Synchronous changes in seawater strontium isotope composition and global climate. *Nature* **363**: 607–10.

Clemens, S. C., Gromet, L. P., and Farrell, J. W., 1995. Artefacts in Sr isotope records. *Nature* **373**: 201.

Clemens, W. A., 2002, Evolution of mammalian fauna across the Cretaceous–Tertiary boundary in northeastern Montana and other areas of the Western Interior. *Geological Society of America Special Paper* 361. Boulder, CO: Geological Society of America, pp. 217–45.

Clement, B. M. and Hailwood, E. A., 1991, Magnetostratigraphy of sediments from Sites 701 and 702. In *Proceedings of the Ocean Drilling Program, Scientific Results*, Vol. 114. College Station, TX: Ocean Drilling Program, pp. 359–65.

Cleveland, W. S., 1979, Robust locally weighted regression and smoothing scatterplots. *Journal of the American Statistical Association* **74**: 829–36.

1981, LOWESS: a program for smoothing scatterplots by robust locally weighted regression. *The American Statistician* **35**: 45.

Cleveland, W. S., Grosse, E., and Shyu, W. M., 1992, Local regression models. In J. M. Chambers and T. Hastie, eds., *Statistical Models in S.* Pacific Grove, CA: Wadsworth and Brooks/Cole, pp. 309–76.

Cleveringa, P., Meijer, T., van Leeuwen, R. J. W., *et al.*, 2000, The Eemian stratotype locality at Amersfoort in the central Netherlands: a re-evaluation of old and new data. *Geologie en Mijnbouw/Netherlands Journal of Geosciences* **79**: 197–216.

Cloud, P., 1972, A working model of the primitive Earth. *American Journal of Science* **272**(6): 537–48.

1987, Trends, transitions, and events in Cryptozoic history and their calibration; apropos recommendations by the Subcommission of Precambrian Stratigraphy. *Precambrian Research* **37**(3): 257–65.

1988, *Oasis in Space: Earth History from the Beginning.* New York: W. W. Norton, p. 508.

Clyde, W. C., Zonneveld, J.-P., Stamatakos, J., Gunnell, G., and Bartels, W. S., 1997, Magnetostratigraphy across the Wasatchian/Bridgerian NALMA boundary (Early to Middle Eocene) in the western Green River Basin, Wyoming. *Journal of Geology* **105**: 657–69.

Clyde, W. C., Sheldon, N. D., Koch, P. L., Gunnell, G. F., and Bartels, W. S., 2001, Linking the Wasatchian/Bridgerian boundary to the Cenozoic Climate Optimum: new magnetostratigraphic and isotopic results from South Pass, Wyoming. *Palaeogeography, Palaeclimatology, Palaeoecology* **167**: 175–99.

Cobban, W. A., 1993, Diversity and distribution of Late Cretaceous ammonites, Western Interior, United States. In W. G. E. Caldwell and E. G. Kauffman, eds., *Evolution of the Western Interior Basin, Geological Association of Canada, Special Paper*, Vol. 39. Geological Association of Canada, pp. 435–51.

Coccioni, R., 2003, Cretaceous anoxic events: the Italian record. *Paléocéanographie du Mésozoique en réponse aux forçages de la paléogéographic et du paléoclimat, Séance spécialisée de la Société Géologique de France,* Paris, June 10–11, 2003 [To appear in Coccioni, R., 2004, *Developments in Paleontology and Stratigraphy*, Vol. 20. Amsterdam: Elsevier].

Coccioni, R., Basso, D., Brinkhuis, H., *et al.*, 2000, Marine biotic signals across a late Eocene impact layer at Massignano, Italy. *Terra Nova* **12**: 258–63.

Cocks, L. R. M., 1985, The Ordovician–Silurian boundary. *Episodes* **8**: 98–100.

1989, The Wenlock Series in the type area. In C. H. Holland and M. G. Bassett, eds., *A Global Standard for the Silurian System, Geological Series*, Vol. 9. Cardiff: National Museum of Wales.

2000, Ordovician and Silurian global geography. *Journal of the Geological Society of London* **158**: 197–210.

Cocks, L. R. M., Holland, C. H., Rickards, R. B., and Strachan, I., 1971, A correlation of the Silurian rocks on the British Isles. *Journal of the Geological Society of London* **127**: 103–36.

Cocks, L. R. M., Woodcock, N. H., Rickards, R. B., Temple, J. T., and Lane, P. D., 1984, The Llandovery Series of the type area. *Bulletin of the British Museum of Natural History (Geology)* **38**: 131–82.

Coe, A. L., 1995, A comparison of the Oxfordian successions of Dorset, Oxfordshire, and Yorkshire. In P. D. Taylor, ed., *Field Geology of the British Jurassic.* London: Geological Society, pp. 151–72.

1996, Unconformities within the Portlandian Stage of the Wessex Basin and their sequence stratigraphical significance. In S. P. Hesselbo and D. N. Parkinson, eds., *Sequence Stratigraphy in British Geology, Geological Society Special Publication,* Vol. 103. London: Geological Society of London, pp. 109–43.

Cohee, G. V., Glassner, M. F., and Hedberg, H. D. (eds.), 1978, *Contributions to the Geologic Time Scale, Special Publication*, Vol. 6. Tulsa, OK: American Association of Petroleum Geologists.

Cohen, A. S. and Coe, A. L., 2002, New geochemical evidence for the onset of volcanism in the Central Atlantic magmatic province and environmental change at the Triassic–Jurassic boundary. *Geology* **30**: 267–70.

Cohen, A. S., Coe, A. L., Bartlett, J. M., and Hawkesworth, C. J., 1999, Precise Re–Os ages of organic-rich mudrocks and the Os isotope composition of Jurassic seawater. *Earth and Planetary Science Letters* **167**: 150–73.

Cohen, A. S., Coe, A. L., and Harding, S. M., 2004, Osmium isotope evidence for the regulation of atmospheric CO_2 by continental weathering. *Geology* **32**: 157–60.

Cohen, B. A., Swindle, T. D., and King, D. A., 2000, Support for the lunar cataclysm hypothesis from lunar meteorite impact melt ages. *Science* **290**: 1754–6.

Cojan, I., Moreau, M. G., and Stott, L. E., 2000, Stable carbon isotope stratigraphy of the Paleogene pedogenic series of southern France as basis for marine–continental correlation. *Geology* **28**: 259–63.

Colalongo, M. L., di Grande, A., D'Onofrio, S., *et al.*, 1979, A proposal for the Tortonian–Messinian boundary. *Annales Géologiques des Pays Helléniques, Tome hors série* **1**: 285–94.

Colin, J.-P., 1998, Ostracodes. [Jurassic portion, Mesozoic and Cenozoic sequence chronostratigraphic framework of European basins, in Hardenbol, J. *et al.*]. In P.-C. de Graciansky, J. Hardenbol., Th. Jacquin, and P. R. Vail, eds., *Mesozoic–Cenozoic Sequence Stratigraphy of European Basins, Society of Economic Paleontologists and Mineralogists Special Publication,* Vol. 60. Tulsa, OK: SEPM, pp. 777–88, Chart 7.

Colin, J.-P. and Babinot, J. F., 1998, Ostracodes. [Cretaceous portion, Mesozoic and Cenozoic sequence chronostratigraphic framework of European basins, in Hardenbol, J. *et al.*]. In P.-C. de Graciansky, J. Hardenbol., Th. Jacquin, and P. R. Vail, eds., *Mesozoic–Cenozoic Sequence Stratigraphy of European Basins, Society of Economic Paleontologists and Mineralogists Special Publication,* Vol. 60. Tulsa, OK: SEPM, pp. 772, Chart 5.

Colloque sur la limite Jurassique–Crétacé, 1975, Summary. *Mémoires du Bureau de Recherches Géologiques et Minières* **86**: 386–93.

Combemorel, R., 1998, Belemnites. [column for Jurassic chart, Mesozoic and Cenozoic sequence chronostratigraphic framework of European basins, in Hardenbol, J. *et al.*]. In P.-C. de Graciansky, J. Hardenbol., Th. Jacquin, and P. R. Vail, eds., *Mesozoic–Cenozoic Sequence Stratigraphy of European Basins, Society of Economic Paleontologists and Mineralogists Special Publication,* Vol. 60. Tulsa, OK: SEPM, Chart 7.

Combemorel, R. and Christen, W. K., 1998, Belemnites. [columns for Cretaceous chart and Mesozoic and Cenozoic sequence chronostratigraphic framework of European basins, in Hardenbol, J. *et al.*]. In P.-C. de Graciansky, J. Hardenbol., Th. Jacquin, and P. R. Vail, eds., *Mesozoic–Cenozoic Sequence Stratigraphy of European Basins, Society of Economic Paleontologists and Mineralogists Special Publication,* Vol. 60. Tulsa, OK: SEPM, Chart 5.

Company, M., Sandoval, J., and Tavera, J. M., 1995, Lower Barremian ammonite biostratigraphy in the Subbetic Domain (Betic Cordillera, southern Spain). *Cretaceous Research* **16**: 243–56.

Compston, W., 2000a, Interpretations of SHRIMP and isotope dilution zircon ages for the geological time-scale: I. The early Ordovician and late Cambrian. *Mineralogical Magazine* **64**(1): 43–57.

2000b, Interpretations of SHRIMP and isotope dilution zircon ages for the Palaeozoic time-scale: II. Silurian to Devonian. *Mineralogical Magazine* **64**(6): 1127–46.

Compston, W. and Pidgeon, R. T., 1986, Jack Hills, evidence of more very old detrital zircons in Western Australia. *Nature* **321**(6072): 766–9.

Compston, W. and Williams, I. S., 1992, Ion probe ages for the British Ordovician and Silurian stratotypes. In B. D. Webby and J. R. Laurie, eds., *Global Perspectives on Ordovician Geology*. Rotterdam: Balkema, pp. 59–67.

Compston, W., Williams, I. W., and Meyer, C., 1984, U–Pb geochronology of zircons from lunar breccia 73217 using a sensitive high mass-resolution ion microprobe. *Journal of Geophysical Research* **89**: B525–34.

Compston, W., Kinny, P. D., Williams, I. S., and Foster, J. J., 1986, The age and Pb loss behaviour of zircons from the Isua supracrustal belt as determined by ion microprobe. *Earth and Planetary Science Letters* **80**: 71–81.

Compston, W., Williams, I. S., Kirschvink, J. L., Zhang Zichao, and Ma Guogan, 1992, Zircon U–Pb ages for the Early Cambrian time-scale. *Journal of the Geological Society of London* **149**: 171–84.

Conil, R. and Lys, M., 1970, Données nouvelles sur les Foraminiferes du Tournaisien inférieur et des couches de passage du Famennien au Tournaisien dans l'Avesnois. In *Colloque sur la stratigraphie du Carbonifère Congrès et Colloques de l'Université de Liège*, Vol. 55. Université de Liege, pp. 241–65.

Conil, R., Austin, R. L., Lys, M., and Rhodes, F. H. T., 1969, La limite des etages Tournaisien et Viseen au stratotype de l'assise de Dinant. *Bulletin de la Société Belge de Géologie* **77**(1): 39–52.

Conil, R., Lys, M., Paproth, E., Ramsbottom, W. H. C., and Sevastopulo, G. D., 1984, Synthesis of biostratigraphic data of the classic Dinantian of western Europe. In S. V. Meyen, V. V. Menner, Y. A. Reytlinger, A. P. Rotay, and M. N. Solov'yeva, eds., *Obshchiye problemy stratigrafii kamennougol'nykh otlozheniy, Compte Rendu Congrès International de Stratigraphie et de Géologie du Carbonifère*, Vol. 8. pp. 170–179.

Conil, R., Groessens, E., Laloux, M., Poty, E., and Tourneur, F., 1990, Carboniferous guide foraminifera, corals and conodonts in the Franco-Belgian and Campine Basins: their potential for widespread correlation. In P. L. Brenkle and W. L. Manger, eds., Intercontinental correlation and division of the Carboniferous System: contributions from the Carboniferous Subcommission meeting. *Courier Forschungsinstitut Senckenberg* **130**: 15–58.

Conte, G., 1994, La limite Bédoulien–Gargasien dans la coupe stratotypique de Cassis–La Bédoule (Bouches du Rhône, France). *Géologie Alpine Mémoire, Hors Série* **20**: 321–6.

Conybeare, W. D. and Phillips, W., 1822, *Outlines of the Geology of England and Wales, with an Introduction Compendium of the General Principles of that Science, and Comparative Views of the Structure of Foreign Countries,* Part 1. London: William Phillips, p. 470.

Cooper, R. A., 1992, A relative timescale for the Early Ordovician derived from depositional rates of graptolitic shales. In B. D. Webby and J. R. Laurie, eds., *Global Perspectives on Ordovician Geology.* Rotterdam: Balkema Press, pp. 3–20.

1999a, Ecostratigraphy, zonation and global correlation of earliest planktic graptolites. *Lethaia* **32**: 1–16.

1999b, The Ordovician time scale – calibration of graptolite and conodont zones. *Acta Universitatis Carolinae Geologica* **43**(1/2): 1–4.

Cooper, R. A. and Lindholm, K., 1990, A precise worldwide correlation of Early Ordovician graptolite sequences. *Geological Magazine* **127**: 497–525.

Cooper, R. A. and Nowlan, G. S., 1999, Proposed global stratotype section and point for base of the Ordovician System. *Acta Universitatis Carolinae Geologica* **43**(1/2): 61 4.

Cooper, R. A. and Stewart, I. R., 1979, The Tremadoc graptolite sequence of Lancefield, Victoria. *Palaeontology* **22**: 767–97.

Cooper, R. A., Fortey, R. A., and Lindholm, K., 1991, Latitudinal and depth zonation of early Ordovician graptolites. *Lethaia* **24**: 199–218.

Cooper, R. A., Jenkins, J. F., Compston, W., and Williams, I. S., 1992, Ion-probe zircon dating of a mid-Early Cambrian tuff in South Australia. *Journal of the Geological Society of London* **149**: 185–92.

Cooper, R. A., Nowlan, G. S., and Williams, H. S., 2001, Global stratotype section and point for base of the Ordovician System. *Episodes* **24**(1): 19–28.

Cope, J. C. W., 1967, The paleontology and stratigraphy of the lower part of the Upper Kimmeridge Clay of Dorset. *Bulletin of the British Museum of Natural History (Geology)* **15**: 3–79.

1993, The Bolonian Stage: an old answer to an old problem. *Newsletters on Stratigraphy* **28**: 151–6.

1995, Towards a unified Kimmeridgian Stage. *Petroleum Geoscience* **1**(4): 351–4.

2003, Latest Jurassic stage nomenclature. *International Subcommission on Jurassic Stratigraphy Newsletter* **30**: 27–9.

Cope, J. C. W., Duff, K. L., Parsons, C. F., *et al.*, 1980a, *A Correlation of Jurassic Rocks in the British Isles. Part Two: Middle and Upper Jurassic*, *Geological Society of London Special Report*, Vol. 15. Oxford: Blackwell Scientific, p. 109.

Cope, J. C. W., Getty, T. A., Howarth, M. K., Morton, N., and Torrens, H. S., 1980b, *A Correlation of Jurassic Rocks in the British Isles. Part One: Introduction and Lower Jurassic*, *Geological Society of London Special Report*, Vol. 14. Oxford: Blackwell Scientific, p. 73.

Copestake, P., Johnson, B., Morris, P. H., Coleman, B. E., and Shipp, D. J., 1989, Jurassic. In D. G. Jenkyns and J. W. Murray, eds., *Stratigraphical Atlas of Fossil Foraminifera*. Chichester: Ellis Horwood/The British Micropalaeontological Society, pp. 125–272.

Coquand, H., 1857a, Notice sur la formation crétacée du départment de la Charente. *Bulletin de la Société Géologique de France, série 2* **14**: 55–98.

1857b, Position des Ostrea columba et biauriculata dans le groupe de la craie inférieure. *Bulletin de la Société Géologique de France, série 2* **14**: 745–66.

1861, Sur la convenance d'établir dans le groupe inférieur de la formation crétacée un nouvel étage entre le Néocomien proprement dit (couches à Toxaster complanatus et à Ostrea couloni) et le Néocomien supérieur (étage Urgonien de d'Orbigny). *Mémoires de la Société d'Emulation de Provence* **1**: 127–39.

1871, Sur le Klippenkalk du département du Var et des Alpes-Maritimes. *Bulletin de la Société Géologique de France* **28**: 232–3.

Cornet, B., 1977, The palynology and age of the Newark Supergroup. Unpublished Ph.D. thesis, Pennsylvania State University.

1993, Applications and limitations of palynology in age, climatic, and paleoenvironmental analyses of Triassic sequences in North America. In S. G. Lucas and M. Morales, eds., The nonmarine Triassic. *New Mexico Museum of Natural History and Science Bulletin* **3**: 75–93.

Cornet, B. and Olsen, P. E., 1985, A summary of the biostratigraphy of the Newark Supergroup of Eastern North America with comments on early Mesozoic provinciality. In R. Weber, ed., *III Congreso Latinoamericano de Paleontologia Mexico, Simposio Sobre Floras del Triasico Tardio, su Fitogeografia y Paleoecologia, Memoria*. Mexico City: University National Auton. Mexico Inst. Geol, pp. 67–81.

Corradini, C. and Serpagli, E., 1999, A Silurian conodont zonation from late Llandovery to end Prídolí in Sardinia. In E. Serpagli, ed., *Studies on Conodonts, Proceedings of the Seventh European Conodont Symposium, Bollettino della Societê Paleontologica Italiana,* Vol. **37**. pp. 255–83.

Correia, A., Laskar, J., and Néron de Surgy, O., 2003, Long term evolution of the spin of Venus – I. Theory. *Icarus* **163**: 1–23.

Cortijo, E., Duplessy, J. C., Labeyrie, L., *et al.*, 1994, Cooling in the Norwegian Sea and North Atlantic before the continental ice sheet growth during the Eemian inter-glacial period. *Nature* **372**: 446–9.

Costa, L. L. and Manum, S. B., 1988, Dinoflagellate cysts: the description of the international zonation of the Paleogene (D1–D15) and the Miocene (D16–D20). *Geologisches Jahrbuch* **A100**: 321–30.

Courtillot, V. E. and Renne, P. R., 2003, On the ages of flood basalt events. *Comptes Rendus Géoscience* **35**: 113–40.

Cowie, J. W., 1992, Two decades of research on the Proterozoic–Phanerozoic transition: 1972–1991. *Journal of the Geological Society of London* **149**: 589–92.

Cowie, J. W. and Bassett, M. G., 1989, International Union of Geological Sciences 1989 global stratigraphic chart with geochronometric and magnetostratigraphic calibration, Supplement. *Episodes* **12**(2): 1 sheet.

Cowie, J. W., Ziegler, W., Boucot, A. J., Bassett, M. G., and Remane, J., 1986, Guidelines and statutes of the International Commission of Stratigraphy (ICS). *Courier Forschungsinstitut Senckenberg* **83**: 1–14.

Cox, A. V., 1968, Lengths of geomagnetic polarity reversals. *Journal of Geophysical Research* **73**: 3247–60.

1975, The frequency of geomagnetic reversals and the symmetry of the nondipole field. *Reviews of Geophysics and Space Physics* **13**: 35–51.

Cox, A. V. and Dalrymple, G. B., 1967, Statistical analysis of geomagnetic reversal data and the precision of potassium–argon dating. *Journal of Geophysical Research* **72**(10): 2603–14.

Cox, A., Doell, R. R., and Dalrymple, G. B., 1964, Reversals of Earth's magnetic field. *Science* **144**: 1537–43.

Cox, B. M., 1990, A review of Jurassic chronostratigraphy and age indicators for the UK. In R. F. P. Hardman and J. R. V. Brooks, eds., *Tectonic Events Responsible for Britain's Oil and Gas Reserves, Geological Society Special Publication*, Vol. 55. London: Geological Society pp. 169–190.

Cox, B. M. and Gallois, R. W., 1981, The stratigraphy of the Kimmeridge clay of the Dorset type area and its correlation with some other Kimmeridgian sequences. *Report of the Institute of Geological Sciences*, Vol. 8014. Nottingham: British Geological Survey, 44 pp.

Craig, L. E., Smith, A. G., and Armstrong, R. L., 1989, Calibration of the geologic time scale: Cenozoic and Late Cretaceous glauconite and non glauconite dates compared. *Geology* **17**: p. 830–2.

Crasquin, S., 1985, Zonation par les Ostracodes dans le Mississippien de l'Ouest canadien. *Revue de Paléobiologie* **4**(1): 45–52.

Creaser, R. A., Sannigrahi, P., Chacko, T., and Selby, D., 2002, Further evaluation of the Re–Os geochronometer in organic-rich sedimentary rocks: a test of hydrocarbon maturation effects in the Exshaw Formation, Western Canada Sedimentary Basin. *Geochimica et Cosmochimica Acta* **66**: pp. 3441–52.

Cresta, S., 1999, A proposal for the Global Boundary Stratotype Section and Point (GSSP) of the Aalenian (Middle Jurassic) and the Toarcian/Aalenian boundary. Internal document of the International Commission on Stratigraphy, p. 27.

Cresta, S., Goy, A., Ureta, S., *et al.*, 2001, The global stratotype section and point (GSSP) of the Toarcian–Aalenian boundary (Lower–Middle Jurassic). *Episodes* **24**: 166–75.

Crick, R. E., Ellwood, B. B., Feist, R., *et al.*, 1997, Magnetosusceptibility Event and Cyclostratigraphy (MSEC) of the Eifelian–Givetian GSSP and associated boundary sequences in north Africa and Europe. *Episodes* **20**: 167–74.

Crick, R. E., Ellwood, B. B., Feist, R., El Hassani, A., and Feist, R., 2000, Proposed Magnetostratigraphy Susceptibility Stratotype for the Eifelian–Givetian GSSP (Anti-Atlas Morocco). *Episodes* **23**: 93–101.

Crick, R. E., Ellwood, B. B., El Hassani, A., *et al.*, 2001, Magnetostratigraphy susceptibility of the Pridoli–Lochkovian (Silurian–Devonian) GSSP (Klonk, Czech republic) and a coeval sequence in Anti-Atlas Morocco. *Palaeogeography, Palaeoclimatology, Palaeoecology* **167**: 73–100.

Crick, R. E., Ellwood, B. B., Feist, R., *et al.*, 2002, Magnetostratigraphy susceptibility of the Frasnian/Famennian boundary. *Palaeogeography, Palaeoclimatology, Palaeoecology* **181**: 67–90.

Croft, J. S., 1978, Upper Permian conodonts and other microfossils from the Pinery and Lamar Limestones Members of the Bell Canyon Formation and from the Rustler Formation, West Texas. Unpublished Master's thesis, The Ohio State University, p. 176.

Croll, J., 1875, *Climate and Time in Their Geological Relations.* New York: Appleton.

Crook, K. A. W., 1989, Why the Precambrian time-scale should be chronostratigraphic: a response to recommendations by the Subcommission on Precambrian Stratigraphy. *Precambrian Research* **43**(1–2): 143–50.

Crouch, E. M., Heilmann-Clausen, C., Brinkhuis, H., *et al.*, 2001, Global dinoflagellate event associated with the late Paleocene thermal maximum. *Geology* **29**: 315–18.

Crowhurst, S. J., 2002, Composite isotope sequence. *The Delphi Project*. www.sc.cam.ac.uk/new/v10/research/institutes/godwin/body.html.

Crowley, J. C., Kim, K.-Y., Mengel, J. G., and Short, D. A., 1992, Modeling 100,000-year climate fluctuations in pre-Pleistocene time series. *Science* **255**: 705–707.

Crux, J. A., 1989, Biostratigraphy and palaeogeographical applications of Lower Cretaceous nannofossils from north-western Europe. In J. A. Crux and S. E. van Heck, eds., *Nannofossils and Their Applications*. Chichester: Ellis Horwood, pp. 143–211.

Crux, J., 1991, Calcareous nannoplankton recovered by Leg 114 in the Subantarctic south Atlantic Ocean. In *Proceedings of the Ocean Drilling Program, Scientific Results*, Vol. 114. College Station, TX: Ocean Drilling Program, pp. 155–77.

Cummins, D. I. and Elderfield, H., 1994, The strontium isotopic composition of Brigantian (late Dinantian) seawater. *Chemical Geology* **118**: 255–70.

Curry, D., 1981. Thanetian. In C. Pomerol, ed., *Stratotypes of Paleogene Stages, Mémoire hors série 2 du Bulletin d'Information des Géologues du Bassin de Paris.* pp. 255–65.

Cushman, J. A. and Stainforth, R. M., 1945, *The Foraminifera of the Cipero Marl Formation of Trinidad, British West Indies,* Special publication, Vol. 14. Ithaca, NY: Cushman Laboratory for Foraminiferal Research, pp. 1–75.

Daams, R., Alcalá, L., Alvarez, M. A., *et al.*, 1998, A stratigraphical framework for the Miocene (MN 4–MN 13) continental sediments of Central Spain. *Comptes rendus de l'Académie des Sciences, Paris, Serie II. Sciences de la Terre et des Planètes* **327**: 625–31.

Daams, R., Van der Meulen, A. J., Álvarez Sierra, M., *et al.*, 1999, Stratigraphy and sedimentology of the Aragonian (Early to Middle Miocene) in its type area (North-Central Spain). *Newsletters on Stratigraphy* **37**: 103–39.

Dachroth, W., 1976, Gesteinsmagnetische Marken im Perm Mitteleuropas. *Geologisches Jahrbuch E* **10**: 3–63.

Daily, B. and Jago, J. B., 1975, The trilobite Leiopyge Hawle and Corda and the middle–upper Cambrian boundary. *Palaeontology* **18**(3): 527–50, pls 62–3.

Dalrymple, G. B. and Lanphere, M. L., 1969, *Potassium– Argon Dating*. San Francisco, CA: Freeman.

1971, $^{40}Ar/^{39}Ar$ technique of K–Ar dating: a comparison with the conventional technique. *Earth and Planetary Science Letters* **12**: 300–8.

Dansgaard, W. and Tauber, H., 1969, Glacier oxygen-18 content and Pleistocene ocean temperatures. *Science* **166**: 499–502.

Dansgaard, W., Johnsen, S. J., Clausen, H. B., *et al.*, 1993, Evidence for general instability of past climate from a 250-kyr ice-core record. *Nature* **364**: 218–20.

Dan'shin, V. M., 1947, Geology and mineral resources of the region of the Moscow Basin. *Transactions of Moscow Society of Natural Studies*. Moscow: Moscow Society of Natural Studies, p. 308 [In Russian].

Dashzeveg, D., 1993, Asynchronism of the main mammalian events near the Eocene–Oligocene boundary. *Tertiary Research* **14**: 141–9.

Davidek, K., Landing, E., Bowring, S. A., *et al.*, 1998, New uppermost Cambrian U–Pb date from Avalonian Wales and age of the Cambrian–Ordovician boundary. *Geological Magazine* **135**: 305–9.

Davies, S. M., Branch, N. P., Lowe, J., and Turney, C. S. M., 2002, Towards a European tephrochronological framework for termination 1 and the early Holocene. *Philosophical Transactions of the Royal Society of London, series A* **360**: 767–802.

Davis, D. W., 1982, Optimum linear regression and error estimation applied to U–Pb data. *Canadian Journal of Earth Sciences* **19**(11): 2141–9.

Davydov, V. I., 1986, Upper Carboniferous and Asselian fusulinids of the southern Urals. In B. I. Chuvashov, E. Y. Leven, and V. I. Davydov, eds., *Carboniferous–Permian Boundary Beds of the Urals, Pre-Urals and Central Asia*. Moscow: Nauka Publishing House, pp. 77–103 [In Russian].

1990, Clarification of the origin and phylogeny of triticitids and of the Middle/Upper Carboniferous boundary. *Paleontological Journal* **24**(2): 13–25 [In Russian]. *Scripta Technica* 1991: 39–51 [In English].

1992, Subdivision and correlation of Upper Carboniferous and Lower Permian deposits in Donets Basin according to fusulinid data. *Soviet Geology* **5**: 53–61 [In Russian].

1996, Fusulinid biostratigraphy and correlation of Moscovian–Guadalupian North American, Tethyan and Boreal (Russian Platform/Uralian) standards. *Permophiles* **29**: 47–52.

1997, Middle/Upper Carboniferous Boundary: the problem of definition and correlation. In M. Podemski, J. S. Dybova, K. Jaworowski, J. Jureczka, and R. Wagner, eds., *Proceedings of the XIII International Congress on the Carboniferous and Permian, Warszawa*, Vol. 157, Part 1: *Prace Panstwowego Instytutu Geologicznego*. Warsaw: Wydawnictwa Geologiczne, pp. 114–122.

1999, Still contradictions: Moscovian–Kasimovian boundary problems. *Newsletter on Carboniferous Stratigraphy* **17**: 18–22.

2001, The terminal stage of the Carboniferous: Orenburgian versus Bursumian. *Newsletter on Carboniferous Stratigraphy* **19**: 58–64.

2002, Carboniferous system and the current status of its subdivisions. In B. I. Chuvashov and E. A. Amon, eds., *Stratigraphy and Paleogeography of Carboniferous of Eurasia*. Ekaterinburg: Uralina Branch of the Russian Academy of Sciences, pp. 92–111 [In English], pp. 72–91 [In Russian].

Davydov, V. I. and Khramov, A. N., 1991, Paleomagnetism of Upper Carboniferous and Lower Permian in the Karachatyr region, southern France, and the problems of correlation of the Kiama hyperzone. In A. N. Kramov, ed., *Paleomagnetism and Paleogeodynamics of the Territory of USSR*. Leningrad: Russian Federation, VNIGRI, pp. 45–53 [In Russian].

Davydov, V. I. and Krainer, K., 1999, Fusulinid assemblages and facies of the Bombaso Formation and basal Meledis Formation (Moscovian–Kasimovian) in the Central Carnic Alps (Austria/Italy). *Facies* **40**: 157–96.

Davydov, V. I. and Nilsson, I., 1999, Fusulinids in the Middle–Upper Carboniferous Boundary Beds on Spitsbergen, Arctic Norway. *Paleontologia Electronica* **2**(1):www-odp.tamu.edu/paleo/index.html.

Davydov, V. I. and Popov, A. V., 1986, Upper Carboniferous and Lower Permian sections of the southern Urals. In B. I. Chuvashov, E. Y. Leven, and V. I. Davydov, eds., *Carboniferous–Permian Boundary Beds of the Urals, Pre-Urals and Central Asia*. Moscow: Nauka, pp. 29–33 [In Russian].

Davydov, V. I., Barskov, I. S., Bogoslovskaya, M. F., *et al.*, 1992, The Carboniferous/Permian boundary in the former USSR and its correlation. *International Geology Review* **34**(90): 889–906.

Davydov, V. I., Glenister, B. F., Spinosa, C., *et al.*, 1995, Proposal of Aidaralash as GSSP for base of the Permian System. *Permophiles* **26**: 1–8.

1998, Proposal of Aidaralash as Global Stratotype Section and Point (GSSP) for base of the Permian System. *Episodes* **21**(1): 11–18.

Davydov, V. I., Anisimov, R. M., and Nilsson, I., 1999a, Upper Paleozoic sequence stratigraphy of the Arctic Region implied from graphic correlation: Spitsbergen and the Barents Sea shelf. In H. Yin and J. Tong, eds., *Pangea and the Paleozoic–Mesozoic Transition*. Hubei: China University Geoscience Press, pp. 94–7.

Davydov, V. I., Snyder, W. W., and Schiappa, T. A., 1999b, Moscovian–Kasimovian transition in Nevada and problems of its intercontinental correlation. *Program with Abstracts XIV ICCP*. Calgary: University of Canada, p. 27.

Davydov, V. I., Chernykh, V. V., Chuvashov, B. I., Northrup, C. J., and Snyder, W. S., 2002, Volcanic ashes in the upper Paleozoic of the southern Urals: new perspectives in the Pennsylvanian time scale calibration. In B. I. Chuvashov and E. A. Amon, eds., *Stratigraphy and Paleogeography of Carboniferous of Eurasia*. Ekaterinburg; pp. 112–22 [In Russian, with extended abstract in English, p. 123].

Dazé, A., Lee, J. K. W., and Villeneuve, M. E., 2003, An intercalibration study of the Fish Canyon sanidine and biotite $^{40}Ar/^{39}Ar$ standards and some comments on the age of the Fish Canyon Tuff. *Chemical Geology* **199**: 111–27.

Dean, W. T., Donovan, D. T., and Howarth, M. K., 1961, The Liassic ammonite Zones and Subzones of the North West European Province. *Bulletin of the British Museum of Natural History (Geology)* **4**: 435–505.

de Beaulieu, J. L. and Reille, M., 1995, Pollen records from the Velay craters: a review and correlation of the Holsteinian Interglacial with isotopic stage 11. *Mededelingen van de Rijks Geologische Dienst* **52**: 59–70.

Debrenne, F., 1992, Diversification of Archaeocyatha. In J. H. Lipps and P. W. Signor, eds., *Origin and Early Evolution of the Metazoa*. New York: Plenum Press, pp. 425–43.

Debrenne, F. and Rozanov, A. Y., 1983, Paleogeographic and stratigraphic distribution of regular Archaeocyatha (Lower Cambrian fossils). *Géobios* **16**(6): 727–36.

De Bruijn, H., Daams, R., Daxner-Höck, G., *et al.*, 1992, Report of the RCMNS working group on fossil mammals, Reisensburg 1990. *Newsletters on Stratigraphy* **26**: 65–118.

De Coninck, J., 1990, Ypresian organic-walled phytoplankton in the Belgian Basin and adjacent areas. *Bulletin de la Société belge de Géologie* **97**: 287–319.

Deflandre, G. and Fert, C., 1954, Observations sur les coccolithophoridés actuels et fossiles en microscopie ordinaire et éléctronique. *Annales de Paléontologie* **40**: 115–76.

de Graciansky, P.-C., Hardenbol, J., Jacquin, Th., and Vail, P. R., 1998, *Mesozoic–Cenozoic Sequence Stratigraphy of European Basins, Society of Economic Paleontologists and Mineralogists Special Publication,* Vol. 60. Tulsa, OK: SEPM, 786 pp.

de Grossouvre, A., 1901, Recherches sur la Craie supérieure. 1: Stratigraphy générale. *Mémoires pour servir à l'explication de la carte géologique détaillée de la France.* p. 1013.

de Jong, J., 1988, Climatic variability during the past three million years, as indicated by vegetational evolution in northwest Europe and with emphasis on data from The Netherlands. *Philosophical Transactions of the Royal Society of London B* 318: 603–17.

de Kaenel, E., Bergen, J. A., and von Salis Perch-Nielsen, K., 1996, Jurassic calcareous nannofossil biostratigraphy of western Europe. Compilation of recent studies and calibration of bioevents. *Bulletin de la Société géologique de France* **167**: 15–28.

de Lapparent, A., 1900, *Traité de Géologie Paris*, p. fasc. 2, pp. 591–1237.

Delepine, G., 1911, *Recherches sur le Calcaire Carbonifère de Belgique.* Lille: Université de Lille, pp. 1–419.

de Lumley, H., 1976, La Préhistoire Française, Vol. 1. Paris: Editions Centre National de la Recherche Scientifique, pp. 5–23.

Demarcq, G. and Carbonnel, G., 1975, Burdigalian (stratotype rhodanian). In F. F. Steininger and L. A. Nevesskaya, eds., *Stratotypes of Mediterranean Neogene Stages* Vol. 2. Bratislava: CMNS, pp. 51–6.

DeMenocal, P., Archer, D., and Leth, P., 1997, Pleistocene variations in deep Atlantic circulation and calcite burial between 1.2 and 0.6 Ma: a combined data–model approach. In *Proceedings of the Ocean Drilling Program, Scientific Results*, Vol. 154. College Station, TX: Ocean Drilling Program, pp. 285–98.

Denison, R. E., Koepnick, R. B., Fletcher, A., Dahl, D. A., and Baker, M. C., 1993, Reevaluation of early Oligocene, Eocene and Paleocene seawater strontium isotope ratios using outcrop samples from the U.S. Gulf Coast. *Paleoceanography* **8**: 101–26.

Denison, R. E., Koepnick, R. B., Burke, W. H., Hetherington, E. A., and Fletcher, A., 1994, Construction of the Mississippian, Pennsylvanian and Permian seawater $^{87}Sr/^{86}Sr$ curve. *Chemical Geology* **112**: 145–67.

1997, Construction of the Silurian and Devonian seawater $^{87}Sr/^{86}Sr$ curve. *Chemical Geology* **140**: 109–21.

Denison, R. E., Koepnick, R. B., Burke, W. H., and Hetherington, E. A., 1998, Construction of the Cambrian and Ordovician seawater $^{87}Sr/^{86}Sr$ curve. *Chemical Geology* **152**: 325–40.

Denison, R. E., Miller, N. R., Scott, R. W., and Reaser, D. F., 2003, Strontium isotope stratigraphy of the Comanchean Series in north Texas and southern Oklahoma. *Geological Society of America Bulletin* **115**: 669–682.

DePaolo, D. J. and Finger, K. L., 1991, High-resolution strontium-isotope stratigraphy and biostratigraphy of the Miocene Monterey Formation, central California. *Geological Society of America Bulletin* **103**: 112–24.

DePaolo, D. J. and Ingram, B., 1985, High-resolution stratigraphy with strontium isotopes. *Science* **227**: 938–41.

Depéret, M., 1892, Note sur la classification et le parallélisme du Système Miocène: *Comptes Rendus Sommaires de la Société Géologique de France, serie 3* **20**(13): 145–56.

Derry, L. A., Brasier, M. D., Corfield, R. M., Rozanov, A. Y., and Zhuravlev, A. Y., 1994, Sr and C isotopes in Lower Cambrian carbonates from the Siberian craton: a paleoenvironmental record during the "Cambrian explosion." *Earth and Planetary Science Letters* **128**: 671–81.

DesMarais, D. J., Strauss, H., Summons, R. E., and Hayes, J. M., 1992, Carbon isotope evidence for the stepwise oxidation of the Proterozoic environment. *Nature* **359**: 605–9.

Desnoyers, J., 1829, Observations sur un ensemble de dépôts marins plus récents que les terrains tertiaires du Bassin de la Seine et constituant une formation géologique distincte: précédées d'un aperçu de la non-simultanéité des bassins tertiaires. *Annales scientifiques naturelles* **16**: 171–214, 402–19.

Desor, E., 1847, Sur le terrain danien, nouvel étage de la Craie. *Bulletin de la Société Géologique de France, serie 2* **4**: 179–82.

1854, Quelques mots sur l'étage inférieur du groupe néocomien (étage Valanginien). *Bulletin de la Société des Sciences Naturelles de Neuchâtel* **3**: 172–80.

Desor, E. and Gressly, A., 1859, Etudes géologiques sur le Jura neuchâtelois. *Bulletin de la Société des Sciences Naturelles de Neuchâtel* **4**: 1–159.

Destombes, P. and Destombes, J. P., 1965, Distribution zonale des ammonites dans l'Albien du bassin de Paris. *Mémoire Bureau de Recherches Géologiques et Minières*, Colloque sur le Crétacé Inférieur, 1963 **34**: 255–70.

Detmers, J., Brüchert, V., Habicht, K. S., and Kuever, J., 2001, Diversity of sulfur isotope fractionations by sulfate-reducing prokaryotes. *Applied Environmental Microbiology* **67**: 888–94.

de Verteuil, L. and Norris, G., 1996, Miocene dinoflagellate stratigraphy and systematics of Maryland and Virginia. *Micropaleontology* **42** (Supplement): 172.

Devuyst, F.-X., Hance, L., Hou, H., *et al.*, 2003, A proposed Global Stratotype Section and Point for the base of the Visean Stage (Carboniferous): the Pengchong section, Guangxi, South China. *Episodes* **26**(2): 105–15.

De Wever, P., 1982, *Radiolaires du Trias et du Lias de la Téthys, Systématique, Stratigraphie, Societé Géologique du Nord*, Vol. 7. Lille: Université de Lille, p. 599.

1998, Radiolarians. [for Cretaceous and Jurassic portions of Mesozoic and Cenozoic sequence chronostratigraphic framework of European basins, in Hardenbol, J. *et al.*]. In P.-C. de Graciansky, J. Hardenbol., Th. Jacquin, and P. R. Vail, eds., *Mesozoic–Cenozoic Sequence Stratigraphy of European Basins, Society of Economic Paleontologists and Mineralogists*

Special Publication, Vol. 60. Tulsa, OK: SEPM, p. 778, Charts 5 and 7.

De Zanche, V., Gianolla, P., Mietto, P., Siorpaes, C., and Vail, P. R., 1993, Triassic sequence stratigraphy in the Dolomites. *Memorie di Scienze Geologische* **45**: 1–27.

D'Hondt, A. V., 1998, Inoceramids. [column for Cretaceous chart, Mesozoic and Cenozoic sequence chronostratigraphic framework of European basins, in Hardenbol, J. *et al.*]. In P.-C. de Graciansky, J. Hardenbol., Th. Jacquin, and P. R. Vail, eds., *Mesozoic–Cenozoic Sequence Stratigraphy of European Basins, Society of Economic Paleontologists and Mineralogists Special Publication,* Vol. 60. Tulsa, OK: SEPM, Chart 5.

Dickens, G. R., Castillo, M. M., and Walter, J. C. G., 1997, A blast of gas in the latest Paleocene: simulating first-order effects of massive dissociation of oceanic methane hydrate. *Geology* **25**: 259–62.

Dickey, J. O., Bender, P. L., Faller, J. E., *et al.*, 1994, 1994 Lunar laser ranging: a continuing legacy of the Apollo program. *Science* **265**: 182–90.

Dickin, A. P., 1997, *Radiogenic Isotope Geology*. Cambridge: Cambridge University Press.

Diener, A., Ebneth, S., Veizer, J., and Buhl, D., 1996, Strontium isotope stratigraphy of the Middle Devonian: brachiopods and conodonts. *Geochimica et Cosmochimica Acta* **60**: 639–52.

Diener, C., 1912, The Trias of the Himalayas. *Geological Survey of India Memoirs* **36**(3): 159.

Dietl, G., 1981, Zur systematischen Stellung von *Ammonites subfurcatus* ZIETEN und dessen Bedeutung für die *subfurcatum*-Zone (Bajocium, Mittlerer Jura). *Stuttgarter Beiträge zur Naturkunde, Serie B (Geologie und Paläontologie)* **81**: 1–11.

Dietl, G., 1994, Der hochstetteri-Horizont – ein Ammonitenfaunen-Horizont (Discus-Zone, Ober-Bathonium, Dogger) aus dem Schwäbischen Jura. *Stuttgarter Beiträge zur Naturkunde, Serie B (Geologie und Paläontologie)* **202**: 1–39.

Dietl, G. and Callomon, J. H., 1988, Der Orbis-Oolith (Ober-Bathonium, Mittl. Jura) von Sengenthal/Opf. Fränk. Alb, und seine Bedeutung für Korrelation und Gliederung der Orbis-Zone. *Stuttgarter Beiträge zur Naturkunde, Serie B (Geologie und Paläontologie)* **142**: 1–31.

Dietl, G. and Etzold, A., 1977, The Aalenian at the type locality. *Stuttgarter Beiträge zur Naturkunde, Serie B (Geologie und Paläontologie)* **30**: 13.

DiVenere, V. J. and Opdyke, N. D., 1991a, Magnetic polarity stratigraphy and Carboniferous paleopole position from the Joggins Section, Cumberland Basin, Nova Scotia. *Journal of Geophysical Research* **96**: 4051–64.

1991b, Magnetic polarity stratigraphy in the uppermost Mississippian Mauch Chunk Formation, Pottsville, Pennsylvania Series/Source. *Geology* **19**(2): 127–30.

Doderlein, P., 1870–1872, Note illustrative della Carta Geologica del Modenese e del Reggiano 1870. Atti Regia Accademie Scienze, Lettere, Arti Modena, tomo IX-1870 e segg., 1–114 (Mem. I e II) (1870, Tip. erede Soliani Modena) e 1–74 (Mem. III) (1872 L. Gaddi Modena).

Dodson, M. H., 1973, Closure temperature in cooling geochronological and petrological systems. *Contributions to Mineralogy and Petrology* **40**: 259–74.

Dollfus, G. F., 1880, Essai sur l'étendue des terrains tertiaires dans le bassin anglo-parisien. *Bulletin de la Société Géologique de Normandie*, Compte rendu de l'éxposition de 1887.

1909, Essai sur l'étage Aquitanien. *Bull. Serv. Carte Géol. Fr., Paris, 19*, **124**: 379–495.

d'Omalius d'Halloy, J. G. J., 1822, Observations sur un essai de carte géologique de la France, des Pays-Bas, et des contrées voisines. *Annales des Mines* **7**: 353–76.

1868, *Précis élémentaire de Géologie*. Paris: Bruxelles-Paris, p. 470.

Dommergues, J.-L. and Meister, C., 1992, Late Sinemurian and Early Carixian ammonites in Europe with cladistic analysis of sutural characters. *Neues Jahrbuch für Geologie und Paläontologie Abhandlungen* **185**: 211–37.

Dong, X. and Bergström, S. M., 2001, Middle and Upper Cambrian protoconodonts and paraconodonts from Hunan, South China. *Palaeontology* **44**: 949–85; pls 1–6.

Donnelly, T. H., Shergold, J. H., and Southgate, P. N., 1988, Anomalous geochemical signals from phosphatic Middle Cambrian rocks in the southern Georgina Basin, Australia. *Sedimentology* **35**: 549–70.

d'Onofrio, S. 1964, I Foraminiferi del neostratotypo del Messiniano. *Giornale di Geologia* **32**: 400–61.

d'Orbigny, A., 1840, *Paléontologie francaise. Terrains crétacés. I. Céphalopodes*. Paris: published by author, p. 622.

1842, *Paléontologie francaise. Terrains crétacés.* Vol. 2, *Gastropodes*. Paris: Editions Masson, p. 456.

1842–1851, Description zoologique et géologique de tous les animaux mollusques et rayonnés fossiles de France comprenant leur application à la reconnaissance des couches. *Terrains oolitiques ou Jurassiques*, Vol. I, *Céphalopodes, Paléontologie Française*. Paris: Editions Masson, 642 pp, 234 pl. (published in segments through 1851).

1847, *Paléontologie francaise. Terrains crétacés*. Vol. 4, *Brachiopodes.* Paris: Editions Masson, p. 390.

1852, *Cours élémentaire de paléontologie et de géologie stratigraphique*, Vol. 2. Paris: Editions Masson, pp. 383–847.

Douglas, R. G. and Savin, S. M., 1975, Oxygen and carbon isotope analyses of Tertiary and Cretaceous microfossils from Shatsky Rise and other sites in the North Pacific Ocean. In *Initial Reports of the Deep Sea Drilling Project*, Vol. 32. Washington, DC: US Govt. Printing Office, pp. 509–20.

Doyle, E. N., Höey, A. N., and Harper, D. A. T., 1991, The rhynchonellide brachiopod *Eocoelia* from the upper

Llandovery of Ireland and Scotland. *Palaeontology* **34**: 439–54.

Driever, B. W. M., 1988, Calcareous nannofossil biostratigraphy and paleoenvironmental interpretation of the Mediterranean Pliocence. *Utrecht Micropaleontological Bulletin* **36**: 245.

Drooger, C. W., 1960, Miogypsina in Northwester Germany. *Proceedings of the Koninklijke Nederlandse Akademie van Wetenschappen, Series B* **63**: 38–50.

Drooger, C. W., Meulenkamp, J. E., Schmidt, R. R., and Zachariasse, W. J., 1976, The Paleogene–Neogene boundary. *Koninklijke Nederlandse Akademie van Wetenschappen, Amsterdam, Series B* **79**(5): 317–29.

Druce, E. C. and Jones, P. J., 1971, Cambro–Ordovician conodonts from the Burke River Structural Belt, Queensland. *Bulletin of the Bureau of Mineral Resources of Australia* **110**: 158, 20 pls.

Dumont, A., 1849, Rapport sur la carte géologique du Royaume. *Bulletin de l'Académie royale des Sciences et des Lettres et des Beaux-Arts de Belgique* **16**(11): 351–73.

1851, Sur la position géologique de l'argile rupelienne et sur le synchronisme des formations tertiaires de la Belgique, de l'Angleterre et du Nord de la France. *Bulletin de l'Académie royale des Sciences et des Lettres de la Belgique* **18**: 179–95.

Dunbar, C. O., 1940, The type Permian; its classification and correlation. *Bulletin of the American Association of Petroleum Geologists* **24**(2): 237–81.

Dunn, D. L., 1970, Conodont zonation near the Mississippian–Pennsylvanian boundary in western United States. *Geological Society of America Bulletin* **81**(10): 2959–74.

Dunning, G. R. and Hodych, J. P., 1990, U/Pb zircon and baddeleyite ages for the Palisades and Gettysburg sills of the northeastern United States: implications for the age of the Triassic/Jurassic boundary. *Geology* **18**: 795–8.

Dunning, G. R. and Krogh, T. E., 1991, Stratigraphic correlation of the Appalachian Ordovician using advanced U–Pb zircon geochronology techniques. In *Geological Survey of Canada Paper* 90–9. Ottawa: Geological Survey of Canada, pp. 85–92.

Dyer, R., 1998, North Sea Central and Viking Graben radiolarians. [column for Jurassic chart, Mesozoic and Cenozoic sequence chronostratigraphic framework of European basins, in Hardenbol, J. *et al.*]. In P.-C. de Graciansky, J. Hardenbol., Th. Jacquin, and P. R. Vail, eds., *Mesozoic–Cenozoic Sequence Stratigraphy of European Basins, Society of Economic Paleontologists and Mineralogists Special Publication,* Vol. 60. Tulsa, OK: SEPM, Chart 7.

Dyer, R. and Copestake, P., 1989, A review of Late Jurassic to earliest Cretaceous radiolaria and their biostratigraphic potential to petroleum exploration in the North Sea. In D. J. Batten and M. C. Keen, eds., *Northwest European Micropalaeontology and Palynology*. Chichester: Ellis Horwood/British Micropalaeontological Society, pp. 214–35.

Eardley, A. J., Shuey, R. T., Gvodetsky, V., *et al.*, 1973, Lake cycles in the Bonneville Basin, Utah. *Geological Society of America Bulletin* **84**: 211–15.

Ebneth, S., Shields, G. A., Veizer, J., Miller, J. F., and Shergold, J. H., 2001, High resolution strontium isotope stratigraphy across the Cambrian–Ordovician transition. *Geochimica et Cosmochimica Acta* **65**(14): 2273–92.

Eddy, J. D. and McCollum, L. B., 1998, Early Middle Cambrian Albertella Biozone trilobites of the Pioche Shale, southeastern Nevada. *Journal of Paleontology* **72**(5): 864–87.

Edwards, D., Fairon-Demaret, M., and Berry, C. M., 2000, Plant megafossils in Devonian stratigraphy: a progress report. *Courier Forschungsinstitut Senckenberg* **220**: 25–37.

Edwards, L. E., 1984, Insights on why graphic correlation (Shaw's method) works. *Journal of Geology* **92**: 583–97.

Edwards, R. L., Beck, J. W., Burr, G. S., *et al.*, 1993, A large drop in atmospheric $^{14}C/^{12}C$ and reduced melting in the Younger Dryas, documented with ^{230}Th ages in corals. *Science* **260**: 962–7.

Edwards, R. L., Cheng, H., Murrell, M. T., and Goldstein, S. J., 1997, Protactinium-231 dating of carbonates by thermal ionozation mass spectrometry: implications for Quaternary climate change. *Science* **276**: 782–86.

Eerola, T. T., 2001, Climate change at the Neoproterozoic–Cambrian Transition. In A. Yu. Zhuravlev and R. Riding, eds., *The Ecology of the Cambrian Radiation*. New York: Columbia University Press, pp. 90–106.

Ehrenberg, S. N., Nielsen, E. B., Svana, T. A., and Stemmerik, L., 1998, Depositional evolution of the Finnmark carbonate platform, Barents Sea: results from wells 7128/6-1 and 7128/4-1. *Norsk Geologisk Tidsskrift* **78**: 185–224.

Einsele, G., Ricken, W., and Seilacher, A., 1991, *Cycles and Events in Stratigraphy*. Berlin: Springer-Verlag.

Elles, G. L., 1925, The characteristic assemblages of the graptolite zones of the British Isles. *Geological Magazine* **62**: 337–47.

Elles, G. L. and Wood, E. M. R. 1901–1918, *A Monograph of British Graptolites*. London: Palaeontographical Society, p. clxxi + 539 pp., 52 plates.

Ellwood, B. B., Crick, R. E., El Hassani, A., Benoist, S., and Young, R., 2000, Magnetosusceptibility event and cyclostratigraphy (MSEC) in marine rocks and the question of detrital input versus carbonate productivity. *Geology* **28**: 1135–8.

Ellwood, B. B., Crick, R. E., Garcia-Alcalde Fernandez, J. L., *et al.*, 2001, Global correlation using magnetic susceptibility data from Lower Devonian rocks. *Geology* **29**: 583–6.

Elmi, S., 1999, Report of the Pliensbachian–Toarcian Boundary Working Group. *International Subcommission on Jurassic Stratigraphy Newsletter* **26**: 43–6.

2003, Toacian working group. *International Subcommission on Jurassic Stratigraphy Newsletter* **30**: 14.

Elrick, M., 1995, Cyclostratigraphy of Middle Devonian carbonates in the eastern Great Basin. *Journal of Sedimentary Research* **B65**: 61–79.

Embry, A. F., 1988, Triassic sea-level changes: evidence from the Canadian Arctic Archipelago. In C. Wilgus, ed., *Sea Level Changes – An Integrated Approach. Society of Economic Paleonologists and Mineralogists Special Publication*, Vol. 42. Tulsa, OK: SEPM, pp. 249–59.

1997, Global sequence boundaries of the Triassic and their identification in the Western Canada Sedimentary Basin. *Canadian Society of Petroleum Geologists Bulletin* **45**: 415–533.

Emeis, K.-C., Sakamoto, T., Wehausen, R., and Brumsack, H.-J., 2000, The sapropel record of the eastern Mediterranean Sea – results of Ocean Drilling Program Leg 160. *Palaeogeography, Palaeoclimatology, Palaeoecology* **158**(3–4): 371–95.

Emiliani, C., 1955, Pleistocene temperatures. *Journal of Geology* **63**: 538–78.

1966a, Isotopic paleotemperatures. *Science* **154**: 851–7.

1966b, Palaeotemperature analysis of Caribbean cores P6304-8 and P6304-9 and a generalized temperature curve for the past 425 000 years. *Journal of Geology* **74**: 109–26.

Encarnación, J., Rowell, A. J., and Gronow, A. M., 1999, A U–Pb age for the Cambrian Taylor Formation, Antarctica: implications for the Cambrian time scale. *Journal of Geology* **107**: 497–504.

Endt, P. M., 1990, Energy levels of A = 21–44 (VII) nucleii. *Nuclear Physics A* **521**: 1–830.

Endt, P. M. and van der Leun, C., 1973, Energy levels of A = 21–44 (V) nuclei. *Nuclear Physics A* **214**: 1–625.

England, G. L., Rasmussen, B., McNaughton, N. J., and Fletcher, I. R., 2001, SHRIMP U–Pb ages of diagenetic and hydrothermal xenotime from the Archean Witwatersrand Supergroup of South Africa. *Terra Nova* **13**: 360–7.

Enos, P. and Samankassou, E., 1998, Lofer cyclothems revisited (Late Triassic, Northern Alps, Austria). *Facies* **38**: 207–28.

Erba, E., 1994, Nannofossils and superplumes: the Early Aptian nannoconid crisis. *Paleoceanography* **9**: 483–501.

Erba, E., with contributions from Avram, R., Barboschkin, E., Bergen, E. J., *et al.*, 1996, The Aptian Stage. *Bulletin de l'Institut Royal des Sciences Naturelles de Belgique, Sciences de la Terre*, Supplément **66**: 31–43.

Erba, E., Channell, J. E. T., Claps, M., *et al.*, 1999, Integrated stratigraphy of the Cismon Apticore (Southern Alps, Italy): a "reference section" for the Barremian–Aptian interval at low latitudes. *Journal of Foraminiferal Research* **29**: 371–91.

Erbacher, J., and Thurow, J., 1997, Influence of oceanic anoxic events on the evolution of Mid-Cretaceous Radiolaria in the North Atlantic and western Tethys. *Marine Micropaleontology* **30**:139–58.

Erbacher, J., Thurow, J., and Littke, R., 1996, Evolution patterns of radiolaria and organic matter variations: a new approach to identify sea-level changes in mid-Cretaceous pelagic environments. *Geology* **24**: 499–502.

Erwin, D. H., 1993, *The Great Paleozoic Crisis: Life and Death in the Permian*. New York: Columbia University Press, p. 327.

1995, The end-Permian mass extinction. In P. A. Scholle, T. M. Peryt, and D. S. Ulmer-Scholle, eds., *The Permian of Northern Pangea*, Vol. I, *Paleogeography, Paleoclimates, Stratigraphy*. Berlin: Springer-Verlag, pp. 20–34.

Erwin, D. H., Bowring, S. A., and Yugan, J., 2002, End-Permian mass extinctions: a review. In C. Koeberl and K. G. MacLeod, eds., *Catastrophic Events and Mass Extinctions: Impacts and Beyond*, *Geological Society of America Special Paper*, Vol. 356. Boulder, CO: Geological Society of America, pp. 363–83.

Esaulova, N. K., Lozovsky, V. R., and Rozanov, A. Y., 1998, Stratotypes and reference sections of the Upper Permian in the regions of the Volga and Kama Rivers. In *International Symposium on the Upper Permian Stratotypes of the Volga Region*, Moscow, p. 300.

Evans, D. A. D., 2000, Stratigraphic, geochronological, and paleomagnetic constraints upon the Neoproterozoic climatic paradox. *American Journal of Science* **300**: 347–433.

Evans, P., 1971, Towards a Pleistocene time-scale. In W. B. Harland, H. Francis, *et al.*, eds., *The Phanerozoic Time-Scale: A Supplement*, *Geological Society of London Special Publication*, Vol. 5. London: Geological Society, pp. 123–356.

Evenchick, C. A. and McNicoll, V. J., 1993. U–Pb age for the Jurassic McEwan Creek pluton, north-central British Columbia: regional setting and implications for the Toarcian stage boundary. In *Radiogenic Age and Isotopic Studies: Report 7. Geological Survey of Canada Paper* 93–2. Ottawa: Geological Survey of Canada, pp. 91–7.

Eyles, N. and Young, G. M., 1994, Geodynamic controls on glaciation in Earth history. In M. Deynoux, J. M. G. Miller, E. W. Domack, *et al.*, eds., *Earth's Glacial Record*. Cambridge: Cambridge University Press, pp. 1–28.

Fahlbusch, V., 1976, Report on the International Symposium on Mammalian Stratigraphy of the European Tertiary (München, April 11–14, 1975). *Newsletters on Stratigraphy* **5**: 160–7.

Farquhar, J., Bao, H., and Thiemens, M., 2000, Atmospheric influence on Earth's earliest sulfur cycle. *Science* **289**: 756–8.

Farrell, J. W., Clemens, S. C., and Gromet, L. P., 1995, Improved chronostratigraphic reference curve of late Neogene seawater $^{87}Sr/^{86}Sr$. *Geology* **23**: 403–6.

Faure, G., 1986, *Principles of Isotope Geology*. London: Wiley.

Fearnsides, W. G., 1905, On the geology of Arenig Fawr and Moel Llyfnant. *Quarterly Journal of the Geological Society of London* **61**: 608–37.

1910, The Tremadoc Slates of south-east Carnarvonshire. *Quarterly Journal of the Geological Society of London* **66**: 142–88.

Feist, R. and Flajs, G., 1987, La limite devonien–carbonifère dans la Montagne Noire (France), biostratigraphie et environnement.

Comptes rendus de l'Académie des Sciences, Paris, Série II. Sciences de la Terre et des Planètes **305**(20): 1537–44.

Feist, R., Flajs, G., and Girard, C., 2000, The stratotype section of the Devonian–Carboniferous boundary. *Courier Forschungsinstitut Senckenberg* **225**: 77–82.

Fensome, R. A., Taylor, F. J. R., Norris, G., *et al.*, 1993, *A Classification of Fossil and Living Dinoflagellates*. New York: American Museum of Natural History Micropaleontology Press, 351 p.

Fernández-López, S. R., 2003, Bathonian working group. *International Subcommission on Jurassic Stratigraphy Newsletter* **30**: 15.

Fiet, N., Beaudoin, B., and Parize, O., 2001, Lithostratigraphic analysis of Milankovitch cyclicity in pelagic Albian deposits of central Italy: implications for the duration of the stage and substages. *Cretaceous Research* **22**: 265–75.

Finney, S. C. and Bergström, S. M., 1986, Biostratigraphy of the Ordovician *Nemagraptus gracilis* Zone. In C. P. Hughes and R. B. Rickards, eds., *Palaeoecology and Biostratigraphy of Graptolites, Geological Society of London Special Publication* Vol. 20. London: Geological Society, pp. 47–59.

Finney, S. C. and Berry, W. B. N., 1997, New perspectives on graptolite distributions and their use as indicators of platform margin dynamics. *Geology* **25**(10): 919–22.

Fischer, A. G., 1964, The Lofer cyclothems of the Alpine Triassic. In D. F. Merriam, ed., *Symposium on Cyclic Sedimentation. Kansas Geological Survey Bulletin* **169**: 107–49.

1995, Cyclostratigraphy, Quo Vadis? In M. R. House, and A. S. Gale, eds., *Orbital Forcing Timescales and Cyclostratigraphy, Geological Society of London Special Publication*, Vol. 85. London Geological Society, pp. 199–204.

Fisher, M. J. and Dunay, R. E., 1981, Palynology and the Triassic/Jurassic boundary. *Review of Palaeobotany and Palynology* **34**: 129–35.

Fisunenko, O. P., 1979, Korrelyatsiya karbona Donetskogo basseyna i Zapadnoy Evropy (po dannym paleobotaniki). [The correlation of the Carboniferous between the Donets Basin and Western Europe, by paleontological data.] In S. V. Meyen, V. V. Menner, Y. A. Reytlinger, A. P. Rotay, and M. N. Solov'yeva, eds., *Compte Rendu Congrès International de Stratigraphie et de Géologie du Carbonifère* **8**: pp. 235–43.

Floris, S. and Thomsen, E., 1981, Danian. In C. Pomerol, ed., *Stratotypes of Paleogene Stages, Mémoire hors série 2 du Bulletin d'Information des Géologues du Bassin de Paris*. pp. 77–81.

Flower, B. P. and Kennett, J. P., 1994, The middle Miocene climatic transition: East Antarctic ice sheet development, deep ocean circulation and global carbon cycling. *Palaeogeography, Palaeoclimatology, Palaeoecology* **108**: 537–55.

Flynn, J. J. and Swisher, C. C., III, 1995, Cenozoic South American Land Mammal Ages: correlation to global geochronologies. *Society of Economic Paleontologists and Mineralogists Special Publication*, Vol. 54. Tulsa, OK: SEPM, pp. 317–33.

Flynn, J. J. and Tauxe, L., 1998, Magnetostratigraphy of Upper Paleocene–Lower Eocene marine and terrestrial sequences. In M.-P. Aubry, S. G. Lucas, and W. A. Berggren, eds., *Late Paleocene–Early Eocene Climatic and Biotic Events in the Marine and Terrestrial Records*. New York: Columbia University Press, pp. 67–90.

Fontugne, M. R. and Calvert, S. E., 1992, Late Pleistocene variability of the carbon isotopic composition of organic matter in the eastern Mediterranean: monitor of changes in carbon sources and atmospheric CO_2 concentrations. *Palaeoceanography* **7**: 1–20.

Forbes, E., 1846, On the connection between the distribution of the existing fauna and flora of the British Isles, and the geological changes which have affected their area, especially during the epoch of the Northern Drift. *Great Britain Geological Survey Memoir*, Vol. 1. Kenilworth, BGS: pp. 336–42.

Fordham, B. G., 1992, Chronometric calibration of mid-Ordovician to Tournaisian conodont zones: a compilation from recent graphic correlation and isotope studies. *Geological Magazine* **129**: 709–21.

1998, Silurian time: how much of it was Pridoli? In J. C. Gutiérrez-Marco and I. Rábano, eds., *Proceedings of the Sixth International Graptolite Conference on the GWG (IPA) and the 1998 Field Meeting of the International Subcommission on Silurian Stratigraphy (ICS–IUGS), Temas Geologico-Mineros*, Vol. 23. Madrid: Instituto Technológico Geominero de España Instituto Geologico e Mineiro PRT/ Consejo Superior de Investigaciones Cientificas, pp. 80–4.

Foreman, H., 1973, Radiolaria of Leg 10 with systematics and ranges for the families Amphipyndacidae, Artostrobiidae and Theoperidae. In *Initial Reports of the Deep Sea Drilling Project*, Vol. 10. Washington, DC: US Govt. Printing Office, pp. 407–74.

Foresi, L. M., Bonomo, S., Caruso, A., *et al.*, 2002, Calcareous plankton high resolution biostratigraphy (Foraminifera and nannofossils) of the uppermost Langhian-lower Serravallian Ras il-Pellegrin section (Malta). In S. M. Iaccarino, ed., *Integrated Stratigraphy and Paleoceanography of the Mediterranean Middle Miocene*, Vol. 108(2). Milan: Rivista Italiana di Paleontologia e Stratigrafia, pp. 195–210.

Foresi, L. M., Iaccarino, S., Mazzei, R., and Salvatorini, G., 1988, New data on calcareous plankton biostratigraphy of the Middle–Upper Miocene of the Mediterranean area. *Rivista Italiana di Paleontologia e Stratigrafia* **104**: 95–114.

Fornaciari, E., 1996, Biocronologia a nannofossili calcarei e stratigrafia ad eventi nel Miocene italiano. Unpublished Ph.D. thesis, University of Padua, Italy.

Fornaciari, E. and Rio, D., 1996, Latest Oligocence to early Miocene quantitative calcareous nannofossil biostratigraphy in the Mediterranean region. *Micropaleontology* **42**: 1–36.

Fornaciari, E., Di Stefano, A., Rio, D., and Negri, A., 1996, Middle Miocene quantitative calcareous nannofossil biostratigraphy in the Mediterranean region. *Micropaleontology* **42**: 37–64.

Fornaciari, E., Iaccarino, S., Mazzei, R., *et al.*, 1997, Calcareous plankton biostratigraphy of the Langhian historical stratotype. In A. Montanari, G. S. Odin, and R. Coccioni, eds., *Miocene Stratigraphy: An Integrated Approach*, Vol. 15, *Developments in Paleontology and Stratigraphy 15*. Amsterdam: Elsevier, pp. 89–96.

Forte, A. M. and Mitrovica, J. X., 1997, A resonance in the Earth's obliquity and precession over the past 20 Myr driven by mantle convection. *Nature* **390**: 676–80.

Fortey, R. A., Bassett, M. G., Harper, D. A. T., *et al.*, 1991, Progress and problems in the selection of stratotypes for the bases of series in the Ordovician System of the historical type area in the U.K. In C. R. Barnes and S. H. Williams, eds., *Advances in Ordovician Geology, Geological Survey of Canada Paper* 90–9. Ottawa: Geological Survey of Canada, pp. 5–25.

Fortey, R. A., Harper, D. T., Ingham, J. K., Owen, A. W., and Rushton, A. W. A., 1995, A revision of Ordovician series and stages from the historical type area. *Geological Magazine* **132**(1): 15–30.

Fortey, R. A., Harper, D. T., Ingham, J. K., *et al.*, 2000, *A Revised Correlation of the Ordovician Rocks of the British Isles, Geological Society of London, Special Report*, Vol. 24. London: Geological Society.

Fortwengler, D. and Marchand, D., 1994, Nouvelles unités biochronologiques de la zone à Mariae (Oxfordien inférieur). In E. Cariou and P. Hantzperque, eds., *Third International Symposium on Jurassic Stratigraphy,* Poitiers, September 22–29, 1991, *Géobios Mémoire Spéciale*, Vol. 17. Lyon: Université Claude Bernard, pp. 203–9.

Foucher, J.-C. and Monteil, E., 1998, Dinoflagellate cysts. [for Cretaceous portion of Mesozoic and Cenozoic sequence chronostratigraphic framework of European basins, in Hardenbol, J. *et al.*]. In P.-C. de Graciansky, J. Hardenbol., Th. Jacquin, and P. R. Vail, eds., *Mesozoic–Cenozoic Sequence Stratigraphy of European Basins, Society of Economic Paleontologists and Mineralogists Special Publication,* Vol. 60. Tulsa, OK: SEPM, pp. 770–2, Chart 5.

Fowell, S. J. and Olsen, P. E., 1993, Time calibration of Triassic/Jurassic microfloral turnover, eastern North America. *Tectonophysics* **222**: 361–9.

1995, Time calibration of Triassic/Jurassic microfloral turnover, eastern North America – Reply. *Tectonophysics* **245**: 96–9.

Frakes, L. A., Francis, J. E., and Syktus, J. L., 1992, *Climate Modes of the Phanerozoic: The History of the Earth's Climate Over the Past 600 Million Years.* Cambridge: Cambridge University Press.

Frankel, C., 1999, *The End of the Dinosaurs, Chicxulub Crater and Mass Extinctions.* Cambridge: Cambridge University Press.

Franz, S. O. and Tiedemann, R., 1997, Physical and chemical changes of deep-water masses between 3000 and 4400 m water depth in the equatorial W. Atlantic during the Pliocene (ODP Leg 154, Ceara Rise). *Development of Paleoceanography as a New Field of Science, 50th Anniversary of the Albatross Expedition 1947–1948, Royal Swedish Academy of Science, Stockholm, Sweden, Abstract Volume,* 50. Stockholm: Royal Swedish Academy of Science.

Franzen, J. L. and Haubold, H., 1987, The biostratigraphic and palaeoecologic significance of the Middle Eocene locality Geiseltal near Halle (German Democratic Republic). In N. Schmidt-Kittler, ed., International Symposium on Mammalian Biostratigraphy and Paleoecology of the European Paleogene, Mainz, February 18–21. *Münchner Geowissenschaftliche Abhandlungen A* **10**.

Fritz, W. H., 1992, Walcott's Lower Cambrian olenellid trilobite collection 61K, Mount Robson area, Canadian Rocky Mountains. *Geological Survey of Canada Bulletin* **432**: 65.

Froude, D. O., Ireland, T. R., Kinny, P. D., *et al.*, 1983, Ion microprobe identification of 4,100–4,200-Myr-old terrestrial zircons. *Nature* **304**(5927): 616–18.

Fuchs, T., 1894, Tertiärfossilien aus den kohlenführenden Miozänablagerungen der Umgebung von Krapina und Radobog und über die Stellung der sogenannten Stufe. *Mitteilungen und Jahrbuch der königlichen Ungarischen Geologischen Anstalt* **10**: 161–75.

Fuller, W. A., 1987, *Measurement Error Models*. New York: Wiley.

Funnell, B. M., 1964, The Tertiary Period. *Quarterly Journal of the Geological Society of London* **1205**: 179–91.

Funnell, B. W., 1996, Plio–Pleistocene palaeogeography of the southern North Sea Basin (3.75–0.60 Ma). *Quaternary Science Reviews* **15**: 391–405.

Furnish, W. M., 1973, Permian Stage names. In A. Logan and L. V. Hills, eds., *The Permian and Triassic Systems and their Mutual Boundary, Canadian Society of Petroleum Geologists Memoir* 2. Calgary: Canadian Society of Petroleum Geologists, pp. 522–49.

Gabilly, J., 1976, Le Toarcien à Thouars et dans le centre-ouest de la France. *Biostratigraphie-Evolution de la faune (Harpoceratinae–Hildoceratinae), Les Stratotypes Français 3.* Paris: Centre National Recherche Scientifique.

Gaetani, M., 1993, Anisian/Ladinian boundary field workshop Southern Alps–Balaton Highlands, 27 June–4 July, *Field-guide book*. IUGS Subcommission of Triassic Stratigraphy.

1993, Anisian/Ladinian boundary field workshop Southern Alps–Balaton Highlands, 27 June–4 July 1993. *Albertiana* **12**: 5–9.

1994, Annual report of the Working Group on the Anisian, Ladinian and Carnian stage boundaries. *Albertiana* **14**: 51–3.

Gaetani, M., Gnaccolini, M., Jadoul, F., and Garzanti, E., 1998, Multiorder sequence stratigraphy in the Triassic system of the western Southern Alps. In P.-C. de Graciansky, J. Hardenbol, T. Jacquin, and P. R. Vail, eds., *Mesozoic–Cenozoic Sequence Stratigraphy of European Basins; Society of Economic Paleontologists and Mineralogists Special Publication,* Vol. 60. Tulsa, OK: SEPM, pp. 701–17.

Galbrun, B., 1984, Magnétostratigraphie de la limitè Jurassique/Crétacé. Proposition d'une échelle de polarité à partir du stratotype du Berriasien (Berrias, Ardèche, France) et la Sierra de Lugar (Province de Murcie, Espagne). *Mémoires des Sciences de la Terre*, Vol. 38. Paris: Université Pierre et Marie Curie.

Galbrun, B., Gabilly, J., and Rasplus, L., 1988, Magnetostratigraphy of the Toarcian stratotype sections at Thouars and Airvault (Deux-Sevres, France). *Earth and Planetary Science Letters* **87**: 453–62.

Gale, A. S., 1995, Cyclostratigraphy and correlation of the Cenomanian Stage in Western Europe. In M. R. House and A. S. Gale, eds., *Orbital Forcing Timescales and Cyclostratigraphy, Geological Society of London Special Publication,* Vol. 85. London: Geological Society, pp. 177–97.

1996, Turonian correlation and sequence stratigraphy of the Chalk in southern England. In S. P. Hesselbo and D. N. Parkinson, eds., *Sequence Stratigraphy in British Geology, Geological Society Special Publication*, Vol. 103. London: Geological Society, pp. 177–95.

Gale, A. S., Kennedy, W. J., Burnett, J. A., Caron, M., and Kidd, B. E., 1996a, The Late Albian to Early Cenomanian succession at Mont Risou near Rosans (Drôme, SE France): an integrated study (ammonites, inoceramids, planktonic foraminifera, nannofossils, oxygen and carbon isotopes). *Cretaceous Research* **17**: 515–606.

Gale, A. S., Montgomery, P., Kennedy, W. J., *et al.*, 1996b, Definition and global correlation of the Santonian–Campanian boundary. *Terra Nova* **7**: 611–22.

Gale, A. S., Young, J. R., Shackleton, N. J., Crowhurst, S. J., and Wray, D. S., 1999, Orbital tuning of Cenomanian marly chalk successions: towards a Milankovitch time-scale for the Late Cretaceous. *Philosophical Transactions of the Royal Society of London, series A* **357**: 1815–29.

Gale, N. H., 1985, Numerical calibration of the Paleozoic time-scale: Ordovician, Silurian and Devonian periods. In N. J. Snelling, ed., *The Chronology of the Geological Record, Geological Society of London Memoir* 10. Oxford: Geological Society, pp. 81–8.

Galeotti, S., Coccioni, R., Angori, E., *et al.*, 2000, The Paleocene/Eocene transition in a classical Tethyan setting: the Contessa Road Section. *Bulletin de la Société géologique de France, Serie 9* **3**: 355–65.

Gallet, Y., Besse, J., Krystyn, L., Marcoux, J., and Théveniaut, H., 1992, Magnetostratigraphy of the late Triassic Bolücektasi Tepe section (southwestern Turkey): implications for changes in magnetic reversal frequency. *Physics of the Earth and Planetary Interiors* **73**: 85–108.

Gallet, Y., Vandamme, D., and Krystyn, L., 1993, Magnetostratigraphy of the Hettangian Langmoos section (Adnet, Austria): evidence for time-delayed phases of magnetization. *Geophysical Journal International* **115**: 575–85.

Gallet, Y., Krystyn, L., and Besse, J., 1998, Upper Anisian to Lower Carnian magnetostratigraphy from the Northern Calcareous Alps (Austria). *Journal of Geophysical Research* **103**: 605–21.

Gallet, Y., Krystyn, L., Besse, J., Saidi, A., and Ricou, L.-E., 2000, New constraints on the Upper Permian and Lower Triassic geomagnetic polarity time scale from the Abadeh section (central Iran). *Journal of Geophysical Research* **105**: 2805–15.

Gallup, C. D., Edwards, R. L., and Johnson, R. G., 1994, The timing of high sea levels over the past 200,000 years. *Science* **263**: 796–800.

Gao, G. and Land, L. S., 1991, Geochemistry of the Cambro–Ordovician Arbuckle Limestone, Oklahoma: implications for diagenetic δ^{18}O alteration and secular δ^{13}C and ^{87}Sr/^{86}Sr variation. *Geochimica et Cosmochimica Acta* **55**: 2911–20.

Garner, E. L., Murphy, T. J., Gramlich, J. W., Paulsen, P. J., and Barnes, I. L., 1975, Absolute isotopic abundances and the atomic weight of a reference sample of potassium. *Journal of Research of the National Bureau of Standards* **79A**: 713–25.

Gartner, S., 1977, Calcareous nannofossil biostratigraphy and revised zonation of the Pleistocene. *Marine Micropalaeontology* **2**: 1–25.

Gavrilov, Y. O., Shcherbinina, E. A., and Muzylöv, N. G., 2000, Paleogene sequence in central North Caucasus: a response to paleoenvironmental changes. *Geologiska Föreningen i Stockholm Förhandlingar* **122**: 51–3.

Gehling, J. G., Jensen, J. S., Droser, M. L., Myrow, P. M., and Narbonne, G. M., 2001, Burrowing below the basal Cambrian GSSP, Fortune Head, Newfoundland. *Geological Magazine* **138**: 213–18.

Gelati, R., Gnaccolini, M., Falletti, P., and Catrullo, D., 1993, Stratigrafia sequenzionale della successione oligo-miocenica delle Langhe, Bacinoterziario ligure-piemontese. *Rivista Italiana di Paleontologia e Stratigrafia* **98**: 425–52.

Geluk, M. C. and Röhling, H.-G., 1997, High-resolution sequence stratigraphy of the Lower Triassic 'Buntsandstein' in the Netherlands and northwestern Germany. *Geologie en Mijnbouw* **76**: 227–46.

George, T. N., Johnson, G. A. L., Mitchell, M., *et al.*, 1976, A correlation of Dinantian rocks in the British Isles. *Geological Society of London Special Publication*, Vol. 7. London Geological Society, pp. 1–87.

Gerasimov, L. P., 1937, The Uralian series of the Permian System: scientific reports of Kazan University. *Geology* **97** (3–4): fasc. 8–9, 68.

German Stratigraphic Commission, 2002, *Stratigraphic Table of Germany 2002*. Pottsdam: Druckerei E. Stein Gmbtt.

Gervais, P., 1867–1869, *Zoologie et paleontology générales*. Nouvelles recherches sur les animaux vertétébrés et fossiles, Paris.

Geyer, O. F. and Gwinner, M. P., 1979, Die Schwäbische Alb und ihr Vorland. *Sammlung Geologischer Führer* **67**: 271.

Geyer, G. and Shergold, J. H., 2000, The quest for internationally recognised divisions of Cambrian time. *Episodes* **23**(3): 188–95.

Gheerbrant, E., Sudre, J., Sen, S., *et al.*, 1998, Nouvelles données sur les mammifères du Thanétien et de l'Yprésien du Bassin d'Ouarzazate (Maroc) et leur contexte stratigraphique. *Palaeovertebrata* **27**: 155–202.

Gialanella, P. R., Heller, F., Haag, M., *et al.*, 1997, Late Permian magnetostratigraphy on the eastern Russian Platform. In M. J. Dekkers, C. G. Langereis, and R. Van der Voo, eds., Analysis of Paleomagnetic Data: a tribute to Hans Zijderveld. *Geologie en Mijnbouw* **76**: 145–54.

Gianolla, P. and Jacquin, T., 1998, Triassic sequence stratigraphic framework of western European basins. In P.-C. de Graciansky, J. Hardenbol, T. Jacquin, and P. R. Vail, eds., *Mesozoic–Cenozoic Sequence Stratigraphy of European Basins, Society of Economic Paleontologists and Mineralogists Special Publication*, Vol. 60. Tulsa, OK: SEPM, pp. 643–50.

Gianolla, P., De Zanche, V., and Mietto, P., 1998, Triassic sequence stratigraphy in the Southern Alps (northern Italy): definition of sequences and basin evolution. In P.-C. de Graciansky, J. Hardenbol, T. Jacquin, and P. R. Vail, eds., *Mesozoic–Cenozoic Sequence Stratigraphy of European Basins, Society of Economic Paleontologists and Mineralogists Special Publication* Vol. 60. Tulsa, OK: SEPM, pp. 719–47.

Gianotti, A., 1953, Microfaune della serie tortoniana del Rio Mazzapiedi–Catellania (Tortona–Alessandria). *Rivista Italiana di Paleontologia e Stratigrafia Mem.* **VI**: 167–308.

Gibbard, P. L., 2003, Definition of the Middle–Upper Pleistocene boundary. *Global and Planetary Change* **36**: 201–8.

Gibbard, P. L. and West, R. G., 2000, Quaternary chronostratigraphy: the nomenclature of terrestrial sequences. *Boreas* **29**: 329–36.

Gibshman, N. B., 2001, Foraminiferal biostratigraphy of the Serpukhovian Stage stratotype (Zaborie quarry, Moscow Basin). *Newsletter on Carboniferous Stratigraphy* **19**: 31–4.

Gignoux, M., 1910, Sur la classification du Pliocène et du Quaternaire dans Italie du Sud. *Comptes rendus de l'Academie des Sciences, Paris* **150**: 841–4.

1913, Les formations marines du Pliocène et Quaternaire dans Italie du Sud et la Sicile. *Annales Université Lyon*, n.s., I-36. Impr. A. Rey Lyon.

1950, *Géologie Stratigraphique*, 4th edition. Paris: Masson and Cie.

Gilbert, G. K., 1895, Sedimentary measurement of geological time. *Journal of Geology* **3**: 121–7.

Gilder, S., Chen, Y., Cogne, J. P., *et al.*, 2003, Paleomagnetism of Upper Jurassic to Lower Cretaceous volcanic and sedimentary rocks from the western Tarim basin and implications for inclination shallowing and absolute dating of the M-0 (ISEA?) chron. *Earth and Planetary Science Letters* **206**: 567-600.

Gingerich, P. D., 1976, Cranial anatomy and evolution of early Tertiary Plesiadapidae (Mammalia, Primates). *University of Michigan Papers on Paleontology* **15**: 1–141.

1992, Marine mammals (Cetacea and Sirenia) from the Eocene of Gebel Mokattam and Fayum, Egypt: stratigraphy, age, and paleoenvironments. *University of Michigan Papers on Paleontology* **30**: 1–84.

2000, Paleocene/Eocene boundary and continental vertebrate faunas of Europe and North America. *Geologiska Föreningen i Stockholm Förhandlingar* **122**: 57–9.

Girard, C. and Lecuyer, C., 2002, Variation in Ce anomalies of conodonts through the Frasnian/Famennian boundary of Poland (Kowala –Holy Cross Mountains): implications for the redox state of seawater and biodiversity. *Palaeogeography, Palaeoclimatology, Palaeoecology* **181**: 209–311.

Giraud, F., 1995, Recherche des périodicités astronomiques et des fluctuations du niveau marin à partir de l'étude du signal carbonaté des séries pélagiques alternantes. *Documents des Laboratoires de Géologie de Lyon* **134**: 279.

Giraud, F., Beaufort, L., and Cotillon, P., 1995, Periodicities of carbonate cycles in the Valanginian of the Vocontian Trough: a strong obliquity control. In M. R. House and A. S. Gale, eds., *Orbital Forcing Timescales and Cyclostratigraphy: Geological Society of London Special Publication* Vol. 85. London Geological Society, pp. 143–64.

Glass, B. P. and Pizzuto, J. E., 1994, Geographic-variation in Australasian microtektite concentrations – implications concerning the location and size of the source crater. *Journal of Geophysical Research* **99**(E9): 19 075–81.

Glass, B. P., Kent, D. V., Schneider, D. A., and Tauxe, L., 1991, Ivory coast microtektite strewn field: description and relation to the Jaramillo geomagnetic event. *Earth and Planetary Science Letters* **107**: 182–96.

Glenister, B. F. and Furnish, W. M., 1961, The Permian ammonoids of Australia. *Journal of Paleontology* **35**: 673–736.

Glenister, B. F., Boyd, D. W., Furnish, W. M., *et al.*, 1992, The Guadalupian: proposed international standard for a Middle Permian Series. *International Geology Review* **34**(9): 857–88.

Glenister, B. F., Wardlaw, B. R., Lambert, L. L., *et al.*, 1999, Proposal of Guadalupian and component Roadian, Wordian and Capitanian Stages as International Standards for the Middle Permian Series. *Permophiles* **34**: 3–11.

Glowniak, E., 1997, Middle Oxfordian ammonites. *International Subcommission on Jurassic Stratigraphy Newsletter* **25**: 45–6.

Goddéris, Y. and Veizer, J., 2000, Tectonic control of chemical and isotopic composition of ancient oceans: the impact of continental growth. *American Journal of Science* **300**: 434–61.

Godwin, H., 1975, *The History of the British Flora*. Cambridge: Cambridge University Press.

Goerges, J., 1957, Die Mollusken der oberoligozänen Schichten des Doberges bei Bünde in Westfalen. *Paläontologische Zeitschrift* **31**: 116–34.

Goldhammer, R. K., Dunn, P. A., and Hardie, L. A., 1987, High frequency glacio-eustatic oscillations with Milankovitch characteristics recorded in northern Italy. *American Journal of Science* **287**: 853–92.

1990, Depositional cycles, composite sea level changes, cycle stacking patterns, and the hierarchy of stratigraphic forcing: examples from the Alpine Triassic platform carbonates. *Geological Society of America Bulletin* **102**: 535– 62.

Goldhammer, R. K., Oswald, E. J., and Dunn, P. A., 1994, High-frequency glacio-eustatic cyclicity in the Middle Pennsylvanian of the Paradox Basin: an evaluation of Milankovitch forcing. In P. L. De Boer and D. G. Smith, eds., *Orbital Forcing and Cyclic Sequence, International Association of Sedimentologists Special Publication*, Vol. 19. Oxford: Blackwell.

Gonzales, C. R., 1990, Development of the late Paleozoic glaciation of the South American Gondwana in western Argentina. *Palaeogeography, Palaeoclimatology, Palaeoecology* **79**: 275–87.

Gorak, S. V., 1977, *Carboniferous ostracods of the Great Donets Basin (Paleoecology, Paleozoogeography and Biostratigraphy)*. Kiev: Naukova Dumka pp. 1–148.

Goreva, N. V. and Alekseev, A. S., 2001, Conodonts. In A. S. Alekseev and S. M. Shik, eds., *Middle Carboniferous of the Moscow Synclise,* Vol. 2, *Paleontologic Characteristic*. Moscow: Scientific World, pp. 33–54 [In Russian].

Goreva, N. V., Gorjunova, R. V., Isakova, T. N., *et al.*, 2001, Paleontologic characteristics. In A. S. Alekseev and S. M. Shik, eds., *Middle Carboniferous of the Moscow Syneclise*, Vol. 2. Moscow: Scientific World, pp. 1–230, 48 plates [In Russian].

Gourmelon, F., 1987, Les Radiolaires tournaisiens des nodules phosphatés de la Montagne Noire et des Pyrénées centrales; systematique; biostratigraphie, paleobiogeographie. *Biostratigraphie du Paleozoique* **6**: 172.

Goy, A., Ureta, M. S., Arias, C., *et al.*, 1994, The Fuentelsaz section (Iberian Range, Spain), a possible stratotype for the base of the Aalenian Stage. In S. Cresta and G. Pavia, eds., Proceedings of the Third International Meeting on Aalenian and Bajocian stratigraphy, Vol. 5. Rome: Institute Poligrafics e Zecca deuo Stato, pp. 1–31.

1997, Die Toarcium/Aalenium-Grenze im Profil Fuentelsaz (Iberische Ketten, Spanien). In W. Ohmert, ed., *Die Grenzziehung Unter-/Mitteljura (Toarcium/Aalenium) bei Wittnau und Fuentelsaz. Beispiele interdisziplinärer geowissenschaftlicher Zusammenarbeit, Geologisches Landesamt Baden-Württemberg, Informationen* Vol. 8. pp. 43–52.

Gradstein, F. M., 1970, Foraminifera from the type Sicilian at Ficarazzi, Sicily (Lower Pleistocene). *Proceedings of the Koninklijke Nederlandse Akademie van Wetenschappen, Series B* **4**: 1–30.

Gradstein, F. M., 1996, *STRATCOR – Graphic Zonation and Correlation: Software-Users Guide, Authors Edition* Version 4.

Gradstein, F. M. and Ogg, J., 1996, A Phanerozoic time scale. *Episodes* **19**: 3–5, with insert.

Gradstein, F. M., Agterberg, F. P., Brower, J. C., and Schwarzacher, W. S., 1985, *Quantitative Stratigraphy*. Dordrecht: D. Reidel and Paris: UNESCO.

Gradstein, F., Kristiansen, I. L., Loemo, L., and Kaminski, M. A., 1992, Cenozoic foraminiferal and dinocyst biostratigraphy of the central North Sea. *Micropaleontology* **38** 111–37.

Gradstein, F. M., Agterberg, F. P., Ogg, J. G., Hardenbol, J., van Veen, P., Thierry, T., and Huang, Z., 1994a, A Mesozoic time scale. *Journal of Geophysical Research* **99**(12): 24 051–74.

Gradstein, F. M., Kaminsiki, M. A., Berggren, W. A., Kristiansen, I. L., and D'Iorio, M. A., 1994b, Cenozoic biostratigraphy of the North Sea and Labrador Sea. *Micropaleontology* **40**: 152.

Gradstein, F. M., Agterberg, F. P., Ogg, J. G., Hardenbol, J., van Veen, P., Thierry, T., and Huang, Z., 1995, A Triassic, Jurassic and Cretaceous time scale. In W. A. Berggren, D. V. Kent, M.-P. Aubry, and J. Hardenbohl, eds., *Geochronology, Time Scales and Global Stratigraphic Correlations: A Unified Temporal Framework for a Historical Geology, Society of Economic Paleontologists and Mineralogists Special Publication*, Vol. 54: Tulsa, OK: SEPM, pp. 95–128.

Gradstein, F. M., Kaminski, M. A., and Agterberg, F. P., 1999, Biotratigraphy and paleoceanography of the Cretaceous seaway between Norway and Greenland. *Earth Science Reviews* **46**(1–4): 27–98.

Grant, S. F., Coe, A. L., and Armstrong, H. A., 1999, Sequence stratigraphy of the Coniacian succession of the Anglo–Paris Basin. *Geological Magazine* **136**: 17–38.

Grant, S. W. F., 1990, Shell structure and distribution of *Cloudina*, a potential index fossil for the terminal Proterozoic. *American Journal of Science* **290A**: 261–94.

Grayson, Jr., R. C., 1984, Morrowan and Atokan (Pennsylvanian) conodonts from the north-eastern margin of the Arbuskle Mountains, southern Oklakhoma. In P. K. Sutherland and W. L. Manger, eds., *The Atokan Series (Pennsylvanian) and its Boundaries: a Symposium, Bulletin Oklahoma Geological Survey* Vol. 136. Oklahoma Geological Survey, pp. 41–62.

Grayson, Jr., R. C., Merrill, G. K., Lambert, L. L., and Turner, J., 1989, Phylogenetic basis for species recognition within the conodont genus Idignathodus: applicability to correlation and boundary placement. In D. R. Boardman II, J. E. Barrick, J. Cocke, and M. K. Nestell, eds., *Middle and Late Pennsylvanian Chronostratigraphic Boundaries in North-central Texas: Glacial-eustatic Events, Biostratigraphy, and Paleoecology, Texas Technical University Studies Geology*, Vol. 2. Texas Technical University, pp. 75–94.

Graziano, S. and Ogg, J. G., 1994, Lower Triassic magnetostratigraphy in the Dolomites region (Italy) and correlation to Arctic ammonite zones. *EOS Transactions American Geophysical Union* **75**: 203.

Green, K. A. and Brecher, A., 1974, Preliminary paleomagnetic results for sediments from Site 263, Leg 27. In J. J. Veevers, J. R. Heirtzler, *et al.*, eds., *Initial Reports of the Deep Sea Drilling Project,* Vol. 27. Washington, DC: US Govt. Printing Office, pp. 405–13.

Griesbach, C. L., 1880, Paleontological notes on the Lower Trias on the Himalayas. *Records of the Geological Survey of India* **13**(2): 94–113.

Grieve, R. A. F., 2001, The terrestrial cratering record. In E. B. Peucker and B. Schmitz, eds., *Accretion of Extraterrestrial MatterThroughout Earth's History*. New York: Kluwer Academic, pp. 379–402.

Grimm, E. C., Jacobson, G. L., Watts, W. A., Hansen, B. C. S., and Maasch, K. A., 1993, A 50,000-year record of climate oscillations from Florida and its temporal correlation with the Heinrich events. *Science* **261**: 198–200.

Grippo, A., Fischer, A., Hinnov, L. A., Herbert, T. D., and Premoli Silva, I., 2004, Cyclostratigraphy and Chronology of the Albian stage (Piobbico core, Italy). In B. D'Argenio, A. G. Fischer, I. Premoli Silva, H. Weissert, and V. Ferreri, eds., *Cyclostratigraphy: An Essay of Approaches and Case Histories, Society of Economic Paleontologists and Mineralogists Special Publication*, Vol. 81. Tulsa, OK: SEPM.

Groessens, E. and Noël, B., 1977, Etude litho- et biostratigraphique du Rocher du Bastion et du Rocher Bayard Dinant. *Symposium Namur 1974* **15**: 1–17.

Groos-Uffenorde, H., Lethiers, F., and Blumenstengel, H., 2000, Ostracodes and Devonian stage boundaries. *Courier Forschungsinstitut Senckenberg* **220**: 99–111.

Grossman, E. L., Mii, H.-S., and Yancey, T. E., 1993, Stable isotopes in Late Pennsylvanian brachiopods from the United States: implications for Carboniferous paleoceanography. *Geological Society of America Bulletin* **105**(10): 1284–96.

Grossman, E. L., Mii, H.-S., Zhang, C., and Yancey, T. E., 1996, Chemical variations in Pennsylvanian brachiopod shells: effect of diagenesis, taxonomy, microstructure, and paleoenvironments. *Journal of Sedimentary Research* **66**: 1011–22.

Grossman, E. L., Bruckschen, P., Mii, H.-S., *et al.*, 2002, Carboniferous paleoclimate and global changes: isotopic evidence from the Russian Platform. In B. I. Chuvashov and E. O. Amon, eds., *Carboniferous Stratigraphy and Paleogeography in Eurasia*. Ekaterinburg: pp. 61–71.

Grotzinger, J. P. and Knoll, A. H., 1999, Proterozoic stromatolites: evolutionary mileposts or environmental dipsticks? *Annual Review of Earth and Planetary Science* 27: 313–58.

Grotzinger, J. P., Bowering, S. A., Saylor, B. Z., and Kaufman, A. J., 1995, Biostratigraphic and geochronologic constraints on early animal evolution. *Science* **270**: 598–604.

Grotzinger, J. P., Watters, W. A., and Knoll, A. H., 2000, Calcified metazoans in thrombolite–stromatolite reefs of the terminal Proterozoic Nama Group, Namibia. *Paleobiology* **26**: 334–59.

Group, I.-C. R., 1995, Field work on the Lopingian stratigraphy in Iran. *Permophiles* **27**: 5–6.

Groves, J. R., 1986, Foraminiferal characterization of the Morrowan–Atokan (lower Middle Pennsylvanian) boundary. *Geological Society of America Bulletin* **97**: 346–53.

Groves, J. R., Nemyrovska, T. I., and Alekseev, A. S., 1999, Correlation of the type Bashkirian Stage (Middle Carboniferous, south Urals) with the Morrowan and Atokan series of the midcontinental and western United States. *Journal of Paleontology* **73**: 529–39.

Guerstein, G .R., Chiesa, J. O., Guler, M. V., and Camacho, H. H., 2002, Bioestratigrafia basada en quistes de dinoflagelados de la Formacion Cabo Pena (Eoceno terminal–Oligoceno temprano), Tierra del Fuego. *Argent. Rev. Esp. Micropal.*, **34**: 105–16.

Guest, J. E. and Greeley, R., 1977, *Geology on the Moon*. The Wykeham Science Series. New York: Crane, Russak and Co.

Guex, J., 1995, Ammonites hettangiennes de la Gabbs Valley Range, Nevada. *Mémoires de Géologie Lausanne* **23**: 130.

Guex, J., Rakus, M., Taylor, D., and Bucher, H., 1997, Proposal for the New York Canyon area, Gabbs Valley Range (Nevada) USA. *International Subcommission on Jurassic Stratigraphy Newsletter* **24**: 26–30.

Gümbel, C. W., 1861, *Geognostische Beschreibung des bayerischen Alpengebirges und seines Vorlands, Perthes Gotha*, p. 950.

Gygi, R. A., Coe, A. L., and Vail, P. R., 1998, Sequence stratigraphy of the Oxfordian and Kimmeridgian stages (Late Jurassic) in northern Switzerland. In P. C. de Graciansky, J. Hardenbol, T. Jacquin, and P. R. Vail, eds., *Mesozoic–Cenozoic Sequence Stratigraphy of European Basins, Society of Economic Paleontologists and Mineralogists Special Publication*, Vol. 60. Tulsa, OK: SEPM, pp. 527–44.

Haag, M. and Heller, F., 1991, Late Permian to Early Triassic magnetostratigraphy. *Earth and Planetary Science Letters* **107**(1): 42–54.

Habib, D. and Drugg, W. S., 1983, Dinoflagellate age of Middle Jurassic–Early Cretaceous sediments in the Blake–Bahama Basin. In R. E. Sheridan, F. M. Gradstein, *et al.*, eds., *Initial Reports of the Deep Sea Drilling Project*, Vol. 76, Washington, DC: US Govt. Printing Office, pp. 623–38.

Haflidason, H., Eiriksson, J., and Van Kreveld, S., 2000, The tephrochronology of Iceland and the North Atlantic region during the Middle and Late Quaternary: a review. *Journal of Quaternary Science* **15**: 3–22.

Hagelberg, T. K., Bond, G., and DeMenocal, P., 1994, Milankovitch band forcing of sub-Milankovitch climate variability during the Pleistocene. *Palaeoceanography* **9**: 545–58.

Hailwood, E. A., 1979, Paleomagnetism of late Mesozoic to Holocene sediments from the Bay of Biscay and Rockall

Plateau, drilled on IPOD Leg 48. In L. Montadert, *et al.*, eds., *Initial Reports of the Deep Sea Drilling Project*, Vol. 48. Washington, DC: US Govt. Printing Office, pp. 305–39.

Hall, T. S., 1895, The geology of Castlemaine, with a subdivision of part of the Lower Silurian Rocks of Victoria, and a list of minerals. *Proceedings of the Royal Society of Victoria* **7**: 55–88.

Hallam, A., 1978, Eustatic cycles in the Jurassic. *Palaeogeography, Palaeoclimatology, Palaeoecology* **23**: 1–32.

1981a, The end-Triassic bivalve extinction event. *Palaeogeography, Palaeoclimatology, Palaeoecology* **35**: 1–44.

1981b, A revised sea-level curve for the Early Jurassic. *Journal of the Geological Society of London* **138**: 735–43.

1986, The Pliensbachian and Tithonian extinction events. *Nature* **319**: 765–8.

1988, A re-evalation of the Jurassic eustasy in the light of new data and the revised Exxon curve. In C. K. Wilgus, B. S. Hastings, G. S. C. Kendall, *et al.*, eds., *Sea Level Changes – an Integrated Approach, Society of Economic Paleonologists and Mineralogists Special Publication*, Vol. 42. Tulsa, OK: SEPM, pp. 261–73.

1990, Correlation of the Triassic–Jurassic boundary in England and Austria. *Journal of the Geological Society of London* **147**: 421–4.

1992a, *Great Geological Controversies*. Oxford: Oxford University Press.

1992b, *Phanerozoic Sea-Level Changes*, The Perspectives in Paleobiology and Earth History Series. New York: Columbia University Press.

1996, Major bio-events in the Triassic and Jurassic. In O. H. Walliser, ed., *Global Events and Event Stratigraphy in the Phanerozoic*. Berlin: Springer Verlag, pp. 265–83.

2001, A review of the broad pattern of Jurassic sea-level changes and their possible causes in the light of current knowledge. *Palaeogeography, Palaeoclimatology, Palaeoecology* **167**: 23–37.

Hallam, A. and Wignall, P. B., 1999, Mass extinctions and sea-level changes. *Earth Science Reviews* **48**: 217–50.

Hallam, A., Hancock, J. M., LaBreque, J. L., Lowrie, W., and Channell, J. E. T., 1985, Jurassic to Paleogene: Part I. Jurassic and Cretaceous geochronology and Jurassic to Palaeogene magnetostratigraphy. In N. J. Snelling, ed., *The Chronology of the Geological Record, Geological Society of London Memoir* 10. London: Geological Society, pp. 118–40.

Hald, A., 1952, *Statistical Theory with Engineering Applications*. New York: Wiley.

Hames, W. E., Renne, P. R., and Ruppel, C., 2000, New evidence for geologically instantaneous emplacement of earliest Jurassic Central Atlantic magmatic province basalts on the North American margin. *Geology* **28**: 859–62.

Hance, L. and Muchez, P., 1995, Study of the Tournaisian–Visean transitional strata in South China (Guangxi). *Abstracts of XIII International Congress on Carboniferous and Permian*. Krakow: Panstwowy Instytut Geologiczny, p. 51.

Hance, L., Brenckle, P. L., Coen, M., *et al.*, 1997, The search for a new Tournaisian–Visean boundary stratotype. *Episodes* **20**(3): 176–80.

Hancock, J. M., 1991, Ammonite scales for the Cretaceous System. *Cretaceous Research* **12**: 259–91.

2000, Names are better than numbers: nomenclature for sequences. *Geology* **28**: 567–8.

2001, A proposal for a new position for the Aptian/Albian boundary. *Cretaceous Research* **22**: 677–83.

Hancock, J. M., Peake, N. B., Burnett, J., *et al.*, 1993, High Cretaceous biostratigraphy at Tercis, SW France. *Bulletin de l'Institut Royal des Sciences naturelles de Belgique, Sciences de la Terre* **63**: 133–48.

Hancock, J. M., Gale, A. S., Gardin, S., *et al.*, 1996, The Campanian Stage. *Bulletin de l'Institut Royal des Sciences Naturelles de Belgique, Sciences de la Terre, Supplément* **66**: 103–9.

Handschumacher, D. W., Sager, W. W., Hilde, T. W. C., and Bracey, D. R., 1988, Pre-Cretaceous evolution of the Pacific plate and extension of the geomagnetic polarity reversal time scale with implications for the origin of the Jurassic "Quiet Zone." *Tectonophysics* **155**: 365–80.

Haq, B. U. and van Eysinga, F. W. B., 1998, *Geological Time Table*, 5th Revised Edition. Amsterdam: Elsevier.

Haq, B. U., Berggren, W. A., and van Couvering, J. A., 1977, Corrected age of the Plio/Pleistocene boundary. *Nature* **269**: 483–8.

Haq, B. U., Hardenbol, J., and Vail, P. R., 1987, Chronology of fluctuating sea levels since the Triassic. *Science* **235**: 1156–67.

1988, Mesozoic and Cenozoic chronostratigraphy and eustatic cycles. In C. K. Wilgus, B. S. Hastings, G. S. C. Kendall, *et al.*, eds., *Sea Level Changes – an Integrated Approach, Society of Economic Paleontologists and Mineralogists Special Publication*, Vol. 42. Tulsa, OK: SEPM, pp. 71–108.

Hardenbol, J., 1968, The Priabon type section (Italy). *Mémoires du Bureau de Recherches géologiques et minières* **58**: 629–35.

Hardenbol, J. and Berggren, W. A., 1978, A new Paleogene numerical time scale. In G. V. Cohee, M. F. Glaessner, and H. D. Hedberg, eds., *Contributions to the Geologic Time Scale, Studies in Geology*, Vol. 6. Tulsa, OK: American Association of Petroleum Geologists, pp. 213–34.

Hardenbol, J., Thierry, J., Farley, M. B., *et al.*, 1998, Mesozoic and Cenozoic sequence chronostratigraphic framework of European basins. In P.-C. de Graciansky, J. Hardenbol, T. Jacquin, and P. R. Vail, eds., *Mesozoic–Cenozoic Sequence Stratigraphy of European Basins, Society of Economic Paleontologists and Mineralogists Special Publication*, Vol. 60. Tulsa OK: SEPM, pp. 3–13, 763–781, and chart supplements.

Hardie, L. A. and Hinnov, L. A., 1997, Biostratigraphic and radiometric age data question the Milankovitch characteristics of the Latemar cycles (Southern Alps, Italy) – Comment. *Geology* **25**: 470–71.

Harland, W. B. and Francis, H., 1971, The Phanerozoic time-scale: a supplement, *Geological Society of London, Special Publication*, Vol. 5. London: Geological Society, p. 356.

Harland, W. B. and Pudsey, C. J., 2002, Protoperidiniacean dinoflagellate cyst taxa from the Upper Miocene of ODP Leg 178, Antarctic Peninsula. *Review of Palaeobotany and Palynology* **120**: 263–84.

Harland, W. B., Smith, A. G., and Wilcock, B., 1964, The Phanerozoic time-scale. (A symposium dedicated to Professor Arthur Holmes.) *Quarterly Journal of the Geological Society of London* **120s**: 458.

Harland, W. B., Cox, A. V., Llewellyn, P. G., *et al.*, 1982, *A Geologic Time Scale*. Cambridge: Cambridge University Press.

Harland, W. B., Armstrong, R. L., Cox, A. V., *et al.*, 1990, *A Geologic Time Scale 1989*. Cambridge: Cambridge University Press.

Harland, R., Hine, N. M., and Wilkinson, I. P., 1992, Paleogene biostratigraphic markers. In R. W. O. B. Knox and S. Holloway, eds., *Paleogene of the Central and Northern North-Sea*. Nottingham: British Geological Survey, pp. A1–A5.

Harries, P. J., Kauffman, E. G., Crampton, J. S., *et al.*, 1996, Lower Turonian Euramerican Inoceramidae: a morphologic, taxonomic, and biostratigraphic overview. A report from the First Workshop on Early Turonian Inoceramids, October 5–8, 1992, Hamburg, Germany. In C. Spaeth, ed., *Mitteilung aus dem Geologisch-Paläontologischen Institut der Universität Hamburg 77* (Jost Wiedmann Memorial Volume); *Proceedings of the Fourth International Cretaceous Symposium*, Hamburg, 1992, pp. 641–71.

Harris, W. J. and Keble, R. A., 1932, Victorian graptolite zones, with correlations and description of species. *Proceedings of the Royal Society of Victoria* **44**: 25–48.

Harris, W. J. and Thomas, D. E., 1938, A revised classification and correlation of the Ordovician graptolite beds of Victoria. *Mining and Geological Journal Department of Mines, Victoria* **1**(3): 62–72.

Harrison, C. G. A. and Funnell, B. M., 1964, Relationship of paleomagnetic reversals and micropaleontology in two Late Cainozoic cores from the Pacific Ocean. *Nature* **204**: 566.

Harrison, T. M., 1981, Diffusion of ^{40}Ar in hornblende. *Contributions to Mineralogy and Petrology* **78**: 324–31.

Harrison, T. M., Duncan, I., and McDougall, I., 1985, Diffusion of ^{40}Ar in biotite: temperature, pressure and compositional effects. *Geochimica et Cosmochimica Acta* **49**: 2461–8.

Hart, C. J. R., 1997, A transect across northern Stikinia: geology of the northern Whitehorse map area, Yukon Territory (105D/13-16). *Yukon Geology Program Bulletin* **8**.

Hart, M., Amédro, F., and Owen, H., 1996, The Albian stage and substage boundaries. *Bulletin de l'Institut Royal des Sciences Naturelles de Belgique, Sciences de la Terre, Supplement* **66**: 45–56.

Hartmann, W. K. and Davis, D. R., 1975, Satellite-sized planetesimals and lunar origin. *Icarus* **24**: 504–15.

Haug, E., 1908–1911, Les Périodes géologiques. *Traité de Géologie* **11**: 1766–76.

1910, *Traité de Géologie. II. Les périodes géologiques, Fascicule 2, Jurassique et Crétacé*. Paris: Armand Collin, pp. 929–1396.

1911, *Traité de Géologie. II. Les périodes géologiques. Fascicule 3, Période Nummulitique*. Paris: Armand Colin, pp. 1397–598.

Hay, W. S. W., 1964, Utilisation stratigraphique des discoasteridés pour la zonation du Paléocène et de l'Eocène inférieur. *Mémoires du Bureau de Recherches géologiques et minières* **28**: 885–9.

Hay, W. W. and Mohler, H. P., 1967, Calcareous nanoplankton from early Tertiary rocks at Pont Labau, France, and Paleocene–Eocene correlations. *Journal of Paleontology* **41**: 1505–41.

Hay, W. W., Mohler, H. P., Roth, H. P., Schmidt, R. R., and Boudreaux, J. E., 1967, Calcareous nanoplankton zonation of the Cenozoic of the Gulf Coast and Caribbean–Antillan area, and trans-oceanic correlation. *Transactions of the Gulf-Coast Association of Geological Societies* **17**: 428–80.

Hayes, J. M., Strauss, H., and Kaufman, A. J., 1999, The abundance of ^{13}C in marine organic matter and isotopic fractionation in the global biogeochemical cycle of carbon during the past 800 Ma. *Chemical Geology* **161**: 103–25.

Hays, J. D., Imbrie, J., and Shackleton, N. J., 1976, Variations in the Earth's orbit: pacemaker of the ice ages. *Science* **194**: 1121–32.

Heaman, L. and Parrish, R., 1991, U–Pb geochronology of accessory minerals. In L. Heaman, and J. N. Ludden, eds., *Applications of Radiogenic Isotope Systems to Problems in Geology: Short Course Handbook, Mineralogical Association of Canada* 19, pp. 59–102.

Heath, R. J., Brenchley, P. J., and Marshall, J. D., 1998, Early Silurian carbon and oxygen stable-isotope stratigraphy of Estonia: implications for climate change. In E. Landing, and M. E. Johnson, eds., *Silurian Cycles: Linkages of Dynamic Stratigraphy with Atmospheric, Oceanic and Tectonic Changes, New York State Museum Bulletin*, Vol. 491. Albany, NY: State Education Department, pp. 313–27.

Heckel, P. H., 1986, Sea-level curve for Pennsylvanian eustatic marine transgressive–regressive depositional cycles along the midcontinent outcrop belt, North America. *Geology* **14**: 330–4.

1989, Updated Middle–Upper Pennsylvanian eustatic sea-level curve for the Midcontinent North America and preliminary biostratigraphic characterization. In *Compte Rendu Onieme Congrès International de Stratigraphie et de Géologie du Carbonifère 1987*, Vol. 4. Nanjing: Nanjing University Press, pp. 160–85.

1999, Overview of Pennsylvanian (Upper Carboniferous) stratigraphy in Midcontinent region of North America. In P. H. Heckel, ed., *Guidebook, Fieldtrip No. 8, XIV International Congress on the Carboniferous–Permian, Kansas Geological*

Survey Open File Report 99–27. Lawrence, KS: Kansas Geological Survey, pp. 68–102.

2002, Overview of Pennsylvanian cyclothems in Midcontinent North America and brief summary of those elsewhere in the world. In L. V. Hills, M. Charles, and E. W. Bamber, eds., *Carboniferous and Permian of the World, XIV ICCP Proceedings Memoir* 19. Calgary: Canadian Society of Petroleum Geologists, pp. 79–98.

Heckel, P. H., Alekseev, A. S., and Nemyrovska, T. I., 1998, Preliminary conodont correlations of late Middle to early Upper Pennsylvanian rocks between North America and eastern Europe. *Newsletters on Carboniferous Stratigraphy* **16**: 8–12.

Heckel, P. H., Boardman, D. R., and Barrick, J. E., 2002, Desmoinesian–Missourian regional stage boundary reference position for North America. In L. V. Hills, C. M. Henderson, and E. W. Bamber, eds., *Carboniferous and Permian of the World: XIV ICCP Proceedings, Memoir of the Canadian Society of Petroleum Geologists*, Vol. 19. Calgary: Canadian Society of Petroleum Geologists, pp. 710–24.

Heckert, A. B. and Lucas, S. G., 1999, Global correlation and chronology of Triassic theropods (Archosauria: Dinosauria). *Albertiana* **23**: 22–35.

Hedberg, H. D., 1976, *International Stratigraphic Guide. A Guide to Stratigraphic Classification, Terminology, and Procedure*. New York: Wiley.

Heilmann-Clausen, C., 1985, Dinoflagellate stratigraphy of the uppermost Danian to Ypresian in the Viborg 1 borehole, central Jylland, Denmark. *Danske Geologiske Undersøgelse* **A7**: 1–89.

1988, The Danish Sub-basin. Paleogene dinoflagellates. *Neues Jahrbuch für Geologie und Paläontologie Abhandlungen* **A101**: 339–43.

1994, Review of Paleocene dinoflagellates from the North Sea region. *Geologiska Föreningen i Stockholm Förhandlingar* **116**: 51–3.

Heinrich, H., 1988, Origin and consequences of cyclic ice rafting in the northeast Atlantic Ocean during the past 130,000 years. *Quaternary Research* **29**: 143–52.

Heinzelin, J. de and Tavernier, R., 1957, Flandrien. In P. Pruvost, ed., *Lexique stratigraphique international*, Vol. 1, *Europe*. Paris: Centre National de la Recherche Scientifique, p. 32.

Heirtzler, J. R., Dickson, G. O., Herron, E. M., Pitman, W. C., and Le Pichon, X., 1968, Marine magnetic anomalies, geomagnetic field reversals, and motions of the ocean floor and continents. *Journal of Geophysical Research* **73**: 2119–39.

Helby, R., Morgan, R., and Partridge, A. D., 1987, A palynological zonation of the Australian Mesozoic. In P. A. Jell, ed., *Studies in Australian Mesozoic Palynology, Memoir Association Australasian Paleontologists*, Vol. 4. Association of Australasian Paleontologists, pp. 1–94.

Heller, F., Chen, H., Dobson, J., and Haag, M., 1995, Permian–Triassic magnetostratigraphy: new results from South China. *Physics of the Earth and Planetary Interiors* **89**(3–4): 281–95.

Heller, F., Haag, M., Gialanella, P. R., *et al.*, 1998, Late Permian magnetostratigraphy on the eastern part of the Russian Platform. *Abstracts, Upper Permian Stratotypes of the Volga region*. Kazan: Kazan State University, pp. 67–68.

Hellman, K. N. and Lippolt, H. J., 1981, Calibration of the Middle Triassic time scale by conventional K–Ar and $^{40}Ar/^{39}Ar$ dating of alkali feldspars. *Journal of Geophysics* **50**: 73–88.

Henderson, C. M., 1988, Conodont paleontology and biostratigraphy of the Upper Carboniferous to Lower Permian Canyon Fiord, Belcher Channel, Nansen, unnamed, and Van Hauen Formations, Canadian Arctic Archipelago. Unpublished Ph.D. thesis, University of Calgary.

2002, Kungurian to Lopingian correlations along western Pangea. *Geological Society of America Abstracts with Programs* **34**(3): A–30.

Henderson, C. M., Martel, D. J., O'Nions, R. K., and Shackleton, N. J., 1994, Evolution of seawater ^{87}Sr / ^{86}Sr over the last 400 ka: absence of glacial/interglacial cycles. *Earth and Planetary Science Letters* **128**: 643–51.

Henderson, C. M., Wardlaw, B., and Mei, S., 2001, New conodont definitions at the Guadalupian–Lopingian Boundary. *Permophiles* **38**: 35–6.

Henderson, C. M., Mei, S. L., and Wardlaw, B. R., 2002, New conodont definitions at the Guadalupian–Lopingian boundary. In L. V. Hills, C. M. Henderson, and W. B. E., eds., *Carboniferous and Permian of the World, Canadian Society of Petroleum Geologists Memoir* 19. Calgary: Canadian Society of Petroleum Geologists, pp. 725–35.

Henderson, G. and Slowey, N. C., 2000, Evidence against northern-hemisphere forcing of the penultimate deglaciation from U–Th dating. *Nature* **402**: 61–6.

Henderson, R. A., 1976, Upper Cambrian (Idamean) trilobites from western Queensland, Australia. *Palaeontology* **19**(2): 325–64, pls 47–51.

1977, Stratigraphy of the Georgina Limestone and a revised zonation for the early Upper Cambrian Idamean Stage. *Journal of the Geological Society of Australia* **23**: 423–33.

Henningsmoen, G., 1957, The trilobite family Olenidae. *Skrifter Utgitt av det Norske Videnskaps-Akademi i Oslo, 1, Matematisk-naturvidenskapelig Klasse* **1**: 303, 31 plates.

1972, The Cambro–Ordovician boundary. *Lethaia* **6**: 423–39.

Henriques, M. H., 1992, Biostratigrafia e paleontologia (Ammonoidea) do Aaleniano em Portugal (Sector Setentrional da Bacia Lusitaniana). Unpublished Ph.D. thesis, Centro de Geociências da Universidade de Coimbra, Instituto Nacional de Investigacao Cientifica.

Henriques, M. H., Sadki, D., and Mouterde, R., 1988, Graphoceratidés (*Ammonitina*) de la base du Bajocien portugais. In R. Rocha and A. F. Soares, eds., *Second International Symposium on Jurassic Stratigraphy*, Lisbon, September 1987, pp. 243–54.

Henriques, M. H., Gardin, S., Gomes, C. R., *et al.*, 1994, The Aalenian–Bajocian boundary at Cabo Mondego (Portugal). In S. Cresta and G. Pavia, eds., *Proceedings of the Third International Meeting on Aalenian and Bajocian Stratigraphy. Miscellanea del Servizio Geologico Nazionale*, Vol. 5. Rome: Instituto Poligrafico e Zecca dello Stato, pp. 63–77.

Herbert, T. D., 1992, Paleomagnetic calibration of Milankovitch cyclicity in Lower Cretaceous sediments. *Earth and Planetary Science Letters* **112**: 15–28.

1994, Reading orbital signals distorted in sedimentation. In De Boer, P. and Smith, D. G., eds. *Orbital Forcing and Cyclic Sequences, International Association of Sedimentologists Special Publication*, Vol. 19. Oxford: Blackwell Scientific, pp. 483–508.

1999, Toward a composite orbital chronology for the Late Cretaceous and Early Paleocene GPTS. *Philosophical Transactions of the Royal Society of London, series A* **357**: 1891–905.

Herbert, T. D., D'Hondt, S. L., Premoli-Silva, I., Erba, E., and Fischer, A. G., 1995, Orbital chronology of Cretaceous–Early Palaeocene marine sediments. In W. A. Berggren, D. V. Kent, and J. Hardenbol, eds., *Geochronology, Time Scales and Global Stratigraphic Correlations: A Unified Temporal Framework for a Historical Geology, Society of Economic Paleontologists and Mineralogists Special Publication*, Vol. 54. Tulsa, OK: SEPM, pp. 81–94.

Herschel, J. F. W., 1830, On the geological causes which may influence geological phenomena. *Geological Transactions* **3**: 293–9.

Heslop, D., Langereis, C. G., and Dekkers, M. J., 2000, A new astronomical time scale for the loess deposits of Northern China. *Earth and Planetary Science Letters* **184**: 125–39.

Hess, J. C. and Lippolt, H. J., 1986, $^{40}Ar/^{39}Ar$ ages of tonstein and tuff sanidines: new calibration points for the improvement of the Upper Carboniferous time scale Series/Source: calibration of the Phanerozoic time scale. *Chemical Geology* **59**(2–3): 143–54.

Hess, J., Stott, L. D., Bender, M. L., Kennet, J. P., and Schilling, J.-G., 1989, The Oligocene marine microfossil record: age assessments using strontium isotopes. *Paleoceanography* **4**: 655–79.

Hess, J. C., Lippolt, H. J., and Burger, K., 1999, High-precision $^{40}Ar/^{39}Ar$ spectrum dating on sanidine from the Donets Basin, Ukraine: evidence for correlation problems in the Upper Carboniferous. *Journal of the Geological Society of London* **156**: 527–33.

Hesselbo, S. P. and Jenkyns, H. C., 1995, A comparison of the Hettangian to Bajocian successions of Dorset and Yorkshire. In P. D. Taylor, ed., *Field Geology of the British Jurassic*. London: Geological Society, pp. 105–50.

1998, British Lower Jurassic sequence stratigraphy. In P.-C. de Graciansky, J. Hardenbol, T. Jacquin, and P. R. Vail, eds., *Mesozoic–Cenozoic Sequence Stratigraphy of European Basins, Society of Economic Paleontologists and Mineralogists Special Publication*, Vol. 60. Tulsa, OK: SEPM, pp. 562–81.

Hesselbo, S. P., Coe, A. L., and Jenkyns, H. C., 1990, Recognition and documentation of depositional sequences from outcrop: an example from the Aptian and Albian on the eastern margin of the Wessex Basin. *Journal of the Geological Society of London* **147**: 549–59.

Hesselbo, S., Meister, C., and Gröcke, D. R., 2000a, A potential global stratotype for the Sinemurian–Pliensbachian boundary (Lower Jurassic), Wine Haven, Robin Hood's Bay, Yorkshire, UK: ammonite faunas and isotope stratigraphy. *Geological Magazine* **137**: 601–07.

Hesselbo, S. P., Grocke, D. R., Jenkyns, H. C., *et al.*, 2000b, Massive dissociation of gas hydrate during a Jurassic oceanic anoxic event. *Nature* **406**: 392–5.

Hesselbo, S. P., Robinson, S. A., Surlyk, F., and Piasecki, S., 2002, Terrestrial and marine extinction at the Triassic–Jurassic boundary synchronized with major carbon-cycle perturbation: a link to initiation of massive volcanism? *Geology* **30**: 251–4.

Hesselbo, S. P., Morgans-Bell, H. S., McElwain, J. C., *et al.*, 2003, Carbon-cycle perturbation in the Middle Jurassic and accompanying changes in the terrestrial paleoenvironment. *Journal of Geology* **111**: 259–76.

Hicks, H., 1881, The classification of the Eozoic and Lower Paleozoic rocks of the British Isles. *Popular Science Review* **5**: 289–308.

Hicks, J. F. and Obradovich, J. D., 1995, Isotopic age calibration of the GRTS from C33N to C31N: Late Cretaceous Pierre Shale, Red Bird section, Wyoming, USA. *Geological Society of America Abstracts with Programs* **27**: A-174.

Hicks, J. F., Obradovich, J. D., and Tauxe, L., 1995, A new calibration point for the Late Cretaceous time scale: the $^{40}Ar/^{39}Ar$ isotopic age of the C33r/C33n geomagnetic reversal from the Judith River Formation (Upper Cretaceous), Elk Basin, Wyoming, USA. *Journal of Geology* **103**: 243–56.

1999, Magnetostratigraphy, isotopic age calibration and intercontinental correlation of the Red Bird section of the Pierre Shale, Niobrara County, Wyoming, USA. *Cretaceous Research* **20**: 1–27.

Higgins, A. C., 1975, Conodont zonation of the late Visean–early Westphalian strata of the south and central Pennines of northern England. *Geological Survey of Great Britain Bulletin*, Vol. 53. Keyworth: British Geological Survey, pp. 1–90.

Higgs, K. T., McPhilemy, B., Keegan, J. B., and Clayton, G., 1988, New data on palynological boundaries within the Irish Dinantian. *Review of Palaeobotany and Palynology* **56**(1–2): 61–8.

Higgs, K. T., Streel, M., Korn, D., and Paproth, E., 1993, Palynological data from the Devonian–Carboniferous boundary beds in the new Stockum trench II and the Hasselbach borehole, Northern Rhenish Massif, Germany. *Annales de la Société Géologique de Belgique* **115**: 551–7.

Hildebrand, A. R., Penfield, G. T., Kring, D. A., *et al.*, 1991, Chicxulub Crater: a possible Cretaceous–Tertiary boundary impact on the Yucatan Peninsula, Mexico. *Geology* **19**: 867–71.

Hilgen, F. J., 1991a, Extension of the astronomically calibrated (polarity) time scale to the Miocene–Pliocene boundary. *Earth and Planetary Science Letters* **107**: 349–68.

1991b, Astronomical calibration of Gauss to Matuyama sapropels in the Mediterranean and implication for the Geomagnetic Polarity Time Scale. *Earth and Planetary Science Letters* **104**: 226–44.

Hilgen, F. J. and Krijgsman, W., 1999, Cyclostratigraphy and astrochronology of the Tripoli diatomite formation (pre-evaporite Messinian, Sicily, Italy). *Terra Nova* **11**: 16–22.

Hilgen, F. J. and Langereis, C. G., 1988, The age of the Miocene–Pliocene boundary in the Capo Rossello area (Sicily). *Earth and Planetary Science Letters* **91**: 214–22.

Hilgen, F. J., Krijgsman, W., Langereis, C. G., *et al.*, 1995, Extending the astronomical (polarity) time scale into the Miocene. *Earth and Planetary Science Letters* **136**: 495–510.

Hilgen, F. J., Krijgsman, W., and Wijbrans, J. R., 1997, Direct comparison of astronomical and $^{40}Ar/^{39}Ar$ ages of ash beds: potential implications for the ages of mineral dating standards. *Geophysical Research Letters* **24**: 2043–46.

Hilgen, F. J., Bissoli, L., Iaccarino, S., *et al.*, 2000a, Integrated stratigraphy and astrochronology of the Messinian GSSG at Oued Akrech (Atlantic Morocco). *Earth and Planetary Science Letters* **182**: 237–51.

Hilgen, F. J., Iaccarino, S., Krijgsman, W., *et al.*, 2000b, The Global Boundary Stratotype Section and Point (GSSP) of the Messinian Stage (uppermost Miocene). *Episodes* **23**: 172–78.

Hilgen, F. J., Krijgsman, W., Raffi, I., Turco, E., and Zachariasse, W. J., 2000c, Integrated stratigraphy and astronomical calibration of the Serravallian/Tortonian boundary section at Monte Gibliscemi, Sicily. *Marine Micropaleontology* **38**: 181–211.

Hilgen, F. J., Abdul-Aziz, H., Krijgsman, W., Raffi, I., and Turco, E., 2003, Integrated stratigraphy and astronomical tuning of Serravallian and lower Tortonian at Monte dei Corvi (Middle–Upper Miocene, Northern Italy). *Palaeogeography, Palaeoclimatology, Palaeoecology* **199**: 229–64.

Hilgen, F. J., Iaccarino, S., Krijgsman, W., *et al.* in press, The global stratotype section and point (GSSP) of the Tortonian stage (Upper Miocene): a proposal. Submitted to the International Union of Geological Sciences, International Commission on Stratigraphy, Subcommission on Neogene Stratigraphy.

Hinnov, L. A., 2000, New perspectives on orbitally forced stratigraphy: *Annual Review of Earth and Planetary Sciences* **28**: 419–75.

Hinnov, L. A. and Goldhammer, R. K., 1991, Spectral analysis of the Middle Triassic Latemar Limestone. *Journal of Sedimentary Petrology* **61**: 1173–93.

Hinnov, L. A. and Park, J. J., 1999, Strategies for assessing Early–Middle (Pliensbachian–Aalenian) Jurassic cyclochronologies. *Philosophical Transactions of the Royal Society of London, series A* **357**: 1831–59.

Hinnov, L. A., Park, J. J., and Erba, E., 1999, Lower–middle Jurassic rhythmites from the Lombard Basin, Italy: a record of orbitally-forced cycles modulated by long-term secular environmental changes in west Tethys. In R. L. Hall and P. L. Smith, eds., *Advances in Jurassic Research* 2000, Proceedings of the Fifth International Symposium on the Jurassic System, 1998, GeoResearch Forum, Vol. 6. Zurich: Transtech Publications, pp. 437–53.

Hochuli, P. A., 1998, Dinoflagellate systs. Spore pollen [for Triassic portion, Mesozoic and Cenozoic sequence chronostratigraphic framework of European basins, in Hardenbol, J. *et al.*]. In P.-C. de Graciansky, J. Hardenbol., Th. Jacquin, and P. R. Vail, eds., *Mesozoic–Cenozoic Sequence Stratigraphy of European Basins, Society of Economic Paleontologists and Mineralogists Special Publication,* Vol. 60. Tulsa, OK: SEPM, pp. 781–2, Chart 8.

Hochuli, P. A., Menegatti, A. P., Weissert, H., *et al.* 1999, Episodes of high productivity and cooling in the early Aptian Alpine Tethys. *Geology* **27**: 657–60.

Hodell, D. A. and Woodruff, F., 1994, Variations in strontium isotopic ratio of sea water during the Miocene: stratigraphic and geochemical implication. *Paleoceanography* **9**: 405–26.

Hodell, D. A., Mueller, P. A., and Garrido, J. R., 1991, Variations in the strontium isotopic composition of seawater during the Neogene. *Geology* **19**: 24–7.

Hodell, D. A., Benson, R. H., Kent, D. V., Boersma, A., and Rakic-El Bied, K., 1994, Magnetostratigraphic, biostratigraphic, and stable isotope stratigraphy of an Upper Miocene drill core from the Salé Briqueterie (northwestern Morocco): a high-resolution chronology for the Messinian stage. *Paleoceanography* **9**: 835–55.

Hodych, J. P. and Dunning, G. R., 1992, Did the Manicouagan impact trigger end-of-Triassic mass extinction? *Geology* **20**: 51–4.

Hoedemaeker, P. J. and Bulot, L., 1990, Preliminary ammonite zonation for the Lower Cretaceous of the Mediterranean region: report. *Géologie Alpine* **66**: 123–7.

Hoedemaeker, P. J. and Leereveld, H., 1995, Biostratigraphy and sequence stratigraphy of the Berriasian–lowest Aptian (Lower

Cretaceous) of the Río Argos succession, Caravaca, SE Spain. *Cretaceous Research* **16**: 195–230.

Hoedemaeker, P. J., Company, M. R., Aguirre Urreta, M. B., *et al.*, 1993, Ammonite zonation for the Lower Cretaceous of the Mediterranean region: basis for the stratigraphic correlation within IGCP. Project 262. *Revista Española de Paleontología* **8**: 117–20.

Hoedemaeker, P. J., Reboulet, S., *et al.*, 2003, Report on the 1st International Workshop of the IUGS Lower Cretaceous Ammonite Working Group, the "Kilian Group" (Lyon, July 11, 2002). *Cretaceous Research* **24**: 89–94.

Hofmann, H. J., 1990, Precambrian time units and nomenclature – the geon concept. Geology **18**: 340–1.

1991, Letter: Precambrian time units; geon or geologic unit? *Geology* **19**(9): 958–9.

1994, Problematic carbonaceous compressions ("metaphytes" and "worms"). In S. Bengtson, ed., *Early Life on Earth.* New York: Columbia University Press, pp. 342–58.

1999, Geons and geons. Geology **27**: 855–6.

Hofmann, H. J., Narbonne, G. M., and Aitken, J. D., 1990, Ediacaran remains from intertillite beds in northwestern Canada. *Geology* **18**: 199–1202.

Holdsworth, B. K. and Murchey, B. L., 1988, Paleozoic radiolarian biostratigraphy of the northern Brooks Range, Alaska. In *Geology and Exploration of the National Petroleum Reserve in Alaska, 1974 to 1982, US Geological Survey Professional Paper.* US Geological Survey, pp. 777–97.

Holland, C. H., 1989, History of classification of the Silurian System. In C. H. Holland and M. G. Bassett, eds., *A Global Standard for the Silurian System. Geological Series* Vol. 9. Cardiff: National Museum of Wales.

Holland, C. H. and Bassett, M. G., 1989, *A Global Standard for the Silurian System, Geological Series* Vol.9. Cardiff: National Museum of Wales.

Holland, C. H., Lawson, J. D., and Walmsley, V. G., 1963, The Silurian rocks of the Ludlow District, Shropshire. *Bulletin of the British Museum of Natural History* (*Geology*) **8**: 93–171.

Hollingsworth, J. S., 1997, Cambrian trilobite extinctions, biomeres and stages. *Mid-America Paleontological Society Digest* **20**(4): 17–27.

Hollis, C. J., 1993, Latest Cretaceous to Late Paleocene radiolarian biostratigraphy: a new zonation from the New Zealand region. *Marine Micropaleontology* **21**: 295–327.

1997, Cretaceous–Paleocene radiolaria from eastern Marlborough, New Zealand. *Institute of Geological and Nuclear Sciences, Monograph* **17**: 152.

Hollis, C. J., Waghorn, D. P., Strong, C. P., and Crouch, E. M., 1997, *Integrated Paleogene Biostratigraphy of DSDP Site 277 (Leg 29), Foraminifera, Calcareous Nannofossils, Radiolaria and Palynomorphs, Science Report* 7. Lower Hutt: New Zealand Institute of Geological and Nuclear Sciences, p. 87.

Holmes, A., 1913, *The Age of the Earth.* London: Harper.

1937, *The Age of the Earth.* London: Nelson.

1947, The construction of a geological time-scale. *Transactions Geological Society of Glasgow* **21**: 117–52.

1960, A revised geological time-scale. *Transactions of the Edinburgh Geological Society* **17**: 183–216.

Holroyd, P. A. and Ciochon, R. L., 1994, Relative ages of Eocene primate-bearing deposits of Asia. In J. G. Fleagle and R. F. Kay, eds., *Anthropoid Origins.* New York: Plenum, pp. 123–41.

Holser, W. T. and Magaritz, M., 1987, Events near the Permian–Triassic boundary. *Modern Geology* **11**: 155–80.

Hood, K. C., 1997, GRAPHCOR–Interactive Graphic Correlations Software, Private Edition.

Hooker, J. J., 1986, Mammals from the Bartonian (middle/late Eocene) of the Hampshire Basin, southern England. *Bulletin of the British Museum of Natural History (Geology)*, **39**(191): 478.

1987, Mammalian faunal events in the English Hampshire Basin (late Eocene–early Oligocene) and their application to European biostratigraphy. In N. Schmidt-Kittler, ed., International Symposium on Mammalian Biostratigraphy and Paleoecology of the European Paleogene, Mainz, February 18–21. *Münchner Geowissenschaftliche Abhandlungen A* **10**.

1991, The sequences of mammals in the Thanetian and Ypresian of the London and Belgian basins. Location of the Paleocene–Eocene boundary. *Newsletters on Stratigraphy* **25**: 75–90.

1996, Mammalian biostratigraphy across the Paleocene–Eocene boundary in the Paris, London and Belgian basins. *Geological Society of London Special Publication,* Vol. **101**. London: Geological Society, pp. 205–18.

1998, Mammalian faunal change across the Paleocene–Eocene transition in Europe. In M.-P. Aubry, S. G. Lucas, and W. A. Berggren, eds., *Late Paleocene–Early Eocene Climatic and Biotic Events in the Marine and Terrestrial Records.* New York: Columbia University Press, pp. 428–50.

Hooker, J. J. and Millbank, C., 2001, A Cernaysian mammal from the Upnor Formation (Late Paleocene, Herne Bay, UK) and its implications for correlation. *Proceedings of the Geologists' Association* **89**: 331–38.

Hopwood, A. T., 1935, Fossil elephants and Man. *Proceedings of the Geologists' Association* **46**: 46–60.

Horner, F., 1983, Paläomagnetismus von Karbonatsedimenten der Südlichen Tethys: Implikationen für die Polarität des Erdmagnetfeldes im untern Jura und für die Tektonik der Ionischen Zone Griechenlands. Unpublished Ph.D. thesis, ETH, Zurich, p. 139.

Horner, F. and Heller, F., 1983, Lower Jurassic magnetostratigraphy at the Breggia Gorge (Ticino, Switzerland) and Alpi Turati (Como, Italy). *Geophysical Journal of the Royal Astronomical Society.***73**: 705–18.

Hörnes, M., 1853, Mitteilung an Prof. Bronn gerichtet: Wien, 3. Okt., 1853. *Neues Jahrbuch für Mineralogie, Geologie, Geognosie und Petrefaktenkunde*, pp. 806–10.

1864, Die fossilen Mollusken des Tertiärbeckens von Wien. *Jahrbuch der Geologischen Reichsanstalt* **14**: 509–14.

Horng, C.-S., Lee, M.-Y., Pälike, H., *et al.*, 2002, Astronomically calibrated ages for geomagnetic reversals within the Matuyama chron. *Earth Planets Space* **54**: 679–90.

Hornibrook, N. D., Brazier, R. C., and Strong, P. C., 1989, Manual of New Zealand Permian to Pleistocene foraminiferal biostratigraphy. *New Zealand Geological Survey Paleontological Bulletin* **56**: 175.

Hottinger, L., 1960, Recherches sur les Alvéolines du Paléocène et de l'Eocène. *Mémoires Suisses de Paléontologie* **75–76**: 243.

1977, Foraminifères operculiniformes. *Mémoires du Museum National d'Histoire Naturelle de Paris* **C40**: 159.

Hottinger, L. and Schaub, H., 1960, Zur Stufeneinteilung des Paläoz äns und des Eozäns. Einführung der Stufen Ilerdien und Biarritzien. *Eclogae Geologicae Helvetica* **53**: 453–80.

Hounslow, M., Mørk, A., Peters, C., and Weitschat, W., 1996, Boreal Lower Triassic magnetostratigraphy from Deltadalen, central Svalbard. *Albertiana* **17**: 3–10.

House, M. R., 1983, Devonian eustatic events. *Proceedings of the Ussher Society* **5**: 396–405.

1985, Correlation of mid-Palaeozoic ammonoid evolutionary events with global sedimentary perturbations: *Nature* **313**: 17–22.

1991, Devonian Period. *Encyclopedia Brittanica* **19**: 804–14.

1995, Devonian precessional and other signatures for establishing a Givetian timescale. *Geological Society of London Special Publication,* Vol. 85. London: Geological Society, pp. 37–49.

2000, Chronostratigraphic framework for the Devonian and Old Red Sandstone. In P. F. Friend and B. P. J. Williams, eds., *New Perspectives on the Old Red Sandstone.* London: Geological Society, pp. 23–7.

2002, Strength, timing, setting and cause of mid-Palaeozoic extinctions. *Palaeogeography, Palaeoclimatology, Palaeoecology* **181**: 5–25.

House, M. R. and Ziegler, W., 1997, Devonian eustatic fluctuation in North Eurasia. *Courier Forschunginstitut Senckenberg* **199**: 13–23.

House, M. R., Feist, R., and Korn, D., 2000, The Middle/Upper Devonian boundary at Puech de la Suque, Southern France. *Courier Forschungsinstitut Senckenberg* **225**: 49–58.

Howarth, R. J. and McArthur, J. M., 1997, Statistics for strontium isotope stratigraphy: a robust LOWESS fit to the marine strontium isotope curve for the period 0 to 206 Ma, with look-up table for the derivation of numerical age. *Journal of Geology* **105**: 441–56.

Huang, T. K., 1932, The Permian formations of Southern China. *Memoirs of the Geological Survey of China, Series A* **10**: 1–40.

Huang, Z., Ogg, J. G., and Gradstein, F. M., 1993, A quantitative study of Lower Cretaceous cyclic sequences from the Atlantic Ocean and the Vocontian Basin (SE France). *Paleoceanography* **8**: 275–91.

Huber, B. T., Hodell, D. A., and Hamilton, C. P., 1995, Middle–Late Cretaceous climate of the southern high latitudes: stable isotopic evidence for minimal equator-to-pole thermal gradients. *Geological Society of America Bulletin* **107**: 1164–91.

Huber, B. T., Norris, R. D., and MacLeod, K. G., 2002, Deep sea paleotemperature record of extreme warmth during the Cretaceous. *Geology* **30**: 123–6.

Huff, W. D., Bergström, S. M., and Kolata, D. R., 1992, Gigantic Ordovician volcanic ash fall in North America and Europe: biological, tectonomagmatic, and event-stratigraphic significance. *Geology* **20**: 875–8.

2000, Silurian K-bentonites of the Dnestr Basin, Podolia, Ukraine. *Journal of the Geological Society* **157**: 493–504.

Huff, W., Davis, D., Bergström, S. M., e*t al.* 1997, A biostratigraphically well-constrained K-bentonite U–Pb zircon age of the lowermost Darriwilian Stage (Middle Ordovician) from the Argentine Precordillera. *Episodes* **20**(1): 29–33.

Hughen, K. A., Overpeck, J. T., Lehman, S. J., *et al.*, 1998, Deglacial changes in ocean circulation from an extended radiocarbon calibration. *Nature* **391**: 65–68.

Hughes, C. P., Jenkins, C. J., and Rickards, R. B., 1982, Abereiddy Bay and the adjacent coast. In M. G. Bassett, ed., *Geological Excursions in Dyfed, South-West Wales.* Cardiff: National Museum of Wales, pp. 51–63.

Hunter, M. A., Bickle, M. J., Nisbet, E. G., Martin, A., and Chapman, H. J., 1998, Continental extensional setting for the Archean Belingwe greenstone belt, Zimbabwe. *Geology* **26**(10): 883–6.

Hutt, J. E., 1974, A new group of Llandovery biform monograptids. In R. B. Rickards, D. E. Jackson, and C. P. Hughes, eds., *Graptolite Studies in Honour of O. M. B. Bulman, Special Papers, Paleontology* **13**: 189–203.

Hyvärinen, H., 1978, Use and definition of the term Flandrian. *Boreas* **7**: 182.

Iaccarino, S., 1985, Mediterranean Miocene and Pliocene planktic foraminifera. In H. M. Bolli, J. B. Saunders, and K. Perch-Nielsen, eds., *Plankton Stratigraphy.* Cambridge: Cambridge University Press, pp. 283–314.

Iannuzzi, R. and Pfefferkorn, H. W., 2002, A pre-glacial, warm-temperate floral belt in Gondwana (Late Visean, Early Carboniferous). *Palaios* **17**: 571–90.

Idnurm, M., Klootwijk, C., Théveniaut, H., and Trench, A., 1996, Magnetostratigraphy. In G. C. Young and J. R. Laurie, eds., *An Australian Phanerozoic Timescale.* Oxford: Oxford University Press, pp. 22–51.

Imbrie, J. and Imbrie, K., 1979, *Ice Ages: Solving the Mystery.* Short Hills: Harvard University Press.

Imbrie, J., Hays, J. D., Martinson, D. G., *et al.*, 1984, The orbital theory of Pleistocene climate: support from a revised chronology of the marine $\delta^{18}O$ record. In A. L. Berger, J. Imbrie, J. Hays, G. Kukla, and B. Saltzman, eds., *Milankovitch and Climate*, Part I. Dordrecht: D. Riedel, pp. 269–305.

Imlay, R. W., 1967. Twin Creek Limestone (Jurassic) in the Western Interior of the United States. *US Geological Survey Professional Paper* 540. Reston, VA: US Geological Survey, 105 pp.

1980. Jurassic paleobiogeography of the conterminous United States in its continental setting. *US Geological Survey Professional Paper* 1062. Reston, VA: US Geological Survey, 134 pp.

Innocenti, M., Mangold, C., Pavia, G., and Torrens, H. S., 1988, A proposal for the formal ratification of the boundary stratotype of the Bathonian stage based on a Bas Auran section (S. E. France). In R. B. Rocha and A. F. Soares, eds., *2nd International Symposium on Jurassic Stratigraphy*, Vol. I. Lisbon: International Subcommission on Jurassic Stratigraphy Instituto Nacional de Investigação Científica, pp. 333–46.

Inosova, K. I., Kruzina, A. K., and Shvartsman, E. G., 1976, *Atlas of the Miospores and Pollen of the Upper Carboniferous and Lower Permian of the Donets Basin Moscow*. Leningrad: Nedra, pp. 1–154 [In Russian].

International Subcommission on Stratigraphic Classification, 1979, Magnetic polarity units, a supplemental chapter of the International Subcommission on Stratigraphic Classification, International Stratigraphic Guide. *Geology* 7: 578–83.

INTERRAD Jurassic–Cretaceous Working Group, 1995, Middle Jurassic to Lower Cretaceous radiolaria of Tethys: occurrences, systematics, biochronology. *Mémoires de Géologie Lausanne* 23: 1172.

Ioannides, N., Riding, J., Stover, L. E., and Monteil, E., 1998, Dinoflagellate cysts. [columns for Jurassic chart, Mesozoic and Cenozoic sequence chronostratigraphic framework of European basins, in Hardenbol, J. *et al.*]. In P.-C. de Graciansky, J. Hardenbol., Th. Jacquin, and P. R. Vail, eds., *Mesozoic–Cenozoic Sequence Stratigraphy of European Basins, Society of Economic Paleontologists and Mineralogists Special Publication,* Vol. 60. Tulsa, OK: SEPM, Chart 7.

Iranian–Chinese Research Group, 1995, Field work on the Lopingian stratigraphy in Iran. *Permiophiles* 27: 5–6.

Irving, E., 1966, Paleomagnetism of some Carboniferous rocks from New South Wales and its relation to geological events. *Journal of Geophysical Research* 71: 6025–51.

Irving, E. and Parry, L. G., 1963, The magnetism of some Permian rocks from New South Wales. *Geophysical Journal of the Royal Astronomical Society* 7(4): 395–411.

Irving, E. and Pullaiah, G., 1976, Reversals of the geomagnetic field, magnetostratigraphy, and relative magnitude of paleosecular variation in the Phanerozoic. *Earth Science Reviews* 12: 35–64.

Isachsen, C. E., Bowring, S. A., Landing, E., and Samson, S. D., 1994, New constraint on the division of Cambrian time. *Geology* 22: 496–8.

Italian Commission on Stratigraphy, 2002, Quaternary chronostratigraphy and establishment of related standards. Forum. *Episodes* 25(4).

Ivanov, A. P., 1926, Middle–Upper Carboniferous deposits of the Moscow province. *Bulletin of Moscow Society of Natural Studies, Geological Series* 4: 133–80 [In Russian].

Ivanova, E. A. and Khvorova, I. V., 1955, Stratigraphy of the Middle and Upper Carboniferous of the western part of the Moscow syneclise. *Transactions of the Paleontological Institute of the Academy of Sciences of the USSR* 53: 3–279 [In Russian].

Ivanova, E. A. and Rosovskaya, C. E., 1967, Biostratigraphy of the Upper Carboniferous of the Russian Platform in the light of the type sections study. *Moscow Society of Natural Studies Bulletin, Geological Series* 42(5): 89–99 [In Russian].

Izart, A., Briand, C., Vaslet, D., *et al.*, 1998, Stratigraphy and sequence stratigraphy of the Upper Carboniferous and Lower Permian in the Donets Basin. *Memoires du Museum National d'Histoire Naturelle* 177: 9–33.

Izart, A., Kossovaya, O., Vachard, D., and Vaslet, D., 1999, Stratigraphy, sedimentology and sequence stratigraphy of the Permian along the Kosva River (Gubakha area, Central Urals, Russia). *Bulletin de la Société Géologique de France* 170(6): 799–820.

Izett, G. A., 1982, Stratigraphic succession, isotopic ages, partial chemical analyses and sources of certain silicic volcanic ash beds (4.0 to 0.1 my) of the western United States. *US Geological Survey Open File Report* 82-0582. Reston, VA: US Geological Survey, p. 47.

1990, The Cretaceous/Tertiary boundary interval, Raton Basin, Colorado and New Mexico, and its content of shock metamorphosed minerals. *Geological Society of America Special Paper* 249. Boulder, CO: Geological Society of America, pp. 1–100.

Jaanusson, V., 1960, On series of the Ordovician System. *Reports of the 21st International Geological Congress*, Copenhagen, Vol. 7, pp. 70–81.

Jacquin, Th. and de Graciansky, P.-C., 1998, Major transgressive–regressive cycles: the stratigraphic signature of European basin development. In P.-C. de Graciansky, J. Hardenbol., Th. Jacquin, and P. R. Vail, eds., *Mesozoic–Cenozoic Sequence Stratigraphy of European Basins, Society of Economic Paleontologists and Mineralogists Special Publication,* Vol. 60. Tulsa, OK: SEPM, pp. 15–29.

Jacquin, Th. and Vail, P. R., 1998, Sequence chronostatigraphy [columns for Triassic chart, Mesozoic and Cenozoic sequence chronostratigraphic framework of European basins, in Hardenbol, J. *et al.*]. In P.-C. de Graciansky, J. Hardenbol., Th. Jacquin, and P. R. Vail, eds., *Mesozoic–Cenozoic Sequence*

Stratigraphy of European Basins, Society of Economic Paleontologists and Mineralogists Special Publication, Vol. 60. Tulsa, OK: SEPM, Chart 8.

Jaeger, H., 1977, Graptolites. In A. Martinsson, ed., The Silurian–Devonian boundary. *IUGS Series A* **5**: 337–45.

Jaffey, A. H., Flynn, K. F., Glendenin, L. E., Bentley, W. C., and Essling, A. M., 1971, Precision measurement of half-lives and specific activities of ^{235}U and ^{238}U. *Physical Review C: Nuclear Physics* **4**: 1889–906.

Jago, J. B. and Haines, P. W., 1998, Recent radiometric dating of some Cambrian rocks in southern Australia: relevance to the Cambrian time scale. *Revista Española de Paleontología, no. extr. Homenaje al Prof. Gonzalo Vidal*, pp. 115–22.

Jagodzinski, E. A. and Black, L. P., 1999, U–Pb dating of sodic lavas, sills and syneruptive resedimented volcaniclastic deposits of the Lower Devonian Crudine Group, Hill End Trough, New South Wales. *Australian Journal of Earth Sciences* **46**: 749–64.

Janis, C. M. E., 1998, *Evolution of Tertiary Mammals of North America*, Vol. 1, *Terrestrial Carnivores, Ungulates and Ungulatelike Mammals*. Cambridge: Cambridge University Press.

Jarrard, R. D., 1974, Paleomagnetism of some Leg 27 sediment cores. In J. J. Veevers, J. R. Heirtzler, *et al.*, eds., *Initial Reports of the Deep Sea Drilling Project*, Vol. 27. Washington, DC: US Govt. Printing Office, pp. 415–23.

Javaux, E., Knoll, A. H., and Walter, M. R., 2001, Ecological and morphological complexity in early eukaryotic ecosystems. *Nature* **412**: 66–9.

Jeans, C. V., Merriman, R. J., Mitchell, J. G., and Bland, D. J., 1982, Volcanic clays in the Cretaceous of southern England and northern Ireland. *Clay Minerals* **17**: 105–56.

Jell, P. A., 1974, Faunal provinces and possible planetary reconstruction of the Middle Cambrian. *Journal of Geology* **82**: 319–50.

1978, Asthenopsis Whitehouse, 1939 (Trilobita), Middle Cambrian in northern Australia. *Memoirs of the Queensland Museum* **18**(2): 219–31, pls 31–38.

Jell, P. A. and Robison, R. A., 1978, Revision of a late Middle Cambrian faunule from northwestern Queensland. *The University of Kansas Paleontological Contributions Paper*, Vol. 90. Lawrence, KS: University of Kansas, p. 21, 4 pls.

Jenkins, D. G., 1967, Planktonic foraminiferal zones and new taxa from the lower Miocene to the Pleistocene of New Zealand. *New Zealand Journal of Geology and Geophysics* **10**: 1064–78.

1985, Southern mid-latitude Paleocene to Holocene planktic foraminifera. In H. M. Bolli, J. B. Saunders, and K. Perch-Nielsen, eds., *Plankton Stratigraphy*. Cambridge: Cambridge University Press, pp. 263–82.

1993, Cenozoic southern middle and high latitude biostratigraphy and chronostratigraphy based on planktonic foraminifera. In J. P. Kennet and D. A. Warnke, eds., *The Antarctic Paleoenvironment: A Perspective on Global Change, Antarctic Research Series*, Vol. 60. Washington, DC: American Geophysical Union, pp. 125–44.

1995, Southern mid-latitude Paleocene to Holocene planktic foraminifera. In H. M. Bolli, J. B. Saunders, and K. Perch-Nielsen, eds., *Plankton Stratigraphy*. Cambridge: Cambridge University Press, pp. 263–82.

Jenkins, D. G. and Luterbacher, H. P., 1992, Paleogene stages and their boundaries: introductory remarks. *Neues Jahrbuch für Geologie und Paläontologie Abhandlungen* **186**: 1–5.

Jenkins, D. G. and Murray, J. W., 1989, *Stratigraphical Atlas of Fossil Foraminifera*, British Micropalaeontological Society Publication Series. Chichester: Ellis Horwood.

Jenkins, T. B. H., Crane, D. T., and Mory, A. J., 1993, Conodont biostratigraphy of the Visean Series in eastern Australia. *Alcheringa* **17**: 211–83.

Jenkyns, H. C., 1988, The early Toarcian (Jurassic) anoxic event: stratigraphic, sedimentary and geochemical evidence. *American Journal of Science* **288**: 101–51.

1996, Relative sea-level change and carbon isotopes: data from the Upper Jurassic (Oxfordian) of central and southern Europe. *Terra Nova* **8**: 75–85.

1999, Mesozoic anoxic events and palaeoclimate. *Zentralblatt für Geologie und Paläontologie (Stuttgart)* **I**(7–9): 943–9.

Jenkyns, H. C. and Clayton, C. J., 1986, Black shales and carbon isotopes in pelagic sediments from the Tethyan Lower Jurassic. *Sedimentology* **33**: 87–106.

1997, Lower Jurassic epicontinental carbonates and mudstones from England and Wales: chemostratigraphic signals and the early Toarcian anoxic event. *Sedimentology* **44**: 687–706.

Jenkyns, H. C., Gale, A. S., and Corfield, R. M., 1994, Carbon- and oxygen-isotope stratigraphy of the English Chalk and the Italian Scaglia and its palaeoclimatic significance. *Geological Magazine* **131**: 1–34.

Jenkyns, H. C., Paull, K., Cummins, D. I., and Fullagar, P. D., 1995, Strontium-isotope stratigraphy of Lower Cretaceous atoll carbonates in the mid Pacific Mountains. In *Proceedings of the Ocean Drilling Program, Scientific Results*, Vol. 143. College Station, TX: Ocean Drilling Program, pp. 89–97.

Jenkyns, H. C., Jones, C. E., Gröcke, D. R., Hesselbo, S. P., and Parkinson, D. N., 2002, Chemostratigraphy of the Jurassic system: applications, limitations and implications for palaeoceanography. *Journal of the Geological Society of London* **159**: 351–78.

Jensen, S., Gehling, J. G., and Droser, M. L., 1998, Ediacara-type fossils in Cambrian sediments. *Nature* **393**(6685): 567–9.

Jeppsson, L., 1997, A new latest Telychian, Sheinwoodian and Early Homerian (Early Silurian) Standard Conodont Zonation. *Transactions of the Royal Society of Edinburgh, Earth Sciences* **88**: 91–114.

1998, Silurian oceanic events: summary of general characteristics. In E. Landing and M. E. Johnson, eds., *Silurian Cycles: Linkages of Dynamic Stratigraphy with Atmospheric, Oceanic and Tectonic Changes*, *New York State Museum Bulletin*, Vol. 491. Albany, NY: New York State Museum, pp. 239–57.

Jessen, K. and Milthers, V., 1928, Stratigraphical and palaeontological studies of interglacial freshwater deposits in Jutland and north-west Germany. *Danmarks Geologisk Undersogelse, II Raekke, No. 48.*

Jiménez, A. P., Jiménez de Cisneros, C., Rivas, P., and Vera, J. A., 1996, The early Toarcian anoxic event in the westernmost Tethys (Subbetic): paleogeographic and paleobiogeographic significance. *Journal of Geology* **104**: 399–416.

Jin, Y. G., 1994, Two phases of the end-Permian mass extinction. *Canadian Society of Petroleum Geologists Memoir* **17**: 813–22.

1996, A global chronostratigraphic scheme for the Permian System – Two decades of the Permian Subcommission. *Permophiles* **28**: 4–9.

Jin, Y. G., Mei, S. L., and Zhu, A. L., 1993, The potential stratigraphic levels of Guadalupian/Lopingian boundary. *Permophiles* **23**: 17–20.

Jin, Y. G., Glenister, B. F., Kotlyar, G. V., and Sheng, J.-Z., 1994, An operational scheme of Permian Chronostratigraphy. In Y. Jin, J. Utting, and B. R. Wardlaw, eds., Permian Stratigraphy, Environments and Resources. *Palaeoworld* **4**: 1–14.

Jin, Y. G., Wardlaw, B. R., Glinister, B. F., and Kotlyar, G. V., 1997, Permian chronostratigraphic subdivisions. *Episodes* **20**(1): 10–15.

Jin, Y. G., Shang, Q. H., Wang, X. D., Wang, Y., and Sheng, J. Z., 1999, Chronostratigraphic subdivision and correlation of the Permian of China. *Acta Geologica Sinica* **73**(2): 127–38.

Jin, Y. G., Wang, Y., Wang, W., *et al.*, 2000, Pattern of marine mass extinction near the Permian–Triassic boundary in South China. *Science* **289**: 432–6.

Jin, Y. G., Henderson, C. M., Wardlaw, B. R., *et al.*, 2001, Proposal for the Global Stratotype Section and Point (GSSP) for the Guadalupian–Lopingian boundary. *Permophiles* **39**: 32–42.

Jinnan, T., Jianjun, Z., and Laishi, Z., 2001, Report on the Lower Triassic of Chaohu, Anhui Province, China. *Albertiana* **25**: 23–7.

Joachimski, M. and Buggisch, W., 1993, Anoxic events in the late Frasnian: Causes of the Frasnian–Famennian faunal crisis? *Geology* **21**: 657–78.

1996, The upper Devonian reef crisis: insights from the carbon isotope record. *Göttinger Arbeiten zur Geologie und Paläontologie, Sonderband* **2**: 365–70.

2002, Conodont apatite $\delta^{18}O$ signatures indicate climatic cooling as a trigger of the Late Devonian mass extinction. *Geology* **30**(8): 711–14.

Joachimski, M. M., Pancrost, R. D., Freeman, K. A., Ostertag-Henning, C., and Buggisch, W., 2002, Carbon isotope geochemistry of the Frasnian–Famennian transition. *Palaeogeography, Palaeoclimatology, Palaeoecology* **181**: 91–109.

Johannson, G. G. and McNicoll, V. J., 1997, New U–Pb data from the Laberge Group, northwest British Columbia: implications for Stikinian arc evolution and Lower Jurassic time scale calibrations. In *Current Research, Part F, Geological Survey of Canada Paper* 1997-1F. Ottawa: Geological Survey of Canada, pp. 121–9.

Johnsen, S. J., Dansgaard, W., Clausen, H. B., and Langway, C. C., 1972, Oxygen isotope profiles through Antarctic and Greenland ice sheets. *Nature* **235**: 429–33.

Johnson, D. A., Schneider, D. A., Nigrini, C. A., Caulet, J. P., and Kent, D. V., 1989, Pliocene–Pleistocene radiolarian events and magnetostratigraphic calibrations for the tropical Indian Ocean. *Marine Micropaleontology* **14**: 33–66.

Johnson, J. G., Klapper, G., and Sandberg, C. A., 1985, Devonian eustatic fluctuations in Euramerica. *Geological Society of America Bulletin* **96**: 567–87.

Johnson, M. E., 1996, Stable cratonic sequences and a standard for Silurian eustasy. In B. J. Witzke, G. A. Ludvigson, and J. Day, eds., *Paleozoic Sequence Stratigraphy: Views from the North American Craton, Geological Society of America Special Paper*, Vol. 306. Boulder, CO: Geological Society of America, pp. 203–11.

Johnson, M. E., Rong, J.-Y., and Kershaw, S., 1998, Calibrating Silurian eustasy against the erosion and burial of coastal paleotopography. In E. Landing and M. E. Johnson, eds., *Silurian Cycles: Linkages of Dynamic Stratigraphy with Atmospheric, Oceanic and Tectonic Changes, New York State Museum Bulletin*, Vol. 491. Albany, NY: New York State Museum, pp. 3–13.

Johnson, R. G., 1982, Brunhes–Matuyama magnetic reversal dated at 790 000 yr BP by marine astronomical correlations. *Quaternary Research* **17**: 135–147.

Jones, C. E. and Jenkyns, H. C., 2001, Seawater strontium isotopes, oceanic anoxic events, and seafloor hydrothermal activity in the Jurassic and Cretaceous. *American Journal of Science* **301**: 112–49.

Jones, C. E., Jenkyns, H. C., and Hesselbo, S. P., 1994a, Strontium isotopes in Early Jurassic seawater. *Geochimica et Cosmochimica Acta* **58**: 1285–301.

Jones, C. E., Jenkyns, H. C., Coe, A. L., and Hesselbo, S. P., 1994b, Strontium isotope variations in Jurassic and Cretaceous seawater. *Geochimica et Cosmochimica Acta* **58**: 3061–74.

Jones, P. J., 1995, *Timescales: 5. Carboniferous, Australian Geological Survey Organisation Record*, No. 1995/34. Canberra: Australian Geological Survey, pp. 3–45.

Jones, P. J., Shergold, J. H., and Druce, E. C., 1971, Late Cambrian and Early Ordovician Stages in western Queensland. *Journal of the Geological Society of Australia* **18**: 1–32.

Juárez, M. T., Osete, M. L., Meléndez, G., Langereis, C. G., and Zijderveld, J. D. A., 1994, Oxfordian magnetostratigraphy of Aguilón and Tosos sections (Iberian Range, Spain) and evidence of a pre-Oligocene overprint. *Physics of the Earth and Planetary Interiors* **85**: 195–211.

Juárez, M. T., Osete, M. L., Meléndez, G., and Lowrie, W., 1995, Oxfordian magnetostratigraphy in the Iberian Range. *Geophysical Research Letters* **22**: 2889–92.

Kaesler, R. L., 1997, *Treatise on Invertebrate Paleontology, Part O, Arthropoda 1, Trilobita, Revised,* Vol. 1. Boulder, CO: The Geological Society of America/Lawrence, KS: The University of Kansas.

Kah, L. C., Lyons, T. W., and Chesley, J. T., 2001, Geochemistry of a 1.2 Ga carbonate–evaporite succession, northern Baffin and Bylot Islands: implications for Mesoproterozoic marine evolution. *Precambrian Research* **111**: 203–34.

Kaljo, D., Kiipli, T., and Martma, T., 1998, Correlation of carbon isotope events and environmental cyclicity in the East Baltic Silurian. In E. Landing and M. E. Johnson, eds., *Silurian Cycles: Linkages of Dynamic Stratigraphy with Atmospheric, Oceanic and Tectonic Changes, New York State Museum Bulletin*, Vol. 491. Albany, NY: New York State Museum, pp. 297–312.

Kamber, B. S., Moorbath, S., and Whitehouse, M. J., 2001, The oldest rocks on Earth: time constraints and geological controversies. *Geological Society of London Special Publication*, Vol. 190. London: Geological Society, pp. 177–203.

Kaminski, M. A. and Gradstein, F. M., 2004, *Atlas of Paleogene Cosmopolitan Deep-water Agglutinated Foraminifera, Special Publication* 4. Grzybowski Foundation.

Kamo, S. L., Czamanske, G. K., and Drogh, T. E., 1996, A minimum U–Pb age for Siberian flood-basalt volcanism. *Geochimica et Cosmochimica Acta* **60**: 3505-11.

Kamo, S. L., Czamanske, G. K., Amelin, Y., *et al.*, 2003, Rapid eruption of Siberian flood-volcanic rocks and evidence for the coincidence with the Permian–Triassic boundary. *Earth and Planetary Science Letters* **214**: 75–91.

Kampschulte, A., Bruckschen, P., and Strauss, H., 2001, The sulphur isotopic composition of trace sulphates in Carboniferous brachiopods: implications for coeval seawater, correlation with other geochemical cycles and isotope stratigraphy. *Chemical Geology* **175**(1–2): 149–73.

Kaplan, I. R. and Rittenberg, S. C., 1964, Microbiological fractionation of sulphur isotopes. *Journal of General Microbiology* **34**: 195–212.

Kappelman, J., Simons, E. L., and Swisher, C. C. I., 1992, New age determinations for the Eocene–Oligocene boundary sediment in the Fayum Depression, northern Egypt. *Journal of Geology* **100**: 647–68.

Karner, D. B. and Renne, P. R., 1998, ^{40}Ar/^{39}Ar geochronology of Roman Volcanic Province tephra in the Tiber River Valley: age calibration of Middle Pleistocene sea-level changes. *Bulletin of the Geological Society of America* **110**(6): 740–7.

Karner, D. B., Renne, P., Alvarez, W., Ammerman, A. J., and Marra, F., 1995, ^{40}Ar/^{39}Ar geochronology of the Tiber River delta, Italy: how well does Milankovitch predict the timing of middle Pleistocene climate cycles? *Abstracts Geological Society of America Annual Meeting 27*. Boulder, CO: Geological Society of America, pp. 175.

Karpinsky, A. P., 1874, Geological investigation of the Orenburg area. *Berg-Jour., II, Verhandl. Miner. Gesellsch.* **IX**: 13–31.

1889, Über die Ammoneen der Artinsk-Stufe und einige mit denselben verwandte carbonische Formen. *Memoir Imperial Academy of Science, St. Petersburg ser. 7* **37**(2): 104.

1891, On Artinskian ammonites and some similar Carboniferous forms. *Zap. Imper. Sakt-Peterburskogo Mineral. O-va*, No. 27

Kauffman, E. J., Kennedy, W. J., Wood, C. J., with contributions by Dhondt, A. V., Hancock, J. M., Kopaevich, L. F., and Walaszczyk, I., 1996, The Coniacian stage and substage boundaries. *Bulletin de l'Institut Royal des Sciences Naturelles de Belgique, Sciences de la Terre, Supplement*, **66**: 81–94.

Kaufman, A. J. and Knoll, A. H., 1995, Neoproterozoic variation in the C-isotopic composition of seawater: stratigraphic and biogeochemical implications. *Precambrian Research* **73**: 27–49.

Kay, R. F., Madden, R. H., Vucetich, M. G., *et al.*, 1999, Revised geochronology of the Casamayoran South American Land Mammal Age: climatic and biotic implications. *Proceedings of the National Academy of Sciences, USA* **96**: 13 235–40.

Keating, B. H. and Helsley, C. E., 1978a, Magnetostratigraphic studies of Cretaceous age sediments from Sites 361, 363, and 364. In H. M. Bolli, W. B. F., *et al.*, eds., *Initial Reports of the Deep Sea Drilling Project*, Vol. 40. Washington, DC: US Govt. Printing Office, pp. 459–67.

1978b, Magnetostratigraphic studies of Cretaceous age sediments from DSDP Site 369. In Y. Lancelot, E. Seibold, *et al.*, eds., *Initial Reports of the Deep Sea Drilling Project*, Vol. 41 supplement. Washington, DC: US Govt. Printing Office, pp. 983–986.

1978c, Paleomagnetic results from DSDP Hole 391C and the magnetostratigraphy of Cretaceous sediments from the Atlantic Ocean floor. In W. E. Benson, R. E. Sheridan, *et al.*, eds., *Initial Reports of the Deep Sea Drilling Project*, Vol. 44. Washington, DC: US Govt. Printing Office, pp. 523–8.

Keigwin, L. D. and Shackleton, N., 1980, Uppermost Miocene carbon isotope stratigraphy of a piston core in the equatorial Pacific. *Nature* **284**: 613–14.

Keigwin, L. D. Curry, W. B., Lehman, S. J., and Johnsen, S., 1994, The role of the deep ocean in North Atlantic climate change between 70 and 130 kyr ago. *Nature* **371**: 323–5.

Keller, G., 1988, Extinction, survivorship and evolution of planktonic foraminifera across the Cretaceous–Tertiary boundary at El Kef, Tunisia. *Marine Micropaleontology* **13**: 239–63.

Kelley, S., 1995, Ar–Ar dating by laser microprobe. In J. Potts, J. F. W. Bowles, S. J. B. Reed, and M. R. Cave, eds., *Microprobe Techniques in the Earth Sciences*. Chapman and Hall, pp. 327–58.

Kelly, D. C., Bralower, T., J., and Zachos, J. C., 2000, On the demise of the early Paleogene *Morozovella vlascoensis* lineage: terminal progenesis in the planktic foraminifera? *Geologiska Föreningen i Stockholm Förhandlingar* **122**: 86–7.

Kemp, E. M., Balme, B. E., Helby, R. J., *et al.*, 1977, Carboniferous and Permian palynostratigraphy in Australia and Antarctica: a review. *Bureau of Mineral Resources Journal of Australian Geology and Geophysics* **2**: 177–208.

Kemper, E., Mutterlose, J., and Wiedenroth, K., 1987, Die Grenze Unter-/Ober-Hauterive in Nordwestdeutschland. Beispiel eines stratigraphisch zu nutzenden Klima-Umschwunges. *Geologisches Jahrbuch* **A96**: 209–18.

Kemple, W. G., Sadler, P. M., and Strauss, D. J., 1995, Extending graphic correlation to many dimensions: stratigraphic correlation as constrained optimisation. In K. O. Mann, H. R. Lane, and P. A. Scholle, eds., *Graphic Correlation, Society of Economic Paleontologists and Mineralogists Special Publication*, Vol. 53. Tulsa, OK: SEPM, pp. 65–82.

Kennedy, M. J., Runnegar, B., Prave, A. R., Hoffmann, K.-H., and Arthur, M. A., 1998, Two or four Neoproterozoic glaciations. *Geology* **26**: 1059–63.

Kennedy, W. J., 1984, Ammonite faunas and the "standard zones" of the Cenomanian to Maastrichtian Stages in their type areas, with some proposals for the definition of the stage boundaries by ammonites. *Geological Society of Denmark Bulletin* **33**: 147–61.

Kennedy, W. J. and Cobban, W. A., 1991, Stratigraphy and inter-regional correlation of the Cenomanian–Turonian transition in the Western Interior of the United States near Pueblo, Colorado. A potential boundary stratotype for the base of the Turonian Stage. *Newsletters on Stratigraphy* **24**: 1–33.

Kennedy, W. J., Walaszczyk, I., and Cobban, W. A., 2000, Pueblo, Colorado, USA, candidate global boundary stratotype section and point for the base of the Turonian Stage of the Cretaceous and for the Middle Turonian substage, with a revision of the Inoceramidae (bivalve). *Acta Geologica Polonica* **50**: 295–334.

Kennedy, W. J., Gale, A. S., Bown, P. R., *et al.*, 2002, Integrated stratigraphy across the Aptian–Albian boundary in the Marnes Bleues at the Col de Pré-Guittard, Arnayon (Brôme), and at Tartonne (Alpes-de-Haute-Provence), France: a candidate global boundary stratotype section and boundary point for the base of the Albian Stage. *Cretaceous Research* **21**: 591–720.

Kennedy, W. J., Gale, A. S., Lees, J. A., and Caron, W. A., in press, Definition of a global boundary stratotype section and point (GSSP) for the base of the Cenomanian Stage, Mont Risou, Hautes-Alpes, France. *Episodes*.

Kennett, J. P. and Srinivasan, M. S., 1983, *Neogene Planktonic Foraminifera: A Phylogenetic Atlas*. Stroudsburg, PA: Hutchinson Ross.

Kennett, J. P. and Stott, L. D., 1991, Abrupt deep sea warming, paleoceanographic changes and benthic extinctions at the end of the Paleocene. *Nature* **353**: 225–9.

Kent, D. and Gradstein, F. M., 1983, A Jurassic and Cretaceous geochronology. *Bulletin of the Geological Society of America* **96**: 1419–27.

1986, A Jurassic to Recent Geochronology. *The Western North Atlantic Region: The Geology of North America*. Vol. M Boulder, CO: Geological Society of America, pp. 45–50.

Kent, D. V., 1999, Orbital tuning of geomagnetic polarity time-scales: *Philosophical Transactions of the Royal Society, Series A* **357**: 1995–2007.

Kent, D. V. and Olsen, P. E., 1999, Astronomically tuned geomagnetic polarity timescale for the Late Triassic. *Journal of Geophysical Research* **104**: 12 831–41.

Kent, D. V., Olsen, P. E., and Witte, W. K., 1995, Late Triassic–earliest Jurassic geomagnetic polarity sequence and paleolatitudes from drill cores in the Newark rift basin, eastern North America. *Journal of Geophysical Research* **100**: 14 965–98.

Kerr, A. C., 1998, Oceanic plateau formation: a cause of mass extinction and black shale deposition around the Cenomanian–Turonian boundary? *Journal of the Geological Society of London* **155**: 619–26.

Kerr, R. A., 2000, Did volcanoes drive ancient extinctions? *Science* **289**: 1130–1.

Keyes, C. R., 1892, The principal Mississippian section. *Geological Society of America Bulletin* **3**: 283–300.

Khramov, A. N., 1958, *Palaeomagnetism and Stratigraphic Correlation*. Leningrad: Gostoptechizdat 8.

1963, *Paleomagnetism of Paleozoic. Transactions of VNIGRI*. Leningrad: Nedra, p. 283. [In Russian].

1987, *Paleomagnetology*. Berlin: Springer-Verlag, pp. 1–308.

2000, The general magnetostratigraphic scale of the Phanerozoic. In A. I. Zhamoida, ed., *Supplements to the Stratigraphic Code of Russia. Transactions of VSEGEI*. St. Petersburg: VSEGEI, pp. 24–45 [In Russian].

Khramov, A. N. and Davydov, V. I., 1984, Paleomagnetism of Upper Carboniferous and Lower Permian in the South of the U.S.S.R. and the problems of structure of the Kiaman hyperzone. *Transactions of VNIGRI*. Leningrad: Nedra, pp. 55–73. [In Russian].

1993, Results of paleomagnetic investigations. *Permian System: Guides to Geol. Excursions in the Uralian Type Localities, Occasional Publications ESRI* Vol. 10. Earth Sciences and Resources Institute, pp. 34–42.

Khramov, A. N. and Rodionov, V. P., 1981, The geomagnetic field during Palaeozoic time. In M. W. McElhinny, A. N. Khramov, M. Ozima, and D. A. Valencio, eds., *Global Reconstruction and the Geomagnetic Field during the Palaeozoic, Advances in Earth and Planetary Science*, Vol. 10, pp. 99–115.

Khramov, A. N., Goncharov, G. I., Komisssarova, R. A., *et al.*, 1974, *Paleozoic Paleomagnetism, Transactions of VNIGRI*. Leningrad: Nedra, pp. 1–238 [In Russian].

Kimura, H. and Watanabe, Y., 2001, Oceanic anoxia at the Precambrian–Cambrian boundary. *Geology* **29**: 995–8.

Kiparisova, L. D. and Popov, Y. N., 1956, Subdivision of the Lower series of the Triassic system into stages. *Doklady Akademiya Nauk USSR, Seriya Geologicheskaya* **109**: 842–5 [In Russian].

1964, The project of subdivision of the Lower Triassic into stages. *XXII International Geological Congress, Reports of Soviet Geologists, Problem* 16a, pp. 91–9 [In Russian].

Kireeva, G. D., 1964, The taxonomical analysis of the wall structure of some fusulinacea at the Middle–Upper Carboniferous transition. *Problems of Micropaleontology*. Publishing Office of the Academy of Science of the USSR, pp. 53–56 [In Russian].

Kirnbauer, T. and Reischmann, T., 2001, Pb/Pb zircon ages from the Hunsrück Slate Formation (Bundenbach, Rhenish Massif): a contribution to the age of the Lower/Middle Devonian boundary. *Newsletters on Stratigraphy* **38**(2–3): 185–200.

Kirschvink, J. L., 1978a, The Precambrian–Cambrian boundary problem: magnetostratigraphy of the Amadeus Basin, central Australia. *Geological Magazine* **115**(2): 139–50.

1978b, The Precambrian–Cambrian boundary problem: palaeomagnetic directions from the Amadeus Basin, central Australia. *Earth and Planetary Science Letters* **40**: 91–100.

Kirschvink, J. L. and Raub, T. D., 2003, A methane fuse for the Cambrian explosion: carbon cycles and true polar wander. *Geoscience* **335**(1): 65–78.

Kirschvink, J. L. and Rozanov, A. Y., 1984, Magnetostratigraphy of Lower Cambrian strata from the Siberian Platform: a palaeomagnetic pole and a preliminary polarity time-scale. *Geological Magazine* **121**: 189–203.

Kirschvink, J. L., Margaritz, M., Ripperdan, R. L., Zhuravlev, A. Y., and Rozanov, A. Y., 1991, The Precambrian–Cambrian boundary: magnetostratigraphy and carbon isotopes resolve correlation problems between Siberia, Morocco and South China. *GSA Today* **1**(4): 61–91.

Kirschvink, J. L., Ripperdan, R. L., and Evans, D., 1997, Evidence for a large scale reorganisation of Early Cambrian continental masses by inertial interchange true polar wander. *Science* **227**: 541–5.

Kirwan, R., 1799, *Additional Observations on the Proportion of Real Acid in the Three Ancient Known Mineral Acids and on the Ingredients in Various Neutral Salts and Other Compounds*. Dublin: George Bonham, p. 137.

Kitagawa, H. and Van der Plicht, J., 1998, Atmospheric radiocarbon calibration to 45,000 yr BP: late glacial fluctuations and cosmogenic isotope production. *Science* **279**(5354): 1187–90.

Klapper, G., 1989, The Montagne Noire Frasnian (Upper Devonian) conodont succession. In N. J. McMillan, A. F. Embry, and D. J. Glass, eds., *Devonian of the World, Canadian Society of Petroleum Geology, Memoir* 14. Calgary: Canadian Society of Petroleum Geology, pp. 449–68.

2000a, Species of Spathiognathodontidae and Polygnathidae (Conodonta) in the recognition of Upper Devonian stage boundaries. *Courier Forschungsinstitut Senckenberg* **220**: 153–9.

2000b, A flawed Devonian timescale. *The Australian Geologist* **114**: 12.

Klapper, G. and Becker, R. T., 1999, Comparison of Frasnian (Upper Devonian) conodont zonations. *Bolletino della Società Paleontologica Italiana* **37**: 339–47.

Klapper, G., Feist, R., and House, M. R., 1987, Decision on the Boundary Stratotype for the Middle/Upper Devonian Series Boundary. *Episodes* **10**: 97–101.

Klapper, G., Feist, R., Becker, R. T., and House, M. R., 1993, Definition of the Frasnian/Famennian Stage boundary. *Episodes* **16**: 433–41.

Kleffner, M. A., 1989, A conodont-based Silurian chronostratigraphy. *Geological Society of America Bulletin* **101**(7): 904–12.

Klootwijk, C. T., 1980, Early Palaeozoic magnetism in Australia. *Tectonophysics* **64**: 249–332.

Klug, C., 2001, Early Emsian ammonoids from the eastern Anti-Atlas (Morocco) and their succession. *Paläontologische Zeitschrift* **74**: 479–515.

Knoll, A. H., 1994, Proterozoic and Early Cambrian protists: evidence for accelerating evolutionary tempo. *Proceedings of the National Academy of Sciences, USA* **91**: 6743–50.

1996, Archean and Proterozoic paleontology. In J. Jansonius and D. C. McGregor, eds., *Palynology: Principles and Applications, Special Publication*, Vol. 1. Tulsa, OK: American Association of Stratigraphic Palynologists Foundation 1, pp. 51–80.

2001, Learning to tell Neoproterozoic time. *Precambrian Research* **100**: 3–20.

Knoll, A. H. and Sergeev, V. N., 1995, Taphonomic and evolutionary changes across the Mesoproterozoic–Neoproterozoic boundary. *Neues Jahrbuch für Geologie und Paläontologie Abhandlungen* **195**: 289–302.

Knoll, A. H. and Walter, M. R., 1992, Stratigraphy and latest Proterozoic Earth history. *Nature* **356**: 673–8.

Koch, P. L., Zachos, J. C., and Gingerich, P. D., 1992, Correlation between isotope records in marine and continental carbon reservoirs near the Paleocene/Eocene boundary. *Nature* **358**: 319–22.

Koeberl, C., Armstrong, R. A., and Reimold, W. U., 1997, Morokweng, South Africa: a large impact structure of Jurassic–Cretaceous boundary age. *Geology* **25**: 731–4.

Koepnick, R. B., Denison, R. E., Burke, W. H., Hetherington, E. A., and Dahl, D. A., 1990, Construction of the Triassic and

Jurassic portion of the Phanerozoic curve of seawater $^{87}Sr/^{86}Sr$. *Chemical Geology* **80**: 327–349.

Köhler, M. and Moyà-Solà, S., 1999, A finding of Oligocene primates on the European continent. *Proceedings of the National Academy of Sciences, USA* **96**: 14 664–14 667.

Kolesov, E. V., 1984, Paleomagnetic stratigraphy of the Devonian–Carboniferous boundary beds in the Soviet North-East and in the Franco–Belgian Basin. *Annales de la Société Géologique de Belgique* **107**: 135–6.

2001, The Paleomagnetism of the Upper Paleozoic volcanogenic–sedimentary sequences of Prikolym uplift. In K. K. Simakov, ed., *Paleomagnetic Studies of Geological Rocks on the North-Eastern of Russia*. Magadan. For East Branch of the Russian Academy of Sciences, pp. 32–44 [In Russian].

Königer, S., Lorenz, V., Stollhofen, H., and Armstrong, R. A., 2002, Origin, age and stratigraphic significance of distal fallout ash tuffs from the Carboniferous–Permian continental Saar–Nahe Basin (SW Germany). *International Journal of Earth Sciences* **91**: 341–56 [formerly *Geologische Rundschau*].

Köppen, W. and Wegener., A., 1924, *Die Klimate der Geologischen Vorzeit*. Berlin: Borntraeger.

Koppers, A. A. P., Staudigel, H., and Duncan, R. A., 2003, High resolution $^{40}Ar/^{39}Ar$ dating of the oldest oceanic basement basalts in the western Pacific basin. *Geochemistry, Geophysics, Geosystems* **4**(11) = doi 10.1029/2003 GC000574.

Koren', T. N., Lenz, A. C., Loydell, D. K., *et al.*, 1996, Generalized graptolite zonal sequence defining Silurian time intervals for global paleogeographic studies. *Lethaia* **29**: 59–60.

Korte, C., Kozur, H. W., Bruckschen, P., and Veizer, J., 2003, Strontium isotope evolution of Late Permian and Triassic seawater. *Geochimica et Cosmochimica Acta* **67**: 47–62.

Kossovaya, O. L., 1996, Correlation of Uppermost Carboniferous and Lower Permian rugose coral zones from the Urals to Western North America. *Palaios* **11**(1): 71–82.

1997, Middle and Upper Carboniferous composite zonal sequence based on Rugosa corals (western part of Russia). *Proceedings of the XIII International Congress on the Carboniferous and Permian*, Prace Panstwowego Instytutu Geologicznego, Vol. 157, pp. 85–98.

Kossovaya, O. L., Vevel', Y. A., and Zhuravlev, A. V., 2001, Fauna and sedimentation near the Visean/Serpukhovian boundary (Izyaya River section, Tchernyshev Swell, Subpolar Urals). *Newsletter on Carboniferous Stratigraphy* **19**: 29–31.

Köthe, A., 1990, Paleogene dinoflagellates from northwest Germany: biostratigraphy and paleoenvironment. *Geologisches Jahrbuch* **A118**: 3–111.

1996, Dinoflagellatenzysten- und Kalknannoplankton-Untersuchungen im Grenzbereich Eozän/Oligozän am Doberg bei Bünde ("Piepenhagen-Profil", Westfalen). *Newsletters on Stratigraphy* **33**: 145–55.

Kotlyar, G. V. and Stepanov, D. L., 1984, *Main Features of the Stratigraphy of the Permian System in USSR*. VSEGEI n. ser. 286, Leningrad: Nedra, p. 233.

Kovács, S., Dosztály, L., Góczán, F., Oravecz-Scheffer, A., and Budai, T., 1994, The Anisian/Ladinian boundary in the Balaton Highland, Hungary: a complex microbiostratigraphic approach. *Albertiana* **14**: 53–64.

Kowallis, B. J., Christiansen, E. H., Deino, A. L., *et al.*, 1998, The age of the Morrison Formation. *Modern Geology* **22**: 235–60.

Kozlova, G. E., 1983, Distribution of radiolarian zones of the Atlantic in the Paleogene of the Volga region. *Doklady Akademiya Nauk SSSR, Nov. Seriya* **3**: 46–51 [In Russian].

1984, Zonal subdivision of the boreal Paleogene by radiolarians. In M. G. Petrushevskaya and S. D. Stepanjants, eds., *Morfologiy, ekologiya i evolutsiya radiolyrii. Materialy IV simpoziuma evropeiskikh radilaristov Eurorad IV*. Leningrad: Akademyia Nauk SSSR, Zoologicheskii Insitut, pp. 196–210 [In Russian].

1993, Radiolarian zonal scale of the boreal Paleogene. *Micropaleontology Special Publication* **6**: 90–3.

Kozur, H., 1977, Beiträge zur Stratigraphie des Perm. Teil 1: Probleme der Abgrenzung und Gliederung des Perm. *Freiberger Forschungshefte* **319**: 79–121.

1998, Some aspects of the Permian–Triassic boundary (PTB) and of the possible causes for the biotic crisis around this boundary. *Palaeogeography, Palaeoclimatology, Palaeoecology* **143**: 227–72.

1999, Remarks on the position of the Norian–Rhaetian boundary. (Proceedings of the Epicontinental Triassic International Symposium, Halle, Germany, September 21–23, 1998). *Zentralblatt für Geologie und Paläontologie (Stuttgart)* **I**: 523–35.

Kozur, H. W. and Mostler, H., 1994, Anisian to Middle Carnian radiolarian zonation and description of some stratigraphically important radiolarians. *Geologisch-Paläontologische Mitteilungen Innsbruck, Sonderband* **3**: 39–255.

Krasheninnikov, V. A. and Akhmeteiev, M. A., 1998, Late Eocene–early Oligocene geological and biotical events on the territory of the former Soviet Union. Part II. The geological and biotical events. 507, Rossiiskaya Akademiya Nauk, Geologicheskii Institut, Trudy, p. 250 [In Russian].

Krasheninnikov, V. A. and Ptukhian, A. E., 1986, Stratigraphical subdivisions of Armenian Paleogene deposits by planktonic microfossils and nummulites. (Regional stratigraphy, zonal scales by planktonic and benthonic microfossils, their correlation). *Voprosy Mikropaleontologii* **28**: 60–98 [In Russian].

Krasheninnikov, V. A., Muzylov, N. G., and Ptukhian, A. E., 1985, Stratigraphical subdivision of Paleogene deposits of Armenia by planktonic foraminifera, nannoplankton and nummulites. Part I. Reference Paleogene sections of Armenia. *Voprosy Mikropaleontologii* **27**: 130–69 [In Russian].

Krasnenkov, R. V., Iossifova, Y. I., and Semenov, V. V., 1997, The Upper Don drainage basin: an important stratoregion for climatic stratigraphy of the early Middle Pleistocene (the early Neopleistocene) of Russia. *Quaternary Geology and Paleogeography of Russia*. Moscow: Geosynthos, pp. 82–96 [In Russian with abstract in English].

Krijgsman, W., Hilgen, F. J., Langereis, C. G., Santarelli, A., and Zachariasse, W. J., 1997a, Late Miocene magnetostratigraphy, biostratigraphy and cyclostratigraphy in the Mediterranean. *Earth and Planetary Science Letters* **136**: 475–99.

Krijgsman, W., Hilgen, F. J., Negri, A., Wijbrans, J. R., and Zachariasse, W. J., 1997b, The Monte del Casino section: a potential Tortonian–Messinian boundary stratotype? *Palaeogeography, Palaeoclimatology, Palaeoecology* **133**: 27–48.

Krijgsman, W., Hilgen, F. J., Raffi, I., Sierro, F. J., and Wilson, D. S., 1999, Chronology, causes and progression of the Messinian salinity crisis. *Nature* **400**: 652–5.

Krijgsman, W., Langereis, C. G., Daams, R., and Van der Meulen, A., 1994, Magnetostratigraphic dating of the middle Miocene climate change in the continental deposits of the Aragonian type area in the Calatayud–Teruel basin (Central Spain). *Earth and Planetary Science Letters* **128**: 513–26.

Kříz, J., 1989, The Prídolí Series in the Prague Basin (Barrandian area, Bohemia). *In* C. H. Holland and M. G. Bassett, eds., *A Global Standard for the Silurian System: Geological Series*, Vol. 9. Cardiff National Museum of Wales.

Kříz, J., Jaeger, H., Paris, F., and Schönlaub, H.-P., 1986, Prídolí: - the Fourth Series of the Silurian. *Jahrbuch der Geologischen Bundesanstalt, Wein* **129**: 291–360.

Krogh, T. E., 1979, Advances in U–Pb dating of zircon in the Canadian Precambrian Shield with case histories. *Program with Abstracts* – Geological Association of Canada, Mineralogical Association of Canada, Canadian Geophysical Union, Joint Annual Meeting 4, 61 pp.

1982a, Improved accuracy of U–Pb zircon ages by the creation of more concordant systems using an air abrasion technique. *Geochimica et Cosmochimica Acta* **46**(4): 637–49.

1982b, Improved accuracy of U–Pb zircon dating by selection of more concordant fractions using a high gradient magnetic separation technique. *Geochimica et Cosmochimica Acta* **46**(4): 631–5.

Kroon, D., Alexander, I., Little, M., *et al.*, 1998, Oxygen isotope and sapropel stratigraphy in the eastern Mediterranean during the last 3.2 million years. In *Proceedings of the Ocean Drilling Program, Scientific Results*, Vol. 160. College Station, TX: Ocean Drilling Program, pp. 181–9.

Krull, E. S. and Retallack, G. J., 2000, δ^{13}C depth profiles from paleosols across the Permian–Triassic boundary: evidence for methane release. *Geological Society of America Bulletin* **112**: 1459–72.

Kruse, P. and Shi, G. R., 2000, Archaeocyaths and radiocyaths. In G. Brock, ed., *Palaeobiogeography of Australian Cambrian Faunas, Memoir of the Association of Australasian Palaeontologists*. Association of Australasian Palaeontologists.

Krymholts, G. Ya., Mesezhnikov, M. S., and Westermann, G. E. G., 1982, Zony iurskoi sistemy v SSSR. *Interdepartmental Stratigraphic Committee of the USSR Transactions*, Vol. 10. Leningrad: Nauka [Revised 1986. Trans. Vassiljeva, T. I, 1988, *The Jurassic Ammonite Zones of the Soviet Union. Geological Society of America Special Paper 223*. Boulder, CO: Geological Society of America].

Krystyn, L., 1980, Triassic conodont localities of the Salzkammergut region (northern Calcareous Alps). In H. P. Schonlaub, ed., *Second European Conodont Symposium, Guidebook and Abstracts*. Abhandlungen der Geologischen Bundesanstalt, pp. 61–98.

1990, A Rhaetian Stage chronostratigraphy, subdivisions and their intercontinental correlations. *Albertiana* **8**: 15–24.

Krystyn, L. and Gallet, Y., 2002, Towards a Tethyan Carnian–Norian boundary GSSP. *Albertiana* **27**: 12–19.

Krystyn, L. and Orchard, M. J., 1996, Lowermost Triassic ammonoid and conodont biostratigraphy of Spiti, India. *Albertiana* **17**: 10–21.

Krystyn, L., Gallet, Y., Besse, J., and Marcoux, J., 2002, Integrated Upper Carnian to Lower Norian biochronology and implications for the Upper Triassic magnetic polarity time scale. *Earth and Planetary Science Letters* **203**: 343–51.

Kuiper, K. F., in press, Intercalibration of radio-isotopic and astronomical time in the Mediterranean Neogene. *Geologica Ultraiectina*.

Kukla, G. J., 1977, Pleistocene land–sea correlations. I, Europe. *Earth Science Reviews* **13**: 307–74.

Kulagina, E. I. and Sinytsina, Z. A., 1997, Foraminiferal zonation of the lower Bashkirian in the Askyn section, southern Urals, Russia. In C. A. Ross, J. Ross, and P. L. Brenckle, eds., *Late Paleozoic Foraminifera: Their Biostratigraphy, Evolution, and Paleoecology,* and *the Mid-Carboniferous Boundary*, *Special Publication*, *Cushman Foundation for Foraminiferal Research*, Vol. 36. Ithaca, NY: Cushman Foundation for Foraminiferal Research, pp. 83–8.

Kulagina, E. I., Rumyantseva, Z. S., Pazukhin, V. N., and Kochetkova, N. N., 1992, Lower and Middle Carboniferous boundary in the Southern Urals and Tien Shan. *Transactions of Geological Institute of Bashkortostan*. Bashkirian Branch of Russian Academy of Sciences. Moscow: Nauka, p. 110 [In Russian].

Kulagina, E. I., Pazukhin, V. N., Kochetkova, M. N., Sinytsina, Z. A., and Kochetkova, N. N., 2001, *Stratotypical and Key-sections of the Bashkirian Stage of the Carboniferous in South Urals Ufa*. Ufa: Gilem, Bashkirian Branch of Russian Academy of Sciences, p. 138 [In Russian].

Kullmann, J., Korn, D., and Weyer, D., 1990, Ammonoid zonation of the Lower Carboniferous subsystem. In P. L. Brenkle and W. L. Manger, eds., Intercontinental correlation and division of the Carboniferous System: contributions from the Carboniferous Subcommission meeting. *Courier Forschungsinstitut Senckenberg* **130**: 127–31.

Kulp, J. L., 1961, Geologic time-scale. *Science* **133**: 1105–14.

Kump, L. R., Arthur, M. A., Patzkowsky, M. E., *et al.*, 1999, A weathering hypothesis for glaciation at high atmospheric pCO_2 during the Late Ordovician. *Palaeogeography, Palaeoclimatology, Palaeoecology* **152**: 173–87.

Kunk, M. J. and Rice, C. L., 1994, High-precision $^{40}Ar/^{39}Ar$ age spectrum dating of sanidine from the Middle Pennsylvanian fire clay tonstein of the Appalachian Basin. Elements of Pennsylvanian stratigraphy, central Appalachian Basin. *Geological Society of America Special Paper*, Vol. 294. Boulder, CO: Geological Society of America. pp. 105–13.

Kunk, M. J. and Sutter, J. F., 1984, $^{40}Ar/^{39}Ar$ age spectrum dating of biotite from Middle Ordovician bentonites, eastern North America. In D. L. Bruton, ed., *Aspects of the Ordovician System.* Oslo: Universitets for laget. Also *Paleontologic Contributions from University of Oslo* **295**: 11–22.

Kunk, M. J., Sutter, J., Obradovitch, J. D., and Lanphere, M. A., 1985, Age of biostatigraphic horizons within the Ordovician and Silurian Systems. In N. J., Snelling, ed., *The Chronology of the Geological Record, Geological Society of London Memoir* 10. Geological Society of London, pp. 89–92.

Kuríz, J., 1989, The Prídolí Series in the Prague Basin (Barrandian area, Bohemia). In C. H. Holland and M. G. Bassett, eds., *A Global Standard for the Silurian System: Geological Series,* Vol. 9. Cardiff: National Museum of Wales.

Kuríz, J., Jaeger, H., Paris, F., and Schönlaub, H.-P., 1986, Prídolí: -the Fourth Series of the Silurian. *Jahrbuch der Geologischen Bundesanstalt, Wien* **129**: 291–360.

Kürschner, W., Becker, R. T., Buhl, D., and Veizer, J., Strontium isotopes in conodonts: Devonian–Carboniferous transition, the northern Rhenish Slate Mountains, Germany. *Annales de la Société Géologique de Belgique* **115**: 595–621.

Kwon, J., Min, K., Bickel, P. J., and Bickel, R. R. P., 2002, Statistical methods for jointly estimating the decay constant of ^{40}K and the age of a dating standard. *Mathematical Geology* **34**(4): 457–74.

LaBrecque, J. L., Kent, D. V., and Cande, S. C., 1977, Revised magnetic polarity time scale for Late Cretaceous and Cenozoic time. *Geology* **5**: 330–5.

Lambert, L. L., Barrick, J. E., and Heckel, P. H., 2001, Provisional lower and middle Pennsylvanian conodont zonation in Midcontinent North America. *Newsletter on Carboniferous Stratigraphy* **19**: 50–5.

Lambert, L. L., Wardlaw, B. R., Nestell, M. K., and Nestell, G. P., 2002, Latest Guadalupian (Middle Permian) conodonts and foraminifers from West Texas. *Journal of Micropaleontology* **48**(6): 111–34.

Lamolda, M. A. and Hancock, J. M., with contributions from Burnett, J. A., Collom, C. J., Christensen, W. K., *et al.*, 1996, The Santonian Stage and substages. *Bulletin de l'Institut Royal des Sciences Naturelles de Belgique, Sciences de la Terre*, Supplément **66**: 95–102.

Land, L. S., 1980, The isotopic and trace element geochemistry of dolomite: the state of the art. *Society of Economic Paleontologists and Mineralogists Special Publication*, Vol. 28. Tulsa, OK: SEPM, pp. 87–110.

Landing, E. L., 1994, Precambrian–Cambrian boundary global stratotype ratified and a new perspective of Cambrian time. *Geology* **22**: 179–82.

Landing, E., Bowring, S. A., Fortey, R. A., and Davidek, K., 1997, U–Pb zircon date from Avalonian Cape Breton Island and geochronological calibration of the Early Ordovician. *Canadian Journal of Earth Sciences* **34**: 724–30.

Landing, E., Bowring, S. A., Davidek, K. L., *et al.*, 1998, Duration of the Early Cambrian: U–Pb ages of volcanic ashes from Avalon and Gondwana. *Canadian Journal of Earth Sciences* **35**: 329–38.

Landing, E., Bowring, S. A., Davidek, K. L., *et al.*, 2000, Cambrian–Ordovician boundary age and duration of the lowest Ordovician Tremadoc Series based on U–Pb zircon dates from Avalonian Wales. *Geological Magazine* **137**(5): 485–94.

Landman, N. H. and Waage, K. M., 1993, Scaphitid ammonites of the Upper Cretaceous (Maastrichtian) Fox Hills Formation in South Dakota and Wyoming. *Bulletin of the American Museum of Natural History* **215**: 257.

Lane, H. R., 1977, Morrowan (Early Pennsylvanian) conodonts of northwestern Arkansas and northeastern Oklahoma. In P. K. Sutherland and W. L. Manger, eds., *Mississippian–Pennsylvanian Boundary in Northeastern Oklahoma and northwestern Arkansas. Guidebook* 18. Norman: Oklahoma Geological Survey: pp. 177–80.

Lane, H. R. and Brenckle, P. L., 2001, Type Mississippian subdivisions and biostratigraphic succession. In P. H. Heckel, ed., *Stratigraphy and Biostratigraphy of the Mississippian Subsystem (Carboniferous System) in its Type Region, the Mississippi Valley of Illinois, Missouri, and Iowa.* IUGS Subcommission on Carboniferous Stratigraphy Guidebook for Field Conference, September 8–13, 2001, pp. 83–107.

Lane, H. R., Merrill, G. K., Straka II, J. J., and Webster, G. D., 1971, North American Pennsylvanian conodont biostratigraphy. *Conodont Biostratigraphy, Memoir of the Geological Society of America* Vol. 127. Boulder, CO: Geological Society of America, pp. 395–414.

Lane, H. R., Sandberg, C. A., and Ziegler, W., 1980, Taxonomy and phylogeny of some Lower Carboniferous conodonts and preliminary standard post-Siphonodella zonation. *Geologica et Palaeontologica* **14**: 117–64.

Lane, H. R., Brenckle, P. L., Baesemann, J. F., and Richards, B., 1999, The IUGS boundary in the middle of the Carboniferous: Arrow Canyon, Nevada, USA. *Episodes* **22**(4): 272–83.

Lang, W. D., 1913, The Lower Pliensbachian–"Carixian"–of Charmouth. *Geological Magazine* **5**(10): 401–12.

Langereis, C. G. and Hilgen, F. J., 1991, The Rossello composite: a Mediterranean and global standard reference section for the Early to early Late Pliocene. *Earth and Planetary Science Letters* **104**: 211–25.

Langereis, C. G., van Hoof, A. A. M., and Hilgen, F. J., 1994, Steadying the rates. *Nature* **369**: 615.

Langereis, C. G., Dekkers, M. J., De Lange, G. J., Paterne, M., and Van Santvoort, P. J. M., 1997, Magnetostratigraphy and astronomical calibration of the last 1.1 Myr from an eastern Mediterranean piston core and dating of short events in the Brunhes. *Geophysical Journal International* **129**: 75–94.

Lanphere, M. A. and Baadsgaard, H., 2001, Precise K–Ar, ^{40}Ar/^{39}Ar, Rb–Sr and U/Pb mineral ages from the 27.5 Ma Fish Canyon Tuff reference standard. *Chemical Geology* **175**: 653–71.

Lanphere, M. A., Churkin, M. J., and Erberlein, G. D., 1977, Radiometric age of the *Monograptus cyphus* graptolite zone in south-eastern Alaska: an estimate of the age of the Ordovician–Silurian boundary. *Geological Magazine* **114**: 15–24.

Lapparent, A. D., 1883, *Traité de Géologie Paris*. Paris: Savy.

Lapworth, C., 1878, The Moffat Series. *Quarterly Journal of the Geological Society of London* **34**: 240–346.

1879a, On the tripartite classification of the Lower Paleozoic rocks. *Geological Magazine, new series* **6**: 1–15.

1879b, On the geological distribution of the Rhabdophora. *Annals and Magazine of Natural History, series 5* **3**: 245–57, 449–55; **4**: 333–41, 423–31.

1880, On the geological distribution of the Rhabdophora. *Annals andMagazine of Natural History, series 5* **5**: 45–62, 273–85, 359–69; **6**: 16–29, 185–207.

Larson, R. L., 1991, Latest pulse of the Earth: evidence for a mid Cretaceous superplume. *Geology* **19**: 547–50.

Larson, R. L. and Erba, E., 1999, Onset of the mid-Cretaceous greenhouse in the Barremian–Aptian: igneous events and the biological, sedimentary, and geochemical responses. *Paleoceanography* **14**: 663–78.

Larson, R. L. and Hilde, T. W. C., 1975, A revised time scale of magnetic reversals for the Early Cretaceous and Late Jurassic. *Journal of Geophysical Research* **80**: 2586–94.

Laskar, J., 1990, The chaotic motion of the solar system: a numerical estimate of the size of the chaotic zones. *Icarus* **88**: 266–91.

1999, The limits of Earth orbital calculations for geological time-scale use. *Philosophical Transactions of the Royal Society of London, series A* **357**: 1785–95.

Laskar, J. and Robutel, P., 2001, High order symplectic integrators for perturbed Hamiltonian systems. *Celestial Mechanics* **80**: 39–62.

Laskar, J., Joutel, F., and Boudin, F., 1993, Orbital, precessional and insolation quantities for the Earth from −20 Myr to +10 Myr. *Astronomy and Astrophysics* **270**: 522–33.

Laskar, J., Robutel, P., Joutel, F., *et al.*, in prep., A numerical solution for the insolation quantities for the Earth from −100 Ma to +50 Ma.

Laurin, B., 1998. Brachiopods. [column for Jurassic chart, Mesozoic and Cenozoic sequence chronostratigraphic framework of European basins, in Hardenbol, J. *et al.*]. In P.-C. de Graciansky, J. Hardenbol., Th. Jacquin, and P. R. Vail, eds., *Mesozoic–Cenozoic Sequence Stratigraphy of European Basins, Society of Economic Paleontologists and Mineralogists Special Publication,* Vol. 60. Tulsa, OK: SEPM, Chart 7.

Lawson, J. D. and White, D. E., 1989, The Ludlow Series in the type area. In C. H. Holland and M. G. Bassett, eds., *A Global Standard for the Silurian System, Geological Series* Vol. 9. Cardiff: National Museum of Wales.

Layer, P. W., Kroener, A., and McWilliams, M., 1996, An Archean geomagnetic reversal in the Kaap Valley Pluton, South Africa. *Science* **273**: 943–46.

Leckie, R. M., Bralower, T. J., and Cashman, R., 2002, Oceanic anoxic events and plankton evolution: biotic response to tectonic forcing during the mid-Cretaceous. *Paleoceanography* **17**: doi 10.1029/2001PA000623.

Lee, D. C., Halliday, A. N., Snyder, G. A., and Taylor, L. A., 1997, Age and origin of the Moon. *Science* **278**(5340): 1098–103.

Leereveld, H., 1995, Dinoflagellate cysts from the Lower Cretaceous Río Argos succession (SE Spain). *LPP Contributions Series*, Vol. 2. Utrecht: LPP Foundation, p. 175.

1997a, Upper Tithonian–Valanginian (Upper Jurassic–Lower Cretaceous) dinoflagellate cyst stratigraphy of the western Mediterranean. *Cretaceous Research* **18**: 385–420.

1997b, Hauterivian–Barremian (Lower Cretaceous) dinoflagellate cyst stratigraphy of the western Mediterranean. *Cretaceous Research* **18**: 421–56.

Legrand-Blain, M., 1990, Brachiopods as potential boundary-defining organisms in the Lower Carboniferous of Western Europe; recent data and productid distribution. In P. L. Brenkle and W. L. Manger, eds., Intercontinental correlation and division of the Carboniferous System: contributions from the Carboniferous Subcommission meeting. *Courier Forschungsinstitut Senckenberg* **130**: 157–71.

Le Hérissé, A., Servais, T., and Wicander, R., 2000, Devonian acritarchs and related forms. *Courier Forschungsinstitut Senckenberg* **220**: 195–204.

Lerbekmo, J. F. and Braman, D. R., 2002, Magnetostratigraphic and biostratigraphic correlation of late Campanian and Maastrichtian marine and continental strata from the Red Deer Valley to the Cyprus Hills, Alberta, Canada. *Canadian Journal of Earth Sciences* **39**: 539–57.

Leslie, S. A. and Bergström, S. M., 1995, Revision of the North American Late Middle Ordovician standard stage classification and timing of the Trenton transgression based on K-bentonite bed correlation. In J. D. Cooper, M. L. Droser, and S. F. Finney, eds., *Ordovician Odyssey: Short Papers for the Seventh International Symposium on the Ordovician System, Society of Economic Paleontologists and Mineralogists Special Publication,* Vol. 77. Tulsa, OK: Pacific Section, Society for Sedimentary Geology (Society of Economic Paleontologists and Mineralogists), pp. 49–54.

Less, G., 1987, Paleontology and stratigraphy of the European Orthophragminae. *Geologica Hungarica (Paleontologia)* **51**: 373.

Less, G., 1998, The zonation of the Mediterranean Upper Paleocene and Eocene by Orthophragminae. *Slovenska Akademija Znanosti in Umetnosti, Razred za Naravoslovne Vede Dela* **34**(2): 21–43.

Less, G. and Kovács, O., 1996, Age-estimates by European Paleogene Orthophragminae using numerical evolutionary correlations. *Géobios* **29**: 261–85.

Leupold, W. and van der Vlerk, I. M., 1936, De stratigraphie van Nederlandsch Ostindiè: 20. The Tertiary. *Leidse Geologische Mededelingen (Feestbundel Prof. Dr. K. Martin)* **5**: 611–48.

Leven, E. Y. and Davydov, V. I., 2001, Stratigraphy and fusulinids of the Kasimovian and Lower Gzhelian (Upper Carboniferous) in the Southwestern Darvas (Pamir). *Revista Italiana di Paleontologia e Stratigrafia* **107**(1): 3–46.

Levrard, B. and Laskar, J., 2003, Climate friction and the Earth's obliquity. *Geophysical Journal International* **154**(3): 970–90.

Li, L.-Q. and Keller, G., 1999, Variability in Late Cretaceous climate and deep waters: evidence from stable isotopes. *Marine Geology* **161**: 171–90.

Likharew, B. K., 1966, *The Stratigraphy of USSR, The Permian System.* Moscow: Nedra, p. 536.

Lindsay, E. and Tedford, R., 1990, Development and application of land mammal ages in North America and Europe, a comparison. In E. H. Lindsay, V. Fahlbusch, and P. Mein, eds., *European Neogene Mammal Chronology*. New York: Plenum Press, pp. 601–24.

Lini, A., Weissert, H., and Erba, E., 1992, The Valanginian carbon isotope event: a first episode of greenhouse climate conditions during the Cretaceous. *Terra Nova* **4**: 374–84.

Lipina, O. A. and Reitlinger, E. A., 1970, Stratigraphie zonale et paléozoogéographie du carbonifére inférieur d'aprés les foraminiféres. In *Compte Rendu Congrès International de Stratigraphie et de Géologie du Carbonifère*, Vol. 3. pp. 1101–12.

Lirer, F., Caruso, A., Foresi, L. M., *et al.*, 2002, Astrochronological calibration of the upper Serravallian–lower Tortonian sedimentary sequence at Tremiti Islands (Adriatic Sea, Southern Italy). In S. Iaccarino, ed., Integrated stratigraphy and paleoceanography of the Mediterranean Middle Miocene. *Milano, Rivista Italiana di Paleontologia e Stratigrafia* **108**: 241–56.

Litt, T. and Turner, C., 1993, Arbeitsergebnisse der Subkommission für Europäische Quartärstratigraphie: Die Saalesequenz in der Typusregion (Berichte der SEQS 10). *Eiszeitalter und Gegenwart* **43**: 125–8.

Litt, T., Brauer, A., Goslar, T., *et al.*, 2001, Correlation and synchronisation of late glacial continental sequences in northern central Europe based on annually laminated lacustrine sediments. *Quaternary Science Reviews* **20**: 1233–49.

Litwin, R. J., Traverse, A., and Ash, S. R., 1991, Preliminary palynological zonation of the Chinle Formation, southwestern USA, and its correlation to the Newark Supergroup (eastern USA). *Review of Palaeobotany and Palynology* **68**: 269–87.

Litwin, R. J., Peterson, F., and Turner, C. E., 1998, Palynological evidence on the age of the Morrison Formation, Western Interior US: a preliminary report. *Modern Geology* **22**: 297–319.

Löfgren, A., 1993, Conodonts from the lower Ordovician at Hunneberg, south-central Sweden. *Geological Magazine* **130**(2): 215–32.

1996, Lower Ordovician conodonts, reworking, and biostratigraphy of the Orreholmen quarry, Vastergötland, south-central Sweden. *Lethaia* **118**: 169–83.

Longacre, S. A., 1970, Trilobites of the Upper Cambrian Ptychaspid Biomere, Wilberns Formation, central Texas. *Journal of Paleontology Supplement* **44**(2), *The Paleontological Society Memoir* 4: The Paleontological Society, 70 pp., 6 plates.

Lonsdale, W., 1840, Notes on the age of the Limestones of South Devonshire. *Transactions of the Geological Society of London, Series* 2 **5**: 721–38.

López-Martínez, N. and Peláez-Campomanes, P., 1999, New mammals from south-central Pyrenees (Tremp Formation, Spain) and their bearing on late Paleocene marine–continental correlations. *Bulletin de la Société Géologique de France* **170**: 681–96.

Lourens, L. J. and Hilgen, F. J., 1997, Long-periodic variations in the Earth's obliquity and their relation to third-order eustatic cycles and late Neogene glaciations. *Quaternary International* **40**: 43–52.

Lourens, L. J., Hilgen, F. J., Gudjonsson, L., and Zachariasse, W. J., 1992, Late Pliocene to early Pleistocene astronomically forced sea surface productivity and temperature variations in the Mediterranean. *Marine Micropaleontology* **19**: 49–78.

Lourens, L. J., Hilgen, F. J., Raffi, I., and Vergnaud-Grazzini, C., 1996a, Early Pleistocene chronology of the Vrica section (Calabria, Italia). *Palaeoceanography* **11**: 797– 812.

Lourens, L. J., Hilgen, F. J., Zachariasse, W. J., *et al.*, 1996b, Evaluation of the Plio–Pleistocene astronomical time scale. *Paleoceanography* **11**: 391–413.

Lourens, L. J., Hilgen, F. J., and Raffi, I., 1998, Base of large Gephyrocapsa and astronomical calibration of early Pleistocene sapropels in ODP sites 967 and 969: solving the chronology of the Vrica section. In *Proceedings of the Ocean Drilling Program, Scientific Results*, Vol. 160. College Station, TX: Ocean Drilling Program, pp. 191–7.

Lourens, L. J., Wehausen, R., and Brumsack, H.-J., 2001, Geological constraints on tidal dissipation and dynamical ellipticity of the Earth over the past three million years. *Nature* **409**: 1029–33.

Louwye, S., 1999, New species of organic-walled dinoflagellates and acritarchs from the upper Miocene Diest Formation, northern Belgium (southern North Sea Basin). *Review of Palaeobotany and Palynology* **107**(1–2): 109–23.

2000, Dinoflagellate cysts and acritarchs from the Miocene Zonderschot Sands, northern Belgium; stratigraphic significance and correlation with contiguous areas. *Geologica Belgica* **3**(1–2): 55–65.

Lovera, O. M., Richter, F. M., and Harrison, T. M., 1989, The $^{40}Ar/^{39}Ar$ geothermometry for slowly cooled samples having a distribution of diffusion domain sizes. *Journal of Geophysical Research* **94**: 17 917–35.

Lowe, J. J. and Walker, M. J. C., 1997, *Reconstructing Quaternary Environments*. London: Longmans.

Lowe, J. J., Birks, H. H., Brooks, S. J., *et al.*, 1999, The chronology of palaeoenvironmental changes during the last glacial –Holocene transition: towards an event stratigraphy for the British Isles. *Journal of the Geological Society of London* **156**: 397–410.

Lowell, T. V., Heusser, C. J., Andersen, B. G., *et al.*, 1995, Interhemispheric correlation of late Pleistocene glacial events. *Science* **269**: 1541–9.

Lowrie, W. and Alvarez, W., 1977, Late Cretaceous geomagnetic polarity sequence: detailed rock and palaeomagnetic studies of the Scaglia Rossa limestone at Gubbio, Italy. *Geophysical Journal of the Royal Astronomical Society* **51**: 561–81.

1981, One hundred million years of geomagnetic polarity history. *Geology* **9**: 392–7.

Lowrie, W., Alvarez, W., Premoli-Silva, I., and Monechi, S., 1980, Lower Cretaceous magnetic stratigraphy in Umbrian pelagic carbonate rocks. *Geophysical Journal, Royal Astronomical Society* **60**: 263–81.

Loydell, D. K., 1991, The biostratigraphy and formational relationships of the upper Aeronian and lower Telychian (Llandovery, Silurian) formations of western mid-Wales. *Geological Journal* **26**: 209–44.

1998, Early Silurian sea-level changes. *Geological Magazine* **135**: 447–71.

Loydell, D. K. and Cave, R., 1996, The Llandovery–Wenlock boundary and related stratigraphy in eastern mid-Wales with special reference to the Banwy River section. *Newsletters on Stratigraphy* **34**: 39–64.

Loydell, D. K., Štorch, P., and Melchin, M. J., 1993, Taxonomy, evolution and biostratigraphical importance of the Llandovery graptolite *Spirograptus*. *Palaeontology* **36**: 909–26.

Loydell, D. K., Männik, P., and Nestor, V., 2003, Integrated biostratigraphy of the lower Silurian of the Aizpute-41 core, Latvia. *Geological Magazine* **140**: 205–29.

Lucas, S. G., 1997, *Dinosaurs, The Textbook*, 2nd edition. Dubuque, IA: W. C. Brown.

1998a, Global Triassic tetrapod biostratigraphy and biochronology. *Palaeogeography, Palaeoclimatology, Palaeoecology* **143**: 345–82.

1998b, Fossil mammals and the Paleocene/Eocene series boundary in Europe, North America, and Asia. In M.-P. Aubry, S. G. Lucas, and W. A. Berggren, eds., *Late Paleocene–Early Eocene Climatic and Biotic Events in the Marine and Terrestrial Records*. New York: Columbia University Press, pp. 451–500.

1999, A tetrapod-based Triassic timescale. *Albertiana* **22**: 31–40.

Lucas, S. G., Wilde, G. L., Robbins, S., and Estep, J. W., 2000, Lithostratigraphy and fusulinaceans of the type section of the Bursum Formation, Upper Carboniferous of south-central New Mexico. *New Mexico Museum of Natural History and Science Bulletin* **16**: 1–13.

Ludden, J., 1992, Radiometric age determinations for basement from Sites 765 and 766, Argo Abyssal Plain and Northwestern Australia. In *Proceedings of the Ocean Drilling Program, Scientific Results*, Vol. 123. College Station, TX: Ocean Drilling Program, pp. 557–9.

Ludvigsen, R. and Westrop, S. R., 1985, Three new Upper Cambrian stages for North America. *Geology* **13**: 139–43.

Ludvigsen, R., Westrop, S. R., Pratt, B. R., Tuffnell, P. A., and Young, G. A., 1986, Paleoscene No. 73. Dual biostratigraphy: zones and biofacies. *Geoscience Canada* **13**: 139–54.

Ludwig, K. R., 1980, Calculation of uncertainties of U–Pb isotope data. *Earth and Planetary Science Letters* **46**(2): 212–20.

1984, Minimizing the effect of variable initial ratios on U–Pb isochrons. *Abstracts Geological Society of America annual meeting* **16**: 579.

1993, *Pbdat: A Computer Program for Processing Pb–U–Th Isotope Data*, version 1.23. Open File Report 88–542. Denver: US Geological Survey.

1998, On the treatment of concordant uranium–lead ages. *Geochimica et Cosmochimica Acta* **62**(4): 665–76.

2000, Decay constant errors in U–Pb Concordia–intercept ages. *Chemical Geology* **166**(3–4): 315–18.

Ludwig, K. R., Mundil, R., and Renne, P. R., 1999, How well can we really do timescale geochronology with zircon U–Pb? *Journal of Goldschmidt Conference Abstracts* **243**: 178–179.

Lumbers, S. B. and Card, K. D., 1991, Chronometric subdivision of the Archean. *Geology* **20**: 56–7.

Luterbacher, H., 1973, La sección-tipo del Ilerdiense. In E. Perconig, ed., *XIII Coloquio Europeo de Micropaleontología*. Madrid: ENADIMSA, pp. 113–40.

Luterbacher, H. P., 1998, Sequence stratigraphy and the limitations of biostratigraphy in the marine Paleogene strata of the Tremp

Basin (central part of the Southern Pyrenean Foreland Basin, Spain). In P.-C. de Graciansky, J. Hardenbol, T. Jacquin, and P. R. Vail, eds., *Mesozoic–Cerozoic Sequence Stratigraphy of European Basins, Society of Economic Paleontologists and Mineralogists Special Publication*, Vol. 60. Tulsa, OK: SEPM, pp. 303–9.

Lyell, C., 1833, *Principles of Geology: Being an Inquiry How Far the Former Changes of the Earth's Surface are Referable to Causes Now in Operation*, Vol. III. London: John Murray, p. 398.

1839, *Nouveaux éléments de géologie*. Paris: Pitois-Levranet, p. 648.

1857, *Principles of Geology*, Supplement to the fifth edition. London: John Murray.

1867, *Principles of Geology*, Vol. 1. London: John Murray.

1873, *The Geological Evidences of the Antiquity of Man: With an Outline of Glacial and Post-Tertiary Geology and Remarks on the Origin of Species; With Special Reference to Man's First Appearance on the Earth*. London: John Murray.

Lyle, M., Wilson, P. A., Janecek, T. R., *et al.*, 2002, Site 1218. In *Proceedings of the Ocean Drilling Program, Initial Reports*, Vol. 199. College Station, TX: Ocean Drilling Program [Online: www-odp.tamu.edu/publications/199_IR/ 199ir.htm., 25 March 2004].

Mabillard, J. E. and Aldridge, R. J., 1985, Microfossil distribution across the base of the Wenlock Series in the type area. *Palaeontology* **28**: 89–100.

Magniez-Jannin, F., 1995, Cretaceous stratigraphic scales based on benthic foraminifera in West Europe (biochronohorizons). *Bulletin de la Société Géologique de France* **166**: 565–72.

Magny, M., 1993, Solar influences on Holocene climatic changes illustrated by correlations between past lake-level fluctuations and the atmospheric ^{14}C record. *Quaternary Research* **40**: 1–9.

Magyar, I., Geary, D. H., and Müller, P., 1999a, Palaeogeographic evolution of the Late Miocene Lake Pannon in Central Europe. *Palaeogeography, Palaeoclimatology, Palaeoecology* **147**: 151–67.

Magyar, I., Geary, D. H., Suto-Szentai, M., Lantos, M., and Muller, P., 1999b, Integrated bio-, magneto- and chronostratigraphic correlation of the Late Miocene Lake Pannon deposits. *Acta Geologica Hungarica* **42**: 5–31.

Mahoney, J. J., Storey, M., Duncan, R. A., Spencer, K. J., and Pringle, M., 1993, Geochemistry and geochronology of the Ontong Java Plateau. In M. Pringle, W. W. Sager, W. V. Sliter, and S. Stein, eds., *The Mesozoic Pacific: Geology, Tectonics, and Volcanism, Geophysical Monographs*, Vol. 77. Washington, DC: American Geophysical Union, pp. 233–61.

Mahoney, J. J., Fitton, G., Wallace, P. J., *et al., 2001, Proceedings of the Ocean Drilling Program, Initial Reports,* Vol. 192. College Station, TX: Ocean Drilling Progam [Online: www-odp.tamu.edu/publications/192_IR].

Maiorano, P., Marino, M., Di Stefano, E., and Ciaranfi, N., in press, Calcareous nannofossil events in the Lower–Middle Pleistocene transition at Montalbano Jonico section (Southern Italy) and ODP Site 964 (Ionian Sea), and their calibration with oxygen isotope and sapropel stratigraphy.

Makhlina, M. K., Kulikova, A. M., and Nikitina, T. A., 1979, Stratigraphy, biostratigraphy and paleogeography of Upper Carboniferous of the Moscow Syneclise. In M. K. Makhlina and C. M. Shik, eds., *Stratigraphy, Paleontology and Paleogeography of Carboniferous of the Moscow Syneclise, Transactions of Geological Foundation of Ministry of Geology of Russian Federation*. Moscow: Ministry of Geology of the USSR, pp. 25–69 [In Russian].

Makhlina, M. K., Vdovenko, M. V., Alekseev, A. S., *et al.*, 1993, Lower Carboniferous of the Moscow syneclise and Voronezh anteclise. *Moscow Society of Natural Studies Bulletin, Geological Series*. Moscow: Nauka, pp. 3–220 [In Russian].

Makhlina, M. K., Alekseev, A. S., Goreva, N. V., *et al.*, 2001, Stratigraphy. In A. S. Alekseev and S. M. Shik, eds., *Middle Carboniferous of the Moscow Syneclise*, Vol. I. Moscow: Scientific World 4, pp. 1–244.

Maletz, J., Löfgren, A., and Bergström, S. M. N., 1996, The base of the *Tetragraptus approximatus Zone* at Mt. Hunneberg, SW Sweden: a proposed global stratotype for the base of the second series of the Ordovician System. *Newsletters on Stratigraphy* **34**: 129–59.

Mallory, V. S., 1959, *Lower Tertiary Biostratigraphy of the California Coast Ranges*. Tulsa, OK: American Association of Petroleum Geologists.

Mamet, B. L., 1974, Une zonation par Foraminifères due Carbonifère Inférieur de la Téthys Occidentale. In *Compte Rendu 7ième Congrès International de Stratigraphie et de Géologie du Carbonifère*, Vol. 3, Krefeld, 1971. pp. 391–408.

Mamet, B. L. and Skipp, B., 1970, Lower Carboniferous calcareous foraminifera: preliminary zonation and stratigraphic implications for the Mississippian of North America. *Compte Rendu Congrès International de Stratigraphie et de Géologie du Carbonifère* **6**: 1129–46.

Mamet, B. L., Bamber, E. W., and Macqueen, R. W., 1986, Microfacies of the Lower Carboniferous Banff Formation and Rundle Group, Monkman Pass map area, northeastern British Columbia. *Geological Survey of Canada Bulletin* **353**: 93.

Mangerud, J., Andersen, S. T., Berglund, B. E., *et al.*, 1974, Quaternary stratigraphy of Norden, a proposal for terminology and classification. *Boreas* **3**: 109–28.

Mangerud, G., Dreyer, T., Søyseth, L., *et al.*, A., 1999, High-resolution biostratigraphy and sequence development of the Paleocene succession, Grane Field, Norway. *Geological Society of London Special Publication*, London: Geological Society, Vol. 152. pp. 167–84.

Mangin, J.-P., 1957, Remarques sur le terme Paléocène et sur la limite Crétacé–Tertiaire. *Comptes rendus des séances de la Société Géologique de France* **14**: 319–21.

Mangold, C., 1997, Report of the Bajocian–Bathonian Boundary Working Group. *International Subcommission on Jurassic Stratigraphy Newsletter* **25**: 42–3.

1999, Report of the Bajocian–Bathonian Boundary Working Group. *International Subcommission on Jurassic Stratigraphy Newsletter* **26**: 50–2.

Mankinen, E. A. and Dalrymple, G. B., 1972, Electron microprobe evaluation of terrestrial basalts for whole-rock K–Ar dating. *Earth and Planetary Science Letters* **17**: 89–94.

1979, Revised geomagnetic polarity time scale for the interval 0–5 my BP. *Journal of Geophysical Research* **84**: 615–26.

Mann, K. O. and Lane, R. H., 1995, *Graphic Correlation, Special Publication 53*. Tulsa, OK: Society for Sedimentary Geology, p. 263.

Marcour, J., 1859, Dyas et Trias ou le Nouveau grès rouge en Europe, en l'Amerique du nord et en l'Inde. *Archives Sciences* **5**: 5–37, 116–46.

Margaritz, M., 1989, δ^{13}C minima follow extinction events: a clue to faunal radiation. *Geology* **17**: 337–40.

Margaritz, M., Kirschvink, J. L., Latham, A. J., Zhuravlev, A. Y., and Rozanov, A. Y., 1991, Precambrian/Cambrian boundary problem: carbon isotope correlations for Vendian and Tommotian time between Siberia and Morocco. *Geology* **19**: 847–50.

Mark-Kurik, E., Blieck, A., Lobaziak, S., and Candilier, A.-M., 1999, Miospore assemblage from the Lode member (Gauja Formation) in Estonia and the Middle–Upper Devonian boundary problem. *Proceedings of the Estonian Academy of Sciences, Geology* **48**: 86–98.

Marks, P., 1971a, Turolian. *Giornale di Geologia* **37**: 209–13.

1971b, Vallasian. *Giornale di Geologia* **37**: 215–19.

Marr, J. E., 1883, *The Classification of the Cambrian and Silurian Rocks*. Cambridge: Deighton Bell.

1905, The classification of the sedimentary rocks. *Quarterly Journal of the Geological Society of London* **61**: Lxi–Lxxxvi.

1913, The Lower Paleozoic rocks of the Caughtley District. *Quarterly Journal of the Geological Society of London* **69**: 1–17.

Märss, T., Fredholm, D., Talimaa, V., *et al.*, 1995, Silurian vertebrate biozonal scheme. In H. Lelievre, S. Wenz, A. Blieck, and R. Cloutier, eds., *Premiers Vertébrés et Vertebres Inferieurs, Géobios Mémoire Spécial 19*. Lyon: Université Claude Bernard, pp. 369–72.

Martin, E. E. and Macdougall, J. D., 1991, Seawater Sr isotopes at the Cretaceous/Tertiary boundary. *Earth and Planetary Science Letters* **104**: 166–80.

1995, Sr and Nd isotopes at the Permian/Triassic boundary: a record of climate change. *Chemical Geology* **125**(1–2): 73–99.

Martin, E. E., Macdougall, J. D., Herbert, T. D., Paytan, A., and Kastner, M., 1995, Strontium and neodymium isotopic analyses of marine barite separates. *Geochimica et Cosmochimica Acta* **59**: 1353–61.

Martin, E. E., Shackleton, N. J., Zachos, J. C., and Flower, B. P., 1999, Orbitally-tuned Sr isotope chemostratigraphy for the late middle to late Miocene. Palaeoceanography **14:** 74–83.

Martin, M. W., Grazhdankin, D. V., Bowring, S. A., *et al.*, 2000, Age of Neoproterozoic bilatarian body and trace fossils, White Sea, Russia: implications for metazoan evolution. *Science* **288**: 841–5.

Martini, E., 1958, Discoasteriden und verwandte Formen im NW-deutschen Eozän (Coccolithophorida). 1. Taxonomische Untersuchungen. *Senckenbergiana Lethaea* **39**: 353–88.

1959a, Discoasteriden und verwandte Formen im NW-deutschen Eozän (Coccolithophorida). 2. Stratigraphische Auswertung. *Senckenbergiana Lethaea* **40**: 137–57.

1959b, Der stratigraphische Wert von Nanno-Fossilien im norwestdeutschen Tertiär. *Erdöl und Kohle* **12**: 137–40.

1971, Standard Tertiary and Quaternary calcareous nannoplankton zonation. In A. Farinacci, ed., Proceedings of the *II Planktonic Conference*, Roma 1970. Vol. 2 Rome: Edizioni Tecnoscienza, pp. 739–85.

1988, Late Oligocene and Early Miocene calcareous plankton (Remarks on French and Moroccan sections). *Newsletters on Stratigraphy* **18**(2): 75–80.

Martini, E. and Müller, C., 1971, Das marine Alttertiär in Deutschland und seine Einordung in die Standard Nannoplankton Zonen. *Erdöl Erdgas Kohle* **24**: 381–4.

Martini, E., von Benedek, N. P., and Müller, C., 1976, Calcareous nannofossils, silicoflagellates, dinoflagellates and related forms in the NW-German Tertiary basin. *IGCP Project 124, Report* 1. Mainz: International Geological Correlation Programme, pp. 73–81.

Martinson, D. G., Pisias, N. G., Hays, J. D., *et al.*, 1987, Age dating and the orbital theory of the ice ages: development of a high-resolution 0 to 300 000 year chronostratigraphy. *Quaternary Research* **27**: 1–29.

Martinsson, A., 1977, The Silurian–Devonian boundary: final report of the Committee of the Siluro–Devonian Boundary within IUGS Commission on Stratigraphy and a state of the art report for Project Ecostratigraphy. *International Union of Geological Sciences, Series A5*. Stuttgart: Schweizerbartsche Verlagsbuchhandlung.

Marzoli, A., Renne, P. R., Piccirillo, E. M., *et al.*, 1999, Extensive 200-million-year-old continental flood basalts of the Central Atlantic Magmatic Province. *Science* **284**: 616–18.

Masse, J.-P. and Philip, J., 1998, Rudists. [column for Cretaceous portion of Mesozoic and Cenozoic sequence chronostratigraphic framework of European basins, in Hardenbol, J. *et al.*]. In P.-C. de Graciansky, J. Hardenbol., Th. Jacquin, and P. R. Vail, eds., *Mesozoic–Cenozoic Sequence Stratigraphy of European Basins, Society of Economic Paleontologists and Mineralogists Special Publication,* Vol. 60. Tulsa, OK: SEPM, pp. 774–7, Chart 5.

Matsuoka, A., 1992, Jurassic and Early Cretaceous radiolarians from Leg 129, Sites 800 and 801, Western Pacific Ocean. In *Proceedings of the Ocean Drilling Program, Scientific Results*, Vol. 129. College Station, TX: Ocean Drilling Program, pp. 203–11.

Matsuoka, A. and Yao, A., 1986, A newly proposed radiolarian zonation for the Jurassic of Japan. *Marine Micropaleontology* **11**: 91–105.

Mattinson, J. M., 1994a, Real and apparent concordance and discordance in the U–Pb systematics of zircon: limitations of "high precision" U/Pb and Pb/Pb ages. *Abstracts of the American Geophysical Union,* fall meeting. p. 691.

1994b, Uranium decay constant uncertainties, and their implications for high-resolution U–Pb geochronology. Abstracts with Programs, Geological Society of America, annual meeting 26(7), p. 221.

Matyja, B. A. and Wierzbowski, A., 1997, The quest for a unified Oxfordian/Kimmeridgian boundary: implications of the ammonite succession at the turn of the Bimammatum and Planula Zones in the Wielun Upland, Central Poland. *Acta Geologica Polonica* **47**(1–2): 77–105.

Maubeuge, P. L., 1964, *Colloque du Jurassique à Luxembourg 1962*, Luxembourg: Institut Grand-Ducal, Section des Sciences Naturelles, Physiques et Mathématiques.

Maurasse, F. and Glass, B. P., 1976, Radiolarian stratigraphy and North American mikrotektites in Caribbean RC9-58: implications concerning Late Eocene radiolarian chronology and the age of the Eocene–Oligocene boundary. *Transactions de la VII Conférence géologique des Caraïbes*, pp. 205–12.

Mayer-Eymar, K., 1858, Versuch einer neuen Klassifikation der Tertiär-Gebilde Europa's. *Verhandlungen der Schweizer Naturforschenden Gessellschaft für gesammt. Naturwissensch. Trogen* **42**: 165–99.

1864, *Tableau synchronistique des terrains jurassiques*. Zurich.

1868, *Tableau synchronistique des terrains tertiaires supérieurs*, IV edn. Zurich.

McArthur, J. M., 1994, Recent trends in strontium isotope stratigraphy. *Terra Nova* **6**: 331–58.

McArthur, J. M. and Morton, N., 2000, Strontium isotope stratigraphy of the Aalenian/Bajocian auxiliary stratotype point at Bearreraig, Isle of Skye, NW Scotland. GeoResearch Forum, Vol. 6. Zurich: Transtech Publications, pp. 137–44.

McArthur, J. M., Kennedy, W. J., Gale, A. S., 1992, Strontium isotope stratigraphy in the late Cretaceous: international correlation of the Campanian/Maastrichtian boundary. *Terra Nova* **4**: 332–45.

McArthur, J. M., Thirlwall, M. F., Gale, A. S., Chen, M., and Kennedy, W. J., 1993, Strontium isotope stratigraphy in the Late Cretaceous: numerical calibration of the Sr isotope curve and intercontinental correlation for the Campanian. *Paleoceanography* **8**: 859–73.

McArthur, J. M., Kennedy, W. J., Chen, M., Thirlwall, M. F., and Gale, A. S., 1994, Strontium isotope stratigraphy for the Late Cretaceous: direct numerical age calibration of the Sr-isotope curve for the US Western Interior Seaway. *Palaeogeography, Palaeoclimatology, Palaeoecology* **108**: 95–119.

McArthur, J. M., Thirlwall, M. F., Engkilde, M., Zinsmeister, W. J., and Howarth, R. J., 1998, Strontium isotope profiles across K/T boundary sequences in Denmark and Antarctica. *Earth and Planetary Science Letters* **160**: 179–92.

McArthur, J. M., Donovan, D. T., Thirlwall, M. F., Fouke, B. W., and Mattey, D., 2000, Strontium isotope profile of the early Toarcian (Jurassic) Oceanic Anoxic Event, the duration of ammonite biozones, and belemnite paleotemperatures. *Earth and Planetary Science Letters* **179**: 269–85.

McArthur, J. M., Howarth, R., and Bailey, T. R., 2001, Strontium isotope stratigraphy: LOWESS Version 3. Best-fit line to the marine Sr-isotope curve for 0 to 509 Ma and accompanying look-up table for deriving numerical age. *Journal of Geology* **109**: 155–69.

McArthur, J. M., Mutterlose, J., Price, G. D., *et al.*, 2004, Belemnites of Valanginian, Hauterivian and Barremian age: Sr-isotope stratigraphy, composition ($^{87}Sr/^{86}Sr$, $\delta^{13}C$, $\delta^{18}O$, Na, Sr, Mg), and palaeo-oceanography. *Palaeogeography, Palaeoclimatology, Palaeoecology* **202**: 253–72.

McArthur, J. M., Bailey, T. R., Houghton, S., and Thirlwall, M., in review, A new $^{87}Sr/^{86}Sr$ date for the interglacial Cockburn Island Formation, northern Antarctic Peninsula. *Geological Society of London Special Publication*. London: Geological Society.

McDougall, I., 1985, K–Ar and $^{40}Ar/^{39}Ar$ dating of the hominid-bearing Pliocene–Pleistocene sequence at Koobi Fora, Lake Turkana. *Geological Society of America Bulletin* **96**: 159–75.

McDougall, I. and Harrison, T. M., 1999, *Geochronology and Thermochronology by the $^{40}Ar/^{39}Ar$ Method*. Oxford: Oxford University Press.

McDowell, F. W., 1983, K–Ar dating: incomplete extraction of radiogenic argon from alkali feldspar. *Chemical Geology* **1**: 119–26.

McElhinny, M. W. and McFadden, P. L., 2000, *Paleomagnetism: Continents and Oceans*. San Diego, CA: Academic Press.

McIntyre, A. and Molfino, B., 1996, Forcing of Atlantic equatorial and subpolar millennial cycles by precession. *Science* **274**: 1867–70.

McKay, W. and Green, R., 1963, *Mississippian Foraminifera of the Southern Canadian Rocky Mountains, Alberta. Research Council of Alberta Bulletin*. Edmonton: Research Council of Alberta.

McKerrow, W. S., Lambert, R. S. J., and Cocks, L. R. M., 1985, The Ordovician, Silurian and Devonian Periods. In N. J. Snelling, ed., *The Chronology of the Geological Record, Geological Society of London Memoir* 10. London: Geological Society, pp. 73–80.

McManus, J. F., Bond, G. C., Broecker, W. S., *et al.*, 1994, High-resolution climate records from the North Atlantic during the last interglacial. *Nature* **371**: 326–29.

McMillan, N. J., Embry, A. F., and Glass, D. J., 1988, *Devonian Geology of the World*, Vols. 1–3. Calgary: Canadian Society of Petroleum Geologists.

McMinn, A., 1992, Neogene dinoflagellate distribution in the eastern Indian Ocean from Leg 123, Site 765. In F. Gradstein, *et al.*, eds., *Proceedings of the Ocean Drilling Program, Scientific Results*, Vol. 123. College Station, TX: Ocean Drilling Program, pp. 429–41.

1993, Neogene dinoflagellate cyst biostratigraphy from sites 815 and 823, Leg 133, northeastern Australian margin. In J. A. McKenzie, *et al.*, eds., *Proceedings of the Ocean Drilling Program, Scientific Results*, Vol. 133. College Station, TX: Ocean Drilling Program, pp. 97–105.

1995, Why are there no post-Paleogene dinoflagellate cysts in the Southern Ocean? *Micropaleontology* **41**(4): 383–6.

McMinn, A. and Wells, P., 1997, Use of dinoflagellate cysts to determine changing Quaternary sea-surface temperature in southern Australia. *Marine Micropaleontology* **29**: 407–22.

McMinn, A., Howard, W. R., and Roberts, D., 2001, Late Pliocene dinoflagellate cyst and diatom analysis from a high-resolution sequence in DSDP Site 594, Chatham Rise, southwest Pacific. *Marine Micropaleontology* **43**: 207–21.

McNaughton, N. J., Rasmussen, B., and Fletcher, I. R., 1999, SHRIMP uranium–lead dating of diagenetic xenotime in siliciclastic sedimentary rocks. *Science* **285**: 78–80.

Mead, G. A. and Hodell, D. A., 1995, Controls on the $^{87}Sr/^{86}Sr$ composition of seawater from middle Eocene to Oligocene: Hole 689B, Maud Rise, Antarctica. *Paleoceanography* **10**: 327–46.

Mearon, S., Paytan, A., and Bralower, T., 2003, Cretaceous strontium isotope stratigraphy using marine barite. *Geology* **31**: 15–18.

Meert, J. G. and McPowell, C. M., 2001, Assembly and break-up of Rodinia: introduction to the special volume. *Precambrian Research* **110**: 1–8.

Mei, S. L., Jin, Y. G., and Wardlaw, B. R., 1994a, Succession of conodont zones from the Permian "Kufeng Formation," Xuanhan, Sichuan and its implications in global correlation. *Acta Palaeontologica Sinica* **33**(1): 1–23.

1994b, Succession of Wuchiapingian conodonts from northeastern Sichuan and its worldwide correlation. *Acta Micropalaeontologica Sinica* **11**(2): 121–39.

1994c, *Clarkina* species succession for the Wuchapingian and its worldwide correlation. *Acta Micropalaeontologica Sinica* **11**(2): 121–39.

1998a, Conodont succession of the Guadalupian–Lopingian boundary strata in Laibin of Guangxi, Chian and West Texas, USA. *Palaeoworld* **9**: 53–76.

Mei Shilong, Zhang Kexian, and Wardlaw, B. R., 1998b, A refined succession of Changhsingian and Griesbachian neogondolellid conodonts from the Meishan section, candidate of the global stratotype section and point of the Permian–Triassic boundary. *Palaeogeography, Palaeoclimatology, Palaeoecology* **143**(4): 213–26.

Mei Shilong, Henderson, C., and Wardlaw, B., 2001, Progress on the definition for the base of the Changhsingian. *Permophiles* **38**: 36–7.

2002, Evolution and distribution of the conodonts *Sweetognathus* and *Iranognathus* and related genera during the Permian, and their implications for climate change. *Palaeogeography, Palaeoclimatology, Palaeoecology* **180**: 57–91.

Mein, P., 1975, Résultats du Groupe de Travail des Vertébrés. In J. Senes, ed., *Report on Activity of the Regional Committee on Mediterranean Neogene Stratigraphy (RCMNS) Working Groups (1971–1975)*. Bratislava: RCMNS, pp. 78–81.

1999, European Miocene mammal biochronology. In G. E. Rössner and K. Heissig, eds., *The Miocene Land Mammals of Europe*. München: Pfeil, pp. 25–38.

Mein, P., Moissenet, E., and Adrover, R., 1990, Biostratigraphie du Néogène supérieur de Teruel. *Paleontologia i Evolució* **23**: 121–39.

Meister, C., 1999a, Report of the Sinemurian–Pliensbachian Boundary Working Group. *International Subcommission on Jurassic Stratigraphy Newsletter* **26**: 33–42.

1999b. Report of the Sinemurian–Pliensbachian Boundary Working Group. *International Subcommission on Jurassic Stratigraphy Newsletter* **27**: 25–6.

2001, Pliensbachian Working Group. *International Subcommission on Jurassic Stratigraphy Newsletter* **28**: 8.

Meister, C., Blau, J., Dommergues, J.-L., *et al.*, 2003, A proposal for a stratotype of the Pliensbachian Stage. *Eclogae Geologicae Helvetiae* **96**: 275–98.

Melchin, M. J. and Holmden, C., 2000, Carbon isotope stratigraphy of the mid-Ashgill (Late Ordovician) to Lower Telychian (Early Silurian) of the Cape Phillips Formation, Central Canadian Arctic Islands. *Palaeontology Down Under 2000, Geological Society of Australia, Abstracts* **61**: 166.

Melchin, J. M. and Mitchell, C. E., 1991, Late Ordovician extinction in the Graptoloidea. In C. R. Barnes and S. H. Williams, eds., *Advances in Ordovician Geology, Geological Survey of Canada, Paper* 90–9. Ottawa: Geological Survey of Canada, pp. 143–56.

Melchin, M. J. and Williams, S. H., 2000, A restudy of the akidograptine graptolites from Dobb's Linn and a proposed redefined zonation of the Silurian Stratotype. *Palaeontology Down Under 2000, Geological Society of Australia, Abstracts* **61**: 63.

Melchin, M. J., Koren', T. N., and Štorch, P., 1998, Global diversity and survivorship patterns of Silurian graptoloids. In E. Landing and M. E. Johnson, eds., *Silurian Cycles: Linkages of*

Dynamic Stratigraphy with Atmospheric, Oceanic and Tectonic Changes, New York State Museum Bulletin, Vol. 491. Albany, NY: State Education Department, pp. 165–82.

Meléndez, G.,1999, Report of the Callovian–Oxfordian Boundary Working Group. *International Subcommission on Jurassic Stratigraphy Newsletter* **26**: 61–7.

2002, Oxfordian Working Group. *International Subcommission on Jurassic Stratigraphy Newsletter* **29**: 1–10.

2003. Oxfordian Working Group. *International Subcommission on Jurassic Stratigraphy Newsletter* **30**: 16–18.

Meléndez, G. and Atrops, F., 1999, Report of the Oxfordian–Kimmeridgian Boundary Working Group. *International Subcommission on Jurassic Stratigraphy Newsletter* **26**: 67–74.

Melezhik, V. A., Fallick, A. E., Medvedev, P. V., and Makarikhin, V. V., 1999, Extreme ^{13}C enrichment in ca. 2.0 Ga magnesite–stromatolite–dolomite–'red beds' association in a global context: a case for the world-wide signal enhanced by a local environment. *Earth Science Reviews* **48**: 71–120.

Menegatti, A. P., Weissert, H., Brown, R. S., *et al.*, 1998, High resolution δ^{13}C-stratigraphy through the early Aptian "Livello Selli" of the Alpine Tethys. *Paleoceanography* **13**: 530–45.

Meng, J. and McKenna, M. C., 1998, Faunal turnovers of Paleogene mammals from the Mongolian Plateau. *Nature* **395**: 364–67.

Menning, M., 1987, Problems of stratigraphic correlation: Magnetostratigraphy. In *Sedimentary and volcanic Rotliegendes of the Saale Depression*, excursion guidebook. Potsdam: Akademie der Wissenschaften für Physik der Erde pp. 92–6.

1989, A synopsis of numerical timescales 1917–1986. *Episodes* **12**: 3–4 (wall chart).

1995, A numerical time scale for the Permian and Triassic Periods: an integrated time analysis. In P. A. Scholle, T. M. Peryt, and D. S. Ulmer-Scholle, eds., *The Permian of Northern Pangea*, Vol. I, *Paleogeography, Paleoclimates, Stratigraphy*. Berlin: Springer Verlag, pp. 77–97.

2000, Magnetostratigraphic results from the Middle Permian type section, Guadalupe Mountains, West Texas. *Permophiles* **37**: 16.

Menning, M. and Hendrich, A., 2002, *Stratigraphische Tabelle von Deutschland 2002*. Potsdam: Deutsche Stratigraphische Kommission, GeoForschungsZentrum.

Menning, M., Weyer, D., Drozdzewski, G., and van Amerom, H. W. J., 1997, Carboniferous time scales revised 1997 Time Scale A (min. ages) and Time Scale B (max. ages) use of geological time indicators. *Carboniferous Newsletter* **15**: 26–8.

Menning, M., Weyer, D., Drozdzewski, G., van Amerom, H. W. J., and Wendt, I., 2000, A Carboniferous time scale 2000. Discussion and use of geological parameters as time indicators from Central and Western Europe. *Geologische Jahresberichte* **A156**: 3–44.

Menning, M., Weyer, D., Drozdzewski, G., and Wendt, I., 2001, More radiometric ages for the Carboniferous. *Newsletter on Carboniferous Stratigraphy* **19**: 16–19.

Menning, M., Schneider, J. W., Chuvashov, B. I., *et al.*, 2003, A Devonian–Carboniferous-Permian correlation chart. *Abstracts of the International Congress on Carboniferous and Permian Stratigraphy*, Utrecht. pp. 350–1.

Merrihue, C. and Turner, G., 1966, Potassium–argon dating by activation with fast neutrons. *Journal of Geophysical Research* **71**: 2852–2857.

Merrill, G. K., 1973, Pennsylvanian conodont biostratigraphy and paleoecology of northwestern Illinois. *Geological Society of America Special Paper*, Vol. 141. Geological Society of America, pp. 239–74.

Merrill, R. T., McElhinny, M. W., and McFadden, P. L., 1996, *The Magnetic Field of the Earth: Paleomagnetism, the Core, and the Deep Mantle*. San Diego, CA: Academic Press.

Mertz, D. F., Swisher, C. C., Franzen, J. L., Neuffer, F. O., and Lutz, H., 2000, Numerical dating of the Eckfeld Maar fossil site, Eifel, Germany: a calibration mark for the Eocene time scale. *Naturwissenschaften* **87**: 270–4.

Metcalfe, I. and Mundil, R., 2001, Age of the Permian–Triassic boundary and mass extinction. *Permophiles* **39**: 11–12.

Metcalfe, I., Nicoll, R. S., Black, L. P., *et al.*, 1999, Isotope geochronology of the Permian–Triassic boundary and mass extinction in South China. In H. Yin and J. Tong, eds., *Pangea and the Paleozoic–Mesozoic Transition*. Wuhan: China University Geosciences Press, pp. 134–7.

Metcalfe, I., Ludwig, K. R., Renne, P. R., Oberli, F., and Nicoll, R. S., 2001, Timing of the Permian–Triassic biotic crisis: implications from new zircon U/Pb age data (and their limitations). *Earth and Planetary Science Letters* **187**(1–2): 131–45.

Meyen, S. V., Afanasieva, G. A., Betekhtina, O. A., *et al.*, 1996, The former USSR: Angara and surrounding marine basins. In C. Martinez Diaz, R. H. Wagner, C. F. Winkler Prins, and L. F. Granados, eds., *The Carboniferous of the World*, Vol. III *The former USSR, Mongolia, Middle Eastern platform, Afghanistan, and Iran*. Ottawa: International Union of Geological Sciences, pp. 180–237.

Miall, A. D., 1992, EXXON global cycle chart – an event for every occasion. *Geology* **20**: 787–90.

Michel, J.-P., Fairbridge, R. W., and Carpenter, M. S. N., 1997, *Dictionnaire des Sciences de la Terre*, 3rd edition. Paris: Masson/Chichester: Wiley.

Michelsen, O., Thomsen, E., Danielsen, M., *et al.*, 1998, Cenozoic sequence stratigraphy in the eastern North Sea: *Society of Economic Paleontologists and Mineralogists Special Publication*, Vol. 60. Tulsa, OK: SEPM, pp. 91–118.

Mietto, P. and Manfrin, S., 1995, A high resolution Middle Triassic ammonoid standard scale in the Tethys Realm. A preliminary

report. *Bulletin de la Société Géologique de France* **166**(5): 539–63.
1999, A debate on the Ladinian–Carnian boundary. *Albertiana* **22**: 23–7.
Mii, H.-S., Grossman, E. L., and Yancey, T. E., 1999, Carboniferous isotope stratigraphies of North America: implications for Carboniferous paleoceanography and Mississippian glaciation. *Geological Society of America Bulletin* **111**(7): 960–73.
Mii, H.-S., Grossman, E. L., Yancey, T. E., Chuvashov, B. I., and Egorov, A., 2001, Isotopic records of brachiopod shells from the Russian Platform evidence for the onset of mid-Carboniferous glaciation. *Chemical Geology* **175**(1–2): 133–47.
Miklukho-Maclay, A. D., 1963, *Upper Paleozoic of Central Asia*. Leningrad: Leningrad State University, pp. 1–329 [In Russian].
Milankovitch, M. M., 1920, *Théorie mathématique des phénomènes thermiques produits par la radiation solaire*. Académie Yougoslave des Sciences et des Arts de Zagreb, Gauthier-Villars.
1941, *Kanon der Erdbestrahlung und seine Anwendung auf das Eiszeitenproblem. Royal Serbian Academy, Section of Mathematical and Natural Sciences, Special Publication* 32. Belgrade: Royal Serbian Academy, p. 633.
1998, Canon of insolation and the ice-age problem. *Serbian Academy of Sciences and Arts, Special Publication* 132, Section of Mathematical and Natural Sciences, Belgrade **39**(3/4): 634.
Miller, J. F., 1980, Taxonomic revisions of some Upper Cambrian and Lower Ordovician conodonts with comments on their evolution. *The University of Kansas Paleontological Contributions Paper 99*. Lawrence, KS: Kansas, Paleontological Institute, p. 39, 2 pls.
1988, Conodonts as biostratigraphic tools for redefinition and correlation of the Cambrian–Ordovician boundary. *Geological Magazine* **125**: 349–62.
Miller, K. G., Aubry, M.-P., Khan, M. J., *et al.*, 1985, Oligocene–Miocene biostratigraphy, magnetostratigraphy and isotopic stratigraphy of the western North Atlantic. *Geology* **13**: 257–61.
Miller, K. G., Feigenson, M. D., Kent, D. V., and Olson, R. K., 1988., Upper Eocene to Oligocene isotope ($^{87}Sr/^{86}Sr$, $\delta^{18}O$, $\delta^{13}C$) standard section, Deep Sea Drilling Project Site 522. *Paleoceanography* **3**: 223–33.
Miller, K. G., Kent, D. V., Brower, A. N., *et al.*, 1990, Eocene–Oligocene sea-level changes on the New Jersey Plain linked to the deep-sea record. *Geological Society of America Bulletin* **102**: 331–9.
Miller, K. G., Feigenson, M. D., Wright, J. D., and Clement, B. M., 1991, Miocene isotope reference section, Deep Sea Drilling Project Site 608: an evaluation of isotope and biostratigraphic resolution. *Paleoceanography* **6**: 33–52.
Miller, K. G., Thomason, P. R., and Kent, D. V., 1993, Integrated late Eocene–Oligocene stratigraphy of the Alabama coastal plain: correlation of hiatuses and stratal surfaces to glacioeustatic lowerings. *Paleoceanography* **8**: 313–31.
Miller, K. G., Mountain, G. S., and Leg 150 Shipboard Party and Members of the New Jersey Coastal Plain Drilling Project, 1996. Drilling and dating New Jersey Oligocene–Miocene sequences: ice volume, global sea level, and EXXON records. *Science* **271**: 1092–5.
Miller, K. G., Barrera, E., Olsson, R. K., Sugarman, P. J., and Savin, S. M., 1999, Does ice drive early Maastrichtian eustasy? *Geology* **27**: 783–6.
Miller, K. G., Sugarman, P. J., Browning, J. V., *et al.*, 2003, Late Cretaceous chronology of large, rapid sea-level changes: glacioeustasy during the greenhouse world. *Geology* **31**: 585–8.
Min, K., Mundil, R., Renne, P. R., and Ludwig, K. R., 1998, Evaluating the decay constants of K: implications from high precision Ar/Ar and U–Pb dating of 1.1 Ga rhyolite. In *American Geophysical Union, Fall Meeting Abstracts 80*.
2000, A test for systematic errors in $^{40}Ar/^{39}Ar$ geochronology through comparison with U/Pb analysis of a 1.1-Ga rhyolite. *Geochimica et Cosmochimica Acta* 64(1): 72–98.
Missarzhevsky, V. V., 1973, Conodont-shaped organisms from the Precambrian–Cambrian boundary strata of the Siberian Platform and Kazakhstan. In I. T. Zhuravleva, ed., *Palaeontological and Biostratigraphic Problems in the Lower Cambrian of Siberia and the Far East*. Novosibirsk: Nauka, pp. 53–57, pls 9–10 [In Russian].
Mitchell, C. E., Brussa, E. D., and Astini, R. A., 1998, A diverse Daz fauna preserved within an altered volcanic ash fall, Eastern precordillera, Argentina: implications for graptolite paleoecology. In J. C. Gutiérrez-Marco and I. Rábano, eds., *Proceedings of the Sixth International Graptolite Conference of the GWG (IPA) and the SW Ibera Field Meeting 1998 of the Commission on Silurian Stratigraphy (ICS–IUGS), Temas Geológico-Mineros* 23. Madrid: Instituto Technológico Geominero de España, pp. 222–3.
Mitchell, C. E., Chen Xu, Bergstrom, S. M., *et al.*, 1997, Definition of a global boundary stratotype for the Darriwilian Stage of the Ordovician System. *Episodes* **20**(3): 158–66.
Mitchell, J. G., 1968, The argon-40/argon-39 method for potassium–argon age determinations. *Geochimica et Cosmochimica Acta* **32**: 781–90.
Mix, A. C., Le, J. and Shackleton, N. J., 1995a, Benthic foraminifer stable isotope stratigraphy of Site 846: 0–1.8 Ma. In N. G. Pisias, L. Mayer, T. Janecek, A. Palmer-Julson, and T. H. van Andel, eds., *Proceedings of the Ocean Drilling Program, Scientific Results,* Vol. 138. College Station, TX: Ocean Drilling Program, pp. 839–56.
Mix, A. C., Pisias, N. G., Rugh, W., *et al.*, 1995b, Benthic foraminiferal stable isotope record from Site 849: 0–5 Ma. In N. G. Pisias, L. A. Mayer, T. R. Janecek, A. Palmer-Julson, and T. H. van Andel, eds., *Proceedings of the Ocean Drilling*

Program, Scientific Results, Vol. 138. College Station, TX: Ocean Drilling Program, pp. 371–412.

Moczydlowska, M., 1991, Acritarch biostratigraphy of the Lower Cambrian and the Precambrian–Cambrian boundary in southwestern Poland. *Fossils and Strata* **29**: 1–127.

1999, The Lower–Middle Cambrian boundary recognised by acritarchs in Baltica and at the margin of Gondwana. *Bolletino della Società Paleontologica Italiana* **38**.

Moeller, V., 1878, Spirally coiled foraminifers in the Carboniferous limestone of Russia. *Materials on the Geology of Russia* **8**: 1–129.

1880, The foraminifers of the Carboniferous limestone of Russia. *Materials or the Geology of Russia* **9**: 1–182.

Mojsisovics, E., von, 1869, Über die Gliederung der oberen Triasbildungen der östlichen Alpen. *Jahrbuch der Geologischen Reichsanstalt* **19**: 91–150.

Mojsisovics, E., von, 1893, Faunistische Ergebnisse aus der Untersuchung der Ammoneen-faunen der Mediterranen Trias. *Abhandlungen der Geologischen Reichsanstalt* **6**: 810.

Mojsisovics, M. E., von and Diener, C., 1895, Entwurf einer Gliederung der pelagischen Sedimente des Trias-Systems. *Sitzungsberichte der Akademie der Wissenschaften Wien* **104**: 1271–302.

Molina, E., Angori, E., Arenillas, I., *et al.*, 2000, Integrated stratigraphy across the Paleocene/Eocene boundary at Campo, Spain. In B. Schmitz, B. Sundquist, and F. P. Andreasson, eds., Early Paleogene warm climates and biosphere dynamics. *Geologiska Föreningen i Stockholm Förhandlingar* **122**: 106–7.

Molina Garza, R. S. and Zijderveld, J. D. A., 1996, Paleomagnetism of Paleozoic strata, Brabant and Ardennes massifs, Belgium: implications of prefolding and postfolding Late Carboniferous secondary magnetizations for European apparent polar wander. *Journal of Geophysical Research* **101**(7): 15 799–818.

Molyneux, S. G., Le Hérissé, A., and Wicander, R., 1996, Paleozoic phytoplankton. In J. Jansanius and D. C. McGregor, eds., *Palynology: Principles and Applications, Special Publication* Vol. 2. Dallas, TX: American Association of Stratigraphic Palynologists Foundation, pp. 493–529.

Monechi, S., 1986, Calcareous nannofossil events around the Eocene–Oligocene boundary in the Umbrian Apennines (Italy). *Palaeogeography, Palaeoclimatology, Palaeoecology* **57**: 61–9.

Monechi, S. and Thierstein, H. R., 1985, Late Cretaceous–Cretaceous nannofossil and magnetostratigraphic correlations near Gubbio. *Marine Micropaleontology* **9**: 419–40.

Montañez, I. P., Banner, J. L., Osleger, D. A., Borg, L. E., and Bosserman, P. J., 1996, Integrated Sr isotope variations and sea level history of Middle to Upper Cambrian platform carbonates: implications for the evolution of Cambrian seawater ^{87}Sr/^{86}Sr. *Geology* **24**: 917–20.

Montañez, I. P., Osleger, D. A., Banner, J. L., Mack, L. E., and Musgrove, M., 2000, Evolution of the Sr and C isotope composition of Cambrian oceans. *GSA Today* **10**: 1–7.

Montgomery, P., Hailwood, E. A., Gale, A. S., and Burnett, J. A., 1998, The magnetostratigraphy of Coniacian–Late Campanian chalk sequences in southern England. *Earth and Planetary Science Letters* **156**: 209–24.

Moore, R. C. and Thompson, M. L., 1949, Main divisions of Pennsylvanian period and system. *American Association of Petroleum Geologists Bulletin* **33**: 275–301.

Moore, T. C., 1971, Radiolaria. In *Initial Reports of the Deep Sea Drilling Project*, Vol. 8. Washington, DC: US Govt. Printing Office, pp. 727–75.

1995, Radiolarian stratigraphy, Leg 138. In *Proceedings of the Ocean Drilling Program, Scientific Results*, Vol. 138. College Station, TX: Ocean Drilling Program, pp. 191–232.

Moorkens, T., Steurbaut, E., Jutson, D., and Dupuis, C., 2000, The Knokke borehole of northwestern Belgium re-analysed: new data on the Paleocene–Eocene transitional strata in the southern North Sea Basin. *Geologiska Föreningen i Stockholm Förhandlingar* **122**: 111–14.

Moorkhoven, F. P. C. M. van, Berggren, W. A. and Edwards, A. S., 1986, Cenozoic cosmopolitan deep-water benthic foraminifera. *Bulletin des Centres Recherches Exploration-Production de Elf-Aquitaine, Mémoires* **11**: 420.

Morante, R. and Hallam, A., 1996, Organic carbon isotopic record across the Triassic–Jurassic boundary in Austria and its bearing on the cause of the mass extinction. *Geology* **24**: 391–4.

Morbey, S. J., 1975, The palynostratigraphy of the Rhaetian stage, Upper Triassic in the Kendelbachgraben, Austria. *Palaeontographica Abteilung B* **152**: 1–75.

Moreau, M.-G., Bucher, H., Bodergat, A.-M., and Guex, J., 2002, Pliensbachian magnetostratigraphy: new data from Paris Basin (France). *Earth and Planetary Science Letters* **203**: 755–67.

Morettini, E., Santantonio, M., Bartolini, A., *et al.*, 2002, Carbon isotope stratigraphy and carbonate production during the Early–Middle Jurassic: examples from the Umbria–Marche–Sabina Apennines (central Italy). *Palaeogeography, Palaeoclimatology, Palaeoecology* **184**: 251–73.

Mørk, A., Embry, A. F., and Weitschat, W., 1989, Triassic transgressive–regressive cycles in the Sverdrup Basin, Svalbard and the Barents Shelf. In J. D. Collinson, ed., *Correlation in Hydrocarbon Exploration*. London: Norwegian Petroleum Society, Graham and Trotman, pp. 113–30.

Mörner, N. A., 1976, The Pleistocene/Holocene boundary: proposed boundary stratotypes in Gothenburg, Sweden. *Boreas* **5**: 193–275.

Morton, N., 1974, The definition of standard Jurassic Stages. [Paper prepared by "Jurassic Working Group" of Mesozoic Era Subcommittee of the Geological Society of London], *Colloque du Jurassique à Luxembourg 1967, Mémoires du Bureau de Recherches Géologiques et Minières*, Vol. 75. Paris, Bureau de Rescherches Géologiques et Minières, pp. 83–93.

Moullade, M., Masse, J.-P., Tronchetti, G., *et al.*,1998, Le stratotype historique de l'Aptien inférieur (région de Cassis-La Bédoule, SE France): synthèse stratigraphique. *Géologie Méditerranéenne* **25**: 289–98.

Mudge, D. C. and Bujak, J. P., 1996, An integrated stratigraphy for the Paleocene and Eocene of the North Sea. *Geological Society of London Special Publication*, Vol. 101. London: Geological Society, pp. 91–113.

Mudie, P. J. and Harland, R., 1996, Aquatic Quaternary. In J. Jansonius and D. C. McGregor, eds., *Palynology: Principles and Applications Special Publication*, Vol. 2. Dallas, TX: American Association of Stratigraphic Palynologists Foundation, pp. 843–77.

Muehlenbachs, K., 1998, The oxygen isotopic composition of the oceans, sediments and the seafloor. *Chemical Geology* **145**: 263–73.

Mukasa, S. B., Wilson, A. H., and Carlson, R. W., 1998, A multielement geochronologic study of the Great Dyke, Zimbabwe: significance of the robust and reset ages. *Earth and Planetary Science Letters* **164**(1–2): 353–69. Also at: www.elsevier.nl/locate/epsl (full text); www.elsevier.com/locate/epsl (mirror site).

Müller, C. and Pujol, C., 1979, Etude du nannoplancton calcaire et des Foraminifères planctoniques dans l'Oligocène et le Miocène en Aquitaine (France). *Géologie Méditerranéenne, Marseille, IV* **2**: 357–68.

Müller, K. J. and Hinz, I., 1991, Upper Cambrian conodonts from Sweden. *Fossils and Strata* **28**: 153 pp, 45 pls.

Mullins, G. L., 2000, A chitinozoan morphological lineage and its importance in Lower Silurian stratigraphy. *Palaeontology* **43**: 359–73.

Mullins, G. L. and Aldridge, R. J., in press, Chitinozoan biostratigraphy of the basal Wenlock Series (Silurian) Global Stratotype Section and Point. *Palaeontology.*

Mundil, R., Brack, P., Meier, M., Rieber, H., and Oberli, F., 1996, High resolution U–Pb dating of Middle Triassic volcaniclastics: time-scale calibration and verification of tuning parameters for carbonate sediments. *Earth and Planetary Science Letters* **141**: 137–51.

Mundil, R., Ludwig, K. R., and Renne, P. R., 1999, High resolution U/Pb data for the Permian–Triassic transition at Meishan, China. *Geological Society of America, Abstracts with Programs* **31**(7): 289.

2000, New U/Pb single-crystal age data for the Permo–Triassic transition. *Abstracts 31st International Geological Congress* (CD-ROM). Rio de Janeiro: International Geological Congress.

Mundil, R., Ludwig, K. R., Renne, P. R., and Metcalfe, I., 2001a, Constraints on the timing of the Permian–Triassic biotic crisis: new U/Pb zircon-ages. *EOS Transactions, American Geophysical Union* **82**(47, Suppl.): abstract V42F-03.

Mundil, R., Ludwig, K. R., Renne, P. R., and Min, K., 2001b, How reliable are "high-resolution" radio-isotopic ages? *Geological Society of America, Abstracts with Programs*, p. A-322.

Mundil, R., Metcalfe, I., Ludwig, K. R., *et al.*, 2001c, Timing of the Permian–Triassic biotic crisis: implications from new zircon U/Pb age data (and their limitations). *Earth and Planetary Science Letters* **187**: 131–45.

Munier Chlamas, E. and De Lapparent, A., 1893, Note sur la Nomenclature des Terrains sedimentaires. *Bulletin de la Société Géologique de France 3* **21**: 438–93.

Murchison, R. D., 1839, *The Silurian System Founded on Geological Researches in the Counties of Salop, Hereford, Radnor, Montgomery, Caermarthen, Brecon, Pembroke, Monmouth, Glouster, Worcestor, and Stafford: with Descriptions of the Coal-fields, and Overlying Formations*, Vols. 1 and 2. London: John Murray.

Murchison, R. I., 1841, First sketch of the principal results of a second geological survey of Russia. *Philosophical Magazine, Series 3* **19**: 417–22.

Murchison, R. I., de Verneuil, E., and von Keyserling, A., 1845, *The Geology of Russia in Europe and the Ural Mountains*, Vol. 1, *Geology*. London: John Murray/Paris: P. Bertrand, 700 pp.

Murphy, M. A. and Salvador, A., 1999, International Stratigraphic Guide – An abridged version. *Episodes* **22**: 255–71.

Murray, B., Malin, M. C., and Greeley, R., 1981, *Earthlike Planets: Surfaces of Mercury, Venus, Earth, Moon, Mars*. San Francisco, CA: W. H. Freeman.

Mutterlose, J. and Böckel, B., 1998, The Barremian–Aptian interval in NW Germany: a review. *Cretaceous Research* **19**: 539–68.

Mutterlose, J. C., Autran, G., Baraboschkin, E. J., *et al.*, 1996, The Hauterivian Stage. *Bulletin de l'Institut Royal des Sciences Naturelles de Belgique, Sciences de la Terre, Supplement* **66**: 19–24.

Mutterlose, J., Bornemann, A., Luppold, F. W., *et al.*, 2003, The Vöhrum section (northwest Germany) and the Aptian/Albian boundary. *Cretaceous Research* **24**: 203–52.

Muttoni, G., Kent, D. V., and Gaetani, M., 1995, Magnetostratigraphy of a Lower-Middle Triassic boundary section from Chios (Greece). *Physics of the Earth and Planetary Interiors* **92**: 245–60.

Muttoni, G., Kent, D. V., Nicora, A., Rieber, H., and Brack, P., 1996, Magneto-biostratigraphy of the 'Buchenstein Beds' at Frötschbach (western Dolomites, Italy). *Albertiana* **17**: 51–6.

Muttoni, G., Kent, D. V., Meço, S., *et al.*, 1998, Towards a better definition of the Middle Triassic magnetostratigraphy and biostratigraphy in the Tethyan realm. *Earth and Planetary Science Letters* **164**: 285–302.

Muttoni, G., Kent, D. V., DiStephano, P., *et al.*, 2001, Magnetostratigraphy and biostratigraphy of the

Carnian/Norian boundary interval from the Pizzo Mondello section (Sicani Mountains, Sicily). *Palaeogeography, Palaeoclimatology, Palaeoecology* **166**: 383–99.

Naish, T. R., Abbott, S. T., Alloway, B. V., *et al.*, 1998, Astronomical calibration of a southern hemisphere Plio–Pleistocene reference section, Wanganui Basin, New Zealand. *Quaternary Science Reviews* **17**: 695–710.

Naish, T. R., Woolfe, K. J., Barrett, P. J., *et al.*, 2001, Orbitally induced oscillations in the East Antarctic ice sheet at the Oligocene/Miocene boundary. *Nature* **413**: 719–23.

Narcisi, B. and Vezzoli, L., 1999, Quaternary stratigraphy of distal tephra layers in the Mediterranean – an overview. *Global and Planetary Change* **21**: 31–50.

Naumann, C. F. 1866, *Lehrbuch der Geognosie*, Vol. 3. Leipzig: Engelmann.

Nawrocki, J., 1997, Permian to Early Triassic magnetostratigraphy from the Central European Basin in Poland: implications on regional and worldwide correlations. *Earth and Planetary Science Letters* **152**: 37–58.

Nawrocki, J. and Szulc, J., 2000, The Middle Triassic magnetostratigraphy from the Peri-Tethys basin in Poland. *Earth and Planetary Science Letters* **182**: 77–92.

Nazarov, B. B., 1988, Radiolaria of the Paleozoic. In A. I. Zhamoida, ed., *Practical Guidebook in Microfauna of the USSR*. Leningrad: Nedra, pp. 1–232.

Nazarov, B. B. and Ormiston, A. R., 1985, Radiolaria from the late Paleozoic of the Southern Urals, USSR, and West Texas, USA. *Micropaleontology* **31**(1): 1–54.

Neal, J. E., 1996, A summary of Paleogene sequence stratigraphy in northwest Europe and the North Sea. *Geological Society of London Special Publication*, Vol. 101. London: Geological Society. pp. 15–42.

Neal, J. E. and Hardenbol, J., 1998, Introduction to the Paleogene. *Society of Economic Paleontologists and Mineralogists Special Publication*, Vol. 60. Tulsa, OK: SEPM, pp. 87–93.

Neff, U., Burns, S. J., Mangini, A., *et al.*, 2001, Strong coherence between solar variability and the Monsoon in Oman between 9 and 6 kyrs ago. *Nature* **411**: 290–3.

Negri, A. and Villa, G., 2000, Calcareous nannofossil biostratigraphy, biochronology and paleoecology at the Tortonian/Messinian boundary of the Faneromeni section (Crete). *Palaeogeography, Palaeoclimatology, Palaeoecology* **156**: 195–209.

Negri, A., Giunta, S., Hilgen, F. J., Krijgsman, W., and Vai, G. B., 1999, Calcareous nannofossil biostratigraphy of the Monte del Casino section (northern Apennines, Italy) and paleoceanographic conditions at times of late Miocene sapropel formation. *Marine Micropaleontology* **36**: 13–30.

Nemirovskaya, T. I., 1999, Bashkirian conodonts of the Donets Basin. *Scripta Geologica* **119**: 115 p.

Nemirovskaya, T. I. and Alekseev, A. S., 1995, The Bashkirian conodonts of the Askyn Section, Bashkirian mountains, Russia. *Bulletin de la Société Belge de Géologie* **103**(1–2): 109–33.

Nemirovskaya, T. I. and Nigmadganov, I. M., 1994, The mid-Carboniferous event. *Courier Forschungsinstitut Senckenberg* **168**: 319–35.

Nemirovskaya, T. I., Perret, M. F., and Meischner, D., 1994, *Lochrea ziegleri* and *Lochrea senckenbergica* new conodont species from the latest Visean and Serpukhovian in Europe. *Courier Forschungsinstitut Senckenberg* **168**: 311–19.

Nemirovskaya, T. I., Perret-Mirouse, M.-F., and Alekseev, A. S., 1999, On Moscovian (Late Carboniferous) conodonts of the Donets Basin, Ukraine. *Neues Jahrbuch für Geologie und Paläontologie Abhandlungen* **214**(1/2): 169–94.

Néron de Surgy, O. and Laskar, J., 1996, On the long term evolution of the spin of the Earth. *Astronomy and Astrophysics* **318**: 975–89.

Nestor, H., Einasto, R., Nestor, V., Marss, T., and Viira, V., 2001, Description of the type section, cyclicity and correlation of the Riksu Formation (Wenlock, Estonia). *Proceedings of the Estonian Academy of Sciences, Geology* **50**: 149–73.

Neumayr, M., 1873, Die Fauna des Schichten mit *Aspidoceras acanthicum*. Abhandlungen der Kais.-Königl. *Geologischen Reichsanstalt* **5**(6): 141–257.

Newell, A. J., 2000., A sequence stratigraphy of the Coralline Formation, Upper Jurassic, Wessex–Weald Basin. *Journal of the Geological Society of London* **157**: 83–92.

Newell, N. D., 1994, Is there a precise Permian–Triassic boundary? *Permophiles* **24**: 46–8.

Nicoll, R. S., 1990, The genus Cordylodus and a latest Cambrian–earliest Ordovician conodont biostratigraphy. *BMR Journal of Australian Geology and Geophysics* **11**: 529–58.

1991, Differentiation of Late Cambrian–Early Ordovician species of Cordylodus (Conodonta) with biapical basal cavities. *BMR Journal of Australian Geology and Geophysics* **12**: 223–44.

Nicoll, R. S. and Shergold, J. H., 1992, Revised Late Cambrian (pre-Payntonian–Datsonian) conodont stratigraphy at Black Mountain, Georgina Basin, western Queensland, Australia. *BMR Journal of Australian Geology and Geophysics* **12**: 93–118.

Nicoll, R. S., Thorshoej Nielsen, A., Laurie, J. R., and Shergold, J. H., 1992, Preliminary correlation of latest Cambrian to Early Ordovician sea level events in Australia and Scandinavia. In B. D. Webby and J. R. Laurie, eds., *Global Perspectives on Ordovician Geology*. Rotterdam: Balkema Press, pp. 381–94.

Nielsen, A. T., 1992, Ecostratigraphy and the recognition of Arenigian (Early Ordovician) sea-level changes. In B. D. Webby and J. R. Laurie, eds., *Global Perspectives on Ordovician Geology*. Rotterdam: A. A. Balkema, pp. 355–66.

2003a, Ordovician sea-level changes: potential for global event stratigraphy. In *Abstracts ISOS*, Argentina.

2003b, Sea-level changes – a Baltoscandian perspective. In B. D. Webby, M. Droser, and F. Paris, eds., *The Great Ordovician Biodiversification Event*, Part II, *Conspectus of the Ordovician World*. New York: Columbia University Press.

Nigrini, C., 1971, Radiolarian zones in the Quaternary of the equatorial Pacific Ocean. In B. M. Funnell and W. R. Riedel, eds., *The Micropaleontology of the Oceans*. Cambridge: Cambridge University Press, pp. 443–61.

1974, Cenozoic Radiolaria from the Arabian Sea, DSDP Leg 23. In *Initial Reports of the Deep Sea Drilling Project*, Vol. 23. Washington, DC: US Govt. Printing Office, pp. 1051–121.

Nikitin, S. N., 1881, Jurassic formations between Rogbinsk, Mologa, and Myshkin: Materials on the Geology of Russia. *Mineralogical Society Special Publication*, Vol. X. p. 194 [In Russian].

1890, The Carboniferous of the Moscow Basin and artesian water in the region of the Moscow Basin. *Transactions of Geological Committee* **5**(5): 1–138 [In Russian], 139–82 [In French].

Nikolaeva, S. V., 1995, Ammonoids of the late Lower and early Upper Carboniferous of Central Asia. *Courier Forschungsinstitut Senckenberg* **179**: 1–106.

Nikolaeva, S. V. and Kullmann, J., 2001, Problems in Lower Serpukhovian ammonoid biostratigraphy. *Newsletter on Carboniferous Stratigraphy* **19**: 35–7.

Nikolaeva, S. V., Kulagina, E. I., Pazukhin, V. N., and Kochetkova, N. N., 2001, Integrated Serpukhovian biostratigraphy in the Southern Urals. *Newsletter on Carboniferous Stratigraphy* **19**: 38–42.

Nikolaeva, S. V., Gibshman, N. B., Kulagina, E. I., Barskov, I. S., and Pazukhin, V. N., 2002, Correlation of the Visean–Serpukhovian boundary in its type region (Moscow Basin) and the South Urals and a proposal of boundary markers (ammonoids, foraminifers, conodonts). *Newsletter on Carboniferous Stratigraphy* **20**: 16–21.

Nilsson, T., 1983, *The Pleistocene*. Dordrecht: Reidel.

Nisbet, E. G., 1991, Of clocks and rocks; the four aeons of Earth. *Episodes* **14**(4): 327–30.

Nishimura, A., 1987, Cenozoic radiolaria in the western North Atlantic, Site 603, Leg 93 of the Deep Sea Drilling Project. In *Initial Reports of the Deep Sea Drilling Project*, Vol. 93. Washington, DC: US Govt. Printing Office, pp. 713–37.

1992, Paleocene radiolarian biostratigraphy in the northwest Atlantic at Site 384, Leg 43 of the Deep Sea Drilling Project. *Micropaleontology* **38**: 317–62.

Nocchi, M., Parisi, G., Monaco, P., Monechi, S., and Madile, M., 1988, Eocene and early Oligocene micropaleontology and paleoenvironments in SE Umbria, Italy. *Palaeogeography, Palaeclimatology, Palaeoecology* **67**: 181–244.

Norris, R. D. and Röhl, U., 1999, Carbon cycling and chronology of climate warming during the Paleocene/Eocene transition. *Nature* **401**: 775–8.

2001, Astronomical chronology for the Paleocene–Eocene transition. *Geologiska Föreningen i Stockholm Förhandlingar (GFF)* **122**: 117–118.

North American Commission on Stratigraphic Nomenclature, 1983, North American stratigraphic code. *American Association of Petroleum Geologists Bulletin* **67**: 841–75.

NUNA, 2003, New frontiers in the fourth dimension: generation, calibration and application of geological timescales. In *NUNA 2003*. Mont Tremblant, Quebec, Canada, March 15–18, 2003.

Nutman, A. P., Friend, C. R. L., Kinny, P. D., and McGregor, V. R., 1993, Anatomy of an early Archean gneiss complex: 3900 to 3600 Ma crustal evolution in southern West Greenland. *Geology* **21**(5): 415–18.

Nutman, A. P., McGregor, V. R., Friend, C. R. L., Bennet, V. C., and Kinny, P. D., 1996, The Itsaq Gneiss Complex of southern West Greenland: the world's most extensive record of early crustal evolution (3900–3600 Ma). *Precambrian Research* **78**: 1–39.

Nutman, A. P., Bennett, V. C., Friend, C. R. L., and Rosing, M. T., 1997, Approximately 3710 and $\geq$3790 Ma volcanic sequences in the Isua (Greenland) supracrustal belt: structural and Nd isotope implications. *Chemical Geology* **141**(3–4): 271–87.

Nutman, A. P., Bennet, V. C., Friend, C. R. L., and Norman, M. D., 1999, Meta-igneous (non-gneissic) tonalites and quartz-diorites from an extensive ca. 3800 Ma terrain south of the Isua supracrustal belt, southern West Greenland: constraints on early crust formation. *Contributions to Mineralogy and Petrology* **137**: 364–88.

Nystuen, J. P., 1986, Regler og råd for navnsetting av geologiske enheter i Norge. *Norsk Geologisk Tidsskrift, Supplement* **66**: 1.

Oakly, K. P., 1950, The Pliocene–Pleistocene Boundary. *Proceedings of the International Geological Congress, London, 1948*. Section H.

Oberli, F., Bachmann, O., Meier, M., and Dungan, M. A., 2002, The Fish Canyon Tuff: Ar–Ar versus U–Pb age discrepancy re-assessed. *Geochimica et Cosmochimica Acta* **66**(15A): A565.

Oberthuer, T., Davis, D. W., Blenkinsop, T. G., and Hoehndorf, A., 2002, Precise U–Pb mineral ages, Rb–Sr and Sm–Nd systematics for the Great Dyke, Zimbabwe: constraints on late Archean events in the Zimbabwe Craton and Limpopo Belt. *Precambrian Research* **113**(3–4): 293–305.

Obradovich, J. D., 1984, An overview of the measurement of geologic time and the paradox of geologic time scales. *Proceedings of 27th International Geological Congress,* Vol. 1, *Stratigraphy.* pp. 11–30.

1988, A different perspective an glauconite as a chronometer. *Paleoceanography* **3**: 757–70.

1993, A Cretaceous time scale. In W. G. E. Caldwell and E. G. Kauffman, eds., *Evolution of the Western Interior Basin, Geological Association of Canada Special Paper* 39. Ottawa: Geological Association of Canada, pp. 379–96.

Obradovich, J. D., Sutter, J. F., and Kunk, M. J., 1986, Magnetic polarity chron tie points for the Cretaceous and early Tertiary. *Terra cognita* **6**: 140.

Obradovich, J. D., Matsumoto, T., Nishida, T., and Inoue, Y., 2002, Integrated biostratigraphic and radiometric scale on the Lower Cenomanian (Cretaceous) of Hokkaido, Japan. *Proceedings of the Japan Academy, Series B-Physical and Biological Sciences* **78**(6): 149–53.

O'Brien, S. R., Mayewski, P. A., Meeker, L. D., *et al.*, 1995, Complexity of Holocene climate as reconstructed from a Greenland ice core. *Science* 270: 1962–64.

Odin, G. S., 1982, *Numerical Dating in Stratigraphy*, Vols. 1 and 2. Chichester: Wiley-Interscience.

1985, Remarks on the numerical scale of Ordovician to Devonian times. In N. J. Snelling, ed., *The Chronology of the Geological Record, Geological Society of London Memoir* 10. London: Geological Society, pp. 93–7.

1992, New stratotypes for the Paleogene: the Cretceous/Paleogene, Eocene/Oligocene and Paleogene/Neogene boundaries. *Neues Jahrbuch für Geologie und Paläontologie Abhandlungen* **186**: 7–20.

1994, Geologic time scale (1994). *Comptes rendus de l'Académie des Science de Paris, série II* **318**: 59–71.

1996, Le site de Tercis (Landes). Observations stratigraphiques sur le Maastrichtien. Arguments pour la localisation et la corrélation du Point Stratotype Global de la limite Campanien–Maastrichtien. *Bulletin de la Société Géologique de France* **167**: 637–43.

2001, *The Campanian–Maastrichtian Boundary: Characterisation at Tercis les Bains (France) and Correlation with Europe and other* Continents, *IUGS Special Publication Series,* Vol. 36, *Developments in Palaeontology and Stratigraphy Series,* Vol. 19. Amsterdam: Elsevier, 910 pp. [Online: www.elsevier.com/locate/isbn/0-444-50647-0].

Odin, G. S. and Curry, D., 1985, The Paleogene time scale: radiometric dating versus magnetostratigraphic approach. *Journal of the Geological Society of London* **142**: 1179–88.

Odin, G. S. and Lamaurelle, M. A., 2001, The global Campanian–Maastrichtian stage boundary. *Episodes* **24**: 229–38.

Odin, G. S. and Luterbacher, H. P., 1992, The age of the Paleogene stage boundaries. *Neues Jahrbuch für Geologie und Paläontologie Abhandlungen* **186**: 21–48.

Odin, G. S. and Montanari, A., 1988, The Eocene–Oligocene boundary at Massignano (Ancona, Italy): a potential boundary stratotype. International Subcommission on Paleogene Stratigraphy. In S. Premoli-Silva, R. Coccioni, and A. Montanari, eds., *The Eocene–Oligocene Boundary in the Marche-Umbria Basin (Italy): Interdisciplinary Studies Presented at the Eocene–Oligocene Boundary ad hoc Meeting*, Vol. 2. Ancona: International Union of the Geological Sciences, Committee on Stratigraphy, pp. 253–63.

Odin, G. S. and Odin, C., 1990, Échelle numérique des temps géologiques. *Géochronologie* **35**: 12–20.

Odin, G. S., Montanari, A., Deino, A., *et al.*, 1991, Reliability of volcano–sedimentary biotite ages across the E–O boundary. *Chemical Geology* **86**: 203–24.

Odin, G. S., Baadsgaard, H., Hurford, A. J., and Riccardi, A. C., 1992, U–Pb and fission track geochronology of Bathonian–Callovian 'tuffs' from Argentina. In G. S. Odin, ed., *Phanerozoic Time Scale, Bulletin de Liaison et Information* (the newsletter of IUGS Subcommission on Geochronology) 11. IUGS, pp. 11–17.

Odin, G. S., with contributions by Hancock, J. M., Antonescu, E., Bonnemaison, M., *et al.*, 1996, Definition of a Global Boundary Stratotype Section and Point for the Campanian/Maastrichtian boundary. *Bulletin de l'Institut Royal des Sciences Naturelles de Belgique, Sciences de la Terre, Supplément*, **66**: 111–17.

Odin, G. S., Montanari, A., and Coccioni, R., 1997, Chronostratigraphy of Miocene stages: a proposal for the definition of precise boundaries. In A. Montanari, G. S. Odin, and R. Coccioni, eds., *Miocene Stratigraphy: an Integrated Approach. Developments in Palaeontology and Stratigraphy,* Vol. 15. Amsterdam: Elsevier, pp. 597–629.

Oeschger, H., Beer, J., Dansgaard, W., *et al.*, 1984, Late Glacial climate history from ice cores. *Geophysical Monograph* **29**: 299–306.

Oeschger, H., Beer, J., Dansgaard, W., *et al.*, 1984, Late Glacial climate history from ice cores. In *Geophysical Monograph*, Vol. 29. Washington, DC: American Geophysical Union, pp. 299–306.

Ogg, J. G., 1987, Early Cretaceous magnetic polarity time scale and the magnetostratigraphy of DSDP Sites 603 and 534, western Central Atlantic. In *Initial Reports of the Deep Sea Drilling Project*, Vol. 93. Washington, DC: US Govt. Printing Office, pp. 849–88.

1988, Early Cretaceous and Tithonian magnetostratigraphy of the Galicia margin (Ocean Drilling Program Leg 103). In *Proceedings of the Ocean Drilling Program, Scientific Results*, Vol. 103. College Station, TX: Ocean Drilling Program, pp. 659–82.

1995, Magnetic polarity time scale of the Phanerozoic. In T. J. Ahrens, ed., *Global Earth Physics: A Handbook of Physics Constants*, AGU Reference Shelf. Washington, DC: American Geophysical Union, pp. 240–70.

2004, Introduction to concepts and proposed standardization of the term "Quaternary." *Episodes* **27**(2): 126.

Ogg, J. G. and Bardot, L., 2001, Aptian through Eocene magnetostratigraphic correlation of the Blake Nose Transect (Leg 171B), Florida Continental Margin. In D. Kroon, R. D. Norris, and A. Klaus, eds., *Proceedings of the Ocean Drilling Program, Scientific Results,* Vol. 171B. College Station, TX: Ocean Drilling Program, Chapter 9

[Online: www-odp.tamu.edu/publication/171B_SR/chap_09/chap_09.htm].

Ogg, J. G. and Coe, A. L., 1997, Oxfordian magnetic polarity time scale. *EOS Transactions American Geophysical Union* **78** (1997 Fall Meeting Supplement): F186.

Ogg, J. G. and Gutowski, J., 1996, Oxfordian and lower Kimmeridgian magnetic polarity time scale. In A. C. Riccardi, ed., *Advances in Jurassic Research, GeoResearch Forum* 1–2. Aedermannsdorf: Transtech Publications, pp. 406–414.

Ogg, J. G. and Lowrie, W., 1986, Magnetostratigraphy of the Jurassic–Cretaceous boundary. *Geology* **14**: 547–50.

Ogg, J. G. and Steiner, M. B., 1991, Early Triassic magnetic polarity time scale – Integration of magnetostratigraphy, ammonite zonation and sequence stratigraphy from stratotype sections (Canadian Arctic Archipelago). *Earth and Planetary Science Letters* **107**: 69–89.

Ogg, J. G., Steiner, M. B., Oloriz, F., and Tavera, J. M., 1984, Jurassic magnetostratigraphy: 1. Kimmeridgian– Tithonian of Sierra Gorda and Carcabuey, southern Spain. *Earth and Planetary Science Letters* **71**: 147–62.

Ogg, J. G., Steiner, M. B., Company, M., and Tavera, J. M., 1988, Magnetostratigraphy across the Berriasian–Valanginian stage boundary (Early Cretaceous) at Cehegin (Murcia Province, southern Spain). *Earth and Planetary Science Letters* **87**: 205–15.

Ogg, J. G., Hasenyager, R. W., Wimbledon, W. A., Channell, J. E. T., and Bralower, T. J., 1991a, Magnetostratigraphy of the Jurassic–Cretaceous boundary interval – Tethyan and English faunal realms. *Cretaceous Research* **12**: 455–82.

Ogg, J. G., Wieczorek, J., Steiner, M. B., and Hoffmann, M., 1991b, Jurassic magnetostratigraphy, 4. Early Callovian through Middle Oxfordian of Krakow Uplands (Poland). *Earth and Planetary Science Letters* **104**: 289–303.

Ogg, J. G., Karl, S. M., and Behl, R. J., 1992, Jurassic through Early Cretaceous sedimentation history of the central Equatorial Pacific and of Sites 800 and 801. In *Proceedings of the Ocean Drilling Program, Scientific Results*, Vol. 129. College Station, TX: Ocean Drilling Program, pp. 571–613.

Ogg, J. G., Hasenyager II, R. W., and Wimbledon, W. A., 1994, Jurassic–Cretaceous boundary: Portland–Purbeck magnetostratigraphy and possible correlation to the Tethyan faunal realm. *Géobios Mémoire Spécial* **17**: 519–27.

Ohde, S. and Elderfield, H., 1992, Strontium isotope stratigraphy of Kita-daito-jima Atoll, North Philippine Sea: implications for Neogene sea-level change and tectonic history. *Earth and Planetary Science Letters* **113**: 473–86.

Ohmert, W., 1996, *Die Grenzziehung Unter-/Mitteljura (Toarciuim/Aalenium) bei Wittnau und Fuentelsaz. Beispiele interdisziplinarer geowissenschaftlicher Zusammenarbeit*, Vol. 8. Geologisches Landesamt, Baden-Württemberg: p. 53.

Okada, H. and Bukry, D., 1980, Supplementary modification and introduction of code numbers to the low-latitude coccolith biostratigraphic zonation (Bukry, 1973; 1975). *Marine Micropaleontology* **51**: 321–25.

Okulitch, A. V., 2002, Geological time chart. *Geological Survey of Canada Open File* 3040. Ottawa: Geological Survey of Canada, 1p.

Olausson, E., 1982, *The Pleistocene/Holocene Boundary in South-western Sweden. Sveriges Geologiska Undersökning, Serie C794, Avhondlingar och Uppsatser*, Vol. 76. Uppsala: Sveriges Geologiska Undersökning.

Olsen, P. E., 1999, Giant lava flows, mass extinctions, and mantle plumes. *Science* **284**: 604–5.

Olsen, P. E. and Kent, D. V., 1999, Long-period Milankovitch cycles from the Late Triassic and Early Jurassic of eastern North America and their implications for the calibration of the Early Mesozoic time-scale and the long-term behaviour of the planets. *Philosophical Transactions of the Royal Society of London, series A* **357**: 1761–86.

2001, Cyclostratigraphic and magnetostratigraphic constraints on the duration of the CAMP. *EOS Transactions American Geophysical Union* **82**: S276.

Olsen, P. E., Kent, D. V., Cornet, B., Witte, W. K., and Schlische, R. W., 1996, High-resolution stratigraphy of the Newark rift basin (early Mesozoic, eastern North America). *Geological Society of America Bulletin* **10**: 40–77.

Olsen, P. E., Kent, D. V., Et Touhami, M., and Puffer, J. H., 2003, Cyclo-, magneto-, and bio-stratigraphic constraints on the duration of the CAMP event and its relationship to the Triassic–Jurassic boundary. In *The Central Atlantic Magmatic Province: insights from the Fragments of Pangea, AGU Geophysical Monograph* 136. Washington, DC: American Geophysical Union, pp. 7–32.

Olsson, R. K., 1970, Planktonic foraminifera from the base of Tertiary, Millers Ferry, Alabama. *Journal of Paleontology* **44**: 598–604.

Olsson, R. K., Hemleben, C., Berggren, W. A., and Huber, B. T., 1999, *Atlas of Paleocene Planktonic Foraminifera. Smithsonian Contributions to Paleobiology*, Vol. 85. Washington, DC: Smithsonian Institution.

Opdyke, N. D. and Channell, J. E. T., 1996, *Magnetic Stratigraphy*. New York: Academic Press.

Opdyke, N. D., Khramov, A. N., Gurevitch, E., Iosifidi, A. G., and Makarov, I. A., 1993, A paleomagnetic study of the middle Carboniferous of the Donets Basin, Ukraine (abstract). *EOS Transactions American Geophysical Union*, Spring Meeting, San Francisco, p. 118.

Opdyke, N. D. Roberts, J., Claoue-Long, J., Irving, E., and Jones, P. J., 2000, Base of Kiaman: its definition and global significance. *Geological Society of America Bulletin* **112**(9): 1315–41.

Öpik, A. A., 1958, The Cambrian trilobite Redlichia: organisation and generic concept. *Bureau of Mineral Resources of Australia Bulletin* **42**: 50, 6 pls.

1961, Alimentary caeca of agnostids and other trilobites. *Palaeontology* **3**(4): 410–38, pls 68–70.

1963, Early Upper Cambrian fossils from Queensland. *Bureau of Mineral Resources of Australia Bulletin* **64**: 133, 9 pls.

1966, The Early Upper Cambrian crisis and its correlation. *Journal and Proceedings of the Royal Society of New South Wales* **100**: 9–14.

1967, The Mindyallan fauna of north-western Queensland. *Bureau of Mineral Resources of Australia Bulletin* **74**(1): 404; (2): 167, 67 pls.

1968, The Ordian Stage of the Cambrian and its Australian Metadoxididae. *Bureau of Mineral Resources of Australia Bulletin* **92**: 133–70, pls 19–20.

1970, Nepeiid trilobites of the Middle Cambrian of northern Australia. *Bureau of Mineral Resources of Australia Bulletin* **113**: 48, 17 pls.

1979, Middle Cambrian agnostids: systematics and biostratigraphy. *Bureau of Mineral Resources of Australia Bulletin* **172**(1): 188; and (2): 67 pls.

1982, Dolichometopid trilobites of Queensland, Northern Territory and New South Wales. *Bureau of Mineral Resources of Australia Bulletin* **175**: 85, 32 pls.

Oppel, C. A., 1856–1858, *Die Juraformation Englands, Frankreichs und des südwestlichen Deutschlands, Württemberger Naturforschende Jahreshefte* 12–14. Stuttgart: Ebner and Seubert, p. iv + 857.

1865, Die Tithonische Etage. *Zeitschrift der Deutschen Geologischen Gesellschaft, Jahrgang* **17**: 535–58.

Orbell, G., 1983, Palynology of the British Rhaeto–Liassic. *Bulletin of the Geological Survey of Great Britain* **44**: 1–44.

Orchard, M. J. and Tozer, E. T., 1997a, Triassic conodont biochronology and intercalibration within the Canadian ammonoid sequence. *Albertiana* **20**: 33–44.

1997b, Triassic conodont biochronology, its intercalibration with the ammonoid standard, and a biostratigraphic summary for the Western Canada Sedimentary Basin. *Canadian Society of Petroleum Geologists Bulletin* **45**: 675–92.

Orchard, M. J., Carter, E. S., and Tozer, E. T., 2000, Fossil data and their bearing on defining a Carnian–Norian (upper Triassic) boundary in western Canada. *Albertiana* **24**: 43–50.

Orchard, M. J., Zonneveld, J. P., Johns, M. J., *et al.*, 2001, Fossil succession and sequence stratigraphy of the Upper Triassic of Black Bear Ridge, northeast British Columbia, a GSSP prospect for the Carnian–Norian boundary. *Albertiana* **25**: 10–22.

Osleger, D. A., 1995, Depositional sequences on Upper Cambrian carbonate platforms: variable sedimentologic responses to allogenic forcing. In B. U. Haq, ed., *Sequence Stratigraphy and Depositional Responses to Eustatic, Tectonic and Climate Forcing*. Dordrecht: Kluwer Academic, pp. 247–76.

Oslick, J. S., Miller, K. G., and Feigenson, M. D., 1994, Oligocene–Miocene strontium isotopes: stratigraphic revisions and correlations to an inferred glacioeustatic record. *Paleoceanography* **9**(3): 427–43.

Over, D. J., 1999, Notes. *Subcommission on Devonian Stratigraphy Newsletter* **16**: 13.

Overpeck, J., Rind, D., Lacis, A., and Healy, R., 1996, Possible role of dust-induced regional warming in abrupt climate change during the last glacial period. *Nature* **384**: 447–9.

Owen, H. G., 1971, Middle Albian stratigraphy in the Anglo–Paris Basin. *Bulletin of the British Museum of Natural History (Geology), Supplement* **8**: 164.

1979, Ammonite zonal stratigraphy in the Albian of North Germany and its setting in the Hoplitinid Faunal Province. In J. Wiedmann, ed., *Aspekte der Kreide Europas, 6 IUGS series A*. Stuttgart: International Union of Geological Science, pp. 563–88.

1996a, Boreal and Tethyan late Aptian to late Albian ammonite zonation and palaeobiogeography. In C. Späth, ed., *Mitteilung aus dem Geologisch-Paläontologischen Institut der Universität Hamburg 77* (Jost Wiedmann Memorial Volume), Proceedings of the 4th International Cretaceous Symposium, Hamburg, 1992, pp. 461–81.

1996b, "Uppermost Wealden facies and Lower Greensand Group (Lower Cretaceous) in Dorset, southern England: correlation and palaeoenvironment by Ruffell and Batten (1994)" and "The Sandgate Formation of the M20 Motorway near Ashford, Kent, and its correlation" by Ruffell and Owen (1995):" reply. *Proceedings of the Geologists' Association* **107**: 74–6.

Owen, H. G., 1999, Correlation of Albian European and Tethyan ammonite zonations and the boundaries of the Albian Stage and substages: some comments. *Scripta Geologica, Special Issue* **3**: 129–49.

2002, The base of the Albian Stage: comments on recent proposals. *Cretaceous Research* **23**: 1–13.

Owens, B., 1984, Miospore zonation of the Carboniferous. In P. K. Sutherland and W. L. Manger, eds., *Compte Rendu IX Congrès International de Stratigraphie et de Géologie du Carbonifère, 1979*, Vol. 2. Washington and Champaign–Urbana: pp. 90–102.

Padden, M., Weissert, H., and de Rafelis, M., 2001, Evidence for Late Jurassic release of methane from gas hydrate. *Geology* **29**: 223–26.

Padden, M., Weissert, H., Funk, H., Schneider, S., and Gansner, C., 2002, Late Jurassic lithological evolution and carbon-isotope stratigraphy of the western Tethys. *Eclogae Geologicae Helvetiae* **95**: 333–46.

Page, K. N. and Bloos, G., 1998, The base of the Jurassic System in west Somerset, south-west England – new observations on the succession of ammonite faunas of the lowest Hettangian Stage:

Geoscience in south-west England. *Proceedings of the Ussher Society* **9**: 231–5.

Page, K. N., Bloos, G., Bessa, J. L., *et al.*, 2000, East Quantoxhead, Somerset: a candidate Global Stratotype Section and Point for the base of the Sinemurian Stage (Lower Jurassic). *GeoResearch Forum* Vol. 6. Zurich: Transtech Publication, pp. 163–71.

Paillard, D., Labeyrie, L., and Yiou, P., 1996, Macintosh program performs time-series analysis. *EOS Transactions American Geophysical Union* **77**: 379.

Pálfy, J., 1995, Development of the Jurassic geochronologic scale. *Hantkeniana*, Géczy Jubilee Volume **1**: 13–25.

Pálfy, J. and Smith, P. L., 2000, Synchrony between Early Jurassic extinction, oceanic anoxic event, and the Karoo–Ferrar flood basalt volcanism. *Geology* **28**: 747–50.

Pálfy, J. and Vörös, A., 1998, Quantitative ammonoid biochronological assessment of the Anisian–Ladinian (Middle Triassic) stage boundary proposals. *Albertiana* **21**: 19–26.

Pálfy, J., Parrish, R. R., and Smith, P. L., 1997, A U–Pb age from the Toarcian (Lower Jurassic) and its use for time scale calibration through error analysis of biochronologic dating. *Earth and Planetary Science Letters* **146**: 659–75.

Pálfy, J., Smith, P. L., Mortensen, J. K., and Friedman, R. M., 1999, Integrated ammonite biochronology and U–Pb geochronometry from a basal Jurassic section in Alaska. *Geological Society of America Bulletin* **111**: 1537–49.

Pálfy, J., Mortensen, J. K., Carter, E. S., *et al.*, 2000b, Timing the end-Triassic mass extinction: first on land, then in the sea? *Geology* **28**: 39–42.

Pálfy, J., Mortensen, J. K., Smith, P. L., *et al.*, 2000c, New U–Pb zircon ages integrated with ammonite biochronology from the Jurassic of the Canadian Cordillera. *Canadian Journal of Earth Sciences* **37**: 549–67.

Pálfy, J., Smith, P. L., and Mortensen, J. K., 2000a, A U–Pb and $^{40}Ar/^{39}Ar$ time scale for the Jurassic. *Canadian Journal of Earth Sciences* **37**: 923–44.

Pálfy, J., Demeny, A., Haas, J., *et al.*, 2001, Carbon isotope anomaly and other geochemical changes at the Triassic–Jurassic boundary from a marine section in Hungary. *Geology* **29**: 1047–50.

Pálfy, J., Parrish, R. R., and Vörös, A., 2002a, Integrated U–Pb geochronology and ammonoid biochronology from the Anisian/Ladinian GSSP candidate at Felsors. *Proceedings of the STS/IGCP 467 Field Meeting*, Vesprem, Hungary, September 5–8, 2002, p. 28.

Pálfy, J., Smith, P. L., and Mortensen, J. K., 2002b, Dating the end-Triassic and Early Jurassic mass extinctions, correlative large igneous provinces, and isotopic events. In C. Koeberl and K. G. MacLeod, eds., *Catastrophic Events and Mass Extinctions: Impacts and Beyond. Geological Society of America Special Paper*, Vol. 356. Boulder, CO: Geological Society of America, pp. 523–32.

Pálfy, J., Parrish, R. R., David, K., and Vörös, A., 2003, Middle Triassic integrated U–Pb geochronology and ammonoid biochronology from the Balaton Highland (Hungary). *Journal of the Geological Society of London* **160**(2): 271–84.

Pälike, H. and Shackleton, N. J., 2000, Constraints on astronomical parameters from the geological record for the last 25 Myr. *Earth and Planetary Science Letters*. **182**: 1–14.

Pälike, H., Shackleton, N. J., and Röhl, U., 2001, Astronomical forcing in late Eocene marine sediments. *Earth and Planetary Science Letters* **193**: 589–602.

Palmer, A. R., 1965a, Biomere – a new kind of biostratigraphic unit. *Journal of Paleontology* **39**(1): 149–53.

1965b, *Trilobites of the Late Cambrian Pterocephaliid Biomere in the Great Basin, United States, US Geological Survey Professional Paper* 493. Reston; VA: US Geological Survey, p. 105, 23 pls.

1973, Cambrian trilobites. In A. Hallam, ed., *Atlas of Palaeobiogeography*. New York: Elsevier, pp. 3–18.

1979, Biomere boundaries re-examined. *Alcheringa* **3**: 33–41.

1981, Subdivision of the Sauk Sequence. In M. E. Taylor, ed., *Short Papers for the Second International Symposium on the Cambrian System, US Geological Survey Open-file Report* 81–743. Reston, VA: US Geological Survey, pp. 160–162.

1983, *Geologic Time Scale, Decade of North American Geology (DNAG)*. Boulder, CO: Geological Society of America.

1984, The biomere problem: evolution of an idea. *Journal of Paleontology* **58**(3): 599–611.

1998, A proposed nomenclature for stages and series for the Cambrian of Laurentia. *Canadian Journal of Earth Sciences* **35**: 323–8.

Palmer, A. R. and Halley, R. B., 1979, *Physical Stratigraphy and Trilobite Biostratigraphy of the Carrara Formation (Lower and Middle Cambrian) in the Southern Great Basin. US Geological Survey Professional Paper* 1047. Reston, VA: US Geological Survey, p. 131, 16 pls.

Palmer, A. R. and Repina, L. N., 1993, Through a glass darkly: taxonomy, phylogeny, and biostratigraphy of the Olenellina. *University of Kansas Paleontological Contributions, New Series* **3**: 35, 13 figs.

Palmer, J. A., Perry, S. P. G., and Tarling, D. H., 1985, Carboniferous magnetostratigraphy. *Journal of the Geological Society of London* **142**: 945–55.

Paproth, E., 1996, Minutes of the SCCS meeting, Krakow, 1995. *Newsletter on Carboniferous Stratigraphy* **14**: 4–6.

Paproth, E., Feist, R., and Flajs, G., 1991, Decision on the Devonian–Carboniferous boundary stratotype. *Episodes* **14**(4): 331–6.

Parandier, H., 1891, Notice géologique et paléontologique sur la nature des terrains traversés par le chemin de fer entre Dijon at Châlons-sur-Saône. *Bulletin de la Société Géologique de France, série 3* **19**: 794–818.

Pareto, L., 1865, Note sur les subdivisions que l'on pourrait établir dans les terrains tertiaires de l'Apennin septentrional. *Bulletin de la Société Géologique de France, serie* 2 **22**: 210–17.

Paris, F., 1996, Chitinozoan biostatigraphy and paleoecology. In J. Jansonius and D. C. McGregor, eds., *Palynology: Principles and Applications, American Association of Stratigraphic Palynologists Foundation, Special Publication* 2. Dallas, TX: American Association of Stratigraphic Palynologists Foundation, pp. 531–552.

Park, J., D'Hondt, S. L., King, J. W., and Gibson, C., 1993, Late Cretaceous precessional cycles in double time: a warm-Earth Milankovitch response. *Science* 261: 1431–4.

Parrish, R. R., 1987, An improved micro-capsule for zircon dissolution in uranium–lead geochronology. *Chemical Geology* **66**: 99–102.

1990, U–Pb dating of monazite and its application to geological problems. *Canadian Journal of Earth Sciences* **27**(11): 1431–50.

Parrish, R. R. and Krogh, T. E., 1987, Synthesis and purification of ^{205}Pb for U–Pb geochronology. *Chemical Geology* **66**: 103–110.

Parsons, I., Brown, W. L., and Smith, J. V., 1999, ^{40}Ar/^{39}Ar thermochronology using alkali feldspars: real thermal history or mathematical mirage of microtexture? *Contributions to Mineralogy and Petrology* **136**: 92–110.

Partington, M. A., Copestake, P., Mitchener, B. C., and Underhill, J. R., 1993, Biostratigraphic calibration of genetic stratigraphic sequences in the Jurassic–lowermost Cretaceous (Hettangian to Ryazanian) of the North Sea and adjacent areas. In J. R. Parker, ed., *Petroleum Geology of Northwest Europe, Proceedings of the 4th Conference*. London: The Geological Society, pp. 371–86.

Partridge, T. C., 1997a, The Plio–Pleistocene boundary, *Quaternary International* **40**: 1–100.

1997b, Reassessment of the position of the Plio–Pleistocene boundary: is there a case for lowering it to the Gauss–Matuyama Palaeomagnetic reversal? In T. C. Partridge, ed., The Plio–Pleistocene boundary. *Quaternary International* **40**: 5–10.

Patterson, C., 1956, Age of meteorites and the earth. *Geochimica et Cosmochimica Acta* **10**: 230–7.

Paul, H. A., Zachos, J. C., Flower, B. P., and Tripati, A., 2000, Orbitally induced climate and geochemical variability across the Miocene–Oligocene boundary. *Paleoceanography* **15**: 471–85.

Paull, R. K., 1997, Observations on the Induan–Olenekian boundary based on conodont biostratigraphic studies in the Cordillera of the western United States. *Albertiana* **20**: 31–2.

Pavia, G. and Enay, R., 1997, Definition of the Aalenian–Bajocian Stage boundary. *Episodes* **20**: 16–20.

Pavia, G., Chandler, R., Fernandez Lopez, S., *et al.*, 1995, A proposal for the global boundary stratotype section and point (GSSP) of the Bajocian (Middle Jurassic) and the Aalenian–Bajocian boundary. *International Subcommission on Jurassic Stratigraphy*. Torino: IUGS, 29 pp.

Paytan, A., Kastner, M., Martin, E. E., Macdougall, J. D., and Herbert, T., 1993, Marine barite as a monitor of seawater strontium isotope composition. *Nature* **366**: 445–9.

Pearson, P. N. and Chaisson, W. P., 1997, Late Paleocene to middle Miocene planktonic foraminifer biostratigraphy of the Ceara Rise. In N. J. Shackleton, W. B. Curry, C. Richter, and T. J. Bralower, eds., *Proceedings of the Ocean Drilling Program, Scientific Results*, Vol. 154. College Station, TX: Ocean Drilling Program, pp. 33–68.

Pechersky, D. M., 1997, Some properties of the geomagnetic field over 1700 million years. *Izvestiya*, *Physics of the Solid Earth* **33**: 341–57.

Pechersky, D. M. and Khramov, A. N., 1973, Mesozoic paleomagnetic scale of the USSR. *Nature* **244**: 499–501.

Peláez-Campomanes, P., López-Martinez, N., Álvarez-Sierra, D. M., and Daams, R., 2000, The earliest mammal of the European Paleocene: the multituberculate Hainina. *Journal of Paleontology* **74**: 701–11.

Penck, A. and Brückner, E., 1909, *Die Alpen in den Eiszeiten*. Leipzig: Tauchnitz, p. 1199.

Peng, S. and Robison, R. A., 2000, Agnostid biostratigraphy across the Middle–Upper Cambrian boundary in Hunan, China. *Journal of Paleontology* **74**(4): 1–104.

Peng Shanchi, Babcock, L. E., Robison, R. A., *et al.*, 1996, Cambrian to Silurian magnetostratigraphy. In G. C. Young and J. Laurie, eds., *An Australian Phanerozoic Timescale*. Oxford: Oxford University Press, pp. 23–9.

Peng Shanchi, Babcock, L. E., Lin Huan-ling, Yong-An, C., and Xue-Jian, Z., 2001, Potential global stratotype section and point for the base of an Upper Cambrian series defined by the first appearance of the trilobite *Glyptagnostus reticulatus*, Hunan Province, China. *Acta Palaeontologica Sinica*, suppl. **40**: 157–66, 6 pls.

Peppers, R. A., 1996, Palynological correlation of major Pennsylvanian (Middle and Upper Carboniferous) chronostratigraphic boundaries in the Illinois and other coal basins. *Geological Society of America Memoir* **188**. Boulder, CO: Geological Society of America, pp. 1–111.

Perch-Nielsen, K., 1981, New Maastrichtian and Paleocene calcareous nannofossils from Africa, Denmark, the USA and the Atlantic, and some Paleocene lineages. *Eclogae Geologicae Helvetiae* **74**: 831–863.

1985, Cenozoic calcareous nannofossils. In H. M. Bolli, J. B. Saunders, and K. Perch-Nielsen, eds., *Plankton Stratigraphy*. Cambridge: Cambridge University Press, pp. 427–554.

Perch-Nielsen, K. and Hansen, J. M., 1981. Selandian. In C. Pomerol, eds., *Stratotypes of Paleogene Stages, Mémoire hors série 2 du Bulletin d'Information des Géologues du Bassin de Paris*. pp. 219–28.

Perkins, C. and Walshe, J. L., 1993, Geochronology of the Mount Read Volcanics, Tasmania, Australia. *Economic Geology* **88**: 1176–97.

Permanent Interdepartmental Stratigraphic Commission, 1963, Decision of the Permanent Interdepartmental Stratigraphic Commission on the Paleogene of the USSR. *Sovietskaya Geologiya* **6**: 145–51 [In Russian].

Pessagno, E. A., Jr. and Meyerhoff Hull, D., 1996, "Once upon a time in the Pacific:" Chronostratigraphic misinterpretation of basal strata at ODP Site 801 (Central Pacific) and its impact on geochronology and plate tectonic models: *GeoResearch Forum*, Vols. 1 and 2. Zurich: Transtech Publications, p. 79–92.

Pessagno, E. A., Jr., Blome, C. D., Hull, D. M., and Six, W. M., 1993, Jurassic Radiolaria from the Josephine ophiolite and overlying strata, Smith River subterrane (Klamath Mountains), northwestern Californian and southwestern Oregon. *Micropaleontology* **39**(2): 93–166.

Pestiaux, P., Van der Mersch, I., Berger, A., and Duplessy, J. C., 1988, Paleoclimatic variability at frequencies ranging from 1 cycle per 10 000 years to 1 cycle per 1000 years: evidence for nonlinear behaviour of the climate system. *Climate Change* **12**: 9–37.

Peterson, D. N. and Nairn, A. E. M., 1971, Palaeomagnetism of Permian redbeds from the South-western United States. *Geophysical Journal of the Royal Astronomical Society* **23**(2): 191–205.

Peterson, L. C., Haug, G. H., Hughen, K. A., and Röhl, U., 2000, Rapid changes in marine and terrestrial climate in the tropical Atlantic during the last glacial. *Science* **290**: 1947–51.

Petit, J. R., Jouzel, J., Raynaud, D., *et al.*, 1999, Climate and atmospheric history of the past 420 000 years from the Vostok ice core, Antarctica. *Nature* **399**: 429–36.

Petrushevskaya, M. G., 1975, Cenozoic radiolarians of the Antarctic, Leg 29, DSDP. In *Initial Reports of the Deep Sea Drilling Project*, Vol. 29. Washington, DC: US Govt. Printing Office, pp. 541–675.

Peybernes, B., 1998a, Larger benthic foraminifers [column for Triassic chart, Mesozoic and Cenozoic sequence chronostratigraphic framework of European basins, in Hardenbol, J. *et al.*]. In P.-C. de Graciansky, J. Hardenbol., Th. Jacquin, and P. R. Vail, eds., *Mesozoic–Cenozoic Sequence Stratigraphy of European Basins, Society of Economic Paleontologists and Mineralogists Special Publication,* Vol. 60. Tulsa, OK: SEPM, Chart 8.

1998b, Larger benthic foraminifers. [column for Jurassic chart, Mesozoic and Cenozoic sequence chronostratigraphic framework of European basins, in Hardenbol, J. *et al.*]. In P.-C. de Graciansky, J. Hardenbol., Th. Jacquin, and P. R. Vail, eds., *Mesozoic–Cenozoic Sequence Stratigraphy of European Basins, Society of Economic Paleontologists and Mineralogists Special Publication,* Vol. 60. Tulsa, OK: SEPM, Chart 7.

Peybernes, B., Fondecave-Wallez, M.-J., Gourinard, Y., and Eichène, P., 1997, Stratigraphie séquentielle, biozonation par les foraminifères planctoniques, grade-datation et évaluation des taux de sédimentation dans les calcaires crayeux campano-maastrichtiens de Tercis (SW de la France). *Bulletin de la Société Géologique de France* **168**: 143–53.

Phillips, J., 1840, Palaeozoic Series. In G. Long, ed., *The Penny Cyclopaedia of the Society for the Diffusion of Useful Knowledge*, Vol. 17. London: Knight, pp. 153–4.

1841, *Figures and Descriptions of the Palaeozoic Fossils of Cornwall, Devon and East Somerset*. London: Longman, Brown, Green, and Longmans.

Pia, J., 1930, *Grundbegriffe der Stratigraphie mit ausführlicher Anwendung auf die europäische Mitteltrias*. Leipzig und Wien: Deuticke.

Pickard, A. L., 2002, SHRIMP U–Pb zircon ages of tuffaceous mudrocks in the Brockman Iron Formation of the Hamersley Range, Western Australia. *Australian Journal of Earth Sciences* **49**: 491–507.

Pickford, M., 1981, Preliminary Miocene mammalian biostratigraphy for Western Kenya. *Journal of Human Evolution* **10**: 73–97.

Pignatti, J. S., 1998, *Paleogene Larger Foraminifera*. Reference List, Slovenska Akademija Znanosti in Umetnosti, Razred za Naravoslovne Vede Dela 34/1, p. 298.

Pillans, B., 1991, New Zealand Quaternary stratigraphy: an overview. *Quaternary Science Reviews* **10**: 405–18.

in press, Proposal to redefine the Quaternary. *Episodes*.

Pillans, B., 2004, Proposal to redefine the Quaternary. *Episodes* **27**(2): 127.

Pitman, W. C., Herron, E. M., and Heirtzler, J. R., 1968, Magnetic anomalies in the Pacific and seafloor spreading. *Journal of Geophysical Research* **73**: 2069–85.

Playford, G., 1985, Palynology of the Australian Lower Carboniferous: a review. In *Compte Rendu 7ème Congrès International de Stratigraphie et de Géologie du Carbonifère 1983*, Vol. 4. Madrid: pp. 247–65.

1991, Australian miospores relevant to extra-Gondwanic correlations: an evaluation. *Courier Forschunginstitut Senckenberg* **130**: 85–125.

Playford, G., Rigby, J., Roberts, J., Turner, S., and Webb, G. E., 2000, Carboniferous biogeography of Australasia. In A. J. Wright, G. C. Young, J. A. Talent, and J. R. Laurie, eds., *Palaeobiogeography of Australasian Faunas and Floras, Association of Australasian Palaeontologists Memoir* **23**. Canberra: Association of Australian Palaeontologists, pp. 259–86.

Plumb, K. A., 1991, New Precambrian time scale. *Episodes* **14**(2): 139–40.

Plumb, K. A. and James, H. L., 1986, Subdivision of Precambrian time: recommendations and suggestions by the

SubCommission on Precambrian Stratigraphy. *Precambrian Research* **32**(1): 65–92.

Pnev, V. P., Polozova, A. N., Pavlov, A. M., and Faddeyeva, I. Z., 1975, Stratotype section of the Orenburgian Stage, southern Urals (Nikol'skoe village). *News of Academy of Sciences of USSR, Geology Series* **6**: 100–9 [In Russian].

Poignant, A. and Pujol, C., 1978, Nouvelles données micropaléontologiques (foraminifères planctoniques et petits foraminifères benthiques) sur le stratotype bordelais du Burdigalien. *Géobios* **11**: 655–712.

Poignant, A., Pujol, C., Ringeade, M., and Londeix, L., 1997a, The Aquitinian historical stratotype. In A. Montanari, G. S. Odin, and R. Coccioni, eds., *Miocene Stratigraphy: An Integrated Approach, Development in Paleontology and Stratigraphy*, Vol. 15. Amsterdam: Elsevier, pp. 10–16.

1997b, The Burdigalian historical stratotype. In A. Montanari, G. S. Odin, and R. Coccioni, eds., *Miocene Stratigraphy: An Integrated Approach, Development in Paleontology and Stratigraphy*, Vol. 15. Amsterdam: Elsevier, pp. 17–24.

Poletaev, V. I. and Lazarev, S. S., 1994, General stratigraphic scale and brachiopod evolution in the Late Devonian and Carboniferous subequatorial belt. *Bulletin de la Société Belge de Géologie* **103**(1–2): 99–107.

Pomerol, C., 1973, *Stratigraphie et Paléogéographie–Ére cénozoïque (Tertiaire et Quaternaire)*. Paris: Doin, p. 269.

1981, *Stratotypes of Paleogene Stages, Bulletin d'Ínformation des Géologues du Bassin de Paris, Mémoire* hors Série 2, p. 301.

Popov, A. V., 1992, Gzhelian ammonoids of Central Asia. *Problems of Paleontology* **10**: 52–62 [In Russian].

Popov, A. V., Davydov, V. I., Donakova, L. M., and Kossovaya, O. L., 1985, On the Gzhelian stratigraphy of the South Urals. *Sovietskaya Geologiya* **3**: 57–67 [In Russian].

Popp, B. N., Anderson, T. F., and Sandberg, P. A., 1986, Brachiopods as indicators of original isotopic compositions in some Paleozoic limestones. *Geological Society of America Bulletin* **97**(10): 1262–9.

Porter, S. C. and Zhisheng, A., 1995, Correlation between climate events in the North Atlantic and China during the last glaciation. *Nature* **375**: 305–8.

Porter, S. M., Meisterfeld, R., and Knoll, A. H., 2003, Vase-shaped microfossils from the Neoproterozoic Chuar Group, Grand Canyon: a classification guided by modern testate amoebae. *Journal of Paleontology* **77**: 409–29.

Posen, P., Hounslow, M. W., and Warrington, G., 1998, The magnetostratigraphy of the Rhaetian in St Audrie's Bay section, north Somerset, UK. *Hallesches Jahrbuch Geowissenschaften, Reihe B, Beiheft* **5**: 143 (summary figure available from Mark Hounslow).

Postuma, J. A., 1971, *Manual of Planktonic Foraminifera*. Amsterdam: Elsevier, p. 421.

Poty, E., 1985, A rugose coral biozonation for the Dinantian of Belgium as a basis for a coral biozonation of the Dinantian of Eurasia. In *Compte Rendu VII Congrès International de Stratigraphie et de Géologie du Carbonifère*, Vol. 10. pp. 29–31.

Pouyet, S., Carbonnel, G., and Demarcq, G., 1997, The Burdigalian historical stratotype in the Rhodanian area. In A. Montanari, G. S. Odin, and R. Coccioni, eds., *Miocene Stratigraphy: An Integrated Approach, Development in Paleontology and Stratigraphy*, Vol. 15. Amsterdam: Elsevier, pp. 25–32.

Powell, A. J., 1992a, Dinoflagellate cysts of the Tertiary System. In A. J. Powell, ed., *A Stratigraphic Index of Dinoflagellate Cysts*, British Micropalaeontological Society, Publication Series. London: Chapman and Hall, pp. 155–251.

1992b, *A Stratigraphic Index of Dinoflagellate Cysts*, British Micropalaeontological Society Publication Series. London: Chapman and Hall, p. 290.

Powell, A. J., Brinkhuis, H., and Bujak, J. P., 1996, Upper Paleocene–Lower Eocene dinoflagellate cyst sequence of southeast England. *Geological Society of London Special Publication*, Vol. 101. London: Geological Society, pp. 145–83.

Poyato-Ariza, F. J., Talbot, M. R., Fregenal-Martinez, M. A., Melendez, N., and Wenz, S., 1998, First isotopic and multidisciplinary evidence for nonmarine coelacanths and pycnodontiform fishes: palaeoenvironmental implications. *Palaeogeography, Palaeoclimatology, Palaeoecology* **144**: 65–84.

Prasad, N. and Roscoe, S. M., 1996, Evidence of anoxic to oxic atmospheric change during 2.45–2.22 Ga from lower and upper sub-Huronian Paleosols, Canada. *Catena* (Giessen) **27**(2): 105–21.

Pratt, B. R., 1992, Trilobites of the Marjuman and Steptoean stages (Upper Cambrian), Rabbitkettle Formation, southern Mackenzie Mountains, northwest Canada. *Palaeontographica Canadiana* **9**: 179, 34 pls.

Prell, W. L., Imbrie, J., Morley, J. J., *et al.*, 1986, Graphic correlation of oxygen isotope stratigraphy applications to the Late Quaternary. *Paleoceanography* **1**(2): 137–62.

Premoli-Silva, I. and Jenkins, D. G., 1993, Decision on the Eocene–Oligocene boundary stratotype. *Episodes* **16**: 379–82.

Premoli-Silva, I. and Sliter, W. V., 1999, Cretaceous paleoceanography: evidence from planktonic foraminiferal evolution. In E. Barrera and C. C. Johnson, eds., *Evolution of the Cretaceous Ocean–Climate System, Society of Economic Paleontologists and Mineralogists Special Publication,* Vol. 332. Tulsa, OK: SEPM, pp. 301–28.

Premoli-Silva, I., Coccioni, R., and Montanari, A., 1988, The Eocene–Oligocene boundary in the Marche–Umbria Basin (Italy). *Proceedings of the Eocene–Oligocene Boundary Meeting,* Ancona, October 1987. Ancona: Annibali/International Union of Geological Sciences.

Premoli-Silva, I., Erba, E., Salvini, G., Locatelli, C., and Verga, D., 1999, Biotic changes in Cretaceous anoxic events of the Tethys. *Journal of Foraminiferal Research* **29**: 352–70.

Premoli-Silva, I., Coccioni, R., Monechi, S., Montanari, A., and Nocchi, M., 2000, In search of the Early Oligocene/Late Oligocene boundary stratotype in Central Italy. In *Abstracts of the 31st International Geological Congress*, Rio de Janeiro, August 2000, pp. 6–13.

Press, W. H., Teukolsky, S. A., Vettering, W. T., and Flannery, B. R., 1992, *Numerical Recipes in FORTRAN*. Cambridge: Cambridge University Press, p. 963.

Pringle, M. S., 1992, Radiometric ages of basaltic basement recovered at Sites 800, 801, and 802, Leg 129, western Pacific Ocean. In *Proceedings of the Ocean Drilling Program, Scientific Results* Vol. 129. College Station, TX: Ocean Drilling Program, pp. 389–98.

Pringle, M. S. and Duncan, R. A., 1995, Radiometric ages of basaltic lavas recovered at Lo-En, Wodejebato, MIT, and Takuyo-Daisan Guyots, northwestern Pacific Ocean. In *Proceedings of the Ocean Drilling Program, Scientific Results*, Vol. 144. College Station, TX: Ocean Drilling Program, pp. 547–57.

Pringle, M. S., Chambers, L., and Ogg, J. G., 2003, Synchronicity of volcanism on Ontong Java and Manihiki plateaux with global oceanographic events? In *American Geophysical Union and European Geosciences Union Joint Assembly 2003*, Nice, May, *Geophysical Research Abstracts*, Vol. 5.

Prokopenko, A. A., Karabanov, E. B., Williams, D. F., *et al.*, 2001, Biogenic silica record of the Lake Baikal response to climatic forcing during the Brunhes. *Quaternary Research* **55**: 123–32.

Prothero, D. R., 1995, Geochronology and magnetostratigraphy of Paleogene North American land mammal "ages." *Society of Economic Paleontologists and Mineralogists Special Publication*, Tulsa, OK: SEPM, pp. 305–15.

Prothero, D. R. and Emry, R. J., 1996, *The Terrestrial Eocene–Oligocene Transition in North America*. Cambridge: Cambridge University Press, p. 688.

Pucéat, E., Lécuyer, C., Sheppard, S. M. F., *et al.*, 2003, Thermal evolution of Cretaceous Tethyan marine waters inferred from oxygen isotope composition of fish tooth enamels. *Paleoceanography* **18**: doi 10.1029/2002PA00823.

Pujalte, V., Baceta, J.-I., Dinares-Turell, J., *et al.*, 1995, Biostratigraphic and magnetostratigraphic intercalibration of the latest Cretaceous and Paleocene depositional sequences from the deep-water Basque Basin, western Pyrenees, Spain. *Earth and Planetary Science Letters* **136**: 17–30.

Purdy, J. W. and Jäger, E., 1976, K–Ar ages on rock-forming minerals from the Central Alps, Institute of Geology and Mineralogy Memoirs 30. Padova: University of Padova, pp. 1–31.

Qing, H., Barnes, C. R., Buhl, D., and Veizer, J., 1998, The strontium isotopic composition of Ordovician and Silurian brachiopods and conodonts: Relationships to geological events and implications for coeval seawater. *Geochimica et Cosmochimica Acta* **62**: 1721–3.

Qiu, Z., Wu, W., and Qiu, Z., 1999, Miocene mammal faunal sequence of China: paleozoogeography and Eurasian relationships. In G. E. Rössner and K. Heissig, eds., *The Miocene Land Mammals of Europe*: München: Friedrich Pfeil, pp. 443–55.

Quenstedt, F., 1848, *Der Jura*. Tübingen: H. Laupp, p. 103, 43 pls.

Quinn, T. R., Tremaine, S., and Duncan, M., 1991, A three million year integration of the Earth's orbit. *Astronomical Journal* **101**: 2287–305.

Rabien, A., 1954, Zur Taxionomie und Chronologie der Oberdevonischen Ostracoden. *Abhandlungen des Hessischen Landesamtes für Bodenforschung*, **9**: 1–268.

Racki, G. and House, M. R., 2002, The Frasnian/Famennian boundary extinction event. *Palaeogeography, Palaeoclimatology, Palaeoecology*, p. 181 (1–3), 181, v + 374 pp.

Raffi, I. 1999, Precision and accuracy of nannofossil biostratigraphic correlation. *Philosophical Transactions of the Royal Society of London, series A* **357**: 1975–1993.

Raffi, I. and Flores, J. A., 1995, Pleistocene through Miocene calcareous nannofossils from eastern Equatorial Pacific Ocean (Leg 138). In *Proceedings of the Ocean Drilling Program, Scientific Results,* Vol. 138. College Station, TX: Ocean Drilling Program, pp. 233–86.

Raffi, I. and Rio, D., 1979, Calcareous nannofossil biostratigraphy of DSDP Site 132, Leg 13 (Tyrrhenian Sea, Western Mediterranean). *Rivista Italiana di Paleontologia e Stratigrafia* **85**: 127–72.

Raffi, I., Backman, J., Rio, D., and Shackleton, N. J., 1993, Plio–Pleistocene nannofossil biostratigraphy and calibration to oxygen isotope stratigraphies from Deep Sea Project Drilling Site 607 and Ocean Drilling Program Site 677. *Paleoceanography* **8**(3): 387–408.

Raffi, S., Rio, D., Sprovieri, R., *et al.*, 1989, New stratigraphic data on the Piacenzian stratotype. *Bolletino della Società Geologica Italiana* **108**: 183–96.

Raffi, I., Mozzato, C., Fornaciari, E., Hilgen, F. J., and Rio, D., 2003, Late Miocene calcareous nannofossil biostratigraphy and astrobiochronology for the Mediterranean region. *Micropaleontology* **49**(1): 1–26.

Rainaud, C., Master, S., Robb, L. J., and Armstrong, R. A., 2001, A fertile Palaeoproterozoic magmatic arc beneath the central African Copperbelt. In C. J. Stanley, ed., *Mineral Deposits: Processes to Processing*. Rotterdam: Balkema, pp. 1427–30.

Ramezani, J., Davydov, V. I., Northrup, C. J., *et al.*, 2003, Volcanic ashes in the Upper Paleozoic of the Southern Urals: opportunities for high-precision calibration of the Upper Carboniferous Cisuralian Interval. *Abstract of the International Congress on Carboniferous and Permian Stratigraphy,* Utrecht. pp. 431–2.

Rampino, M., Prokoph, R. A., and Adler, A. C., 2000, Tempo of the end-Permian event: high-resolution cyclostratigraphy at the Permian–Triassic boundary. *Geology* **28**: 643–6.

Rampino, M., Prokoph, R., Adler, A. C., and Schwindt, D. M., 2002, Abruptness of the end-Permian mass extinction as determined from biostratigraphic and cyclostratigraphic analysis of European western Tethyan sections. In C. Koeberl and K. G. MacLeod, eds., *Catastrophic Events and Mass Extinctions: Impacts and Beyond, Geological Society of America Special Paper* 356. Boulder, CO: Geological Society of America, pp. 415–27.

Ramsbottom, W. H. C., 1973, Transgressions and regressions in the Dinantian: a new synthesis of British Dinantian stratigraphy. *Proceedings of the Yorkshire Geological Society* **39**(4): 567–607.

1977, Major cycles of transgressions and regressions (mesothems) in the Namurian. *Proceedings of the Yorkshire Geological Society* **41**(24): p. 261 91.

1981, Eustasy, sea level and local tectonism, with examples from the British Carboniferous. *Proceedings of the Yorkshire Geological Society* **43**(4): 473–82.

1984, The founding of the Carboniferous System. In G. Mackenzie, ed., *Compte Rendu 9ème Congrès International de Stratigraphie et de Géologie du Carbonifère 1979*, Vol. 1. Carbondale, IL: South Illinois University, Press, pp. 109–12.

Ramsbottom, W. H. C. and Saunders, W. B., 1985, Evolution and evolutionary biostratigraphy of Carboniferous ammonoids. *Journal of Paleontology* **59**(1): 123–39.

Ramsbottom, W. H. C., Claver, M. A., Eager, R. M. C., *et al.*, 1978, *Silesian (Upper Carboniferous): A Correlation of Silesian Rocks in the British Isles, Geological Society of London Special Report,* Vol. 10. Edinburgh: Scottish Academic Press.

Rasbury, E. T., Hanson, G. N., Meyers, W. J., and Saller, A. H., 1997, Dating of the time of sedimentation using U–Pb ages of paleosol calcite. *Geochimica et Cosmochimica Acta* **61**(7): 1525–9.

Rasbury, E. T., Hanson, G. N., Meyers, W. J., *et al.*, 1998, U–Pb dates of paleosols: constraints on late Paleozoic cycle durations and boundary ages. *Geology* **26** (5): 403–6.

Rauser-Chernousova, D. M., 1937, About fusulinids and stratigraphy of Upper Carboniferous and Artinskian of the western slope of the Urals. *Bulletin Moscow Society of Nature Studies, new series, vol. 45, geological series* **15**(5): 478 [In Russian].

1941, New Upper Carboniferous stratigraphic data from the Oksko–Tsninskyi Arch. *Reports of the Academy of Sciences of the USSR* **30**(5): 434–6 [In Russian].

1949, Stratigraphy of Upper Carboniferous and Artinskian deposits of Bashkirian Preurals. In D. M. Rauser-Chernousova, ed., *Foraminifers of Upper Carboniferous and Artinskian deposits of Bashkirian Pre-Urals, Transactions of Geological Institute of Academy of Sciences of USSR* **105**(35): 3–21 [In Russian].

1958, The experience of super-detailed subdivision of Upper Carboniferous section in the Kuibyshev's Hydro-electric PowerStation/HEPS/area. *Transactions of the Geological Institute of Academy of Science of USSR* **13**: 121–38 [In Russian].

Rauser-Chernousova, D. M. and Reitlinger, E. A., 1954, Biostratigraphic distribution of foraminifers in Middle Carboniferous deposits of southern limb of Moscow Syneclise. In D. V. Nalivkin and V. V. Menner, eds., *Regional Stratigraphy of the USSR Vol. 2 Stratigraphy of Middle Carboniferous deposits of Central and Eastern parts of Russian Platforms (based on foraminifers study), Vol. 1. Moscow syneclise, Transactions of the Geological Institute of Academy of Sciences of USSR*. pp. 7–120, 21 pls [In Russian].

Rauser-Chernousova, D. M. and Schegolev, A. K., 1979, The Carboniferous–Permian boundary in the USSR. In R. H. Wagner, A. C. Higgins, and S. V. Meyen, eds., *The Carboniferous of the USSR, Yorkshire Geological Society Occasional Publication No. 4*. Leeds: Yorkshire Geological Society, pp. 175–91.

Rauser-Chernousova, D. M., Bensh, F. R., Vdovenko, M. V., *et al.*, 1996, *Guidebook on the Systematics of Foraminifers of Paleozoic*, Academy of Sciences of the USSR. Moscow: Nauka, pp. 3–202 [In Russian].

Ravizza, G. and Turekian, K. K., 1989, Application of ^{187}Re–^{187}Os system to black shale geochronometry. *Geochimica et Cosmochimica Acta* **53**: 3257–62.

Rawson, P. F., 1983, The Valanginian to Aptian stages: current definitions and outstanding problems. *Zitteliania* **10**: 493–500.

1990, Event stratigraphy and the Jurassic–Cretaceous boundary. *Transactions of the Institute of Geology and Geophysics, Academy Sciences USSR, Siberian branch* **699**: 48–52 [In Russian with English summary].

Rawson, P. F., Curry, D., Dilley, F. C., *et al.*, 1978, *A Correlation of Cretaceous Rocks in the British Isles, Geological Society of London Special Report* 9. London: Geological Society, p. 70.

Rawson, P. F., Avram, E., Baraboschkin, E. J., *et al.*, 1996a, The Barremian Stage. *Bulletin de l'Institut Royal des Sciences Naturelles de Belgique, Sciences de la Terre, Supplément* **66**: 25–30.

Rawson, P. F., Dhondt, A. V., Hancock, J. M., and Kennedy, W. J., 1996b, Proceedings of Second International Symposium on Cretaceous Stage Boundaries, Brussels, September 8–16, 1995. *Bulletin de l'Institut Royal des Sciences Naturelles de Belgique, Sciences de la Terre Supplément* **66**: 117.

Raymo, M. E., Ruddiman, W. F., and Froelich, P. N., 1988, Influence of late Cenozoic mountain building on ocean geochemical cycles. *Geology* **16**: 649–53.

Raymo, M. E., Ruddiman, W. F., Backman, J., Clement, B. M., and Martinson, D. G., 1989, Late Pliocene variation in northern Hemisphere ice sheets and North Atlantic deep water circulation. *Paleoceanography* **4**: 413–46.

Raymo, M. E., Ganley, K., Carter, S., Oppo, D. W., and McManus, J., 1998, Millennial-scale climate instability during the early Pleistocene epoch. *Nature* **392**: 699–702.

Reboulet, S., 1995, L'évolution des ammonites du Valanginien–Hauterivien inférieur du bassin vocontien et de la plate-forme provençale (sud-est de la France): relations avec la stratigraphie séquentielle et implications biostratigraphiques. *Documents des Laboratoires de Géologie de Lyon* **137**: 137.

2001, Limiting factors on shell growth, mode of life and segregation of Valanginian ammonoid populations: evidence from adult size populations. *Géobios* **34**(4): 423–35.

Reboulet, S. and Atrops, F., 1999, Comments and proposals about the Valanginian–Lower Hauterivian ammonite zonation of south-eastern France. *Ecologae Geologicae Helvetiae* **92**: 183–97.

Reboulet, S., Atrops, F., Ferry, S., and Schaaf, A., 1992, Renouvellement des ammonites en fosse vocontienne à la limite Valanginien–Hauterivien. *Géobios* **25**(4): 469–76.

Reboulet, S., Mattioli, E., Pittet, B., *et al.*, 2000, Ammonoid and nannoplankton abundance in Valanginian limestone–marl alternations: carbonate dilution or productivity? *Sixth International Cretaceous Symposium, Abstracts,* Vienna, August 27–September 4, 2000, p. 110.

in press, Ammonoid and nannoplankton abundance in Valanginian (early Cretaceous) limestone-marl alternations: carbonate dilution or productivity? *Palaeogeography, Palaeoclimatology, Palaeoecology*.

Reimond, W. U., Koerberl, C., Brandstater, F., *et al.*, 1999, Morokweng impact structure, South Africa: geologic, petrographic, and isotopic results, and implications for the size of the structure. In. B. O. Dressler and V. L. Sharpton, eds., *Large Meteorite Impacts and Planetary Evolution II, Geological Society of America Special Paper,* Vol. 339. Boulder, CO: Geological Society of America, pp. 61–90.

Remane, J., 1985, Calpionellids. In H. M. Bolli, J. B. Saunders, and K. Perch Nielsen, eds., *Plankton Stratigraphy.* Cambridge: Cambridge University Press, pp. 555–72.

1990, The Jurassic–Cretaceous boundary: problems of definition and procedure. *Transactions of the Institute of Geology and Geophysics, Academy Sciences USSR, Siberian branch* **699**: 7–17 [In Russian with English summary].

1997, Foreword: chronostratigraphic standards: how they are defined and when they should be changed. In T. C. Partridge, ed., The Plio–Pleistocene boundary. *Quaternary International* **40**: 3–4.

1998, Calpionellids. [for Cretaceous and Jurassic portions of Mesozoic and Cenozoic sequence chronostratigraphic framework of European basins, in Hardenbol, J. *et al.*]. In P.-C. de Graciansky, J. Hardenbol., Th. Jacquin, and P. R. Vail, eds., *Mesozoic–Cenozoic Sequence Stratigraphy of European Basins, Society of Economic Paleontologists and Mineralogists Special Publication,* Vol. 60. Tulsa, OK: SEPM, pp. 773–4, Charts 5 and 7.

2000, International stratigraphic chart, with explanatory note. Sponsored by ICS, IUGS and UNESCO. 31st International Geological Congress, Rio de Janeiro 2000, p. 16.

2003, Chronostratigraphic correlations: their importance for the definition of geochronologic units. *Palaeogeography, Palaeoclimatology, Palaeoecology* **196**: 7–18.

Remane, J. and Michelsen, O., 1999, Report – On the vote to demand to lower the Plio–Pleistocene boundary. *Neogene Newsletter* (Subcommission of Neogene Stratigraphy under IUGS) **6**: 9–14.

Remane, J., Bakalova-Ivanova, D., Borza, K., *et al.*, 1986, Agreement on the subdivision of the Standard Calpionellid Zones defined at the IInd Planktonic Conference Roma 1971. *Acta Geologica Hungarica* **29**: 5–14.

Remane, J., Bassett, M. G., Cowie, J. W., *et al.*, 1996, Revised guidelines for the establishing of global chronostratigraphic standards by the International Commission on Stratigraphy (ICS). *Episodes* **19**(3): 77–81.

Remizova, S. T., 1992, Micropaleontological definition of the boundary between the Middle and Late Carboniferous. *Scientific Report of KOMI section of Russian Academy of Sciences 285*. Syktyvkar: Russian Academy of Sciences, pp. 3–19 [In Russian].

Renevier, E., 1864, Notices géologique et paléontologiques sur les Alpes Vaudoises, et les régions environnantes. I. Infralias et Zone à Avicula contorta (Ét. Rhaetien) des Alpes Vaudoises. *Bulletin de la Société Vaudoise des Sciences Naturelles, Laussane* **8**: 39–97.

1873, Tableau de terrains sédimentaires formés pendant les époques de la phase organique du globe terrestre. *Bulletin de la Société Vaudoise des Sciences Naturelles, Laussane* **12**: 218–52.

1874, Tableau des terrains sédimentaires. *Bulletin de la Société Vaudoise des Sciences Naturelles, Laussane* **13**: 218–52.

Renne, P. R. and Basu, A. R., 1991, Rapid eruption of the Siberian Traps flood basalts at the Permo–Triassic boundary. *Science* **253**(5016): 176–9.

Renne, P. R., Deino, A. L., Walter, R. C., *et al.*, 1994, Intercalibration of astronomical and radioisotopic time. *Geology* **22**(9): 783–6.

Renne, P. R., Zhang, Z., Richards, M. A., Black, M. T., and Basu, A. R., 1995, Synchrony and causal relations between Permian–Triassic boundary crises and Siberian flood volcanism. *Science* **269**(5229): 1413–16.

Renne, P. R., Sharp, W. D., Deino, A. L., Orsi, G., and Civetta, L., 1997, $^{40}Ar/^{39}Ar$ dating into the historical realm: calibration against Pliny the Younger. *Science* **277**: 1279–80.

Renne, P., Karner, D. B., and Ludwig, K. R., 1998a, Radioisotope dating: absolute ages aren't exactly. *Science* **282**: 1840–1.

Renne, P. R., Deino, A. L., and Sharp, W. D., 1998b, $^{40}Ar/^{39}Ar$ Dating of the 79 AD eruption of Vesuvius and some implications. In *Joint Annual Meeting of the Geological Association of Canada and the Mineralogical Association of Canada with the Society of Economic Geologists Program with Abstracts*, Vol. 23. Ottawa: Geological Association of Canada.

Renne, P. R., Swisher III, C. C., Deino, A. L., *et al.*, 1998c, Intercalibration of standards, absolute ages and uncertainties in $^{40}Ar/^{39}Ar$ dating. *Chemical Geology* **145**: 117–52.

Renne, P. R., Mundil, R., Min, K., and Ludwig, K. R., 1999, Progress report on high resolution comparison of argon/argon and uranium/lead systems. In *9th Goldschmidt Conference, Volume LPI Contribution 971*. Houston, TX: Lunar and Planetary Institute, pp. 243–4.

Renne, P. R., Ludwig, K. R., and Karner, D. B., 2000, Progress and challenges in geochronology. *Science Progress* **83**: 107–21.

Resolution, 1955, *Resolution of the All-Union Meeting on the Unified Scheme of the Mesozoic Stratigraphy for the Russian Platform*. Leningrad: Gostoptechnizdat, 30 pp. [In Russian].

Rhodes, G. M., Ali, J. R., Hailwood, E. A., King, C., and Gibson, T. C., 1999, Magnetostratigraphic correlation of Paleogene sequences from northwest Europe and North America. *Geology* **27**: 451–4.

Riba, O. and Reguant, S., 1986, Una taula dels temps geològics. *Institut d'Estudis Catalans, Arxius Secció Ciéncies* **81**:127.

Riccardi, A. C., Westerrmann, G. E. G., and Elmi, S., 1991, Biostratigraphy of the upper Bajocian–middle Callovian (Middle Jurassic), South America. *Journal of South American Earth Sciences* **4**(3): 149–57.

Rice, C. L., Belkin, H. E., Henry, T. W., Zartman, R. E., and Kunk, M. J., 1994, The Pennsylvanian fire clay tonstein of the Appalachian Basin; its distribution, biostratigraphy, and mineralogy. In C. L. Rice, ed., *Elements of Pennsylvanian Stratigraphy, Central Appalachian Basin, Geological Society of America Special Paper 294*. Boulder, CO: Geological Society of America, pp. 87–104.

Richardson, J. B. and McGregor, D. C., 1986, Silurian and Devonian spore zones of the Old Red Sandstone Continent and adjacent areas. *Geological Survey of Canada Bulletin* **364**: 1–79.

Richardson, J. B., Rasul, S. M., and Al-Ameri, T., 1981, Acritarchs, miospores and correlation of the Ludlovian–Downtonian and Silurian–Devonian boundaries: *Reviews of Palaeobotany and Palynology* **34**: 209–24.

Richardson, J. B., Rodriguex, R. M., and Sutherland, J. E., 2000, Palynology and recognition of the Silurian/Devonian boundary in some British terrestrial sediments by correlation with other European marine sequences – a progress report. *Courier Forschungsinstitut Senckenberg* **220**: 1–7.

Richmond, G. M., 1996, The INQUA-approved provisional Lower–Middle Pleistocene boundary. In C. Turner, ed., *The Early Middle Pleistocene in Europe*. Rotterdam: Balkema, pp. 319–26.

Rickards, R. B., 1976, The sequence of Silurian graptolite zones in the British Isles. *Geological Journal* **11**: 153–88.

Riding, J. B. and Ioannides, N. S., 1996, Jurassic dinoflagellate cysts. *Bulletin de la Société géologique de France* **167**: 3–14.

Riding, J. B. and Thomas, J. E., 1997, Marine palymorphs from the Staffin Bay and Staffin Shale formations (Middle–Upper Jurassic) of the Trotternish Peninsula, NW Skye. *Scottish Journal of Geology* **33**: 59–74.

Rieber, H., 1984, Report of the Aalenian Working Group. Aalenian, present status and open problems. In O. Michelsen and A. Zeiss, eds., *Proceedings of 1st International Symposium on Jurassic Stratigraphy*, Vol. 1. Copenhagen: Geological Society of Denmark, pp. 45–54.

Riedel, W. R., 1957, Radiolaria: a preliminary stratigraphy. In H. Petterson, ed., *Reports of the Swedish deep-Sea Expedition, 1947–1948*. Goteborg: Elanders Boktryckeri Aktiebolag, pp. 59–96.

Riedel, W. R. and Sanfilippo, A., 1970, Radiolaria, Leg 4, Deep Sea Drilling Project. In *Initial Reports of the Deep Sea Drilling Project*, Vol. 4. Washington, DC: US Govt. Printing Office, pp. 503–75.

1971, Cenozoic Radiolaria from the western tropical Pacific, Leg 7. In *Initial Reports of the Deep Sea Drilling Project*, Vol. 7. Washington, DC: US Govt. Printing Office, pp. 1529–672.

1978, Stratigraphy and evolution of tropical Cenozoic radiolarians. *Micropaleontology* **24**: 61–96.

Riley, N. J., 1996, Dinantian ammonoids from the Craven Basin, north-west England. *Special Papers in Palaeontology* **53**: 1–87, 8 plates, 27 tables, sketch map.

2000, New project group proposal: a GSSP close to the Visean–Namurian/Serpukhovian boundary. *Newsletter on Carboniferous Stratigraphy* **18**: 7.

Riley, N. J., Razzo, M. J., and Owens, B., 1985, A new boundary stratotype section for Westphalian B/C in northern England. In *Tenth International Congress of Carboniferous Geology and Stratigraphy, Abstracts*. Madrid: Com. Nac. Geol., p. 14.

Riley, N. J., Claoué-Long, J., Higgins, A. C., Owens, B., Spears, A., Taylor, L., and Varker, W. J., 1994, Geochronometry and geochemistry of the European Mid Carboniferous boundary global stratotype proposal, Stonehead Beck, North Yorkshire, UK. In M. Street and E. Paproth, eds., *Carboniferous Biostratigraphy, Annales de la Société Géologique de Belgique*, Vol. 116. Liege: Société géologique de Belgique, pp. 275–89.

Rio, D., Sprovieri, R., Raffi, I., and Valleri, G., 1988, Biostratigrafia e paleoecologia della sezione stratotipo del Piacenziano. *Bollettino della Società Paleontologica Italiana* **27**: 213–38.

Rio, D., Raffi, I., and Villa, G., 1990a, Plio–Pleistocene calcareous nannofossil distribution patterns in the western Mediterranean. In K. A. Kastens, J. Mascle, *et al.*, eds., *Proceedings of the Ocean Drilling Program, Scientific Results*, Vol. 107. College Station, TX: Ocean Drilling Program, pp. 513–33.

Rio, D., Fornaciari, E., and Raffi, I., 1990b, Late Oligocene through early Pleistocene calcareous nannofossils from the western equatorial Indian Ocean. In *Proceedings of the Ocean Drilling Program, Scientific Results*, Vol. 115. College Station, TX: Ocean Drilling Program, pp. 175–235.

Rio, D., Sprovieri, R., and Thunell, R., 1991, Pliocene–lower Pleistocene chronostratigraphy: a re-evaluation of Mediterranean type sections. *Geological Society of America Bulletin* **103**: 1049–58.

Rio, D., Sprovieri, R., and Di Stefano, E., 1994, The Gelasian Stage: a proposal of a new chronostratigraphic unit of the Pliocene Series. *Rivista Italiano di Paleontologia e Stratigrafia* **100**: 103–24.

Rio, D., Cita, M. B., Iaccarino, S., Gelati, R., and Gnaccolini, M., 1997, Langhian, Serravallian, and Tortonian historical stratotypes. In A. Montanari, G. S. Odin, and R. Coccioni, eds., *Miocene Stratigraphy: An Integrated Approach, Developments in Paleontology and Stratigraphy*, Vol. 15. Amsterdam: Elsevier, pp. 57–87.

Rio, D., Sprovieri, R., Castradori, D., and Di Stefano, E., 1998, The Gelasian Stage (Upper Pliocene): a new unit of the global standard chronostratigraphic scale. *Episodes* 21: 82–7.

Rio, D., Castradori, D., and Van Couvering, J. A., 2000, The Pliocene/Pleistocene boundary-stratotype at Vrica (Calabria, Italy) survived the last challenge. *Subcommission on Neogene Stratigraphy Newsletter 7*.

Rioult, M., 1964, Le stratotype du Bajocien. In P. L. Manbenge, ed., *Colloque du Jurassique à Luxembourg 1962*. Luxembourg: l'Institut Grand-Ducal, Section des Sciences Naturelles, Physiques et Mathématiques, pp. 239–58.

Ripley, B. D. and Thompson, M., 1987, Regression techniques for the detection of analytical bias. *Analyst* **112**: 377–83.

Ripperdan, R. L. and Kirschvink, J. L., 1992, Paleomagnetic results from the Cambrian–Ordovician boundary section at Black Mountain, Georgina Basin, western Queensland, Australia. In B. D. Webby and J. R. Laurie, eds., *Global Perspectives on Ordovician Geology*. Rotterdam: Balkema, pp. 93–103.

Ripperdan, R. L. and Miller, J. F., 1995, Carbon isotope ratios from the Cambrian–Ordovician boundary section Lawson Cove, Ibex area, Utah. In J. D. Cooper, M. L. Droser, and S. F. Finney, eds., *Ordovician Odyssey: Short Papers for the Seventh International Symposium on the Ordovician System*, Fullerton, CA: The Pacific Section for the Society for Sedimentary Geology (Society of Economic Paleontologists and Mineralogists), pp. 129–32.

Ripperdan, R. L., Margaritz, M., Nicoll, R. S., and Shergold, J. H., 1992, Simultaneous changes in carbon isotopes, sea level and conodont biozones within the Cambrian–Ordovician boundary interval at Black Mountain, Australia. *Geology* **20**: 1039–42.

Ripperdan, R. L., Margaritz, M., and Kirschvink, J. L., 1993, Magnetic polarity and carbon isotope evidence for non-depositional events within the Cambrian–Ordovician boundary section at Dayangcha, Jilin Province, China. *Geological Magazine* **130**: 443–52.

Ritter, S. M., 1986, Taxonomic revision and phylogeny of post-Early Permian crisis *bisselli–whitei* Zone conodonts with comments on late Paleozoic diversity. *Geologica et Palaeontologica* **20**: 139–65.

1995, Upper Missourian–lower Wolfcampian (upper Kasimovian–lower Asselian) conodont biostratigraphy of the midcontinent, USA. *Journal of Paleontology* **69**: 1139–54.

Ritter, S. M., Barrick, J. E., and Skinner, R. M., 2002, Conodont sequence biostratigraphy of the Hermosa Group (Pennsylvanian) at Honaker Trail, Paradox Basin, Utah. *Journal of Paleontology* **76**(3): 495–17.

Riveline, J., 1998, Charophytes. [column for Cretaceous chart, Mesozoic and Cenozoic sequence chronostratigraphic framework of European basins, in Hardenbol, J. *et al.*]. In P.-C. de Graciansky, J. Hardenbol., Th. Jacquin, and P. R. Vail, eds., *Mesozoic–Cenozoic Sequence Stratigraphy of European Basins, Society of Economic Paleontologists and Mineralogists Special Publication,* Vol. 60. Tulsa, OK: SEPM, Chart 5.

Robaszynski, F., 1998, Planktonic foraminifera. [columns for Cretaceous chart, Mesozoic and Cenozoic sequence chronostratigraphic framework of European basins, in Hardenbol, J. *et al.*]. In P.-C. de Graciansky, J. Hardenbol., Th. Jacquin, and P. R. Vail, eds., *Mesozoic–Cenozoic Sequence Stratigraphy of European Basins, Society of Economic Paleontologists and Mineralogists Special Publication,* Vol. 60. Tulsa, OK: SEPM, Chart 5.

Robaszynski, F. and Caron, M., 1995, Foraminiféres planctoniques du Crétacé: commentaire de la zonation Europe-Méditérranée. *Bulletin de la Société Géologique de France* **166**: 681–92.

Robaszynski, F., Caraon, M., Amédro, F., *et al.*, 1993, Sequence stratigraphy in a distal environment: the Cenomanian of the Kalaat Senan region (central Tunisia). *Bulletin des Centres de Recherches Exploration–Production Elf–Aquitaine* **17**: 395–433.

Robb, L. J., Davis, D. W., and Kamo, S. L., 1990, U–Pb ages on single detrital zircon grains from the Witwatersrand Basin, South Africa: constraints on the age of sedimentation and on the evolution of granites adjacent to the basin. *Journal of Geology* **98**(3): 311–28.

Roberts, J., Claoué-Long, J. C., and Jones, P. J., 1995a, Australian Early Carboniferous time. In W. A. Berggren, D. V. Kent, M.-P. Aubry, and J. Hardenbol, eds., *Geochronology, Time Scales and Global Stratigraphic Correlations: A Unified Temporal Framework for a Historical Geology, Society of Economic Paleontologists and Mineralogists Special Publication*, Vol. 54. Tulsa, OK: SEPM, pp. 23–40.

Roberts, J., Claoué-Long, J. C., Jones, P. J., and Foster, C. B., 1995b, SHRIMP zircon age control of Gondwanan sequences in Late Carboniferous Australia. In R. Dunay, and E. Hailwood, eds.,

Non-Biostratigraphical Methods of Dating and Correlation, Geological Society of London Special Publication, Vol. 89. London: Geological Society, pp. 145–74.

Roberts, J., Claoué-Long, J. C., and Foster, C. B., 1996, SHRIMP zircon dating of the Permian System of eastern Australia. *Australian Journal of Earth Sciences* **43**: 401–21.

Robinson, E. and Wright, R. M. J., 1993, Jamaican Paleogene larger foraminifera. *Geological Society of America Memoir* 182. Boulder, CO: Geological Society of America, pp. 283–545.

Robinson, S. G., Maslin, M. A., and McCave, I. N., 1995, Magnetic susceptibility variations in Upper Pleistocene deep-sea sediments of the NE Atlantic: implications for ice rafting and paleocirculation at the last glacial maximum. *Paleoceanography* **10**(2): 221–50.

Robison, R. A., 1994, Agnostid trilobites from the Henson Gletscher and Kap Stanton formations (Middle Cambrian), North Greenland. *Grønlands Geologiske Undersøgelse Bulletin* **169**: 25–77.

Rochon, A., De Vernal, A., Turon, J. L., Mathiessen, J., and Head, M. J., 1999, *Distribution of Recent Dinoflagellate Cysts in Surface Sediments from the North Atlantic Ocean and Adjacent Seas in Relation to Sea-surface Parameters, AASP Contributions Series*, Vol. 35. Dallas, TX: American Association of Stratigraphic Palynologists Foundation.

Roddick, J. C., 1983, High precision intercalibration of ^{40}Ar–^{39}Ar standards. *Geochimica et Cosmochimica Acta* **47**: 887–98.

1987, Generalized numerical error analysis with applications to geochronology and thermodynamics. *Geochimica et Cosmochimica Acta* **51**: 2129–35.

1988, The assessment of errors in ^{40}Ar/^{39}Ar dating. *Geological Survey of Canada Paper* 88–02. Ottawa: Geological Survey of Canada, pp. 3–8.

1996, Efficient mass calibration of magnetic sector mass spectrometers. *Current Research, Geological Survey of Canada* 1995-F. Ottawa: Geological Survey of Canada, pp. 1–9.

Roddick, J. C. and van Breemen, O., 1994, U–Pb zircon dating: a comparison of ion microprobe and single grain conventional analyses. *Current Research, Geological Survey of Canada* 1994-F. Ottawa: Geological Survey of Canada, pp. 1–9.

Roddick, J. C., Cliff, R. A., and Rex, D. C., 1980, The evolution of excess argon in Alpine biotites: a ^{40}Ar–^{39}Ar analysis. *Earth and Planetary Science Letters* **48**: 185–208.

Roddick, J. C., Loveridge, W. D., and Parrish, R. R., 1987, Precise U/Pb dating of zircon at the sub-nanogram Pb level. *Chemical Geology* **66**: 111–21.

Roden, M. K., Parish, R. R., and Miller, D. S., 1990, The absolute age of the Eifelian Tioga ash bed. *Journal of Geology* **98**: 282–5.

Rodionov, V. P., 1966, Dipole character of geomagnetic field in the Late Cambrian and Ordovician in the south of the Siberian Platform. *Geologiya i Geofizika* **1**: 94–101.

Roe, H. M., 2001, The Late Middle Pleistocene biostratigraphy of the Thames Valley, England: new data from eastern Essex. *Quaternary Science Reviews* **20**(16–17): 1603–19.

Rogers, R. R., Swisher III, C. C., Sereno, P. C. *et al.*, 1993, The Ischigualasto tetrapod assemblage (Late Triassic, Argentina) and ^{40}Ar/^{39}Ar dating of dinosaur origins. *Science* **260**: 794–7.

Röhl, U. and Ogg, J. G., 1996, Aptian–Albian sea level history from guyots in the western Pacific. *Paleoceanography* **11**: 595–624.

Röhl, U., Bralower, T. J., Norris, R. D., and Wefer, G., 2000, New chronology for the late Paleocene thermal maximum and its environmental implications. *Geology*. **28**: 927–30.

Röhl, U., Ogg, J. G., Geib, T., and Weber, G., 2001, Astronomical calibration of the Danian time scale. In D. Kroon, R. D. Norris, and A. Klaus, eds., *Western North Atlantic Paleogene and Cretaceous Palaeoceanography, Geological Society of London Special Publication*, Vol. 183. London: Geological Society, pp. 163–83.

Röhl, U., Norris, R. D., and Ogg, J. D., 2003, Cyclostratigraphy of upper Paleocene and lower Eocene sediments at Blake Nose site 1051 (western North Atlantic), *Geological Society of America Special Paper* 369 pp. 567–588. Boulder, CO: Geological Society of America.

Romein, A. J. T., 1979, Lineages in early Paleogene calcareous nannoplankton. *Utrecht Micropaleontological Bulletin* **22**: 1–231.

Ropolo, P., Conte, G., Gonnet, R., *et al.*, 1998, Les faunes d'Ammonites du Barrémien supérieur/Aptien inférieur (Bédoulien) dans la region stratotypique de Cassis-Le Bédoule (SE France): état des connaissances et propositions pour une zonations par Ammonites du Bédoulien-type. *Géologie Méditerranéenne* **25**: 167–75.

Roscoe, S. M., 1973, The Huronian Supergroup, a paleoaphebian succession showing evidence of atmospheric evolution, Huronian stratigraphy and sedimentation. *Geological Association of Canada Special Paper 12*. Geological Association of Canada, pp. 31–47.

Rosenkrantz, A., 1924, De Københavnske Grøbsandslag og deres placering i dem Danske lagrække. *Meddedelinger danske geologiske Foreningen* **6**: 3–39.

Rosovskaya, S. E., 1950, Triticites genus, its development and stratigraphic significance. *Transactions of the Paleontological Institute of the Academy of Sciences of the USSR* **26**: 5–78 [In Russian].

1975, Composition, phylogeny and system of the order Fusulinida, *Transactions of the Paleontological Institute of the Academy of Sciences of the USSR* **149**: 1–267 [In Russian].

Ross, C. A., 1979, Carboniferous. In R. A. Robinson and C. Teichert, eds., *Introduction, Part A, Treatise on Invertebrate Paleontology*. Boulder, CO: Geological Society of America/Lawrence, KS: Kansas University Press, pp. 254–290.

Ross, C. A. and Ross, J. R. P., 1985, Late Paleozoic depositional sequences are synchronous and worldwide. *Geology* **13**: 194–7.

1987, Biostratigraphic zonation of late Paleozoic depositional sequences. In C. A. Ross and H. Diew, eds., *Special Publication* 24. Ithaca, NY: Cushman Foundation for Foraminiferal Research, pp. 151–68.

1988, Late Paleozoic transgressive–regressive deposition. In C. K. Wilgus, B. S. Hastings, H. Posamentier, *et al.*, eds., *Sea Level Changes – an Integrated Approach, Society of Economic Paleontologists and Mineralogists Special Publication*, Vol. 42. Tulsa, OK: SEPM, pp. 227–47.

1994, Permian sequence stratigraphy and fossil zonation. In A. F. Embry, B. Beauchamp, and D. J. Glass, eds., *Pangea, Global Environments and Resources, Canadian Society of Petroleum Geologists Memoir* 17. Alberta: Canadian Society of Petroleum Geologists, pp. 219 231.

1995a, North American depositional sequences and correlations. In J. D. Cooper, M. L. Droser, and S. F. Finney, eds., *Ordovician Odyssey: Short Papers for the Seventh International Symposium on the Ordovician System*. Fullerton, CA: The Pacific Section for the Society for Sedimentary Geology (Society of Economic Paleontologists and Mineralogists), pp. 309–14.

1995b, Foraminiferal zonation of late Paleozoic depositional sequences. In M. R. Langer, J. H. Lipps, J. C. Ingle, and W. V. Sliter, eds., *Forams' 94* - selected papers from the Fifth international symposium of Foraminifera. *Marine Micropaleontology* **26**(1–4): 469–78.

1996, Silurian sea-level fluctuations. In B. J. Witzke, G. A. Ludvigson, and J. Day, eds., *Paleozoic Sequence Stratigraphy: Views from the North American Craton, Geological Society of America Special Paper 306*. Boulder, CO: Geological Society of America, pp. 187–92.

Ross, R. J., Naeser, C. W., Izett, G. A., *et al.*, 1982, Fission track dating of British Ordovician and Silurian stratotypes. *Geological Magazine* **119**(2): 135–53.

Ross, R. J., Hintze, L. F., Ethington, R. L., *et al.*, 1997, The Ibexian, lowermost series in the North American Ordovician. In M. E. Taylor, ed., *Early Paleozoic Biochronology of the Great Basin, Western United States, US Geological Survey Professional Paper* 1579. Reston, VA: US Geological Survey, pp. 1–50.

Rossignol-Strick, M. and Paterne, M., 1999, A synthetic pollen record of the eastern Mediterranean sapropels of the last 1 Ma: implications for the time-scale and formation of sapropels. *Marine Geology* **153**: 221–37.

Rossignol-Strick, M., Paterne, M., Bassinot, F., Emeis, K.-C., and De Lange, G. J., 1998, An unusual mid-Pleistocene monsoon period over Africa and Asia. *Nature* **392**: 269–72.

Rostovtsev, K. O. and Prozorowsky, V. A., 1997, Information on resolutions of standing commissions of the Interdepartmental Stratigraphic Committee (ISC) on the Jurassic and Cretaceous systems. *International Subcommission on Jurassic Stratigraphy Newsletter* **24**: 48–9.

Roth, P. H., 1970, Oligocene calcaieous nannoplankton biostratigraphy. *Ecologae Geological Helvetiae* **63**: 799–881.

Roth, P. H., Baumann, P., and Bertolino, V., 1971. Late Eocene–Oligocene calcareous nannoplankton from central and northern Italy. *Proceedings, II Planktonic Conference*, Vol. 2. Rome: International, pp. 1069–97.

Roveda, V., 1961, Contributo allo studio di alcune macroforaminiferi di Priabona. *Rivista Italiana di Paleontologia e Stratigrafia* **67**: 201–7.

Royer, D. L., Berner, R. A., and Beerling, D. J., 2001, Phanerozoic atmospheric CO_2 change: evaluating geochemical and paleobiological approaches. *Earth Science Reviews* **54**(4): 349.

Royer, J. Y. and Sandwell, D. T., 1989, Evolution of the Eastern Indian Ocean since the late Cretaceous: Constraints from GEOSAT altimetry. *Journal of Geophysical Research* **94**: 13 755–82.

Ruan, Y., 1996, Zonation and distribution of the early Devonian primitive ammonoids in South China. In H. Wang and X. Wang, eds., *Centennial Memorial Volume of Prof. Sun Yunzhu. Palaeontology and Stratigraphy*. China University of Geoscience Press, pp. 104–12.

Rubincam, D. P., 1994, Insolation in terms of Earth's orbital parameters. *Theoretical and Applied Climatology* **48**: 195–202.

1995, Has climate changed the Earth's tilt? *Paleoceanography* **10**: 365–72.

Ruddiman, W. F. and McIntyre, A., 1984, Ice-age thermal response and climatic role of the surface Atlantic Ocean, 40° N to 63° N. *Geological Society of America Bulletin* **95**: 381–96.

Ruddiman, W. F., Cameron, D., and Clement, B. M., 1987, Sediment disturbance and correlation of offset holes drilled with the hydraulic piston corer: Leg 94. In W. F. Ruddiman, R. B. Kidd, E. Thomas, *et al.*, eds., *Initial Reports of the Deep Sea Drilling Project*, Vol. 94. Washington, DC: US Govt. Printing Office, pp. 615–34.

Rudwick, M. J. S., 1985, *The Great Devonian Controversy: the Shaping of Scientific Knowledge among Gentlemanly Specialists*. Chicago, IL: University of Chicago Press, p. 194.

Ruedemann, R., 1904, *Graptolites of New York*, Part 1, *New York State Museum Memoir*, Vol. 7. Albany, NY: New York State Museum.

Ruget, C. and Nicollin, J.-P., 1998, Smaller benthic foraminifers. [column for Jurassic chart, Mesozoic and Cenozoic sequence chronostratigraphic framework of European basins, in Hardenbol, J. *et al.*]. In P.-C. de Graciansky, J. Hardenbol., Th. Jacquin, and P. R. Vail, eds., *Mesozoic–Cenozoic Sequence Stratigraphy of European Basins, Society of Economic Paleontologists and Mineralogists Special Publication,* Vol. 60. Tulsa, OK: SEPM, Chart 7.

Ruggieri, G. and Sprovieri, R., 1979, Selinunitiano, nuovo superpiano per il Pleistocene inferiore. *Bollettino della Societa Geologica Italiana* **96**: 797–802.

Ruggieri, G., Rio, D., and Sprovieri, R., 1984, Remarks on the chronostratigraphic classification of Lower Pleistocene. *Bollettino della Societa Geologica Italiana* **103**(2): 251–9.

Runnegar, B. and Saltzman, M. R., 1998, Global significance of the Late Cambrian Steptoean positive carbon isotope excursion (SPICE). In E. Landing and S. R. Westrop, eds., *Avalon 1997: The Cambrian Standard, New York State Museum Bulletin* 492. Albany, NY: State Education Department, p. 89.

Ruppel, S. C., James, E. W., Barrick, J. E., Nowlan, G., and Uyeno, T. T., 1996., High-resolution $^{87}Sr/^{86}Sr$ chemostratigraphy of the Silurian: implications for event correlation and strontium flux. *Geology* **24**: 831–4.

1998, High-resolution $^{87}Sr/^{86}Sr$ record: evidence for eustatic control of seawater chemistry? In E. Landing and M. E. Johnson, eds., *Silurian Cycles: Linkages of Dynamic Stratigraphy with Atmospheric, Oceanic and Tectonic Changes, New York State Museum Bulletin* 491. Albany, NY: State Education Department, pp. 285–95.

Rushton, A. W. A., 1982, The biostratigraphy and correlation of the Merioneth–Tremadoc Series boundary in North Wales. In M. G. Bassett and W. T. Dean, eds., *The Cambrian–Ordovician Boundary: Sections, Fossil Distributions, and Correlations, National Museum of Wales Geological Series*, Vol. 3. Cardiff: National Museum of Wales, pp. 41–59.

Rutten, M. G., 1959, Paleomagnetic reconnaissance of mid-Italian volcanoes. *Geologie en Mijnbouw* **21**: 373–4.

Ruzhenzev, V. E., 1936, New data on the stratigraphy of the Carboniferous and Lower Permian deposits of the Orenburg and Aktyubinsk regions. *Problems of Soviet Geology* **6**: 470–506.

1945, Suggestions for subdividing Upper Carboniferous into Stages. *Reports of the Academy of Sciences of the USSR* **46**(7): 314–17 [In Russian].

1950a, Upper Carboniferous ammonoids of the Urals. *Transactions of the Paleontological Institute of the Academy of Sciences of the USSR* **29**: 223 [In Russian].

1950b, Type section and biostratigraphy of the Sakmarian Stage. *Doklady Akademiya Nauk USSR, Nov. Ser.* **71**(6): 1101–4.

1954, Asselian Stage of the Permian System. *Doklady Akademiya Nauk USSR, Seriya Geologicheskaya* **99**(6): 1079–82.

1955, Main stratigraphic assemblages of ammonoids in the Permian System. *News of the Academy of Sciences of USSR, Series Biological* **4**: 120–32 [In Russian].

1965, The major ammonoid assemblages of the Carboniferous Period. *Paleontological Journal* **2**: 3–17 [In Russian]. Transliterated: Osnovnye kompleksy ammonoidey kamennougol'nogo perioda.

1974, Late Carboniferous ammonoids of the Russian Platform and Preurals. *Paleontological Journal* **8**(3): 311–323 [In Russian].

Ruzhenzev, V. E. and Bogoslovskaya, M. F., 1971, The Namurian Stage in the evolution of the Ammonoidea: early Namurian Ammonoidea. *Transactions of the Paleontological Institute of the Academy of Sciences of the USSR* **133**: 1–382 [In Russian].

1978, The Namurian Stage in the evolution of the Ammonoidea: late Namurian Ammonoidea. *Transactions of the Paleontological Institute of the Academy of Sciences of the USSR* **167**: 1–338 [In Russian].

Ryan, W. B. F., Bolli, H. M., Foss, G. N., *et al.*, 1978, Objectives, principal results, operations, and explanatory notes of Leg 40, South Atlantic. In H. M. Bolli, W. B. F. Ryan, *et al.*, eds., *Initial Reports of the Deep Sea Drilling Project*, Vol. 40. Washington, DC: US Govt. Printing Office, pp. 5–20.

Sadler, P. M., 1981, Sediment accumulation rates and the completeness of stratigraphic sections. *Journal of Geology* **89**: 569–84.

1999, *Constrained Optimization Approaches to the Stratigraphic Correlation and Seriation problems. A User's Guide and Reference Manual to the CONOP Program Family* (copyright 1998–1999 Peter M. Sadler). Riverside, CA: University of California, p. 142.

Sadler, P. M. and Cooper, R. A., 2003, Best-fit intervals and consensus sequences. In P. Harries, ed., *High Resolution Approaches in Stratigraphy*. New York: Plenum Press.

2004, Calibration of the Ordovician timescale for the IGCP Project 410. In B. D. Webby, M. L. Droser, and F. Paris, eds., *The Great Ordovician Biodiversfication Event*. New York: Columbia University Press.

2004b, Best-fit intervals and consensus sequences: comparison of the resolving power of traditional biostratigraphy and computer-assisted correlation. In P. J. Harris, ed., *High-Resolution Approaches in Stratigraphic Paleontology*. Dordrecht: Kluwer Academic, pp. 50–94.

Sadler, P. M. and Strauss, D. J., 1990, Estimation of completeness of stratigraphical sections using empirical data and theoretical models. *Journal of the Geological Society of London*. **147**: 471–85.

Sager, W. W., Weiss, C. J., Tivey, M. A., and Johnson, H. P., 1998, Geomagnetic polarity reversal model of deep-tow profiles from the Pacific Jurassic Quiet Zone. *Journal of Geophysical Research* **103**: 5269–86.

Sahagian, D., Pinous, O., Olferiev, A., and Zakharov, V., 1996, Eustatic curve for the middle Jurassic through Cretaceous based on Russian platform and Siberian stratigraphy: zonal resolution. *American Association of Petroleum Geologists Bulletin* **80**: 1433–58.

Salfeld, H., 1913, Certain Upper Jurassic strata of England. *Quarterly Journal of the Geological Society of London* **69**: 423–30.

1914, Die Gliederung des Oberen Jura in Nordwest Europa. *Neues Jahrbuch für Mineralogie, Geologie und Paläontologie, Beilage-Band* **32**: 125–246.

Salter, J. W., 1866, On the fossils of North Wales. In A. C. Ramsay, ed., *The Geology of North Wales, Memoirs of the Geological*

Survey of Great Britain 3. London: His Majesty's Stationery Office, pp. 239–363, 372–81, pls 1–26.

Saltzman, M. R., 2002, Carbon and oxygen isotope stratigraphy of the Lower Mississippian (Kinderhookian–lower Osagean), Western United States: implications for seawater chemistry and glaciation. *Geological Society of America Bulletin* **114**(1): 96–108.

2003, Late Paleozoic ice age: oceanic gateway or pCO_2? *Geology* **31**(2): 151–4.

Saltzman, M. R., Runnegar, B., and Lohmann, K. C., 1998, Carbon isotope stratigraphy of Upper Cambrian (Steptoean Stage) sequences of the eastern Great Basin: record of a global oceanographic event. *Geological Society of America Bulletin* **110**(3). 285–97.

Saltzman, M. R., Ripperdan, R. L., Brasier, M. D., *et al.*, 2000, A global carbon isotope excursion (SPICE) during the Late Cambrian: relation to trilobite extinctions, organic-matter burial and sea level. *Palaeogeography, Palaeoclimatology, Palaeoecology* **162**: 211–23.

Salvador, A. E., 1994. *International Stratigraphic Guide:– A Guide to Stratigraphic Classification, Terminology and Procedure*. Boulder, CO: Geological Society of America.

Samson, S. D. and Alexander, E. C., 1987, Calibration of the interlaboratory ^{40}Ar–^{39}Ar dating standard MMhb-1. *Chemical Geology* **66**: 27–34.

Samtleben, C., Munnecke, A., Bickert, T., and Pätzold, J., 1996, The Silurian of Gotland (Sweden): facies interpretation based on stable isotopes in brachiopod shells. *Geologische Rundschau* **85**: 278–92.

Samtleben, C., Munnecke, A., and Bickert, T., 2000, Development of facies and C/O-isotopes in transects through the Ludlow of Gotland: evidence for global and local influences on a shallow-marine environment. *Facies* **43**: 1–38.

Samuelsberg, T. J. and Pickard, N. A. H., 1999, Upper Carboniferous to Lower Permian transgressive–regressive sequences of central Spitsbergen, Arctic Norway. *Geological Journal* **34**(4): 393–411.

Sanchez-Goñi, M. F., Eynaud, F., Turon, J. L., and Shackleton, N. J., 1999, High resolution palynological correlation off the Iberian margin: direct land–sea correlation for the last interglacial complex. *Earth and Planetary Science Letters* **171**: 123–37.

Sandberg, C. A., 1979, Devonian and Lower Mississippian conodont zonation of the Great Basin and Rocky Mountains. In C. A. Sandberg and D. L. Clark, eds., *Conodont Biostratigraphy of the Great Basin and Rocky Mountains*, Vol. 26, *Geology Studies*. Provo, UT: Department of Geology, Brigham Young University, pp. 87–103.

Sandberg, C. A., Ziegler, W., Leuteritz, K., and Brill, S. M., 1978, Phylogeny, speciation, and zonation of Siphonodella (Conodonta, Upper Devonian and Lower Carboniferous). *Newsletters on Stratigraphy* **7**(2): 102–20.

Sando, W. J., 1990, Global Mississippian coral zonation. In P. L. Brenkle and W. L. Manger, eds., Intercontinental correlation and division of the Carboniferous System: contributions from the Carboniferous Subcommission meeting. *Courier Forschungsinstitut Senckenberg* **130**: 179–87.

Sanfilippo, A. and Blome, C. D., 2001, Biostratigraphic implications of mid-latitude Paleocene–Eocene radiolarian faunas from Hole 1051A, Ocean Drilling Program Leg 171B, Blake Nose, western North Atlantic. *Geological Society of London Special Publication 183*. London: Geological Society, pp. 185–224.

Sanfilippo, A. and Hull, D. M., 1999, Upper Paleocene–lower Eocene radiolarian biostratigraphy of the San Francisco de Paula Section, western Cuba: Regional and global comparisons. *Micropaleontology*, *Supplement 2*, **45**: 57–82.

Sanfilippo, A. and Nigrini, C., 1995, Radiolarian stratigraphy across the Oligocene/Miocene transition. *Marine Micropaleontology* **24**: 239–85.

1996, Radiolarian biomarkers at the Oligocene/Miocene boundary. In A. Moguilevsky and R. Whatley, eds., *Microfossils and Oceanic Environments, Proceedings of the 'ODP and the Marine Biosphere' International Conference*, Aberystwyth, April 19–21, 1994. Aberystwyth: University of Wales Press, pp. 317–26.

1998a, Upper Paleocene–Lower Eocene deep-sea radiolarian stratigraphy and the Paleocene/Eocene series boundary. In W. A. Berggren, S. G. Lucas, and M.-P. Aubry, eds., *Late Paleocene–Early Eocene Climatic and Biotic Events in the Marine and Terrestrial Records*. New York: Columbia University Press, pp. 244–76.

1998b, Code numbers for Cenozoic low latitude radiolarian biostratigraphic zones and GPTS conversion tables. *Marine Micropaleontology* **33**: 109–56.

Sanfilippo, A., Westberg-Smith, M. J., and Riedel, W. R., 1985, Cenozoic radiolaria. In H. M. Bolli, J. B. Saunders, and K. Perch-Nielsen, eds., *Plankton Stratigraphy*. Cambridge: Cambridge University Press, pp. 631–712.

Santarelli, A., 1997, Dinoflagellate cysts and astronomical forcing in the Mediterranean Upper Miocene. Unpublished Ph.D. thesis, Utrecht University, p. 141.

Santarelli, A., Brinkhuis, H., Hilgen, F. J., *et al.*, 1998, Orbital signatures in a late Miocene dinoflagellate record from Crete (Greece). *Marine Micropaleontology* **33**(3/4): 273–97.

Sanvoisin, R., D'Onofrio, S., Lucchi, R., Violanti, D., and Castradori, D., 1993, 1 Ma paleoclimatic record from the eastern Mediterranean–Marflux project: first results of a micropaleontological and sedimentological investigation of a long piston core from the Calabrian Ridge. *Il Quaternario* **6**(2): 169–88.

Satterley, A. K., 1996, The interpretation of cyclic successions of the Middle and Upper Triassic of the Northern and Southern Alps. *Earth Science Reviews* 40: 181–207.

Saunders, J. B., Bernoulli, D., Müller-Mertz, E., *et al.*, 1985, Stratigraphy of the late middle Eocene to early Oligocene in the Bath Cliff Section, Barbados, West Indies. *Micropaleontology* **30**: 390–425.

Savage, D. E. and Russell, D. E., 1977, Comments on mammalian paleontologic stratigraphy and geochronology: Eocene stages and mammal ages of Europe and North America. *Géobios Mémoires Spéciales* **1**: 47–56.

Scaillet, S., 2000, Numerical error analysis in $^{40}Ar/^{39}Ar$ dating. *Chemical Geology* **162**: 269–98.

Schärer, U., 1984, The effect of initial ^{230}Th disequilibrium on young U–Pb ages: the Makalu case, Himalaya. *Earth and Planetary Science Letters* **67**(2): 191–204.

Schaub, H., 1981, Nummulites et Assilines de la Téthys paléogène. *Mémoires Suisses de Paléontologie* **236**: 104–6.

Schaub, H., Benjamini, C., and Moshkovitz, S., 1995, The biostratigraphy of the Eocene of Israel. *Mémoires Suisses de Paléontologie* **117**: 58.

Schiappa, T. A., 1999, Lower Permian (Asselian–Sakmarian stratigraphy and biostratigraphy (ammonoid and conodont) of Novogafarovo and Kondurovsky, southern Ural Mountains, Russia Moscow. Unpublished Ph.D. thesis, University of Idaho, p. 295.

Schidlowski, M., 1993, The initiation of biological processes on Earth: summary of empirical evidence. In M. H. Engel and S. A. Macko, eds., *Organic Geochemistry*. New York: Plenum Press.

Schidlowski, M., Eichmann, R., and Junge, C. E., 1975, Precambrian sedimentary carbonates: carbon and oxygen geochemistry and implications for the terrestrial oxygen budget. *Precambrian Research* **2**: 1–69.

Schimper, W. P., 1874, *Traité de Paléontologie végétale III*. Paris: J. P. Baillière, p. 896.

Schindewolf, O. H., 1937, Zur Stratigraphie und Paläontologie der Wocklemer Schichten (Oberdevon). *Abhandlungen der Preussischen geologischen Landesanstalt, Neue Folge* **178**: 1–132.

Schindler, E., 1993, Event-stratigraphic markers within the Kellwasser crisis near the Frasnian/Famennian boundary (Upper Devonian) in Germany. *Palaeogeography, Palaeoclimatology, Palaeoecology* **104**: 115–25.

Schlanger, S. O. and Jenkyns, H. C., 1976, Cretaceous oceanic anoxic events: causes and consequences. *Geology en Mijnbouw* **55**: 179–84.

Schlanger, S. O., Arthur, M. A., Jenkyns, H. C., and Scholle, P. A., 1987, The Cenomanian–Turonian oceanic anoxic event. I. Stratigraphy and distribution of organic carbon-rich beds and the marine $\delta^{13}C$ excursion. In J. Brooks and A. J. Fleet, eds., *Marine Petroleum Source Rocks, Geological Society of London Special Publication 26*. London: Geological Society, pp. 371–99.

Schlüchter, C., 1992, Terrestrial Quaternary stratigraphy. *Quaternary Science Reviews* **11**: 603–7.

Schmidt, H., 1925, Die carbonischen Goniatiten Deutschlands. *Jarbuch preuß. geologischen Landesanst* **45**: 489–609.

Schmidt-Kittler, N., 1987, International Symposium on Mammalian Biostratigraphy and Paleoecology of the European Paleogene, Mainz, February 18–21. *Münchner Geowissenschaftliche Abhandlungen A* **10**: 1–312.

Schmitz, B., Aberg, G., Werdelin, L., Forey, P., and Bendix-Almgreen, S. E., 1991, $^{87}Sr/^{86}Sr$, Na, F, Sr and La in skeletal fish debris as a measure of the paleosalinity of fossil-fish habitats. *Geological Society of America Bulletin* **103**: 786–94.

Schmitz, B., Ingram, S. L., Dockery III, D. T., and Åberg, G., 1997, Testing $^{87}Sr/^{86}Sr$ as a paleosalinity indicator on mixed marine, brackish-water and terrestrial vertebrate skeletal apatite in late Paleocene–early Eocene near-coastal sediments, Mississippi. *Chemical Geology* **140**: 275–87.

Schmitz, B., Molina, E., and von Salis, K., 1998, The Zumaya section: a possible stratotype for the Selandian and Thanetian stages. *Newsletters on Stratigraphy* **36**: 35–42.

Schmitz, B., Pujalte, V., and Nuñez-Betelu, K., 2001, Climate and sea-level perturbations during the initial Eocene thermal maximum: evidence from siliciclastic units in the Basque Basin (Ermua, Zumaia and Trabakua Pass). *Palaeogeography, Palaeoclimatology, Palaeogeography* **165**: 299–320.

Schmitz, M. D. and Bowring, S. A., 2001, U–Pb zircon and titanite systematics of the Fish Canyon Tuff: an assessment of high-precision U–Pb geochronology and its application to young volcanic rocks. *Geochimica et Cosmochimica Acta* **65**(15): 2571–87.

Schneer, C. J., 1969, Introduction. In C. J. Schneer, ed., *Towards a History of Geology*. Cambridge, MA: Massachusetts Institute of Technology Press, pp. 1–18.

Schneider, D. A., 1995, Paleomagnetism of some Leg 138 sediments: detailing Miocene magnetostratigraphy. In N. Pisias, L. Mayer, T. Janecek, *et al.*, eds., *Proceedings of the Ocean Drilling Program, Scientific Results*, Vol. 138. College Station, TX: Ocean Drilling Program, pp. 59–72.

Schneider, D. A. and Kent, D. V., 1990, Ivory-coast microtektites and geomagnetic reversals. *Geophysical Research Letters* **17**: 163–66.

Schneider, J. W. and Roscher, M., 2002, Charts for the non-marine/marine stratigraphic correlation of the Carboniferous and Permian. Oberkarbon–*Untertrias in Zentraleuropa: Prozesse und ihr Timing Workshop*. Freiburg: TU Bergakademie Freiburg, Geologisches Institut, pp. 39–41.

Scholger, R., Mauritsch, H. J., and Brandner, R., 2000, Permian–Triassic boundary magnetostratigraphy from the Southern Alps (Italy). *Earth and Planetary Science Letters* **176**: 495–508.

Schönfeld, J., Schulz, M.-G. C., McArthur, J. M., *et al.*, 1996, New results on biostratigraphy, geochemistry and correlation from the standard section for the Upper Cretaceous white chalk of northern Germany (Lägerdorf–Kronsmoor–Hemmoor). In C. Spaeth, ed., *Mitteilung aus dem Geologisch-Paläontologischen Institut der Universität Hamburg 77* (Jost Wiedmann Memorial Volume); *Proceedings of the 4th International Cretaceous Symposium*, Hamburg, 1992, pp. 545–75.

Schönlaub, H. P., Attrep, M., Boeckelmann, K., *et al.*, 1992, The Devonian/Carboniferous boundary in the Carnic Alps – a multidisciplinary approach. *Jahresberbericht der Geologischen Bundesanstalt, Wien* **135**: 57–98.

Schouten, S., Van Kaam-Peters, H. M. E., Rijpstra, W. I. C., Schoell, M., and Sinninghe Damste, J. S., 2000, Effects of an oceanic anoxic event on the stable carbon isotopic composition of Early Toarcian carbon. *American Journal of Science* **300**: 1–22.

Schuuarman, W. M. N., 1979, Aspects of Late Triassic palynology. 3. Palynology of the latest Triassic and earliest Jurassic deposits of the Northern Limestone Alps in Austria and southern Germany, with special reference to a palynological characterization of the Rhaetian stage in Europe. *Review of Palaeobotany and Palynology* **27**: 53–75.

Schudack, M. E., Turner, C. E., and Peterson, F., 1998, Biostratigraphy, paleoecology, and biogeography of charophytes and ostracods from the Upper Jurassic Morrison Formation, Western Interior, USA. *Modern Geology* **22**: 379–414.

Schulz, H., von Rad, U., and Erlenkeuser, H., 1998, Correlation between Arabian Sea and Greenland climate oscillations of the past 110 000 years. *Nature* **393**: 54–7.

Schulz, H., Emeis, K.-C., Erlenkeuser, H., von Rad, U., and Rolf, C., 2002, The Toba volcanic event and interstadial/stadial climates at the marine isotopic stage 5 to 4 transition in the northern Indian Ocean. *Quaternary Research* **57**: 22–31.

Schwarz, W. H. and Lippolt, H. J., 2002, Coeval argon-40/argon-39 ages of moldavites from the Bohemian and Lusatian strewn fields. *Meteoritics and Planetary Science* **37**(12): 1757–63.

Schwarzacher, W., 1947, *Über die sedimentäre Rhythmik des Dachsteinkalkes von Lofer*. Wien: Verhandlungen der geologischen Bundesanstalt, pp. 175–88.

1954, Die Grossrhythmik des Dachsteinkalkes von Lofer. *Tschermaks mineralogische und petrographische Mitteilungen* **4**: 44–54.

1993, *Cyclostratigraphy and the Milankovitch Theory*, *Developments in Sedimentology*, Vol. 54. Amsterdam: Elsevier, p. 225.

Schweigert, G. and Callomon, W. J., 1997, Der bauhini-Faunenhorizont unde seine Bedeutung für die Korrelation zwischen tethyalem und subborealem Oberjura. *Stuttgarter Beiträge zur Naturkunde, Serie B (Geologie und Paläontologie)* **247**: 1–69.

Scotese, C. R., Boucot, A. J., and McKerrow, W. S., 1999, Gondwanan paleogeography and paleoclimatology. *Journal of African Earth Sciences* **28**(1): 99–144.

Sedgwick, A., 1838, A synopsis of the English series of stratified rocks inferior to the Old Red Sandstone – with an attempt to determine the successive natural groups and formations. *Proceedings of the Geological Society of London*, **1**(20): 270–9.

1847, On the classification of the fossiliferous slates of North Wales, Cumberland, Westmoreland and Lancashire. *Quarterly Journal of the Geological Society of London* **3**: 133–64.

1852, On the classification and nomenclature of the Lower Palaeozoic rocks of England and Wales. *Quarterly Journal of the Geological Society of London* **8**: 136–68.

Sedgwick, A. and Murchison, R. I., 1835, On the Cambrian and Silurian Systems exhibiting the order in which the older sedimentary strata succeeded each other in England and Wales. *London and Edinburgh Philosophical Magazine* **7**: 483–5.

1839, Stratification of the older stratified deposits of Devonshire and Cornwall. *Philosophical Magazine*, *Series 3* **14**: 241–60.

1840, On the physical structure of Devonshire, and on the subdivisions and geological relations of its older stratified deposits. *Transactions of the Geological Society of London*, *Series 2* **5**: 633–704.

Seguenza, G., 1868, La formation zancléenne, ou recherches sur une nouvelle formation tertiaire. *Bulletin de la Société Géologique de France, series 2* **25**: 465–86.

1879, Le formazioni terziarie nella provincia di Reggio Calabria. *Mem. R. Acc. Lincei, Cl. Sc. Fis. Mat. Nat., Ser. 3* **VI**: 1–446, tav. I–XVII, Roma.

Selby, D., Nesbitt, B. E., Muehlenbachs, K., and Prochaska, W., 2000, Hydrothermal alteration and fluid chemistry of the Endako porphyry molybdenum deposit, British Columbia. *Economic Geology* **95**: 183–201.

Selli, R., 1960, Il Messiniano Mayer-Eymar 1867. Proposta di un neostratotipo. *Giornale di Geologia* **28**: 1–34.

Semikhatova, S. V., 1934, Moscovian deposits in the lower and upper Povolzhie and position of the Moscovian Stage within general stratigraphic scale of the Carboniferous in the USSR. *Problems of Soviet Geology* **3**(8): 73–92 [In Russian].

Semikhatova, S. V., Einor, O. L., Kireeva, G. D., *et al.*, 1979, Bashkirian Stage in its type area of the Urals. In R. H. Wagner, A. C. Higgins, and S. V. Meyen, eds., *The Carboniferous of the USSR, Occasional Publications of Yorkshire Geological Society 4*. Leeds: Yorkshire Geological Society, pp. 83–98.

Sen, S., 1997, Magnetostratigraphic calibration of the European Neogene mammal chronology. *Palaeogeography, Palaeoclimatology, Palaeoecology* **133**: 181–204.

Sephton, M. A., Amor, K., Franchi, I. A., *et al.*, 2002a, Carbon and nitrogen isotope disturbances and an end-Norian (Late Triassic) extinction event. *Geology* **30**: 1119–22.

Sephton, M. A., Looy, C. V., Veefkind, R. J., *et al.*, 2002b, Synchronous record of $\delta^{13}C$ shifts in the oceans and atmosphere at the end of the Permian. In C. Koeberl and K. G. MacLeod, eds., *Catastrophic Events and Mass Extinctions: Impacts and Beyond, Geological Society of America Special Paper 356*. Boulder, CO: Geological Society of America, pp. 455–62.

Sepkoski, J. J., 1995, The Ordovician radiations: diversification and extinction shown by global genus-level taxonomic data. In J. D. Cooper, M. L. Droser, and S. F. Firrey, eds., *Ordovician Odyssey: Short Papers for the Seventh International Symposium on the Ordovician System*. Fullerton, CA: The Pacific Section for the Society for *Sedimentary Geology* (Society of Economic Paleontologists and Mineralogists), pp. 393–6.

Sepkoski, J. J. and Sheehan, P., 1983, Diversification, faunal change, and community replacement during the Ordovician radiations. In M. J. S. Tevesz and P. L. McCall, eds., *Biotic Interactions in Recent and Fossil Benthic Communities*. New York: Plenum Press, pp. 673– 717.

Séronie-Vivien, M., 1972, Contribution à létude de Sénonien en Aquitaine septentrionale, ses stratotypes. *Coniacien, Santonien, Campanien, Les Stratotypes Français*, Vol. 2. Centre National de la Recherche Scientifique, p. 195.

Serra-Kiel, J., Hottinger, L., Caus, E., *et al.*, 1998, Larger foraminiferal biostratigraphy of the Tethyan Paleocene and Eocene. *Bulletin de la Société Géologique de France* **169**: 281–99.

Seslavinski, K. B. and Maidanskaya, I. D., 2001, Global facies distributions from Late Vendian to Mid-Ordovician. In A. Yu. Zhuravlev and R. Riding, eds., *The Ecology of the Cambrian Radiation*. New York: Columbia University Press, pp. 47–68.

Sevastopulo, G. and Hance, L., 2000, Report of the Working Group to establish a boundary close to the existing Tournaisian–Visean boundary within the lower Carboniferous. *Newsletter on Carboniferous Stratigraphy* **18**: 6.

Sevastopulo, G., Devuyst, F. X., Hance, L., *et al.*, 2002, Progress report of the Working Group to establish a boundary close to the existing Tournaisian–Visean boundary within the Lower Carboniferous. *Newsletter on Carboniferous Stratigraphy* **20**: 6–7.

Shackleton, N. J., 1977, The oxygen isotope stratigraphic record of the Late Pleistocene. *Philosophical Transactions of the Royal Society of London B* **318/** *B* **280**: 169–82.

1989, The Plio–Pleistocene ocean: stable isotope history. In J. Rose and C. Schlüchter, eds., *Quaternary Type Sections: Imagination or Reality?* Rotterdam: Balkema, pp. 11– 24.

1995, New data on the evolution of Pliocene climatic variability. In E. S. Vrba, H. H. Denton, T. C. Partridge, and L. H. Burckle, eds., *Paleoclimate and Evolution with Emphasis on Human Origins*. New Haven, CT: Yale University Press, pp. 242–8.

1997, The deep-sea record and the Pliocene–Pleistocene boundary. In T. C. Partridge, ed., The Plio–Pleistocene boundary. *Quaternary International* **40**: 33–6.

Shackleton, N. J. and Crowhurst, S., 1997, Sediment fluxes based on an orbitally tuned time scale 5 Ma to 14 Ma, Site 926. In *Proceedings of the Ocean Drilling Program, Scientific Results*, Vol. 154. College Station, TX: Ocean Drilling Program, pp. 69–82.

Shackleton, N. J. and Hall, M. A., 1989, Stable isotope history of the Pleistocene at ODP site 677. In *Proceedings of the Ocean Drilling Program, Scientific Results*, Vol. 111. College Station, TX: Ocean Drilling Program, pp. 295– 316.

1997, The late Miocene stable isotope record, Site 926. In *Proceedings of the Ocean Drilling Program, Scientific Results*, Vol. 154. College Station, TX: Ocean Drilling Program, pp. 367–74.

Shackleton, N. J. and Opdyke, N. D., 1973, Oxygen isotope and paleomagnetic stratigraphy of equatorial Pacific core V28–238: oxygen isotope temperatures and ice volumes on a 10^5 and 10^6 year scale. *Journal of Quaternary Research* **3**: 39–55.

Shackleton, N. J., Backman, J., Zimmerman, H., *et al.*, 1984, Oxygen-isotope calibration of the onset of ice-rafting in DSDP site 552A: history of glaciation in the North Atlantic region. *Nature* **307**: 620–33.

Shackleton, N. J., Berger, A., and Peltier, W. A., 1990, An alternative astronomical calibration of the lower Pleistocene time scale based on ODP Site 677. *Transactions of the Royal Society of Edinburgh, Earth Sciences* **81**: 251–61.

Shackleton, N. J., Crowhurst, S., Hagelberg, T. K., Pisias, N. G., and Schneider, D. A., 1995a, A new late Neogene time scale: application to Leg 138 sites. In N. G. Pisias, L. A. Mayer, T. R. Janecek, A. Palmer-Julson, and T. H. van Andel, eds., *Proceedings of the Ocean Drilling Program, Scientific Results*, Vol. 138. College Station, TX: Ocean Drilling Program, pp. 73–101.

Shackleton, N. J., Hall, M. A., and Pate, D., 1995b, Pliocene stable isotope stratigraphy of Site 846. In N. G. Pisias, L. A. Mayer, T. R. Janecek, A. Palmer-Julson, and T. H. van Andel, eds., *Proceedings of the Ocean Drilling Program, Scientific Results*, Vol. 138. College Station, TX: Ocean Drilling Program, pp. 337–55.

Shackleton, N. J., Crowhurst, S. J., Weedon, G. P., and Laskar, J., 1999, Astronomical calibration of Oligocene–Miocene time. *Philosophical Transactions of the Royal Society of London, series A* **357**: 1907–29.

Shackleton, N. J., Hall, M. A., Raffi, I., Tauxe, L., and Zachos, J., 2000, Astronomical calibration age for the Oligocene–Miocene boundary. *Geology* **28**: 447–50.

Shackleton, N. J., Raffi, I., and Rohl, U., 2001, Astronomical age calibration in the Middle Miocence. *EOS, Transactions, American Geophysical Union*, Spring Meeting, San Francisco, Abstract.

Shackleton, N. J., Sanchez-Goni, M. F., Pailler, D., and Lancelot, Y., 2002, Marine isotope substage 5e and the Eemian interglacial. *Global and Planetary Change* **36**(3): 151–55.

Shanchi, P. and Robison, R. A., 2000, Agnostoid biostratigraphy across the Middle–Upper Cambrian boundary in Hunan, China. *Journal of Paleontology* **74**(4): 1–104.

Shangsi, P., Babcock, L. E., Robison, R. A., *et al.*, 2002, *Proposed Global Standard Stratotype Section and Point for the Paibian Stage and Furongian Series (Upper Cambrian)*. International Subcommission on Cambrian Stratigraphy, p. 19.

Shaw, A. B., 1964, *Time in Stratigraphy*: New York: McGraw-Hill, p. 365.

Sheehan, P. M., 2001, The Late Ordovician mass extinction. *Annual Reviews of Earth and Planetary Sciences* 29: 331–64.

Shen, Y., Buick, R., and Canfield, D. E., 2001, Isotopic evidence for microbial sulphate reduction in the early Archaean era. *Nature* **410**: 77–81.

Sheng, J. Z. and Jin, Y. G., 1994, Correlation of Permian deposits of China. In Y. G. Jin, J. Utting, and B. R. Wardlaw, eds., Permian stratigraphy, environments and resources. 1, Palaeontology and stratigraphy. *Palaeoworld* **4**: 14–113.

Shergold, J. H., 1969, Oryctocephalidae (Trilobita: Middle Cambrian) of Australia. *Bureau of Mineral Resources of Australia Bulletin* **104**: 66.

1972, Late Upper Cambrian trilobites from the Gola Beds, western Queensland. *Bulletin of the Bureau of Mineral Resources of Australia* **112**: 126.

1973, *Meneviella viatrix* sp. nov., a new conocoryphid trilobite from the Middle Cambrian of western Queensland. *Bulletin of the Bureau of Mineral Resources of Australia* **126**: 19–26.

1975, Late Cambrian and Early Ordovician trilobites from the Burke River Structural Belt, western Queensland, Australia. *Bureau of Mineral Resources of Australia Bulletin* **153** (1): 251; (2): 58 pls.

1980, Late Cambrian trilobites from the Chatsworth Limestone, western Queensland. *Bulletin of the Bureau of Mineral Resources of Australia* **186**: 111.

1982, Idamean (Late Cambrian) trilobites, Burke River Structural Belt, western Queensland. *Bulletin of the Bureau of Mineral Resources of Australia* **187**: 69.

1993, The Iverian Stage (Late Cambrian) and its subdivision in the Burke River Structural Belt, western Queensland. *BMR Journal of Australian Geology and Geophysics* **13**(4): 345–58.

1995, *Timescales* 1. *Cambrian*, Canberra: Record 1995/30. Australian Geological Survey p. 32.

1996, Cambrian. In G. C. Young, and J. R. Laurie, eds., *An Australian Phanerozoic Timescale*. Melbourne: Oxford University Press, pp. 63–76.

Shergold, J. H. and Laurie, J. R., 1997, Suborder Agnostina. In R. L. Kaesler, ed., *Treatise on Invertebrate Paleontology, Part O, Arthropoda 1, Trilobita, Revised*. Boulder, CO: Geological Society of America; Lawrence: KS, The University of Kansas, pp. 331–383.

Shergold, J. H. and Nicoll, R. S., 1992, Revised Cambrian–Ordovician boundary biostratigraphy, Black Mountain, western Queensland. In B. D. Webby and J. R. Laurie, eds., *Global Perspectives on Ordovician Geology*. Rotterdam: Balkema, pp. 81–92.

Shields, G. A., 1999, Towards a new calibration scheme for the Terminal Proterozoic. *Eclogae Geologicae Helvetiae* **92**: 221–233.

Shields, G. A. and Veizer, J., 2002, The Precambrian marine carbonate isotope database: version 1.1. *Geochemistry, Geophysics, Geosystems* 3(6):doc. 101029/2001GC00026.

Shields, G. A., Strauss, H., Howe, S. S., and Siegmund, H., 1999, Sulphur isotope compositions of sedimentary phosphorites from the basal Cambrian of China: implications for Neoproterozoic–Cambrian biogeochemical cycling. *Journal of the Geological Society of London* **156**: 943–55.

Shields, G. A., Carden, G. A. F., Veizer, J., *et al.*, 2003, Sr, C, and O isotope geochemistry of Ordovician brachiopods: a major isotopic event around the Middle–Late Ordovician transition. *Geochimica et Cosmochimica Acta* **67**: 2005–25.

Shik, S. M., Borisov, B. A., and Zarrina, E. P., 2002, About the project of the interregional stratigraphic scheme of the Neopleistocene of East European Platform and improving regional stratigraphic schemes, The Third All Russian Meeting on Quaternary Research. *Abstracts*. Smolensk: Geological Institute RAN/Smolensky Pedagogical University, pp. 125–9 [In Russian].

Shipboard Scientific Party, 1988, Sites 677 and 6.78. In K. Becker, *et al.*, eds., *Proceedings of the Ocean Drilling Program, Initial Reports,* Vol. 111. College Station, TX: Ocean Drilling Program, pp. 253–348.

Shipboard Scientific Party, 1990, Site 801: Pigafetta Basin, western Pacific. In *Initial Reports of the Deep Sea Drilling Project*, Vol. 129. Washington, DC: US Govt. Printing Office, pp. 91–170.

Shipboard Scientific Party, 1995, Leg 54 synthesis. In W. B. Curry, N. J. Shackleton, C. Richter, *et al.*, eds., *Proceedings of the Ocean Drilling Program, Initial Reports,* Vol. 154. College Station, TX: Ocean Drilling Program, pp. 421–2.

Shipboard Scientific Party, 1998, Site 1049: paleomagnetism section. In D. Kroon, R. D. Norris, A. Klaus, *et al.*, eds., *Proceedings of the Ocean Drilling Program, Initial Reports,* Vol. 171B. College Station, TX: Ocean Drilling Program, pp. 70–5.

Shipboard Scientific Party, 2002, Ocean Drilling Project Site 1218. In M. Lyle, P. A. Wilson, T. R. Janecek, *et al.*, eds., *Initial Reports of the Deep Sea Drilling Project* [Online] 199, www-odp.tamu.edu/publications/199_IR/chap_11/chap_11.htm.

Shipboard Scientific Party, 2004, Explanatory notes: biostratigraphy. In *Proceedings of the Ocean Drilling Program,*

Initial Reports, Vol. 207. College Station, TX: Ocean Drilling Program [Online: www-odp.tamu.edu/publications/207_IR].

Short, D. A., Mengel, J. G., Crowley, T. J., Hyde, W. T., and North, G. R., 1991, Filtering Milankovitch cycles by Earth's geography. *Quaternary Research* **35**: 157–73.

Sibrava, V., 1978, Isotopic methods in Quaternary geology. In G. V. Cohee, M. F. Glaessner, and H. D. Hedberg, eds., *Contributions to the Geologic Time Scale, Studies in Geology*, Vol. 6. Tulsa, OK: American Association of Petroleum Geologists, pp. 165–9.

Sierro, F. J., Hilgen, F. J., Krijgsman, W., and Flores, J. A., 2001, The Abad composite (SE Spain): a Messinian reference section for the Mediterranean and the APTS. *Palaeogeography, Palaeoclimatology, Palaeoecology* **168**: 141–69.

Sikora, P. and Bergen J., in press, Lower Cretaceous biostratigraphy of Ontong Java sites from DSDP Leg 30 and ODP Leg 192. In G. Fitton, J. Mahoney, P. Wallace, and S. Saunders, eds., *Origin and Evolution of the Ontong Java Plateau, Geological Society of London Special Publication.* London: Geological Society.

Sikora, P. J., Howe, R. W., Gale, A. S., and Stein, J. A., 2003, Comparative microfossil chronostratigraphy of the Coniacian–Santonian boundary from the Western Interior, USA (Austin and Niobara formations) and the UK Chalk. In M. Lamola, J. M. Pons, and A. Dhondt, eds., *Meeting on the Coniacian–Santonian Boundary, Abstracts,* Bilboa, September 13–17, 2002 [Online: ww.ehu.es/~gpplapam/congresos/csb/abstract. html#sikora. Submitted to *Cretaceous Research*].

Silberling, N. J. and Tozer, E. T., 1968, Biostratigraphic classification of the marine Triassic in North America. *Geological Society of America Special Paper* 10. Boulder, CO: Geological Society of America, pp. 1–63.

Simmons, M. D., Williams, C. L., and Hancock, J. M., 1996, Planktonic foraminifera across the Campanian/Maastrichtian boundary at Tercis, south-west France. *Newsletters on Stratigraphy* **34**: 65–80.

Sinemurian Boundary Working Group, 2000, Submission of East Quantoxhead (West Somerset, SW England) as the GSSP for the base of the Sinemurian Stage. In G. Bloos, ed., International Commssion on Stratigraphy Internal Document, 14 pp. and 15 figures.

Sinitsyna, Z. A. and Synitsin, I. I., 1987, *Biostratigraphy of the Bashkirian Stage at its Type Section*. Ufa: Bashkirian Branch of Academy of Sciences of USSR, Bashkirian Institute of Geology, p. 72 [In Russian].

Sircombe, K. N., Bleeker, W., and Stern, R. A., 2001, Detrital zircon geochronology and grain-size analysis of ~2800 Ma Mesoarchean proto-cratonic cover succession, Slave Province, Canada. *Earth and Planetary Science Letters* **189**(3–4): 207–220, www.elsevier.com/locate/epsl.

Sirocko, F., Garbe-Schönberg, D., McIntyre, A., and Molfino, B., 1996, Teleconnections between the subtropical monsoons and high-latitude climates during the last deglaciation. *Science* **272**: 526–9.

Skevington, D., 1963, A correlation of Ordovician graptolite-bearing sequences. *Geologiska Föreningens i Stockholm Förhandlingar* **85**: 298–319.

Skipp, B., 1969, Foraminifera. In E. D. McKee and R. C. Gutschick, eds., History of the Redwall limestone of northern Arizona. *Geological Society of America Memoir 114*. Boulder, CO: Geological Society of America, pp. 173–229.

Skjold, L. J., Van Veen, P. M., Kristensen, S.-E., and Rasmussen, A. R., 1998, Triassic sequence stratigraphy of the southwestern Barents Sea. In P.-C. de Graciansky, J. Hardenbol, T. Jacquin, and P. R. Vail, eds., *Mesozoic–Cenozoic Sequence Stratigraphy of European Basins, Society of Economic Paleontologists and Mineralogists Special Publication*, Vol. 60. Tulsa, OK: SEPM, pp. 651–66.

Skompski, S., Alekseev, A., Meischner, D., *et al.*, 1995, Conodont distribution across the Visean/Namurian boundary. *Courier Forschungsinstitut Senckenberg* **188**: 177– 209.

Sliter, W. V. and Leckie, R. M., 1993, Cretaceous planktonic foraminifers and depositional environments from the Ontong Java Plateau with emphasis on Sites 803 and 807. In W. H. Berger, L. W. Kroenke, L. A. Mayer, *et al.*, *Proceedings of the Ocean Drilling Program, Scientific Results*, Vol. 130. College Station, TX: Ocean Drilling Program, pp. 63–83.

Smith, A. G., 1993, Methods for improving the chronometric time-scale. In E. Hailwood and R. B. Kidd, eds., *High Resolution Stratigraphy*, *Geological Society of London Special Publication*, Vol. 70. London: Geological Society, pp. 9–25.

2001, Paleomagnetically and tectonically based global maps for Vendian to mid-Ordovician time. In R. Riding and A. Y. Zhuravlev, eds., *The Ecology of the Cambrian Radiation*. New York: Columbia University Press, pp. 11–46.

Smith, A. G., Smith, D. G., and Funnell, B. M., 1994, *Atlas of Mesozoic and Cenozoic Coastlines*. Cambridge: Cambridge University Press, p. 112.

Smith, P. E., Evensen, N. M., and York, D., 1993, First successful ^{40}Ar–^{39}Ar dating of glauconies: argon recoil in single grains of cryptocrystalline material. *Geology* **21**: 41–4.

Smith, P. E., Evensen, N. M., York, D., and Odin, G. S., 1998, Single-grain ^{40}Ar–^{39}Ar ages of glauconies: implications for the geologic time scale and global sea level variations. *Science* **279**: 1517–19.

Smith, P. E., Evensen, N. M., York, D., Szatmari, P., and Oliveira, D. C., 2001, Single-crystal ^{40}Ar–^{39}Ar dating of pyrite: no fool's clock. *Geology* **29**: 403–6.

Snelling, N. J., 1985, *The Chronology of the Geological Record*, *Geological Society of London Memoir* 10. Oxford: Blackwell Scientific Publications.

Sohl, L. E., Christie-Blick, N., and Kent, D. V., 1999, Paleomagnetic polarity reversals in Marinoan (ca. 600 Ma) glacial deposits of Australia: implications for the duration of low-latitude glaciation in Neoproterozoic time. *Geological Society of America Bulletin* **111**: 1120–39.

Solovieva, M. N., 1977, Zonal fusulinid stratigraphy of Middle Carboniferous of the USSR. *Questions of Micropaleontology* **19**: 43–67 [In Russian].

1986, Zonal fusulinid scale of Moscovian stage based on restudy of the type sections of intrastage subdivisions. *Questions of Micropaleontology* **28**: 3–23 [In Russian].

Sorgenfrei, T., 1964, *Report of the 21st International Geological Congress*, Norden, 1960, Vol. 28. pp. 254–277.

Southgate, P. N. and Shergold, J. H., 1991, Application of sequence stratigraphic concepts to Middle Cambrian phosphogenesis, Georgina Basin, Australia. *BMR Journal of Australian Geology and Geophysics* **12**(2): 119–44.

Spell, T. L. and McDougall, I., 2003, Characterization and calibration of ^{40}Ar–^{39}Ar dating standards. *Chemical Geology* **198**(3–4): 189–211.

Sprenger, A. and ten Kate, W. G., 1993, Orbital forcing of calcilutite–marl cycles in southeast Spain and an estimate for the duration of the Berriasian stage. *Geological Society of America Bulletin* **105**: 807–18.

Sprovieri, R., Di Stefano, E., Howell, M., *et al.*, 1998, Integrated calcareous plankton biostratigraphy and cyclostratigraphy at Site 964. In *Proceedings of the Ocean Drilling Program, Scientific Results*, Vol. 160. College Station, TX: Ocean Drilling Program, pp. 155–66.

Sprovieri, M., Caruso, A., Foresi, L. M., *et al.*, 2002a, Astronomical calibration of the upper Langhian/lower Serravallian record of Ras Il–Pellegrin section (Malta Island, central Mediterranean). In S. Iaccarino, ed., *Integrated Stratigraphy and Paleoceanography of the Mediterranean Middle Miocene, Rivista Italiana di Paleontologia e Stratigrafia, Milano 108*. Milan: Rivista Italiana di Paleontologia e stratigrafia, pp. 183–193.

Sprovieri, R., Bonomo, S., Caruso, A., *et al.*, 2002b, Integrated calcareous plankton biostratigraphy and biochronology of the Mediterranean Middle Miocene. In S. Iaccarino, ed., *Integrated Stratigraphy and Paleoceanography of the Mediterranean Middle Miocene, Rivista Italiana di Paleontologia e Stratigrafia, Milano 108*. Milan: Rivista Italiana di Paleontologia e Stratigrafia, pp. 337–53.

Spudis, P. D., 1999, The Moon. In J. K. Beatty, C. Collins Petersen, and A. Chaikin, eds., *The New Solar System*. Cambridge: Cambridge University Press, pp. 125– 40.

Srinivasan, M. S. and Kennett, J. P., 1981, Neogene planktonic foraminiferal biostratigraphy and evolution: equatorial to subantarctic, South Pacific. *Marine Micropaleontology* **6**: 499–533.

Stacey, J. S. and Kramers, J. D., 1975, Approximation of terrestrial lead isotope evolution by a two stage model. *Earth and Planetary Science Letters* **27**: 201–21.

Staerker, T. S., 1998, Quantitative calcareous nannofossil biostratigraphy of Pliocene and Pleistocene sediments from the Eratosthenes Seamount region in the Eastern Mediterranean. In *Proceedings of the Ocean Drilling Program, Scientific Results*, Vol. 160. College Station, TX: Ocean Drilling Program, pp. 83–98.

Stainforth, R. M., Lamb, J. L., Luterbacher, H. P., Beard, J. H., and Jeffords, R. M., 1975, Cenozoic planktonic foraminiferal zonation and characteristics of index forms. *The University of Kansas Paleontological Contributions Paper* 62. Lawrence, KS: University of Kansas, p. 425.

Standish, E. M., 1998, JPL planetary and Lunar ephemerides, DE405/LE405. JPL IOM 312. F-98–048, Pasadena.

Steemans, P., 1989, Etude palynostratigraphique du Devonien Inferieur dans l'ouest de l'Europe. *Mémoire Explication des Cartes Géologiques et Mineralogiques de Belgique* **27**: 1–453.

Steenbrink, J., Van Vugt, N., Hilgen, F. J., Wijbrans, J. R., and Meulenkamp, J. E., 1999, Sedimentary cycles and volcanic ash beds in the Lower Pliocene lacustrine succession of Ptolemais (NW Greece): discrepancy between $^{40}Ar/^{39}Ar$ and astronomical ages. *Palaeogeography, Palaeoclimatology, Palaeoecology* **152**: 283–303.

Steenbrink, J., Van Vugt, N., Kloosterboer-van Hoeve, M. L., and Hilgen, F. J., 2000, Refinement of the Messinian APTS from sedimentary cycle patterns in the lacustrine Lava section (Servia Basin, NW Greece). *Earth and Planetary Science Letters* **181**: 161–73.

Steenbrink, J., Kloosterboer van Hoeve, M. I., and Hilgen, F. J., 2003, Millennial-scale climate variations recorded in early Pliocene colour reflectance time series from the lacustrine Ptolemais Basin (NW Greece).*Global and Planetary Change* **36**(1–2): 47–75.

Steiger, R. H., and Jäger, E., 1977, Subcommission on Geochronology: convention on the use of decay constants in geo- and cosmo-chronology. *Earth and Planetary Science Letters* **36**: 359–62.

1978, Subcommission on geochronology: convention on the use of decay constants in geochronology and cosmochronology. In G. V. Cohee, M. F. Glaessner, and H. D. Hedberg, eds.,*Contributions to the Geologic Time Scale, American Association of Petroleum Geologists 6*. Tulsa. OK: American Association of Petroleum Geologists, pp. 67–71.

Steiner, M. B., 1988, Paleomagnetism of the Late Pennsylvanian and Permian, a test of the rotation of the Colorado Plateau. *Journal of Geophysical Research* **93**(3): 2201–15.

1994, Die neoproterozoischen Megalgen Südchinas. *Berliner geowissenschaftliche Abhandlungen* **E15**: 1–146.

Steiner, M. B. and Ogg, J. G., 1988, Early and Middle Jurassic magnetic polarity time scale. In R. B. Rocha and A. F. Soares, eds., *Second International Symposium on Jurassic Stratigraphy*,

Lisbon, September 1987. Lisbon: Instituto Nacional de Investigação Científica, pp. 1097–11.

Steiner, M. B., Ogg, J. G., Melendez, G., and Sequieros, L., 1986. Jurassic magnetostratigraphy. 2. Middle–Late Oxfordian of Aguilon, Iberian Cordillera, northern Spain. *Earth and Planetary Science Letters* **76**: 151–66.

Steiner, M. B., Ogg, J. G., and Sandoval, J., 1987, Jurassic magnetostratigraphy. 3. Bajocian–Bathonian of Carcabuey, Sierra Harana and Campillo de Arenas (Subbetic Cordillera, southern Spain). *Earth and Planetary Science Letters* **82**: 357–72.

Steiner, M. B., Ogg, J. G., Zhang, Z., and Sun, S., 1988, Paleomagnetism of the Permian/Triassic boundary in south China and the Early Triassic polarity time scale. *Journal of Geophysical Research* **94**: 7343–63.

Steiner, M. B., Morales, M., and Shoemaker, E. M., 1993, Magnetostratigraphy, biostratigraphic, and lithostratigraphic correlations in Triassic strata of the western United States: applications of paleomagnetism to sedimentary geology. *Society of Economic Paleontologists and Mineralogists Special Publication*, Vol. 49. Tulsa, OK: SEPM, pp. 41–57.

Steiner, M. B., Lucas, S. G., and Shoemaker, E. M., 1994, Correlation and age of the Upper Jurassic Morrison Formation from magnetostratigraphic analysis. In M. V. Caputo, J. A. Peterson, and K. J. Franczyk, eds., *Mesozoic Systems of the Rocky Mountain Region*. Denver: Society of Economic Paleontologists and Mineralogists Rocky Mountain Section, pp. 315–30.

Steininger, F. F., 1999, Chronostratigraphy, geochronology, and biochronology of the Miocene "European Land Mammal Mega-Zones (ELMMZ) and the Miocene Mammal-Zones (MN-Zones)." In G. E. Rössner and K. Heissig, eds., *The Miocene Land Mammals of Europe*. München: Pfeil, pp. 9–24.

Steininger, F. F., Berggren, W. B., Kent, D. V., *et al.*, 1996, Circum Mediterranean Neogene (Miocene and Pliocene) marine–continental chronologic correlations of European mammal units and zones. In R., Bernro, V. Fahlbusch, and H. W. Mittmann, eds., *The Evolution of Western Eurasian Neogene Mammal Faunas*. New York: Columbia University Press, pp. 64–77.

Steininger, F. F., Aubry, M. P., Berggren, W. A., *et al.*, 1997a, The global stratotype section and point (GSSP) for the base of the Neogene. *Episodes* **20**: 23–8.

Steininger, F. F., Aubry, M. P., Biolzi, M., *et al.*, 1997b, Proposal for the global stratotype section and point (GSSP) for the base of the Neogene (the Palaeogene/Neogene boundary). In A. Montanari, G. S. Odin, and R. Coccioni, eds., *Miocene Stratigraphy: An Integrated Approach*, *Development in Paleontology and Stratigraphy*, Vol. 15. Amsterdam: Elsevier, pp. 125–47.

Stemmerik, L., Nilsson, I., and Elvebakk, G., 1995, Gzhelian–Asselian depositional sequences in the western Barents Sea and North Greenland. In R. J. Steel, V. L. Felt, E. P. Johannessen, and C. Mathieu, eds., *Sequence Stratigraphy of the Northwest European Margin*, *NPF Special Publication* 5. Amsterdam: Elsevier, pp. 529–44.

Stepanov, D. L., 1973, The Permian System in the USSR. In A. Logan and L. V. Hills, eds., *The Permian and Triassic Systems and their Mutual Boundary*, *Canadian Society of Petroleum Geologists Memoir 2*. Calgary: Canadian Society of Petrolium Geologists, pp. 120–36.

Stephenson, F. R., 1997, *Historical Eclipses and Earth's Rotation*. Cambridge: Cambridge University Press.

Stern, R. A., 2001, A new isotopic and trace-element standard for the ion microprobe: preliminary thermal ionization mass spectrometry (TIMS), U–Pb and electron microprobe data. *Radiogenic Age and Isotopic Studies*, *Report 14*, *Current Research* 2001-F1. Ottawa: Geological Survey of Canada, p. 11.

Stern, R. A. and Amelin, Y., 2003, Assessment of errors in SIMS zircon U–Pb geochronology using a natural zircon standard and NIST SRM 610 glass. *Chemical Geology* **197**: 111–42.

Stern, R. A. and Bleeker, W., 1998, Age of the world's oldest rocks refined using Canada's SHRIMP: the Acasta Gneiss Complex, Northwest Territories, Canada. *Geoscience Canada* **25**(1): 27–31.

Steurbaut, E., 1992, Integrated stratigraphic analysis of lower Rupelian deposits (Oligocene) in the Belgian Basin. *Annales de la Société Géologique de Belgique* **115**: 287–306.

1998, High-resolution holostratigraphy of the Middle Paleocene to Early Eocene strata in Belgium and adjacent areas. *Palaeontographica Abteilung A* **247**: 91–156.

Stewart, K., Turner, S., Kelley, S., *et al.*, 1996, 3–D ^{40}Ar–^{39}Ar geochronology in the Parana continental flood basalt province. *Earth and Planetary Science Letters* **143**: 95–109.

Stitt, J. H., 1975, Adaptive radiation, trilobite paleoecology and extinction, Ptychaspidid Biomere, Late Cambrian of Oklahoma. *Fossils and Strata* **4**: 381–90.

Stoffler, D., Artemieva, N. A., and Pierazzo, E., 2002, Modeling the Ries–Steinheim impact event and the formation of the moldavite strewn field. *Meteoritics and Planetary Science* **37**(12): 1893–907.

Štorch, P., 1994, Graptolite biostratigraphy of the Lower Silurian (Llandovery and Wenlock) of Bohemia. *Geological Journal* **29**: 137–165.

Stott, L. D., Sinha, A., Thiry, M., Aubry, M.-P., and Berggren, W. A., 1996, Global $\delta^{13}O$ changes across the Paleocene–Eocene boundary: criteria for terrestrial–marine correlations. *Geological Society of London Special Publication*, Vol. 101. London: Geological Society, pp. 381–99.

Stover, L. E. and Hardenbol, J., 1993, Dinoflagellates and depositional sequences in the Lower Oligocene (Rupelian)

Boom Clay Formation, Belgium. *Bulletin de la Société belge de Géologie* **102**: 5–77.

Stover, L. E., Brinkhuis, H., Damassa, S. P., *et al.*, 1996, Mesozoic–Tertiary dinoflagellates, acritarchs and prasinophytes. In J. Jansonius and D. C. McGregor, eds., *Palynology: Principles and Applications*. Dallas, TX: American Association of Stratigraphic Palynologists Foundation, pp. 641–750.

Stradner, H., 1958, Die fossilen Discoasteriden Österreichs: 1. Teil. *Erdöl-Zeitschrift* **74**: 178–88.

1959a, First report on the discoasters of the Tertiary of Austria and their stratigraphic use. *Proceedings of the Fifth World Petroleum Congress*, Vol. 1. Chichester: Wiley, pp. 1080–95.

1959b, Die fossilen Discoasteriden Österreichs. 2.Teil. *Erdöl-Zeitschrift* **75**: 472–88.

Strauss, H., 1993, The sulfur isotopic record of Precambrian sulfates: new data and a critical evaluation of the existing record. *Precambrian Research* **63**: 225–46.

2001a, The sulfur isotopic composition of Precambrian sedimentary sulfides – seawater chemistry and biological evolution. In W. Altermann and P. Corcoran, eds., *Precambrian Sedimentary Environments: A Modern Approach to Ancient Depositional Systems, Special Publication of the International Association of Sedimentologists*, Vol. 33. Oxford: Blackwell, pp. 67–105.

2001b, 4 Ga of seawater evolution – evidence from sulfur isotopes. *Geological Society of America, Abstracts with Programs* **33**(6): 95.

Strauss, H., Banerjee, D. M., and Kumar, V., 2001, The sulfur isotopic composition of Neoproterozoic to early Cambrian seawater – evidence from the cyclic Hanseran Evaporites, NW India. *Chemical Geology* **175**: 17–28.

Streel, M., Higgs, K., Loboziak, S., Riegel, W., and Steemans, P., 1987, Spore stratigraphy and correlation with faunas and floras in the type marine Devonian of the Ardenne – Rhenish regions. *Review Palaeobotany and Palynology* **50**: 211–29.

Streel, M., Loboziak, S., Steemanns, P., and Bultynck, P., 2000a, Devonian miospore stratigraphy and correlation with the global stratotype sections and points. *Courier Forschungsinstitut Senckenberg* **220**: 9–23.

Streel, M., Caputo, M. V., Loboziak, S., and Melo, J. H. G., 2000b, Late Frasnian–Famennian climates based on palynomorph analyses and the question of the Late Devonian glaciations. *Earth Science Reviews* **52**(1–3): 121–73.

Stuckenberg, A. A., 1890, General geological map of Russia, Sheet 138: geological investigations in the northwestern area of the sheet. *Trudy Geol. Kom-ta* **4**(2).

Stucky, R. K., 1992, Mammalian faunas in North America of Bridgerian to early Arikareean "ages" (Eocene and Oligocene). In D. R. Prothero and W. A. Berggren, eds., *Eocene–Oligocene Climatic and Biotic Evolution*. Princeton, NJ: Princeton University Press, pp. 464–93.

Stuiver, M. and Braziunas, T. F., 1993, Modeling atmospheric ^{14}C influences and ^{14}C ages of marine samples back to 10 000 BC. *Radiocarbon* **35**: 137–89.

Stuiver, M. and Reimer, P. J., 1993, Extended ^{14}C database and revised CALIB 3.0 ^{14}C age calibration program. *Radiocarbon* **35**: 215–30.

Stuiver, M., Braziunas, T. F., Becker, B., and Kromer, B., 1991, Climatic, solar, oceanic and geomagnetic influences on late-glacial and Holocene atmospheric ^{14}C/^{12}C change. *Quaternary Research* **35**: 1–24.

Stuiver, M., Reimer, P. J., Bard, E., *et al.*, 1998, INTCAL98 radiocarbon age calibration, 24 000–0 cal BP. *Radiocarbon* **40**(3): 1041–84.

Stukalina, G. A., 1988, Studies in Paleozoic crinoid columnals and stems. *Palaeontographica Abteilung A, Palaeozoologie–Stratigraphie* **204**(1–3): 1–66.

Sturani, C., 1967, Ammonites and stratigraphy of the Bathonian in the Digne–Barrême area (SE France). *Bolletino della Società Paleontologica Italiana* **5**: 1–55.

Subcommission on Silurian Stratigraphy, 1995, Left hand column for correlation charts. *Silurian Times* **2**: 7–8.

Subbotina, N. N., 1953, Fossil Foraminifera of the USSR. Globigerinids, Hantkeninids and Globorotaliids. *Trudy Vsesoyuznovo Nauchno-Issledovatelskoo Geologo-Razvedochnovo Insituta (VNIGRI)* **76**: 296 [In Russian].

Suc, J. P., Bertini, A., Leroy, S. A. G., and Suballyova, D., 1997, Towards a lowering of the Pliocene/Pleistocene boundary to the Gauss/Matuyama Reversal. In T. C. Partridge, ed., The Plio–Pleistocene boundary. *Quaternary International* **40**: 37–42.

Sudbury, M., 1958, Triangulate monograptids from the *Monograptus gregarius* Zone (Lower Llandovery) in the Rheidol Gorge (Cardiganshire). *Philosophical Transactions of the Royal Society of London, series B* **241**: 485–554.

Sugarman, P. J., Miller, K. G., Bukry, D., and Feigenson, M. D., 1995, Uppermost Campanian–Maastrichtian strontium isotopic, biostratigraphic, and sequence stratigraphic framework of the New Jersey Coastal Plain. *Geological Society of America Bulletin* **107**: 19–37.

Suggate, R. P., 1974, When did the Last Interglacial end? *Quaternary Research* **4**: 246–52.

Suggate, R. P. and West, R. G., 1969, Stratigraphic nomenclature and subdivision in the Quaternary. Working group for Stratigraphic Nomenclature, INQUA Commission for Stratigraphy, unpublished discussion document.

Sugiyama, K., 1997, Triassic and Lower Jurassic radiolarian biostratigraphy in the siliceous claystone and bedded chert units of the southeastern Mino Terrane, Central Japan. *Bulletin of the Mizunami Fossil Museum* **24**: 79–193.

Sullivan, F. R., 1964, Lower Tertiary nannoplankton from the California Coast Ranges. I. Paleocene. *University of California Publications in Geological Sciences* **44**: 163–227.

1965, Lower Tertiary nannoplankton from the California Coast Ranges. II. Eocene. *University of California Publications in Geological Sciences* **53**: 1–74.

Sundberg, F. A., 1994, *Corynexochida and Ptychopariida (Trilobita, Arthropoda) of the Ehmaniella Biozone (Middle Cambrian), Utah and Nevada. Natural History Museum of Los Angeles County, Contributions in Science 446*. Los Angeles, CA: Natural History Museum of Los Angeles 137 pp., 93 figs.

Sutherland, P. K. and Manger, W. L., 1983, The Morrowan–Atokan (Pennsylvanian) boundary problem. *Geological Society of America Bulletin* **94**(4): 543–8.

Sutter, J. G., 1988, Innovative approaches to the dating of igneous events in the early Mesozoic basins of the Eastern United States. *US Geological Survey Bulletin 1776*. Boulder, CO: US Geological Survey, pp. 194–200.

Sweet, W. C., 1984, Graphic correlation of upper Middle and Upper Ordovician rocks, North American Midcontinental Province, USA. In D. L. Bruton, ed., *Aspects of the Ordovician System, Paleontological Contributions of the University of Oslo* 295. Oslo: University of Oslo, pp. 23–35.

1988, Mohawkian and Cincinnatian chronostratigraphy. *Bulletin of the New York State Museum* **462**: 84–90.

1995, A conodont-based composite standard for the North American Ordovician: progress report. In J. D. Cooper, M. L. Droser, and S. F. Finney, eds., *Ordovician Odyssey: Short Papers for the Seventh International Symposium on the Ordovician System*. Fullerton, CA: The Pacific Section for the Society for Sedimentary Geology (Society of Economic Paleontologists and Mineralogists), pp. 15–20.

Sweet, W. C. and Bergström, S. M., 1976, Conodont biostratigraphy of the Middle and Upper Ordovician of the United States Midcontinent. In M. G. Bassett, ed., *The Ordovician System: Proceedings of a Palaeontological Association Symposium*, Birmingham, September 1974. Cardiff: University of Wales Press and National Museum of Wales, pp. 121–51.

1984, Conodont provinces and biofacies of the Late Ordovician. *Geological Society of America Special Paper* 196. Boulder, CO: Geological Society of America, pp. 69–87.

1986, Conodonts and biostratigraphic correlation. *Annual Reviews of Earth and Planetary Sciences* **14**: 85–112.

Swisher, C. C., III, Grajales, N. J. M., Montanari, A., *et al.*, 1993, Coeval Ar–Ar ages of 65 million years ago from Chixculub crater melt-rock and Cretaceous–Tertiary boundary tektites. *Science* **257**: 954–8.

Takahashi, M. and Saito, K., 1997, Radiometric ages of the first occurrence of Globigerina nepenthes in the Tomioka sequence, central Japan. In A. Montanari, G. S. Odin, and R. Coccioni, eds., *Miocene Stratigraphy: An Integrated Approach, Development in Paleontology and Stratigraphy 15*. Amsterdam: Elsevier, pp. 381–93.

Takemura, A., 1992, Radiolarian Paleogene biostratigraphy in the southern Indian Ocean, Leg 120. In *Proceedings of the Ocean Drilling Program, Scientific Results*, Vol. 120. College Station, TX: Ocean Drilling Program, pp. 735–56.

Takemura, A. and Ling, H.-Y., 1997, Eocene and Oligocene radiolarian biostratigraphy from the Southern Ocean: correlation of ODP Legs 114 (Atlantic Ocean) and 120 (Indian Ocean). *Marine Micropaleontology* **30**: 97– 116.

Talent, J. A. and Yolkin, E. A., 1987, Transgression patterns for the Devonian of Australia and Southwestern Siberia. *Courier Forschungsinstitut Senckenberg* **92**: 235–49.

Talent, J. A., Mawson, R., Andrew, A. S., Hamilton, P. J., and Whitford, D. J., 1993, Middle Palaeozoic extinction events: faunal and isotopic data. *Palaeogeography, Palaeoclimatology, Palaeoecology* **104**: 139–52.

Tamaki, K. and Larson, R. L., 1988, The Mesozoic tectonic history of the Magellan Microplate in the western Central Pacific. *Journal of Geophysical Research* **93**: 2857–74.

Tanguy, J. C., Condomines, M., and Kieffer, G., 1997, Evolution of the Mount Etna magma: constraints on the present feeding system and eruptive mechanism. *Journal of Volcanology and Geothermal Research* **75**(3–4): 221–50.

Tankard, A. J., Eriksson, K. A., Hunter, D. R., *et al.*, 1982, *Crustal Evolution of Southern Africa: 3.8 Billion Years of Earth History*. Berlin: Springer-Verlag.

Tantawy, A. A., Ouda, K., Von Salis, K., and Saad El-Din, A., 2000, Biostratigraphy of Paleocene sections in Egypt. *Geologiska Föreningen i Stockholm Förhandlingar* **122**: 163–5.

Tarduno, J. A., 1990, A brief reversed polarity interval during the Cretaceous Normal Polarity Superchron. *Geology* **18**: 638–86.

1992, Magnetic susceptibility cyclicity and magnetic dissolution in Cretaceous limestones of the Southern Alps (Italy). *Geophysical Research Letters* **19**: 1515–18.

Tarduno, J A., Sliter, W. V., Kroenke, L. *et al.*, 1991, Rapid formation of Ontong Java Plateau by Aptian mantle volcanism. *Science* **254**: 399–403.

Tauxe, L. and Hartl, P., 1997, 11 million years of Oligocene geomagnetic field behaviour. *Geophysical Journal International* **128**: 217–29.

Tauxe, L., Opdyke, N. D., Pasini, G., and Elmi, C., 1983, Age of the Pliocene–Pleistocene boundary in the Vrica section, southern Italy. *Nature* **304**: 125–9.

Tauxe, L., Tucker, P., Peterson, N. P., and LaBrecque, J. L., 1984, Magnetostratigraphy of Leg 73 sediments. In *Initial Reports of the Deep Sea Drilling Project*, Vol. 73. Washington, DC: US Govt. Printing Office, pp. 609–21.

Tauxe, L., Gee, J., Gallet, Y., Pick, T., and Bown, T., 1994, Magnetostratigraphy of the Willwood Formation, Bighorn Basin, Wyoming: new constraints on the location of the Paleocene/Eocene boundary. *Earth and Planetary Science Letters* **125**: 159–172.

Tavera, J. M., Aguado, R., Company, M., and Oloriz, F., 1994, Integrated biostratigraphy of the *Durangites* and *Jacobi* Zones (J/K boundary) at the Puerto Escano section in southern Spain (Province of Cordoba). In E. Cariou and P. Hantzperque, eds., *Third International Symposium on Jurassic Stratigraphy*, Poitiers, September 22–29, 1991, *Géobios Mémoire Spéciale 17*, Vol. 1. Lyon: Université Claude Bernard, p. 469–76.

Taylor, J. F., 1997, Upper Cambrian biomeres and stages, two distinctly different and equally vital stratigraphic units. Abstracts with Program. In *2nd International Trilobite Conference*, August 1997, St. Catharine's, p. 47.

Taylor, S. P., Sellwood, B., Gallois, R., and Chambers, M. H., 2001, A sequence stratigraphy of Kimmeridgian and Bolonian stages, Wessex–Weald Basin. *Geological Magazine* **158**: 179–92.

Tedford, R. H., Galusha, T., Skinner, M. F., *et al.*, 1987, Faunal succession and biochronology of the Arikareean through Hemphillian interval (late Oligocene through earliest Pliocene epochs) in North America. In M. O. Woodburne, ed., *Cenozoic Mammals of North America*. Berkeley: University of California Press, pp. 152–210.

Teodorovich, G., 1949, On subdividing of the Upper Carboniferous into stages. *Reports of the Academy of Sciences of the USSR*, **67**(3): 537–540 [In Russian].

Thaler, L., 1965, Une échelle de zônes biochronologique pour les mammifères du Tertiaire de l'Europe. *Comptes Rendus Sommaires de la Société Géologique de France*. Paris: La Société Géologique de France p. 118.

Thiede, J., Winkler, A., Wolf-Welling, T., *et al.*, 1998, Late Cenozoic history of the polar north Atlantic: results from ocean drilling. *Quaternary Science Reviews* **17**: 185–208.

Thieuloy, J.-P., 1977, La zone à Callidiscus du Valanginien supérieur vocontien (Sud-Est de la France). Lithostratigraphie, ammonitofaune, limite Valanginien–Hautérivien, corrélations. *Géologie Alpine* **53**: 83–143.

Thirwall, M. F., 1988, Geochronology of late Caledonian magmatism in northern Britain. *Journal of the Geological Society of London* **145**: 951–67.

Thisted, R. A., 1988, *Elements of Statistical Computing*. New York: Chapman and Hall, p. 427.

Thomas, D. E., 1960, The zonal distribution of Australian graptolites. *Journal and Proceedings of the Royal Society of New South Wales* **94**: 1–58.

Thomas, E., 1998, Biogeography of the Late Paleocene benthic foraminiferal extinction. In M.-P. Aubry, S. G. Lucas, and W. A. Berggren, eds., *Late Paleocene–Early Eocene Climatic and Biotic Events in the Marine and Terrestrial Records*. New York: Columbia University Press, pp. 214–43.

Thomas, E. and Shackleton, N. J., 1996, The Paleocene–Eocene benthic foraminiferal extinction and stable isotope anomalies. *Geological Society of London Special Publication*, Vol. 101. London: Geological Society, pp. 401–41.

Thompson, L. G., Mosley-Thompson, E., Davis, M. E., *et al.*, 1995, Late glacial stage and Holocene tropical ice core records from Huascaran, Peru. *Science* **269**: 46–50.

Thomsen, E. and Heilmann-Clausen, C., 1985, The Danian–Selandian boundary at Svejstrup with remarks on the biostratigraphy of the boundary in western Denmark. *Bulletin of the Geological Society of Denmark* **33**: 341–62.

Thomsen, E., 1994, Calcareous nannofossil stratigraphy across the Danian–Selandian boundary in Denmark. *Geologiske Föreningen Förhandlingar* **116**: 65–7.

Thomsen, E. and Vorren, T. O., 1986, Macrofaunal paleoecology and stratigraphy in late Quaternary shelf sediments off northern Norway. *Palaeogeography, Palaeoclimatology, Palaeoecology* **56**: 103–50.

Thomson, D. J., 1990, Quadratic-inverse spectrum estimates: applications to palaeoclimatology. *Philosophical Transactions of the Royal Society of London, series A* **332**: 539–97.

1982, Spectrum estimation and harmonic analysis. *Proceedings of the IEEE* **70**: 1055–96.

Thorne, A. M. and Trendall, A. F., 2001, Geology of the Fortescue Group, Pilbara Craton, Western Australia. *Geological Survey of Western Australia Bulletin* **144**: 249.

Thorshoej Neilsen, A. T., 1992, Ecostratigraphy and the recognition of Arenigian (Early Ordovician) sea-level changes. In B. D. Webby and J. R. Laurie, eds., *Global Perspectives on Ordovician Geology*. Rotterdam: Balkema Press, pp. 355–66.

Thurmann, J., 1836, *Bulletin de la Société Géologique de France, Serie 1* **7**: 209.

Tiedemann, R. and Franz, S. O., 1997, Deep-water circulation, chemistry, and terrigenous sediment supply in the equatorial Atlantic during the Pliocene, 3.3–2.6 Ma and 5–4.5 Ma. In N. J. Shackleton, W. B. Curry, C. Richter, and T. J. Bralower, eds., *Proceedings of the Ocean Drilling Program, Scientific Results*, Vol. 154. College Station, TX: Ocean Drilling Program, pp. 299–318.

Tiedemann, R., Sarnthein, M., and Shackleton, N. J., 1994, Astronomical timescale for the Pliocene Atlantic $\delta^{18}O$ and dust flux records of ODP Site 659. *Palaeoceanography* **9**: 619–38.

Timofeev, B. V., 1966, *Mikropaleofitologicheskoe issledovanie drevnikh svit*. Moskva: Akademiya Nauk SSSR, Isdatelskvo Nauka Moskva, pp. 1–147 [Transl. 1974 British Library-Lending Division, Yorkshire].

Tipper, H. W., Carter, E. S., Orchard, M. J., and Tozer, E. T., 1994, The Triassic–Jurassic (T–J) boundary in Queen Charlotte Islands, British Columbia, defined by ammonites, conodonts and radiolarians. In E. Cariou and P. Hantzpergue, eds., *Third International Symposium on Jurassic Stratigraphy*, Poitiers, September 22–29, 1991, *Géobios Mémoire Spéciale 17*. Vol. 1. Lyon: Université Claude Bernard, pp. 485–92.

Titus, A. L., Webster, G. D., Manger, W. L., and Dewey, C. P., 1997, Biostratigraphic analysis of the Mid-Carboniferous Boundary,

South Syncline Ridge Section, Nevada Test Site (NTS), southern Nye County, Nevada, United States. In M. Podemski, S. Dybova-Jachowicz, J. Jureczka, and R. Wagner, eds., *Proceedings of the XIII International Congress on the Carboniferous and Permian*, Warszawa, Part 2, 157, Part 3. Warsaw: Wydawnictwa Prace Panstwowego Instytutu Geologicznego. Geologiczne, pp. 207–13.

Tjalsma, R. C. and Lohmann, G. P., 1983, Paleocene–Eocene bathyal and abyssal benthic foraminifera from the Atlantic Ocean. *Micropaleontology Special Publication* 4. p. 90.

Toghill, P., 1968, The graptolite assemblages and zones of the Birkhill Shales (Lower Silurian) at Dobb's Linn. *Palaeontology* **11**: 654–68.

Torrens, H. S., 1965, Revised zonal scheme for the Bathonian stage of Europe: *Reports of the Seventh Carpato–Balkan Geological Association Congress*, Sofia, Part 2, pp. 47–55.

Torsvik, T. H. and Trench, A., 1991, Ordovician magnetostratigraphy: Llanvirn–Caradoc limestones of the Baltic Platform. *Geophysical Journal International* **107**: 171–84.

Toumarkine, M. and Luterbacher, H. P., 1985, Paleocene and Eocene planktic foraminifera. In H. M. Bolli, J. B. Saunders, and K. Perch-Nielsen, eds., *Plankton Stratigraphy*. Cambridge: Cambridge University Press, pp. 87–154.

Townsend, H. A. and Hailwood, E. A., 1985, Magnetostratigraphy of Paleogene sediments in the Hampshire and London basins, southern UK. *Journal of the Geological Society of London* **142**: 957–82.

Tozer, E. T., 1965, Lower Triassic stages and ammonoid zones of Arctic Canada. *Geological Survey of Canada Paper* 65. Ottawa: Geological Survey of Canada, p. 12.

1967, A standard for Triassic time. *Geological Survey of Canada Bulletin* **156**: 104, 10 pls.

1984, *The Trias and its Ammonites: The Evolution of a Time Scale*, *Geological Survey of Canada Miscellaneous Report* 35. Ottawa: Geological Survey of Canada, p. 171.

1986, Definition of the Permian–Triassic (P–T) boundary: the question of the age of the Otoceras beds. *Memorie della Società Geologica Italiana* **34**: 291–301.

1990, How many Rhaetians? *Albertiana* **8**: 10–13.

1994a, Age and correlation of the Otoceras beds at the Permian–Triassic boundary. *Albertiana* **14**: 31–7.

1994b, Canadian Triassic ammonoid faunas. *Geological Survey of Canada Bulletin* **467**: 1–663.

Trapp, E. and Kaufmann, B., 2002, *Hochpräzise U–Pb Datierungen von Pyroklastika im Jungpaläozoikum*. Neustadt: Schwerpunktprogramm der deutschen Forschungsgemeinschaft DFG (SPP 1054), pp. 18–19.

Trapp, E., 2001, U-Pb–Geochronologie an unterkarbonischen Pyroklastika des Harzes: Der Schlüssel zu einer verbesserten Zeitskala des Unterkarbons. TU Braunschweig [Diss.]. v. 106 S.

Trapp, E., Mezger, K., Baumann, A., Wachendorf, H., and Scherer, E., in press, Time scale calibration by precise U–Pb single-zircon dating of silicified volcanic ashes in the Lower Carboniferous (Dinantian) of the Harz Mountains, Germany. *Journal of Geology*.

Traverse, A., 1988, *Paleopalynology*. Boston: Unwin Hyman, p. 600.

Trench, A., 1996, Magnetostratigraphy: Cambrian to Silurian. In G. C. Young and J. R. Laurie, eds., *An Australian Phanerozoic Timescale*. Melbourne: Oxford University Press, pp. 23–9.

Trench, A., McKerrow, W. S., Torsvik, T. H., Li, X., and McCracken, S. R., 1993, The polarity of the Silurian magnetic field: indications from global data compilation. *Journal of the Geological Society of London* **150**: 823– 31.

Trendall, A. F., 1991, Opinion: The "geological unit" (g.u.) – a suggested new measure of geologic time. *Geology* **19**(3): 195.

Trendall, A. F., de Laeter J. R., Nelson, D. R., and Hassler, S. W., 1998, Precise zircon U–Pb ages from the Marra Mamba Iron Formation and Wittenoom Formation, Hamersley Group, Western Australia. *Australian Journal of Earth Sciences* **45**(1): 137–42.

Tröger, K.-A. and, Kennedy, W. J., with contributions by Burnett, J. A., Caron, M., Gale, A. S., and Robaszynski, F., 1996, The Cenomanian Stage. *Bulletin de l'Institut Royal des Sciences Naturelles de Belgique, Sciences de la Terre* Supplement **66**: 57–68.

Truswell, E. M., 1997, Palynomorph assemblages from marine Eocene sediments on the west Tasmanian continental margin and the South Tasman Rise. *Australian Journal of Earth Science* **44**: 633–54.

Tucker, R. D., 1992, U–Pb dating of plinian-eruption ashfalls by the isotope dilution method: a reliable and precise tool for time-scale calibration and biostratigraphic correlation. *Abstracts and Programs* **24**: A198.

Tucker, R. D. and McKerrow, W. S., 1995, Early Paleozoic chronology: a review in light of new U–Pb zircon ages from Newfoundland and Britain. *Canadian Journal of Earth Sciences* **32**(4): 368–79.

Tucker, R. D., Krogh, T. E., Ross, R. J., and Williams, S. H., 1990, Time-scale calibration by high-precision U–Pb zircon dating of interstratified volcanic ashes in the Ordovician and lower Silurian stratotypes of Britain. *Earth and Planetary Science Letters* **1000**: 51–8.

Tucker, R. D., Bradley, D. C., Ver Straeten, C. A., *et al.*, 1998, New U–Pb zircon ages and the duration and division of Devonian time. *Earth and Planetary Science Letters* **158**(3–4): 175–86.

Turco, E., Hilgen, F. J., Lourens, L. J., Shackleton, N. J., and Zachariasse, W. J., 2001, Punctuated evolution of global climate cooling during the late Middle to early Late Miocene: high-resolution planktonic foraminiferal and oxygen isotope records from the Mediterranean. *Paleoceanography* **16**: 405–23.

Turco, E., Bambini, A. M., Foresi, L., *et al.*, 2003, Middle Miocene high-resolution calcareous plankton biostratigraphy at Site 926 (Leg 154, equatorial Atlantic Ocean): palaeoecological and palaeobiogeographical implications. *Géobios* **35**(1): 257–76.

Turner, C., 1998, Volcanic maars, long Quaternary sequences and the work of the INQUA Subcommission on European Quaternary Stratigraphy. *Quaternary International* **47/48**: 41–9.

Turner, C. and West, R. G., 1968, The subdivision and zonation of interglacial periods. *Eizeitalter und Gegenwart* **19**: 93–101.

Turner, G., Miller, J. A., and Grasty, R. L., 1966, The thermal history of the Bruderheim meteorite. *Earth and Planetary Science Letters* **1**: 155–7.

Turner, G., Huenke, J. C., Podosek, F. A., and Wasserburg, G. J., 1971, ^{40}Ar–^{39}Ar ages and cosmic ray exposure age of Apollo 14 samples. *Earth and Planetary Science Letters* **12**: 19–35.

Tzedakis, P. C., Andrieu, V., de Beaulieu, J. L., *et al.*, 1997, Comparison of terrestrial and marine records of changing climate of the last 500 000 years. *Earth and Planetary Science Letters* **150**: 171–6.

Underwood, C. J., 1993, The position of graptolites within Lower Palaeozoic planktic ecosystems. *Lethaia* **26**: 189–202.

Vai, G. B., 1997, Twisting or stable Quaternary boundary? A perspective on the glacial Late Pliocene concepts. *Quaternary International* **40**: 11–22.

Vai, G. B., Villa, I. M., and Colalongo, M. L., 1993, First direct radiometric dating of the Tortonian–Messinian boundary. *Comptes rendus de l'Académie des Sciences, Paris, Serie II, Sciences de la Terre et des Planétes* **316**: 1407–14.

Vail, P. R., Mitchum Jr., R. M., Todd, R. G., *et al.*, 1977, Seismic stratigraphy and global changes of sea level. In C. E. Payton, ed., *Seismic Stratigraphy – Applications to Hydrocarbon Exploration, AAPG Memoir* 26. Tulsa, OK: American Association of Petroleum Geologists, pp. 49–212.

Vakarcs, G., Hardenbol, J., Abreu, V. S., Vail, P. R., Várnai, P., and Tari, G., 1998, Oligocene–Middle Miocene depositional sequences of the Central Paratethys and their correlation with regional stages. *Society of Economic Paleontologists and Mineralogists Special Publication*, Vol. 60. Tulsa, OK: SEPM, pp. 209–231.

Van Buchem, F. S. P., McCave, I. N., and Weedon, G. P., 1994, Orbitally induced small-scale cyclicity in a siliciclastic epicontinental setting (Lower Lias, Yorkshire, UK). In P. L. de Boer and D. G. Smith, eds., *Orbital Forcing and Cyclic Sequences, International Association of Sedimentologists Special Publication,* Vol. 19. Oxford: Blackwell, pp. 345–66.

Van Couvering, J. A., 1995, Setting Pleistocene marine stages. *Geotimes* **40**: 10–11.

1997, The new Pleistocene. In J. A. Van Couvering, ed., *The Pleistocene Boundary and the Beginning of the Quaternary*. Cambridge: Cambridge University Press, pp. Preface, ii–xvii.

Van Couvering, J. A., Castradori, D., Cita, M. B., Hilgen, F. J., and Rio, D., 2000, The base of the Zanclean Stage and of the Pliocene Series. *Episodes* **23**(3): 179–87.

Van Dam, J. A., Alcalá, L., Alonso Zarza, A. M., *et al.*, 2001, The Upper Miocene mammal record from the Teruel–Alfambra region (Spain): the MN system and continental stage/age concepts discussed. *Journal of Vertebrate Paleontology* **21**(2): 367–85.

Van Dam, J. A., in press, European Neogene mammalian chronology: past, present and future. *Deinsia*.

VandenBerg, A. H. M. and Cooper, R. A., 1992, The Ordovician graptolite sequence of Australasia. *Alcheringa* **16**: 33–65.

VandenBerg, J. and Wonders, A. A. H., 1980, Paleomagnetism of Late Mesozoic pelagic limestones from the Southern Alps. *Journal of Geophysical Research* **85**(B7): 3623–7.

VandenBerg, J., Klootwijk, C. T., and Wonders, A. A. H., 1978, Late Mesozoic and Cenozoic movements of the Italian peninsula: further paleomagnetic data from the Umbrian sequence. *Geological Society of America Bulletin* **89**: 133–50.

Vandenberghe, N., Laenen, B., Van Echelpoel, E., and Lagrou, D., 1997, Cyclostratigraphy and climatic eustasy. Example of the Rupelian stratotype. *Comptes rendus de l'Académie des Sciences, Paris, Serie II, Sciences de la Terre et des Planètes* **325**: 305–15.

Vandenberghe, N., Laga, P., Steurbaut, E., Hardenbol, J., and Vail, P. R., 1998, Tertiary sequence stratigraphy at the southern border of the North Sea Basin in Belgium. *Society of Economic Paleontologists and Mineralogists Special Publication*, Vol. 60. Tulsa, OK: SEPM, pp. 119–54.

van den Boogaard, C., Dorfler, W., Glos, R., *et al.*, 2002, Two tephra layers bracketing late Holocene paleoecological changes in northern Germany. *Quaternary Research* **57**: 314–24.

van der Boogaard, M. and Bless, M. J. M., 1985, Some conodont faunas from the Aegiranum Marine Band. *Proceedings of the Koninklijke Nederlandse Akademie van Wetenschappen, Series B* **88**(2): 133–54.

Van Donk, J., 1976, A record of the Atlantic Ocean for the entire Pleistocene Epoch. *Geological Society of America Memoir* **145**: 147–63.

Van Echelpoel, E., 1994, Identification of regular sedimentary cycles using Walsh spectral analysis with results from the Boom Clay Formation, Belgium. In *Orbital Forcing and Cyclic Sequences, Special Publication of the International Association of Sedimentologists,* Vol. 19. Oxford: Blackwell, pp. 63–76.

van Heck, S. E. and Prins, B., 1987, A refined nannoplankton zonation for the Danian of the Central North Sea. *Abhandlungen der Geologischen Bundesanstalt* **39**: 353–88.

van Hinte, J. E., 1965, The type Campanian and its planktonic Foraminifera. *Proceedings of the Koninklijke Nederlandse Akademie van Wetenschappen, Series B* **68**: 8–28.

1976, A Jurassic time scale. *American Association of Petroleum Geologists Bulletin* **60**: 489–97.

van Hoof, A. A. M., 1993, Geomagnetic polarity transitions of the Gilbert and Gauss chrons recorded in marine marls from Sicily. *Geologica Ultraiectina* **100**: 125.

Van Kolfschoten, T. and Gibbard, P. L., 2000a, Introduction. *Geologie en Mijnbouw/Netherlands Journal of Geosciences* **79**.

2000b, The Eemian: local sequences, global perspectives – introduction. *Geologie en Mijnbouw/Netherlands Journal of Geosciences* **79**: 129–33.

van Kranendonk, M. J., Hickman, A. H., Smithies, R. H., *et al.*, 2002, Geology and tectonic evolution of the Archean North Pilbara Terrain, Pilbara Craton, Western Australia. *Economic Geology and the Bulletin of the Society of Economic Geologists* **97**(4): 695–732.

van Leckwijck, W. P., 1960, Report of the Subcommission on Carboniferous Stratigraphy. In *Compte Rendu 4ième Congrès International de Stratigraphie et de Géologie du Carbonifère 1958*, Heerlen, pp. 24–6.

van Leeuwen, R. J. W., Beets, D. J., Bosch, J. H. A., *et al.*, 2000, Amsterdam-Terminal borehole, the Netherlands. *Geologie en Mijnbouw/Netherlands Journal of Geosciences* **79**: 161–96.

Van Morkhoven, P. C. M., Berggren, W. A., and Edwards, A. S., 1986, Cenozoic cosmopolitan deep-water benthic Foraminifera. *Bulletin des Centers de Recherches Exploration-Production Elf-Aquitaine,* memoil **11**. 421 pp.

Van Simaeys, S., in press, The Rupelian–Chattian boundary in the North Sea Basin and its calibration to the international time-scale. *Netherlands Journal of Geosciences.*

Van Simaeys, S., De Man, E., Vandenberge, N., Brinkhuis, H., and Steurbaut, E., in press, Stratigraphic and paleoenvironmental analysis of the Rupelian–Chattian transition in the type region: evidence from dinoflagellate cysts, foraminifera and calcareous nannofossils. *Palaeogeography, Palaeoclimatology, Palaeoecology*.

Van Veen, P. M., 1995, Time calibration of Triassic/Jurassic microfloral turnover, eastern North America - Comment. *Tectonophysics* **245**: 93–5.

Van Veen, P., 1998, Ammonoids and pelecypods [columns for Triassic chart, Mesozoic and Cenozoic sequence chronostratigraphic framework of European basins, in Hardenbol, J. *et al.*]. In P.-C. de Graciansky, J. Hardenbol, Th. Jacquin, and P. R. Vail, eds., *Mesozoic–Cenozoic Sequence Stratigraphy of European Basins, Society of Economic Paleontologists and Mineralogists Special Publication,* Vol. 60. Tulsa, OK: SEPM, Chart 8.

Van Veen, P., Hochuli, P. A., and Vigran, J. O., 1998, Arctic spores/pollen [column for Triassic chart, Mesozoic and Cenozoic sequence chronostratigraphic framework of European basins, in Hardenbol, J. *et al.*]. In P.-C. de Graciansky, J. Hardenbol, Th. Jacquin, and P. R. Vail, eds., *Mesozoic–Cenozoic Sequence Stratigraphy of European Basins, Society of Economic Paleontologists and Mineralogists Special Publication,* Vol. 60. Tulsa, OK: SEPM, Chart 8.

Van Vugt, N., Steenbrink, J., Langereis, C. G., Hilgen, F. J., and Meulenkamp, J. E., 1998, Magneto-stratigraphy-based astronomical tuning of the early Pliocene lacustrine sediments of Ptolemais (NW Greece) and bed-to-bed correlation with the marine record. *Earth and Planetary Science Letters* **164**: 535–51.

Varol, O., 1989, Paleocene calcareous nannofossil biostratigraphy. In J. A. Crux and S. van Heck, eds., *Nannofossils and their Applications, British Micropaleontological Society Publication Series*. Leeds: British Micropaleontological Society, pp. 267–310.

1997, Paleogene. In P. R. Bowen, ed., *Calcareous Nannofossil Biostratigraphy, British Micropaleontological Society Publication Series*. Leeds: British Micropaleontological Society, pp. 200–24.

Veevers, J. J. and Powell, M., 1987, Late Paleozoic glacial episodes in Gondwanaland reflected in transgressive–regressive depositional sequences in Euramerica. *Geological Society of America Bulletin* **98**(4): 475–87.

Veizer, J., 1989, Strontium isotopes in seawater through time. *Annual Reviews of Earth and Planetary Sciences* **17**: 141– 67.

Veizer, J. and Compston, W., 1976, $^{87}Sr/^{86}Sr$ in Precambrian carbonates as an index of crustal evolution. *Geochimica et Cosmochimica Acta* **40**(8): 905–14.

Veizer, J., Compston, W., Hoefs, J., and Nielsen, H., 1982, Mantle buffering of the early oceans. *Naturwissenschaften* **69**: 173–80.

Veizer, J., Buhl, D., Diener, A., *et al.*, 1997, Strontium isotope stratigraphy: potential resolution and event correlation. *Palaeogeography, Palaeoclimatology, Palaeoecology* **132**: 65–77.

Veizer, J., Ala, D., Azmy, K., *et al.*, 1999, $^{87}Sr/^{86}Sr$, $\delta^{13}C$ and $\delta^{18}O$ evolution of Phanerozoic seawater. *Chemical Geology* **161**: 59–88, datasets posted at www.science.uottawa.ca/geology/isotope_data/.

Veizer, J., Godderis, Y., and François, L. M., 2000, Evidence for decoupling of atmospheric CO_2 and global climate during the Phanerozoic eon. *Nature* **408**: 698–701.

Veldkamp, A. and van den Berg, M. W., 1993, 3-dimensional modeling of Quaternary fluvial dynamics in a climo–tectonic dependent system – a case history of the Maas record (Maastricht, The Netherlands). *Global and Planetary Change* **8**: 203–18.

Verniers, J., Nestor, V., Paris, F., *et al.*, 1995, A global Chitinozoa biozonation for the Silurian. *Geological Magazine* **132**: 651–66.

Versteegh, G. J. M., 1994, Recognition of cyclic and non-cyclic environmental changes in the Mediterranean Pliocene: a palynological approach. *Marine Micropaleontology* **23**: 147–71.

1997, The onset of major northern hemisphere glaciations and their impact on dinoflagellate cysts and acritachs from the Singa section, Calabria (southern Italy) and DSDP Holes 607/607A (North Atlantic). *Marine Micropaleontology* **30**: 319–43.

Versteegh, G. J. M. and Zonneveld, C. A. F., 2002, Use of selective degradation to separate preservation from productivity. *Geology* **30**(7): 615–18.

Vervloet, C. C., 1966, *Stratigraphical and Micropaleontological Data on the Tertiary of Southern Piedmont (Northern Italy)*. Utrecht: Schotanus, pp. 1–88.

Vidal, G., 1976, Late Precambrian microfossils from the Visingsö beds in southern Sweden. *Fossils and Strata* **9**: 1–57.

Villa, E., 2001, Report of the Working Group to define a GSSP close to the Moscovian/Kasimovian boundary. *Newsletter on Carboniferous Stratigraphy* **19**: 8–11.

Villeneuve, M. E., Sandeman, H., and Davis, W. J., 2000, Intercalibration of U–Pb and ^{40}Ar/^{39}Ar ages in the Phanerozoic. *Geochimica et Cosmochimica Acta* **64**: 4017–30.

Vinken, R., von Daniels, C., Graham, F., *et al.*, 1988, The Northwest European Tertiary Basin: results of the International Geologic Correlation Programme Project No. 124. *Geologisches Jahrbuch* **A100**: 508.

Voigt, S. and Hilbrecht, H., 1997, Late Cretaceous carbon isotope stratigraphy in Europe: correlation and relations with sea level and sediment stability. *Palaeogeography, Palaeoclimatology, Palaeoecology* **134**: 39–59.

Vojacek, H. J., 1979, UNESCO geological world atlas. *Cartography* **11**: 32–9.

von Alberti, F. A., 1834, *Beitrag zu einer Monographie des Bunten Sandsteins, Muschelkalks und Keupers und die Verbindung dieser Gebilde zu einer Formation*. Stuttgart and Tübingen: Verlag der J. G. Cottaishen Buchhandlung, p. 326.

von Buch, L., 1839, *Über den Jura in Deutschland*. Berlin: Der Königlich Preussischen Akademie der Wissenschaften.

von Hillebrandt, A., 1994, The Triassic/Jurassic boundary and the Hettangian biostratigraphy in the area of the Utcubamba Valley (Northern Peru). In E. Cariou and P. Hantzpergue, eds., *Third International Symposium on Jurassic Stratigraphy*, Poitiers, September 22–29, 1991, *Géobios Mémoire Spéciale* 17. Lyon: Université Claude Bernard, pp. 297–307.

1997, Proposal for the Utcubamba Valley sections in northern Peru. (A proposal for base-Jurassic GSSP.) *International Subcommission on Jurassic Stratigraphy Newsletter* **24**: 21–5.

von Humboldt, F. W. H. A., 1799, *Über die unterirdischen Gasarten und die Mittel ihren Nachtheil zu vermindern*. Wiewag: Ein Beitrag zur Physik der Praktischen Bergbaukunde Braunschweig, p. 384.

von Salis, K., 1998, Calcareous nannofossils [column for Triassic and Cretaceous charts, Mesozoic and Cenozoic sequence chronostratigraphic framework of European basins, in Hardenbol, J. *et al.*]. In P.-C. de Graciansky, J. Hardenbol., Th. Jacquin, and P. R. Vail, eds., *Mesozoic–Cenozoic Sequence Stratigraphy of European Basins, Society of Economic Paleontologists and Mineralogists Special Publication*, Vol. 60. Tulsa, OK: SEPM, p. 779, Charts 5 and 8.

von Salis, K., Bergen, J., and de Kaenel, E., 1998, Calcareous nannofossils. [columns for Jurassic chart, Mesozoic and Cenozoic sequence chronostratigraphic framework of European basins, in Hardenbol, J. *et al.*]. In P.-C. de Graciansky, J. Hardenbol., Th. Jacquin, and P. R. Vail, eds., *Mesozoic–Cenozoic Sequence Stratigraphy of European Basins, Society of Economic Paleontologists and Mineralogists Special Publication,* Vol. 60. Tulsa, OK: SEPM, Chart 7.

von Zittel, K. A., 1901, *History of Geology and Paleontology*. London: W. Scott, 562 pp.

Vonhof, H. B., Smit, J., Brinkhuis, H., Montanari, A., and Nederbragt, A. J., 2000, Global cooling accelerated by early late Eocene impacts? *Geology* **28**: 687–90.

Vörös, A., Szabó, I., Kovács, S., Dosztály, L., and Budai, T., 1996, The Felsöörs section: a possible stratotype for the base of the Ladinian Stage. *Albertiana* **17**: 25–40.

Vrielynck, B., 1998, Conodonts [column for Triassic chart, Mesozoic and Cenozoic sequence chronostratigraphic framework of European basins, in Hardenbol, J. *et al.*]. In P.-C. de Graciansky, J. Hardenbol., Th. Jacquin, and P. R. Vail, eds., *Mesozoic–Cenozoic Sequence Stratigraphy of European Basins, Society of Economic Paleontologists and Mineralogists Special Publication,* Vol. 60. Tulsa, OK: SEPM, p. 780, Chart 8.

Waagen, W. and Diener, C., 1895, Untere Trias. In E. van Mojsisovics, W. von Waagen, and C. Diener, eds., *Entwurf einer Gliederung der pelagischen Sedimente des Trias-Systems*, Vol. 104. Vienna: Sitzungsberichtes der Akademie der Wissenschaften, pp. 1271–302.

Wagner, R. H., 1984, Megafloral zones of the Carboniferous. In *Compte Rendu Congress Internationale Stratigraphie et du Géologie du Carbonifère,* Vol. 9. pp. 109–34.

Wagner, R. H. and Winkler Prins, C. F., 1985, Stratotypes of the two lower Stephanian stages, Cantabrian and Barruelian Series. In *Compte Rendu Congrès International de Stratigraphie et de Géologie du Carbonifère* Vol. 4. pp. 473–85.

1994, General overview of Carboniferous stratigraphy. *Annales de la Société Géologique de Belgique* **116**(1): 163–74.

1997, Carboniferous chronostratigraphy: quo vadis? In M. Podemski, S. Dybova-Jachowicz, J. Jureczka, and R. Wagner, eds., *Proceedings of the XIII International Congress on the Carboniferous and Permian*, Warszawa, Vol. 157, Part 1, Prace Panstwowego Instytutu Geologicznego. Warsaw: Wydawnictwa Geologiczne, pp. 188–96.

Wahlman, G. P., Verville, G. J., and Sanderson, G. A., 1997, Biostratigraphic significance of the fusulinacean Protriticites in the Desmoinesian (Pennsylvanian) of the Rocky Mountains, Western USA. In C. A. Ross, J. Ross, and P. L. Brenckle, eds., *Late Paleozoic Foraminifera: Their Biostratigraphy, Evolution, and Paleoecology, and the Mid-Carboniferous Boundary. Special Publication, Cushman Foundation for Foraminiferal Research 36.*

Ithaca, NY: Cushuman Foundation for Foraminiferal Research, pp. 163–168.

Walliser, O. H., 1984a, Geological processes and global events. *Terra Cognita* **4**: 17–20.

1984b, Pleading for a natural D/C-boundary. In E. Paproth and M. Streel, eds., The Devonian–Carboniferous boundary. *Courier Forschungsinstitut Senckenberg*: 241–6.

1996, Global events in the Devonian and Carboniferous. In O. D. H. Walliser, ed., *Global Events and Event Stratigraphy in the Phanerozoic*. Berlin: Springer-Verlag, pp. 225– 50.

2000, The Eifelian–Givetian Stage boundary. *Courier Forschungsinstitut Senckenberg* **225**: 37–47.

Walliser, O. H., Bultynck, P., Weddige, K., Becker, R. T., and House, M. R., 1996, Definition of the Eifelian–Givetian Stage boundary. *Episodes* **18**: 107–15.

Wallmann, K., 2001, The geological water cycle and the evolution of marine δ^{18}O values. *Geochimica et Cosmochimica Acta* **65**: 2469–85.

Walsh, S. L., 2001, Notes on geochronologic and chronostratigraphic units. *Geological Society of America Bulletin* **113**: 704–13.

2003, Notes on geochronologic and chronostratigraphic units: reply. *Geological Society of America Bulletin* **115**: 1017–19.

Walter, M. R., Veevers, J. J., Calver, C. R., Gorjan, P., and Hill, A. C., 2000, Dating the 840–544 Ma Neoproterozoic interval by isotopes of strontium, carbon, and sulfur in seawater, and some interpretative models. *Precambrian Research* **100**: 371–433.

Wang, C. Y., 1993, Auxiliary stratotype sections for the global stratotype section and point (GSSP) for the Devonian–Carboniferous boundary: Nanbiancun. In M. Streel, G. Sevastopulo, and E. Paproth, eds., Devonian–Carboniferous boundary. *Annales de la Societe Geologique de Belgique* **115**(2): 707–8.

2000, The base of the Permian System in China defined by *Streptognathodus isolatus*. *Permophiles* **36**: 14–5.

Ward, P. and Orr, W., 1997, Campanian–Maastrichtian ammonite and planktonic foraminiferal biostratigraphy from Tercis, France: implications for defining the stage boundary. *Journal of Paleontology* **71**: 407–18.

Ward, P. D., Haggart, J. W., Carter, E. S., *et al.*, 2001, Sudden productivity collapse associated with the Triassic–Jurassic boundary mass extinction. *Science* **292**: 1148–51.

Wardlaw, B. R., 2000, Guadalupian conodont biostratigraphy of the Glass and Del Norte Mountains. In B. R. Wardlaw, R. E. Grant, and D. M. Rohr, eds., *The Guadalupian Symposium, Smithsonian Contributions to the Earth Sciences* 32. Washington, DC: Smithsonian Institution, pp. 37–87.

Wardlaw, B. R. and Davydov, V. I., 2000, Preliminary placement of the International Lower Permian Working Standard to the Glass Mountains, Texas. *Permophiles* **36**: 11–14.

Wardlaw, B. R. and Grant, R. E., 1987, Conodont biostratigraphy of the Cathedral Mountain and Road Canyon Formations, Glass Mountains, West Texas. In D. Cromwell and L. Mazzullo, eds., *The Leonardian Facies in W. Texas and S. E. New Mexico and Guidebook to the Glass Mountains, West Texas Permian Basin Section, Special Publication of the Society of Economic Geologists and Paleontologists* 87–27. Tulsa, OK: Society of Economic Geologists and Paleontologists, pp. 63–6.

1988, The Word–Capitan boundary (Permian, US regional standard section) records a eustatic sea level fall. *Geological Society of America Abstracts with Programs* **20**: A268.

Wardlaw, B. R. and Lambert, L. L., 1999, Evolution of *Jinogondolella* and the definition of Middle Permian stage boundaries. *Permophiles* **33**: 12–13.

Wardlaw, B. R. and Mei, S., 1998, *Clarkina* (conodont) zonation for the Upper Permian of China. *Permophiles* **31**: 3–4.

Wardlaw, B. R. and Pogue, K. R., 1995, The Permian of Pakistan. In P. A. Scholle, T. A. Peryt, and D. S. Ulmer-Scholle, eds., *The Permian of Northern Pangea*, Vol. 2. Heidelberg: Springer Verlag, pp. 215–24.

Wardlaw, B. R., Leven, E. Y., Davydov, V. I., Schiappa, T. A., and Snyder, W. S., 1999, The Base of the Sakmarian Stage: call for discussion (possible GSSP in the Kondurovsky Section, Southern Urals, Russia). *Permophiles* **34**: 19–26.

Wardlaw, B. R., Rudine, S. F., and Nestell, M. K., 2000, Conodont biostratigraphy of the Permian beds at Las Delicias, Coahuila, Mexico. In B. R. Wardlaw, R. E. Grant, and D. M. Rohr, eds., *The Guadalupian Symposium, Smithsonian Contributions to the Earth Sciences* **32**. Washington, DC: Smithsonian Institution, pp. 381–92.

Wardlaw, B. R., Boardman, D. R., II, and Nestell, M. K., 2003, Conodonts from the uppermost Carboniferous and lowermost Permian of Kansas. *Kansas Geological Survey Bulletin*, *Part A*. Lawrence, KS: Kansas Geological Survey.

Warrington, G., 1999, Triassic/Jurassic boundary WG, report by the secretary. *International Subcommission on Jurassic Stratigraphy Newsletter* **27**: 19–23.

2003, Triassic/Jurassic boundary Task Group, report by the secretary. *International Subcommission on Jurassic Stratigraphy Newsletter* **30**: 8–13.

Warrington, G. and Bloos, G., 2001, Triassic/Jurassic boundary Working Group. *International Subcommission on Jurassic Stratigraphy Newsletter* **28**: 3–4.

Warrington, G., Cope, J. C. W., and Ivimey-Cook, H. C., 1994, St Audrie's Bay, Somerset, England: a candidate global stratotype section and point for the base of the Jurassic System. *Geological Magazine* **131**: 191–200.

Waterhouse, H. K., 1995, High-resolution palynofacies investigation of Kimmeridgian sedimentary cycles. In M. R. House and A. S. Gale, eds., *Orbital Forcing Timescales and Cyclostratigraphy*, *Geological Society Special Publication*,

Vol. 85. London: Geological Society of London, pp. 75–114.

Waterhouse, J. B., 1976, World correlations for marine Permian faunas. *Queensland University Department of Geology Papers* 7(2): 232.

1982, An early Djulfian (Permian) brachiopod faunule from Upper Shyok Valley, Karakorum Range, and the implications for dating of allied faunas from Iran and Pakistan. *Contribution to Himalayan Geology* **2**: 188–233.

Watts, W. A., Allen, J. R. M., and Huntley, B., 1995, Vegetation history and palaeoclimate of the last glacial period at Lago Grande di Monticchio, southern Italy. *Quaternary Science Reviews* **15**: 133–53.

Webby, B. D., 1995, Towards an Ordovician timescale. In J. D. Cooper, M. L. Droser, and S. F. Finney, eds., *Ordovician Odyssey: Short Papers for the Seventh International Symposium on the Ordovician System*. Fullerton, CA: The Pacific Section for the Society for Sedimentary Geology, Society of Economic Paleontologists and Mineralogists. Tulsa, OK: SEPM, pp. 5–9.

1998, Steps towards a global standard for Ordovician stratigraphy. *Newsletters on Stratigraphy* **36**(1): 1–33.

Webby, B. D., Young, G. C., Talent, J. A., and Laurie, J. R., 2000, Palaeobiogeography of Australasian faunas and floras. *Association of Australasian Palaeontologists Memoir* **23**: 63–126.

Webby, B. D., Cooper, R. A., Bergström, S. M., and Paris, F., in press, Global, regional and zonal subdivisions and time slices. In B. D. Webby, M. Droser, and F. Paris, eds., *The Great Ordovician Biodiversification Event*. New York: Columbia University Press.

Webby, B. D., Droser, M. L., and Paris, F., in press, *The Great Ordovician Biodiversification Event*. New York: Columbia University Press.

Wedekind, D., 1918, Trilobiten und Ammonoideen aus der Entogonites nasutus-Zone (Unterkarbon) des Büchenbergsattels (Elbingeröder Komlex, Harz). Teil 2. *Geologie* **21**: 318–49.

Weedon, G. P., 1989, The detection and illustration of regular sedimentary cycles using Walsh power spectra and filtering, with examples from the Lias of Switzerland. *Journal of the Geological Society of London* **146**: 133–44.

2001, An astronomical time scale for the type Kimmeridge Clay, Late Jurassic, S. England, 2001. In S. Amodio, F. P. Buonocunto, and R. Sandulli, eds., *Abstract Volume of the Multidisciplinary Approach to Cyclostratigraphy*, Society of Economic Paleontologists and Mineralogists International Workshop, Sorrento, Italy, May 26–28, 2001. Napoli: Istituto di Ricerca Geomare Sud/National Research Council (CNR), pp. 54–5.

Weedon, G. P. and Jenkyns, H. C., 1999, Cyclostratigraphy and the Early Jurassic timescale: data from the Belemnite Marls, southern England. *Geological Society of America Bulletin* **111**: 1823–40.

Weedon, G. P., Shackleton, N. J., and Pearson, P. N., 1997, The Oligocene time scale and cyclostratigraphy on the Ceara Rise, western Equatorial Atlantic. In *Proceedings of the Ocean Drilling Program, Scientific Results*, Vol. 154. College Station, TX: Ocean Drilling Program, pp. 101–14.

Weedon, G. P., Jenkyns, H. C., Coe, A. L., and Hesselbo, S. P., 1999, Astronomical calibration of the Jurassic time-scale from cyclostratigraphy in British mudrock formations. *Philosophical Transactions of the Royal Society of London, series A* **357**: 1787–813.

Weedon, G. P., Coe, A. L., and Gallois, R. W., 2004, Cyclostratigraphy, orbital tuning and inferred productivity for the type Kimmeridgian Clay (Late Jurassic), Southern England. *Journal of the Geological Society London* **161**: 655–66.

Wei, W., 1994, Age conversion table for different time scales. *Journal Nannoplankton Research* **16**: 71–3.

1995, Revised age calibration points for the geomagnetic polarity time scale. *Geophysical Research Letters* **22**: 957–60.

Wei, W. and Pospichal, J. J., 1991, Danian calcareous nannofossil succession at Site 738 in the southern Indian Ocean. In *Proceedings of the Ocean Drilling Program, Scientific Results*, Vol. 119. College Station, TX: Ocean Drilling Program, pp. 495–512.

Wei, W. and Wise, S. W., 1990a, Middle Eocene to Pleistocene calcareous nannofossils recovered by Ocean drilling Program Leg 113 from the Weddell Sea. In *Proceedings of the Ocean Drilling Program, Scientific Results*, Vol. 113. College Station, TX: Ocean Drilling Program, pp. 639–66.

1990b, Biogeographic gradients of Middle Eocene–Oligocene calcareous nannoplankton in the South Atlantic Ocean. *Palaeogeography, Palaeoclimatology, Palaeoecology* **79**: 29–61.

Weissert, H. and Bréheret, J. G., 1991, A carbonate-isotope record from Aptian–Albian sediments of the Vocontian Trough (SE France). *Bulletin de la Société Géologique de France* **162**: 1133–40.

Weissert, H. and Channell, J. E. T., 1989, Tethyan carbonate carbon isotope stratigraphy across the Jurassic–Cretaceous boundary: an indicator of decelerated global carbon cycling? *Paleoceanography* **4**: 483–94.

Weissert, H. and Lini, A., 1991, Ice Age interludes during the time of Cretaceous greenhouse climate? In D. W. Muller, J. A. McKenzie, and H. Weissert, eds., *Controversies in Modern Geology*. London: Academic Press, pp. 173–91.

Weissert, H., Lini, A., Fölmi, K. B., and Kuhn, O., 1998, Correlation of Early Cretaceous carbon isotope stratigraphy and platform drowning events: a possible link? *Palaeogeography, Palaeoclimatology, Palaeoecology* **137**: 189–203.

Wenzel, B. and Joachimski, M. M., 1996, Carbon and oxygen isotopic composition of Silurian brachiopods

(Gotland/Sweden): palaeoceanographic implications. *Palaeogeography, Palaeoclimatology, Palaeoecology* **122**: 143–66.

West, R. G., 1968, *Pleistocene Geology and Biology* (2nd edition, 1977). London: Longmans.

1979, Further on the Flandrian. *Boreas* **8**: 126.

1984, Interglacial, interstadial and oxygen isotope stages. *Dissertationes Botaniceae* **72**: 345–57.

West, R. G. and Turner, C., 1968, The subdivision and zonation of interglacial periods. *Eiszeitalter und Gegenwart* **19**: 93–101.

Westergård, A. H., 1946, Agnostidea of the Middle Cambrian of Sweden. *Sveriges Geologiska Undersökning Series C 477, Årsbok* **40**(1): 140, 16 pls.

Westermann, G. E. G., 1988, Duration of Jurassic stages based on averaged and scaled subzones. In F. P. Agterberg and C. N. Rao, eds., *Recent Advances in Quantitative Stratigraphic Correlation*. Delhi: Hindustan Publ. Co., pp. 90–100.

Westermann, G. E. G. and Riccardi, A. C., 1972, Middle Jurassic ammonite fauna and biochronology of the Argentine–Chilean Andes. 1, Hildoceratacea. *Palaeontographica Abteilung A* **140**: 1–116.

Westrop, S. R., 1986, Trilobites of the Upper Cambrian Sunwaptan Stage, southern Canadian Rocky Mountains, Alberta. *Palaeontographica Canadiana* **3**: 179, 41 pls.

1995, Sunwaptan and Ibexian (Upper Cambrian–Lower Ordovician) trilobites of the Rabbitkettle Formation, Mountain River region, northern Mackenzie Mountains, northwest Canada. *Palaeontographica Canadiana* **12**: 75, 15 pls.

White, E. I., 1950, The vertebrate faunas of the Lower Old Red sandstone of the Welsh Borders. *Bulletin of the British Museum of Natural History (Geology)* **1**: 51–67.

Whitehouse, F. W., 1936, The Cambrian faunas of northeastern Australia. Part 1, Stratigraphic outline; Part 2, Trilobita (Miomera). *Memoirs of the Queensland Museum* **11**(1): 59–112, pls 8–10.

1939, The Cambrian faunas of northeastern Australia. Part 3, The Polymerid trilobites (with supplement No. 1). *Memoirs of the Queensland Museum* **11**(3): 179–282, pls 19–25.

1945, The Cambrian faunas of northeastern Australia. Part 5, The trilobite genus Dorypyge. *Memoirs of the Queensland Museum* **12**(3): 117–23, pl. 11.

Whittaker, A. and Green, G. W., 1983, *Geology of the Country around Weston-super-Mare*, Memoir for 1 : 50 000 geological sheet 279, New Series, with parts of sheets 263 and 295. Keyworth: Geological Survey of Britain; England and Wales.

Whittard, W. F., 1961, *Lexique Stratigraphique Internationale*, Vol. 1: *Europe. Fasc. 3a, Pays de Galles, Ecosse, V., Silurien*. Paris: Centre National de la Recherche Scientifique, p. 273.

Whittington, H. B., 1992, Trilobites, Vol xi. Rochester, NY: The Boydell Press, p. 145, 120 plates.

1997, Mode of life, habits and occurrence. In R. L. Kaesler, ed., *Treatise on Invertebrate Paleontology, Part O, Arthropoda 1, Trilobita, Revised*. Boulder, CO: Geological Society of America/Lawrence, KS: University of Kansas, pp. 137–69.

Whittington, H. B., Dean, W. T., Fortey, R. A., *et al.*, 1984, Definition of the Tremadoc series and the series of the Ordovician System in Britain. *Geological Magazine* **121**(1): 17–33.

Wiedenbeck, M., Allé, P., Corfu, F., *et al.*, 1995, Three natural zircon standards for U–Th–Pb, Lu–Hf and trace element and REE analyses. *Geostandards Newsletter* **19**: 1–23.

Wierzbowski, A., 2001, Kimmeridgian Working Group. *International Subcommission on Jurassic Stratigraphy Newsletter* **28**: 15–16.

2002, Kimmeridgian Working Group. *International Subcommission on Jurassic Stratigraphy Newsletter* **29**: 18–19.

2003, Kimmeridgian Working Group. *International Subcommission on Jurassic Stratigraphy Newsletter* **30**: 15–16.

Wiese, F. and Kaplan, U., 2001, The potential of the Lengerich section (Münster Basin, northern Germany) as a possible candidate global boundary stratotype section and point (GSSP) for the Middle/Upper Turonian boundary. *Cretaceous Research* **22**: 549–63.

Wignall, P. B., 2001, Large igneous provinces and mass extinctions. *Earth Science Reviews* **53**: 1–33.

Wignall, P. B. and Hallam, A., 1992, Anoxia as a cause of the Permo–Triassic mass extinction: facies evidence from northern Italy and the western United States. *Palaeogeography, Palaeoclimatology, Palaeoecology* **93**: 21–46.

Wilde, S. A., Valley, J. W., Peck, W. H., and Graham, C. M., 2001, Evidence from detrital zircons for the existence of continental crust and oceans on the Earth 4.4 Gyr ago. *Nature* **409**(6817): 175–78.

Willems, W., Bignot, G., and Moorkens, T., 1981, Ypresian. In C. Pomerol, ed., *Stratotypes of Paleogene Stages, Mémoire hors série 2 du Bulletin d'Information des Géologues du Bassin de Paris.* 267 pp.

Williams, D. F., Peck, J., Karabanov, E. B., *et al.*, 1997, Lake Baikal record of continental climate response to orbital insolation during the past 5 million years. *Science* **278**(5340): 1114–17.

Williams, E. A., Friend, P. F., and Williams, P. J., 2000, A review of Devonian time scales: databases, construction and new data. In P. F. Friend and B. P. J. Williams, eds., *New Perspectives on the Old Red Sandstone, Geological Society of London Special Publication,* Vol. 180. London: Geological Society, pp. 1–21.

Williams, G. L. and Bujak, J. P., 1985, Mesozoic and Cenozoic dinoflagellates. In H. M. Bolli, J. B. Saunders, and K. Perch-Nielsen, eds., *Plankton Stratigraphy*. Cambridge: Cambridge University Press, pp. 847–964.

Williams, G. L., Brinkhuis, H., Bujak, J. P., *et al.*, 1998b, Dinoflagellate cysts [Cenozoic portion of Mesozoic and Cenozoic sequence chronostratigraphic framework of European basins (Hardenbol, J., *et al.*)]. In P.-C. de Graciansky,

J. Hardenbol, Th. Jacquin, and P. R. Vail, eds., *Mesozoic–Cenozoic Sequence Stratigraphy of European Basins, Society of Economic Paleontologists and Mineralogists Special Publication,* Vol. 60. Tulsa, OK: SEPM, pp. 764–5, Chart 3.

Williams, G. L., Brinkhuis, H., Bujak, J. P., *et al.*, 1998a, Cenozoic Era – dinoflagellates. Basins. *Society of Economic Paleontologists and Mineralogists Special Publication,* Vol. 60. Tulsa, OK: SEPM, pp. 764–65.

Williams, G. L., Brinkhuis, H., Pearce, M. A., Fensome, R. A., and Weegink, J. W., in press, Southern Ocean and global dinoflagellate cyst events compared: index events for the Late Cretaceous–Neogene. In N. F. Exon, J. P. Kennett, M. J. Halone, *et al.*, eds., *Proceedings of the Ocean Drilling Program, Scientific Results*, Vol. 189. College Station, TX: Ocean Drilling Program.

Williams, H. S., 1891, *Correlation Papers: US Geological Survey Bulletin. Devonian and Carboniferous.* Reston, VA: US Geological Survey, pp. 1–279.

Williams, S. H., 1983, The Ordovician–Silurian boundary graptolite fauna of Dobb's Linn, southern Scotland. *Palaeontology* **26**: 605–39.

Williams, S. H. and Bruton, D. L, 1983, The Caradoc–Ashgill boundary in the central Oslo Region and associated graptolite faunas. *Norsk Geologisk Tidsskrift* **63**: 147–191.

Williamson, M. A., 1987, Quantitative biozonation of the Late Jurassic and Early Cretaceous of the East Newfoundland Basin. *Micropaleontology* **33**: 37–65.

Williamson, T. E., 1996, The beginning of the Age of Mammals in the San Juan Basin, New Mexico: biostratigraphy and evolution of Paleocene mammals of the Nacimiento Formation. *Bulletin of the New Mexico Museum of Natural History and Science* **8**: 1–141.

Wilpshaar, M., 1995, Direct stratigraphic correlation of the Vercors carbonate platform in SE France with the Barremian stratotype by means of dinoflagellate cysts. *Cretaceous Research* **16**: 273–81.

Wilpshaar, M., Santarelli, A., Brinkhuis, H., and Visscher, H., 1996, Dinoflagellate cysts and mid-Oligocene chronostratigraphy in the central Mediterranean region. *Journal of the Geological Society London* **153**: 553–61.

Wilson, D. S., 1993, Confirmation of the astronomical calibration of the magnetic polarity time scale from sea-floor spreading rates. *Nature* **364**: 788–90.

Wilson, G. J., 1984, New Zealand Late Jurassic to Eocene dinoflagellate biostratigraphy. *Newsletters on Stratigraphy* **13**: 104–17.

Wilson, G. S., Roberts, A. P., Verosub, K. L., Florindo, F., and Sagnotti, L., 1998, Magnetobiostratigraphic chronology of the Eocene–Oligocene transition in the CIROS-1 core, Victoria Land margin, Antarctica: implications for Antarctic glacial history. *Geological Society of America Bulletin* **110**: 35–47.

Wilson, G. S., Lavelle, M., McIntosh, W. C., *et al.*, 2002, Integrated chronostratigraphic calibration of the Oligocene–Miocene boundary at 24.0 ± 0.1 Ma from the CRP-2A drill core, Ross Sea, Antarctica. *Geology* **30**: 1043–46.

Windley, B. F., 1984, The Archaean–Proterozoic boundary. *Tectonophysics* **105**(1–4): 43–53.

1995, *The Evolving Continents.* New York: Wiley, p. 526.

Windley, B. F., Naqvi, S. M., Gupta, H. K., and Balakrishna, S., 1984, The Archaean–Proterozoic boundary. Lithosphere: structure, dynamics and evolution. *Tectonophysics*, **105**(1–4): 43–53.

Wing, S. L. and Sues, H.-D. R., 1992, Mesozoic and early Cenozoic terrestrial ecosystems. In A. K. Behrensmeyer, J. D. Damuth, W. A. DiMichele, *et al.*, eds., *Terrestrial Ecosystems through Time: Evolutionary Paleoecology of Terrestrial Plants and Animals.* Chicago: University of Chicago Press, pp. 324–416.

Wing, S. L., Bown, T. M., and Obradovitch, J. D., 1991, Early Eocene biotic and climatic change in interior western North America. *Geology* **19**: 1189–92.

Wing, S. L., Bao, H., and Koch, P. L., 2000, An early Eocene cool period? Evidence for continental cooling during the warmest parts of the Cenozoic. In B. T. Huber, *et al.*, eds., *Warm Climates in Earth History.* Cambridge: Cambridge University Press, pp. 197–237.

Wingate, M. T. D., 2000, Ion microprobe U–Pb zircon and baddeleyite ages for the Great Dyke and its satellite dykes, Zimbabwe. *South African Journal of Geology* **103**(1): 74–80.

Winograd, I. J., Coplen, T. B., Landwehr, J. M., *et al.*, 1992, Continuous 500 000-year climate record from vein calcite in Devils Hole, Nevada. *Science* **258**: 255–60.

Wise, S. W., 1983, Mesozoic and Cenozoic calcareous nannofossils recovered by Deep Sea Drilling Project Leg 71 in the Falkland Plateau region, Southwest Atlantic Ocean. In *Initial Reports of the Deep Sea Drilling Project,* Vol. 71. Washington, DC: US Govt. Printing Office, pp. 481–550.

Woillard, G. M., 1978, Grande Pile peat bog: a continuous pollen record for the past 140 000 years. *Quaternary Research* **9**: 1–21.

Woldstedt, P., 1962, Über die Bennenung einiger Unterabteilungen des Pleistozäns. *Eiszeitalter und Gegenwart* **3**: 14–18.

Wolf-Welling, T. C. W., Cremer, M., O'Connell, S., Winkler, A., and Thiede, J., 1996, Cenozoic Arctic Gateway paleoclimate variability. Indications from changes in coarse-fraction composition (ODP Leg 151). In J. Thiede, A. M. Myhre, J. Firth, *et al.*, eds., *Proceedings of the Ocean Drilling Program, Scientific Results,* Vol. 151. College Station, TX: Ocean Drilling Program, pp. 515–67.

Won, M.-Z., 1998, A Tournaisian (Lower Carboniferous) radiolarian zonation and radiolarians of the *A. pseudoparadoxa* Zone from Oese (Rheinisches Schiefergebirge),

Germany. *Journal of the Korean Earth Science Society* **19**(2): 216–59.

Wood, C. J., Ernst, G., and Rasemann, G., 1984, The Turonian–Coniacian stage boundary in Lower Saxony (Germany) and adjacent areas: the Salzgitter–Salder Quarry as a proposed international standard section. *Bulletin of the Geological Society of Denmark* **33**: 225–38.

Wood, C. J., Morter, A. A., and Gallois, R. W., 1994, Upper Cretaceous stratigraphy of the Trunch Borehole (TG 23 SE 8). In R. S. Arthurton, S. J. Booth, A. N. Morigi, *et al.*, eds., *Geology of the Country around Great Yarmouth*. Memoir for 1 : 50 000 geological sheet 162 (England and Wales). London: British Geological Survey/Her Majesty's Stationery Office.

Wood, H. E., II., Chaney, R. W., Clark, J., *et al.*, 1941, Nomenclature and correlation of the North American continental Tertiary. *Geological Society of America Bulletin* **52**: 1–48.

Woodburne, M. O., 1987, *Cenozoic Mammals of North America*. Berkeley: University of California Press, p. 336.

Woodburne, M. O. and Swisher, C. C., 1995, Land mammal high-resolution geochronology, intercontinental overland dispersals, sea level, climate, and vicariance. *Society of Economic Paleontologists and Mineralogists Special Publication*, Vol. 54. Tulsa, OK: SEPM, pp. 335–364.

Woodruff, F. and Savin, S. M., 1991, Mid-Miocene isotope stratigraphy in the deep sea: high resolution correlations, paleoclimate cycles, and sediment preservation: *Paleoceanography* **6**: 755–806.

Woollam, R. and Riding, J. B., 1983, Dinoflagellate cyst zonation of the English Jurassic. *Report of the Institute of Geological Sciences* **83**(2): 41.

Work, D. M., 2002, Secretary report. *Newsletter on Carboniferous Stratigraphy* **20**: 3.

Wrenn, J. H. and Hart, G. F., 1988, Paleogene dinoflagellate cyst biostratigraphy of Seymour Island, Antarctica. *Geological Society of America Memoir* **169**: 321–447.

Wrenn, J. H., Hannah, M. J., and Raine, J. J., 1998, Diversity and palaeoenvironmental significance of Late Cainozoic marine palynomorphs from the CRP-1 core, Ross Sea, Antarctica. *Terra Antarctica* **5**(3): 553–570.

Wright, J. K., 1973, The Middle and Upper Oxfordian and Kimmeridgian Staffin Shales at Staffin, Isle of Skye. *Proceedings of the Geologists' Association* **84B**: 447–457.

1989, The Early Kimmeridgian ammonite succession at Staffin, Isle of Skye. *Scottish Journal of Geology* **25**: 263–72.

Wyborn, D., Owen, N., Compston, W., and McDougall, I., 1982, The Laidlaw Volcanics: a late Silurian point on the geological time scale. *Earth and Planetary Science Letters* **59**: 99–100.

Xiao, S. and Knoll, A. H., 2000, Phosphatized animal embryos from the Neoproterozoic Doushantuo Formation at Weng'an, Guizhou Province, South China. *Journal of Paleontology* **74**: 767–88.

Xiao, S., Knoll, A. H., Zhang, Y., and Yin, L., 1997, Neoproterozoic fossils in Mesoproterozoic rocks? Chemostratigraphic resolution of a biostratigraphic conundrum from the North China Platform. *Precambrian Research* **84**: 197–220.

Xiao, S., Yuan, X., and Knoll, A. H., 2000, Eumetazoan fossils in terminal Proterozoic phosphorites? *Proceedings of the National Academy of Sciences, USA* **97**: 13 684–9.

Xiao, S., Yuan, X., Steiner, M., and Knoll, A. H., 2001, Carbonaceous macrofossils in a terminal Proterozoic shale: a systematic reassessment of the Miaohe biota, South China. *Journal of Paleontology* **76**: 345–74.

Yang, W., Harmsen, F., and Kominz, M. A., 1995, Depositional cyclicity of the Middle and Late Devonian Lost Burro Formation, Death Valley, California – a possible record of Milankovitch climatic cycles. *Journal of Sedimentary Research* **B65**: 306–22.

Yang, Z., Moreau, M.-G., Bucher, H., Dommergues, J.-L., and Trouiller, A., 1996, Hettangian and Sinemurian magnetostratigraphy from the Paris Basin. *Journal of Geophysical Research* **101**: 8025–42.

Yin, H., 1996, *The Palaeozoic–Mesozoic Boundary, Candidates of Global Stratotype Section and Point of the Permian–Triassic Boundary*. Wuhan: China University of Geosciences Press, p. 137.

Yin, H., Sweet, W. C., Glenister, B. F., *et al.*, 1996, Recommendation of the Meishan section as Global Stratotype Section and Point for basal boundary of Triassic System. *Newsletters on Stratigraphy* **34**(2): 81–108.

Yin, H., Zhang, K., Tong, J., Yang, Z., and Wu, S., 2001, The Global Stratotype Section and Point (GSSP) of the Permian–Triassic boundary. *Episodes* **24**: 102–14.

Yolkin, E. A., Kim, A. I., Weddige, K., Talent, J. R., and House, M. R., 1998, Definition of the Pragian/Emsian Stage Boundary. *Episodes* **20**: 235–40.

Young, G. C., 2000, Flawed timescale, or flawed logic. *The Australian Geologist* **115**: 6.

Young, G. C. and Laurie, J. R., 1996, *An Australian Phanerozoic Timescale*. Melbourne: Oxford University Press, 279 pp. + charts.

Yu Changmin, 1988, *Devonian–Carboniferous Boundary in Nanbiancum, Guflin, China: Aspects and Records*. Beijing: Science Press, 379 pp.

Yudina, A. B., 1995, Genus Auncyrodella succession in the earliest Frasnian (?) of the Northern Chernyschev Swell. *Geoloines, Praha* **3**: 17–20.

Zachariasse, W. J., Zijderveld, J. D. A., Langereis, C. G., Hilgen, F. J., and Verhallen, P. J. J. M., 1989, Early Late Pliocene biochronology and surface water temperature variations in the Mediterranean. *Marine Micropaleontology* **14**: 339–55.

Zachos, J. C. and Arthur, M. A., 1986, Paleoceanography of the Cretaceous/Tertiary boundary event: inferences from stable isotopic and other data. *Paleoceanography* **1**: 5–26.

Zachos, J. C., Berggren, W. A., Aubry, M.-P., and Mackenson, A., 1992, Isotope and trace element geochemistry of Eocene and Oligocene foraminifers from Site 748, Kerguelen Plateau. In S. W. Wise, R. Schlich, *et al.*, eds., *Proceedings of the Ocean Drilling Program, Scientific Results*, Vol. 120. College Station, TX: Ocean Drilling Program, pp. 839–854.

Zachos, J. C., Lohmann, K. C., Walker, J. C. G., and Wise, S. W., 1993, Abrupt climate change and transient climates during the Paleogene: a marine perspective. *Journal of Geology* **101**(1): 91–213.

Zachos, J. C., Opdyke, B. N., Quinn, T. M., and Halliday, A. N., 1999, Early Cenozoic glaciation, Antarctic weathering, and seawater $^{87}Sr/^{86}Sr$: is there a link? *Chemical Geology* **161**: 165–80.

Zachos, J. C., Shackleton, N. J., Revenaugh, J. S., Pälike, H., and Flower, B. P., 2001b, Periodic and non-periodic climate response to orbital forcing across the Oligocene–Miocene boundary. *Science* **292**: 274–77.

Zachos, J., Pagani, M., Sloan, L., Thomas, E., and Billups, K., 2001a, Trends, rhythms, and aberrations in global climate 65 Ma to present. *Science* **292**: 686–93.

Zagwijn, W. H., 1992, The beginning of the Ice Age in Europe and its major subdivisions. *Quaternary Science Reviews* **11**: 583–91.

Zakharov, Y. D., 1994, Proposals on revision of the Siberian standard for the Lower Triassic and candidate stratotype section and point for the Induan–Olenekian boundary. *Albertiana* **14**: 44–51.

Zakharov, V. A., Bown, P., and Rawson, P. F., 1996, The Berriasian Stage and the Jurassic–Cretaceous boundary. *Bulletin de l'Institut Royal des Sciences Naturelles de Belgique, Sciences de la Terre*, Supplement **66**: 7–10.

Zakharov, Y. D., Shigata, Y., Popov, A. M., *et al.*, 2000, The candidates of global stratotype of the boundary of the Induan and Olenekian stages of the Lower Triassic in Southern Primorye. *Albertiana* **24**: 12–26.

Zakharov, Y. D., Shigeta, Y., Popov, A. M., *et al.*, 2002, Triassic ammonoid succession in South Primorye. 1, Lower *Olenekian Hedenstroemia bosphorensis* and *Anasibirites nevolini* Zones. *Albertiana*: **27**: 42–64.

Zalasiewicz, J., Williams, M., Verniers, J., and Jachowicz, M., 1998, A revision of the graptolite biozonation and calibration with the chitinozoa and acritarch biozonations for the Wenlock succession of the Builth Wells district, Wales, UK. In J. C. Gutiérrez-Marco and I. Rábano, eds., Proceedings of the Sixth International Graptolite Conference on the GWG (IPA) and the 1998 Field Meeting of the International Subcommission on Silurian Stratigraphy (ICS-IUGS), Instituto Tecnológico Geominero de España, *Temas Geológico-Mineros*, Vol. 23, p. 141.

Zalasiewicz, J., Smith, A. G., Brenchley, P., *et al.*, 2004, Simplifying the stratigraphy of time. *Geology* **32**(1): 1–4.

Zang, W. and Walter, M. R., 1992, Late Proterozoic and Cambrian microfossils and biostratigraphy, Amadeus Basin, central Australia. *Association of Australasian Palaeontologists Memoir* **12**: 1–132.

Zapfe, H., 1974, Die Stratigraphie der Alpin-Mediterranen Trias: Schriftenreihe der Erdwissenschaftlichen Kommissionen. *Österreichische Akademie der Wissenschaften* **2**: 137–44.

Zegers, T. E. and Dekkers, M. J., 1999, The relationship between tectonics and remagnetisation in the Variscan Ardennes Massif Series/Source. *European Union of Geosciences conference abstracts*; *EUG 10*: *Journal of Conference Abstracts* **4**(1): 628.

Zeuner, F. E., 1935, The Pleistocene chronology of central Europe. *Geological Magazine* **72**: 350–76.

1959, *The Pleistocene Period*. London: Hutchinson, p. 447.

1945, The Pleistocene period: its climate, chronology and faunal successions. *Ray Society Monograph* **130**.

Zhang, Y., Yin, L., Xiao, S., and Knoll, A. H., 1998, *Permineralized Fossils from the terminal Proterozoic Doushantuo Formation, China*. *Paleontolological Society Memoir*, Vol. 50. Bridgewater, MA: Paleontological Society.

Zhao, J. K., Sheng, J. Z., Yao, Z. Q., *et al.*, 1981, The Changhsingian and Permian–Triassic boundary of South China. *Bulletin of the Nanjing Institute of Geology and Palaeontology, Academica Sinica* **2**: 1–112.

Zhou, L. P. and Shackleton, N. J., 1999, Misleading positions of geomagnetic reversal boundaries in Eurasian loess and implications for correlation between continental and marine sedimentary sequences. *Earth and Planetary Science Letters* **168**: 117–30.

Zhou, Z., Glenister, B. F., and Spinosa, C., 1996, Multi-episodal extinction and ecological differentiation of Permian ammonoids. *Permophiles* **29**: 52–62.

Zhuravlev, A. Y., 1995, Preliminary suggestions on the global Early Cambrian zonation. In E. Landing and G. Geyer, eds., *Morocco '95: The Lower–Middle Cambrian Standard of Western Gondwana, Beringeria, Special Issue* 2. Wuerzburg: Freunde der Würzburger Geowissenschaften, pp. 147–60.

Zhuravlev, A. Yu., 2001, Biotic diversity and structure during the Neoproterozoic–Ordovician transition. In A. Yu. Zhuravlev and R. Riding, eds., *The Ecology of the Cambrian Radiation*. New York: Columbia University Press, pp. 173–99.

Zhuravlev, A. Y. and Gravestock, D. I., 1994, Archaeocyaths from Yorke Peninsula, South Australia and archaeocyathan Early Cambrian zonation. *Alcheringa* **18**: 1–54.

Zhuravlev, A. and Riding, R., 2001, *The Ecology of the Cambrian Radiation*, *Perspectives in Paleobiology and Earth History*, Vol. 7. New York: Columbia University Press, 525 pp.

Ziegler, W., 1962, Taxionomie und Phylogenie Oberdevonischer Conodonten und ihre stratigraphische Bedeutung. *Abhandlungen des Hessischen Geologischen Landesamtes für Bodenforschung* **38**: 1–166.

Ziegler, W., 2000, The Lower Eifelian boundary. *Courier Forschungsinstitut Senckenberg* **225**: 27–36.

Ziegler, W. and Sandberg, C. A., 1990, The Late Devonian standard zonation. *Courier Forschungsinstitut Senckenberg* **121**: 115.

Ziegler, W. and Werner, R., 1982, On Devonian stratigraphy and palaeontology of the Ardenno–Rhenish mountains and related Devonian matters. *Courier Forschungsinstitut Senckenberg* **55**: 1–505.

Zielinski, U. and Gersonde, R., 2002, Plio–Pleistocene diatom biostratigraphy from ODP Leg 177, Atlantic sector of the Southern Ocean. *Marine Micropaleontology* **45**: 309–56.

Zijderveld, J. D. A., Hilgen, F. J., Langereis, C. G., Verhallen, P. J. J. M., and Zachariasse, W. J., 1991, Integrated magnetostratigraphy and biostratigraphy of the upper Pliocene–lower Pleistocene from the Monte Singa and Crotone areas in Calabria (Italy). *Earth and Planetary Science Letters* **107**: 697–714.

Zimmerman, M. K., Noble, P. J., Holmden, C., and Lenz, A. C., 2000, Geochemistry of Early Silurian (Wenlock) graptolite extinctions: Cape Phillips Formation, Arctic Canada. *Palaeontology Down Under 2000, Geological Society of Australia, Abstracts* **61**: 183.

Stratigraphic Index

General Index